Studies in Surface Science and Catalysis 120A

# ADSORPTION AND ITS APPLICATIONS IN INDUSTRY AND ENVIRONMENTAL PROTECTION
# VOL. I: APPLICATIONS IN INDUSTRY

**Studies in Surface Science and Catalysis**

**Advisory Editors:** B. Delmon and J.T. Yates

**Vol. 120A**

# ADSORPTION AND ITS APPLICATIONS IN INDUSTRY AND ENVIRONMENTAL PROTECTION

## Vol. I: Applications in Industry

Editor

**A. Dąbrowski**
*Faculty of Chemistry, Department of Theoretical Chemistry, Maria Curie-Skłodowska University, Lublin 20-031, Poland*

**1999**
**ELSEVIER**
**Amsterdam — Lausanne — New York — Oxford — Shannon — Singapore — Tokyo**

ELSEVIER SCIENCE B.V.
Sara Burgerhartstraat 25
P.O. Box 211, 1000 AE Amsterdam, The Netherlands



First edition 1999

Library of Congress Cataloging in Publication Data

Adsorption and its applications in industry and environmental
protection / editor, A. Dąbrowski.
p. cm. -- (Studies in surface science and catalysis ; vol.
120 A-B)
Includes bibliographical references and index.
Contents: pt. A. Applications in industry -- pt. B. Applications
in environmental protection.
ISBN 0-444-82828-1 (set). -- ISBN 0-444-50165-7 (pt. A : alk.
paper). -- ISBN 0-444-50166-5 (pt. B : alk. paper)
1. Adsorption. 2. Adsorption--Environmental aspects.
I. Dąbrowski, A. II. Series: Studies in surface science and
catalysis ; 120 A-B.
TP156.A35A27 1999
660'.284235--dc21 98-48805
CIP

ISBN Vol 120A: 0-444-50165-7
ISBN Set: 0-444-82828-1

∞ The paper used in this publication meets the requirements of ANSI/NISO Z39.48-1992 (Permanence of Paper).

Printed in The Netherlands.

## PREFACE

Justification for a book concerned with practical applications of adsorption is hardly neccssary. Adsorption has a great cognitive and utilitarian importance and is widely used in industry and environmental protection. Moreover, adsorption makes stability of colloidal systems possible, initiates many biological phenomena, also plays a significant role in agricultural chemistry, biology, physics, biophysics and several interdisciplinary areas. The surface science literature has presently reached such an extent that its fundamental advantage cannot be disputed. On the other hand, it is widely known that many non-academic projects or industrial applications are connected with the adsorption phenomena occurring at various interfaces. A few specific examples deal with heterogeneous catalysis, biocatalysis, chromatography, membrane separation technology, corrosion, energy conversion, friction and lubrication, solid state electronics, nucleation and crystal growth, flotation and detergency, isolation and purification on a large scale in chemical and pharamceutical industries, etc.

Adsorption utility has a very long history and its first applications were noted in ancient times. However, the rational use of the phenomenon under consideration for industrial purposes started at the end of the 18th century when the Swedish chemist and apothecary Carl Wilhelm Scheele discovered the uptake of gases by porous solids in 1773. Later, the Rusian academician Lowitz found that charcoal when immersed in the tartaric acid solution, decolorizes it by adsorbing the organic contaminants present. This discovery led to the first industrial application of charcoal in the sugar industry in England in 1794. The discovery of adsorption process selectivity by the Russian scientist Tswett in 1903 originated a new analytical technique which is adsorption chromatography. Tswett recommended this process for separation of various mixtures. Extensive adsorption investigations were continued during World War I in order to protect man's respiratory tracts from toxic warfare agents introduced into the air. The world war experiences and researches conducted in 1930's led to the development of new technologies for obtaining active carbons which were found as the most effective sorbents. The revolutionary progress in this field was made owing to the discovery in mid 50's the synthetic zeolites, unique solids with the wide - spread applications. In adsorption and related domains, carbon molecular sieves, fullerenes and recently heterofullerenes, the so-called nanomaterials, both carbonaceous and inorganic ones became more and more important. At present, the production of new adsorbents, catalysts and ion-exchangers is strictly connected with the computational material sience which may be considered as a strategic technology for the 21st century.

The adsorption theory initiated by Langmuir's fundamental works, has been developed extensively during the last seventy years and presently comprises very advanced approaches, including the wide spectrum of modern surface chemistry. The revolutionary progress in the growth of computer power produced a new revolution: molecular modelling of adsorption phenomena which gives a quite new insight into a molecular picture of the surface phenomena. In terms of computer modelling and simulations methods it is possible to investigate systems not subjected to the analytical description. These methods have emerged as an efficient tool for improving the fundamental understanding of basic microscopic occurences and for helping to solve industrial and environmental relevant problems involving wide fields of chemical and physicochemical processes.

As follows from the above considerations, the subject of utility of adsorption science has enormous technological, environmental and economic value and continues to present significant challanges with prospects for further significant development. As the themes to be considered refer to such a wide spectrum, no single author could write a critical review on more than a very limited number of problems. That is why this book includes chapters written approximately by 140 authoritative specialists from more than forty countries of the world. I would like to present a study including papers and critical reviews describing selected but most important tasks which deal with practical aspects of adsorption and related domains.

The book is divided into two volumes, consisting of chapters arranged in a consistent order, though some chapters could be connected with the industrial (volume I) or environmental (volume II) fields. In order to highlight for potential readers all topics and considerations each volume of the monograph comprises the complete contents and the complete list of authors, but includes its own subject index only. It should be emphasized that all contributions were subjected to a rigorous review process, with almost all papers receiving two reviews from a panel of approximately fifty reviewers.

The first aim of volume I is a brief review of adsorption history and its development for practical purposes until the up to date stage. The second aim deals with the presentation of some important fundamental information on the adsorbents and catalysts as well as on the methods of their characterization. Among the methods for characterizing their bulk and surface structural properties the classical techniques as adsorption, chromatography, calorimetric, spectroscopy as well as adsorption - thermodesorption methods are presented. This section includes also a fascinating paper on adsorption with soft adsorbents or adsorbates, which gives a preview of what might be a theory of adsorption beyond the traditional paradigm of this surface science branch. The part of this volume connected with practical industrial applications includes the chapters presenting advanced technical tools for high capacity adsorption separation of liquid and gas mixtures, development of new adsorbents for removal of hazardous contaminants from combustion flue gases and waste waters, degasification of coal seams, fabrication of inorganic membranes and their applications, the comprehensive review on contemporary utility of self-assembled monolayers, adsorption proteins and its role in modern industry, adsorption methods in technology of optical fibre glasses, sol-gel technology, solid dessicant dehumidification systems, etc. The presented articles give both the scientific backgrounds of the phenomena discussed and bring out in strong relief their practical aspects. From this point of view, the book may be addressed to a large group of research workers both in academic institutions and industrial laboratories, whose professional activity is related to the fields of surface chemistry, adsorption, catalysis, ion-exchange, chromatography or spectroscopy techniques, etc. This book is also meant for the above mentioned readers and for university teachers, who could use the presented material for introductory, graduate or post graduate courses and those who have a general interest in the subject find the overviews offered practically interesting.

Moreover, the chapters give not only brief current knowledge about the studied problems, but are also a source of topical literature on it. Thus each chapter can constitute a literature guide for a given topic and encourage the reader to get to know a problem and for further detailed studies. At the end of the book the comprehensive bibliography on adsorption principles, design data and adsorbent materials for industrial applications are presented which deal with the period 1967-1997. I hope that the broad aim of this volume to review the current status of knowledge concerning a wide spectrum of utility of adsorption phenomena will provide a stimulus for future work in this exciting area of applied surface chemistry.

I am quite aware that the panorama of the researches presented here is incomplete. Nevertheless, I believe that this monograph is a substantial step toward presenting the actual trends and state of the art. I would like to express my warmest thanks to all the authors for their efforts contributing to the scientific problems. Finally, I wish to acknowledge the great help I had from my wife, Mrs. Iwonna Dąbrowska, during all stages of the growth of the monograph. Her patience, encouragment and support made it possible for this book to appear in the present form.

*Lublin, September, 1998.* *A.Dąbrowski (ed.)*

## Complete List of Authors

1. **Alexandratos S.D.**
Department of Chemistry, University of Tennessee at Knoxville, Knoxville, TN 37996-1600, USA

2. **Andrushkova O.V.**
Department of Total and Bioorganic Chemistry, Novosibirsk Medical Institute, Krasny Prospekt 52, Novosibirsk 630091, Russia

3. **Baldini F.**
Instituto di Ricerca sulle Onde Elettromagnetiche „Nello Carrara", CNR, Via Panciatichi 64, 50127 Firenze, Italy

4. **Bandosz T.J.**
Department of Chemistry, City College of New York, New York, NY 10031, USA

5. **Blom J.**
Tauw Milieu P.O.Box 133, 7400 AC Deventer, The Netherlands

6. **Błądek J.**
Institute of Chemistry, Military University of Technology, Kaliskiego 2, 01-489 Warsaw, Poland

7. **Boere J.A.**
NORIT N.V., Research & Development, Nijverheidsweg - Noord 72, P.O.Box 105, 3800 AC Amersfoort, The Netherlands

8. **Bogillo V.I.**
Institute of Surface Chemistry, National Academy of Sciences, Prospekt Nauki 31, 252022 Kiev, Ukraine

9. **Bracci S.**
Centro di Studio sulle Cause di Deperimento e Metodi di Conservazione Opere d'Arte, CNR, Via G.Capponi 9, 50121 Firenze, Italy

10. **Bülow M.**
The BOC Group Gases Technical Center, 100 Mountain Ave., Murray Hill, NJ 07974, USA

11. **Buczek B.**
Faculty of Fuels and Energy, University of Mining and Metallurgy, 30-059 Cracow, Poland

12. **Burke M.**
University of Arizona, Old Chemistry Bldg., Tucson, AZ 85721, USA

13. **Cacciola G.**
National Council of Research, Institute for Research on Chemical Methods and Processes for Energy Storage and Transformation, S.Lucia sopra Contesse, 98126 Messina, Italy

14. **Carey T.R.**
Radian International, LLC, 8501 N.Mopac Blvd., Austin, TX 78759, USA

15. **Cerofolini G.F.**
SGS-THOMSON Microelectronics, 20041 Agrate MI, Italy

16. **Chang R.**
Electric Power Research Institute, 3412 Hillview Ave., Palo Alto, CA 94403, USA

17. **Chen J.**
Georgia Institute of Technology, School of Civil and Environmental Engineering, Atlanta, GA 30332-0512, USA

18. **Chen S.**
Illinois State Geological Survey, 615 E. Peabody Dr. Champaign, IL 61820, USA

19. **Dabou X.**
Chemical Process Engineering Laboratory, Department of Chemical Engineering, Aristotle University of Thessaloniki and Chemical Process Engineering Research Institute, PO Box 1520, Thessaloniki 54006, Greece

20. **DallBauman L.A.**
NASA Johnson Space Center , Houston, TX 77058 , USA

21. **Dąbrowski A.**
Faculty of Chemistry, M.Curie-Skłodowska University, 20031 Lublin, Poland

22. **Deka R.C.**
India Catalysis Division, National Chemical Laboratory, Pune - 411008, India

23. **Deng S.G.**
USA Department of Chemical Engineering, University of Cincinnati, Cincinnati, Ohio 45221, USA

24. **Dobrowolski R.**
Faculty of Chemistry, M.Curie-Sklodowska University, 20031 Lublin, Poland

25. **Domingo-Garcia M.**
Grupo de Investigación en Carbones, Dpto. de Quimica Inorganica, Fac. de Ciencias, Universidad de Granada, 18071 Granada, Spain

26. **Dybko A.**
Department of Chemistry, Warsaw University of Technology, Noakowskiego 3, 00-664 Warsaw, Poland

27. **Fadoni M.**
Department of Physical Chemistry and Electrochemistry, University of Milan, Via Golgi 19, 20133 Milan, Italy

28. **Fernandez-Morales I.**
Grupo de Investigación en Carbones, Dpto. de Quimica Inorganica, Fac. de Ciencias, Universidad de Granada, 18071 Granada, Spain

29. **Finn J.E.**
NASA Ames Research Center, Moffett Field CA, USA

30. **Fleming H.**
Cochrane Inc., 800 3$^{nd}$ Avenue, King of Prussia, 19406 PA, USA

31. **Ghosh T.K.**
Particulate Systems Research Center, Nuclear Engineering Program, E 2434 Engineering Building East, University of Missouri-Columbia, Columbia, MO 65211, USA

32. **Ghzaoui A.El.**
UM II LAMMI ESA 5079, Case 015, Place Eugene Bataillon, 34095 Montpellier Cedex 5, France

33. **Golden T.C.**
Air Products and Chemicals, Inc., 7201 Hamilton Boulevard, Allentown, PA 18195-1501, USA

34. **Groszek A.J.**
MICROSCAL LTD, 79 Southern Row, London W 10 5 AL, UK

35. **Haukka S.**
Microchemistry Ltd., P.O.Box 132, FIN-02631 Espoo, Finland

36. **Heijman S.G.J.**
KIWA Research and Consultancy, P.O.Box 1072, 3430 BB Nieuwegein, The Netherlands

37. **Hines A.L.**
Honda of America Mfg.Inc., 24 000 Honda Parkway, Marysville, OH 43040, USA

38. **Hopman R.**
KIWA Research and Consultancy, P.O.Box 1072, 3430 BB Nieuwegein, The Netherlands

39. **Horvath G.**
University of Veszprem, H-8201 Veszprem, P.O.Box 158, Egyetem u.10, Hungary

**40. Hsi H-C.**
University of Illinois, Environmental Enegineering Program,
205 N.Mathews Ave., Urbana, IL 61801, USA

**41. Hubicki Z.**
Faculty of Chemistry, M.Curie-Skłodowska University, 20031 Lublin, Poland

**42. Isupov V.P.**
Institute of Solid State Chemistry and Raw Mineral Processing
Kutateladze-18, 630128, Novosibirsk, Russia

**43. Iverson I.**
Department of Chemistry, University of Nevada, Reno, NV 89557, USA

**44. Izmailova V.N.**
Moscow State University, Department Colloid Chemistry, Vorob'evy Gory, 119899 Moscow, Russia

**45. Jakowicz A.**
Faculty of Chemistry, M.Curie-Skłodowska University, 20031 Lublin, Poland

**46. Janusz W.**
Faculty of Chemistry, M.Curie-Skłodowska University, 20031 Lublin, Poland

**47. Kalvoda R.**
J.Heyrovsky Inst.Phys.Chem., Czech Acad. Scis, Dolejskova 3,
18223 Prague 8, Czech Republic

**48. Kaneko K.**
Chiba University, Department of Chemistry, Faculty of Science, 1-33 Yayoi, Inage, Chiba 263, Japan

**49. Kanellopoulos N.**
Institute of Physical Chemistry NCSR „DEMOKRITOS", Aghia Paraskevi Attikis, GR-153 10, Athenes, Greece

**50. Kikkinides E.S.**
Institute of Physical Chemistry NCSR „DEMOKRITOS", Aghia Paraskevi Attikis, GR-153 10, Athenes, Greece

**51. Kirichienko O.A.**
Institute of Solid State Chemistry, SB RAS, Kutateladze 18,
Novosibirsk 630128, Russia

**52. Kleut D.v.d.**
NORIT N.V., Research & Development, Nijverheidsweg - Noord 72,
P.O.Box 105, 3800 AC Amersfoort, The Netherlands

**53. Kobal I.**
Department of Physical and Environmental Chemistry, J.Stefan Institute, 61000 Ljubljana, Slovenia

54. **Kotsupalo N.P.**
Ekostar - Nautech Company, B.Chmielnitsky 2, 630075 Novosibirsk, Russia

55. **Krebs K.-F.**
Merck KGaA, LAB CHROM Synthese, D-64271 Darmstadt, Germany

56. **Kubo M.**
Department of Molecular Chemistry and Engineering, Faculty of Engineering, Tohoku University, Sendai 980-77, Japan

57. **Lakomaa E.-L.**
Neste Oy, Technology Center, P.O.Box 310, FIN-06101 Porvoo, Finland

58. **Lemcoff N.O.**
The BOC Group, 100 Mountain Avenue, Murray Hill, NJ 07974, USA

59. **Lin Y.S.**
USA Department of Chemical Engineering, University of Cincinnati, Cincinnati, Ohio 45221, USA

60. **Liu Y.**
Department of Chemical Engineering, Swearingen Engineering Center, University of South Carolina, Columbia, SC 29208, USA

61. **Long R.**
Department of Chemical Engineering, The University of Michigan, Ann Arbor, Michigan 48109-2136, USA

62. **Lopez-Cortes A.**
Center for Biological Research, P.O. Box 128, La Paz 23000, BCS, Mexico

63. **Lopez-Garzon F.J.**
Grupo de Investigación en Carbones, Dpto. de Quimica Inorganica, Fac. de Ciencias, Universidad de Granada, 18071 Granada, Spain

64. **Lucarelli L.**
ThermoQuest Italy S.p.A., Strada Rivoltana, 20090 Rodano (Milan), Italy

65. **Luo R.G.**
Department of Chemical Engineering, Chemistry and Environmental Science, New Jersey Institute of Technology, University Heights, Newark, NJ 07102-1982, USA

66. **Lutz W.**
Holzmarktstrasse 73, D-10179 Berlin, Germany

67. **Łodyga A.**
Fertilizers Research Institute, 24110 Puławy, Poland

68. **Łukaszewski Z.**
Poznań University of Technology, Institute of Chemistry and Technical Electrochemistry, Piotrowo 3, 60-965 Poznań, Poland

69. **MacDowall J.D.**
NORIT United Kingdom Ltd., Clydesmill Place, Cambuslang Industrial Estate, Glasgow G32 8RF, Scotland

70. **Matyska M.**
Department of Chemistry, San Jose State University, San Jose, CA 95192 USA

71. **Matijevic E.**
Center for Advanced Materials Processing, Clarkson University, P.O.Box 5814, Potsdam, New York 13699-5814, USA

72. **Meda L.**
EniChem - Istituto Guido Donegani, 28100 Novara NO, Italy

73. **Menzeres L.T.**
Ekostar - Nautech Company, B.Chmielnitsky 2, 630075 Novosibirsk, Russia

74. **Meyer K.**
Bundesanstalt für Materialforschung und -prüfung (BAM), Zweiggelande Adlershof, Rudower Chaussee 5, D-12489 Berlin, Germany

75. **Mitropoulos A.Ch.**
Institute of Physical Chemistry NCSR „DEMOKRITOS", Aghia Paraskevi Attikis, GR-153 10, Athenes, Greece

76. **Miyamoto A.**
Department of Molecular Chemistry and Engineering, Faculty of Engineering, Tohoku University, Sendai 980-77, Japan

77. **Mizukami K.**
Department of Molecular Chemistry and Engineering, Faculty of Engineering, Tohoku University, Sendai 980-77, Japan

78. **Moon H.**
Department of Chemical Technology, Chonnam National University, Kwangju 500-757, Korea

79. **Moreno-Castilla C.**
Grupo de Investigación en Carbones, Dpto. de Quimica Inorganica, Fac. de Ciencias, Universidad de Granada, 18071 Granada, Spain

80. **Neffe S.**
Institute of Chemistry, Military University of Technology, Kaliskiego 2, 01-489 Warsaw, Poland

81. **Nemudry A.P.**
Institute of Solid State Chemistry and Raw Mineral Processing, Kutateladze-18, 630128, Novosibirsk, Russia

82. **Nijdam D.**
Tauw Milieu, P.O.Box 133, 7400 AC Deventer, The Netherlands

83. **Ochoa J.L.**
Center for Biological Research, P.O.Box 128, La Paz 23000, BCS, Mexico

84. **Pan G.**
Department of Earth Sciences, University of Leeds, Leeds LS2 9JT, UK

**85. Partyka S.**
UM II LAMMI ESA 5079, Case 015, Place Eugene Bataillon, 34095 Montpellier Cedex 5, France

**86. Patel D.C.**
Department of Chemical Engineering, Chemistry and Environmental Science, New Jersey Institute of Technology, University Heights, Newark, NJ 07102-1982, USA

**87. Pesek J.**
Department of Chemistry, San Jose State University, San Jose, CA 95192, USA

**88. Pokrovskiy V.A.**
Institute of Surface Chemistry, National Academy of Sciences, Prospekt Nauki 31, 252022 Kiev, Ukraine

**89. Raisglid M.**
University of Arizona, Old Chemistry Bldg., Tucson, AZ 85721, USA

**90. Ramarao B.V.**
Syracuse University, Faculty of Paper Science and Engineering and Engineering, SUNY, College of Environmental Science and Forestry, Syracuse, NY 13210, USA

**91. Rao M.B.**
Air Products and Chemicals, Inc., 7201 Hamilton Boulevard, Allentown, PA 18195-1501, USA

**92. Ray M.S.**
Department of Chemical Engineering, Curtin University of Technology, GPO Box U1987, Perth 6845, Western Australia

**93. Reimerink W.M.T.M.**
NORIT N.V., Research & Development, Nijverheidsweg - Noord 72, P.O.Box 105, 3800 Ac Amersfoort, The Netherlands

**94. Restuccia G.**
National Council of Research, Institute for Research on Chemical Methods and Processes for Energy Storage and Transformation, S.Lucia sopra Contesse, 98126 Messina, Italy

**95. Richardson C.F.**
Radian International, LLC, 8501 N.Mopac Blvd., Austin, TX 78759, USA

**96. Ripperger K.P.**
Department of Chemistry, University of Tennessee at Knoxville, Knoxville, TN 37996-1600, USA

**97. Ritter J.A.**
University of South Carolina, Department of Chemical Engineering, Swearingen Engineering Center, Columbia, South Carolina 29208, USA

**98. Robens E.**
Institut für Anorganische Chemie und Analytische Chemie der J.Gutenberg-Universitat D-55099 Mainz, Germany

**99. Rodrigues A.E.**
Laboratory of Separation and Reaction Engineering, University of Porto, 4099 Porto Codex, Portugal

**100. Rood M.**
University of Illinois, Environmental Engineering Program, 205 N.Mathews Ave., Urbana, IL 61801, USA

**101. Rosenhoover W.**
CONSOL, 4000 Brownsville Rd., Library, PA 15129, USA

**102. Rostam-Abadi M.**
Illinois State Geological Survey, 615 E. Peabody Dr. Champaign, IL 61820, USA

**103. Rule J.**
College of Sciences, Old Dominion University, Norfolk, VA 23529-0163, USA

**104. Saba J.**
Faculty of Chemistry, M.Curie-Skłodowska University, 20031 Lublin, Poland

**105. Sakellaropoulos G.P.**
Chemical Process Engineering Laboratory, Department of Chemical Engineering, Aristotle University of Thessaloniki and Chemical Process Engineering Research Institute, PO Box 1520, Thessaloniki 54006, Greece

**106. Samaras P.**
Chemical Process Engineering Laboratory, Department of Chemical Engineering, Aristotle University of Thessaloniki and Chemical Process Engineering Research Institute, P.O. Box 1520, Thessaloniki 54006, Greece

**107. Shintani H.**
National Institute of Hygienic Sciences, 18-1 Kamiyoga 1-Chome, Setagaya-ku, Tokyo 158, Japan

**108. Silva da F.A.**
Laboratory of Separation and Reaction Engineering, University of Porto, 4099 Porto Codex, Portugal

**109. Silva J.A.C.**
Laboratory of Separation and Reaction Engineering, University of Porto, 4099 Porto Codex, Portugal

**110. Sircar S.**
Air Products and Chemicals, Inc., 7201 Hamilton Boulevard, Allentown, PA 18195-1501, USA

**111. Sivasanker S.**
Catalysis Division, National Chemical Laboratory, Pune - 411008, India

112. **Stubos A.K.**
Institute of Nuclear Technology and Radiation Protection, NCSR „DEMOKRITOS", 15310 Aghia Paraskevi Attikis, GR-15310, Athenes, Greece

113. **Subramanian D.**
University of South Carolina, Department of Chemical Engineering, Swearingen Engineering Center, Columbia, South Carolina 29208, USA

114. **Suckow M.**
Fachhochschule Lausitz, Grossenhainer Strasse, D-01968 Senftenberg, Germany

115. **Suntola T.**
Microchemistry Ltd., P.O.Box 132, FIN-02631 Espoo, Finland

116. **Suzuki M.**
Institute of Industrial Science, University of Tokyo, 7-221 Roppongi, Minato-ku, Tokyo 106, Japan

117. **Szczypa J.**
Faculty of Chemistry, M.Curie-Skłodowska University, 20031 Lublin, Poland

118. **Sykut K.**
Faculty of Chemistry, M.Curie-Skłodowska University, 20031 Lublin, Poland

119. **Świątkowski A.**
Institute of Chemistry, Military Technical Academy, Kaliskiego 2, 01-489 Warsaw, Poland

120. **Takaba H.**
Department of Molecular Chemistry and Engineering, Faculty of Engineering, Tohoku University, Sendai 980-77, Japan

121. **Tam-Chang S.-W.**
Department of Chemistry, University of Nevada, Reno, NV 89557, USA

122. **Tarasevich Yu.I.**
Institute of Colloid Chemistry and Chemistry of Water, 42 Vernadsky avenue, Kiev 252680, Ukraine

123. **Tóth J.**
Hungarian Academy of Sciences, Research Laboratory for Mining Chemistry, 3515 Miskolc-Egyetemvaros, P.O. Box 2, Hungary

124. **Tzevelekos K.P.**
Institute of Physical Chemistry NCSR „DEMOKRITOS", Aghia Paraskevi Attikis, GR-153 10, Athenes, Greece

125. **Unger K.K.**
Institut für Anorganische Chemie und Analytische Chemie der J.Gutenberg-Universitat, D-55099 Mainz, Germany

**126. Ushakov V.A.**
Institute of Solid State Chemistry, SB RAS, Kutateladze 18, Novosibirsk 630128, Russia

**127. Vansant E.F.**
Laboratory of Inorganic Chemistry, University of Antwerpen (U.I.A.), Universiteitsplein 1, 2610 Wilrijk, Belgium

**128. Vetrivel R.**
Catalysis Division, National Chemical Laboratory, Pune - 411008, India

**129. Vigneswaran S.**
University of Technology, Sydney, Faculty of Engineering, Building 2, Level 5 P.O.Box 123 Broadway, NSW 2007, Australia

**130. Waghmode S.B.**
Catalysis Division, National Chemical Laboratory, Pune - 411008, India

**131. Wróblewski W.**
Department of Chemistry, Warsaw University of Technology, Noakowskiego 3, 00-664 Warsaw, Poland

**132. Yampolskaya G.P.**
Moscow State University, Department Colloid Chemistry, Vorob'evy Gory, 119899 Moscow, Russia

**133. Yang R.T.**
Department of Chemical Engineering, The University of Michigan, Ann Arbor, Michigan 48109-2136, USA

**134. Yiacoumi S.**
Georgia Institute of Technology, School of Civil and Environmental Engineering, Atlanta, GA 30332-0512, USA

# Contents of Volume I

## Contents of Volume II

# FUNDAMENTALS OF ADSORPTION

Adsorption and its Applications in Industry and Environmental Protection
Studies in Surface Science and Catalysis, Vol. 120
A. Dabrowski (Editor)



# Adsorption - its development and application for practical purposes

A.Dąbrowski

Faculty of Chemistry, M.Curie-Skłodowska University, 20031 Lublin, Poland

## 1. INTRODUCTION

Many physical and chemical processes occur at the boundary between two phases, while others are initiated at that interface. The change in concentration of a given substance at this interface as compared with the neighbouring phases is referred to as adsorption. Depending on the type of phases in contact, we can consider this process in the following systems: liquid/gas, liquid/liquid, solid/liquid and liquid/liquid. Adsorption is also possible even between solid and solid materials [1].

The major development of adsorption processes on a large, industrial scale deals mainly with the solid/gas [2-6] and solid/liquid [7-9] boundries, but in various laboratory separation techniques all types of interfaces are applied [10-13]. The term "fluid" is commonly used to denote gas or liquid in contact with the boundary surface of solids.

A basic conception in adsorption occurring at every interface is the real adsorption system. Let us consider this conception towards the solid/gas interface. The real adsorption system is an equilibrium one including the adsorbent being in contact with the bulk phase and the so-called interfacial layer. This layer consists of two regions: the part of gas residing in the force field of the solid surface (adsorption space) and the surface layer of the solid. In the light of the above definition of adsorption system, the term "adsorption" deals with the process in which molecules accumulate in the interfacial layer, but desorption denotes the converse process. Adsorption hysteresis is said to occur when the adsorption and desorption curves deviate from one another. In such a case the isotherm possesses a hysteresis loop, the shape of which varies from one adsorption system to another. Hysteresis loops are connected with mesoporous solids, where the so-called capillary condensation occurs [14]. The material in the adsorbed state is defined as the "adsorbate", but that in the bulk gas or vapour phase prior to being adsorbed is called the "adsorptive". The penetration by the adsorbate molecules into the bulk solid phase is determined as "absorption". The term "sorption" - together with the terms "sorbent", "sorbate" and "sorptive" - is also used to denote both adsorption and absorption, when both occur simultaneously or cannot be distinguished [15].

The above considerations about the real adsorption system can be extended to the adsorption at the solid/liquid interface [16-22]. With regard to nonporous solids or solids with large pores the concept of specific surface area (the surface area divided by the mass of the solid) has a real physical meaning. Therefore, in such solid/fluid systems, the interfacial layer denotes the surface phase formed on the surface of a solid adsorbent. The macroscopic concept of the specific surface area loses physical meaning for microporous solids i.e., for the solids with pores of molecular dimensions [23]. The microporous adsorbents saturated by adsorbate molecules are rather considered as a uniphase system [24-25]. Consequently, the aforementioned concept of interfacial layer should be used most adequately with respect to uptake of fluids by microporous solids.

The most important conception in adsorption science is that named as the "adsorption isotherm". It is the equilibrium relation between the quantity of the adsorbed material and the pressure or concentration in the bulk fluid phase at constant temperature.

Apart from the results of the calorimetric measurements, the adsorption isotherm is the fundamental source of information on the adsorption process. Its strict meaning can be derived from the so-called adsorption excess concept which was introduced to the surface science by Gibbs more than one hundred years ago [26] and forms an important tool for describing adsorption phenomena occurring at various types of interfaces. In terms of the original Gibbs treatment, the real interface is assumed to be an imaginary, geometrical plane, which is called the Gibbs Dividing Surface (GDS). This surface divides two neighbouring, noninteracting homogeneous bulk phases of a hypothetical reference system. Moreover, these phases are supposed to maintain constant concentration of adsorbed component(s) up to GDS. All the extensive properties of a real adsorption system - among them adsorption - are referred to the above hypothetical model. The Gibbs approach, one of the most fundamental in adsorption at all interfaces, has an elegant but a formal character and requires no models of the structure of the interface layer. Misunderstanding of the Gibbs formalism was common for many years, especially at the liquid/liquid and liquid/solid interfaces. Yet in the 1940's, this method was criticized as physically unrealistic [27]. Moreover, the surface tension (or interface tension) parameter in the Gibbs adsorption equation [26] is not directly measurable for the liquid/solid system. The anisotropic force field acting on particles at the liquid/vapor and liquid/liquid interfaces is revealed by the surface tension which can be measured directly. For the gas/solid interface, it appears in the accumulation of the gas molecules at the solid surface (internal or external). In the case of the liquid/solid interface, the foregoing force field is revealed by changes in concentration of the distinguished component in the bulk solution. It should be stated that in the light of the above considerations, adsorption of pure liquids is not considered, although the anisotropic force field mentioned above is revealed by the exothermic heat of immersion and slight changes in the liquid density at the interface. Structural changes of a liquid close to a solid surface and of structures similar to two-dimensional liquid crystals are sometimes possible [28].

As follows from the above considerations, the thermodynamic method of Gibbs surface excess does not take into account the conception of real surface phase, its structure and thickness. Moreover, in a quantitative sense, the Gibbs surface excess depends on the position of GDS. The conception of GDS is intuitive for flat or only slightly curved surfaces of the solids, but for the surfaces whose radii of curvature are comparable with molecular dimensions, its meaning becomes ambiguous. Thus this conception seems to be obscure for porous adsorbents that is those most frequently found in the industrial and laboratory practice. Moreover, the Gibbs dividing surface can be situated arbitrary within the interfacial layer in such a way that at every point the constituent matter is in an identical thermodynamic condition [21]. There is an infinite number of surfaces satisfying this requirement. However, the ambiguities of the Gibbs treatment are only apparent and constitute a main advantage of this theory. It is possible to replace the Gibbs equilibrium excesses by operational definitions of the surface excesses which do not depend on the position of GDS [29-30]. Moreover, these operational quantities can be easily connected - for adsorption at any interface - with the experimental quantities describing the phenomenon of adsorption quantitatively [31]. Strict operational definitions of the surface excesses were introduced by Wagner [32]. They are very useful to clarify the abstract character of Gibbs definitions. Considering the adsorption at the solid/gas interface, where the adsorption of any component is not too weak and its equilibrium pressure is sufficiently small, the identification between the surface excess amount and the total amount of adsorbed substance can be justified [14]. The same remark is in force for adsorption of solutes from dilute solutions on solids [33]. However, it should be remembered that taking into account adsorption from the solutions in the full range of their concentrations or adsorption from the gas phase (mono- or multicomponent) under very high pressure, we should take advantage of the surface excess concept whose fundamental validity cannot be disputed.

An alternative convention for the thermodynamic treatment of adsorption phenomena was proposed by Verschaffeldt [34] and Guggenheim [35]. In their formulation an interface is treated as a separate surface phase located between two adjacent bulk phases and which has a finite thickness and volume. This phase can be essentially described thermodynamically in a way analogous to bulk phases. By the use of this approach, the adsorption values and all extensive thermodynamic functions are total - not excess quantities as in the Gibbs method.

The choice of one of the above conventions for the treatment of adsorption systems is a matter of convenience, although the thermodynamics of excesses is the most rigorous mode of interfacial phenomena description. It should be stressed at this point that the Gibbs method does not require any distinct model of the surface phase and its popularity in the chemistry of interfaces is due to its great utility in experimental studies [36-37].

Many years ago Hill wrote [38]: ..."there are two uncertainties, both connected with a precise definition of the surface phase: (1) the exact location of dividing surface between the bulk phases and the surface phase, and (2) the meaning and experimental measurements of the "volume" of the pure adsorbent in some cases (e.g., highly - porous solids). These difficulties can be avoided by the use of Gibbs surface thermodynamics or by considering the bulk phase plus the adsorbed phase as the thermodynamic system. The Gibbs method can be applied to uncertainty (1), at least in some instances, while the second method (full statistical description) is completely general and takes care of both (1) and (2)."

Irrespective of the convention used for description of the solid/fluid interface, the formalism of the phenomenological [39-43] as well as statistical thermodynamics [44-49] may be applied.

The most advanced theories of adsorption are statistical ones and should make it possible to calculate ab initio the profile of component concentration at the solid/fluid interface using standard methods of statistical thermodynamics [50-51]. Simultaneously, it should also enable determination of orientation of the adsorbed molecules and calculation of the thermodynamic functions characteristic for interface. However, the ab initio methods require knowledge of the analytical form of the intermolecular interaction potentials in the fluid and at the interface space. Taking into consideration the actual knowledge about these interactions, it is impossible to use these methods for studying the most real adsorption systems. Thus, the ab initio methods, though very promising, at present can be used merely in the studies of adsorption of noble gases and simple molecules on well defined surfaces [52-53].

It follows from the above considerations, that at present and in the near future theoretical descriptions requiring simple but realistic models of the adsorption process will be of great importance in the studies of adsorption at the solid/fluid interface. In the generally accepted model of the adsorption system, the real concentration profile is replaced by a step function which "divides" the fluid phase between the surface and bulk phases. These phases are at the thermodynamic equilibrium with the thermodynamically inert adsorbent which creates a potential energy field above the surface. The inertness of the solid is believed to be true in the case of physical adsorption, but there are several instances when it can be questioned [54].

A complete statistical description is furthermore complicated by the heterogeneity of the solid materials which include porous adsorbents, i.e., the most industrial adsorbents.

Assuming thermodynamic equilibrium between the surface and bulk phases we can derive various adsorption isotherms by utilizing the equality of the chemical potentials of a given component in both phases. The analytical forms of these equations depend on the assumed models for the surface bulk phases. The surface phase may be considered as a monolayer or multilayer, and as either localized, mobile or partially mobile. The analytical forms of adsorption isotherms are complicated due to structural and energetic heterogeneity of the solid surfaces, which is characteristic for a great number of adsorbents used in practice [19-21,54].

The equilibrium between a bulk phase and the surface layer may be established with regard to neutral or ionic particles. If the adsorption process of one or several ionic species is accompanied by the simultaneous desorption of an equivalent amount of ionic species, this process is considered as an ion exchange.

Adsorption can result either from the van der Waals universal interactions (physical adsorption, physisorption) or it can have a character of the chemical process (chemical adsorption or chemisorption). Contrary to physisorption, the chemisorption only occurs as a monolayer [14]. Physical adsorption can be compared to the condensation process of the adsorptive. As a rule, it is a reversible process occurring at a temperature lower or close to the critical temperature of an adsorbed substance.

Physical adsorption is very effective particularly at a temperature close to the critical temperature of a given gas. Chemisorption usually occurs at temperatures much higher than the critical temperature and - by contrast to physisorption - is a specific process which can occur only on some solid surfaces for a given gas. In other words, chemisorption only occurs if the fluid is capable of forming a chemical bond with the adsorbent. Under favourable conditions both processes can occur simultaneously or alternately. Adsorption of oxygen on the active carbon can serve as an example. At a temperature of liquid air it has a character of the physical process, at temperatures 100 - 200°C it has a mixed character and over 200°C the phenomenon of oxygen uptake by active carbon is complicated due to the dominant oxygen chemisorption [55].

So, the distinction between physical and chemical adsorption is sometimes blurred, and many principles of physisorption apply to both kinds of adsorption.

The physical adsorption is always accompanied by a decrease in free energy and entropy of the adsorption system. It results from decreasing of freedom degrees of adsorbed molecules. Consequently, the physical adsorption is always an exothermic process and the extent of gas adsorption increases with decreasing temperature.

Experimental studies concerning the development in the measurements of gas adsorption isotherms on solid adsorbents and the various experimental techniques have been reviewed and summarized in detail (see reviews [56-60] and references therein). With regard to the solid/liquid interface the comprehensive literature on the subject was presented by Kipling [7], Everett [61], Dabrowski and Jaroniec [21].

Most of the industrial solid adsorbents of important industrial applications possess a complex porous structure which consists of pores of different sizes and shapes. If the pores are slit shaped we can speak of their "width" but for the pores with a cylindrical shape the term "diameter" is frequently used.

In terms of experiences of adsorption science, total porosity is usually classified into three groups. According to the IUPAC recommendation [62], the micropores are defined as pores of a width not exceeding 2 nm, mesopores are pores of a width between 2 and 50 nm, but macropores represent pores of a width greater than 50 nm. The above classification is widely accepted in the adsorption literature. On the other hand, this classification is to some extent arbitrary because the adsorption mechanism depends not only on the pore size, but also on the size and structure of the adsorbate molecules. However, taking into account the typical sizes of adsorbed molecules (approximately 0.3 nm) this classification has great practical meaning, since the boundary values of pore sizes below which the adsorbed molecules cannot exhibit the bulk phase properties is about 2 nm.

The significance of pores in the adsorption processes largely depends on their sizes. Because sizes of micropores are comparable to those of adsorbate molecules, all atoms or molecules of the adsorbent can interact with the adsorbate species. That is the fundamental difference between adsorption in micropores and larger pores like meso- and macropores. Consequently, the adsorption in micropores is essentially a pore-filling process in which their volume is the main controlling factor. Thus as the essential parameter characterizing micropores is their volume usually referred to a unit of the solid and characteristics of their sizes. This characteristics is expressed by the so-called micropore - distribution function evaluated mainly from the low concentration adsorption data [25]. Determination of microporous adsorbent specific area from generally accepted adsorption equations is only of a formal character.

In case of mesopores whose walls are formed by a great number of adsorbent atoms or molecules the boundary of interphases has a distinct physical meaning. That means that the adsorbent surface area has also a physical meaning. In micropores the action of adsorption forces does not occur throughout their volume but at a close distance from their walls. Therefore the mono- and multilayer adsorption takes place successively on the surface of mesopores and their final fill proceeds according to the mechanism of adsorbate capillary condensation [14]. The basic parameters characterizing mesopores are then: specific surface area, pore volume and pore-size or pore-volume distribution. Mesopores, like macropores play also an essential role in the transport of adsorbate molecules inside the micropores volume.

The mechanism of adsorption on the macropores surface does not differ from that on flat surfaces. The specific surface area of macroporous solids is very small, that is why adsorption on this surface is usually neglected [57]. For obvious reasons, the adsorbate capillary condensation does not occur in macropores.

Dubinin [63] proposed to classify the pores as follows: micropores - width less than 2 nm, mesopores - width between 2 and 200 nm, and macropores - width greater than 200 nm. It is also useful to subdivide micropores further into ultra - (width less than 0.5 nm) and super - (width between 1 and 2 nm) micropores [64].

As mentioned earlier, a main source of information about adsorption and its mechanism is, besides the colorimetric measurements of adsorption heats [65], the adsorption isotherm. It should be pointed out that the fact that the adsorption isotherm is integral characteristic of a concrete adsorption system is rather rarely highlighted in literature. It means that all information derived from the adsorption isotherm deals only with a concrete adsorbent and adsorbate. By means of various adsorbates and the same adsorbent, we can obtain quite different information (characteristics). It is inappropriate to consider that some of them is correct, but other is incorrect (on condition that the adsorption isotherm is reliably determined). Both kinds of information are truly valid and suitable characteristics should be chosen according to what is required to be measured. The foregoing remarks are true regarding all real adsorption systems, but they are particularly important in the case of porous adsorbents which are mostly industrial ones containing a great variety of pores of different sizes belonging to those of a mixed structure.

The complete and actual terminology, symbols and definitions dealing with physical adsorption at various interfaces - among them those appropriate for adsorption at the solid/gas and solid/liquid interfaces - were prepared by the International Union of Pure and Applied Chemistry [15,62,66].

## 2. HISTORICAL PERSPECTIVE OF PHYSICAL ADSORPTION

The aim of this chapter is to present briefly the development of physical adsorption (physisorption) from the earliest works on the subject until the up to date state of theory as well as basic principles of this process. In many cases the concepts and results - and usually terminology - date back several decades. Therefore, the state of the art of physisorption will be preceded by a brief review of the main events of the history of physical adsorption, with some emphasis on theory. We will dwell mainly on the physisorption at the solid/gas interface.

Justification for giving prority to this interface is the fact that both experimental and theoretical studies of adsorption at the solid/liquid interface preceded those from the gaseous phase. Moreover, some equations of isotherms for adsorption at the solid/liquid interface, particularly, those referring to adsorption from the diluted solutions, are derived from the theoretical description of single gases and their mixtures on solid surfaces.

The comprehensive reviews of physical adsorption at the solid/liquid boundry can be found in several books and monographs [7,19-21,39-43,67] as well as references therein.

Certain phenomena which we now associate with adsorption were known in ancient times, but the first quantitative observations were carried out by Scheele [68] in 1773 and Fontana [69] in 1777 who reported some experiments of the uptake of gases by charcoal and clays.

Modern application of adsorption is connected with Lowitz's observation [70,71] who used charcoal for decolourization of the tartaric acid solution as result of organic impurities uptake. Systematic studies of adsorption dating from the works by de Saussuere [72,73] started in 1814. He came to the conclusion that all kinds of gases are taken up the by porous substances (sea - foam, cork, charcoal, asbestos) and this process is accompanied by evolution of heat. Thus he discovered the exothermic character of adsorption processes and as the first drew the attention to the commonness of adsorption. Moreover, de Saussure observed that the gases undergoing condensation readily are taken up to the greatest extent by porous substances.

Of very few papers from the 19th century, there should be mentioned those by Chappuis [74-76], Joulin [77] and Kayser [78,79].

Chappuis measured at a constant temperature adsorption of ammonia on charcoal and asbestes and that of sulphur dioxide, carbon dioxide and air on charcoal depending on the pressure of the gas taken up. He also made the first calorimetric measurements of heat evolved during wetting of adsorbents with liquids. This problem was also studied by Pouillet [80], Junck [81], Fitzgerald [82], Lagergren [83], Gaudechon [84] and Dewar [85].

The term "adsorption" was proposed by E. du Bois-Reymond but introduced into literature by Kayser [78,79] and during the next few years the terms "isotherm" and "isothermal curve" were used for describing the results of adsorption measurements at a constant temperature. Kayser also developed some theoretical concepts which became a basis for the monomolecular adsorption theory. He assumed that the molecules on the solid surface possess free unsaturated attractive forces. Due to that they are able to attract and accumulate molecules of gas or liquid which form a monomolecular layer on the solid surface. According to Kayser, a pressure increase is favourable for adsorption but a temperature increase causes thermal vibrations which disturb and decrease adsorption.

McBain [86] in 1909 introduced the term "absorption" to determine a much slower uptake of hydrogen by carbon than adsorption. He proposed the term "sorption" for adsorption and absorption. It is not always possible to distinguish these two phenomena and to define them precisely. In doubtful cases, the term "sorption" and consequently the terms "sorbent", "sorbare" and "sorptive" are used.

As it is known adsorption can be considered as a physical or chemical process. The forces responsible for physical adsorption are van der Waals or London dispersion forces. The hydrogen bonds are also responsible for this kind of adsorption. Physical adsorption is a rule, a reversible process, i.e., the physicosorbed layer is removed by reducing pressure or by increasing temperature.

Chemisorption required most drastic conditions to break the chemical bonds. The heat of physisorption is much lower than the heat of chemisorption and is comparable to the liquefaction of the adsorbate. The heat of chemisorption is of the same order as the heat of the relevant chemical reaction.

The theory of adsorption forces was developed by London [87,88], de Boer and Custers [89], Lenel [90] and de Boer [91,92]. The studies by Lennard-Jones [93], who proposed the Lennard-Jones attraction - repulsion (6-12) potential used up till now, were of significant importance. Development of mechanics and quantum chemistry (e.g. [94]) had a great effect on understanding the character of interactions in the physical and chemical adsorption processes. An excellent review of developments on adsorption forces has been presented by Young and Crowell [95].

Practical application of adsorption processes is based mainly on selective uptake of individual components from their mixtures with other substances. Selective adsorption was discovered by the Russian scientist Tswett in 1903 [96]. He took advantage of this phenomenon for separation of chlorophyl and other plant pigments by means of silica materials. This separation was possible owing to various adsorption affinity of silica gel with reference to various pigments. The technique proposed by Tswett has been termed as "column solid/liquid adsorption chromatography". This discovery was not only the beginning of a new analytical technique, but also the origin of a new field of surface science.

The fact that selective adsorption was found experimentally by Dewar in 1904 [97] is little known. Studying adsorption of gas mixtures by charcoal he found greater adsorption of oxygen than that of nitrogen from the air uptake by charcoal. The reverse process i.e. oxygen desorption was slower than that of nitrogen. The separation of neon and hellium mixture was made using adsorption on active carbon and next fractional desorption.

After Tswett's death, his method was forgotten and the experiments on adsorption separation of substances were very rare. However, in 1941 Martin and Synge [98] published their famous article where they proposed the so-called liquid/liquid partition chromatography, both in columns and a planar form. These scientists also provided a theoretical basis of chromatographic processes. In 1952 Martin and Synge were awarded the Nobel Prize in chemistry for their work which had a great influence especially in biochemistry.

At present chromatography is a separate well developed field of knowledge derived from adsorption [99] and one of the most important analytical methods [100-102]. It is commonly applied for industrial separation of complex mixtures [103-105]. The historical development of chromatography is presented in the book [106] that describes some important milestones of its theory and practice and the scientists who deserve thanks for developing this technique.

The isotherms dealing with physical adsorption of gases and vapours give most important characteristics of industrial sorbents which include, among other things, pore volume, pore size or energy distribution and surface area. Moreover, these very specific curves can be interpreted in order to obtain information about the adsorption mechanism strictly connected with interactions between adsorbent and adsorbate molecules and give the possibility to assess the efficiency of industrial adsorbents applied in separation, purification and other utilitarian processes.

It should be clearly emphasized that the correct interpretation of experimental adsorption isotherms should be realized in terms of mathematical adsorption equations, i.e., in terms of adsorption isotherms. As mentioned in Introduction such equations are derived in close connection with the assumptions concerning a physical model of adsorption system. Such a model should include basic parameters which determine or can determine the adsorption process. The model assumptions usually are a result of the experiment observation.The experimental results allow for formation of hypothesis about character of the adsorption process. This hypothesis can be tested experimentally. If a hypothesis is not disapproved by repeated experiments, it develops into a theory, i.e., a suitable adsorption equation. Thus, a theory is tested in order to explain the behavior of the adsorption system investigated. A theory always serves as a guide to new experiment and is constantly tested. If a theory appears to be incorrect to describe the experiment, it must either be discarded in favour of a new one or should be modified so that all experimental observations can be accounted for. Adsorption science has been developed by a constant interplay between theory and experiment.

There was not any theory enabling interpretations of adsorption isotherms till 1914. The Freundlich equation [107] was used but is was not theoretically justified. According to McBain [108], the empirical equation mentioned above was proposed by J.M. van Bemmelen in 1888.

The so-called Freundlich adsorption isotherm was also proposed by Boedecker in 1895 [109] as an empirical equation. This equation is known in literature as the Freundlich's equation, because Freundlich [110] assignated great importance to it and popularized its use.

In 1914 - 1918 there were proposed two independent descriptions of adsorption phenomena. These descriptions are associated with such names as Langmuir [111] and Euckena [112] as well as Polanyi [113,114].

The Langmuir equation initially derived from the kinetic studies was based on the assumption that on the adsorbent surface there is a definite and equivalent energetically number of adsorption sites, at each of which one molecule of a perfect gas may be adsorbed. The bonding to the adsorption sites can be either chemical or physical, but must be sufficiently strong to prevent displacement of adsorbed molecules along the surface. So, the localized adsorption has been assumed as distinct from non-localized adsorption, where the adsorbed molecules can move along the adsorbed surface. Because the bulk phase is constituted by a perfect gas, the lateral interactions among the adsorbate molecules were neglected. On the energetically homogeneous surface of the adsorbent a monolayer surface phase is thus formed. So, Langmuir, for the first time, introduced a clear concept of the monomolecular adsorption.

There are very interesting statements which Langmuir wrote in his earlier paper [115]: "If any molecules impinging on the surface are condensed, a certain time interval must elapse before they can evaporate. This time lag will bring about the accumulation of molecules in the surface layer, and may thus be looked upon as the cause of adsorption. ..... The forces acting between two layers of adsorbed molecules will usually be very much less than those between the crystal surface and the first layer of molecules. The rate of evaporation in the second layer will, therefore generally be so much more rapid than in the first, that the number of molecules in the second layer will be negligible". However, in this paper Langmuir stated that as the external pressure became the saturation vapour pressure "...there will be a tendency for the adsorbed film to become several molecules deep" and later he expressed the view: "According to this theory it is very improbable that films more than one on two molecules deep would ever be held on a surface by adsorption, except with nearly saturated vapors".

The statement proposed by Langmuir applied to chemisorption and with some restrictions to physical adsorption. The constant parameters of the Langmuir equation have a strictly defined physical meaning (contrary to the parameters of the Freundlich empirical equation). Its meaning became particularly clear later while deriving this equation by Volmer [116] within the phenomenological thermodynamics formalism and while deriving this equation by Fowler within the statistical thermodynamics [117]. The Langmuir equation describes relatively well physical (or chemical) adsorption on the solid surface with one type of adsorption active centres.

Langmuir found, however, that his isotherm equation "...which apply to adsorption by plane surfaces, could not apply to adsorption by charcoal" [115]. Moreover, he stated in the same paper that for porous sorbents like charcoals "...it is impossible to know definitely the area on which the adsorption takes place". In the light of the above comments, the Sing's critical statement [118] that Langmuir equation is still widely used in order to analyse the experimental isotherms measured on activated carbons is extraordinarily accurate.

Langmuir tried also to extend his theoretical approach by taking into account the heterogeneity of solid adsorbent and the multilayer character of adsorption. In this later papers Langmuir noticed that one of the fundamental assumptions of his theory about the heterogeneity of the adsorbent surface is not satisfed in many cases. Surfaces of most solids are energetically heterogeneous because the adsorption sites are distributed on energetically different levels [115].

As the first, Langmuir proposed description of adsorption on the energetically heterogeneous polycrystalline surface by means of a finite sum of monolayer isotherms. He assumed that the surface of such an adsorbent is composed of a finite number of crystal planes of the same energy falling to one active site. For adsorption on amorphous surfaces, addition can be replaced by integration.

It ought to be stressed at this point, that the Langmuir studies of the adsorption of gases surfaces, led to the formulation of a general treatment of the kinetics of reactions at surfaces. Langmuir stated that the surface catalysis is usually preceded by chemisorption and he interpreted the kinetics of surface reactions in terms of his monolayer equation. He also showed how the adsorption isotherm can be applied to interpret the kinetics of a variety of surface reactions. These fundamental investigations in the field of surface catalysis provided the background for the detailed and exhaustive studies that are carried out today as a result of the industrial importance of catalysis. In 1932 Langmuir was awarded the Nobel Prize in chemistry for "his discoveries and researches in the realm of surface chemistry" [119].

Later some attempts were made to generalize the Langmuir equation taking into account lateral interactions among adsorbed molecules, their mobility and the surface heterogeneity of the solids. The Langmuir equation itself should be considered, however, as the useful equation which coresponds with the so-called ideal localized monolayer. The ideal localized monolayer model, despite its obvious imperfections, occupies a central position in surface and adsorption science. It started comprehensive theoretical studies whose aim was and still is a search for more and more perfect and real descriptions of experimental adsorption systems.

The milestone towards development of the adsorption science was the multilayer isotherm equation proposed by Brunauer, Emmett and Teller in 1938 [120]. The multilayer adsorption theory was preceded by two significant works by Brunauer and Emmett which appeared in 1935 [121] and 1937 [122].

In 1935 Brunauer and Emmett [121] carried out the first successful attempt to determine - by means of isotherm adsorption of six different gases - the surface area of an iron synthetic ammonia catalyst. Later [122], in 1937, these authors determined the surface area of two different silica gels measuring adsorption isotherms of seven different gases.. In the above mentioned works the surface area was determined by extrapolating the middle linear sections of experimental isotherms to zero pressure in order to obtain the amount of gas required to cover the adsorbent surface with a monomolecular layer. On condition that the monomolecular layer was in a close-packed stage, the surface area was then evaluated from the monolayer adsorbed amount. Brunauer and Emmett [121,122] also proposed to determine the monolayer adsorption amount from the so-called point B of the experimental isotherm. It was assumed that this point corresponds to the inflection point and can be obtained from the beginning of the linear section of adsorption isotherms.

At first the BET equation was derived from the kinetic considerations analogous to those proposed by Langmuir while deriving the monomolecular adsorption isotherm. First, statistical thermodynamic derivation was carried out by Cassie [123]. Lately, a slightly modified derivation has been proposed by Hill [124-126], Fowler and Guggenheim [127].

The principal assumption of the BET theory is that the Langmuir equation applies to every adsorption layer. Like the Langmuir theory, the first adsorption layer is formed on an array of surface sites of uniform energy. The major simplifying assumptions deal with the assumption, that beginning from the second adsorbed layer the condensation heat is equal to the evaporation heat of gas and that the ratio between the rate constant of uptake from one layer and the condensation of a lower situated layer for all the layers starting from the second one is the same. Later these assumptions were extensively discussed in literature [128-131]. The BET isotherm is generalization of the Langmuir isotherm to some extent.

The BET adsorption isotherm was originated for a definite model of adsorption layer [120], and next it was extended to the finite, n-number of layers [132,133].

In the former treatment the restriction was placed on the number of layers of adsorbate which is adsorbed on an open and plane surface of the solid. However, in pores the adsorption is clearly restricted. If it is supposed that adsorption takes place on the two parallel plane walls of a capillary, the maximum n-number of adsorbed layers of adsorbate can be accommodated on each of the walls.

An alternative well - known approach to multilayer adsorption was proposed by Frankel, Halsey and Hill, usually called the FHH slab theory [95]. Later, several modifications of the BET equation were proposed [95,134,135], but those did not find wide acceptance for studying gas and vapour adsorption processes [14].

The n-layer treatment mentioned above takes an account of the forces of capillary condensation. In 1940, Brunauer, Deming, Deming and Teller [132] were able to extend the BET theory by introducing an additional contribution to the energy of adsorption resulting from these forces. The so-called BDDT equation, contrary to the BET isotherm, can be applied over a wider range of relative pressures. This equation, however, is rather complex and contains four adjustable parameters which cannot be assessed independently. For this reason, this is seldom applied and only a few attempts have been made to use the equation under consideration [95,118].

Another important contribution by Brunauer, Deming, Deming and Teller [132] deals with the identification of five principal types of adsorption isotherms for gases and vapours. This identification is known as the BDDT classification and is recommended as the basis for a more complete classification introduced by IUPAC [15].

The BET theory, despite many restrictions, was the first attempt to create an universal theory of physical adsorption. It describes the whole course of the isotherm including the areas of monomolecular adsorption, polymolecular adsorption and capillary condensation. Unfortunately, the last area is described in an unsatisfactory way, particularly, when the solid is of a heterogeneous porous structure including pores of various capillary widths. This theory cannot be applied at temperatures higher than the critical temperature because it is not possible to assume equality of heats of adsorption and condensation as well as stability of the adsorption coefficients ratio with regard to individual adsorption layers.

None the less the so-called BET low temperature nitrogen method is still widely applied as the standard procedure for determining the surface area of fine powders and porous materials. It was stated by Sing [118], that there are two principal causes for its popularity. First, under favourable conditions, the BET plot can provide a reliable estimation of the monolayer capacity - especially by means of the nitrogen adsorption at 77K - and secondly, the BET approach is easy both to apply and to comprehend.

The Langmuir and BET theories, originating from one common assumption, assume existence of the interface geometrical surface on which mono- or multilayer adsorption takes place. The basic geometrical parameter applied to flat macroscopic adsorbents is their surface area.

As mentioned earlier, already Langmuir [115] supposed that adsorption on the adsorbents possessing narrow pores proportional to the sizes of adsorbed molecules must proceed according to a different mechanism. With any character of adsorption interactions causing physical adsorption in the whole area of micropores, a strong adsorption field appears in them. A limited space of micropores causes that next molecules adsorbing in micropores do not form adsorption layers but fill up micropores in respect of volume according to the mechanism of micropore filling. Therefore a basic geometrical parameter characterizing microporous adsorbents is the volume of micropores but not their "surface".

Adsorption theory of the volume filing of micropores (TVFM theory) has been proposed by Dubinin and Radushkievich [136], but this approach has originated from the potential theory of adsorption introduced by Eucken [112] and Polanyi [113,114].

According to their conception, the adsorbed layer has a multilayer character, therefore it is not two-dimensional as follows from the Langmuir theory but possesses a definite volume and applies to the van der Waals equation. The assumptions underlying the so-called potential theory of adsorption, commonly named the Polanyi theory originates from the earlier works by de Saussuere [72,73]. According to this assumption the adsorbed layer is considered as the thick, multilayer film of decreasing density and increasing distance from the solid surface. The basic concept of the Polanyi theory includes the adsorption potential and the characteristic adsorption curve. This characteristic curve presents a simple relationship between the adsorption potential and the distance from the solid surface. This relationship was termed as the characteristic adsorption equation. The aforementioned distance may be expressed in terms of volume units of the adsorbed phase. Polanyi assumed [113,114] that the adsorption potential is a temperature independent one over a wide range of temperatures. It means that the characteristic curve of adsorption is temperature independent, too. Such a statement follows from the fact, that the van der Waals forces are also independent of temperature. However, for polar adsorbates this does not always hold true.

Having calculated the curve characteristic for a given temperature it is possible to determine adsorption isotherms at other temperatures. The Polanyi potential theory does not give a definite equation of adsorption isotherm which, to some extent, replaces the characteristic adsorption equation.

Experimental verification of the Polanyi theory deals with the calculation of the characteristic adsorption curve by means of an experimental isotherm and hence determination of the isotherms at different temperatures. The first practical examinations of the potential theory of adsorption carried out by Titoff [137] and Berenyi [138] showed the good agreement with experiment.

The Polanyi theory assuming the description of three-dimensional layer adsorbed with the van der Waals equation implies not only the adsorbate concentration increase at a suitably low temperature but also its condensation into the liquid which takes place on the adsorbent flat surface. In contrast to the condensation which can proceed only on the flat surface of the solid, the capillary condensation which takes place if the adsorbent has a porous structure should be distinguished.

The capillary condensation phenomenon was discovered by Zsigismody [139], who investigated the uptake of water vapour by silica materials. Zsigismody proved that the condensation of physicosorbed vapours can occur in narrow pores below the standard saturated vapour pressure. The main condition for the capillary condensation existence is the presence of liquid meniscus in the adsorbent capillaries. As it is known, the decrease of saturated vapour pressure takes place over the concave meniscus. For cylindrical pores, with the pore width in the range 2-50 nm, i.e., for the mesopores, this phenomenon is relatively well described by the Kelvin equation [14]. This equation is still widely applied for the pore size analysis, but its main limitations remain unresolved. Capillary condensation is always preceded by mono- and/or multilayer adsorption on the pore walls. It means that this phenomenon plays an important, but secondary role in comparison with the physical adsorption of gases by porous solids. Consequently, the true pore width can be assessed if the adsorbed layer thickness is known.

Zsigismody, as the first one, drew attention to adsorption on the capillary inner walls which is primary in relation to the capillary condensation [139]. This unusually correct observation corresponds to modern views about the processes of gases and vapours uptake by porous (i.e. industrial) adsorbents. Such a process usually includes mono- and multilayer adsorption followed by the capillary condensation in the final stage of uptake. A quantitative part of capillary condensation in the uptake of a definite vapour changes for different adsorbents depending on their porous structure. This process is dominant for the adsorbents where mesopores constitute a larger part [140].

A very important characteristics of capillary condensation is the so-called hysteresis loop occurring on many experimental adsorption isotherms. According to Foster [141] and Cohan [142] the adsorption branch of hysteresis loop is caused by the polymolecular adsorption and capillary condensation but the desorption branch appears only by the condensation phenomenon.

Zsigismody stated [139] that hysteresis phenomenon is not immanent feature of any adsorbate, but depends on impurities found on the capillary walls. This view was confirmed in some cases [143].

The occurrence of reversible and irreversible hysteresis [144] was proved experimentally and there was different interpretation of these phenomena [141,142,145-148].

De Boer [149] classified the capillary hysteresis loops according to their shapes, relating the latter to the occurence of pores of given types. He distinguished five types of hysteresis loops with one vertical or steep branch. Owing to the non-homogeneous porous structure of most adsorbents, the experimental hysteresis loops of adsorption - desorption isotherms are a combination of two or more types described by de Boer.

An important contribution to explanation of the hysteresis feature was made by Everett and coworkers [150-155].

Up to this day, various attempts have been made to assess the mesopore size distribution function by means of capillary condensation part of adsorption isotherms. Folowing the IUPAC recommendations [15,140], these procedures should be carefully applied in order to obtain reliable results.

At present the Polanyi theory has rather historial meaning. However, the theory of volume filling micropores (TVFM) also called the Dubinin-Radushkevich theory [136] which is generally accepted, though always improved, originates from the Polanyi theory. TVFM has significant importance for the characteristic of most industrial adsorbents which have a well developed porous structure.

Dubinin and his co-workers [156-158] as well as Radushkevich [159] found that the characteristic adsorption curve is related to the porous structure of the adsorbent. Radushkevich [159] proved theoretically the equations of the characteristic adsorption curves for the two extreme types of adsorbents: with narrow and wide pores. Based on this Dubinin proposed the expression known in literature as the Dubinin-Radushkevich (DR) isotherm equation [136].

It was shown [136,160], that the values of constant "B" in the DR adsorption isotherm are directly associated with the porous structure of the adsorbent, i.e. related to the adsorption potential of the micropores. Its value is a qualitative measurement of the relative average sizes of the adsorbent. The second constant in the DR equation, "$W_o$", is the so-called limiting value of the total volume of micropores. Dubinin [160] assumed the slit-shaped model of pores in his theory and pore size distribution in the gaussian distribution form. So, by means of DR equation the following information can be obtained [64]: extent of micropores volume capacity, relative pore-size or pore-energy distributions and deviations of adsorption energies from a gaussian or related type of distributions. The Dubinin-Radushkevich equation was further developed by many scientists, e.g.[161-166]. The main advantage of the DR equation is that in can be used to interpret the experimental isotherms measure for porous adsorbents over low ranges of relative pressures (below 0.03).

As follows from the above considerations the DR equation, unlike the Langmuir and BET equations, is not based on a defined model process to describe physisorption of gases. It is rather based on considerations of adsorption energies.

The excellent reviews of this equation, its applicability and limitations were presented in quite recent monographs [60,167].

The Dubinin-Radushkevich equation with its numerous modifications is very important for the adsorption methods of characteristics of most industrial adsorbents. These adsorbents have a complex and well developed porous structure including pores of different shapes and widths but micropores play the most significant role in the structure. Industrial porous adsorbents, like active carbons or silica gels, have pores of different sizes, from the smallest - molecular sizes, to the large ones visible under the microscope or even with a naked eye. In the pores of molecular dimensions, the capillary condensation phenomenon does not occur because a meniscus cannot be formed in them. The conception "meniscus" can lose its physical meaning with regard to micropores. The micropores are filled with the uptaken gas or vapour in terms of the mechanism of micro-pore filling and capillary condensation does not occur in them at all or in a small range of relative pressures if a meniscus can be formed. The mechanism of micropore filling depends on the ratio of pore width/molecular diameter of adsorbate as well as the nature of adsorption system [166].

Discussing and comparing the basic adsorption theories presented so far, it is difficult to say which one is fundamentally right. The Polanyi thermodynamic theory neither determines a definite adsorption isotherm equation nor gives a detailed mechanism of the process. Development of this theory for description of adsorption on microporous substances leads to the analytical DR equation but its character is semiempirical. The Langmuir and BET theories introduce the concept of localized mono- and multilayer formed on the energetically homogeneous solid surface.

The Langmuir equation and the method of solid surface area determination based on it can be applied for the systems in which the adsorption process is not complicated by formation of a multilayer as well as by adsorption in micropores and capillary condensation. Adsorption of gases at the temperatures higher than the critical temperature on nonporous or wideporous adsorbents is an example of such cases. Despite this limitation, the Langmuir equation is used in technical adsorption for calculations of kinetics and dynamics of impurities uptake from the gas medium or diluted solutions.

The BET equation reducing in the area of low relative pressures to the Langmuir equation describes relatively well adsorption in the area of relative pressures 0.05 to 0.35. It means that the agreement of experimental data with this equation is included rather in a small area. The more typical deviations resulting from application of the BET equation consist in the fact that it predicts too small adsorption under low pressures and too large under high pressures. Using the BET equation it is possible to calculate correctly the specific surface area of macroporous adsorbents and transiently porous ones lacking a great number of micropores. The presence of micropores volume filling with adsorbate molecules in the adsorbent leads to incorrect results, though in a definite range of relative pressures the experimental data can correspond to a linear form of the BET equation. In case of water and nitrogen adsorption on zeolites of various types the upper boundary of the relative pressure is about 0.1 [57]. Thus the BET theory can be applied only in the area of average relative pressures i.e. after formation of a monolayer. Then heterogeneity of the adsorbent surface does not appear in the second adsorption layer and the energy of adsorbate condensation is a good approximation of the adsorption heat in the second and successive layers. Moreover, in the area of relative pressures under consideration the capillary condensation does not generally appear. In some cases the BET equation gives satisfactory results in the range of relative pressures up to 0.5 [14]. In the capillary condensation area the BET theory should be used cautiously particularly if the adsorbent is the characterized by a large spectrum of capillary sizes. In this area the Kelvin equation is usually applied. Finally, the BET theory cannot be used in description of adsorption on microporous adsorbents where the DR isotherm is a proper expression.

The three equations discussed above are used as linear graphs. It should be stressed that obtaining a correct straight line based on the experimental data is a necessary condition but not sufficient to accept applicability of a given isotherm. Another condition is obtaining real and reliable values of adsorption parameters based on the linear graphs, which having a definite physical meaning, should be thermodynamically verified.

The least we demand from every theory is that it describes the free energy change accurately. This is related to differing enthalpy and entropy effects at varying degrees of coverage of the adsorbent surface. The enthalpy and entropy contributions arising from the BET and also the Langmuir theory agree at best semiquantitatively with experiment [95]. The calculated enthalpy values are too small and the entropy values too high. There are, therefore several doubts as to the applicability of the physical picture given by both BET and Langmuir theories. The above remarks demonstrate that the extension of the theory of gas or vapour adsorption to cover the whole range of pressures has not been successful in the first period of adsorption science. So, the next step in the studies of adsorption will be continued to lead to even better results, providing a more accurate description of the true mechanism of adsorption from the gaseous phase.

Anticipating the considerations presented further, it should be stressed that the general theory of adsorption including all factors affecting a given process can lead to such a complicated equation of isotherm that it can describe correctly each experimental isotherm independently of this shape. Such an equation would be totally useless because none of the constants occurring in it could be experimentally determined. It often happens that simplicity of a given theory can be its greatest advantage at the same time.

In the final part of considerations about early adsorption science it should be stated that only the most important conceptions and equations of adsorption isotherms have been discussed. However, the isotherms including the lateral interactions between molecules in the surface monolayer as well as the equations concerning mobile and mobile-localized adsorption have been omitted. These equations can be derived in a simple way by assuming that molecules in the surface phase produce the surface film whose behaviour is described by the so-called surface equation of state. This equation is a two-dimensional analogue of the three-dimensional equation of state and relates the surface pressure (spreading pressure) of the film to the adsorption. This adsorption can be expressed by the Gibbs adsorption isotherm [26]. Consequently, it is possible to interconvert the adsorption isotherm and a surface equation of state by means of the Gibbs adsorption equation. Assuming various pictures for the two-dimensional adsorbed gas, the various adsorption isotherms were developed. Among them, the Henry's law for ideal non-localized monolayer [14], the Langmuir equation for ideal localized monolayer [22], the Volmer [95] and Hill - de Boer [144,168] equations for the non-ideal and non-localized monolayers, and the Fowler - Guggenheim [169] equation for the non-ideal, but localized monolayers. On the other hand, the above equations can be derived by means of statistical thermodynamics formalism [168,169]. The two-dimensional equations of state were very useful for investigating the phase transitions in the adsorbed layers [10,95].

The above considerations did not include basic conception of adsorption kinetics and dynamics on porous adsorbents and in the heterogeneous catalysis process. The fundamental field of science dealing with adsorption i.e. adsorption thermodynamics was also omitted. Even a brief discussion of these problems goes beyond the limits of this chapter. The reader interested in these problems can find them in numerous, generally accessible textbooks and monographs [6,10-12,25,35,40,45,47,54,57, 58,144,168-172] as well as references therein.

On the other hand, the equations and definitions presented above are the most fundamental ones for adsorption science although more than fifty years have elapsed since their publication. As it can be seen in the further considerations a large part of modern adsorption theory is derived directly from the equations discussed above. Moreover, these equations play a fundamental role in the studies of structure of most industrial adsorbents and solid catalysts.

Catalysis, particularly heterogeneous catalysis, is closely connected with adsorption, though contrary to chromatography, it was a separate field of knowledge in the initial, empirical period and its development. At present, it is known, that the action of solid catalysts is inseparably connected with their abilities for adsorption of reacting substances and the requirements from industrial adsorbents and catalysts are very close or in line. For these reasons it is advisable to discuss briefly this very important branch of surface science.

## 3. CATALYSIS

Catalysis plays an extremely important role in modern industry, environmental protection and our everyday life. Moreover, its importance in sustainable development is beyond discussion [173-176].

Approximately, its accounts for as much as 90% of chemicals and materials manufactured throughout the world. Catalysis, as a vital process, is technology of 21st century.

To give a comprehensive description of practical applications in which catalysis is an essential feature, it requires a book or books itself (see references [10,176-180] and references therein). Besides, this section is not an attempt to review applications of catalytic technology in field of every modern life. Rather, the aim is to highlight the relation between catalysis and adsorption, which should be considered as the most important domains of the surface science.

Catalysis is not a new phenomenon, but its wilful applications by humans has really begun in this century. Enzymatic catalysis is necessary for all living mater. Most essential of all catalysis processes is photosynthesis which underlay most of the simplest and earliest evolved life forms.

One of the first catalytic processes was probably fermentation of fruits in order to obtain alcoholic drinks [177]. Here the natural catalysts in yeast are enzymes which convert sugar into alcohol. In fact, ancient Sumerians described the beer preparation on clay plates many millenia ago. Systematic studies of the phenomenon under consideration began at the beginning of 19th century. In 1815 Davy developed the experiments dealing with catalytic combustion with platinum gauzes. The term "catalysis", however, was introduced by J.J.Berzelius in 1836. Berzelius argued that [181]: "Many bodies have the property of exerting on other bodies an action which is very different from chemical affinity. By means of this action they produce decomposition in bodies, and from new compounds into the composition of which they do not enter. This new power, hitherto unknown, I shall call it catalytic power. I shall also call catalysis the decomposition of bodies by this force".

The term catalyst denotes a body or a material which accelerates but does not appear in the chemical equation for this reaction. It enhances the rate of the reaction, and finally it is regenerated at the end of it. Usually, catalysts are classified both as homogeneous and heterogeneous catalysts. Homogeneous catalysts occur in the same phase as the reactants, but heterogeneous catalyst is in a different phase. Heterogeneous catalysis includes heterogeneous catalysts which are typically solids. Besides, the important subdisciplines of the process under consideration are biological, enzymatic and photo-catalysis [177].

The condition for hetergeneous catalysis to occur is adsorption (usually chemical one) of molecule of the reacting substances on the inner or outer surface of the adsorbent or of the catalyst spread there, then molecular dissociation of at least one or two reacting components, usually preceded by the surface diffusion [175].

The next is a surface reaction whose step is often the rate determining in a catalytic reaction. Then the product desorption occurs because the surface bond is broken and the final product enters the bulk phase, diffuses through the catalyst pores. This very simple scheme highlights the basic idea of heterogeneous catalysis. Moreover, this scheme points out clearly that development of adsorption science - including theory, practice and achievements of preparation and production of new solid sorbents - is a main factor conditioning development of heterogeneous catalysis.. As mentioned above the important advances in the field of heterogeneous catalysis were made by Langmuir, who showed how his adsorption isotherm can be used to interpret the kinetics of surface reactions.

Nowadays, a broad range of advanced catalysts is used to facilitate numerous chemical reactions. Among others, the role of catalysts is played by [182]: metals or metal complexes grafted individual sites on high area micro- or mesoporous solid supports, enzymes, antibodies, ribozymes, membranes - ceramic or biological and nanoparticles.

Moreover, there is a great variety of adsorbents which, depending on the chemical structure of their internal or external surfaces, play the role of solid catalysts or their supports. The basic function of catalyst supports is to keep the catalytically active phase in a highly dispersed state. For this reason, catalyst supports are usually porous adsorbents characterized by a well developed inner structure and possessing a large surface area. A large surface area is, however, not always desirable.

At present catalysis similarly to chromatography is a separate field of surface science. However, contrary to chromatography it is not derived directly from the science about adsorption. But it cannot be considered apart from adsorption, particularly from adsorption at the soldid/gas and solid/solution interfaces. As it is generally known, most adsorbents play a role of catalysts or their supports. In consequence, methods of preparation and characterization of adsorbents and catalysts are very similar or identical. Physical structure of catalysts is investigated by means of both adsorption methods and various instrumental techniques derived for estimating their porosity and surface area [177].

The most fundamental characteristics of any industrial catalyst is its chemical composition. The other factors such as surface area, distribution of pore volumes, pore sizes, stability and mechanical properties are also very important. Such catalysts as metals or oxides of various kinds (pure or mixed) are not thermally stable in the high surface area form in which they have to be applied. So, they are prepared as small particles bonded to the support material, usually oxides, such as alumina and silica gels.

The activated carbon, owing to its porosity and chemical surface composition, both of which may be appropriately controlled, is also recommended as a suitable catalyst support [179]. On the other hand, both silica and alumina oxides as well as natural amorphous aluminosilicates and zeolites are widely used as heterogeneous catalysts. These adsorbents having acid or (and) base sites are named as solid acid-base catalysts [183].

In the mid 1960s an important group of catalysts has been developed utilizing synthetic zeolites which have open, well controlled structure of interconnected cavities [184].

Especially, microporous aluminium phosphates and metal substituted aluminium phosphates can be routinely manufactured [185]. Synthetic microporous zeolites are nowadays of central importance industrially, as they are the powerful acid-base catalysts. These microporus materials have usually pore diameters in the range from 0.4 to 1.4 nm. By means of the ion-exchange of alkali metals for proton compounds acid catalysts whose acidity can be many times higher than that of sulfonic acid can be obtained [182]. An important feature of zeolite catalysts deals with their three-dimensional framework of channels and cavities, to give possibility for selection of reactants and products due to the dissimilarity of their molecular sizes and shapes. Other important perspectives of the greatest practical application are connected with the so-called mesoporus catalysts, exhibiting unique properties and prepared by a revolutionary synthesis method. These so-called MCM-41 zeolites (Mobile Catalytic Material, number 41) first reported in 1992 [186], are of quite new possibilities in a generation of shape - selective catalysts and as supports for the metal - based catalytic sites. Their typical pore dimensions of the channels are 2 - 10 nm, with the internal surface areas exceeding 1000 $m^2/g$. The mesoporous MCM materials may be easily modified by incorporating various heteroatoms into their framework [187].

A new class of solid acid-base catalysts is developed from mineral clays [183]. They have a two-dimensional layer lattice in which oxyanions are exchanged by hydrated cations. These intercalated cations with acid properties give the so-called pillared clays [188] having structural cavities - whose sizes can be controlled - similarly to those in zeolites. Both three and two-dimensional solid acid catalysts are very useful for selecting reactants and products of catalytic processes, among them some isomers of profound petrochemical significance [189].

The role of special catalysts is played by well known Keggin heteropoly acids. The catalysts of a regular structure (zeolites, clays and heteropoly acids) as well as mechanism of the reactions in them,, can be designed and investigated by means of the Monte Carlo, molecular dynamics and quantum mechanics methods [190-194].

Rationally designed, both microporous and mesoporous inorganic catalysts for many industrial and environmentally compatible technologies are already the facts [177].

From the above remarks about catalysis there can be drawn the following conclusions:

1. Adsorption and catalysis are closely related to each other ;
2. Action of solid catalysts results from their capability of reacting substances adsorption;
3. The same porous solids can be used as adsorbents, catalysts supports and catalysts;
4. Chemical character and size of solid surface areas, their porous structure, mechanical properties and thermal stability play an essential role in adsorption and catalysis;
5. Development of theoretical studies on adsorption, design and production of new adsorbents affects heterogeneous catalysis development.

The excellent insight into the current problems of heterogeneous catalysis under symptomatic title: "Tales of tortured ecstasy: probing the secrets of solid catalysts", was recently published by Sir John Meuring Thomas, the Fullerian Professor of Chemistry at the Royal Institution of Great Britain [187]. Readers interested in this exciting area of surface chemistry are recommended to consult this article and also Ref. [177,182,195,196].

## 4. CURRENT STATE OF PHYSICAL ADSORPTION ON SOLIDS

The first period of adsorption theory development and developing adsorptiom experiment in a wide range (mainly measurements of adsorption isotherms and calorimetric studies of adsorption heats) was based mainly on Langmuir and BET equations, capillary condensation theory, Polanyi potential theory and DR equation related to it. Advantages and disadvantages of the above mentioned equations were discussed earlier. This first period of adsorption development is termed as the "pioneering age" of the adsorption science [197]. It can be accepted by mutual agreement that this period ended on the turn of the forties and fifties. At that time the above theories and equations of adsorption isotherms were extensively verified experimentally and improved theoretically. The improvements included mainly interactions between adsorbed molecules, localized and mobile-localized adsorption as well as the studies of phase transitions in the adsorbed layers. Then the Freundlich and Langmuir - Freundlich empirical adsorption equations were used. The latter equation was proposed for the first time by Bradley [198]. Among other empirical equations the Temkin equation should be mentioned [199].

Another new period of adsorption development is connected mainly with intensive theoretical and experimental studies of physical adsorption of gases, their mixture as well as liquid solutions on heterogeneous solid surfaces. In spite of many improvements of the Langmuir and BET equations, distinct deviations from experiment were observed particularly in the range of small and great relative pressures. The attemps to fit the experimental isotherms by theoretical equations showed negative deviations from the experimental points at low adsorbate pressure and positive deviations at higher adsorbate pressure. The most dramatic deviations were observed in relation to the measurements of the enthalpy changes upon adsorption. On the other hand, the theoretically predicted isosteric heats of adsorption should be an increasing function of the surface coverage, but almost all the reported experimental heats showed an opposite trend [197]. The above mentioned disagreements between theory and experiment pointed out that there must exist an additional physical factor which influences the adsorption processes, which can be compared with the effect resulting from interactions in the adsorbed layer and this factor had not be taken into account in the existing theories. This missing factor deals with the energetic heterogeneity of the most real solid adsorbents.

The concept of solid surface heterogeneity and existence of active centres on it was introduced into the adsorption science by Taylor in the twenties [200], based on the results of works by various authors, among others, Pease [201]. The concept of surface heterogeneity is in agreement with general concepts regarding the solid theory. Real solids have a polycrystalline and amphorous structure. Fine crystals of various sizes create grains of different forms and sizes. The polycrystalline structure causes an increased adsorption activity of the solid. Rhodin [202] measured adsorption of nitrogen on the individual walls of copper monocrystal (110), (100) and (111) as well as on polycrystalline copper by means of very precise measurements using a vacuum torsion balance. The author stated that the copper monocrystal walls are homogeneous. They do not possess adsorption sites with the increased activity, but the polycrystalline surfaces are characterized by a significantly increased adsorption activity showing great values of adsorption heats at small extents of filling.

Besides cracks, fissures and other defects on the solid surface whose existence can be detected by means of various observation methods, heterogeneity is caused by disturbances in the crystal lattice corresponding to the difference between the real structure and the ideal crystal lattice. Besides dislocation of surface atoms, another type of disturbances can result, among others, from the lattice defects of the Frenkl or Schottky type. Finally, heterogeneity can be caused by impurities whose presence can affect significantly the surface properties of adsorbents, partcularly of catalysts. The surface heterogeneity generally plays an important role in the adsorption on crystalline and noncrystalline, nonporous, mesoporous and macroporous solids. However, the main source of heterogeneity for microporous solids is their complex porous structure which contains micropores of different dimensions and forms. This distribution of micropores makes the main source of the heterogeneity of a microporous structure.

To discover the surface heterogeneity Roginsky [203] proposed the so-called differential-isotope method. On the studied surface he adsorbed successively two portions of hydrogen differing in the isotope composition. In case of heterogeneous surface, the adsorption centres are filled in the order of decreasing adsorption energy. In desorption, the part of gas which was taken up as the last, would desorb as the first one. If the adsorption centres possess the same energy i.e. the surface is energetically homogeneous, then the isotope composition of gas remains unchanged throughout the desorption period.

Defects and disturbances in the solid structure and thus in its surface cause that the interaction between adsorbing molecules and this surface varies in different points. It means that the adsorption energy value depends which area of the surface adsorption takes place on. In a given area the adsorption energy value is constant and elementary theories of adsorption are applied. Thus the process occurring on the whole heterogeneous surface can be considered in the first approximation as a result of summing up independent processes taking place on individual areas of the surface or active centres.

Based on this assumption, Langmuir [111], as the first, observed that for adsorption from the gas phase on polycrystal surfaces, his equation with suitably chosen constants can be applied to describe adsorption on a certain kind of sites and that the experimental adsorption data can be approximated by the sum of Langmuir equations, each multiplied by the fraction of a given kind of site on the solid surface. In case when a number of areas with different values of energy is very large, energy changes in a continuous way and addition is replaced by integration. Walker and Zettlemoyer [204],McMillan [205], Smith i Pierce [206] made some attempts to extend applicability of the BET equation introducing some corrections including heterogeneity of the adsorbent surface.

The conception of integral adsorption equation was introduced for the first time by Zeldowitsch [207] in 1934, who, based on it, derived the overall adsorption isotherm being the Freundlich equation. However, the first wilful apllication of the integral adsorption equation is associated with Schuchowitzky [208] and Roginsky et al. [203,209-212]. Roginsky assumed, that the differential distribution of the number of adsorption sites over the values of the adsorption energy can be represented by the so-called contiuous energy distribution function. Moreover, he proposed to use the Stieltjes transform method for solving the adsorption integral equation. The next important step in terms of this approach was firstly proposed by Roginsky and co-workers [203] and further by Todes and Banareva [213], who introduced the concept of the condensation approximation (CA). Halsey and Taylor also studied this problem [214]. This concept of CA - very fruitful in further studies of adsorption on heterogeneous surfaces - deals with the assumption that adsorption process proceeds gradually when adsorbate pressure increases on adsorption sites with decreasing adsorption energies. At a given temperature and pressure, the adsorption sizes having the so-called critical adsorption energy are completly covered, but others are totally empty. Thus the critical adsorption energy corresponds to some critical temperature $T_c$ and critical preassure $p_c$ .

By means of the integral adsorption equation introduced by Roginsky one may obtain essentially what follows: (1) analytical forms of the overall isotherm for various analytical forms of the adsorption energy distribution function assumed a priori, (2) analytical forms of the energy distribution function for the overall adsorption isotherms assumed a priori, and (3) numerical values of the adsorption energy distribution for an overall experimental adsorption data. Consequently, the integral adsorption equation gives possibility to determine the mutual dependence between the shape of the overall adsorption isotherm and the energy distribution function. This distribution gives a quantitative characterization of global adsorbent heterogeneity and provides no information with regard to distribution topography of the adsorption sites. At the end of 1940's and at the beginning of 1950's the works on physical adsorption of gases on energetically heterogeneous solid surfaces became an object of interest of many American scientists [214-220]. Among others Sips [215], using the Stieltjes transformat method, derived overall adsorption isotherm in the form of Langmuir-Freundlich equation and the so-called generalized Freundlich equation. In the same way the Temkin [199] and Tóth [221] empirical equations were justified.

Initially the first two possibilities of integral adsorption equation were intensively expolored by investigators who used almost exclusively the Langmuir adsorption isotherm [223-231]. Then, the type of the topography of adsorption sites is of free choice. The mathematical forms of the overall adsorption isotherms depend only on the shape of the energy distribution functions, which characterize the global heterogeneity of the adsorbent surface. In terms of this approach, the Freundlich and Dubinin-Radushkevich equations have been derived as the overall adsorption isotherms for adsorption on heterogeneous solid surfaces. As mentioned earlier, the DR equation is a fundamental equation for adsorption from the gas phase on microporous solids. However, Hobson [232] realized its great applicability for description of gas adsorption on real flat surfaces obeying non-porous, porous as well as mesoporous materials. Hobson [233] was also the first who used the condensation approximation method to find the analytical form of energy distribution related to the DR equation. The most advanced studies, giving theoretical foundations for the wide-spread DR equation applicability, were made by Cerofolini [234,235]. Cerofolini [237-237] launched a well-documented hypothesis that the so-called DR behaviour of numerous experimental systems is related to the rules governing the formation of the real surfaces. The same conclusions were true of the Freundlich equation. The papers of Hobson [232,233] and Cerofolini [234,235] as well as exhaustive experimental investigations (see [54] and references therein) were the reasons for reexamination of the DR equation for description of gas adsorption on microporous solids. This reexamination was the basis for developing new methods for determining the distribution of micropore sizes [238-241].

The further development of extensive researches in the seventies and the eighties deals with local isotherm equations that took into account the interactions between the adsorbed molecules [242]. In such a case, the model of heterogeneous solids that assume a definite distribution of adsorption sites on the surface must be exactly defined.

Two main models of heterogeneous surface are frequently used in adsorption. The patchwise model was suggested by Langmuir [111] and popularized by Ross and Olivier [242]. In this model the adsorption sites of equal adsorption energies are assumed to be grouped together into patches; these patches are so large that the interactions between two molecules adsorbed into different patches can be neglected. The random model of adsorbent surfaces, in which the adsorption sites of equal adsorption energies are assumed to be distributed fully at random over a heterogeneous surface was introduced by Hill [217]. This last model seems to be more realistic for majority of real adsorbents, especially towards the adsorption from the gas phase [20]. Jaroniec and co-workers considered also the models with medial and regular distributions of adsorption sites on the surface [243,244]. In recent years, the new kinds of heterogeneous surfaces have emerged that cannot be classified in either of the basic models (patchwise or random) discussed above. These models, including also fractal surfaces were discussed by Cerofolini and Re [245]. Later, an alternative concept of adsorbent heterogeneity was proposed for adsorption on the microporous solids with micropores of different dimensions and shapes [246,247]. Such micropores are a source of the structural heterogeneity, which may be described by the micropore-size distribution function, but surfaces of meso- and macropores generate the surface heterogeneity. The total adsorbent heterogeneity is a simple sum of the structural and surface heterogeneities and is usually characterized by the adsorption energy distribution function. This distribution function may be evaluated on the basis of the overall experimental adsorption isotherm. However, homogeneous microporus adsorbents have micropores of identical shapes and dimensions. Consequently, a homogeneous microporous adsorbent is energetically heterogeneous. Thus, the concept of energetic heterogeneity is more universal and may be applied to describe the heterogeneity effects in adsorption on non-porous and porous solids [245].

It was mentioned above that if the local adsorption isotherm is described by an equation assuming lateral interactions then the simultaneous consideration of the effects arising from these interactions and the effects of the energetic heterogeneity of adsorption sites must be taken into account. It means that a definite model of heterogeneous solids is of great importance. At the end of the last section, the isotherm equations describing non-localized and localized monolayer adsorption with lateral interactions on homogeneous surfaces were considered briefly. These equations such as the Hill - de Boer [144,168] and Fowler - Guggenheim [169] ones were used as the local isotherms in the integral equation of adsorption. This equation was solved analytically - in exact or approximate ways - or numerically for various types of energy distribution functions. The Kiselev equation describing the localized monolayer with lateral interactions [248] was used also as the local isotherm. The results obtained by means of the above equations used as the local isotherms are summarized by Jaroniec and Madey [249].

Another important progress in physical adsorption deals with description of multilayer adsorption of gases on heterogeneous surfaces. In this case most considerations are based on the following assumptions: (1) energetic heterogeneity of the adsorbent surface is restricted to the first layer only, and (2) lateral interactions in each adlayer are neglected. Assumption (2) means that the kind of topography of adsorption sites is of no significance. As the local equation there were used both the classical BET equation [250], n-layer BET equation [250,251], and the so-called Jovanovic isotherm [252,253]. A new set of papers on multilayer adsorption on heterogeneous surfaces was initiated by Cerofolini [245,255]. In contrast to numerous analytical solutions of the integral equation for multilayer adsorption, the numerical studies were scanty. Typical papers on this subject were published by Jaroniec and Rudziński [250] and Hsu et al. [256]. Both the n-layer and infinite layer models of the surface phase were taken into account. The results published in the above papers were successful compared with the experimental data.

Up to this point we reported the papers which retained the Langmuir's assumption that each adsorbed molecule occupies one site on the surface. While this assumption may be justified for small molecules, it is unlikely to be realistic for large ones, and often more industrally important molecules such as the n-alkanes. It ought to be stressed, however, that for homogeneous solid surfaces the Langmuir isotherm became a starting point for derivation of the equation, which was reported independently by Williams in 1918 [257,258] and Henry in 1922

[259]. According to the assumption underlying the Williams-Henry equation, the adsorbed molecule occupies more than one active site.

Contemporary theory on the multi-site occupancy adsorption on heterogeneous solid surfaces is rather scanty. The reason is that such description involves a new degree of complexity and only a few papers have been published on this subject [260,261]. Rudziński and co-workers [262-265] accepted the Everett's equation for the multisite occupancy adsorption from liquid mixtures [266] on solid surfaces as a starting expression for description of adsorption on heterogeneous surfaces with random and patchwise topography. The short review on the subject is presented in reference [267].

In 1970s and 1980s several numerical methods were proposed in order to find the distribution energy functions of adsorption on the basis of tabulated data of experimental adsorption isotherm. From a mathematical point of view the integral adsorption equation is the Fredholm integral equation of the first kind. The particular nature of this equation poses severe difficulties to its solution and strict limits to the range of numerical methods that can be used in such a task.

Several numerical algorithms have been developed in order to solve the Fredholm integral equation and many of them have been applied for determining the adsorption energy distribution function from the experimental adsorption data. The following list includes the most popular and useful numerical methods:

1. discretization methods, in which the solution of the integral equation is reduced to the inversion of a square matrix [268,269],
2. optimization methods based on the choice of an analytical form for the distribution function containing some parameters which are subsequently determined by a best fit of the calculated overall isotherm to the experimental data [270-272],
3. regularization methods, in which the ill-posed problem is replaced by a stable minimization problem in various ways [273-276],
4. iterative methods, which use various iterative algorithms in order to improve an initial guess for the distribution function [277-279],
5. expansion methods, which are based on the expansion of the all functions appearing in integral equations in series of a complete orthonormal set [280-282], and
6. integral transform methods, which are based on analytical methods in which the experimental isotherm is approximated by a suitable interpolating expression and the analytical inversion formula is expressed in an easily computable form.

The aforementioned methods should be considered as an important tool for characterizing most adsorbents of great practical importance. The readers interested in mathematical features, applicability and limits of the above numerical methods are recommended to consult the reviews [283-285].

As follows from the hitherto existing considerations, the adsorption integral equation can be solved considering the distribution function in an analytical and numerical way. Approximate methods are another set of methods used for solution of this equation. These methods are based on the above mentioned condensation approximation (CA) approach which was first proposed by Roginsky and co-workers [203,213]. Later this formalism was extended by Harris [286] and Cerofolini [234,287,288]. The CA method is based on the replacement of the true local isotherm by a step function which has two values: 0 below the critical pressure $p_c$ and the value 1 for $\geq p_c$ . The critical pressure $p_c$ is usually interpreted as a condensation pressure and its value depends on the critical adsorption energy belonging to given adsorption sites. This method leads to simple analytical expressions for the distribution functions in terms of the derivative of the global isotherm. The CA method was next improved by Hobson [289] and Cerofolini [287] who proposed the so-called asymptotically correct approximation (ACA). It is based on the replacement of the true local isotherm by an approximate kernel of the integral equation which shows the correct asymptotic behaviours both at low and high pressures. In the ACA method the distribution function is evaluated in terms of the second order derivatives of the global isotherm. The comparison between the CA and ACA methods was presented in the paper [237]. Another approximation method, originating from the CA approach has been proposed by Rudziński and co-workers [290-292]. This so-called third-order approximation was developed without assuming a local isotherm approximation. Rudziński found that his method gives the results remarkably better than those obtained by the CA and ACA approaches. The other approximation method for solving the adsorption integral equation with regard to the energy distribution function was proposed by Nederlof and co-workers [293]. The main advantage of the above approximated methods deals with [245]: (1) the stability of the obtained solutions, (2) the simplicity of the expressions for the distribution function and (3) the local character of the solutions i.e., the fact that the calculation of the distribution function in one point does not need the knowledge of the whole isotherm.

Another method for estimating surface heterogeneity is based on the calorimetric measurements of effects of adsorption which are more sensitive to the nature of a particular adsorption system than adsorption isotherms [292,294-298]. This method is promising for characterization of adsorbent heterogeneity, but it needs accurate calorimetric data for a given adsorption system [299].

Recently, thermal-desorption [300,301] and adsorption kinetics measurements on heterogeneous surfaces [302-304] have been proposed for evaluating the energy distribution functions. These methods have been tested extensively during the last few years.

In 1974 W.Steele published his famous book "The Interaction of Gases with Solid Surfaces" [46] which should be considered as important quidelines for scientists investigating the nature of solid/gas interactions in terms of the so-called virial formalism. This trend in the theoretical studies of physical adsorption is very fruitful and numerous papers have been published on this subject. Virial formalism originated from statistical mechanics and statistical thermodynamics [168,169,305,306].

In 1974 Bergman [307] and Morrison and Ross [308] improved the Hill - de Boer model of mobile surface phase formed on homogeneous surfaces in terms of virial description formalism adopted for the two-dimensional systems. Based on some simplifications they were able to assess the numerical values of the second, third and fourth two-dimensional virial coefficients which give qualitative information on the interactions between adsorbed molecules and adsorption sites as well as on the interactions among suitable numbers of admolecules. Next, the above homogeneous two-dimensional mobile adsorption approach has been extended to the case of the three-dimensional systems, i.e., to the case of the three-dimensional model of mobile adsorption in which the adsorptive molecules are supposed to move in a three-dimensional potential field created by the presence of a solid phase. The numerical values of the second, third and other virial coefficients were obtained from the low-pressure adsorption or the solid gas chromatographic measurements, neglecting the gas imperfections [309,310]. The necessary condition for the application of the virial description formalism is that sufficiently low-pressure can be achieved in an adsorption experiment [46]. Thus, the virial formalism turned out to be a valuable method to investigate the effect of surface heterogeneity at low surface coverages [311].

The virial expansions in activity or density have the same form regardless if the adsorbent surface is homogeneous or heterogeneous. On the other hand, the values of the virial coefficients are changed. In terms of suitably selected virial formalism it is possible to obtain quite new insight into the adsorption energy distribution [312] and the topography of the solid surface [313,314]. The second and higher virial coefficients are defined by the integrals which contain the distribution function. That is why the main problem in virial formalism on heterogeneous surfaces is to omit the numerical difficulties in evaluating the virial coefficients [315-317]. However, the main information on the surface topography is included in the higher solid/gas virial coefficients and a few approaches have been proposed in literature for their evaluation [315,316,318,319]. Rudziński and Jagiełło [320] presented a careful virial description of adsorption on strongly heterogeneus surfaces in the Henry's low region. This work proved that the behaviour of gas at low pressures is determined not only by maximum adsorption energies, but also the spread energies. A promising way refers to the application of gas/solid chromatography data to study low-pressure adsorption [321,322] in order to determine the virial coefficients. Some papers were published presenting the possibility of evaluating energy distribution functions from the gas chromatographic measurements [323,324].

The above brief discussion shows that the virial description of adsorption on heterogeneous surfaces is very promising, but further studies in this field are expected to lead to even better results, providing a more accurate description of the true mechanism of adsorption from the gaseous phase [53].

In the seventies and eighties there were developed also the studies of adsorption from liquid solutions [14,19,20] and gas mixtures [325] on the solid heterogeneous surfaces. Then adsorption both from diluted solutions [325] and from the solutions in the full range of their concentrations [21] was studied. Such studies are of important practical application in the industrial separation of gas and liquid mixtures as well as in the environmental protection processes. Moreover, adsorption from multicomponent solutions is a basis for the theory of adsorption liquid chromatography [327].

The simplest description refers to the single-solute adsorption [328-331]. Assuming great dilution of the solution, the interactions between the molecules of the dissolved substance and the solvent can be neglected and the process can be described as in the case of single -gas adsorption. The description of multi-solute adsorption is more complex [332]. Some important expressions were developed and used widely for predicting the multi-solute adsorption equilibria by means of single-solute adsorption parameters [333,334].

Considering adsorption from undiluted solutions and gaseous mixtures under high pressure, the competition among adsorbing molecules towards adsorption centres should be taken into account. In this case, the difference of adsorption potentials of adsorbate molecules plays a significant role and experimental isotherms have a character of excess isotherms [19]. A competetive adsorption process makes the theoretical description complicated and physical interpretation of adsorption energy distribuction function difficult. Moreover, for obvious reasons in the system in which the adsorption contains more than one kind of molecules, there increases a number of parameters which must be taken into account in the theoretical description of the phenomena. In case of the relatively simple adsorption from the two-component solutions in the full range of their concentrations, the process is determined by the following parameters at constant temperaure [335]: (1) molecular interactions in the bulk phase, (2) molecular interactions in the surface phase, (3) interactions between the bulk and surface phases, and (4) molecular interactions at the solid/solution interface. In the case of adsorption from the multi-component solutions and adsorption from gas mixtures the description is largely complicated [336,337]. It should be emphasized that the effects resulting from the surface heterogeneity are sometimes compared quantitatively with those caused by the intermolecular interactions of the adsorptive [335]. The character and value of interactions mentioned in point (4) depend on structural and chemical heterogeneity of the solid surface, which is characteristic for a great number of adsorbents used in practice.

In the majority of papers devoted to liquid adsorption on solids, the adsorbent surface was explicitly assumed to be homogeneous. Some scientists such as Everett [266] or Zettlemoyer and Micale [338], while not attacking the essence of the problem, suggested that the surface heterogeneity of the solid might cause a change in the sign of the adsorption excess as well as solution non-ideality in the surface phase. Therefore, energetic heterogeneity should be regarded as one of the main physical factors governing the adsorption process at the solid/solution interface. According to Zettlemoyer and Micale [338]: "The major problem in the development of the thermodynamic theory of adsorption at the solid/liquid interface is concerned with the definition of the usually uncertain heterogeneous solid surface in terms of a mathematical model", and „Although extensive experimental work on solution adsorption has been in progress for over 50 years, there is no satisfactory physical model available which allows for a rigorous thermodynamic treatment of the process. The main reason for the lack of theoretical development in the field is the complexity of the process due to the heterogeneity of most solid surfaces and the lack of knowledge of the various interactions both between two components and adsorbent". In 1973 Coltharp and Hackerman [339,340] showed in an experimental way that adsorbent heterogeneity might determine the type and range of competitive adsorption from solutions of solids. In the same year, Rudziński and co-workers [341] published the first paper in the world literature in which the heterogeneity effects were described in a quantitative way for benzene adsorption from cyclohexane on silica gel. Their description was based on the patchwise model of heterogeneous solid surfaces. This paper was the basis for further investigations dealing with the determination of the distribution function from the liquid adsorption data. Numerical methods of determination of this distribution function were next elaborated by several authors [342-344]. Extensive studies of heterogeneity effects in adsorption from solutions on solids are still being developed [345-347] and they include investigations of the influence of different types of distribution functions on the shape of the excess adsorption isotherms [348,349] and the study of correlation between adsorption from gaseous and liquid phases [350,351]. Some authors considered the role of adsorbent heterogeneity and molecular interactions in the adsorption process at the solid/liquid interface [335,352,353].

Up till now, despite the above mentioned difficulties connected with the complex system of liquid adsorption on heterogeneous surfaces of solids, satisafctory results were obtained in regard to:

1. single and multisolute adsorption on heterogeneous solid surfaces [325,326] ,
2. adsorption from two- and three component solutions of non-electrolytes on homogeneous and heterogeneous solid surfaces [335,354-356],
3. methods for determination of capacity of surface phases formed on heterogeneous solid surfaces [357,358],
4. possibilities of prediction of adsorption from liquid and gas mixtures based on the measurements of individual isotherms of the components from the gaseous phase [350,351,359,360].

In 1982 Everett wrote as follows [67]: "But most surfaces of practical importance are not homogeneous and it is important to assess the influence which heterogeneity of the surface has on adsorption phenomena. It turns out that this problem is not easily resolved, for there is no unambigous way of separating out the influence of non-ideal behaviour in the surface region (arising from intermolecular forces in the surface region) from that of surface heterogeneity". The above problem was solved by Dąbrowski and co-workers [361], who proposed the concept of the global surface activity coefficients to separate the non-ideality of the surface phase caused by intermolecular interactions from that generated by surface heterogeneity of the solid.

From the technological point of view, besides adsorption from liquid mixtures, that from gas mixtures is also important. In spite of the impressive progress that has been made in improving new techniques for the measurement of adsorption from gas mixtures, they are still time consuming experiments. On the other hand, the measurements of single gas isotherms can be made easily and quickly by means of various adsorption techniques. Thus, predicting mixed-gas adsorption equilibria from knowledge of the single adsorption isotherms is of the great, most practical importance and is the main goal of the theories of adsorption from gas mixtures on solid surfaces. In many important technological processes it is essential to know the mixed-gas adsorption equilibria which correspond to a variety of different technological regimes. Practically it is impossible to get the necessary knowledge of mixed-gas adsorption equilibria from mixed-gas adsorption measurements. It results from the fact that in the above mentioned technological processes the physical regimes may change very quickly. So, it is important to use a fast method for calculating the mixed gas adsorption equilibria. Thus, it is not suprising that in the industrial separation of gases by adsorption processes very simple isotherm equations have been used in the systems of differential equations describing the dynamic separation of gases. However, the isotherm equations which has been used are so crude that they give rise to the following question: would it be possible in the present state of theory, to propose equations that are more accurate, yet simple enough to calculate fast mixed-gas adsorption equilibria? To make the calculations accurate, the theoretical considerations, have to take into account all the most important physical factors which govern the behaviour of adsorption systems. At present there are the following two factors: (1) the gas-solid interactions and their dispersion at various points on the real gas-solid interfaces called surface energetic heterogeneity, (2) the interactions between the adsorbent molecules. The most advanced theories should introduce both the above mentioned factors. Theoretical approach of adsorption of mixed-gas on heterogeneous solid surfaces is based either on the integral adsorption equation [362] or on the ideal adsorbed solution (IAS) theory proposed by Myers and Prausnitz [363]. For the same model of single-gas adsorption various expressions for mixed-gas adsorption were developed in terms of various theoretical approaches [364]. The excellent review on the subject under consideration has been presented by Rudziński and co-workers [359,360].

In the eighties a new trend of studies on adsorption started from the work by Mandelbrot "The Fractal Geometry of Nature" [365]. This work gave a new possibility for a quantitative characterization of a geometric irregularity at the surface. In the classical meaning the surface was assumed to be a two-dimensional matter whose geometrical irregularities present small deviations from the ideal arrangement. But when these geometrical irregularities have spatial extensions which are comparable with the size of the adsorbate molecules, we can speak of fractal surfaces [365,367]. This situation is adequate for several amorphous solid adsorbents and leads to numerous consequences of their adsorbing behaviour. The strictest consequence is the loss of meaning of the concept of surface area which is replaced by the definition of fractal surface itself. For a series of adsorbate molecules of different sizes, but with geometrically similar cross-sections, the apparent surface area of several solids (silica gels, active carbons, activated charcoals, crushed glasses, etc.) is connected with the effective cross-sectional area of the adsorbate molecule, $a_m$ , by the relation: $A \propto \rho^{2-D}$ , where $\rho = a_m^{1/2}$ . Here D is a non-integer number between 2 and 3, known as the fractal dimension. Its value varies from D = 2 for an ideal planar surface to D = 3 for a hypothetical porous solid with neglegibly thin walls. Many methods have been developed in order to obtain D on the basis of mercury porosimetry, scanning electron microscopy and small angle X- ray diffraction as well as neutron-scattering measurements [366-368]. On the other hand, several methods have been proposed to understand the influence of fractal geometry of the solid adsorbents on the adsorption isotherms [369-375]. Some of them, are simple and convenient since they require only one complete adsorption isotherm far a given solid to calculate the value of D [376,377]. The arguments were put forward against the assesement of the fractal dimension from the experimental adsorption data [378]. Later their relevance was questioned [379].

It follows from the above considerations that the fractal analysis perfomed in terms of adsorption measurements, both from gaseous and liquid phases, is probably a next important tool for characterizing the surface irregularities. However, this way of researches is new and still being developed. It is too early to summarize its achievements now. Undoubtedly, the fractal analysis as applied to porous materials, can provide important additional information for the characterization of adsorbent surfaces.

In the industrial usage of adsorbents, the time dependence of adsorption on solid surfaces, is named as adsorption kinetics. The more general term "adsorption dynamics" deals with the time evolution adsorption processes. With the development of the theory of equilibria of adsorption on heterogeneous solid surfaces, the theory of adsorption/desorption kinetics on the heterogeneous surfaces was also developed. Adsorption kinetics is determined by the following stages: (1) diffusion of molecules from the bulk phase towards the interface space; it is the so-called external diffusion, (2) diffusion of molecules inside the pores; it is the internal diffusion, (3) diffusion of molecules in the surface phase named as the surface diffusion, and (4) adsorption/desorption elementary processes. The total rate of the kinetic process is determined by the rate of the slowest process. The diffusion processes in the porous solids can be complicated by molecular sieve and activated diffusion effects. In the former case, the adsorption into the smaller pores is excluded, because the adsorbate molecules is too large to enter these smaller volume elements. The activated diffusion takes place when the dimensions of the adsorbate molecules are only slightly smaller than the pore diameter. Consequently, at low temperatures the adsorbate molecules have insufficient kinetic energy to enter the pores. The activated diffusion is confirmed when by measuring the isotherm at higher temperature, the equilibration time is reduced. For non-porous and macroporous solids the internal diffusion may be neglected. In this case, the adsorption kinetics is determined by external diffusion and molecular adsorption/desorption processes. The Langmuirian kinetics, based on the ideal monolayer adsorbed model, proved to be deceptive for most real adsorption systems, which include structurally (high porous) and energetically heterogeneous solids. On the other hand, the adsorption/desorption kinetics theories are technologically extremely important, because the diffusion of adsorbed particles on solid surfaces is the phenomenon of great importance in catalysis, metallurgy, microelectronics, material science and other numerous scientific and technological applications. As for the matter of catalysis, it should be pointed out, that since the adsorption of reactants and desorption of products are fundamental steps of heterogeneous catalysis, there is need to understand the kinetics of adsorption/desorption phenomena on heterogeneous surfaces.

In 1970s and 1980s in the majority of papers devoted to the adsorption kinetics, the models of mass transfer kinetics were examined [6,47,380]. These models took into account external, internal and surface diffusion of adsorbate molecules. However, in terms of mass-transfer kinetics, some authors investigated the influence of the energetic [381,382] and structural heterogeneity [383,384] of solid adsorbents. Jaroniec [385] discussed the adsorption and desorption rates for kinetics of localized adsorption of single gases and their mixtures on energetically heterogeneous solid surfaces, assuming discrete, continuous distribution of energies among adsorption sites. In terms of simple considerations he derived both the kinetic and various equilibrium adsorption isotherms. Czarniecki and Jaroniec proposed the stochastic modelling of adsorption kinetics of localized and mobile adsorption on homogeneous and heterogeneous surfaces [386-388]. Their approach based on the Monte-Carlo technique, allowed to take into account the following aspects [389]: (1) the monolayer and multilayer character of surface phase, (2) the localized and mobile character of the surface phase, (3) the association of molecules in the surface phase, (4) energetic heterogeneity of the adsorbent surface, (5) the kind of topography of adsorption sites, and (6) surface diffusion in the adsorbed phase. The above investigations were carried out for adsorption gas, gas mixtures and liquid mixtures.

The other way of investigations of adsorption/desorption kinetics on solids is connected with the famous and commonly used empirical expression known, in literature as the Elovich equation [390,391]. The excellent work on application of this formula for describing single-component adsorption/desorption kinetics on heterogeneous surfaces was presented recently by Cerofolini [392].

Quite new perspectives for investigations of kinetics and surface diffusion processes on flat (energetically heterogeneous) and porous (structurally heterogeneous) solid surfaces are connected with the molecular modelling of adsorption phenomena. Moreover, the invention of Atomic Force Microscopy

(AFM) and Scanning Tunneling Microscopy (STM) gave the excellent source of information about the mechanistic models which can be accepted in computer simulation. The comprehensive overview of the state of the art of researches on this subject was presented in the monograph: "Equilibria and Dynamics of Gas Adsorption on Heterogeneous Solid Surfaces" [393]. The relevant states of the art articles were recently published by Ruthven [394], Nicholson [395] and Gubbins [396].

During last two decades the methods of molecular modelling of adsorption phenomena have developed extensively. The term "molecular modelling" refers to both approximate methods of statistical thermodynamics and the molecular compter simulations methods. These methods have emerged as an efficient tool for improving the fundamental understanding of basic microscopic phenomena and for helping to solve industrial relevant problems involving wide fields of chemical and physico-chemical processes. Starting from the atomic scale, molecular modelling is the appropriate tool to develop a qualitative and quantitative knowledge of structure - properties relationships in a wide range of systems like molecules, bulk phases (gas and fluid) and interfaces [53,397-403]. One of the great advantages of the molecular modelling of surface phenomena includes recognizing and retrieving useful or even predictive information about the system investigated. When regarding the complex molecular systems like porous and heterogeneous solids being in contact with fluids, the experimental studies have been preceded by the theoretical ones [404,405]. Computer modelling is also a very promising method for fast development of industrial separation processes [406,407]. Gubbins showed [406] that the molecular modelling methods are useful for determining the effects of pore size and shape, temperature and other variables on the selectivity adsorption and its maximum effects. As programming technologies improve and computational power increases, the methods under consideration are being increasingly used as an addendum to experiment.

The classical models of adsorption processes like Langmuir, BET, DR or Kelvin treatments and their numerous variations and extensions, contain several uncontrolled approximations. However, the classical theories are convenient and their usage is very widespread. On the other hand, the aforementioned classical theories do not start from a well - defined molecular model, and the result is that the link between the molecular behaviour and the macroscopic properties of the systems studied are blurred. The more developed and notable descriptions of the condensed systems include lattice models [408] which are solved by means of the mean - field or other non-classical techniques [409]. The virial formalism of low - pressure adsorption discussed above, integral equation method and perturbation theory are also useful approaches. However, the state of the art technique is the density functional theory (DFT) introduced by Evans [410] and Tarazona [411]. The DFT method enables calculating the equilibrium density profile, $\rho(\mathbf{r})$, of the fluid which is in contact with the solid phase. The main idea of the DFT approach is that the free energy of inhomogeneous fluid which is a function of $\rho(\mathbf{r})$, can be expressed as a sum of a hard sphere short range contribution and a longer range attractive force contribution. The equilibrium density profile ρ ( r) is obtained by minimization of the free energy. Once ρ (r)is known, it is possible to obtain all macroscopic information - like adsorption isotherm, density and relevant thermodynamic functions - about the adsorption systems studied. DFT can be considered as a useful method for investigating the systems of simple geometry as well as pure and mixed fluids composed of spherical molecules [412]. For these systems it is possible to explore a large number of system variables relatively quickly, such as pore sizes and pore shapes, solid/fluid interaction parameters, temperatures, pressures, etc. The most advanced form of this theory is the so-called non-local density functional theory (NDFT) [413]. NDFT has been used for describing fluids confined in narrow pores [414,415] and for analysing pore-size distribution of microporous carbons [416]. The main disadvantage of the density functional theory is that this approach is difficult to apply to the complex systems including fluids consisting of non-spherical molecules, pores of mixed geometry or heterogeneous surfaces. In this case, the other way of molecular modelling, namely computer simulation methods are much more adequate [412].

One of the main reasons for using computer simulations is the fact that they eliminate inaccuracies resulting from the approximate statistical thermodynamic methods. In terms of computer simulations methods, it is possible to investigate the systems not subjected to the analytical description. As the methods under investigations can be used to study the complex systems, they provide the standard data for verification of approximate theories. Moreover, they make it possible to compare the molecular models with the experimental data as well as to form the correctness criteria for their choice. Thus, the molecular simulation methods are useful for testing the exactness of assumed intermolecular potentials, to validate approximate statistical thermodynamical theories and to explore systems under conditions and with a level of detail which is hardly to achieve by conventional experiment.

As an example there can serve the internal diffusion phenomenon which cannot be “observed” in the direct laboratory experiment but can be observed and assessed quantitatively using the computer simulation methods. In the last few years the revolutionary progress in the field is connected with developing new algorithms and with striking increases in computing speed.

The starting point of molecular simulation methods is - as in the density functional theory - the well-defined microscopic description of the system studied. This macroscopic (molecular) specification includes: (1) the equations of statistical thermodynamics describing the fluid/fluid and solid/fluid interactions, and (2) the molecular model of solid adsorbent. This model should take into account all possible and reliable information on the solids, most of which can be developed from various modern surface science techniques [417]. For instance, some important data on the bulk crystalline structures are given by the X-ray diffraction or neutron diffraction, but the scanning tunelling microscopy is a valuable source of information on the topography of a surface solid. For solving the equations of statistical mechanics corresponding to the molecular model of the system, two methods are usually used: Monte Carlo (MC) and Molecular Dynamics (MD). The MC method was developed by Metropolis in 1957 [418] but the first molecular simulations were performed by Alder and Wainwright [419] in the Livermore Laboratory in California [420]. The main idea of the MC approach refers to estimating the configurational contributions to the thermodynamic quantities of a given statistical mechanical system, which are calculated by averaging over all accepted configuraions. The MC method allows the generation of different types of statistical ensembles according to the equilibrium properties of the system which are searched [396]. For instance, isosteric heats of adsorption are developed from the canonical ensemble, but adsorption isotherms and fluid phase equilibria from the grand canonical and Gibbs ensembles, respectively. As follows from the above considerations, the MC method gives configurational equilibrium properties of the system investigated. In the MD method, the molecules are dislocated owing to their own intermolecular forces. The trajectories of the molecules are calculated by direct solution of Newton's equations of motion, but desired properties of a system are assesed by averaging over the trajectories. Thus, contrary to the MC approach, the MD method allows to obtain both equilibrium and dynamic properties of the system. Therefore, MD simulations are very useful for studying diffusion effects and phase separation phenomena. The detail discussion of advantages and disadvantages of MC and MD methods for solving specifing problems of the surface science has been recently presented by Gubbins [412]. He also showed the relevant point of view on applications of the molecular modelling methods toward selective adsorption of mixtures of great industrial importance.

To give a short description of numerous applications in which the molecular modelling of surface phenomena is a significant feature, would require a book in itself. Some examples were included in the section dealing with the heterogeneous catalysis, but others have been presented above. Let us attempt to summarize the considerations on the subject as follows:

- simulation molecular methods are of great importance for the computational studies on the design of new synthetic sorbents and catalysts like zeolites [193,421], aluminophosphates [192,422,423], aluminosilicates [424], nano- and mesotubes [425,426], fullerenes and heterofullerenes [427-429], pillared-clays [430,431] and other disordered porous solids [432-435]. The above solid materials are very important in selective adsorption, catalysis and separation technology;
- in terms of molecular modelling of adsorption phenomena the following problems were examined: adsorption simple fluids and (or) their mixtures on (and in) heterogeneous solid phases like zeolites [436], porous carbons [394,395,437], carbon nanotubes [192,438] and other porous materials [439]. Quite recently, the results were reported on the molecular simulations of adsorption phenomena on new types of heterogeneous surfaces including porous structures [440,441]. Adsorption of fluids on the amorphous solid

surfaces was also investigated [402,442]. An important part of the aforementioned studies deals with the dynamic aspects like kinetics or diffusion and phase transitions;

- simulation methods are very efficient for studying the effects of surface heterogeneity on equilibrium and kinetics of fluid adsorption, thermal desorption, surface diffusion and surface reactions [442,443] (see also Ref. [393] and references therein).

To mantain a compact character of this chapter, we restricted ourselves only to the selected problems in the molecular modelling of adsorption phenomena, but it should be stressed that the topic is vast.

Finally, we should point out some difficulties connected with the practical aspects of molecular modelling approach. The main difficulty refers to a lack of definite knowledge of the structure of solid adsorbents and inadequate or improper experimental measurements which can be used to refine the models and to compare the calculations. In the case of crystalline materials, like zeolites or aluminophosphates or aluminosilicates, their structure can be exactly determined from the X-ray or neutron diffraction. However, the morphology of amorphous materials, such as activated carbons, oxides or silica gels is still poorly recognized. In this case, the simple model structures in which the pores are represented as slits or cylinders are usually applied. The second main disadvantage of molecular simulation is that our knowledge about intermolecular forces is still inadequate [53]. In the article on atomic calculations of Henry's constant in A zeolite which was published in 1975, Derrah and Ruthven wrote: [445] ..." the practical limitations of this approach arise not from the difficulty of performing such calculations but rather from the present unsatisfactory state of knowledge of intermolecular forces". Unfortunately, after 23 years the subject still needs better understanding. Evidently, the advances have been made in this field as well as in the simulation methods and at present more complex systems can be simulated. It is clear that quantum chemistry ab initio methods should be developed to increase the accuracy of calculations of the solid/fluid interaction potentials for all kinds of adsorption systems. The third disadvantage of molecular simulations is the fact that long computing time may be needed for complex systems. For this reason, the dynamical processes that are slow, e.g. desorption from strongly binding sites, activated diffusion effects or phase separations at high density in the associated pore structures, are difficult to study. This problem will be solved by further improvements in computing power. And the final remark: the principal difficulty in the field of development of molecular modelling method is connected with the poor contact between modelling and experiment [412]. The more complex and realistic models arising from the molecular modelling should be tested by means of more accurate experimental data interpreted in terms of well-posed theories. Thus there is a need for close cooperation between theoretical and experimental groups, in which the experiments and models are designated to complement each other.

It follows from the above considerations that at present and in the near future the development of adsorption will be stimulated by:

- development and application of widely understood modern methods to study adsorption experiment; these methods should be applied for precise measurements of both adsorption isotherms and heats of adsorption; to this end the structural techniques such as STM, AFM, XRD, FTIR, various kinds of NMR, etc. should be also used;
- development of molecular modelling methods of adsorption which depends on progress in software, hardware and in increasing the accuracy of quantum chemistry ab initio approaches for calculations of the solid/fluid interaction potentials for all adsorption systems;
- one can suppose that for the complex systems that is: fluid/solid including amorphous solids with complex porous structures and complex fluids or their mixtures as H-bonding or associated mixtures, mixtures composed of non-spherical species, simple and realistic phenomenological models of adsorption systems will be still of great practical importance; such models cannot lead to too complicated equations because their practical applicability is doubtful;
- the other area for future work is making careful comparison of the model results with experiment, particularly for the microscopic details of adsorption.

## 5. PRACTICAL APPLICATION OF ADSORPTION

Adsorption has a very long history. Its first practical applications were noted in ancient times. Hipocrates recommended dusting wounds with powdered charcoal in order to remove their unpleasant odour. However, the rational use of adsorption for industrial purposes started at the end of the 18th century. The Swedish chemist Carl Wilhelm Scheele, an apothecary by profession, was the first to discover the phenomenon of adsorption of gases on charcoal in 1773. A dozen years later the Russian academician Lowitz found that charcoal when immersed in the taric acid solution, decolorizes it by adsorbing the organic contaminants present. This discovery led to the first industrial application of charcoal in the sugar industry in England in 1794, where it was used as a decolorizating agent for sugar syrup. This event initiated the research on adsorption from the liquid phase. The discovery of adsorption process selectivity by the Russian scientist Tswett in 1903 originated a new analytical technique which is adsorption chromatography. Tswett recommended this process for separation of various mixtures [106].

In the year 1901 R.V. Von Ostrejko [446] set the basis for the commercial development of activated carbons through the processes involving the incorporation of metallic chlorides with carbonaceous materials before carbonization and the mild oxidation of charred materials with carbon dioxide or steam at increased temperatures. Based on the Ostrejko's patents, in 1909 in Raciborz on the then German territory, the plant, called the Chemische Werke was built to manufacture, for the first time on a commercial scale, the powdered, active carbon called eponit. In 1911 a new kind of active carbon, known as norit and purit obtained from peat by activation with steam was produced in this plant. About the same time as the Raciborz factory was founded, a wood distillation plant was built in Hajnowka (East Poland), initially manufacturing active carbon solely from wood. In 1911 the NORIT factory in Amsterdam was founded, now one of the most advanced international manufacturers of active carbons. The process of chemical activation of sawdust with zinc chloride was carried out for the first time in 1914 in the Austrian plant in Aussing, and also in 1915 in the dye plant of Bayer. The powdered carbons were used at that time mainly for decolorizing solutions in the chemical and food industries.

World War I introduced the problem of protecting man's respiratory tracts from toxic warfare agent introduced intentionally into the air. In April 1915 in France and in May of the same year in the neighbourhood of Warsaw, the German army used warfare gases for the first time against the British and French in the West, and against Russian soldiers in the East. This gave rise to a hasty search for means of protection. Professor N.Zelinsky of Moscov University was the first to suggest the use of active carbon as the adsorption medium in gas masks. Such masks, of course with many modifications are the basis for protecting the respiratory tracts of soldiers throughout the world to the present day. During World War I, coconut shells provided the raw material for production of active carbon. These world war experiences and researches conducted in 1930's led to the development of new technologies for obtaining granulated active carbons of supersorbon and of benzosorbon types. These carbons have found commercial application in the adsorption of gases and vapours. The possibility of purifying municipal gas by removing benzene using active carbon, and other recuperative methods in which this adsorbent was used, has extended to commercial, wide-spread utilization of active carbon.

As follows from the above considerations active carbon was the first widely used adsorbent. In different periods of time development of the adsorption technique was based on various types of adsorbents: before World War I on carbon adsorbents, in the period between World War I and World War II on active carbons, silicic acid gels and aluminium oxides, but after World War II revolutionary progress was made owing to discovery and application of synthetic zeolites. The method of zeolite synthesis was worked out by the English physicst Barrer [447] and the American investigator Breck [448].

Zeolites are the only existing crystalline aluminosilicates with a well defined pore structure in the microporous range. They exhibit unique features: (1) due to the presence of aluminium in the structure, most zeolites have strong acid sites at their surface, making them superior cracking catalysts. A product selectivity in the catalytic reaction is ensured by the microporous matrix of the catalysts; steric phenomena are very important in zeolite catalysis, and a term "shape selective catalysis", is coined to describe these effects; (2) because cations are free to migrate in and out of zeolite structures, these solids are good ion-exchangers. This property can be used to introduce different cations in the structure, creating selective sites for adsorption purposes or catalysis; (3) their narrow pore size and tunable affinity for certain molecules make them ideal adsorbents for selective purification of gas mixtures or to encapsulate hazardous compounds. Zeolites are characterized not only by a large selectivity (selective separation mechanism) but by the ability to separate substances based on the difference in sizes and shapes of separated molecules (steric separation mechanism). Consequently, the species with a molecular diameter which makes them too large to pass through a zeolite pore are effectively sieved. Adsorption processes based on molecular sieving and selectivity are always reversible in theory and usually reversible in practice. This allows the zeolite to be reused many times, cycling between adsorption and desorption. This accounts for the considerable economic value of zeolites in adsorption applications. Modification of the zeolites changes permanently their pore structure. The pore size reduction can be tuned by changing the modification conditions. Modification of zeolites can also be applied for encapsulating gas molecules in zeolite channels. The encapsulates are homogeneous and stable towards acids, mechanical grinding and $\gamma$-irradiation. By controlling the pore size reduction, the thermal stability of zeolites can be controlled. In order to control the pore-opening size without affecting the internal pore systems of the zeolite, modification can be performed using modifying agents with a molecular size larger than that of the zeolite pore so that they cannot enter the pores and interact only with the external surface. The deposition of silicon dioxide at the external surface of the zeolite reduces the size of the pore opening without changing the internal properties of the zeolite. Depending on the degree of the silicon dioxide deposition at the zeolite crystals the adsorption behaviour can be influenced.

Based on the Barrer's and Breck's patents the North-American Linde Company started production of synthetic zeolites on a commercial scale in 1955. At present, besides 40 natural zeolites, there are known over 150 synthetic zeolites.

As follows from the above remarks, development and application of adsorption cannot be considered separately from development of technology of adsorbents applied both on the laboratory and industrial scales. These sorbents can take a broad range of chemical forms and different geometrical surface structures. This is reflected in the range of their applications in industry, or helpfulness in the laboratory practice. It is comparable to the variety of adsorbents one finds in various environmental applications as well. A fundamentally important feature of industrial sorbents is their high porosity and usually high surface. That is why their most important characteristics deal with total pore volume, pore size distribution over the pore diameter and the specific surface area. Other features of practical importance like bulk density, crush strength and erosion resistance have been presented elsewhere [140]. A greater part of solid adsorbents of signifacant industrial applications possesses a complex porous structure which consists of pores of different sizes and shapes. Of the greatest importance are micropores which give a source of a considerable increase of the adsorption capacity because their whole accessible volume can be considered as the adsorption space. Adsorption in micropores is essentially a pore-filling process in which their volume is the controlling factor. In contrast to micropores, for macropores and mesopores the layer-by-layer adsorption mechanism is accepted.

Application of solid adsorbents and catalysts requires their many-sided characterization which comprises determination of their chemical composition, crystallographical and geometrical structure, surface and mechanical properties, and the energy distribution functions as well as the shape and size distribution of the porous materials. The former functions characterize a global energetic heterogeneity of solids. Their physical interpretation is rather complex, but when associated with additional independent measurements as calorimetric, spectroscopic and other ones, gives valuable details about the correlation between the energy distribution of adsorption sites and their chemical nature [243]. Various modern techniques provide direct information about physicochemical properties of these solids [140]. It is worthy to mention electron and scanning microscopy, X-ray diffraction and X-ray spectroscopy, Auger electron spectroscopy (AES), Raman spectroscopy, small angle X-ray spectroscopy (SAXS), NMR, TPD spectra, etc. Although the importance of these techniques is constantly increasing, the adsorption/desorption data are widely used because they provide information on the behaviour of a solid with regard to definite adsorbate. The experimental adsorption/desorption isotherms of nitrogen at 77 K are applied as the standard in order to measure the specific surface area and the distribution of pore sizes which are most important properties of an adsorbent that determine how helpful it is. According to the IUPAC recommendation, the BET equation has been accepted as the conventional method for determining the adsorbent surface area [15]. This area can also be determined in terms of the point B method [15,64]. In both methods adsorption isotherms are measured, assuming the $N_2$ molecular size to be 0.162 $nm^2$. On the other hand, from the adsorption / desorption isotherms of nitrogen at 77 K, the suitable pore distribution can be calculated by means of the Kelvin equation. The specific surface area evaluation requires only a monolayer part of the adsorption isotherm, but the pore size - distribution is measured by means of either the multilayer part of isotherm or in terms of mercury porosimetry [140]. The latter method deals with the penetration of larger pores only and the suitable distribution function does not include micropores. Low concentration adsorption data are very useful for characterization of the microporous structure of the solids and give the excellent basis for evaluating both the micropore-size distribution and the adsorption potential distribution which is a funadmental source of information on the adsorbate - adsorbent interactions.

One of the alternative methods for assessment of the porosity of solids within the mesopore range was proposed at the beginning of 90s (see reference [449] and references therein). The method consists of the measurements of temperature programmed desorption of liquids wetting the porous solid perfectly. Experimentally obtained desorption curves representing the weight loss of the sample against temperature may be converted into the curves volume loss against the pore/core radius using the Kelvin equation. The specific conditions of thermal desorption experiment cause that the problems of the surface film remaining on the walls of pores after their emptying may be neglected. The method was tested for several inorganic sorbents e.g. silica gels, aluminum oxides, solis and activated carbons. On the basis of trasformed desorption curves it is possible to obtain the total pore volume of porous materials and pore size distribution curve (PSD). Pore size distributions, mean pore radii and total pore volumes derived from thermogravimmetric (TG) data are very similar to those calculated from adsorption isotherms measured at static conditions especially from low temperature nitrogen adsorption. Parameters characterizing the porosity of investigated adsorbents are also very similar to the parameters calculated from the mercury intrusion data. The temperature programmed desorption method was also tested for organic porous polymers and nuclear membranes which swell in contact with wetting liquids. Pore size distributions derived from these data differ considerably in comparison to PSD calculated from the adsorption isotherms measured in dry conditions at very low temperature. However, the pore dimensions for nuclear membranes derived from the TG data and small angle scattering data are close to one another. It follows from the data existing in literature that temperature programmed desorption may be a complementary and alternative method for estimation of the porosity of solids in relation to widely used adsorption and mercury intrusion techniques. By means of this approach the pore size distribution, pore volume and specific surface area of adsorbents and catalysts may be obtained.

Carbon and mineral adsorbents are commonly applied in practice. The first one includes active carbons [59,60,167,450], the other - silica gels [451-453], activated alumina [447,448,454-458], oxide- and hydroxides of metals [459-462], zeolites [463,464], clay minerals [188,465] and developed from them the pillared clays [189,190]. In adsorption and catalysis carbon molecular sieves [466], fullerenes and heterofullerenes [427-429] and the so-called nanomaterials, both carbonaceous [425,426] and inorganic ones [467-469] become more and more important. A comprehensive review related to adsorption on new and modified inorganic sorbents was presented in the monograph published recently in this series [470]. It should be emphasized that production of new adsorbents and catalysts and also ion-exchangers is connected with the computational material science, which may be considered as a strategic technology for the 21-st century [173,176,471]. Technology of optical fiber glasses is based on the sol-gel technique which originates directly from the technique of preparation of commonly applied adsorbents and catalysis i.e. silica gels. The same content is true of ceramic membranes [472,473]. Moreover, almost all of the important inorganic oxides can be prepared by the sol-gel process [474,475]. In adsorption processes a significant role is played also by microporous glasses, whose sorptive properties are similar to those of gels and silica zeolites, however, the area of their application is wider. Microporus glasses may be used as a semi-permeable membranes for the separation or liquid and gaseous mixtures, as well as gel filling in chromatography. Because of the high silica dioxide content they may find application in chemical, metallurgical, electrotechnical and other industries [476]. There exists also the possibility to make use of these glasses in biochemistry, e.g. for long-lasting storing of enzymes or as protein testers capturing the conformation anomalies and in this way finding clinical and diagnostic applications. On account of their sorptive properties, microporous glasses represent an excellent material for storing high-energy radioactive waste products in nuclear power engineering and for bounding toxins in natural environmental [477]. There are two main reasons accounting for the aforementioned statement regarding the computational material science as the perspective and modern technology for the 21st century [471]: (1) theoretical methods make it possible to calculate a great number of material properties prior to their synthesis, and (2) the performance of computer hardware increases exponentially while its cost diminishes. Computer simulations of adsorbents, catalysts and ion-exchangers will thus become more accurate, faster, more reliable and cheaper compared with experiment. Since an analysis of the number of known compounds shows that the majority of combination of elements have yet to be made, computational approaches will be a strategic tool to discover the innovative and marked materials. In fact, new developments in this field are the key for development of other areas of science and technology.

The fundamental practical application of adsorption and related domains are following: separation and purification of liquid and gas mixtures, drying gases and liquids before loading them into the industrial system as well as removal of impurities from the liquid and gas media. In many cases these impurities are dangerous for biological life. The most important adsorption separation methods are: pressure swing adsorption (PSA) technology and its various variants [6,478,479], membrane processes [480-482], chromatographic techniques applied in the process scale [99-105] and ion-exchange [483,484]. All the above mentioned separation methods play a very important role both in industry and environmental protection. Moreover, the adsorption processes are alternative to distillation ones [6] owing to the costs which for adsorption separation are generally lower than those for distillation. Heterogeneous catalysis connected closely with adsorption plays a significant role in modern industry and environmental protection. Its numerous applications were discussed briefly earlier. The ion exchange also related to adsorption plays an important part in the effective removal of impurities including heavy metals and radioactive wastes from the liquid media. On the laboratory and industrial scales quantitative separation and qualitative determination of numerous substances are carried out using various chromatographic methods. At present chromatography is one of the most modern analytical methods. The other important application deals with the adsorption of charged species at the solid/liquid interfaces. This gives rise to the formation of electrical double layers, which have a topical use in the environment of the soil systems [485,486]. Also, the corrosion inhibition is connected with the formation of double layers [487]. These layers also play an important role in electroanalysis [488]. Adsorption of various solutes on dispersed colloids in liquid (aqueous) solutions can alter the charge of the solids, their dispersion stability, solubility and adhesion on solid surfaces. These changes can be manipulated by the choice of solutes in contact with a given solid and then employed in numerous applications, including catalysis, medical diagnostics, cosmetics, inks, agricultural products, pigment preparations, corrosion, etc. [489]. There is an abundance of systems between those containing a macroscopic solid/liquid interface and those which are colloidal in nature. Dyes, soaps or detergents are adsorbed from solution onto the surfaces of fibres in the dyeing and washing processes. Coal cleaning, mineral extraction and soil science are the fields where adsorption process and solid/liquid interface are of paramount importance. Adsorption is one route to surface modification of the solids in order to design new, reliable, highly-specified adsorbents, selective catalysts, polymer extenders, thickners of dispersive media and efficient chromatographic packings. Surface modification is also a main element in flotation, a process in the field of ore refinement [470]. Adsorption can have a dramatic impact on the crystal growth if some crystal planes are blocked by trace amounts of impurities. In this way, their selective adsorption determines the eventual crystal shape [490]. Another practical area of great importance deals with the adsorption of surfactants [491,492]. This process has wide applications in detergency, flotation, numerous pharmaceutical purposes, cosmetics, paints and the stabilization of suspensions in general [493].

Recently, the extensive studies have been carried out on the self-assembled monolayers (SAMs) formed by spontaneous adsorption of a long-chain, terminally functionalized molecules on various solid films. These molecules form stable monolayers on solids. The long-chains are oriented away from the surface and are present at the air or liquid interfaces. The interfacial properties of SAMs depend on the nature of the terminal functional groups. The SAMs have stability in air and under solvents for the period of months, but can be destroyed under extreme conditions. At present, SAMs have wide applications in numerous problems [494].

Adsorption of a protein from an aqueous solution on various interfaces is one of the numerous natural properties of such substances and investigations of physicochemical principles of adsorption phenomena in this case are of importance for biology, modern technology and environmental protection. Adsorption of proteins has been used in food production for a long time. New applications of adsorbed proteins push forward successful development of biotechnology, pharmacology and medicine, determining the usefulness of new drugs and the control of drug delivery. Regulation of the selective adsorption on solid surfaces form mixtures of proteins is a main problem of biocompatibility of synthetic materials for medical purposes. Applications of immobilized or adsorbed enzymes as a specific catalyst open new routes in modern applied chemistry [495,496].

Adsorption technologies have also been developed to overcome the degradation of environmental quality. They play a significant role both in environmental control (water, ground water and air treatment) and in prevention from a global warming and ozone layer depletion. The neccessity to reduce the ozone depletion gases such as chlorofluorocarbons (CFCs) and the request of primary energy diversification in the air conditioning sector, are the main reasons for the increasing interest in adsorption devices considered as alternative to the traditional compressor heat pumps and cooling systems. Adsorption processes are the heart of several new energy technologies which can find suitable applications in the domestic sector as reversible heat pumps, and in the industrial sector as refrigerating systems and heat transformers using industrial waste heat as the primary energy source. They can also be used for technologies to be applied in the transportation sector, for automobile air conditioning or for food preservation in trucks. The use of environmental friendly materials like zeolite as an adsorbent and water as a refrigerant makes this system very acceptable in any sector. Obviously the concrete possibility of economic and efficient machines depends on the solution to the problems still open today [497]. The adsorption dessicant dehumidification technology is also emerging as an alternative to the vapour compression systems for cooling and conditioning air for a space. The dessicant based systems can improve indoor air quality and remove air pollutants due to coadsorption by the dessicant materials. Moreover, a number of microorganisms is removed or killed by the dessicant [498]. Other problems, like production of drinking water [499,500], removal of industrial pollutants from air [501], soils [502] and removal microorganisms from indoor air [503] and other important problems are solved in terms of adsorption technologies.

Discussing even a part of adsorption application and that connected directly or indirectly with related domains goes beyond the limits of this chapter. The reader can find the examples of numeorus and well documented applications of adsorption phenomena in industry and environmental protection in this monograph. However, to maintain compact and overall character of this chapter one should mention also importance and role of adsorption processes in many fields of modern industry, technique and everyday life.

Adsorbents are used for drying and purifying gases and liquids, uptake of volatile solvents, harmful industrial wastes contaminating atmosphere and water basins, evolution of valuable components from mixtures of gases and vapours for their further chemical and biological treatment.

Adsorption processes are used for very thorough purification of monomers - the initial substances for preparation of multimolecular materials. Owing to such purification, polymers of new and desirable exploitation properties are obtained.

Adsorption processes are very important in purification of various petroleum products: fuels, oils, extraction benzines, etc. Profound drying of natural gas by means of adsorbents prevents from formation of congestion in the gas piping, particularly in the regions of low temperatures. Owing to the adsorption process there can be achieved the most effective drying of air before its low-temperature separation into nitrogen, liquid oxygen and noble gases. In the same way, dry atmosphere in the technology of semiconductors production is obtained. Adsorbents make it possible to create protective atmospheres during welding and thermal treatment of some important elements of machinery construction. Effectiveness of catalytic processes in the chemical industry depends on the extent of initial gases drying and purifying for the contact synthesis which can be also achieved using adsorbents. Their application makes it possible to capture vapours of precious solvents from the atmosphere, to turn them for production, thus decreasing the environmental pollution to a large extent. Adsorbents are widely applied in chemical, food and pharmaceutical industries. Some of them are effective catalyst supports. A large number of adsorbents is also used in the rubber industry as fillers for rubber mixtures, often with vulcanization accelerating agents which improves the quality of rubber products and removes the defect caused by too early vulcanization whose sign are listers. Adsorbents are also used for conservation of machinery because they protect from corrosion of metal parts.

Hydrocarbons which are raw materials used for the chemical treatment and in biological processes of valuable protein preparation can be removed from petroleum where heavy paraffins are separated from aromatic and cyclic hydrocarbons and their isomers.

Protection of the environment from the impurities originating from industrial wastes and transportation means is a very important and current problem. Adsorbents play a significant role in neutralization of waste gases and sewages and at the same time in capturing valuable components found in wastes. Compared with other methods, adsorbents allow for the most thorough purification of raw materials with relatively low costs.

Adsorbents are widely applied in medicine, among others, to take up poisons found in living organisms and in case of some diseases of the alimentary canal. Lately adsorbents have been used for purification of blood from noxious substances using chemisorption.

In the pharmaceutical industry adsorbents are used for purification of anaesthetics, removal and purification of vitamins, antibiotics and others. To-day more than fifty per cents of the pharmaceuticals are enantiomers. They cannot be often obtained by means of stereoselective synthesis. Then, the only solution of the problem is to apply the so-called chiral adsorbents and to perform separation.

Of particular importance in the laboratory practice are the adsorbents used in chromatography for analysis and separation of mixtures with simultaneous evolution of high purity components. Adsorption gas chromatography is used in the industrial laboratories for periodical inspection of technological processes and in the system of automatic control and steering of various production processes.

Using adsorbents it is possible to obtain high vacuum, among others, in large size machinery and to ensure long and stable work of semiconducting, cooling and other equipment. Adsorbents make it possible for a man to work in closed spaces, among others, in spaceships. Adsorption can also be expected to play a significant role in the environmental control and life support systems on planetary bases, where sorbents may be used to process habitat air or to recover useful substances from the local environment [504].

Catalysts, on which about 90% processes in the chemical plants are based, are as common as adsorbents. Catalysts are applied in production of fertilizers, fuels, polymers as well as modelling pastes, paints and varnishes, synthetic fibers and washing products. At present catalysts and adsorbents are used in almost all fields of science and technology. As mentioned before, adsorption and catalysis are interrelated because most adsorbents are applied in both adsorption and catalytic processes. In the latter case adsorbents either play a function of catalyst supports or are catalysts themselves. Solution of many theoretical and practical tasks depends on the choice of adsorbents and catalysts of optimal surface and porous structure. As follows from the above, it is essential to work out rules of formation of porous structure of adsorbents, catalysts and supports of catalytically active substances of beforehand determined properties and to elaborate the methods for their preparation. It should be stressed that preparation of new adsorbents and catalyst supports of desirable porous structure and chemical structure of the surface is one of the most important problems cammon in adsorption technique and catalysis, but determination of structural characteristics of adsorbents and catalysts is based largely on adsorption methods.

Adsorption, catalysis, membrane separation, ion-exchange or process scale chromatography deal with technologies for reducing the ecological load, introduction of renewable energy sources, strategies for selection of and search for ecology - friendly processes, formulation of criteria for estimation of acceptability of current chemical technologies and for design and production of new revolutionary solid materials. It is widely known, that broadly-understood adsorption science has gained a dominating role in modern industry under environmental, economical and energy saving aspects. Doubtless, the adsorption technologies are rapidly improved and adopted to contemporary tasks of mankind. Both industrial as well as recent environmental problems require wide body scientists and engineers to develop the theory of adsorption science and to produce of new adsorbents, catalysts and other advanced solids of great practical importance. Nowadays, only such technologies which give the possibility of sustainable development of the people and society are justified. Adsorption, catalysis and aforementioned related domains have a major impact in many areas central to the question of our future. In this context they may be esteemed as the technologies of 21st century.

## REFERENCES:

1. J.H.S.Haggin, Chemistry, 44 (4) (1971) 6.
2. D.A.Degras, in: Physics in Industry, E.O'Mongain and C.O'Toole (eds.), Pergamon Press, Oxford, 1976, 85.

3. C.J.King, Separation Processes, McGraw-Hill, New York, 1980.
4. G.E.Keller, in: Industrial Gas Separation, T.E.Whyte, Jr., C.M.Yon and E.H.Eagener (eds.), ACS Symp.Ser., 223 (1983) 145.
5. B.L.Karger, L.R.Snyder and C.Horvath, An Introduction to Separation Science, Wiley, New York, 1973.
6. R.T.Yang, Gas Separation by Adsorption Processes, Imperial College Press, London, 1997.
7. J.J.Kipling, Adsorption from Solutions of Non-Electrolytes, Academic Press, London, 1965.
8. Adsorption from Solution at the Solid/Liquid Interface, G.D.Parfitt and C.H.Rochester (eds.), Academic Press, London, 1983.
9. Adsorption from Solution, R.H.Ottawil, C.H.Rochester and A.L.Smith (eds.), Academic Press, London, 1983.
10. A.W.Adamson, Physical Chemistry of Surfaces, Wiley, New York, 1990.
11. J.Lyklema, Fundamentals of Interface and Colloid Science, vol. 1, Academic Press, London, 1991.
12. J.Lyklema, Fundamentals of Interface and Colloid Science, vol. 2, Academic Press, London, 1995.
13. Preprints of Topical Conference on Separation Science and Technologies, AJChE Annual Meeting, Los Angeles, California, November, 1997.
14. J.Ościk, Adsorption, Ellis Horwood, Chichester, PWN, Warsaw, 1982.
15. K.S.W.Sing, D.H.Everett, R.A.W.Haul, L.Moscow, R.A.Pierotti, J.Rouquerol and T.Siemieniewska, Pure and Appl.Chem., 57 (1985) 603.
16. R.S.Hansen, J.Phys.Chem., 66 (1962) 410.
17. F.C.Goodrich, in: Surface and Colloid Science, vol. 1, E.Matijevic (ed.), Wiley, New York, 1969, 1.
18. G.Schay, in: Physical Chemistry, Engineering Topics from Colloid and Surface Science, H.van Olphen and K.J.Mysels (eds.), Theorex, La Jolla, California, 1975, 229.
19. A.Dąbrowski, M.Jaroniec and J.Ościk, in: Surface and Colloid Science, vol. 14, E.Matijevic (ed.), Plenum Press, New York, 1987, 83.
20. A.Dąbrowski and M.Jaroniec, Adv.Colloid Interface Sci., 27 (1987) 211.
21. A.Dąbrowski and M.Jaroniec, Adv.Colloid Interface Sci., 31 (1990) 155.
22. R.Aveyard and D.A.Haydon, An Introduction to the Principles of Surface Chemistry, Cambridge University Press, 1973.
23. M.M.Dubinin, J.Colloid Interface Sci., 23 (1967) 487.
24. M.M.Dubinin, in: Characterization of Porous Solids, S.J.Gregg, K.S.W.Sing and H.F.Stoeckli (eds.), Soc.Chem.Ind., London, 1979, 1.
25. S.J.Gregg and K.S.W.Sing, Adsorption, Surface Area and Porosity, Academic Press, London, 1982.
26. J.W.Gibbs, Collected Works, vol. 1, Longmans, London, 1928.
27. E.A.Guggenheim, Trans.Faraday Soc., 36 (1940) 397.
28. D.H.Everett, Israel J.Chem., 14 91975) 267.

29. E.A.Guggenheim and N.K.Adam, Proc.R.Soc., London, ser. A, 139 (1933) 218.
30. R.Defay, J.Prigogine, A.Bellemans and D.H.Everett, Surface Tension and Adsorption, Longmans, London, 1966.
31. G.Schay, Pure Appl.Chem., 48 (1976) 393.
32. C.Wagner, Nachr.Akad.Wiss. Göttingen Math.Phys., 3 (1973) 37.
33. A.Deryło-Marczewska and M.Jaroniec, in: Surface and Colloid Science, vol. 14, E.Matijević (ed.), Plenum Press, 1987, 301.
34. J.E.Verschaffeldt, Bull.Ac.Roy.Belg., 22 91937) 373.
35. E.A.Guggenheim, Thermodynamics, North Holland, Amsterdam, 1967, 46.
36. D.K.Chattoray and K.S.Birdi, Adsorption and the Gibbs Surface Excess, Plenum Press, New York, 1984.
37. J.Tóth, Adv.Coloid Interface Sci., 55 (1995) 1.
38. T.L.Hill, Adv.Catal., Academic Press, New York, 4 (1952) 231.
39. D.H.Everett, in: Specialist Periodical Reports, vol. 1, D.H.Everett (ed.), Chemical Society, London, 1973.
40. C.E.Brown and D.H.Everett, in: Specialist Periodical Reports, vol. 2, D.H.Everett (ed.), Chemical Society, London, 1975.
41. D.H.Everett and R.T.Podoll, in: Specialist Periodical Reports, vol. 3, D.H.Everett (ed.), Chemical Society, London, 1979.
42. J.H.Davis and D.H.Everett, in: Specialist Periodical Reports, vol. 4, D.H.Everett (ed.), Chemical Society, London, 1979.
43. G.Schay, in: Surface and Colloid Science, vol. 2, E.Matijević (ed.), Wiley, New York, 1970, 155.
44. M.Jaroniec, A.Patrykiejew and M.Borówko, in: Progress in Surface and Membrane Science, vol. 14, Academic Press, New York, 1981, 2.
45. R.A.Pierotti, Physical Adsorption. The Interaction of Gases and Solids, Wiley, New York, 1971.
46. W.A.Steele, The Interaction of Gases with Solid Surfaces, Pergamon Press, Oxford, 1974.
47. D.M.Ruthven, Principles of Adsorption and Adsorption Processes, J.Wiley, New York, 1984.
48. A.V.Bogdanov, G.V.Dubrovskiy, M.P.Kriutkov, D.V.Kulginov and V.M.Streclchenya, Interaction of Gases with Surfaces, Springer, 1996.
49. L.W.Bruch, M.W.Cole and E.Zaremba, Physical Adsorption: Forces and Phenomena, Clarendon Press, Oxford, 1997.
50. A.R.Altenberger and J.Stecki, Chem.Phys.Lett., 5 (1970) 29.
51. J.E.Lane, Austr.J.Chem., 21 (1968) 927.
52. S.Sokołowski, Adv.Colloid Interface Sci., 15 (1981) 71.
53. W.A.Steele, Chem.Rev., 93 (1993) 2355.
54. M.Jaroniec and R.Madey, Physical Adsorption on Heterogeneous Solids, Elsevier, Amsterdam, 1988, chap. 1.
55. W.E.Garner and D.McKie, J.Chem.Soc., (1927) 2451.

56. N.D.Parkyns and K.S.W.Sing, in: Specialist Periodical Reports, vol. 2, D.H.Everett (ed.), Chemical Society, London, 1975, 1.
57. N.V.Keltsev, Principles of Adsorption Technology (in Russian), WNT, Warsaw, 1980.
58. T.Paryjczak, Gas Chromatography in the Study of Adsorption and Catalysis, Ellis Harwood, Chichester, PWN, Warsaw, 1985.
59. H.Jankowska, A.Świątkowski and J.Choma, Active Carbon, Ellis Horwood, Chichester, PWN, Warsaw, 1991.
60. Porosity in Carbons, J.W.Partick (ed.), Edward Arnold, London, 1995.
61. D.H.Everett, in.: Specialist Periodical Reports, vol. 4, D.H.Everett (ed.), Chemical Society, London, 1983.
62. IUPAC Manual Appendix II, Part I, D.H.Everett, Pure Appl.Chem., 31 (1973) 579; Part II: Heterogeneous Catalysis, Pure and Applied Chemistry, 31 (1976) 71.
63. M.M.Dubinin, J.Colloid Interface Sci., 46 (1974) 351.
64. J.F.Byrne and H.Marsh, in: Porosity in Carbons, J.W.Patrick (ed.), Edward Arnold, London, 1995, chap. 1.
65. A.Groszek, this book, vol. 1.
66. J.Rouquerol, D.Avnir, C.W.C.Faibridge, D.H.Everett, J.H.Haynes, N.Pernicone, J.D.F.Ramsay, K.S.W.Sing and K.K.Unger, Pure Appl.Chem., 66 (1994) 1793.
67. D.H.Everett, in: Colloidal Disspersions, Special Publication Royal Soc.Chem., 43 (1982) 71.
68. C.W.Scheele, Chemische Abhandlung von der Luft und dem Feuer, 1777; see: Ostwald's Klassiker der exakten Wiss., 58 (1894).
69. F.Fontana, Mem.Mat.Fis.Soc.Ital., 1 (1777) 679.
70. T.Lowitz, Crell's Chem.Ann., 1 (1786) 211.
71. T.Lowitz, Crell's Chem.Ann., 2 (1788) 36.
72. T.de Saussure, Gilbert's Ann.der Physik, 47 (1814) 113.
73. T.de Saussure, Ann.Phil., 6 (1915) 241.
74. P.Chappuis, Wied.Ann., 8 (1879) 1.
75. P.Chappuis, Wied.Ann., 12 (1881) 161.
76. P.Chappuis, Wied.Ann., 19 (1883) 21.
77. L.Joulian, Ann.Chim.Phys., 22 (1881) 398.
78. H.Kayser, Wied.Ann.der Physik, 12 (1881) 526.
79. H.Kayser, Wied.Ann.der Physik, 14 (1881) 450.
80. C.M.Pouillet, Ann.Chim.Phys., 20 (1822) 141.
81. C.G.Junck, Pogg.Ann., 125 (1865) 292.
82. D.Fitzgerald, Nature, 49 (1894) 293.
83. S.Lagergren, Kgl.Vetenskaps Akad., 24B (1898).
84. H.Gaudechon, Compt.Rend., 157 (1913) 207.
85. J.Dewar, Proc.Roy.Soc., 74 (1904) 122.
86. J.W.McBain, Phil.Mag., 18 1909) 916.
87. F.London, Z.Physik, 63 (1930) 245.

88. F.London, Z.physik.Chem., (B) 11 (1931) 222.
89. J.H.De Boer and J.F.H.Custers, Z.physik.Chem., (B) 25 (1934) 225.
90. F.V.Lenel, Z.physik.Chem., (B) 23 (1933) 379.
91. J.H.de Boer, Atomic Forces and Adsorption, in: Advances in Colloid Science, Interscience Publishers, New York, 3 (1950) 1.
92. J.H.de Boer, Advances in Catalysis, vol. 8, Academic Press, New York, 1956.
93. J.E.Lennard-Jones, Trans.Farady Soc., 28 (1932) 333.
94. e.g. P.W.Atkins, Physical Chemistry, 6th ed., Oxford University Press, Oxford, 1997, Part 2.
95. D.M.Young and A.D.Crowell, Physical Adsorption of Gases, Butterworth & Co. Ltd., London, 1962.
96. K.Sakodynskii, The Life and Scientific Works of Michael Tswett, J.Chromatogr., 73 (1972) 303.
97. J.Dewar, Compt.Rend., 133 (1904) 261.
98. A.J.P.Martin and B.L.M.Synge, Biochem.J., 35 (1941) 1358.
99. L.R.Snyder, Principles of Adsorption Chromatography, Marcel Dekker, New York, 1968.
100. J.H.Knox, High-Performance Liquid Chromatography, Edinburg University Press, Edinburg, 1978.
101. L.R.Snyder, J.L.Glajch and J.J.Kirkland, Practical HPLC Method Development, Wiley, New York, 1938.
102. G.Guiochon, S.G.Shirazi and A.M.Katti, Fundamentals of Preparative and Non-Linear Chromatography, Academic Press, Boston, 1994.
103. G.Subramanian, Process Scale Chromatography, VCH, Weinheim, 1995.
104. G.K.Sofer and L.E.Nyström, Process Chromatography - a Practical Guide, Academic Press, London, 1989.
105. K.K.Unger, in: Physical Adsorption: Experiment, Theory and Applications, J.Fraissard and C.W.Conner (eds.), NATO ASI Series, Kluwer Academic Publ., Dordrecht, 1996, 407.
106. 75 Years Chromatography - A Historical Dialogue, L.S.Ettre and A.L.Zlatkis (eds.), Elsevier, Amsterdam, 1979.
107. H.Freundlich, Kapilarchemie, Leipzig, 1930.
108. J.W.McBain, The Sorption of Gases and Vapours by Solids, London, 1932.
109. C.Boedecker, J.Landw., 7 (1985) 48.
110. H.Freundlich, Colloid and Capillary Chemistry, Methuen, London, 1926.
111. J.Langmuir, J.Am.Chem.Soc., 40 (1918) 1461.
112. A.Eucken, Verh.deutsch.phys.Ges., 16 (1914) 345.
113. M.Polanyi, Verh.deutsch.phys.Ges., 16 (1914) 1012.
114. M.Polanyi, Verh.deutsch.phys.Ges., 18 (1916) 55.
115. J.Langmuir, J.Am.Chem.Soc., 38 (1916) 2221.
116. M.Volmer, Z.phys.Chem., 115 (1925) 253.
117. R.H.Fowler, Proc.Cambridge Phil.Soc., 31 (1935) 260.

118. K.S.W.Sing, in: Physical Adsorption: Experiment, Theory and Applications, J.M.Fraissard and C.W.Conner (eds.), NATO ASI Series, Kluwer Academic Publ., Dordrecht, 1996, 3.
119. K.J.Laidler, The World of Physical Chemistry, Oxford University Press, Oxford, 1993, 309.
120. S.Brunauer, P.H.Emmett and E.Teller, J.Am.Chem.Soc., 60 (1938) 309.
121. S.Brunauer and P.H.Emmett, J.Am.Chem.Soc., 57 (1935) 1754.
122. S.Brunauer and P.H.Emmett, J.Am.Chem.Soc., 59 (193) 2682.
123. A.B.D.Cassie, Trans.Farady Soc., 41 (1945) 450.
124. T.L.Hill, J.Chem.Phys., 14 (1946) 263.
125. T.L.Hill, J.Chem.Phys., 15 (1947) 767.
126. T.L.Hill, Adv.Catalysis, 4 (1952) 211.
127. R.H.Fowler and E.A.Guggenheim, Statistical Thermodynamics, Cambridge University Press, Cambridge, 1952.
128. G.Pickett, J.Am.Chem.Soc., 67 (1945) 1958.
129. R.B.Anderson, J.Am.Chem.Soc., 68 (1946) 686.
130. G.F.Hüttig, Monats.Chem., 78 (1948) 177.
131. G.F.Hüttig and G.Pietzka, Monats.Chem., 78 (1948) 175.
132. S.Brunauer, L.S.Deming, W.E.Deming and E.Teller, J.Am.Chem.Soc., 62 (1940) 1723
133. S.Brunauer, The Adsorption of Gases and Vapours, vol. 1, Oxford University Press, Oxford, 1944, 154.
134. J.S.Gregg and K.S.W.Sing, Adsorption Surfaces Area and Porosity, Academic Press, New York, 1967.
135. R.Aveyard and D.A.Hayden, An Introduction to Principles of Surface Chemistry, Cambridge University Press, Cambridge, 1973.
136. M.M.Dubinin and L.W.Radushkevich, Compt.Rend.Acad.Sci.URSS, 55 (1947) 327.
137. A.Titoff, Z.physik.Chem., 74 (1910) 641.
138. L.Berenyi, Z.physik.Chem., 94 (1920) 628.
139. R.Zsigismody, Z.anorg.Chem., 71 (1911) 356.
140. J.Rouquerol, D.Avnir, D.H.Everett, C.Faibridge, M.Haynes, N.Pernicone, J.D.Ramsay, K.S.W.Sing and K.K.Unger, in: Characteristic of Porous Solids III, Proceedings of the IUPAC Symposium (COPS III), France, May, 1993, Elsevier, Amsterdam 1994, 1.
141. A.G.Foster, Trans.Faraday Soc., 28 (1932) 645.
142. L.H.Cohan, J.Am.Chem.Soc., 60 (1938) 433.
143. A.W.Partick, Colloid Symp.Ann., 7 (1930) 129.
144. J.H.de Boer, The Dynamic Character of Adsorption, Oxford University Press, Oxford, 1953.
145. S.Brunauer, The Physical Adsorption of Gases and Vapours, Clarendon Press, Oxford University Press, Princeton, 1945.
146. J.W.McBain, J.Am.Chem.Soc., 57 (1935) 699.
147. M.B.Coelingh, Kolloid Z., 87 (1939) 251.

148. R.M.Barrer, N.McKenzie and J.S.S.Reay, Colloid Sci., 11 (1956) 479.
149. J.H.de Boer, The Structure and Properties of Porous Materials, in: X Symp.Colston Research Soc.Univ.Bristol, Butterworths Sci.Publ., D.H.Everett and F.S.Stone (eds.), London, 1958, 68.
150. D.H.Everett, The Structure and Properties of Porous Materials, in: X Symp.Colston Research Soc.Univ.Bristol, Butterworths Sci.Publ., D.H.Everett and F.S.Stone (eds.), London, 1958, 95.
151. D.H.Everett and W.J.Whitton, Trans.Faraday Soc., 48 (1952) 749.
152. D.H.Everett and F.W.Smith, Trans.Faraday Soc., 50 (1953) 187.
153. D.H.Everett, Trans.Faraday Soc., 50 (1954) 1077.
154. D.H.Everett, Trans.Faraday Soc., 57 (1955) 1557.
155. D.H.Everett, in: The Solid-Gas Interface, E.A.Flood (ed.), Marcel-Dekker, New York, 1967, 1055.
156. M.M.Dubinin, E.D.Zaverina and L.W.Radushkevich, Zh.Fiz.Khim., 21 (1947) 1351.
157. M.M.Dubinin and E.D.Zaverina, Dokl.Akad.Nauk SSSR, 72 (1950) 319.
158. M.M.Dubinin and E.D.Zaverina, Dokl.Akad.Nauk SSSR, 72 (1950) 327.
159. L.W.Radushkevich, Zh.Fiz.Khim., 23 (1949) 1410.
160. M.M.Dubinin, J.Colloid Interface Sci., 23 (1967) 487.
161. A.W.House, in: Specialist Periodical Reports, vol. 4, D.H.Everett (ed.), Chemical Society, London, 1979.
162. H.F.Stoeckli and F.Kraehenbuehl, Carbon, 19 (1981) 353.
163. H.Marsh, Carrbon, 25 (1987) 49.
164. H.Marsh and B.Rand, J.Colloid Interface Sci., 33 (1970) 101.
165. M.Jaroniec and R.Madey, Carbon, 27 (1989) 77.
166. M.Jaroniec and J.Choma, in: Chemistry and Physics of Carbon, P.A.Thrower (ed.), vol. 22, Marcel Dekker, New York, 1989, 197.
167. Introduction to Carbon Technology, H.Marsh, E.A.Heintz and F.Rodriguez-Reinoso (eds.), University of Alicante, Alicante, 1997.
168. T.L.Hill, An Introduction to Statistical Thermodynamics, Addison-Wesley Publ.Co., Reading, Mass., London, 1960.
169. R.H.Fowler and E.A.Guggenheim, Statistical Thermodynamics, Cambridge University Press, Cambridge, 1943.
170. D.Nicholson and K.S.W.Sing, in: Specialist Periodical Reports, vol.3, D.H.Everett (ed.), Chemical Society, London, 1979.
171. S.Yiacoumi and Ch.Tien, Kinetics of Metal Ion Adsorption from Aqueous Solutions, Kluwer Academic Publ., Boston, 1995.
172. G.F.Cerofolini, in: Adsorption on New and Modified Inorganic Sorbents, A.Dąbrowski and V.K.Tertykh (eds.), Elsevier, Amsterdam, 1996, 435.
173. IUPAC CHEMRAWN IX, World Conference on the Role of Advanced Materials in Sustainable Development, September, 1996, Seul, Korea (Proceedings).
174. K.I.Zamaraev, Chemistry for Sustainable Development, 1 (1993) 133.
175. J.E.Maxwell and J.E.Naber, Catalysis Letters, 12 (1997) 105.

176. Perspectives in Catalysis: A Chemistry for the 21st Century, J.M.Thomas and K.I.Zamaraev (eds.), IUPAC, Blackwells, 1992.
177. M.Bowker, The Basis and Applications of Heterogeneous Catalysis, Oxford University Press, Oxford, 1998.
178. Catalysis in Petroleum Reffining and Petrochemical Industries, M.Absi-Halabi, J.Beshara, H.Qbazard and A.Stanislaus (eds.), Elsevier, Amsterdam, 1996.
179. F.Rodriguez-Reinoso, in: Porosity in Carbons, J.W.Patrick (ed.), Edward Arnold, 1995, 253.
180. K.I.Zamaraev, Topics in Catalysis, 3 (1996) 1.
181. J.J.Berzelius, Edinburg New Phylosophical J., 21 (1836) 223.
182. J.M.Thomas, in: Proceedings IUPAC CHEMRAWN IX, September, 1996, Seul, Korea, 33.
183. J.M.Thomas and K.I.Zamaraev, Agnew.Chem., 106 (1994) 603.
184. M.J.S.Dewar and W.Thiel, J.Am.Chem.Soc., 99 (1977) 4899.
185. P.A.Barretr, R.H.Jones, J.M.Thomas, G.Sankar, J.J.Shanon and C.R.A.Catlow, J.Chem.Soc.Chem.Comm., (1996) 2001.
186. C.T.Kresge, M.E.Leonowicz, W.J.Roth, J.C.Vartuli and J.S.Beck, Nature, 359 (1992) 710.
187. J.M.Thomas, Faraday Discuss., 100 (1995) C9.
188. I.Dekany, in: Adsorption on New and Modified Inorganic Sorbents, A.Dąbrowski and V.A.Tertykh (eds.), Elsevier, Amsterdam, 1996.
189. J.M.Thomas, Agnew.Chem., Int.Ed.Engl., 33 (1994) 913.
190. D.W.Lewis, C.R.A.Catlow, J.M.Thomas, D.J.Wilcock and G.J.Hutchings, Nature, 382 (1996) 604.
191. C.M.Freeman, C.Richard, C.R.A.Catlow, J.M.Thomas and S.Brode, Chem.Phys.Lett., 186 (1991) 137.
192. R.Vetrivel, H.Takaba, M.Katagiri, M.Kubo and A.Miyamoto, in: Adsorption on New and Modified Inorganic Sorbents, A.Dąbrowski and V.A.Tertykh (eds.), Elsevier, Amsterdam, 1996.
193. R.Vetrivel, R.C.Deka, S.B.Waghmode, S.Sivasanker, K.Mizukami, H.Takaba, M.Kubo and A.Miyamoto, this book, vol. 2.
194. D.W.Lewis, C.R.A.Catlow and C.M.Freeman, J.Phys.Chem., 99 (1995) 194.
195. A.Clark, The Theory of Adsorption and Catalysis, Academic Press, New York, 1970.
196. T.I.Hill, Advances in Catalysis, Academic Press, vol. 4, New York, 1952.
197. Equilibric and Dynamics of Gas Adsorption on Heterogeneous Solids Surfaces, W.Rudziński, W.Steele and G.Zgrablich (eds.), Elsevier, Amsterdam, 1997, V.
198. H.Bradley, Nature, 120 (1927) 82.
199. M.I.Temkin, Zhur.Fiz.Khim., 4 (1933) 573.
200. H.S.Taylor, Proc.Roy.Soc., A110 (1925) 105.
201. R.N.Pease, J.Am.Chem.Soc., 45 (1923) 1196, 2235.
202. T.N.Rhodin, Jr., Advances in Catalysis, vol. X, New York, 1953.

203. S.Z.Roginsky and O.M.Todes, Acta Physicochim., URSS, 21 (1946) 519.
204. W.C.Walker and A.C.Zettlemoyer, J.Phys.&Colloid Chem., 52 (1948) 47.
205. W.G.McMillan, J.Chem.Phys., 15 (1947) 390.
206. R.N.Smith and C.Pierce, J.Phys.&Colloid Chem., 52 (1948) 1115; 54 (1950) 354.
207. J.Zeldowitsch, Acta Physicochim., URSS, 1 (1934) 961.
208. A.Schuchowitzky, Acta Physicochim., URRS, 8 (1938) 531.
209. S.Z.Roginsky, Dokl.Akad.Nauk SSSR, 45 (1944) 61.
210. S.Z.Roginsky, Dokl.Akad.Nauk SSSR, 45 (1944) 194.
211. S.Z.Roginsky and N.I.Janowsky, Izv.Akad.Nauk SSSR, Ser.Khim., 52 (1952) 59.
212. S.Z.Roginsky, Adsorption and Catalysis on Heterogeneous Surfaces, Izd.AN SSSR, Moscov, 1949.
213. M.Todes and A.K.Boundareva, Zh.Prikl.Chimii, 21 (1948) 693.
214. G.D.Halsey and H.S.Taylor, J.Chem.Phys., 15 (1947) 624.
215. R.Sips, J.Chem.Phys., 16 (1948) 490.
216. R.Sips, J.Chem.Phys., 18 (1950) 1024.
217. T.L.Hill, J.Chem.Phys., 17 (1949) 762.
218. F.C.Tompkins, Trans.Faraday Soc., 46 (1949) 569.
219. J.M.Honig and T.L.Hill, J.Phys.Chem., 22 (1954) 851.
220. J.M.Honig, J.Chem.Phys., 23 (1955) 1024.
221. J.Tóth, Acta Chim.Hung., 32 (1962) 31.
222. F.C.Tompkins, Trans.Faraday Soc., 46 (1950) 569.
223. P.Brauer and G.Heller, Wiss.Z.F.S.Univ.Jena Math.Naturwiss., R., 26 (1977) 719.
224. D.N.Misra, Surface Sci., 18 (1969) 367.
225. D.N.Misra, J.Chem.Phys., 52 (1970) 5499.
226. D.N.Misra, Indian J.Pure and Appl.Phys., 9 (1971) 1358.
227. P.J.Cirkmore and B.W.Wojciechowski, J.Chem.Soc.Faraday I, 73 (1977) 1216.
228. B.P.Bering and V.V.Serpinsky, Izv.Akad.Nauk SSSR, Ser.Khim., 74 (1974) 2227.
229. C.J.Radeke and J.M.Prausnitz, Ind.Eng.Chem.Fundam., 11 (1972) 946.
230. M.Jaroniec and A.W.Marczewski, Monats.Chem., 115 (1987) 997.
231. M.Jaroniec, Surface Sci., 50 (1975) 1553.
232. J.P.Hobson, J.Chem.Phys.Chem., 34 (1961) 1850.
233. J.P.Hobson and R.A.Armstrong, J.Phys.Chem., 67 (1963) 2000.
234. G.F.Cerofolini, Thin Solid Films, 23 (1974) 129.
235. G.F.Cerofolini, in: Specialist Periodical Reports, vol. 4, D.H.Everett (ed.), London, Chemical Society, 1982, 59.
236. G.F.Cerofolini, Surface Sci., 51 (1975) 423.
237. G.F.Cerofolini, Surface Sci., 61 (1976) 678.
238. M.M.Dubinin and H.F.Stoeckli, J.Colloid Interface Sci., 75 (1980) 34.
239. B.McEnaney, Carbon, 25 (1987) 69.

240. M.Jaroniec, Langmuir, 3 (1987) 673.
241. M.Jaroniec and J.Choma, in: Equilibria and Dynamics of Gas Adsorption on Heterogeneous Solid Surface, W.Rudziński, W.A.Steele and G.Zgrablich (eds.), Elsevier, 1997, 715.
242. S.Ross and J.P.Olivier, On Physical Adsorption, Wiley, New York, 1964.
243. M.Jaroniec, Advances in Colloid Interface Sci., 18 (1983) 149.
244. M.Jaroniec and P.Brauer, Surf.Sci.Reports, 6 91986) 65.
245. G.F.Cerofolini and N.Re, The Mathematical Theory of Adsorption on Non-Ideal Surfaces, Rivista del Nuovo Cimento, 16 (7) (1993) 1.
246. T.I.Izotova and M.M.Dubinin, Zh.Fiz.Khim., 39 (1978) 362.
247. H.F.Stoeckli, J.Colloid Interface Sci., 53 (1977) 184.
248. A.W.Kiselev, Koll.Zh., 20 (1958) 338.
249. M.Jaroniec and R.Madey, Physical Adsorption on Heterogeneous Surfaces, Elsevier, Amsterdam, 1988, 82.
250. M.Jaroniec and W.Rudziński, Acta Chim.Hung., 88 (1976) 351.
251. G.F.Hüttig and O.Theimer, Monats.Chem., 83 (1952) 650.
252. T.Sasaki, J.Chem.Soc.Japan, 73 (1952) 217.
253. D.S.Jovanović, Colloid Polymer Sci., 235 (1969) 1203.
254. D.S.Jovanović, Colloid Polymer Sci., 253 (1969) 1214.
255. G.F.Cerofolini, J.Low Temperature Phys., 6 (1972) 473.
256. C.C.Hsu, W.Rudziński and B.Wojciechowski, J.Chem.Soc., Faraday I, 72 (1976) 453.
257. A.M.Williams, Proc.Roy.Soc. Edinburg, 38 (1918) 23; 39 (1919) 48.
258. A.M.Williams, Proc.Roy.Soc., A96 (1919) 287.
259. D.C.Henry, Phill.Mag., 44 (1922) 689.
260. A.W.Marczewski, A.Deryło-Marczewska and M.Jaroniec, J.Colloid Interface Sci., 109 (1986) 310.
261. T.Nitta, M.Kuro-Oka and T.Katayama, J.Chem.Eng. Japan, 17 (1984) 45.
262. W.Rudziński, L.Łajtar, J.Zając, E.Wolfram and I.Paszli, J.Colloid Interface Sci., 96 (1983) 334.
263. W.Rudziński and J.Zając, Acta Ged.Geoph.Montan.Acad.Sci.Hung., 20 (1985) 1337.
264. W.Rudziński, J.Zając and C.C.Hsu, J.Colloid Interface Sci., 103 (1985) 528.
265. W.Rudziński, J.Zając. I.Dekany and F.Szanto, J.Colloid Interface Sci., 112 (1986) 473.
266. D.H.Everett, Trans.Farady Soc., 61 (1965) 2478.
267. W.Rudziński and D.H.Everett, Adsorption of Gases on Heterogeneous Surfaces, Academic Press, London, 1992, chap. 12.
268. G.F.Cerofolini, G.Forda and G.Spadini, Thin Solid Films, 68 (1980) 315.
269. P.Cavalotti, G.F.Cerofolini and A.Casarico, in: Principles and Applications of Pore Structural Characterization, J.M.Haynes and P.Rossi-Doria (eds.), J.W.Arrowsmith, Bristol, 1985, 171.
270. S.Ross, in: The Solid-Gas Interface, E.A.Flood (ed.), Marcel Dekker, New York, 1967, chap. 15.

271. S.E.Hoory and J.M.Prausnitz, Surf.Sci., 67 (1977) 195.
272. B.Kindl, E.Negri and G.F.Cerofolini, Surf.Sci., 23 (1970) 299.
273. S.Ross and I.D.Morrison, Surf.Sci., 52 (1975) 103.
274. R.S.Sacher and I.D.Morrison, J.Colloid Interface Sci., 70 (1979) 153.
275. C.H.Vos and L.K.Koopal, Colloid Surfaces, 14 (1988) 87.
276. C.H.Vos and L.K.Koopal, J.Colloid Interface Sci., 105 (1985) 183.
277. A.W.Adamson and J.Ling, Adv.Chem.Ser., 33 (1961) 51.
278. I.D.Morrison and S.Ross, Surf.Sci., 62 (1977) 331.
279. W.A.Hause and M.J.Jaykock, Colloid Polymer Sci., 256 (1978) 52.
280. G.F.Cerofolini, Thin Solid Films, 23 (1974) 129.
281. S.Sokołowski, M.Jaroniec and G.F.Cerofolini, Surf.Sci., 47 (1975) 429.
282. W.Rudziński, M.Jaroniec, S.Sokołowski and G.F.Cerofolini, Czech.J.Phys.B, 25 (1975) 891.
283. P.Brauer and M.Jaroniec, J.Colloid Interface Sci., 108 (1986) 50.
284. P.Brauer, M.Fassler and M.Jaroniec, Chem.Phys.Lett., 123 (1985) 245.
285. P.Brauer, M.Fassler and M.Jaroniec, Chem.Phys.Lett., 125 (1986) 241.
286. L.B.Harris, Surf.Sci., 10 (1968) 129.
287. G.F.Cerofolini, Surf.Sci., 24 (1971) 391.
288. G.F.Cerofolini, Chem.Phys., 33 (1978) 423.
289. J.P.Hobson, Can.J.Phys., 43 (1965) 1934.
290. W.Rudziński, J.Narkiewicz, B.W.Wojciechowski and C.C.Hsu, J.Colloid Interface Sci., 67 (1978) 292.
291. W.Rudziński, J.Narkiewicz and A.Patrykiejew, Z.physik.Chem., Leipzig, 260 (1979) 1097.
292. W.Rudziński, J.Jagiełło and Y.Grillet, J.Colloid Interface Sci., 87 (1982) 478.
293. M.M.Nederlof, W.H.Van Riemsdijk and L.K.Koopal, J.Colloid Interface Sci., 135 (1990) 410.
294. K.Tsutsumi, Y.Mitani and H.Takahashi, Colloid Polym.Sci., 263 (1985) 838.
295. K.Tsutsumi, Y.Matsushima and A.Matsumoto, Langmuir, 9 (1993) 2665.
296. B.Fubini, V.Bolis, A.Cavanego, E.Garrone and P.Ugliengo, Langmuir, 9 (1993) 2712.
297. A.J.Groszek and S.Partyka, Langmuir, 9 (1993) 2721.
298. V.Bolis, G.Magnacca, C.Cerrato and C.Morterra, Langmuir, 13 (1997) 888.
299. W.Rudziński and D.H.Everett, Adsorption of Gases on Heterogeneous Surfaces, Academic Press, London, 1992, chap. 11.
300. W.Rudziński, T.Pańczyk and A.Dominko, in: Proceedings of the ISSHAC III, Toruń, Poland, August, 1998, 168.
301. M.Jaroniec, in: Proceedings, 14th IUPAC Conf.on Chemical Thermodynamics, ICCT-96, Osaka, Japan, August, 1996.
302. G.F.Cerofolini and N.Re, Langmuir, 13 (1997) 990.
303. M.Jaroniec, Advanced Colloid Interface Sci., 18 (1983) 149.
304. F.Villieras, L.J.Michot, J.M.Cases, I.Bardot, M.Fracois, G.Gerard and J.Ivon, in: Equilibria and Dynamics on Gas Adsorption on Hetergeneous

Solids Surfaces, W.Rudziński, D.H.Everett and G.Zgrablich (eds.), Elsevier, 1997, 573.
305. T.L.Hill, Statistical Mechanics, McGraw-Hill, New York, 1956.
306. G.S.Rushbrooke, Introduction to Statistical Mechanics, Oxford University Press, Oxford, 1949.
307. E.Bergman, J.Phys.Chem., 78 (1974) 405.
308. S.Ross and I.D.Morrison, Surface Sci., 39 (1973) 21.
309. J.F.Hanalan and M.P.Freeman, Can.J.Phys., 37 (1959) 843.
310. W.Rudziński, Z.Suprynowicz and J.Rayss, J.Chromatography, 66 (1972) 1.
311. W.Rudziński and D.H.Everett, Adsorption of Gases on Heterogeneous Solids, Academic Press, London, 1992, chap. 7.
312. A.Waksmundzki, W.Rudziński and Z.Suprynowicz, J.Gas Chromatgraphy, 2 (1966) 93.
313. M.P.Freeman and K.Kolb, J.Phys.Chem., 67 (1963) 217.
314. D.C.Hinman and D.D.Halsey, J.Chem.Phys., 64 (1975) 3353.
315. W.A.Steele, J.Phys.Chem., 67 (1963) 2016.
316. R.A.Pierotti and H.E.Thomas, Trans.Faraday Soc., 70 (1974) 1725.
317. S.E.Hoory and J.M.Prausnitz, Surface Sci., 6 (1967) 377.
318. R.Ripan and G.Zgrablich, J.Phys.Chem., 79 (1975) 2118.
319. W.Rudziński, Phys.Lett., 42A (1972) 519.
320. W.Rudziński and J.Jagiełło, Vacuum, 32 (1982) 577.
321. 321.R.J.Laub and R.L.Pecsok, Physicochemical Applications of Gas Chromatography, Wiley, New York, 1978.
322. D.H.Everett, Gas Chromatogr., 16 (1964) 219.
323. R.Leboda and S.Sokołowski, J.Colloid Interface Sci., 61 (1977) 365.
324. Z.Suprynowicz, M.Jaroniec and J.Gawdzik, J.Chromatogr., 9 (1976) 61.
325. M.Jaroniec and R.Madey, Physical Adsorption on Heterogeneous Solids, Elsevier, 1988, chap. 4.
326. A.Deryło-Marczewska and M.Jaroniec, in: Surface and Colloid Science, E.Matijević (ed.), Plenum Press, 1987.
327. M.Jaroniec, D.E.Martire and M.Borówko, Adv.Colloid,Interface Sci., 22 (1985) 177.
328. S.Sircar, A.L.Myers and M.C.Molstad, Trans.Faraday Soc., 66 (1970) 2354.
329. R.W.Walters and R.G.Luthy, Environ.Sci.Technol., 18 (1984) 395.
330. M.S.Abdo, Indian J.Technol., 17 (1979) 201.
331. M.Jaroniec, in: Fundamentals of Adsorption, A.L.Myers and G.Belford (eds.), Amer.Inst.Chem.Eng., New York, 1984, 239.
332. M.Jaroniec, A.Deryło and A.W.Marczewski, Monats.Chem., 114 (1983) 393.
333. A.Deryło and M.Jaroniec, Environment Protection Eng., 8 (1982) 75.
334. M.Jaroniec, Thin Solid Films, 81 (1981) L97.
335. A.Dąbrowski, Monats.Chem., 117 (1986) 139.
336. M.Borówko and M.Jaroniec, Adv.Colloid Interface Sci., 19 (1983) 137.
337. J.Goworek and R.Kusak, Colloid & Polymer Sci., 267 (1989) 539.
338. A.C.Zettlemoyer and F.Micale, Croat.Chem.Acta, 42 (1970) 247.

339. M.T.Coltharp and N.Hackerman, J.Colloid Interface Sci., 43 (1973) 176.
340. M.T.Coltharp and N.Hackerman, J.Colloid Interface Sci., 43 (1973) 186.
341. W.Rudziński, J.Ościk and A.Dąbrowski, Chem.Phys.Lett., 20 (1973) 5.
342. W.A.House, Chem.Phys.Lett., 60 (1978) 169.
343. J.Pappenhuijzen and L.K.Koopal, in: Adsorption from Solutions, R.H.Ottewill, C.H.Rochester and A.L.Smith (eds.), Academic Press, New York, 1983, 211.
344. J.Ościk, A.Dąbrowski, M.Jaroniec and W.Rudziński, J.Colloid Interface Sci., 56 (1976) 403.
345. S.Sircar, J.Chem.Soc., Faraday I, 79 (1983) 2085.
346. S.Sircar, Surface Sci., 148 (1984) 478.
347. J.Ościk, A.Dąbrowski, M.Jaroniec and S.Sokołowski, J.Res.Inst.Catalysis, Hokkaido Univ., 23 (1976) 91.
348. A.Dąbrowski, J.Ościk, M.Jaroniec and W.Rudziński, J.Colloid Interface Sci., 69 (1979) 287.
349. A.Patrykiejew, M.Jaroniec and A.Dąbrowski, Croat.Chem.Acta, 53 (1980) 9.
350. A.W.Marczewski and A.Deryło-Marczewska, Chem. Scripta, 28 (1988) 173.
351. K.V.Chmutov and O.G.Larionov, Progress in Surface and Membrane Sci., 14 (1981) 237.
352. W.Rudziński and S.Partyka, J.Colloid Interface Sci., 86 (1982) 25.
353. A.Dąbrowski, Chem.Scripta, 25 (1985) 182.
354. H.Minka and A.L.Myers, AIChE J., 19 (1973) 453.
355. D.Schiby and E.Ruckenstein, Colloid and Surfaces, 15 (1985) 17.
356. J.S.C.Hsieh, R.M.Turian and Ch.Thien, AIChE J., 23 (1977) 263.
357. A.Dąbrowski and P.Podkościelny, Langmuir, 13 (1997) 3464.
358. A.Dąbrowski and P.Podkościelny, Colloids and Surfaces (in press).
359. W.Rudziński, K.Nieszporek, H.Moon and H-K.Rhee, Heterogeneous Chem.Rev., 1 (1994) 275.
360. W.Rudziński, K.Nieszporek, H.Moon and H.K.Ree, Chemical.Ing.Sci., 50 (1995) 2641.
361. A.Dąbrowski, M.Jaroniec and J.K.Garbacz, Chem.Scripta, 29 (1989) 21.
362. M.Jaroniec and W.Rudziński, Phys.Lett., 53A (1975) 59.
363. A.L.Myers and J.M.Prausnitz, AIChE J., 11 (1965) 212.
364. R.Danner and E.C.F.Choi, Ind.Eng.Chem.Fundamentals, 17 (1978) 248.
365. B.B.Mandebort, The Fractal Geometry of Nature, Freeman, New York, 1982.
366. The Fractal Approach to Heterogeneous Chemistry, D.Avnir (ed.), Wiley, New York, 1989.
367. P.Pfeifer and D.Avnir, J.Chem.Phys., 79 (1983) 1990.
368. D.Avnir and D.Farin, New J.Chem., 14 (1990) 197.
369. D.Avnir, D.Farin and P.Pfeifer, J.Chem.Phys., 79 (1983) 3566.
370. D.Avnir, D.Farin and P.Pfeifer, J.Coolid Interface Sci., 103 (1985) 112.
371. D.Farin, A.Volpert and D.Avnir, J.Am.Chem.Soc., 107 (1985) 3368.

372. J.J.Fripiat and H.Van Damme, Bul.Soc.Chim.Belg., 94 (1985) 825.
373. Ch.C.Jacquin and P.M.Adler, J.Cooloid Interface Sci., 107 (1985) 405.
374. D.Farin and D.Avnir, in: Characterization of Porous Solids, K.K.Unger, J.Rouquerol, K.S.W.Sing and H.Kral (eds.), Elsevier, Amsterdam, 1988, 421.
375. J.M.Drake, P.Levitz and J.Klafter, Israel J.Chem., 31 (1991) 135.
376. A.V.Neimark, Russ.J.Phys.Chem., 64 (1990) 1398.
377. D.Avnir and M.Jaroniec, Langmuir, 5 (1989) 1431.
378. J.M.Drake, P.Levitz and J.Klafter, New J.Chem., 14 (1990) 77.
379. D.Avnir, D.Farin and P.Pfeifer, New J.Chem., 16 91992) 439.
380. Kinetics and Dynamics of Physical Adsorption, M.M.Dubinin and L.V.Raduskevich (eds.), Nauka, Moscow, 1973 (in Russian).
381. P.Ditl, R.W.Coughlin and H.E.Iera, J.Colloid Interface Sci., 63 (1978) 410.
382. J.Czarniecki and M.Jaroniec, Thin Solid Films, 75 (1981) 347.
383. M.M.Dubinin, Carbon, 13 (1975) 193.
384. M.M.Dubinin, Pure Appl.Chem., 48 (1976) 407.
385. M.Jaroniec, Thin Solid Films, 71 (1980) 273.
386. J.Czarniecki and M.Jaroniec, Thin Solid Films, 75 (1981) L11.
387. J.Czarniecki and M.Jaroniec, Thin Solid Films, 17 (1981) 231.
388. J.Czarniecki and M.Jaroniec, Thin Solid Films, 94 (1982) 365.
389. J.Czarniecki and M.Jaroniec, Surface Sci. Reports, 6 (1986) 65.
390. F.Characorin and S.Y.Elovich, Acta Physicochim.URSS, 5 (1936) 325.
391. S.Y.Elovich and G.M.Zhabrova, Zh.Fiz.Khim., 13 (1939) 1761.
392. G.F.Cerofolini, in: Adsorption on New and Modified Inorganic Sorbents, A.Dąbrowski and V.A.Tertykh (eds.), Elsevier, Amsterdam, 1996.
393. Equilibria and Dynamics of Gas Adsorption on Heterogeneous Solid Surfaces, W.Rudziński, W.A.Steele and G.Zgrablich (eds.), Elsevier, Amsterdam, 1997.
394. D.M.Ruthven, in: Physical Adsorption: Experiment, Theory and Applications, J.Fraissard and C.W.Conner (eds.), NATO ASI Ser., Kluwer Academic Publ., Dordrecht, 1997, 241.
395. D.Nicholson, in: Physical Adsorption: Experiment, Theory and Applications, J.Fraissard and C.W.Conner (eds.), NATO ASI Ser., Kluwer Academic Publ., Dordrecht, 1997, 105.
396. K.E.Gubbins, in: Physical Adsorption: Experiment, Theory and Applications, J.Fraissard and C.W.Conner (eds.), NATO ASI Ser., Kluwer Academic Publ., Dordrecht, 1997, 65.
397. G.C.Maitland, M.Rigby, E.B.Smith and W.A.Wakeham, Intermolecular Forces, Clarendon Press, Oxford, 1981.
398. M.Allen and D.J.Tildesley, Computer Simulation of Liquids, Clarendon Press, Oxford, 1987.
399. G.Ciccotti, D.Frenkel and J.R.McDonald, Simulation of Liquids and Solids, North Holland, 1987.
400. D.Nicholson and N.G.Parsonage, Computer Simulation and the Theory of Physical Adsorpion, Academic Press, London, 1982.

401. C.G.Gray and K.E.Gubbins, Theory of Molecular Fluids, Clarendon Press, Oxford, 1984.
402. W.Rudziński and D.H.Everett, Adsorption of Gases on Heterogeneous Surfaces, Academic Press, London, 1992, chap. 13.
403. Computer Simulations of Polymers, E.Colbourne (ed.), Longman, London, 1994.
404. C.Mellot and J.Lignieres, in: Physical Adsorption: Experiment, Theory and Applications, NATO ASI Ser., J.Fraissard and C.W.Conner (eds.), Kluwer Academic Publ., Dordrecht, 1997, 429.
405. Computer Simulation in Chemical Physics, M.P.Allen and D.J.Tildesley (eds.), Kluwer, The Netherlands, 1992.
406. Molecular Simulation and Industrial Process, K.E.Gubbins and N.Quirke (eds.), Gordon & Breach, London, 1996.
407. e.g., Preprints of the Topical Conference on Separation Science and Technologies, Part II, Section 6: New Development in Adsorption Process Design and Simulations, AIChE Annual Meeting, Los Angeles, California, November, 1997.
408. e.g., Yu.K.Tovbin, Theory of Physical Chemistry Processes at Gas-Solid Interface, CRC Press, Boca Raton, 1991.
409. G.L.Aronovich and M.D.Donohue, J.Colloid Interface Sci., 189 (1997) 101.
410. R.Evans, Adv.Phys., 28 (1979) 143.
411. P.Tarazona, M.B.Narconi and U.Evans, Mol.Phys., 60 (1987) 573.
412. K.E.Gubbins, in: Preprints of the Topical Conference on Separation Science and Technologies, Part II, Section 6: New Development in Adsorption Process Design and Simulations, AIChE Annual Meeting, Los Angeles, California, November, 1997, 1266.
413. R.Evans and P.Tarazona, Phys.Rev.Lett., 52 (1984) 557.
414. B.K.Paterson, J.P.R.B.Walton and K.E.Gubbins, J.Chem.Soc., Faraday Trans., 2, 82 (1986) 1789.
415. P.B.Balbuena and K.E.Gubbins, Langmuir, 9 (1993) 1801.
416. C.M.Lasotskie, K.E.Gubbins and N.Quirke, J.Phys.Chem., 97 (1993) 4786.
417. E.F.Vansant, P.van der Voort and K.C.Vranckan, Characterization and Chemical Modification of the Silica Surface, Elsevier, Amsterdam, 1995, Appendix D, 511.
418. N.Metropolis, A.W.Rosenbluth, M.N.Rosenbluth, A.Teller and E.Teller, J.Chem.Phys., 21 (1957) 1087.
419. B.J.Alder and T.E.Wainwright, J.Phys.Chem., 27 (1957) 1208.
420. W.W.Wood, Early History of Computer Simulation in Statistical Mechanics, in: Proceedings of the Enrico Fermi Summer School, Varenna, 1986, 3.
421. Modelling of Structure and Reactivity in Zeolites, C.A.Catlow (ed.), Academic Press, London, 1992.
422. M.D.Daris, C.Saldarriage, C.Ontes, J.Graces and C.Crasder, Nature, 331 (1988) 6158.
423. K.Maeda, Y.Kiyozumi and F.Mizukami, Angew.Chem., 33 (1994) 2334.

424. T.Schultz, in: Preprints of the Topical Conference on Separation Science and Technologies, Part II, Section 6: New Development in Adsorption Process Design and Simulations, AIChE Annual Meeting, Los Angeles, California, November, 1997, 1494.
425. S.Iijima, Nature, 354 (1991) 56.
426. T.Kyoutuchi, Li-fu Tsai and A.Tomita, Chem.Mater., 7 (1995) 1427.
427. T.Guo, J.Changmin and R.E.Smalley, J.Phys.Chem., 95 (1991) 4948.
428. W.Adreoni, F.Gygi and M.Painello, Chem.Phys.Lett., 190 (1992) 159.
429. S.-H.Wang, F.Chen, Y.-C.Faan, M.Kashani, M.Malaty and S.A.Jansen, J.Phys.Chem., 99 (1995) 6801.
430. K.S.Page and P.A.Monson, Phys.Rev.E., 54 (1996) 6557.
431. E.Lomba, Phys.Rev.E., 48 (1993) 283.
432. E.Kiesak, M.L.Rosinberg and G.Trajdos, J.Phys.-Cond.Matter, 8 (1996) 9621.
433. L.Belloni, J.Chem.Phys., 98 (1993) 8080.
434. E.Pittard, M.L.Rosinberg and G.Trajdos, Molec.Simul., 17 (1996) 399.
435. A.Meroni, D.Levesque and J.J.Weiss, J.Chem.Phys., 105 (1996) 1101.
436. A.T.Bell, E.J.Maginn and D.N.Theodorou, in: Hanbook of Heterogenous Catalysis, G.Ertl, H.Knözinger and J.Weitkamp (eds.), VCH, Weinheim, 1997.
437. T.Suzuki, K.Kaneko and K.E.Gubbins, Langmuir, 13 (1997) 2545.
438. K.Kaneko, this book, vol. 2.
439. M.Borówko, P.Borówko and W.Rżysko, Ber.Bunsenges.Phys.Chem., 101 (1997) 1050.
440. M.Schoen and D.J.Diestler, Phys.Rev., E56 (1997) 4427.
441. M.Schoen and D.J.Diestler, Chem.Phys.Lett., 273 (1997) 296.
442. J.L.Riccardo, W.A.Steele, A.J.R.Cuesta and G.Zgrablich, Langmuir, 13 (1997) 1064.
443. V.A.Bakaev and W.A.Steele, in: Adsorption on New and Modified Inorganic Sorbents, A.Dąbrowski and V.A.Tertykh (eds.), Elsevier, Amsterdam, 1996, 335.
444. A.Patrykiejew J.Sawczak, S.Sokołowski and P.Szabelski, in: Proceedings ISSHAC III, Toruń, Poland, August, 1998, 271.
445. R.I.Derrah and D.M.Ruthven, Can.J.Chem., 53 (1975) 996.
446. R.V.von Ostrejko, German Patent No. 136 792 (1901).
447. R.M.Barrer, Zeolite and Clay Minerals, Academic Press, London, 1978.
448. D.W.Breck, Zeolite Molecular Sieves, Wiley, New York, 1974.
449. A.Dąbrowski, R.Leboda, J.Goworek and J.K.Garbacz, in: Adsorption on New and Modified Inorganic Sorbents, Elsevier, Amsterdam, 1996, 664.
450. 450 R.Ch.Bansal, J-B.Donnet and F.Stoeckli, Active Carbon, Marcel Dekker, New York and Basel, 1988.
451. E.F.Vansant, P.Van Der Voort and K.C.Vrancken, Characterisation and Chemical Modification of the Silica Surface, Elsevier, Amsterdam, 1995.

452. The Colloid Chemistry of Silica, H.E.Bergna (ed.), Amer.Chem.Soc., Washington, D.C., 1995.
453. R.P.W.Scott, Silica Gel and Bonded Phases, Wiley, Chichester, 1993.
454. K.P.Goodbay and H.L.Fleming, Chem.Eng.Prog., (1984) 63.
455. B.C.Lippens, Structure and Texture of Aluminas, Delft University of Technology, Delft, 1961.
456. J.B.Peri, J.Phys.Chem., 69 (211) 1965.
457. K.Wefers and G.M.Bell, Oxides and Hydroxides of Aluminum, Technical Paper No. 19, Alcoa, Pittsburgh, 1972.
458. H.L.Fleming, this book, vol. 1.
459. K.K.Unger and U.Trudinger, Anal.Chem., 98 (1989) 145.
460. U.Trudinger, G.Muller and K.K.Unger, J.Chromatogr., 535 (1990) 111.
461. J.Nawrocki, M.P.Rigney and P.W.Carr, J.Chromatogr., A, 657 (1993) 229.
462. J.J.Pesek and V.H.Tang, Chromatographia, 39 (1994) 649.
463. Zeolites: Facts, Figures, Future, Elsevier, Amsterdam, 1989, parts A and B.
464. Recent Advances and New Horizons in Zeolite Science and Technology, Elsevier, Amsterdam, 1996 (see also other volumes on the subject - this series).
465. Yu.I.Tarasevich, this book, vol. 2.
466. P.Samaras, X.Dabou, G.P.Sakellaropoulos, this book, vol. 1.
467. B.E.Yoldas, US Patent No. 3 941 719 (1979).
468. Z.Jaworska-Galas, S.Janiak, W.Mista, J.Wrzyszcz and M.Zawadzki, J.Mater.Sci., 28 (1993) 2075.
469. J.Wrzyszcz, Z.Jaworska-Galas, S.Janiak and M.Zawadzki, PL Patent No. 165 026 (1994).
470. Adsorption on New and Modified Inorganic Sorbents, A.Dąbrowski and V.A.Tertykh (eds.), Elsevier, Amsterdam, 1996.
471. E.Wimer in: Proccedings of the Int.Conf., The World of New Technologies, Cracow, May 1996, 95.
472. J.Lin and S.Deng, this book, vol. 1.
473. E.S.Kikkinides, A.K.Stubos, K.P.Tzevelekos, A.Ch.Mitropoulos, N.K.Kanellopoulos, this book, vol. 1.
474. Chemistry for Innovative Materials, G.F.Cerofolini, R.M.Mininni and P.Schwartz, EniChem, Milan, 1991.
475. C.J.Brinker and G.W.Scherer, Sol-Gel Science, Academic Press, New York, 1990.
476. A.L.Dawidowicz, in: Adsorption on New and Modified Inorganic Sorbents, A.Dąbrowski and V.A.Tertykh (eds.), Elsevier, Amsterdam, 1996, 31.
477. B.Procyk, L.Stoch, A.Kielski and K.Wodnicka, in: Proceedings of the Int.Conf., The World of New Technologies, Cracow, May, 1996, 165.
478. D.M.Ruthven, S.Farooq and K.S.Knaebel, Pressure Swing Adsorption, VCH Publishers, New York, 1994.
479. The Special Issue of Adsorption Journal, Adsorption, 4 (2) (1998).
480. R.Ash, R.M.Barrer and C.G.Pope, Proc.Roy.Soc., London, A 271 (1963) 19.

481. R.Ash, R.M.Barrer and R.T.Lowson, J.Chem.Soc., Faraday Trans. I, (1969) 2166.
482. Membrane Technology: Applications to Industrial Wastewater Treatment, A.Caetano, M.N.de Pinho, E.Drioli and H.Muntau (eds.), Kluwer Academic Publ., Dordrecht, 1994.
483. P.Simon, Ion-Exchange Training Manual, Chapman and Hall, London, 1996.
484. Z.Hubicki, A.Jakowicz and A.Łodyga, this book, vol. 2.
485. J.Szczypa, I.Kobal and W.Janusz, this book, vol. 2.
486. J.Rule, this book, vol. 2.
487. K.Sykut, J.Saba, this book, vol. 2.
488. R.Kalvoda, this book, vol. 2.
489. E.Matijevic, this book, vol. 1.
490. A.Lopez-Cortes and J.L.Ochoa, this book, vol. 2.
491. Z.Łukaszewski, this book, vol. 2.
492. A.El Ghzaoui and S.Partyka, this book, vol. 2.
493. D.H.Everett, Basic Principles of Colloid Science, Royal Soc.Chem., Letcherworth, 1989.
494. S-W. Tam Chang and I.Iverson, this book, vol. 1.
495. V.N.Izmailova and G.P.Yamploskaya, this book, vol. 1.
496. D.C.Patel and R.G.Luo, this book, vol. 1.
497. G.Cacciola and G.Restuccia, this book, vol. 2.
498. T.K.Ghosh and A.L.Hines, this book, vol. 1.
499. G.Pan, this book, vol. 2.
500. S.G.Heijman and R.Hopman, this book, vol. 2.
501. M.Rostam-Abadi, H-C.Hsi, S.Chen, M.Rood, R,Chang, T.R.Carey, C.F.Richardson and W.Rosenhoover, this book, vol. 1.
502. Yu.I.Tarasevich, this book, vol. 2.
503. T.K.Ghosh and A.L.Hines, this book, vol. 2.
504. L.DallBauman and J.Finn, this book, vol. 2.

Adsorption and its Applications in Industry and Environmental Protection
Studies in Surface Science and Catalysis, Vol. 120
A. Dabrowski (Editor)



# Industrial carbon adsorbents

A. Świątkowski

Institute of Chemistry, Military Technical Academy, Kaliskiego 2, 01-489 Warsaw, Poland

Industrial carbon adsorbents are complex products difficult to classify on the basis of their behaviour, surface characteristics, properties or utility. However, they are usually categorised according to their particle shapes and size into powdered, granulated, spherical, pelletised or fibrous (in the form of felt or cloth) activated carbons. These categories can be further sub-divided according to their various properties, e.g. porous structural or surface chemical structural characteristics.

The various forms of activated carbon cover a wide range of amorphous, carbon-based materials produced for their high degree of porosity and extensive specific surface area. Activated carbons belong to a group of carbonaceous materials whose structure and related properties are roughly similar to those of graphite. The properties of such materials, the structural units of which are based on hexagonal layers of carbon atoms, are closely dependent on their structure (size, thickness and stacking order of the hexagonal layers) and texture (aggregation of the layers on a nanometric scale), as characterised by X-ray diffraction and electron microscopy. Activated carbons are unique and excellent adsorbents because they are porous. The different types of pores are defined by their width, the distance between the walls of a slit-shaped pore or the radius of a cylindrical pore. The porous structure of an activated carbon is the result of a wide range of pore sizes. Their linear dimensions range from about 0.35 to above one thousand nanometers, so for practical reasons they are classified into three groups: micropores, mesopores and macropores. The suitability of an activated carbon for a particular application depends on the ratio in which pores of different sizes are present. Since activated carbon can be manufactured with various pore structures, it has been widely used as an adsorbent. Its use has expanded greatly in an attempt to counteract environmental problems due to increasing amounts of hazardous pollutants in water supplies and flue gases. Several authors [1-12] have extensively reviewed the production, properties and uses of activated carbons.

# 1. ACTIVATED CARBONS IN THE FORM OF PARTICLES OF VARIOUS SHAPES AND SIZES

About 55% of such activated carbons are produced in powder form, and about 35% in granular form; the remainder is manufactured as pellets or extrudates [11]. About 80% of the total production (powder, granular and formed carbon) is used in liquid-phase applications, the remaining 20% (granular and formed carbons only) in gaseous-phase applications [7]. While granular or formed carbons are more desirable for continuous or cycling processes, powdered carbons find a use in once-through processes.

## 1.1. Raw materials for the production of activated carbons

The principal adsorptive and physical properties of manufactured activated carbons depend on the type and properties of the raw material used. Any cheap substance with a high carbon and low inorganic content can be used as a starting material for the production of activated carbon. Important considerations to be made when choosing a commercial carbon precursor include cost, availability, quality, workability, and, particularly in the case of coals, peat and lignite, the mineral matter and sulphur contents. In practice, the principal raw materials used in the industrial production of activated carbons include wood, coal, lignite, coconut shells and peat.

Where wood is a precursor, some 55% of activated carbons are produced from pine alone; the other 45% is derived from other woods. Where coal is the raw material, more than 90% of it consists of the bituminous and sub-bituminous varieties [7].

The carbons obtained from wood have a large pore volume but a low hardness and density and are therefore generally used in liquid (aqueous) phase applications. These carbons are not very suitable for vapour adsorption applications. The carbons obtained from hard coal have a large pore volume and a high hardness and density, and are used in gas/vapour adsorption applications. Activated carbons obtained from soft coal have a medium pore volume, hardness and density and can therefore be used in both vapour phase and liquid phase applications. Lignite produces hard carbons with a small micropore volume and is generally preferred in aqueous applications. Coconut shells or other nut shells produce hard carbons with large micropore volumes suitable mainly for vapour phase applications.

Some activated carbons are obtained by initially mixing the carbonaceous raw material with a binding agent, which usually is wood or coal tar, or pitch. Attempts have also been made to use resins and products from the petroleum industry for this purpose, but so far it has not been possible to provide clear principles for selecting a binder for extruding formed carbons.

## 1.2. The manufacture of activated carbons

The manufacture of activated carbons involves two main processes: the carbonization of a suitable carbonaceous raw material at temperatures below 800°C in an inert atmosphere and the activation of the carbonized product thereby obtained. The carbonization and activation steps are sometimes carried out simultaneously. The most general commercial processing route to producing activated carbons can be shown schematically:

Raw Materials<br>
↓ ↓<br>
Reconstitution → Grinding/Classifying<br>
↓ ↓<br>
Carbonization<br>
↓<br>
Activation<br>
↓<br>
Sizing - Crushing, Grinding,<br>
Sieving, Classifying<br>
↓<br>
Product

The preparation of the raw material depends to a large extent on its nature and the desired form of the activated carbon product. Before carbonization, precursors are first crushed or ground into lumps or granules of the proper size. Reconstitution involves pulverizing the starting material and then agglomeration by extrusion, briquetting or tabletting. Precursors that tend to form soft carbons have to be reconstituted at this stage, often with the use of a binding agent, in order to produce strong granular or formed carbons. As a result of carbonization, volatiles are eliminated, hydrogen is lost, and a fixed carbon mass with a rudimentary porosity is formed. This pore structure is developed further during activation. Carbon activation procedures can be broadly divided into two main types: physical and chemical. Physical activation is a heterogeneous, solid-gas reaction, involving the gasification of the more reactive portions of the carbon skeleton of the carbonized precursor by oxidation with water vapour or carbon dioxide in the 850-1000°C temperature range. In the chemical activation process, carbonization and activation are carried out in a single step by the thermal decomposition of the starting material impregnated with reagents such as phosphoric acid, zinc chloride or sulphuric acid. These activating agents act as both dehydrating agents and oxidants. Following activation, a final sizing operation takes place. There may be additional size reduction or de-dusting of the product.

Carbonization

This process is usually conducted in rotary kilns or multiple hearth furnaces at temperatures below 800°C in a stream of an inert gas, and leads to the elimination of the non-carbon species as volatiles during the thermal decomposition of the precursor, loss of hydrogen, and the formation of free radicals, which condense to form a rigid cross-linked solid char. Carbon atoms group themselves into sheets of condensed aromatic ring systems with a certain degree of planar structure. The mutual arrangement of these aromatic sheets is irregular and therefore leaves free interstices between them. These interstices may become filled with tarry matter or decomposition products, or at least partially blocked by the disorganised carbon. Carbonization thus produces a fixed carbon mass with a rudimentary pore structure. Carbonized products with the requisite properties are obtained by suitably adjusting the conditions under which the carbonaceous material is pyrolysed. The important parameters of the process that determine the quality and the yield of the carbonized material are on the one hand the rate of heating, the final temperature achieved, and the soaking time at this temperature, and on the other, the type and properties of the precursor. The process occurs in two important stages that significantly determine the properties of the final product. The first one is the softening period, during which the temperature control has an important bearing on the type of char obtained. After the softening phase the char begins to harden and shrink. The shrinkage of the char also plays a role in the development of porosity in the carbonized product.

Physical activation

Physical activation converts the carbonized raw material into a product that contains an extremely high surface area and a porous structure of molecular dimensions. The aim of this process is to enhance the volume and to enlarge the diameters of the pores formed during carbonization and to create some new porosity. Physical activation is usually carried out at temperatures between 800 and 1000°C in the presence of suitable oxidizing gases such as steam, $CO_2$ or air, or any mixture of these gases.

The process is usually carried out in rotary kilns, vertical multiple hearth furnaces or fluidized-bed reactors. Rotary kilns are the most versatile and widely used furnaces. They can be used for carbonization and/or activation and are available in a wide range of sizes. In this type of reactor the feed material is fed in granular or extrudate form. The vertical multiple hearth furnace consists of separate circular hearths, each of which can be maintained under different conditions of temperature and steam flow. Such reactors are usually used for activating carbonized material in granular form. The fluidised-bed reactor can be used for carbonization and/or activation. This type of reactor was used in a process developed to produce activated carbon from anthracite.

The heat is supplied by the combustion of coke oven or natural gas, as it is the most economical: both the required heat and the activation agent are supplied simultaneously. In directly fired reactors, extra steam is added to moderate the temperature.

Gasification of the carbonized raw material with steam and carbon dioxide occurs by the following endothermic reactions:

$$C + H_2O \rightarrow CO + H_2 \qquad \Delta H = +117\ kJ\ mol^{-1} \qquad (1)$$

$$C + CO_2 \rightarrow 2CO \qquad \Delta H = +159\ kJ\ mol^{-1} \qquad (2)$$

Activation with carbon dioxide requires a higher temperature than that with steam. The heterogeneous reaction of carbon with steam is accompanied by the water-gas shift reaction catalysed by the carbon surface:

$$CO + H_2O \rightarrow CO_2 + H_2 \qquad \Delta H = -42\ kJ\ mol^{-1} \qquad (3)$$

The rate of reaction of carbon with steam is retarded by the product $H_2$. The inhibiting effect due to hydrogen may be attributed to the fact that the active centres of the carbon surface are blocked by its adsorption.

In actual industrial processes, the activating agent is generally flue gas, to which a certain amount of steam has been added so that a combined activation with carbon dioxide and steam can occur.

Where oxygen acts as activating agent, both the reactions

$$C + O_2 \rightarrow 2CO_2 \qquad \Delta H = -406\ kJ\ mol^{-1} \qquad (4)$$

and

$$2C + O_2 \rightarrow 2CO \qquad \Delta H = -226\ kJ\ mol^{-1} \qquad (5)$$

are exothermic; combustion is excessive and the process is difficult to control. As the reaction is extremely violent, the burning is not just restricted to the pores but also occurs on the external surface of the carbon grains, causing excessive loss of mass. Thus, activation with oxygen as gasifying agent is rarely used.

Although the exact mechanism of physical activation is not completely understood, it can be visualised as a reaction of the activating agent with the carbon atoms to form the structure of the carbonized raw material. The activation reaction involves two stages [7]. In the first stage the disorganised carbon is burned out preferentially when the extent of burnoff (gasification) does not exceed 10%. This results in the opening of the blocked pores. In the second step, the carbon of the aromatic ring system starts burning, producing active sites and wider pores. At higher burnoffs the significant effect is the widening of the existing pores or the formation of the large-sized pores by the complete burnout of the walls between the adjacent pores. Activation with $CO_2$ promotes external oxidation and development of larger pores compared to activation with steam. Generally, the differences in porosity created by the different activating agents depend on the extent of burnoff.

Chemical activation

Chemical activation involves carbonization and activation in a single step in which the raw material, impregnated with certain chemical agents, is thermally decomposed. This process is usually applied where the precursor is of wood. The reagents most commonly used in industry are zinc chloride, phosphoric acid or sulfuric acid. These act both as dehydrating agents and as oxidants, so that carbonization and activation take place simultaneously. Chemical activation involves the reaction of the precursor with the activating agents at temperatures between 500 and 800°C. The raw material is impregnated with the given chemical reagent in the form of a concentrated solution by mixing and kneading: this procedure causes the cellulose to break down. The impregnated material is then extruded and pyrolyzed in the absence of air. In this stage, the chemical activator dehydrates the raw material, as a result of which the carbon skeleton is charred and aromatised, and a porous structure is created. The product is then cooled and washed to remove the activating agent, which is recycled.

Up to 1970 the most commonly used activating agent was zinc chloride [11]. With this compound, the optimum temperature is around 600-800°C. Studies of the activation of different woods, almond shells, nut shells and agricultural waste have shown that the main factors controlling porosity development are the lignin-to-cellulose ratio in the raw material, the activating agent-to-precursor ratio, the calcination temperature and heating rate [13]. However, because of limited reagent recovery efficiency and environmental disadvantages, the use of $ZnCl_2$ is decreasing in favour of phosphoric acid.

In the phosphoric acid process, the activation temperature is comparatively low, between 350 and 500°C. This activating agent is most commonly used when sawdust is the raw material. The dried sawdust is mixed with a concentrated solution of phosphoric acid into a paste, which is then calcined. The acid can be recovered at a high concentration by means of multi-stage extraction. The activated carbon is washed with water, dried and finely ground.

The chemical activation of coals has had only a limited application [11].

Recently, activated carbons with very high surface areas have been prepared by chemical activation with alkaline reagents. High-surface-area activated carbon obtained from a mixture of various carbonaceous materials and excess KOH has attracted attention because of its extremely large surface area (over 3000 $m^2/g$ BET surface area) and vast adsorptive capacity. The production of high-surface-area activated carbon by using a mixture of petroleum coke or coal and excess KOH was originally developed in the USA by AMOCO [14,15]; the production of this type of carbon (MAXSORB) was commercialised by the Kansai Coke and Chemicals Co. Ltd. (Japan), in 1992 [16,17]. The continuous process consists of a two-stage heat treatment of the mixture in an inert atmosphere [18]. In the first step the mixture of dried petroleum coke and excess KOH is dehydrated at 400°C. In the second stage activation is carried out at temperatures between 500 and 900°C under nitrogen. The remaining KOH and salts formed during the activation are removed by washing thoroughly with water, which contains added acid if necessary. The product is then dried and processed to obtain typical forms such as powders or pellets in various sizes.

The activation mechanism involves the concept of the potassium cycle [19]. The possible potassium cycle occurs by the following reactions:

$2KOH \rightarrow K_2O + H_2O$ (dehydration) (6)

$C + H_2O \rightarrow H_2 + CO$ (water-gas reaction) (7)

$CO + H_2O \rightarrow H_2 + CO_2$ (water-gas shift reaction) (8)

$K_2O + CO_2 \rightarrow K_2CO_3$ (carbonate formation) (9)

A considerable amount of metallic potassium is formed above 700°C. This is thought to be formed by reduction with carbon or hydrogen at high temperatures:

$K_2O + H_2 \rightarrow 2K + H_2O$ (reduction with hydrogen) (10)

$K_2O + C \rightarrow 2K + CO$ (reduction with carbon) (11)

It was therefore, concluded that since the metallic potassium was mobile at the activation temperatures, it must be intercalated within the carbon matrix. As a result, several atomic layers of carbon were widened and formed cage-like pores.

### 1.3. Particle size and shapes of commercial activated carbons

Commercial activated carbons are produced in various forms. They are therefore classified on the basis of their particle size and shapes into powdered, granular and formed.

Powdered activated carbons (PAC) have very small particle sizes (usually less than 100 μm in diameter). Their advantage over large particles is that there is less diffusional resistance to adsorption and, hence, much higher adsorption rates are attained. Powdered carbons are generally prepared by chemical activation from sawdust. They are preferably used for adsorption from the liquid phase and their application is simple. PAC is added to the solution directly, agitated, left in contact for a short time, and subsequently separated by filtration.

Granular activated carbons (GAC) have a relatively larger size of particles compared to PAC and consequently present a smaller external surface. Diffusion of the adsorbate is thus an important factor. Although granular or formed carbons are more expensive than powdered ones, they may be more cost-effective if the usage rate is high, since they can be regenerated and re-used. Granular or formed carbons are more desirable for continuous or cyclic processing in fixed-bed operations. The pressure drop over the bed can be controlled by adjustment of the particle-size distribution. The grain size is selected depending upon the height of the bed to be used. GACs can be prepared by physical activation methods using a variety of raw materials such as bituminous, sub-bituminous and lignite coals, petroleum coke, and peat.

Extruded activated carbons are obtained by preliminary mixing of the carbonaceous raw material with a binding agent, which usually is a wood or coal tar or else pitch. However, attempts have also been made to use resins, products from the petroleum industry or other binding materials for this purpose. The granulation process involves the preparation of a paste composed of finely ground coal dust and tar, extrusion of the granules and drying. The most common method of producing cylindrical granules consists of extruding the plastic paste including the coal-binding agent through a nozzle. The filaments obtained are cut into cylindrical granules of required dimensions. The formed granules after drying are carbonized. Many methods of granulating activated carbons have been developed that allow the production of spherical granules. Such granules have considerable advantages over cylindrical or irregularly shaped ones, the latter being particularly liable to crumbling or abrasion.

## 1.4. Porous structure of activated carbons

Activated carbon is generally considered to consist of turbostratic crystallites. These basic structural units of activated carbon are similar to graphite. The graphite crystal is composed of layers of fused hexagons held approximately 0.335 nm apart by weak van der Waals forces. The carbon-carbon bond distance within each layer is 0.142 nm. The carbon layers are so arranged that half of the carbon atoms in any plane lie above the centre of the hexagons in the layer immediately below it. The level of structural imperfections in activated carbon crystallites is very high. The ringed structures at the edges of the planes are often heterocyclic, owing to the nature of either the raw material or the preparation process. The turbostratic character of the orientation between adjacent planes in the crystallites is produced partly by the functional groups terminating the crystallite. Carbons are distinguished from each other on the basis of crystallite size. The crystallites are composed of a few (about three [2]) parallel plane graphite layers, the diameter of which is estimated to be about 15 nm [20] or about nine times the width of one carbon hexagon [2]. The regular array of carbon bonds in the surface of the crystallites is disrupted during the activation process, yielding free valences which are very reactive. The crystallites are randomly oriented and extensively interconnected. The structure of an activated carbon may be visualised as stack of poorly developed aromatic sheets (crystallites), distributed and cross-linked in a random manner, separated by disorganised carbonaceous matter and inorganic matter (ash) dependent on the raw material. The anisotropic crystallite alignment is associated with the presence of voids.

The type of the raw material, the specific additives, and the carbonization and activation temperatures are of great importance in the preparation of activated carbons in relation to their porous structure. During carbonization, most of the non-carbon elements (hydrogen, oxygen, traces of sulfur and nitrogen) are removed in gaseous form by the pyrolytic decomposition of the precursor; the carbonized product is then formed by more or less disordered elementary graphitic crystallites with a poorly developed porous structure. During activation the spaces between the crystallites become cleared of less organised carbonaceous matter and, at the same time, part of the carbon is also removed from the crystallites. Selective oxidation of intercrystalline material and planes of the crystallites may give rise to an extensive porous system. The resulting channels through the graphitic regions and the spaces between the crystallites of activated carbon together with the fissures within and parallel to the graphitic planes constitute the porous structure, with a large internal surface area (usually in the range 500 to 1500 $m^2/g$). The total porous structure is formed by a wide range of pore sizes.

For practical reasons the pores of activated carbons can be divided into three basic classes: micropores, mesopores and macropores. A schematic representation of these types of pores is shown in Figure 1.

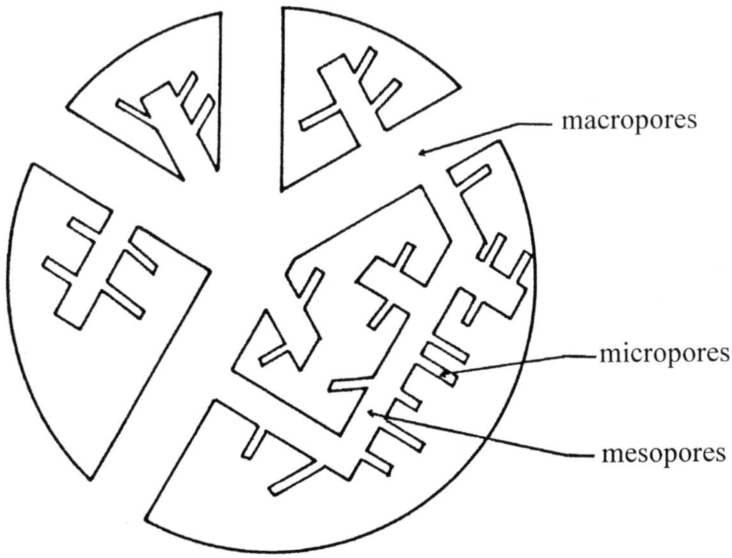


Figure 1. Schematic representation of the different types of pores in a particle of activated carbon (From Rodriguez-Reinoso and Linares-Solano [9]. Reproduced with permission from Marcel Dekker, Inc.).

Several standards for the grouping of pore size ranges have been used in the past. According to Dubinin [21], adsorbent pores can be classified into micropores with linear size up to x< 0.6-0.7 nm, supermicropores with 0.6-0.7< x <1.5-1.6 nm, mesopores with 1.5-1.6< x <100-200 nm, and macropores with x >100-200 nm. The linear size of a pore is the half-width in the slit-like pore model, and the radius in cylindrical or spherical pores. The classification proposed by Dubinin is based on the difference in the mechanisms of adsorption and capillary phenomena occurring in adsorbent pores. The finest pores, i.e. the micropores, are commensurate with the molecules adsorbed. When the dispersion force fields of the opposite micropore walls are superimposed, the adsorption energies in micropores are greatly increased. While the curvature of the mesopore surface hardly affects adsorption, mono- and polymolecular adsorption does occur there, which acquires a clear-cut physical significance, which causes the mesopore volume to be filled by the capillary condensation mechanism. The supermicropores form the transitional porosity region above which the characteristic features of the micropores gradually degenerate and the mesopore properties begin to manifest themselves. Finally, macropores remain practically unfilled by capillary condensation because of the extremely low sorption rate at x >100-200 nm, and act as large transport arteries in the adsorption process.

Porosity in carbon materials can be conventionally classified using a scheme recently proposed by the International Union of Pure and Applied Chemistry (IUPAC) and reported by Sing et al. [22]:
micropores - pore width < 2 nm,
mesopores - pore width between 2 and 50 nm,
macropores - pore width > 50 nm.

The definition of the different types of pores is based on their width, which represents the distance between the walls of a slit-shaped pore or the radius of a cylindrical pore. This classification, which is not entirely arbitrary, is now widely accepted and used. It takes into account differences in the behaviour of molecules adsorbed in micropores and in mesopores. It appears that for pore widths exceeding 1.5-2.0 nm, the gaseous adsorbate condenses in a liquid-like state and a meniscus is formed. As a consequence, a hysteresis loop appears on desorption and its interpretation can lead to the distribution of the mesopores in the adsorbent [23]. The limit between mesopores and macropores at 50 nm is more artificial, and corresponds to the practical limit of the method for pore-size determination based on the analysis of the hysteresis loop. As a rule, the porous structure of the usual types of activated carbons is tridisperse, i.e. they contain micropores, mesopores and macropores. Micropores are of the greatest significance for adsorption owing to their very large specific surface area, and their large specific volume. At least 90-95% of the total surface area of an activated carbon can correspond to micropores.

The porous structure of activated carbons can be characterised by various techniques [1,2,5-12,21,23] such as:
- adsorption of gases ($N_2$, Ar, $CO_2$) or vapours (benzene, water) by static (volumetric or gravimetric) or dynamic methods,
- adsorption from liquid solutions of solutes which have a limited solubility (i.e. from dilute solutions) and of solutes which are completely miscible with the solvent in all proportions,
- gas chromatography,

- immersion calorimetry,
- flow microcalorimetry,
- temperature-programmed desorption,
- mercury porosimetry,
- transmission electron microscopy (TEM) and scanning electron microscopy (SEM),
- small-angle X-ray scattering (SAXS),
- X-ray diffraction (XRD).

## 1.5. Chemical structure of the surface of activated carbon

Although the adsorption capacity of activated carbons is determined mainly by their porous structure, it is also influenced by the chemical structure of their inner surface. Imperfections in the elementary microcrystalline structure due, for example, to the presence of partially burnt off graphite layers, alter the arrangement of the electron clouds in the carbon skeleton. As a result, unpaired electrons and incompletely saturated valencies appear, and this condition influences the adsorptive properties of activated carbons, especially towards polar substances. Furthermore, activated carbons are almost invariably associated with appreciable numbers of heteroatoms such as oxygen and hydrogen. In addition, they can be associated with atoms of nitrogen, sulphur or chlorine. These elements may have originated from the raw material and remain as a result of imperfect carbonisation, or they may become chemically bonded to the surface during activation. These heteroatoms are bonded at the edges of the aromatic sheets and form surface compounds (surface functional groups). They can also be incorporated within the carbon layers forming heterocyclic ring systems. In general, 5 to 20 percent by weight of activated carbon consists of non-carbon elements. The presence of oxygen and hydrogen has a great effect on the properties of the carbon. Carbon-oxygen surface compounds, often containing hydrogen, are by far the most important structures in influencing the surface characteristics and surface behaviour of activated carbons. Attempts have been made to characterise (identify and estimate) the surface oxygen chemical structures (surface functional groups) using several chemical, physicochemical and physical methods, which include neutralisation of bases, direct analysis by specific chemical reactions, potentiometric and thermometric titrations, thermal decomposition of the surface compounds, enthalpy of immersion measurements, polarography, infrared spectroscopy and X-ray photoelectron spectroscopy. The functional groups more frequently suggested as being present on the surface of activated carbon include carboxyl groups, phenolic hydroxyl groups, quinone-type carbonyl groups, lactones (normal and fluorescein-type) groups, carboxylic acid anhydride groups and cyclic peroxide groups. Fig. 2 shows the diagrammatic representations of different oxygen-containing functional groups existing on carbon surfaces (identified as IR-active functionalities) recently proposed by Fanning and Vannice [24].

However, the surface functional groups cannot be treated as ordinary organic compounds. They behave as combined structures presenting numerous mesomeric forms depending on their location on the same polyaromatic frame.

It is common practice to classify surface oxygen groups as acidic or basic. Steenberg [25] examined the acid-base properties of activated carbons as a function of the activation temperature. He used the uptake of inorganic acid and base as a method of characterising the carbons activated and oxidized at different temperatures. He further proposed a nomenclature for the carbons he studied, classifying those oxidized at low temperature which adsorbed primarily hydroxide ions as L carbons and those activated at high temperature by adsorbing hydrogen ion as H carbons. The borderline activation-oxidation temperature between L and H carbons was about 500-600°C. Electrokinetic studies have indicated that L carbons exhibit a negative surface potential and H carbons a positive surface potential. Although the variety of activation procedures results in both acidic and basic surface oxides on the activated carbon, this classification is still useful to describe the overall nature of a carbon.

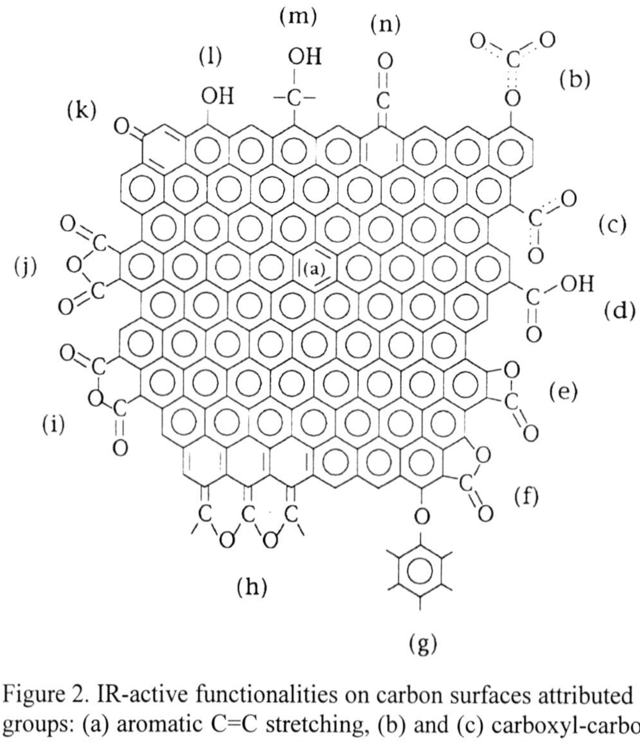


Figure 2. IR-active functionalities on carbon surfaces attributed to different oxygen functional groups: (a) aromatic C=C stretching, (b) and (c) carboxyl-carbonates, (d) carboxylic acid, (e) lactone (4-membered ring), (f) lactone (5-membered ring), (g) ether bridge, (h) cyclic ethers, (i) cyclic anhydride (6-membered ring), (j) cyclic anhydride (5-membered ring), (k) quinone, (l) phenol, (m) alcohol, and (n) ketone (Fanning and Vannice [24]. Reprinted with permission from Pergamon Press Ltd.).

Acidic carbons defined as those carbons that adsorb appreciable amounts of bases but very small quantities of acids can be produced by oxidizing as-received carbons. There are several techniques by which acidic oxygen surface groups can be formed: methods utilising oxidizing gases such as oxygen, water vapour, carbon dioxide or nitrogen oxides, and those involving oxidizing solutions such as nitric acid, a nitric-sulfuric acid mixture, ammonium persulfate, chlorine water or acidified potassium permanganate. Carboxyl, lactone and phenolic are functional groups generally regarded as producing acidic surface oxides. The neutralization of bases is one of the earliest and simplest methods used to determine the nature of the acidic surface groups on carbons. Boehm [26] differentiated between acidic groups by selective neutralization with bases of different basicities: $NaHCO_3$, $Na_2CO_3$, NaOH, and $C_2H_5ONa$. He suggested that the neutralization of $NaHCO_3$ was due to carboxyl groups, that of $Na_2CO_3$ to carboxyl + lactone groups, that of NaOH to carboxyl + lactone + phenolic groups, and that of $C_2H_5ONa$ to carboxyl + lactone + phenolic + carbonyl groups. The neutralization capacity of the four bases with respect to the oxidized carbons was in the simple ratio 1:2:3:4. Puri et al. [27] examined a large number of sugar and coconut charcoals before and after outgassing and extensive oxidation, and attempted to correlate the base neutralization capacity of the charcoal with the oxygen evolved as $CO_2$ on evacuation at 1200°C. It was found that in each case the amount of alkali neutralized was close to the amount of $CO_2$ evolved on evacuation (they termed this 'the $CO_2$ complex'). They hold the view that in charcoals the same surface group which evolves $CO_2$ on evacuation is involved in the neutralization of alkalis.

Basic carbons are carbons exhibiting basic behaviour, and adsorbing acids but few bases. The fact that activated carbons can adsorb strong acids such as HCl has been known for a long time, but the nature of the basic surface functional groups has not yet been elucidated entirely satisfactorily. Steenberg [25] was of the opinion that acid adsorption resulted largely from physical forces. The protons were held close to the surface by primary adsorption forces and the anions by secondary electrostatic forces forming the outer part of the diffuse double layer. Garten and Weiss [28] suggested the existence of chromene-type structures on basic carbons, which were readily oxidised at room temperature in the presence of acid to the corresponding benzopyrylium structures with the adsorption of the acid anion and the formation of hydrogen peroxide. However, Voll and Boehm [29] later proposed the existence of pyrone-type structures with the oxygen atoms, in general located in two different rings of a graphitic layer, the positive charge being stabilised by resonance. The existence of pyrone-type structures is supported by the studies of Papirer et al. [30] and Leon Y Leon et al. [31]. The former noted that accurate detection of basic oxides in the presence of acidic surface functional groups is difficult because of the reduced activity towards acid (HCl) caused either by internal neutralization by acidic groups on the same aromatic system, or by a different electron distribution caused by the presence of acidic groups. The latter presented data which were interpreted as favouring electron donor interaction as the source of the basic behaviour of activated carbons. Thus, it has not been clearly established that oxide structures are responsible for the basic reactions shown by some carbon surfaces.

The surface chemistry of activated carbons is extensively studied (recently e.g. [32,33]) and discussed in several reviews [3,4,7,10,25,26,34,35].

## 1.6. Applications of activated carbons

Typically, activated carbons have an extensive internal surface area, implied by high specific surface areas (commonly in the range of 1000-1500 $m^2/g$), a highly developed porosity, a high degree of surface reactivity and hence a large capacity for adsorbing chemicals from liquids or gases. Consequently, they are extremely versatile adsorbents of major industrial significance. Activated carbons are used in a wide range of applications concerned with the removal of species by adsorption from the liquid or gas phase, in order to effect purification or the recovery of chemicals.

Liquid-phase adsorption

One of the most widespread uses of activated carbons for liquid-phase adsorption is in water treatment. Recent years have seen an increase in the level of synthetic organic chemicals (SOC) in public water supplies. Hundreds of SOCs, such as pesticides, herbicides, detergents, polycyclic aromatic hydrocarbons, nitrosamines, phenolic compounds, trihalomethanes and other pollutants, have been identified in drinking water supplies. On the other hand, natural organic material (NOM) is found in varying concentrations in all natural water sources. It is a complex mixture of compounds formed as a result of the breakdown of animal and plant material in the environment. Most NOM consists of a range of compounds, from small hydrophilic acids, proteins and amino-acids to larger humic and fulvic acids. The reactions between NOM and disinfectants such as chlorine can produce disinfectant by-products, e.g. the reaction of chlorine with humic acids in groundwaters can produce chlorophenols and halomethanes, and these are indeed the most common products of chlorination. Many of these organic chemicals are carcinogenic. Several methods have been used with varying degrees of success for the control of organic pollutants in water. However, the use of activated carbons is perhaps the best broad-spectrum technology available at the present moment. As a consequence, the use of activated carbons in water treatment has increased throughout the world. GAC adsorption is an effective treatment technology for the removal of organics from drinking water supplies and for improving taste and odour. It is also a practicable technique for the removal of trace (heavy) metals such as Cd, Cr, Hg, Cu, Fe, V, Zn, Ni. Activated carbons are now being used on a much larger scale than ever before. Important properties of GAC for water treatment are their adsorptive capacity and selectivity, ability to withstand thermal regeneration and resistance to attrition losses during transport and handling.

The use of GAC for the treatment of municipal and industrial wastewaters has developed rapidly in the last 25 years. Moving beds, downflow fixed beds and upflow expanded beds have all been used in industrial wastewater applications. In most wastewater applications, the cost of virgin carbon usually precludes its use on a throwaway basis, so the thermal reactivation of hard, coal-based carbons has proved to be both economical and practicable. Chemical regeneration is generally limited to applications where partial recovery of capacity is acceptable and regenerant disposal is not a problem.

A second important use of activated carbons is in the removal of colour from sugars and in the purification of various foods and beverages. At present the sugar industry uses powdered and/or granular activated carbons as decolourants. Sugar solutions contain different colouring matter, such as caramels, melanoidines and iron-containing polyphenolic complexes. In part, the colour originates from the raw material, and in part it is formed during the refining process. The most easily adsorbed components are usually the melanoidines (nitrogeneous brown polymers formed by reactions between amino compounds and sugars). Activated carbons also help to remove surface-active agents and colloidal substances by raising their surface tension and decreasing their viscosity. This leads to higher rates of sugar crystallisation. In sugar processing, a typical procedure is to treat the sugar liquor with PAC. The carbon is subsequently separated by filtration. GAC is also used in various configurations of the decolourizing system (e.g. in countercurrent pulsed-bed columns).

Another, similar, application of activated carbons is in the treatment of edible oils and fats to remove undesirable components. Here, they are used in conjunction with certain bleaching clays. Activated carbon are also used in treating wines and spirits to remove any traces of fusel oil. In the production of brandies, they are used to remove undesirable flavours and to reduce the amount of aldehydes in the raw distillate. In the case of beers, activated carbons are used to improve their colour, and to remove flavours attributed to phenol and colouring matter.

A relatively recent use of activated carbons is in the recovery of gold from its ores. The subject has been extensively reviewed by Bansal et al. [7]. The process usually involves the treatment of finely ground ore with a very dilute solution of NaCN and oxygen. The gold and other metallic impurities present in the ore are oxidized and form cyanide complexes. After this stage they are leached together. The gold is recovered from the cyanide pulp by adsorption on activated carbons. In the carbon-in-pulp (CIP) process developed in the early 1950s [36] the carbon granules were directly added to the cyanide pulp and moved countercurrent to it. The gold-loaded carbon is removed by screening and the gold is recovered from it by elution using either solutions of metal salts, such as $K_2CO_3$-KOH, or mixtures of organic solvents [7]. After regeneration the carbon is recycled.

Gas-phase adsorption

Activated carbon adsorption is widely used in solvent recovery. The solvent properties and volatility of organic compounds make them suitable for a great many industrial processes, such as the manufacture of paints, polymers, rayon, adhesives and explosives, the extraction of oil from seeds, and dry cleaning. However, the high volatility of many spent solvents creates unacceptable problems if they are emitted into the atmosphere, as they constitute health, fire and explosion hazards. These problems can be resolved by collecting the vapour and recovering the solvent. The pores of the activated carbons used in solvent recovery should have a higher adsorption capacity in the supermicro- and mesopore range. The activated carbons are usually used in a cyclic process. The solvent is recouped by desorption in the regeneration stage by passing low-pressure steam through the bed in a direction counter to the air flow. For instance, flat-bed and annular-bed adsorbers, which use granular or extruded carbons, operate on a cyclic basis [11]. A relatively recent development is the continuous rotary adsorber [11].

An another industrial use of activated carbon is in hazardous gases adsorption. Activated carbon filters, combined in series with ventilation units are commonly used for the removal of toxic contaminants such as $SO_2$, $CS_2$, $H_2S$ from effluent air.

Also for separation of radioactive matter from nuclear power plants activated carbons are used [11,41]. Activated carbon filters are applied to prevent the release of radioactive iodine (in elemental form and as methyl iodide) and noble gases such as krypton and xenon into the atmosphere. Filter units of that kind are used to adsorb accidental leakages of radioactive gases and vapours. Activated carbons are usually impregnated with potassium iodide or triethylenediamine. Nuclear-grade carbon adsorbents are narrow-pored (highly microporous) and a process consists of a fixed bed.

Activated carbons have long been used to protect against toxic gases and vapours. Because of the wide range of offensive gases that can be used, the gas masks and larger protection units required for military purposes demand impregnated activated carbons. Whetlerite, activated carbon impregnated with a complex mixture of metal compounds that include copper, chromium, silver and sometimes organic species is the most commonly used active agent in protective devices. The impregnator has a high reactivity with some of toxic chemicals (e.g. cyanogen chloride or hydrogen cyanide). In the case of the others (e.g. nerve gases) physical adsorption on the carbon adsorbents occurs. Protection against these gases require the use of activated carbons in a form that allows rapid and effective adsorption (fine granules or fibres).

The technology used by the military has been adopted by industry (mainly chemical industry) to provide protection to workers against hazardous gases and vapours.

Activated carbons are also used in inhabited spaces, such as hospitals, laboratories, offices and food processing plants, where clean air is required. Air purification in inhabited spaces operates at low pollutant concentrations (below 10 ppm), and panel-type carbon filters can be used in such systems. The activated carbon should be highly microporous to effect greater adsorption at lower concentrations [7].

Particular attention is given to separation $SO_2$ and $NO_x$ from power station exhausts. Environmental regulations pertaining to emissions from stationary fossil fueled power sources have focused on $SO_2$ because it was implicated as an important component in acid deposition. Nitrogen oxides, including NO and $NO_2$, have also been implicated as precursors to acid precipitation, whereas $N_2O$ is considered a potentially important greenhouse gas. Activated carbon is used as an adsorbent for the removal of $SO_2$, and also as a catalyst for the reduction (and hence removal) of $NO_x$ from combustion flue gases.

## 2. ACTIVATED CARBON FIBRES

Activated carbon fibre (ACF) is a relatively new kind of adsorption material which has attracted increasing attention because of its advantages over conventional activated carbons, for example, its large adsorption capacity, rapid adsorption/desorption rate and multi-use forms.

ACFs in the form of felt or cloth have come under the spotlight in recent years as an adsorbent for purifying water. The first studies on this new material were carried out mainly in the gaseous phase and showed it to be very effective in removing contaminants from air. Its high specific surface area and microporosity endow ACF with good adsorptive properties both in the gaseous and the liquid phase.

ACFs have a characteristically higher adsorption rate and larger capacity than conventional GAC. Initial adsorption rates are 2.5-10 times larger with fibres than with granules, this improved efficiency being attributed to a more uniform pore-size distribution and a more efficient contact surface. ACF adsorption performance is superior to conventional activated carbon (AC) on a weight basis. On a volume basis, however, the adsorption performance of ACF is no better than that of AC, because the bulk density of ACF (~0.1 $g/cm^3$) is much smaller than that of AC (~0.5 $g/cm^3$). In order to demonstrate the merits of ACF effectively, the bulk density of ACF has to be increased by some means. Many attempts to do this have been made, including the preparation of formed ACFs with binders. These, however, tend to plug the micropores of ACFs, thereby significantly reducing the efficiency of ACF. Miura et al. [37] described a method of producing an ACF of high bulk density (HD-ACF) without any binder by the use of hot briquetting. A stabilised carbon fibre was heated under mechanical pressure to prepare successfully formed carbon fibres (CF) with elevated bulk densities. The so-formed CFs were carbonized at 950°C and activated with steam at 900°C to prepare HD-ACFs.

ACF adsorption appears to be selective towards low-molecular-weight compounds. The pores of ACFs are classified into sub-micropores (pore size

< 0.8 nm), micropores (0.8-2 nm), mesopores (2-50 nm) and macropores (> 50 nm). The adsorption capabilities of ACFs depend mainly on their pore size and specific surface area. They have quite uniform micropores, as compared with conventional activated carbons. As the micropore walls are composed mainly of micrographitic crystallites, the micropores are slit-like. ACFs with fixed pore widths are available in the 0.7 - 1.5 nm range, so the graphitic micropores of ACF are excellent hydrophobic nanospaces. As a microporous carbon, ACF strongly adsorbs gases at low pressure due to the overlap of force fields from opposite pore walls, so ACF is capable of removing a trace impurity from a contaminated gas very rapidly. While microporous ACFs have a high adsorption affinity for relatively small molecules, mesoporous ACFs are suitable for the adsorption of larger molecules. When ACFs are used to remove macromolecular impurities from solutions, such as waste water and biological fluids, only the surface area corresponding to the mesopores and macropores is effective; the micropores are inaccessible to these large molecules. So work is now intensifying on developing novel ACFs with extensive mesoporosity, but so far there have been few reports on the preparation of such fibres.

Research in ACF has attracted increasing attention in the last few years in terms of their synthesis, and their suitability in different applications that include solvent recovery, molecular sieving, gas storage and catalysis. Activated carbon fibres are usually prepared from precursors of low or intermediate crystallinity: such raw materials include polyacrylonitrile (PAN) fibres, cellulose fibres, phenolic resin fibres, pitch fibres, cloth or felts made from them, and viscose rayon cloth. They are first pyrolysed and then activated at a temperature of 700-1000°C in an atmosphere of steam or carbon dioxide. Both the processing costs and the properties of the fibre products are dependent on the nature of the starting material.

## 3. CARBON MOLECULAR SIEVES

Carbon molecular sieves (CMS) are highly microporous materials having a preponderance of pores of < 1 nm. Among the various types of carbon, CMS materials represent one member of the family of activated carbons. CMS differ from activated carbons in the actual surface composition and the pore size distribution. Unlike CMS, activated carbons display far better detectable surface functionalities. CMS are finding a number of possible uses for the separation of air or other gases and in catalysis. CMS for use as air separation sorbents are usually made from activated carbons by a post-treatment that narrows the pore-size distribution to produce a material with a biomodal pore distribution having a predominance of pores < 0.6 nm [38]. Key to the performance of these materials is their size specific selectivity. CMS are similar to zeolites in that their porous structures have dimensions sized close to the critical dimensions of small to medium sized molecules, that is, the range between 0.3 and 1 nm. As a result, separations can be made on the basis of differences in molecular sizes and shapes. Although CMS can be classified as part of a broader grouping of materials described as molecular sieves, the similarities between the zeolites and CMS ends there. The primary difference is that the CMS materials are globally amorphous, whereas the zeolites have extended long-range order. More specifically, on the length scale of X-ray coherence, approximately 2.5 nm, the CMS materials do not display any distinct, sharp diffraction pattern. Despite the rather significant difference between an amorphous and a crystalline network, it is remarkable that the physical characteristics of CMS and zeolites are at all similar. The ranges of pore volumes, apparent surface areas, and the pore sizes are quite similar. Carbon molecular sieves have some advantages over molecular sieve zeolites in terms of shape selectivity for planar molecules, high hydrophobicity, high resistance to both alkaline and acid media and thermal stability under inert atmosphere at higher temperatures.

CMS materials are commercially prepared by the pyrolysis of polymers (natural or synthetic) or the decomposition of organic molecules (deposition of carbonaceous material) onto microporous carbon hosts in order to narrow their porosity. Although polymer pyrolysis chemistry seems quite simple, it is its complexity that gives rise to the richly structured carbons displaying molecular sieving properties [39]. Most polymers will degrade upon being heated above their decomposition temperatures, but relatively few will meet the necessary criterion of generating a high yield of carbon. With some polymer precursors, such as coal tars, polyvinylchloride, phenol-formaldehyde resin, polyfurfuryl alcohol or polyacrylonitrile, high yields of carbon residue (coke or char) are obtained. The pyrolysis of any polymer can be generalised and broken down into stages (precarbonization, carbonization, dehydrogenation, annealing and graphitization) that form a continuum of chemistries with rising temperature. The combined physical processes that follow the initial chemical transformations of the polymer into the CMS are a highly activated kind of annealing. The total process can be viewed as one that transforms amorphous $sp^2$ carbon into aromatic domains which enlarge with time and temperature. As the aromatic domains grow at the expense of the amorphous carbon, the disorder in the structure drops; so too does the tendency to form open domains with a high free volume. The system always tends towards graphite. Most notable is that the micropore sizes of CMS, made from quite different precursors and by different processes, fall into essentially the same narrow range.

The micro- and nanostructures proposed by Armor [38] for air separation of CMS commercial materials is illustrated in Figure 3. The CMS pellets are composed of ~ 100 μm domains of carbon material with size selective gates capable of distinguishing the 0.02 nm size difference between $O_2$ and $N_2$.

Commonly used methods for preparing CMS include heat treatment and carbonaceous material deposition from the pyrolysis of hydrocarbons such as propene and benzene on previously prepared activated carbon in order to narrow down its porosity. The efficiency of these treatments depends on the pore-size

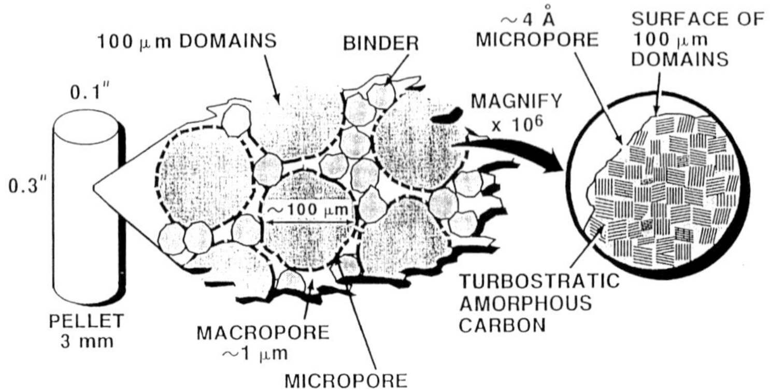


Figure 3. Proposed micro- and nanostructure for air separation of CMS materials (From Armor [38]).

distribution of the starting material and on the extent and conditions of the treatment. Juntgen et al. from Bergbau-Forschung [40-43] demonstrated and patented the large-scale, commercial production of high-quality CMS for air separation by the use of pyrolyzed coal as a host. They gave a general description of a number of routes employing the controlled pyrolysis of coal followed by cracking with organic vapours under an inert atmosphere to produce a material with the desired pore size. In one of their early patents, the workers at Bergbau-Forschung [40] provide a detailed route to $O_2$-selective CMS from mineral-grade coal. The coal is first activated in 12% $O_2$ at 230°C, then treated with soft pitch/water. This mixture is extruded into pellets and carbonized under an inert atmosphere at 10°C/min to 800°C. Finally, the product is treated with 5-12% benzene vapour at 800°C for 20 min to yield hard, 2 mm extrudates. A number of Japanese companies have patented [44-46] detailed routes to CMS materials based on coconut shell material. For example, Kuraray Chemical patented [44] a process using coal tar pitch to pelletise coconut shell carbon powder. The pellets were carbonised at 600-900°C, immersed in mineral acid (to remove alkaline metal compounds), washed and dried, impregnated with creosote, heated again at 600-900°C for 10-60 min and cooled in an inert gas. General methods of manufacturing CMS materials include carbonizing coal or nut hulls in an inert atmosphere to produce chars. These can then be activated in a reactive atmosphere, e.g. air or steam, to develop the necessary porosity. Non-$O_2$-selective materials are made selective to $O_2$ by depositing a gate-keeping layer at the pore entrances. This layer is often formed by pyrolysis of carbonaceous materials such as pitch, benzene or furfuryl alcohol. Through proper matching of hydrocarbon size to the particular carbon support and employing a two-step process for the deposition of the carbonaceous residue [38], one can convert less expensive, non-selective commercial microporous carbons into materials of potential use for the kinetic separation of $O_2$ from air. A two-step pyrolysis with a single hydrocarbon such as isobutylene is superior to any single-step treatment, since it allows a very high percentage of the pore openings to be selectively narrowed for oxygen adsorption relative to nitrogen, without becoming so narrow that capacity is lost, adsorption becomes too slow, or pores become blocked. The same hydrocarbon can be used by varying the pyrolysis conditions in two independent steps. Figure 4 illustrates the deposition of the carbonaceous residue [38].

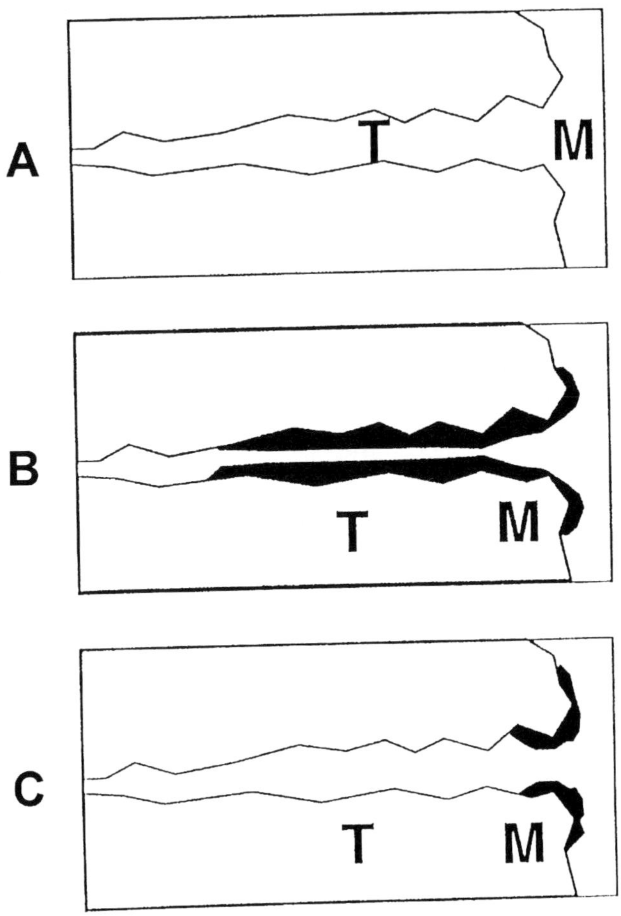


Figure 4. Deposition of carbon onto/into a pore: A - uncoated micropore of 0.5 nm; B - deposition of carbon via pyrolysis of isobutylene into the throat (T) of the pore; C - deposition of carbon via the pyrolysis of isobutylene into some of the throat (T), but mainly at the pore mouth (M) (From Armor [38]).

The last twenty years have seen a tremendous growth in research activities and commercial applications of CMS-based gas separation processes [38] such as pressure swing adsorption (PSA). PSA [47] is basically a regenerable adsorption process in which the adsorbent beds are operated cyclically between high and low pressure gas flows. The increased usage of CMS has been attributed to the development of more selective adsorbents and improved theoretical understanding of adsorption from gaseous mixtures. Although many processes have been proposed in patents which produce a variety of products, the major application of CMS in large number of units and high capacity is still the PSA process for the production of nitrogen from air. In the mid-seventies, when CMS became available, nitrogen separation by PSA quickly established itself on the market as an alternative to adsorption-based inert gas generators. In just a few years, more nitrogen PSA units were installed than all other PSA units since their introduction. The introduction of CMS caused many more manufacturers to enter the market place in the eighties, mainly in the low capacity area. The number of technical and scientific papers, the number of research groups and the number of patents filed have increased dramatically in the last decade.

CMS materials are kinetic adsorbents that separate $N_2$ from air by the faster sorption of $O_2$ by the carbon. They separate $O_2$ from $N_2$ by allowing the smaller $O_2$ molecule to fill the micropores of the carbon before the $N_2$. The kinetic diameters of $O_2$ and $N_2$ are 0.346 nm and 0.364 nm respectively. The carbon molecular sieves used for air separation are capable of distinguishing the 0.02 nm difference between $O_2$ and $N_2$. The rejection of nitrogen at the pore mouth need not be absolute; simply reducing the number of trajectories for nitrogen that successfully lead to its translation through the pore compared to those of oxygen will ensure adequate separation. This is indeed the case, because it can be shown that the equilibrium loading levels of oxygen and nitrogen on these commercial CMS are nearly equivalent; it is only the rates of approach to the equilibrium points that differ. A number of important parameters govern the performance of CMS materials as sorbents for air separation. They include $O_2$ capacity (volume of $O_2$ sorbed), the $O_2/N_2$ selectivity (the ratio of their rates of adsorption), the rate of $O_2$ sorption and the pore size distribution. Carbon molecular sieves were succesfully applied to the commercial recovery of nitrogen from air in the 1970s. Other examples of kinetic separation on CMS are the $CH_4$-$CO_2$ [48], $CH_4$-$N_2$ [49], $CH_4$-$C_2H_6$ and $N_2$-$C_2H_6$ systems [50]. The choice of adsorbent for PSA or temperature swing adsorption (TSA) applications is governed by its adsorption capacity and selectivity for the gas to be separated. Ruthven [51] has reported that for a PSA process to be economical, the minimum acceptable intrinsic separation factor for the desired component is approximately 3. With separation factors of less than 2, it is difficult to design a satisfactory process. Moreover, increases in the separation factor beyond 4-5 are of only minimal benefit. Three factors, applied singly or in combination, may be employed to achieve the desired selectivity: steric factors, such as the difference in shape and size of the adsorbate molecules; equilibrium effects, i.e. when the adsorption isotherms of the components of the gas mixture differ appreciably; and kinetic effects, when the components have substantially different diffusion or adsorption rates. Commonly used adsorbents usually take the form of granules or extruded pellets (extrudates). During gas separation in a PSA system, granular CMS can settle, resulting in channel formation which allows the gas stream to by-pass the adsorbent. Moreover, attrition of the adsorbent occurs due to abrasive wear of the granules. The carbon dust produced in this way can block the system or cause unacceptable pressure drops across the adsorption beds. The height of PSA vessels is limited by the relatively low crush strength of granular carbons.

A monolithic adsorbent carbon material has been developed that is not only strong and rigid, but can also overcome the problems associated with unit separation operations using granular adsorbents. This material is a carbon fibre composite molecular sieve (CFCMS), which consists of a petroleum pitch-derived carbon fibre and a phenolic resin-derived carbon binder [52]. Routinely, CFCMS with surface areas >1000$m^2/g$ are obtained using isotropic pitch derived fibres. The binder phase content is quite low and provides a monolithic structure by bonding the fibres at their contact points only. Even though the material can be fabricated in large sizes, it retains the beneficial gas adsorption-separation properties of its precursor carbon fibres. CFCMS is unique in that it is monolithic, potentially eliminating the attrition, dust and settling problems associated with granular carbons. The CFCMS material structure contains large voids (>100 μm) between the 10-20 μm diameter fibres, which allows for free flow of fluids through the material. A crush strength of >1 MPa is retained after activation to 40 wt % burn-off. A potentially large application for CFCMS material is gas separation using the PSA process. Current restrictions on PSA bed size and orientation will be removed because CFCMS is a strong monolith. The CFCMS allows certain equipment variations not possible with granular carbons. The use of the CFCMS would permit the employment of, for example, horizontally-oriented vessels with controlled flow of the gas mixture through the adsorbent without risk of channelling and bypass flows.

As a result of the continuity of the fibre-matrix unit, the CFCMS material is an electrical conductor. Advantage has been taken of this property to develop a novel desorption process which the authors [53] have termed 'electrical swing adsorption' (ESA). On saturation of the CFCMS with, for example, $CO_2$, immediate desorption can be accomplished by the application of a low voltage across the adsorbent. There are many important operational implications of this desorption process. ESA is very fast in comparison to pressure or temperature swing adsorption, and is an inherently low-energy process compared to the energy requirements of PSA or TSA. The major operating cost associated with gas processing by adsorption-desorption techniques is the energy requirement for system bed pressurisation (in PSA) and system heating (in TSA). The CFCMS and the ESA technique may be the basis of a new unit operation separation system for energy and environmentally relevant gas separations. Potential applications include the recovery of $CO_2$ from flue gas [54], the removal of acid gases such as $N_2O$ and $SO_x$ from flue gases, and the separation of $CO_2$ and $H_2S$ from natural gas [55].

## REFERENCES


1. M.M. Dubinin, in: Chemistry and Physics of Carbon, P.L. Walker Jr. (ed.), vol. 2, Marcel Dekker, Inc., New York, 1966, 51.
2. M. Smíšek and S. Černý, Active Carbon, Elsevier Publishing Company, Amsterdam- London- New York, 1970.
3. J.S. Mattson and H.B. Mark, Jr., Activated Carbon, Marcel Dekker, Inc., New York, 1971.
4. P.N. Cheremisinoff and F. Ellerbusch (eds.), Carbon Adsorption Handbook, Ann Arbor Science Publishers, Inc., Ann Arbor, 1978.
5. H. von Kienle and E. Bader, Aktivkohle und ihre industrielle Anwendung, Ferdinand-Enke Verlag, Stuttgart, 1980.
6. A. Capelle and F. de Vooys (eds.), Activated Carbon ... A Fascinating Material, Norit N.V., Amersfoort, 1983.
7. R.C. Bansal, J.-B. Donnet, F. Stoeckli, Active Carbon, Marcel Dekker, Inc., New York and Basel, 1988.
8. H. Sontheimer, J.C. Crittenden and R.S. Summers, Activated Carbon for Water Treatment, DVGW-Forschungsstelle, Karlsruhe, 2nd edn, 1988.
9. F. Rodriguez-Reinoso and A. Linares-Solano, in: Chemistry and Physics of Carbon, P.A. Thrower (ed.), vol. 21, Marcel Dekker, Inc., New York and Basel, 1989, 1.
10. H. Jankowska, A. Świątkowski and J. Choma, Active Carbon, Ellis Horwood, New York, 1991.
11. J.W. Patrick (ed.), Porosity in Carbons, Edward Arnold, London, 1995.
12. H. Marsh, E.A. Heintz and F. Rodríguez-Reinoso (eds.), Introduction to Carbon Technologies, Publicaciones de la Universidad de Alicante, Alicante 1997.
13. F. Caturla, F. Molina-Sabio and F. Rodríguez-Reinoso, Carbon, 29 (1991) 999.
14. H. Marsh, D.C. Crawford, T.M. O'Grady and A. Wennerberg, Carbon, 20 (1982) 419.
15. A.N. Wennerberg and T.M. O'Grady, US Patent No. 4 082 694 (1978).
16. T. Otowa, M. Shiraishi, R. Tanibata and N. Tanaka, CARBON'92, Essen, Proceedings, 944.
17. T. Otowa, R. Tanibata and M. Itoh, Gas Separation & Purification, 7 (1993) 241.
18. T. Otowa, Y. Nojima and M. Itoh, CARBON'94, Granada, Extended Abstracts, 808.
19. T. Otowa, M. Yamada, R. Tanibata and M. Kawakami, Gas Separation Technology, E.F. Vansant and R. Dewolfs (eds.), Elsevier Science Publishers B.V., Amsterdam 1990, 263.
20. W.F. Wolff, J. Phys. Chem., 63 (1959) 653.

21. M.M. Dubinin, in: Progress in Surface Membrane Science, J.F. Danielli (ed.), vol. 9, Academic Press, New York, 1975, 1.
22. K.S.W. Sing, D.H. Everett, R.A.W. Haul, L. Moscou, R.A. Pierotti, J. Rouquerol and T. Siemieniewska, Pure and Appl. Chem., 57 (1985) 603.
23. S.J. Gregg and K.S.W. Sing, Adsorption, Surface Area and Porosity, Academic Press, London, 2nd ed., 1982.
24. P.E. Fanning and M.A. Vannice, Carbon, 31 (1993) 721.
25. B. Steenberg, Adsorption and Exchange of Ions on Activated Charcoals, Almquist and Wiksell, Uppsala, 1944.
26. H.P. Boehm, Advances in Catalysis, D.D. Eley, H. Pines and P.B. Weisz, (eds), vol. 16, Academic Press, New York, 1966, 179.
27. B.R. Puri, Y.P. Myer and L.R. Sharma, J. Indian Chem. Soc., 33 (1956) 781.
28. V.A. Garten and D.E. Weiss, Australian Journal of Chemistry, 10 (1957) 309.
29. M. Voll and H.P. Boehm, Carbon, 9 (1971) 481.
30. E. Papirer, S. Li and J. Donnet, Carbon, 25 (1987) 243.
31. C.A. Leon, Y. Leon, J.M. Solar, V. Calemma and L.R. Radovic, Carbon, 30 (1992) 797.
32. B.R. Puri, in: Chemistry and Physics of Carbon, P.L. Walker Jr. (ed.), vol. 6, Marcel Dekker, Inc., New York, 1970, 191.
33. J. Zawadzki, in: Chemistry and Physics of Carbon, P.A. Thrower (ed.), vol. 21, Marcel Dekker, Inc., New York and Basel, 1989, 147.
34. J. Kazmierczak, S. Biniak, A. Świątkowski and K.H. Radeke, J. Chem. Soc., Faraday Trans., 87 (1991) 3557.
35. S. Biniak, G. Szymański, J. Siedlewski and A. Świątkowski, Carbon, 35 (1997) 1799.
36. J.B. Zandra, Process for recovery of gold from activated carbon by leaching and electrolysis, R.I. No 11672, Washington, DC, US Bureau of Mines, 1950.
37. K. Miura, H. Nakagawa, H. Okamoto and A. Shudo, 23rd Biennial Conference on Carbon, The Pennsylvania State University, 1997, Extended Abstracts, vol. 2, 78.
38. J.N. Armor, Carbon molecular sieves for air separation, in: Separation Technology, E.F.Vansant (ed.), Elsevier, Amsterdam, 1994.
39. H.C. Foley, M.S. Kane and J.F. Goellner, Access in Nanropoous Materials, T.J.Pinnavaia and M.F. Thorpe (eds.), Plenum Press, New York, 1995, 39.
40. H. Munzer, H. Heimbach, W. Korbacher, W. Peters, H.Juntgen, K. Knoblauch and D. Zundorf, U.S. Patent 3 801 513 (1974).
41. H. Juntgen, Carbon, 15 (1977) 273.
42. H. Juntgen, K. Knoblauch and K. Harder, Fuel, 60 (1981) 817.
43. H. Munzer, H. Heimbach, W. Korbacher, W. Peters, H. Juntgen, K. Knoblauch and D. Zundorf, US Patent 3 960 769 (1976).
44. T. Ohsaki and S. Abe, US Patent No. 4 742 040 (1988).
45. Y. Eguchi, K. Itoga, T. Sawada and H. Nishino, Jpn. Patent Sho 49-37036 (1974).
46. C. Marumo, E. Hayata and N. Shiomi, US Patent 4 933 314 (1990).

47. G.V. Baron, Industrial gas separation using PSA, in: Separation Technology, E.F. Vansant (ed.), Elsevier, Amsterdam, 1994.
48. A. Kapoor and R.T. Yang, Chem. Eng. Sci., 44 (1989)1723.
49. M.W. Ackley and R.T. Yang, AIChE J., 36 (1990) 1229.
50. A. Kapoor and R.T. Yang, AIChE J., 33 (1987) 1215.
51. D.L. Ruthven, Principles of adsorption and adsorption processes, Wiley Interscience, New York, 1984.
52. T.D. Burchell, C.E. Weaver, F. Derbyshire, Y.Q. Fei and M. Jagtoyen, in: Proc. CARBON'94, Granada, Spain, 1994, 650.
53. T.D. Burchell, R.R. Judkins, M.R. Rogers and A.M. Williams, Carbon, 35 (1997) 1279.
54. T.D. Burchell and R.R. Judkins, Energy Convers. Mgmt., 37 (1996) 947.
55. T.D. Burchell, in: Proc. CARBON'96, Newcastle upon Tyne, UK, 1996, vol. II, 749.

Adsorption and its Applications in Industry and Environmental Protection
Studies in Surface Science and Catalysis, Vol. 120
A. Dabrowski (Editor)



# Standardization of sorption measurements and reference materials for dispersed and porous solids

E. Robens[a], K.-F. Krebs[b], K. Meyer[c], K.K. Unger[a]

[a]Institut für Anorganische Chemie und Analytische Chemie der Johannes Gutenberg-Universität, D-55099 Mainz, Germany

[b]Merck KGaA, LAB CHROM Synthese, D-64271 Darmstadt, Germany

[c]Bundesanstalt für Materialforschung und -prüfung (BAM),
Division I.1: Inorganic Chemical Analysis; Reference Materials,
Rudower Chaussee 5, D-12489 Berlin, Germany

In this chapter a survey is presented on standardised methods for the characterization of the surface structure of dispersed or porous solids and to determine its sorption properties. The survey includes also reference materials which can be used as a comparative sample and to adjust and control the measuring equipment.

## 1. INTRODUCTION

To make industrial products more easily comparable and to facilitate the exchange of measuring results two routes have been taken:

- standardizing of measuring methods, procedures and instruments and
- certification of reference materials.

In practice, no absolute parameters for the characterization of dispersed or porous materials exist. With few exceptions, e.g. single crystals of molecular sieves, dispersed materials exhibit a fractal nature within a more or less extended region of size. As a consequence, the results of investigating the geometric structure depend strongly on the unit size of area of the probe used to investigate the surface. In general different conditions of sample preparation or different measuring methods lead to different results. Measuring method and procedure, therefore, need to be standardized. For its practical application they are supplemented in addition by field-experienced tests. Disregarding the large number of standards listed below, only a few basic adsorption measuring methods and some instruments are standardized. Efforts are made to create a standard column and to certify adsorbing materials for high performance liquid chromatography (HPLC).

Round robin tests using candidate reference materials revealed that the surface of dispersed solids can alter remarkably during storage and may be seriously affected during preparation for its measurement. Nevertheless, a variety of certified reference materials for the particle size and for the specific surface area is available from national and international standardization administrations and from industrial distributors. Recently, also reference materials for porous materials have been certified.

## 2. STANDARDISATION

The most important parameters for the characterization of the surface texture of dispersed solids are

- density
- specific surface area
- specific pore volume
- pore size distribution
- particle size distribution.

Table 1 covers standards which describe methods of any kind to characterize the surface structure.

Table 1
Standards for characterization of the surface structure of dispersed or porous solids

| | | |
|---|---|---|
| **AFNOR** | | **Association Française de Normalisation**, Tour Europe, F-92080 Paris la Défense Cedex, France |
| X 11-601 | 79 | Tamisage et granulométrie - Détermination de l'aire massique ou volumique des poudres par perméabilimétrie - Méthode de Lea et Nurse |
| X 11-602 | 77 | Détermination de l'aire massique des poudres par divers méthodes de perméabilimétrie à l'air |
| X-11-620 | 94 | Détermination de la surface spécifique des solides par adsorption de gaz à l'aide de la méthode BET |
| X-11-621 | 75 | Détermination de l'aire massique (surface spécifique) des poudres par adsorption de gaz - Méthode B.E.T.: Mesure volumétrique par adsorption d'azote à basse température |
| X 11-622 | 77 | - Variantes de la méthode de base |
| X 11-630 | 81 | Granulométrie – Vocabulaire |
| X 11-632 | 83 | - Expression des résultats expérimentaux d'analyse granulométrique |

Table 1
Standards for characterization of the surface structure of dispersed or porous solids

| | | |
|---|---|---|
| X 11-634 | 88 | - Caractérisation de la taille et de la forme des éléments d'une population granulaire |
| X 11-635 | 85 | - Représentation des distributions granulométriques - Partie 1: Modèles de référence |
| X 11-636 | 85 | - - Partie 2: Ajustement d'une courbe expérimentale à un modèle de référence. Cas du tamisage |
| X 11-640 | 79 | - Analyse granulométrique des poudres fines sur tamiseuse à dépression d'air |
| X 11-642 | 82 | - Tamisage en milieu liquide des poudres de granulométrie inférieure à 200 micromètres |
| X 11-660 | 83 | - Analyse granulométrique par microscopie optique - Généralités sur le microscope |
| X 11-661 | 84 | - Détermination de la taille des particules d'une poudre - Méthode par microscopie optique |
| X 11-666 | 84 | - Analyse granulométrique des poudres - Méthode par diffraction |
| X 11-667 | | - Méthode optique par laser - Mesurage du temps de transition |
| X 11-670 | 89 | - Analyse granulométrique de particules en suspension dans un électrolyte par utilisation du compteur à variation de résistance |
| X 11-680 | 80 | - Triage par fluides - Analyse granulométrique par sédimentation par gravité dans un liquide |
| X 11-681 | 82 | - Analyse granulométrique par sédimentation par gravité dans un liquide - Méthode de la pipette |
| X 11-682 | 84 | - - Méthode par occultation - Photosédimentomètre |
| X 11-683 | 81 | - Analyse granulométrique d'une poudre par sédimentation par gravité à hauteur variable dans un liquide - Méthode par mesure d'absorption de rayons X |
| X 11-684 | 85 | - Analyse granulométrique par sédimentation cumulative dans un liquide immobile - Méthode de la balance de sédimentation |
| X 11-685 | 83 | - Analyse granulométrique par sédimentation centrifuge dans un liquide immobile par rapport à l'axe de centrifugation |
| X 11-690 | 84 | - Triage par gravité dans un fluide en mouvement (Lévigation - Elutration) |
| X 11-693 | 83 | - Analyses granulométriques - Liquides de suspension et agents dispersants |
| X 11-695 | 87 | - Caractérisation des séparations granulométriques |
| X 11-696 | 89 | Granulométrie par analyse d'images |

Table 1
Standards for characterization of the surface structure of dispersed or porous solids

| | | |
|---|---|---|
| **AIA** | | **Asbestos International Association**, 68 Gloucester Place, GB - London W1H 3HL, UK |
| RTM1 | 82 | Reference method for the determination of the concentration of asbestos fibres |
| RTM2 | 84 | Method airborne asbestos and other inorganic fibres by scanning electron microscopy |
| **ASTM** | | **American Society for Testing and Materials**, 1916 Race Street, Philadelphia 3, PA, USA |
| C 115 | | Wagner Turbidimeter. Ditto: AASHTO T 98, ANSI A 1.7 |
| C 204 | 68 | Standard for the Determination of the Fineness of Portland Cement by Air Flow. Blaine Method. Ditto: Federal Test Method Standard 158 + Method 2101, AASHTO T 153 |
| C 373 | 88 | Determination of Apparent Density, Dry Density, Water Absorption and Porosity of Burnt Porcelain and Flint Ware |
| D 1193 | | Specification for Reagent Water |
| D 2355 T | 65 | Determination of Isotherms of Decolorization |
| D 3663 | 84 | Test Method for Surface Area of Catalysts |
| D 3766 | 86 | Definitions of Terms Relating to Catalysts and Catalysis |
| D 3906 | 85 | Test Method for Relative Zeolite Diffraction Intensities |
| D 3908 | 88 | Test Method for Hydrogen Chemisorption on Supported Platinum on Alumina Catalysts by Volumetric Vacuum Method |
| D 3942 | 85 | Test Method for Determination of the Unit Cell Dimension of a Faujasite-Type Zeolite |
| D 4222 | 83 | Test Method for Determination of Nitrogen Adsorption and Desorption Isotherms of Catalysts by Static Volumetric Measurements |
| D 4284 | 88 | Standard Test Method for Determining Pore Volume Distribution of Catalysts by Mercury Intrusion Porosimetry |
| D 4365 | 85 | Test Method for Determining Zeolite Area of a Catalyst |
| D 4567 | 86 | Test Method for Single-Point Determination of the Specific Surface Area of Catalysts Using Nitrogen Adsorption by the Continuous Flow Method |
| D 4641 | 88 | Practice for Calculation of Pore Size. Distributions of Catalysts from Nitrogen Desorption Isotherms |
| D 4824 | 88 | Test Method for Determination of Catalyst Acidity by Ammonia Chemisorption |

Table 1
Standards for characterization of the surface structure of dispersed or porous solids

| | | |
|---|---|---|
| **BSI** | | **British Standards Institution**, BS House, 2 Park Street, GB – London W1A 2BS / BSI Sales Department, Linford Wood, Milton Keynes MK14 6LE, UK |
| 2955 | 58 | Glossary of terms |
| 3406 | | Methods for the determination of particle size distributions |
| 3406-1 | 86 | Guide to powder sampling |
| 3406-2 | 84 | Recommendations for gravitational liquid sedimentation methods for powders and suspensions |
| 3406-3 | 83 | Air elutriation methods |
| 3406-4 | 85 | Optical microscope methods |
| 3406-5 | 83 | - Part 5: Recommendations for electrical sensing zone method (the Coulter principle) |
| 3406-5 | 85 | - Part 6: Recommendations for centrifugal liquid sedimentation methods for powders and suspensions |
| 3406-7 | 88 | Recommendations for single particle light interaction methods |
| 4359 | | Determination of the specific surface area of powders |
| 4359-1 | 84 | Recommendations for gas adsorption (BET) methods |
| 4359-2 | 87 | Recommended air permeability methods |
| 4359-4 | 95 | Recommendations for methods of determination of metal surface area using gas adsorption techniques |
| | | Recommended methods for the evaluation of porosity and pore size distribution. - Part 1. Mercury porosimetry |
| | | - Part 2. Gas adsorption |
| | | - Part 3. Challenge test |
| | | - Part 4. Liquid expulsion |
| **DIN** | | **Deutsches Institut für Normung e.V.**, D-10772 Berlin, Germany |
| 4760 | 82 | Gestaltabweichung, Begriffe, Ordnungssystem |
| 4761 | 78 | ≡ ISO 8785 Oberflächencharakter; Geometrische Oberflächentextur-Merkmale, Begriffe, Kurzzeichen |
| 4762 | 89 | ≡ ISO 4287-1 Oberflächenrauheit; Begriffe; Oberfläche und ihre Kenngrößen |
| 4763 | 81 | Stufung der Zahlenwerte für Rauheit |
| 7726 | 82 | ≡ ISO 1382 Schaumstoffe, Begriffe und Einteilung |
| 51 005 | 96 | Thermische Analyse (TA) – Begriffe |
| 51 006 | 90 | Themische Analyse (TA) - Thermogravimetrie (TG) |

Table 1
Standards for characterization of the surface structure of dispersed or porous solids

| | | |
|---|---|---|
| 51 918 | 86 | Bestimmung der Rohdichte nach der Auftriebsmethode und der offenen Porosität durch Imprägnieren mit Wasser |
| 51 957 | | Scheindichte |
| 52 102 | 88 | Bestimmung von Dichte, Trockenrohdichte, Dichtigkeitsgrad und Gesamtporosität |
| 53 108 | 75 | Prüfung von Papier und Pappe: Bestimmung der Rauhigkeit nach Bendtsen |
| 53 193 | | Scheindichte |
| 66 100 | 78 | Körnung; Korngrößen zur Kennzeichnung von Kornklassen und Korngruppen |
| 66 111 | 83 | Partikelgrößenanalyse; Sedimentationsanalyse; Grundlagen |
| 66 111 B1 | 83 | Partikelgrößenanalyse; Auswertegleichungen zur Mengenmessung im Fliehkraftfeld |
| 66 115 | 83 | Partikelgrößenanalyse; Sedimentationsanalyse im Schwerefeld; Pipetteverfahren |
| 66 116 | 73 | Korn-(Teilchen-)größenanalyse; Sedimentationsanalyse im Schwerefeld; Sedimentationswaage |
| 66 118 | 84 | Partikelgrößenanalyse; Sichtanalyse; Grundlagen |
| 66 119 | 83 | Partikelgrößenanalyse; Sichtanalyse mit Schwerkraft-Gegenstromsichter |
| 66 120 | 83 | Partikelgrößenanalyse; Sichtanalyse mit Fliehkraftsichter |
| 66 126 T 1 | 89 | Bestimmung der spezifischen Oberfläche disperser Feststoffe mit Durchströmungsverfahren. Grundlagen |
| 66 127 | 89 | - Verfahren und Gerät nach Blaine |
| 66 131 | 93 | Bestimmung der spezifischen Oberfläche von Feststoffen durch Gasadsorption nach Brunauer, Emmett und Teller (BET); Grundlagen |
| 66 132 | 75 | Bestimmung der spezifischen Oberfläche von Feststoffen durch Stickstoffadsorption. Einpunkt-Differenzverfahren nach Haul und Dümbgen |
| 66 133 | 93 | Bestimmung der Porenvolumenverteilung und der spezifischen Oberfläche von Feststoffen durch Quecksilberintrusion |
| 66 134 | 97 | Bestimmung der Porengrößenverteilung und der Oberfläche mesoporöser Feststoffe durch Stickstoffsorption. Verfahren nach Barrett, Joyner und Halenda (BJH) |
| 66 135 T1 | | Mikroporenanalyse mittels Gasadsorption, Grundlagen und Meßverfahren |

Table 1
Standards for characterization of the surface structure of dispersed or porous solids

| | | |
|---|---|---|
| 66 135 T2 | | - Auswerteverfahren: Isothermenvergleichsverfahren |
| 66 135 T3 | | - Berechnung des Mikroporenvolumens nach Dubinin und Radushkevich |
| 66 135 T4 | | - Berechnungsmethode nach Horvath-Kaeazoe |
| 66 136 | | Bestimmung der Oberfläche von Metallen durch Verfahren der Gasadsorption |
| 66 141 | 74 | Darstellung von (Korn-) Teilchengrößenverteilungen; Grundlagen |
| 66 142 T1 | 81 | Darstellung und Kennzeichnung von Trennungen disperser Güter; Grundlagen |
| 66 142 T2 | 81 | - Anwendung bei analytischen Trennungen |
| 66 142 T3 | 82 | - Auswahl und Ermittlung von Kennwerten bei betrieblichen Trennungen |
| 66 143 | 74 | Darstellung von (Korn-) Teilchengrößenverteilungen; Potenznetz |
| 66 144 | 74 | - Logarithmisches Normalverteilungsnetz |
| 66 145 | 76 | - RRSB-Netz |
| 66 160 | 85 | Partikelgrößenanalyse – Begriffe |
| 66 161 | 85 | Partikelgrößenanalyse- Formelzeichen, Einheiten |
| 66 165 T1 | 83 | Partikelgrößenanalyse; Siebanalyse – Grundlagen |
| 66 165 T2 | 84 | Partikelgrößenanalyse; Siebanalyse – Durchführung |
| **EN** | | **Comité Européenne de Normalisation - CEN**, Rue de Stassart 36, B-1050 Bruxelles, Belgium |
| 623 | | Hochleistungskeramik; Monolithische Keramik: Allgemeine und strukturelle Eigenschaften |
| 623-1 | 95 | Prüfung auf Anwesenheit von Oberflächenfehlern durch Farbstoffeindringtests |
| 623-2 | 93 | Bestimmung von Dichte und Porosität |
| 623-3 | 93 | Bestimmung der Korngröße |
| 623-4 | 93 | Bestimmung der Oberflächenrauheit |
| 725 | | Hochleistungskeramik; Prüfverfahren für keramische Pulver |
| 725-5 | 96 | Bestimmung der Teilchengröße |
| 725-6 | 96 | Bestimmung der spezifischen Oberfläche |
| 725-7 | 95 | Bestimmung der absoluten Dichte |
| 725-8 (pr) | | Bestimmung der geklopften Schüttdiche |
| 725-9 (pr) | | Bestimmung der Schüttdichte |
| 725-10(pr) | | Bestimmung der Verdichtungseigenschaften |

Table 1
Standards for characterization of the surface structure of dispersed or porous solids

| | | |
|---|---|---|
| 993-1 | 95 | Prüfverfahren für dichte geformte feuerfeste Erzeugnisse. - Bestimmung der Rohdichte, offene Porosität und Gesamtporosität |
| **GOST R** | | **Committee of the Russian Federation for Standardization, Metrology and Certification**, Leninsky Prospekt 9, Moskva 117049, Russia |
| | 79 | Density, specific surface area (BET) and pore size distribution (BJH, Dollimore and Heal) |
| **ISO** | | **International Organization for Standardization**, Central Secretariat, 1 rue de Varembé, BP 56, CH- 1211 Genève 20 |
| 1382 | 82 | Cellular materials; definitions of terms and classification |
| 1953 | 72 | Hard coal - Size analysis |
| 4287-1 | | Surface roughness; terminology; surface and its parameters |
| 8785 | 94 | Surface character; geometrical characteristics of surface texture terms, definitions, symbols |
| 9276 | 90 | Representation of results of particle size analysis. - Part 1: Graphical representation |
| 9277 | 96 | Determination of specific surface area by BET gas adsorption method |
| 10070 | 89 | Metallic powders. - Determination of envelope-specific surface area from measurements of the permeability to air of a powder bed under steady-state flow conditions |
| 10076 | 89 | Determination of particle size distribution by gravitational sedimentation in a liquid and attenuation measurement |
| Draft | 97 | Pore size distribution and porosity of solid materials - Evaluation by mercury porosimetry and gas adsorption |
| **IUPAC** | | **International Union of Pure and Applied Chemistry**, Bank, 2, Pound Way, GB - Oxford, UK |
| Appendix II, Part I | 72 | Manual of symbols and terminology for physico-chemical quantities and units. Terminology and symbols in colloid and surface chemistry |
| Part II | 76 | Terminology in heterogeneous catalysis |
| | 85 | Reporting physisorption data for gas/solid systems with special reference to the determination of surface area and porosity. Pure and Applied Chemistry 57 (1985) 4, 603-619 |
| | 86 | Reporting data on adsorption from solution at the solid/solution interface. Pure and Applied Chemistry 58 (1986) 968-984 |

Table 1
Standards for characterization of the surface structure of dispersed or porous solids

| | | |
|---|---|---|
| | 94 | Recommendations for the characterization of porous solids. Pure and Applied Chemistry (8.1994) |
| **SEV-ASE** | | **Schweizerischer Elektrotechnischer Verein,** Seefeldstrasse 301, CH - 8034 Zürich, Switzerland |
| 3621-1-3 | 91 | Allgemeine Prüfung für Isolier- und Mantelwerkstoffe für Kabel und isolierte Leitungen; Teil 1. Allgemeine Prüfverfahren; Hauptabschnitt 3: Prüfverfahren zur Dichtebestimmung; Wasseraufnahmeprüfung; Schrumpfungsprüfng |
| **VDI** | | **Verein Deutscher Ingenieure e.V.**, Postfach 101045, D - 40001 Düsseldorf, Germany |
| 2031 | 62 | Feinheitsbestimmungen an technischen Stäuben |
| 2066 B1 | 75 | Messen von Partikeln. Staubmessungen in strömenden Gasen. Gravimetrische Bestimmung der Staubbeladung. – Übersicht |
| 2066 B2 | 81 | - Manuelle Staubmessung in strömenden Gasen. Gravimetrische Bestimmung der Staubbeladung, Filterkopfgerät |
| 2066 B4 E | 80 | - Staubmessung in strömenden Gasen. Bestimmung der Staubbeladung durch kontinuierliches Messen der optischen Transmission |
| 2119 B1 E | 72 | Messung partikelförmiger Niederschläge. – Übersicht |
| 2119 B2 E | 72 | - Bestimmung des partikelförmigen Niederschlags mit dem Bergerhoff-Gerät (Standardverfahren) |
| 2119 B3 E | 72 | - Bestimmung des partikelförmigen Niederschlags mit dem Hibernia- und Löbner-Liesegang-Gerät |
| 2119 B4 E | 72 | - Bestimmung des partikelförmigen Niederschlags mit Haftfolien |
| 2265 | 80 | Feststellen der Staubsituation am Arbeitsplatz zur gewerbehygienischen Beurteilung |
| 2266 B1 | 68 | Messung der Staubkonzentration am Arbeitsplatz. Messung der Teilchenzahl. Messen mit dem Thermalpräzipitator |
| 2266 B2 | 68 | - Messen mit dem Konimeter |
| 2266 B2 | 71 | - Messen unter Benutzung von Membranfiltern |
| 2267 B2 | 83 | Stoffbestimmung an Partikeln in der Außenluft. Messen der Blei-Massenkonzentration mit Hilfe der Röntgenfluoreszensanalyse |
| 2267 B3 | 83 | - Messen der Blei-Massenkonzentration mit Hilfe der Atomabsorptionsspektrometrie |
| 2267 B11 | 83 | - Messen der Blei-Massenkonzentration mit Hilfe der energiedispersiven Röntgenfluoreszenzanalyse |

Table 1
Standards for characterization of the surface structure of dispersed or porous solids

| | | |
|---|---|---|
| 2269 B1 E | 72 | Mikroskopische Untersuchungsverfahren feiner Teilchen. Übersicht |
| 2463 B1 | 74 | Messen von Partikeln in der Außenluft. Übersicht |
| 2463 B2 E | 77 | Messen von Partikeln. Messen der Massenkonzentration von Partikeln in der Außenluft. High Volume Sampler - HV 100 |
| 2463 B3 E | 76 | - - TBF-Filterverfahren |
| 2463 B4 E | 76 | - - LIB-Filterverfahren |
| 2463 B5 | 86 | - - Filterverfahren, Automatisches Filtergerät FH 621 |
| 2463 B6 | 86 | - - - Automatisiertes Filtergerät BETA-Staubmeter F 703 |
| 2463 B7 | 82 | - Messen der Massenkonzentration (Immission), Filterverfahren, Kleinfiltergerät GS 050 |
| 2463 B8 | 82 | - - Basisverfahren für den Vergleich von nichtfraktionierenden Verfahren |
| 2463 B9 E | 81 | - Messen der Massenkonzentration von Partikeln in der Außenluft. Filterverfahren, LIS/P-Filtergerät |
| 3491 B1 | 80 | - Kennzeichnung von Partikeldispersionen in Gasen. Begriffe und Definitionen |
| 3491 B2 | 80 | - Herstellungsverfahren für Prüfaerosole, Grundlagen und Übersicht |
| 3491 B3 | 80 | - Herstellung von Latex-Aerosolen unter Verwendung von Düsenzerstäubern |
| 3491 B4 | 80 | - Herstellungsverfahren für Prüfaerosole, Aerosolgenerator nach Sinclair und Lamer |
| 3491 B5 | 80 | - Herstellung von Prüfaerosolen aus Farbstofflösungen mit Düsenzerstäubern |
| 3491 B6 | 80 | - Herstellungsverfahren für Prüfaerosole, Platinoxid-Aerosolgenerator |
| 3491 B7 E | 80 | - - Aerosolgenerator nach Rapaport und Weinstock |
| 3492 B1 E | 86 | Messen anorganisch, faserförmiger Partikel in der Außenluft. Rasterelektronenmikroskopische Verfahren |

Standardized density determinations are based on the displacement of a liquid by the sample in a calibrated volume. A standard for the measurement of the displacement of a gas is in preparation. Other methods apply differences of the buoyancy in gas and in a liquid by means of a hydrostatic balance. When preparing a sample for a thermogravimetric or gravimetric sorption investigation the pressure dependence of the buoyancy may be used to calculate the density difference to the counterweight.

Important methods for the determination of the specific surface area and of the pore size distribution are based on the measurement of the gas adsorption isotherm [1,2]. The gas adsorption method and the evaluation according to Brunauer, Emmett and Teller using the two-parameter BET equation has been standardized in several countries for a number of years and an ISO standard just appeared. To establish the pore size distribution the method of Barrett, Joyner and Halenda (BJH) is generally accepted. Other methods for this purpose make use of the flow resistance of air through the compressed sample. The Blaine test and other flow tests used to characterize building materials are standardized world-wide.

Thermoanalytical methods comprise calorimetry (Differential Scanning Calorimetry, DSC) and thermogravimetry (TG) [3]. Mostly, the temperature of the sample is increased linearly and peaks of the sample temperature in comparison to a reference sample are recorded (Differential Thermal Analysis, DTA). In thermogravimetry the mass loss is observed. In this way the content of adsorbed liquids can be determined and the desorption kinetics can be observed. For special tasks also a quasi-isothermal treatment may be applied [4]. The most important standards on such methods are summarized in Table 2.

Adsorption from solution may be observed as a function of concentration of solved species. Because its adsorption competes with that of the solvent, resulting isotherms are rather complicated. Nevertheless a number of tests, notably discoloration tests, exist. Specifications are widely scattered. For activated carbons tests are summarized by the CEFIG [5].

Particle size analysis include separation by sieving, sedimentation in a liquid or in air, sedimentation by means of a centrifuge, air separation, optical and X-ray observation of particles moved through a slot by means of a gas or liquid stream. In the latter case also the change of the electrical resistance may be observed. Image analysis of a powder distributed on a plain can be made.

A great number of standard test methods are concerned with ad- and absorption of water and other vapors by various materials. They cover the measurement of the content of the liquid and the sorption kinetics, e.g. when drying materials. Table 3 summarizes a number of standards on this item.

German DIN standards on the characterization of dispersed or porous solids are collected in ref. [6]. The most comprehensive description of the adsorption method is found in an IUPAC recommendation [7]. Nowadays national standards are being harmonized either in the framework of the European Communities or at the international level. A list of standardization committees working in this field is appended (Table 4). Different methods of particle counting and characterization are collected in a VDI manual [8].

Basis for the choice of symbols and units are ISO 31 and ISO 1000. Correspondingly the IUPAC "Green Book" [9] may be used. In both, indices are not employed in elevated position to avoid confusion with exponents.

Table 2
Most important standards for thermoanalytical methods (TA)

| | | |
|---|---|---|
| ASTM D 2766 | 91 | Test method for specific heats of liquids ad solids |
| ASTM D 3417 | 89 | Test method for heats of fusion and crystallization of polymers by thermal analysis |
| ASTM D 3418 | 88 | Test method for transition temperatures of polymers by thermal analysis |
| ASTM E 472 | 91 | Reporting thermoanalytical data |
| ASTM E 473 | 88 | Definition of terms relating to thermal analysis |
| ASTM E 537 | 86 | Test method for assessing the thermal stability of chemicals by methods of thermal analysis |
| ASTM E 698 | 84 | Test method for Arrhenius kinetic constants for thermally unstable materials |
| ASTM E 793 | 89 | Test method for heats of fusion and crystallization by differential scanning calorimety |
| ASTM E 794 | 89 | Test method for melting and crystallization temperatures by thermal analysis |
| ASTM E 831 | 86 | Test method for linear thermal expansion of solid materials by thermomechanical analysis |
| ASTM E 914 | 87 | Standard practice for evaluating temperature scale for thermogravimetry |
| ASTM E 928 | 89 | Test method for mol percent impurity by differential scanning calorimety |
| ASTM E 967 | 92 | Standard practice for temperature calibration of differential scanning calorimeters and differential thermal analyzers |
| ASTM E 968 | 87 | Standard practice for heat flow calibration of differential scanning calorimeters |
| ASTM E 1131 | 86 | Test method for compositional analysis by thermogravimetry |
| ASTM E 1296 | 90 | Test method for determining specific heat capacity by differential scanning calorimety |
| ASTM E 1256 | 91 | Test method for glass transition temperatures by differential scanning calorimetry or differential thermal analysis |
| ASTM E 1582 | 93 | Standard practice for calibration of temperature scale for thermogravimetry |
| ASTM E 1641 | 93 | Standard test method for decomposition kinetics by thermogravimetry |
| DIN 51 004 | 94 | - Bestimmung der Schmelztemperatur kristalliner Stoffe mit der Differenzthermoanalyse (DTA). |
| DIN 51 005 | 93 | Thermische Analyse (TA) - Begriffe. |
| DIN 51 006 | 90 | - Thermogravimetrie (TG) - Grundlagen |
| DIN 51 007 | 94 | - Differenzthermoanalyse (DTA) - Grundlagen |
| DIN 51045-1 | 89 | Determination of the thermal expansion of solids. Basic rules |
| JIS K 0129 | 94 | General rules for thermal analysis |

Table 3
Standards on determination of water sorption, water content and on drying

ASTM C 121, C 272, C 373, D 570, D 1815, D 2842, D 3285, D 3816/M, D 5455, E 946
DIN 1230-3, 1996-8, 4051, 7708-2/3, 7746-2, 16890, 16911, 16913-2, 16955, 16956, 16958, 18132, 18180, 19538, 40634-1, 52103, 52251-2, 52351, 52364, 52459, 52617, 53129, 53330, 53338/-2, 53433, 53434, 53495, 53184, 53923, 54540-4, 57291-268750, 68752,
EN 99, 121, 15 9, 176, 177, 178, 186-1/2, 187-1/2, 188, 532, 382-2, 2155-2, 2378, 20535
IEC 811-1-3
ISO 62, 535, 2417, 2508, 2896, 5635, 5637, 6783, 8361-1/2, 8787
LN (Luftfahrt-Norm) 29820, 65336
OENORM B 3006, B 3122/-2, 3234, 7013, 7017, 7018, 7735-1/2, 7873-7, 53495
SEV-ASE 3621-1-3
VDE 0291-2

Table 4
Institutes which work on standards for the characterization of solid surfaces and of adsorption properties.

| | |
|---|---|
| AFNOR | Association Française de Normalisation, Tour Europe - Cedex 7, F-92080 Paris la Défence |
| AIA | Asbestos International Association, 68 Gloucester Place, GB - London W1H 3HL, UK |
| ASTM | American Society for Testing Materials, 1916 Race Street, Philadelphia PA 19103, USA |
| AWWA | American Water Works Association, 6666 West Quincy Avenue, Denver CO 80235, USA |
| BSI | British Standards Institution, BS House, 2 Park Street, GB - London W1Y 4AA, UK |
| CEFIG | European Council of Chemical Manufacturers' Federations, Avenue Louise, Bte. 71, B - 1050 Brussels |
| DIN | Deutsches Institut für Normung, e.V., D-10772 Berlin, Germany |
| EN | Comité Européenne de Normalisation - CEN, Rue de Stassart 36, B-1050 Bruxelles, Belgium |
| GOST R | Committee of the Russian Federation for Standardization, Metrology and Certification, Leninsky Prospekt 9, Moskva 117049, Russia |
| JIS | Japanese Industrial Standards Institution |
| ISO | International Organization for Standardisation. B.P. 56, CH -1211 Genève |
| IUPAC | International Union of Pure and Applied Chemistry, Bank, 2, Pound Way, GB - Oxford, UK |
| SEV-ASE | Schweizerischer Elektrotechnischer Verein, Seefeldstrasse 301, CH - 8034 Zürich |
| VDI | Verein Deutscher Ingenieure e.V., Postfach 101045, D - 40001 Düsseldorf, Germany |

## 3. SAMPLING AND SAMPLE PREPARATION

The analysis error includes the error by the variance of the sample from the average and the measuring error. Based on the law of error propagation, the total error in the value of results is

$$S_{total} = \sqrt{S^2_{measurement} + S^2_{sample}} \tag{1}$$

The standard deviation of the value of the results in terms of the analyzed material is determined by the analyzer to a significant extent only if the variance for the sample preparation is markedly less than the variance of the measurement. Therefore, the results are only reproducible if the sample to be analyzed is fully representative of the material to be tested. To minimize the errors, as many samples as possible must therefore be taken from random locations in the basic whole and then combined. This may be done by a laboratory sample divider [10], using a procedure as described in DIN 53 803 (1991): Sampling, Part 1: Statistical basis, one-way layout; Part 2: Practical execution.

Another sample error may happen during sample preparation. Degassing at elevated temperature can easily alter the surface properties of dispersed or porous samples. Appropriate conditions may be found by trial experiments variegating pressure, temperature and time. We recommend to keep strictly to the rules included in the standards for the different measuring methods.

## 4. VALIDATION

Modern manufacturing methods in the area of biology, pharmacy, semiconductors etc. are accompanied by quantitative analysis of very high precision and accuracy. In some cases a detection limit down to the molecular region is required. On the other hand, the measuring methods discussed here are charged with remarkable errors. Therefore, the suitability of methodology for providing useful analytical data should be determined by a validation process [11]. By the validation a known accuracy and precision of the method will be obtained, but most important its limitations will be revealed. As a final conclusion of the validation an approximate system suitability test for the daily use of an analytical method should be outlined. Some general guidelines for validation are included in EN 45001: General requirements of the competence of test and calibration laboratories and ISO/DIS 5725-1 - 6: Accuracy (trueness and precision) of measurement methods and results.

## 5. HPLC REFERENCE COLUMN

High-Performance Column Liquid Chromatography (HPLC) is the most widely employed separation technique for the analysis of complex mixtures. The method is based on the adsorption on a adsorbent filled in a column as the stationary phase, when a sample solution flows through this column. Differences in binding forces and the kinetics of adsorption of the components allow for the determination of the chemical composition of the sample [12]. On the other hand using macromolecular solutes of different but known size the pore size distribution of the column material can be determined by means of inverse size exclusion chromatography.

Although HPLC methods are validated large difficulties exist in batch-to-batch and column-to-column reproducibility. In addition the life-time of the columns is not guaranteed. Furthermore, changes in the column characteristics are not traceable to the manufacturing process due to proprietary reasons. Coordinated by the Institut für Anorganische Chemie und Analytische Chemie der Johannes Gutenberg-Universität, a consortium was formed to develop a HPLC reference column (EU-project SMTCT 95 No. 2026). The consortium is composed of industrial, academic and research laboratories including HPLC column manufacturers. The overall objectives of the project are:

- to develop a manufacturing process of a reversed phase, C-18 bonded silica column [13],
- to asses the physicochemical and chromatographic properties, the homogeneity and stability,
- to standardize and validate the manufacturing process,
- to perform the certification of the reference column, the reference tests and reference solution.

## 6. COMPARATIVE MEASUREMENTS

Comparative measurements at different laboratories using candidate reference materials revealed that surfaces of highly dispersed materials can be significantly affected during storing and sample preparation. This resulted from investigations of SCI, IUPAC and NPL [14] on surface area standards and the measurements in the context of the choice of particle size and surface area reference materials by the BCR [15], both using the adsorption method. For the characterization of building materials the mercury intrusion method was tested [16]. A new project was created within the 4th framework of the BCR program "Standards, Measurement and Testing" (SMT) 1994-1998: SMT4-CT 95-2025: Certification of pore size reference materials (1996-98). It is concerned on materials with micro- and mesopores.

## 7. REFERENCE MATERIALS

Reference material (RM) is defined as a material or substance one or more of whose property values are sufficiently homogeneous and well established to be used for the calibration of an apparatus, the assessment of a measurement method or for assigning values to materials. To control their products and to calibrate their instruments many manufacturers of powders or porous materials produce such materials for their own needs [17]. After a careful selection and evaluation by official working groups RMs may be certified. For a CRM one or more properties are certified by a procedure that establishes its traceability to an accurate realization of the unit in which the property values are expressed, and for which each certified value is accompanied by an uncertainty at a stated level of confidence. All CRMs are supplied with a certificate, the content of which and the intended use of the CRM are defined by ISO REMCO Guides 31 to 35. Disregarding the difficulties in storage a variety of particle size and surface area reference materials are certified (Table 5) and more are under consideration [18]. For public use RMs are stored and offered by national and international standardization institutions and by some commercial distributors (Table 6).

Table 5
Certified reference materials for dispersed matter

| Material | Order number (see Table 1) | Distributor | Particle diameter μm | Spec. surface area*) $m^2g^{-1}$ | Pore region nm |
|---|---|---|---|---|---|
| Alpha-Alumina | CRM-PM-102 | BAM | | 5,41 | |
| Alpha-Alumina | CRM 169 | BCR | | 0.104 | |
| Alpha-Alumina | CRM 170 | BCR | | 1.05 | |
| Alpha-Alumina | M 11-05/09 | NPL | | 2.1/0.7* | |
| Alpha-Alumina | M 11-06/10 | NPL | | 0.3/0.1* | |
| Alpha-Alumina | M 11-07/11 | NPL | | 0.1/0.04* | |
| Alpha-Alumina | M 11-08/12 | NPL | | 1.0/0.3* | |
| Alumina | CRM 171 | BCR | | 2.95 | |
| Alumina | 8571 | NIST | | 158.7 | |
| Alumina/Silica | CRM-PM-103 | BAM | | 156.0 | 1.93 |
| Alumina/Silica | CRM-PM-104 | BAM | | 79.9 | 2.23 |
| Bronze | CRM 174 | BCR | | 0.06 | |
| Graphitized carbon black-Vulcan FT-G | M 11-01 | NPL | | 11 | |
| - Sterling 3-G | M 11-02 | NPL | | 71 | |
| Glass spheres | 1003a | NIST | 8 - 58 | | |
| Glass spheres | 1004a | NIST | 50 - 100 | | |
| Glass spheres | 1017a | NIST | 100 - 310 | | |
| Glass spheres | 1018a | NIST | 225 - 780 | | |
| Glass spheres | 1019a | NIST | 760 -2160 | | |
| Caolin, calcined | 8570 | NIST | | 10.89 | |
| Latex | CRM 165 | BCR | 2.223 | | |
| Latex | CRM 166 | BCR | 4.821 | | |
| Latex | CRM 167 | BCR | 9.475 | | |

Table 5
Certified reference materials for dispersed matter

| Material | Order number (see Table 1) | Distributor | Particle diameter µm | Spec. surface area*) $m^2g^{-1}$ | Pore region nm |
|---|---|---|---|---|---|
| Polystyrene spheres | 1690 | NIST | 0.895 | | |
| Polystyrene spheres | 1691 | NIST | 0.269 | | |
| Polystyrene spheres | 1660 | NIST | 9.89 | | |
| Polystyrene spheres | 1661 | NIST | 29.64 | | |
| Polystyrene spheres | 1665 | NIST | 9.94 | | |
| Silica | CRM-PM-101 | BAM | | 0.177 | mesoporous |
| Silica | M 12-01 | NPL | | | mesoporous |
| Silica | M 12-02 | NPL | | | mesoporous |
| Silica Tk 800 | M 11-03 | NPL | | 152 | non-porous |
| Silica Gasil | M 11-04 | NPL | | 260 | mesoporous |
| Silica/Alumina | 8572 | NIST | | 291.2 | |
| Quartz | CRM 066 | BCR | 0.35-3.50 | | |
| Quartz | CRM 067 | BCR | 2.4- 32 | | |
| Quartz | CRM 068 | BCR | 160 - 630 | | |
| Quartz | CRM 069 | BCR | 14 - 90 | | |
| Quartz | CRM 070 | BCR | 1.2- 20 | | |
| Quartz | CRM 130 | BCR | 50 - 220 | | |
| Quartz | CRM 131 | BCR | 480 -1800 | | |
| Quartz | CRM 132 | BCR | 1400 -5000 | | |
| Quartz | M 13-02 | NPL | 0.1 - 3 | | |
| Quartz | M 13-03 | NPL | 3 - 40 | | |
| Quartz | M 13-04 | NPL | 40 -1000 | | |
| Quartz | M 13-05 | NPL | 10 - 100 | | |
| Quartz | M 13-06 | NPL | 1 - 10 | | |
| Quartz | 66 | NPL | 0.35- 2.5 | | |
| Quartz | 67 | NPL | 3 - 20 | | |
| Quartz | 68 | NPL | 140 - 650 | | |
| Quartz | 69 | NPL | 12 - 90 | | |
| Quartz | 70 | NPL | 0.5- 90 | | |
| Quartz | CRM 172 | BCR | | 2.50 | |
| Titania | M 13-01 | NPL | 0.1- 3 | | |
| Titania-Rutile | CRM 173 | BCR | | 8.23 | |
| Tungsten | CRM 175 | BCR | | 0.181 | |
| Portland-cement | 114n | NIST | | $0.202^{+}$<br>$0,346^{*}$ | |
| Zirconia | M 13-07 | NPL | 1 - 2 | | |
| Zirconia | M 13-08 | NPL | 5 | | |
| Zirconia | M 13-09 | NPL | 15 | | |

Table 5
Certified reference materials for dispersed matter

| Material | Order number (see Table 1) | Distributor | Particle diameter µm | Spec. surface area*) $m^2g^{-1}$ | Pore region nm |
|---|---|---|---|---|---|
| Zirconia | M 13-10 | NPL | 30 | | |
| Zirconia | M 13-11 | NPL | 60 | | |
| Test dust | | Loughborough University | | | |
| Test dust | | Carborundum | | | |
| Test dust | | Delco | | | |
| Test dirt | | Spark Plug | | | |
| Test dirt | | Wäscherei-forschung | | | |

*) Specific surface areas are measured using the nitrogen adsorption method at 77 K, except those marked by *) with the permeation method and +) using the Wagner turbidimeter.

Table 6
Distributors of reference materials for dispersed products

BAM - Bundesanstalt für Materialforschung und - prüfung, Referat 10.01, Rudower Chaussee 5, D - 12489 Berlin, Germany

BAS - Bureau of Analyse Samples Ltd., Newham Hall, Newby, GB - Middlesbrough, Cleveland TS8 9EA, U.K.

CTIF - Centre Technique des Industries de la Fonderie, 44 Avenue de la Division Leclerc, F - 92310 Sèvres, France

Duke Standards Co., 445 Sherman Avenue, Palo Alto, CA 94306, USA

Eidgenössische Materialprüfungs - und Versuchsanstalt für Industrie Bauwesen und Gewerbe, Unterstr. 11, CH - 900 St.Gallen

BCR - Community Bureau of Reference, 200 rue de la Loi, B - 1049 Brüssel

IIII - International Trade and Industry Inspection Institute, 49-10, 2 Nishihara Shibuya-ku, Tokyo 151, Japan

IRSID - Institut de Recherches de la Sidérurgie Française, B.P. 64, F - 57210 Maizières-les-Metz

LGC - Laboratory of the Government Chemist, Office of Reference Materials, Bldg. 95 RMAI, Queen's road, Teddington, Middlesex TW11 OLY, UK

MBH Analytical Ltd., Holland House, Queens Road, GB - Barnet, Hertfordshire, EN5 4DJ, U.K.

NBL - U.S. Department of Energy, New Brunswick Laboratory, Reference Materials Sales, 9800 S. Cass Avenue, Bldg. 350, Argonne, IL 60439, USA

Table 6
Distributors of reference materials for dispersed products

| | |
|---|---|
| NIST - | Office of Standard Reference Materials, US Department of Commerce, National Institute of Standards and Technology, Rm. B311 Chemistry Bldg., Gaithersburg, MD 20899, USA |
| NPL - | National Physical Laboratory, Teddington, Middlesex, TW11 OLY, UK. Distributor: LGC |
| NRC - | CRM - National Research Centre for CRM, No. 7 District 11, Hepingjie, Chaoyangqu, 100013, Peoples Republic of China |
| Particle | Information Service, Inc., P.O.Box 792, 222 Granite Hill Road, Grands Pass, Oregon 97526, USA |
| EMC - | Council Committee on Reference Materials, International Organization for Standardization, 1, rue de Varembe, B.P. 56, CH - 1211 Genève 20 |
| RBS - | Regine Brooks, Pariser Str. 5, D - 53117 Bonn, Germany |
| SMR - | LNE - Service des Materiaux de Référence, 1, rue Gaston Boissier, F - 75015 Paris |
| Silikose | -Forschungsinstitut, Hunscheidtstr. 12, D - 44789 Bochum, Germany |
| Staubforschungsinstitut des Hauptverbandes der gewerblichen Berufsgenossenschaften e.V., Langwartweg 103, D - 53129 Bonn, Germany | |
| Steinkohlenbergbauverein, Hauptstelle für Staub- und Silikosebekämpfung, Frillendorfer Str. 351, D - 45139 Essen-Kray, Germany | |
| Testfabrics Inc., 55 Vandam Str., New York 13, N.Y., USA | |
| UNIIM | - Ural Research Institute for Metrology, Krasnoarmeiskaya Street, 4, 620219 Ekaterinburg, GSP-824, Russia |
| Wäschereiforschung Krefeld e.V., Adlerstr. 44, D - 47798 Krefeld, Germany | |
| Wirtschaftsverband Asbestzement e.V., Arbeits- und Umweltschutz, Kölner Str. 102-104, D - 41464 Neuss, Germany | |

With more than 200 producers of RMs throughout the world, it is often difficult to find the best one for a specific application. The database COMAR has been developed to assist chemists in finding the RM they need. COMAR is available on discs by LGC, LNE, BAM, NIST, UNIIM, NRC-CRM, IIII (see Table 6).

The Institute for Inorganic and Analytical Chemistry of the University Mainz has developed single crystals of zeolites and aluminophosphates which may be used as reference materials in the micropore range and as an internal standard for the calibration of isotherms [19]. New porous/finely dispersed reference materials (CRM) have been developed in the German Federal Institute for Materials Research and Testing (BAM) in 1996. It was certified with respect to the specific surface area, specific pore volume and mean pore radius by means of the gas adsorption method using krypton and nitrogen [20,21]. The CRMs were produced in compliance with BCR guidelines [22], ISO guidelines [23] and are available by BAM [24].

With regard to gaseous and liquid sorptivs a great number of chemicals in any degree of purity are offered by the chemical industry. Organic compounds of extremely high purity can be obtained by separation using HPLC techniques as applied by specialized contract laboratories [25]. Such substances may be used as a reference for pharmaceuticals.

## 8. OUTLOOK

More national and international standardization procedures for mercury porosimetry and the derivation of pore size distributions from adsorption isotherms are in preparation. Regarding the weakness of the two-parameter BET model for surface area determination in addition the three-parameter BET equation or improved approximations [26] should be considered. Competitive evaluation methods, like the method of Dubinin, Horwath-Kawazoe, Kaganer and Radushkevich are being discussed.

Standard procedures for adsorption measurements and for the characterization of surface area and pore size distribution are formulated in two IUPAC recommendations [7,27]. Guidelines for measurements have been published in the COPS III proceedings [28]. These documents provide a basis for pure and applied research. Besides, national and international standardization authorities issue official rules for instrumentation and measuring procedures which allow for the comparison of industrial products. A survey on British activities has been published recently [29] and German standards are collected in a paperback [30]. Future plans of British and German working groups are concerned with the determination of micropores [31,32].

The ISO working group ISO/TC 24: Sieves, Sieving, and Other Sizing Methods is engaged in elaborating methods for the surface characterization of dispersed materials and just issued an ISO standard on the BET method.

Recently, at the University of Austin, Texas, comparative sorption measurements have begun with the aim of standardizing a procedure for the determination of the fractal dimension [33] of dispersed materials [34]. Using the method of Neimark [35] the fractal dimension can be obtained by evaluating a single isotherm. It turned out, however, that most adsorbents are fractal only in a restricted region of size of pores or particles, respectively [36]. Thus, the calculation of the fractal dimension does not provide additional information [37,38].

An adsorption database containing several thousands equilibrium values of pure materials was established recently at the Carl von Ossietzky-University, Oldenburg (Germany) [39]. The database allows for the supplementation by own results.

## REFERENCES


1. S.J. Gregg and K.S.W. Sing, Adsorption, Surface Area and Porosity, 2nd ed., Academic Press, London 1982.

2. R. Sh. Mikhail and E. Robens, Microstructure and Thermal Analysis of Solid Surfaces, Wiley, Chichester 1983.
3. F. Paulik, Special Trends in Thermal Analysis, Wiley, Chichester 1995.
4. P. Staszczuk, J. Thermal Analysis, 46 (1996) 1821.
5. Testmethoden für Aktivkohlen. CEFIG 1986.
6. DIN-Taschenbuch 133, Partikelmeßtechnik, 3. ed., Beuth, Berlin 1990.
7. IUPAC Recommendations 1984, K.S.W. Sing, D.H. Everett, R.A.W. Haul, L. Moscou, R.A. Pierotti, J. Rouquérol and T. Siemieniewska, Pure & Appl. Chem., 57 (1985) 4, 603.
8. VDI-Handbuch Reinhaltung der Luft, VDI Verlag, Düsseldorf, 1984.
9. I. Mills, T. Cvitas, K. Homann, N. Kallay and K. Kuchitsu (eds.), Quantities, Units and Symbols in Physical Chemistry, Blackwell, Oxford, 1989.
10. W. Mutter, Int. Lab., 27 (6.1997) 3A, 8.
11. A.-M. Olsson, Pharmaceutical Analysis R&D, Ferring AB, Malmö, 1996.
12. A.I. Liapis and K.K. Unger, in: G. Street (ed.), Highly Selective Separations in Biotechnology. Chapman & Hall, London 1994, 121.
13. M. Grün, A.A. Kurganov, S. Schacht, F. Schüth and K.K. Unger, J. Chromatogr. A, 740 (1996) 1.
14. D.H. Everett, G.D. Parfitt, K.S.W. Sing and R. Wilson, J. Appl. Chem. Biotechnol., 24 (1974) 199.
15. Reference Materials and Methods. Status Report 1979, ECSC-EEC-EAEC, Brussels-Luxembourg, 1979.
16. K. Hinrichsmeyer, S. Abdul-Maula, U. Diederichs and F.S. Rostásy, Quecksilberporosimetrie, Ringversuche an erhärtetem Zementstein. Bericht des Instituts für Baustoffe, Massivbau u. Brandschutz der Technischen Universität Braunschweig, 3,1988.
17. E. Robens and R. Meyer, Powder Metallurgy International, 13 (1981) 1, 44.
18. P.J. Jenks, Analysis Europe (April 1997), 22.
19. U. Müller, A. Tissler and K.K. Unger, GIT Fachz. Lab., 32 (1988) 635.
20. K. Meyer, P. Klobes and B. Röhl-Kuhn, Cryst. Res. Technol., 32 (1997) 1, 173.
21. B. Röhl-Kuhn, K. Meyer and Th. Fritz, Fresenius J. Anal. Chem., (1997).
22. BCR/48/93: Guidelines for the Production and Certification of BCR Reference Materials, Brussels, 1994.
23. ISO Guide 30: Terms & Definitions Used in Connection with Reference Materials, 1992.
ISO Guide 31: Contents of Certificates of Reference Materials, 1981.
ISO Guide 32: (draft): Calibration of Chemical Analysis and Use of Certified Reference Materials, 1994.
ISO Guide 33: Uses of Certified Reference Materials, 1989.
ISO Guide 34: Quality System Guidelines for the Production of Reference Materials, 94.
ISO Guide 35: Certification of Reference Materials - General and Statistical Principles,1989.

24. Bundesanstalt für Materialforschung und -prüfung (BAM), Unter den Eichen 87, D - 12205 Berlin.
25. Res pura GmbH, Institut für Anorganische Chemie und Analytische Chemie der Johannes Gutenberg-Universität, J.J. Becherweg 24, D - 55099 Mainz.
26. G.L. Aranovich, Russian J. Phys. Chem., 62 (1988) 11, 1561.
27. IUPAC Recommendations 1994: J. Rouquérol, D. Avnir, C.W. Fairbridge, D.H. Everett, J.H. Haynes, N. Pernicone, J.D.F. Ramsay, K.S.W. Sing and K.K. Unger, Pure & Appl. Chem., 66 (1994) 8, 1739.
28. J. Rouquerol, D. Avnir, D.H. Everett, C. Fairbridge, M. Haynes, N. Pernicone, J.D.F. Ramsay, K.S.W. Sing and K.K. Unger, in: J. Rouquerol, F. Rodriguez-Reinoso, K.S.W. Sing and K.K. Unger (eds.), Characterization of Porous Solids III. Elsevier, Amsterdam 1994, 1.
29. R.W. Lines, in: N.G. Stanley-Wood and R.W. Lines (eds.), Particle Size Analysis. Royal Society of Chemistry, Cambridge 1992, 40.
30. DIN-Taschenbuch 133: Partikelmeßtechnik, 3. ed., Beuth, Berlin 1990.
31. J. Schröder, GIT Fachz. Lab., (1986) 10, 978 and 11, 1095.
32. J. Seifert and G. Emig, Chem.-Ing.-Techn., 59 (1987) 6, 475.
33. D. Avnir, The Fractal Approach to Heterogeneous Chemistry, Wiley, Chichester, 1989.
34. Prof. D. Avnir (Dpt. of Chemistry, University of Austin, Texas, Texas 78712-1167), pers. communication.
35. A.V. Neimark, Ads. Sc. Techn., 7 (1990) 4, 210.
36. A.V. Neimark, E. Robens and K.K. Unger, Z. Phys. Chem., 187 (1994) 265.
37. F. Gottsleben and D. Hesse, Hung. J. Industrial Chem., Vesprém 20 (1992) 121.
38. C. Meyer and D. Hesse, Chem.-Ing.-Techn., 65 (1993) 12, 1478.
39. Prof. Dr. J. Gmehling, Fachbereich 9, Technische Chemie, Carl von Ossietzky-Universität Oldenburg, Postach 2503, D-26111 Oldenburg, Germany.

Adsorption and its Applications in Industry and Environmental Protection
Studies in Surface Science and Catalysis, Vol. 120
A. Dabrowski (Editor)



# Spectroscopic characterization of chemically modified oxide surfaces

J.J. Pesek and M.T. Matyska

Department of Chemistry, San Jose State University, One Washington Square, San Jose, CA 95192, USA

## 1. INTRODUCTION

Chemically modified oxide surfaces are found in many industrial processes and in environmental analysis and remediation. Among the most frequently encountered of these uses are chemically modified separation materials and catalysts. Whatever the ultimate application of the modified oxide material is, it is the surface that determines the properties which are exploited for the intended use. Therefore a survey of spectroscopic techniques which can study the nature of surfaces and chemically modified surfaces will be applicable to most materials regardless of their specific use. This paper presents a selection of the most important and most frequently used spectroscopic methods for the characterization of chemically modified surfaces. In most cases, the examples are related to materials used in separation processes. The most commonly used separation support material is silica which is available in a wide range of physical formats, including porous and nonporous substances. When a porous matrix is required, silica can be obtained in a broad range of particle sizes with a well-defined pore structure and specific surface areas from a few square meters to hundreds of square meters per gram.

## 2. SPECTROSCOPIC CHARACTERIZATION METHODS

### 2.1. Diffuse Reflectance Fourier Transform Infrared Spectroscopy (DRIFT)

Infrared spectroscopy is a classical instrumental method long used by chemists to characterize inorganic and organic molecules by absorption frequencies that are characteristic of certain functional groups in the compound. With respect to surface analysis, it is possible to obtain spectra of molecules that are either adsorbed on or chemically bonded to surfaces which have a reasonable area. For high surface area materials, an ordinary transmission spectrum can be obtained if the sample is mixed with KBr to improve light throughput. As the surface area decreases, it is necessary to signal average which can be most efficiently accomplished by acquiring the spectrum in the Fourier transform mode. Further increases in sensitivity can be achieved if the spectrum is obtained via diffuse reflectance. In this mode the infrared light beam from the source is directed to the sample, usually a mixture of the material to be studied and KBr, which is contained in a small cup. The light which is reflected from the surface is collected by a mirror and focused onto the infrared detector. The reflected infrared light contains the spectrum of the surface including any molecules bonded or adsorbed. Signal averaging improves the sensitivity so that materials with relatively small surface areas (a few square meters per gram) or a low concentration of bonded or adsorbed species can be characterized. The choice of detector also has an influence on the ultimate sensitivity. The liquid nitrogen cooled mercury cadmium telluride (MCT) detector is at least a factor of five more sensitive that the standard deuterated triglycine sulfate (DTGS) detector which is commonly used in the mid-infrared range [1].

The surface nature of the DRIFT technique can be easily illustrated by some simple examples. Figure 1 shows the DRIFT spectra of two inorganic oxides: silica and titania. In Figure 1A the most prominent features of the spectrum of silica include the very broad band between 4000 and 3000 $cm^{-1}$ which is the result of water adsorbed on the surface. A sharp peak near 3750 $cm^{-1}$ is superimposed on this broad band and is the result of isolated silanol groups on the surface (≡Si-OH) [2-4]. Since silica can be considered a polymeric

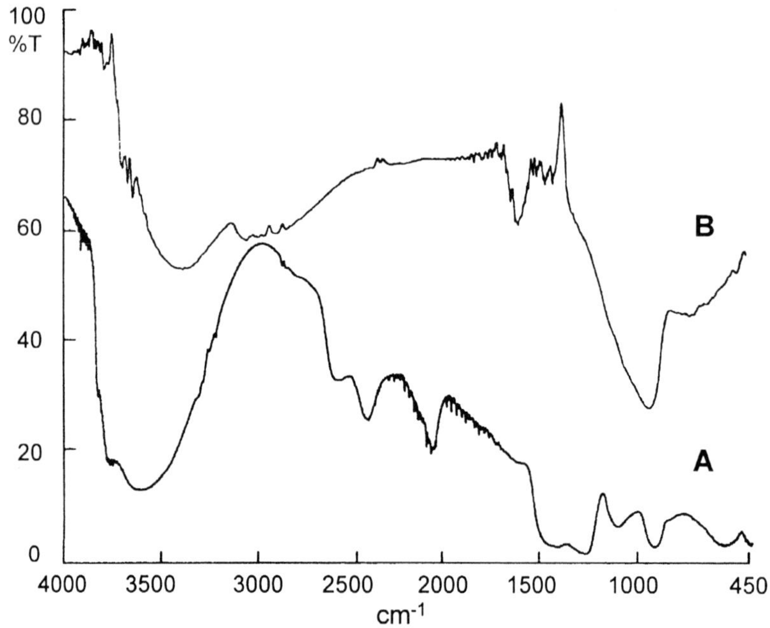


Figure 1. DRIFT spectra of native oxides. A) silica and B) titania.

condensation product of silicic acid, each silicon atom on the surface is bonded to three other silicon atoms via siloxane linkages (Si-O-Si) and the terminal silanol (Si-OH). A variety of silicon-oxygen bending and stretching modes are responsible for the other features in the spectrum. A similar result can be seen for titania (Figure 1B). Since most oxides are very adsorptive materials with respect to water, a broad band between 4000 and 3000 $cm^{-1}$ is observed for the unmodified titania surface. Some sharp peaks near 3700 $cm^{-1}$ are due to the titanols (=Ti-OH) which are present on the surface [5]. Other features in the spectrum are the result of various bending and stretching modes associated with the titania polymer matrix (=Ti-O-Ti=). There is a great deal of similarity in the essential features of the DRIFT spectra of these two oxides and indeed similar infrared bands can be identified in the spectra of alumina, zirconia and thoria [6]. In all of these inorganic oxides, the reactive group with respect to potential chemical modification or as a site of adsorption is the surface hydroxyl moiety.

The power of DRIFT as a surface analytical tool is evident from the spectra obtained after modification of most oxide surfaces such as those mentioned above. Figure 2A is the DRIFT spectrum of silica after it has been reacted with triethoxysilane.

The main features of the spectrum are the decrease in the intensity of the silanol band at 3750 $cm^{-1}$ and the appearance of a band at 2250 $cm^{-1}$ which represents the Si-H stretching mode [7]. Since a chemical reaction can only take place at the silanols and the Si-OH groups are only at the surface, these distinct

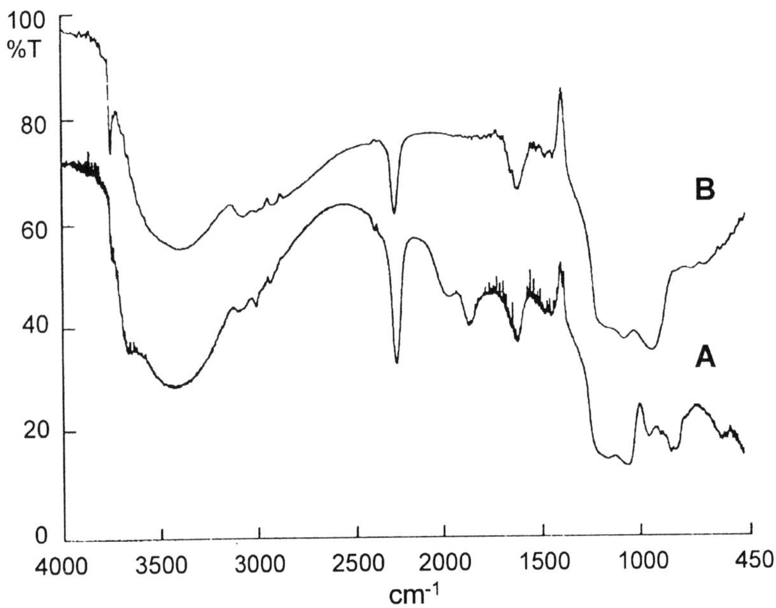


Figure 2. DRIFT spectra of oxides modified with TES. A) silica and B) titania.

changes in the infrared spectrum are proof that the DRIFT technique is a useful tool for characterizing surface chemistry and structures. A similar result is obtained when the same reaction takes place on a titania surface (Figure 2B). The intensity of the Ti-OH peaks decrease and a peak near 2250 $cm^{-1}$ from the Si-H stretch appears in the spectrum [6]. The same reasoning used for silica to document the surface nature of the spectral features observed can be applied to the changes observed in the titania DRIFT spectrum.

In many cases, particularly for chromatographic materials, it is desirable to modify oxide surfaces so that they possess significant hydrophobic properties [8]. Reacting the silica hydride surface of the material shown in Figure 2A with 1-octadecene will produce such a hydrophobic product. The DRIFT spectrum of this new surface is shown in Figure 3A. Again there are two essential features in the spectrum which verify the successful addition of 1-octadecene to the hydride surface. First, the intensity of the Si-H peak at 2250 $cm^{-1}$ decreases which indicates that this group has undergone a chemical reaction. Second, new peaks appear in the spectrum between 2800 and 3000 $cm^{-1}$ which represent the C-H stretching modes from the methylene and methyl carbons of the attached alkyl chain ($\equiv Si\text{-}CH_2\text{-}(CH_2)_{16}\text{-}CH_3$). Since the hydride groups exist only at the surface, then the bonded organic moiety is also only present at the silica surface [9]. The same result can also be observed for the modification of the hydride surface on titania shown in Figure 2B. The spectrum of the product obtained from the

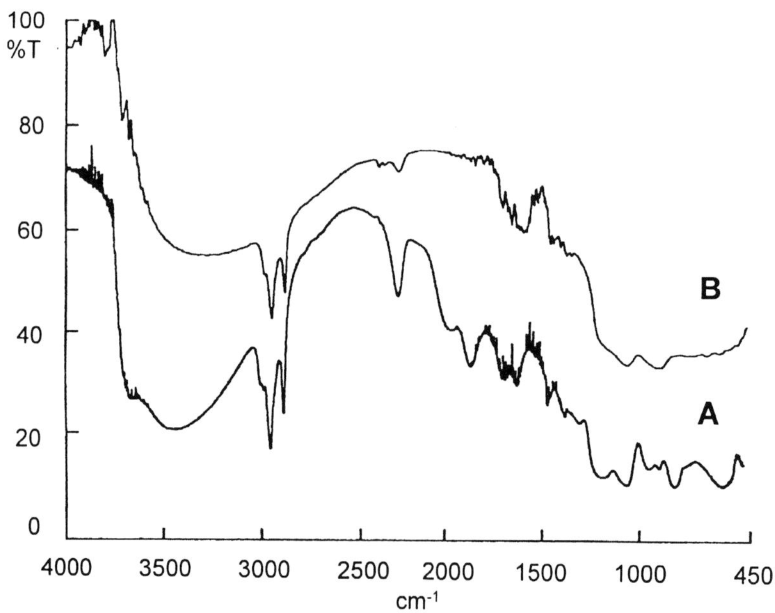


Figure 3. DRIFT spectra of hydride oxides modified with 1-octadecene. A) silica and B) titania.

reaction of 1-octadecene with the titania hydride (Figure 3B) material again displays a decrease in the Si-H stretching band at 2250 $cm^{-1}$ and a concomitant appearance of C-H stretching bands between 2800 and 3000 $cm^{-1}$ [5]. Similar results have been obtained with other oxides such as alumina, zirconia and thoria [6].

While the presence of hydrocarbon chains is almost always found for most typical bonded phase chromatographic materials, other functional groups on the surface can also be identified by the FTIR spectra. For example, the compound 4-phenyl-1-butene attached via hydrosilation results in the presence of aromatic carbon-hydrogen stretching bands above 3000 $cm^{-1}$ [10] - Figure 4A. Another

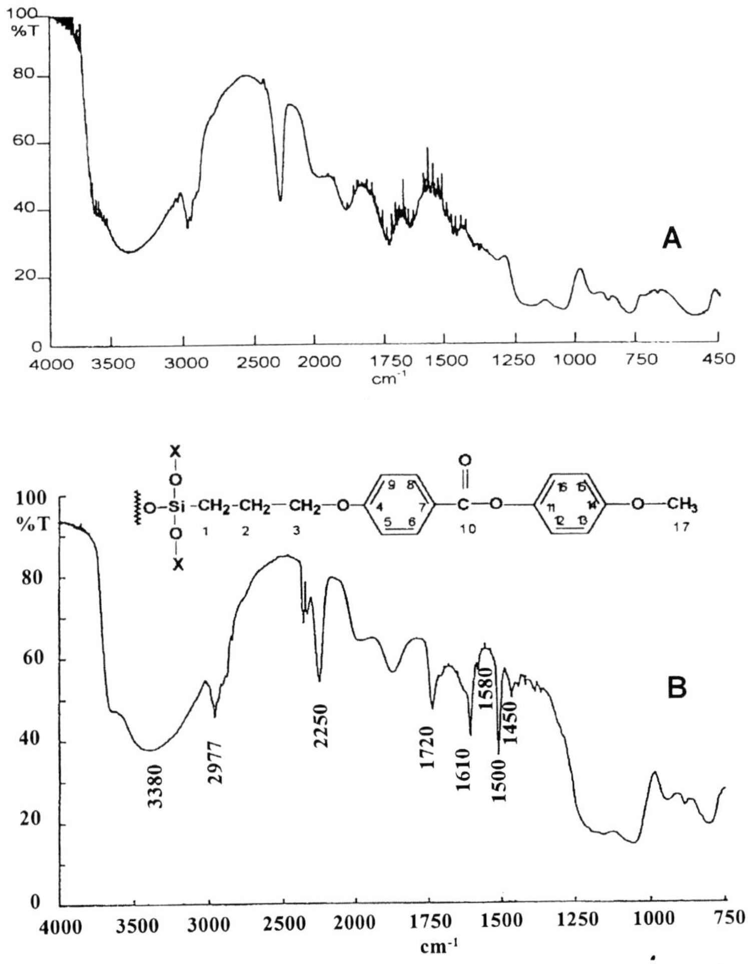


Figure 4. DRIFT spectra of chemically modified silicas. A) 4-pheny-1-butene and B) 4-methoxyphenyl-4-allyloxy benzoate (MPAB).

example is the liquid crystal whose structure and spectrum are shown in Figure 4B. In particular the ester carbonyl group can be identified by the peak at 1720 $cm^{-1}$ [11]. Other groups which have been bonded to silica surfaces and characterized by FTIR spectroscopy include cyano with an absorption at 2256 $cm^{-1}$[12], amine with an absorption near 1595 $cm^{-1}$ [13], and a ketone carbonyl which appears near 1720 $cm^{-1}$ [14]. This is not a complete list but is representative of some typical stretching and bending vibrational modes which have been used to confirm the bonding or adsorption of molecules on silica surfaces.

While silica is probably the most frequently encountered oxide surface, other materials particularly alumina, titania and zirconia also have considerable use and spectroscopic characterization is beneficial. One study mentioned above explored the potential for modifying alumina, zirconia, titania and thoria surfaces in a manner similar to silica [6]. While bonding of the moiety is usually the method of choice, in some cases adsorption is sufficient. For example, polyacrylates adsorbed on alumina are a useful dispersent in the production of certain ceramic products. The successful adsorption of these compounds on aluminia has been monitored by FTIR using the carbonyl stretching frequency for the acrylate species which appears between 1602-1606 $cm^{-1}$ [15]. Polybutadiene which has been adsorbed on alumina for use as a chromatographic phase can be detected by FTIR [16]. A similar adsorption process has also been tested on zirconia [17,18]. It has also been shown that FTIR can be used to detect Langmuir-Blodgett layers on the metal surfaces of thin-film devices [19].

While DRIFT spectroscopy is probably the most frequently used technique for studying surfaces by IR, other options are also available. Among these are photoacoustic spectroscopy (PAS) where the IR absorbances are excited by a typical infrared source but detection is accomplished via a microphone-like system [13]. The frequency dependent perturbations in the microphone are caused by reflections off the surface which is being studied. Another reflectance technique, attenuated total reflectance (ATR), relies on multiple reflections off the surface which is being studied and a crystal placed against the surface [1]. Sensitivity can be improved through signal averaging and Fourier transformation techniques. This method is often preferred for surfaces which are smoother than typical porous oxide materials such as silica or zeolites. However, it is applicable to the study of polymer surfaces, semiconductor materials and coated fused silica surfaces. Another possibility is reflection-absorption spectroscopy (RAIRS) where the incident light is plane polarized and detection of the polarization of the reflected beam can be used to determine the molecular orientation of the functional group with respect to the surface [1]. These measurements are most practical for determining the surface nature of a thin film in the study of deposition processes.

Infrared spectroscopy is an extremely useful technique for the study of both native and modified surfaces. Several modes of acquisition are available each with its own inherent advantages and disadvantages. In combination with signal averaging and Fourier transform signal analysis, a high degree of sensitivity can be achieved. The different modes also make it possible to characterize a broad range of surfaces from highly porous materials such as oxides to very smooth surfaces such as semiconductor devices and polymers. The variety of FTIR techniques will continue to be one of the primary means for acquiring qualitative (functional group) information about a broad range of surfaces on many different types of materials.

### 2.2. Nuclear Magnetic Resonance (NMR) spectroscopy

NMR has proven to be one of the most valuable spectroscopic techniques for the identification and characterization of a wide range of organic and inorganic species. It can provide unparalleled information about the chemical composition and structure of molecules. The newer two-dimensional pulse techniques [20] are expanding the capabilities of NMR even further so that in some cases structural prediction can approach that of x-ray crystallography. The use of NMR has traditionally been associated with chemical structure determination or characterization of new compounds produced by synthetic chemists or as an aid in the identification of unknown compounds in the solving of analytical problems. In most instances the compound is dissolved in an appropriate solvent and the spectrum is obtained in the solution state. For materials analysis or surface characterization, it is often impossible or impractical to dissolve the sample. Therefore, the ability to obtain NMR spectra in the solid state is crucial to the extension of this most important spectral technique to the study of surfaces.

#### 2.2.1. Solid state NMR

During the past 10-15 years solid state NMR spectroscopy [21] has been developed and applied to the study of a broad range of materials characterization problems. Several instrumental factors allow the acquisition of NMR spectra of solids so that useful information, comparable to that from spectra of compounds in solution, can be obtained.

Under the conditions used for solutions, the NMR spectrum of a solid material most often consists of a single broad line for each nucleus. This is due to the fact that nuclei from the same element, such as carbon-13, are fixed in a variety of orientations with respect to the applied magnetic field. This effect, known as chemical shift anisotropy (CSA), accounts for the broad range and continuous nature of chemical shifts observed in solids. While theoretically this same phenomena is present in solutions, the fact that molecules are free to tumble randomly averages out the orientations to a single chemical shift for each chemically different type of a particular nucleus. Spinning the sample at a moderate rate (50-60 Hz) further averages these chemical shift variations so that a single sharp line is obtained for each chemically different nucleus of the same type (carbon-13). In order to simulate this same effect in solid materials, the sample is spun at a much higher speed (several kHz) at an angle of 55° (the "magic angle") with respect to the applied magnetic field. The "magic angle" is the diagonal of a cube and spinning a solid very rapidly in this orientation closely approximates the same effect produced by freely tumbling molecules in solution. Magic angle spinning (MAS) results in nuclei experiencing a more uniform magnetic field environment. Additional broadening results in solids from dipolar coupling between the nucleus observed (spin = ½ and not hydrogen) and hydrogen. So while MAS will improve the line width to a great extent, it still remains fairly broad until the hydrogen dipolar coupling is removed. This is accomplished by broad band proton decoupling where the sample is subjected to a high power rf field at the hydrogen frequency. This process simultaneously decouples all protons from the nucleus being observed and narrows the absorption line even further. This same type of experiment is often performed on solution samples in order to simplify the spectrum so that each chemically different carbon atom for example will give a single line. While not quite as narrow as those obtained for solution samples, the spectrum of a solid obtained with MAS and broad band proton decoupling will consist of lines where individual nuclei of the same type in different chemical environments can be distinguished.

While MAS and proton decoupling can solve the problems associated with broad lines in solids, further complications can occur in the analysis of surfaces and other solid materials due to the low sensitivity of the nucleus to be observed. This is especially true of nuclei such as carbon-13 and silicon-29. Under normal circumstances this lack of sensitivity would greatly inhibit the ability to study surfaces and some materials by NMR. However, it is possible to utilize a technique, cross-polarization (CP), developed for solution samples to alleviate this problem. By appropriate matching of the precessional frequencies between the insensitive nucleus and hydrogen, some of the sensitivity associated with protons can be transferred to a nucleus like carbon-13. Use of CP significantly shortens the time needed to acquire the spectrum of solid samples containing carbon-13 or silicon-29 so that native or modified surfaces can be characterized. Some specific examples of surfaces characterized by NMR using three different nuclei are given below.

*Carbon-13*

For chemically modified surfaces like those found on separation materials, carbon-13 CP-MAS NMR spectroscopy is one of the most useful methods for characterization. In most cases it is possible to positively identify the bonded organic moiety on the surface. An example of a carbon-13 CP-MAS NMR spectrum of an organic group bonded to silica is shown in Figure 5A. For this particular bonded moiety, allyl glycidyl ether (AGE), it is possible to identify resonances for each of the six carbon atoms in the compound [22]. The peak at 7 ppm represents the carbon directly bonded at the silica surface and the peak at 22 ppm is due to the methylene group next to the surface bonded carbon. The peaks at 71 ppm and 65 ppm are due to carbons 3,4 and 5, and carbon 6 respectively. Two additional peaks in the spectrum are labeled **a** from the methyl group and **b** from the methylene group of the ethoxy moiety which remains after

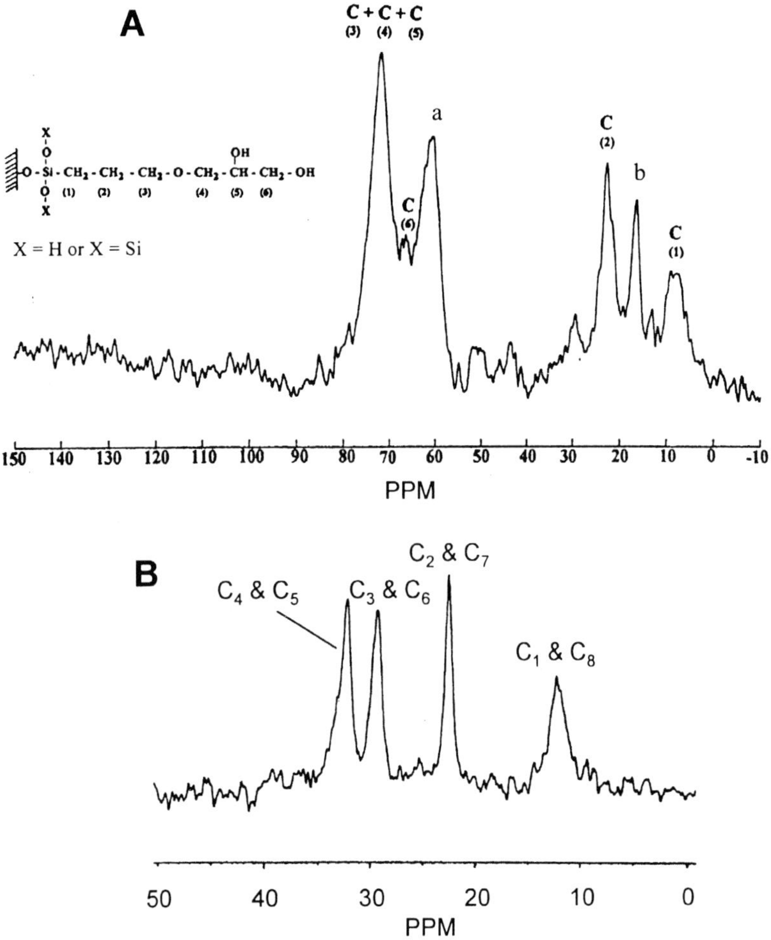


Figure 5. Carbon-13 CP-MAS spectra of bonded organic groups on silica. A) allyl glycidyl ether (AGE) and B) octyl.

the TES silanization process to create the hydride surface necessary for bonding AGE. The spectrum of another bonded group, n-octyl, is shown in Figure 5B [23]. In this case only 4 peaks are observed since there is coincidental overlap of four pairs of carbon atoms. This spectrum is a good example of the limits of resolution encountered in solid state NMR spectroscopy. The line widths are not quite as narrow as those encountered in typical solution spectra so while each carbon in the alkyl chain has a different chemical shift, the difference between each of four pairs is so small that they are not resolved. However, the correct identification of the bonded group is not difficult in this case and there is no doubt that the appropriate moiety is attached to the surface. In some cases the complete

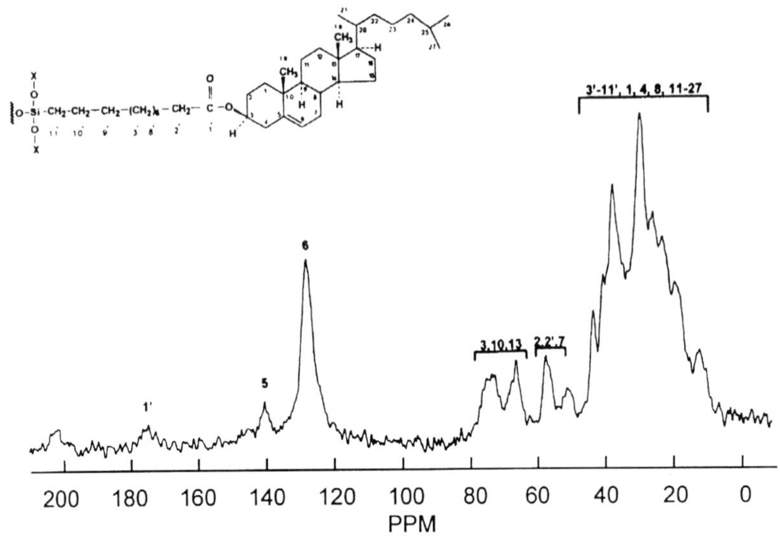


Figure 6. Carbon-13 CP-MAS spectrum of cholesterol bonded to silica.

assignment of all carbons to peaks in the spectrum is not possible such as the example shown in Figure 6 where a large molecule, a liquid crystal cholesterol compound, is bonded to silica [24]. By comparison to the unbonded compound and reasonable expectations of chemical shifts, it is possible to give general ranges for particular carbons and to specifically identify one or more individual carbons so that verification of the structure of the bonded moiety is quite certain. When other oxides such as zirconia, alumina and titania are first converted to a hydride intermediate and then further modified by hydrosilation to attach an organic group, the carbon-13 CP-MAS NMR spectrum of the product can be used to confirm the success of the bonding process and to identify the structure of the bonded group [4,5]. These examples illustrate the usefulness of carbon-13 CP-MAS NMR for the identification or verification of the chemical structure of organic compounds that are either bonded to or adsorbed on oxide materials having a reasonable surface area..

In addition to the information about structure, the carbon-13 CP-MAS NMR spectrum has also been shown to be useful for determining relative mobilities of various bonded moieties. The motion of the alkyl chain is reflected in the relaxation times of the various carbons. As the mobility in a certain part of the alkyl chain is increased, the relaxation time of the carbons in this part of the bonded organic moiety will increase due to less efficient dipolar interactions. In an early study it was concluded that motion increases up the alkyl chain from the surface to about the eighth carbon but after that point it remains relatively constant [25]. At uniform surface coverage it was determined that alkyl chains of C-6 or C-8 had the highest mobility [26]. Similar chain mobility studies were used to correlate the retention of fullerenes on various C-18 columns [27]. Another variable temperature study was used to identify molecular motion and phase transitions in bonded liquid crystal stationary phases [24].

*Silicon-29*

Silicon-29 CP-MAS NMR is probably most useful for the study of oxide surfaces used in chemically modified stationary phases for HPLC and catalytic materials which contain a considerable fraction of silica. The spectrum of silica in a variety of physical forms has been well-documented in many studies. An example of a typical silica spectrum is shown in Figure 7 [28]. The three peaks observed are readily understandable in terms of the expected surface structure of

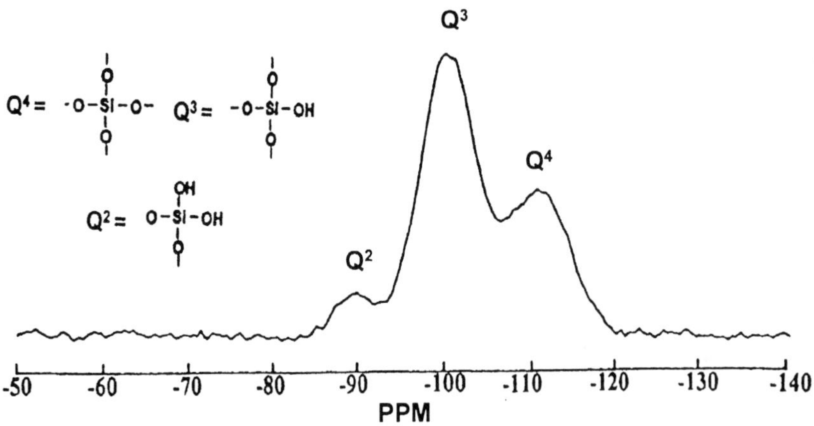


Figure 7. Silicon-29 CP-MAS NMR spectrum of native silica.

silica. The highest field peak (-110 ppm) is due to silicon atoms which have four siloxane linkages, i.e. the backbone structure of silica. The peak near -100 ppm is assigned to silicon atoms having three siloxane linkages and a hydroxyl group. This is a surface species and its intensity is greater that the -110 ppm peak because the higher field peak has no hydrogens nearby so there is considerably less cross-polarization even though there are significantly more backbone silicon atoms than silanols (≡Si-OH). The final peak near -90 ppm represents the few surface silicon atoms that contain two siloxane linkages and two hydroxyl groups, referred to as geminal silanols.

The spectra of modified silica surfaces can be used to determine the type of linkages formed between the modifying group and the oxide substrate. For example, silanization of silica with triethoxysilane (TES) results in two new peaks appearing in the spectrum [29]: at -85 ppm due to the silicon atom having three siloxane linkages and a hydride and at -75 ppm for silicon atoms having two siloxane linkages, a hydride and a hydroxyl as shown in Figure 8A. When

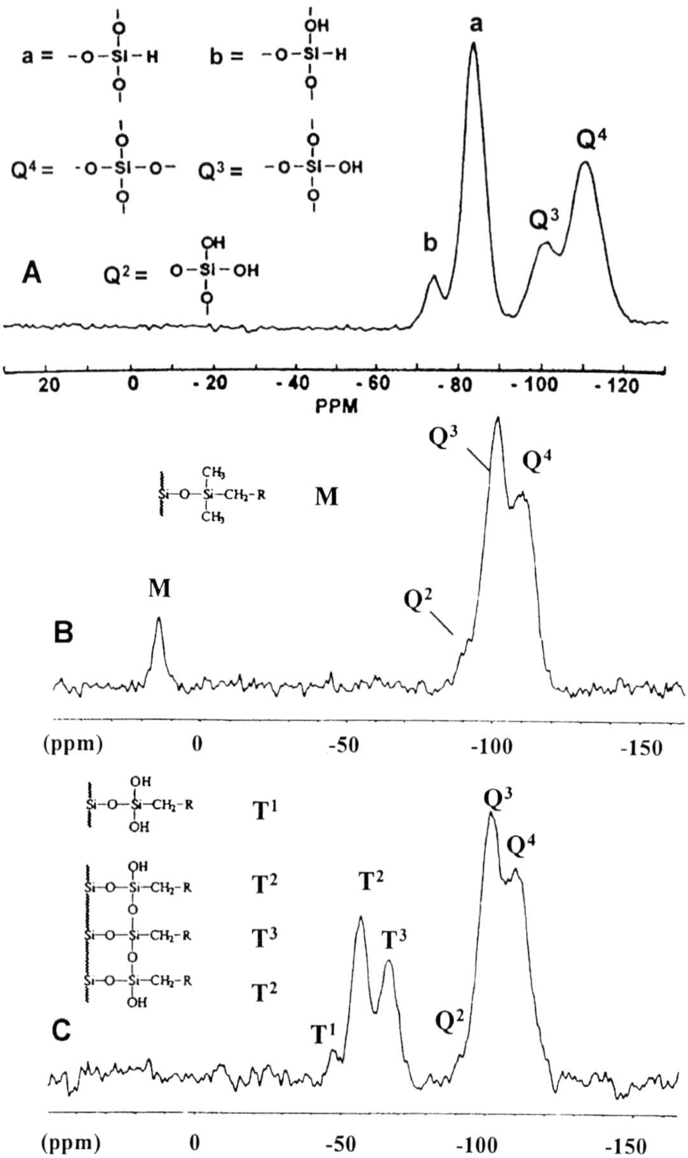


Figure 8. Silicon-29 CP-MAS spectra of modified silica surfaces. A) hydride from TES, B) monomeric phase from organosilanization [31] and C) polymeric phase from organosilanization [31].

this new hydride surface is modified further through a hydrosilation reaction (≡Si-H + $CH_2$=CH-R → ≡Si-$CH_2$-$CH_2$-R) two additional peaks appear in the spectrum. These new peaks represent silicon atoms on the surface with three siloxanes and a carbon (-65 ppm) [30] and silicon atoms with two siloxane linkages, a carbon and a hydroxyl (-55 ppm). For bonded phases made by traditional organosilanization, the type of bonding reaction can be distinguished from the silicon-29 CP-MAS NMR spectrum [31]. Figure 8B is the spectrum of the product from a monofunctional organosilanization reaction (≡Si-OH + X-Si(RR'R") → ≡Si-O-Si(RR'R") ) on a silica surface. This bonded material is identified by the peak at 13 ppm which represents the silicon atom on the surface with three Si-C bonds. The spectrum of a polymeric phase is shown in Figure 8C which is the product from the reaction of silica with a trifunctional silane (≡Si-OH + $X_3$Si-R → ≡Si-O-Si$Z_2$R where Z = -O-Si≡ or -OH). The peak at -66 ppm represents the material where the two Z groups = -O-Si≡ while the peak at -58 ppm appears when the surface contains a Z group of each type. While the carbon-13 CP-MAS spectrum is a direct means of identifying a bonded organic group on an oxide surface, the silicon-29 CP-MAS spectrum is most useful as a means of characterizing the types of silicon atoms which are present on native and modified silica surfaces. However, when various silane reagents are used to modify other oxide surfaces besides silica, the silicon-29 CP-MAS NMR spectrum can be used to characterize the surface species. For example when alumina, titania, zirconia or thoria are modified with TES, the same two peaks at -85 ppm and -75 ppm that are observed on silica appear in the spectra of these other oxides [4,5]. When these oxides with hydride surfaces undergo a hydrosilation reaction, peaks appear at -65 ppm and -55 ppm indicating bonding of the organic group via formation of an Si-C bond.

*Proton*

Proton NMR spectra have been an essential tool for the structural identification of organic compounds for more than 40 years. Spectra for dilute solutions are quite easy because of the high sensitivity of the hydrogen nucleus. With improved instrumentation the limit of detection is continually moving to lower levels. However, in solids the presence of strong homonuclear dipole-dipole interactions results in very broad lines. In addition, the chemical shift range of protons (≈10 ppm) is considerably less than the 200 ppm for carbon-13 and 150 ppm for silicon-29.

Two methods have been developed to overcome the effects of homonuclear dipole-dipole interactions in order to obtain line widths narrow enough to distinguish between protons with reasonably different chemical shifts. The first technique involves the use of magic angle spinning at 1.5-2.0 kHz and multiple-pulse line narrowing [32]. The multiple-pulse approach results in rapid orientation of the proton magnetic vector to many positions with respect to the externally applied magnetic filed so that the net effect is similar to the random movement of a freely tumbling molecule in solution. In conjunction with MAS, the technique has been given the name Combined Rotation and Multiple-Pulse Spectroscopy (CRAMPS). In the CRAMPS spectrum of a carefully dried silica hydride sample, sharp peaks are observed at 4.5 ppm (Si-H) and 2.0 ppm (isolated silanols) and a broad peak between 2-8 ppm is assigned to hydrogen bonded silanols.

More recently it has been demonstrated that for systems such as bonded organic groups on silica surfaces, the use of high rotational frequencies (12-14 kHz) alone is sufficient to provide reasonably narrow line widths in the proton spectrum [32,33]. An interesting example is shown in Figure 9 which is a series of $^{1}H$ MAS NMR spectra for various C-18 phases at different bonding densities. It is quite easy to distinguish the peak for the methyl group at 0.8 ppm from peak for the methylene groups at 1.2 ppm. However, what can be clearly seen is that as the bonding density increases the peaks become broader due to the stronger dipolar interactions which result from the closer proximity of

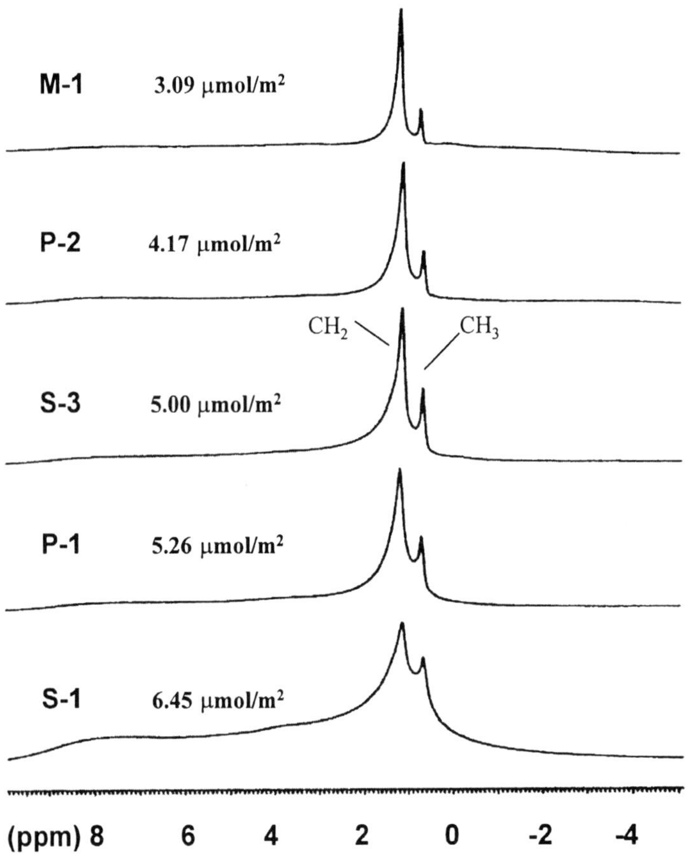


Figure 9. Proton MAS NMR Spectrum of C-18 modified silica at different bonding densities [33].

protons to each other. Fortunately, very few phases have bonding densities greater than 5.0 μmol/m$^2$ so that proton MAS NMR spectra should be a viable technique for characterizing most chromatographic separation materials.

### 2.2.2. Solution phase (slurry) NMR

While solid state NMR offers many advantages for obtaining information about organic moieties bonded to oxide surfaces, for separation materials it lacks one essential component of a real system: the liquid mobile phase. Therefore, an NMR technique that could characterize the stationary phase in the presence of typical mobile phase liquids is highly desirable. For reversed-phase materials the appropriate liquids would be water and organic solvents such as methanol, acetonitrile and tetrahydrofuran. These experiments involve placing the stationary phase material and an appropriate solvent combination inside an NMR sample tube to simulate the situation inside of an HPLC column [34,35]. There are two approaches that can be taken to gain insight into the dynamics of stationary phase/mobile phase interactions. The first method is to study the stationary phase in a manner similar to the solid state experiments. The second technique would involve studying the solvent in the system, particularly that portion of the liquid that is in the immediate vicinity of the bonded organic moiety.

As in the solid state, the solution phase characterization of bonded materials is best studied by carbon-13 NMR measurements with the relaxation time providing the information about the nature of the organic group under various solvent conditions. For example, the C-18 bonded moiety has been characterized over the temperature range 5-70°C in an 80:20 methanol/water solvent system [36]. The van't Hoff plot of the relaxation time for the methylene carbons 4-15 vs. 1/T is a line with a small negative slope in agreement with the theoretical prediction based on dipolar interactions. However, similar plots for the $C_{17}$ methylene and $C_{18}$ methyl groups have a measurable curvature with a change in slope near 35°C. The interpretation of this data is that hindered movement of this segment of the molecule occurs below 35°C which results in a change of the dominant relaxation mechanism. Therefore, it is likely that major changes in the retention behavior of this bonded material will occur around this "transition" temperature as the stationary phase becomes more ordered below 35°C and less ordered above this temperature.

To study the solvent behavior in the presence of a bonded moiety, it is necessary to use deuterated solvents since deuterium is a quadrupolar nucleus whose relaxation is controlled by the molecular correlation time, a measure of molecular motion [37]. The C-18 bonded phase has been studied in water/methanol and water/acetonitrile mixtures of varying composition [38]. Figure 10 illustrates one part of this study where the change in the deuterium relaxation time in $D_2O$ is monitored as a function of % organic modifier in the mobile phase for methanol and acetonitrile using two different bonding density C-18 stationary phases. These results show that little change in the deuterium

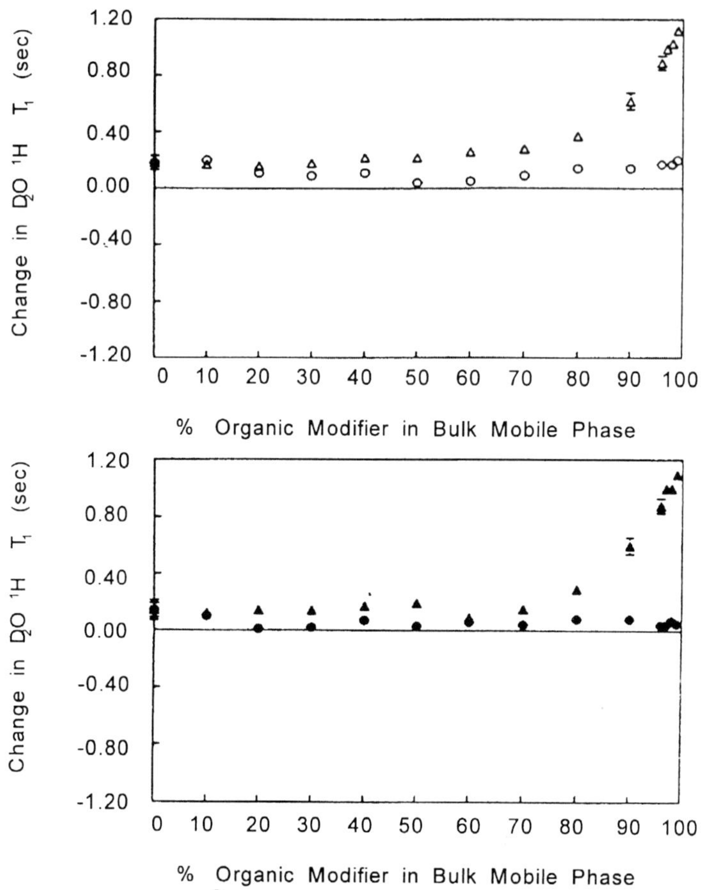


Figure 10. Change in $^2H$ relaxation in $D_2O$ in binary solvents in contact with C-18 moieties bonded to a silica surface. Top: low bonding density. Bottom: high bonding density. Circles = methanol. Triangles = acetonitrile [38].

relaxation time occurs when methanol is the organic component for both stationary phases. However, when the organic constituent is acetonitrile, a relatively large change occurs at high percentages of the organic component. These results are interpreted in terms of $D_2O$ interactions with residual silanols on the surface. Since water and acetonitrile form relatively weak hydrogen bonds in comparison to methanol, $D_2O$ is more likely to associate with the surface silanols in the presence of $CH_3CN$. In addition, acetonitrile is better able to solvate the alkyl moiety which opens the bonded phase structure allowing easier penetration of water to the surface.

Despite the fact that these solution studies have been used to investigate stationary phases for more than 10 years, there are still relatively few reports utilizing this technique. In addition, virtually all of the results reported to date involve the characterization of C-18 as the bonded organic moiety.

## 2.3. X-ray Photoelectron Spectroscopy (XPS)

In XPS, the sample surface to be studied is irradiated with high energy x-rays using a collimated beam with a diameter between 100 μm and a few millimeters. The sample is placed in an ultrahigh vacuum (UHV) since adsorption of any species on the surface can complicate the interpretation of the spectra obtained [39]. Elemental information about surface species, up to a depth of about 70 Å, is obtained by analysis of the energy of electrons which are emitted upon irradiation by x-rays. All elements except H and He can be identified by XPS. If a narrow range of electron energies around a particular element is scanned (high resolution spectrum), information of the oxidation state of that species can be obtained. For carbon, this often translates into the type of hybridization and in some instances can provide a clue about different functional groups on the surface. However, the extent of structural information available from XPS spectra is limited in comparison to either DRIFT or carbon-13 CP-MAS NMR.

An example of typical XPS spectral information is shown in Figure 11A for hexadecanoic acid covalently immobilized on quartz [40]. The two major peaks in

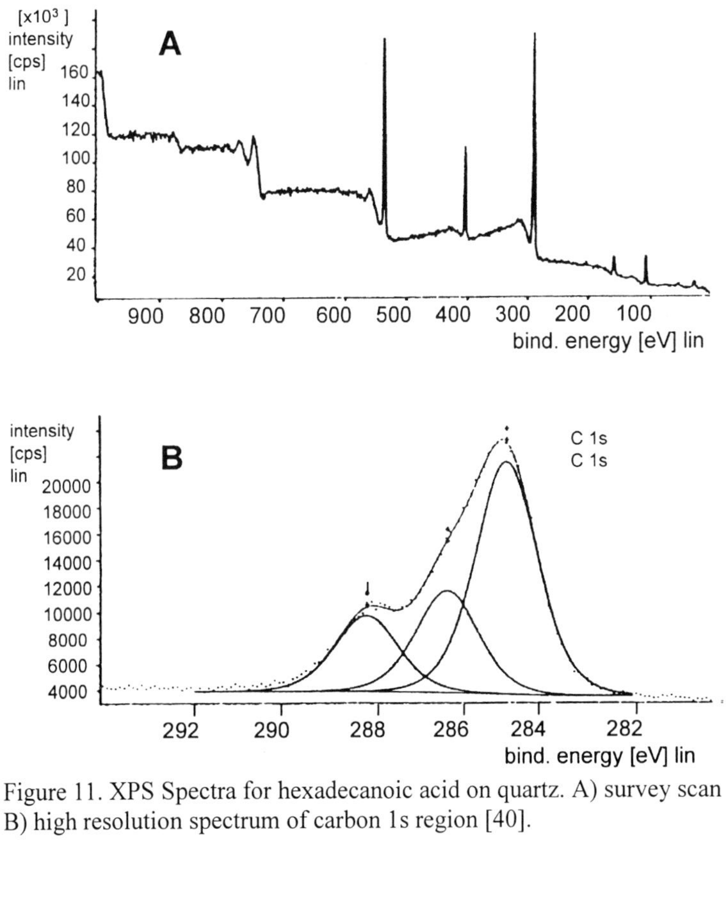


Figure 11. XPS Spectra for hexadecanoic acid on quartz. A) survey scan and B) high resolution spectrum of carbon 1s region [40].

the spectrum are at energies which are characteristic of oxygen and carbon. The high resolution spectrum (Figure 11B) for the carbon peak near 300 eV can be deconvoluted into three components. The three individual components are assigned as follows: 285 eV is due to the methylene carbons of the alkyl chain; 286.7 eV results from the methyl groups on the silane reagent used to attach the hexadecanoic acid to the surface; and 288.7 eV comes from the carbonyl carbon of the acid group. In addition to the elemental information provided by the survey scan and the functional group data contained in the high resolution scan, peak intensities can also be used to estimate the thickness of an adsorbed or chemically bonded layer. By comparison to the expected intensity from a monolayer at a fixed angle of electron emission, the thickness of the sample shown in Figure 11 is estimated to be about 5.3 nm. Generally a correction must be made for the organic material which adsorbs on the surface even under UHV conditions. The thickness of this randomly adsorbed carbon containing material is on the order of 1nm.

For chemically bonded separation materials, XPS can be used to characterize the organic moiety on any type of oxide surface. However, in general other methods such as DRIFT and solid state NMR provide more molecular information about the organic species on the surface. For quantitative information, ordinary elemental (combustion) analysis allows the determination of the surface coverage in terms of $\mu mol/m^2$ if the structure of the starting compound is known. However, XPS can confirm the results of the other spectroscopic methods. The thickness of the bonded layer is not possible to determine by XPS on porous oxide materials since a relatively smooth surface is necessary for this type of measurement. One unique application for XPS is in the bonding of organic groups to hydride oxide surfaces via hydrosilation reactions. These reactions utilize metal complexes (particularly platinum compounds) as catalysts for the bonding of the unsaturated functional group (usually an olefin) to the hydride surface. Appropriate reaction conditions can be developed which minimize or eliminate any deposition of metal on the surface [41]. The presence of the reduced species, such as Pt(0), can have detrimental effects on chromatographic performance. So far XPS has been shown to be most useful for the characterization of smooth surfaces such as semiconductor materials or quartz plates.

### 2.4. Scanning Electron Microscopy (SEM)

SEM has been a primary tool for characterizing the fundamental physical properties of oxide materials for some time. For example, SEM is particularly useful for determining the particle shape and approximate size distribution of various silica materials used as supports in chemically bonded stationary phases for chromatography [8]. The visual images provide resolution at the micron to in some cases the submicron level so that surface morphology can be determined. This information is especially useful when evaluating a new synthetic approach to the formation of oxide materials. For example, a recently developed method involved the synthesis of both silica and titania by the sol-gel approach using alkoxides as the starting materials [42]. The shape and size distribution of the oxide particles was studied by SEM at various stages during the formation process and under several conditions. Therefore, it was possible to determine the best conditions for the sol-gel process and to establish the optimum time for the most uniform distribution of particles at the desired size, i.e. particle diameter. Another example is the evaluation of coating procedures for catalysts used in emission control [43]. From SEM photos, it was possible to determine the nature (thickness and uniformity) of the active coating on a support surface for automotive catalytic converters. Another use of SEM involved the evaluation of various surfactants on the structure of coal produced by a liquefaction process [44].

For characterization of separation materials, a more recent application of SEM has been the evaluation of different surfaces on the inner walls of fused silica tubes used in capillary electrophoresis. SEM photos were particularly useful for verifying the formation of extended polymer networks in 100 μm capillaries [45]. The goal was to produce coatings of controllable thickness. Therefore, from the micrographs it was possible to determine the coating thickness under various reaction conditions so that an appropriate experimental protocol could be developed. Even the surfaces of uncoated capillaries have been evaluated by SEM [46]. In particular SEM was used to determine the extent of adsorption of proteins on uncoated surfaces and the extent of their removal by various washing processes. A new approach to capillary electrochromatography (CEC) has recently been developed which involves first etching the inner wall of the fused silica tube with a reagent that releases HF [47]. The goal is to increase the area of the inner surface by a factor of 100-1000 so that subsequent attachment of various organic groups will facilitate interactions between solutes to improve separation. An example of an SEM photo for an etched inner capillary surface is shown in Figure 12.

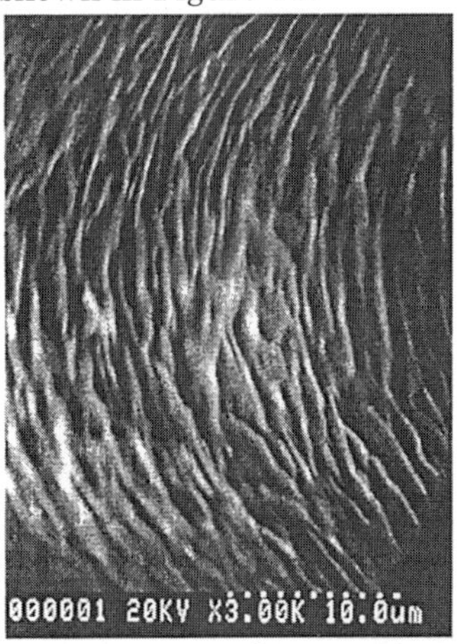


Figure 12. SEM of the chemically etched inner wall of a fused silica capillary [47].

The degree of roughness on the surface varies with reagent concentration, reaction time and reaction temperature.

SEM is an extremely useful method for visual confirmation of the physical state of many different types of surfaces. It provides morphology information that can be used to evaluate or predict performance based on surface characteristics. It will continue to be an essential technique for the evaluation of materials used in catalytic reactions, semiconductor devices and separation processes.

## 2.5. Atomic Force Microscopy (AFM)

AFM is one of the newest techniques for the characterization of surface morphology. The information provided by AFM does not duplicate that of SEM but is generally quite complementary. SEM photos can be used to study surface features that are several tenths of a nanometer while the resolution of AFM is less than 0.1 nm. Therefore, AFM has the ability to distinguish objects on smooth surfaces of molecular dimensions. In fact, under optimum conditions atomic force microscopy has been able to attain resolution on the atomic scale. The highest resolution is achieved for surfaces such as pure metals used for electrodes or for silicon materials used in semiconductor devices.

The principle of AFM is based on the design originally developed for scanning tunneling microscopy (STM) [48]. In STM, a current develops between an electrode and a cantilever as a result of deflection caused by interaction between the probe tip and the surface. In order for a response to be measured, the surface must be conductive. AFM utilizes coulombic or van der waal's forces between the surface and the probe tip on the cantilever as a means of investigating surface features. The movement of the cantilever is converted into images as the probe is rastered across the surface. The choice of probe tip material is important so that a suitable response can be generated as the surface is scanned. Considerable effort has also gone into techniques which can make the probe as sharp as possible in order to obtain the highest resolution.

For separation materials, AFM has recently proved to be an effective method for studying the inner walls of fused silica capillaries used in capillary electrophoresis. Under the high magnification of AFM, the apparently smooth surface of a bare unmodified capillary displays a variety of defects [49]. There is much interest in coating silica capillaries in order to control electroosmotic flow or to prevent the adsorption of certain species, usually peptides and proteins, on the inner wall. AFM has been used to monitor the homogeneity of the coating under various reaction conditions [50]. It was possible to correlate the best images with respect to the uniformity of the coating in the capillaries to the deposited layer of highest stability and the greatest reduction in electroosmotic flow. As with SEM, AFM has also proved useful for detecting the presence of adsorbed proteins on capillary surfaces [51]. An important aspect of AFM that distinguishes it from SEM is its ability to determine the depth of surface features. An example of an AFM photo of a bare fused silica capillary that reveals the surface morphology is shown in Figure 13 [46]. From a depth profiling analysis on this micrograph, it was determined that some of the grooves visible on this surface were as deep as 500 nm. An examination of a polyacrylamide capillary revealed a fairly smooth surface at a maximum resolution of 10 nm [46]. Some irregularities were seen but depth profiling revealed that changes in vertical elevation were not greater than 50 nm. The cause of these irregularities was attributed to the roughness of the original fused silica surface. While it has been possible to study fairly smooth structures by AFM, the large irregularities of porous silica material have prevented to date any high resolution images of these surfaces. However, the inner walls of etched fused silica capillaries have been imaged by AFM [52] so that it seems possible to characterize materials with reasonably rough surfaces.

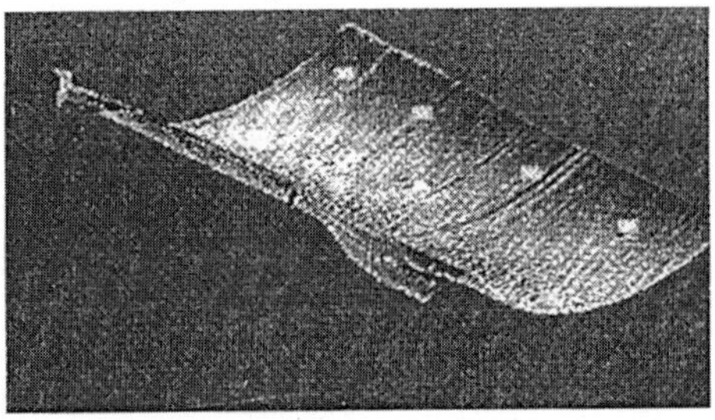

Figure 13. AFM of fused silica capillary surface [46].

The use of Atomic Force Microscopy for the characterization of oxide materials is still limited but the high resolution capabilities of the method show promise for the study of a wide variety of surfaces. Both the principles of the technique and the instrumentation must be advanced so that a wider range of morphologies can be studied with resolution approaching that achieved at present on relatively smooth surfaces.

## 2.6. Miscellaneous spectroscopic methods

The methods discussed above have been the most extensively used or have unique advantages for the study of modified oxide surfaces. A number of other techniques have been applied to the characterization of oxide materials. A few of these will be described in this section.

One method used for the study of monolayer structures on glass or quartz surfaces is fluorescence microscopy [53]. A fluorescent dye is placed on the modified surface and the intensity of the light observed is used to distinguish the uniformity of the organic coating. Raman spectroscopy has been used to characterize catalysts with respect to the type of vanadium oxide species deposited on alumina and titania supports [54]. An interesting study to determine if ions like sodium can migrate into the structure of fused silica quartz capillaries utilized secondary ion mass spectroscopy (SIMS) [46]. After exposure of the inner wall of the capillary to a typical high pH buffer, sodium borate, the SIMS determination revealed a significant amount of $Na^+$ on the surface.

However, after ion sputtering to remove about 5 nm of the surface material, there was little evidence of sodium ions indicating relatively little penetration due to the presence of the buffer.

Spectroscopic techniques are used to probe the degree and rate of interaction between a molecule and a modified surface, most often as a model for chromatographic processes. For example, fluorescence spectroscopy has been used to determine the extent of interaction between a charged complex, ruthenium bipyridine, and both the silanols and a bonded organic moiety on a silica surface [55]. An example of the results obtained for bare silica, C-2 silica and C-18 silica at two pH values are shown in Figure 14. The data shows that

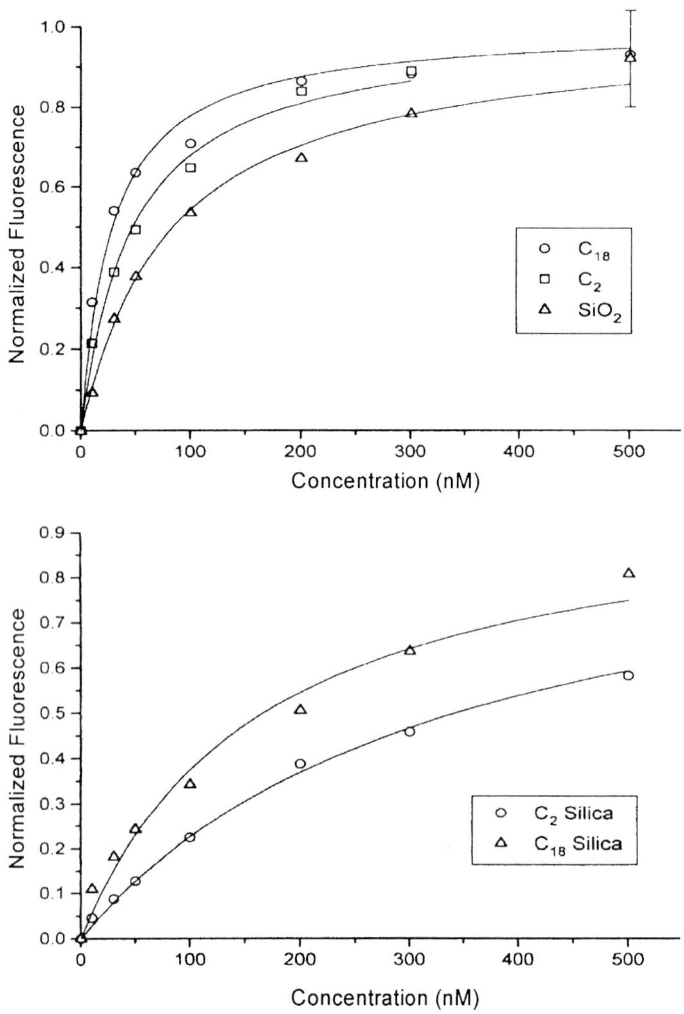


Figure 14. Normalized fluorescence of ruthenium bipyridine vs. its concentration on various silica surfaces. Top: pH = 8.0 Bottom: pH = 2.0 [55].

while there is significant negative charge on silica at pH 8, hydrophobic interactions between the bonded alkyl group and the bipyridine moiety of the complex are also present since increased adsorption over that observed on bare silica is measured when the C-2 and C-18 ligands are present. As expected, C-18 has more hydrophobic interactions than C-2. At pH 2 (Figure 14B), only hydrophobic interactions are observed since the silanols are protonated and therefore no measurements for bare silica are possible. Surface-enhanced Raman spectroscopy (SERS) has also been used to monitor the adsorption of molecules onto silica surfaces [56]. The intensity measurements in the Raman can be used to obtain absorption isotherms like those shown in Figure 14. In addition, the high sensitivity allows the measurement of kinetic processes on the surface. An example of such a determination is the adsorption of pyridine on silica shown in Figure 15. Fluorescent probes have also been used to monitor the diffusion of molecules within bonded alkyl groups [57]. It was found that diffusion decreased

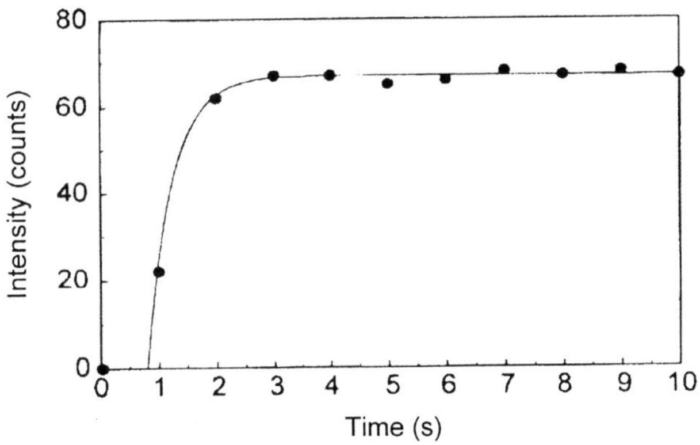


Figure 15. Raman intensity vs. time for adsorption of pyridine on silica [56].

with decreasing surface coverage and with decreasing bonded alkyl chain length. Conditions that provide the most continuity between the solute and bonded phase such as the presence of an organic modifier, high alkyl chain density and longer alkyl chains result in higher rates of diffusion. Further modifications to the fluorescent probe method which involve temperature jump measurements have been used to measure adsorption/desorption kinetics between the bonded phase and the bulk solution [58]. The results of these measurements under different mobile phase compositions are shown in Figure 16. These data indicate that as the amount of water in the mobile phase increases, the rate at which the hydrophobic probe will adsorbed on the alkyl bonded moiety increases and the rate at which it desorbs decreases.

The above examples represent several of the different methods which have been used to characterize various aspects of oxide and chemically modified oxide surfaces. Some of these may eventually be more important than the techniques described earlier but at present have provided specialized information often using equipment not available in the average laboratory. A search of the literature will reveal other methods not described here, often having even more limited utility.

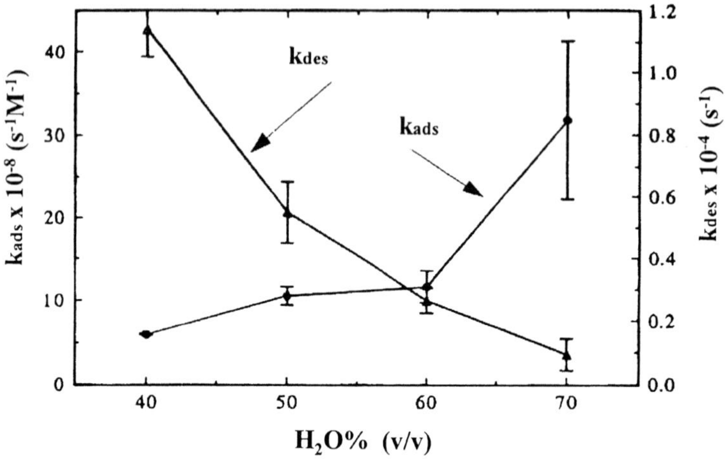


Figure 16. Rates of adsorption and desorption for fluorescent probe on silica as a function of percent water in the solvent [58].

## 3. CONCLUSIONS

A wide variety of methods have been developed to determine the chemical and physical nature of oxide and modified oxide surfaces. For chemical characterization, infrared spectroscopy, solid state nuclear magnetic resonance spectroscopy and x-ray photoelectron spectroscopy have proved to be the most useful. For physical measurements, in particular surface morphology, scanning electron microscopy is the method of choice. However, atomic force microscopy is now being used more often because no surface preparation is required and resolution can be better in some cases. AFM is still limited to generally smooth surfaces but advances in experimental techniques and instrumentation are expanding the degree of roughness that can be accommodated. Overall surface characterization is still a growing field with numerous developments in recent years due to its importance for catalysts, semiconductor devices and separation materials.

## REFERENCES

1. P.R. Griffiths and J.A. de Haseth, Fourier Transform Infrared Spectroscopy, Wiley Interscience, New York, 1986.
2. J. Nawrocki, Chromatographia, 31 (1991) 177.
3. J. Nawrocki, Chromatographia, 31 (1991) 193.
4. J. Nawrocki, J. Chromatogr., 779 (1997) 29.
5. J.J. Pesek, M.T. Matyska and J. Ramakrishnan, Chromatographia, 44 (1997) 538.

6. J.J. Pesek and V.H. Tang, Chromatographia, 39 (1994) 649.
7. C.H. Chu, E. Jonsson, M. Auvinen, J.J. Pesek and J.E. Sandoval, Anal. Chem., 65 (1993) 808.
8. E.F. Vansant, P. Van Der Voort and K.C. Vrancken, Characterization and Chemical Modification of the Silica Surface, Elsevier, Amsterdam, 1995.
9. J.E. Sandoval and J.J. Pesek, Anal. Chem., 63 (1991) 2634.
10. J.J. Pesek, M.T. Matyska, E. Soczewinski and P. Christensen, Chromatographia, 39 (1994) 520.
11. J.J. Pesek, M.T. Matyska, E.J. Williamsen, R. Tam and Z. Wang, J. Liq. Chromatogr. & Rel. Technol., in press.
12. K.G. Proctor, S.K. Ramirez, K.L. McWilliams, J.L. Huerta and J.J. Kirkland, in: Chemically Modified Surfaces: Recent Developments, J.J. Pesek, M.T. Matyska and R. Abuelafiyia (eds.), Roy.Soc.Chem., Cambridge, 1996, 45.
13. K.C. Vrancken, P. Van Der Voort, K. Possemiers, P. Grobet and E.F. Vansant, in: Chemically Modified Surfaces, J.J. Pesek and I.E. Leigh (eds.), Roy.Soc.Chem., Cambridge, 1994, 46.
14. J.M. Berquier, in: Chemically Modified Surfaces, J.J. Pesek and I.E. Leigh (eds.), Roy.Soc.Chem., Cambridge, 1994, 210.
15. D.H. Lee, R.S. Condrate and J.S. Reed, in: Chemically Modified Surfaces: Recent Developments, J.J. Pesek, M.T. Matyska and R. Abuelafiya (eds.), Roy.Soc.Chem., Cambridge, 1996, 98.
16. U.B. Vogelsang, A. Deege, H. Figge, J. Kohler and G. Schomberg, Chromatographia, 19 (1984) 170.
17. J.A. Blackwell and P.W. Carr, Anal. Chem., 64 (1992) 853.
18. J.A. Blackwell and P.W. Carr, Anal. Chem., 64 (1992) 863.
19. B. Berno, R. Aroca and A. Nazri, in: Chemically Modified Surfaces, J.J. Pesek and I.E. Leigh (eds.), Roy.Soc.Chem., Cambridge, 1994, 91.
20. J.K.M. Sanders and B.K. Hunter, Modern NMR Spectroscopy, Oxford Press, Oxford, 1990.
21. C.A. Fyfe (ed.), Solid State NMR for Chemists, CFC Press, Guelph, Ontario, 1983.
22. J.J. Pesek and M.T. Matyska, J. Chromatogr., 687 (1994) 33.
23. J.E. Sandoval and J.J. Pesek, Anal. Chem., 63 (1991) 2634.
24. J.J. Pesek, M.T. Matyska, E.J. Williamsen and R. Tam, Chromatographia, 41 (1995) 301.
25. D.W. Sindorf and G.E. Maciel, J. Amer. Chem. Soc., 105 (1983) 1848.
26. B. Pfleiderer, K. Albert, K.D. Lark, K.K. Unger, H. Brukner and E. Bayer, Angew. Chem., Int. Ed. Engl., 28 (1989) 327.
27. H. Ohta, Y. Saito, K. Jinno, J.J. Pesek, M.T. Matyska, Y.-L. Chen, J. Archer, J.C. Fetzer and W.R. Biggs, Chromatographia, 40 (1995) 507.
28. J.E. Sandoval and J.J. Pesek, Anal. Chem., 61 (1989) 2067.
29. C.-H. Chu, E. Jonsson, M. Auvinen, J.J. Pesek and J.E. Sandoval, Anal. Chem., 65 (1993) 808.
30. J.J. Pesek, M.T. Matyska, J.E. Sandoval and E.J. Williamsen, J. Liq. Chromatogr. & Rel. Technol., 19 (1996) 2843.

31. M. Pursch, S. Strohschein, H. Handel and K. Albert, Anal. Chem., 68 (1996) 386.
32. C.C. Liu and G.E. Maciel, Anal. Chem., 68 (1996) 1401.
33. M. Pursch, L.C. Sander and K. Albert, Anal. Chem., 68 (1996) 4107.
34. R.K. Gilpin and M.E. Gangoda, Anal. Chem., 56 (1984) 1470.
35. E.C. Kelusky and C.A. Fyfe, J. Am. Chem. Soc., 108 (1986) 1746.
36. K.B. Sentell, D.M. Bliesner and S.T. Shearer, in: Chemically Modified Surfaces, J.J. Pesek and I.E. Leigh (eds.), Roy.Soc.Chem., Cambridge, 1994, 190.
37. D.B. Marshall and W.P. McKenna, Anal. Chem., 56 (1984) 2090.
38. D.M. Bliesner and K.B. Sentell, Anal. Chem., 65 (1993) 1819.
39. D. Briggs and M.P. Seah, Practical Surface Analysis, John Wiley and Sons, New York, 1993.
40. J.B. Brennan, R.F. De Bono, K.M.R. Kallury and U.J. Krull, in: Chemically Modified Surfaces, J.J. Pesek and I.E. Leigh (eds.), Roy.Soc.Chem., Cambridge, 1994, 72.
41. J.J. Pesek, M.T. Matyska, H. Hemphala and P. Christensen, J. Liq. Chromatogr. & Rel. Technol., 18 (1995) 1445.
42. K. Tani and Y. Suzuki, Chromatographia, 38 (1994) 291.
43. R. M. Heck and R.J. Farrauto, in: Chemically Modified Surfaces, J.J. Pesek and I.E. Leigh (eds.), Roy.Soc.Chem., Cambridge, 1994, 120.
44. G.S. Hickey and P.K. Sharma, in: Chemically Modified Surfaces, J.J. Pesek and I.E. Leigh (eds.), Roy.Soc.Chem., Cambridge, 1994, 139.
45. O. Bruggemann, R. Freitag, M.J. Whitcombe and E.N. Vulfson, J. Chromatogr. A, 781 (1997) 43.
46. S. Kaupp and H. Watzig, J. Chromatogr. A, 781 (1997) 55.
47. J.J. Pesek and M.T. Matyska, J. Chromatogr. A, 736 (1996) 255.
48. G. Binning, H. Rohrer, Ch. Gerber and E. Weibel, Phys. Rev. Lett., 50 (1985) 120.
49. R. Barberi, M. Giocondo, R. Bartolino and P.G. Righetti, Electrophoresis, 16 (1995) 1445.
50. R. Barberi, J.J. Bonvent, R. Bartolino, J. Roeraade, L. Capelli and P.G. Righetti, J. Chromatogr. B, 683 (1996) 3.
51. J.J. Bonvent, R. Barberi, R. Bartolino, L. Capelli and P.G. Righetti, J. Chromatogr. A, 756 (1996) 233.
52. J.J. Pesek, M.T. Matyska and E. Pullen, unpublished results.
53. V. Tscharner and H.M. McConnell, Biophys. J., 36 (1981) 409.
54. D.D. Rostrup-Nielsen, P.F. Schubert and R.K. Sato, in: Chemically Modified Surfaces: Recent Developemnts, J.J. Pesek, M.T. Matyska and R. Abuelafiya (eds.), Roy.Soc.Chem., Cambridge, 1996, 1.
55. X. Huang, J.M Kovaleski and M.J. Wirth, Anal. Chem., 68 (1996) 4119.
56. W.B. Lavy, J.M. Williams, L.A. Wenzler, T.P. Beebe and J.M. Harris, Anal. Chem., 68 (1996) 1003.
57. R.L. Hansen and J.M. Harris, Anal. Chem., 68 (1996) 2879.
58. F.Y. Rem, S.W. Waite and J.M. Harris, Anal. Chem., 67 (1995) 3441.

Adsorption and its Applications in Industry and Environmental Protection
Studies in Surface Science and Catalysis, Vol. 120
A. Dabrowski (Editor)



# Advances in characterisation of adsorbents by flow adsorption microcalorimetry

A. J. Groszek

Microscal Ltd., 79 Southern Row, London W10 5AL, United Kingdom

## 1. INTRODUCTION

One of the most striking phenomena accompanying adsorption at solid–liquid and solid–gas interfaces is the evolution of heat. The existence of this thermal effect has been recognised at least from the early nineteenth century. Such effects were mostly confined to wetting or gas saturation processes of various finely divided solids, such as cements, active carbons, clays and metal oxides subjected to drying, evacuation, or heating [1].

In many cases the heat effects were caused by a combination of adsorption and surface reactions and were often not completely reversible, the wetting of cements with water being a case in point [2]. The heat of wetting measurements were often not carried out under well defined conditions and even as late as 1968 A. C. Zettlemoyer concluded that the experimental techniques used by many research workers in this area were deficient [3] and that the meaning of the heat of wetting (or the heat of immersion) is not very clear. However in some cases the heats of wetting were found to be reliable and were used for the evaluation of specific surface areas of powders following the work of V. D. Harkins and G. Jura [4, 5]. Thermodynamic parameters were defined for relating the heats of immersions to specific surface areas and their absolute values obtained independently of the estimates obtained from gas adsorption measurements and the application of the BET equation. This could only be done for systems in which the immersion liquids formed physically adsorbed monolayers, the forces binding the adsorbed molecules being confined to a monolayer in direct contact with the solid surface, with little or no significant adsorbate–adsorbate interaction. It was also assumed that all of the surface was accessible for the adsorption and that the attractive forces between the adsorbed monolayer and the bulk liquid were closely similar to those between the molecules in the bulk liquid. In this situation the heats of immersion were directly proportional to the specific surface areas.

Since the early nineteen forties there has been rapid development of chromatographic techniques which involved displacement of weakly adsorbed carrier liquids or gases percolating through adsorbent columns by more strongly adsorbed substances introduced into the carrier fluids and eluted by them from the adsorbent.

It was observed by the author during his work on the chromatographic separation of liquid mixtures that the movement of bands of separated components of the mixtures was accompanied by heat waves confined to the regions in which the separated components were moving through the column. It was concluded that the heat bands represented the heats of preferential adsorption of the components separated from the carrier fluid on the adsorbent filling the column. Following these observations the author pioneered the development of flow methods of determining the heats of adsorption and desorption from liquids and gases percolating through powdered or granulated solids [6].

This paper reviews the work of the author and that of other users of flow microcalorimetry (FMC), which led to discoveries of new phenomena at solid–liquid and solid–gas interfaces [7, 8] including the behaviour of adsorbates at different surface coverages [9, 10], and many, previously unknown, interactions of surfaces of metals, metal oxides, graphites [11], active carbons, zeolites, clays, and other minerals with organic and inorganic adsorptives. The work led to development of new selective adsorption processes [12], prediction of the performance of liquid lubricants and its components [13], surface reactions [14, 15], dispersion technology [16] and determinations of specific areas of surface sites with different chemical properties [17, 18].

The progress achieved in the above areas is summarised with emphasis on the most recent developments in the flow techniques, instrumentation and new results in the studies of microporous solids, the properties of individual surface sites and studies of new selective adsorption processes and catalyst supports.

## 2. FLOW ADSORPTION MICROCALORIMETRY – APPARATUS AND PROCEDURES

### 2.1. Solid–liquid interfaces

Description of the first flow microcalorimeter and its initial applications was published by the author some 40 years ago [19, 20]. A novel feature in this microcalorimeter was the use of a single calorimetric cell for the measurement of temperature changes in the adsorbent bed due to adsorption/desorption phenomena employing the cold junction thermocouples as the reference sensors. Heat effects produced during surface interactions were rapidly dissipated in the heat sink constituted by a water jacket in direct contact with the glass walls of the calorimetric cell. The reference junctions were placed in water adjoining the glass walls, whilst the measuring heat junctions were in direct contact with the adsorbent. Heat loss from the cell occurred within a few minutes after surface interactions took place and in this way the heats of adsorption/desorption on granulated solids could be measured under effectively isothermal conditions.

This type of microcalorimetry has been referred to as diathermal calorimetry and has been adopted by most adsorption calorimeters of the Tian–Calvet type [21].

The latter calorimeters usually employed two cells, one containing the adsorbent, and the other an empty reference cell. Both were inserted into a massive metal block which acted as a heat sink. Heat produced in the measuring cell flows into the heat sink, which is the reason why such calorimeters are sometimes referred to as belonging to the *heat flow* category. This must not be confused with the *flow microcalorimetry* developed by the author, which involves the use of carrier fluids passing continuously through the adsorption cell with adsorption events taking place in competition with the carrier fluid.

A single adsorption cell was used in the first commercial version of a flow microcalorimeter (FMC) developed by Microscal Ltd. [22]. In this instrument thermocouples were replaced by thermistors and the water jacket by a metal block surrounding a small PTFE cell accommodating an adsorbent bed having a volume of 0.17 cc [22, 23].

The current version of the FMC, as represented in Figure 1, provides reliable data on heats of adsorption and desorption and the corresponding amounts of adsorption/desorption for virtually all types of liquids and solutions. The heats can be determined at a range of temperatures, from ambient to 200°C, with different flow rates of the carrier liquid (usually between 1 $\mu lmin^{-1}$ and 100 $\mu lmin^{-1}$), and using adsorbents in the form of powders, filaments, ribbons, pellets, and compacts completely filling the adsorption cell. In most cases the experiments are carried out in the presence of a small calibration coil encapsulated in PTFE, producing in situ heat effects electrically. Effects as small as a few microjoules evolved over a period of 10 to 200 seconds can be determined as well as much larger and longer effects lasting several hours. Even better long term thermal stability is obtained if the calorimeter is situated in an air thermostat, permitting the determination of heat evolutions lasting several days.

The adsorbent in the adsorption cell can be evacuated and wetted with the carrier liquid generating the heats of immersion which can approximate the heats of wetting obtained in static calorimeters used under the same conditions (degree of evacuation, temperature of the adsorbent, and the wetting liquid). The evacuation and wetting are carried out under static conditions. After the establishment of thermal equilibrium following the evolution of the heat of wetting, the carrier liquid is percolated through the adsorbent at a constant rate using a precision syringe pump. Individual solutions are supplied to the adsorbent by another syringe pump operating at the same flow rate as that supplying the carrier liquid. The solution pump is activated some ten minutes before its output is substituted for the flow of carrier liquid. The back pressures in the solvent (carrier liquid) and the solution lines have to be adjusted to a value not differing by more than 0.05 bar g to avoid disturbance of the base line during the changeover between the two streams.

An example of a series of evacuation, wetting and adsorptions/desorptions from solution is shown in Figure 2 for an active carbon immersed in n–heptane and n–butanol solutions.

The adsorptions from volatile solvents can be studied at elevated temperatures by applying appropriate back pressures at the outlet of the down stream detector (DSD). For example adsorptions from aqueous solutions of n–butanol can be carried out at 120°C when the applied back pressure is 2.5 bar g. Surprisingly the heats of adsorption of n–butanol on carbon surfaces are much higher at 120°C than at room temperature, 57 $Jg^{-1}$ and 20 $Jg^{-1}$ respectively, which is probably due to a reduced affinity of water for the carbon surfaces at the higher temperature.

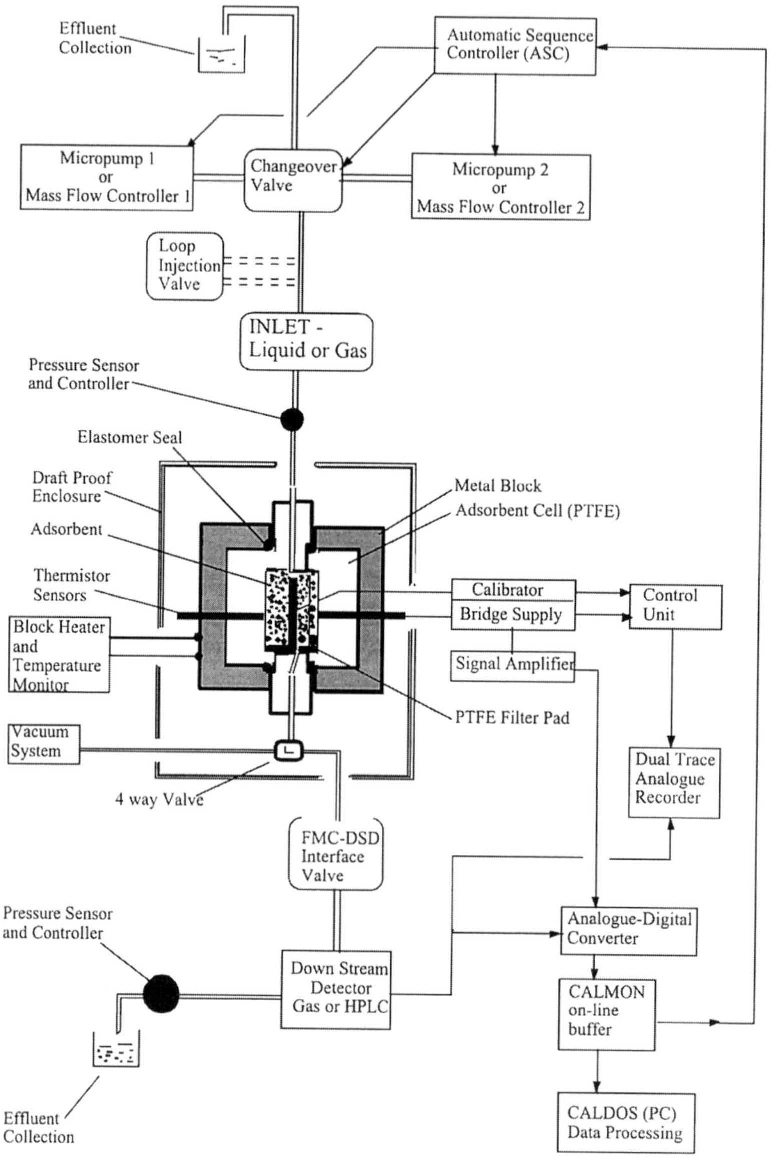


Figure 1. System diagram of the MICROSCAL flow adsorption microcalorimeter, Mark 4, for the simultaneous determination of the heats and amounts of adsorption at solid-liquid and solid-gas interfaces.

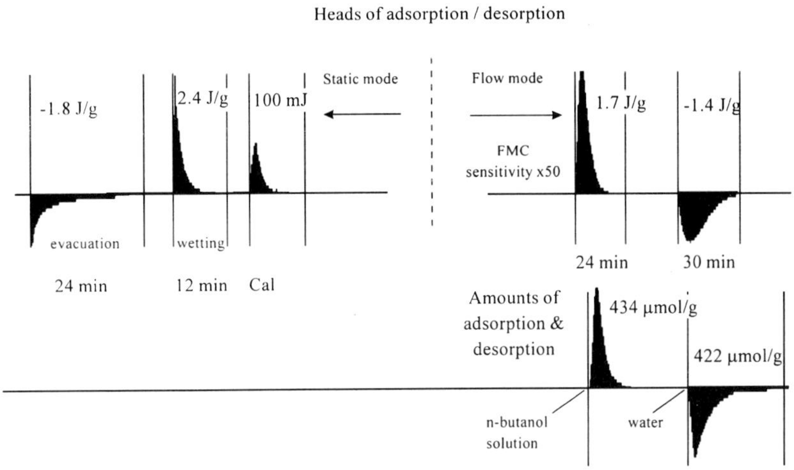


Figure 2. Simultaneous determination of heats and amounts of adsorption/desorption of n-butanol from 10g/l solution in water on 0.0696g graphitized carbon black at 25°C after evacuation and wetting under static conditions.

## 2.2. Determination of heat and adsorption/desorption isotherms and reversibility – Flow Equilibrium Adsorption Thermodynamics (FEAT)

One of the important advantages of flow adsorption microcalorimetry is that it can determine changes of enthalpy at solid–liquid interfaces due to adsorption from solutions with precisely predetermined concentration over the whole range of molar fractions from zero (pure solvent) to unity (pure solute, if it is a liquid). The initial change of enthalpy is that due to immersion of adsorbent in one of the pure liquid components, which usually generates a sharp evolution of heat, the heat of wetting, or the heat of wetting and swelling if the adsorbent absorbs the liquid. Subsequently the adsorbent can be successively equilibrated with a series of solutions of increasing concentration in a stepwise fashion leading to progressive displacement of solvent molecules from the interface. At each successive step the achievement of equilibrium is indicated by cessation of heat evolution and the corresponding absence of change in the response of the DSD which monitors the concentration of solute in the effluent from the adsorption cell.

The amount of adsorption that takes place at each step is evaluated by integrating the difference in the detector response to changes in the effluent composition after it leaves the cell filled with the adsorbent, and a parallel response obtained when the cell is filled with a non–adsorbing solid. This is illustrated in Figure 3 by the concentration profiles obtained for adsorption of n–butanol from n–heptane on a graphitised carbon black (Graphon).

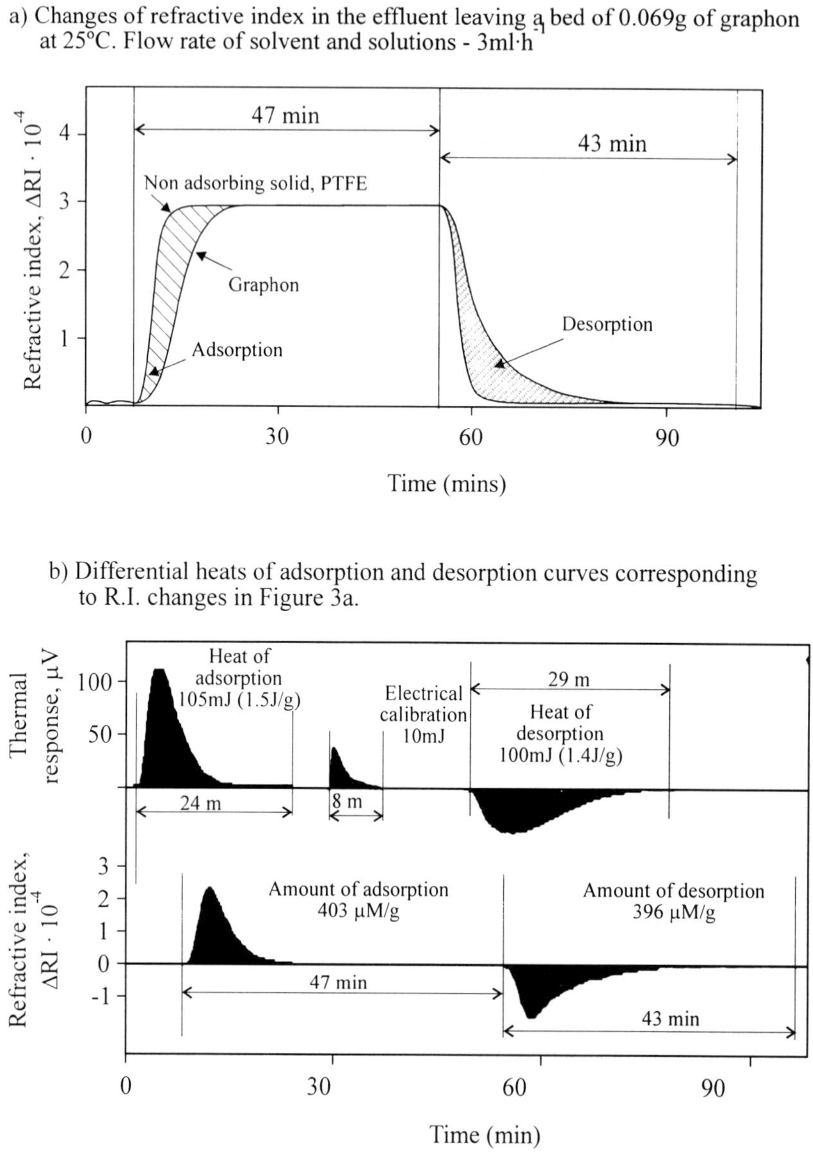


Figure 3. Computer graphics of: (a) changes in the refractive index of the effluent from the adsorption cell filled with the adsorbent, graphitised carbon black, and a non–adsorbing solid (PTFE); and, (b) thermal output of the thermistors measuring heat effects due to interactions at solid–water interface caused by adsorption and desorption of n–butanol from its 10 $gl^{-1}$ solution in water on 0.0696 g of GRAPHON at 25°C. Flow rate of water carrier – 3 $mlh^{-1}$.

Desorption can be effected by a stepwise dilution of the solution percolating through the adsorbent. For reversible adsorption the heats of adsorption and desorption are equal, but opposite in sign. However, the kinetics of desorptions are usually slower than those of adsorptions especially for high molecular weight adsorptives and microporous adsorbents.

A flow method that produces a complete saturation of surfaces with solutions or with gas mixtures having a certain *fixed* composition and determination of both the associated heat and the amount of adsorption is named by the author as *Flow Equilibrium Adsorption Thermodynamics*. The method can be used for adsorption, desorption, and the determination of adsorption reversibility.

A series of heat of adsorption runs which can be used to determine adsorption isotherms is illustrated in Figure 4 by the record of sequential adsorptions of n–butanol from n–heptane on an active carbon. In each case the adsorptions reach a well defined plateau and conform with Langmuir's equation.

| Soln $g\,l^{-1}$ | Heat of adsorption $Jg^{-1}$ |
|---|---|
| 2 | 9.3 |
| 5 | 3.0 |
| 12.5 | 1.1 |
| 25.0 | 0.3 |

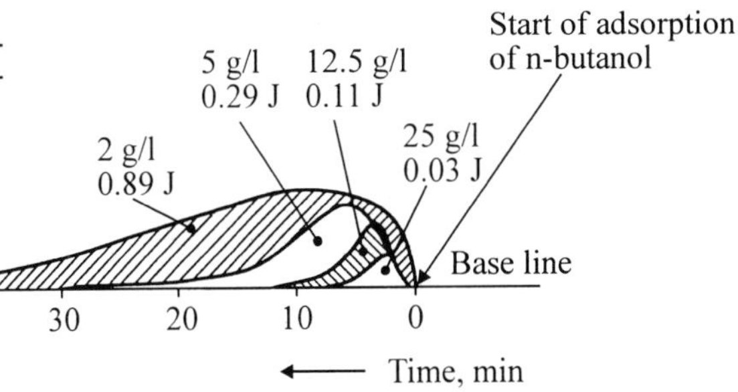


Figure 4. Heat evolution on adsorption of n-butanol from n-heptane onto 96 mg active carbon Chemviron BPL.

For dilute solutions the heats of successive adsorptions are often proportional to the amounts of adsorption (a necessary condition for conformity with Langmuir's equation). The equilibrium constants can be determined directly from the heat isotherms, as suggested by T. Allen and R. M. Patel [10], writing the relationship

$$\frac{C}{q} = \frac{1}{kq_0} + \frac{C}{q_0} \tag{1}$$

where C is the solute concentration producing the heat of adsorption q, $q_0$ is the heat of adsorption corresponding to monolayer coverage and k is the equilibrium constant.

The free energy, ΔG, and the entropy, ΔS, of adsorption can be evaluated using the well known equations 2 and 3 below:

$$-\Delta G = RT \ln k \tag{2}$$

$$-\Delta G = \Delta H - T\Delta S \tag{3}$$

The values of $\Delta G$ evaluated in this way could be substituted into equation 3 to obtain the entropy change $\Delta S$, as was shown by the adsorption of n–paraffins and n–alcohols on graphitic carbons [11]. Furthermore, if the equilibrium constants are determined at two different temperatures $T_1$ and $T_2$ and the corresponding heats of adsorption are also determined, the change of entropy of adsorption occurring between these two temperatures can be calculated from the individual values of enthalpy–changes $\Delta H_1$ and $\Delta H_2$ occurring at temperatures $T_1$ and $T_2$ by using equation 3.

The heats of adsorption from dilute solutions forming well defined plateaux at relatively low solute concentrations were used by the author for the determination of specific surface areas of metal oxides [26]. The surface areas were determined by saturating the powdered solids placed in the calorimetric cell with a solution of n–butanol in n–heptane, known to form a close packed monolayer on the oxide surfaces, and determining the resulting integral heat of a monolayer formation. The heats proved to be proportional to the specific surface areas for a large number of polar solids and correlated well with their BET($N_2$) surface areas. The levels of the heat adsorption plateaux corresponded to the individual surface areas with the correlating factor of 1 $Jg^{-1}$ being equivalent to 11.3 $m^2g^{-1}$ of the surface.

Adsorption reversibility can be evaluated by carrying out several adsorption–desorption cycles with a solution which is sufficiently concentrated to produce a high degree of saturation of the adsorbent surface. The difference between the amounts of adsorption obtained in the first two adsorption cycles can be taken as the irreversible adsorption which is usually chemisorption. After the first cycle further reversible adsorption cycles take place on the chemically modified surface. Chemisorption is often confined to individual active sites forming only a small proportion of the total surface. The reversible adsorption may take place on top of the chemisorbed layer as well as on the sites where chemisorption does not take place. In some cases the initial chemisorption leads to further reactions which may produce soluble reaction products. Typical cases of adsorption from solution producing a significant proportion of irreversible interactions is provided by selective adsorption of transition metal salts from aqueous solutions on activated carbons, and other adsorbents possessing basic surfaces sites. This is illustrated by adsorption of potassium chloroplatinate on active carbon shown in Figure 5. Similar results are produced by adsorbents with strong acid sites which irreversibly adsorb bases such as ammonia and pyridine.

Percolation of carrier liquids through the adsorbent may be associated with the dissolution of some of the surface material on its surface. Quartz and $\gamma$–alumina adsorbents often produce extended endothermic heat effects when water is percolated through them, as illustrated for $\gamma$–alumina in Figure 6. The dissolution process in this case is completed in 84 minutes of percolation (22 bed volumes). Heat of adsorption determinations can then be carried out on the adsorbents equilibrated with the carrier liquid. In some cases the carrier liquids are absorbed by the solid in the cell causing an increase in volume (its swelling). This often occurs, for example, with ion–exchangers when wetted by water. It is essential, of course, that the volume of the adsorbent in the cell does not exceed the volume of the adsorption cell itself and, therefore, the degree of swelling must be determined before the adsorption of such solids is undertaken.

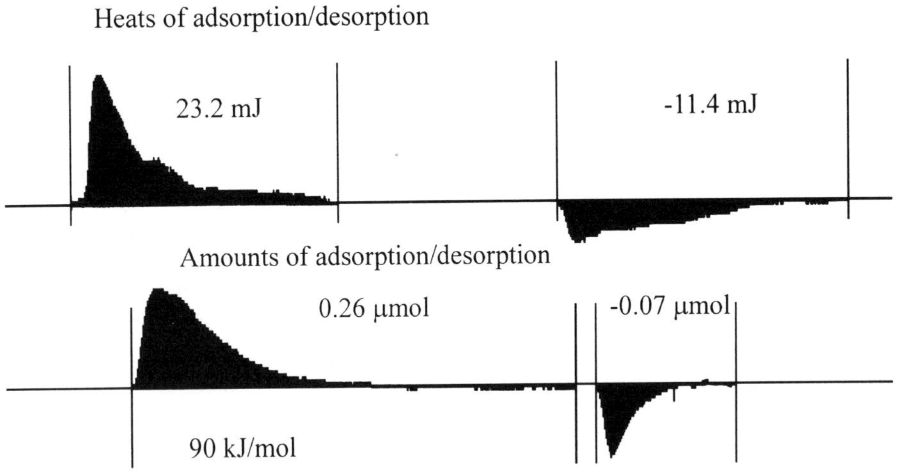


Figure 5. Simultaneous determination of heats and amount of adsorption/desorption of $K_2PtCl_6$ from water on 10.9 mg graphitised carbon black mixed with 90 mg of PTFE at 25°C.

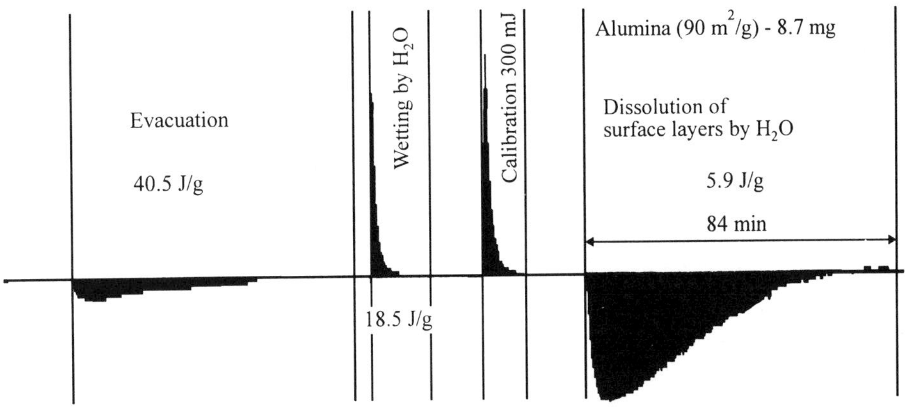


Figure 6. A sequence of evacuation, wetting and start of water percolation through 0.087 g of alumina adsorbent causing dissolution of its surface impurities and absorption of heat.

An endothermic heat effect produced after the initiation of carrier liquid percolation may also be caused by displacement of surface impurities. In that case the use of an appropriate DSD may indicate the nature of the impurity by passing the effluent from the adsorption cell through it.

### 2.3. Heats of dissolution

As indicated above, percolation of carrier liquids may cause dissolution of the surface impurities in adsorbents causing substantial heat effects. If a known amount of a solid which is completely soluble in the carrier liquid is added to an inert granulated material filling the adsorption cell, the dissolution of the added solid will produce a heat effect. This effect may be positive or negative depending on the chemical nature of the carrier liquid and the dissolved solid. For example, dissolution of 22 mg of NaCl added to PTFE powder in a water carrier fluid produced a negative heat effect 57.2 $Jg^{-1}$ which gave a molar heat of dissolution of 3.4 $KJmol^{-1}$ against 3.7 $KJmol^{-1}$ reported in the literature [24] for infinite dilution of NaCl.

### 2.4. Heats of mixing

The changes of solution concentration required to obtain adsorption isotherms produce heat effects which may be large enough in relation to the heats of adsorption to affect accuracy. Usually, changes of concentration of the order of 1 – 2 percent, by weight, produce heats of mixing which are too small to lead to significant errors in the determination of the much larger heats of adsorption.

A *heat of mixing unit*, in which two fluids are merged, can be used to establish permissible concentration steps if required.

Alternatively, for more concentrated solutions it is possible to determine the heats of mixing in the flow microcalorimeter used under static conditions. Appropriate corrections could then be introduced to the heats of adsorption. In the static mode the calorimeter cell is filled with a known volume of the carrier liquid (150 μl) using a microsyringe and microlitre quantities of the second liquid (or solution) injected directly into the liquid in the cell. Since the volume of the cell is small, the injected solute diffuses quickly into the liquid filling the cell and the heat due to mixing is measured by the same thermistors as those used for adsorption work. Each injection gives a differential heat of mixing of the injected liquid with the liquid in the cell forming a dilute solution.

For example, the addition of 1 μl quantities of pure n–butyl alcohol to 150 μl of water placed in the cell produces heat effects shown in Figure 7. The heats of mixing decrease to a value close to zero as the saturation concentration of n–butanol in water (7 percent by weight at 25°C) is approached. Additions of 1 percent n–butanol solutions to water produce no measurable heat of mixing effects and can, therefore, be ignored when the heat of n–butanol adsorption is measured from its 1 percent solution in water is measured on various hydrophobic adsorbents.

This method can also be adapted for adsorption on very finely dispersed solids which would produce excessive back pressure if percolation of the liquid in which the solid particles are dispersed was attempted, or would pass through the filter

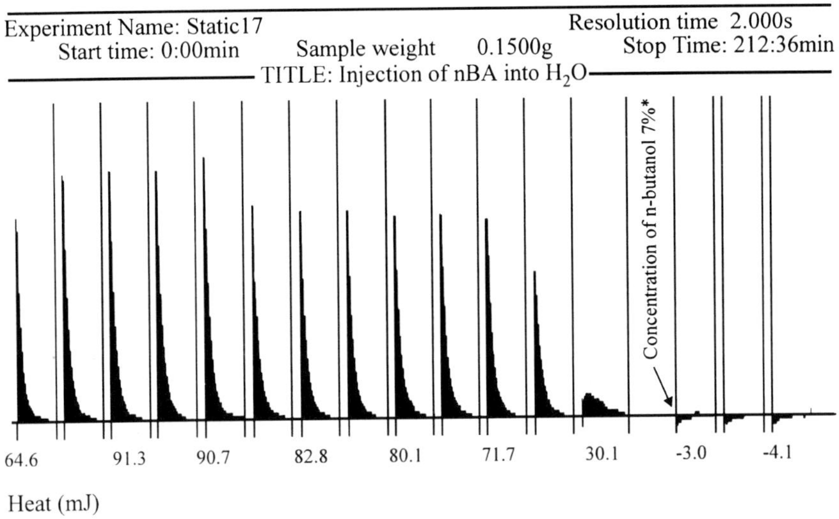


*Literature volume for solubility of n-butanol in water at 25°C is 7.0% wt.

Figure 7. Heats of mixing of n-butanol in water produced by direct injection of 1 μl quantities of the pure alcohol into 150 μl of water in the adsorption cell. Determinations under static conditions at 25°C.

retaining the solid in the adsorption cell, can also be studied under static conditions in the same way as the determination of the heats of mixing. Direct injections of relatively concentrated solutions into the cell filled with 150 μl of a stable dispersion generates heat effects due to adsorption of the injected solute combined with its mixing with the dispersion medium. The heats of mixing can then be subtracted from the gross heat effects produced by mixing *and* adsorption. Injections can be continued until the volume of liquid in the cell reaches 170 μl.

## 2.5. Pulse adsorption and differential heats of adsorption – Flow Injection Adsorption Thermodynamics (FIAT)

One of the methods involves injection of small volumes of solutions into the stream of carrier liquid percolating through the adsorbent in the calorimetric cell and is designed for the determination of the heats of adsorption of irreversibly adsorbed solutes at increasing degrees of surface coverage. This technique is based essentially on flow injection analysis techniques, but it is confined to strong interactions between small amounts of active components of fluid mixtures and adsorbents and has been named *Flow Injection Adsorption Thermodynamics*. The downstream detector determines the concentration of that part of the injected solute in the effluent from the cell which is not retained by the adsorbent. The heat effect obtained following the injection corresponds only to the strongly (irreversibly) adsorbed component. The relevant differential molar heat of adsorption can, therefore, be accurately determined for a given degree of surface coverage.

The injections can be continued until there is no more irreversible adsorption, i.e. until all the injected solute passes with the effluent to the DSD. At that point all the active sites on the adsorbent are saturated and further adsorption is fully reversible. To speed up the saturation process and reduce the number of injections to saturate the active surface, it is possible to dilute the adsorbent bed in the calorimeter with an inert powder. In such a case 5 – 30 mg of adsorbent can be mixed with about 90 – 65 mg of inert PTFE powder, making up the volume required to fill the cell (0.17 cc).

A typical example of the FIAT method applied to the determination of differential enthalpies of adsorption is illustrated in Figure 8 for adsorption of potassium chloroplatinate on a graphitised carbon black. The first injection of 0.2 μmol of $K_2PtCl_6$ produced a molar heat of adsorption of 128 $kJmol^{-1}$, and the fourth injection yielded 31 $kJmol^{-1}$. The individual adsorptions occur on surface sites of different activity and reflect the heterogeneity of the surface. On the other hand, similar adsorption experiments on alumina produced relatively low heats of adsorptions emphasising the high affinity for transition metal salts of the polar sites present in graphitised carbons.

The high sensitivity of the latest versions of flow microcalorimeter permits accurate determination of the heats of adsorption produced by the injection of nanomole quantities of solutes into the carrier liquids. This means that the heats of adsorption can be determined on solids having very low specific surface areas, or, on the other hand, the differential heats can be obtained at very low surface coverages for high surface area adsorbents.

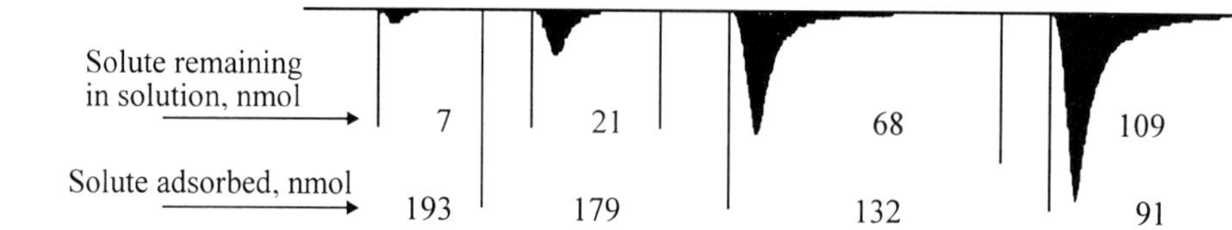


Figure 8. Pulse adsorptions following sequential injections of 20 μl of 0.01 molar aqueous solution of $K_2PtCl_6$ (0.2 μmol) into water percolating throught 25.8 mg of graphitised carbon black mixed with PTFE powder. Flow rate of water – 3 $mlh^{-1}$, temperature – 20°C.

Simultaneous determination of the heats and the amounts of adsorption is essential in the pulse adsorption experiments to establish what proportion of the injected solute is irreversibly adsorbed. In most cases the capacity of solid surfaces for irreversible adsorption is limited to a part of the total surface as determined by the BET method. The other part of the surface may continue to selectively adsorb the injected solutes, but the adsorptions become irreversible after saturation of the surface sites absorbing irreversibly. The injections of solutes which are reversibly adsorbed produce heat effects composed of overlapping positive and negative portions prohibiting an accurate determination of the heats of adsorption, in which case the saturation technique of section 2.2 may be required.

For reversible adsorption the differential heats can be obtained by slicing the heat evolution peak obtained by use of an extended saturation of the adsorbent with a solution establishing a state of interfacial equilibrium. This is exemplified by the work of F. Fowkes and T. Lloyd who applied this technique to the adsorption of acids and bases on metal oxide pigments [25].

## 2.6. Adsorption of gases and vapours

As in the case of adsorption from liquids, the flow adsorption of gases is carried out after evacuation or purging of the adsorbent, e.g. with helium, at an appropriate temperature. The adsorbent is then equilibrated at a desired temperature with the carrier gas, such as helium, which can then be exchanged for another gas or a gas mixture. The gas flows are regulated by mass flow controllers and thermal effects caused by displacement of the carrier gas are measured in the same way as for the adsorption from solution. The amounts of adsorption are measured by passing the gaseous effluent through a detector employed in gas chromatography, such as a thermal conductivity detector (TCD). Continuous monitoring of changes in mixture composition during the adsorption process (decreasing molar fraction of the adsorbed component) or the desorption process (increasing molar fraction of the adsorbed component) produces adsorption/desorption profiles which can be used for the evaluation of adsorption kinetics.

A schematic flow diagram used in FMC gas adsorption work is shown in Figure 9. An important feature of the Microscal microcalorimeter is its suitability for the measurements of the heats of adsorption or desorption under substantial pressures. Thus after saturation of an adsorbent with a single gas or a gas mixture the outlet from the adsorption cell can be closed and pressure increased or decreased, which causes adsorptions or desorptions. In this case the heats of adsorption cannot be measured simultaneously with the amounts of adsorption which have to be measured separately. Illustrations of adsorption/desorption cycles of $CO_2$ from He and $N_2$ on an active carbon are shown in figures 10 and 11. In Figure 11 the peaks represent heat effects produced by increasing/decreasing the pressure of $O_2$ after its adsorption from He at atmospheric pressure at 25°C. Cumulative heats of adsorption of $N_2$ and $O_2$ on the carbon molecular sieve are subjected to a pressure cycle reaching 20 bar g is shown in Figure 12. The results clearly indicate the relatively high heats of adsorption of $O_2$ compared with those of $N_2$.

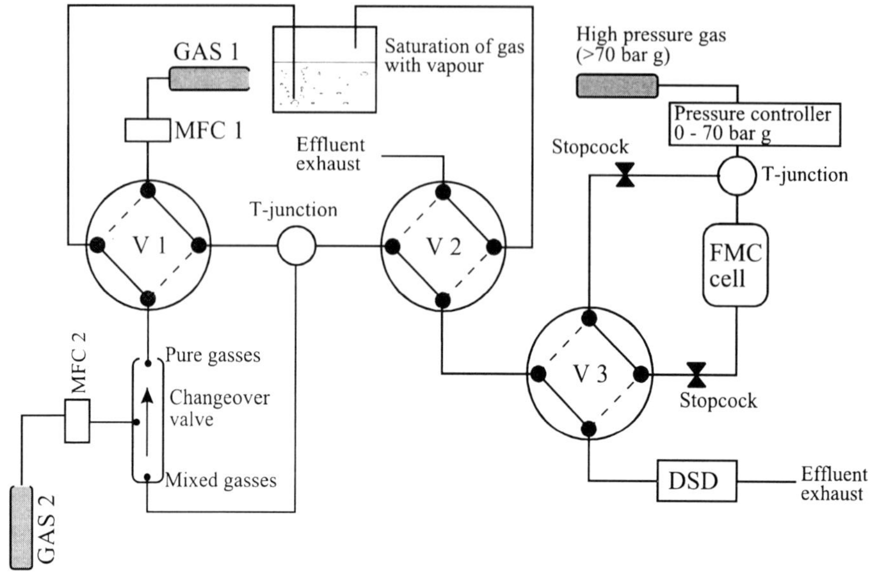


Figure 9. Schematic flow diagram of Microscal flow adsorption microcalorimeter used for gas adsorption/desorption studies.

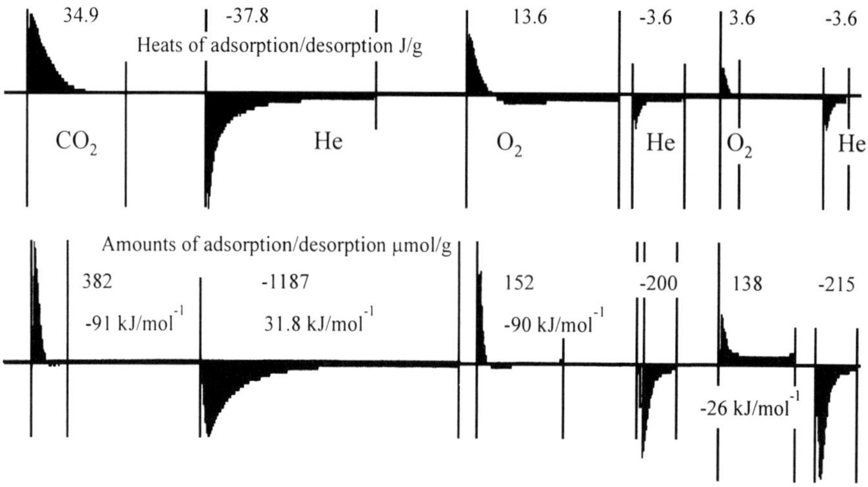


Figure 10. Heats of displacement of He by sequential adsorption of $O_2$ and $CO_2$ on 0.0697 g of active carbon CHEMIVIRON BPL at 25°C.

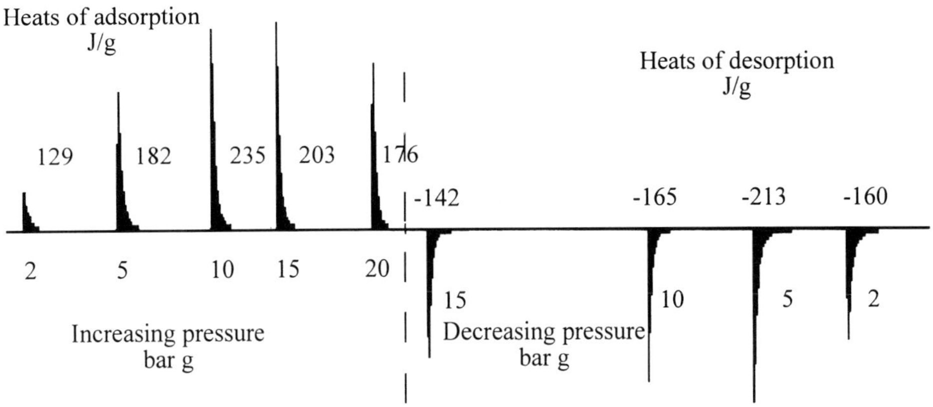


Figure 11. Effect of pressure on the heats of adsorption and desorption of oxygen on 0.0852 g of Bergbau carbon molecular sieve at 25°C using helium carrier gas.

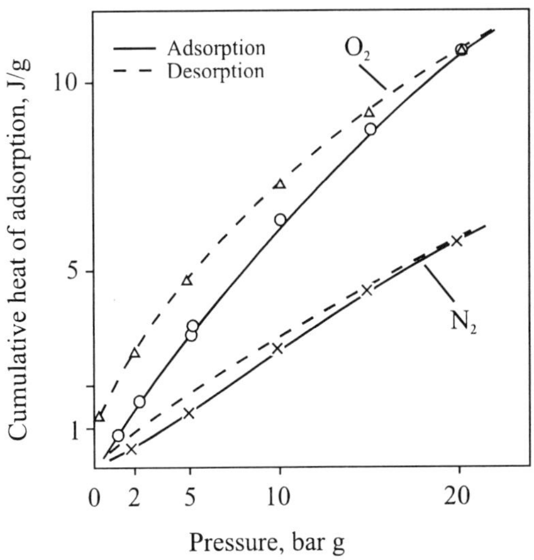


Figure 12. Cumulative heats of adsorption/desorption of oxygen and nitrogen at 25°C on a carbon molecular sieve caused by pressure changes between 0 and 20 bar g.

The heats of adsorption of vapours can be carried out by passing a carrier gas through a saturator in which the gas is bubbled in a liquid kept at a constant temperature. The rate of flow of the carrier gas is adjusted to achieve its complete saturation with the vapour of the liquid. Usually a gas flow of 1 $mlmin^{-1}$ is sufficient. For low partial pressures of vapour, such as that of water at 20°C, and its adsorption on adsorbents having high surface areas, it is desirable to reduce the amount of adsorbent in the FMC cell. This can be done by mixing the adsorbent with a low surface area solid such as glass balls, purified quartz sand, or PTFE powder.

An example of such a practice is shown in Figure 13 for a mixture of 4.5 mg of an active carbon and 0.230 mg of sand, the volume of the mixture being sufficient to fill the FMC adsorption cell (0.17 cc). Even with this small amount of the active carbon the adsorption of water vapour could only be completed in about 20 hours, the bulk of the heat evolution taking some 7 hours. The integral molar heat of water adsorption amounted to 70 kJmol$^{-1}$, much higher than the heat of water vapour liquefaction, indicating that the state of the adsorbed water is very different from that of bulk water.

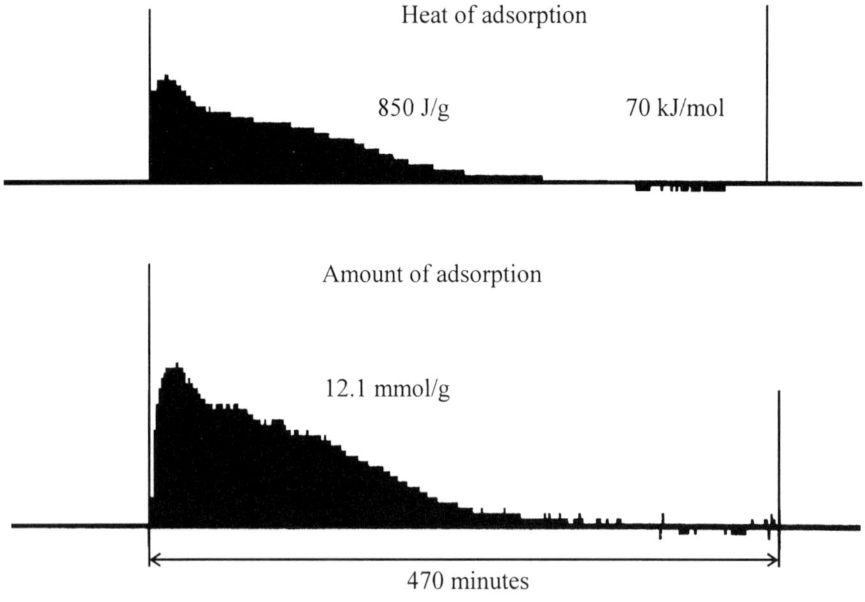


Figure 13. The heat of adsorption of water vapour from helium carrier on 0..0045 g of active carbon CHEMIVIRON BPL mixed with sand. The amount of adsorption is measured simultaneously using a thermal conductivity detector. Helium flow rate – 1 mlh$^{-1}$. Temperature – 23°C. The heat of adsorption of water on sand is 0.03 % of that on the carbon.

## 2.7. Data collection, storage, and processing

The latest model of Microscal's FMC incorporates the new Calorimetric Digital Output and Sequencer, CALDOS 3, which provides simple access to the output data from the instrument and any downstream devices employed with it. In addition, CALDOS 3 offers control of the fluid delivery devices, the valves which run the experiment, and automated calibration facilities for both flow and for static studies.

The processing options are designed to enhance the quality and accessibility of FMC data and so extend its applicability in research. Automated experimentation is performed by using a set of Sequence Elements which permit the control of fluid delivery and of data collection accommodating large departures from anticipated output signals. CALDOS 3 will also permit an inexperienced operator to repeatedly perform the same experiment as many times as is required, automatically.

So as to ensure that operators can reliably obtain the most from the FMC data without losing or corrupting it, the system provides for: automatic calibration; data marking; blank and live data matching; very long run protocols; multiple user options; and a data security cascade. Also provided are data presentation graphics as well as the options to prepare stored data for use in other software packages.

## 3. CHARACTERISATION OF ADSORBENTS

### 3.1. Determination of specific surface areas

In a number of instances adsorption of polar compounds from dilute solutions passes through a region in which the surface is covered by a closely packed monolayer of the polar solute. The author discovered in 1966 that most non–porous metal oxides adsorb preferentially n–butanol from n–heptane with the formation of such a monolayer. The integral heat produced during the formation of the monolayer correlated very well with the specific areas of the metal oxides measured by the BET($N_2$) method [26]. The specific surface areas of these solids could be effectively measured by a „single point" method, in which a sample of the solid immersed in n–heptane was flooded with a 2 $gl^{-1}$ solution of n–butanol producing a heat of n–heptane displacement which was proportional to the total surface area of the sample.

Later work with various binary liquid mixtures and different types of solids indicated that the flow calorimeter methods of specific surface determination can also be applied to homogenous surfaces represented by graphitised carbon blacks. In this case strong preferential adsorption of long–chain paraffins from n–heptane was used to produce a monolayer and the corresponding heat of adsorption measured the total graphitic basal plane in such carbons [11].

The long–chain normal paraffins were also strongly adsorbed on the hydrophobic surfaces of layered metal disulphides and it was found that the method used for the determination of the basal graphitic planes could be extended to these layered solids, as well as to boron nitride [7, 17]. Each solid surface had its own calibration factor relating the heat of monolayer formation to a specific surface area and the results derived in this way could also be used for solids having mixed hydrophobic–polar character as shown below in section 3.2.

Most alumino–silicates, including clays and pillared clays, quartz minerals and oil reservoir sands, possess surfaces which are completely hydrophilic. A common property of such surfaces is their decreasing affinity for alcohols with increasing size of the alkyl group. Thus for kaolinite and an oil reservoir sand the heats of adsorption of methanol from n–heptane are much higher than those of n–butanol as illustrated in Figure 14. Methanol tends to interact with the clay surfaces exceptionally strongly and appears to alter their structure, but does not form a monolayer of the type observed on metal oxides such as $Fe_2O_3$ and $Al_2O_3$. The high methanol heats suggest that it intercalates in kaolinite and clays present on the surface of oil reservoir sand, unlike n–butanol which forms close packed monolayers.

Normal butanol is also superior to longer alkyl alcohols which tend to produce adsorbate–adsorbate interactions causing irregular increases in the heat of adsorption and preventing the establishment of reliable correlations between the heats and the surface areas. The unique suitability of n–butanol as the adsorbate for the determination of polar surface areas of inorganic solids has been confirmed by P. Saluja et al who examined a wide range of minerals with the use of the flow calorimeter method developed by the author [27]. Saluja found that the method is fast, less than two hours per run, and can be adapted for routine operation. Only a small amount of a sample is required, generally about 0.1 g, even if its surface area is well below 1 $m^2g^{-1}$. As can be seen below, the method is especially useful for following changes in the surface chemistry which are difficult to detect with other analytical methods.

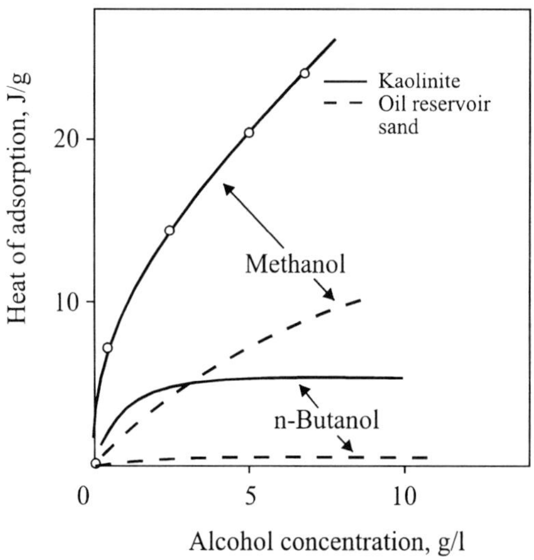


Figure 14. Cumulative heats of adsorption of n-butanol and n-heptane on kaolinite and oil reservoir sand at 25°C.

Another flow calorimetric approach to the determination of specific surface area is based on the preferential adsorption of $CO_2$ from He, which is used as a carrier gas. Carbon dioxide produces close packed monolayers on most solids at room temperatures and the author found that the integral heat of adsorption of $CO_2$ from He correlates well with the surface areas determined by the BET($N_2$) method, as illustrated in Table 1.

Table 1
Heats of $CO_2$ adsorption and estimates of surface areas

| Adsorbent | Heat of $CO_2$ Adsorption $Jg^{-1}$ | Estimated Surface Area $M^3g^{-1}$ | BET($N_2$) Surface Area $m^3g^{-1}$ |
|---|---|---|---|
| Silica | 0.22 | Standard | 10.0 |
| $Fe_2O_3$ | 0.09 | 4.10 | 3.7 |
| Kaolin | 0.20 | 9.00 | 9.1 |
| Alumina 1 | 1.60 | 72.70 | ca 80 |
| Alumina 2 | 3.50 | 159.10 | ca 180 |

This method is even faster than that involving adsorption of n–butanol from n–heptane, as the FMC used for gas adsorption studies contains a ceramic cell reducing the time constant of the calorimeter by a factor of three compared to the FMC used for liquid work. Thus saturating an adsorbent sample with $CO_2$ takes only a few minutes permitting the determination of low surface areas in less than one hour, which includes the filling of the adsorption cell with adsorbent, purging with He, adsorbing the $CO_2$, and processing the results.

There is, of course, a large number of possible combinations of carrier fluids and adsorbable solutes that could be used for the determination of surface areas. A necessary condition for candidates suitable for this application is the conformity with Langmuir's isotherm, i.e. no change in the heat of adsorption of the adsorptive with surface coverage, and the ability to saturate the total fluid–solid interface at low solution or low gas mixture concentrations. Other important conditions are the completeness of the displacement of carrier fluid from the interface and the formation of a close packed monolayer by the adsorbate. The complete displacement of carrier fluid may not occur on microporous solids for which the determinations of total surface areas by adsorption are always difficult to interpret irrespective of the adsorptive used in such determinations.

Thus adsorption microcalorimetry continues to provide valuable information on the surface properties of solids and retains its potential for further applications in research work and routine industrial work on catalysts, pigments, fillers, cements, clays, other minerals, and particulates in general.

### 3.2. Polar and hydrophobic sites

The work on the determination of surface areas by using the heats of adsorption of n–butanol from n–heptane was extended to carbonaceous solids and it became apparent that their calorimetric surface areas were much lower than those determined by the BET($N_2$) method. The author's study of various types of graphite has shown that graphites invariably have a small proportion of their total surface in the form of polar groups which strongly adsorb alcohols, such as n–butanol, from solutions in n–heptane producing adsorption plateaux corresponding to close packed monolayers [11]. The adsorption was Langmuirian in character, but the factor correlating the heats of n–butanol adsorption and the surface area of the sites from which n–butanol displaced n–heptane was lower than that obtained for metal oxide surfaces with 1 J of the heat of adsorption corresponding to 6.7 $m^2g^{-1}$ of the polar area.

The work was later extended to the study of adsorption of n-butanol from water and it was found that, again, as in the case of the adsorption from n–heptane, n–butanol forms closed packed monolayers but this time exclusively on the hydrophobic sites, which for graphitic carbons are basal planes. The work on carbons was reported by the author in 1987 [18]. Subsequently the work on carbons was continued with other partly hydrophobic solids, such as heat treated silica gels, ZSM–5 zeolites, coals, and carbon blacks [28]. All these solids possessed various proportions of hydrophilic sites, the surface areas of which could be independently estimated from the integral heats of adsorption of n–butanol from water.

The methods proved to be very useful for the evaluation of the surface properties of various solids with mixed surface characteristics and changes in the surface character following oxidation, heat reduction, and comminution treatments. For example the effect of heat treatments on the amount of polar sites in a high surface area carbon black is shown in Figure 15, and the effect of graphite comminution in air and in liquid hydrocarbons is given in Table 2.

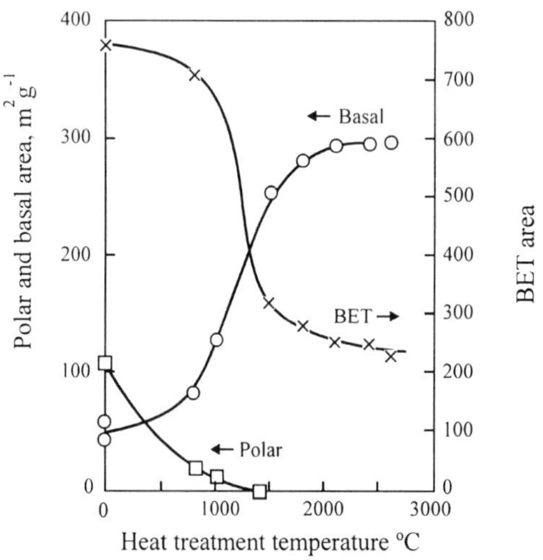


Figure 15. Changes in BET($N_2$) surface area, graphitic basal plane area, and graphitic polar area in carbon BLACK PEARLS 2 heated in argon to 2600°C.

Table 2
Surface properties of graphites

| Type of graphite | Time of grinding Mins | BET surface area $m^2g^{-1}$ | Basal plane area* $m^2g^{-1}$ | Polar area* $m^2g^{-1}$ |
|---|---|---|---|---|
| Synthetic, as received. Ground in air | 0 | 5 | 4.7 | 1.4 |
| | 2 | 14 | 13.3 | 1.4 |
| | 35 | 68 | 58.0 | 17.0 |
| | 65 | 307 | 181.0 | 39.0 |
| | 120 | 449 | 217.0 | – |
| | 180 | 658 | 246.0 | 87.0 |
| | 285 | 565 | 240.0 | – |
| Ground in vacuum | 180 | 436 | 243.0 | 76.0 |
| Ground in n–heptane | 480 | 92 | 114.0 | 2.2 |
| Graphon | – | 81 | 81 | 1.1 |

* Calculated as described in reference 11.

In many cases the increases in surface areas of carbons by oxidation or by comminution were accompanied by the creation of microporosity which influenced the correlation between the heats of adsorption and surface area estimates as shown below in section 3.4. For non–porous purely hydrophobic solids there is an excellent agreement between the determinations of surface areas by flow adsorption microcalorimetry and the BET($N_2$) method as can be seen for Graphon in Table 2.

### 3.3. Acid and basic sites

The polar surface areas determined by n–butanol adsorptions from n–heptane are not homogenous and contain a number of functional groups with electron donating or electron accepting properties. The author has studied the heats of adsorption of ammonium hydroxide on various solids and found that, like n-butanol, it tends to form adsorption plateaux on most solids having hydrophilic surface properties as illustrated, for zeolites and silica gels, in Figure 16. The data indicate the presence of acidity in all the above solids with the heats being relatively low for silica but very high, as expected, for an acid treated mordenite. The work was extended to the study of adsorption reversibility and it was found that a proportion of $NH_4OH$ was adsorbed irreversibly, the proportions being very different for individual adsorbents as shown in Table 3.

As can be seen the strongly acidic mordenite gave a high proportion of irreversible adsorption, in contrast to the zeolites, ZMS–5 and a sodium form of 13–X, which showed relatively low heats of adsorption and a high degree of reversibility. No doubt the heats of adsorption and the degree of irreversibility would strongly increase after acid or heat treatment of the zeolites. The low surface area silica gave complete reversibility of the adsorption indicating its very weak acidity.

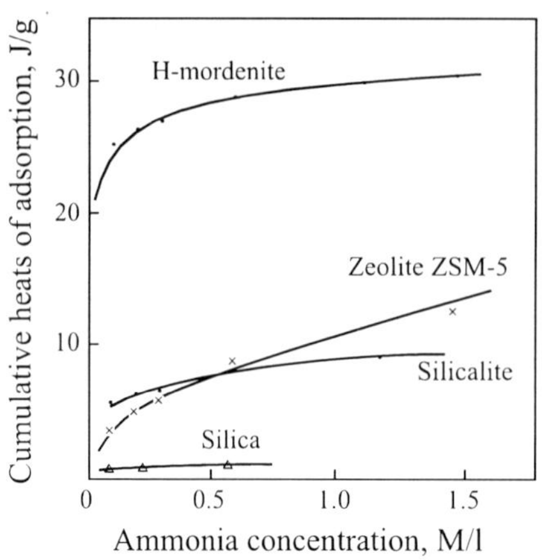


Figure 16. Cumulative heats of adsorption of $NH_4OH$ from aqueous solutions on zeolites and silica.

Table 3
Reversibility of adsorption of ammonia from water solutions onto silicas and zeolites

| Adsorbent | Heat of adsorption of 0.1 N solution in $Jg^{-1}$ | | | |
|---|---|---|---|---|
| | Adsorption 1 | Desorption | Adsorption 2 | Reversibility % |
| Silica, 25 $m^2g^{-1}$ | 1.17 | 1.34 | 1.39 | 119* |
| Silica, 300 $m^2g^{-1}$ | 14.60 | 10.38 | 12.67 | 87 |
| Silicalite 0.2 % $Al_2O_3$ | 5.56 | 3.61 | 3.76 | 94 |
| ZSM–5 3 % $Al_2O_3$ | 3.76 | 2.90 | 3.54 | 68 |
| H–Mordenite | 25.30 | 28.73 | 9.22 | 36 |
| Zeolite 13–X (Na) | 4.07 | 2.08 | 4.07 | 100 |

* This result was probably due to the removal of an unidentified surface impurity during the first cycle

An interesting index of total acidity of solid surfaces is provided by dividing the heat of ammonia adsorption from its 0.1 N solution by the total surface area estimated from the heats of adsorption of n–butanol from water as given in Table 4.

H–Mordenite gives the highest heat of adsorption mainly due to the low estimate of its surface area by n–butanol adsorption which does not displace the water in its micropores. The displacement is apparently achieved by the $NH_4^+$ ions. Otherwise, the pillared clay gives high values followed by kaolinite and the Linde sieve in its Ca form.

Table 4
Adsorption of $NH_4OH$ onto clays and zeolites

| Adsorbent | Heat of adsorption | | Surface area |
|---|---|---|---|
| | $Jg^{-1}$ | $Jm^{-2}$ | $M^2g^{-1}$ * |
| Pillared clay | 22.6 | 0.094 | 241 |
| Kaolinite | 1.5 | 0.068 | 22 |
| Silica gel | 6.1 | 0.039 | 155 |
| Zeolite 13–X (Na) | 12.1 | 0.027 | 443 |
| H–Mordenite | 48.2 | 1.60 | 30 |
| Linde 5A sieve (Ca) | 7.2 | 0.72 | 10 |
| Silicalite 0.2 % $Al_2O_3$ | 4.4 | 0.038 | 116 |
| ZMS–5 5 % $Al_2O_3$ | 8.2 | 0.021 | 385 |

* Estimated from the heat of adsorption of n–butanol from n–heptane

An extensive study of the acidity and basicity of minerals by flow adsorption calorimetry over the past decade has been carried out by F. Fowkes and his group [25, 29]. His conclusion was that flow adsorption microcalorimetry was the only method that can correctly evaluate the distribution and strength of acid and base sites in solids, and predict the effect of these parameters on their catalytic and adhesive properties. This was fully supported by later work done by D. P. Ashton and D. Briggs [30].

Evaluation of surface acidity can also be carried out in the gas phase by contacting adsorbents saturated with a carrier gas, such as He, with gaseous ammonia or another base. The adsorption and the determination of the heats of adsorption can be carried out at high pressures which may produce results altogether different to the measurements of adsorption at very low surface coverages and the associated low partial pressures of gaseous bases at which they may totally saturate the adsorbent surface.

A recently obtained example of pulse adsorption of $NH_3$ on ZMS–5 zeolite, activated at 200°C for 8 hours, from a He carrier at atmospheric pressure is shown in Figure 17. The pulses of ammonia were irreversibly adsorbed producing constant heats of adsorption ranging from 110 kJmol to 130 kJmol. These values are in the lower range of the heats of adsorption obtained by A. Auroux et al on a different ZMS–5 zeolite, activated at 500°C [31]. Similar work with the flow adsorption microcalorimeter at higher pressures is likely to reveal new aspects of $NH_3$ interactions with a variety of adsorbents and catalysts which will probably be more closely related to their uses in industrial practice.

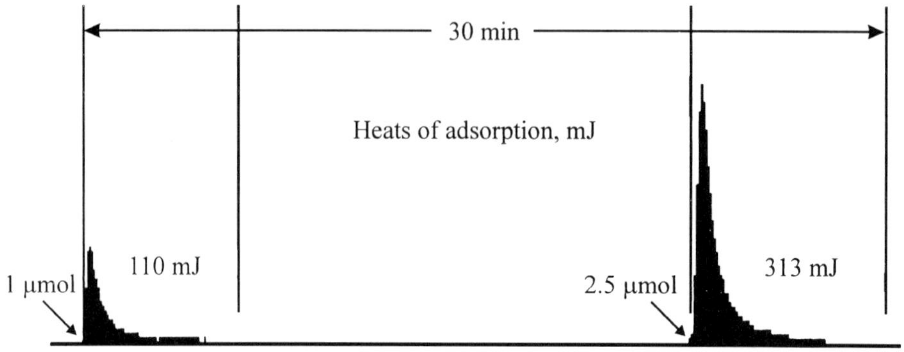


Figure 17. Pulse adsorption of $NH_3$ on 0.0756 g of zeolite ZSM-5 from He carrier at 25°C.

### 3.4. Microporosity

Displacement of carrier fluids by probes possessing different molecular dimensions and kinetic diameters is ideally suited for the determination of pore size distribution in adsorbents. The proportion of adsorption occurring in micropores as opposed to the external surface could also be determined if the amounts of adsorptions could be determined independently.

In relatively simple cases of non–admission into the pores that are too small for the adsorptive, flow adsorption microcalorimetry can produce striking results as exemplified by the heats of adsorptions of normal and tertiary butyl alcohols shown in Figure 18 indicating a major reduction in adsorptive accessibility due to a

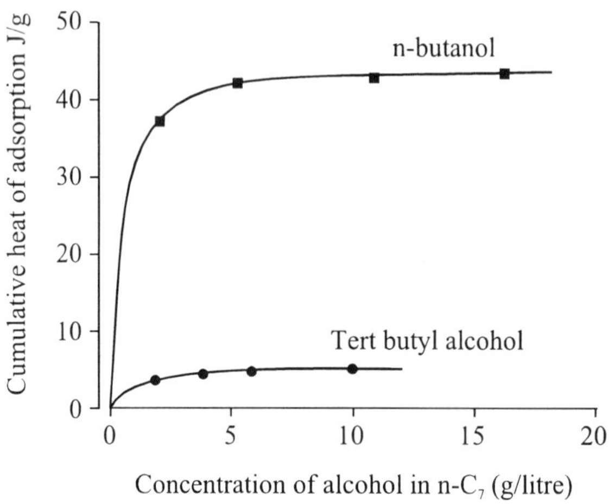


Figure 18. Cumulative heats of adsorption of primary and tertiary butyl alcohols on zeolite ZSM-5 at 25°C.

relatively small change in kinetic diameters. Further increases in kinetic diameter produced relatively small changes as reported by L. T. Canham and the author [32, 33].

Ultra–micropores, with diameters less than 0.5 nm, present special problems for strongly adsorbed carriers, even for supposedly inert gases such as He [34]. The separation of individual components of gas mixtures in pores with diameters between 0.3 nm and 0.5 nm may strongly depend on the kinetics of adsorption governed by the shape of adsorptive molecules, as well as that of the pores, as shown by M. L. Sykes et al [35], but the differences in adsorbate affinity for the surface may also play a role as shown recently by the author for a carbon molecular sieve [36].

For pores having hydrophilic internal surfaces, the adsorption of water does not seem to produce any problems and it can apparently easily displace most of the He used as a carrier gas as shown in Figure 13. In this case the rate of water adsorption may well depend on the chemical properties of micropore surfaces as the heat of preferential adsorption measurements clearly indicate two steps in the He displacement process shown in the above figure.

For pores with hydrophobic internal surface properties the kinetics of their penetration by hydrophobic liquids, such as n–heptane, has been dramatically demonstrated by the author for PVDC chars subjected to different degrees of oxidation. On wetting with n–heptane the unoxidised chars produced drastically longer heat evolution patterns than the wetting after oxidation, a process which increased the pore sizes to 1 nm to 2 nm from less than 1 nm [37]. There were also interesting differences in the heats of wetting of microporous ZSM zeolites which gave much lower heats of wetting with isooctane compared with those produced by n–heptane [33]. This paralleled the work with the adsorption of straight–chain and branched–chain alcohols referred to above [32].

The determination of graphitic basal planes in microporous carbons by n–dotriacontane is confined only to the pores with diameters exceeding 10 nm. It seems that the n–dotriacontane molecule is unable to penetrate the finer micropores filled with n–heptane, which, incidentally, is very difficult to remove completely from the micropores even at temperatures as high as 200°C (the adsorption of n–heptane vapour from helium carrier is also partly irreversible at room temperature [36]). The adsorption of n–butanol from water is much more complete, however, and returns high values of hydrophobic surface areas which are generally about 5 – 10 times higher than those determined by n–dotriacontane adsorption [28].

The adsorption of n–butanol and its heats of adsorption are enhanced by pore filling phenomena, and the values of surface area obtained from these adsorption measurements are excessive. The same applies to the estimates of surface areas by the BET methods, irrespective of the nature of the gas used in those determinations. The problem extends to the use of He which, according to the author's recent studies of its adsorption on active carbons, gives heats of adsorption as high as 70 $kJmol^{-1}$ at room temperature [37]. This adsorption is confined, of course, to a small part of the ultra–micropores which are present in certain high surface active carbons adsorbing 0.15 $\mu molg^{-1}$ of He [34].

# 4. AFFINITY OF SURFACES FOR SPECIFIC COMPOUNDS

## 4.1. Active carbons and graphites

Flow microcalorimetry has revealed a number of unique surface properties of the basal planes of graphite and, separately, the properties of edges of the graphene layers – which are usually combined with oxygen.

An important discovery by the author in the early sixties was the strong affinity of the basal planes for n–paraffins and the fact that the affinity increases markedly with their chain length. This work was fully documented by the author [8, 11, 12] and led to a number of applications in selective adsorption and lubrication of metal surfaces. The normal paraffins were postulated to lie flat in the basal plane surface with their methylene groups falling neatly into the centres of carbon hexagons in the basal planes [11]. This was later confirmed by STM microscopy [38] and x–ray diffraction studies [39].

Another important property of the basal planes is their affinity for polycyclic aromatics. This was found by pulse adsorption of dilute solutions of anthracene and pyrene in cyclohexane which produced very high molar heats of adsorption at low surface coverages. Later work by the author indicated that in spite of the strong initial adsorption, the polycyclic compounds do not form closely packed layers on the basal planes but are localised on certain more active parts of the graphene surface without the formation of plateaux in their adsorption isotherms [40].

Experiments with graphitised carbon blacks and transition metal compounds indicated a strong affinity of a small part of the surface for transition metal ions such as Cu, Ag, and Au [41]. The affinity increased regularly with the molecular weight of the cations or anions containing the metals, such as $Hg^{++}$ or $PtCl_6^-$. The affinities were certainly not caused by the oppositely charged ions without any transition metals such as $K^+$ in $K_2PtCl_6$ or $NO_3^-$ in $AgNO_3$, as both KCl and $KNO_3$ have very little tendency to be preferentially adsorbed on the graphitic basal planes from aqueous solutions. The irreversible adsorption of $PtCl_6^-$ ions on one of the graphitised carbon blacks amounted to 26 $\mu$molg$^{-1}$, as shown in figure 5, which corresponds to 1.3 % of the weight of the carbon. The integral heat of water displacement by the platinum complex is 90 kJmol$^{-1}$ with the values reaching 300 kJmol$^{-1}$ at surface coverages below 10 % [42].

Platinum ions are therefore extremely strongly adsorbed on certain sites of the carbon surface which is likely to render these ions resistant to agglomeration and the consequential loss of platinum surface area at high temperatures. This is, of course, important in catalytic applications. The strongly adsorbed Pt occupies about 550 Å per molecule, i.e. much less than the amount required to form a close packed monolayer on the total basal plane surface, but much more than the amount that could be accommodated on the polar sites available on the carbon surface.

It has to be accepted, therefore, that the strong adsorption cannot take place exclusively on the polar sites, but occurs predominantly on a small part of the basal plane surface showing more affinity for transition metal ions than the rest of that surface.

Recent work published by the author reveals another property of the basal planes which has not been previously reported, namely their strong affinity for sulphur [43]. Experiments with urea and thiourea, molecules that have a similar structure but differing only in respect of an O atom linked to carbon in urea being replaced by sulphur, show clearly very much stronger adsorption of thiourea compared with urea. An illustration of this effect is given in figures 19 and 20, showing cumulative heats of adsorption of urea and of thiourea from water on a graphitised carbon and an active carbon. The affinity for $Na_2S$ is also much greater than that for NaOH or NaCl, confirming the same effect.

The molar heat of adsorption of thiourea on Graphon from its 0.01 molar solution in water is quite high at 112 $kJmol^{-1}$. The corresponding heat on the active carbon is only 43 $kJmol^{-1}$, indicating the preference of the sulphur atom for extended but accessible basal plane surfaces in the graphitised carbon.

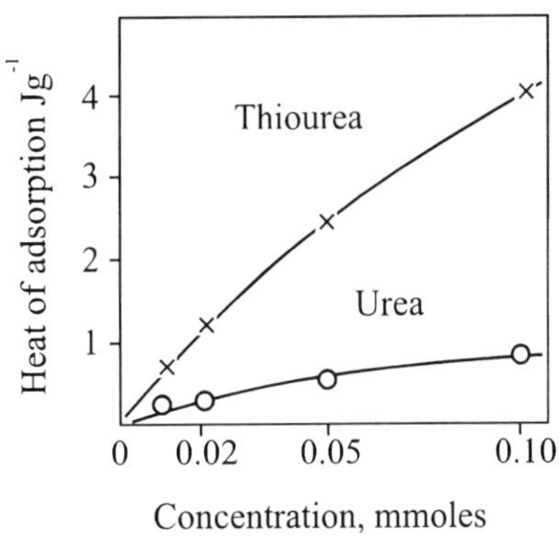


Figure 19. Cumulative heats of adsorption on 0.0704 g of graphitised carbon black, from water at 25°C.

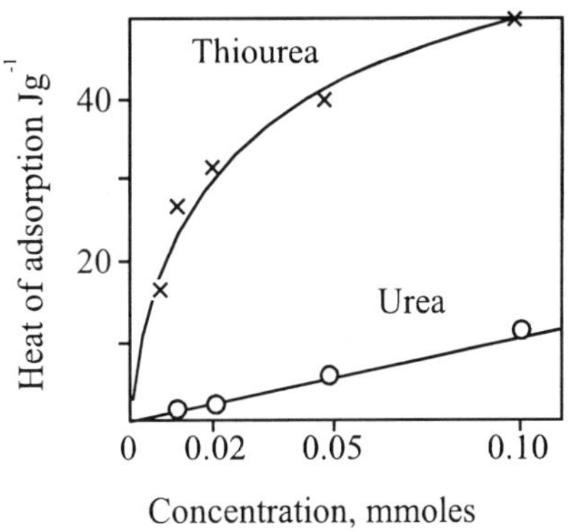


Figure 20. Cumulative heats of adsorption on 0.0067 g of active carbon CHEMIVIRON BPL mixed with 0.090 g of PTFE powder, from water at 25°C.

## 4.2. Metals and metal oxides – Effects of adsorption on resistance to wear and corrosion

One of the first applications of flow adsorption microcalorimetry has been the study of interactions of n–heptane solutions of long chain alcohols, carboxylic acids, and amines *with* metal surfaces and metal oxides. Strong correlation was found between the heats of adsorption of their long chain compounds and anti–wear action. Later work was extended to the study of adsorption on freshly formed metal surfaces, which behaved quite differently from the metal oxides [46]. Freshly formed iron surfaces were protected by layers of adsorbed n–heptane, the liquid hydrocarbon in which the surfaces were produced. The adsorbed hydrocarbon was easily displaced by stearic acid, producing much higher heats of adsorption than that measured for adsorption on iron oxides. The adsorption of stearic acid was, in fact, to a large extent chemical in nature with the heat of desorption being only 50 % of that produced on adsorption.

Anti–wear performance of adsorbates is only one step removed from their action as corrosive agents. This is exemplified by the action of stearic acid on copper surfaces which rapidly produce reaction products after initial absorption. This also applies to alcohols which adsorb strongly on an iron surface activated by milling in a dilute solution of stearic acid in n–heptane in the absence of $O_2$. The adsorption is followed by a chemical reaction characterised by a steady heat evolution after completion of the adsorption process [15].

The behaviour of specific strongly adsorbed compounds which do not react after adsorption but render the surface resistant to the action of corrosive agents, such as acetic acid, is exemplified by the interaction of benzotriazole with copper surfaces [47]. Pulse adsorption of acetic acid on a pure copper powder produced a strong irreversible adsorption which was substantially reduced after saturation of the metal powder with a dilute benzotriazole solution. A return to high reactivity with acetic acid was obtained after the desorption of benzotriazole.

Corrosion inhibition can thus be evaluated conveniently by sequential pulse experiments with surfaces in contact with or in the absence of corrosion inhibitors. Similar work can be carried out with gaseous corrosive agents and metal surfaces treated with corrosion inhibitor vapours such as butylamine.

Effective boundary lubrication greatly depends of the formation of strong films at the metal/liquid lubricant interface which can be formed by rapid adsorption from the liquids. The stronger the films the stronger the resistance that will be offered to the metal to metal contact which leads to lubrication failure. Most of the factors contributing to the strength of the adsorbed films, and their ability to reduce wear and friction, such as the thickness of adsorbed films, their lowering of the surface energy, and resistance to detachment from the metal surface can be investigated by FMC adsorption–desorption experiments, and the determination of the heats of adsorption and adsorption kinetics. The latter can be readily evaluated from the length of the adsorption–time profiles and the time required to desorb the protective film from the interface. Successful work of this type has been reported by the author [48], and other investigators [49, 50].

### 4.3. Solid lubricants

The solids in the form of fine powders or dispersions of fine powders in minerals, which lower friction and wear when added to lubricating oils, possess certain surface properties in common which are responsible for their anti–wear action. Flow adsorption microcalorimetry was instrumental in discovering some important surface properties of the solid powders that make them either lubricating or abrasive when placed in contact with metal surfaces in close relative movement, i.e. rolling or sliding contact.

The author discovered that the key surface property of effective solid lubricants is the *oleophilic* nature of their surface properties. This surface character is related to low surface energy of the majority of surface sliding sites, especially those that come into contact with the sliding metal surfaces. The FMC test applied to a candidate for a solid lubricant is to establish whether a normal paraffin, such as dotriacontane is strongly adsorbed on the surface of the solid. If so, it would probably make a good solid lubricant – with one proviso: *the polar surface sites which are present, in addition to the low–energy oleophilic sites,* must *not be abrasive nor form a high proportion of the total surface.* Layered solids which are composed of particles having a high aspect ratio, i.e. a high ratio of basal planes to edge sites, fulfill this requirement. Another important requirement is that the basal planes must be capable of forming a strong bond with the metal surface which it is lubricating.

Examples of effective solid lubricants are provided by layered metal disulphides such as $MoS_2$, $WS_2$ and graphite. The basal planes of these solids have a high affinity for mineral oils but can also adhere strongly to steel surfaces which they have to lubricate and are, therefore, capable of forming strongly adhering films on the metal surfaces which markedly reduce wear and friction under dry or wet conditions. However, the graphite or $MoS_2$ powders which have excessive amounts of the polar edge surface are abrasive and cannot be used as lubricants.

Tests for the quality of the solid lubricants can be conveniently carried out by flow adsorption microcalorimetry whereby the ratio of the basal planes to the polar sites can be determined, and the affinity of the basal planes for n–paraffins, as well as the affinity for transition metal ions can both be confirmed. The samples with excessive amounts of edge sites are not likely to be acceptable for lubrication [51, 52].

## 4.4. Ion exchange

Sorption on ion exchangers produces marked heat effects when they sorb ions from aqueous solutions. The author used flow microcalorimetry to determine the heat effects caused by the sorption and desorption of ions and discovered that the heats of cation exchange on strongly acidic exchangers were generally very large, but very much dependant on the ion's valency [52].

Care had to be taken to load the ion exchange resin into the adsorption cell of the FMC in such a way as to allow for its swelling when wetted by water. For resin used in the author's work, Amberlite IR120, the swelling amounted to one third of the original volume of the resin. The adsorption of water by the dry resin in its acid form caused a sharp heat of wetting which was similar to those produced by high surface area activated carbons and zeolites.

The heats of sorption from 0.15 M solutions of $NH_4^+$ ions and alkali chlorides were positive but were slow to reach equilibrium in the first adsorption/desorption cycles. Generally, the sorption effects have been much slower on the resin than the adsorptions on active carbons. The rate of the exchange process on the resins was clearly governed by the diffusion of the ions into and out of the resin which changed with the number of adsorption/desorption cycles. The heats of adsorption of $Na_2^+$ ions generally decreased with the number of cycles and there was evidence for irreversible sorption for most of the ions investigated.

Sorption of $Cu^{++}$ ions produced endothermic heat effects indicating the strong effect of dehydration of the ions. Sorption of $NH_4Cl$ produced exceptionally high heat effects, 174 $Jg^{-1}$ compared to 75.4 $Jg^{-1}$ for NaCl. There was evidence that $NH_4^+$ ions are irreversibly blocking some of the sites in the resin, reducing the heats of sorption of Ca and Cu ions.

Further work by the author [42] compared the adsorption of $PtCl_6^-$ and $AuCl_4^-$ ions on a basic ion exchanger and a graphitised carbon adsorbent and has confirmed that flow adsorption microcalorimetry can rapidly determine the capacity and the kinetics of sorption of the ions on such adsorbents under dynamic conditions. The rate of sorption of the Pt and the Au ions was limited by their diffusion into the resin matrix and the high affinity of the resin for water. High surface area graphitic carbons suffer much less from these effects and appeared to be much more effective adsorbents.

The results of this work indicated that sequential determination of the heats of ion exchange on sorbent or adsorbent surfaces give a unique insight into the performance of individual resin sorbents, which offers a good potential for a prediction of the performance of the resins in practical applications.

### 4.5. Fillers and pigments

This area has recently been extensively investigated by flow adsorption microcalorimetry, mostly in respect of the strength of adsorption of surfactants and resins on the surfaces of the above materials and the effect on the stability of pigment dispersion and adhesion properties as a function of the surface chemistry of solids. A major contribution in this field has been made by the late Prof. F. Fowkes whose work was recently summarised by D. P. Ashton and D. Briggs [30].

Prof. Fowkes's flow calorimetric work led to important advances in the understanding of the interactions between mineral fillers, polymers, and polymer additives. The work of S. T. Joslin and F. M. Fowkes [29] in which they time–sliced the outputs from an FMC and its accompanying DSD enabled them to determine differential molar heats of adsorption as a function of progress in the adsorption process, and gave an indication of the heterogeneity of strength of the acid or base adsorption sites on the filler or pigment surfaces.

Similar determinations can be carried out by using the pulse adsorption FIAT methodology described in section 2.5.

## 5. CONCLUSIONS

Flow adsorption microcalorimetry has now been developed to a point at which it provides accurate and reliable adsorption and desorption data for events occurring at solid–liquid and solid–gas interfaces within a wide range of temperatures, pressures, and/or solution concentrations.

The data can be used to obtain thermodynamic parameters characterising the nature of adsorbate–solid interactions which are important in gaining a fundamental understanding of selective adsorption mechanisms, the kinetics of adsorption, and associated processes such as: catalysis, lubrication, dispersion technology, corrosion, adhesion, and the determination of surface areas of chemically different sites on solid surfaces.

However, the considerable potential of flow microcalorimetric studies is still under used at present. It is hoped that this paper will promote the wider use of flow techniques and their application to more efficiently adjust and control industrial processes. Much can be achieved in the future by studying the heats of adsorption at high temperatures and pressures, which is of especial importance in the development of new catalysts.

## REFERENCES

1. W. Swietosławski, Reinhold Publishing Corporation, 92 (1946) 127.
2. Ibid, 173.
3. A. C. Zettlemoyer, J. Colloid Interface Sci., 28 (1968) 343.
4. V. D. Harkins and G. Jura, J. Am. Chem. Soc., 66 (1944) 1365.
5. T. L. Hill and G. Jura, J. Am. Chem. Soc., 74 (1952) 1598.
6. A. J. Groszek, J. Chromatography, 3 (1960) 454.
7. A. J. Groszek, Nature, 204 (1964) 680.
8. A. J. Groszek, Nature, 196 (1962) 531.
9. A. J. Groszek, ASLE Transactions, 13 (1971) 278.
10. T. Allen and R. M. Patel, J. Applied Chemistry, (1970) 20.
11. A. J. Groszek, Proc. Royal Society, A314 (1970) 473.
12. A. J. Groszek, Adsorption at the Solid – Gas & Solid – Liquid Interface, Elsevier Science Corporation, 1982, 55.
13. A. J. Groszek, Liquid Lubrication Technology, 1973, NASA, 184.
14. H. Stack and R. P. Fletcher, Symp. on Thermal Analysis & Calorimetry, 1977, 35.
15. A. J. Groszek, Wear, 18 (1971) 279.
16. A. J. Groszek, 4th International Conference on Surfactants, 1964, 2B, 1015.
17. A. J. Groszek, ASLE Transactions, 9 (1966) 67.
18. A. J. Groszek, Carbon, 25 (1987) 717.
19. A. J. Groszek, Nature, 182 (1958) 1152.
20. G. H. Bell and A. J. Groszek, Nature, 191 (1961) 1184.
21. S. Cerny, V. Ponec and L. Hładek, J. Chem. Thermodynamics, 2 (1970) 391.
22. C. E. Templer, Proc. Particle Size Analysis, 26 (1970) 4.

23. C. E. Templer, Nature - Physical Sciences, 235 (1972) 158.
24. R. C. Weast (ed), Handbook of Chemistry and Physics, CRC Press, 1974, D81.
25. F. M. Fowkes, T. B. Lloyd, J. R. Brand and L. J. Dizikes, J. Coatings Tech., 64 (1992) 91.
26. A. J. Groszek, Chemistry and Industry, (1966) 1754.
27. P. P. S. Saluja, D. W. Oscarsou, H. G. Miller and J. C. LeBlanc, Proc. 8th International Clay Conference (1985), Denver Co. USA.
28. A. J. Groszek and S. Partyka, Langmuir, 9 (1993) 2721.
29. S. T. Joslin and F. M. Fowkes, Ind. Eng. Prod. Dev., 24 (1985) 369.
30. D. P. Ashton and D. Briggs, Partially Filled Polymer Composites, 1995, 99 (chapter 3 entire).
31. A. Auroux, N. Muscas, D. J. Coster and J. T. Pripiat, Catalysis Letters, 28 (1994) 179.
32. A. J. Groszek, Preprint of IVDAC Symposium on Characterisation of Porous Solids (COPS), Bad Soden, Germany, 115 (April 1987).
33. L. T. Canham and A. J. Groszek, J. Appl. Phys., 72 (1992) 1558.
34. R. Staudt, F. Dreisbach and J. V. Keller, 23rd Biennial Conference on Carbon, Penn State, 178 (July 1997).
35. M. L. Sykes, H. Chaqqer and K. M. Thomas, Carbon, 31 (1993) 827.
36. A. J. Groszek, Carbon, 35 (1997) 1399.
37. A. J. Groszek, Carbon, 27 (1989) 33.
38. G. C. McGonigal, R. H. Bernhard and D. J. Thomson, Appl. Phys. Letters, 57 (1990) 28.
39. K. Morishe, Y. Takami and Y. Yokata, Physical Review B, 48 (1993) 11.
40. A. J. Groszek, Chem. Soc. Faraday Discussions, 59 (1975) 109.
41. A. J. Groszek, Carbon, 35 (1997) 1329.
42. A. J. Groszek and M. J. Templer, Proc. of IEX'92, M. J. Slater (ed.), Elsevier Applied Science, London, 1992, 175.
43. A. J. Groszek, Proc. of 22nd Biennial Conference on Carbon, University of California, San Diego, 448 (1995).
44. A. J. Groszek, ASLE Transactions, 5 (1962) 105.
45. A. J. Groszek, Chemistry and Industry, (1965) 482.
46. A. J. Groszek, ASLE Transactions, 13 (1970) 278.
47. G. Steinberg, Chemtech, 11 (1981) 730.
48. A. J. Groszek, Liquid Lubricant Technology, 1973, NASA, 477.
49. S. Hirousha, Y. Yohagi and T. Sakurai, Bulletin of the Japan Petroleum Inst., 17 (1975) 201.
50. H. Kondo, A. Seki, H. Watanabe and I. Seto, Colloids and Surfaces, 26 (1990) 2691.

51. G. L. Andrews, A. J. Groszek and N. Heirs, ASLE Transactions, 15 (1971) 184.
52. A. J. Groszek and R. Witheridge, ASLE Transactions, 14 (1971) 254.
53. A. J. Groszek and M. J. Templer, Ion Exchange Advances, M. J. Slater (ed.), Elsevier Applied Science, London, 1992, 175.

Adsorption and its Applications in Industry and Environmental Protection
Studies in Surface Science and Catalysis, Vol. 120
A. Dabrowski (Editor)




# Temperature programmed desorption, reduction, oxidation and flow chemisorption for the characterisation of heterogeneous catalysts. Theoretical aspects, instrumentation and applications

M. Fadoni[a] and L. Lucarelli[b]

[a] State University of Milan, Chemical Physical Department, via Golgi 19, 20133 Milan, Italy
[b] CE Instruments (ThermoQuest S.p.A.), Strada Rivoltana 20090 Rodano (Milan), Italy

Some aspects related to catalysts characteristic and behaviour will be treated such as determination of metal surface area and dispersion, spillover effect and synterisation. A detailed description of the available techniques will follow, taking in consideration some aspects of the gas-solid interactions mechanisms (associative/dissociative adsorption, acid-base interactions, etc.). Every technique will be treated starting from a general description of the related sample pre-treatment, due to the fundamental importance of this step prior to catalysts characterisation. The analytical theories will be described in relation to static and dynamic chemisorption, thermal programmed desorption and reduction/oxidation reactions. Part of the paper will be dedicated to the presentation of the experimental aspects of chemisorption, desorption and surface reaction techniques, and the relevant calculation models to evaluate metal surface area and dispersion, energy distribution of active sites, activation energy and heat of adsorption.

The combination of the described techniques and the integration of the experimental results produce a detailed picture of the investigated catalyst, allowing a better comprehension of the reaction mechanisms in complicated processes and a detailed characterisation of catalyst activity and selectivity. Most of the experimental results shown in the present paper have been obtained in the application lab of CE Instruments (ThermoQuest S.p.A.), Milan – Italy. All the graphs related to static volumetric chemisorption have been obtained by the adsorption apparatus Sorptomatic 1990, while the graphs related to TPD, TPR/O and pulse chemisorption analyses with the dynamic apparatus TPDRO 1100.

## 1. INTRODUCTION

A catalyst can be defined in many ways but generally it is a substance that, when added in the balance of a chemical reaction, accelerates the achievement of the chemical equilibrium between reactants and products without influencing the thermodynamic equilibrium of the process. Usually catalysts are not consumed during the reaction and they could be found unchanged after the reaction. In reality, catalysts are submitted to a slow transformation with use, causing a general decrease of the activity and/or selectivity. The first main distinction between catalysts depends on the catalyst nature in relation to the reactants. A homogeneous catalyst is in the same physical state of the reactants (liquid, solid or gaseous) while heterogeneous ones are in a different state. In this paper only the heterogeneous solids catalysts will be taken into consideration. Among solid catalysts, we can identify three main groups: metal supported, acid sites and/or basic sites. The metal supported catalysts are prepared by supporting a metal (usually a noble one) onto a porous material, such as a gamma alumina or silica, featuring a suitable pore size distribution and specific surface area. Examples of acid catalyst are, for instance, zeolites. It is very important to characterise these materials to classify carefully the catalysts in function of the chemical reaction, to improve reactivity, selectivity and/or the production technique in order to understand better the role of the catalyst in a chemical reaction. Furthermore, it is possible to analyse the reasons for a catalyst poisoning or deactivation after use. Catalysts can be characterised by different techniques giving a wide range of information:

- activity is defined as the rate at which a chemical reaction reaches the equilibrium. From the industrial point of view activity is also defined as the amount of reactant transformed into product per unit of time and unit of reactor volume,
- selectivity is defined as the rate of reactant conversion into the desired products. Selectivity usually depends on reaction parameters such as temperature, pressure, reactants composition and also on the catalyst nature.

Activity, selectivity and other parameters can be measured by performing the chemical reaction in a pilot reactor but a basic characterisation of the catalyst surface is necessary to correlate the catalyst nature to its performance. Considering as an example a metal supported catalyst, there are two main aspects that should be investigated: the porous nature of the support and the active sites nature/distribution. Considering as an example a homogeneous reaction profile in comparison with the same reaction performed by using a catalyst, in the second process the catalyst action is to decrease substantially the total time to reach the thermodynamic equilibrium, that is to speed up the conversion process. Therefore the main effect of a catalyst is to provide an alternative reaction path that permits to decrease the activation energies of the different reaction steps, reaching therefore the equilibrium in an easier and faster way. The two different reaction paths (without and with catalyst) are represented in Figure 1. In Figure 1, Enc is the activation energy that is necessary for the reaction without the use of catalyst, Eads is the adsorption energy of reactants on the catalyst surface, Ecat is the energy related to the activation of the chemical reaction between reactants on the catalyst surface, Edes is the desorption energy of the products of reaction.

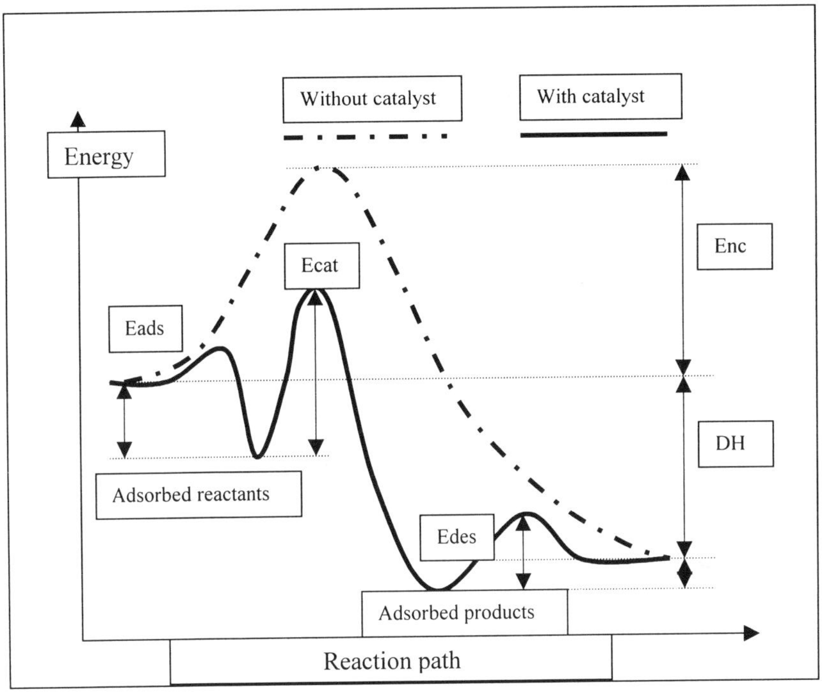


Figure 1. Reaction profile for a chemical reaction with and without catalyst.

In the above example the process of reactants adsorption is considered as an exothermic process, while the products desorption is considered as endothermic. Finally DH is the total heat of reaction that will be the same for the homogeneous and the catalytic process. Usually the activation energy of a catalytic reaction is lower than the one related to the homogeneous reaction. Sometimes by increasing the temperature of the process there is a limit where the homogeneous reaction becomes faster than the catalytic reaction. Therefore, the use of a catalyst should also be evaluated according to the energy profile of a certain chemical reaction.

In a catalytic reaction besides the knowledge of the energy profile, it is of extreme importance the study of the reaction kinetic profile. This permits to identify the slower stage of the reaction. Usually in heterogeneous reactions, we can divide the catalytic process in five main steps:

1- diffusion of reactants from the fluid into the catalyst support porous structure, reaching the internal surface,
2- adsorption of reactants on the catalyst active sites,
3- chemical reaction between the reactants,
4- desorption of the reaction products from the active sites,
5- diffusion of products through the catalyst support, reaching the external fluid.

One or more of the above stages could be the rate-determining step, influencing the total speed of reaction. The speed related to the steps 1 and 5 is mainly due to the porous nature of the support and the reactants/products geometrical parameters. In fact, it is necessary that the limiting step of the reaction should not be a diffusion problem. The pore size of the catalyst support should be chosen in relation to the reactant molecule volume and geometry. Steps 2 and 4 are related to the nature of the reactants/products and the active sites deposited on the catalyst surface. If the diffusion is not the limiting effect, the speed of reaction is directly function of the active surface area of the catalyst: the higher is the number of active sites available for the adsorption process of reactants the faster is the speed of reaction. A very high active surface can be achieved by using high surface area supports and optimising the deposition process of the metal. On the contrary, the specific surface area of a solid porous support is inversely proportional to the pore size of the support itself: at parity of pore specific volume, the smaller the pores, the higher the specific surface. Therefore, the characterisation of the support in terms of pore size distribution and specific surface area is of fundamental importance in the choice of a suitable catalyst. While a very high surface is advisable, a correct pore dimension should fit with the reactants/products geometry.

## 2. ASPECTS RELATED TO CATALYSTS CHARACTERISATION

As described above, the basic catalyst characterisation involves two main steps: the investigation on the porous nature of the catalyst support (physical properties) and on the properties of the active sites that are dispersed on the support surface (see Table 1).

Therefore, the physical characterisation of the support and of the supported catalyst is related to the measurement of the parameters as:

- geometry of the catalyst (solid shape or powder)
- specific surface area - square meters per mass
- pore specific volume – volume per mass
- pore size distribution – volume versus width
- mean pore size
- pore shape
- real density - weight per volume
- apparent density - weight per volume
- bulk density - weight per volume
- percent porosity
- particle size distribution (in case of powders) – relative percentage versus diameter.

Table 1
General scheme of catalysts characterisation

| Catalyst texture | | | |
|---|---|---|---|
| Physical properties | | Chemical properties | |
| Result | Technique | Result | Technique |
| Geometry and shape | | Chemical composition | Electron spectroscopy<br>Atomic adsorption |
| Total specific surface area | Gas physisorption<br>Mercury porosimetry | Active site surface area | Selective chemisorption (static or dynamic) |
| True density | X-ray analysis<br>Neutron diffraction | Degree of dispersion | Selective chemisorption<br>X-ray<br>Electron microscopy<br>Magnetisation analysis |
| Bulk and apparent density | Helium pycnometry<br>Mercury porosimetry<br>Liquid displacement | Surface energy | Thermal analysis tests<br>Temperature-programmed desorption and reaction<br>Calorimetry |
| Pore specific volume<br>Porosity | Mercury porosimetry<br>Gas adsorption | Acid-base sites | Selective chemisorption<br>Temp. programmed desorption |
| Pore size and mean pore size | Mercury porosimetry<br>Gas adsorption | Redox sites | Spectroscopic methods<br>Temp. programmed reduction<br>Temp. programmed oxidation |
| Particle size | Sieves<br>Laser scattering<br>Sedimentation<br>Electrical sensing zone<br>Etc. | Catalytic properties<br>Activity<br>Selectivity | Reactor tests and simulation |
| Surface structure | Optical microscopy<br>Electron microscopy<br>X-ray analyses | | |
| Surface charge | Z potential | | |

There are two main techniques available to determine the above parameters, mercury pressure porosimetry and gas physisorption. These two techniques should be chosen according to the pore size. In fact, pores are classified according to three main groups:
Micropores: up to 2 nm
Mesopores: between 2 and 50 nm
Macropores: over 50 nm.

The gas physisorption technique permits to obtain parameters as:
Specific surface area: generally from 0.0005 $m^2/g$, theoretically no upper limit
Pore size distribution: generally covering the range of micropore and mesopore
Pore specific volume: in the range of validity.

Mercury porosimetry completes the information above with regard to the pore size:
Pore size distribution: from 3.6 nm up to 600000 nm.

The detailed description of the above methods is not matter of the present paper. The characterisation of the active nature of a catalyst can be split into two main types: basic textures of active sites and reactivity tests. In Table 2 a general overview of the methods involved in catalyst characterisation is represented.

Table 2
Adsorption/desorption techniques in catalysts characterisation

| Analytical technique | Information |
|---|---|
| Static volumetric chemisorption | Active sites surface area<br>Degree of dispersion<br>Distinction of weak/strong gas-solid interaction<br>Acid/base surface properties<br>Isosteric heat of adsorption |
| Pulse chemisorption in flow | Active sites surface area<br>Degree of dispersion<br>Determination of strong gas-solid interaction<br>Acid/base surface properties<br>Isostheric heat of adsorption |
| Temperature programmed desorption | Active sites surface area<br>Degree of dispersion<br>Activation energy as function of metal saturation degree<br>Surface reactions<br>Kinetic and thermodynamic parameters of surface reactions |
| Temperature programmed reduction | Reduction degree of active sites<br>Activation energy related to reduction |
| Temperature programmed desorption | Oxidation degree of active sites<br>Activation energy related to oxidation |

The analytical methods reported in Table 2 are of particular interest not only in the research and in development of catalysts but also from the industrial point of view. The industry requirements regarding analytical methods are based on two levels:

- research and development:
  - accuracy and precision
  - flexibility
- quality control
  - precision
  - reproducibility
  - speed of analysis (productivity)
  - ease of use
  - certification of the method and of the instrument.

For the above reasons not only the analytical instrumentation should be developed in a way to meet the industrial demand but also the analytical method should be relatively easy to handle by different operators and be fast. The techniques described in this paper meet the above requirements, providing essential catalyst parameters with high precision at a limited cost.

### 2.1. Selective chemisorption techniques

The chemisorption techniques are very well established analytical methods to evaluate the free metal specific surface area and metal dispersion degree. These methods consist in performing a real chemical reaction between a reactive gas and the catalyst that has been previously prepared in a suitable way. Different types of chemisorption techniques can be used, the main ones are gas chemisorption on metals which are in zero oxidation degree, hydrogen/oxygen titration and acid/base reaction. The pre-treatment procedures must be chosen therefore according to the catalyst nature and to the technique that will be applied. In all the above cases, a common procedure to be performed before the real pre-treatment is to clean the catalyst surface. The cleaning generally consists in degassing the sample at a suitable temperature to remove water or other vapours eventually adsorbed on the surface, even if the catalyst has already been calcinated. The degassing can be done under vacuum or under a flow of inert gas. If the catalyst comes from a reactor it is necessary to remove eventual reaction residual that can block the catalytic surface (i.e. carbon derived by cracking) by a forced oxidation using air or oxygen. After the preliminary cleaning, the pre-treatment procedure should be differentiated according to the required analysis. In the first case the sample preparation has the task to oxidise or reduce the metal deposited on the support surface to zero oxidation state. This procedure activates the catalyst to the chemisorption measurement with a suitable reactive gas. A common sequence is oxidation – reduction – removal of hydrogen chemisorbed by flowing an inert gas. At this point the sample is activated to chemisorb a reactive gas. The second type of pre-treatment should produce an oxidised or reduced status of the metal. Therefore, it will be possible to perform a chemical reaction between hydrogen injected into the sample holder and oxygen bound to the metal active sites (or viceversa). In the last case, acid/base reaction, the catalyst surface should be only free from pollutant vapours and the gas used for the analysis must have acid (i.e. carbon dioxide) or base (i.e. dry ammonia) properties to react selectively with the base or acid sites of the sample. It is of extreme importance that the sample after the pre-treatment should not have any contact with the environment otherwise the reliability of the measurement could be seriously compromised.

### 2.2. Active surface area and metal dispersion

It is commonly used and convenient to define in a catalyst the surface area of the free active sites. Considering as example a metal supported catalyst, we can define as total surface area the surface of the support that can have contact with the external fluid mass. When an active phase (i.e. noble metal) is deposited on the support, only part of the available support surface can be covered. The chemisorption techniques permit to evaluate selectively the surface area of the active phase that is usually smaller than the total catalyst surface area.

Furthermore, only a small part of the active phase is physically free to react with the measuring gas due to the formation of metal aggregates. The metal atoms that are contained inside the aggregate cannot have contact with external fluids therefore they have no influence on the chemical reaction. In case of metal supported catalysts the total amount of metal fixed on the support can be conveniently determined by techniques such as atomic adsorption giving as result the total metal percentage present in the sample. The chemisorption techniques evaluate the free metal surface in square meter per gram by counting the number of surface metal atoms available on the metal aggregates. Finally, the degree of dispersion is defined as the ratio between the free metal atoms and the total number of metal atoms that are fixed on the support surface (in other words, the fraction of metal exposed to an external fluid phase).

### 2.3. Acid-base sites

The catalyst surface may contain acid and base sites that can interact together. On a certain surfaces the acid or base behaviour may prevail even if both sites are always present. In a catalyst characterisation, it is very useful to define the nature (Lewis or Brönsted, see par. 4.3) of these sites, their density, location, distribution and strength. Generally, an acid site is defined as a site that can react with a base and, on the contrary, a base site is one that can react with an acid. The above information (acid-base sites density) can be obtained by performing a chemisorption measurement using an acid (such as $CO_2$, $SO_2$) or base gas (such as dry ammonia) while their strength could be measured by temperature programmed techniques (desorption).

### 2.4. Spillover effect

Spillover is a phenomenon that involves the migration of an active chemisorbed species, formed on a first active phase (metal) onto a second phase that usually could not react if present alone in the same conditions. The phenomenon of spillover is not desired in the determination of adsorption stoichiometry because it always involves an increase of the amount of adsorbed gas. In case spillover takes place the free metal surface area and dispersion are always overestimated therefore spillover is not desired in catalyst characterisation but well accepted in a catalytic reaction because the number of active sites greatly increases. In the following picture are reported three cases in which the spillover-effects can take place [1].

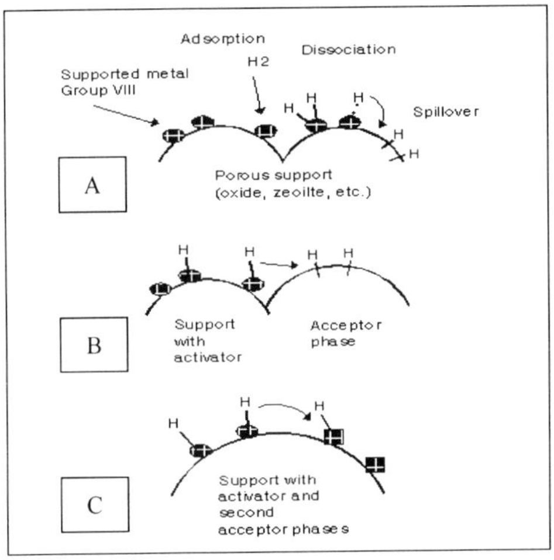


Figure 2. Different mechanisms of spillover: a) the first active phase is directly supported on the second phase, the acceptor; b) the first active phase is supported on an activated support, mixed with a non-activated support; c) the first active phase is fixed on a support that is activated by another active phase.

The physical conditions (temperature and pressure) causing the spillover effect depend on the catalyst (metal and support) and on the reactants. General conditions in which spillover might be avoided are:

Temperature: between 5 and 40 °C

Pressure: between 0 and 100 torr.

Higher temperature and pressure values during the experiment could promote the spillover effect therefore influencing negatively the estimation of parameters such as metal surface area and dispersion.

### 2.5. Synterization

Synterization of a metal catalyst is a process that consists of a migration of the supported metal atoms to form larger metal aggregates. Synterisation is a direct consequence of temperature, time and ageing. During the catalyst activation (pre-treatment) prior to the analysis, there are several phases in which the catalyst should be heated at very high temperatures (i.e. to remove, hydrogen after the reduction process). The catalyst nature and history should be very well known in order not to overtake the maximum conditions of temperature and time used for the catalyst preparation (i.e. calcination). In fact, the metal is finely dispersed on the support in order to maximise the metal surface area in relation to the minimum amount of noble metal. By heating the catalyst during the pre-treatment procedures, the mobility of the metal particles is increased. If the temperature overtakes certain limits, depending on the catalyst nature, the metal particles migrate to form larger aggregates decreasing therefore the metal dispersion. This effect reduces directly the number of active sites exposed to the fluid reactants, dramatically reducing the catalytic activity. Synterisation is a non-reversible phenomenon and the original metal surface area and dispersion cannot be restored.

### 2.6. Poisoning

A poison, when referred to catalysts, is an impurity that is present in the fluid phase and that reacts selectively with some active sites, stopping their activity. Usually the poisoning effect should be always avoided but sometimes could be useful to stop the formation of undesired secondary products. Poisoning could occur by chemical reaction (chemical poisoning) or fouling (physical poisoning). An example of chemical poisoning is the reaction between sulphur and some noble metals. The chemical poisoning is non-reversible if the product of reaction is stable while sometimes it is possible to remove the poison by a suitable chemical reaction. For instance, in case of some sulphur compounds, it is possible to remove the poison by hydrogenation to produce $H_2S$. The physical poisoning takes place when an external substance blocks directly the access of the fluid to the active surface. This effect could be caused by encrusting of powders, carbon coke or pitches on the catalyst surface or inside the pores (fouling). In case the physical poisoning is due to carbon coke formation, an oxidation process can remove the poison.

## 3. EVALUATION OF CATALYST SURFACE PROPERTIES BY CHEMISORPTION

As described above, the main purpose of the chemisorption methods is to evaluate the number of active sites that can be reached or that can interact with a fluid phase. These techniques are based on a chemical reaction between a suitable reactive gas and the surface reactive site. There are different methods to perform the above operation, the static volumetric, the static gravimetric or the flow methods. In the volumetric method, the sample is kept under high vacuum before the analysis. The analytical instrument then introduces known doses of reactive gas into the sample holder, measuring afterwards the equilibrium pressure that will be established between the sample and the gaseous phase. The static volumetric method requires the calibration of the system dead volume, being an absolute method (all the parameters related to the analysis are absolute values: temperature, pressure and dead volumes). The dead volume calibration is usually defined as "blank measurement" and it consists in performing a run in the same analytical conditions but using an inert gas instead of a reactive one. The pressure range should be chosen in order to complete the adsorption isotherm covering a monolayer of reacted molecules over the sample. The pressure range therefore depends on the catalyst/adsorbate nature and the analysis temperature. Most of catalytic systems analysed around room temperature show the monolayer covering below 100 torr equilibrium pressure.

The advantage of the static method is that the system catalyst/adsorbate reaches the real equilibrium conditions. Furthermore, it is possible to distinguish in a quantitative way the amount of gas strongly bound to the active sites and weakly bound to the support. On the contrary, the static volumetric method requires high vacuum system, long lasting measurements, and generally higher instrumentation costs. The gravimetric technique is in principle the same as the volumetric one; the difference is that the amount of gas adsorbed is measured by the weight change of the sample during the adsorption process. This technique is very precise but also very expensive due to the high cost of the necessary microbalance. Flow methods are carried out at atmospheric pressure. An inert carrier gas is passing through the sample in a suitable sample holder. A suitable injection system, typically a loop valve, can pulse in the gas stream before the sample, the reactive gas. The detector in this case can be a thermal conductivity detector, a microbalance or a mass spectrometer. Using the thermal conductivity detector the amount of gas adsorbed is calculated by integrating the peaks generated by the detector signal in function of time. This method requires the calibration by injecting the gas by a calibrated loop and measuring afterward the generated peak area. The advantages of this analytical method are based to fast analysis, relative low cost of the apparatus, relatively low risk of leaks. The flow method can be used mainly for systems catalyst/adsorbate that show fast equilibrium time. Furthermore, this method can only evaluate the strong interactions between the gas and the active sites, while the weak interactions cannot be measured as the gas is inmediately removed by the carrier flow.

### 3.1. Choice of reactive gas

The reactive gas used for chemisorption textures should be chosen mainly according to the active phase of the catalyst. The knowledge of the stoichiometric relation between the gas and the metal, therefore the mechanism of chemisorption, is of extreme importance to determine correctly the number of free active sites. In fact the evaluation of the number of free active sites by chemisorption and thermal programmed techniques is based on the accurate measurement of the gas amount that is adsorbed or desorbed during the analytical cycle. The purpose of the chemisorption methods is to evaluate the amount of gas used to cover a monolayer over the free active phase. The monolayer might be expressed in moles or volumes (NTP) of gas referred to sample mass or to the amount of metal present in the catalyst.

The correct choice of the adsorptive therefore should take into consideration the following factors:
- reaction with the metal
- short equilibrium time
- minimum effect of adsorption on the support (spillover)
- minimum effect of gas–gas interaction (multi-layer formation)
- minimum effect of gas dissolution in the metal (i.e. hydrogen and palladium)
- the stoichiometry must be known
- gas purity (minimum of 99.99 %).

In Table 3, we report a general classification of some metals and gases in function of the adsorption type [2].

Table 3
Classification of metals according to adsorption type

| Metals | Dissociative form | | | | | Associative form | | | | |
|---|---|---|---|---|---|---|---|---|---|---|
| Gases | $H_2$ | $O_2$ | $N_2$ | NO | CO | $H_2$ | $O_2$ | $N_2$ | NO | CO |
| Hg, Ta, Zr, Nb, W, Ti, V, Mn, Cr, Mo | + | + | + | + | + | - | - | - | - | - |
| Fe, Re | + | + | + | + | + | - | - | - | + | + |
| Ni, Co, Tc | + | + | - | + | + | - | - | - | + | + |
| Os, Ir, Ru, Pt, Rh, Pd | + | + | - | + | - | - | - | - | + | + |

(+) = Possible form
(-) = Impossible form

Hydrogen is the most commonly used gas for chemisorption measurements. It can be used with ruthenium (very slow equilibrium time), rhodium and iridium. Anyway, it is frequently used to analyse also the noble metals of the Group VIII. Hydrogen cannot be used to characterise silver and it forms with palladium a solution that is function of the temperature and pressure (absorption) [3]. Generally, hydrogen is adsorbed in a dissociative form (stoichiometric factor 0.5), but many other mechanism of interaction with active surfaces are known (mobile, subsurface, hydride, adsorbed in the bulk, partially reversible, strong residual and spilled over). Oxygen can be used with metals showing easy oxidation properties. The disadvantage of using this gas is that oxygen might change the surface-active structure by corrosive chemisorption. Furthermore, it can penetrate beneath the surface forming an oxide layer. Such effects are undesired for surface area characterisation and they could be minimised by using low temperatures (sub-ambient). Oxygen is the only convenient gas to characterise silver catalysts. Carbon monoxide is a very reactive gas and, in principle, it might be used with most of metals. The problem related to this gas is the proper evaluation of the stoichiometric factor of the chemical reaction. Furthermore, smaller particles of metal can dissociate in presence of this gas to form carbonyl complexes. Although the reacted gas can be removed by evacuation in flow or in vacuum, the metal particles form again, but it is not certain if the structure will be the same. Carbon monoxide can also form volatile carbonyls, especially with iron and nickel. Nitrogen oxide is a very reactive oxidant and can be conveniently used to characterise cobalt and silver catalyst. The main issue in the use of NO is relative to its toxicity.

### 3.2. Metal surface area calculation

The calculation of the metal specific surface area is based on the simple following relation:

$$MSS = (Vm\ Na) / (Sf\ Sd) \qquad (1)$$

MSS = metal surface area (square meters per gram of sample)
Vm = gas adsorbed at monolayer (moles per gram of sample)
Na = Avogadro number (molecule per mole of gas)
Sf = Stoichiometric factor of the reaction (number of molecule of adsorbate per surface metal atom)
Sd = Metal surface density (number of metal atoms per square meter)

The stoichiometric factor Sf of a chemical reaction is therefore of fundamental importance in the calculation of the correct specific surface area. Sf should be investigated also by other analytical methods (i.e. for CO with infrared spectroscopy). The most common gases used in chemisorption techniques are hydrogen, carbon monoxide, oxygen and di-nitrogen oxide. In the following table, we report some examples of possible reactions between the above gases and a generic metal M [4].

For carbon monoxide, two adsorption stoichiometries are commonly found (linearly and bridged bonded) and infrared measurements of adsorbed carbon monoxide can be used to distinguish between the two possibilities. With platinum, it is chiefly the linear species that is formed. Various proportions of the bridged species have also been noted. However, the carbon monoxide to accessible metal atom ratio (CO/M) is generally similar to the H/M ratio. With palladium, larger fractions of the bridged species are usually observed. In the following table, we report some examples of stoichiometric factors and metal atomic densities per unit of polycrystalline surface [5]. The values of stoichiometric factors are expressed in moles of adsorbing gas per metal surface atom.

Table 4
Stoichiometric factors for different types of reaction

| Method | Reaction (M = accessible metal atom) | Stoichiometry |
|---|---|---|
| Hydrogen (or oxygen) chemisorption | $M + ½ H_2 = M\text{-}H$ | 0.5 |
| Carbon monoxide chemisorption | $M + CO = M\text{-}CO$<br>$2 M + CO = M\text{-}CO\text{-}M$ | 1<br>0.5 |
| Hydrogen/oxygen titration | $M\text{-}O + 3/2 H_2 = M\text{-}H + H_2O$ | 1.5 |
| Oxygen/hydrogen titration | $M\text{-}H + ¾ O_2 = M\text{-}O + ½ H_2O$ | 0.75 |
| Nitrous oxide reaction | $2 M + N_2O = M\text{-}O\text{-}M + N_2$ | 0.5 |

Table 5
Parameters related to some metals and stoichiometric factors with different gases

| Metal (M) | $H_2$/M | CO/M | $O_2$/M | Atomic weight | Sd x $10^{20}$ (atoms/$m^2$) | Note |
|---|---|---|---|---|---|---|
| Pt | 0.5 | 1<br>1.15 | 0.5 | 195.09 | 0.125 | Spillover<br>T > 200 C |
| Pd | 0.5 | 0.6 | - | 106.4 | 0.127 | Absorption with $H_2$ |
| Ir | | | 0.5 | 192.22 | 0.130 | |
| D > 1.5 nm | 0.5 | 1 | | | | |
| D < 1.5 nm | 1 | 1 | | | | |
| Rh | | | 0.5 | 102.905 | 0.133 | |
| D > 2 nm | 0.5 | 1 | | | | |
| D < 2 nm | 0.5 - 1 | 2 | | | | |
| Ru | 0.5 | | | 101.07 | 0.163 | Spillover with $H_2$ |
| D > 2 nm | | 1 | 0.5 | | | T > 25 C |
| D < 2 nm | | > 1 | > 0.5 | | | P > 100 torr |
| Os | 0.5 | - | - | 190.2 | 0.159 | |
| Ni | 0.5 | - | - | 58.71 | 0.154 | Carbonyl formation |
| Co | 0.5 | - | - | 55.993 | 0.151 | |
| Fe | 0.5 | 0.5 | - | 55.847 | 0.163 | |
| Ag | - | - | 0.4 | 107.868 | 0.115 | At 420 K |
| Au | - | - | 0.25<br>0.5 | 196.9665 | 0.115 | At 470 K<br>At 570 K |

### 3.3. Metal dispersion calculation

The metal dispersion is another important parameter related to the nature or the status of a catalyst. The metal dispersion is a non-dimensional number representing the fraction of free metal atoms over the total number of metal atoms in the catalyst. The calculation of metal dispersion does not require the knowledge of the metal surface density.

$$MD = (Vm\ Aw\ 10^4) / (W\%\ Sf) \quad (2)$$

MD = metal dispersion percentage
Vm = monolayer volume (moles of gas per gram of sample)
Aw = metal atomic weight (gram of metal per mole)
W% = metal percentage in the sample
Sf = stoichiometric factor (molecule of gas per metal atom)

### 3.4. Average size of catalytic aggregate calculation

A further information that can be calculated by chemisorption techniques is the average diameter of the metal aggregates.

$$Da = (10^4\ F) / (MSS\ Dm) \quad (3)$$

Da = average diameter of metal aggregate (Angstrom)
MSS = metal surface area (square meters per gram of metal)
Dm = metal density (gram per metal volume unit)
F = shape factor

The shape factor depends on the geometric shape of the metal aggregate. It is a delicate parameter because requires that we consider the metal aggregate as having a regular shape. The shape factor is defined according to the supposed geometry and it is equal to the ratio between the surface of a solid and its volume. For a generic solid with a non-regular geometry:

$$S / V = F / d \quad (4)$$

where S is the external surface, V is the solid volume, F is the shape factor and d is the average diameter. The equation for the average diameter is calculated using for S the metal specific surface area measured by the chemisorption technique and for V the metal density. The shape factor for regular solids is:

- for a cube with all the sides exposed or for a sphere F = 6
- for a cube with one side non-exposed F = 5 (i.e. a metal aggregate deposited on the catalyst support).

## 4. GAS SOLID INTERACTION: MECHANISM OF CHEMISORPTION

All heterogeneously catalysed processes must be preceded by gas adsorption on the surface of the solid catalyst before the reaction. There are two fundamental kinds of mechanistic situations that can arise in the surface-catalysed transformation (see Figure 3) [6].

A) Langmuir-Hinshelwood

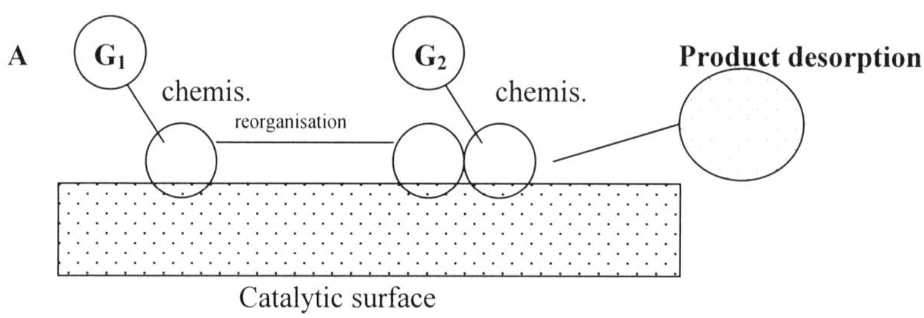


B) Eley-Rideal

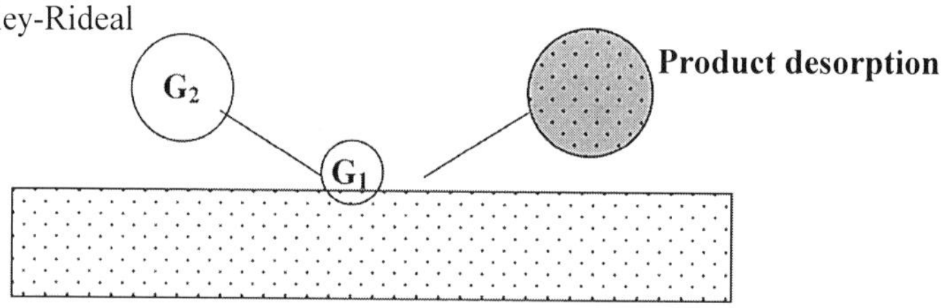


Figure 3. Different mechanisms of chemisorption.

For a generic reaction as:

$$G_1 + G_2 \longrightarrow P$$

it is possible to have different mechanisms: according to Langmuir-Hinshelwood type mechanism, both species ($G_1$ and $G_2$) are attached to the surface, and an atomic reorganisation takes place on the surface converting the gaseous reactants in the product (P); according to the Eley-Rideal mechanism, only one of the reactants ($G_1$) is bound to the surface of the catalyst and it is converted to a product only when the second reactant ($G_2$) has an interaction with $G_1$.

In effect, the distinction between the two methods is not always easy. Although this is only a simplification of the catalytic processes, it confirms that it is very important to know the definition and the role of the surface gas adsorption. The amount of gas uptake by a solid at a fixed pressure and temperature, is proportional to its surface area and not to its total volume as the chemisorption is not a bulk phenomenon but a surface phenomenon. It is also helpful to distinguish physical adsorption from chemical adsorption. The fundamental difference between the two kinds of adsorption is that the physical adsorption involves a molecular interaction force including permanent dipole, induced dipole and quadrupole attraction (forces of Van der Waals). In chemical adsorption (chemisorption), the forces involved are very strong, with consequential formation and/or rupture of chemical bonds. Physical adsorption is characterised by small enthalpy changes, typically in the range -10 to -40 $kJmol^{-1}$ with heat of adsorption of 10-40 $kJmol^{-1}$, whereas chemisorption heat is rarely less than 80 $kJmol^{-1}$ and often exceeds 400 $kJmol^{-1}$. Another difference between these two extreme types of adsorption concerns the temperatures at which adsorption takes place and its specificity. The physical adsorption occurs, in general, only at low temperature, whereas chemisorption has no restrictions. Finally: chemisorption is a chemical reaction upon the surface of the solid and it is therefore specific. This mechanism is extensively studied because it is strictly related to the catalytic process.

### 4.1. Energy of adsorption

In general, the interaction of a molecule of reactant with a surface involves a sequence of three reactions: (A) adsorption, (B) surface reaction and (C) desorption:

A $R_g \longrightarrow R_{ad}$

B $R_{ad} \longrightarrow I_{ad} \longrightarrow P_{ad}$

C $P_{ad} \longrightarrow P_g$

Every reaction described above may consist in various reaction steps: adsorption (A) may involve the formation of a weakly adsorbed initial state. Subsequently surface diffusion or reorganisation takes place leading to an adsorption state with a chemical bond. Desorption can occur in the reverse sequence. In a catalysed reaction the study of the reaction kinetic permits to identify the rate-limiting steps: in this case the other reaction steps are assumed to be in equilibrium. It is important to estimate the binding strength of the species that are adsorbed upon the surface [7] . As reported in Figure 1, it is possible to have quite different values of the activation energy barriers (Eads, Ecat, Edes) or of the energy minima (adsorbed Reactants and Products). These energies depend on the catalyst and its exposed crystallographic phases.

The degree of efficiency gained in the following catalytic path is controlled by the energetic steps of intermediates, by the activation energy required to convert the reactants into a surface intermediate, by the activation energy needed to yield adsorbed products and finally by the activation energy for products desorption (see Figure 1).

### 4.2. Associative and dissociative chemisorption

The measurement of the metallic surface area in a multi-component system as a bimetallic supported catalyst or an alloy is feasible by selective chemisorption on the metallic phase. The chemisorption stoichiometry is defined with reference to the adsorbate related to the metallic element [8]. Therefore, the chemisorption process is very different if the adsorbed gas molecule is dissociated or not. The two kinds of chemisorption involve different energetic behaviours and different theoretical models define them: associative and dissociative adsorption. In the first case, the gas is adsorbed without fragmentation; in the second case, the gas molecule is adsorbed after its decomposition in one or more fragments. Hydrogen, for example, is always adsorbed in its dissociated form.

$H_2 \longrightarrow 2H$, whereas $N_2$ is adsorbed in the two forms (see Table 3).

The gas-metal interaction (see par. 4) involves 3 or 4 steps:
1) Adsorption of molecular gas system
2) Dissociation of gas molecule (if the process is dissociative)
3) Migration of atoms on the surface and into the bulk
4) Formation of gas-metal bond.

In general, in case of associative mechanism, the chemisorption process follows the first order; in the other case it is of the second order. As already reported (Table 3, par. 3.1), the adsorption mechanism depends upon the energetic of the process and it is different for each pair of metal-gas. The knowledge of the chemisorption mechanism permits to calculate the activation energies. Thermal programmed desorption is commonly used to estimate activation energy:

$$Q_{ad} = -\Delta H = E_d - E_a \tag{5}$$

where $Q_{ad}$ is the heat of adsorption; $E_d$ and $E_a$ are the energy of the desorption and adsorption process respectively. In the case of dissociative chemisorption, the activation energy $E_a$ for the adsorption on metal is zero. Therefore, the heat of adsorption is equal to the activation energy for the desorption process. The binding energies, BE, for the bond M-H are obtained from the equation:

$$E_{M\text{-}H} = ½\,(E_{diss,H2} - E_a) \tag{6}$$

where $E_{diss,H2}$ is the dissociation energy of $H_2$.

If the catalyst system is not able to provide enough energy for the gas dissociation, the adsorption process can be activated by a temperature increase. Cobalt on $Al_2O_3$ is a typical system able to adsorb dissociated $H_2$ only at high temperature (100°C). The presence of a second metal (i.e. Ru, Pt) can activate the adsorbate dissociation even at lower temperatures [8]. Atomic hydrogen can be therefore chemisorbed on cobalt atoms.

### 4.3. Acid-base interaction

The investigation of the surface acidity is helpful to explain the behaviour of some kind of catalyst. The acid surfaces are known to be able to generate, for example, carbon ions that are useful intermediates in catalytic reforming, cracking and isomerisation. In general, the evaluation of the acidity, or more specifically of the surface acidity, needs the determination of the nature, the strength and the number of acid sites. There are many techniques able to determine some aspects of acidity, to characterise the acid properties of solid surfaces used as catalysts in industrial processes or to find some potential solid candidates for new developments. The characterisation of acid sites can be performed by the chemisorption of a basic gas which, adsorbed by an acid site, is transformed into its conjugated acid form. The Brönsted definition is used to describe an acid or a base: an acid is an electron-pair acceptor and is able to transfer a proton from the solid to the adsorbed molecule. Whereas, according to Lewis definition, the acid site is an electron pair acceptor from the adsorbed molecule and implies a co-ordinated bond with the surface [9]. The surface acid nature depends on the co-ordination of a metal cation with oxygen in the support structure. For example, alumina and alumina-silicates show tetrahedral co-ordinations between the Al-Si cations and oxygen, and negative charges are created. For cations with charges lower than 4 a Brönsted acid site is generated. To create a Lewis type acid site it is necessary to have the formation of co-ordinated unsaturated sites. Generally, the acid properties of a give solid are due to the presence of a wide distribution of acid site type. Sometimes both Brönsted and Lewis sites are simultaneously present on the surface. The investigation on the acid surface properties is therefore of great importance to foresee the catalytic activity and selectivity. In any case it is very important, in the catalyst characterisation, to carry out each determination in analytical conditions as close as possible to the operative conditions of the catalytic reaction. Common methods for surface acidity measurement include: aqueous methods (titration or ion exchange); indicators methods without solvent, with spectroscopic instruments; calorimetric methods by determination of the heat of adsorption, temperature programmed methods (in particular the temperature programmed desorption, TPD), where a sample previously saturated with a reactive gas, as ammonia or an acid molecule in a static or dynamic system, is submitted to a linear ramp of temperature in a flow of inert gas. Finally, other spectroscopic methods are: ultraviolet spectroscopy (UV), nuclear magnetic resonance (NMR), electron spin resonance (ESR) and x-ray photoelectron spectroscopy (XPS).

## 5. CHEMISORPTION TECHNIQUES

As mentioned above, the methods to perform chemisorption analyses are mainly two: the static and the dynamic adsorption. In the case of static adsorption, the sample is generally pre-treated in high vacuum, and, after the catalyst activation, known doses of reactive gas are injected in the sample holder. Modern analytical instruments detect the amount of gas adsorbed for each introduction by determining accurately the time necessary to establish the correct equilibrium between the gaseous phase and the adsorbed one. Once equilibrium has been reached, the equilibrium pressure is measured systematically, until the complete chemisorption isotherm is collected and a monolayer of gas molecule is covering the active surface. The dynamic method consists in flowing continuously (constant flow) an inert gas through the sample, and by injecting known doses of reactive gas by a loop system into the gas stream. When the reactive gas pulses reach the catalytic surface, the gas molecules react with the sample being adsorbed. In the case of dynamic adsorption, it is not possible to state the correct equilibrium time for the chemical reaction. Therefore, the only parameter that can be adapted to allow sufficient contact time between the gas and the solid is the flow rate. Using the dynamic method, the lower the flow rate the longer the contact time between the gas pulse and the sample. Operative conditions during the chemisorption analysis are of extreme importance. Temperature of analysis, equilibrium time and pressure for static techniques, flow rates for dynamic techniques should be carefully chosen to avoid phenomena such as spillover or solutions between the adsorbate and the solid (i.e. absorption of hydrogen in palladium). The static technique allows to perform an analytical procedure called "back sorption". During a first analytical cycle the reactive gas can interact with the catalytic surface producing gas-solid interactions showing different bounding energies. This leads to different species of gas adsorbed: strong and weak chemisorption. While the gas that is strongly bounded to the surface can be removed only by increasing the temperature and applying vacuum or a flow of inert gas (thermal desorption), the gas molecules weakly bound can be removed at the same analytical temperature only applying vacuum or inert flow. Therefore the back sorption procedure consists in measuring a first gas adsorption isotherm (strong + weak adsorption) followed by a sample degassing at the same temperature, moving only the weakly adsorbed molecules. Finally, a second run is performed with the same reactive gas, measuring only the amount of gas adsorbed by the weak interactions. The two collected adsorption isotherms can be now subtracted one from the other providing a third curve related only to the strong interactions. In Figure 4 the isotherm labelled as "A" represents the first run of hydrogen chemisorption on 6.3% platinum supported on silica, "B" is the second adsorption analysis and "C" is the curve resulting from the subtraction of the two analyses. Separate calculation of the gas uptake permits the distinction of the hydrogen molecules adsorbed with different energies. The calculation of metal specific surface area and dispersion should be carried out on isotherm “C”.

The dynamic method cannot provide the above information as during the adsorption process the inert gas flow continuously moves the molecules weakly adsorbed. On the contrary, a complete static analysis takes many hours (typically 6 to 12 hours for the first run and additional 6 – 8 hours for the second run) and it is operatively more complicated while the dynamic method is extremely fast and easy to be performed.

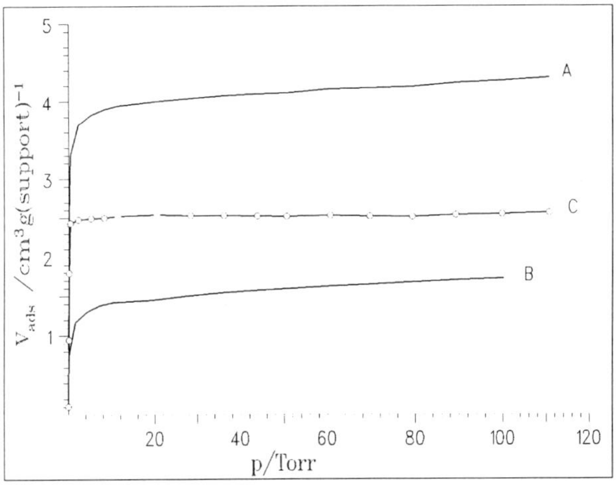


Figure 4. Chemisorption of hydrogen on 6.3% platinum on silica catalyst at 25°C performed by the Sorptomatic 1990 (CE Instruments).

### 5.1. Sample preparation and catalyst activation

Sample preparation is a fundamental step before any type of chemisorption measurement. This phase has the purpose of activating the catalyst to the chemical reaction that will take place. Therefore, the sample pre-treatment should be carefully studied to obtain later on the most accurate experimental data. A typical procedure for chemisorption measurements consists in the following steps:

1. Accurate cleaning of the catalyst surface before the pre-treatment procedure.

The surface cleaning consists in removing water vapour or other pollutants adsorbed on the catalyst. This step should be chosen according to the catalyst nature and catalyst life cycle. For instance, the cleaning conditions should be determined in function of the treatments performed on the catalyst. In case of a new catalyst the promoters of metal deposition should be removed to avoid any kind of interaction with the reactive gas used for the measurement. If the catalyst has been used, it is important to clean the surface, if possible, from pollutants or poisons eventually blocking the active sites. The cleaning phase can be performed under high vacuum degree or flowing an inert gas through the sample at the highest temperature possible without modifications of the catalyst structure (i.e. synterisation). The cleaning time has also a great importance because, even if the selected temperature for a short time is not affecting any sample modification, it can happen that if the catalyst is kept at this temperature for long periods the metal atoms can migrate, increasing the aggregate size.

2. Choice of proper pre-treatment procedure.

This is a very delicate step; the nature of the catalyst should be carefully investigated by a bibliographic search, if available, to chose the best preparation procedure. Sample pre-treatment usually consists in several steps that must be repeated carefully for successive analyses to assure the best reproducibility in the analytical results. For instance, it is sufficient a small difference in temperature rates, vacuum degree or final temperatures to obtain complete different results on the same catalyst. Pre-treatment is used to activate the catalyst surface to the chemical reaction with the measuring gas, therefore it is difficult to generalise a procedure for it but most of catalytic systems are well studied and a bibliographic search is always advisable. As the most common gas used for chemisorption analyses are hydrogen, oxygen, carbon monoxide or di-nitrous oxide, a generic pre-treatment procedure can be performed according to the following steps:

a. Sample oxidation, by flowing air or, better, pure oxygen over the sample. This step oxides all the metal atoms on the surface, eventually removing also carbon residual.
b. Sample reduction, by flowing pure hydrogen over the catalyst. Hydrogen will react with the metal oxide forming hydrides and water.
c. Hydrogen removal. This step can be carried out in high vacuum or in flow of inert gas obtaining a surface metal in zero valence state, ready to react with the measuring gas in the analytical cycle. Usually it is better to run first the oxidation and later the reduction because for most metal systems the hydride can be dissociated more easily during the final degassing procedure. Metal oxides are generally very stable. Of course, in the case of oxygen/hydrogen titration analyses, the final step will leave a layer of oxygen or hydrogen bound to the surface active sites, in order to perform the chemical reaction between the two gases during the analysis (one adsorbed and one in the gaseous phase).

Temperature is one of the main parameters to be carefully chosen. In fact, during the sample reduction by flowing hydrogen, water vapour is produced by the chemical reaction. The formed water is in part carried away by the gas stream but in part can be adsorbed by the porous structure of the support during step b. here above. When performing hydrogen removal (step c.) sample temperature should usually be a little lower then the previous step. In fact, during the sample reduction, the stream of hydrogen while reducing the metal also removes the formed water on the metal surface and from the porous support, but this happens at the reduction temperature. If in the next phase the temperature is slightly higher, the water adsorbed by the support begins to desorb slowly. When reaching the catalytic aggregates, which are now activated, water reacts again with the supported metal and a non-controlled oxidation takes place. The bad consequence is the non-reproducibility of the adsorbate uptake measurement. It is sufficient to keep the temperature of the last step few degree lower than the reduction step just to be sure that the overshooting or the temperature oscillation of the oven stay below the selected temperature for reduction.

### 5.2. Static adsorption

To perform static adsorption the analytical instrument must be equipped as follow:

- high vacuum pumping system, able to generate a vacuum degree over the sample of at least $10^{-4}$ torr.
- Stainless steel plumbing with high vacuum fittings to ensure a correct tightening. Leaks are the most common problems related to static adsorption. In fact, the experiment is usually carried out below atmospheric pressure and a leak of air coming inside the system will affect the results. Equilibrium pressure will be higher than the real ones and the catalyst may be oxidised by the oxygen contained in the air.
- Well calibrated injection system. The injection system could be a fixed calibrated volume, where only the adsorbate loading pressure can be varied, or a moving calibrated piston, where also the volume injection can be changed. The advantage of using this last system is that after the gas injection in the sample holder the piston will fill completely the calibrated volume. In this way the calibrated volume for injection will not increase the dead space over the sample. On the contrary, when using a fixed injection volume, during the equilibrium pressure measurement the calibrated loading volume is added to the system dead space, decreasing the precision in pressure measurement.
- Detection method. Usually two main methods are used to detect the amount of gas adsorbed during the analysis: the volumetric system and the gravimetric one. Anyway the equilibrium pressure is measured by a suitable pressure transducer chosen according to the pressure range that is established during the experiment. Typical pressure range for chemisorption measurement is up to 100 torr (most of catalytic systems show the monolayer formation below this pressure). In a volumetric apparatus the raw experimental data are the equilibrium pressures and the amount of gas adsorbed for each step. The gas uptake is calculated directly from the equilibrium pressures values but a proper dead volume calibration has to be performed before or after the measurement by a "blank run" (that is an analysis by using an inert gas not reacting with the sample). In case of gravimetric systems the dead volume

calibration is not necessary because a suitable microbalance directly detects the weight changes of the sample when adsorbing the measuring gas. Static volumetric instruments are very commonly used because they provide a reasonable compromise between degree of precision, ease of use and price, while the gravimetric systems are very precise but also very expensive due to the high cost of the microbalance.

- Suitable manifold dead spaces. The system dead volume in theory should be as small as possible but in reality it should be carefully evaluated, especially for static volumetric systems. The volumes of adsorbed gas are calculated by the pressure difference between the experiment (with a reactive gas) and the blank (dead volume calibration with an inert gas): the smaller the dead volume, the higher the difference in pressure and more precise is the adsorbed volume calculation. On the contrary, by decreasing the system dead space the gas dose to be injected should be decreased accordingly to avoid the risk of injecting too much gas that might overtake the necessary amount to form a monolayer or, in the best case, to produce an isotherm with few experimental data points. In fact, when the injection volume is too small it is very difficult to calibrate it with the required precision.
- Manifold and sample temperature. The manifold, injection system and pressure transducers temperature should be fixed at a value slightly higher than room temperature (i.e. 35 – 40 °C) for better thermal conditioning (it is very difficult to keep temperature constant around room temperature). On the other hand the temperature should not be too high because this might badly affect the transducers linearity. Sample temperature should be as stable as possible (maximum oscillation +/- 0.1 °C) and should be chosen according to the type of experiment that is required. Typical chemisorption isotherms are performed around room temperature (20 – 25 °C) but for some systems might be necessary higher or lower values (i.e. analysing palladium with hydrogen, higher temperature shifts to higher pressure the gas dissolution in the metal or measuring ruthenium with hydrogen higher temperature shortens the equilibrium time).

### 5.2.1. Experimental aspects

A typical scheme of a volumetric apparatus for chemisorption is reported in Figure 5.

A minimum of two gas inlets is required, one for the inert gas used in the calibration of the dead volume and one for the reactive gas. The reference volume for gas injection might be a fixed volume type or a moving piston. In the second case it is possible to change the injection volume in addition to the loading pressure according to the sample adsorption rate and the required number of experimental points. One or two separate transducers might perform the measure of the loading pressure and the equilibrium one. In case that only one transducer is used, the pressure transducer volume must be added as dead volume in the reference volume.

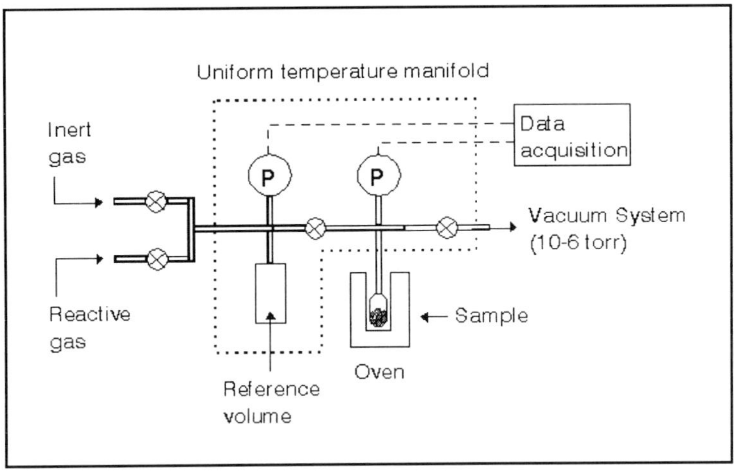


Figure 5. Typical diagram of a static volumetric apparatus for gas adsorption.

## 5.3. Dynamic adsorption

In chemisorption measurements performed by static volumetric systems, the sample is submitted during the analysis in a vacuum system. In effect, this method has the advantage of separating the contribution of physical and chemical adsorbed gas (weak and strong bounding). On the other hand, the analysis conditions are very different from the working conditions of the catalyst during the reaction in the reactor (high temperature and pressure, very long analysis time). When the catalyst characterisation have to be performed in analytical conditions which are very similar to the catalytic process, the flow chemisorption method (dynamic adsorption) might be very useful. In flow technique the sample is placed in a tubular reactor and it is analysed in operative conditions as close as possible to real work conditions, as already cited. There are several advantages using the dynamic method:

1) analysis is very fast,
2) during the analysis there is no physical adsorption and therefore the second adsorption run is not necessary, obtaining immediately the data related the strongly chemisorbed gas,
3) this technique allows an easier characterisation of samples with a small dispersion or a very low percentage of active phase,
4) manual operations are very easy.

The sample preparation is similar to the ones already described for static adsorption (see par. 5.1) the catalyst must be cleaned and pre-treated opportunely in order to perform a correct quantification of the active sites.

### 5.3.1. Experimental aspects

The chemisorption analysis by flow system is named Pulse Chemisorption. The typical apparatus to perform pulse chemisorption analyses is very similar to the one used for TPD/TPR analyses. In Figure 6 is represented the schematic diagram of a typical apparatus performing most of the dynamic analytical techniques. The gas to be adsorbed is introduced as a pulse, by an opportune system such as a multi-port loop valve, into the stream of the inert gas used as carrier. The injection system is placed before the sample reactor. The choice of the probe gas and of the carrier gas must assure the best answer of the detector if a TCD is used. In this case the two gases should have different thermal conductivity to have the best sensitivity of the filament. It is important also to use the carrier and reactive gases with the highest purity; both gases should flow through a water trap or a suitable filter to completely retain eventual vapours. The detector calibration should be performed before starting the pulse chemisorption analysis by using the same analytical conditions such as temperature, types of gas and flow rate, in order to estimate the answer of the detector when the calibration pulse is introduced. The peak obtained in the chromatogram without adsorption must be then integrated and therefore the relation between the peak area and quantity of gas injected can be easily calculated.

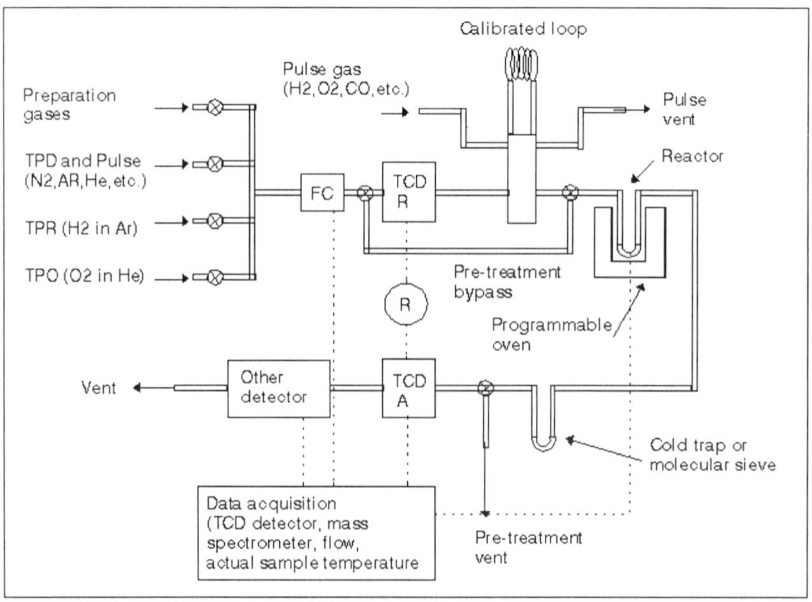


Figure 6. Diagram of a multipurpose apparatus for pulse chemisorption, TPD, TPR and TPO.

After the quantitative calibration, it is possible to start the analysis by introducing a number N of pulses of gas that will be adsorbed by the activated catalyst. The amount of gas adsorbed for each pulse is given by the difference between the area of the peak given for the reference pulse and the area under the peak acquired during the adsorption process (see Figure 7). In general, the injection loop volume should be chosen to provide an analysis with several experimental points to define better the saturation curve. The total amount of gas adsorbed Vt during the analysis is given by:

$$V_t = \sum_{i=1}^{N} V_{ai} \tag{7}$$

where N is the number of pulses necessary to saturate the active sites and $V_{ai}$ is the volume of gas adsorbed from the pulse i given by the difference: Vref - $V_{pi}$ where $V_{pi}$ is the volume of gas non adsorbed from the pulse i.
Once the reactive gas saturates the sample, the peaks areas acquired by the TCD (b, Figure 7) are equal to the reference calibration peak (a, Figure 7) and the analysis is over. In general, as for static adsorption analyses, the dynamic chemisorption is performed at constant temperature conditions (isotherms). The choice of the temperature depends on many factors such as type of sample and gas, type of information required, but it is very common to perform chemisorption at room temperature. Also the pre-treatments of the sample are effected in order to obtain a clean surface able to chemisorb the probe gas as described in 5.1.

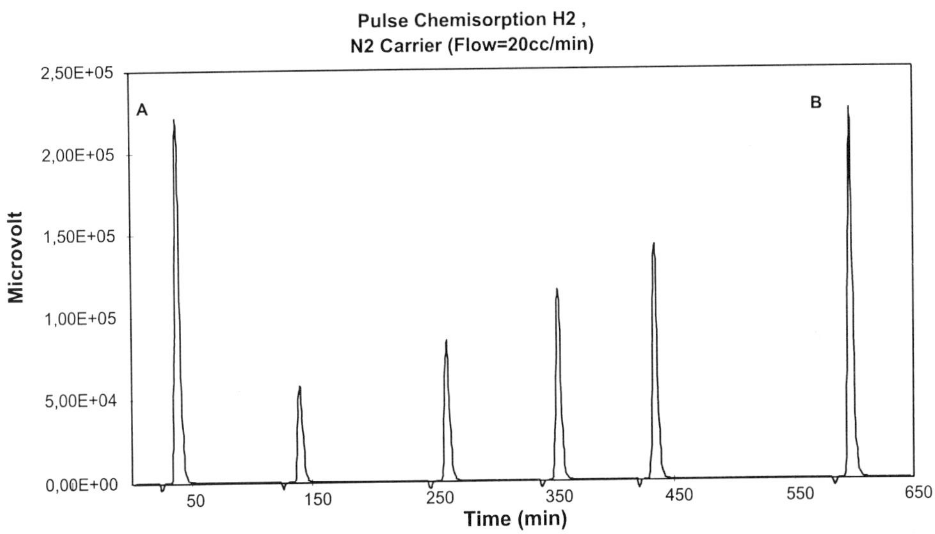


Figure 7. Pulse chemisorption analysis performed by the TPDRO 1100 (CE Instruments).

## 5.4. Calculation of monolayer volume of chemisorbed gas

The experimental data collected by a static chemisorption apparatus are the equilibrium pressures and the adsorbed volumes. Using these two sets of values it is possible to draw the chemisorption isotherm (adsorbed amount versus pressure). The mathematical interpretation of these curves leads to the calculation of the gas uptake covering the monolayer. In case of static adsorption, the isotherm resulting from the subtraction between the first and second run (only strongly chemisorbed gas) must be considered, since, the first run is also comprehensive of weak and reversible effects that depend mainly from the system pressure. In the following paragraph two most commonly used models will be described, the Langmuir model and the extrapolation to zero pressure, taking into consideration that other models can be applied such as the Freundlich and Temkin ones [11]. The main difference among these models is related to the heat of adsorption. Whereas according to Langmuir the heat of adsorption is constant in relation to the covering degree θ, according to the others the heat of adsorption changes in function of the surface saturation degree. In the following table are reported the three models equations.

Table 6
Isotherms of adsorption

| Model | Heat of adsorption (q) | Mathematical equation |
|---|---|---|
| Associative Langmuir | $q = q_0 = const.$ | $V/V_m = bP/(1+bP)$ |
| Dissociative Langmuir | $q = q_0 = const.$ | $V/V_m = (bP)^{1/2}/((1+(bP)^{1/2})$ |
| Freundlich | $q = -q_0 \ln(\theta)$ | $V/V_m = (bP)^{RT/Q0}$ |
| Temkin | $q = q_0 (1 - \alpha \theta)$ | $V/V_m = RT\ln(A_0P) / q_0\theta$ |

Where $q_0$ is the heat of adsorption for θ close to zero.

In case of pulse chemisorption technique, the best way to evaluate the monolayer volume is to take into account the total volume of gas adsorbed during the experiment. In fact, in this case it is not necessary any volume correction as the carrier flow during the run removes continuously the physisorbed or weak chemisorbed probe gas.

### 5.4.1. Langmuir isotherm

The Langmuir model [12] is commonly used to process chemisorption isotherms because it is relatively simple and it can be used in developing several kinetic equations. It is based on two fundamental assumptions:

- The adsorption energy is the same for every active site and there are no interactions between adsorbed molecules,
- Adsorption is localised.

From these assumptions derives that every active site has the same probability to interact with an adsorbing gas molecule, independently from the fact that it is surrounded or not by other occupied sites. Furthermore the maximum uptake that can be adsorbed corresponds to a monolayer of molecules (or, better, of gas atoms). The Langmuir model can be conveniently used to describe the associative or dissociative adsorption.

*Associative adsorption*

By indicating as A the adsorbate gas molecule and S the free active site, the associative adsorption follows the following mechanism: S + A = SA. By indicating as $\theta$ the covering degree:

$$\theta = (\text{occupied sites} / \text{total available sites})$$

therefore the speed of adsorption during the process is proportional to (1- $\theta$) (free sites) and to the gas pressure P.

The speed of adsorption, Uads = [adsorbed moles / time], is calculated as follow:

$$Uads = Kd\ Pa\ (1- \theta) \quad (8)$$

While the speed of desorption is:

$$Udes = Ki\ \theta \quad (9)$$

All the experimental points in a chemisorption isotherm are at the equilibrium, therefore for every point the two speeds are equal, Uads = Udes, that is:

$$Kd\ Pa\ (1 - \theta) = Ki\ \theta \quad (10)$$

$$\theta = (Kd\ Pa) / (Ki + Kd\ Pa) = K\ Pa / (1 + K\ Pa) \quad (11)$$

where K = Kd / Ki is the adsorption equilibrium constant that depends on the absolute temperature according to the Van't Hoff equation:

$$K = a\ e^{-dH / R\ T} \quad (12)$$

where - dH is the heat of adsorption. According to the Langmuir model assumptions, the heat of adsorption is independent from the covering degree, but this is true only if the active surface is energetically homogeneous and if there are no interactions between adsorbed molecules. On the other hand, by considering this assumption as true the mistake appears to be minimal.

*Dissociative adsorption*

When the adsorbent molecule covers two or more active sites, it is possible to derive the dissociative Langmuir model in the following way:

$$\theta = (K\ Pa)^{1/n} / (1 + (K\ Pa)^{1/n}) \quad (13)$$

where "n" corresponds to the stoichiometry of the reaction Sn + nA = n(S A). Defining the covering degree as

$$\theta = (Vads / Vm) \quad (14)$$

where Vads = adsorbed gas volume for each experimental point and Vm = monolayer uptake, we obtain:

$$(Vads / Vm) = (K\ P)^{1/n} / (1 + (K\ P)^{1/n}) \quad (15)$$

and after several passages:

$$P^{1/n} / Vads = (1 / (Vm\ K)^{1/n})\ (P^{1/n} / Vm) \quad (16)$$

This is the equation of a straight line whose slope is the inverse of the monolayer uptake and from the intercept, it is possible to calculate the parameter K, better known as "B" parameter of the Langmuir model. The "n" parameter should be optimised according to the correlation coefficient calculated by the least squared method. In case of associative adsorption n = 1.

### 5.4.2. Extrapolation to zero pressure

In chemisorption isotherms the value of adsorbed volume of gas is not constant and, even after the monolayer has been formed, the volume of gas continuously increases due to weak adsorption effects that are mainly pressure dependent. Usually to obtain a real horizontal plateau after the monolayer covering it is necessary to perform the back sorption procedure and the consequent isotherm subtraction. Some models (i.e. Freundlich) can be applied also to the first run only but an empirical method is available giving consistency in the monolayer volume calculation in comparison to the Langmuir model when applied to the subtracted isotherm. The method of back extrapolation [13] consists in identifying a linear part in the isotherm, even if the slope is still higher than zero, and, by applying a linearisation method, it is possible to extrapolate from these data a straight line. The intercept value of the resulting line is the monolayer uptake as the theory behind this method supposes that by extrapolating the linear part of the isotherm to a zero pressure, the reversible effects of adsorption can be excluded (see Figure 8). An interesting comparison is reported in Table 7, where the system palladium supported on carbon has been characterised by oxygen chemisorption.

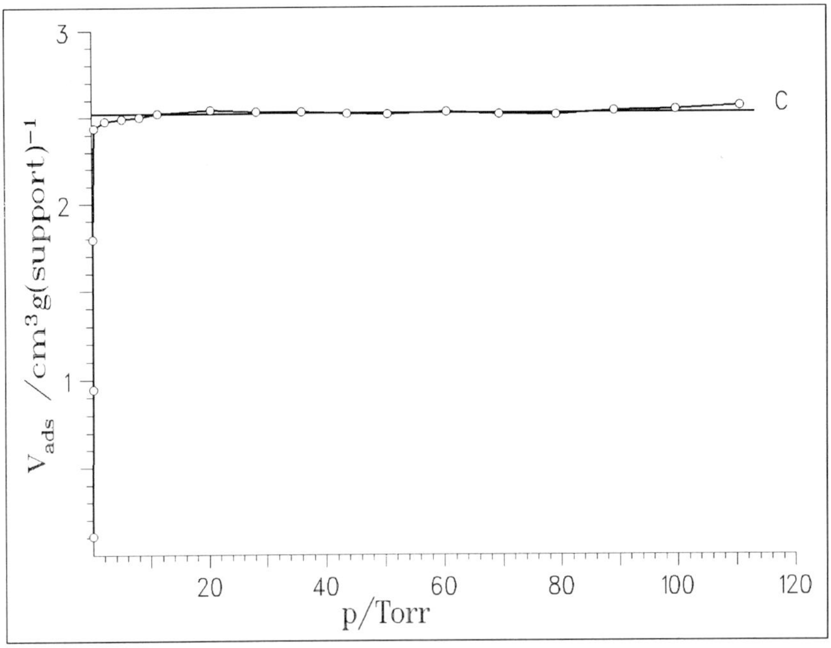


Figure 8. Extrapolation to zero pressure to a corrected isotherm (Sorptomatic 1990).

Table 7
Results by different calculation methods of Pd 5% supported on carbon (Sorptomatic 1990)

| Palladium 5% on activated carbon | | Vm (Ncc/g support) | Surface area ($m^2$/g metal) | Metal dispersion (%) |
|---|---|---|---|---|
| First run | Back extrap. (1) | 2.089 | 249.52 | 56.51 |
| isotherm | Langmuir (2) | 2.877 | 343.76 | 77.85 |
| Isotherm after | Back extrap. (3) | 1.855 | 221.67 | 50.20 |
| subtraction | Langmuir (4) | 2.228 | 266.17 | 60.28 |

The results calculated by the Langmuir model on the first run isotherm (2) are not reliable as the first run is taking into account also reversible adsorption. The Langmuir model is not applicable here. However, a comparison between the other results leads to very similar values. In fact, the results at point (1) and (3) are quite close each other as the back extrapolation theory supposes that at zero pressure the weak effects can have no influence. Therefore the results obtained by the application of this model to both isotherms (first run and subtracted one) should give similar values. The values obtained by applying the Langmuir model to the corrected isotherm are consistent with the previous results.

# 6. THERMAL DESORPTION TECHNIQUE

The mechanism of catalytic processes is directly correlated to the surface behaviour of the solid catalyst. The characterisation of catalytic system requires the detailed description of the preparation method, the specification of the catalyst physical properties and detailed information about its structural and chemical characteristics. A complete description of a catalyst should be comprehensive of information about the rate of chemical change, about the product composition and about the catalyst modification in time, due to poisoning, inhibition, sintering and chemical reactions. The thermal analysis methods are very helpful to know all of these catalytic properties. The thermal analysis methods may be used with different objectives: the reaction or desorption profile may be used qualitatively to fingerprint a given system. Then it is possible to make a quantitative considerations about the nature of chemical processes, the amount of gas involved in the chemical reaction and finally to calculate the number of the active sites, the reducibility degree of the sample related to the catalytic activity. All the methods relating some characteristic properties of a catalyst to its temperature during a programmable heating ramp are named commonly in the field of the thermal analysis.

As the thermal change at a given temperature depends on many factors such as the nature of the system, the type of gas used, the flow rate, pressure and several kinetics factors, therefore all the experimental aspects should be taken in consideration. The thermal analyses are often used to investigate surface modifications and bulk reactivity by varying the surface composition, the catalyst preparation method, the pre-treatment for catalyst activation and the analytical conditions. The fundamental differences between temperature programmed desorption and temperature programmed reaction (reduction, TPR, and oxidation, TPO), is that the first analysis involves a surface process, while the TPR/O involve a bulk reaction.

## 6.1. Theory of thermal desorption

During the Temperature Programmed Desorption analysis (TPD), the sample, adequately pre-treated, is submitted at an increasing temperature with constant rate and is swept by an inert gas such as helium, argon or nitrogen. The sample surface desorbs the gas that has been previously chemisorbed and a suitable detector monitors the process. Most commonly used detectors are thermal conductivity (as described in 6.2) or a mass spectrometer. In the TPD studies the solid system is previously equilibrated until saturation with a probe gas in isothermal conditions and at a given partial pressure (in general atmospheric pressure if the dynamic method is used). The Langmuir adsorption model may be used as well for the TPD spectra interpretation as it describes both gas adsorption and desorption in the two cases of associative and dissociative adsorption (see 4.2) [14]. The enthalpy of adsorption is considered as independent from the fraction of occupied active sites and the number of the available sites is fixed. Furthermore, both quantities are supposed to be independent from the temperature. The two types of possible mechanism (associative and dissociative), as already mentioned are the following:

1) $S + A \leftrightarrows S\text{-}A$ non dissociative

2) $2S + A_2 \leftrightarrows 2S\text{-}A$ dissociative

According to the above hypotheses, in case of non dissociative adsorption it is possible to derive the number of occupied sites $N_o$ at a given time t by:

$$dN_o / dt = pk_a (N_{tot} - N_o) - k_d N \qquad (17)$$

where:
$k_a$ and $k_d$ are respectively the kinetic constant of adsorption and desorption process,
p is the partial pressure of reactive gas and
$N_{tot}$ is the total number of active sites.

In general, the constant $k_a$ and $k_d$ are depending on the Arrhenius equation:

$$k_i = A_i \exp(-E_i/RT) \qquad i = a, d \qquad (18)$$

where $E_i$ is the activation energy of the process. Moreover, as already described in equation (5), $\Delta H_a = E_a - E_d$. In case of dissociative adsorption, the process is of the second order and the adsorbate molecule dissociates in two or more parts. Accordingly, in the equation (17), the terms $(N_{tot} - N_O)$ and N are at the second power $(N_{tot} - N_O)^2$, $N^2$. The TPD analysis profile, in both cases, the process rate is given by the difference between the rate of desorption $r_d$ and the rate of adsorption $r_a$. When $r_a \cong r_d$ the regime is in dynamic equilibrium, while when $r_d >> r_a$ the TPD profile depends by the heating rate $\beta$ (expressed in K/s). The correlation between the energy of desorption and the factor $\beta$ is given by:

$$\ln (T_m^2 / \beta) = E_d/RT_m + \ln (E_d / k_d R) \qquad (19)$$

as shown by Anderson et al. [14]. Equation (19) shows that the activation energy Ed for the desorption process is an experimental quantity, easy to be obtained from the temperature programmed desorption data. The activation energy for desorption can be estimated from the temperature of the maximum desorption rate, $T_m$, from the heating rate parameter $\beta$ and from kinetic constant of the desorption reaction $k_d$.

### 6.2. Experimental aspects

Suitable characterisation techniques permit to determine the characteristic of the catalyst such as the surface area, the metal dispersion, the type of the deactivation or the structural modifications during and after catalysed reactions. Therefore, catalyst characterisation is essential for evaluating and improving the preparation methods or the reaction parameters. The techniques available for this purpose are often not very helpful to characterise catalysts under working conditions. In general, the analytical methods based on a flow system, as thermal programmed desorption, reaction/oxidation, reaction and pulse chemisorption, are the best methods to characterise the adsorption and reaction energetic, the bulk or surface active phase and the site distribution of the supported catalyst. In fact, it is possible to approach the analytical conditions used in these methods to the real reaction conditions. The flow-based techniques (TPD, TPR/O and pulse) use essentially the same equipment. A typical flow system diagram is represented schematically in Figure 6. The TPD/R/O analyses are carried out by flowing a suitable reactive gas or gas mixture through the catalyst placed in a tubular sample holder (flow-through or flow-over types). In case of TPD analysis, the sample is previously saturated with the chosen adsorbate by flowing the reactive gas or executing a pulse chemisorption analysis. Gas mixture is typically used to perform TPR and TPO analyses, as a small percentage of hydrogen or oxygen (about 5%) diluted in argon or nitrogen for TPR and helium for TPO. In fact, when the detector is a thermal conductivity one, the thermal conductivities of the carrier gas and the detected gas must be different. If the detector is a quadrupole mass spectrometer, the desorbing gases should present typical and unique mass fragments, to be distinguished from the carrier mass fragments. In temperature programmed techniques the temperature increase must be linear, therefore the oven must be able to perform a wide range of temperature rates (typically from 1 to 20 K/min) in a wide temperature range (from ambient up to 1373 K). Moreover, the best furnace type is the anti-magnetic one, that is the heating coils should not generate any magnetic field on the sample holder to avoid signal oscillation when using carriers with a polar moment (i.e. nitrogen). The real sample temperature must be monitored continuously by a suitable sensor placed inside the catalyst bed to detect possible endo or exothermic reactions. The temperature sensor should be opportunely protected by an inert material sheath (i.e. quartz made) to avoid chemical reactions of the sensor itself (typically a thermocouple) with the reactive gases. It is important, from the experimental point of view, to use gases of the highest available purity and also to remove traces of water impurity by using a suitable cold trap or molecular sieve trap. The gas flow must be very stable to optimise the detection sensibility (an electronic mass flow controller is the best system for this purpose). The choice of the carrier gas depends on the reactive gas to be detected: in Table 8 are reported some typical gas coupling when using a thermal conductivity detector.

Table 8
Detection of some gases in relation to different carriers by TCD

| Gas | Main use | Thermal conductivity (*) | Detectable reactive gases |
|---|---|---|---|
| He | Carrier | 3363 | $O_2$, $CH_4$, CO, $CO_2$, $SO_2$, $H_2S$,$NH_3$, NO, $N_2O$ |
| Ar | Carrier | 406 | $H_2$ |
| $N_2$ | Carrier | 580 | $H_2$ |
| $H_2$ | React./Carrier | 4130 | CO, $CO_2$ |
| $O_2$ | React. | 583 | - |
| $CH_4$ | React. | 720 | - |
| CO | React. | 540 | - |
| $CO_2$ | React./Carrier | 343 | $H_2$ |
| $SO_2$ | React. | 195 | - |
| $H_2S$ | React. | 327 | - |
| $NH_3$ | React. | 514 | - |
| NO | React. | 555 | - |
| $N_2O$ | React. | 374 | - |

(*) Determined at 273 K, values $.10^7$ (cal / cm.s.K)

In a TCD detector, two sets of filaments are mounted in a Wheatstone bridge circuit, one set is immersed in the pure carrier gas stream (reference) while the other in the stream exiting from the reactor (measure). The filaments are made of suitable metals (i.e. tungsten or gold) having high temperature coefficient of electrical resistance:

$$R(T) = R_0 (1+aT)$$

where T is the filament temperature, $R_0$ is the metal electrical resistivity at 293 K and "a" is the resistance temperature coefficient $K^{-1}$. When the bridge circuit is power supplied, the filaments temperature changes according to the thermal conductivity of the gas, the flow rate and the environment temperature. It is necessary to have only one variable in the system, that is the change of the gas thermal conductivity between the reference branch and the measure one, therefore all other parameters must be absolutely constant: the flow rate, the environment temperature and the current supply. If the reference and measure filaments are immersed in the same gas type (no gases are adsorbed or desorbed by the sample) the bridge is in equilibrium (same resistance). When the sample, due to temperature increase, begins to adsorb or desorb other species, the bridge is unbalanced, and the detector generates a positive or negative current. Usually the sensitivity by using tungsten filaments is higher, while gold filaments are advisable when the reactive gas is corrosive. The thermal conductivity of the gas flowing over the TCD determines the filaments temperature and consequently also their resistance. It is possible to correlate the measured potential (V) to the change in the gas composition d[C] by V = s d[C], where “s” is the detector sensitivity factor. The thermal conductivity detector is extremely sensitive and it is able to reveal gas quantities in the order of 1 μl. The TCD sensitivity is more efficient when the flow is low (20-50 Ncc/min) and constant.

### 6.2.1. Sample preparation

All samples are pre-treated before the analysis by using various procedures in order to obtain a clean surface and to eliminate undesired contaminants. A typical scheme is first to heat the sample at high temperature in flow of inert gas to effect a complete elimination of physisorbed water and/or other pollutants. After the preliminary cleaning, the sample can be saturated with a suitable gas probe (i.e. hydrogen, oxygen, NO, CO, $NH_3$) at a given temperature. Once the saturation is over, the excess gas is removed by flowing an inert gas at the same temperature of the saturation. In this way the reactive gas weakly chemisorbed and present in the piping can be removed. It is also possible to cool the sample at a room temperature while keeping the flow of the same probe gas. The inert gas used as carrier for the TPD analysis can be introduced once the room temperature has been reached. Sometimes, some substances as carbonaceous residues of calcination or residues of precursors used to prepare the catalyst are present in the catalyst. These “pollutants” can react with the gas probe. In this case, the pre-treatment procedure can be more complex and comprehensive of more steps as pre-oxidation in oxygen, followed by cleaning with an inert gas, reduction with hydrogen and finally another purge of he system. At this point it is possible to saturate the catalyst whit the probe gas. The purpose of the above complex procedure is to assure that the analysis profile takes into consideration only the probe gas desorption that was adsorbed on the investigated active phase.

### 6.2.2. Analytical method

TPD analyses can be performed by the apparatus described in 5.3.1. As already anticipated, the sample submitted to a linear temperature rate, releases the adsorbed gas in the carrier stream. The thermal conductivity detector will measure the current generated by the bridge unbalance. The data acquired are reported in a TPD profile relating the amount of gas desorbed versus the sample temperature and time (the rate is linear). To assure a precise quantitative calculation of the desorbed gas it is necessary to remove from the stream possible vapours that are produced during the analysis. For this reason is rather common the use of cold trap or a molecular sieve trap placed before the detector. Sometimes a mass detector might be very useful to verify which types of gases are desorbing together with the probe gas (detection of surface reactions). The experimental conditions should be prepared in a standardised manner to obtain reproducible results. Same factors to be considered include the nature of the gas, the analysis pressure and the flow rate in order to avoid possible phenomena of re-adsorption, that should be avoided because otherwise the resulting peaks will be too large. Best condition for TPD analyses is a low temperature rate to separate the peaks related to different active sites. In the case of ammonia desorption from molecular sieves, the rate of 10 K/min assures that there is not significant re-adsorption, while the data obtained a 2 K/min show undesired peaks enlargement, giving evidence of free ammonia re-adsorption.

### 6.3. Calculation of total desorbed volume

The limitation of the thermal conductivity detector used in dynamic techniques is related to its inability to identify the species desorbed. That is the reason why a mass spectrometer is largely used after the TCD detector. In fact it is not possible (or, anyway, very difficult) to effect a correct quantitative calculation of desorbed gas only using quadrupole system, therefore the combination of the two detection systems is very appreciated. In the dynamic techniques, the amount of desorbed gas (generally expressed in μmol) is directly proportional to the peak area. Modern acquisition software can easily perform the integration of the resulting spectra if a proper system calibration has been previously carried out. It is also very useful to apply de-convolution models to TPD spectra to identify the contribution of different energetic sites that are dispersed on the catalyst surface. The calibration of an apparatus fitted with a TCD detector consists in a "blank", which is an analysis without sample using the same analytical conditions as flow rate and gas type. Known gas doses are injected by a syringe or by a calibrated loop in the carrier flow stream and the obtained peak is integrated. The resulting area is correlated to the injected gas dose by a linear relationship. Of course, the response of the TCD detector must be linear in a wide range of injected volumes; otherwise, a non-linear correlation must be performed collecting additional data points in the blank. In case of TPR and TPO analyses, another calibration procedure is commonly used. In this case, the reference peak is obtainable effecting a real analysis on cupric oxide (TPR) or metallic cupric (TPO) previously weighed. The redox stoichiometry of the reaction between hydrogen or oxygen with the above materials is well known: therefore it is possible to correlate directly the sample weight used in the calibration with the amount of reacted gas.

### 6.4. Energy distribution of active sites and isosteric heat of adsorption

The temperature at which species are desorbed from the catalyst surface reflects the strength of the surface bond. The higher is the temperature, the stronger is the bond. Temperature programmed desorption data permit to estimate the heat of adsorption of a given species or the formed surface by using the equation (5). When different species are adsorbed $N_a$ it is interesting to evaluate the energy distribution of the active sites on the surface of the sample. For a desorption process that occurs with a kinetic order x, the relation between the activation desorption energy $E_d$ and the number of adsorbed species $N_a$ is given by:

$$-dN_a/dt = kN_a^x \exp(-E_d/RT) \qquad (20)$$

During the TPD analysis the temperature is increased linearly by:

$$T_t = T_0 + \beta t \qquad (21)$$

where $\beta = dT/dt$ and $T_0$ is the initial temperature.
Thus:

$$(-dN_a/dt)\beta = kN_a^x \exp(-E_d/RT) \qquad (22)$$

Reporting 1/T versus $\ln(dxN_a/dt)\ \beta$ it is possible to estimate the strength of the binding energy put in evidence by the peak presence. The de-convolution and the integration of the peaks in a TPD spectra permit to evaluate the energy distribution of the active sites as each peak is produced by different types of desorbing sites. Note that, in the equation (22), the desorption energy is supposed to be independent from coverage degree. If $T_M$ is maximum temperature of a given TPD spectrum, we are able to set:

$$d/dt\left[N_a^x (k/\beta)\exp(-E_d/RT)\right]_{TM} = 0 \qquad (23)$$

$$\beta E_d = RT_M^2 k \exp(-E_d/RT_M) \qquad (24)$$

For a first order desorption (x=1) $T_M$ is independent from the initial coverage degree, while in the second order desorption (x=2), the maximum of the peak shifts to lower temperatures as the coverage increases. This demonstrates that TPD data are very helpful and valuable source of information on mechanistic features of catalysed reaction. Another important application of TPD analyses is related to the thermal desorption of ammonia to characterise the acid nature of some supports or catalysts as zeolites, alumina and molecular sieve [16]. The isosteric heat of adsorption can be easily calculated by performing various analyses at different saturation temperatures in order to obtain different degrees of coverage. The analytical temperatures should be chosen in a range that will not modify the surface structure. After several measurements the average adsorption enthalpies can be calculated with respect to the isosteric heat of adsorption Qst. Qst should be evaluated at the same surface covering degree $\theta$ for the different isotherms. The Clausius-Clapeyron equation permits to calculate the heats of adsorption from the isotherm data. This equation puts in relation the vapour pressure of a condensed compound with the temperature. Considering the gas as ideal and that the liquid molar volume is negligible with respect to the gas molar volume, the Clausius-Clapeyron equation can be written in the following way:

$$d \ln p/dT = \Delta H_{ev} / RT^2 \quad (25)$$

where $\Delta H_{ev}$ is the evaporation enthalpy. The relation (25) is applicable to the adsorption/desorption processes of gases and vapours on solid surfaces:

$$(\delta p / \delta T)\theta = Qst / RT^2 \quad (26)$$

where Qst is the isosteric heat of adsorption. For every covering degree, equation (26) must be evaluated. The equilibrium pressure, during the adsorption process, is function of θ according the resulting adsorption isotherm. To evaluate the relation between the pressure and the temperature the adsorption conditions must be related to a constant saturation degree θ. These conditions are named isosteric. The isosteric heat of adsorption coincides with the average adsorption enthalpy (ΔH), unless the negative sign according to the heat convention. ΔH is calculated for a close system at constant pressure and temperature, whereas the unique form of labour is the one of the volume. For a given saturation degree θ, by integrating the equation (26), it is possible to obtain:

$$\ln p = (Qst/R)(1/T) + \text{cost} \quad (27)$$

To apply the equation (27), it is necessary to collect various adsorption isotherms at different temperatures. For a given saturation degree and for each isotherm, we obtain a pair of values of pressure and temperature. The linear regression of lnp versus 1/T permits to calculate the values of Qst/R from the slope of the straight line that is obtained. By drawing more lines of this type for different saturation degrees, it is possible to study the dependence of Qst form θ. This relation is very helpful to investigate the catalyst surface homogeneity.

### 6.5. Analytical examples

In Figure 9 is reported the TPD profile obtained on a commercial supported catalyst (5% $Ru/Al_2O_3$, Engelhard) saturated at 100°C in a flow of pure hydrogen. After saturation at 100 °C, the sample was cooled down to room temperature in flow of $N_2$ to clean the reactor and the piping from the hydrogen in excess. Then a thermal ramp of 10 °C /min was started in flow of nitrogen. The two resulting peaks can be correlated to two different types of active sites. In Figure 10 is reported the desorption profile of mordenite saturated with ammonia. The carrier gas in this case is helium at 30cc/min, with a temperature rate of 10°C/min.

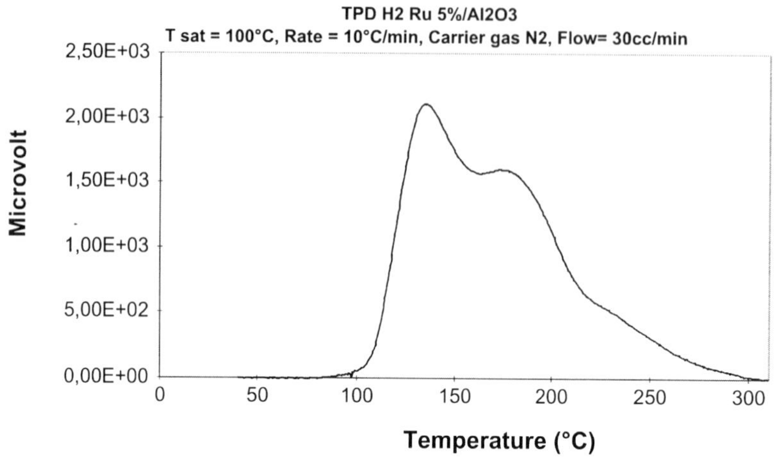


Figure 9. TPD profile of hydrogen adsorbed at 100 °C on 5% Ru on $Al_2O_3$ (TPDRO 1100).

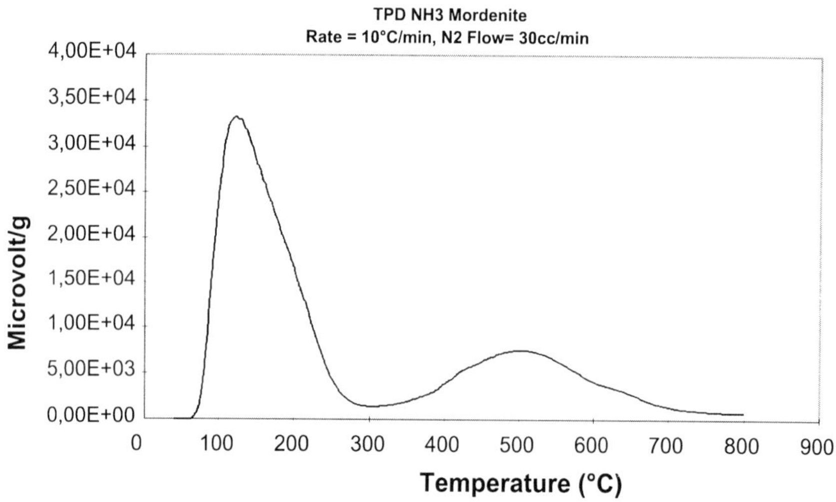


Figure 10. TPD profile of ammonia adsorbed on mordenite (TPDRO 1100).

## 7. TEMPERATURE PROGRAMMED REDUCTION AND OXIDATION

The objectives of this technique are essentially the following:

1. To find the most efficient reduction conditions
2. To identify the supported precursor phases and their interactions with the support

3. To characterise complex systems, as bimetallic or doped catalyst, to determine the role of the second component and to establish alloy formation or promotion effects.

There are several interesting studies about this technique: Robertson et al. [17] first reported TPR profile of nickel and nickel-copper catalysts and since then many catalysts have been investigated. In the TPR technique an oxidised catalyst precursor is submitted to a programmed temperature rise, while a reducing gas mixture is flowed over it (usually, hydrogen diluted in some inert gas as nitrogen or argon). In the TPO technique, the catalyst is in the reduced form and is submitted to a programmed temperature increase, but in this case, an oxidising mixture of gas (oxygen in helium) is flowed over the sample. The reduction or oxidation rates are continuously measured by monitoring the change in composition of the reactive mixture of after the reactor. The decrease in $H_2$ or $O_2$ concentration in the effluent gas with respect to the initial percentage monitors the reaction progress. An interesting application of this technique is that the TPR/O analysis may be used to obtain evidence for the interaction between the atoms of two metallic components, in the case of bimetallic system or alloy as already cited. In general, TPR/TPO studies are carried out under low partial pressure of the reactive gas. In this way it is possible to observe the intermediate reactions, depending from analytical conditions such as temperature rate, flow rate and concentration of reactive gas. The TPR/TPO methods are used for quantitative and quantitative analysis. In effect, the spectra produced are characteristic of a given solid. TPO is less commonly used than TPR, but the quantitative considerations for this type of analysis are more correct, in particular if the two analyses are performed in succession (hydrogen/oxygen titration). When used in combination, the two techniques can provide useful information in the study of the reactivity and redox behaviour of catalysts.

### 7.1. Reduction and oxidation reactions

The reaction between a metal oxide $M_xO_y$ and hydrogen, reducing the system to produce the pure metal M is represented by the equation:

$$M_xO_{y\,(solid)} + H_2 \longrightarrow M_{(solid)} + H_2O$$

From the thermodynamic point of view, the reduction of a solid oxide is feasible if the standard free energy change $\Delta G^0$ is negative. If $\Delta G^0$ is positive, the second term of the equation (28) must be sufficiently negative to make also negative $\Delta G$:

$$\Delta G = \Delta G^\circ + RT \log (P_{H2O} / P_{H2}) \qquad (28)$$

The reduction process is a bulk phenomenon and the degree of reduction ($\alpha$ as a function of time or temperature and hydrogen pressure) is interpreted in terms of mechanism by which the reduction occurs. Two different models can interpret the reduction processes: the nucleation model and the contracting sphere model. In the first case, according to nucleation mechanism, the reduction begins after some time and at a given temperature bringing to the formation of a solid product nucleus. During the nucleation, oxygen ions are removed from the lattice with progressive formation of solid metal and hydrogen and oxygen molecules diffuse at the interface oxide/metal/atmosphere. If the nucleation process is very fast, the real formation of separated and independent nuclei cannot be distinguished and the second mechanism takes place (contracting sphere model). The result during the reduction process, in this case, is a total coverage of the solid oxide particle with a thin layer of metallic product as an eggshell. In fact, the distinction between the two models is not only theory, but it has a consequence in the rate of reduction that is very different. In Figure 11 is reported a graphic comparison between the different dependence of the degree of the reduction from the time [18]. The A diagram is relative to metal oxide reduction by a nucleation mechanism, while the diagram B reports the case of contracting sphere model.

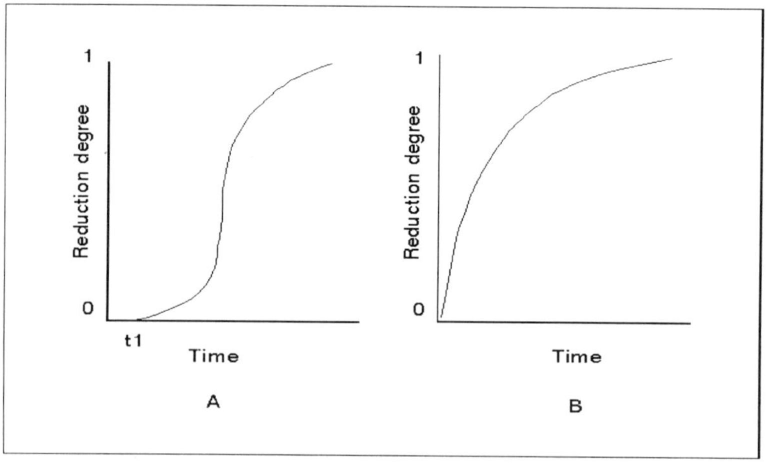


Figure 11. Dependence of reduction degree from time.

In the first case, it is possible to identify a maximum rate; this profile is typical of auto-catalysed reactions. In the second case, the rate of reaction decreases continuously until the reaction process is completed as there is a continuous decrease of the metal /oxide interface. It is common, in catalysis, to have a supported system that may exhibit a different reductive behaviour in comparison to unsupported metal oxides due to possible interactions between the metal and the support. The metal/support interactions may modify the reaction mechanism, promoting the atom diffusion on the surface of supported metal oxides or inhibiting the reduction process. This last is the case of cobalt supported on alumina, where cobalt aluminate, that is a system very difficult to be reduced, is formed. Similar possibility occurs in the case of bimetallic systems, where the second metallic compound (the doping species) may have a promoting effect by increasing the number of nucleation sites or providing a higher concentration of dissociated hydrogen that is transferred trough the support by the spillover effect. In the case of the TPO analysis, the reaction involved is an oxidation of a pre-reduced system:

$$xM + y/2\ O_2 \longrightarrow M_xO_y$$

In the above reaction, water is not produced and the oxidation degree can be interpreted according to the same model of TPR. TPO analyses are often performed in combination with TPR. In this way, it is possible to obtain additional information about the metallic compounds in the catalyst active phase and it is possible to separate the contribution of different metallic species in multi-metallic systems. The combination of the two reactions is a real titration of the hydrogen/oxygen consumption, permitting the calculation of the metal phase percentage in the catalyst (of course if the stoichiometric factor of the reaction is known). Another advantage of combining the two analyses is that the TPO permits to remove undesired contaminants then to concentrate the attention on the characterisation of the catalyst active phase.

### 7.2. Experimental aspects

The experimental apparatus for TPR/TPO analyses is usually the same as the one used for TPD measurement (see 5.3.1). The fundamental difference is the type of carrier gas flowing trough the sample (see par. 6.2) and the pre-treatment procedure. Moreover, it is important to underline again that the TPR/TPO analyses investigate the bulk system while TPD gives information about the surface behaviour of the catalyst.

#### 7.2.1. Sample preparation

The procedure to collect the TPR/TPO/TPD data is also comprehensive of the sample pre-treatment. Several types of procedure can be chosen in relation to the sample nature and type of information required. In fact, the diversification of the pre-treatment permits to obtain a wider range of parameters on a given catalyst. Generally, before starting a TPR analysis, the sample should be in its oxide form. The pre-treatment, in this case, consists in oxidising the catalyst in flow of pure oxygen or air, then flowing an inert gas to purge the product formed as water or carbon residues. Both pre-treatments must be effected at a given temperature to assure that the two processes are feasible. In case of TPO analysis, the sample must be previously reduced to obtain the active metal in zero valence form. The standard pre-treatment is a reducing procedure effected at a given temperature (isothermal or increased by a constant rate). The pre-treatment procedure permits also to remove undesired compounds as residual solvent traces or products resulting from the precursor decomposition. Alternatively, it is possible to remove only the physisorbed water to obtain information on the efficiency of the activation procedure or on the poisoning phenomena of exhaust catalysts. The calcination operation is effected at high or medium temperature in flow of air to decompose the precursor compound. The precursor presence in fact can negatively influence the reducibility of the catalyst. In the case of cobalt supported on alumina, for example, if the calcination temperature necessary to decompose the precursor (generally cobalt nitrate) is too high cobalt aluminate is formed. The consequence is a decrease of the metal active surface. By changing the pre-treatment methods before the TPR or TPO analyses, it is possible to investigate other catalyst behaviours related to the temperature. For example, modifications of analytical profiles due to temperature variations in the pre-treatment permit to estimate effects as synterisation or other metal/support interactions. In the example of cobalt/alumina catalyst, this type of studies permitted to state the best pre-treatment procedure to avoid the formation of cobalt aluminate. The best reducibility of this supported metal is achieved by pre-treating the catalyst at temperatures below 375°C and by performing the calcination process in flow of pure oxygen.

### 7.2.2. Analytical method

During the TPR/TPO analyses, several products as water, CO or $CO_2$ are formed. It is important to remove all undesired gas molecules that can interfere in the signal output. A correct pre-treatment and the use of suitable traps to stop secondary products are therefore necessary. The choice of the analytical parameters, in particular temperature and flow rates, is fundamental to obtain significant reaction profiles. The problem related to the difficulty in comparing different analyses has received little attention in literature because the conditions of sample preparation, pre-treatment and acquisition of experimental data are often omitted. Delanay G. [19], for example, reported the demonstration that the experimental conditions affect the temperature at which the reduction occurs. In any case, all the experimental parameters as hydrogen or oxygen concentration in the gas mixture, temperature increasing rate, total flow rate, sample weight and contact time can make influence the analytical profiles. These parameters have effect also on the detector sensitivity (i.e. the flow rate). Monti et al. [20] proposed a method to standardise the TPR/TPO data defining a number k, given by:

$$k = S_0 / (V^* C_0) \tag{29}$$

where $S_0$ is the hypothetical amount of initial reducible species in the sample expressed in μmol, $V^*C_0$ is the molar flow rate (μmol/s) of the reactive gas. This number should be in the range 55-150 s to have accurate and reliable results from the TPR/TPO analysis and above all to have comparable data. A typical example is the TPR analysis of cupric oxide: changing the temperature and the flow rates of the analysis, two reaction profiles will result: the resolution of the analysis is changed and it is possible either to distinguish the two phases of the reduction process identified by two peaks ($Cu^{II} \longrightarrow Cu^{I} \longrightarrow Cu^{0}$) or to obtain only one peak comprehensive of the total hydrogen consumption that is involved in the two processes. In the second case, the advantage is to calculate more easily the total quantity of reacted gas. In general, when the sample contains only one component it is useful to perform the analysis with a low temperature rate to observe the mechanism of the reaction process. In the case of multi-metallic catalyst higher temperature rate permits to separate the different contribution of the reactive components [21].

### 7.3. Quantitative calculation of reduced/oxided sites

When a reduction process is considered (similarly in the oxidation process), it is possible to express the rate of the reaction by the equation:

$$r = -d[M_xO_y]/dt = -d[H_2]/dt = k[M_xO_y]^p[H_2]^q \quad (30)$$

where k is a constant given by the Arrhenius equation $k = A\ e^{-E/RT}$ and $dT = \beta\ dt$, T is the temperature (K) and t is the time (min).
As temperature is increased linearly it is possible for both TPR and TPO, it is possible to correlate the concentration variation of the reactive gas by:

$$dH_2/dt = -\beta d[H_2]/dt \quad (31)$$

The possibility to correlate the parameters determining the reaction process ($H_2$ concentration, temperature rate and time) and the kinetic-thermodynamic parameters confirms that the TPR/TPO data are very useful characterisation techniques. Experimental TPR/TPO data offer important information about the change rate of some parameters in function of the temperature. The system can be described as a rector by correlating reduction/oxidation profiles to kinetic/thermodynamic parameters. The consumption rate of the reactive gas r, is correlated to the flow rate $\phi$, to the reactor element dx and the fraction of conversion df by the following expression:

$$r = \phi\ df/dx \quad (32)$$

### 7.4. Evaluation of average metal oxidation degree

Temperature programmed reaction permits to estimate exactly the amount of reactive gas consumed during the reaction. This quantity is correlated to the oxidised form of the sample, but it is necessary to follow several conditions:

1. A suitable pre-treatment of the sample must be carefully chosen to avoid secondary and undesired reactions.
2. The detection system must be correctly calibrated with standard samples or blank analysis to estimate exactly the amount of gas involved in the reaction.

3. Analytical parameters used during the measurement must guarantee that the reaction is thermodynamically feasible.

If all the above conditions are respected, the average metal oxidation degree can be measured if the metal percentage and the reaction stoichiometry are known. The degree of sample oxidation is given by the ratio:

$$\alpha = n_H / ( n_m S_F ) \tag{33}$$

where $n_H$ is the number of detected hydrogen atoms that are proportional to the peak area, $N_m$ is the total number of metal atoms contained in the sample, $S_f$ is the stoichiometric factor depending by the initial oxidation state and by the final product.

### 7.5. Analytical examples

In Figure 12 is reported the overlay of TPR analyses carried out on four catalysts containing the 5%(wt) of cobalt supported on alumina. They have been prepared by wetness impregnation ad then doped with different percentage of iridium [22]. The pre-treatment procedure is the same for all the samples: the catalysts, pre-calcinated in air at 350°C, have successively been cleaned in $N_2$ flow at 150°C and finally cooled at room temperature. The TPR was carried out with a temperature rate of 10°C/min and a flow rate of 30 cc/min of a mixture of

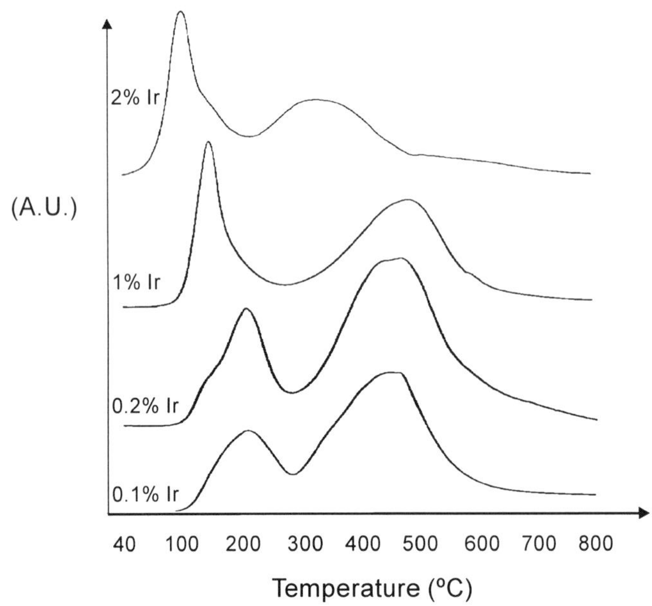


Figure 12. TPR profiles on 5% cobalt on alumina with different doping percentage of iridium. Gas used 5% hydrogen in nitrogen, flow 30 cc/min, rate 10 C/min (TPDRO 1100).

5% $H_2/N_2$. There are two evidences in the TPR profiles: the $H_2$ consumption increases when the percentage of doping metal (Ir) is increased while and the maximum temperature, related to the maximum consumption of gas, decreases accordingly. This example is a clear demonstration that the TPR analysis offers information about the reducibility of metallic samples and that it is possible to estimate quantitatively the effect due to the presence of a second metallic species. Multi-metallic systems are known for the difficulty in their characterisation.

In Figure 13 is reported a typical reduction profile of pure cupric oxide. Cupric oxide can be conveniently used to calibrate the detector signal. Sharp reduction peaks permits a better integration and a correct calculation of the reacted hydrogen. This result can be achieved by using a relatively high temperature rate (15 °C/min) and a small amount of sample (20-30 mg).

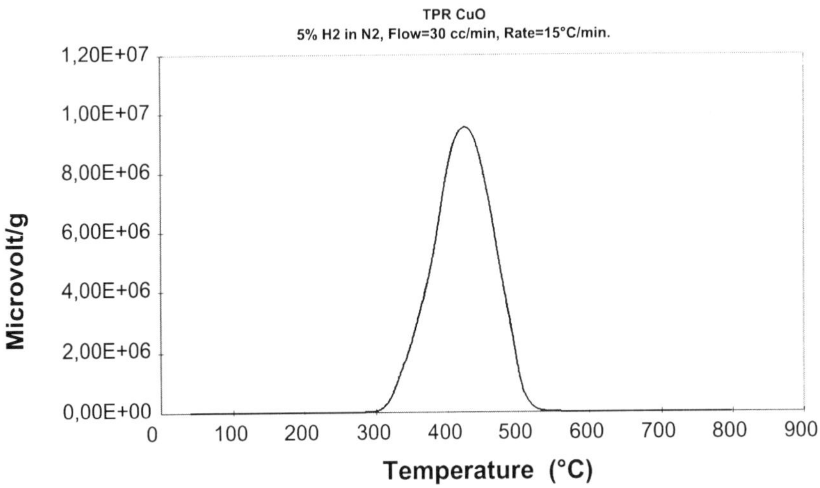


Figure 13. TPR profile on CUO using 5% hydrogen in nitrogen (TPDRO 1100).

## 8. CONCLUSIONS

The analytical methods described in detail in this paper can be considered as the basic texture methods for supported catalysts. The complete investigation on such complex structures and on the role of a given catalyst in an industrial process should take into account also tests on the reactivity.

The choice of the most suitable method should take into consideration the main task of the method itself. Static volumetric chemisorption provides the most reliable data from the scientific point of view. In fact, this is the only technique assuring the correct equilibrium time between the gaseous and adsorbed phases, that is the basic hypothesis to apply most of the thermodynamic equations we described in this paper. On the contrary, the main disadvantage related to static chemisorption consists in the long analysis time required. Furthermore, the handling of a static apparatus requires a certain basic knowledge about vacuum systems and their use (long degassing times, risk of leaks, etc.).

Dynamic methods are very fast and relatively easy to handle, even by inexperienced users. The analytical results, especially for pulse chemisorption, can be compared to the static methods ones only taking into consideration the basic differences between the two systems. In fact, in pulse chemisorption, the weak chemisorbed species are removed as they forms and it is not possible to state that there is a real equilibrium between the probe gas phases (gaseous and bound).

Temperature programmed methods provide very useful results on kinetic and thermodynamic aspects that are related to solid/gaseous interactions. In this case, the method of adsorption used to saturate the catalyst before the analysis (i.e. TPD) might have an influence on the results.

Anyway, an extremely important aspect, common to all the above techniques, is the sample preparation to the analysis. Very often, the pre-treatment procedures are not sufficiently described in scientific publications, making sometimes impossible to reproduce experimental data. It is also important to underline that the analytical reproducibility should be always verified. Analytical reproducibility is influenced not only by the experimental parameters but mainly by the pre-treatment procedures. For this reason, fully automatic equipment performing the catalyst preparation before the analysis is highly recommended, especially when these type of measurements are performed in industrial quality control laboratories.

## REFERENCES


1. P. A. Sermon and G.C. Bond, Catalysis Rev., 8 (1974) 211.
2. E. Miyazaki, J.Catal., 65 (1980) 84.
3. D.P. Smith, Hydrogen in Metals, Un.of Chicago Press.
4. British Standard, BS 4359: Part 4, (1995) 13.
5. Specialist Periodical Report Catalysis vol.11, Royal Soc. of Chemistry (1994).
6. J.M. Thomas and W.J. Thomas, Principle and practice of heterogeneous catalysis, Weinheim, New York (1997).
7. J.B. Benziger in: Metal-surface reaction energetics, E. Shustorovich (ed.), VCH Publishers Inc., 1991.
8. C.H. Bartholomew, in: Catalysis, Royal Soc. of Chemistry, vol. 11, (1994) 3.
9. J.M. Zowtiak and C.H. Bartolomew, J.Catal., 83 (1983) 107.
10. G.A. Olah, J.K.S. Praksh and J. Sommer, SUPERACIDS, J. Wiley, New York (1985).

11. D.O. Hayward and B.M.W. Trapnell, in: Chemisorption, Butterworths, London (1964).
12. I. Langmuir, J.Chem. Soc., 40 (1918) 1361.
13. J.R. Anderson, Structure of metallic catalysts, Academic Press (1975).
14. J.L. Lemaitre, P.G. Menon and F. Delannay, in: The measurements of catalyst dispersion, M.Decker Inc., New York, 1984, 266.
15. Characterisation of heterogeneous catalysts, F. Delannay, M.Dekker Inc., New York, Cap. 7 (1984) 299.
16. J.R. Anderson and K.C. Pratt, Introduction to characterisation and testing of catalysts, Academic Press, Harcourt Brace Jovanovich, Publishers, Australia, (1985).
17. L. Forni, F.P. Vatti and E. Ortoleva, Micro.Mat., 3 (1995) 367.
18. S.D. Robertson, B.D. McNicol, J.H. De Baas, S.C. Kloet and J.W. Jenkins, J. Catal., 37 (1975) 424.
19. N.W. Hurts, S.J. Gentry, A. Jones and B.D. McNicol, Catal. Rev-Sci. Eng., 24 (1982) 233.
20. G. Delahay, Ph.D. Thesis, Université de Toulouse, France, 1991.
21. D.A.M. Monti and A. Baiker, J. Catal., 83 (1983) 323.
22. B. Jouguet, A. Gervasini and A. Auroux, Chem. Eng. Technol. 18 (1995) 243.
23. C.L. Bianchi, L.Aina, M.Fadoni and V. Ragaini, J.Catal., subbmitted for publication.

Adsorption and its Applications in Industry and Environmental Protection
Studies in Surface Science and Catalysis, Vol. 120
A. Dabrowski (Editor)



# Adsorption with soft adsorbents or adsorbates. Theory and practice

G. F. Cerofolini[a], L. Meda[b] and T. J. Bandosz[c]

[a] SGS-Thomson Microelectronics, 20041 Agrate MI, Italy

[b] EniChem - Istituto G. Donegani, 28100 Novara NO, Italy

[c] Department of Chemistry, City College of New York, New York, NY 10031, USA

Physical adsorption on solid surfaces has in most cases been described in the hypotheses that: the adsorbate is not modified during adsorption, the adsorbate internal degrees of freedom are not excited during adsorption, and each adsorbed molecule fills one and only one adsorption site. This set of assumption constitutes the *paradigm of adsorption*, quite irrespective of the detailed adsorption model considered. This work is intended to give a preview of what might be a theory of adsorption beyond the paradigm. The emphasis will mainly be posed on special systems taken from practical situations rather than on general abstract cases.

## 1. THE UBIQUITOUS OCCURRENCE OF ADSORPTION

The unsaturated forces which result when a solid is terminated with a surface may be relieved in a lot of ways, like relaxation, reconstruction, and - in the presence of an atmosphere - adsorption.

According to the nature of the gas-solid interaction one speaks of chemisorption, for specific forces involving chemical bonds, or physisorption, for non-specific forces of electrostatic or Van der Waals nature[1] [1]. Since these secondary forces are manifested among all atoms, physisorption is an ubiquitous phenomenon. However, since they are usually weak, physisorption is an important phenomenon only at low temperature.

Since the forces which keep a molecule in liquid phase are of electrostatic or Van der Waals type, when a solid surface at a temperature T is in contact with a

[1] How coordination adducts can be arranged in this classification is not clear - the relatively high binding energies stabilizing the adducts suggest that they are classified as chemisorbed species, while the poor specificity of the involved forces allows them to be considered as physisorbed species. The very fact that an equilibrium between the gas phase and adsorbed phase is often attained in a short time (in the laboratory time scale) suggests they are classified as physisorbed. Numerous adducts of extreme biological relevance run in this situation: $O_2$-haemoglobin, $CO_2$-haemoglobin, $CO_2$-chlorophyll, etc. are familiar examples.

vapour, whose critical temperature $T_c$ is higher than T, the system undergoes multilayer adsorption. This phenomenon is manifested with the formation of a relatively thick film, whose thickness depends on the relative pressure of the vapour, but is usually in the interval 3 -15 Å (the relative pressure x is defined as the ratio of the partial pressure p to saturated vapour pressure $p_0$ at the considered temperature).

Most of systems with which the mankind is in contact are immersed in an air atmosphere at an average ground temperature of 290 K with a dispersion around 10%. With 'air' one intends a gas mixture formed by 78% of $N_2$, 21% of $O_2$, 0.9% of Ar, 0.08% of $CO_2$, a variable amount (however around 0.2%) of $H_2O$ [1], and small percentages of $O_3$, NO, $NO_2$, CO, $SO_2$, saturated and unsaturated hydrocarbons, etc. Though most of these minor substances (e.g., $O_3$, NO, $NO_2$, CO, $SO_2$, etc.) are contaminants produced by the human activity, they may be produced even in `natural' conditions (e.g. $SO_2$, in volcanic activity; NO and $NO_2$ in thunderbolts; $O_3$ in upper atmosphere during solar irradiation, etc.).

Many vapours, though present as traces in the atmosphere, are able to produce multilayer adsorption. For instance, all hydrocarbons except the lightest ones run in this situation, and their adsorption on stones, clays, inorganic and organic dyes, etc. is one of the major causes of injury to artcrafts (see ref. [2] for a short review of the material problems involved in frescos, temperas, eucastics and oils).

Of all the gases contained in ponderal amounts in the natural atmosphere, only water has a critical temperature higher than room temperature. Since the water relative pressure may be high, most of materials exposed to air must be thought of as covered by a multilayer film of adsorbed water. Typically, the relative humidity (= water relative pressure) of the atmosphere is around 0.5, though values even as low as 0.1 may be attained in desert atmospheres and values above 1 are attained in supersaturated air resulting from the adiabatic expansion of warm water-saturated air.

The film of adsorbed water plays an important role in many systems. For instance, water is adsorbed on the polar sides of lipids constituting the outer surface of cells as well as inside pores of oxides, silicates and carbonates forming soils; adsorbed water promotes metal oxidation by forming hydroxyls groups via cleavage of passivating oxo bridges at metal surface; and adsorbed water can revert to liquid water, thus providing a medium which allows the anion-hydron dissociation via solvation.

That insoluble porous oxides may have had a fundamental role in prebiotic conditions was first hypothesized by Bernal [3]. He proposed that the organic compounds dissolved in the primordial soup had been concentrated by adsorption on the surfaces of clay mineral particles suspended in the early oceans, and that the subsequent prebiotic reactions synthesizing the organic substrates were catalyzed at active sites on the clay surfaces. It is interesting to note that water adsorbed in restricted geometries (in which the organic compounds had concentrated) has properties different from those of liquid water. For instance, in particular situations (water adsorbed in porous silica) $K^+$ is preferentially adsorbed with respect to $Na^+$ [4], so some features which are observed in living cells ($K^+$ accumulation and $Na^+$ depletion) might have been characteristic of the primordial niche of the prebiotic system.

A cell may be viewed as a very dense emulsion of macromolecules in water, embedded in a semipermeable lipoproteic membrane. Water is expected to be adsorbed on the polar sites of the macromolecules. Since in biological cells there are approximately three water molecules only per polar site, it seems that in biosystems all water must be viewed in an adsorbed, rather than liquid, phase. This picture has been able to account for a lot of seemingly uncorrelated phenomena: the $Na^+/K^+$ ratio [5,6], anomalous viscosity [6], and inert-gas anaesthesia [7].

There is a sharp separation between adsorbed water and liquid water, in which an adsorbent may be suspended and dissolved. In adsorbed water the structure is imposed by the adsorption field, while in liquid water the (dis)order is the one characteristic of the bulk phase. This state of affairs suggests that the adsorbed-to-liquid transition is a phase transition. On another side, water on many surfaces continuously undergoes adsorbed-to-liquid transition and *vice versa*. Familiar examples of adsorbents where the state of water frequently cycles between the adsorbed and liquid phases are soils and the skin of terrestrial mammalians. It is also noted that the formation of liquid water in clouds may occur via heterogeneous nucleation (i.e., via multilayer adsorption), either spontaneously on dust particles or artificially on AgI crystals formed by condensation of sublimated AgI.

In spite of this, there is no accurate description accounting for the formation of a liquid film at the surface of a solid starting from an adsorbed film, or conversely of the appearance of an adsorbed film in the final stages of evaporation of a liquid film. In fact, the two major models hitherto developed for multilayer adsorption, namely the model originally proposed by Brunauer, Emmett and Teller (BET) and the one proposed separately by Frenkel, Halsey and Hill (FHH), apply to the description of a thin adsorbed phase the former and of a liquid film the latter.

The gas-solid interaction energies which are responsible for physisorption are not very different from vapour-phase interaction energies which are responsible for condensation to the liquid state. Because of this, multilayer adsorption (i.e., adsorption on the top of already adsorbed molecules) is manifested especially when the temperature of a solid surface is lower than the critical temperature of the adsorbate. In this situation multilayer adsorption results in an increase with pressure of the adsorbed molecules up to bulk condensation as the adsorption pressure approaches the bulk vapour pressure, in this way providing an example of heterogeneous nucleation of the saturated vapour.

## 2. THE PARADIGM - ADSORPTION OF HARD ADSORBATES ON HARD ADSORBENTS

For a long time the possibility that internal degrees of freedom of both the adsorbent and the adsorbate can be modified after adsorption was ignored in most theories. The `paradigm' of the theory of adsorption was based on the assumptions of a ideally flat surface or of a two-dimensional Ising lattice in equilibrium with a gas formed by structureless molecules (hence 'adatoms'), at most endowed with a co-area. Only later was the paradigm extended to account for the energetic heterogeneity of the surface (see, for instance, the monographs of Jaroniec and Madey [8] and Rudziński and Everett [9] and the review of Cerofolini and Rudziński [10]) and for the polyatomic structure of the admolecule (see, for instance, the review by Evans [11]).

Most of adsorption models have been constructed by postulating (often implicitly) that: (i) the adsorbing surface generates the adsorption field and remains unchanged after adsorption, (ii) the internal degrees of freedom of the adsorbate are not modified during the process, and (iii) each adsorbed molecule fills one and only one adsorption site.

The one-molecule/one-site assumption manifestly does not hold true for adsorption in the multilayer regime or of large molecules. The first extensions of the paradigm were just performed trying to remove assumption (iii).

### 2.1. Beyond one-molecule/one-site assumption - multilayer adsorption

Simplified models for multilayer adsorption are formulated within the paradigm by explicitly allowing the possibility of multilayer formation. Among these models we shall consider only the BET and FHH ones. They can be considered as complementary, since the BET isotherm gives generally a satisfactory description of the real systems at coverages lower than 2 – 3 layers, while the FHH isotherm becomes adequate only at coverages higher than 3 layers.

#### 2.1.1. The early stages of multilayer adsorption - the BET isotherm

The original derivation of the BET isotherm was based on kinetic arguments [12], although statistical derivations are known [13]. The hypotheses upon which the BET theory is built are the following:

(BET1): The gas is perfect.

(BET2): Adsorbed molecules are classical objects localized on their adsorption sites.

(BET3): The surface is characterized by $N_m$ identical sites.

(BET4): Adsorption takes place either on surface sites or on the top of molecules already adsorbed (in-between positions are excluded).

(BET5): The first layer only interacts with the surface; all other layers have interparticle interaction with the same energy as would apply in the liquid state, and involving only nearest neighbours in the vertical stack of adsorbed atoms in each site.

(BET6): Adsorbed molecules do not interact laterally.

Since the first three assumptions are the same as for the Langmuir isotherm, the BET model is essentially an extension of Langmuir model to multilayer adsorption.

The above assumptions lead to the following expression for the surface coverage $\Theta$:

$$\Theta(x) = \frac{1}{1-x} \frac{Cx}{1+(C-1)x} \tag{1}$$

where

$$C = \frac{z}{z_0} \exp\left(\frac{q - q_0}{k_B T}\right) \tag{2}$$

In eq. (2) $k_B$ is the Boltzmann constant, z and $z_0$ are the partition functions for a molecule in the first layer and liquid phase, respectively, while q and $q_0$ are the adsorption energies in the first layer and higher layers, respectively.

If C >> 1 (BET isotherm of II type), higher layers are occupied only when the first layer has been filled almost completely. If $C \lesssim 1$ (BET isotherm of III type), adsorption in the first layer occurs in competition with adsorption in higher layers.

The BET assumptions can be modified by imposing that piles with a maximum of n molecules can be accomodated on the surface. In this case one gets

$$\Theta_n(x) = \frac{Cx}{1-x} \frac{1-(n+1)x^n + nx^{n+1}}{1+(C-1)x - Cx^{n+1}} \tag{3}$$

instead of eq. (1). Formally $\Theta_n(x) \longrightarrow \Theta(x)$ for $n \longrightarrow +\infty$.

The n-form (3) of the BET equation can be approximated by the equation

$$\Theta(x) = \frac{1}{1-\kappa x} \frac{C\kappa x}{1+(C-1)\kappa x} \tag{4}$$

provided that $n \cong 2/(1-\kappa) - 1$. Equation (4) is formally obtained by correcting the $\infty$-form BET equation with the adjustable parameter $\kappa$ (usually $\kappa < 1$). This equation was originally proposed by Anderson [14] and Brunauer, Skalny and Bodor [15] on different theoretical grounds and will henceforth be referred to as ABSB equation, from the initials of its proponents.

A simple inspection of the BET assumptions clearly shows that the BET equation cannot give an adequate physical description of multilayer formation.

However, in the neighbourhood of the B point (i.e., the point in the $x - \Theta$ plot where the experimental isotherm changes its concavity) the BET equation has provided a rationalization of so many experimental data as to have become a standard for the quantification of surface areas (see section 2.2.).

Many treatments have been developed in order to improve the BET theory [16, 17]. Among them we mention the model by Hill [16], essentially based on the same assumptions of BET theory, but accounting for lateral interactions among adatoms in the same layer (within the Bragg - Williams approximation). The obtained improved isotherm is considerably more complicated without leading to better agreement with experimental results [18]; for these reasons it is not frequently utilized in practice - that occurs for other seemingly 'improved' isotherms.

### 2.1.2. The final stages of multilayer adsorption - the FHH isotherm

In the FHH theory [19, 20, 21] the adphase is considered as a liquid phase subjected to an external potential generated by the adsorbing solid surface: the attractive gas-solid interactions are responsible for a stabilization of the adphase with respect to the bulk liquid.

The FHH isotherm is derived by assuming the adsorbate as a uniform thin layer of liquid on a planar, homogeneous, solid surface and considering the effect of the replacement of the solid by the liquid: a molecule in the adsorbed layer will feel different potentials in these two situations. Equating such a potential energy difference to the difference of chemical potentials between the adsorbed layer and the bulk liquid, one obtains the following implicit isotherm

$$k_B T \ln(p/p_0) = u_p(L) \tag{5}$$

where L is the thickness of the adsorbed layer and $u_P(x_3)$, known as perturbation energy, is the difference between the actual potential $U(x_3)$ (supposedly to depend on the distance $x_3$ from the surface only) acting on the adsorbed layer at a distance $x_3$ and the hypothetical potential acting on the same point if the solid adsorbent were substituted with liquid adsorbate: $u_P(x_3) = [U(x_3) - u_{liq}(x_3)]$. Equation (5) is interpreted by stating that a liquid condenses in a volume within a distance L from the surface when the perturbation potential in that volume is less than or equal to $k_B T \ln(p/p_0)$. The quantity

$$\varepsilon = -k_B T \ln(p/p_0) \tag{6}$$

is usually referred to as Polanyi potential. See Steele's treatise for a compact discussion of the physical bases of the FHH theory [13].

An explicit form for the FHH isotherm (5) is obtained by putting $L = \Theta\lambda$ (where $\lambda$ is the thickness of each layer), and assuming that the perturbation energy is attractive and varies as an inverse power s of the distance, $U(x_3) \propto x_3^{-s}$. With this assumption the FHH isotherm becomes

$$\ln\left(\frac{p}{p_0}\right) = -\frac{\alpha_{FHH}}{k_B T \lambda^3}\left(\frac{1}{\Theta}\right)^s \tag{7}$$

where $\alpha_{FHH}$ is a proportionality constant. Considering the long-range part of the gas-solid interaction and the intermolecular interaction in the bulk liquid as due to dispersion interactions, the exponent s should be equal to 3. Of course, the assumption $L \propto \Theta$ provides a 'continuum' description of the adsorption process; this description is realistic only for $\Theta$ high enough, say for $\Theta \gtrsim 3$.

Equation (7) with $s = 3$ is rarely, if ever, observed. Actually, in most cases the adsorption isotherm in the multilayer regime is well described by the FHH equation with $s > 3$ though occasionally the case $s < 3$ is observed too. The first explanation for the deviation from $s = 3$ was provided by Halsey himself, who ascribed the occurrence $s > 3$ to energy heterogeneity effects [22]. This explanation, however, has never been fully convincing - in fact, energy effects are responsible for deviations from the BET behaviour mainly in the submonolayer range and tend to disappear once the monolayer is completed [23], that makes it difficult to understand why they should reappear at even higher coverage. In recent years a few collaborations have proposed explanations of these deviations in terms of geometric heterogeneity [24, 25]. According to the model originally proposed in ref. [24] and reviewed in ref. [26], the occurrence $s \neq 3$ is ascribed to a fractal nature of the surface; the exponent is related to the fractal dimension D of the adsorbing surface and to the adsorption mode:

$$\frac{1}{s} = \begin{cases} (3 - D)/3 & \text{Van der Wals regime,} \\ 3 - D & \text{capillary regime.} \end{cases} \tag{8}$$

The high pressure range of an adsorption isotherm provides therefore information on the fractal dimension of the adsorbent.

## 2.2. Beyond one-molecule/one-site assumption - measuring the adsorbent surface area with scaled molecules

Even surfaces obtained by cleaving single crystals are not atomically smooth, because any surface undergoes relaxation and reconstruction, and contains equilibrium defects like vacancies and self-interstitials. Roughening is a fundamental process for all solid surfaces, and results in terraces, steps and kinks (see, for instance, ref. [27]).

This situation is magnified for highly dispersed solids, which are characterized by strongly non-equilibrium configurations obtained imparting a macroscopic energy over relatively few degrees of freedom. Fracture, supercritical extraction of solvent, burning of templates, etc. are familiar examples.

Highly dispersed solids are characterized by a fraction of atoms exposed to the atmosphere which frequently is of the order of 10% or even higher. They provide examples of systems in which the surface properties do not disappear even in the thermodynamic limit. The techniques which are usually adopted for the characterization of flat surfaces (like low- or high-energy electron diffraction, scanning or transmission electron microscopy, atomic-force or scanning tunnel microscopy) are not suitable for the characterization of highly dispersed solids. Rather, the high number of atoms exposed to the atmosphere suggests the use of adsorption techniques for their characterization.

The measurement of the monolayer amount of adsorbates with the same nature but scaled cross sections (e.g, noble gases, alkanes, or polymers) is based on the BET technique, which provides - even accurately – a measure of the monolayer amount for any adsorbed vapour. The one-molecule/one-site assumption, stating that the adsorbate at the monolayer is so arranged that each adsorbed molecule covers one and only one adsorption site, is plausible only for small molecules but is surely false for large molecules. In the second case, a first estimate of the adsorbent area A may be obtained by multiplying the monolayer amount $N_m$ by the adsorbate cross section $\sigma$: $A = N_m \sigma$. A typical experimental arrangement is based on $N_2$ adsorption at 77 K; the surface area $A°$, determined by measuring in these conditions the monolayer amount $N_m°$ and multiplying it by the $N_2$ cross section $\sigma°$ (with $\sigma° := \sigma_{N_2} = 16.2$ Å$^2$; see ref. [28] for a discussion of this value), is usually referred to as *standard surface area*, $A° = N_m° \sigma°$.

Of course, the standard surface area provides an adequate description of the true surface area only if it remains unchanged when the probe is changed, i.e. if

$$N_m = N_m^o \sigma^o / \sigma \tag{9}$$

The cross section $\sigma$ depends on the number of atoms forming the molecule and on their spatial arrangement. Since this arrangement is modified by the vicinity of other molecules, $\sigma$ is expected to depend on the adsorbent too. Though the unambiguous determination of the molecular cross section is a difficult, likely ill-posed, problem, extended compilations are available, based either on bulk (liquid or gas) phase properties of the adsorbate [29] or on molecular dynamics simulations [30].

Starting from these compilations one usually finds that even considering different adsorption probes with similar chemical nature (to minimize the differences in adsorption forces) the determined monolayer amount $N_m(\sigma)$ and area depend on the probe, in a way for which condition (9) is not satisfied. In such analyses two situations are typically encountered: A increases with $\sigma$, or A decreases with $\sigma$.

### 2.2.1. The area increases with adsorbate cross section - smooth surfaces

This case is the one most frequently observed in the practice. According to the adsorbate diameter $\lambda$ ($\cong \sigma^{1/2}$), two extreme situations are observed. When $\lambda$ is much smaller than the mean nearest-neighbour distance $\xi$ of adsorption sites,

$$\lambda << \xi \tag{10}$$

the measured monolayer amount is independent of $\sigma$, so that

$$A \propto \sigma \tag{11}$$

Since typically $\xi \cong 3$ Å, only few adsorbates are such to guarantee that condition (10) is indeed satisfied; the most noticeable ones are: He, $H_2$ (both requiring freezing in the cryogenic regime), and marginally Ne. Equation (11) shows that these adsorbates do not provide a complete coverage of the surface.

For $\lambda >> \xi$ (that happens for most adsorbates), the measured monolayer amounts vary with $\sigma$ as

$$A = A^* + \delta(\sigma) \tag{12}$$

where $\delta(\sigma)$ is a slowly variable, monotonically increasing function such that $\delta(0) = 0$. Equation (12) suggests that in this case the monolayer amount provides a complete coverage of the surface, the quantity $\delta(\sigma)$ being related to a partial adsorbate-adsorbate overlapping. This interpretation allows $A^*$ to be interpreted as outer measure of the adsorbent surface. The outer measure, however, does not provide yet a measure of the surface area. In the theory of measure one should also determine an inner measure $A_*$ and the surface would be measurable when $A_* = A^*$, in which case the common value $\overline{A} := A_* = A^*$ would be referred to as surface area. However, since there is no practical way for the determination of the inner surface area, the measure theory cannot be applied. In the absence of any direct experimental information on $A_*$, one is therefore forced to use the outer measure $A^*$ only. Surfaces described by eq. (12) with finite $A^*$ are smooth.

### 2.2.2. The area decreases with adsorbate cross section – fractal surfaces

The case of A decreasing with $\sigma$ is less frequently encountered in the practice. In these cases, the monolayer amount is usually found to vary as

$$N_m = N_m^o \left(\sigma^o / \sigma\right)^{D/2} \tag{13}$$

where D is a constant exponent (D > 2). The corresponding area varies as

$$A = A^o \left(\sigma^o / \sigma\right)^{D/2-1} \tag{14}$$

Of course, eqs. (13) and (14) have a quite restricted range-of-validity: these equations are satisfied for $\sigma$ usually varying over less than one order of magnitude. Assuming their validity even in the limit for $\sigma \to 0$, they state that the area diverges in this limit. In abstract analysis any surface satisfying condition (13) is said to be a fractal with Hausdorff fractal dimension D; this suggests that real surfaces satisfying eqs. (13) and (14) may be considered *fractal*, at least in the length scale where the above equations are actually observed experimentally.

The problem of the *experimental* determination of the fractal dimension D of real surfaces was first addressed by Pfeifer and Avnir [31], who proposed four different methods based on adsorption data. In a subsequent paper, Avnir, Farin and Pfeifer applied those methods to real adsorbents and gave evidence for surfaces with $D > 2$ [32].

In conclusion, real surfaces display many pathological behaviours. According to its fractal dimension D, a surface is classified either as *smooth* when $D = 2$ or as *fractal* when $2 < D \leq 3$. According to the analysis of Avnir et al. [32], most of adsorbents are smooth, though fractal adsorbents are occasionally found. Smooth surfaces, in turn, may be either *flat* (e.g., lamellar solids) or *rough* (e.g., zeolites, etc.). In most cases the adsorbent is characterized by a distribution of convex or concave regions. Any concave region puts geometric limits to the unlimited growth of the adsorbed film. When the thickness of the adsorbed film becomes comparable with the modulus of the curvature radius of the region, the concavity is said to be a *pore*.

### 2.3. Extending the paradigm - Soft adsorbents and adsorbates

As stated in section 2.1, both the BET and FHH models are based on the implicit assumptions (i) and (ii) which are summarized in the following compact statements:

(HS): *The surface is hard*

and

(HA): *The adsorbate is hard.*

Understanding the general features of what happens when the surface undergoes reconstruction after adsorption or the adsorbate conformation is modified by the adsorption process is the goal of this work. In particular, section 3 will consider what happens when (HS) does not hold true and must be replaced by

($\neg$HS): *The surface is soft*,

while section 4 will consider the effect of substituting

($\neg$HA): *The adsorbate is soft*

for (HA). What happens when (HS) *and* (HA) do not hold true and the surface and adsorbate are *both* soft, though of practical and conceptual relevance, is beyond the scope of this work.

In last years the attribute 'soft' has become of widespread use in natural sciences: 'soft sciences' (like ecology, meteorology, etc.) have had great attendances, 'soft processing' has become the major indication for sustainable industrial development[2], 'soft computing' has proposed itself as a practically valid alternative to exact numerical computing[3], 'soft matter' has attracted large interest [34, 35], and the methods of 'soft chemistry' have become of widespread use for the preparation of new materials [36].

It is therefore not astonishing that even in specialized parts of `hard' science (like physical chemistry) the attention has recently been focused on soft matter. In the theory of adsorption, an adsorbent is considered hard when its energy landscape has few minima separated by high barriers; it is said soft when its energy landscape has several minima separated by low barriers; 'high' and 'low' take their meaning from the comparison with the adsorption energy. Similar definitions apply to adsorbates too. Before entering the details of the theory, we want to emphasize that though at a first glance 'soft' may appear somewhat synonymous of 'thermally unstable and complex', and 'hard' of 'thermally stable and simple', this is not correct. In section 3 we shall give an example of a thermally stable, inorganic surface which behaves as soft after adsorption of water, while in section 4 we shall give an example of thermally stable, relatively simple, however soft, adsorbates. The terms 'hard' and 'soft' are used in the theory of adsorption with a meaning which resembles the one of the same terms in the hard-soft theory of acidity [37].

In principle, the softest adsorbates are liquids, and adsorption at the mercury-solution interface is at the basis of one of the most common analytical technique - polarography. A theory of adsorption on liquid surfaces is facilitated by the fact that liquid are always characterized by an equilibrium configuration and has actually been developed [38]. No such a theory is known for less extreme situations, in which the adsorbate does reconstruct partially without attaining an equilibrium configuration.

## 3. SOFT ADSORBENTS

In this chapter we shall give an overview of the phenomena which may occur in the adsorption on soft adsorbents. In particular, we shall consider the conformational changes of adsorbents (like silica or proteins) produced by the

---

[2] 'Soft processing' means arranging a set of procedures which allow the production, with a minimum of environmental risk and impact, of a certain material. 'Softness' substitutes 'environmental risk and impact' for 'cost', proper of hard processing. The key factor which allows soft processing is the availability of suitable catalytic systems which, in analogy with biological catalysts, combine size (or even shape) selectivity with catalytic activity.

[3] "Soft computing is an association of computing methodologies centering on fuzzy logic, neurocomputing, genetic computing, and probabilistic computing. The methodologies comprising soft computing are for the most complementary and synergistic rather than competitive. The guiding principle of soft computing is: exploit the tolerance for imprecision, uncertainty, partial truth, and approximation to achieve tractability, robustness, low solution cost and better rapport with reality" [33].

adsorption of small polar molecules thereon. We shall also show that the concepts and methods developed for adsorption on soft adsorbents can be used to solve an age-old, and not solved yet, problem of the theory of adsorption on hard adsorbents - a unified description of adsorption from the Henry range to the formation of an adsorbed liquid.

In the considered cases, the adsorbent undergoes profound reconstruction after adsorption. A difference is however noted: while water adsorption on silica produces an irreversible reconstruction (in such a way that the adsorbent is changed by an adsorption-desorption cycle), water adsorption on proteins is responsible for an almost reversible reconstruction.

### 3.1. Silica hydration as a chemisorption process promoted by a physisorbed precursor

The surface terminations of silica are essentially related to its preparation procedure.

The *fracture* in an inert atmosphere is responsible for the homolytic cleavage of Si-O bonds, with the consequent formation of silicon and siloxyl radicals,

$$(-O)_3Si^\bullet \quad \text{and} \quad (-O)_3SiO^\bullet$$

and of strained siloxanic bridges

$$(-O)_3Si\cdots O\cdots Si(O-)_3$$

These defects are present at a concentration of the 0.1 - 1 % each [39, 40, 41].

*Ion bombardment* is a way to prepare the above radicals in pairs or quartets: the displacement of an oxygen atom forms the oxygen-bridge vacancy and its relaxed configuration,

$$(-O)_3Si^\bullet \;\; {}^\bullet Si(O-)_3 \quad \text{and} \quad (-O)_3Si-Si(O-)_3$$

while the displacement of a silicon atom forms the silicon-link vacancy,

$$\begin{array}{l} \equiv Si-O^\bullet \;\; {}^\bullet O-Si\equiv \\ \equiv Si-O^\bullet \;\; {}^\bullet O-Si\equiv \end{array}$$

and its relaxed configurations:

$$\begin{array}{l} \equiv Si-O - \;\; O-Si\equiv \\ \equiv Si-O^\bullet \;\; {}^\bullet O-Si\equiv \end{array} \quad \text{and} \quad \begin{array}{l} \equiv Si-O - \;\; O-Si\equiv \\ \equiv Si-O - \;\; O-Si\equiv \end{array}$$

The *hydrothermal synthesis* of silica results in the formation of surface silanols: In most cases, together with *isolated* silanols $(-O)_3SiOH$, the surface contains *geminal* silanols $(-O)_3Si(OH)_2$ and *vicinal* silanols $(-O)_3SiOH\ HOSi(O-)_3$. Silanols are the preferred adsorption sites of water vapour. Freshly prepared porous silica obtained by hydrothermal processing appears hygroscopic with respect to water vapour (water molecules being adsorbed on silanol groups), but in most cases is hydrophobic with respect to liquid water. The reason for this difference resides in the fact that adsorption from gas phase is a spontaneous process not requiring any activation, while adsorption from liquid phase requires an energy related to the surface energy of the liquid.

Though hydrothermally grown silica is well known for its chemical inertness, its prolonged exposure to water vapour renders it progressively less hydrophobic and eventually hydrophilic. In view of the strength of the Si–O bond, understanding why silica is unstable once is exposed to water vapour and how surface reconstruction occurs is not a trivial task. However, these facts can be explained in terms of progressive increase of surface silanols following water adsorption.

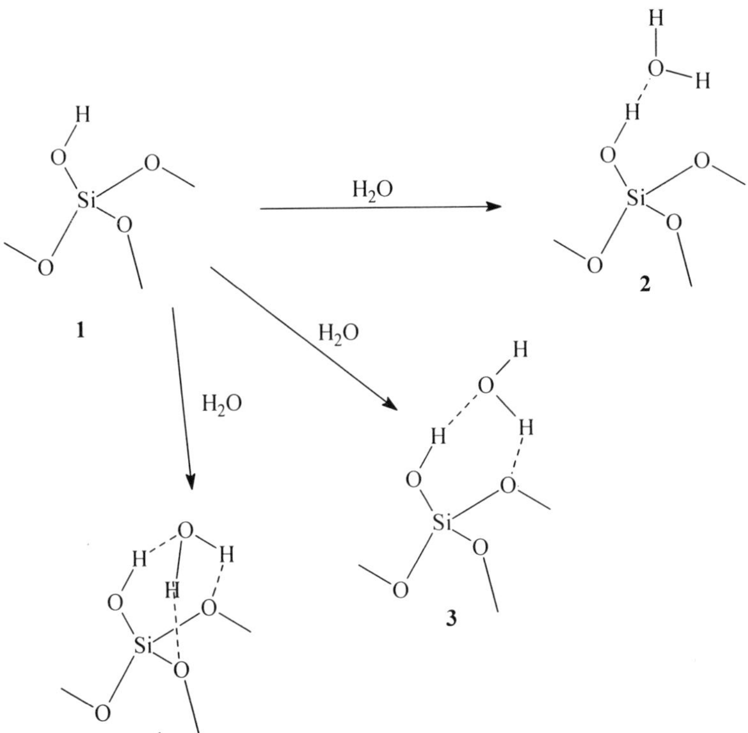


Figure 1. Various configurations of water adsorbed on a silanol group. The number of hydrogen bonds increases progressively from **2** to **4**.

The following pathway seems likely: $H_2O$ physisorption on silanol groups Si –OH is a fast, non-activated phenomenon involving secondary forces only. In most cases water adsorption results simply in the formation of a hydrogen-bonded adduct, as **2** in fig. 1. Occasionally, however, the adsorption will result in more strongly bound configurations (like **3** or even **4** in the same figure), which evolve towards the formation of two new silanols per original sites. Figure 2 sketches a possible pathway involving hydron ($H^+$ ion) transfer. Since the O–H bond energy is the same as the Si – O bond energy, the driving force of the hydroxylation process is only the strain energy of the Si – O – Si bridge. Since this energy is completely relieved after the cleavage of the siloxanic bridge, the system is expected to be stable after the cleavage of the oxygen bridges originally strained.

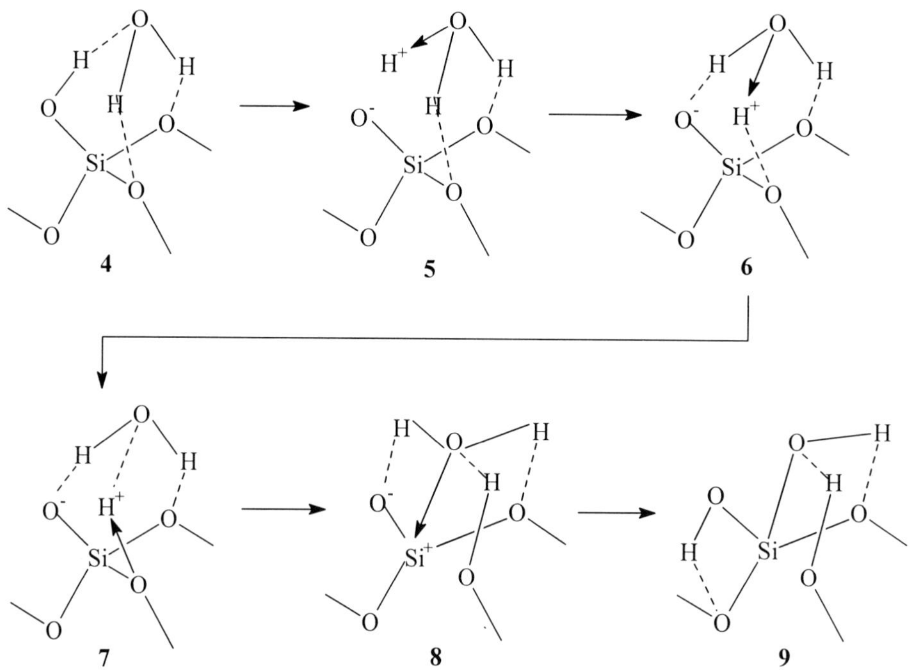

Figure 2. The hypothesized pathway for the cleavage of a siloxanic bridge by adsorbed water.

This conclusion, however, ignores an important factor: namely, the fact that each newly formed silanol is able to adsorb a water molecule. Once completed, this spontaneous process (which brings two OH terminations and two $H_2O$ molecules in a region which originally contained only one O atom) does indeed originate an additional strain on neighbouring Si - O - Si bridge and allows the process to continue. In this way, the physisorption energy, though low, furnishes the hydroxylation reaction with a reaction pathway with low activation energy. Silica surface may therefore be considered a soft adsorbent with respect to water.

Not only does water adsorption on silica cause an irreversible reconstruction of its surface, but also is responsible for creation of acidic centers. On the surface of silica usually two types of oxygen-containing groups such as siloxane, ≡ Si-O-Si ≡, and silanol, ≡ Si-OH, are present. With increasing temperature of heat treatment silanols are converted into very reactive siloxanes [42]:

$$\equiv Si - O\boxed{H + HO}- Si \equiv \longrightarrow \equiv Si \cdots O \cdots Si \equiv + H_2O$$

This reaction is reversible up to 673 K. Rehydration results in silanols, either isolated or in pairs [28], thus suggesting a mechanism like that sketched in fig. 2

The newly formed groups are very weakly acidic in their nature. The determination of their acidic constants $K_a$ using different methods gave $pK_a$ between 6.8 and 8.0 [43, 44, 45, 46]. When potentiometric titration combined with a procedure of solution of the adsorption integral equation using splines [47] was used to study the silica surface, two peaks at $pK_a$ = 6.84 and 8.32 were found. They are shown in fig. 3. [48]. The first peak at $pK_a$ = 6.84 can be assigned to the presence of silanol groups. It is in excellent agreement with the results of Schindler and Kamber who found this value to be 6.8 [46]. The second peak at at $pK_a$ = 8.32 representing weaker acidity can be assigned to proton release caused by breaking of siloxane groups.

Changes in the acidity of silica on various dehydratation levels can be also detected using the analysis of butane and butane sorption isotherms measured at

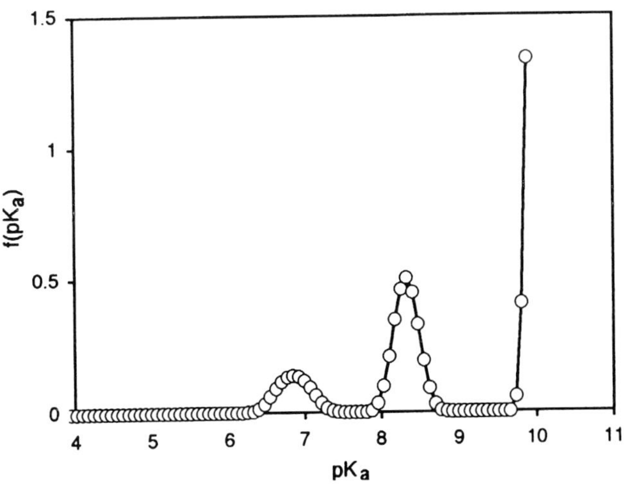


Figure 3. $pK_a$ distribution of a silica surface.

different temperatures [49, 50]. From these isotherms the heats of adsorption and adsorption energy distributions can be calculated. The analysis of results is based on the fact the $\pi$ bond of an alkene molecule interacts with silica acidic groups in a specific way whereas an alkane molecule interacts with the surface only in a dispersive way. This leads to the butene adsorption energy distribution which besides peaks coming from dispersive interactions of molecules with the surface also contains peaks representing acidic centers of the solid. Similar effects are expected to be observed for other oxides.[4]

### 3.2. Surface reconstruction resulting from the adsorption of polar molecules on proteins

It is not necessary that new chemical bonds are formed to have surface reconstruction: the adsorption of polar molecules on biomolecules like proteins or nucleic acids is a familiar example.

The importance of reconstruction phenomena during adsorption, especially in biochemistry, will never be sufficiently stressed: it suffices to mention that one of the most important works in biochemistry - the seminal paper of Monod, Wyman and Changeux (MWC) on allosteric transition [51] - is nothing but an example.

The MWC model is essentially constituted by an adsorbent formed by molecules each with $\nu_m$ identical sites; filling one site produces an equal change in the adsorption properties of each of the remaining $\nu_m - 1$ ones, and so on until a complete filling is obtained. Though this model can be adapted to many biological situations, it can hardly be applied to proteins, which are too complicated to be described by the MWC model.

All proteins are composed of linear chains of aminoacids linked together by peptide bonds. Each individual protein molecule may consist of one or more polypeptide chains, bound together by covalent (disulfide), ionic, hydrophobic, and hydrogen bonds. The side chains of the 20 common aminoacids produce an enormous range of functional characteristics, including charge and hydrophobicity [52].

Most of proteins have a globular structure. Any globular protein "is essentially a one dimensional system folded into a three dimensional structure" [53]. Having the protein been synthesized in an aqueous medium, due to the need of minimizing the protein-water interface energy during synthesis most of polar sites will be at the protein 'surface', while the hydrophobic arms of the chain will be in the protein 'bulk' [54]. The protein bulk has a fractal dimension D slightly less than 3 [55], so that proteins can be viewed as collapsed polymers. This view is consistent with the surface fractal dimension of globular proteins, which is typically in the range 2.1 – 2.2, not much greater than the value D = 2 characteristic of smooth surfaces [56, 57].

---

[4] Water acts as binder in mortar because it allows reticulation among particulates. On another side, the progressive destruction of the oxides by water, suggests that in the long run it is eventually responsible for their dispersion and dissolution. We advace the hypothesis that the spontaneous collapses of several one-thousand-year-old brick-and-mortar buildings like Venice's Bell Tower and Pavia's Cathedral have had this origin.

Usually any protein has proper biological activity in only one of its possible three-dimensional (3D) conformations [58]. When it assumes a different, inactive conformation (which may also be thermodynamically stable) it is called denatured. While some proteins spontaneously refold after denaturation, most require a careful, critical process for refolding (some of them cannot refold at all, thus definitively losing their biological activity). There are proteins which are stable under a wide range of physico-chemical conditions, while others can operate only in very limited conditions. Understanding all factors preserving stability is critical for developing protein purification methods based on specific or non-specific interactions.

The globular structure is preserved by the – S – S - covalent bridges between different segments of the protein backbone and by electrostatic interactions between polar sites on the backbone or in the side-chain groups. In general, only few disulphide bridges are present in a protein so that most of the structure results from a balance between non-specific folding, produced by electrostatic pairing of heteropolar sites, and directional folding, produced either by covalent bonds along the backbone (which impart α-helix configuration to the protein) or by interchain hydrogen bonds (which impart β-helix configuration). The contrast between these driving forces will result in a partial unpairing of heteropolar sites.

Unpaired polar sites are available to adsorb small polar molecules. The amount of water (considered as representative of small polar molecules) bound to haemoglobin (considered as representative of globular proteins) is 0.3 g $H_2O$/g haemoglobin [54]; assuming that one water molecule is bonded to each exposed polar group and that each aminoacid residue contains on the mean 3 polar sites, approximately 65% of the polar sites are covered by water. This amount may be assumed as a measure of the exposed sites. The remaining sites are most presumably present as ionic pairs $B^{\delta+} \cdots A^{\varepsilon-}$ ($\delta, \varepsilon < 1$) and are available to adsorb water only when the solvation energy $-\Delta H_{solv}$ gained in the hydration process,

$$B^{\delta+} \cdots A^{\varepsilon-} \xrightarrow{+H_2O} B^{\delta+} \cdots OH_2 \cdots A^{\varepsilon-} \xrightarrow{+H_2O} B^{\delta+} \cdots OH_2 + OH_2 \cdots A^{\varepsilon-} \qquad (15)$$

is greater than the electrostatic pairing energy, $-\Delta H_{es}$. For sufficiently polar sites ($\delta$ or $\varepsilon$ close to 1), the energy involved in electrostatic pairing is quite high (say, $-\Delta H_{es} \cong 5$ eV), while the hydration energy is much lower (say, $-\Delta H_{solv} \cong 1$eV/$H_2O$ molecule). This implies that reaction (15) is thermodynamically impossible. Elementary electrostatic considerations show however that the reaction becomes possible when a large polar molecule is inserted in between $B^{\delta+}$ and $A^{\varepsilon-}$. This process is facilitated by any increase of the $B^{\delta+} \cdots A^{\varepsilon-}$ distance, so that *the swelling resulting from adsorption softens the adsorbates and renders new sites available to adsorption.*

Hydrogen-exchange experiments on native cytochrome c in low concentrations of denaturing agents have indeed shown the existence of a sequence of metastable, partially unfolded forms holding free-energy levels in between the ones of the native protein and of the fully unfolded state [59].

The complexity of the phenomena undergone by a protein after adsorption of polar molecules explains why this matter is so difficult to model. The first attempts to account for them were tried by one of the present authors, who formulated a model to account for heterogeneity, allostericity and hysteresis in water adsorption on proteins [60, 61], and have been recently reconsidered using a modification of a model proposed by Landsberg [62] to account of the Elovich equation in chemisorption phenomena.

Adsorption of polar molecules takes place on the polar sites of the protein; on another hand, protein folding is stabilized by the electrostatic attraction between heteropolar sites. This means that only a fraction of the existing polar sites is available to adsorption. While adsorption proceeds, however, the protein undergoes a certain unfolding that allows new sites to be exposed to the atmosphere and to adsorb further polar molecules.

Adsorption at room temperature is not exhausted with the formation of a monolayer but proceeds up to multilayer formation. When the relative pressure is varied by an amount $dx$, the number of adsorbed molecules varies because of two factors: (a) the change $d\Theta$ of equilibrium coverage, and (b) the change $dN_m$ of exposed sites. In certain conditions (discussed in ref. [63]) this is equivalent to assume that

$$\frac{dN}{dx} = N_m \frac{d\Theta}{dx} \tag{16}$$

provided that the number of sites is not assigned, but varies because of folding-unfolding. The theory can proceed only specifying how $N_m$ varies during adsorption. In analogy with Landsberg model, the following relationship will be assumed:

$$\begin{aligned} \frac{dN_m}{dx} &= b\frac{dN}{dx}N_m \\ &= b\Theta'(x)N_m^2 \end{aligned} \tag{17}$$

where b is the newly exposed area resulting from the adsorption of one molecule. Equation (17) can be solved by separation of variables:

$$N_m(x) = N_m^0 \frac{1}{1 - bN_m^0\Theta(x)} \tag{18}$$

where $N_m^0$ is the amount of exposed sites at x = 0, $N_m^0 := N_m(0)$. Though formally $N_m$ can increase indefinitely, it is however limited by the total amount $N_m^{tot}$ of polar sites.

Inserting eq. (18) into eq. (16), one has the following differential equation

$$dN = N_m^0 \frac{d\Theta}{1 - bN_m^0\Theta} \tag{19}$$

which applies to adsorption on reconstructable proteins (b ≠ 0) as well as on unreconstructable proteins (b = 0). For the trivial case, b = 0, eq. (19) gives $N(x) = N_m^0\Theta(x)$. Otherwise, i.e. for b ≠ 0, the solution of eq. (16) is

$$N(x) = -\frac{1}{b}\ln\left(1 - bN_m^0\Theta(x)\right) \tag{20}$$

Adsorption on reconstructable proteins occurs in the low coverage limit with the same law as on unreconstructable proteins,

$$\Theta << 1/bN_m^0 \Rightarrow N(x) \cong N_m^0\Theta(x)$$

whatever is the expression of Θ(x). At high coverage, however, different phenomena can occur.

While the above discussion does not require any detailed knowledge of the isotherm Θ(x), understanding the phenomena occurring at high coverage requires a specification of Θ(x). Consistently with the sparse distribution of adsorption sites at protein surface, the BET model seems to provide a natural framework to describe multilayer adsorption. However, since the surface structure of globular proteins is fractal (D = 2.1 – 2.2), adsorption is expected to be described better by eq. (3) than by eq. (1), which is instead expected to describe adsorption either on smooth proteins (D = 2) or on completely unfolded proteins.

The maximum coverage $\Theta_n^m$,

$$\Theta_n^m = \frac{n+1}{2}\frac{C_n}{1+C_n}$$

(attained at x = 1), represents a milestone of the following discussion. A protein will be said *completely reconstructable* when

$$bN_m^0\Theta_n^m \geq 1 - N_n^{tot}/N_m^0 \tag{21}$$

otherwise it will be referred to as *incompletely reconstructable*.

*Incompletely reconstructable proteins* are expected to belong to the class of hard proteins, i.e. globular proteins with a high degree of conformational stability and a low degree of flexibility and able to resist large irreversible changes in conformation upon adsorption at interfaces. When condition (21) is not satisfied, the adsorption produces a progressive exposure of polar sites, but this process is terminated before their exposure has been completed, so that the protein will maintain its original globular structure. The adsorption isotherm will be given by eq. (20) with $\Theta(x)$ given by eq. (3).

*Completely reconstructable proteins* are expected to belong to the class of soft proteins, i.e. relatively flexible globular proteins able to modify their tertiary structure to facilitate adsorption at interfaces [64, 65, 66]. Hard proteins are usually characterized by many intrachain – S – S - bridges, few superficial polar sites, and low foamability; conversely, soft proteins have many superficial polar sites and high foamability [67]. When condition (21) is satisfied, there exists a relative pressure $x_c$ at which all polar sites are exposed to the polar molecule. In this situation unfolding is complete and most presumably the constraint on n disappears. Accordingly, for $x \geq x_c$ the description of adsorption requires the substitution of eq. (1) for eq. (3):

$$N(x)=\begin{cases} -\dfrac{1}{b}\ln\left(1-bN_m^0\Theta_n(x)\right) & \text{for} \quad x<x_c \\ N_m^{tot}\Theta_\infty(x) & \text{for} \quad x\geq x_c \end{cases} \tag{22}$$

As it stands, the model hitherto formulated is too vague for practical applications. Different polar molecules are in fact expected to have different unfolding effects. The effects of some adsorbates can be predicted *a priori*:

**i.** Small polar molecules, like $H_2O$, will be characterized by high n and by a modest ability to reconstruct the adsorbate (low b).

**ii.** Large polar molecules, like $CO(NH_2)_2$, will form multilayers (n as high as for molecules of type i) and unfold the protein (high b); b is expected to increase with the dipole moment of the molecule.

**iii.** Molecules formed by a small polar head and a large nonpolar tail, like $CH_3(CH_2)_2OH$, will form only one layer (n = 1) with a modest surface reconstruction (low b).

### 3.2.1. Small polar adsorbates

According to the sketch of fig. 4, the adsorption of small polar molecules is expected not to produce large unfolding of the protein. It is just because of this reason that the crystalline state of proteins *in vitro* is a good model for the functional behaviour of hydrated proteins in solution *in vivo*. The model could be validated by determining, via X-ray diffraction, the protein structure under different hydration equilibrium levels resulting after exposure to atmosphere at certain relative humidities x. However, we are not aware of such experiments,

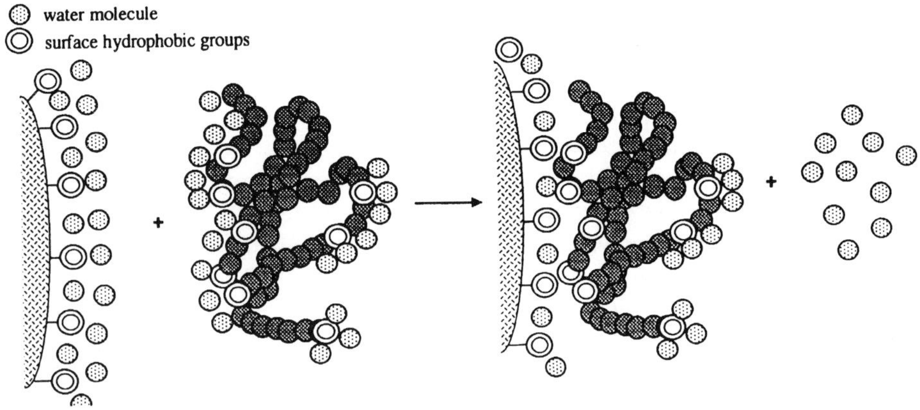


Figure 4. Hydrophobic binding interactions.

except for studies of different crystallographic configurations of proteins (like the triclinic and tetragonal forms of lysozyme [68]) with their proper hydration levels.

Adsorbed water plays a fundamental role in hydrophobic-interaction chromatography. This technique for protein separation and purification exploits the hydrophobicity which results from the organizational structure of the water molecules surrounding the proteins and the binding surface. When hydrophobic areas bind together, water is effectively released from their surrounding, causing a thermodynamically favourable increase of entropy. Figure 4 sketches this process. Salts that induce hydrophobic interactions are those which increase the ordered structure (decrease the entropy) of the water. Since this process is driven by entropy, temperature can have a strong effect; increasing temperature increases the binding strength. On the other hand, temperature can also affect protein conformation and biological activity, causing the resulting binding effects to be quite complex.

### 3.2.2. Large polar adsorbates

The adsorption of large polar molecules is expected to produce large unfolding of the protein, possibly terminating with its denaturation. *Denaturation* can be considered as a phase transition, characterized by a jump discontinuity of the adsorption isotherm, between *native-like state* (with bulk fractal dimension $D \cong 3$ and surface fractal dimension $D = 2.1 - 2.2$) and the *denatured state* (in which the protein can be seen as an excluded-volume polymer with $D = 5/3$ [57]). Denaturation is therefore associated with a sudden variation of the fractal dimension of the protein.

The protein picture hitherto considered is oversimplified and does not consider that different parts of the protein have different hardness with respect to the adsorbate. The tertiary structure of a protein is indeed characterized by a a distribution of α helices and β strands (the so called *folding pattern*, determined by the primary structure) interconnected by loops and endowed with side chains. It was long believed that the protein could admit only two states with deep energy minima: the native-like state and the denatured one. From the above picture, however, it follows that it should be possible to unfold progressively first the side chains, then the loops and eventually the β strands and α helices. In such a way one should obtain, by suitable choice of environment (adsorbate and pressure regime), many folding states.

It is worth noticing that at least one such intermediate state, the *molten globule*, has been identified. The molten globule (first hypothesized as kinetic intermediate and only later discovered to exist as equilibrium configuration) is a protein state characterized by: (a) the same folding of α helices and β strands as in the native state, (b) a complete unfolding of loops, and (c) an open configuration of side chains (see the short review of ref. [69]).

### 3.2.3. Adsorbates with small polar head and large non-polar tail

This situation is interesting because the distribution of polar sites at the surface of the native protein is covered in such a way that, after monolayer completion, the protein is coated with a non polar (hydrophobic) layer, which can be seen as a kind of Langmuir–Blodgett covering of the protein. The adsorption of a second layer is therefore very difficult, its stabilizing energy being negligible. The expected effects of this coating are the following: (a) a reduction of the protein solubility in aqueous ambient, and (b) a reduced interaction among different proteins. The reasons for (a) are trivial; (b) is due to the fact that the strong electrostatic interaction between proteins (dominated by enthalpic effects) is replaced by the weak interaction between the hydrophobic protein envelops (dominated by entropic effects). It is therefore not fortuitous that the addition of alcohols to aqueous solutions of proteins reduces their solubility and that crystalline proteins without mutual interactions are obtained via this route.

### 3.2.4. The effect of pH

Acidity of proteins is the result of various chemical groups present in aminoacid sequence. Changes in a protein conformation and local environment caused by adsorption can also be seen in their pK values determined using titration method [70, 71, 72]. Haynes et al. showed that the pK values of some residues in the folded conformation differ substantially from their intrinsic values indicating that pK for a given residue is intimately linked with the residue local environment [72]. Their titration curves cover about 19 basic groups; they also indicate that intramolecular electrostatic forces are primary responsible for shifting residue pK values during the protein folding process. Moreover, for some proteins they observed pH-induced conformational changes. Based on the titration data obtained for α–lactalbumin they found that pK values for carboxyl groups in the adsorbed proteins was 1.1 unit higher than for the dissolved protein. This suggests that the sorbent (polystyrene) has a dramatic influence on the properties of protein. Similar shifts were reported for other globular proteins adsorbed on polystyrene [70, 71]. These results indicate that some carboxyl groups are close to sorbent surface or that electrostatic repulsion between other carboxylic groups is diminished as a result of the unfolding of the protein during adsorption.

### 3.3. The adsorbate as a soft adsorbent - the C&M theory

Since the forces responsible for multilayer adsorption have the same nature as, and are of comparable intensity with, the ones which stabilize the condensed phase, *multilayer adsorption can be viewed as a kind of adsorption on reconstructable surfaces.* This conclusion became clear to the author immediately after he presented the model of section 3.2 to the 2nd Symposium on the Effects of Surface Heterogeneity in Adsorption and Catalysis on Solids [73]; transforming this idea in a theory, however, was not easy. This goal was achieved only two years later by Cerofolini and Meda[5], who proposed a 'clustering and melting' (C&M) theory of multilayer adsorption [74]. The essentials of the C&M theory are sketched in the following.

#### 3.3.1. Search for the initial hypotheses

The BET theory can be restated in the following equations:

$$dN/dp = N_m \, d\Theta/dp \tag{23}$$

$$N_m = N_m^0 \, (\text{constant}) \tag{24}$$

$$\Theta(p) = \frac{Cp/p_0}{1-p/p_0} \, \frac{1-(n+1)(p/p_0)^n + n(p/p_0)^{n+1}}{1+(C-1)p/p_0 - C(p/p_0)^{n+1}} \tag{25}$$

the last equation being possibly replaced by the ABSB equation.

Though the system {(23),(24),(25)} may seem a complicate way of writing eq. (3), it however clarifies a few striking features of the BET theory: Though the total number of atoms exposed to the vapour increases with p as adsorption proceeds, because of eq. (24) the total number of sites available to adsorb remains constant; rather, what varies is only the average height $\Theta(p)$ of each adsorbed pile as given by eq. (25).

That assumption (24) is a poor description of physical reality may be considered as an understatement. An extension of the theory which takes into account how the number of sites varies as adsorption proceeds is however

---

[5] The formal analogies between the two descriptions are immediately understood comparing (17) with (41), (18) with (42), and (20) with (43).

non-trivial. In fact, the whole paradigm of physical adsorption is based on the assumption that the adsorption sites are fixed [13], and only few works have been devoted to extend the theory to situations where the adsorbing area does not remain constant [60, 63, 73]. The considered situations involved the adsorption of polar molecules on proteins, because these adsorbents are known to undergo deep structural changes during adsorption.

It was however neglected that in the absence of reconstruction[6] *the adsorption energy is also a cohesion energy for the adsorbed film, so that in multilayer adsorption the adsorption energy is always comparable with the energy which stabilizes each adsorbing site*. This implies that *multilayer adsorption represents an ideal arena for the application of models in which the number of sites is not assumed to remain constant*.

Before proceeding with a formal extension of such models (in particular, the one proposed in ref. [63]) to multilayer adsorption, we however intend to discuss two phenomena which play an important role in multilayer adsorption and which are not considered in the BET theory: clustering and melting.

*Clustering*. We continue to describe multilayer adsorption in terms of piles with height increasing with x.

Adjacent piles, however, cannot be considered as mutually non-interacting, because 'vertical forces' have the same nature as 'lateral forces'. While vertical forces are responsible for the persistence of a non-null adsorption energy (whatever large is the pile height), lateral forces are responsible for the formation of new adsorption sites. This phenomenon, explicitly excluded in the BET theory, is referred to as 'clustering'.

Clustering produces a kind of pile branching and is therefore responsible for roughening of the adsorbate surface with a net increase of current adsorption sites as the adsorption proceeds.

In this picture each molecule belongs to a pile only and, at most, contributes to the generation of new adsorption sites via clustering. Because of this situation, each newly adsorbed molecule will be arranged in a low-coordination configuration. Therefore, in the clustering stages of multilayer adsorption the adsorbed molecules may be in one or the other of the following three configurations:

(*) *Molecules at the adsorbent surface*: each of them generates a vertical pile and is held at the surface by an energy $q_1$ related to the secondary (i.e., electrostatic or van der Waals) forces exerted between adsorbent atoms and adsorbate atoms.

---

[6] Reconstruction however plays an important role in adsorption. The 'paradoxical' demonstration that the enthalpic factor of the saturated-vapour pressure is related to a bulk rather than surface cohesion energy requires in an essential way a reconstruction process [75, section 91]. Assuming, as done in the BET theory, that the adsorption energy in upper layers is given by $q_0$ implies that adsorption and complete reconstruction are a unique process - that is inconsistent with a description in terms of piles.

(**) *Molecules at adsorption sites produced by clustering*: each of them generates a lateral pile and is held at the newly born site by an energy $q_l$ related to the secondary forces exerted between adsorbed atoms alone.

(***) *Molecules which are not the generators of new piles*: each of them is held in a pile by an energy $q_2$ related to the secondary forces exerted between adsorbent atoms alone.

Due to the nature of secondary forces (which are pairwise additive and decay rapidly with separation), the adsorption energy is controlled by the nature and number of nearest neighbours of each molecule. Since molecules in configurations (**) and (***) (as well as adsorbate molecules in the liquid phase) are stabilized by forces of the same nature, the adsorption energy is simply related to the number of neighbours, that gives

$$q_2 < q_l < q_0 \tag{26}$$

where $q_0$ is the binding energy of an adsorbate molecule in liquid phase. Of course, even assuming that the surface is truly energetically homogeneous, the intrinsically random nature of clustering will produce a family of new sites characterized by a distribution of adsorption energies; the value $q_l$ in (26) is any value in such a distribution.

It is noted that in almost all cases of practical interest the secondary forces which keep the molecule adsorbed at the surface are stronger than the ones which keep it in the liquid state, so that an inequality can also be established for the adsorption energies of molecules in configurations (*) and (**):

$$q_l < q_1 \tag{27}$$

Moreover, even though the coordination of a molecule in a liquid is higher than at the surface of a solid, the secondary forces are usually strong enough to guarantee the following condition: $q_l < q_0 < q_1$. Adsorption described by isotherms of types III and V of Brunauer's classification does not run in this case.

*Melting.* Since new adsorption sites are continuously created as adsorption proceeds, the adsorbed film will be characterized by a branched structure. This structure will progressively become denser and denser, so that there will eventually be a situation in which any newly adsorbed molecule will be surrounded by as many molecules as in the condensed phase. In other words, while in the early stages of adsorption each newly adsorbed molecule is (always) a *propagator* of its pile and (occasionally) a *generator* of new piles, in the final stages each newly adsorbed molecule belongs to several piles, so that it behaves as a pile *terminator*.

In this case the adsorption of any new molecule will be characterized by an adsorption energy very close to $q_0$ and will result in a kind of local 'melting', eventually responsible for the formation of an adsorbed liquid film. Since this process is characterized by a discontinuous change of the adsorption enthalpy (from a value between $q_2$ and $q_l$ to a value close to $q_0$), it can be viewed as a kind of first order phase transition.

The transition of the adsorption energy from $q_1$ to $q_0$ as N increases does not occur steadily (as does happen for Steele's extension of the BET theory [76] or even for the FHH theory); rather q(N) has a maximum ($q_1$) for adsorption at the original surface, is lower than $q_0$ for adsorption in the second or upper layers, and eventually attains a value close to $q_0$ when N exceeds a characteristic value $\overline{N}$, above which melting starts to occur:

$$q(N) \begin{cases} = q_1 & \text{for} \quad 0 \le N << N_m^0 \\ < q_0 & \text{for} \quad N_m^0 < N \lesssim \overline{N} \\ \cong q_0 & \text{for} \quad \overline{N} \lesssim N \end{cases} \tag{28}$$

Since with this choice q(N) behaves definitively as $q_0$ for large N, the C&M theory can predict a phase transition at $p = p_0$. The final stages of multilayer adsorption are therefore characterized by a kind of local melting produced by filling molecular voids in the adsorbed slab. This process can also be viewed as a kind of reconstruction occurring in the final stages.

It is noted that while eqs. (26) and (27) give the adsorption energy in each particular configuration (irrespective of the value of N), eq. (28) gives the dependence of the average adsorption energy on the adsorbed amount.

### 3.3.2. C&M axioms

The following axioms can be used to formalize the above conclusions keeping the computational difficulties to a minimum:

> (C&M1): Adsorption takes place either in vertical piles, growing on the original adsorption sites, or in lateral piles, growing on sites resulting from clustering in the adsorbed film.

Denoting with $N_v$ the vapour amount adsorbed in vertical piles and wuth $N_l$ the amount adsorbed in lateral piles, the total adsorbed amount N is given by

$$N(p) = N_v(p) + N_l(p) \tag{29}$$

The second and third axioms specify $N_v(p)$ and $N_l(p)$, once the amounts of adsorption sites are known:

> (C&M2): The amount $N_v(p)$ adsorbed in vertical piles is given by

$$N_v(p) = \int_0^p N_{m,v} \frac{d\Theta_v(p)}{dp} dp \tag{30}$$

where $\Theta_v(p)$ is the adsorption isotherm specifying the filling of the $N_{m,v}$ 'vertical' surface sites.

(C&M3): The amount $N_l(p)$ adsorbed in lateral piles is given by

$$N_l(p) = \int_0^p N_{m,l} \frac{d\Theta_l(p)}{dp} dp \tag{31}$$

where $\Theta_l(p)$ is the adsorption isotherm specifying the filling of the $N_{m,l}$ lateral adsorption sites formed because of clustering.

Equations (30) and (31) are just integral forms of eq. (23) without constraint (24). When $N_{m,v}$ or $N_{m,l}$ are not constant with p, eqs. (30) or (31) are not rigorous, but rather provide only an approximate description. The limits of such a description are discussed in ref. [63].

The fourth axiom states that the adsorbent is hard, i.e. its surface does not reconstruct during the process:

(C&M4): The amount $N_{m,v}$ of surface sites remains constant during adsorption,

$$\forall p : N_{m,v}(p) = N_m^0$$

Much more complex is to specify the evolution of $N_{m,l}$ as adsorption proceeds. We assume the following axioms:

(C&M5): In the monolayer range the adsorbed molecules provide the sites for lateral piles:

$$p \to 0 \Rightarrow dN_{m,l}(p) \sim \varepsilon N_m^0 \Theta_v'(p) dp \tag{32}$$

$\varepsilon$ being a constant of proportionality giving the clustering efficiency in a mean field approximation.

(C&M6): In the multilayer regime the adsorption of $dN_l$ molecules increases the current surface area by an amount b $dN_l$, and the newly formed area has the same density of adsorption sites as the pre-existing one:

$$N >> N_m^0 \Rightarrow dN_{m,l} \cong N_{m,l} b \, dN_l = N_{m,l}^2 b \Theta_l'(p) dp \tag{33}$$

The quantity b represents the increase of surface area resulting from the adsorption of one molecule and will be considered independent of N, but possibly dependent on the temperature T (in particular, b is expected to decrease with T).

In view of the ability of eq. (4) to describe adsorption in the monolayer region, we assume that C&M isotherm must reduce to the ABSB equation when $\varepsilon = 0$ or $b = 0$. The following axiom allows this need to be satisfied:

(C&M7): The vertical isotherm is given by

$$\Theta_v(p) = \frac{1}{1-\kappa_v\, p/p_0}\,\frac{C_v \kappa\, p/p_0}{1+(C_v-1)\kappa_v\, p/p_0} \tag{34}$$

where $C_v$ and $\kappa_v$ have the same meaning as C and κ in eq. (4). Consistency reasons suggest the following expression for $\Theta_l(p)$:

(C&M8): The lateral isotherm is given by

$$\Theta_l(p) = \frac{1}{1-\kappa_l\, p/p_0}\,\frac{C_l \kappa_l\, p/p_0}{1+(C_l-1)\kappa_l\, p/p_0} \tag{35}$$

where $C_l$ is given by

$$C_l = \frac{z_l}{z_0}\exp\left(\frac{q_l - q_0}{k_B T}\right) \tag{36}$$

and $\kappa_l$ has the same meaning as κ in eq. (4).

It is noted that in postulating the above axioms, we have retained some axioms (like (BET1), (BET2) and (BET6)) which allow the derivation of eqs. (34) and (35). In principle $\kappa_v \neq \kappa_l$, for simplicity we shall however assume that they have a common value, $\kappa_v = \kappa_l = \kappa$.

On intuitive grounds one realizes that the combination of axioms from (C&M1) to (C&M8) leads to clusters whose surface area increases significantly, for ε, b >0, as the adsorption proceeds. In next part it will be shown that this increase results in a divergence of $N_l(p)$ at a certain pressure $\overline{p}$. The last axiom is a way to deal with this divergence:

(C&M9): When the adsorbate density ρ equals the bulk density $\rho_0$, the adsorbed film melts and reverts to a liquid film.

### 3.3.3. The C&M equation

Though the detailed specification of N(p) requires the knowledge of $\Theta_v(p)$ and $\Theta_l(p)$, the C&M theory can be developed to a large extent without specifying these isotherms. In particular, the following part does require only some asymptotic properties of these isotherms and would hold true even for other choices of $\Theta_v(p)$ and $\Theta_l(p)$.

To develop quantitatively the theory, we have to merge axioms (C&M5) and (C&M6) in a unique differential equation for $N_{m,l}$, whose initial condition is specified by axiom (C&M5):

$$N_{m,l}(0) = 0 \tag{37}$$

Since $\Theta_v(p)$ and $\Theta_l(p)$ behave in the same way for large p ($p \rightarrow p_0/\kappa \Rightarrow \Theta_v(p) \sim \Theta_l(p) \sim (1 - \kappa\, p/p_0)^{-1}$), and $N_{m,l}^2\, b\, \Theta_l'(p)$ vanishes faster than $N_m^0\, \Theta_v'(p)$ for $p \rightarrow 0$, one could simply try a linear composition of (32) and (33):

$$\frac{dN_{m,l}}{dp} = N_m^0 \Theta_v'(p) + N_{m,l}^2 b \Theta_l'(p) \tag{38}$$

This equation, however, has a serious disadvantage - it cannot be solved analytically in closed form.

The composition law (38) can however be modified in such a way as to yield a differential equation which can be solved by separation of variables. For, an integration of eq. (32),

$$p \rightarrow 0 \Rightarrow N_{m,l} \sim \varepsilon N_m^0 \Theta_v(p)$$

allows it to be written in the form

$$p \rightarrow 0 \Rightarrow dN_{m,l} \sim N_{m,l}^2 \frac{\Theta_v'(p)}{\varepsilon N_m^0 \Theta_v^2(p)} dp \tag{39}$$

It is emphasized that eqs. (32) and (39) are equivalent only for the initial condition (37), which is however guaranteed by axiom (C&M5). Since

$$p \rightarrow p_0/\kappa \Rightarrow \Theta_v'(p)/\Theta_v^2(p) \rightarrow 0 \tag{40}$$

and

$$N_{m,l}(p) \rightarrow 0 \Rightarrow bN_{m,l}^2 \Theta_l'(p) \rightarrow 0$$

the descriptions given by axioms (C&M5) and (C&M6) can therefore be merged in a unique differential equation,

$$\frac{dN_{m,l}}{dp} = N_{m,l}^2 \left( \frac{\Theta_v'(p)}{\varepsilon N_m^0 \Theta_v^2(p)} + b\Theta_l'(p) \right) \tag{41}$$

Equation (41) can be solved by separation of variables:

$$N_{m,l}(p) = \frac{\varepsilon N_m^0 \Theta_v(p)}{1 - \varepsilon N_m^0 b \Theta_v(p) \Theta_l(p)} \tag{42}$$

Inserting eq. (42) into eq. (31) one gets a differential equation,

$$dN_l = \varepsilon N_m^0 \Theta_v(p) \frac{d\Theta_l}{1 - \varepsilon N_m^0 b \Theta_v(p) \Theta_l(p)}$$

whose solution is

$$N_l(p) = -\frac{1}{b} \ln\left(1 - \varepsilon N_m^0 b \Theta_v(p) \Theta_l(p)\right) \tag{43}$$

Inserting eqs. (30) and (43) in (29) one gets the general C&M equation:

$$N(p) = N_m^0 \Theta_v(p) - \frac{1}{b} \ln\left(1 - \varepsilon N_m^0 b \Theta_v(p) \Theta_l(p)\right) \tag{44}$$

The derivation of eq. (44) is to a large extent independent of the choice of $\Theta_v(p)$ and $\Theta_l(p)$; this derivation would hold true even for different isotherms. It is however noted that the validity of the above derivation requires that if another 'vertical' isotherm is intended to replace the BET equation, it must satisfy condition (40). If axioms (C&M7) and (C&M8) are accepted, the general C&M equation is specialized substituting eqs. (34) and (35) for $\Theta_v(p)$ and $\Theta_l(p)$, respectively, in eq. (44). The resulting equation, whose explicit expression will be given later (eq. (51)), is referred to as 'special C&M isotherm'.

As already noted, the modified form (4) of the BET equation provides an accurate description of adsorption around the monolayer. Therefore, the C&M theory has two needs: not only must eq. (44) reduce to eq. (4) when the parameters ε and b reduce to 0, but also eq. (44) must be very close to eq. (4) around the monolayer.

The choice of eqs. (34) and (35) for $\Theta_v(p)$ and $\Theta_l(p)$, which has led to the special C&M equation, guarantees that the first need is satisfied.

The second need puts nonholonomic contraints to the parameters ε and b. These constraints are determined by noting that for

$$\varepsilon N_m^0 b \Theta_v(p) \Theta_l(p) << 1 \tag{45}$$

the first order Maclaurin expansion of the logarithm gives

$$N(p) \cong N_m^0 \Theta_v(p)(1 + \varepsilon \Theta_l(p))$$

where the dependence on b has disappeared. The effect of clustering is to produce an overestimate of the monolayer amount when data are analyzed with the BET equation. Since the monolayer region for $\Theta_v(p)$ is observed in the pressure region

$\{ p : 1/C_v << \kappa\, p/p_0 \lesssim x_u\}$, (with $x_u \cong 0.3$), the condition $N(p) \cong N_m{}^0\, \Theta_v(p)$ is surely satisfied for $\varepsilon << 1$, though a less restrictive condition can also be given: $\varepsilon C_l\, x_u << 1$. In such cases condition (45) is satisfied even for

$$N_m^0 b \approx 1 \tag{46}$$

### 3.3.4. Consequences of the C&M theory

*Condensation.* Since both $\Theta_v(p)$ and $\Theta_l(p)$ diverge for $p \to p_0/\kappa$, whatever is the value of $\varepsilon\, N_m{}^0\, b$ there exists a value $\overline{p}$ of pressure such that

$$\varepsilon N_m^0 b \Theta_v(\overline{p}) \Theta_l(\overline{p}) = 1 \tag{47}$$

so that

$$p \to \overline{p} \Rightarrow N_{m,l}(\overline{p}) \to +\infty \tag{48}$$

Even N diverges in the same limit. A divergence of N may also be obtained for vertical isotherms $\Theta_v(p)$ which do not diverge; for the divergence of N(p) it suffices that condition (47) is satisfied. Of course, one may have $\overline{p} < p_0$, $\overline{p} = p_0$, or $\overline{p} > p_0$ according to the values of the parameters contained in eq. (47).

However, *if the divergence of* N *is interpreted in terms of vapour condensation, the intermediate case* $\overline{p} = p_0$ *must hold true*. Putting $\overline{p} = p_0$ in eq. (47) one gets a mutual relationship among the parameters of the theory:

$$\varepsilon N_m^0 b \Theta_v(p_0) \Theta_l(p_0) = 1 \tag{49}$$

For $\Theta_v(p)$ and $\Theta_l(p)$ given by the ABSB equation, condition (49) becomes

$$\varepsilon N_m^0 b / (1-\kappa)^2 = 1 \tag{50}$$

Expressing $\varepsilon$ as a function of the remaining ones, the special C&M equation becomes

$$\frac{N(p)}{N_m^0} = \frac{1}{1-\kappa\frac{p}{p_0}} \frac{C_v \kappa \frac{p}{p_0}}{1+(C_v-1)\kappa\frac{p}{p_0}} - \frac{1}{N_m^0 b} \ln\left(1 - \frac{(1-\kappa)^2}{\left(1-\kappa\frac{p}{p_0}\right)^2} \frac{C_v \kappa \frac{p}{p_0}}{1+(C_v-1)\kappa\frac{p}{p_0}} \frac{C_l \kappa \frac{p}{p_0}}{1+(C_l-1)\kappa\frac{p}{p_0}}\right) \tag{51}$$

*Roughening and flattening.* Equation (42) states that *clustering produces a progressive roughening* of the current surface as adsorption proceeds. The divergence (48) describes quantitatively this phenomenon.

An infinite surface area is consistent with a space-filling surface. Of course, a space-filling surface is a mathematical, rather than physical, concept because of the limitations imposed by the atomic structure of matter (these limitations hold true in other situations too, like for fractal surfaces, whose scale invariance is usually valid in a quite restricted scale range). In view of the atomic nature of matter, the maximum allowed value of $N_{m,l}$ can be estimated by the condition

$$\nu\left(N + N_{m,l}\right) \approx \rho_0 L \tag{52}$$

where $\nu$ is the number of atoms per adsorbed molecule, $\rho_0$ is the adsorbate atomic density in liquid phase ($\rho_0 \approx 10^{23}$ cm$^{-3}$), and L is the thickness of the adsorption slab. The final result of this phenomenon, which we have described as a kind of melting, is the formation of a liquid flat slab at the surface, so that *melting produces a progressive flattening which is completed when condensation has completed.*

In this way C&M have taken into account that the formation of a liquid layer implies the raise of forces like surface tension, not accounted in the BET theory. In a way, the C&M theory merges together seemingly contradictory behaviours (growth in piles and flat surface), so avoiding *ad hoc* corrections, like McMillan and Teller's modification of the BET equation [77].

### 3.3.5. Adsorption in restricted geometries

We assume that the restricted geometry (pore) may be characterized by walls separated by a distance $\delta$. Let $\overline{L}_f := \delta/2$; $\overline{L}_f$ is the maximum slab thickness which allows different zones not to interfere with one another. If $\delta$ is a random variable (i.e., in the presence of a pore-size distribution), $\overline{L}_f$ is in turn a random variable depending on the particular pore considered. To understand the hysteresis due to a restricted geometry, it is however convenient to ignore such a geometric heterogeneity and limit the analysis to the ideal case of a fixed pore size. The case of fixed pore size has a physical counterpart in adsorption in zeolites, though their growth techniques eventually leads to materials with two characteristic feature sizes: the 'inner' pores (empty spaces of zeolite cages) and 'outer' pores (empty spaces in between grains). The existence of an 'inner' surface (with $N^0_{m,in}$ sites), where adsorption is subjected to geometric constraints, and an 'outer' surface (with $N^0_{m,out}$ sites), where adsorption is subjected to more relaxed geometric constraints, characterizes many porous adsorbents. For adsorbents of practical interest $N^0_{m,in} >> N^0_{m,out}$, that allows us to ignore the contribution to the isotherm due to adsorption on the outer surface. In the following $N_m^0$ will be a contracted form for $N_{m,in}^0$.

*Filling the pore.* The basic idea to describe adsorption in restricted geometries is the following: Let L(p) denote the average thickness of the adsorbed slab: $L(p) := N(p)/\rho$, where $\rho$ is the actual adsorbate density. Since $\rho < \rho_0$, a lower bound to L(p) is immediately obtained: $L(p) > N(p)/\rho_0$. When L(p) becomes comparable with $\delta/2$, each newly adsorbed molecule has a chance to belong to two or more piles, so that it behaves as pile terminator. This occurrence will also be expressed with the statement that 'the adsorbate is dense in the pore'.

It does seem reasonable to assume that eq. (44) is not affected by the geometric constraint until the adsorbed film is not dense inside the pore; this occurs when the addition of one other monolayer (involving approximately $N_{m,l}(p)$ molecules) to the pre-existing $N^{\uparrow}(p)$ adsorbed molecules fills the pore:

$$N^{\uparrow}(p) + N_{m,l}(p) \approx N_m^0 n \tag{53}$$

where n is related to the pore-to-adsorbate diameter ratio by the relationship $n = \delta/2\lambda$.

Let $\overline{p}_f$ be the solution of eq. (53) considered as an equation in p, and define the amounts of adsorbed molecules $\overline{N}_f$ and adsorption sites $\overline{N}_m^f$ at $p = \overline{p}_f$ : $\overline{N}_f := N^{\uparrow}(\overline{p}_f)$ and $\overline{N}_m^f := N_{m,l}(\overline{p}_f)$. For $p > \overline{p}_f$ one may consistently assume that the adsorption of one further molecule does not roughen, but rather flattens, the surface. This implies that the newly formed area per adsorbed molecule is negative, so that the adsorption isotherm will eventually display a downward concavity.

Modelling this process does not seem straightforward, so that at this moment we are unable to specify the adsorption isotherm in the whole pressure range:

$$\frac{N(p)}{N_m^0} = \begin{cases} \Theta_v(p) - \dfrac{1}{N_m^0 b}\ln\left(1 - \varepsilon N_m^0 b\Theta_v(p)\Theta_l(p)\right) & \text{for } 0 \le p < \overline{p}_f \\ ? & \text{for } \overline{p}_f \le p \end{cases} \tag{54}$$

*Emptying the pore.* The desorption-branch isotherm is obtained by assuming an additional energy $\overline{q}_f$ stabilizing the film inside the pore:

$$\overline{p}_f = \overline{p}\exp\left(-\frac{\overline{q}_f}{k_B T}\right) \tag{55}$$

Note that if pore filling does indeed occur before heterogeneous condensation (as hypothesized at the beginning of this section), then $\overline{p}_f < \overline{p}$, that implies $\overline{q}_f > 0$, i.e. the existence of a stabilizing energy.

In so doing the adsorption isotherm describing the phenomena occurring while emptying the pore is described by the same equation which describes the adsorption isotherm in the desorption branch provided that b is negative ($b = -|b^{\downarrow}|$) and $\overline{p}^{\downarrow}$ is substituted for $\overline{p}_f$ in eq. (54):

$$\frac{N^{\downarrow}(p)}{N_m^0} = \begin{cases} \Theta_v^{\downarrow}(p) - \dfrac{1}{N_m^0 \left| b^{\downarrow} \right|} \ln\left(1 - \varepsilon^{\downarrow} N_m^0 \left| b^{\downarrow} \right| \Theta_v^{\downarrow}(p) \Theta_l^{\downarrow}(p)\right) & \text{for } 0 \le p < p^{\downarrow} \\ n & \text{for } p^{\downarrow} \le p \end{cases} \qquad (56)$$

where the pressure $p^{\downarrow}$ is the root of the equation

$$\Theta_v^{\downarrow}(p) + \frac{1}{N_m^0 \left| b^{\downarrow} \right|} \ln\left(1 - \varepsilon^{\downarrow} N_m^0 \left| b^{\downarrow} \right| \Theta_v^{\downarrow}(p) \Theta_l^{\downarrow}(p)\right) = n$$

### 3.3.6. The reach of the C&M theory

The C&M theory contains six parameters ($N_m^0, \kappa, b, \varepsilon, C_v$, and $C_l$) against the three parameters ($N_m^0, \kappa$ and C) of the ABSB equation. However, while on one side the ABSB equation requires at least one additional corrective parameter to account for condensation at $p = p_0$, on another side eq. (50) reduces to five the number of free parameters required by the C&M theory. Moreover, $N_m^0$ is a scale parameter; expressing therefore N and 1/b in units of $N_m^0$, the special C&M equation depends on four among the five dimensionless parameters $N_m^0 b$, $\kappa, C_v, C_l$ and $\varepsilon$.

Even limiting the goal to a purely numerical exercise, the study of the behaviours of eq. (51) is manifestly a difficult task, because of the numerous parameters it contains. Such an analysis, carried out in a preliminary form in ref. [74], shows however that eq. (51) is able to account for the Brunauer classification of adsorption isotherms.

Figure 5 shows indeed the graphs drawn by specializing the C&M equation with the following choices of parameters: $C_v = 100$, $C_l = 1$, $\kappa = 0.1$ and $N_m^0 b = 10$ for type I isotherm; $C_v = 100$, $C_l = 1$, $\kappa = 0.6$ and $N_m^0 b = 1$ for type II isotherm; $C_v = 2$, $C_l = 1$, $\kappa = 0.6$ and $N_m^0 b = 1$ for type III isotherm; $C_v = 100$, $C_l = 1$, $\kappa = 0.8$, $N_m^0 b = 0.1$ and $n = 5$ for type IV isotherm; and $C_v = 2$, $C_l = 1$, $\kappa = 0.8$, $N_m^0 b = 0.1$ and $n = 5$ for type V isotherm.

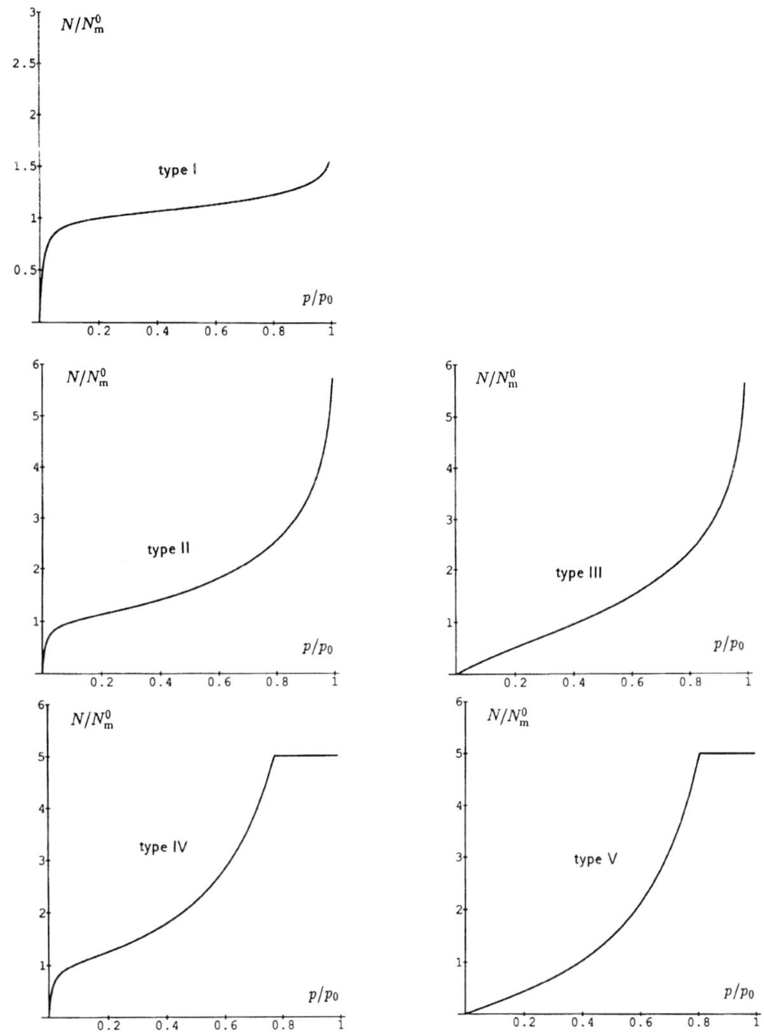


Figure 5. How the C&M equation accounts for Brunauer classification of adsorption isotherms.

The ability of the C&M isotherm to reproduce observed behaviours is not limited to the near monolayer region.

Figure 6 shows that, quite irrespective of the value of s, there exists a choice of parameters which allow the C&M isotherm to reproduce the assigned FHH one. The parameters used for the plots are: $C = C_v = 100$, $C_l = 0.1$, $\kappa_v = \kappa_l$; $\kappa_v = 0.85$

and $N_m^0 b = 0.45$ for $s = 2$; $\kappa_v = 0.6$ and $N_m^0 b = 0.8$ for $s = 3$; and $\kappa_v = 0.1$ and $N_m^0 b = 5.0$ for $s = 15$.

For the isotherms of types IV and V a maximum coverage n has been imposed via an external cut-off and no hysteresis has been considered. The application of eqs. (56) and (54) seems however possible and is expected to lead to a new description of pore filling and emptying in a region difficult to model.

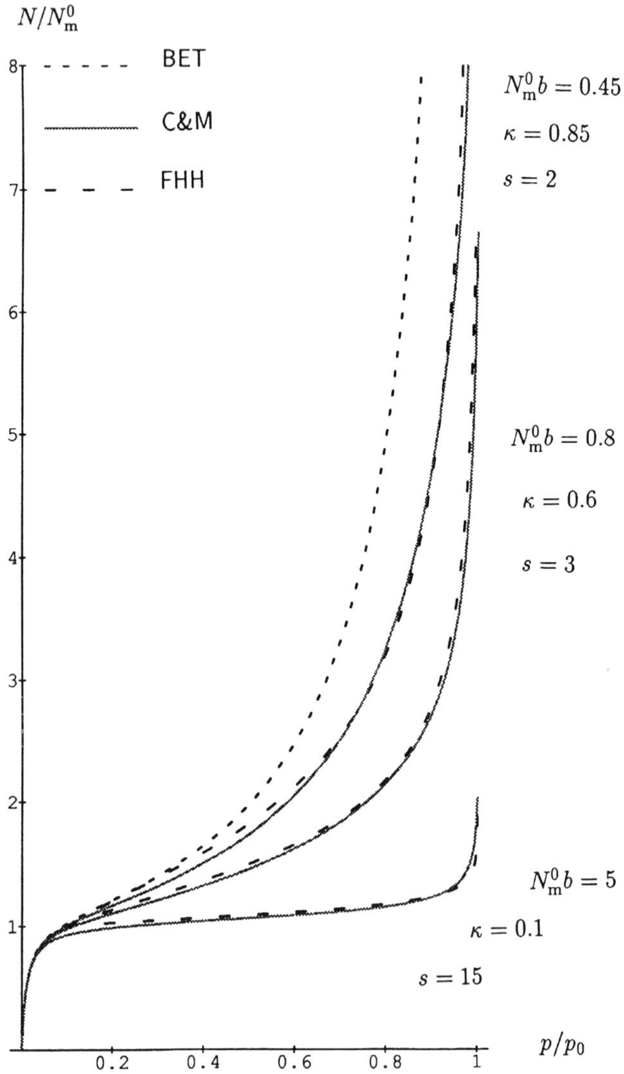


Figure 6. Fitting FHH behaviours with C&M isotherms.

## 4. SOFT ADSORBATES

Many adsorbates of practical interest, like $N_2$ and noble gases, may be considered structureless, since the energy required to excite internal degrees of freedom is much higher than the adsorption energy. Many other adsorbents, however, may assume different conformations whether intrinsically (e.g., the boat-chair conformation of cyclohexane) or in relation to the environment where they are located.

Even when thermally stable, such molecules may behave as soft with respect to adsorption. Thermal stability and hardness, in fact, are not synonymous: compared with alkanes, proteins have a lower thermal stability but a higher hardness. Protein hardness (which is manifested in the constancy of protein shape - a key functional factor) is imparted to proteins by a lot of thermally labile hydrogen bonds. Heating is not the unique way to destroy hardness. As discussed in the previous section, electrostatic interactions cleave the shape-preserving secondary bonds. Sufficiently large molecules with few intrachain hydrogen bonds may be soft because of the possibility to assume a lot of tautomeric configurations. That the adsorbates too may be soft impact severely the determination of surface areas.

### 4.1. How soft adsorbates affect the determination of the fractal dimension of adsorbent surfaces

The analysis of Avnir, Pfeifer and Farin was based on the implicit assumption that the adsorbate conformation remains unchanged after adsorption. The molecular shape, however, is expected to be modified by the intense fields in the vicinity of a surface.

An example showing that adsorbed molecules do indeed undergo deep structural changes after adsorption was given by Tertykh in ref. [78]. That work was dedicated to the search of advantageous routes for the preparation of dense hydrophobic coatings on silica surfaces. In particular, it was found that silica, prepared in such a way as to have $(-O)_3SiH$ terminations, forms hydrophobic surfaces after reaction with alk-1-enes, according to the pathway

$$(-O)_3SiH + CH_2 = CH(CH_2)_{n-3}CH_3 \longrightarrow (-O)_3Si(CH_2)_{n-1}CH_3$$

Modified silica surfaces were prepared in this way by grafting alkane moieties with n = 6,8,10,14,16 or 18. The monolayer amount was found to depend on the adsorbate cross section, $N_m = N_m(\sigma_n)$, as given in table 1. Though the amount of alkane moiety required to complete the coating was found to decrease with n (column 2 of table 1), the hydrophobic nature of the surface was found to increase with n.

Table 1
Chemisorbed amount $N_m(\sigma_n)$ of hydrocarbon moieties on a silica surface as a function of the number n of carbon atoms in the moiety.

| n | $N_m(\sigma_n)$ ($\mu$mol·m$^{-2}$) | $\frac{N_m(\sigma_n)}{N_m(\sigma_6)}\left(\frac{n}{6}\right)^{2/3}$ | $\frac{N_m(\sigma_n)}{N_m(\sigma_6)}\left(\frac{n}{6}\right)$ |
|---|---|---|---|
| 6 | 3.70 | 1 | 1 |
| 8 | 3.01 | 0.98 | 1.08 |
| 10 | 2.05 | 0.78 | 0.92 |
| 14 | 0.99 | 0.47 | 0.62 |
| 16 | 0.88 | 0.46 | 0.63 |
| 18 | 0.71 | 0.40 | 0.58 |

The data for n = 6 and n = 18 have been taken from ref. [78]; the others have been communicated personally by Tertykh, to whom the authors are greatly indebted.

The very fact that the hydrophobicity increases with n faster than the rate of decrease of $N_m(\sigma_n)$ was noted by Tertykh himself, who concluded that hydrocarbon moieties with high n "do not have a brush-like structure, rather these groups are inclined and they pave the silica matrix surface" [78].

However, though column 3 of table 1 shows indeed that the assumption that the hydrocarbon moiety has a globular shape and preserves it after adsorption (so that the number of surface sites covered by the moiety varies as $n^{2/3}$) does not account for the observed hydrophobicity increasing with n, column 4 shows that even the assumption that each atom in the moiety covers a surface site (so that the number of surface sites covered by the moiety varies as n) is insufficient for that.

A fractal analysis (plot of log $N_m(\sigma_n)$ vs. log $\sigma_n$, whose slope gives D/2) of Tertykh data is possible only specifying $\sigma_n$ as a function of n. Taking $\sigma_n \propto n^{2/3}$ this analysis gives $D \cong 4$ (> 3, absurd result which discards the hypothesis), while assuming $\sigma_n \propto n$ gives $D \cong 3$.

Tertykh's data, therefore, give evidence for a space-filling surface (with $D \cong 3$) *and* a completely unfolded moiety (with $\sigma_n \propto n$)[7]. Except for this extreme situation, however, in all other cases *information on the fractal dimension of the surface can be obtained only if the degree of adsorbate unfolding is known.*

Previous determinations of D, which neglected the conformational changes resulting after adsorption, could have been affected by the soft nature of the adsorbate. In other words, when molecules with easily excitable internal degrees of freedom are used as adsorption probes, extreme care must be taken in using

[7] The polar sites on the alkane-coated silica surface are the Si–C grafting sites, because of electron-charge transfer from silicon to carbon. The increasing hydrophobicity with n is therefore understood in terms of decreasing amount of polar sites: an increase of n by a factor of 3 produces a decrease of the grafting sites by a factor of 5.

their bulk (gas- or liquid-phase) data of σ in the fractal analysis of the adsorbent surface.

### 4.2. Adsorption of proteins

Though the adsorption of proteins on surfaces may be considered as a special case of adsorption of polymers [79], the interactions between surfaces and proteins play a special role both in natural and industrial processes: The earliest forms of life might have been formed through assembly of molecules on clays; the adsorption-desorption of proteins play a fundamental role in the regulation of metabolism, control of disease, and repair of tissue damage; and the response of the immuno-system to antigens is expected to be initiated by the adsorption of a protein on a proteic receptor, thus running in a situation of adsorption of a soft adsorbate on a soft adsorbate.

Drug delivery, design of biocompatible materials for prostheses, and protein purification are just a few examples of industrial importance where protein adsorption plays a fundamental role [80, 81, 82, 83, 84]. The 3D conformation of proteins critically affects chromatographic separation. If the protein did not have a stable conformation, all of the aminoacid side chains would potentially influence its interaction with a stationary-phase surface. However, when the protein assumes a 3D conformation, only the aminoacids on the 'outside' of the molecule are available for surface interactions. The aminoacids 'trapped on the inside' of the molecule cannot interact with a surface, and therefore do not affect the adsorption process. However, any change in the experimental conditions (especially mobile-phase parameters such as pH, temperature, etc.) can potentially modify the 3D conformation, exposing new aminoacids to the surface, and altering the binding and elusion characteristics of the molecule.

The 3D conformation also creates binding site for affinity interactions [85]. Affinity results from the same basic physico-chemical interactions (ionic, hydrophobic, hydrogen bonding), but only when certain functional groups are arranged in a particular stereochemical interaction (fig. 7). The arrayed

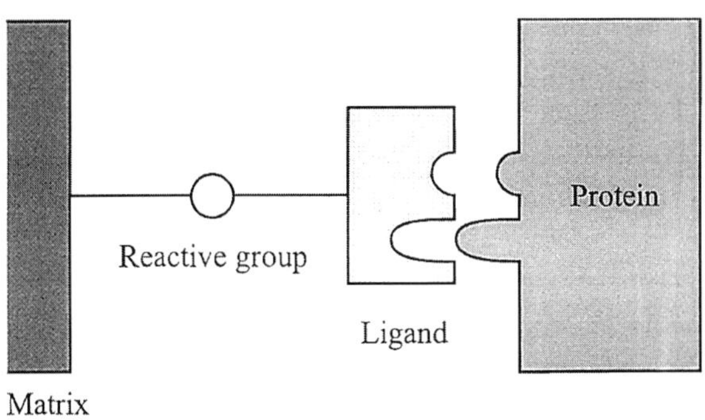


Figure 7. Binding of protein in affinity chromatography.

functional groups in the binding site of one protein fit like a 'lock and key' into complementary groups in the binding site of the other protein [86]. Obviously, correct 3D conformation is necessary for affinity interaction to occur. The ligand can be biologically specific, such as peptide or antibody (glycoproteins), or it can exploit nonspecific interactions with lectines, dyes or hydrophobic moieties. Separation based on specific protein-ligand affinities tend to be unique for different proteins.

One of the proteins which are used in affinity chromatography to adsorb other proteins are lectins, which bind carbohydrates reversibly [87]. They are also useful as affinity ligand for glycoproteins. Lectins are chosen for affinity chromatography according to their specificity and tightness of binding. For example, Concavallin A binds glucose- or mannose-containing proteins, whereas wheat germ lectin binds proteins with N-acetylglucosamine [88]. They can be used as stationary phase but many of them are conjugated to a matrix.

The steps involved in the process of protein adsorption may be listed as follows: (i) protein transport from the solution to the surface, (ii) attachment at the surface, (iii) relaxation of the adsorbed protein to a more stable configuration, (iv) detachment from the surface, and (v) transport away from the surface. The desorption steps (iv) and (v) are not rate-limiting, because a strong driving force (like the use of displacing molecules) is usually required to cleave the numerous bonds linking the protein to the surface. The protein concentration C in the liquid phase is expected to determine the rate-determining step: the overall process is limited by transport at low C or by attachment-relaxation at high C. In the following the latter situation will be investigated.

### 4.2.1. One-state linear model

In the most elementary formulation, there is no relaxation and protein adsorption-desorption is described in terms of Langmuir's kinetics:

$$\frac{dm}{dt} = kC(1 - m\sigma) - \frac{m}{\tau_d}$$

where m is the total number per unit area of adsorbed proteins, $\sigma$ is their cross section, and k is a suitable kinetic coefficient.

This equation predicts that:

- starting from an initial condition m(0) = 0 the amount of adsorbed proteins attains a steady-state value $m_{ss}$,

$$m_{ss} = \frac{kC}{kC\sigma + 1/\tau_d}$$

according to the equation

$$m(t) = m_{ss}\left[1 - \exp(-t/\tau_1)\right] \tag{57}$$

with

$$\frac{1}{\tau_1} = \frac{1}{\tau_d} + \sigma k C$$

while

- starting from an initial condition $m(0) = m_0 \neq 0$ and exposing the surface to the pure solvent, protein molecules are released according to the equation $m(t) = m_0 \exp(-t / \tau_d)$.

None of these behaviours is actually observed: (i) after an initial, almost linear, fast adsorption process there is a slow, monotonic increase of n with t not described by eq. (57); (ii) the adsorption is not completely reversible; (iii) all available space at the surface is filled by irreversibly adsorbed proteins; and (iv) the total adsorbed amount depends weakly on C.

### 4.2.2. Two-state linear model

An improvement of the elementary model of protein adsorption is obtained by allowing for relaxation; that can be done by assuming that the protein may be in one or the other of the following states: *globular*, with cross section $\sigma$ and characteristic of the protein in solution; and *sticked*, in which it holds an area $\Sigma$ at the adsorbing surface. The globular form, i.e. the shape characteristic of the protein in water, is stabilized by the solvation energy of protein polar sites; the energy stabilizing the globular form is related to the strength of the charge-dipole interaction. The sticked form, i.e. the shape resulting after complete relaxation of the protein at the surface, is stabilized by the surface-protein interactions. Very polar surfaces are expected to interact with proteins in a stronger way than water, and the final effect will be to increase $\Sigma$ above the original value $\sigma$, $\Sigma > \sigma$. The more polar are the protein and the surface, the higher is the adhesion energy[8] and thus the ratio $\Sigma / \sigma$. The transition of the protein from the globular form to the sticked form may be described as a first order reaction with lifetime $\tau_s$.

Denoting with m and M the amounts per unit area of adsorbed protein in globular and sticked forms, respectively, the rate equations for the overall adsorption-desorption process are given by

$$\frac{dm}{dt} = kC(1 - \sigma m - \Sigma M) - \frac{m}{\tau_d} - \frac{m}{\tau_s} \qquad (58)$$

$$\frac{dM}{dt} = \frac{m}{\tau_s} - \frac{M}{\tau_D} \qquad (59)$$

[8] The case $\sigma > \Sigma$ is possible too and is expected to occur for non polar surfaces. Even this case is of practical importance for prostheses which must not be felt by the immunitary system.

where $\tau_d$ and $\tau_D$ are the desorption lifetimes for a globular and sticked protein, respectively. Let $\tau_{des}(A)$ denote the desorption lifetime when the contact area is A. If $\tau_{des}$ varies as a function of A as expected for a thermally activated process,

$$\tau_{des}(A) = \tau_{des}(A_0)\exp\left(\frac{\gamma(A - A_0)}{k_B T}\right)$$

(where $\gamma$ is the interface tension), one has

$$\tau_D = \tau_d \exp\left(\frac{\gamma(\Sigma - \sigma)}{k_B T}\right)$$

A higher contact area is thus responsible for a very large increase of the lifetime in the sticked configuration provided that $\gamma(\Sigma - \sigma)/k_B T >> 1$.

If $\tau_d << \tau_s$ and $\tau_d << \tau_D$, the system {(58), (59)} accounts for most of the observed behaviours: (I) Starting from the initial condition $m(0) = M(0) = 0$, the early stages of adsorption are described by the equation $m(t) + M(t) \cong m_{ss} [1 - \exp(- t/\tau_1)]$, with the same parameters as for the linear case. (II) However, $m(t) + M(t)$ increases in the time interval $\tau_1 << t \lesssim \tau_s$ almost linearly with t until attains a steady value only for $t >> \tau_s$. This value is given by

$$m_{ss} = \frac{\tau_s}{\tau_D} \frac{kC}{kC(\Sigma + \sigma\tau_s/\tau_D) + (\tau_s/\tau_D)(1/\tau_d + 1/\tau_s)} \tag{60}$$

$$M_{ss} = \frac{kC}{kC(\Sigma + \sigma\tau_s/\tau_D) + (\tau_s/\tau_D)(1/\tau_d + 1/\tau_s)} \tag{61}$$

In the limit $\tau_D \to + \infty$, $m_{ss} + M_{ss} \sim M_{ss} \sim 1/\Sigma$, independent of C. (III) Starting from a surface covered by a protein amount close to $m_{ss} + M_{ss}$ and exposing the surface to a pure solvent for a time much shorter than $\tau_D$, the total amount decays to a quasi saturation value close to $M_{ss}$ (only adsorbed proteins in globular form are desorbed in the considered time scale).

### 4.2.3. Non-linear effects in the two-state model

Since the two-state model is able to account for so many experimental facts even in its most elementary formulation (the linear model), it is expected to be appreciably improved when non-linear phenomena too are allowed for.

Manifestly enough, many non-linear effects influence the overall adsorption-desorption kinetics. In particular, we mention a possible dependence of $\Sigma$ on $m + M$ ($\Sigma$ is expected to decrease with M for sufficiently high coverage) and of $\tau_D$ on M (because $\tau_D \propto \exp(\gamma\ \Sigma(M)/k_B T)$). If the resulting nonlinearities are sufficiently strong, the considered dependences might result in oscillatory behaviours too.

## 5. CONCLUSIONS

For a long time the rate-determining step for the development of natural sciences has been the possibility to perform numerical calculations. Even when possible, detailed modelling of actual physico-chemical systems was considered useless because of the practical impossibility to perform numerical calculations. The ingenuity of theoreticians was accordingly concentrated more on the search of exactly or approximately solvable models rather than on the search of accurate but unsolvable models. 'Solvable' and 'unsolvable' are however related to the deployed mathematical apparatus - much of what remained unsolvable after centuries of development of calculus and analysis, has become solvable after the advent of electronic computers (even in pure mathematics - think of the four-colour problem).

The enormous progress of electronic computers has allowed physico-chemical systems to be accurately described (by means of new theoretical disciplines like computational physics and theoretical chemistry) via numerical solution of the basic equations. This *hard modelling* of physico-chemical systems is characterized by: extended numerical computations; *ab initio* methods; accurate description of the system; specificity but non extensibility of the results from one system to another.

The counterpart of hard modelling is *soft modelling*. This theoretical approach is characterized by: analytic methods with a minimum of computational efforts (soft theorists consider solved a problem when it has been reduced to quadratures); tailored models obtained by a reduced description of the system (simplification[9]); generality of the results which hold true usually for a large class of systems, but usually lack of specificity.

The progress in numerical computation has progressively extended the domain of systems which can be hard-modelled. Systems which admit only soft modelling have become fewer and fewer, and are usually characterized either by situations too complex to be exactly described (e.g., protein folding) or by a number of particles so high as to make calculations impossible (e.g., phase transitions). At present, the problems involved in adsorption with soft adsorbates or adsorbents admit only soft modelling.

## REFERENCES

1. S. F. Mason, Chemical Evolution, Clarendon Press, Oxford, 1991.
2. W. S. Taft, MRS Bull., 21, No. 12 (1996) 18.
3. J. D. Bernal, The Physical Basis of Life, Rutledge, London, 1951.
4. G. F. Cerofolini and G. Boara, J. Colloid Interface Sci., 161 (1993) 232.
5. G. F. Cerofolini and M. Cerofolini, Specul. Sci. Technol., 3 (1980) 149.

---

[9] "Simplification is an art rather like that of the cartoonist who captures the key features of a familiar face in a few deft strokes to make it instantly recognizable" [89].

6. G. F. Cerofolini, Nuovo Cimento D 2, (1983) 763.
7. G. F. Cerofolini, Nuovo Cimento D 2, (1983) 1156; Fisica in Medicina 2 (1986) 99.
8. M. Jaroniec and R. Madey, Physical Adsorption on Heterogeneous Solids, Elsevier, Amsterdam, 1988.
9. W. Rudziński and D.H. Everett, Adsorption of Gases on Heterogeneous Surfaces, Academic Press, London, 1992.
10. G. F. Cerofolini and W. Rudziński, in: Equilibria and Dynamics of Gas Adsorption on Heterogeneous Solid Surfaces, W. Rudziński, W. A. Steele and G. Zgrablich (eds.), Elsevier, Amsterdam, (1997) 1.
11. J. W. Evans, Rev. Mod. Phys., 65 (1993) 1281.
12. S. Brunauer, P.H. Emmett and E. Teller, J. Am. Chem. Soc., 60 (1938) 309.
13. W. A. Steele, The Interaction of Gases with Solid Surfaces, Pergamon Press, Oxford, 1975.
14. R. B. Anderson, J. Am. Chem. Soc., 68 (1946) 686.
15. S. Brunauer, J. Skalny and E. E. Bodor, J. Colloid. Interface Sci., 30 (1969) 546.
16. T. L. Hill, J. Chem. Phys., 15 (1947) 767.
17. J. C. P. Broekhoff and R. H. Van Dongen, in: Physical and Chemical Aspects of Adsorbents and Catalysis, B.G. Linsen (ed.), Academic Press, London, (1970) ch. 2, 63.
18. W. M. Champion and G. D. Halsey, Jr., J. Phys. Chem., 57 (1953) 646.
19. J. Frenkel, Kinetic Theory of Liquids, Clarendon Press, Oxford, 1946.
20. G. D. Halsey, Jr., J. Chem. Phys., 16 (1948) 931.
21. T. L. Hill, J. Chem. Phys., 17 (1949) 590; 668.
22. G.D. Halsey, Jr., J. Am. Chem. Soc., 73 (1951) 2693.
23. G.F. Cerofolini, J. Low Temp. Phys., 6 (1972) 473; Thin Solid Films, 23 (1974) 129.
24. P. Pfeifer, Y. J. Wu, M. W. Cole and J. Krim, Phys. Rev. Lett., 62 (1989) 1997.
25. D. Avnir and M. Jaroniec, Langmuir, 5 (1989) 1431.
26. P. Pfeifer and K.-Y. Liu, in: W. Rudziński, W. A. Steele and G. Zgrablich (eds.) Equilibria and Dynamics of Gas Adsorption on Heterogeneous Solid Surfaces, Elsevier, Amsterdam, (1997) ch. 12, 625.
27. J. Lapujoulade, Surf. Sci. Rep., 20 (1994) 191.
28. L. Jelinek and E. Sz. Kováts, Langmuir, 10 (1994) 4225.
29. A. L. McClellan and H. F. Harnsberger, J. Colloid Interface Sci., 23 (1967) 577.
30. M. J. Gray, R. C. Mebane, H. N. Womack and T.R. Rybolt, J. Colloid Interface Sci., 170 (1995) 98.
31. P. Pfeifer and D. Avnir, J. Chem. Phys., 79 (1983) 3558.
32. D. Avnir, D. Farin and P. Pfeifer, J. Chem. Phys., 79 (1983) 3566; Nature, 308 (1984) 261.
33. L. A. Zadeh, Soft Computing, 1 (1997) 1.
34. P. G. de Gennes, Rev. Mod. Phys., 64 (1992) 645.

35. P. G. de Gennes and J. Badoz, Fragile Objects, Springer-Verlag, New York, NY, 1996.
36. J. Gopalakrishnan, Chem. Mater., 7 (1995) 1265.
37. R. G. Pearson, J. Am. Chem. Soc., 85 (1963) 3533.
38. O. V. Bychuk and B. O'Shaughnessy, J. Colloid Interface Sci., 167 (1994) 193.
39. V. A. Radtsig and A. V. Bystrikov, Kinet. Katal., 19 (1978) 713.
40. V. A. Radtsig, Kinet. Katal., 20 (1980) 448; 1203; 1206.
41. V. A. Radtsig and V. A. Khalif, Kinet. Katal., 20 (1980) 705.
42. R. K. Iler, The Colloid Chemistry of Silica and Silicates, Cornell Univ. Press, Ithaca, New York, 1955, 233.
43. T. Hiemstra, W. H. Van Riemsdijk and G. H. Bolt, J. Colloid Interface Sci., 133 (1989) 91.
44. T. Hiemstra, J. C. M. de Wit and W. H. van Riemsdijk, J. Colloid Interface Sci., 133 (1989) 105.
45. T. Hiemstra and W. H. van Riemsdijk, J. Colloid Interface Sci., 136 (1990) 132.
46. P. W. Schindler and H. R. Kamber, Helv. Chim. Acta, 51 (1968) 1781.
47. J. Jagiełło, Langmuir, 10 (1994) 2778.
48. T. J. Bandosz, C. Lin and J. A. Ritter, J. Colloid Interface Sci., submitted for publication.
49. J. Jagiełło, T. J. Bandosz, K. Putyera and J. A. Schwarz, in: M. D. LeVan (ed.), Fundamentals of Adsorption, Kluwer, Boston, MA; 1996, 417.
50. T. J. Bandosz, J. Colloid. Interface Sci., 193 (1997) 127.
51. J. Monod, J. Wyman and J. P. Changeux, J. Mol. Biol., 12 (1965) 88.
52. R. K. Scopes, Protein Purification. Principle and Practice, Springer-Verlag, New York, NY, 1982.
53. H. Frauenfelder, in: E. Clementi and R. H. Sarma (eds.), Structure and Dynamics: Nucleic Acids and Proteins, Adeninine Press, Guilderland, NY, 1983, 369.
54. M. F. Perutz, Proteins and Nucleic Acids; Elsevier, Amsterdam, 1962.
55. T. G. Dewey, J. Chem. Phys., 98 (1993) 2250.
56. B. A. Fedorov, B. B. Fedorov and P. W. Schmidt, J. Chem. Phys., 99 (1993) 4076.
57. T. G. Dewey, Het. Chem. Rev., 2 (1995) 91.
58. T. E. Creighton (ed.), Protein Structure: A Practical Approach, IRL Press, Oxford, 1989.
59. Y. Bai, T. R. Sosnick, L. Mayne and S. W. Englander, Science, 269 (1995) 192.
60. G. F. Cerofolini and M. Cerofolini, J. Colloid Interface Sci., 78 (1980) 65.
61. G. F. Cerofolini, Adv. Colloid Interface Sci., 19 (1983) 103.
62. P. T. Landsberg, J. Chem. Phys., 23 (1955) 1079.
63. G. F. Cerofolini, Surf. Sci., 366 (1996) 597.
64. T. Arai and W. Norde, Colloids Surf., 51 (1990) 1.
65. W. Norde and C. I. Anusiem, Colloids Surf., 66 (1992) 73.
66. W. Norde and J. P. Favier, Colloids Surf., 64 (1992) 87.

67. B. C. Tripp, J. J. Magda and J. D. Andrade, J. Colloid Interface Sci., 173 (1995) 16.
68. M. A. Joynson, A. C. T. North, V. R. Sarma, R. E. Dickerson and L. K. Steinrauf, J. Mol. Biol., 50 (1970) 137.
69. O. B. Ptitsyn, Trends Biochem. Sci., 20 (1995) 376.
70. S. Kochwa, R. S. Litwak, R. E. Rosenfield, E. F. Leonard, Ann. N. Y. Acad. Sci., 283 (1977) 37.
71. W. Norde and J. Lyklema, J. Colloid Interface Sci., 66 (1978) 257.
72. C. A. Haynes, E. Sliwinsky and W. Norde, J. Colloid Interface Sci., 164 (1994) 394.
73. G. F. Cerofolini, Langmuir, 13 (1997) 995.
74. G. F. Cerofolini and L. Meda, J. Colloid Interface Sci., (1998) in press.
75. R. W. Gurney, Introduction to Statistical Mechanics, McGraw Hill, New York, NY, 1949.
76. W.A. Steele, J. Chem. Phys., 25 (1956) 819.
77. W. G. McMillan and E. Teller, J. Chem. Phys., 19 (1951) 25; J. Phys. Chem., 55 (1951) 17.
78. V. A. Tertykh, Macromol. Symp., 108 (1996) 55.
79. A. K. Chakraborty and M. Tirrell, MRS Bull., 21, No. 1 (1996) 28.
80. F. MacRitchie, Adv. Prot. Chem., 32 (1978) 283.
81. J. D. Andrade and V. Hladky, Adv. Polym. Sci., 79 (1986) 1.
82. C. A. Haynes and W. Norde, Colloids Surf. B 2, (1994) 517.
83. J. J. Ramsden, Chem. Soc. Rev., 73 (1995).
84. M. Kleijn and W. Norde, Heter. Chem. Rev., 2 (1995) 157.
85. Pharmacia, Affinity Chromatography: Principles and Methods, (1993).
86. G. B. Wisdom, in: J. M. Walker (ed.), New Protein Techniques, Humana Press, Clifton, NJ, 1988, 373.
87. C. Sutton, in: E. L. V. Harris and S. Angal (eds.), Protein Purification Methods: A Practical Approach, IRL Press, Oxford, 1989, 268.
88. R. Lotan and G. L. Nicolson, Biochim. Biophys. Acta, 559 (1979) 329.
89. A. Cottrell, MRS Bull., 21, No. 5 (1997) 15.

# APPLICATION IN INDUSTRY

Adsorption and its Applications in Industry and Environmental Protection
Studies in Surface Science and Catalysis, Vol. 120
A. Dabrowski (Editor)



# Advanced technical tools for the solution of high capacity adsorption separation

G. Horváth[a] and M. Suzuki[b]

[a]Department of Chemical Engineering, University of Veszprém, P.O.Box 158, 8201 Veszprém, Hungary

[b]Institute of Industrial Science, University of Tokyo, 7-221 Roppongi, Minato-ku, Tokyo 106, Japan

## 1. INTRODUCTION

The surface and transport properties of adsorbents have been widely used in different technological fields. The development of compact and large capacity separation processes is one of the most urgent problems in the field of environmental protection. The oxygen production, the purification of stack gases and other polluting off gases demand at least partly new technical solutions.

Among the new geometrical constructions new methods are taken into consideration, like ultra rapid pressure swing adsorption (URPSA), piston driven PSA, parallel passage adsorbers and so on.

The main disadvantages of the traditional packed bed columns are the relatively high pressure drop and the thermal resistance of adsorbent materials leading to a significant power cost.

In order to realise a high capacity adsorptive separator, some fundamental engineering studies have been done during the past years.

## 2. HYDRODYNAMIC ASPECTS

The large-scale "gas-separation" processes may be conveniently divided into two classes, the cryogenic separation systems and the adsorption separation processes. In this chapter we are dealing with adsorption processes only.

Figure 1 shows schematically the pressure swing, thermal swing and combined adsorption operations. In order to estimate the practical adsorption capacities necessary to know the equilibrium relations. The equilibrium relation depending on temperatures between the specific adsorbed amount a (g/g) and the pressure p (bar) is plotted with continous line. In industrial operation this capacity can be as maximum considered and naturally it cannot be fully utilized. All of separation technologies demands energy introduction. From this point of view the form of separation work classifies the technologies. Thermal swing solution in ideal case (ideal $T_{swing}$) shows a difference between isotherms of $T_1$ and $T_2$ at constant pressure. In the reality the ideal differece between the two equilibrium states cannot be reached because the adsorbent warms during the adsorption step and it cools during the desorption step. Thus the dotted isotherms are the virtual equilibrium lines ($T_1'$.and $T_2'$). The mass transfer resistances have to be also taken into consideration, but here we do not deal with them.

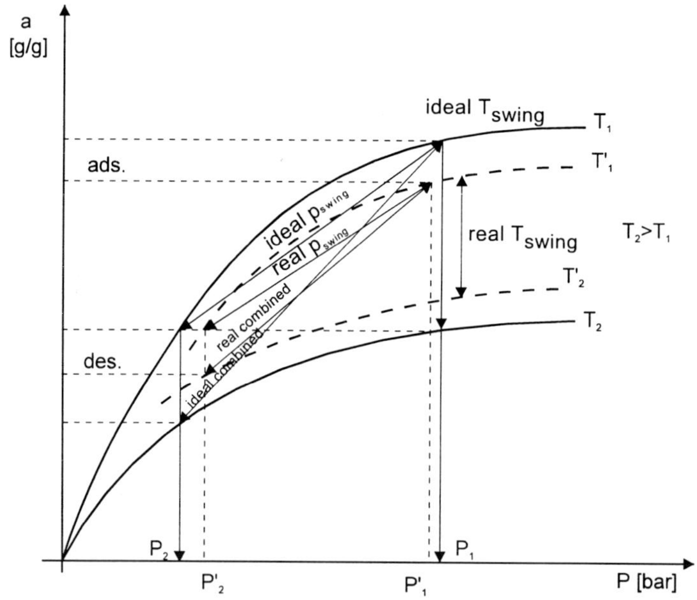


Figure 1. Schematic adsorption swing operations.

As for the ideal pressure swing (ideal $P_{swing}$) adsorption the situation is similar. Usually the main reasons of deviation are the pressure drop in the colums and the adsorbent warming, so the dotted isoterms symbolizes the differences from the ideal operation. Having based on the same logic, the ideal and real combined adsorption processes face to the same limitations.

One of the most important problem of large apparatuses is the solution of the gas transport. The produced gas amount could hardly reach the 1 metric ton/hour. This capacity usually is too small for environmental applications.

**2.1. Method of Ergun-equation**

The hydrodynamic of flow through packed beds has been widely investigated, except the closed end column, which is important from the point of adsorption view. To evaluate the pressurization time from experiments the model of isothermal gas flow in porous materials has been used. The model is based on the following assumptions: [1,4]:

- the gas temperature is constant,
- radial gradients in the pressure and in the macroscopic velocity are negligible (the plug flow condition),
- the inertia forces are negligible,
- the friction forces obey the Ergun equation,
- an ideal gas is assumed,
- the gravitational forces are negligible,
- the dynamic viscosity does not depend on pressure.

In the range of our experimental data the inertia forces and the temperature effects are negligible usually. The mathematical model consists of a mass balance

$$\varepsilon\frac{\partial c}{\partial t}+\frac{\partial}{\partial x}(\varepsilon u c)=0 \tag{1}$$

and a momentum balance which contains only two terms expressing pressure and friction forces.

$$-\frac{\partial p}{\partial x}=\lambda\rho\frac{(1-\varepsilon)}{\varepsilon d_p}u^2 \tag{2}$$

For the friction coefficient λ, the Ergun equation has been used

$$\lambda=\frac{A}{Re}+B \tag{3a}$$

where

$$Re=\frac{d_p\varepsilon\rho u}{(1-\varepsilon)\mu} \tag{3b}$$

Ergun [1] recommends A equal to 150 and B equal to 1.75 based on a correlation of experimental data. The state equation for an ideal gas connects the molar concentration c and pressure p, $c = p/RT$. The density of ·a gas is the product of the molar concentration, c, and the molecular weight, $\rho = Mc$. Instead of the velocity, u, we introduced the mass flux density in the form $M = uc$.. With the constant temperature assumption, the Eqs. (1 ) and (2) can be rearranged to the form

$$\frac{1}{RT}\frac{\partial p}{\partial t}+\frac{\partial M}{\partial x}=0 \tag{4}$$

$$\frac{\partial p}{\partial x}=-A\mu RT\left(\frac{1-\varepsilon}{\varepsilon d_p}\right)^2\frac{M}{p}-B\frac{1-\varepsilon}{\varepsilon d_p}MRT\frac{M^2}{p} \tag{5}$$

There are two dependent variables, p and M, and two independent variables, x and t. Under the assumptions mentioned above, all other quantities are constant. 1n addition to Eqs. (4) and (5), the two initial conditions

$$p(x,0)=p_I \tag{6}$$

$$M(x,0)=0 \tag{7}$$

and two boundary conditions are required. The model describes the pressurization of a fixed bed closed at one end. The pressure at the other end is suddenly fixed at a different pressure. At the closed end of the column, the boundary condition

$$M(x=H,t)=0 \qquad t>0 \tag{8}$$

is applied, and at the open end

$$p(x=0,t)=p_F \qquad t>0 \tag{9}$$

Equations (4) and (5) were transformed into dimensionless form by means of the following dimensionless variables

$$p^*=\frac{p-p_I}{p_F-p_I} \tag{10}$$

$$M^*=M\frac{RT}{p_I^2}\left(\frac{1-\varepsilon}{\varepsilon d_p}\right)^2 HA\mu \tag{11}$$

$$\eta=\frac{x}{H} \tag{12}$$

$$\tau=t\left(\frac{\varepsilon d_p}{1-\varepsilon}\right)^2\frac{p_I}{H^2A\mu} \tag{13}$$

and parameters

$$\Delta p^* = \frac{p_F - p_I}{p_I} \tag{14}$$

$$\delta = \frac{B}{A^2}\left(\frac{\varepsilon d_p}{1-\varepsilon}\right)^3 \frac{p_I^2 M}{H\mu^2 RT} \tag{15}$$

The transformed Eqs.(4), (5) become:

$$\Delta p^* \frac{\partial p^*}{\partial \tau} + \frac{\partial M}{\partial \eta} = 0 \tag{16}$$

$$\Delta p^* \frac{\partial p^*}{\partial \eta} = -\frac{M^*}{1 + \Delta p^* p^*}\left(1 + \delta M^*\right) \tag{17}$$

with the initial conditions

$p^* (\eta, \tau=0)=0$ (18/a)

$M^* (\eta,\tau=0)=0$ (18/b)

and the boundary conditions

$M^* (\eta=1, \tau)=0$ (19)

$p^* (\eta=0,\tau)=1$ (20)

Equations (16) to (20) have been solved numerically. The Figure 2 shows a traditional industrial adsorber.This type of adsorber has a capacity as high as 10 - 50 $m^3$/hr depending on the raw materials. The calculated pressure drop of the former traditional adsorber in a steady state regime can be see on the Figure 3. The parameters are the temperature (T) and the volumetric standard gas velocity (Y). A master sheet for the determination of pressurisation time and dependence of the pressurisation time on the complex parameter $\delta$ can be seen in Figure 4 [20]. For $\delta$ approaching 0, the flow approaches Darcy`s-region, while for $\delta$ >>1 it is in the high velocity range [2-5]. From the results of these studies it can be concluded that the Ergun type equation can be used for the simulation of slower pressurisation processes (i.e. in case of small particles) provided the constants of the equation are determined preliminarily. The very high velocity range like ultra rapid PSA should be studied with more detail.

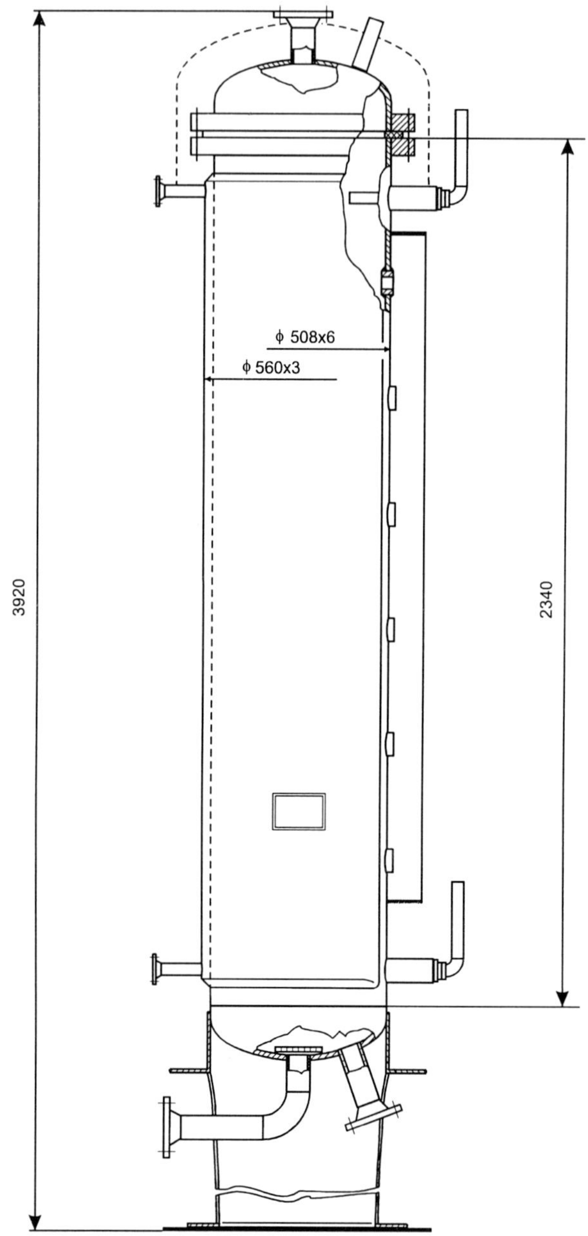


Figure 2. A traditional industrial adsorber.

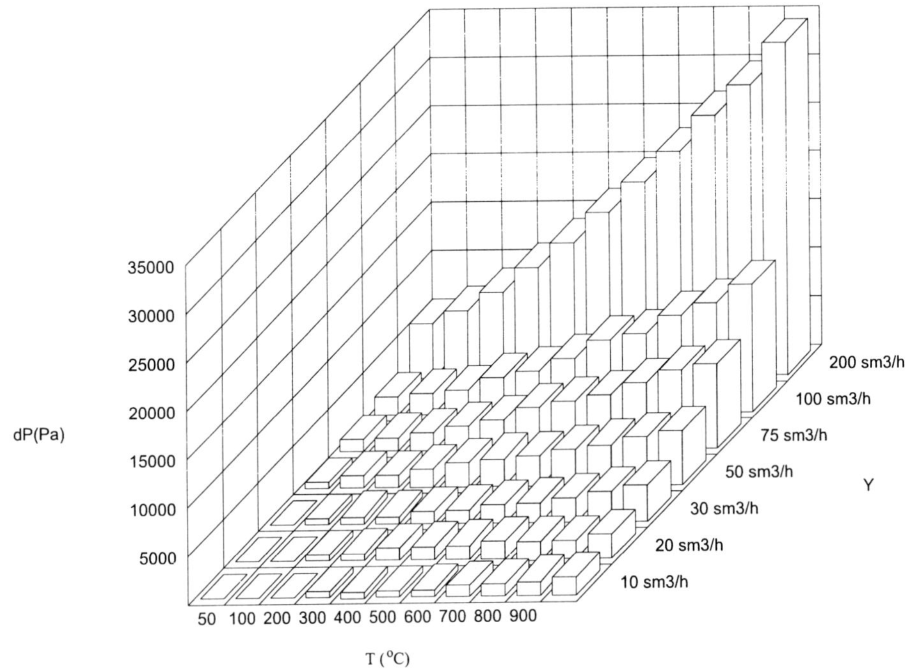


Figure 3. Pressure drop, temperature dependency in steady state.

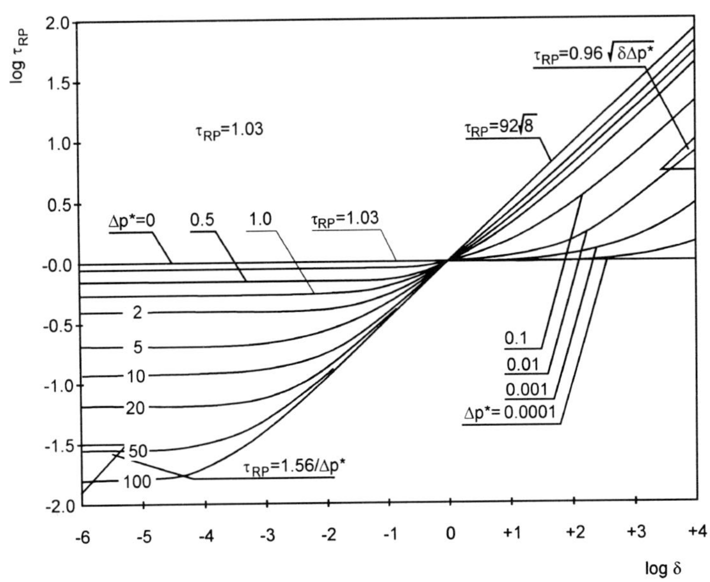


Figure 4. The pressurization time vs. $\delta$ for different inlet pressure differences. (Reprinted with permission from ref. [20]).

## 2.2. The geometry (with the permission of L'AIR LIQUIDE)

Several industrial firms began to construct adsorbers mainly for air separation. So numerous alternatives can be found on the market, but the specific energy in the case of classical column constructions used to be high (0,9-2,1 kWh/Nm$^3$ $O_2$). Therefore the low temperature separation as for the energetic point of wiev is more economic.

The L'AIR LIQUIDE opened a new direction some years ago, and developed the so-called COMPACT VSA family. They could decrease the energy demand into the 0.3-0.45 kWh/Nm$^3$$O_2$ intervall . The gas production lies in a range of 5 metric tons/day to 160 metric tons/day.

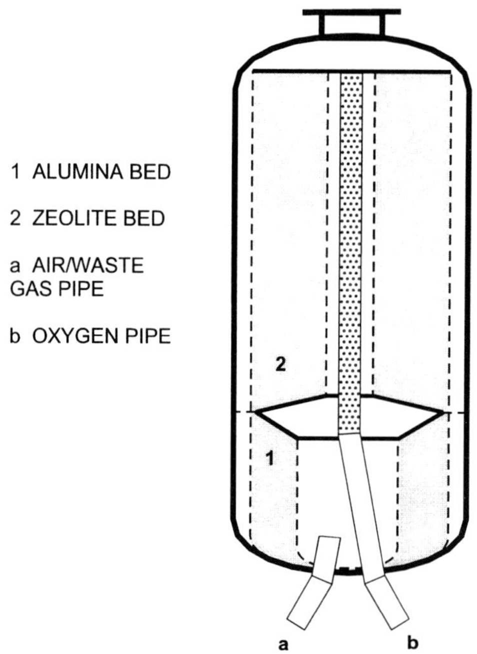


Figure 5. Radial adsorber of L`AIR LIQUIDE.

Among others the secret of this sucsess is the radial adsorber (Figure 5). The air enters into the under part of column, flows through an alumina adsorber, then passes radially the zeolite bed. Evacuation is used for regeneration. Supposing that the adsorber is adiabatic, the temperature change reaches an order of magnitude 50°C.
Such types of constructions open new possibilities in the fied of technological applications.

## 3. FUNDAMENTALLY NEW CONSTRUCTIONS

### 3.1. Ultra rapid PSA

The development of processes is one of the most urgent requirements in such fields as carbon dioxide recovery from stack gases and oxygen enrichment for more efficient engine performance of automobiles. The pressure swing adsorption (PSA) process is a good candidate for achieving these purposes, but the production capacity of PSA must be improved for this purpose. This can be done by the development of more selective adsorbents, operation at low temperature where more adsorption capacity is expected, use of short cycle times where more throughput ratio is expected, etc. Among these approaches, the most significant improvement of capacity is expected to be achieved by rapid cycle operation. However, short cycle operation by conventional PSA processes is difficult bacause the lifetime of the ordinary valves applied is limited, affecting the continuous operation of the process.

The piston-driven ultra rapid pressure swing adsorption (URPSA) equipment was developed and oxygen enrichment from air was examined as an example [6-19]. The adsorbent bed is directly connected to the cylinder where a piston moves at high frequency. Thus pressurization and depressurization in the bed are driven by mechanical piston motion, which can achieve far more rapid cycles compared with the conventional pressure swing operation using valves. The cycle time is usually on the order of seconds or sub seconds. Oxygen enrichment from air up to about 60% or higher of oxygen concentration was achieved by small-scale equipment using zeolite 5A with a oxygen production capacity of 100 $Nm^3$-product gas/$m^3$-zeolite/h. This specific value is about ten times larger than those of commercialized PSAs for the same purpose.

A method to realize rapid operation of PSA, a piston-cylinder module has been applied in a conventional reciprocal engine equipment as the pressurizing and depressurizing mechanism combined with an adsorption unit. Using this equipment, ultra-rapid operation was achieved and oxygen enrichment from air was experimentally investigated. A simplified numerical model has been proposed to describe ultra-rapid pressure swing adsorption (URPSA) operation [21]. Having used this model, the basic characteristics of this operation can be characterised.

#### 3.1.1. Basic steps of piston-driven ultra rapid PSA

The URPSA equipment (Figure 6) consists of a piston, a cylinder, valves and an adsorption bed. The piston is moved by an arm connected to a rotating disc driven by an electric motor. The basic operational steps are suction, adsorption and production, desorption, and exhaust step, as shown in Figure 7. The operation in each step is as follows.

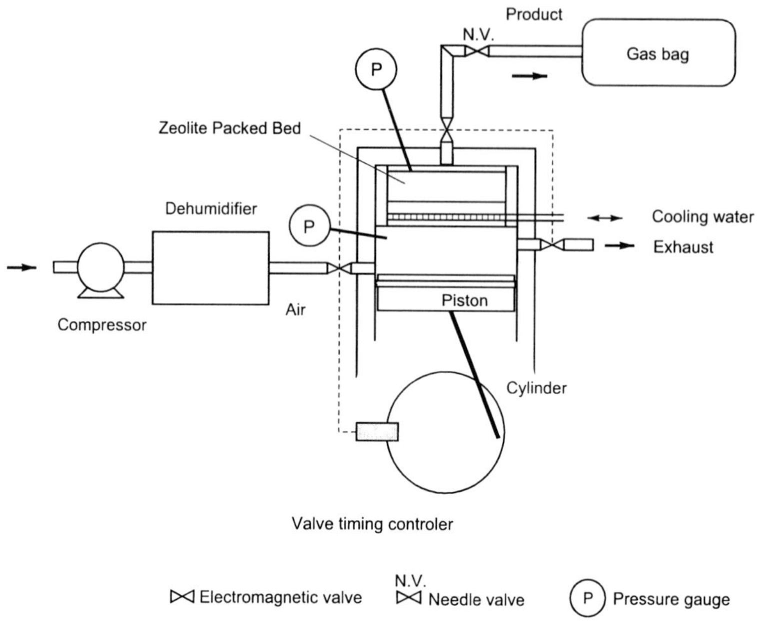


Figure 6. Schematic diagram of experimental apparatus.
(Reprinted with permission from ref. [21].

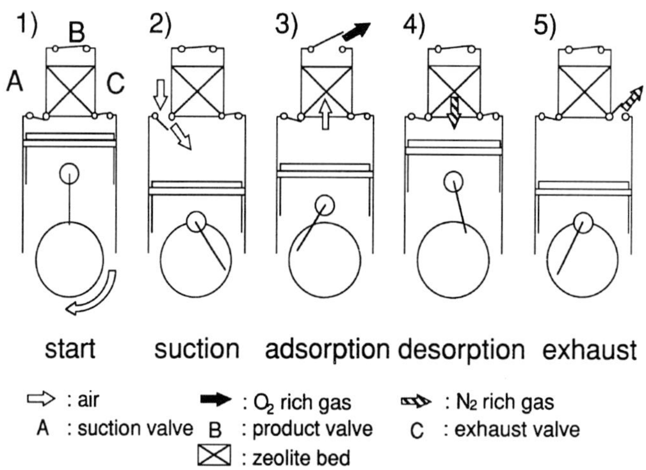


Figure 7. Basic steps of piston-driven ultra rapid pressure swing adsoption.
(Reprinted with permission from ref. [21]).

Step 1: Raw gas is introduced into the cylinder while the piston goes down from the top position to the bottom position.

Step 2: While the piston goes up, gas in the cylinder is compressed and introduced into the adsorbent bed, where adsorptive separation takes place and a less-adsorbable component is enriched in the product gas which is released from the top of the adsorbent bed.

Step 3: The piston goes down with all the valves closed. During this step, the depressurization and evacuation of the adsorbent bed take place. Desorption occurs and the cylinder is filled with the desorbed gas.

Step 4: In this step, the piston goes up again and the desorbed gas in the cylinder is removed as exhaust through the valve at the cylinder wall.

One adsorption-desorption cycle is completed with two up-and-down strokes of the piston. The cycle time in this sequential operation is controlled by the rotation speed of an electric motor and can be adjusted to seconds or sub seconds. Oxygen enrichment by using zeolite as an adsorbent was tried as an example of application.

### 3.1.2. Apparatus

Details of the adsorption unit are shown in Figure 8.

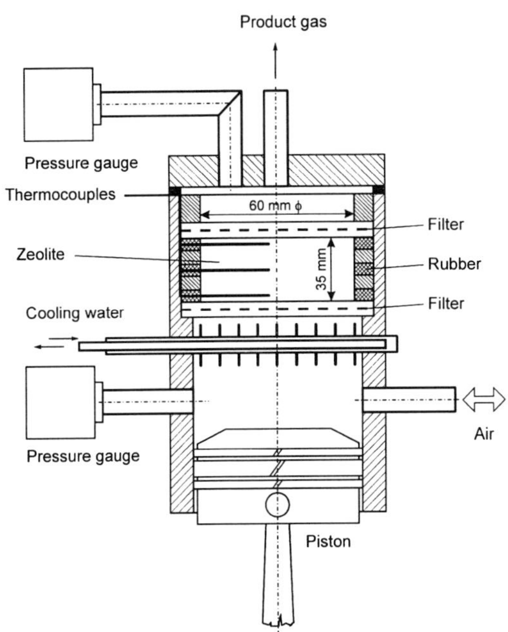


Figure 8. Details of adsorption unit.
(Reprinted with permission from ref. [21]).

The packed bed of zeolite in this case was 35 mm in height and 60 mm in diameter. The zeolite bed was fixed at the top and bottom by a sheet of filter paper (Advantech Toyo Filter Paper, No. SA) and a stainless steel screen of 400 mesh size. Three copper-constantan thermocouples of 1.0 mm in sheath diameter were inserted into the center of the bed at the inlet, middle and outlet sections.

The applied adsorbent was zeolite 5A (Union Carbide Corporation) which was crushed and sieved into 48 to 80, 80 to 150 and 200 to 325 mesh size groups. As a preliminary treatment the zeolite was heated at 653 K for over 3 hours under evacuated conditions. The pre-treated zeolite was carefully packed in the bed to avoid humidity uptake.

The rotation speed of the motor was set to 80, 120, 160, 200 and 240 rpm corresponding, respectively, to cycle times of 1.5,1.0, 0.75, 0.6 and 0.5 seconds. The oxygen concentration in the product gas collected in the gas bag was determined by an oxygen analyzer using zirconia (Toray Corporation, LC-100). The flow rates of the feed air and the product gas were measured with an integration type flowmeter (Shinagawa Manufacturing Co., Dp-2A-1). The temperature of the cooling water was keep constant at 278 K.

### 3.1.3. A simplified mathematical model

The assumptions employed in the modeling of URPSA were as follows (Kowler and Kadlec [11], Jones and Keller [8, 9] and Guan and Ye [6]):

- all gases are ideal,
- air is a mixture of nitrogen and oxygen only, and each gas is independent in terms of the adsorption equilibrium and rate,
- the adsorption equilibrium is linear,
- the adsorption rate is expressed by the linear driving force (LDF) approximation with a correction factor, due to rapid cyclic adsorption and desorption (Nakao and Suzuki, [13]),
- the flow in the packed bed is the plug flow.[6]. The shape of the zeolite particle is spherical,
- the pressure drop in the packed bed is given by Ergun-equation,
- the temperature of the entire system is constant at 303 K.

The model is a box model consisting of the cylinder, the zeolite bed divided into 10 compartments and the head space of the packed bed. The mass balance of each box is written as follows.

The mass balance in the cylinder is,

$$\frac{dP_{cyl}}{dt} V_{cyl} + P_{cyl} \frac{dV_{cyl}}{dt} = \frac{dN_{cyl}}{dt} RT \qquad (21)$$

where

$$\frac{dN_{cyl}}{dt} = F_{feed} + F_{in} + F_{exhaust} \tag{22}$$

$$V_{czl} = S_{czl} + x_{pi} \tag{23}$$

$$x_{pi} = \frac{x_{pi\,max} - x_{pi\,min}}{2} \{\cos(\omega t + \pi) + 1\} \tag{24}$$

$P_{cyl}$, $V_{cyl}$, R and T present the total pressure in the cylinder, the volume, a gas constant and the temperature, respectively. $N_{cyl}$ is the amount of gas present in the cylinder $F_{feed}$, $F_{in}$ and $F_{exhaust}$ the flows of the feed gas, to the packed bed and of the exhaust gas, respectively. $S_{cyl}$ and $x_{pi}$ represent the cross sectional area of the piston and the length between the piston head and the ceiling of the cylinder. $x_{pi\,max}$ and $x_{pi\,min}$ are maximum and minimum of $x_{pi}$. w is the angular velocity.

The mass balance of component *i* in the j-th box of the zeolite bed is given as:

$$\frac{dN_{bed_{i,j}}}{dt} = \left(F_{bed}\right)_{i,j-1} + \left(F_q\right)_{i,j} + \left(F_{bed}\right)_{i,j} \tag{25}$$

$N_{bed\ i,j}$, is the amount of gas present in the bed. $(F_q)_{i,j}$ and $(F_{bed})_{i,j}$ are the fluxes due to adsorption and desorption, and to the adjacent box, respectively.

Adsorption rate and equilibrium relation are written as

$$\gamma \frac{dq_{i,j}}{dt} = \left(K_F a_v\right)_{i,j} \frac{\left(p_{bed}\right)_{i,j} - \left(p^*_{bed}\right)_{i,j}}{RT} \tag{26}$$

$$\left(K_F a_v\right)_{i,j} = \frac{\Omega_{i,j}\left(D_p\right)_{i,j}(1-\varepsilon)}{R_p^2} \tag{27}$$

$$q_{i,j} = \left(\beta_0\right)_i \frac{\left(p_{bed}\right)_{i,j}}{RT} \tag{28}$$

$$\left(\beta_0\right)_i = \left(\beta_0^0\right)_i \exp\left(\frac{Q_{st}}{RT}\right) \tag{29}$$

where $K_F a_v$, $D_p$, $\varepsilon$, $Q_{st}$ and $\left(\beta_0^0\right)$ are the overall mass transfer coefficient, the pore diffusion coefficient, bed void fraction, the heat of adsorption and pre-exponent constant, respectively. $\left(p^{*}_{bed}\right)_{i,j}$ is the pressure in equilibrium with the present amount adsorbed, $q_{i,,j}$, $\Omega_{i,,j}$ is a coefficient depending on $\left(\tau_c\right)_{i,j}$ (Nakao and Suzuki, [13]) given by

$$\left(\tau_c\right)_{i,j} = \frac{\left(D_p\right)_{i,j} t_c}{R_p^2} \tag{30}$$

where $t_C$ is the adsorption time and $R_p$ is the radius of adsorbent particle.

$D_p$ is evaluated as follows.

$$\left(D_p\right)_{i,j} = \frac{\varepsilon_p}{k^2}\left(D_0\right)_{i,j} \tag{31}$$

$$\frac{1}{\left(D_0\right)_{i,j}} = \frac{1}{\left(D_m\right)_{i,j}} + \frac{1}{\left(D_K\right)_i} \tag{32}$$

$$\left(D_K\right)_i = \frac{2}{3} r_a \sqrt{\frac{8RT}{M_i}} \tag{33}$$

where $D_m$ and $D_K$ are the molecular diffusion coefficient and the Knudsen diffusion coefficient, respectively. $\varepsilon_p$ is the particle void fraction, $k^2$ is the tortuosity factor, $r_a$ is a the macropore radius and $M_i$ is the molecular weight of the i-th component.

The mass balance in the head space is given as

$$\frac{dN_{up}}{dt} = \left(F_{bed}\right)_{i,jmax} + F_{product} \tag{34}$$

where $N_{up}$ is the amount of gas present in the head space over the zeolite bed and $F_{product}$ is the flow of the product gas.
The calculation method is as follows.

- The pressure change in the cylinder by the piston motion at time, t, is calculated.
- The amount of gas flowing from the cylinder into the first box of the zeolite bed is obtained.
- The pressure in the first box is calculated.
- The pressure decrease by adsorption is calculated.

- The flow to the next box, the second box, is evaluated according to pressure gradient between the first box and second box.
- The same procedure can be repeated for the following boxes.

Pressure changes at the inlet and outlet of the zeolite bed with cycle time are shown in Figures 9-11. for those cases where particles of 48 - 80, 80 - 150 and 200 - 325 mesh size were employed. The first quarter of the cycle time is the suction step when the piston descends from the upper dead point. The solid and dotted lines respectively represent the pressure swings at the inlet and outlet of the zeolite bed. These were calculated using the model described above.

When smaller particles were used, smaller pressure changes at the outlet of the zeolite bed and larger at the inlet were found. These are caused by the different pressure wave propagations due to different conductance in the beds of different particle sizes.

This empirical fact suggests that the separation performance depends greatly on particle size because the system pressure changes effective for adsorption and desorption are controlled by the pressure drop performances in the packing bed. Having compared with Rapid PSA (Turnock and Kadlec [19]), larger pressure swing over and under the atmospheric pressure is achieved in a shorter cycle time.

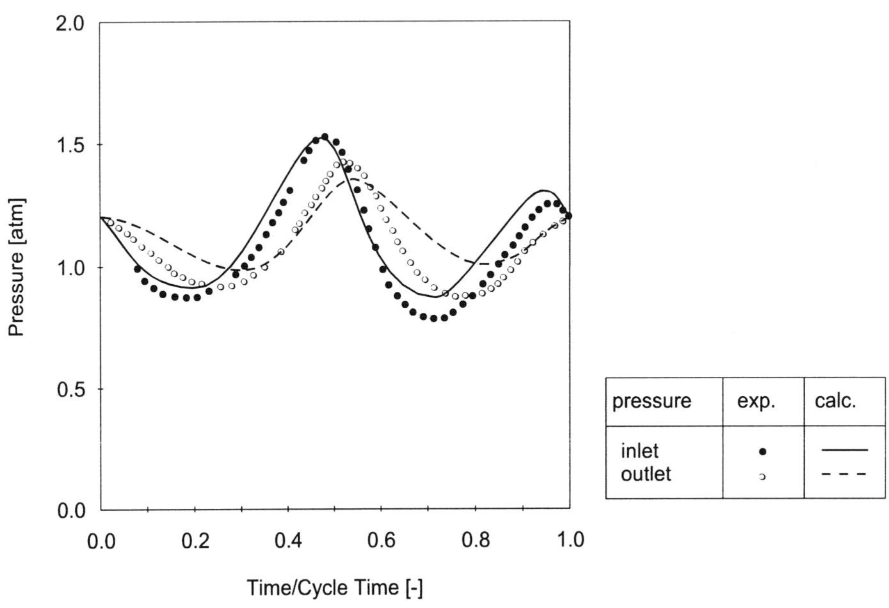


Figure 9. Pressure swing in adsorption bed ( dp:# 48-80, cycle time: 1.0 s.) (Reprinted with permission from ref. [21]).

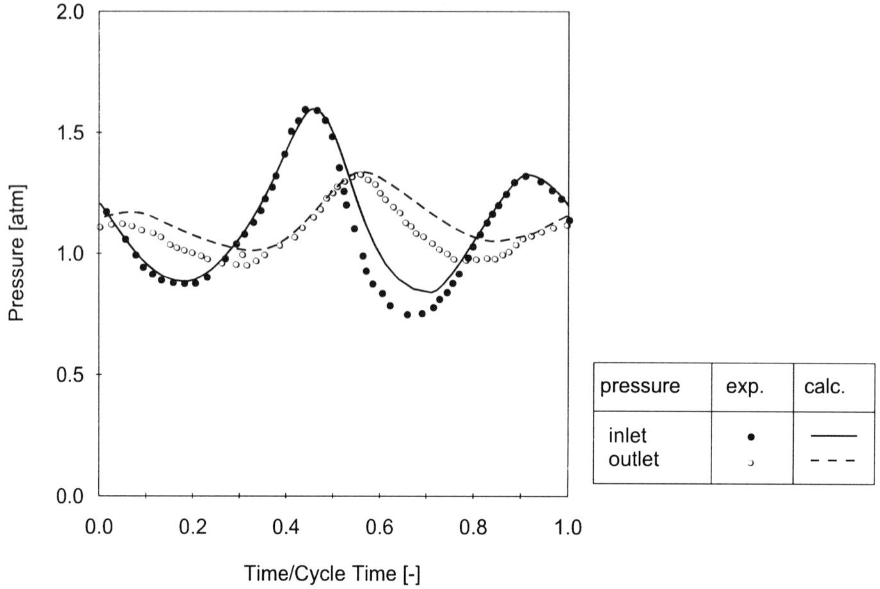


Figure 10. Pressure swing in adsorption bed ( dp: # 80-150, cycle time: 1.0 s.) (Reprinted with permission from ref. [21]).

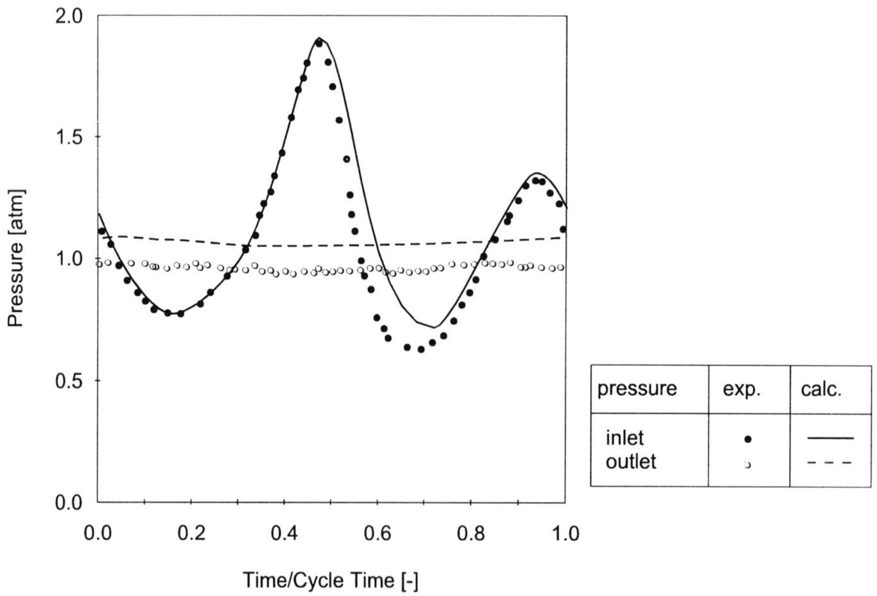


Figure 11. Pressure swing in adsorption bed ( dp: # 200-325, cycle time: 1.0 s ) (Reprinted with permission from ref. [21]).

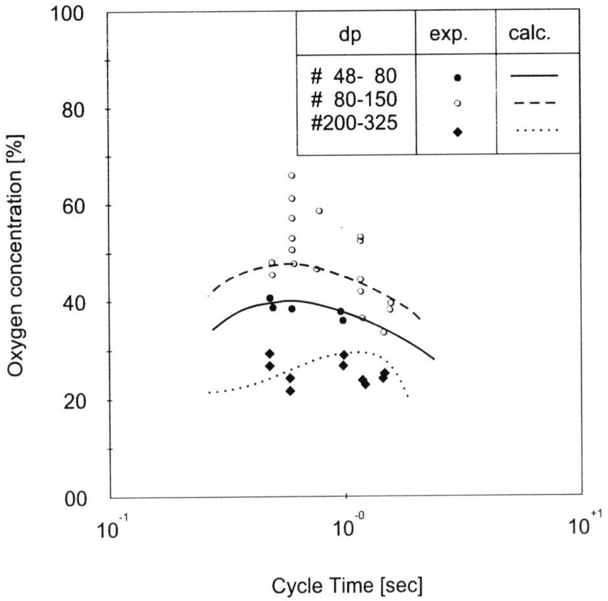


Figure 12. Dependency of oxygen concentration on cycle time.
(Reprinted with permission from ref. [21]).

The pressure swings obtained from experiments and calculations in spite of the simple model are in good agreement. The dependence of the oxygen concentration in the product gas on cycle time is showed in Figure 12. The oxygen concentration depends on the cycle time, since the pressure swing profile is controlled by the cycle time as mentioned before. The oxygen concentration also varies considerably with particle size of zeolite. The highest oxygen concentration was achieved with the particles of 80 to 150 mesh size.
Usually, the production capacity of PSA is defined as:

$$\text{Production Capacity [Nm}^3\text{ product gas/m}^3\text{ bed/hr]} = \frac{G_{product}}{V_{bed}} \tag{35}$$

Since the oxygen production capacity of normal PSA is usually below 10 $Nm^3$ oxygen $m^{-3}$ bed $hr^{-1}$, the oxygen production capacity of the proposed URPSA can be estimated to 10 times more than in the case of normal PSA. The highest oxygen production capacity can be achieved, when the particles of 80 to 150 mesh size were used (Figure 13).

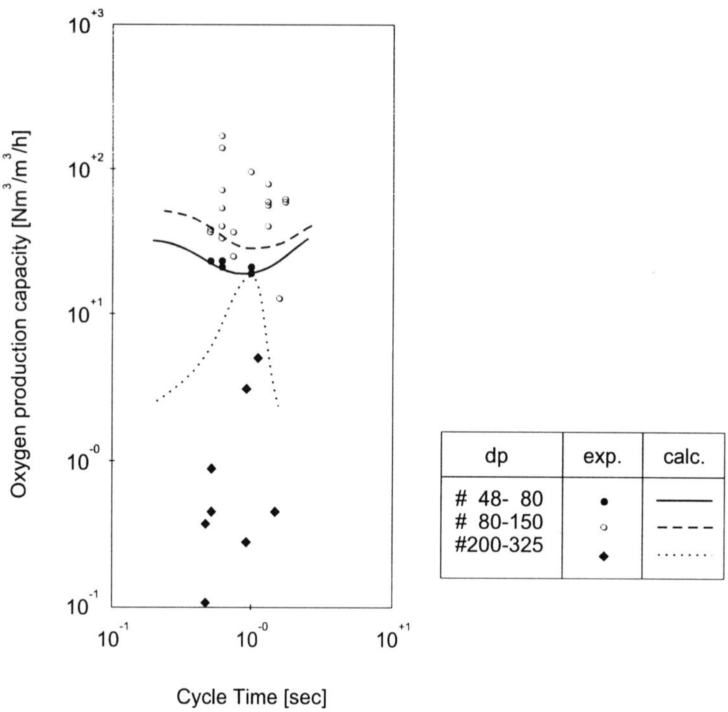


Figure 13. Dependency of oxygen production on cycle time.
(Reprinted with permission from ref. [21]).

Oxygen yield is defined in conventional form as follows:

$$\text{Oxygen yield [\%]} = \frac{\langle \text{Oxygen output} \rangle}{\langle \text{Oxygen input} \rangle} \times 100 = \frac{G_{\text{product}} C_{\text{prouct}}}{G_{\text{feed}} C_{\text{feed}}} \times 100 \qquad (36)$$

The cycle time dependence of the oxygen yield is shown in Figure 14. The oxygen yield of the URPSA is less than 5%. It has become clear that the oxygen enrichment by this URPSA gives extremely high oxygen production capacity and low oxygen yield.

The piston-driven ultra rapid pressure swing adsorption (URPSA) is a new technical solution for oxygen enrichment from air using zeolite. Nearly 60% oxygen-enriched gas was produced with an oxygen production capacity of about 100 $Nm^3$ $m^{-3}$ bed $hr^{-1}$, which is one order of magnitude higher than that obtained by commercialized oxygen-enrichment PSAs. This method has been shown as a variation of the improvement and optimization of cyclic operations and feasibility studies for application of this novel PSA.

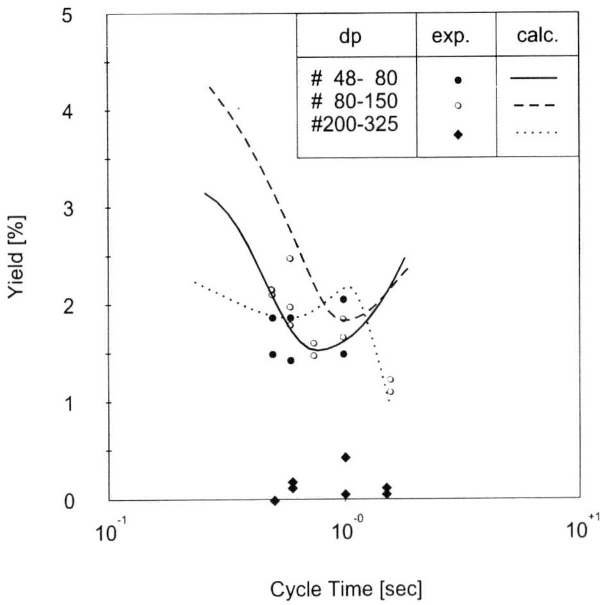


Figure 14. Dependency of oxygen yield on cycle time.
(Reprinted with permission from ref. [21]).

### 3.2. Parallel passage cross-flow adsorber

To measure fast pressurization transients, an experimental apparatus has been constructed [5]. The schematic picture can be seen in Figure 15.

The experimental results of dynamic pressurization show, that this method is suitable for the determination not only the measurement of pressurization velocity, but the real adsorption velocity too. That will bridge over the difficulties originated from the differences among industrial applications and the equilibrium or very slow scientific measurements. With the help of these equipments the RPSA processes can be carefully examined. This system is new in this field in the sense of high frequency as well as the easy data acquition. The maximum velocities of different mass flows show the boundary values of URPSA technologies ( Table 1).

Table 1
The different mass flows for zeolite 5A

| Mass flow Gas | Inlet (mol/s) | Pressurization (mol/s) | Adsorption (mol/s) | Specific adsorption (mol/s/dm$^3$) |
|---|---|---|---|---|
| $H_2$ | 0.29 | 0.216 | 0.074 | 0.056 |
| He | 0.24 | 0.24 | 0 | 0 |
| $CH_4$ | 0.107 | 0.021 | 0.086 | 0.065 |
| $N_2$ | 0.088 | 0.024 | 0.064 | 0.059 |
| Ar | 0.078 | 0.036 | 0.042 | 0.032 |
| $O_2$ | 0.08 | 0.036 | 0.042 | 0.032 |
| $CO_2$ | 0.061 | 0.0008 | 0.053 | 0.04 |

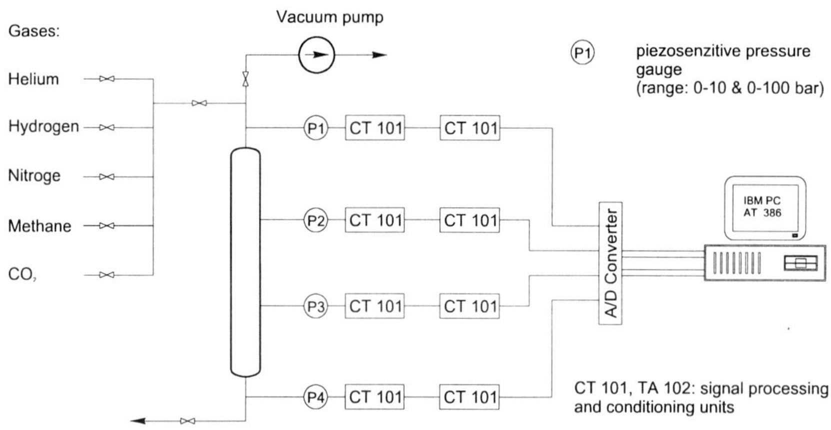


Figure 15. Apparatus for the investigation of fast adsorption processes. (Reprinted with permission from ref. [5]).

### 3.2.1. Method of building cross-flow adsorbers

The schematic picture of a cross-flow parallel passage adsorber-desorber can be seen on the Figure 16.

The structure of the adsorber-desorber was produced by alternate crosswise stacking of square shaped pieces of sheets provided by adsorbent layers on both sides. The technique of manufacturing consists of two parts. One is the method of adsorbent layer production, and the other the manufacturing method of structured metal frame. Because of the necessary high heat conductivity, usually the copper has been chosen as frame material. Taking into consideration, the thin layers, a thermocompression or the so called diffusion bounding can be the suitable technical solution.

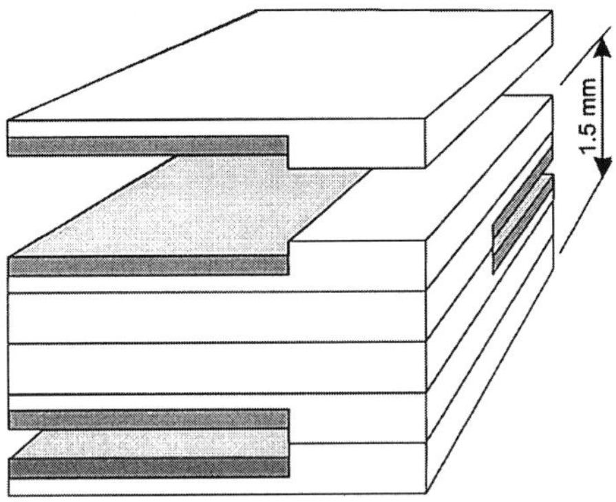


Figure 16. Schematic picture of a cross-flow parallel passage adsorber-desorber.

This process uses both heat and pressure to make a bond. Knowing the equilibrium diagram of a copper-oxygen alloys the bounding process takes place without chemical changes only in inert atmosphere or high vacuum. For the bounding process necessary to decide the pressure, temperature and time. The usual values for the temperature Tb = (0.7 ~ 0.8) Tmp, for the pressure P = 3~100 MPa, the maximal bounding time 1~240 minutes.

On the other hand could be hardly find material into which copper does not diffuse. The graphite seems to be the best. It can bear the relatively high pressure too. Among our circumstances first we pressed the sheets and than they were heated. The applied bonding pressure was 27 MPa.

As for the temperature 1100 K and practically 0 time gave good result. The two 0.1 mm thick sheets were bounded. The 0 time needs a small explanation. It means that the heating period took 3 hours, following directly the cooling period of 10 hours.

An adsorbent layer ( NORIT activated carbon powder d~0.01 mm) in the suspension form with starch was smeared. The layer thickness - after a drying at $150^0$ C temperature - rises as highly as 0.1 ~ 0.2 mm.

### 3.2.2. The adsorption heat and heat transfer

Let us consider how large is the heat transfer capacity of the former adsorber-desorber construction. The heat conductivity of a thin copper layer is larger at least with two magnitudes than the carbon or other adsorbents layers ones. Thus the governing transfer properties are the conductivities of adsorbent layers. The thermal conductivity of a carbon powder ~0.6 kJ/m hr K. The average thickness is ~ 0.1-0.2 mm. Using these data the technically possible heat transfer velocity

~ 800 J/m$^2$ s K. Having based on the proposed construction the transfer surface is 1600 m$^2$/m$^3$. Taking into consideration that the adsorbent amount is 70-100g/m$^2$, and the heat of adsorption (isosteric) in average 28 kJ/mol, the necessary cooling velocity 800-1000 J/m$^2$s. This heat flux seems to be enough for the desorption at a lower pressure.

The former experiments concerning to the adsorption velocities allow an estimated cycle time in the magnitude of 10 s. In this case for a recovery process (600 MW Boiler using liquid natural gas [LNG]) for 1700 000 Nm$^3$/hr an adsorber-desorber system with TCRPSA (thermally coupled rapid PSA) needs ~500 m$^3$ volume of adsorber-desorber apparatus. Comparing with the supposed recovery process (Figure 17), the necessary volume is only a small fraction of them. The proposed technological process can be seen on the Figure 18. The possible adsorption capacity is ~3000 m$^3$ flue gas/hr/m$^3$ bed.

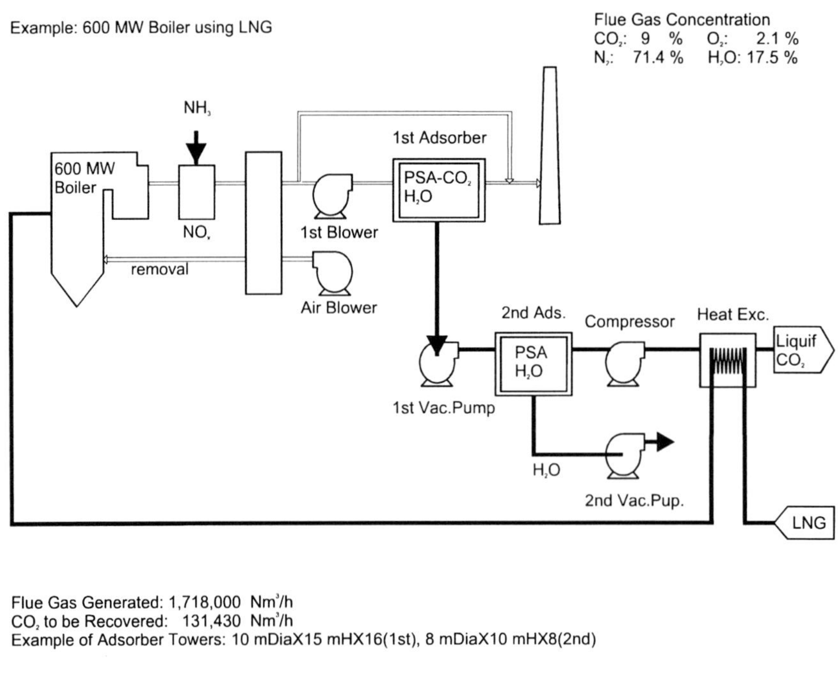


Figure 17. Supposed recovery process of $CO_2$ by PSA (1st adsorbers for $CO_2$, $H_2O$, the 2nd for $CO_2$ drying).

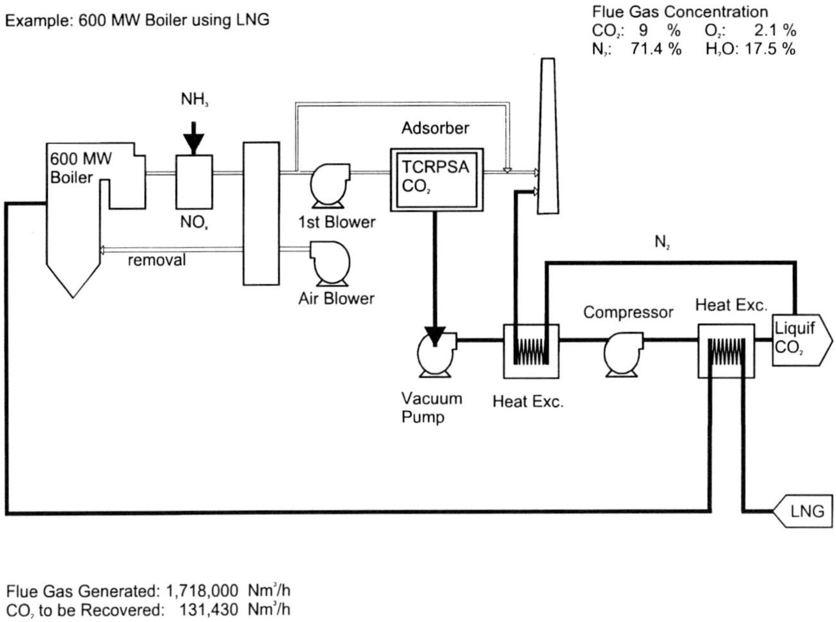


Figure 18. Supposed recovery process of $CO_2$ by TCRPSA.(Necessary adsorber tower 1pc: 10mDiaX15mH).

## CONCLUSION

Results of different experimental and theoretical studies suggest that the traditional mass transfer devices for adsorption processes – in the spite of the fact that the packed beds are now in fashion and they are cheap – cannot answer the new challenges arising from the environmental protection technologies. The main disadvantages of the traditional packed bed are the high pressure drop leading to significant power costs and the low specific capacity resulting telatively large apparatures.

The applications of adsorptive separation on same important fields like the treatment of industrial offgases or large amount of $CO_2$ emmisions were the new technical solutions. In this chapter we introduced some solutions for the decreasing of energy demand originating from the hydrodynamic properties and the application of inner heat coupling, demonstrated with new constructional solutions like parallel passage adsorbers. The velocity barriers for the application of piston driven adsorption methods have been discussed. These basic elements are to be expected for the benefit of the constructors involved in the large capacity adsorber desing.

## NOMENCLATURE

| | |
|---|---|
| A,B | constants of the Ergun-type equation |
| c | gas concentration, kmol/m$^3$ |
| C | molar fraction |
| $D_0$ | constant defined by Eq. (32), m$^2$/s |
| $D_K$ | Knudsen diffusion coefficient, m$^2$/s |
| $D_m$ | molecular diffusion coefficient, m$^2$/s |
| $D_p$ | pore diffusion coefficient, m$^2$/s |
| $d_p$ | particle diameter, m |
| dp | particle size, mesh |
| F | flow rate, mol/s |
| G | flow rate, m$^3$/hr |
| H | bed height, m |
| $k^2$ | tortuosity factor |
| $K_F a_v$ | overall mass transfer coefficient,1/s |
| M | molar weight kg/kmol |
| M(x,t) | mass flux kmol/m$^2$/s |
| $M^*(\eta,t)$ | dimensionless mass flux |
| $M_i$ | molecular weight, kg/mol |
| N | mass, mol |
| p | partial pressure, Pa |
| P | total pressure, Pa |
| p(x,t) | pressure, Pa |
| $p^*(\cdot\eta,t)$ | dimensionless pressure |
| $p^*_{i(t)}$ | dimensionless pressure at the ith gauge |
| $p_F$ | feed pressure, Pa |
| $p_I$ | initial pressure, Pa |
| $\Delta p^*$ | dimensionless pressure difference |
| q | amount adsorbed, mol/kg |
| $Q_{st}$ | heat of adsorption, J/mol |
| R | gas constant, kJ/kmol/K |
| $r_a$ | macropore radius, m |
| Re | Reynolds number |
| $R_p$ | particle radius of adsorbent, m |
| S | cross sectional area, m$^2$ |
| T | temperature, K |
| Tb | bounding temperature, K |
| Tmp | melting point, K |

| | |
|---|---|
| $t$ | time, s |
| $t_c$ | adsorption time, s |
| $u$ | gas velocity, m/s |
| $x$ | axial co-ordinate, m |
| $V$ | volume, $m^3$ |

Greek letters

| | |
|---|---|
| $\beta_0$ | adsorption capacity coefficient, $m^3/kg$ |
| $\beta_c^0$ | pre-exponent constant in Eq. (9) |
| $\gamma$ | packing density, $kg/m^3$ |
| $\delta$ | ratio of lin. and non-lin. friction forces |
| $\varepsilon$ | void fraction |
| $\varepsilon_p$ | particle void fraction |
| $\eta$ | dimensionless co-ordinate |
| $\lambda$ | drag coefficient |
| $\rho$ | gas density, $kg/m^3$ |
| $\tau$ | dimensionless time |
| $\tau_c, \tau_{PR}$ | dimensionless times |
| $\omega$ | angular velocity, rad/s |
| $\Omega$ | constant given by $\tau_c$, |


## ACKNOWLEDGEMENT

We thank the New Energy and Industrial Technology Development Organization and Hungarian National Scientific Fundation for supporting this collaborative work and personally to B. Ferencz and J. Argyelán for drawing figures.


## REFERENCES


1. S. Ergun, Chem. Eng. Prog., 48 (1952) 89.
2. G. Horváth, T. Szánya, P. Szolcsányi, A. László, L. Bilitzky and J. Hüttner, Method for the separation of helium-neon mixtures, Hungarian patent No. HU 195 690 (1990).
3. L. Arányi, J. Hay, G. Horváth, Z. Szabó, T. Szánya and P. Szolcsányi, Magy. Kém. Lapja, 36 (1980) 39.
4. K. Kutics, Hydrodynamic Study of Packed Beds, M.Eng. Thesis, University of Veszprém, 1981.
5. G. Horváth, K. Kutics and M. Suzuki, NATO ASI Series. vol. 491, 511 Kluwer, 1996.
6. J. Guan and Z. Ye, Optimization of the Operation Parameters for Rapid Pressure Swing Adsorption, Fundamentals of Adsorption Proceedings of Fourth International Conference on Fundamentals of Adsorption, ( 1992) 243.

7. J.Hart and W. J. Thomas, Gas Sep. Purif., 5 (1991) 125.
8. R. L. Jones, G. E. Keller and R. C. Wells, Rapid Pressure Swing Adsorption Process with High Enrichment Factor, US Patent No. 4 194 892 (1980).
9. R.L. Jones and G. E. Keller, J. Separ. Proc Technol., 2 ( 3 ) (1981) 17.
10. K. Kawazoe, I. Sugiyama and Y. Fukuda, Kagaku kogaku, 30 (1966) 1007.
11. D.E Kowler and R. H. Kadlec, AIChE Journal,18 ( 6 ) (1972) 1207.
12. G.W Miller, K. S. Knaebel and K. G. Ikels, AIChE Journal, 33 (2) (1987) 194.
13. S. Nakao and M. Suzuki, J. Chem. Eng. Japan, 16(2) (1983) 114.
14. C.L. Pritchard and G. K. Simpson, Chem. Eng. Res. Des., 63 (1986) 467.
15. R. Rota and P C. Wankat, AIChE Journal, 36 (9) (1990) 1299.
16. A. Sakoda and M. Suzuki, Sep. Tech., 1 (1991) 73.
17. C.W. Skarstrom, Recent Developments in Separation Science, N. N. Li (ed.), CRC Press, Ohio, vol. 2, 95 1975.
18. M. Suzuki, Adsorption Engineering, Kodansha, Tokyo, 1990.
19. P.H. Turnock and R. H. Kadlec, AIChE Journal, 17 (2) (1971) 335.
20. L. Rousar, P. Ditl, L. Kotsis and K. Kutics, Chem. Eng. Comm., 112 (1992) 67.
21. M. Suzuki, G. Horvath, D.M. Ruthven and S. Farooq, Research Report, NEDO, Tokyo, 1996.

Adsorption and its Applications in Industry and Environmental Protection
Studies in Surface Science and Catalysis, Vol. 120
A. Dabrowski (Editor)



# The mutual transformation of hydrogen sulphide and carbonyl sulphide and its role for gas desulphurization processes with zeolitic molecular sieve sorbents

M. Bülow [a], W. Lutz [b] and M. Suckow [c]

[a] The BOC Group Gases Technical Center, 100 Mountain Ave., Murray Hill, NJ 07974, USA

[b] Holzmarktstraße 73, D-10179 Berlin, Germany

[c] Fachhochschule Lausitz, Großenhainer Straße, D-01968 Senftenberg, Germany

The physico-chemical fundamentals of gas desulphurization by zeolitic molecular sieves are described. Structure-property relationships are derived for both the selective sorption separations of hydrogen sulphide from gases and the formation of carbonyl sulphide, in the case that carbon dioxide is one of the feed gas components. The formation of carbonyl sulphide under the conditions of a practical desulphurization process for natural gas is interpreted on the basis of fundamental laboratory scale data. A systematic study has been carried out to investigate the influence of zeolite and cation types, cation exchange degree, sorption equilibrium properties of the various materials and gases, presorption of appropriate gases, time, temperature and hydrothermal stability of zeolitic sorbents on the catalytic reaction. A mathematical model to simulate technical desulphurization processes is described and applied to the analysis of a technical purification plant with a utilization of related research data. The breakthrough behaviour of various gas mixture components for a cyclic plant regime is simulated for the cases of sorption purification from hydrogen sulphide without and with formation of carbonyl sulphide. Various consequences that follow from the results, in particular with regard to the use of zeolites for gas desulphurization are discussed.

## 1. INTRODUCTION

The utilization of gases as source of primary chemicals and fuels has increased steadily over the last half of this century. The development of the gas industry in Europe during the thirties and fourties included novel processes in the fuel gas industry that were related mainly to those for liquefaction and gasification of coal, e.g., by Fischer-Tropsch syntheses. With the expansion of petrochemistry after World War II, overwhelming activities were directed globally towards the development of new technological processes and schemes for the utilization of natural gases and those that accompany crude oil. Entirely novel technologies were developed and introduced into practice to satisfy, especially, the rapidly increasing need in carriers of primary energy and to allow for a utilization of many deposits discovered, in particular in the Near East region as well as on the American and Eurasian continents. One of these technologies relates to the adsorption separation of mixtures by means of molecular sieves [1-4] pioneered in the US. This special development had been fostered by chemical industry. Consequently, zeolitic molecular sieves A (LTA) and X (FAU) became industrial products since mid-fifties [5,6].

There are many methods for processing of gases [7], due to a great diversity in their original chemical composition. Gases may stem, for example, from degasification of hard coal, or lignite under conditions of increased pressure and temperature, but also from crude oil processing, natural and biogases. Additional considerations relate to the various purposes of utilization of the gases, e.g., as pure gases of constant composition for industry and households, furthermore, as feed products for chemical processes, etc. The primary treatment of natural and technical gases comprises two major goals, i.e., increase in their energy content and removal of noxious compounds. These ones are mainly sulphur-containing compounds to be separated from gases due to their toxicity, environmental hazardousness, their ability to cause corrosion and precipitation of solids in chemical plants [8]. In modern industrial societies, maximum concentrations of noxious sulphur emissions are limited by law. Upper limits for the content of sulphur compounds in those gases, such as mercaptans, disulphides and carbonyl sulphide (COS), are as low as 100 mg/Nm$^3$. For hydrogen sulphide ($H_2S$), the maximum concentration allowed comprises even much lower values, e.g., 5 mg/Nm$^3$, in Germany. Industrialized countries demand, furthermore, certain dryness of the gases. Both restrictions, viz., "sweetness" and dryness of industrial gases, help actively to prevent the occurrence of stress corrosion cracking in industrial plants and gas transport systems.

Separation of sulphur compounds from gases has been described widely in literature, cf., e.g., [7-19]. A series of papers is specifically dedicated to the removal of $H_2S$. For this particular purpose, several methods were recommended, e.g., physical "gas washing", including absorption/extraction by means of organic solvents, e.g., amines, neutralization of acidic components, oxidation, chemical drying, purification by means of membranes and adsorption by carbonaceous sorbent materials.

Sorption by zeolites offers a powerful means for the development of methods for super purification (super desulphurization with simultaneous drying) of gases with low-level $H_2S$ content (lean gases) and which, at the same time, may have high water ($H_2O$) concentration, up to saturation of the gas by $H_2O$ vapour. This principle, if it uses hydrophilic zeolites, is characteristic of reversibility of the sorption process, high separation performance and long "lifetime" of the sorbent itself. At the same time resulting waste products may not represent environmental hazards. The technology is cheap in terms of capital and operating costs; it can be fully automatted.

Molecular sieve sorption for gas desulphurization has been developed with regard to the utilization of both natural and synthetic zeolites such as the LTA and FAU types in different modifications. The latter can be varied with respect to the type of sulphur compounds to be separated from a given gas but also as a consequence of the principle to be used, viz., "molecular sieving", kinetic or equilibrium separation. Process-specific thermal or hydrothermal conditions may also be accounted for. Zeolites were recommended especially for the treatment of lean gases with an increased but still acceptable content of carbon dioxide ($CO_2$) for fuel gas purposes. This is because utilization of other principles, e.g., various physical and chemical washing methods for $H_2S$ removal may lead to a decrease in their efficiency due to co-absorption of $CO_2$, hydrocarbons and organic sulphur components [20,21]. On the other hand, the formation of COS was reported as a phenomenon that occurs during desulphurization ("sweetening") of $CO_2$-containing gases, by zeolitic molecular sieves [20,22-24], in accordance with the following reaction:

$$H_2S + CO_2 \Leftrightarrow COS + H_2O. \tag{1}$$

A systematic investigation of the sorption of $H_2S$ as a component of gas mixtures resulted in the consideration of COS formation as one of the aspects of the complex process [25,26]. This phenomenon cannot be neglected, although COS itself is not known to cause stress corrosion cracking, especially, because COS is a very volatile compound that is able to leave the gas purification plant together with the "purified" product. However, COS may be as toxic as $H_2S$. A gas purified in a way that neglects the possible occurrence of COS may not be used as town gas.

Appropriate modification of zeolites and other molecular sieves as well as specific process design were claimed to suppress the generation of COS during desulphurization in presence of $CO_2$ [27-29]. A study of related properties of zeolitic sorbents commercialized for gas desulphurization, has shown, however, that these approaches to the problem do not meet entirely those demands [30]. It is, therefore, necessary to optimize zeolites with regard to their chemical composition and related sorption separation efficiency and catalytic activity, on the one hand, and those prerequisites that govern process conditions and management of waste-gas related demands in conjunction with capital and operating costs of a technique to be chosen, on the other hand.

## 2. HISTORICAL BACKGROUND

Technical opportunities for sorptive separation of $H_2O$, $CO_2$, $H_2S$ and organic sulphur compounds from natural gas were considered originally by the Linde Corp. in 1957 [31]. The first plant for drying and sweetening of gases by means of molecular sieve sorption went on stream in the USA in 1958 [32]. Technological processes for treatment of natural gases, liquefied gases and gasoil were reviewed by Sherwood in 1963 [33]. In 1963, some 20 plants existed for propane sweetening [34]. A total of ca. 500 ton of zeolite mass had been in use by US-based desulphurization plants during the mid-sixties [35]. Reports by Schoofs [36] and Thomas and Clark [37] indicated a tendency toward a utilization of zeolites, in particular, the materials 3A (KA) and 4A (NaA) for processes of gas drying, 5A (NaCaA, CaA) for removal of $H_2S$, and 13X (NaX) for removal of organic sulphur compounds from various types of gases. Zeolite sorbents were employed in 2-bed [34,38-40], 3-bed [41-43] and 4-bed sorption plants [32,37,41]. In 3- and 4-bed plants, as usual, several adsorber columns work in the adsorption regime or - as described in [42-44] - in the desorption regime, in a consecutive arrangement. Process schemes were described in detail in [32,36,40-43,45-47]. Modern schemes for related gas processing do not differ significantly from those described therein. As far as the adsorber bed capacity is concerned, a wide spectrum of sizes from 0.5 to 1 ton/bed to 50-70 t/bed is known. The latter is typical of large plants of natural gas industry [32,37]. The length of the adsorption step of the majority of TSA (thermal swing adsorption) processes ranged between 2 and 10 hours [35,40,42,43,48,49]. In a few cases, the adsorption step lasted shorter than 0.5 hours [50] or longer than 36 hours [37]. Feed gas pressure values amounting to ca. 20-60 bar [32,35,37,41,48,51,52] allowed the gas throughput of, e.g., $0.2 \times 10^6$ $Nm^3/d$ [41], and $2.5\text{-}2.6 \times 10^6$ $Nm^3/d$ [32,37,42,43,48]. Zeolites of both 5A and 13X types were used as sorbing medium. These materials were found not to differ significantly in their selectivity towards $H_2S$ [44,53]. Despite this situation, 5A zeolite became a preferred choice [44,48,51,53] because of a certain catalytic activity of 13X with regard to side reactions. However, both types were recommended for the Haines process that includes oxidative regeneration of the zeolite [44,54,55] leading to elemental sulphur. This process did not find broad application due to a decomposition of the zeolite structure during the oxidation. For the removal of organic sulphur compounds, an exclusive utilization of 13X zeolites was reported [45,46]. An analogous situation holds for liquid phases [34,40,48-50,56]. With regard to zeolite 10X (CaX), a decrease in sorption selectivity and increase in catalytic activity towards $H_2S$ was found. (For this reason, its utilization for the Haines process was recommended [42,43,57]). Beside synthetic zeolites, natural species were considered as alternatives, e.g., the types mordenite and clinoptilolite [52,58]. However, the ready availability of clinoptilolite did not diminish efforts needed to optimize those materials to make them competitive compared to synthetic materials that are cheap and homogeneous in properties. For almost all processes described, the operating temperature was in the region of the ambient one [32,35,37,41,44,48,53]. In cases of desulphurization of liquids, separation of sulphur compounds was performed either directly in liquid phase [34,40,49,59], or in vapour phase [50,59] at temperatures ranging from 293 to 313 K and 473 to 623 K, respectively.

For the regeneration of zeolitic sorbents, a series of methods was employed. These include (i) combined thermal/gas replacement desorption; (ii) oxidative regeneration; and (iii) combined pressure/thermal swing desorption. Method (i) has found the most of applications. In plants with a regeneration step under such conditions [32,35,37-40,48,49,51,53,60,61], desorption of sulphur compounds proceeds preferentially at 473-573 K [32,35,37,48,53,60] but also above 573 K [51] by passing ca. 2-10 % of the purified gas [32,37,38,53,62,63] through the bed. In the case of desulphurization of liquefied gas, desorption is described to proceed by an inert gas [56]. In each case, desorption temperature is reached by utilizing hot gases as replacement agents that serve also as purge gases. The desorbing gas loaded with sulphur compounds is either flared [38] or treated by an alkaline absorption process to remove the sulphur compounds and then used as fuel gas [32,38,61] or directed into a Claus process to produce elemental sulphur [64].

The description in the literature of early gas desulphurization processes that utilize zeolites does not mention duly the formation of COS during the removal of $H_2S$, if $CO_2$ is present in the feed gas, except in a few cases, e.g., ref. [20,28,65]. Since modern desulphurization plants work in accordance with the same principles and utilize identical zeolite types, the COS formation reaction may have strong implications for the currently employed processes for desulphurization of gases by means of those sorbents. Therefore, it is necessary to investigate (i) the COS formation as dependence on the zeolite type, the type and content of cations in the sorbent, the concentration and contact time of reactants with the sorbent, the temperature and the conditions of co-adsorption; (ii) the mechanism of that reaction on the sorbent with specific emphasis on its sorption and catalytic properties; and (iii) to develop a mathematical model to simulate dynamic processes that proceed in adsorbers/reactors of technical dimension. This investigation should lead to novel formulations of modified zeolite sorbents and to alternatives with regard to operating conditions of sorption plants with the purpose of either minimization or maximization of the formation of COS.

## 3. FUNDAMENTALS OF GAS DESULPHURIZATION BY MOLECULAR SIEVES

Considerable knowledge of sorption by zeolites was available when sorption processes for gas desulphurization became actual R&D topics. During the reduction of these processes into practice research contributed specifically and continuously to basic understanding of sorption of $H_2S$, $CS_2$ and organic sulphur compounds. However, the formation of COS did not seem to represent any major interest except in [20,28,65] and our work that accounted for both scientific and commercial aspects. As to the scientific aspects, fundamentals of sorption processes and zeolites as metastable systems were main topics. The hydrothermal stability of zeolites became of interest because, if the COS reaction were relevant, preferential sorption of $H_2O$ would take place in the entire sorbent bed.

### 3.1. Zeolitic molecular sieves

Since FAU and LTA zeolites represent sorbents of major interest to gas desulphurization, their structure will be considered briefly; for details, cf., [1-6,27,58]. Zeolites are crystalline aluminosilicates in which silicon and aluminum atoms are interconnected via oxygen atoms and form a 3d-lattice with a micropore or channel system therein. This lattice can be imagined to be composed by $[AlO_4]^{5-}$ and $[SiO_4]^{4-}$ tetrahedrae alternately joined by common O-atoms, at each tetrahedral corner. Such corner sharing creates infinite lattices comprised of identical building blocks or unit cells. In general, the number of $[AlO_4]^{5-}$ tetrahedrae in a zeolitic aluminosilicate causes a defined residual negative charge on the oxygen of the lattice. The negative charge of the lattice is compensated by cations. These cations occupy, as usual, defined sites in the microporous structure and are exchangeable. Relevant numbers and positions of cations in the FAU structure are discussed below. The sizeselective sorption property of zeolites is due to specific dimensions of entrances to the intracrystalline micropore system, so-called zeolitic "windows", and their relation to the molecular cross-sections of sorbing species. Although all of the gas species that comprise a typical natural gas may enter the supercages (LTA zeolite: α-cages) of a zeolite crystal, only $H_2O$ molecules are able to access the sodalite units (LTA zeolite: β-cages), cf., Figure 1.

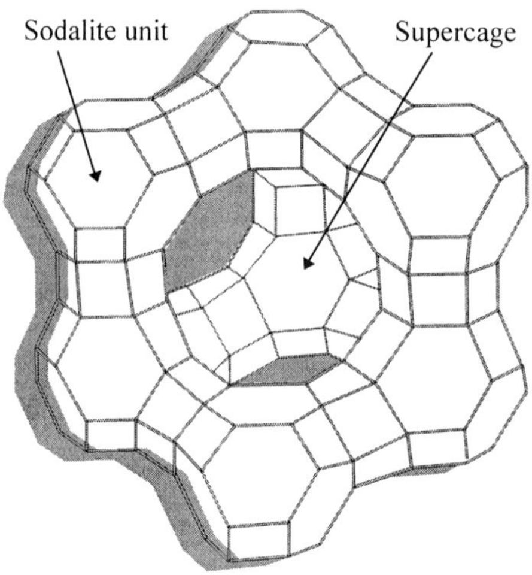


Figure 1. Pseudo-unit cell of faujasite type zeolite (FAU).

This "biporous" structure of FAU and LTA zeolite frameworks is created by an interconnection of basic zeolitic units, viz., sodalite cages, via their double six rings (LTA structure: the sodalite units are interconnected via their double four rings). The FAU zeolite offers a higher micropore volume compared with the LTA zeolite, which leads to a higher saturation capacity with regard to the dipole-carrying sorbing species, i.e., $H_2O$, COS and $H_2S$. However, special properties connected with both non-specific, van der Waals-type, and specific, Coulomb-type, intermolecular interactions - due to the presence of negative and positive charges distributed over the lattice and metal cations, respectively, that compensate those negative charges -, a higher sorption attraction of those sorbing species has often been found for LTA zeolites, at low to moderate partial pressures of those species in the gaseous phase. Except for LSX (low-silicon-X) zeolite [66] with an elemental ratio, Si/Al = 1, as that of LTA zeolite, the LTA zeolite offers the highest number of charges, i.e, sorption centers for specific sorptive interaction. LSX materials were not considered in this work. For the desulphurization of gases, the influence of energetic effects on the sorption selectivity of zeolites with regard to $H_2O$ and $H_2S$, prevails that of steric or sizeselective effects.

### 3.2. Sorption interaction of gas components with molecular sieves

Sorption affinity of fluid components with molecular sieves is governed by the type of their mutual interactions, i.e., fluid-solid interactions, on the microphysical level. For the sorption systems given, besides van der Waals-type atom-atom dispersion interactions of all types of molecules involved, with the solid, zeolitic ion-dipole interactions for $H_2O$, $H_2S$ and COS as well as zeolitic ion-quadrupole interaction for $CO_2$ have to be taken into account. A measure of the overall strength of interaction energy is the isosteric heat of sorption, $q_{isosteric}$. This quantity differs from the differential heat of sorption or sorption enthalpy, $-\Delta H$, usually measured calorimetrically, by the mechanical work term, $|RT|$; $q_{isosteric} = -\Delta H + RT$.

Figure 2 presents a comparison of concentration dependences of sorption enthalpy values for various gases being typical components of natural gas, as measured calorimetrically for their single component sorption by a NaCaA zeolite

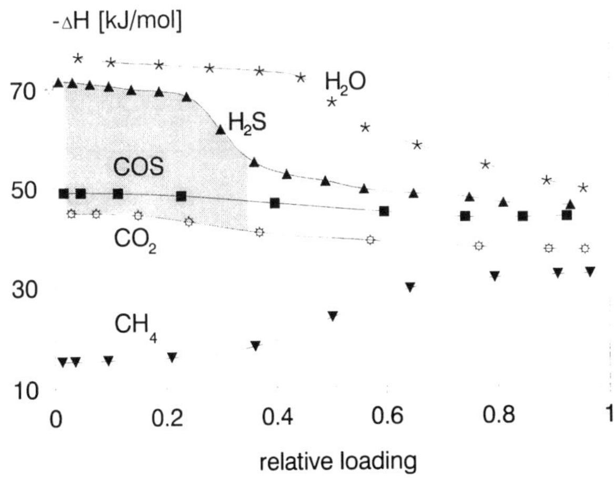


Figure 2. Sorption enthalpies of gases on NaCaA zeolite ($Na^+$ vs. $Ca^{2+}$ exchange: 45 %).

with 45 % of its sodium cations, $Na^+$, exchanged vs. calcium cations, $Ca^{2+}$. The brightly hatched area indicates the strong preferential sorption of $H_2O$ over $H_2S$, COS, $CO_2$ and methane ($CH_4$) by the zeolite considered. At low concentration, $H_2S$ interacts with the sorbent significantly stronger than COS and $CO_2$ (cf., dark hatched area) [30].

For the NaCaA sorbent, the fraction of $H_2S$ concentration which interacts specifically with the zeolite, varies obviously with its cation exchange value. With increasing content of $Na^+$, the sorption enthalpy plateau, at ca. 70 kJ/mol, expands towards higher relative zeolite loading [67]. This is due to a strong binding of $H_2S$ molecules by $Na^+$ cations that leads to a dissociative interaction of $H_2S$ as proven by IR spectroscopy [68,69]. Concordantly, one might assume that NaA zeolite (all cations are $Na^+$ ones) should show outstanding separation properties. However, kinetic reasons, i.e., low intracrystalline mobility of $H_2S$ in NaA [70] (effective cross-sections of windows: ca. 4Å), do not allow for an efficient utilization of this situation. An expected similarity in affinities of $H_2S$ with KA and NaA zeolites may not be relevant because the effective cross-section of KA windows, ca. 3Å, is smaller than that of the $H_2S$ molecules, ca. 3.5 Å.

Obviously, in the case of LTA zeolites as desulphurization sorbents, a material of the general composition, $NaMe^{2+}A$, where $Me^{2+}$ denotes a divalent cation, may be most favourable. The exchange value of $Na^+$ cations against $Me^{2+}$ ones might be envisaged as $\geq$ 33 %, i.e., just above the percolation threshold between the existence of 4A (4Å) and 5A (5Å) species, NaA and $NaMe^{2+}A$, of the basic LTA structure with regard to their cross-sections, and still maintaining a strong interaction between $H_2S$ molecules and $Na^+$ cations [71,72].

The dependences of single component sorption enthalpy on concentration, cf., Figure 2, allow for only a rough estimate of mixture sorption behaviour. In principle, changes of Gibbs free sorption energy for multi-component sorption equilibria should be compared. To gain such information, time-consuming experiments are necessary. Whence such data is known for a few systems only. For the multi-component case, strong competitive sorption effects are expected. For a gas mixture of composition characteristic of natural gas, due to a significant excess of $CH_4$, $N_2$ and $CO_2$ compared with $H_2S$, the preferable specific interaction between $H_2S$ and cations may be suppressed, at least partly. For a gas mixture comprised of 50 % $CH_4$, 35 % $N_2$, 10 % $CO_2$, 0.01 % $H_2O$ and 0.01 % $H_2S$ (all vol. %; residual percentage relates to low-molecular-weight hydrocarbons), the $H_2S$ uptake by NaCaA zeolite at ambient temperature, amounts to one molecule per $\alpha$-cage only, based upon the authors' experience. This corresponds to ca. 8 % of the saturation value for $H_2S$ single component sorption by NaCaA zeolite at ambient temperature.

### 3.3. Cation positions in LTA-type zeolites and base catalysis

Zeolite NaA contains twelve $Na^+$ cations per unit cell - eight strongly coordinated cations on so-called S1 positions and four weakly coordinated cations in positions S2 and S3, i.e., in or in front of the plane of the oxygen-8-rings and oxygen-4-rings of the framework, respectively, (cf., [73], for a definition of extra-framework sites). The weakly coordinated $Na^+$ cations enable strong interaction with $H_2S$ sorbed in α-cages only, its access to β-cages, i.e., to strongly coordinated cations, is impossible for steric reasons. Calorimetric results for single component sorption of $H_2S$ show comparatively high sorption heats for zeolites rich in $Na^+$ cations. These data allows one to conclude [74-76] that the $(SH)^-$ group of a $H_2S$ molecule may be oriented toward cations in positions S2 and S3 and the proton interacts with oxygen that bridges silicon and aluminum in the framework, cf., Figure 3.

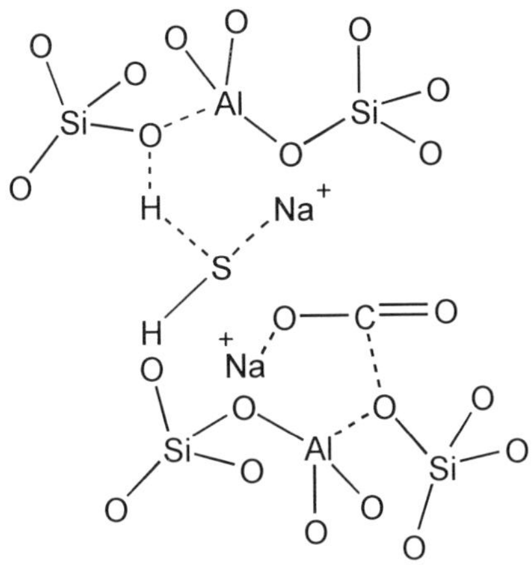

Figure 3. Scheme of possible interaction between $H_2S$ and $CO_2$ and the zeolite framework.

A dissociative destabilization of $H_2S$ that may be considered as a process of base catalysis, seems to cause the high reactivity of $H_2S$ needed for sorption-catalytic reactions such as COS formation under the conditions of gas desulphurization [76]. Besides, basic lattice sites and weakly coordinated $Na^+$ cations were found to catalyze reactions of $H_2S$ with tetrahydrofuran [77] and oxygen [78-80]. Evidence for the existence of those active sites may be obtained by means of IR spectroscopy that detects $(SH)^-$ characteristic bands [68,69,74-77]. If the $Na^+$ cations were replaced by other monovalent metal cations, the catalytic behaviour of the zeolite would change, appropriately. The presence of divalent cations, however, diminishes this type of catalytic activity of the zeolite [76]. For example, if $Ca^{2+}$ cations were exchanged into a NaA zeolite, one $Ca^{2+}$ cation is assumed to replace one weakly and one strongly coordinated $Na^+$ cation, whence, the number of cations per unit cell decreases from twelve to six. After having reached an exchange value, ca. $\geq 67$ %, all weakly coordinated $Na^+$ cations were removed [80,81]. For this reason, values of sorption enthalpy of $H_2S$ for a highly exchanged NaCaA zeolite are lower than those for a NaA zeolite [30,67]. The formation of dissociative $(SH)^-$ species is repressed [74], and the catalytic activity is minimized [27,76]. Alkaline earth cations are strongly coordinated. Either they do not exert specific interaction with $H_2S$ molecules at all, or they do to a lesser extent. If $Na^+$ cations were exchanged by divalent cations of heavy metals such as of cadmium, zinc, lead or manganese, sorption of $H_2S$ leads to an irreversible formation of the metal sulphides.

### 3.4. The reaction of COS formation

Carbonyl sulphide is present in the nature. If $H_2O$ and a catalyst, e.g., a zeolite, are present in a given system, COS decomposes in accordance to eq (1); the value of standard Gibbs free energy is relatively small and amounts to - 33.5 kJ/mol. In gas phase experiments without a catalyst, the COS-to-$H_2S$ reaction does not occur at temperatures below ~ 200 °C, based upon our experience. Therefore, despite natural gases are saturated with $H_2O$, as usual, certain COS concentrations may be detected therein. The equilibrium constant, $K_p$, of reaction (1) [82],

$$K_p = \frac{[COS][H_2O]}{[H_2S][CO_2]}, \qquad (2)$$

is given in Table 1 as dependence on temperature. If reaction (1) takes place in the presence of a hydrophilic zeolite such as those of the LTA- and FAU-types, $H_2O$ is sorbed preferentially at $Na^+$ cations and inside the sodalite units [83], and the reaction is shifted to the right, below 100 °C. Simultaneously, the action of a zeolite as a basic catalyst enhances this process. For the KA zeolite, in particular, that should have a strong activity with regard to reaction (1), formation of COS was observed after a longer time only, due to strong $H_2O$ uptake by α- and β-cages. Since $H_2S$ molecules cannot enter its micropores for size-selectivity reasons, the number of catalytically active sites is relatively restricted and the reaction proceeds unexpectedly slowly [83]. Since the reactants $H_2S$ and $CO_2$ have no/very little access to the intracrystalline void volume due to their comparatively large effective molecular cross-sections, i.e., 3.50 Å and 3.30 Å, respectively, KA zeolite can be used for drying gases that contain $H_2S$ and $CO_2$ ($H_2O$: 2.65 Å) [84].

Table 1
Temperature dependence of the equilibrium constant, $K_p$, of reaction (1)

| Temperature, °C | $K_p$ |
|---|---|
| 20 | $1.38 \times 10^{-6}$ |
| 100 | $3.16 \times 10^{-5}$ |
| 200 | $3.64 \times 10^{-4}$ |
| 300 | $1.8 \times 10^{-3}$ |
| 400 | $5.4 \times 10^{-2}$ |

# 4. FORMATION OF COS UNDER STATIC AND DYNAMIC CONDITIONS

The utilization of a NaCaA zeolite in a production-scale plant (State of Thüringen, Germany) for desulphurization of natural gas showed the necessity to investigate in detail the COS formation under static and dynamic conditions. Besides, models to describe the breakthrough behaviour of single components and mixtures that comprise the natural gas, were developed. The results of their application were compared with the experience collected on the plant.

## 4.1. Sorption catalysis under conditions of technical desulphurization of gases

A process analysis was performed for a 4-bed desulphurization plant with 20 metric tons of NaCaA sorbent (cation exchange: ca. 70 %) in each bed, which served for natural gas processing with the purpose of fuel gas production. The average feed gas composition is as follows (vol %): $CH_4$ (50), $N_2$ (35), $CO_2$ (10), $H_2S$ (0.01), $H_2O$ (0.01), $C_2$-$C_3$ alkanes as residual. Originally, the plant worked according to the following regime. A sequence of three beds is always in the adsorption step with the first bed acting as gas dryer, in general, the fourth bed being regenerated and then cooled. Feed gas enters the plant at pressure, 25 bar, and ambient temperature. Since the gas after purification is supplied directly to a power station, and the demand in fuel gas varies significantly, only an average value, ca. $10 \times 10^3$ $Nm^3/h$, can be used to characterize the throughput. As usual, after a throughput of ca. $10^6$ $Nm^3$, the process is changed by starting desorption of the first bed and utilizing the regenerated bed as third bed in the sequence of those that work in the adsorption regime. The regeneration proceeds by purging the bed countercurrently with regard to the feed gas direction in the adsorption step, with $2.5 \times 10^3$ $Nm^3/h$ of product heated up to a temperature of 340 °C, at a pressure of 22 bar, the gas desorbed being flaked.

According to the analyses of gas streams, the product gas is free of $H_2S$, after the third bed. Instead of $H_2S$, it contains equimolar amounts of COS formed by $H_2S$ and $CO_2$ according to eq. (1), mainly in the second bed. Since $H_2S$ does not react with $CO_2$ instantaneously, there may be large discrepancies between the concentration of $H_2S$ at the inlet of bed 1 and that of COS at the outlet of bed 3. These are caused both by a variation in $H_2S$ content of the feed gas and differences in product gas usage by the customer during a 24 hours' cycle. When the catalytic activity of the zeolite in bed 2 (at the end of which sampling for gas analysis takes place) vanishes, breakthrough of $H_2S$ and zero concentration of COS are observed. At this stage, the plant regime changes: the bed that had been the first until that time, is now regenerated by "purified" product gas that, however, contains COS. Unrecognized, COS reacts with $H_2O$ available in excess in this bed, to reform $H_2S$ and $CO_2$. Finally, a desorption gas loaded by $H_2S$, unrecognized again, is flaked. It may be assumed that a similar scenario could be typical of many zeolite plants employed for desulphurization of $CO_2$ containing gases, in particular, natural gases. If there is not much evidence known with regard to such environmentally hazardous process regime, this might be due to the usual situation of gas stream analysis, viz., the latter is directed onto $H_2S$ only, in most of the practical cases.

### 4.2. Formation of COS under static experimental conditions

Information on COS formation according to eq. (1) in the presence of sorbents is both scarce and contradictive. To chose selectively separating zeolites for gas desulphurization, one has to characterize and systematize their different technical and lab-synthesized modifications with regard to standard conditions. Particular static conditions for COS formation experiments were chosen to allow for a characterization and quantification of catalytic activities of the sorbents.

A binary mixture comprised of 20 % $H_2S$ and 80 % $CO_2$ (vol. %) is brought into contact with a zeolite sample activated at 673 K during ca. 10 h in vacuo, $< 5 \times 10^{-3}$ torr, within a known volume at temperature, 20 °C (5 mmol of mixture with 1 g of zeolite, corrected for binder content if any, in 150 $cm^3$). After exposure of the gas mixture to the zeolite during 5 h, the reaction mixture composed of COS, $CO_2$ and $H_2S$, is frozen out for GC analysis to calculate the conversion of $H_2S$ into COS; during sample collection the zeolite being heated at a rate, 10 K/min, until it reached 433 K. As it follows from single component sorption isotherms, the prevailing amount of the reaction product, $H_2O$, remains in the sorption state, under these conditions. Therefore, the reverse reaction between COS and $H_2O$ is suppressed in the gas phase, cf., $K_p = 1.8 \times 10^{-3}$ at 300 K. The equilibrium value of the $H_2S$ conversion into COS amounts to ca. 75 %. Since technical beds are operated at the condition of a $CO_2$ excess by 2 to 3 orders of magnitudes, a total $H_2S$-to-COS conversion is most probable for almost all zeolitic aluminosilicates. The influence of various typical factors related to zeolites and zeolitic sorption, on the reaction of COS formation is exemplified in the following sections.

#### 4.2.1. Zeolite type

As shown for the pure $Na^+$ cation modifications of A, X and Y zeolites, the catalytic activity decreases with increasing framework Si/Al ratio, i.e., with the decrease in the amount of cations of a given type, cf., Figure 4.

Compared with NaA zeolite, for which at conditions described above an equilibrium value of the $H_2S$-to-COS conversion, ca. 75 %, is achieved, NaX and NaY zeolites show significantly lower values of $H_2S$ conversion, although $H_2S$ is nonetheless able to interact with weakly coordinated $Na^+$ cations in the supercages of these zeolites. The total number of active centers, however, leads to their lower activity, the equilibrium value of $H_2S$-to-COS conversion, viz., 75 %, being reached after extended contact time. The catalytic activity of the LTA and FAU forms as described with regard to reaction (1), should be typical of that for all other synthetic and natural zeolites of other crystallographic structures but with aluminosilicate frameworks.

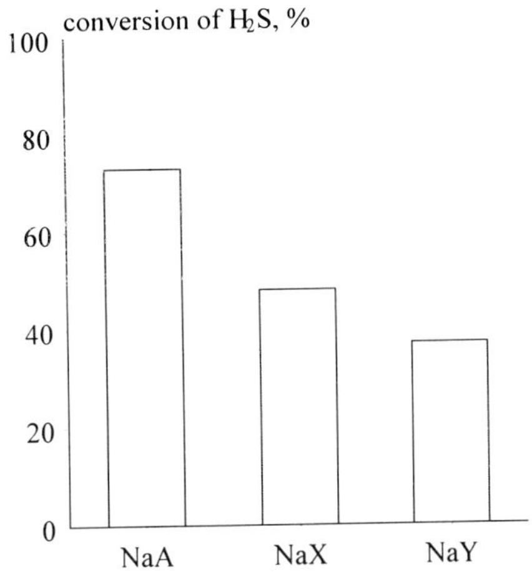


Figure 4. Formation of COS as function of type and Si/Al ratio of zeolites LTA and FAU.

#### 4.2.2. Cation type

The catalytic behaviour of the various alkaline metal cations if present in A, X and Y zeolites is very similar to each other except for NaKA zeolite if the $K^+$ cation concentration therein approaches that of the percolation threshold between the sub-types 4A and 3A, cf., Figure 5.

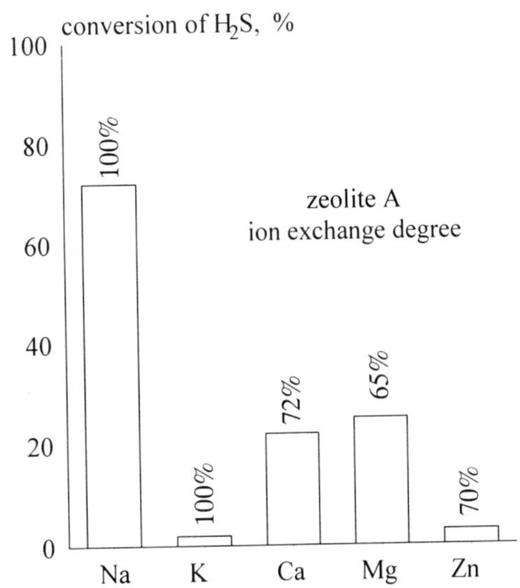


Figure 5. Formation of COS as function of cation type in LTA zeolites.

Since $K^+$ cations block the access of the intracrystalline void volume by $H_2S$ molecules, the rate of their transformation into COS is low; the conversion value given in Figure 5 is correct for an exposure time of 5 h; after ca. 300 h about 65 % of $H_2S$ were transformed into COS. So-called "shrunk" NaA zeolites with narrowed micropore apertures behave similarly.

For the X and Y zeolites, such steric hindrances are irrelevant that leads to the finding of their expected interaction with $H_2S$ [76]. The $H_2S$-to-COS conversion is reduced drastically to ca. 20 % if ca. 70 % of $Na^+$ cations were exchanged vs. $Ca^{2+}$ cations. The removal of weakly coordinated $Na^+$ cations leads to a strong reduction in reaction rate, as already shown for LTA zeolites. An even stronger reduction of the $H_2S$-to-COS conversion is detected for zeolites where $Na^+$ cations were exchanged by $Mn^{2+}$ or $Zn^{2+}$ cations, cf., Figure 6. The reason for this feature is a tendency of those cations to form sulphides [85]. Due to this type of chemisorption of $H_2S$ with a difference in strength, $Mn^{2+} < Zn^{2+}$, $H_2S$ is only partly available to undergo COS formation. With regard to the behaviour of $Zn^{2+}$ containing materials, they are known to be used for a series of purposes related to $H_2S$, including an irreversible utilization of ZnO for gas desulphurization. The chemisorption of $H_2S$ by $Zn^{2+}$-modified zeolites but also for similarly behaving $Cd^{2+}$ species, can hardly be utilized for a suppression of COS formation on large scale due to their environmentally hazardous properties that would cause problems during both manufacturing, handling and regeneration of zeolites loaded by heavy metals. (The behaviour of Zn-LTA species as compared with that of the Ca-LTA ones, is described in [86].) Their relatively low hydrothermal stability [87-92] is another argument against their utilization for gas desulphurization. The hydrothermal stress to which the zeolites are exposed, particularly during regeneration, stems

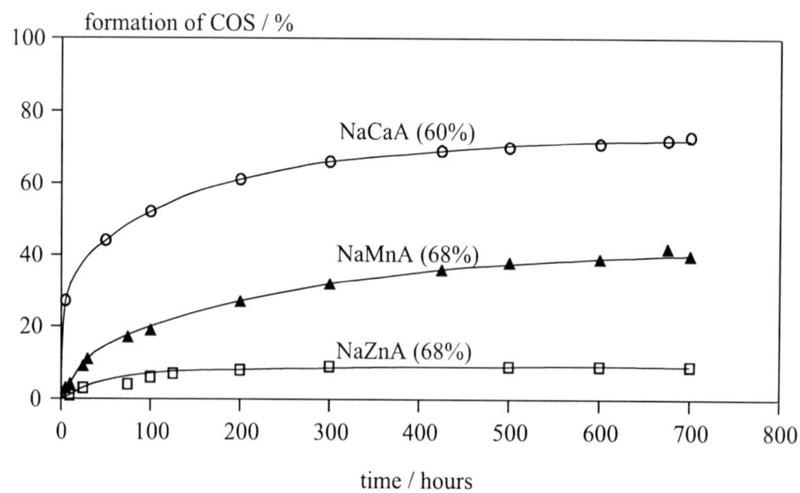


Figure 6. COS formation as dependence on time for various divalent cations in LTA zeolite.

from both $H_2O$ introduced into the bed by the feed gas, and that of the COS reaction, the $H_2O$ being well distributed over the entire bed.

### 4.2.3. Cation exchange value

Differences in the behaviour of cations coordinated either weakly or strongly, are shown for a series of NaCaA zeolites modified at laboratory scale, in Figure 7. The equilibrium value of $H_2S$-to-COS conversion is reached on zeolites with a $Ca^{2+}$ vs. $Na^+$ exchange of ca. 40 %. The zeolite is catalytically active in the usual manner. A drop-down in activity takes place over a region of exchange values between ca. 40 % and 55 %. This is concordant with the removal of weakly coordinated $Na^+$ cations from the α-cages. Towards a complete $Ca^{2+}$ vs. $Na^+$ exchange, the catalytic activity approaches its minimum, i.e., ca. 20 % of $H_2S$-to-COS conversion. Divalent $Ca^{2+}$ cations are strongly coordinated. They conduct dissociative interaction of $H_2S$ to a much lesser extent. The $H_2S$ conversion on highly exchanged zeolites results mainly from a shift in reactant concentrations according to the equilibrium constant, cf., eq. (1). With regard to conditions chosen, the conversion of ca. 20 % of $H_2S$ seems to be a possible limit to the suppression of COS formation, if chemisorption and sorption kinetic effects on $H_2S$ sorption were excluded. The tendency exemplified for the case of LTA zeolites should, in principle, be typical of that for FAU zeolites.

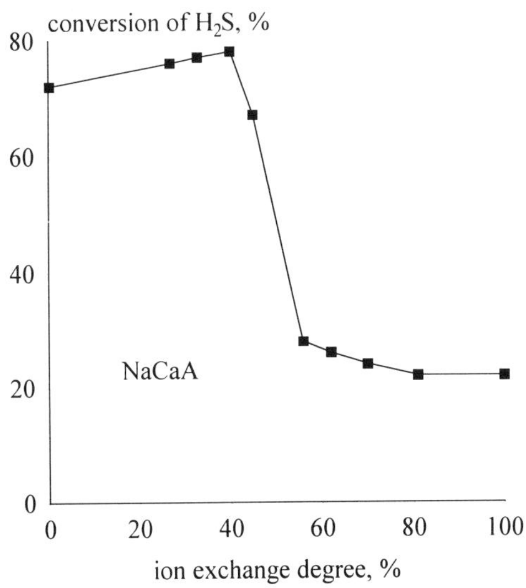


Figure 7. Formation of COS as function of $Ca^{2+}$ exchange value for NaCaA zeolites.

### 4.2.4. Time effect

In contrast to NaA zeolite, for which the COS formation reaches its equilibrium within ca. 5 h, an exposure of the gas mixture during nearly 300 h is necessary for the CaNaA form to lead the reaction to its equilibrium, cf., Figure 8. In this case, the equilibrium is reached due to preferential sorption of $H_2O$ (enrichment in the sorption phase) which shifts the concentration values in accordance with the equilibrium constant, $K_p$, of reaction (1), cf., eq. (2). This tendency being valid similarly for FAU zeolites, such as NaX and NaY, is of great relevance with regard to the layout of technical processes for which cycles with long adsorption and short desorption steps are preferred for economic reasons. If the concentration of $H_2S$ and $CO_2$ in the feed gas varies, this effect may become particularly important. The long-term kinetics of COS formation over CaNaA materials as dependence on the cation exchange value, is shown in Figure 9.

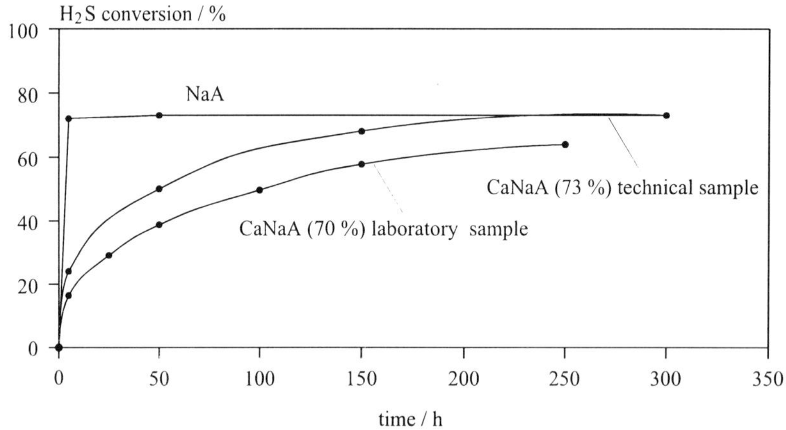


Figure 8. Formation of COS on NaA and NaCaA zeolites as function of time.

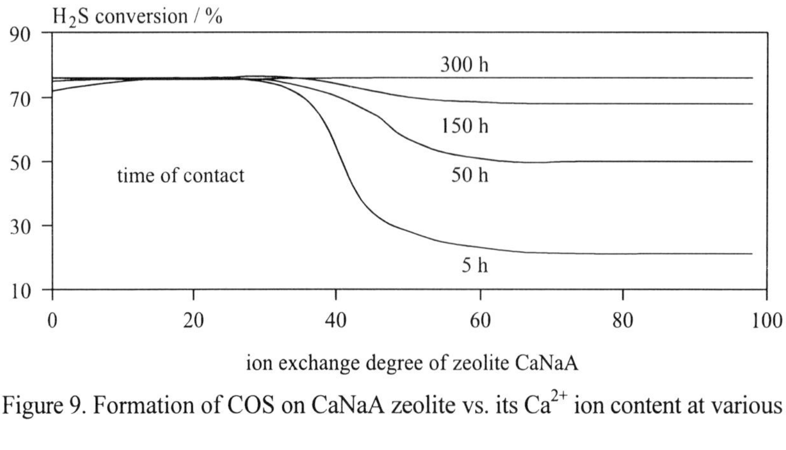


Figure 9. Formation of COS on CaNaA zeolite vs. its $Ca^{2+}$ ion content at various contact times.

### 4.2.5. Presorption effect

The phenomenon of presorption, a specific case of co-adsorption, may play an important role in technical separation and purification processes, cf., [93]. In the presence of $NH_3$ that shows a sorption interaction with zeolitic cations nearly as strong as that of $H_2O$, the COS formation is reduced drastically, cf., Figure 10. Synchronously, the zeolite loses both the selectivity and the capacity. Nonetheless, the $H_2S$ conversion enhances with time because $H_2O$ replaces $NH_3$ partly, according to eq. (1), and occupies sorption sites in the β-cages that cannot be accessed by $NH_3$. This leads, though kinetically retarded, to a shift in concentration values according to the equilibrium constant, $K_p$, of reaction (1), cf., eq. (2). For these reasons, neither $NH_3$ nor $H_2O$ presorption as proposed in the patent literature, seems to be a valid approach to the suppression of COS formation.

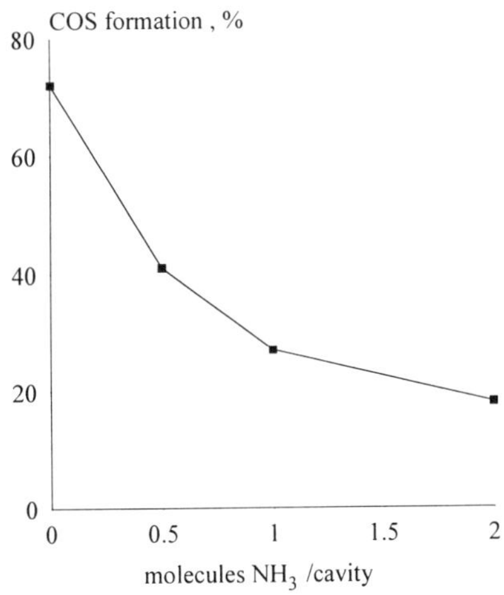


Figure 10. Formation of COS on NaA zeolite as function of presorption of $NH_3$.

### 4.2.6. Temperature effect

The influence of temperature on the catalytic enhancement of reaction (1) over NaA zeolite is found to be less significant, cf., Figure 11. This result contradicts findings for the influence of temperature on reaction (1) in gas phase and absence of a catalyst [94]. A maximum value of $H_2S$ conversion is reached at ambient temperature. Below ambient temperature the conversion of $H_2S$ into COS decreases, probably, because of mobility restrictions of the reactant molecules. With increasing temperature, the decrease in strength of specific dipole-cation charge interaction being typical for both $H_2O$ and $H_2S$ seems to exert its influence on molecular transformation. This effect was confirmed for all hydrophilic zeolites considered. Therefore, suppression of COS formation with simultaneously occurring strong selective sorption of $H_2S$ cannot be achieved by changing the temperature of the purification process. This result contradicts conclusions

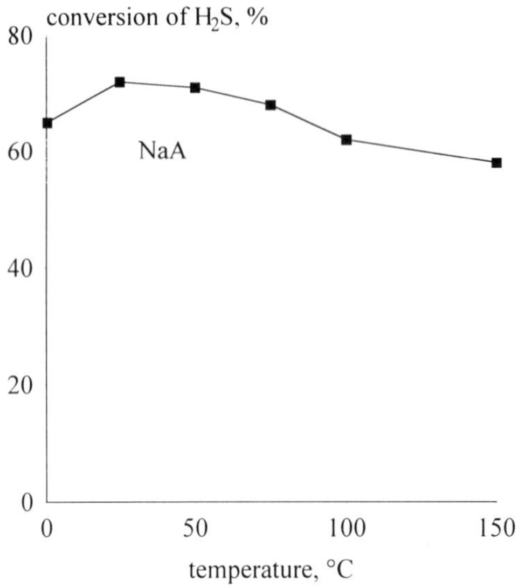


Figure 11. Formation of COS on NaA zeolite as dependence on temperature.

from thermodynamic considerations of eq. (1) in conjunction with single component sorption thermodynamic data [86]; in particular, sorption data for $H_2O$ suggests a suppression of COS formation at temperature > 350 K. The reason for the discrepancy is most probably due to the utilization of single component thermodynamic data for such an assessment of a complex mixture sorption case.

### 4.2.7. Hydrothermal stability of zeolites

The long-term stability of sorbents in industrial plants with an expected life-time of sorbents measured in years, is a major scientific and technological challenge. Structural deterioration caused by one regeneration step as a result of very little differences in chemical properties of various zeolites that were comparable, in principle, may be negligible. However, the accumulation of these damages over a huge number of such steps during practical utilization of the sorbent may strongly affect the duration of its usage. The main damaging agent is $H_2O$ because it causes framework hydrolysis. This effect is directly proportional to temperature [87] and becomes stronger with a decrease in Si/Al atomic ratio of the framework according to the sequence NaY < NaX < NaA with NaY as the most stable form [88,89].

An assessment of hydrolysis effects on different LTA species was made at various conditions. The influence of $H_2O$ on their stability, in the form of either $H_2O$ saturation vapour pressure of 1 bar, or liquid phase in an autoclave at saturation pressures that corresponded to temperatures in the range, 100 to 220 °C, resulted into a zeolite stability sequence, Zn, Co, Mn ≈ Mg < Ni, Cd < Ca, Li < Na, coded via their cations. For LTA zeolites, this sequence is analogous to that of the solubility of hydroxides of those cations in $H_2O$. Due to a formation of hydroxides with low solubility, dissociation of the latter proceeds with progressive generation of protons, $H^+$, that attack the framework. The crystalline state of zeolites transforms gradually into an amorphous one, and the microporous system breaks down. The NaCaA zeolites recommended for desulphurization of natural gas, are characteristic of highest stability that increases with decrease in $Ca^{2+}$ vs. $Na^+$ exchange value [92].

Due to the high solubility of KOH in $H_2O$, one might conclude that the KA zeolite were most stable. However, this is not the case due to a specific mechanism of fast lattice deterioration even at temperatures < ca. 200 °C. The process leads directly to another crystalline but non-microporous aluminosilicate (similarly to that for $Sr^{2+}$-exchanged zeolites) [91,95]. The relatively low hydrothermal stability of the KA zeolite is another reason for its inappropriateness for gas desulphurization, and this restricts its utilization for large-scale gas drying, at elevated temperature.

### 4.3. Suppression of COS formation

The formation of COS cannot be suppressed either by modifying the zeolites by cation exchange, blocking the catalytic centers via presorption of $H_2O$ or $NH_3$, or performing the purification process at increased temperature, cf., Figures 8-11.

A through analysis of reaction (1) over KA zeolite [83] provides evidence for the formation of COS if the contact time of reactants with the zeolite is sufficiently long, cf., Figure 12. The reaction occurs at the interface between zeolite crystals and gas phase together with a space separation of $H_2O$ and COS by selective sorption of $H_2O$ in the β-cages of the KA zeolite. Evidence with regard to reaction (1) was shown by the behaviour of hydrosodalite with a micropore system

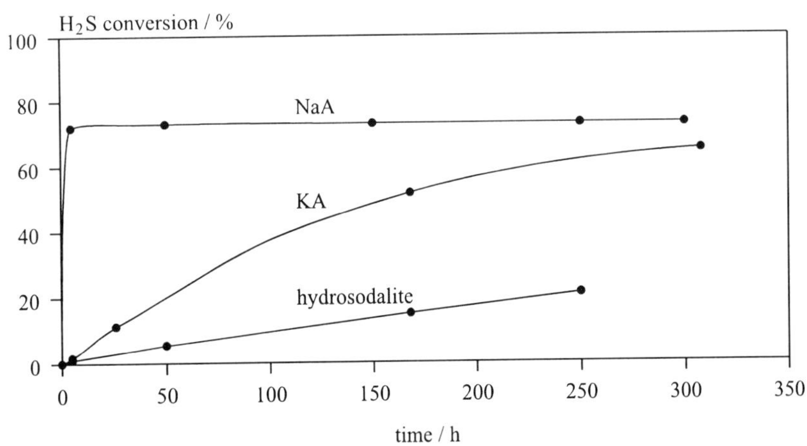


Figure 12. $H_2S$-to-COS conversion on LTA zeolites and hydrosodalite.

(diameter: ca. 0.24 nm) capable of adsorbing only $H_2O$. Hydrosodalite is composed by sodalite units that represent the β-cages in LTA zeolites. The rate of COS formation is limited by the rate of migration of $H_2O$ into the sodalite framework. The particular conversion rates for NaA, KA and hydrosodalite are compared to each other in Figure 12.

This model is supported by the behaviour of sodalites with salts occluded and of an amorphous silica with high $H_2O$ sorption saturation value, respectively ($H_2O$ capacities: 265 mg/g (NaA), 190 mg/g (silica)). On sodalites, the $H_2S$ conversion decreases strongly with increasing pore filling and sizes of salts occluded, cf., Figure 13. This underlines the necessity of space separation for $H_2O$ and COS to foster $H_2S$ conversion. In the silica case, despite strong $H_2O$ sorption, the conversion of $H_2S$ is insignificant because no space separation of $H_2O$ and COS occurs, cf., Figure 14. In addition, no catalytic influence of the silica on reaction (1) is expected. A space separation of $H_2O$ and COS as conversion products of $H_2S$ proceeds not only on sodalites but also on NaA, KA, and CaNaA zeolites if $H_2O$ is sorbed inside the β-cages that cannot be accessed by COS.

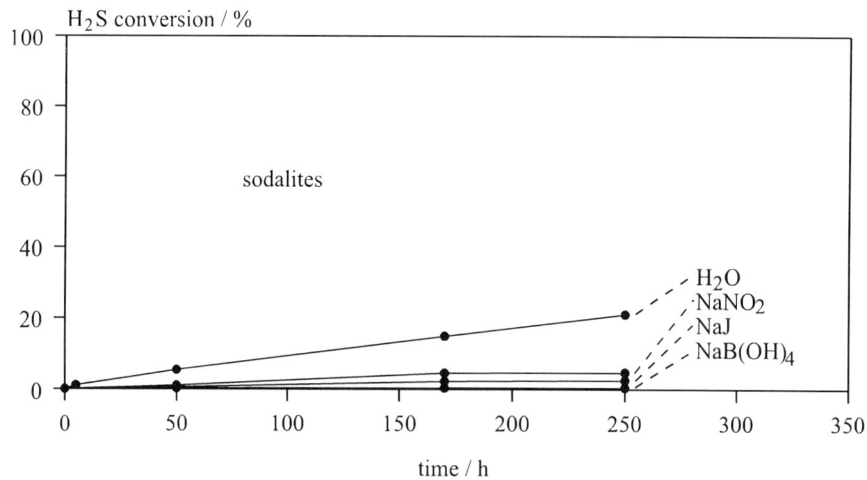


Figure 13. $H_2S$-to-COS conversion on sodalites with salt occlusion.

A blockage of β-cages by $NaNO_2$ performed on NaA zeolite as a result of its primary synthesis, does not prevent entirely the conversion of $H_2S$ to COS. However, if the $Na^+$ vs. $Ca^{2+}$ cation exchange approaches completion, the reaction is suppressed, cf., Figure 15.

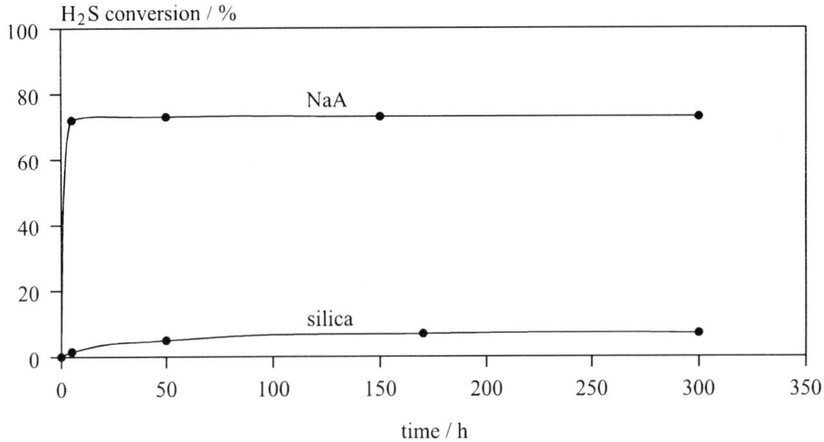


Figure 14. $H_2S$-to-COS conversion on NaA zeolite and hydrophilic silica.

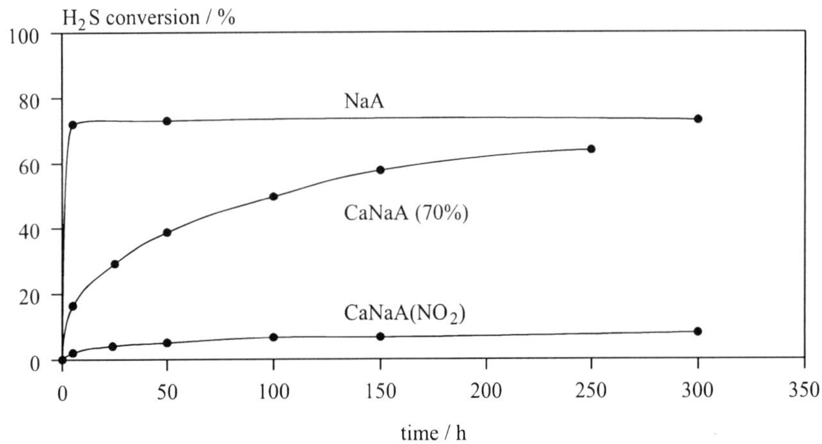


Figure 15. $H_2S$-to-COS conversion on LTA zeolite with $NaNO_2$ occluded in β-cages.

### 4.4. Modeling of dynamic sorption behaviour of natural gas components

Plentiful information on the behaviour of zeolites with regard to components of natural gases gained under conditions of laboratory experiments and technical processes, allowed mathematical modeling of the latter. Further, this work contributed to understanding the behaviour of $H_2S$ during sorption and reaction in sorbent beds. It also allowed for conclusions to be drawn with regard to the layout and operating conditions of plants for sorption desulphurization of gases with respect to both minimization and maximization of the formation of COS [30,67,7172,96].[1]

[1] The authors of these papers were granted about a dozen of related patents that will not be referred to for reasons of space.

### 4.4.1. Mathematical sorbent bed model

#### 4.4.1.1 Basic model

The basic mathematical model for a calculation of concentration vs. time dependences in sorbent beds accounts for non-isothermal sorption in biporous sorbent particles. It considers mass and energy balances in the interparticle void space of the bed and in the macro- and micropores of sorbent particles. Thus, it comprises three spatial coordinates, besides time, ($0 \le t \le \infty$): (i) the height z along the bed, ($0 \le z \le L$); (ii) the radial direction r in macropores of particles, ($0 \le r \le R_p$); and (iii) the radial direction ρ in their micropores, ($0 \le \rho \le R_{zl}$). Different geometries may exist in each single direction but in each of those geometries the transport equations are one-dimensional. For zeolite-based biporous particles, sorption in the macropores is negligible. These pores serve as transport pores, only. Sorption takes place, exclusively, in micropores.

The model comprises the following particular steps: (i) mass transport in void space of column; (ii) mass transfer between fluid phases of the column void space and macropores of particles; (iii) mass transport in macropores of biporous particles; (iv) adsorption; (v) mass transport in micropores of biporous particles; (vi) reaction of sorbed components in the micropores; (vii) heat conduction within particles owing to adsorption heat released; (viii) heat transfer between particles and fluid phase of the column void space; and (ix) heat transfer in void space of column. The model consists of the following equations for a number, i, ($1 \le i \le 6$), of sorbing species.

<u>Mass balance in the void space of sorbent bed:</u>

The mass balance in the void space of the column is described by a plug flow model with axial dispersion; radial gradients are neglected due to convection enforced in axial direction:

$$\frac{\partial g_i}{\partial t} + V_p \cdot \frac{\partial \bar{c}_i}{\partial t} = D_{g,i} \cdot \frac{\partial g_i}{\partial z^2} - u \cdot \frac{\partial g_i}{\partial z}, \qquad (3)$$

where $g_i$ are concentrations in the interparticle volume of column, $c_i$ concentrations in the macropore system of pellet (bar denotes average over a related volume, cf., eq. (4)), V volume fraction, D diffusivity and u flow rate,

$$\bar{c} = \frac{1+\alpha}{R^{1+\alpha}} \cdot \int_0^R r^{\alpha} \cdot c(z,r)\, dr \qquad (4)$$

with the geometry parameters

$$\alpha = \begin{cases} 0 & - \text{ plane geometry} \\ 1 & - \text{ cylindric geometry} \\ 2 & - \text{ spheric geometry.} \end{cases}$$

Mass transfer between fluid phase and pellet surface:

$$\beta \cdot D_{c,i} \cdot \text{grad}_r c_i(t,z,r)\big|_{r=R_p} = k_{g,i} \cdot \left(g_i(t,z) - c_i(t,z,R_p)\right), \tag{5}$$

where $\beta$ and $k$ denote the macropore fraction of pellet and mass transfer coefficient, respectively. By means of eqs. (2) and (3), one receives

$$\begin{aligned} V_p \cdot \frac{\partial \bar{c}}{\partial t} &= \frac{\partial}{\partial t}\left[\frac{1+\alpha}{R_p^{1+\alpha}} \cdot \int_0^{R_p} r^\alpha \left\{V_c \cdot c(z,r) + V_q \cdot \bar{q}(z,r)\right\} dr\right] \\ &= V_c \cdot \frac{1+\alpha}{R_p^{1+\alpha}} \cdot \int_0^{R} r^\alpha \cdot \text{div}(D_c \cdot \text{grad}\, c)\, dr \\ &= V_c \cdot \frac{1+\alpha}{R_p} \cdot k_g \cdot (g - c). \end{aligned} \tag{6}$$

If the sorbent particles were of cylindrical shape, eq. (3) assumes the following form which is often used in literature:

$$\frac{\partial g_i}{\partial t} + V_c \cdot \frac{2}{R_p} \cdot k_{g,i} \cdot (g - c) = D_{ax,i} \cdot \frac{\partial^2 g}{\partial z^2} - u \cdot \frac{\partial g}{\partial z}. \tag{7}$$

Mass transport in macropores of sorbent pellets:

The mass transport in macropores of sorbent pellets is assumed to proceed via intercrystalline diffusion that obeys the 2d Fick law, with $q$ denoting micropore sorption phase concentration:

$$V_c \cdot \frac{\partial c_i}{\partial t} + V_q \cdot \frac{\partial \bar{q}_i}{\partial t} = V_c \cdot \text{div}\left(D_{c,i} \cdot \text{grad}_r c_i\right). \tag{8}$$

Mass transport in micropores of sorbent:

The transport of sorbing species in the micropores of sorbent particles is presumed to proceed as a concentration-dependent diffusional process in accordance with the following equation:

$$V_q \cdot \frac{\partial q_i}{\partial t} = V_q \cdot \mathrm{div}\left( D_{q,i} \cdot q_i \cdot \mathrm{grad}_r \frac{\mu_i}{\Re T_p} \right), \tag{9}$$

where T, $\mu_i$ and $\Re$ denote the absolute temperature, the chemical potential of component i and the universal gas constant, respectively. It is assumed that the driving force of the process is the gradient of the chemical potential of the sorbing species, $\mathrm{grad}_r\, \mu_i$, to take into account both heterogeneities within the primary sorbent particle and, at the same time, to allow for non-constant equilibrium profiles with regard to the spatial coordinate. The chemical potential is calculated by means of eq.(10),

$$\frac{\mu_i}{\Re T_p} = \ln s_i + \mathrm{const.}, \tag{10}$$

in which the gas phase concentration is a fictive one because in micropores, there is a sorption phase only but no gas phase. By means of eq. (10), the term

$$D_{\mu,i} \cdot q_i \cdot \mathrm{grad} \frac{\mu_i}{\Re \cdot T} \tag{11}$$

can be transformed into

$$D_{\mu,i} \cdot \frac{\partial \ln s_i}{\partial \ln q_i} \cdot \mathrm{grad}\, q_i\,, \tag{12}$$

to shape the diffusion equation for micropores as follows:

$$V_q \cdot \frac{\partial q_i}{\partial t} = V_q \cdot \mathrm{div}\left( D_{q,i} \cdot \mathrm{grad}\, q_i \right), \tag{13}$$

where the diffusivity is dependent on concentration in accordance with the Darken equation,

$$D_{q,i} = D_{\mu,i} \cdot \frac{\partial \ln s_i}{\partial \ln q_i}. \tag{14}$$

The advantage of describing the diffusional process by means of eqs. (9)-(10) consists in their more general shape and in their straightforward numerical handling. Eq. (9) can be solved easily, irrespective of the type of sorption isotherm.

The equilibrium at the interface of microporous crystals obeys the condition,

$$c_i = s_i \big|_{r=R_p} \tag{15}$$

and the sorption isotherm is given by

$$q_i = q^* \frac{b_i(T_p) \cdot s_i}{1 + \sum_{j=1}^{n} b_j(T_p) \cdot s_j}, \qquad q^* = \frac{1}{n} \sum_{j=1}^{n} q_{\infty,j} \tag{16}$$

with

$$b_i(T_p) = \frac{\Re T_0}{p_0} \cdot \exp\left( \frac{\Delta S_i}{\Re} - \frac{\Delta E_i}{\Re T_p} \right), \tag{17}$$

where $q_\infty$, $\Delta E$ and $\Delta S$ denote the saturation value of sorbed amount, the sorption energy and sorption entropy, respectively.

Chemical reactions:

The kinetics of the COS formation reaction, cf., eq. (1), is assumed to obey the Langmuir-Hinshelwood model,

$$\frac{\partial \Theta_{CO_2}}{\partial t} = \frac{\partial \Theta_{H_2S}}{\partial t} = -\vec{\kappa} \cdot \Theta_{CO_2} \cdot \Theta_{H_2S} + \overleftarrow{\kappa} \cdot \Theta_{COS} \cdot \Theta_{H_2O} \tag{18}$$

$$\frac{\partial \Theta_{COS}}{\partial t} = \frac{\partial \Theta_{H_2O}}{\partial t} = -\overleftarrow{\kappa} \cdot \Theta_{COS} \cdot \Theta_{H_2O} + \vec{\kappa} \cdot \Theta_{CO_2} \cdot \Theta_{H_2S} \tag{19}$$

assuming that the reaction is of a second order and depends only on temperature and sorbent loading, $\Theta$, of components involved in the reaction.

Heat transfer in interparticle void volume of column:

It is assumed that heat transfer proceeds via convective mass transfer being disturbed by axial heat conduction in accordance with eq. (20), where $\gamma$ and $\lambda_g$ denote a space-specific heat and the heat conductivity, respectively:

$$\gamma_g \cdot \frac{\partial T_g}{\partial t} + V_p \cdot \gamma_p \cdot \frac{\partial T_p}{\partial t} = \lambda_g \cdot \frac{\partial^2 T_g}{\partial z^2} - u \cdot \gamma_g \cdot \frac{\partial T_g}{\partial z}. \tag{20}$$

Heat transfer between gas phase and solid:

It is presupposed that heat transfer between the gas phase and the solid sorbent particles proceeds in accordance with the following balance equation:

$$\lambda_p \cdot \mathrm{grad}_r T_p(t,z,r)\big|_{r=R_p} = k_T \cdot \left(T_g(t,z) - T_p(t,z,R_p)\right). \tag{21}$$

Heat transfer in solid sorbent particles:

Due to Lewis numbers above a value, Le > 10, solid sorbent particles can be considered as uniform material for which a discrimination between gas and solid phase is not necessary. Heat conduction is assumed to obey a Fourier law where the sorption heat of component i, $\Delta Q_i$, generated during the process, is taken into account by a specific term:

$$V_p \cdot \gamma_p \cdot \frac{\partial T_p}{\partial t} = V_p \cdot \mathrm{div}(\lambda_p \cdot \mathrm{grad}_r T_p) - V_q \cdot \sum_{i=1}^{n} \Delta Q_i \cdot \frac{\partial \bar{q}_i}{\partial t} . \tag{22}$$

The model is complemented by appropriate initial and boundary conditions.

### 4.4.1.2. Parameter identification

In the interstitial particle volume of the column, mass transfer is governed by both flow rate and coefficients of axial dispersion. The flow rate, u, is calculated by an expression,

$$u = \frac{\dot{V}}{\mathrm{Quer} \cdot \varepsilon} , \tag{23}$$

with the volume throughput, $\dot{V}$, governed by external conditions (T,p), and the cross-section of the column effective with regard to mass flow, $(\mathrm{Quer} \cdot \varepsilon)$. The flow rate is assumed to be constant with regard to time and space, because throughput and pressure drop over the column height were assumed to be constant during one phase (adsorption or desorption) and negligible, respectively.

To describe axial dispersion in the column, the Bishoff approach [86] that takes into account the tortuosity factors, $\tau_{ax}$ and $\tau_r$, and a Peclet number, $Pe_\infty = 2$, at a technical scale, are utilized:

$$D_{ax} = \tau_{ax} \cdot D_{AB} + \frac{2 \cdot u \cdot R_p \cdot Pe_\infty^{-1}}{1 + \dfrac{\tau_{ax} \cdot \tau_r \cdot D_{AB}}{2 \cdot u \cdot R_p}} . \tag{24}$$

The following expression was also suggested [98]:

$$D_{ax} = \frac{D_{AB}}{\varepsilon} \cdot (20 + 0.5 \cdot Sc \cdot Re) , \tag{25}$$

where Sc and Re are, respectively, the Schmidt number, $Sc = \nu / \rho D_{AB}$, and Reynolds number, $Re = \varepsilon \cdot 2\ u\ R_p\ \rho / \nu$; $\nu$ and $\rho$ denote the dynamic viscosity and density of gas phase, respectively.

The mass transfer through the laminar boundary layer is influenced strongly by a mass transfer coefficient, $k_g$, which can be calculated by means of an empirical correlation [99],

$$Sh = \frac{Sc}{\varepsilon}(0.756\ Re^{0.18} + 0.365\ Re^{0.614}) \qquad \text{for} \qquad 0.01 < Re < 1500, \tag{26}$$

where Sh is the Sherwood number, $Sh = 2\, k_g\, R_p / D_{AB}$. These correlations refer to spherical geometry of particles. For other geometries, the radius, $R_p$, has to be substituted, correspondingly.

The mass transport in the pellets is described by a macropore diffusivity, $D_c$, and a micropore diffusivity, $D_q$. For the macropore diffusivity, it holds:

$$D_c = \frac{D}{\tau} \qquad \text{with} \qquad \frac{1}{D} = \frac{1}{D_{AB}} + \frac{1}{D_K}, \tag{27}$$

and a Knudsen diffusivity that can be calculated as follows:

$$D_K = 9700 \cdot R_m \cdot \left(\frac{T}{M}\right)^{1/2}, \tag{28}$$

where $R_m$ denotes the average macropore radius. If the molecular diffusivity, $D_{AB}$, is not known for a given system, it can be approached by means of the Chapman-Enskog relation:

$$D_{AB} = 0.0018538 \cdot \frac{T^{2/3} \cdot \left(\frac{1}{M_A} + \frac{1}{M_B}\right)^{1/2}}{p \cdot \sigma_{AB}^{2} \cdot \Omega_{AB}}, \tag{29}$$

where $M_A$ and $M_B$ are the molecular weights of components, A and B, respectively; $\sigma_{AB}$ and $\Omega_{AB}$ the collision diameter and collision integral of the two species, respectively. Since $\Omega_{AB}$ depends on temperature the Chen-Othmer relation [62] can be used for a calculation of the diffusivity, $D_{AB}$,

$$D_{AB} = 0.5963 \cdot 10^{-8} \cdot \frac{T^{1.81} \cdot \left(\frac{1}{M_A} + \frac{1}{M_B}\right)^{1/2}}{p \cdot \left(T_{kr,A} \cdot T_{kr,B}\right)^{0.1405} \cdot \left(V_{kr,A}^{0.4} + V_{kr,B}^{0.4}\right)^{2}}. \tag{30}$$

The temperature dependence of micropore diffusion that represents an activated process, is described by an Arrhenius law, with the energy of activation, E:

$$D_{\mu} = D_{\mu,0} \cdot e^{-E/\Re T}, \tag{31}$$

where $D_{\mu,0}$ represents a diffusivity extrapolated for zero concentration of sorbing species in the micropores, at temperature, $T = 273$ K. The concentration dependence of micropore diffusivity is taken into account by the Darken equation. For a single component system that conforms the Langmuir sorption isotherm equation at equilibrium, it holds:

$$D_q = D_{\mu} \cdot \frac{1}{1-\Theta}. \tag{32}$$

Since the diffusion equation is chosen in terms of the gradient of chemical potential, grad $\mu$, an explicit form of eqs. (14) and (32) is not necessary.

<u>Equilibrium relationships:</u>

The calculation of the sorption equilibrium proceeds by means of eqs. (16) and (17). The characteristic parameters, $q_{\infty}$, E and $\Delta S$, are obtained by fitting to data that were either experimentally obtained or taken from literature. Although utilization of eq. (16) with single component sorption equilibrium data may lead to extreme deviations, its applicability had been proven [63].

<u>Heat transfer:</u>

To assess the heat transfer characteristics, a determination of the heat conductivity of the free gas, $\lambda_g$, the heat transfer coefficient, $k_T$, and the effective heat conductivity, $\lambda_P$, in the particle, is necessary, besides that of the above defined gas flow rate. For the heat conductivity, $\lambda_g$, it holds:

$$\lambda_g = \nu \cdot (c_{p,g} + 1.25 \cdot \frac{\Re}{M}). \tag{33}$$

For the dynamic viscosity, $\nu_{mix}$, and the molecular weight, $M_{mix}$, the average values characteristic of the actual gas mixture, are used, to avoid cumbersome calculations of the heat conductivity, $\lambda_g$, by means of relationships for mixtures. The heat transfer is assessed analogously to eq. (26) [97]:

$$Nu = \frac{Pr^{1/3}}{\varepsilon} (2.876 + 0.3023\, Re^{0.65}) \qquad \text{for} \qquad 10 < Re < 10000, \tag{34}$$

with Nu and Pr as Nusselt number, $Nu = 2 R_p k_T / \lambda_g$, and Prandtl number, $Pr = c_p \mu / \lambda_g$, respectively. Literature values [100] of heat conductivity, $\lambda_g$, and specific heat capacity, $c_p$, were used.

### 4.4.1.3. Model simplifications

Due to both common knowledge and related model calculations, at a technical scale that is characterized by $L/2R_p > 200$, axial dispersion as compared with convection contributes little to overall mass and heat transfer. Neglecting both steps in the calculation of breakthrough curves causes deviations within the order of magnitude of numerical errors for the solution of the resulting equations for convection. Further simplification is possible by lumping parameters. A comparison of mass transport resistances in a pellet leads to the following overall resistance [101]:

$$w_p = \left( \frac{R_p}{(1+\alpha) \cdot V_p \cdot k_g} + \frac{R_p^2}{F \cdot V_c \cdot D_c} + \frac{R_{zl}^2}{15 \cdot V_q \cdot D_q \cdot \frac{\partial q}{\partial s}} \right) \cdot \frac{\delta_0^2}{(1+\delta_0)^2}, \tag{35}$$

where $D_q$ is given by eq. (14), and $\delta_0$ is deciphered as follows:

$$\delta_0 = V_c + V_q \cdot \frac{\partial q}{\partial s}. \tag{36}$$

If the intracrystalline diffusional resistance for certain components could be neglected compared to resistances of other types, then the solution to the system of transport equations for a biporous pellet reduces to the macropore diffusion equation (8), in conjunction with the equilibrium relationship, eqs. (15)-(16). For an effective macropore diffusion coefficient, it holds:

$$\frac{R_p^2}{D_{eff}} = \frac{R_p^2}{F \cdot \beta \cdot D_c} + \frac{R_{zl}^2}{15 \cdot (1 - ß) \cdot D_\mu \cdot \frac{q}{s}}, \tag{37}$$

at the assumptions of spherical symmetry of zeolite crystals and of diffusional path ways in macropores that were dependent on pellet geometry (for spheres, $F = 15$, for cylinders, $F = 9.2$).

If even macropore transport is sufficiently quick not to be rate-determining, one may apply a linear driving force model to the overall kinetics, cf., [102]. For this purpose, the molar flux, N, through the pellet interface is correlated with the particular average gas concentration in the pellet that is in equilibrium with the concentration of sorbing species in the zeolite:

$$N = -k_{eff} \cdot (g - \bar{c}^*) . \tag{38}$$

Furthermore, if it is appropriate to consider the molar flux being governed by a diffusional process in the pellet macropores,

$$N = -k_g \cdot (g - c_{of}) \;, \quad c_{of} = c\,|_{r=R_p} \,, \tag{39}$$

an account for the boundary condition of (5) allows to find the following resistance for a generalized effective mass transfer:

$$\frac{1}{k_{eff}} = \frac{1}{k_g} + \frac{c_{of} - \bar{c}^*}{D_{eff} \cdot \left.\dfrac{\partial c}{\partial r}\right|_{r=R_p}} . \tag{40}$$

Assuming a parabolic concentration profile of the gas phase within a pellet,

$$c = c_0 + c_2\, r^2 \qquad \text{with} \qquad c_{of} = c_0 + c_2 \tag{41}$$

and calculating the average concentration therein, in accordance with eq. (4), one finds

$$\frac{1}{k_{eff}} = \frac{1}{k_g} + \frac{R_p}{(3 + \alpha) \cdot D_{eff}} . \tag{42}$$

For a pellet in spherical, $\alpha = 2$, or cylindrical, $\alpha = 1$, symmetry, the second term of eq. (42) assumes the well-known shape, $R_p / (5\, D_{eff})$, and, $R_p / (4\, D_{eff})$, respectively. Since, for the cylindrical shape, only the surface of the radial plane but not that of the axial planes is considered, a shape factor, F, is introduced in eq. (42), analogously to eq. (37):

$$\frac{1}{k_{eff}} = \frac{1}{k_g} + \frac{(1 + \alpha) \cdot R_p}{F \cdot D_{eff}} . \tag{43}$$

In this way, the solution to the mass transport equations reduces to the solution of a linear driving force model with an effective transport coefficient.

Analogously, the heat transfer can be linearised, cf., eq. (44), and eq. (20) reduces to eq. (45) that can be solved directly and easily,

$$\frac{1}{k_{T,eff}} = \frac{1}{k_T} + \frac{(1 + \alpha) \cdot R_p}{F \cdot \lambda_p} , \tag{44}$$

$$V_p \cdot \gamma_p \cdot \frac{\partial T_p}{dt} = V_p \cdot \frac{1+\alpha}{R_p} \cdot k_{T,eff} \cdot (T_g - T_p) - V_q \cdot \sum_{i=1}^{n} \Delta Q_i \cdot \frac{\partial \bar{q}_i}{\partial t} . \tag{45}$$

**4.4.1.4. Numerical solution of model equations**

Before a numerical solution of transport equations is attempted, these were transferred into a dimension-less form to assess more conveniently the relative influence of the various transport steps on the overall transport rate. With normed concentrations in the interparticle column volume, eq. (46), of sorbent particles, eq. (47), and in the micropores of the latter, eq.(48),

$$Y_g = \frac{M \cdot g}{\rho_s}, \tag{46}$$

$$Y_c = \frac{M \cdot c}{\rho_s}, \tag{47}$$

$$X = \frac{(1-\beta) \cdot M \cdot q}{\rho_s}, \tag{48}$$

and with dimensionless time, $\tau$, scaled with respect to the breakthrough time, $t_B$, the mass transport equations read as

$$\frac{\partial Y_{g,i}}{\partial \tau} + V_c \cdot (1+\alpha) \cdot MC \cdot Bi^*_{eff} \cdot \left(Y_{g,i} - Y^s_{c,i}\right) + \gamma \cdot \frac{\partial Y_{g,i}}{\partial \nu} = 0, \tag{49}$$

$$V_c \cdot \frac{\partial Y_{c,i}}{\partial \tau} + V_p \cdot \frac{\partial X_i}{\partial \tau} - V_c \cdot (1+\alpha) \cdot MC \cdot Bi^*_{eff} \cdot \left(Y_{g,i} - Y^s_{c,i}\right) = 0, \tag{50}$$

with the Biot number for mass transport, eq.(51), the effective Biot number, eq.(52),

$$Bi^* = \frac{R_p \cdot k_g}{D_c}, \tag{51}$$

$$\frac{1}{Bi^*_{eff}} = \frac{1}{Bi^*} + \frac{1}{3+\alpha}, \tag{52}$$

and the mass transport coefficient, eq.(53),

$$MC = \frac{t_B \cdot D_c}{R_p}. \tag{53}$$

One yields equations for sorption equilibrium, eq. (54), and heat conduction, eqs. (55)-(56):

$$X_i = X^* \cdot \frac{B_i(\theta_p) \cdot Y_{p,i}}{1 + \sum_j B_j(\theta_p) \cdot Y_{p,j}} \qquad \text{with} \qquad X^* = \frac{1}{n} \cdot \sum_{j=1}^{n} X_{\infty,j}, \tag{54}$$

$$\frac{\partial \theta_g}{\partial \tau} + V_c \cdot (1+\alpha) \cdot HC \cdot Bi_{eff} \cdot \left( \frac{c_p}{c_g} \cdot \theta_g - \theta_p \right) + \gamma \cdot \frac{\partial \theta_g}{\partial \upsilon} = 0, \tag{55}$$

$$\frac{\partial \theta_p}{\partial \tau} - V_c \cdot (1+\alpha) \cdot HC \cdot Bi_{eff} \cdot \left( \frac{c_p}{c_g} \cdot \theta_g - \theta_p \right) + \sum_i H_i \cdot \frac{\partial X_i}{\partial \tau} = 0, \tag{56}$$

with the expressions for normed temperature, eq. (57), the Biot number, eq. (58), and the coefficient of heat conduction, eq. (59),

$$\theta = \frac{M \cdot c_p \cdot T}{Q_0}, \tag{57}$$

$$Bi = \frac{R_p \cdot k_T}{\lambda_p}, \tag{58}$$

$$HC = Le \cdot MC, \qquad \text{with the Lewis number} \qquad Le = \frac{\lambda_p}{\rho_s \cdot c_p \cdot D_c}. \tag{59}$$

The kinetic equations for the COS formation reaction, eq. (1), are scaled as follows:

$$\frac{\partial \Theta_{CO_2}}{\partial \tau} = \frac{\partial \Theta_{H_2S}}{\partial \tau} = -\underline{\vec{\kappa}} \cdot \Theta_{CO_2} \cdot \Theta_{H_2S} + \underline{\overleftarrow{\kappa}} \cdot \Theta_{COS} \cdot \Theta_{H_2O} \tag{60}$$

$$\frac{\partial \Theta_{COS}}{\partial \tau} = \frac{\partial \Theta_{H_2O}}{\partial \tau} = -\underline{\overleftarrow{\kappa}} \cdot \Theta_{COS} \cdot \Theta_{H_2O} + \underline{\vec{\kappa}} \cdot \Theta_{CO_2} \cdot \Theta_{H_2S} \cdot \tag{61}$$

The adsorber model comprises a system of (i) three parabolic partial differential equations for the mass transport of each single component coupled by both sorption isotherm equations and an expression for the temperature dependence of rate coefficients; (ii) two differential equations for chemical reaction; and (iii) two parabolic partial differential equations for heat transfer. Beside time, the model contains three spatial coordinates that refer to the interstitial column volume, the macropore volume and the micropore volume and that may be of different geometry. The solution of the problem for which a module-wise algorithm was developed, is described in detail in refs. [103,104].

### 4.4.2. Model calculations

#### 4.4.2.1. Breakthrough behaviour without formation of COS

The breakthrough time data for $H_2S$ are crucial for a layout of sorption desulphurization plants. A rough estimate of this parameter is possible by a formula,

$$t_B = \frac{L}{u} \cdot \left(1 + V_c + V_p \cdot \frac{X_E^* - X_0}{Y_E - Y_0}\right), \tag{62}$$

that enables to calculate the time needed to fill all the voids (interstitial column volume and macropore volume) and all sorption sites, with regard to sorption equilibrium for a given initial concentration, $X_E$, of sorbing species. Due to the simplifications involved, direct access to the breakthrough time on the basis of breakthrough curves cannot be circumvented. Accordingly, breakthrough curves were determined using typical plant parameters listed in Table 2. The plant was utilized to purify a natural gas of the given composition given in section 4.1.

Table 2
Plant parameters for calculation of breakthrough curves

| Parameter | Value |
|---|---|
| Sorbent type | NaCaA zeolite,<br>Cation exchange value: 60 % |
| Sorbent mass | 20 ton |
| Sorbent bed height | 10 m |
| Throughput | 10,000 $Nm^3/h$ |
| Operating pressure | 2.5 MPa |
| Gas temperature at column inlet | 293 K |

Figure 16 compiles breakthrough curves calculated for each single component with an account for its content in the natural gas. Calculations were performed assuming that in each case only one component is sorbed and all the other ones

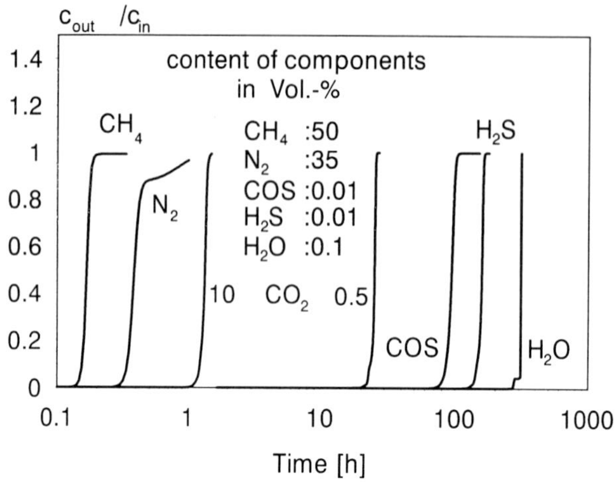


Figure 16. Single component breakthrough curves of natural gas components.

behave inertly with regard to sorption. However, the parameters needed for the calculation of the mass transfer were those of the entire mixture. A calculated $H_2S$ breakthrough time that amounts to ca. 150 h, appears to support the assumption of a successful utilization of both the sorbent material and process for sorption desulphurization. Since the feed gas concentration of $CO_2$ may vary significantly, breakthrough curves were calculated for two different $CO_2$ contents. Although the feed gas does not contain COS, a breakthrough is given for this species that was calculated assuming that the COS concentration in the feed were comparable with that of $H_2S$. Since the sorption properties of COS are close to those of $CO_2$ rather than of $H_2S$, the calculated breakthrough of COS precedes that of $H_2S$. The breakthrough behaviour changes dramatically if competitive sorption is taken into account, i.e., sorption of the natural gas is considered as mixture sorption, cf., Figure 17. Depending on the $CO_2$ content and due to sorption replacement, the $H_2S$ breakthrough shifts towards a shorter time. Although a similar effect becomes obvious for other gas components as well, cf., sorption replacement between $N_2$ and $CH_4$, it is strongest for the most interesting component, $H_2S$.

For a gas that contains 0.5 vol.% $CO_2$ the breakthrough time of $H_2S$ shortens from 150 h to 13 h. The roll-up in concentrations for several components is due to their replacement by the other ones (concentrations at the column outlet are referred to those at the column inlet). First, $N_2$ is replaced by $CH_4$, then $CO_2$ by $H_2S$ and, of course, by $H_2O$, and, finally, $H_2S$ by $H_2O$ of the feed gas. Sorption of $H_2O$ is strongest, due to its large dipole moment, and it is able to replace all the other gas components. For this reason, $H_2O$ breakthrough time changes only a little as compared with single component sorption. If the feed gas $CO_2$ content is increased to 10 vol. %, cf., Figure 18, the result of model calculations becomes even

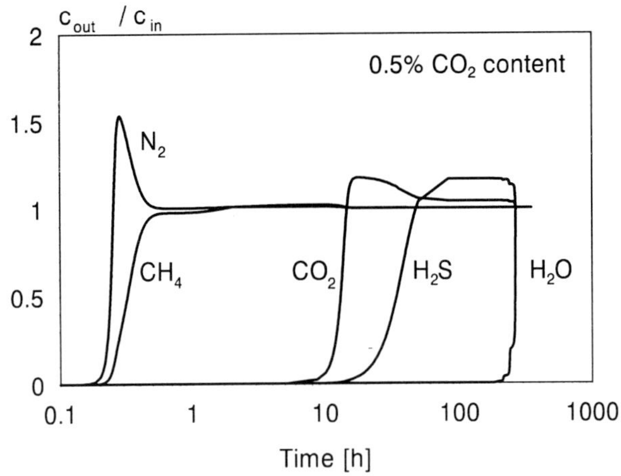


Figure 17. Gas mixture breakthrough curves in presence of 0.5 vol. % $CO_2$.

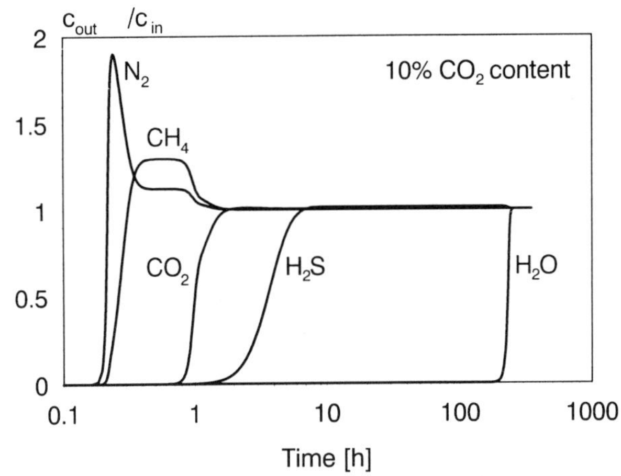


Figure 18. Gas mixture breakthrough curves in presence of 10 vol. % $CO_2$.

more unrealistic with regard to sorption times that stem from "practical experience". This, however, is not a disadvantage of the model but implies that these sorbents are inappropriate for the desulphurization of gases in the presence of $CO_2$ , viz., due to neglecting the process of eq. (1). First of all, the proportion of $CO_2$ and $H_2S$ in the feed gas is responsible for this unfavourable behaviour. Since the sorption properties of the both gases are close to each other with slightly stronger sorption of $H_2S$, cf., Figure 2 for sorption enthalpy values, and since $CO_2$ is always present in excess, an originally empty bed would take up $CO_2$, first, to be replaced by $H_2S$ later. As long as the ratio of concentrations, $[CO_2]$ / $[H_2S]$, is small the roll-up in relative $CO_2$ concentration is significant, cf., Figure 17, but vanishes with increase of the former. Replacement of $CO_2$ by $H_2S$ is negligible at gradient values, ≥ 100, and, hence, breakthrough behaviour of $CO_2$ differs to a small extent for the cases of mixture and single component sorption. In contrast to this effect, a strong replacement of $CH_4$ is proven by the roll-up in relative concentrations, cf., Figure 18. Due to replacement effects, the mass transfer zones broaden, breakthrough curves spread and breakthrough times shorten for the various components. The generation of sorption heat contributes as well to such effects. For unfavourable cases, e.g., for an empty bed and high $CO_2$ content of the feed gas, the increase in temperature of the bed may reach 100 K, which would significantly diminish the sorption attraction of the zeolite for $H_2S$.

#### 4.4.2.2. Breakthrough behaviour with formation of COS

As pointed out above, industrial desulphurization plants work at cycles, ≥ 24 h. The results of model calculations contradict this practical situation, entirely. This discrepancy can be overcome by taking into account the implications of reaction (1) which has been proven to occur on industrial scale. The transformation of $H_2S$ into COS which decreases its concentration in the feed gas to an extent that the loading of the zeolite diminishes significantly and the breakthrough occurs at much later time, was integrated in the mathematical model.

The following additional assumptions are justifiable: (i) for sorption of $H_2O$ as formed by eq. (1), only ca. 10 % of the total number of sorption sites are available (at maximum, ca. 30 $H_2O$ molecules would be sorbed within one α-cage of LTA zeolite); (ii) high $CO_2$ content in the feed gas; it amounts to 10 vol. % which means that the concentration of $CO_2$ exceeds always that of $H_2S$, and the latter transforms completely into COS; (iii) irreversibility of reaction (1) at conditions of zeolitic sorption, i.e., $H_2O$ does not undergo the reverse reaction according to eq. (1) due to strong attraction forces that act on $H_2O$ in zeolite cavities; and (iv) transformation of $H_2S$ into COS depends only on the availability of sorption centers but not on time, which leads to a nearly infinite large value of the reaction constant if a sufficient number of sorption sites were available.

Breakthrough curves calculated under the above assumptions, (i)-(iv), are shown in Figure 19. The COS content at the bed exit is scaled with regard to that of $H_2S$ in the feed gas because this does not contain COS. The entire amount of COS is formed according to reaction (1). Conclusions from Figure 19 conform with the expectation, in particular, of a significant shift, i.e., by factor ca. 10, of the $H_2S$ breakthrough towards large times although the very long breakthrough times in accordance with those obtained from single component sorption behaviour were not reached. On the other hand, the $CO_2$ breakthrough remains practically unchanged. Consumption of $CO_2$ for COS formation stays within the error margins of model calculations. Although COS breakthrough is very close to that of $CO_2$, its replacement by the latter leads to a significant roll-up followed by a phase of quasi-stationary COS

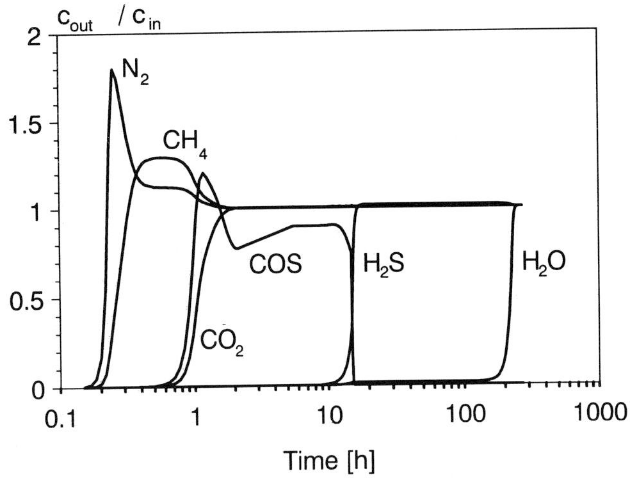


Figure 19. Gas mixture breakthrough curves with COS formation.

formation. Whence sorbent sites for preferential interaction with $H_2O$ were occupied in accordance with the gas inlet concentrations, the driving force vanishes for the direct reaction according to eq. (1), and the reaction stops. Accordingly, the COS concentration in the product gas drops to zero, and $H_2S$ being unable to undergo its transformation breaks through, simultaneously. This breakthrough proceeds much more steeply compared with the case of molecular sieves "without" formation of COS. The breakthrough behaviour of $H_2O$ remains unchanged irrespective of a larger amount of $H_2O$ offered to the zeolite. In summary, the results described are close to those found for a gas desulphurization plant of industrial scale.

#### 4.4.2.3. Influence of $CO_2$ content of feed gas

It follows from Figures 17 and 18 that the $CO_2$ content of the feed gas governs the $H_2S$ breakthrough. Hence, the breakthrough time of $H_2S$ as dependence on the $CO_2$ content in the feed gas is of interest. The breakthrough behaviour was calculated for the cases of both presence and absence of reaction (1). The time at which the $H_2S$ concentration in the product gas reaches half the concentration value in the feed gas, is considered as breakthrough time, $t_B$.

Results of Figure 20 for cases without COS formation prove that small changes in $CO_2$ content cause large shifts in $H_2S$ breakthrough, for a concentration range of $CO_2$ up to ca. 5 %. Since the $CO_2$ feed gas content is subject to changes of an order of ca. 1 %, shortening of the $H_2S$ breakthrough time by ca. 10 h may be possible. In the range of higher $CO_2$ contents, their variation has little influence on $H_2S$ breakthrough times. The realistic case of sorption with COS formation shows an

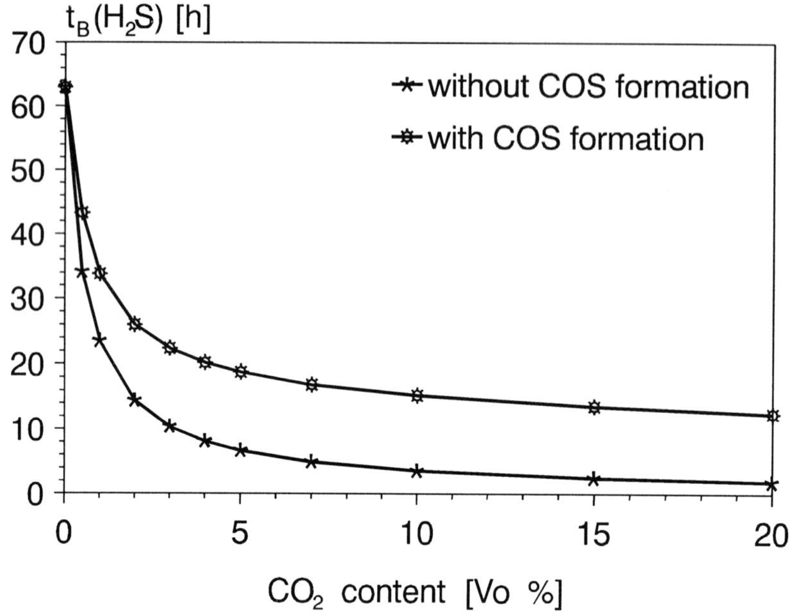


Figure 20. Influence of $CO_2$ content of feed gas on breakthrough time of $H_2S$.

improved but still "insufficient" behaviour. Although the decay in values, $t_B$, becomes less steep, shortening in $H_2S$ breakthrough time by up to ca. 10 h may still occur. For higher $CO_2$ contents, the decay in values, $t_B$, may proceed in parallel for both cases considered. From a certain $CO_2$ feed gas level onwards, breakthrough of $H_2S$ in the "purified" gas is delayed. The difference depends on the $H_2S$ concentration in the feed gas, the operating conditions of the plant and the zeolite type utilized as sorbent. Zeolites with increased sorption capacity for $H_2O$ may show "improved" breakthrough behaviour of $H_2S$, considering the formation of COS.

### 4.4.2.4. Breakthrough curves for cyclic process regime

A cyclic regime exerted over a bed being the typical practical case, excludes naturally the consideration of an "empty" sorbent, which had always been the case at time, t = 0, of the preceding examples. Returning to the practical 4-bed plant regime as described in section 4.1., the first of the three sequentially arranged beds changes from the adsorption step into desorption, after an adsorption time of ca. 100 h. During the subsequent desorption step, hot "purified" product gas is passed through this bed. Next, the regenerated bed is arranged to be at the end of the series of beds in the adsorption regime. Here, this bed serves as fines cleaner. In this regime, it becomes partly loaded with trace amounts of impurities to be removed from the natural gas. As a result of subsequent step shifts in which all beds participate, the considered bed "travels" in the direction of the gas inlet of the plant until it has not reached the position of the first bed and, then, it is brought into the desorption stage, again. Therefore, at the beginning of the adsorption process each of the three beds are loaded with three components, $CO_2$, $N_2$ and $CH_4$. In addition, the first two beds are loaded with fractions of $H_2O$ and $H_2S$, that were not present in the inlet gas. Therefore, the breakthrough behaviour changes in the case of a cyclic regime, cf., Figure 21. At the beginning of the adsorption step, the three components brought into the bed during desorption, leave this bed at the concentration level established during the desorption step. Variations in concentrations were mainly due to differences in conditions for the desorption step, particularly due to the high desorption temperature. Over all the three beds there occur certain "steady-state" concentration profiles with inconsistencies unavoidable at their boundary positions. The breakthrough of $H_2S$ deserves particular attention. For the conditions of the practical case with an adsorption time of ca. 100 h, $H_2S$ breaks through after ca. 2-3 h, i.e., even a cyclic process regime does not allow for a product gas free of $H_2S$, over 100 h.

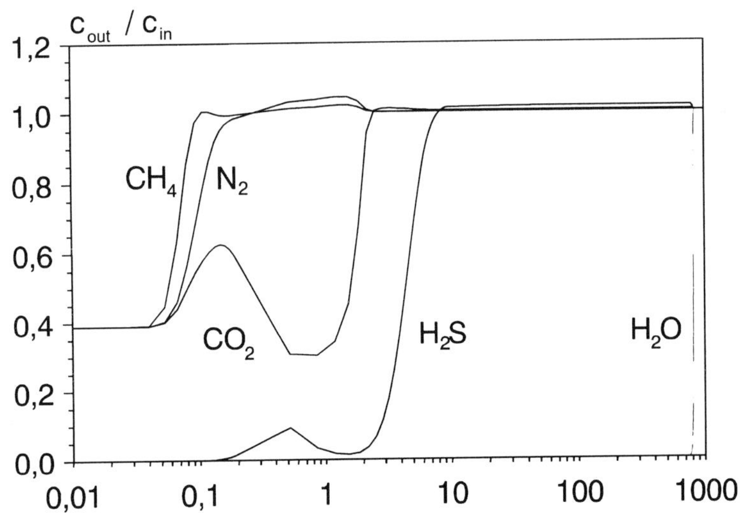


Figure 21. Mixture breakthrough curves for a cyclic regime without formation of COS.

A simulation of the breakthrough behaviour accounting for COS formation results in a picture that corresponds to the plant experience, cf., Figure 22. A relatively high content of the three components, $CO_2$, $N_2$ and $CH_4$, is in line with the results of simulations without COS formation. Breakthrough of COS takes place at a high level from the very beginning because it is present in the product gas used for desorption. As in the case of non-cyclic sorption, the product gas remains free of $H_2S$ during the entire time interval of at least 100 h allocated to the adsorption step. Breakthrough of $H_2S$ that is remarkably sharp, occurs only after all sites specific for sorption of $H_2O$ are saturated. At the same time formation of COS breaks down, and the gas that leaves the bed, is free of COS. Breakthrough of $H_2O$ takes place shortly after that of $H_2S$. Certain deviations of

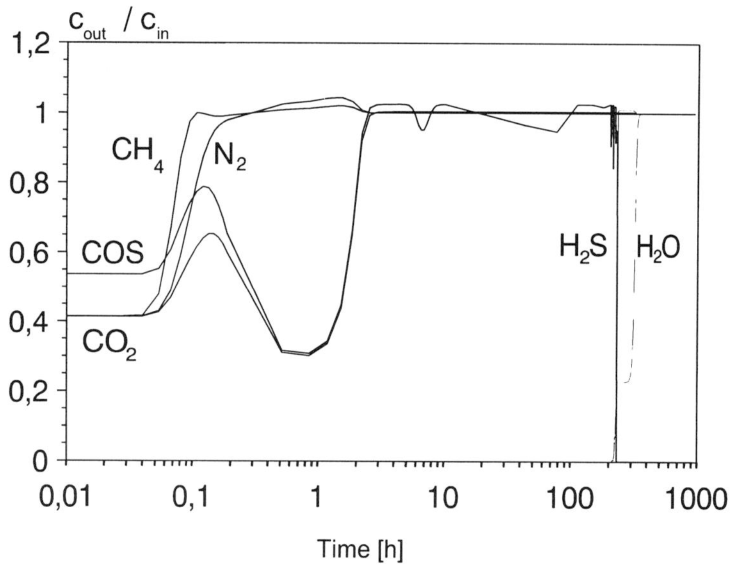


Figure 22. Mixture breakthrough curves for a cyclic regime with formation of COS.

calculated breakthrough times from the experimental data for the various components of natural gas may occur because, in realty, complex features of reaction (1) in zeolites may differ from those, ((i) to (iv)), assumed at the beginning of section 4.4.2.2.

## CONCLUSIONS

Zeolites of the LTA-type are outstanding sorbents for $H_2S$, due to their high capacity, selectivity and long-term stability as well as reversibility of the sorption process. Especially, calcium LTA modifications with any $Ca^{2+}$ cation content up to ca. 100 %, that have, in addition, high thermal and hydrothermal stability, can be utilized successfully for the purification of $CO_2$-free gases. Gases purified in this way are neither toxic nor cause stress corrosion cracking in plants or other installations. However, $CO_2$ is often a component of gases to be sweetened. Since $H_2O$ shows an even stronger adsorbability on zeolites, sorption of $H_2S$ is connected directly with its transformation into COS in accordance with the reaction $H_2S + CO_2 \Leftrightarrow COS + H_2O$. Gases purified in this way, contain COS that is toxic but does not lead to stress corrosion cracking.

Alkaline metal cations at extra-framework sites of the zeolite structure that were coordinated weakly, have the highest catalytic activity with respect to COS formation, in accordance with the process of base catalysis. LTA zeolites with complete alkaline earth cation exchange exhibit the lowest activity due to sufficiently retarded kinetics of COS formation. Nonetheless, even these materials lead the COS formation to its equilibrium after long contact times.

The formation of COS cannot be suppressed either by modifying the zeolites by any cation exchange, blocking the catalytic centers via presorption of $H_2O$ or $NH_3$, or performing gas purification at increased temperature. Potassium LTA modifications exclude $H_2S$ for steric reasons at low temperatures and act non-selectively at high temperatures. COS formation on such zeolites takes place due to sizeselective sorption of $H_2O$ that is generated by the COS formation reaction as a result of the catalytic activity of external crystal surface sites. To date, the only reliable way to the suppression of COS formation on LTA zeolites comprises blocking of their β-cages to prevent adsorption of $H_2O$ therein. This approach has not yet been reduced into practice.

For gases that are free of $CO_2$ or contain a low amount of $CO_2$ compared with that of $H_2S$ (concentration ratio: $[CO_2] / [H_2S] < 10$), desulphurization by means of zeolites as shown by both plant experience and mathematical modeling of various process regimes, represents an effective method. If the gases contain appreciable amounts of $CO_2$, the removal of $H_2S$ by sorption on zeolites becomes possible by short-time techniques, e.g., PSA / VSA processes, only. A cyclic plant regime as usual seems not possible since the mutual replacement of $CO_2$ and $H_2S$, which depends on the content of the former in the feed gas, diminishes significantly the $H_2S$ sorption capacity of the zeolite. In these cases, selective sorption by zeolites could be utilized for the removal of $H_2S$ if only both gases that compete for the sorption sites, undergo a reaction to form $H_2O$ and COS. The first product would remain in the bed and the second would leave it with the "purified" gas. The purpose of gas desulphurization cannot be achieved unless the removal of a corrosive species, i.e., $H_2S$, which would allow for gas transportation over long distances, might be considered as the only but temporary goal of the process.

## ACKNOWLEDGEMENTS

The authors dedicate this paper to Professor Dr. Wolfgang Schirmer, Berlin-Hessenwinkel, on the occasion of his 79th birthday.

M.B. thanks his colleagues, Drs. Sudhakar Jale, Adeola F. Ojo, Dongmin Shen and Qingmin Wang, Murray Hill, for critical remarks and valuable suggestions.

## REFERENCES

1. R.M. Barrer, Zeolites and Clay Minerals as Sorbents and Molecular Sieves, Academic Press, London, 1978.
2. D.W. Breck, Zeolite Molecular Sieves, John Wiley & Sons, New York, 1973.
3. S.P. Zhdanov, S.S. Chvoshchev and N.N. Feoktistova, Synthetic Zeolites, Parts 1 and 2, Gordon and Breach Science Publishers, New York, 1990 (originally published in Russian: Chimija, Moscow, 1981).
4. R.T. Yang, Gas Separation by Adsorption Processes, Butterworth, Boston, 1986.
5. R.M. Milton, US Patent No. 2 882 243 (1953); D.W. Breck, W.G. Eversole, R.M. Milton, T.B. Reed and T.L. Thomas, J. Am. Chem. Soc., 78 (1956) 5963.
6. R.M. Milton, US Patent No. 2 882 244 (1953).
7. J. Schmidt, Verfahren der Gasaufbereitung, VEB Deutscher Verlag für Grundstoffindustrie, Leipzig, 1970.
8. L. Henderson and T. E. Cox, Oil & Gas Journal, 69 (1971) 15.
9. P.G. Stecher, Hydrogen Sulfide Removal Processes, Noyes Data Corp., Park Ridge, 1972.
10. K. Hedden and F. Schnürer, Freiberger Forschungshefte, A 413 (1967) 5.
11. L. Heisler, Erdöl-Erdgas-Zeitschr., 83 (1967) 5.
12. J.J. Collins, Chem. Eng. Progr., 64 (1968) 8.
13. F. Pass and A. Hackl, Erdöl-Erdgas-Zeitschr., 90 (1974) 9.
14. C.D. Swaim, Hydrocarbon Processing, 49 (1970) 3.
15. B.G. Goar, Oil & Gas J., 78 (1980) 18.
16. F. Rosendahl, Gas & Wasserfach (Gas), 106 (1965) 31.
17. F. Rosendahl, Gas & Wasserfach (Gas), 110 (1969) 5.
18. F. Rosendahl, Gas & Wasserfach (Gas), 110 (1969) 13.
19. F. Rosendahl, Erdöl & Kohle-Erdgas-Petrochemie, 34 (1981) 3.
20. P.H. Turnok and K.J. Gustafson, Proc. Gas Condit. Conf. (Abstracts), (1972) B22.
21. R.D. Stoll and S. Röper, Erdöl-Kohle & Erdgas-Petrochemie, 35 (1982) 380.
22. H.A. Larson, M.F. Böhme and J.W. Sheets, US Patent No. 4 522 793 (1985).

23. M.R. Cines, D.M. Huskell and C.G. Houser, Chem. Eng. Prog., 72 (1972) 89.
24. A.M. Aitani, Hydrocarbon Processing, April Issue (1993) 67.
25. K. Hedden and B.R. Rao, GWF-Gas/Erdgas, 123 (1982) 1.
26. J. Burkert, H. Busch, K. Hedden and B.R. Rao, GWF-Gas/Erdgas, 125 (1984) 202.
27. D.M. Ruthven, Principles of Adsorption and Adsorption Processes, John Wiley & Sons, New York, 1984, 360.
28. J.D. Sherman and A.T. Katsaros, US Patent No. 4 329 160 (1982).
29. E.W. Corcoran, D.E.W. Vaughan, P.E. Eberly and K.D. Efird, US Patent No. 5 264 193 (1993).
30. W. Lutz, M. Suckow and B. Schlicht, Erdöl, Erdgas, Kohle, 107 (1991) 413.
31. Linde Co., Molecular Sieve Department, New York, Petroleum Refiner, 36 (1957) 7.
32. T.L. Thomas and E.L. Clark, Oil & Gas Journal, 65 (1967) 12.
33. P.W. Sherwood, Erdöl & Kohle-Erdgas-Petrochemie, 16 (1963) 8.
34. J. Messmer and O. Brunner, Erdöl & Kohle-Erdgas-Petrochemie, 15 (1962) 3.
35. P.N. Kraychy and M. Masuda, Oil Gas Journal, 64 (1966) 32.
36. R.J. Schoofs, Gas, 43 (1966) 11.
37. T.L. Thomas and E.L. Clark, Proc. 46th Ann. Conv., Nat. Gas Process. Assoc., Houston 1967.
38. E.L. Clark, Oil Gas Journal, 57 (1959) 18.
39. C.A. Duval, in: R.A. Jones, Advances Petroleum Chemistry & Refinery, Chapt. 4. J. McKetta (ed.), Dept. Chem. Eng., University of Texas, 1961, 164.
40. K.H. Bacon and A. Henke, Oil Gas Journal, 59 (1961) 12.
41. V.A. Voroncova, Gazovaja promyshlennostj, (Russ.), N1 (1972) 1.
42. Nescio Nomen, Canad. Chem. Processing, 44 (1960) 11.
43. Nescio Nomen, Oil Gas Journal, 58 (1960) 52.
44. H.W. Haines, G.A. van Wielingen and F.B. Mathews, Oil Gas Journal, 59 (1961) 5.
45. S.A. Conviser, Oil Gas Journal, 63 (1965) 49.
46. F.C. Boston, Hydrocarbon Processing & Petroleum Refiner, 43 (1964) 8.
47. E.L. Clark, Oil Gas Journal, 60 (1962) 1962.
48. A.L. Dunbar, Proc. Symp. "The Less Common Means of Separation", Birmingham, April, 1963, London, Inst. Chem. Engineers, 1963.
49. A.M. Henke, H.C. Stauffer and N.L. Carr, Hydrocarbon Processing & Petroleum Refiner, 41 (1961) 5.
50. S.G. Kolesnikova, Poluchenije vodoroda iz uglevodorodnogo syrja, (Russ.), 1979, 36.
51. D.W. Rushton and W. Hays, Oil Gas Journal, 59 (1961) 38.
52. G. Kowshun, Khimicheskij Zhurnal, (Russ.), 43 (1977) 247.
53. J.C. Fails and W.D. Harris, Oil Gas Journal, 59 (1961) 38.
54. H.W. Haines, G.A. van Wielingen and G.H. Palmer, Oil Gas Journal, 59 (1961) 5.

55. H.W. Haines, US Patent No. 3 144 307 (1964).
56. A.M. Henke, H.C. Stauffer and N.L. Carr, Oil Gas Journal, 60 (1962) 4.
57. W.R. Epperly and N.J. Roselle, US Patent No. 3 098 814 (1963).
58. I.N. Dijarov, Prirodnye ceolity: Trudy sovetsko-bolgarskogo simpoziuma po issledovaniju fiziko-khimicheskikh svojstv prirodnykh ceolitov, (Russ.), Mecniereba, Tbilisi, 1979, 269.
59. E.L. Clark, Chem. Eng. Progr., 58 (1960) 9.
60. Nescio Nomen, Nitrogen, 72 (1971), 34; 36.
61. N.V. Kelcev, Gazovaja Promyshlennostj, (Russ.), 9 (1963) 52.
62. N.H. Chen and D.F. Othmer, J. Chem. Eng. Data, 7 (1962) 37.
63. G. Groninger, K. Hedden and B.R. Rao, Chem. Ing. Techn., 10 (1987) 63.
64. Nescio Nomen, Petroleum Proc. Eng., 1 (1960) 9.
65. H. Mineta, S. Nakamura, A. Wakaizumi, T. Tanaka, H. Kawakami and K. Hishinuma, JP 84-235 629 (1984).
66. G. Kühl and H.S. Sherry, Brit. Pat. No. 1 580 929 (1980); G. Kühl, Zeolites, 7 (1987) 451.
67. W. Lutz, M. Suckow and M. Bülow, Gas. Sep. Purif., 4 (1990) 190.
68. H.G. Karge and J. Raskó, J. Coll. Interf. Sci., 64 (1978) 522.
69. H. Förster and M. Schuldt, J. Coll. Interf. Sci., 52 (1975) 380.
70. M. Bülow, Unpublished results.
71. M. Suckow, W. Lutz, K. Fiedler and M. Bülow, Proc. 5th Conf. Appl. Chem, Balatonfüred, Sept., 1989, EFCE Publications Ser., 74, vol. 1 (1989), 182.
72. M. Suckow, W. Lutz and M. Bülow, Proc. Technol. Proc., vol. 8: Gas Sep. Technol., Elsevier, New York, 1990, 289.
73. W.J. Mortier, Compilation of Extra Framework Sites in Zeolites, Butterworth, Guildford, 1982.
74. K. Möller, Dissertationsschrift, Promotion A, Zentralinstitut für Physikalische Chemie, Akademie der Wissenschaften der DDR, Berlin, 1978.
75. H.G. Karge, J. Ladebeck and N.K. Nag, 7th Canad. Symp. Catalysis (Preprints), Edmonton, Canada, 1980, 225.
76. P. Fellmuth, W. Lutz and M. Bülow, Zeolites, 7 (1987) 367.
77. Y. Ono, Stud. Surf. Sci. Catal., 5 (1980) 19.
78. M. Ziolek and Z. Dudzik, Zeolites, 1 (1981) 117.
79. A.B. Verver and W.P.M. van Swaaij, Appl. Catal., 14 (1985) 185.
80. K. Pilchowski, R. Gaedeke, J. Karch and F. Wolf, Krist. & Techn., 13 (1978) 47.
81. W. Lutz, Unpublished results.
82. M.B. Mick, Presentation, Gas Processors Association Meeting, USA, March 24, 1976.
83. W. Lutz, J.-Chr. Buhl and H. Thamm, Erdöl, Erdgas, Kohle, 114 (1998), in print.
84. R.E. Trent, D.F. Craig and R.L. Coleman, Proc. 43d Laurance Reid Gas Cond. Conf., March 3, 1993, 239.

85. M. Wark, G. Schulz-Ekloff, N.I. Jaeger and W. Lutz, Materials Research Society Symp. Proc., vol. 233, MRS, Pittsburgh, 1991, 133.
86. M. Bülow and A. Micke, in: M.D. LeVan and K.S. Knaebel (eds.), Fundamentals of Adsorption, Proc. 5th Int. Conf. Fundamentals of Adsorption, Pacific Grove, May, 1995; Kluwer Publ., Norwell, 1996, 131.
87. M. Suckow, W. Lutz, J. Kornatowski, M. Rozwadowski and M. Wark, Gas. Sep. Purif., 6 (1992) 101.
88. W. Lutz, B. Fahlke, R. Seidel, M. Bülow and W. Wieker, Chem. Techn. (Leipzig), 40 (1988) 121.
89. K. Ehrhardt, M. Suckow and W. Lutz, Stud. Surf. Sci. Catal., 94 (1995) 179.
90. W. Lutz, B. Fahlke, U. Lohse and R. Seidel, Chem. Techn. (Leipzig), 35 (1983) 249.
91. H. Fichtner-Schmittler, W. Lutz, S. Amin, A. Dyer and M. Wark, Zeolites, 12 (1992) 749.
92. W. Lutz, M. Bülow, N.N. Feoktistova, L.M. Vtjurina, S.P. Zhdanov and H. Siegel, Cryst. Res. Technol., 24 (1989) 161.
93. M. Bülow and P. Struve, Stud. Surf. Sci. Catal., 87 (1994) 551.
94. C.W. Chi and W.P. Cummings, Adsorptive Separation - Gases, Encyclopedia of Chemical Technology, vol. 1, John Wiley & Sons, New York, 1978, 544.
95. W. Lutz, H. Fichtner-Schmittler, M. Bülow, E. Schierhorn, Nguyen van Phat, E. Sonntag, I. Kosche, S. Amin and A. Dyer, J. Chem. Soc. Faraday Trans., 86 (1990) 1899.
96. M. Suckow and W. Lutz, in: E. F. Vansant (ed.): Separation Technology, Elsevier, 1994, 143.
97. S. Weiß (ed.), Verfahrenstechnische Berechnungsmethoden, Teil 2, Thermisches Trennen, VEB Deutscher Verlag für Grundstoffindustrie, Leipzig, 1980.
98. N. Wakao and T. Funazkri, Chem. Eng. Sci., 42 (1978) 1375.
99. P.N. Dwiwedi and S.N. Upadhyay, Ind. Eng. Chem., Proc. Des. Dev., 16 (1977) 157.
100. S.N. Gupta, R.B. Chaube and S.N. Upadhyay, Chem. Eng. Sci., 29 (1974) 839.
101. K.-H. Radeke, Chem. Techn. (Leipzig), 39 (1987) 111.
102. R.M. Marutovsky and M. Bülow, Chem. Eng. Sci., 42 (1987) 2745.
103. M. Suckow, Dissertationsschrift, Promotion A, Zentralinstitut für Physikalische Chemie, Akademie der Wissenschaften der DDR, Berlin, 1986.
104. M. Suckow, Gas Sep. Purif., 2 (1988) 151 (part I); 196 (part II).

Adsorption and its Applications in Industry and Environmental Protection
Studies in Surface Science and Catalysis, Vol. 120
A. Dabrowski (Editor)



# Nitrogen separation from air by pressure swing adsorption

N. O. Lemcoff

The BOC Group, 100 Mountain Avenue, Murray Hill, NJ 07974-2064, USA

## 1. ABSTRACT

Pressure swing adsorption is commercially used for the separation of air as an alternative to the conventional cryogenic separation process. A zeolite based process, in which the adsorbent shows preferential adsorption of nitrogen over oxygen under equilibrium conditions, is used for the production of oxygen. On the other hand, the separation of air for the production of nitrogen is carried out by pressure swing adsorption over a carbon molecular sieve. The separation is kinetically based, since the equilibrium adsorption of both oxygen and nitrogen is very similar, but the oxygen is adsorbed faster. In conventional pressure swing adsorption systems, both the productivity and yield decrease significantly as nitrogen purities are increased beyond 99.9 %. Although higher nitrogen purities were achievable, product flows were so low that it was not an economically viable process. Only recently, through both improvements in the manufacture of carbon molecular sieves and new cycle developments, higher performance is achieved in the high purity region. A theoretical dynamic model developed for the kinetic separation of nitrogen from air is presented, and the effect of the different process variables and cycle steps on the process performance is analyzed. A comparison between the experimental performance and the model predictions is discussed.

## 2. INTRODUCTION

Gaseous nitrogen is used in the chemical and petroleum industries for storage tank blanketing and vessel inerting applications. It is also used extensively by the electronics and metals industries for its inert properties. It is also widely used in the food industry to retain food freshness longer. Commercial nitrogen is produced by a variety of air separation processes, including cryogenic liquefaction and distillation, adsorption separation, and membrane separation.

All adsorption separation processes involve two principal steps: adsorption, during which one or more species are preferentially adsorbed, and regeneration, during which these species are removed from the adsorbent. When the regeneration step is carried out through the reduction of the total pressure, the process is called pressure swing adsorption (PSA), and it has become a subject of interest in gas separations because of its low energy requirement and cost. The basic steps involved in a cycle are typically pressurization, high-pressure adsorption, equalization, countercurrent blowdown and low-pressure desorption. PSA processes have found widespread application in hydrogen purification, air drying, and air separation [1,2]. In the first two cases, the separation takes place due to the difference in the equilibrium adsorption isotherms of the components of the gas mixture. On the other hand, the separation of nitrogen from air on a carbon molecular sieve, is kinetically controlled. The adsorption equilibrium isotherms of oxygen and nitrogen are almost identical, and the preferential sorption of oxygen in these adsorbents results from the faster adsorption of that species on the carbon molecular sieve.

The objective of this chapter is to study the effect of different process variables on the performance of a rate induced PSA process, and to compare the experimental results with the predictions of a theoretical model. The effect of process variables and cycle steps on the temperature and concentration profiles is also studied, and the relationship between the profiles and the process performance is discussed.

## 3. CARBON MOLECULAR SIEVES

Carbon molecular sieves (CMS) have played a critical role in the commercialization of the pressure swing adsorption process for the separation of nitrogen from air. They differ from activated carbon mainly in the pore size distribution and surface area. While activated carbons have a broad range of pores, with a typical average pore diameter of 20 Å, carbon molecular sieves have a more narrow pore size distribution, with pore sizes in the range of 3 - 5 Å. A molecular probe method is one of the best approaches to determine the effective micropore size distribution of carbon molecular sieves [3,4]. Typical surface areas for a carbon molecular sieve are in the range of 250-400 $m^2/g$, while the micropore volume is about 0.15-0.25 $cm^3/g$ [2,5].

One of the earliest publications on the preparation of carbon molecular sieves was done by Walker et al. [6]. They started from a Saran copolymer, but currently many different raw materials are being used either for research or commercial purposes. The main sources of carbon molecular sieves for commercial applications are coal and coconut shell. An extensive list of the work done in the preparation of carbon molecular sieves can be found elsewhere [7,8]. The typical steps in the preparation of carbon molecular sieves from coal are shown in Fig. 1 [8]. It can be seen that the main steps are oxidation, carbonization, followed either by steam activation or pyrolysis. During steam activation, pores are opened and the resulting carbon molecular sieve can be used in hydrogen purification, while during pyrolysis carbon deposition takes place, and a narrower pore size distribution is achieved.

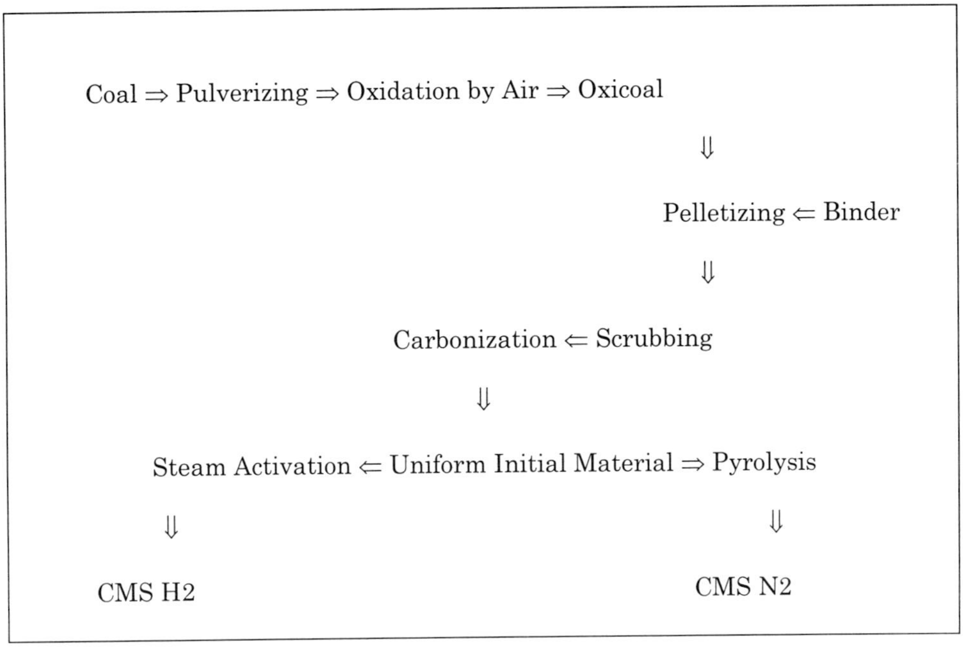


Figure 1. Schematic diagram showing the processes involved in the manufacture of carbon molecular sieve adsorbents.

A recent development was to prepare carbon based molecular sieves which are funcionalized with inorganic oxides and supported metals. The main objective is to combine the molecular sieving properties of the carbon with surface chemical and physical properties of the inorganic oxides in one composite structure. Sharma and Seshan [9] reported copper modified CMS for the selective removal of oxygen at temperatures below 200°C. At higher temperatures, oxygen can be reversible adsorbed only if present in trace amounts.

## 4. ADSORPTION RATE

The separation of nitrogen from air is based on the different uptake rate of nitrogen with respect to oxygen. Typical adsorption isotherms and uptake curves on carbon molecular sieves are shown in Figs. 2 and 3. Selectivities for carbon molecular sieves are in the range between 20 and 50 [7,10,11]. This difference in adsorption kinetics is thought to be related to the different molecular size. The kinetic diameter of oxygen (3.5 Å) is slightly smaller than that of nitrogen (3.6 Å).

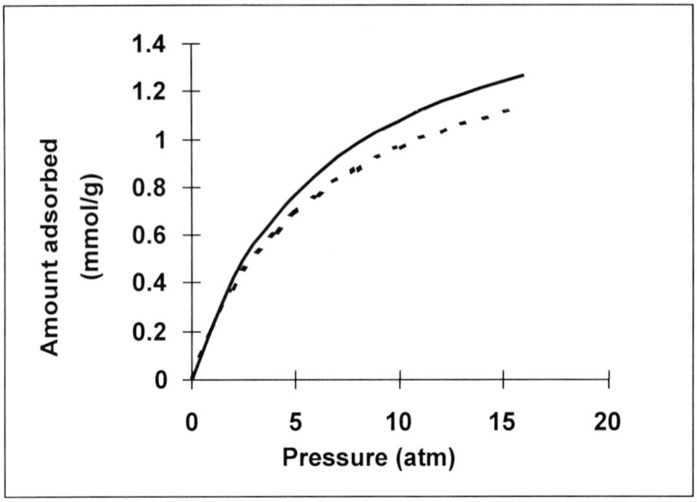


Figure 2. Adsorption isotherms on carbon molecular sieve at 25°C.
———— oxygen — — — — nitrogen.

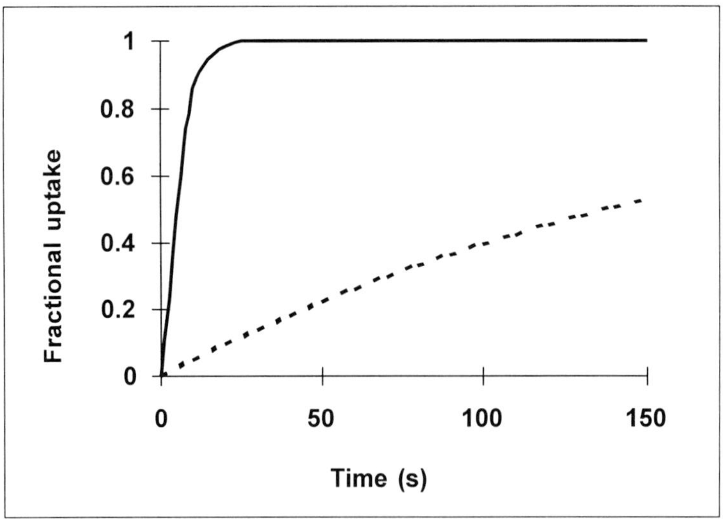


Figure 3. Adsorption kinetics on carbon molecular sieve at 25°C.
———— oxygen — — — — nitrogen.

The adsorption kinetics of oxygen and nitrogen on CMS were studied using different experimental methods: volumetric, gravimetric, chromatographic, thermal desorption, and isotope exchange [11]. From the analysis of the literature it follows that the transport process controlling the uptake rate seem to obey to two different mechanisms, either a surface barrier at the micropore entrances, or a diffusional molecular transport within the bulk of the micropores.

Ruthven [12] analyzed the diffusion of nitrogen and oxygen in carbon molecular sieves that respond to Fickian diffusion. They found that, outside the linear region of the adsorption isotherm, the diffusivity increases strongly with concentration, following the Darken equation. The actual values of the corrected limiting diffusivities for oxygen and nitrogen at 300K are $D_o/r^2 = 2.3 \times 10^{-3}$ and $6 \times 10^{-5}$ 1/s, respectively [12-14]. The corresponding activation energies are 25.1 and 27.2 kJ/mole. An order of magnitude smaller diffusivity values were measured by other researchers [15], and this was attributed to a difference in the activation temperature of the sieve samples.

Fitch et al. [16] studied the uptake of oxygen and nitrogen in a non-Fickian carbon molecular sieve. It is assumed that carbon crystallites are small in size and are loosely cross-linked with other crystallites to form an aperture-cavity structure. The aperture has a slit shape which serves as an entrance to the cavity. It is suggested that the sieving action is produced by the interaction of Van der Waals potential forces in the slit and the molecules. An experimental characteristic of these sieves is that the half-time of the adsorption process depends on the equilibrium coverage, and hence, depends on the adsorption pressure. The slit potential activation energies determined by these studies were 0.10 and 15.5 kJ/mole, respectively. These values are close to the energy barrier values calculated by Rao et al. [17] assuming a two graphite layer as a model for the slit.

Recently, different approaches to describe the kinetics of oxygen and nitrogen sorption in carbon molecular sieves have been presented. It has been shown that they can lead, as particular cases, to the two above mentioned mechanisms. Srinivasan et al. [18] proposed a surface barrier model of diffusion under a chemical potential gradient, while Trifonov and Golden [19] presented a theoretical approach based on a hopping mechanism, in which the molecules jump from occupied to free adsorption sites. Seaton et al. [20] carried out molecular dynamics simulation of diffusion in individual pores and in pore networks. The experimental selectivity can be reproduced both at the level of individual pores, as well as with pore networks having a wide range of pore sizes.

Carbon surfaces exhibit a strong affinity for oxygen even at room temperature. Aging of the carbon molecular sieve has been attributed to the oxidation of the carbon surface, which leads to a reduction in the pore or slit size, and a reduction in the gas uptake rates. Also, sorption of water at or near the pore slits of a CMS can affect the diffusion rate of gases during the adsorption and desorption steps, and therefore significantly change its performance in a pressure swing adsorption system. Passivation of active sites has been suggested to achieve a stable CMS surface [21).

## 5. PRESSURE SWING ADSORPTION PROCESS

In actual operation, pressure swing adsorption systems are composed of two identical beds containing carbon molecular sieve that perform the separation process continuously. Compressed air enters the bottom of the first bed where the sieve adsorbs oxygen and other impurities, letting high-purity nitrogen pass through. When this bed approaches saturation it switches to a regenerative phase, while the second bed automatically begins the separation process. The regeneration is accomplished by reducing the pressure, allowing the pelletized bed material to vent the released gases. After a few minutes, the non-nitrogen components have been vented from the bed and it is ready to separate air again. In this way, high purity nitrogen is produced without an interruption. PSA systems are used primarily for applications that require relatively steady flows of gaseous nitrogen at rates of approximately 30 - 3,000 $Nm^3/h$. Typical operation conditions of nitrogen pressure swing adsorption systems are: adsorption pressure, 8 atm, regeneration pressure, 1 atm, half cycle time, 1-2 min.

The PSA performance is determined in terms of the specific product and yield at a certain nitrogen product purity. The specific product is defined as the product flow rate per unit adsorbent volume, while the yield is the ratio between the net flow rate of nitrogen produced and the flowrate of nitrogen fed to the system. The specific feed, in turn, is defined as the feed flow rate per unit of adsorbent volume.

In order to avoid deterioration of the carbon molecular sieve due to water adsorption, a drying system is generally used. The drying system can be either external or internal to the vessels. In the first case, a chiller is used to reduce moisture. In the second case, a desiccant is used which dries out the gas internally, without the need of bulky external dryers. However, the materials used as desiccants are all polar materials and, specially for zeolites, have a selective adsorption of nitrogen over oxygen. Golden et al. [22] developed a composite adsorbent capable of both dehydrating air and selectively adsorbing oxygen over nitrogen. An increase in the process performance was achieved by using LiCl and silica impregnated carbon molecular sieves. The dehydrating agents reside in the macropores where water adsorption takes place.

Up to recently, nitrogen purities from commercial pressure swing adsorption processes were below 99.95%. In order to get higher purities, an additional step of reduction of the oxygen impurities was required. The low purity nitrogen product is fed together with a fuel, such as hydrogen, to a deoxygenation unit, where the oxygen impurities are converted over a catalyst. Palladium or platinum impregnated alumina are generally used. Purities as high as 99.9999% can be achieved, but the disadvantage is that humidity and excess fuel are introduced into the nitrogen product. Although this may not be a problem in some applications, such as reducing atmospheres, it still requires the availability of the fuel. An alternative way of achieving higher nitrogen purities is the use of vacuum during the regeneration of the adsorbent beds. As the ratio between the adsorption and desorption pressures increases substantially, both productivity and yield increase significantly. However, the addition of another rotating unit is a disadvantage not only from the capital cost, but also from the maintenance point of view.

Oxygen selective sorbents have been developed by researchers in different institutions over the past decades. These metal complexes mimic certain biological oxygen carriers that selectively remove oxygen from air. Cobalt, iron and manganese based materials have been developed with capacities of between 1 and 10% oxygen by weight. However, the most serious drawback is their chemical instability caused by autoxidation reactions. Either the ligand or the central atom is oxidized, or peroxo-bridged dimers are formed, and their oxygen adsorption capacity is lost [23].

Nitrogen PSA performance has improved considerably in the last 20 years. Since 1980, the power consumption has dropped to a half and the productivity doubled, resulting in both smaller air compressors, and a reduction in the size of the units. These advances, which have allowed to commercialize pure pressure swing adsorption systems supplying nitrogen with purities higher than 99.999%, arise from two different areas: molecular sieve developments and process improvements [24].

Probably, the most important advances have come as a result of carbon molecular sieve improvements, but there have been also significant improvements due to the better use of the sieve by modifications to the PSA process. Early improvements comprise the incorporation of additional steps to the basic cycle. In the pressure equalization step, both beds are interconnected to transfer gas from the high pressure bed to the low pressure bed, and therefore save both air feed and compression energy [14,25]. During this step, the gas at the product end of the bed is pushed back, acting like a purge, and therefore improving both purity and recovery. The pressure equalization step plays an important and beneficial role in establishing the shape of the solid mass transfer zone, and it is specially important for cycles with short feed and desorption times [26].

A better regeneration of the beds is achieved through purging with product gas during the desorption step and/or the use of vacuum [27]. Although the purge helps cleaning the bed and a higher purity can be achieved, it can adversely affect the recovery. An increase in the performance is also observed when the adsorbent beds are repressurized by feed gas at an initially slow rate followed by a rapid feed second stage. A combination of product and feed repressurization of the bed before going into production can also improve the process performance [28,29].

The total cycle time is another important variable in the process. As the cycle time increases, the performance goes through a maximum. At short cycle times, both the yield and the specific product go to zero because the nonproductive steps in the cycle have a significant effect. On the other hand, at long cycle times, the bed becomes saturated, and no separation is possible. For fast cycles the system behavior depends on the kinetic selectivity, while for slow cycles it is a function of the equilibrium selectivity [27,30,31].

Other practices that can help achieve a higher productivity are the use of a dense packing technique for the filling of the beds, as well as the use of layers with different quality sieve [24].

The separation of nitrogen from air can also be done using zeolite molecular sieves. Although the kinetic selectivity is similar to that of carbon molecular sieves, the equilibrium selectivity favors oxygen. Shin and Knaebel [32,33] found that there is a trade off between purity and yield, which is characteristic of the diffusion induced separations.

In order to better understand the system behavior, several theoretical models have been developed describing the process. Different mechanisms are considered, and assumptions are made in order to predict the performance of the pressure swing adsorption cycle. These theoretical studies also allow gain an insight into the effect of the cycle steps and operating variables.

One of the advantages of the theoretical studies is the possibility of studying the sensitivity of the process performance to both operating variables and adsorbent properties. For instance, if the gases uptake rates is assumed to increase, while maintaining the same kinetic selectivity, the productivity increases. On the other hand, the effect on the yield depends whether a constant or variable cycle time is assumed [31]. In general it is observed that, while a change in the oxygen uptake rate will mainly affect the purity, a change in the nitrogen uptake rate will mainly affect the recovery [14,34].

Good agreement between experimental data for air separation with results from simulation studies using different gas uptake models has been observed. Although the linear driving force model can correctly predict the trends of the experimentally observed behavior, the pore diffusion model shows a much closer agreement [34].

Several authors have also studied and analyzed the temperature and concentration profiles in the bed during a PSA cycle. Raghavan and Ruthven [35] carried out a numerical simulation of the nitrogen pressure swing adsorption process over a carbon molecular sieve. They run a simple Skarstrom type cycle, with no equalization step, and analyzed the evolution of the gas and solid phase composition profiles using a linear isotherm and a linear driving force model. The mass transfer zones are of an S-shape, but do not show maximum or minimum inside the bed. More realistic simulation studies, as well as experimental measurements, have shown the roll-up effect due to displacement of oxygen by nitrogen, as well as the reduced working capacity between adsorption and pressurization due to the gradual concentration profiles [10,26]. It was also confirmed that the oxygen mass transfer zone is very broad under favorable operating conditions, and spans over a large portion of the bed.

In the low purity region, the purification of the product is relatively independent of cycle time [36]. This could only be so if the shape of the mass transfer zone (MTZ) is insensitive to the cycle time [31]. Except for very long cycle times, it would be expected that, due to the high throughputs and the small amount of $O_2$ adsorbed, the MTZ would span the length of the adsorbent bed.

Recently, Lemcoff et al. [36,37] studied both experimentally and theoretically the nitrogen pressure swing adsorption process in the high purity region, and found that the effect of cycle time on the process performance becomes more significant as the purity increases.

## 6. THEORETICAL MODEL

Numerical simulations were carried out using the Dynamic Adsorption Process Simulator (DAPS), developed by LaCava et al. [38]. The following assumptions were made in the development of the simulator:

- The system is non-isothermal, with the total pressure remaining constant during production and purge. The pressure is assumed to increase linearly during pressurization, and follow an exponential decay during blowdown.
- The flow pattern is described by the axial dispersion model. The axial dispersion is variable point to point along the bed, a function of pressure and velocity.
- In the equalization step, the detailed composition profile for the pressurizing gas is taken from the corresponding profile of the depressurizing gas of the other bed (point by point model).
- The rate of uptake of gases by the non-Fickian carbon molecular sieve is described in terms of the slit-potential rate model [39], which considers the slit-cavity structure of the carbon molecular sieve. Under certain conditions, the Langmuir adsorption rate expression is a good approximation to the slit-potential rate model.

### 6.1. Model equations

The mass balance for component i in the fluid is given by:

$$\varepsilon \frac{\partial c_i}{\partial t} = D_L \frac{\partial^2 c_i}{\partial z^2} - \frac{\partial (v_z c_i)}{\partial z} - (1-\varepsilon) R_i \tag{1}$$

where the rate of adsorption can be expressed in terms of the slit potential rate model:

$$R_i = \frac{\partial q_i}{\partial t} = \frac{k_{s1i} q_{mi}}{K_i (1-\theta_i)} \left[ K_i c_i \left( 1 - \sum \theta_k \right) - \theta_i \right] \tag{2}$$

At very high coverage, or when the slit potential does not control the rate, the rate equation reduces to the classical Langmuir form:

$$R_i = \frac{\partial q_i}{\partial t} = k_{di} q_{mi}\left[K_i c_i\left(1 - \sum \theta_k\right) - \theta_i\right] \quad (3)$$

The heat balance in the adsorber bed is given by:

$$\left[\rho_g C_{pg}\varepsilon + \rho_s C_{ps}(1-\varepsilon)\right]\frac{\partial T}{\partial t} = K_L \frac{\partial^2 T}{\partial z^2} - \rho_g C_{pg}\frac{\partial (v_z T)}{\partial z} + (1-\varepsilon)\sum R_i(-\Delta H_i) - \frac{2h}{r}(T - T_w) \quad (4)$$

The boundary conditions for the production step are:

$$D_L \frac{\partial c_i}{\partial z}(0,t) = -v_z(0,t)\left[c_i(0^-,t) - c_i(0,t)\right] \quad (5a)$$

$$K_L \frac{\partial T}{\partial z}(0,t) = -v_z(0,t)\rho_g C_{pg}\left[T(0^-,t) - T(0,t)\right] \quad (5b)$$

$$\frac{\partial c_i}{\partial z}(L,t) = 0 \qquad \frac{\partial T}{\partial z}(L,t) = 0 \quad (5c)$$

For the pressurization step, the only different boundary condition is the velocity at the top of the bed:

$$v_z(L,t) = 0 \quad (6)$$

During the vent or blowdown step, the dispersion effect is neglected, and the boundary conditions are:

$$\frac{\partial c_i}{\partial z}(L,t) = 0 \qquad v_z(L,t) = 0 \quad (7a)$$

$$\frac{\partial T}{\partial z}(0,t) = 0 \qquad \frac{\partial T}{\partial z}(L,t) = 0 \quad (7b)$$

The initial conditions are:

$$c_i\,(z,0) = 0 \qquad q_i\,(z,0) = 0 \qquad T\,(z,0)= T_0 \qquad \text{for a clean bed} \tag{8a}$$

$$c_i\,(z,0) = c_{io} \qquad q_i\,(z,0) = q_i^* \qquad T\,(z,0)= T_0 \qquad \text{for a saturated bed} \tag{8b}$$

The numerical integration of the partial differential equations is carried out by using the numerical method of lines. Finite differences are used in the spatial discretization procedure, and the Adams multistep method as the time integrator.

Kinetic and equilibrium data were obtained using a batch column device described elsewhere [39,40]. At 25°C, the oxygen and nitrogen monolayer capacity is 1.8 mole/kg CMS, and the adsorption and desorption kinetic constants are, respectively, 76.84 $cm^3$/(mole s) and 0.0176 1/s for oxygen, and 3.11 $cm^3$/(mole s) and 0.000585 1/s for nitrogen.

Shirley and Lemcoff [36] estimated the adsorption half-time from results at different cycle times and product purities. As the oxygen concentration in the product goes to 0 ppm, the feed per cycle goes to some finite, limiting value. These values reflect the amount of air needed to bring the adsorber vessels up to the production pressure, including the amount needed for adsorption as well as that needed to fill the interstitial and pore volumes in the CMS. At short cycle times, the majority of the limiting feed is due to the gas needed to fill the interstitial space in the adsorbent beds, such that very little nitrogen purification occurs, while at long cycle times, the adsorption of both nitrogen and oxygen is near equilibrium levels with little nitrogen purification, again. Therefore, when these values are plotted against the part cycle time, the limiting feed per cycle is a gross measure of the adsorption kinetics of air. Interpolation of the data gives an adsorption half-time of about 1600 sec, which matches closely that for air at the same pressures from batch adsorption experiments.

## 7. EXPERIMENTAL PART

The experiments were carried in a commercial-scale two-bed nitrogen PSA unit built by The BOC Group, Inc. A schematic of the unit is shown in Fig. 4. These units typically run a series of four processing steps in cyclic fashion. In the first step of a cycle, one bed is pressurized with air while cocurrently producing nitrogen gas while the second bed is regenerated by venting countercurrently to the atmosphere or to a vacuum. At the end of this first step, the beds are brought into fluid communication, causing gases from the bed at high pressure to flow into the regenerated bed until pressure equalization occurs. During regeneration, the low-pressure bed is sometimes purged with a small fraction of the purified product stream. Additional steps, such as product backfill can be included in the cycle [41]. In this case, before the regenerated bed is pressurized with air, a fraction of the nitrogen product is used to partially pressurize the bed countercurrently.

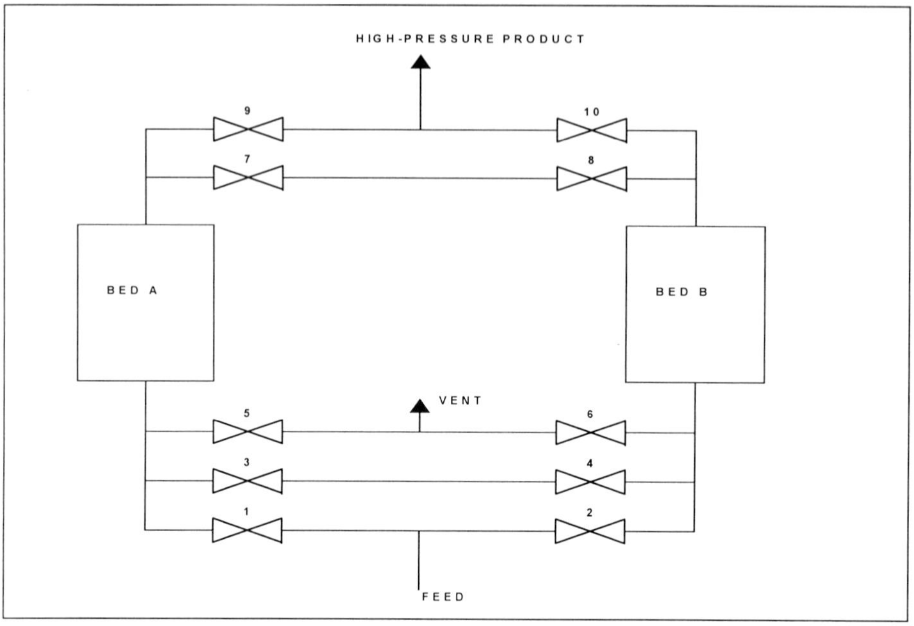


Figure 4. Schematic of nitrogen pressure swing adsorption plant.

Different cycle times (full cycle time between 120 and 1200 s) and product flowrates (specific product between 5 and 60 $Nm^3/m^3/h$) at adsorption pressures between 6.5 and 9.0 atm ($6.6x10^5$ and $9.1x10^5$ Pa) were tested. Regeneration was done at atmospheric pressure in all cases. The change in the cycle time was achieved by changing simultaneously the duration of the production and venting steps. The duration of the remaining steps, pressurization, blowdown, and equalization, were not modified. The purge flowrate was also varied.

Pressurization of the adsorber beds was controlled according to the optimal method disclosed by Shirley and LaCava [29].

## 8. RESULTS

The predictions of the theoretical model have been compared with the experimental results. Oxygen concentration breakthrough curves were measured experimentally by sending to an oxygen analyzer a small fraction of the product gas produced at the exit of the bed. The oxygen analyzer response was about 1 sec, and readings were taken every second. There is a period when the unit is not producing, namely between the end of the production step in one bed, and the end of the pressurization step in the other bed. The experimental results are

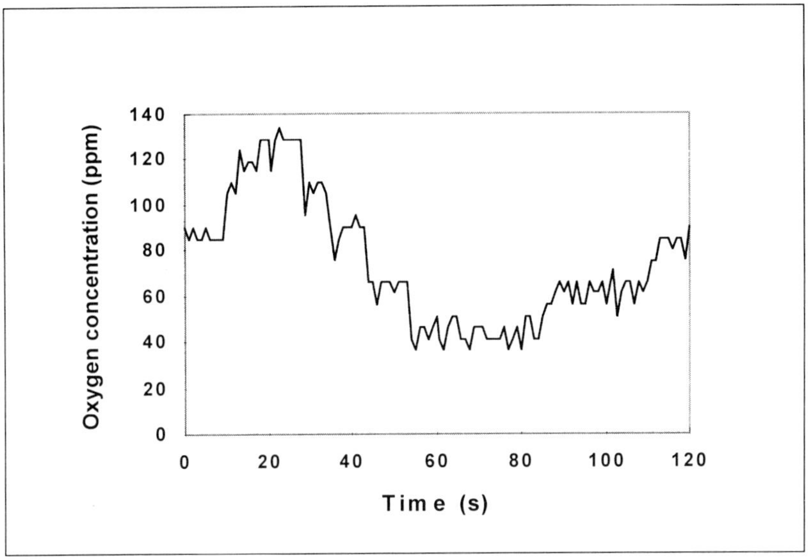


Figure 5. Experimental oxygen breakthrough curve. Product purge during desorption step.

shown in Fig. 5. The peak corresponds to the equalization step, and is related to the concentration of gas being transferred between the beds. As the mass transfer concentration zone shifts, the concentration increases towards the end of the production step.

The predicted oxygen breakthrough curves for the simple 4-step cycle, with and without nitrogen purge, are shown in Fig. 6. A good agreement between the experiments and the simulation is observed. When no purge with nitrogen product is done during the vent step, a more linear response is obtained, and the average oxygen concentration in the half-cycle is higher.

In order to better understand the behavior of the PSA process, oxygen concentration profiles were obtained from the simulation studies for different steps in the cycle. We will consider first the case with nitrogen purge during the vent step. The oxygen mole fraction in the gas phase at different times during the production step are shown in Fig. 7. It can be seen that the shift in the oxygen concentration profile is about one third of the bed length. At the same time it can be seen that, at the beginning of the step, a concave up profile exits, but it immediately changes to an s shape, and to a concave down profile. This rollover effect is due to the fact that oxygen is adsorbed faster than nitrogen, so that initially the adsorbed concentration of oxygen in the carbon molecular sieve is high. As the step proceeds, nitrogen starts to be adsorbed to a greater degree, and displaces the oxygen. Therefore, the oxygen concentration in the gas phase becomes greater than the value in air. Ng et al. [26] studied the shape and movement of the oxygen mass transfer zone at four axial positions in a laboratory N2 PSA unit, and concluded that the concave down profile is achieved only at the end of the production step. The different conclusion can be attributed to the relative large distance between their measurement points, which was about one quarter of the bed length.

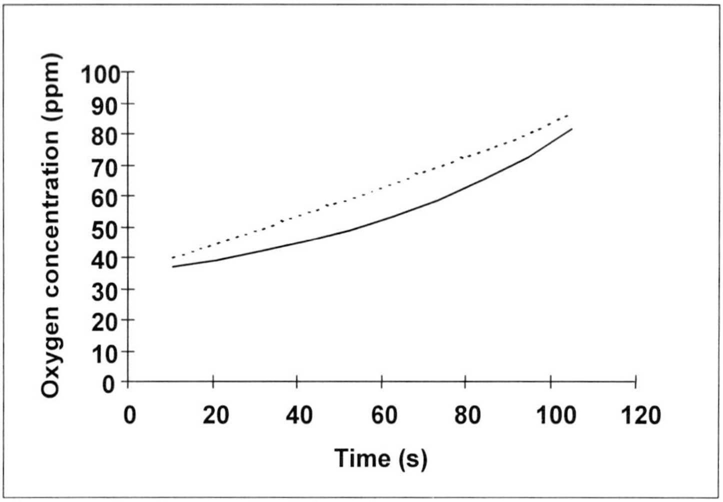


Fig. 6. Predicted oxygen breakthrough curves.
———Purge————No purge.

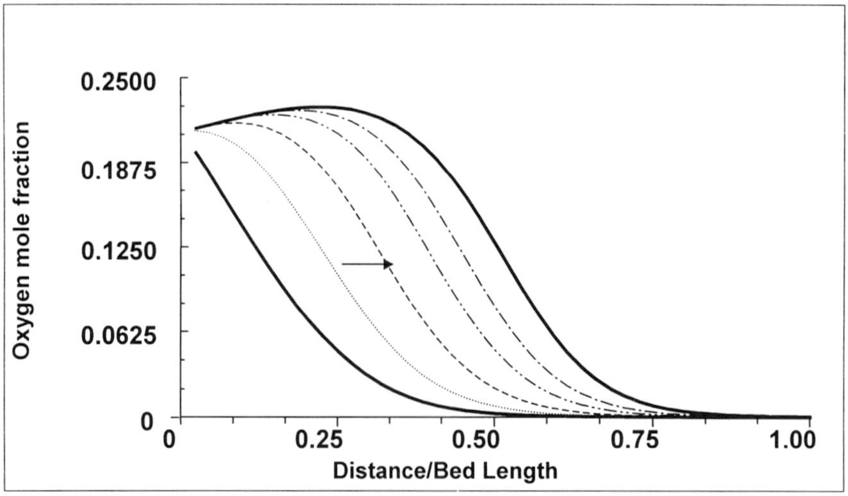


Figure 7. Evolution of gas phase oxygen concentration profiles during the production step. Nitrogen purge during vent step. Oxygen concentration in product: 0.05%.

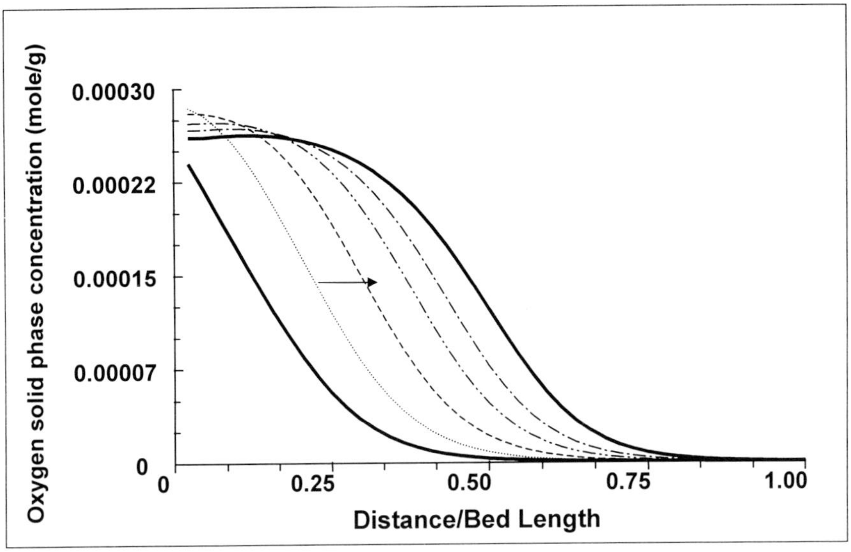


Figure 8. Evolution of solid phase oxygen concentration profiles during the production step. Nitrogen purge during vent step. Oxygen concentration in product: 0.05%.

The corresponding oxygen solid phase concentration profiles are shown in Fig. 8. We can see that the profiles change from concave up to an s-shape. As discussed above, the adsorbed oxygen concentration near the feed end increases at the beginning of the step due to the fast oxygen adsorption, but then decreases due to the displacement by nitrogen. During the production step the solid phase concentration front also shifts by about a third of the bed length.

The predicted bed temperature profiles at different times during the production step are shown in Fig. 9. A maximum temperature difference of about 7°C is observed near the bed center. It can be seen that during the cycle the profiles seem to shift vertically throughout the bed.

The oxygen concentration profiles in the gas and solid phases at different times during the vent step are represented in Figs. 10-11. They span over the whole bed length and are almost proportional to each other. As the step proceeds, both adsorbed gases are being desorbed. In addition, the nitrogen product purge helps the oxygen concentration profiles become flatter. The corresponding temperature profiles show an almost uniform decrease in temperature (Fig. 12).

The concentration and temperature profiles for the case with no nitrogen product purge during the vent step are very similar to the case just described. The major differences are at the bed outlet, during the production step, and at the bed feed end during the vent step. Bed temperatures are slightly higher for the no purge case, while the oxygen concentration profiles during the vent step are slightly higher near the bed feed end.

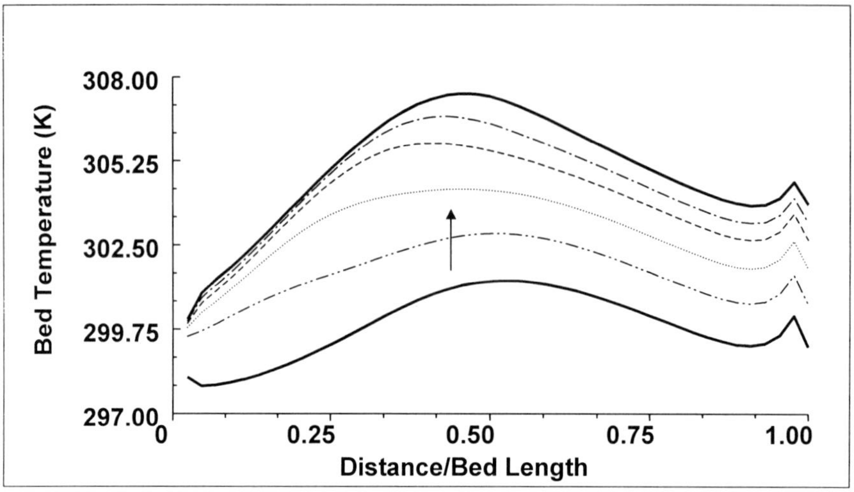


Figure 9. Evolution of bed temperature profiles during the production step. Nitrogen purge during vent step. Oxygen concentration in product: 0.05%.

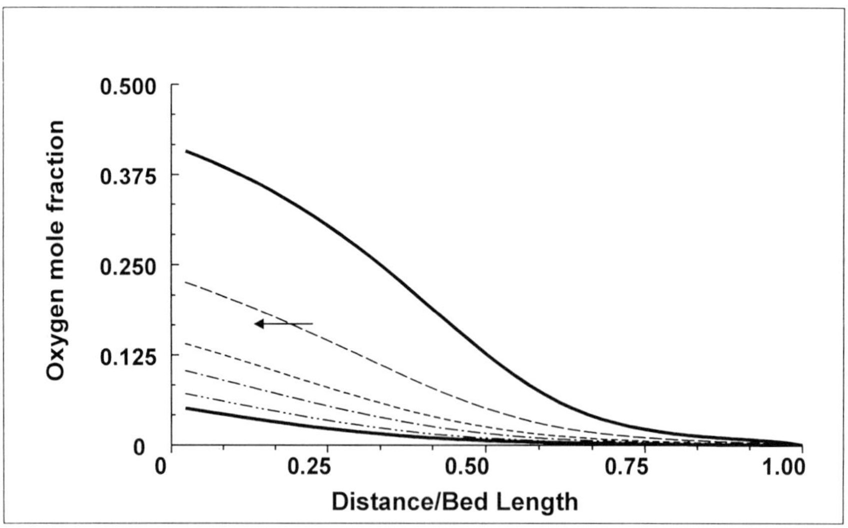


Figure 10. Evolution of gas phase oxygen concentration profiles during the purge step. Nitrogen purge during vent step. Oxygen concentration in product: 0.05%.

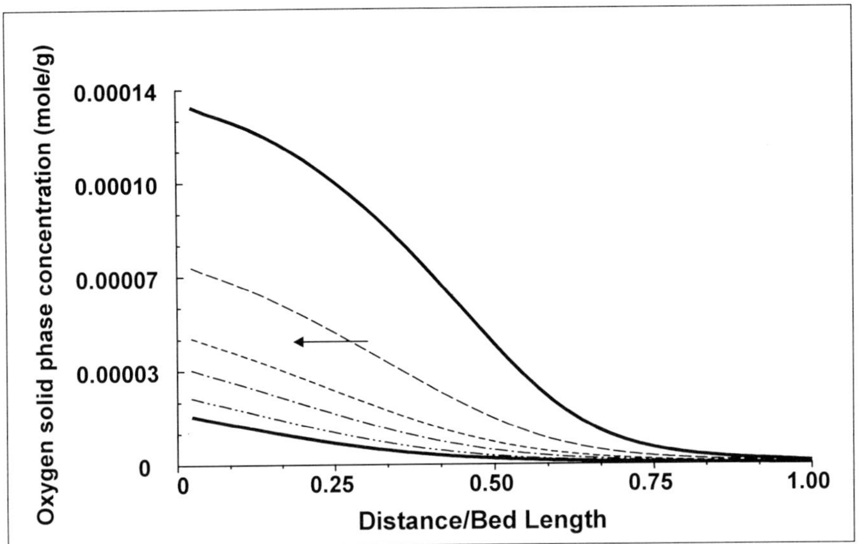


Figure 11. Evolution of solid phase oxygen concentration profiles during the purge step. Nitrogen purge during vent step. Oxygen concentration in product: 0.05%.

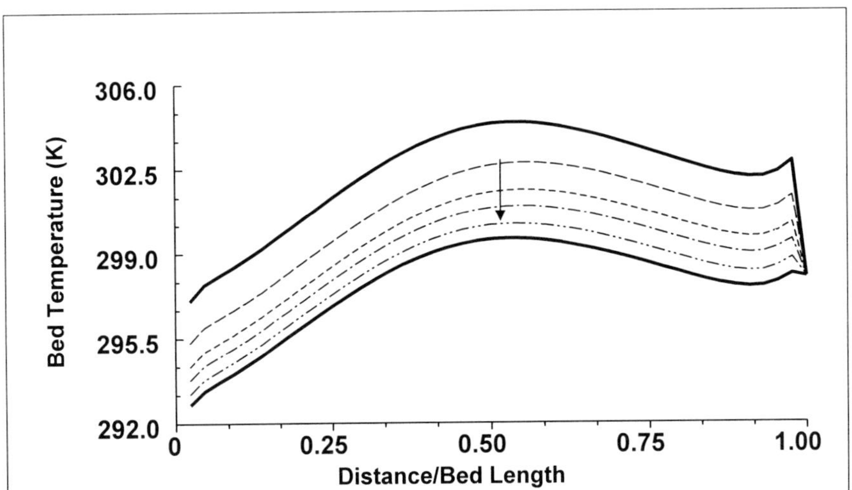


Figure 12. Evolution of bed temperature profiles during the purge step. Nitrogen purge during vent step. Oxygen concentration in product: 0.05%.

The effect of the backfill step on the bed concentration profiles and the process performance is shown in Figs. 13-14. The oxygen concentration profiles in the bed at the end of the production, equalization and purge steps, for a cycle without backfill, are represented in Fig. 13. The rollover effect is observed at the end of

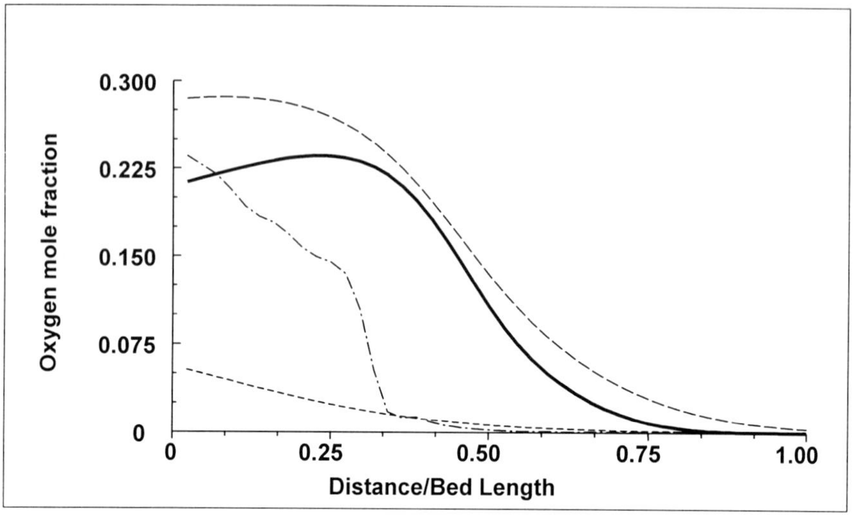


Figure 13. Oxygen concentration profiles at the end of the steps. Cycle with no backfill. Product oxygen concentration: 270 ppm.
——— Production ——— Equalization (down)
- - - - Purge — — — — — — — Equalization (up).

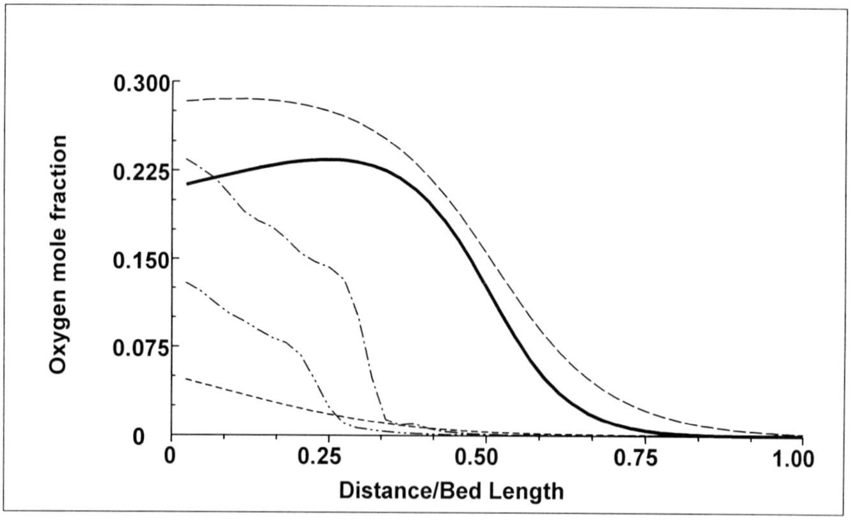


Figure 14. Oxygen concentration profiles at the end of the steps. Cycle with no backfill. Product oxygen concentration: 100 ppm.
——— Production ——— Equalization (down) - - - - Purge
— - — - — Equalization (up) — - - — - - — - - — Backfill.

the production step. During the equalization step, gas is transferred from the high pressure to the low pressure bed through both inlet and outlet. As the decreasing equalization proceeds, the oxygen concentration profile is spread out further and, due to the oxygen desorption, a higher concentration than that in air is observed at the inlet. During the vent step, helped by the countercurrent purge, the oxygen concentration decreases significantly. As the increasing equalization step proceeds, gas enters the bed from both ends, and the mass transfer zone is reduced. At the end of the increasing equalization step, the oxygen mole fraction at the inlet is again higher than its value in air. During the following pressurization step, the oxygen concentration profile shifts to the left, and turns into concave up.

As the backfill is introduced, the oxygen concentration profile shifts towards the feed end, specially in the high concentration region. Now the oxygen mole fraction at the feed end is substantially lower than its value in air. During the following pressurization step, the oxygen concentration profile shifts to the right, and turns into concave up as in the case above. The oxygen concentration at the feed end has its value in air. As the production step proceeds, the profile shifts to the right and becomes steeper, and higher nitrogen purities can be achieved (Fig. 14).

However, there is a limit to the process improvement. The effect of the duration of the backfill step on the purity of the nitrogen product is shown in Fig. 15. As the backfill increases, since more product gas is used, the yield tends to decrease. At the same time, the feed flowrate has to increase to achieve the

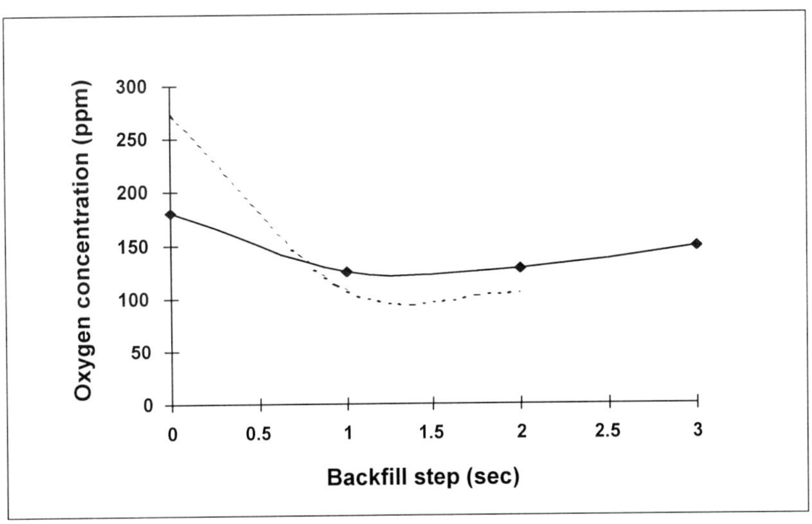


Figure 15. Effect of backfill step on nitrogen purity at constant product flow. ———— Experimental – – – – – Predicted.

same product flow, and an increase in the oxygen concentration in the product is observed for long backfill steps. A good agreement is observed between the experimental results and the model predictions.

Other way of achieving higher nitrogen purities is the use of cryogenic back-up tank vent gas for purging the regenerating bed [42]. The reported results to produce a gas containing 0.1% oxygen are summarized in Table 1. Case I represents the base case, when the purge step is carried out with the same product gas. Cases II and III correspond to the situation when a high purity gas is used to purge the beds during the regeneration step. A significant improvement in the performance, which increases with the amount of purge gas, can be observed.

Table 1
Comparison of purge methods

| | Case I | Case II | Case III |
|---|---|---|---|
| Product impurity (% $O_2$) | 0.1 | 0.1 | 0.1 |
| Purge concentration (% $O_2$) | 0.1 | < 0.001 | < 0.001 |
| Purge to feed ratio (%) | 11.2 | 10.6 | 13.0 |
| Productivity increase (%) | base case | 15.1 | 20.5 |
| Yield increase (%) | base case | 8.1 | 11.8 |

Similar behavior is observed at higher nitrogen purities, as shown in the simulation studies carried out for a system using an external source for the nitrogen purge. The oxygen concentration profiles at the end of the different cycle steps, at constant product flowrate, are shown in Figs. 16-18. From the analysis of the figures it follows that the main differences in the profiles are observed for the purge step. It can be seen that, as the purge to feed ratio increases from 0.05 to 0.10, the oxygen concentration profile at the end of the purge step is flatter, and the oxygen concentration in the product decreases from 500 to 200 ppm (Figs. 16-17). At the higher purge flowrate, a better cleaning of the bed is achieved, and a higher purity product can be obtained. On the other hand, when the purity of the external purge gas is increased, so that the oxygen concentration in the purge gas decreases from 10 ppm to 1 ppm oxygen, no significant change in the product oxygen concentration is observed (Figs. 16 and 18). The profiles also remain unchanged. It can be concluded that the purity of the gas used as external purge has no significant effect of the final product purity, when its purity is more than one order of magnitude higher than the purity of the product gas.

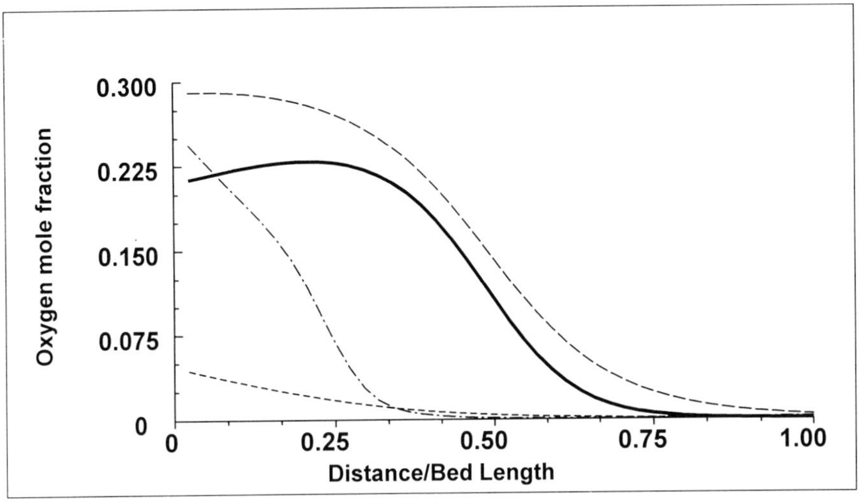


Figure 16. Oxygen concentration profiles at the end of the steps. Cycle time: 240s. Product oxygen concentration: 0.05%. Purge to feed ratio: 0.05. Purge purity: 10 ppm $O_2$.
———— Production — — — — — Equalization down
- - - - - - Purge — — — — — — — — — Equalization up.

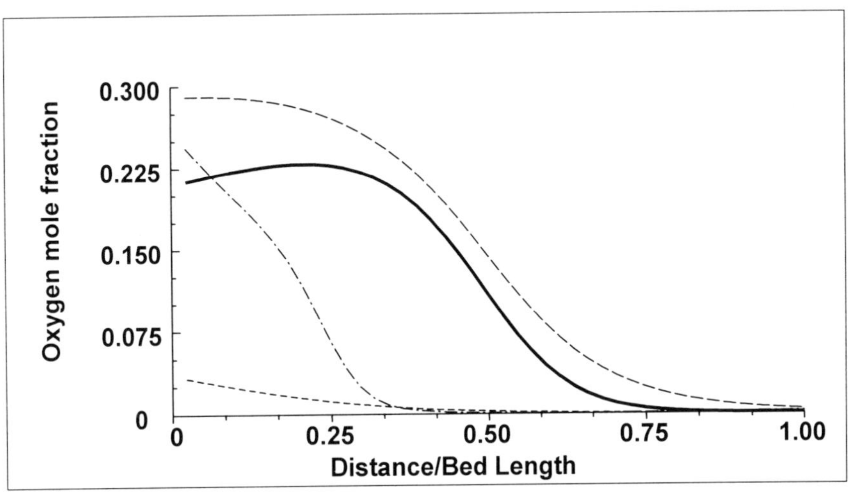


Figure 17. Oxygen concentration profiles at the end of the steps. Cycle time: 240 s. Product oxygen concentration: 0.02%. Purge to feed ratio: 0.10. Purge purity: 10 ppm $O_2$.
———— Production — — — — — Equalization down
- - - - - - Purge — — — — — — — — — Equalization up.

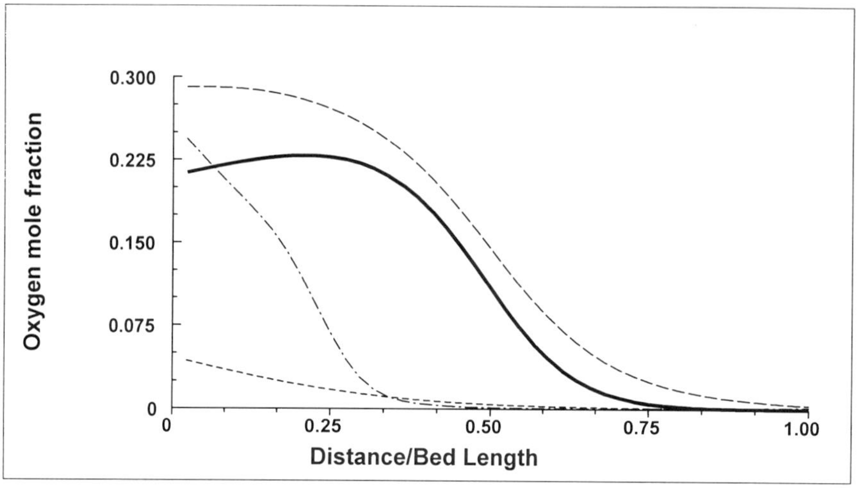


Figure 18. Oxygen concentration profiles at the end of the steps. Cycle time: 240s. Product oxygen concentration: 0.02%. Purge to feed ratio: 0.05. Purge purity: 1 ppm $O_2$.
——— Production — — — — Equalization down
······· Purge — - — - — - — Equalization up.

## 9. CONCLUSIONS

The pressure swing adsorption process for the separation of nitrogen from air over carbon molecular sieves was described, and both experimental and simulation studies were carried out. Two types of adsorbents, with different behavior in the gases uptake are commonly used, and the process characteristics for a non-Fickian type sieve were specially analyzed.

The effects of the different variables and cycle steps on the production of nitrogen were discussed and, in particular, the influence of the purge and backfill steps on the oxygen concentration profiles was analyzed. The shape and movement of these profiles help understand the relationship between the process variables and the pressure swing adsorption performance. A good agreement between the predictions of the theoretical model and experimental results was also observed.

## NOMENCLATURE

$c$ gas phase concentration, mole/m$^3$
$C_p$ specific heat, kJ/kg K
$D_L$ axial dispersion coefficient, m$^2$/s
$\Delta H$ enthalpy change of adsorption, kJ/mole

| | |
|---|---|
| $h$ | heat transfer coefficient, $kJ/m^2$ s K |
| $K$ | Langmuir equilibrium constant, $m^3$/mole |
| $K_L$ | effective thermal conductivity, kJ/m s K |
| $k_d$ | Langmuir desorption constant, 1/s |
| $k_{sl}$ | slit potential rate constant, 1/s |
| $L$ | bed length, m |
| $q$ | adsorbed phase concentration, mole/$m^3$ |
| $q_m$ | monolayer capacity, mole/$m^3$ |
| $R$ | rate of adsorption, mole/($m^3$ s) |
| $r$ | bed radius, m |
| $T$ | temperature, K |
| $t$ | time, s |
| $v_z$ | axial gas velocity, m/s |
| $z$ | axial distance, m |

Greek letters

| | |
|---|---|
| $\varepsilon$ | bed voidage |
| $\rho$ | density, kg/$m^3$ |
| $\theta$ | $q/q_m$ |

Subscripts

| | |
|---|---|
| g | gas |
| i,k | components i,k |
| o | initial |
| s | solid |
| w | wall |

## REFERENCES


1. R.T. Yang, Gas Separation by Adsorption Processes, Butterworths, Boston, 1987.
2. D.M. Ruthven, S. Farooq and K.S. Knaebel, Pressure Swing Adsorption, VCH, New York, 1994.
3. J.E. Metcalfe III, M. Kawahata and P.L. Walker, Fuel, 42 (1963) 233.
4. J.D. Moyer, T.R. Gaffney, J.N. Armor and C.G. Coe, Microporous Materials, 2 (1994) 229.
5. I.P. O'koye, M. Benham and K.M. Thomas, Langmuir, 13 (1997) 4054.
6. P.L. Walker, L.G. Austin and S.P. Nandi, in: Chemistry and Physics of Carbon, P.L. Walker (ed.), Marcel Dekker, NY, vol. 2, 257, 1966.
7. R.V. Jasra, N.V. Choudary and S.G.T. Bhat, Sep. Sci. Tech., 26 (1991) 885.
8. H. Jüntgen, K. Knoblauch and K. Harder, Fuel, 60 (1981) 817.
9. P.K. Sharma and P.K. Seshan, Gas Sep. Purif., 4 (1990) 203.

10. R.M. Thorogood, Gas Sep. Purif., 5 (1991) 83.
11. R.M. Rynders, M.B. Rao and S. Sircar, AIChE Journal, 43 (1997) 2456.
12. D.M. Ruthven, Chem. Eng. Sci., 47 (1992) 4305.
13. D.M. Ruthven, N.S. Raghavan and M.M. Hassan, Chem. Eng. Sci., 41 (1986) 1325.
14. S. Farooq and D.M. Ruthven, Chem. Eng. Sci., 46 (1991) 2213.
15. Y.D. Chen, R.T. Yang, P. Uawithya, AIChE Journal, 40 (1994) 577.
16. F.R. Fitch, M. Bulow and A.I. LaCava, Gas Sep. Purif., 8 (1994) 45.
17. M.B. Rao, R.G. Jenkins and W.A. Steele, Langmuir, 1 (1985) 137.
18. R. Srinivasan, S.R. Auvil and J.M. Shorck, Chem. Eng. J., 57 (1995) 137.
19. Y.Y. Trifonov and T.C. Golden, J. Porous Materials, 3 (1996) 5.
20. N.A. Seaton, S.P. Friedman, J.M.D. MacElroy and B.J. Murphy, Langmuir, 13 (1997) 1199.
21. S.K. Verma and P.L. Walker, Jr., Carbon, 30 (1992) 837.
22. T.C. Golden, P.J. Battavio, Y.C. Chen, T.S. Farris and J.N. Armor, Gas Sep. Purif., 7 (1993) 274.
23. G.Q. Li and R. Govind, Ind. Eng. Chem. Research, 33 (1994) 755.
24. A. Schulte-Schulze Berndt, MUST '96, 185, 1996.
25. N.O. Lemcoff, S.J. Doong and A.I. LaCava, in: Fundamentals of Adsorption, M. Suzuki (ed.), Kodansha, Tokyo, 357, 1993.
26. M. Ng, J.M. Schork and K.R. Fabregas, Gas Sep. Purif., 7 (1993) 159.
27. N.O. Lemcoff and A.I. LaCava, Gas Sep. Purif., 6 (1992) 9.
28. M.M. Hassan, N.S. Raghavan and D.M. Ruthven, Chem. Eng. Sci., 42 (1987) 2037.
29. A.I. Shirley and A.I. LaCava, Ind. Eng. Chem. Research, 32 (1993) 906.
30. M.M. Hassan, N.S. Raghavan and D.M. Ruthven, Chem. Eng. Sci., 41 (1986) 1333.
31. J.M. Schork, R. Srinivasan and S.R. Auvil, Ind. Eng. Chem. Research, 32 (1993) 2226.
32. H.S. Shin and K.S. Knaebel, AIChE Journal, 33 (1987) 654.
33. H.S. Shin and K.S. Knaebel, AIChE Journal, 34 (1988) 1409.
34. S. Farooq and D.M. Ruthven, Chem. Eng. Sci., 45 (1990) 107.
35. S. Raghavan and D.M. Ruthven, AIChE Journal, 31 (1985) 2017.
36. A.I. Shirley and N.O. Lemcoff, AIChE Journal, 43 (1997) 419.
37. A.I. LaCava and N.O. Lemcoff, Gas Sep. Purif., 10 (1996) 113.
38. A.I. LaCava, J.A. Dominguez and J. Cardenas, in: Adsorption: Science and Technology, A.E. Rodrigues, M.D. LeVan and D. Tondeur (eds.), NATO ASI Series, vol. 158, 323, 1989.
39. A.I. LaCava, V.A. Koss and D.A. Wickens, Gas Sep. Purif., 3 (1989) 180.
40. V.A. Koss, D.A. Wickens, P. Cucka and A.I. LaCava, Proc. Carbon 86, 388, Baden-Baden, Germany, July 1986.
41. N.O. Lemcoff, Pressure swing adsorption process, US Patent 5 520 720 (1996).
42. R. Jain, PSA employing high purity purging, US Patent 5 090 973 (1992).

Adsorption and its Applications in Industry and Environmental Protection
Studies in Surface Science and Catalysis, Vol. 120
A. Dabrowski (Editor)



# Methodology of gas adsorption process design. Separation of propane/propylene and n/iso-paraffins mixtures

José A. C. Silva, F. Avelino da Silva and Alirio E. Rodrigues

Laboratory of Separation and Reaction Engineering
Faculty of Engineering, University of Porto
4099 Porto Codex, Portugal

The general guidelines for developing a gas separation process based on adsorption are reviewed. Two important industrial cases based on adsorption processes are selected: the separation of propane/propylene mixtures and n/iso-paraffins mixtures. The 13X zeolite and $Ag^+$-Amberlyst were used as adsorbent for propane/propylene mixture taking into account information from the open literature. The 5A zeolite was selected for n/iso-paraffins system; the adsorption equilibrium and diffusivity data were obtained from gravimetric and ZLC techniques respectively. A mathematical model for the bulk separation in fixed bed upon non-isothermal non-adiabatic conditions is formulated and solved numerically. The simulated results are compared with the available experimental breakthrough curves. Finally, a cyclic process based in the PSA-VSA and TSA concepts is proposed for these systems.

## 1. INTRODUCTION

Separation of mixtures with close boiling points, such as the propane/propylene and n/iso-paraffins mixtures is very energy consuming. Molecular sieve adsorption technology is the modern answer to this separation problem in order to save energy.

One of the first industrial adsorption processes was the separation of n/iso-paraffins in molecular sieves 5A zeolite developed by Union Carbide in the sixties for the octane improvement of gasoline pools and solvent production (Symoniak [1]). Since the seventies, the development of Hysomer and TIP (Total Isomerisation Process) by Shell and UOP respectively, and the demand of environment protection with the restriction of lead in fuel, makes this separation coupled with an isomerisation reactor one of the most licensed adsorption processes in the world. The two major fractions of normal paraffins that exists in the low range naphtha used in TIP for the improvement of the research octane numbers (RON) of gasolines are n-pentane and n-hexane (Holcombe et al. [2]; Minkkinen et al. [3]). In such processes n-hexane and n-pentane which have low octane number are partially converted to high octane branched paraffins in an isomerisation reactor. The unconverted n-paraffins are separated in a selective adsorption unit packed with 5A zeolite pellets and recycled to the isomerisation reactor. The 5A zeolite excludes branched paraffins and adsorbs linear $nC_5$ and $nC_6$.

The propylene/propane gas separation is a well-known energy consuming process, especially when the traditional distillation is used. It requires more than one hundred contact steps since the relative volatility is near to one at temperatures between 244 K and 327 K and total pressures between 1.7 to 22 bar [4]. The production of propylene of high purity (>99.9% mole) is an important issue in the petrochemical industry as the main feed of the polymer production. Common sources of propylene are the refinery gas and steam cracking streams. The composition of these streams is not constant, but the propylene is always mixed with propane and other hydrocarbons. Recently, the interest in reducing the energy costs of the distillation process by improving the energy integration of the overall process has increased [5]. In order to compensate the thermal losses of the traditional distillation, this operation can be combined with other separation steps. For the separation of olefin/paraffin systems other processes have been proposed [6]: Extractive Distillation, Chemical and Physical Adsorption, Chemical and Physical Absorption and Membrane Separation. Physical Adsorption appears as one of the most attractive techniques, based on the experience gained in the last forty years since the Pressure Swing Adsorption concept was applied commercially in the 50's [7]. Also, economic studies indicate that alternative hybrid methods of distillation-adsorption are competitive if compared to the traditional process and they can represent energy savings between 30 and 50% and capital cost savings between 20 and 30% if they were introduced for separating the propylene - propane mixture in the actual refineries [5].

An important tool in the methodology of gas adsorption process design is measurement of adsorption equilibrum isotherms. Sorption isotherms in molecular sieve zeolites are generally type I in IUPAC classification. Several models have been developed in order to interpret equilibrium sorption data (Ruthven, [8]; Yang, [9]). Due to their simplicity, we suggest molecular models like Langmuir [10], Nitta et al. [11] and LRC isotherms. These models allow the predictions of multicomponent equilibria from single component data. This point is of importance since experimental techniques for the measurement of multicomponent sorption data are generally very time consuming and difficult to interpret. Molecular sieve zeolites consist of small microporous crystals formed in a macroporous pellet; kinetics of sorption in these adsorbents generally offers two resistances to mass transfer: the macroporous resistance of the pellet and the microporous resistance of the crystals. Assuming these resistances in series Ruckenstein et al. [12] developed a bidisperse model for the measurement of transient diffusion. Based on that model Ruthven and Loughlin [13] proposed a simple criterion for the relative importance of time constants of diffusion. At that time most of the measurement of transport diffusion was made by volumetric or gravimetric sorption uptake and chromatography in conjunction with the measurement of adsorption isotherms. It is well known that several transport mechanisms can be of importance in such systems making difficult the interpretation of kinetic data (Karger and Ruthven, [14]; Hufton and Ruthven, [15]) and may lead to erroneous results. In order to minimize such effects new techniques have been developed such the ZLC technique of Eic and Ruthven [16] and the Thermal Method of Grenier et al. [17]. With these techniques micropore diffusion in zeolites can be faster in orders of magnitude from the data measured by gravimetry or classical chromatography. Basic concepts needed for the interpretation and modeling of fixed bed adsorption have been given by Rodrigues [18].

In this article we report: i) the measurement of sorption equilibrium data of $nC_5$ and $nC_6$ in 5A zeolite pellets on a flow microbalance; ii) The measurement of intraparticle diffusivity of $nC_5$ and $nC_6$ on 5A zeolite pellets with crystals of different size by ZLC and gravimetry and iii) The development of a mathematical model in order to predict the behavior of fixed bed and cyclic adsorption processes, iv) The prediction of breakthrough curves of propane/propylene and n/iso-paraffins mixtures in a fixed-bed adsorber based on a model including parameters independently measured. iv) Study of cyclic adsorption processes as Pressure Swing Adsorption (PSA) / Vacuum Swing Adsorption (VSA) and Temperature Swing Adsorption (TSA) for the separation of propane /propylene mixtures and n/iso-paraffins mixtures.

## 2. EQUILIBRIUM ISOTHERMS

Figure 1 shows experimental and predicted (Nitta et al., [11]) n-pentane and n-hexane adsorption isotherms between 373-573K obtained in a flow gravimetric system operated at atmospheric pressure (Silva and Rodrigues, [19], [20]).

The governing equation of Nitta et al's. isotherm for homogeneous surfaces neglecting the interaction term between adsorbed molecules is,

$$K_{eq} = \frac{1}{p}\frac{\theta}{(1-\theta)^n}; \theta = \frac{q}{q_{max}}; K_{eq} = k_\infty \exp\left(\frac{-\Delta H_i}{RT}\right) \quad (1)$$

where $\theta = q/q_{max}$ is the degree of filling of sites, p is the partial pressure of sorbate, $K_{eq}$ is an equilibrium constant, $q_{max}$ the maximum concentration of adsorbate at the saturation of the adsorbent, *n* is the number of active sites occupied by an adsorbed molecule, $k_\infty$ is a pre-exponential factor, $\Delta H_i$ is the isosteric heat of

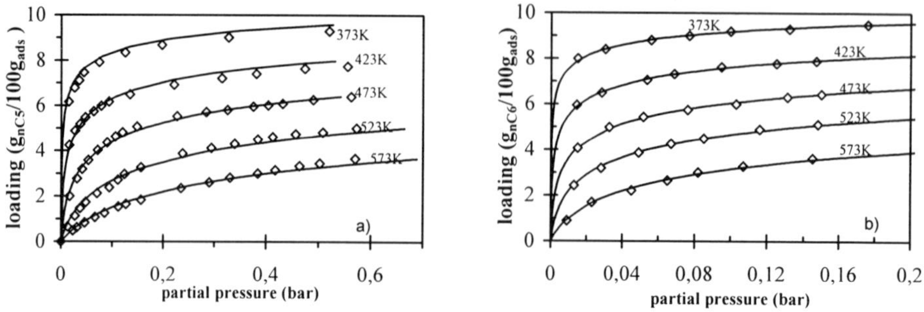


Figure 1. Adsorption equilibrium isotherms on pellets of 5A zeolite: a) n-pentane b) n-hexane. Points are experimental data (Silva and Rodrigues, [19], [20]). Lines are model predictions. Absolute temperatures are quoted in each curve.

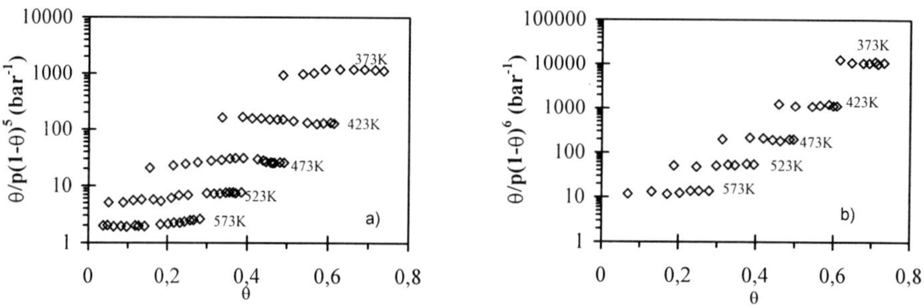


Figure 2. Semi-log plot of $\theta/p(1-\theta)^n$ *versus* $\theta$: a) n=5 for $nC_5$; b)n=6 for $nC_6$.

adsorption and T is the temperature. Experimental validation of this model is simple: a plot of $\theta/p(1-\theta)^n$ *versus* $\theta$ should be a straight line parallel to $\theta$ axis, which in turn multiplied by $q_{max}$ gives the Henry's law coefficient H. Such representation requires the determination of $q_{max}$; Silva and Rodrigues, [19], [20] have shown that $q_{max}$=13g/100$g_{ads}$ is an acceptable value. Figure 2 shows a plot of $\theta/p(1-\theta)^n$ *versus* $\theta$ for $nC_5$ and $nC_6$. Table 1 shows the isotherm model parameters resulting from fitting experimental data. Figure 1 and 2 clearly show that results predicted by Nitta et al. model are in good agreement with experimental data.

Järvelin et al. [21,22] and Huang et al. [23] presented experimental adsorption isotherms for the propylene-propane system over 13X zeolite. They mention that the 13X zeolite presents appropriate properties for both selectivity and adsorption capacity for the propylene-propane system at pressures around the atmospheric. Nevertheless, at higher pressures, Ghosh et al.[24] showed that silica gel has higher selectivity and adsorption capacity than the 13X zeolite.

Table 1
Parameters of Nitta et al´s. model isotherm for adsorption of $nC_5$ and $nC_6$ in pellets of 5A zeolite. (Data of Silva and Rodrigues, [19], [20])

| Equilibrium parameters | n-pentane | n-hexane |
|---|---|---|
| n | 5 | 6 |
| $q_{max}$ (g/100$g_{ads}$) | 13 | 13 |
| $-\Delta H_i$ (kJ/mol) | 55.2 | 59.4 |
| $K_{eq}$ at 573K ($bar^{-1}$) | 2.2 | 12.9 |

Activated carbon has also been used as adsorbent, but in spite of having the highest adsorption capacity of all adsorbents, its selectivity is very poor [25]. More recently Yang and coworkers [26,27] have introduced a new type of ion-exchange resins for paraffin/olefin separations, the $Ag^+$-Amberlyst. In this work, we focus on the adsorption process at the atmospheric pressure using 13X zeolite for the TSA case and the $Ag^+$-Amberlyst for the VSA case. The loading ratio correlation or LRC model was adopted to represent multicomponent equilibrium isotherm over both adsorbents, being the parameters shown in Table 2.

Table 2
Equilibrium parameters for propylene (1) – propane(2) adsorbed over 13X zeolite [23] and $Ag^+$-resins [26,27]

LRC model

$$n_i^* = m_i k_i P_i \chi_i / (1 + \sum_{i=1}^{n} k_i P_i \chi_i)$$

$$m_i = A_i \exp(B_i / T)$$

$$k_i = C_i \exp(D_i / T)$$

Propylene (1) – propane (2) over 13X zeolite

| i | $A_i$ [mol/kg] | $B_i$ [K] | $C_i$ [$kPa^{-1}$] | $D_i$ [K] | $\chi_i$ [-] | $-\Delta H_i$ [kJ/mol] |
|---|---|---|---|---|---|---|
| 1 | 1.31 | 170.7 | $6.437\ 10^{-5}$ | 2895.5 | 13.6 | 52.8 |
| 2 | 0.495 | 353.9 | $3.252\ 10^{-5}$ | 2826.4 | 3.72 | 46.9 |

Propylene (1) – propane (2) over $Ag^+$-Resin

| i | $A_i$ [mol/kg] | $B_i$ [K] | $C_i$ [$kPa^{-1}$] | $D_i$ [K] | $\chi_i$ [-] | $-\Delta H_i$ [kJ/mol] |
|---|---|---|---|---|---|---|
| 1 | 0.351 | 354.2 | $1.28\ 10^{-2}$ | 886.0 | 1.0 | 43.1 |
| 2 | $9.0\ 10^{-5}$ | 2401.8 | $0.268\ 10^{-2}$ | 397.0 | 1.0 | 21.3 |

## 3. KINETICS OF SORPTION

Kinetic studies of adsorption of n-pentane and n-hexane were performed by the chromatographic ZLC (Zero Length Column) technique (Eic and Ruthven, [16]). The ZLC technique consists in a differential bed of porous particles which is first saturated with the fluid mixture containing the adsorbable species; at time zero the carrier gas flows through the ZLC at sufficiently high flowrate and the desorption curve is analyzed in terms of concentration *versus* time. Analysis of experimental data with ZLC technique is straightforward, since modeling of the ZLC leads to a simple equation. For example in a system controlled by macropore diffusion, desorption curves at long times in a semi-log plot of concentration *versus* time are straight lines described by the following equation,

$$\ln\left(\frac{C_{out}}{C_0}\right) = \ln\left(\frac{2L}{\beta_1^2 + L^2}\right) - \frac{\beta_1^2 D_p t}{R_p^2(1+K)} \tag{2}$$

where,

$$\beta_n J_1(\beta_n) - L J_0(\beta_n) = 0 \tag{3}$$

$$L = \frac{1}{2}\frac{\text{Purge flowrate}}{\text{Pellets volume}}\frac{R_p^2}{\varepsilon_p D_p} \tag{4}$$

where $C_{out}$ is the outlet concentration of ZLC cell, $C_{in}$ is the concentration at time zero in the ZLC cell, $\beta_1$ are roots of transcendental Equation 3, $J_1(\beta_1)$ and $J_0(\beta_1)$ are Bessel functions of the first kind, t is the time variable, $D_p$ is the pore diffusivity and $K=(1-\varepsilon_p)H/\varepsilon_p$ is the capacity factor.

In bidisperse porous adsorbents such as zeolite pellets there are two diffusion mechanisms: the macropore diffusion with time constant $R_p^2/D_p$ and the micropore diffusion with time constant $r_c^2/D_c$. Bidisperse porous models for ZLC desorption curves have been recently developed by Brandani [28] and Silva and Rodrigues [29]. In bidisperse porous adsorbents, it is important to carry out experiments in pellets with different sizes but with the same crystal size (different $R_p$, same $r_c$) or pellets with the same size but with different crystals (same $R_p$, different $r_c$). If macropore diffusion is controlling, time constants for diffusion should depend directly on pellet size and should be insensitive to crystal size changes. If micropore diffusion controls the reverse is true. The influence of temperature is also important: when macropore diffusion is dominant the apparent time constant of diffusion defined by $R_p^2(1+K)/D_p$ is temperature dependent in the same order of K (directly related to the heat of adsorption) which is determined independently from the isotherm. The type of purge gas is

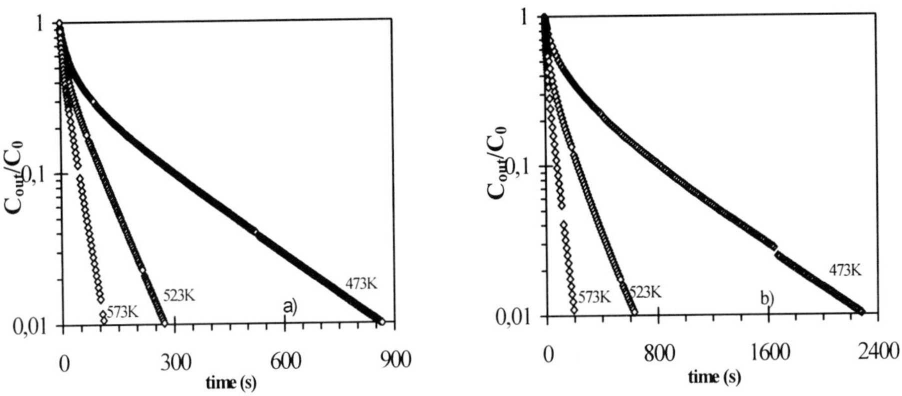


Figure 3. Effect of temperature on ZLC desorption curves in 5A zeolite pellets. a) System He/$nC_5$. b) System He/$nC_6$. Data of Silva and Rodrigues, [19], [20].

also important: if micropore diffusion is dominant, ZLC desorption curves should be independent of the kind of purge gas. To test the consistency of the ZLC model it is important to perform experiments with different purge flowrate: all diffusion parameters obtained should be the same at a given temperature. This strategy has been exploited for the study of diffusion of $nC_5$ and $nC_6$ in 5A zeolite pellets. Details of that study can be found elsewhere (Silva and Rodrigues, [19], [20]). Figure 3 shows the influence of temperature in ZLC desorption curves of the systems He/$nC_5$ and He/$nC_6$. It is clear a strong temperature dependence of desorption time.

The reciprocal of the apparent time constants for diffusion $D_p/R_p^2(1+K)$ of $nC_5$ and $nC_6$ (calculated from ZLC desorption curves) plotted *versus* 1/T are shown in Figure 4. A strongly temperature dependence of time constants exists which is of the order of heat of adsorption. Time constants range from 0.002 $s^{-1}$ at 473K up to 0.03 $s^{-1}$ at 573K in the system He/$nC_5$, and from 0.00035 $s^{-1}$ at 473K up to 0.0053 $s^{-1}$ at 573K in the system He/$nC_6$.

These data indicate that macropore diffusion is the controlling mechanism for diffusion of $nC_5$ and $nC_6$ in the adsorbent. Diffusion mechanisms in macropores are Knudsen diffusivity ($D_K$) in series with molecular diffusion ($D_m$), and so the pore diffusivity can be calculated by, $D_p=1/T_p(1/D_m+1/D_K)$, where $T_p$ is the tortuosity. Table 3 summarizes the values of experimental $\tau_{dif}=R_p^2/\varepsilon_p D_p$; experimental $D_p$, Knudsen $D_K$, molecular $D_m$ diffusivities and tortuosities $T_p$.

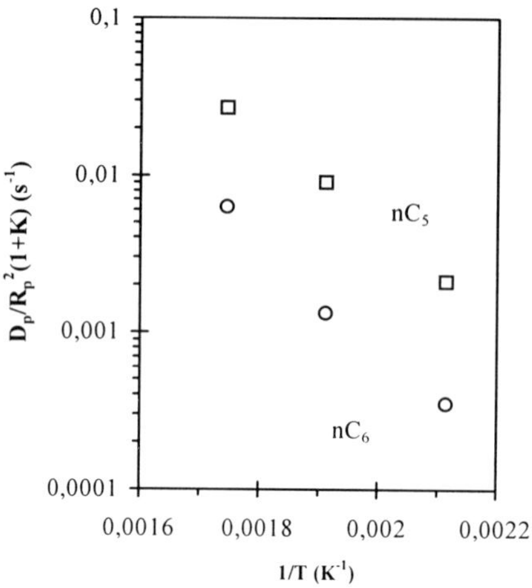


Figure 4. Reciprocal of the apparent time constant of diffusion $D_p/R_p^2(1+K)$ *versus* 1/T for the systems He/$nC_5$ and He/$nC_6$.

Table 3
Experimental pore diffusivities of $nC_5$ and $nC_6$ on 5A zeolite pellets and tortuosities

| T (K) | $\tau_{dif}$ (s) | $D_{p,exp}$ (cm$^2$/s) | $D_m$ (cm$^2$/s) | $D_K$ (cm$^2$/s) | $T_p$ |
|---|---|---|---|---|---|
| | | System He-$nC_5$ | | | |
| 573 | 0.13 | 0.14 | 1.0 | 0.27 | 1.5 |
| 523 | 0.15 | 0.12 | 0.87 | 0.26 | 1.6 |
| 473 | 0.19 | 0.10 | 0.75 | 0.25 | 1.8 |
| | | System He-$nC_6$ | | | |
| 573 | 0.14 | 0.13 | 0.95 | 0.25 | 1.5 |
| 523 | 0.15 | 0.12 | 0.82 | 0.24 | 1.5 |
| 473 | 0.16 | 0.11 | 0.69 | 0.23 | 1.6 |

## 4. FIXED BED ADSORPTION

The performance of fixed-bed adsorbers is governed by equilibrium, kinetics of mass transfer and hydrodynamics. The objective of this part of the work is the prediction of breakthrough curves of mixtures of n/iso-paraffins and propane /propylene in a fixed-bed adsorber including parameters independently measured. The mathematical model for the fixed bed adsorption process is based on the following assumptions:
1) The gas mixture behaves as an ideal gas.

2) Bed radial gradients in velocity, temperature and concentrations are negligible.
3) The temperature inside the pellet is uniform and the gas inside the pellet is in thermal equilibrium with the solid phase.
4) The energy accumulation of the gas phase within the pellet is negligible.
5) The superficial fluid velocity in the bed follows Ergun's relation locally.
6) The gas-solid adsorption equilibrium is represented with a loading ratio correlation or Nitta et al. model isotherms.
7) The linear driving force model (LDF) applies to both the crystal and the pellet.

Table 4
Mass, energy, momentum and equilibrium equations

| Name | Equation |
|---|---|
| Overall mass balance | $\varepsilon\frac{\partial C}{\partial t}=-\frac{\partial uC}{\partial z}-\sum_{i=1}^{n}N_i;N_i=\frac{(1-\varepsilon)(C_i-\bar{c}_i)}{R_{g,i}}$ <br> $R_{g,i}=\left(\frac{1}{aK_{m,i}}+\frac{R_p}{5\varepsilon_p aD_{p,i}}\right)$ |
| Component mass balance in the gas phase | $\varepsilon\frac{\partial C_i}{\partial t}=\frac{\partial}{\partial z}\left(\varepsilon D_{zm,i}C\frac{\partial Y_i}{\partial z}\right)-\frac{\partial uC_i}{\partial z}-N_i$ |
| Component mass balance in the pellet | $\frac{\partial \bar{c}_i}{\partial t}=\frac{15D_{p,i}}{R_p^2\left(1+\frac{5\varepsilon_p D_{p,i}}{R_p K_{m,i}}\right)}(C_i-\bar{c}_i)-\frac{\rho_p w_c}{\varepsilon_p}\frac{\partial \bar{\bar{n}}_i}{\partial t}$ |
| Component mass balance in the crystal | $\frac{\partial \bar{\bar{n}}_i}{\partial t}=\frac{15D_{c,i}}{r_c^2}\left(n_i^*-\bar{\bar{n}}_i\right)$ |
| Gas Energy Balance | $\varepsilon C\ \tilde{C}_V\frac{\partial T_g}{\partial t}=\frac{\partial}{\partial z}\left(\varepsilon D_{zT}\frac{\partial T_g}{\partial z}\right)-uC\ \tilde{C}_p\frac{\partial T_g}{\partial z}+\varepsilon\Re T_g\frac{\partial C}{\partial t}+$ <br> $\frac{2h_w}{R_w}(T_g-T_w)-(1-\varepsilon)ah_f(T_g-T_s)$ |
| Solid Energy Balance | $\rho_b C_{ps}\frac{\partial T_s}{\partial t}=\rho_b w_c\sum_{i=1}^{n}(-\Delta H_i)\frac{\partial \bar{\bar{n}}_i}{\partial t}+(1-\varepsilon)h_f a(T_g-T_s)$ |
| Wall Energy Balance | $\rho_w C_{pw}\frac{\partial T_w}{\partial t}=\alpha_w h_w(T_g-T_w)-\alpha_{ln}U(T_w-T_\infty)$ |
| Ergun´s Equation | $-\frac{\partial P}{\partial z}=\frac{150\mu(1-\varepsilon)^2}{\varepsilon^3 d_p^2}u+\frac{1.75(1-\varepsilon)\rho}{\varepsilon^3 d_p}\lvert u\rvert u$ |

Based on the above assumptions, the model equations are shown in Table 4. The mass balance equations at the pellet and crystal level are based in the double linear driving model equations or bidisperse model[30]. The solution of the set of parabolic partial differential equations showed in Table 4 was performed using the method of lines. The spatial coordinate was discretized using the method of orthogonal collocation in finite elements. For each element 2 internal collocation points were used and the basis polynomial were calculated using the shifted Jacobi polynomials with weighting function $W(x)=1\ (\alpha=0,\beta=0)$ that has equidistant roots inside each element [31]. The set of discretized ordinary differential equations are then solved with DASPK solver [32] which is based on backward differentiation formulas.

It has been shown (Silva and Rodrigues, [19], [20]) that macropore diffusion controls the mass transfer inside the adsorbent for adsorption of n-pentane and n-hexane. The mathematical model shown in Table 4 is used in the limit of macropore diffusion control in order to predict breakthrough curves of mixtures of n/iso-paraffins.

Figures 5 a) and b) show breakthrough curves of single $nC_5$ and $nC_6$ carried in a mixture of isopentane and nitrogen. The temperature history at two locations of the bed is also shown.

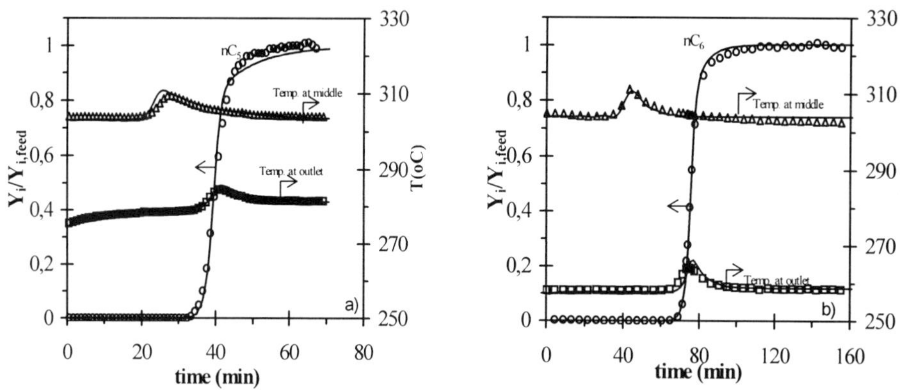


Figure 5. Breakthrough curves of single paraffins carries in a mixture of isopentane and nitrogen: feed mole fraction: a) $nC_5$ 19%; $iC_5$ 30% b) $nC_6$ 10%; $iC_5$ 40% in pellets of 5A zeolite. Temperature excursion at the middle and outlet of the bed is also shown. Points are experimental data. Lines are theoretical values predicted by the model shown in Table 4. Data of Silva and Rodrigues [33].

Experiments of multicomponent adsorption of $nC_5$ /$nC_6$ mixtures carried in nitrogen and isopentane were also performed. For the numerical simulation the extended multicomponent isotherm of Nitta et al.

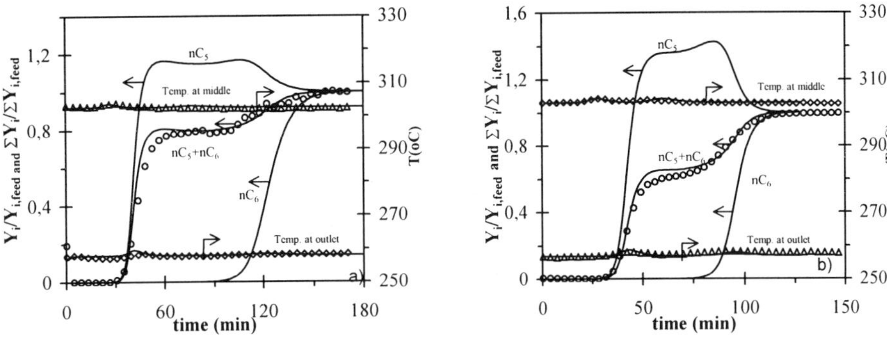


Figure 6. Multicomponent adsorption breakthrough curves of a mixtures of $nC_5/nC_6/N_2/iC_5$ in pellets of 5A zeolite. Feed mole fraction: a) $nC_5$ 2.6%; $nC_6$ 1.2% b) $nC_5$ 3.2%; $nC_6$ 2.7%. Points are experimental data. Lines are model results.

$$\theta_i = K_{eqi} P y_i \left[ 1 - \sum_i \theta_i \right]^{n_i} \tag{5}$$

was used, where $K_{eqi}$ and $n_i$ are pure component parameters. Figures 6 a) and b) show multicomponent breakthrough curves. The technique used to analyze the effluent stream of the column was the frontal chromatography. The prediction of the model is in good agreement with experimental data suggesting that the extended multicomponent adsorption model is representative of this adsorption system.

For the propane/propylene system the bidisperse model presented in Table 4 was tested and compared with the experimental breakthrough curve and temperature history given by Järvelin [22]. Figure 7 shows the breakthrough and outlet temperature predicted by the model compared with the experimental data available for the adsorption step and the predicted values for the desorption step with the hot purge gas (Feed temperature, 296 [K]; Feed pressure, 269 [kPa], Molar feed composition: propylene (1) 1.5%, nitrogen (2) 98.5%; total molar flow: 21.0 mol/m$^2$.s) (see Table 2 and 5 for model parameters). The predicted values compares reasonably with the experimental curves. The desorption step following the adsorption step was performed in countercurrent with a hot purge stream of nitrogen at 206 °C with the same total molar flow introduced in the adsorption step. Experimental data for this desorption step were not available; however, Järvelin [22] shows maximum values in an outlet normalized desorption run for propane near to 2.0-2.5 while here for the desorption of propylene the model predicts 1.51. The desorption experiments with propane reported by Järvelin [22] over 13X zeolite take between 60 to 80 minutes while the predicted desorption time for the propylene takes 100 minutes. Finally, the models predict final outlet

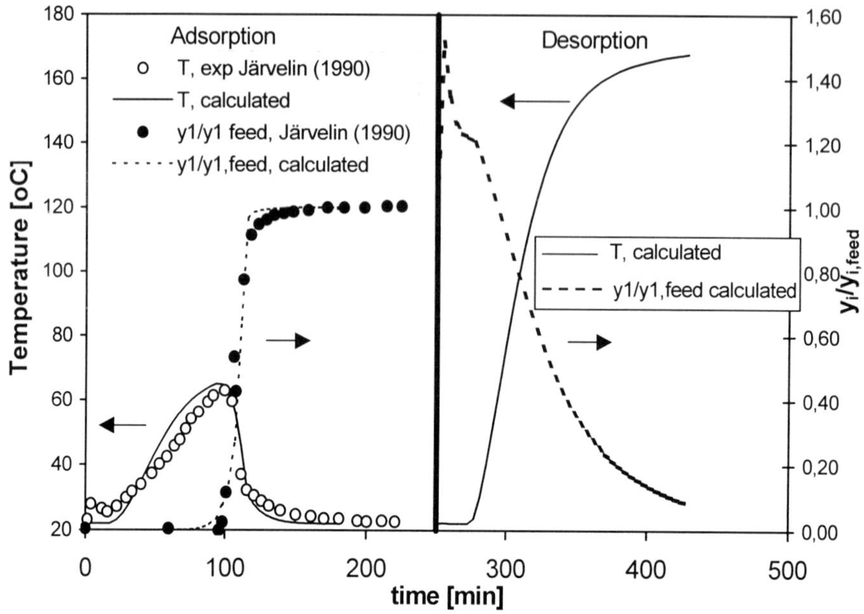


Figure 7. Propylene adsorption and desorption on 13X zeolite and temperature history. Points are experimental data from Järvelin [22]. Lines are model predictions.

Table 5
Fixed-bed parameters of breakthrough curves

| | |
|---|---|
| Bed length, $L = 1.22$ [m] | Specific solid heat, $C_{ps} = 920$ [J/kg K] |
| Column diameter, $D = 8.47$ [cm] | pellet radius, $R_p$=1.59 [mm] |
| Bed porosity, $\varepsilon = 0.395$ | superficial velocity, $u_o = 0.195$ [m/s] |
| Feed temperature, $T_o = 296$ [K] | $\alpha_w = 227$ cm$^{-1}$, $\alpha_{\ln} = 232$ [cm$^{-1}$] |
| Feed pressure, $P_o = 269$ [kPa] | gas viscosity, $\mu_o = 1.72\ 10^{-5}$ [kg/ms] |
| Pore radius, $r_p = 1700$ [Å] | gas thermal conduction, $k_{go} = 2.4410^{-2}$ [W/mK] |
| Crystal radius, $r_c = 1.0$ [μm] | wall heat film coefficient, $h_w = 34.3$ [W/m$^2$ K] |
| Pellet density, $\rho_p = 1140$ [kg/m$^3$] | overall heat coefficient, $U = 2.01$ [W/m$^2$ K] |
| Pellet void, $\varepsilon_p = 0.27$ | |

temperature near to 167°C while the feed temperature is 206°C revealing thermal losses through the wall. Järvelin [22] reports a temperature drop from 221°C to 133°C when testing the column with 4A molecular sieves using nitrogen as drying agent revealing that thermal losses of the actual column are near or higher than predicted with the model. Figure 8 shows the experimental breakthrough of propylene/propane and nitrogen over 13X and the simulated results for the adsorption feed step at similar operating conditions (Feed temperature, 296 [K]; Feed pressure, 269 [kPa], Molar feed composition: propylene (1) 1.161%, propane (2) 1.185% and nitrogen (3) 97.654 %; total molar flow: 22.5 mol/m$^2$.s). The model agrees with the experimental breakthrough given by Järvelin and Fair [21]. Figure 9 shows the simulated temperature and superficial velocity histories for the same breakthrough case where it is seen how the gas superficial velocity follows the gas temperature evolution while the wall temperature shows a small delay with respect to the gas temperature evolution.

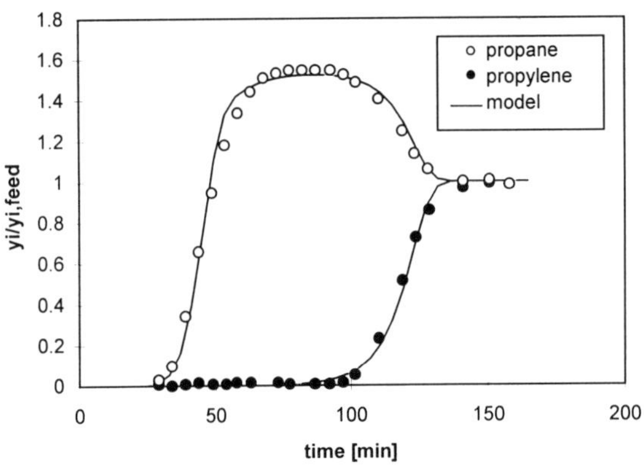


Figure 8. Multicomponent adsorption breakthrough of propylene/propane mixture on 13X zeolite. Points are experimental data from Järvelin and Fair [21]. Lines are model predictions.

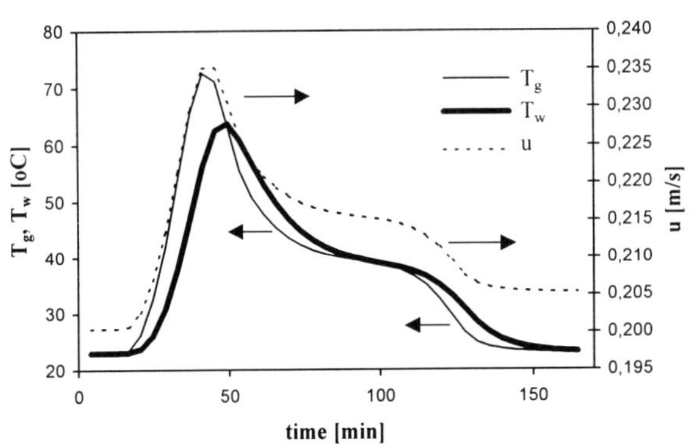


Figure 9. Gas and wall dimensionless temperature history and superficial dimensionless velocity at the outlet end during adsorption step for the propylene / propane /nitrogen system over 13X zeolite under non-isothermal non-adiabatical operating conditions.

The simulated curves obtained with the bidisperse model represents nearly well the breakthrough curve of the propane/propylene system over 13X zeolite and reproduces qualitatively the thermal history behavior of the fixed bed process.

## 5. CYCLIC SEPARATIONS

### 5.1. PSA separation for n/iso-paraffins mixtures

Since basic equilibria, kinetic and fixed-bed data of sorption of $nC_5$ and $nC_6$ in pellets of 5A zeolite were obtained, we are able to simulate a cyclic PSA process for the separation of n/iso-paraffins. The case selected is the patent data shown by Minkkinen et al. [3]. In such process, isomerisation of $C_5/C_6$ normal paraffins with recycling of normal paraffins is described. The recycling is performed in a selective adsorption containing 38Kg of 5A zeolite pellets. In the selective adsorption (lenght=4m; i.d=12.7cm) unit a PSA cycle takes place at 300°C. Adsorption phase occurs at a total pressure of 15 bars with a duration of 6 minutes. Desorption phase is performed in 6 minutes at 2 bars countercurrent to adsorption with a fraction of the $iC_5$ rich product. To obtain continuous operation two columns are used. The effluent of the isomerisation reactor contains approximately 13.9 mole % $nC_5$ and 4.6 mole % $nC_6$. The performance of the unit

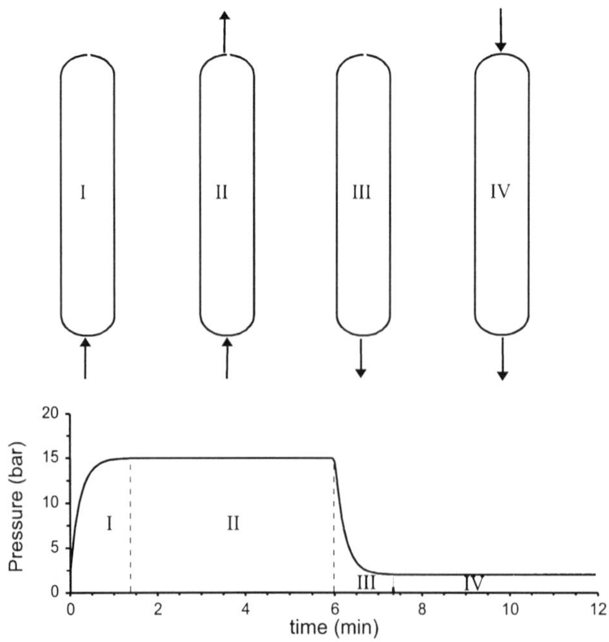


Figure 10. Schematic diagram of cycle used for the separation of n/iso-paraffins. I-pressurization; II- high-pressure feed; III-blowdown; IV- low-pressure purge.

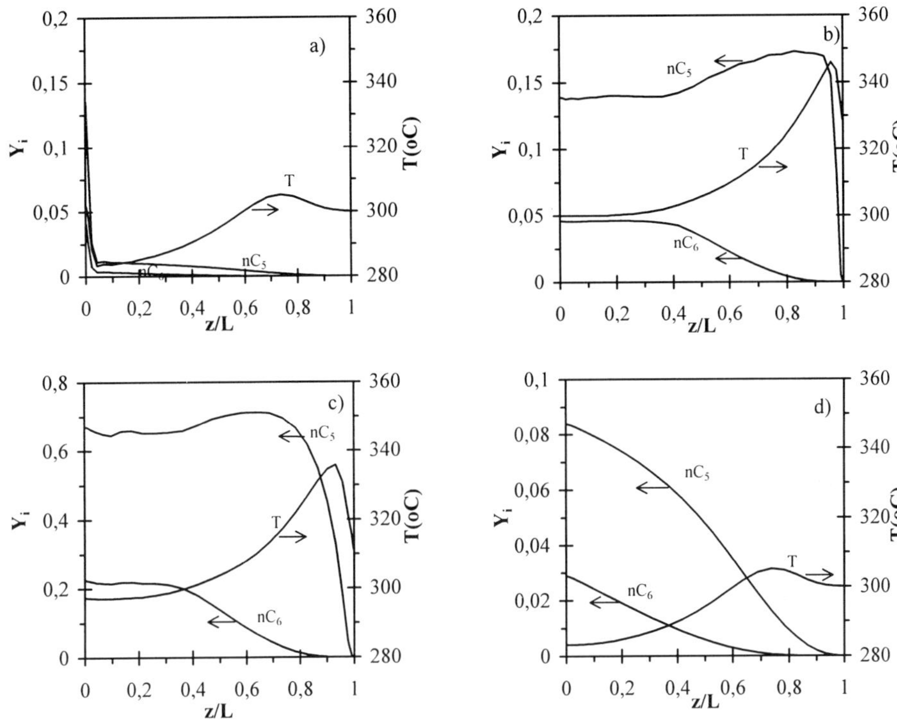


Figure 11. Gas phase concentration profiles of $nC_5$ and $nC_6$ in the bed at cyclic steady state at the end of: a) pressurization; b) high-pressure feed; c) blowdown; d) low-pressure purge step. Temperature profiles are also shown. Feed conditions are 84.6Kg/h of a mixture containing 13.9 % $nC_5$ and 4.6 % $nC_6$ mole fraction. The purge to feed ratio is 0.465. High-pressure feed is 15 bar. Low-pressure purge is at 2 bar.

is a RON product of 88 with less than 1 mole % or normal $C_5/C_6$. The desorption effluent recycled to the reactor contains approximately 27 mole % of $nC_5$ and 7.5 mole % $nC_6$.

To simulate the process we consider that adsorption phase acts in two steps: pressurization with feed plus high-pressure feed. Desorption is performed by blowdown plus contercurrent low-pressure purge step. As a base case we consider that the feed flowrate to the adsorption unit is the same of the isomerisation reactor: 84.6 Kg/h of a liquid containing a mixture of n/iso-paraffins. The $iC_5$ rich product flowrate used in desorption step is 31.8 Kg/h as mentioned by Minkkinen et al. The model used to simulate the process is the one shown in Table 4 with kinetic and equilibria data shown in this work. The PSA cycle simulation starts with the column clean of n-paraffins. A schematic diagram of the process is shown in Figure 10.

Figure 11 shows bed profiles of concentration in the gas phase of $nC_5$ and $nC_6$ at the cyclic steady-state (≈50 cycles) for the base case and at the end of the four steps considered, predicted by the model in adiabatic operating conditions. Temperature profiles are also shown.

Interesting features of bed dynamics are shown in Figure 11. In high-pressure feed (Figure 11b) $nC_5$ concentration wave is near the outlet of the column and it is very sharp. A roll-up phenomenon is clearly seen. A temperature wave exists in the column with a maximum peak of 50 °C relatively to feed inlet temperature. In the blowdown step (Figure 11c) it is clear the increasing concentration of n-paraffins in the bed. $nC_5$ and $nC_6$ mole fraction increases four times relatively to inlet conditions in the feed. At the end of purge step (Figure 11d) mole fractions of $nC_5$ and $nC_6$ are below the feed conditions; two temperature zones exist in the bed: one above feed conditions and other below.

In order to study the performance of the system a plot of purity and recovery of the iso rich fraction versus the purge to feed ratio is shown in Figure 12. Recovery is practically a linear function of purge to feed ratio. Purity decreases significantly when purge to feed ratio is lower than 0.46.

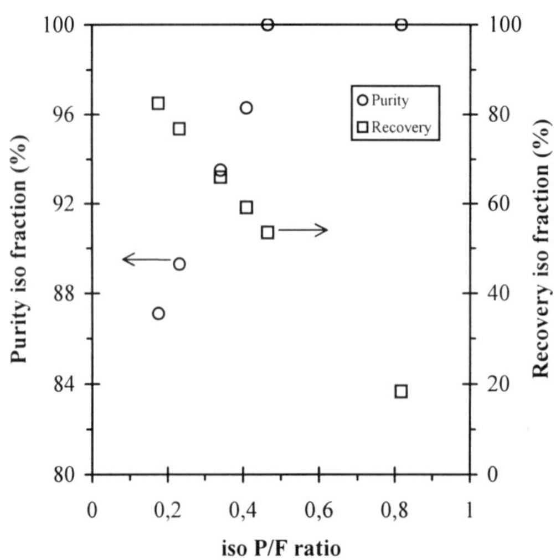


Figure 12. Recovery and purity *versus* the iso fraction purge to feed ratio for a PSA separation of n/iso-paraffins based on patent data of Minkkinen et al. [3]. Simulations are performed with equilibria, kinetic and a fixed bed model shown in this work.

### 5.2. TSA/VSA processes for propane/propylene separation

Various adsorption processes have been proposed for the separation of the propylene-propane system. Kulvaranon et al. [34], describe a variable-temperature stepwise desorption or VSTD process, that is mainly based in two process steps, an adsorption feed and a desorption step where a temperature increment of the adsorption column is used to regenerate the solid phase. From an equimolecular propylene-propane mixture and in one adsorption-desorption complete cycle process a propylene product with 85 mole % is obtained [34]. The main disadvantage is the requirement of longer time cycles than for a PSA process. Kumar et al.[5] suggest a VSA process, where the olefins are first separated from paraffins and the conventional distillation is used to obtain the final high quality product. Ghosh et al. [24] suggest to perform distillation first and then refine the propylene to the desired level of purity by adsorption. Järvelin and Fair [21] propose a three step separation process using an inert gas as a solvent: first, a propylene-propane mixture is diluted with nitrogen (molar concentration of both propane + propylene < 3%), this ternary mixture is separated by an adsorption step in two streams. Propane plus inert gas is obtained in the feed production step while in the desorption step propylene plus inert gas are obtained. Each enriched mixture is further submitted to a second adsorption process where the nitrogen gas is recovered and recycled and high purity streams of propylene and propane are obtained. However, in this TSA process a hot gas purge stream of nitrogen at 200°C is used for regeneration, requiring longer cycle time than for a typical VSA process. In this work two approaches are proposed to perform the propylene/propane separation: One is based on three step Temperature Swing Adsorption but using a lower regeneration temperature with inert gas and the other is based on the VSA concept. The TSA process is simulated with the bidisperse model described in Table 4 with parameters shown in Table 6. Figure 13a shows schematically the steps involved: I-Low temperature feed (25°C) with a equimolar mixture of propylene and propane diluted with nitrogen (10% propylene, 10% propane and

Table 6
Parameters and reference conditions for the TSA process

| | |
|---|---|
| Bed length, $L = 1.0$ [m] | Pellet radius, $R_p$=1.6 [mm] |
| Column diameter, $D = 8$ [cm] | Pellet density, $\rho_p = 1140$ [kg/m$^3$] |
| Bed porosity, $\varepsilon = 0.39$ | Pellet void, $\varepsilon_p = 0.27$ |
| Bulk density, $\rho_b$ 690 [kg/m$^3$] | Gas viscosity, $\mu_o = 1.45\ 10^{-5}$ [kg/(m s)] |
| Specific solid heat, $C_{ps}$ 920 [J/kg K] | Gas thermal conductivity, $k_{go} = 0.015$ [W/m K] |
| Reference Temperature, 298 [K] | Wall heat film coefficient, $h_w = 37$ [W/m$^2$ K] |
| Crystal diffusivities [27]: | Overall heat coefficient, $U = 1.5$ [W/m$^2$ K] |
| $D_{c,propylene} = 4.1\ 10^{-12}$ [m$^2$/s] | $\alpha_w = 488$ cm$^{-1}$, $\alpha_{ln} = 494$ [cm$^{-1}$] |
| $D_{c,propane} = 2.8\ 10^{-12}$ [m$^2$/s] | Reference Pressure, 101.3 [kPa] |

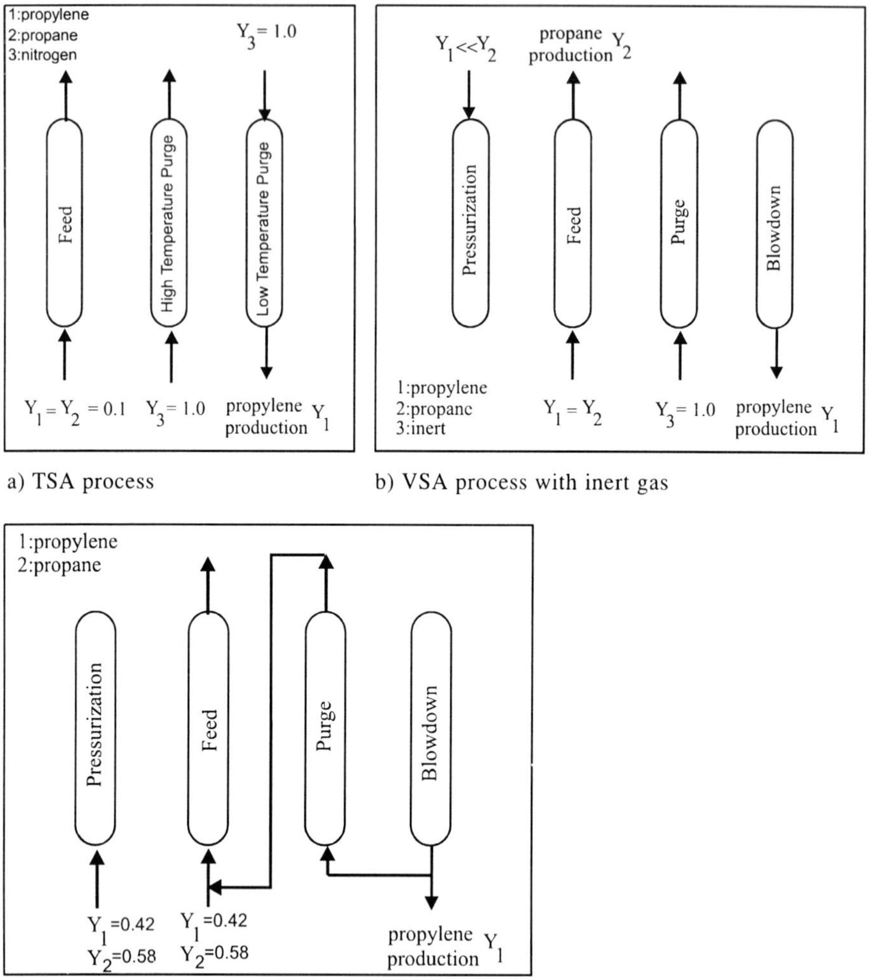


Figure 13. Separating propylene/propane mixture: a) TSA method; b) VSA process with inert gas; c) VSA process without inert gas.

80% of nitrogen); II- A high temperature (85°C) cocurrent purge step with nitrogen; III- A low temperature (25°C) countercurrent purge step with nitrogen, where the propylene product obtained is diluted. All steps are operated at 101 kPa of pressure and each step is during 16 minutes.

Figure 13b shows a VSA process for producing propane/inert and propylene/inert simultaneously: I- Pressurization: with a high purity propane stream at the exit of the column. This step increases the pressure in the system until the level required in the next step and prepares the exit where the propane rich stream is going to be produced; II- Feed: in high-pressure conditions an equimolar propylene-propane mixture plus nitrogen is introduced in the column. In this step the objective is to produce an enriched propane-nitrogen product while at the feed entrance the propylene is being accumulated mainly in the solid phase. This step continues until the mass transfer front of propylene is near the middle of the column. III- Purge: with a pure inert stream. The objective of this step is to push the propylene-propane gas mixture with the feed composition away from the feed entrance and also to reduce the partial pressure of the adsorbed components before the desorption step begins; IV- Blowdown: to a low vacuum pressure. In this step it is attempted to produce enriched propylene - nitrogen mixture and regenerate partially the solid phase where propylene is being desorbed mainly. A similar strategy has been proposed by Sircar [35] for the simultaneous production of nitrogen and oxygen from the air. For both the TSA process and the VSA with inert, after propylene enriched stream is obtained, it would be separated with another adsorption step to recover nitrogen to be recycled and to produce the final high purity propylene as suggested by Järvelin and Fair [21]; however, the regeneration of the solid is performed in the purge and blowdown steps without using hot stream inert gas regeneration step. The VSA proposed here is limited to the easiness of desorbing the propylene from the solid. Preliminary numerical simulations in isothermal conditions at 25°C revealed a strong difficulty to regenerate the solid with the last two steps following the strategy of Figure 13b [36]. The problem found is the irreversible behavior of the propylene adsorption isotherm at low temperatures over the 13X zeolite. Another alternative is shown in Figure 13c, based on the Resin-$Ag^+$ adsorbent proposed by Sikavitsas et al. [26]: I) Pressurization with feed mixture to 101 kPa; II ) High-pressure feed at 101 kPa where some of the purge stream is used as feed; III) Cocurrent-purge with product IV) and countercurrent blowdown to 5 kPa where the propylene product stream is obtained. With this alternative VSA, no inert gas is introduced and a high purity propylene stream is produced in the step IV. Also it is no longer required a recovery step for the inert as in the two previous cycle processes proposed.

Figure 14 compares the recovery vs purity curves for the TSA and the VSA process without inert gas. Table 6 and 7 show the main operating parameters for both processes while Table 2 shows the isotherm data. While the TSA process was simulated with the full bidisperse model in non-isothermal non-adiabatic conditions, the VSA was solved with the micropore control mass transfer model and in adiabatic conditions as suggested by Sikavitsas et al.[26]. For the TSA case, nearly 20 cycles are required until the cycle steady state is achieved while for the VSA more than 100 cycles are required. From Figure 14 it can be seen that both methods are able to predict a high purity propylene stream (>99.5%); however, in both process the recovery obtained is very low (between 2-14%). With the TSA alternative, a high purity propylene product relative to propane is obtained but it is diluted with inert gas (0.23% molar in nitrogen). On the other hand, the simulated VSA process has a long cycle time (40 minutes each step) in order to obtain a high quality propylene product as a consequence of high mass transfer resistance found for propylene and propane on the Resin-$Ag^+$. Sikavitsas et al. [26] obtained similar results using a particle diameter of 0.6 mm with recovery lower than 1%. However, 0.3 mm resin particles as introduced here have the drawback that large pressure drop is obtained (8 kPa with 0.6 mm sorbent while with 0.3 mm sorbent particles a pressure drop of 24 kPa is obtained on the adsorption step) limiting its industrial applicability when used for VSA operations.

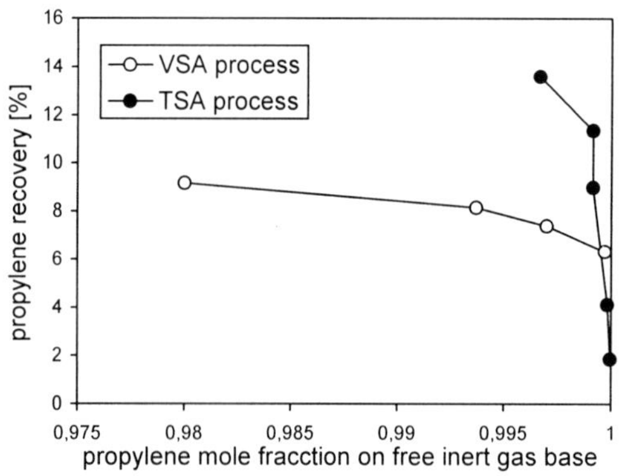


Figure 14. Simulated propylene recovery vs propylene mole fraction on free inert gas base for the TSA and the VSA without inert gas.

Table 7
Parameters and reference conditions for the VSA without inert process

| | |
|---|---|
| Bed length, $L = 3.5$ [m] | Gas viscosity, $\mu_o = 8.6\ 10^{-6}$ [kg/ms] |
| Column diameter, $D = 200$ [cm] | Gas thermal conductivity, $k_{go} = 2.16\ 10^{-2}$ [W/mK] |
| Bed porosity, $\varepsilon = 0.40$ | Wall heat film coefficient, $h_w = 0.0$ [W/m$^2$ K] |
| Bulk density, $\rho_b$ 670 [kg/m$^3$] | Overall heat coefficient, $U = 0.0$ [W/m$^2$ K] |
| Specific solid heat, $C_{ps}$ 1170 [J/kg K] | $\alpha_w = 499.5$ [cm$^{-1}$], $\alpha_{ln} = 499.75$ [cm$^{-1}$] |
| Reference Temperature, 298 K | Reference Pressure, 101.3 [kPa] |
| Diffusion time constants [26]: | Pellet radius, $R_p$=0.15 [mm] |
| Propylene, $15D_c/r_c^2$=1.2 $10^{-5}$ [s$^{-1}$] | Pellet density, $\rho_p = 670$ [kg/m$^3$] |
| Propane, $15D_c/r_c^2$ =4.0 $10^{-5}$ [s$^{-1}$] | Pellet void, $\varepsilon_p = 0.51$ |
| Superficial velocity, $u_o$=0.069[m/s] | |

## 6. CONCLUSIONS

Basic equilibrium and kinetic data were experimentally measured in view of adsorber design for the separation of n/iso-paraffins. Based on those data a fixed-bed model was set-up and experimentally validated for the separation of n/iso-paraffins mixtures. A cyclic PSA process was successfully simulated based on data from patent literature [3].

The fixed bed numerical simulation in non-isothermal non-adiabatic operating conditions also reproduces fairly well the experimental breakthrough of propane/propylene mixtures from the open literature[21,22]. Upon this model, simulated predictions for the propylene/propane system using two different approaches: a 3 step TSA process and a 4 step VSA without inert gas dilution establishes that it is possible to obtain a high purity propylene product from an equimolar mixture but with a low recovery using the commercial adsorbents available. The VSA schemes proposed here are able to perform the propylene-propane separation, having the advantage of shorter time cycles than heating regeneration process, but its success is based in the easiness of the solid regeneration in the blowdown/purge steps.

### Acknowlegments

Financial support from Junta Nacional de Investigação Científica e Tecnológica (research fellowships Praxis XXI/BD/3148/94 and Praxis XXI/BD/5772/95) and JOUE-CT92-0076 are gratefully acknowledged.

## REFERENCES

1. M. F. Symoniak, Hydrocarbon Processing, (May-1980) 110.
2. T. C. Holcombe, T. C. Sager; W. K. Volles and A. Zarchy, US Patent No. 4929799 (1990).
3. A. Minkkinen, L. Mank and J. Sophie, US Patent No. 5 233 120 (1993).
4. D.B. Manley and G.W. Swift, J. Chem. Eng. Data, 16(3) (1971) 301.
5. R. Kumar, T. C. Golden, T. R. White and A. Rokicki, Sep. Sci. and Technol., 27(15) (1992) 2157.
6. R. Eldridge and Bruce, Ind. Eng. Chem. Res., 32 (1993) 2208.
7. D. M. Ruthven, S. Farooq and K. S. Knaebel, Pressure Swing Adsorption, VCH Publishers, Inc., USA, 1994.
8. D. M. Ruthven, Principles of Adsorption and Adsorption Process, John Willey & Sons, New York, 1984.
9. R. T. Yang, Gas Separation by Adsorption Process, Butterworth, Stoneham, 1987.
10. I. Langmuir, J. Am. Chem. Soc. 17 (1918) 39.
11. T. Nitta, T. Shigetomi, M. Kuro-Oka and T. Katayama, J. Chem. Eng. Japan, 17 (1984) 39.

12. E. Ruckenstein, A. S. Vaidynathan and G. R. Youngquist, Chem. Eng. Sci., 26 (1971) 1305.
13. D. M. Ruthven and K. F. Loughlin, Chem Eng. Sci., 26 (1971) 1145.
14. J. Karger and D. M. Ruthven, Diffusion in Zeolites and Other Microporous Solids, John Willey & Sons, New York, 1992.
15. R. J. Hufton and D. M. Ruthven, Ind. Eng. Chem. Res., 32 (1993) 2379.
16. M. Eic and D. M. Ruthven, Zeolites, 8 (1988) 41.
17. Ph. Grenier, V. Bourdin, L. M. Sun and F. Meunier, AIChE J., 41 (1995) 2047.
18. A. E. Rodrigues, Modeling of percolation processes. In Percolation Processes; Theory and Applications; A.E. Rodrigues and D. Tondeur (eds.), Sijthoff & Noordhoff, Alphen aan den Rijn, The Netherlands, 31, 1981.
19. J. A. Silva and A. E. Rodrigues, Eng. Chem. Res., 36 (1997) 493.
20. J. A. Silva and A. E. Rodrigues, AIChE J, 43 (1997) 2524.
21. H. Järvelin and, J. R. Fair, Ind. Eng. Chem., 32 (1993) 2201.
22. H. Järvelin, Separations Research Program, The University of Texas at Austin, Austin, TX 78712, USA, (1990).
23. Y.-H. Huang, J. W. Johnson, A.I. Liapis and O.K. Crosser, Sep. Technol., 4 (1994) 156.
24. T. K. Ghosh, Lin Hon-Da and A. L. Hines, Ind. Eng. Chem. Res., 32 (1993) 2390.
25. W.K. Lewis, C. R. Gilliland, B. Chertow and W. H. Hoffman, AIChE J., 72 (1950) 1153.
26. V.I. Sikavitsas, R.T Yang, M.A. Burns and E.J. Langenmayr, Ind. Eng. Chem. Res., 34 (1995) 2873.
27. R. T. Yang and E. S. Kikkinides, AIChE J., 41(3) (1995).
28. S. Brandani, Chem. Eng. Sci., 51 (1996) 3283.
29. J. A. Silva and A. E. Rodrigues, Gas Separation & Purification, 10(4) (1996) 207.
30. P. L. Cen and R. T. Yang, Chem. Eng. Comm., 78 (1989) 139.
31. J. Villadsen and M. L. Michelsen, Solution of Differential Equation Models by Polynomial Approximation, Printice-Hall, Inc. Englewood Cliffs, N.J., 1978.
32. P.N. Brown, A.C. Hindmarsh and L.R. Petzold, SIAM, J. Sci. Comput., 15 (6) (1994)1467.
33. J. A. Silva and A. E. Rodrigues, Ind. Eng. Chem. Res., 36 (1997) 3769.
34. S. Kulvaranon, M. E. Findley and A.I. Liapis, Ind. Eng. Chem. Res., 29 (1990) 106.
35. S. Sircar and B.F. Hanley, Sep. Sci. & Technol., 28 (1993) 2553.
36. F. Da Silva, A. Macedo and A. E. Rodrigues, JOU2-CT93-0337, Final Report, January - July (1996).
37. S. Brandani, J. Hufton and D. Ruthven, Zeolites, 15 (1995) 624.

## Nomenclature

| | |
|---|---|
| a | specific pellet area, [$m^{-1}$] |
| C | total bulk molar concentration, [$mol/m^3$] |
| $\bar{c}_i$ | average molar concentration of "i" component in the pellet, $c\bar{y}_i$, [$mol/m^3$] |
| $C_i$ | molar concentration of "i" component in the bulk, $CY_i$, [$mol/m^3$] |
| $\tilde{C}_p$ | gas molar specific heat at constant pressure, [J/mol K] |
| $C_{ps}$ | specific heat of the pellet, [J/kg K] |
| $C_{pw}$ | wall specific heat, [J/kg K] |
| $\tilde{C}_v$ | gas molar specific heat at constant volume, [J/mol K] |
| $D_{c,i}$ | diffusion coefficient inside crystal for "i" component, [$m^2/s$] |
| $D_{o,i}$ | crystal diffusivity coefficient at zero coverage, [$m^2/s$] |
| $D_{p,i}$ | pore diffusion coefficient, [$m^2/s$] |
| $D_{zT}$ | axial dispersion heat coefficient, [W/m K] |
| $D_{zm,i}$ | dispersion axial mass coefficient for "i" component, [$m^2/s$] |
| $D_K$ | Knudsen diffusivity, [$m^2/s$] |
| $E_i$ | diffusional activation energy for "i" component, [J/mol] |
| $h_f$ | film transfer coefficient between gas and solid, [$W/m^2$ K] |
| $\Delta H_i$ | isosteric heat of adsorption of the "i" component, [J/mol] |
| $K_{eq}$ | equilibrium constant, [$kPa^{-1}$] |
| $K_{m,i}$ | external mass transfer coefficient for "i" component, [m/s] |
| n | coefficient of Nitta et al. model isotherm |
| $\bar{\bar{n}}_i$ | average adsorbed concentration. for "i" comp. in the pellet, [mol/kg] |
| $n_i^*$ | equilibrium concentration for "i" component, [mol/kg] |
| N | total molar flux , [$mol/m^3$ s] |
| $N_i$ | component molar flux, [$mol/m^3$ s] |
| p | partial pressure, [kPa] |
| P | gas pressure, [kPa] |
| q | amount adsorbed [$g/g_{ads}$] |
| $q_{max}$ | maximum amount adsorbed [$g/g_{ads}$] |
| $\Re$ | ideal gas constant (= 8.3144 [J/mol K]) |
| t | time, [s] |
| $T_g$ | gas temperature, [K] |
| $T_p$ | tortuosity |
| $T_s$ | solid temperature, [K] |
| $T_w$ | wall temperature, [K] |
| u | superficial gas velocity, [m/s] |
| $w_c$ | weight fraction of crystal in the pellet |
| $\bar{y}_i$ | averaged mol fraction for "i" component within the pellet |

| | |
|---|---|
| $Y_i$ | mole fraction of "i" component in the bulk |
| z | axial position, [m] |

greek symbols

| | |
|---|---|
| $\varepsilon$ | interparticle void fraction of bed |
| $\varepsilon_p$ | pellet void fraction |
| $\rho_p$ | pellet density, [kg/m$^3$] |
| $\rho_w$ | wall density, [kg/m$^3$] |
| $\mu$ | gas viscosity of the mixture,[kg/m s] |
| $\chi_i$ | equilibrium mixture parameter for "i" component |
| $\theta$ | coverage of the adsorbent ($=q/q_{max}$) |

Adsorption and its Applications in Industry and Environmental Protection
Studies in Surface Science and Catalysis, Vol. 120
A. Dabrowski (Editor)




# Fractionation of air by zeolites

S. Sircar, M. B. Rao and T. C. Golden

Air Products and Chemicals, Inc.
7201 Hamilton Boulevard, Allentown, PA 18195-1501, USA

## ABSTRACT

**Fractionation of air by selective adsorption of $N_2$ on zeolites has become a common industrial practice. Many different zeolites and air separation processes have been developed for this purpose.**

Pure gas isotherms for adsorption of $N_2$ and $O_2$ on five commercial zeolites (NaX, 5A, Na- Mordenite, CaX and CaLSX) at two different temperatures are reported. The isotherms can be described by the Langmuir model in the range of the data. Mixed gas Langmuir model is used to evaluate the relative $N_2$ adsorption and desorption characteristics for these zeolites in connection with air separation application by the pressure swing adsorption (PSA) concepts.

Nine different PSA processes for air separation using zeolites are reviewed and their process performances are compared. These processes can be designed to produce low (23–50 mole%) and medium purity (90–95 mole%) $O_2$-enriched air and high purity (98+ mole%) $N_2$-enriched air. **Processes can be tailor made to match the adsorptive properties of the zeolite for a given separation need or vice versa.**

## 1. INTRODUCTION

Production of oxygen enriched air containing 23-95mole% $O_2$ and nitrogen enriched air containing 98+ mole% $N_2$ from ambient air by pressure swing adsorption (PSA) processes have become a major unit operation in the chemical industries during the last thirty years. **A survey of Derwent Patent index on adsorptive air separation processes showed that 150 basic patents were granted to 60 corporations around the world between 1970 and 1992 [1].** A large number of these PSA processes use synthetic zeolites as the adsorbent for air fractionation.

**Barrer and Robbins first demonstrated in 1953 that $N_2$ can be selectively adsorbed over Ar from a mixture of these gases on natural**

**chabazite [2].** The $N_2$ is selectively adsorbed (thermodynamic) on polar adsorbents, like cation-containing zeolites, from mixtures with $O_2$ and Ar because of its larger quadrupole moment [$N_2$=1.52x$10^{-26}$esu.cm$^2$, $O_2$=0.30x$10^{-26}$esu.cm$^2$, Ar=0.0x$10^{-26}$esu.cm$^2$] compared to the other two gases [5,7]. On the other hand, there is practically no thermodynamic selectivity of adsorption between $O_2$ and Ar on most practical zeolites under normal conditions due to their weakly or non-polar nature and comparable polarizabilities. The pole-pole interactions between the $N_2$ molecule and the cations within the zeolite structure can, however, be very different depending on (a) the basic structure of the zeolite frame work, (b) the types of cations (single or multicomponent) in the frame-work, (c) the location, the accessibility, and the concentration of the cations, and (d) the presence of strongly polar trace molecules like water in the zeolite, etc. **Academic and industrial scientists are continuously searching for better zeolitic adsorbents for air separation [3].**

The adsorptive air separation processes using zeolites are based on the concept that (a) an oxygen-enriched product gas can be generated by flowing dry air over a packed column of the zeolite which selectively retains the $N_2$ from air (adsorption step) and, (b) a nitrogen-enriched product gas can be generated by removing the adsorbed gas from the zeolite column (desorption step). The adsorption step is carried out in a PSA process by flowing the air over the zeolite at a relatively higher $N_2$ partial pressure (total gas pressure is ambient or super-ambient) while the desorption step is carried out by reducing the partial pressure of $N_2$ over the zeolite (total gas pressure is ambient or sub-ambient). These two steps are carried out in a sequential cyclic fashion using a single or multiple columns of zeolite so that the adsorbent is repeatedly used.

**PSA processes for air separation differ by (a) the type of zeolite used, (b) the modes of execution of the adsorption and desorption steps and (c) the incorporation of a variety of complementary steps which improve separation performance [4].** The key process performance variables include (a) the amount of adsorbent required and the energy consumed to produce unit amount of the product gas, (b) rate of production of the product gas, and (c) purity, pressure and the recovery of the product gas (from the ambient air).

The processes are often designed to best utilize the adsorptive properties of a given zeolite so that the product specifications and separation performance can be met efficiently. Alternatively, zeolites are often tailor-made to obtain the best separation performance from a given PSA process. **The marriage between process and material provides immense flexibility in the design and operation of PSA processes for air fractionation.** That is also the likely reason for vast research and development efforts in this area. The purpose of this report is to give several examples of how different zeolites have been integrated with different PSA air separation process designs in order to produce a variety of $O_2$ and $N_2$ enriched product gases.

## 2. ZEOLITES FOR AIR SEPARATION

**The key adsorptive properties of the zeolites for air separation applications are (a) specific nitrogen adsorption capacity and selectivity of adsorption of $N_2$ over $O_2$ and Ar as functions of gas phase pressure, temperature and composition, (b) isosteric heats of adsorption of $N_2$, and $O_2$ as functions of adsorbate loadings, (c) desorption characteristics of $N_2$ and $O_2$ from the zeolite under various conditions of operation, and (d) kinetics of ad(de)sorption of $N_2$ and $O_2$ on the zeolite.**

Most practical zeolitic adsorbents are used in a pellet form (with or without binders) where a network of meso-macro pores provide the access of the gases to the adsorption sites (inside the micropores of crystalline zeolites). The zeolite crystal and the pellet radii are typically in the range of 0.5-2.0 μm and 0.5-2.0 mm, respectively. Consequently, the kinetics of ad(de)sorption of $N_2$ and $O_2$ are often controlled by the transport of these gases through the mesoporous network, and the ad(de)sorption kinetic (Knudsen, molecular and Poiseuille flow) time constants are large (>0.5 $seconds^{-1}$). Thus, the kinetics of ad(de)sorption processes may not be critical. The thermodynamic adsorptive properties (a,b) and the desorption characteristics (c) under local equilibrium conditions often determine the separation performance of a zeolite.

We consider three different zeolite frame-works (Type A, Type X and Mordenite) and two different cations in the same framework (NaX and CaX). Additionally, we consider a low silica (high alumina) variety of the X type zeolite (CaLSX). The type A zeolite contains a mixture of Na and Ca ions (5A) and the mordenite is in the Na form.

Pure $N_2$ and $O_2$ adsorption isotherms were measured on the zeolites in our laboratory at approximately 30 and 70°C and in the gas pressure range of 0-1.5 atmospheres except for CaLSX where the isotherms were measured at 23 and 45°C. A conventional volumetric adsorption apparatus was used for these measurements [5]. The experimental isotherms are given in Tables 1-5 for easy access. Figures 1 and 2 respectively show the pure $N_2$ and $O_2$ isotherms (amount adsorbed $n_i^0$ vs gas pressure P) on the zeolites at ~30°C. The isotherms for CaLSX were interpolated by using the Langmuir model for pure gas adsorption, which adequately described all isotherms in the range of the data. The solid lines in Figures 1 and 2 show the best fit of the data by the model which is given below:

$$n_i^0 = \frac{mbP}{1 + bP} \tag{1}$$

$$b_i = b_i^0 \exp\left[q_i^0/RT\right] \tag{2}$$

Table 1
Pure gas nitrogen and oxygen adsorption isotherms on NaX zeolite

| $N_2$ T=30.3°C | | $O_2$ T=30.2°C | |
|---|---|---|---|
| P, atm | n, mmol/g | P, atm | n, mmol/g |
| 0.1188 | 0.0303 | 0.1243 | 0.0114 |
| 0.1339 | 0.0340 | 0.5025 | 0.0440 |
| 0.4788 | 0.1195 | 1.1405 | 0.0993 |
| 1.1020 | 0.2614 | 1.2224 | 0.1049 |
| 1.1557 | 0.2894 | | |
| $N_2$ T=70.1°C | | $O_2$ T=69.9°C | |
| P, atm | n, mmol/g | P, atm | n, mmol/g |
| 0.0843 | 0.0099 | 0.1249 | 0.0064 |
| 0.3032 | 0.0379 | 0.2392 | 0.0109 |
| 0.3382 | 0.0404 | 0.5018 | 0.0244 |
| 0.3557 | 0.0437 | 1.1447 | 0.0550 |
| 0.8667 | 0.1048 | | |
| 1.2122 | 0.1306 | | |

Table 2
Pure gas nitrogen and oxygen adsorption isotherms on 5A zeolite

| $N_2$ T=29.8°C | | $O_2$ T=29.8°C | |
|---|---|---|---|
| P, atm | n, mmol/g | P, atm | n, mmol/g |
| 0.1800 | 0.0931 | 0.0888 | 0.0114 |
| 0.3151 | 0.1491 | 0.2420 | 0.0330 |
| 0.7125 | 0.2879 | 0.5039 | 0.0661 |
| 0.8647 | 0.3451 | 0.7679 | 0.0979 |
| 1.0888 | 0.3949 | 0.8175 | 0.0978 |
| | | 1.0313 | 0.1298 |
| $N_2$ T=71.8°C | | $O_2$ T=73.7°C | |
| P, atm | n, mmol/g | P, atm | n, mmol/g |
| 0.1272 | 0.0201 | 0.3707 | 0.0239 |
| 0.4742 | 0.0720 | 0.7175 | 0.0469 |
| 0.9978 | 0.1457 | 0.7653 | 0.0469 |
| 1.1822 | 0.1706 | 1.1657 | 0.0686 |

Table 3
Pure gas nitrogen and oxygen adsorption isotherms on Na-mordenite

| $N_2$ T=29.7°C | | $O_2$ T=29.7°C | |
|---|---|---|---|
| P, atm | n, mmol/g | P, atm | n, mmol/g |
| 0.1881 | 0.1512 | 0.5717 | 0.1138 |
| 0.4134 | 0.2942 | 0.6378 | 0.1255 |
| 0.4184 | 0.2940 | 0.6451 | 0.1383 |
| 0.4850 | 0.3169 | 1.4020 | 0.2553 |
| 0.5628 | 0.3666 | 1.4620 | 0.2639 |
| 0.8757 | 0.5087 | | |
| 1.1141 | 0.6000 | | |
| $N_2$ T=69.5°C | | $O_2$ T=69.5°C | |
| P, atm | n, mmol/g | P, atm | n, mmol/g |
| 0.5967 | 0.1411 | 0.5057 | 0.0415 |
| 0.7668 | 0.1770 | 0.8723 | 0.0746 |
| 1.4558 | 0.2991 | 1.5054 | 0.1239 |
| 3.7347 | 0.6093 | 3.5442 | 0.2811 |

Table 4
Pure gas nitrogen and oxygen adsorption isotherms on CaX zeolite

| $N_2$ T=30.0°C | | $O_2$ T=30.1°C | |
|---|---|---|---|
| P, atm | n, mmol/g | P, atm | n, mmol/g |
| 0.0434 | 0.1410 | 0.0290 | 0.0082 |
| 0.1720 | 0.3950 | 0.0870 | 0.0290 |
| 0.4590 | 0.6650 | 0.2910 | 0.0850 |
| 1.0210 | 0.9030 | 0.7510 | 0.2050 |
| $N_2$ T=70.8°C | | $O_2$ T=71.6°C | |
| P, atm | n, mmol/g | P, atm | n, mmol/g |
| 0.0726 | 0.0590 | 0.0300 | 0.0042 |
| 0.2350 | 0.1700 | 0.0795 | 0.0113 |
| 0.5730 | 0.3450 | 0.3050 | 0.0399 |
| 1.0960 | 0.5220 | 0.7850 | 0.1030 |

Table 5
Pure gas nitrogen and oxygen adsorption isotherms on CaLSX zeolite

| $N_2$ T=23°C | | $O_2$ T=23°C | |
|---|---|---|---|
| P, atm | n, mmol/g | P, atm | n, mmol/g |
| 0.0343 | 0.171 | 0.184 | 0.067 |
| 0.0790 | 0.343 | 0.364 | 0.128 |
| 0.1370 | 0.524 | 0.551 | 0.187 |
| 0.4100 | 0.950 | 0.742 | 0.246 |
| 0.6820 | 1.141 | 0.930 | 0.298 |
| 1.0220 | 1.282 | | |
| **$N_2$ T=45°C** | | **$O_2$ T=45°C** | |
| P, atm | n, mmol/g | P, atm | n, mmol/g |
| 0.068 | 0.151 | 0.211 | 0.047 |
| 0.362 | 0.558 | 0.412 | 0.094 |
| 0.489 | 0.682 | 0.833 | 0.175 |
| 0.769 | 0.879 | 1.054 | 0.213 |
| 1.121 | 1.053 | | |

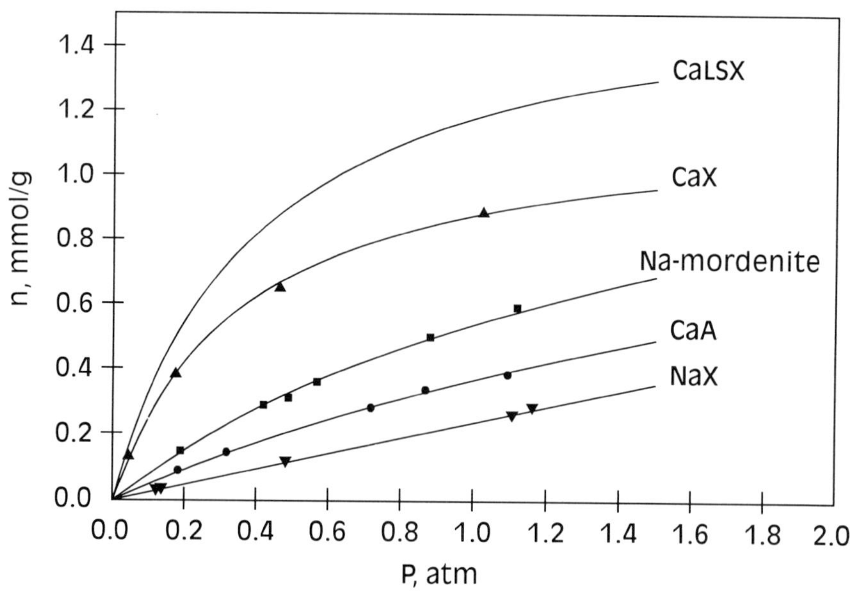


Figure 1. Isotherms for adsorption of pure $N_2$ on various zeolites at ~ 30°C.

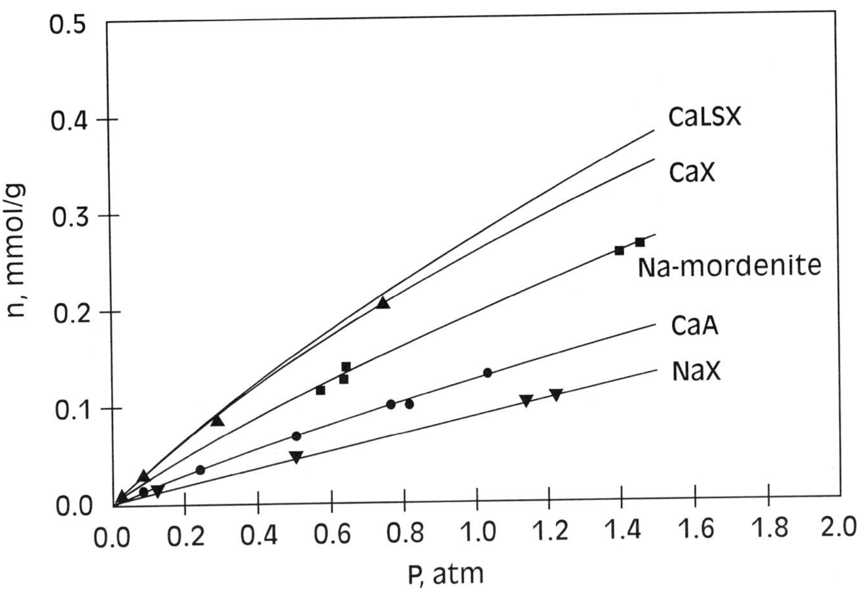


Figure 2. Isotherms for adsorption of pure $O_2$ on various zeolites at ~ 30°C.

$n_i^0$ is the specific amount (moles/g) of a pure gas i adsorbed on the zeolite at a gas pressure of P (atmospheres), m (moles/g) is the saturation adsorption capacity of both $N_2$ and $O_2$ on the zeolites. The variable $b_i$ (atmosphere$^{-1}$) is the corresponding Langmuirian pure gas-solid interaction parameter. The temperature dependence of $b_i$ is given by Equation (2) where $q_i^0$ (Kcal/mole) is the isosteric heat of adsorption of the pure gas i, and $b_i^0$ (atmosphere$^{-1}$) is a constant. T(K) is the system temperature and R is the gas constant.

The ability of the Langmuir model to fit the pure gas isotherms indicates that the zeolites are energetically homogenous [$q_i^0$ constant] for adsorption of both $N_2$ (component 1) and $O_2$ (component 2) in the range of the data. The Langmuirian selectivities of adsorption between $N_2$ and $O_2$ ($S_{12}$) by the zeolites at any temperature (T) are given by ($b_1/b_2$). The variable $S_{12}$ is a function of temperature only, $\left(S_{12} = S_{12}^0 \exp\left\{\left(q_1^0 - q_2^0\right)/RT\right\}\right)$, where $S_{12}^0$ is given by $\left(b_1^0 / b_2^0\right)$. Table 6 reports the Langmuir parameters for adsorption of $O_2$ and $N_2$ by the zeolites.

Table 6
Adsorption characteristics of pure $N_2$ and $O_2$ and their mixtures on zeolites.

| | Pure Gas Langmuir Parameters | | | | | | Binary Gas Adsorption | | |
|---|---|---|---|---|---|---|---|---|---|
| | $N_2$ | | | $O_2$ | | | at P=1.0atm, T=30°C, $y_1$, =0.79 | | |
| Zeolites | $b_i^0$ (atm$^{-1}$) | $q_i^0$ (kcal/mole) | m (mmoles/g) | $b_i^0$ (atm$^{-1}$) | $q_i^0$ (kcal/mole) | M (mmoles/g) | $n_1$ (mmoles/g) | $n_2$ (mmoles/g) | $S_{12}$ |
| NaX(13X) | 7.54x10$^{-5}$ | 4.3 | 3.12 | 15.9x10$^{-5}$ | 3.1 | 3.12 | 0.197 | 0.018 | 2.9 |
| Ca-Na A(5A) | 2.87x10$^{-5}$ | 5.7 | 1.41 | 24.3x10$^{-5}$ | 3.6 | 1.41 | 0.311 | 0.022 | 3.8 |
| Na-Mordenite | 2.26x10$^{-5}$ | 6.2 | 1.47 | 10.9x10$^{-5}$ | 4.4 | 1.47 | 0.462 | 0.031 | 4.0 |
| CaX | 2.28x10$^{-5}$ | 7.1 | 1.20 | 27.2x10$^{-5}$ | 4.2 | 1.20 | 0.821 | 0.020 | 10.9 |
| CaLSX | 2.57x10$^{-5}$ | 6.9 | 1.66 | 15.7x10$^{-5}$ | 4.3 | 1.66 | 1.087 | 0.023 | 12.7 |

The mixed gas Langmuir model can be used to calculate the amounts of component $i$ adsorbed $(n_i)$ from their mixtures at a total gas pressure (P), temperature (T), and a gas phase mole fraction $y_i$ for component $i$:

$$n_i = \frac{mb_iPy_i}{1+\sum_i b_iPy_i} \qquad i = 1,2 \tag{3}$$

Table 6 shows the calculated amounts of $N_2$ and $O_2$ adsorbed on the zeolites from synthetic air ($y_1$=0.79, $y_2$=0.21) at a gas pressure of 1.0 atmosphere and at 30°C. It may be seen that the $N_2$ adsorption capacity from air increases in the order CaLSX>CaX>Na-Mordenite>5A>NaX. There is about five fold difference in the $N_2$ adsorption capacities between NaX and CaSLX zeolites.

The $O_2$ adsorption capacities from air also increases on these zeolites in the same order but the differences are relatively smaller. This results in about four-fold increase in the selectivity of adsorption of $N_2$ over $O_2$ between NaX and CaLSX zeolites.

The strength of adsorption of $N_2$ on these zeolites follow the same order as indicated by the $N_2$ isosteric heats of adsorption. $N_2$ is more strongly adsorbed on the zeolites as its capacity and selectivity of adsorption over $O_2$ increases. Consequently, the difficulty of desorption of $N_2$ from these zeolites decreases in the order NaX<5A<Na-Mordenite<CaX<CaLSX.

We estimated the $N_2$ desorption characteristics from these zeolites under two idealized but common concepts of operation of PSA processes. They are (a) isothermal evacuation of an adsorbent column which is initially equilibrated with a binary gas mixture of $N_2$ ($y_1$=0.79) and $O_2$ ($y_2$=0.21), and (b) isothermal and isobaric desorption of pure $N_2$ from an adsorbent column by flowing a stream of pure $O_2$ (called purging) through the column. The adsorbers are initially at a pressure of one atmosphere and at a temperature of 30°C in both cases. Analytical model solutions are available for the above described desorption processes when they are carried out under local equilibrium conditions and when the adsorbates follow Langmuir isotherms [1,6].

Figure 3 shows the fraction of $N_2$ removed from the column (ratio of the amount of $N_2$ removed to the total amount of $N_2$ initially present in the column as adsorbed and void gases) as a function of column pressure *(P)* during the evacuation process. It is much easier to remove the $N_2$ from the column by evacuation when it is held less tightly (e.g. NaX). A much deeper vacuum is necessary to remove a substantial quantity of $N_2$ from the zeolites when its selectivity of adsorption over $O_2$ is large (e.g. CaX, CaLSX).

Figure 4 shows the mole fraction of $N_2$ in the column effluent gas during the above described evacuation process as a function of fraction desorbed (between pressure levels of 1.0-0.10 atmospheres). It also gives the specific quantities of the total desorbed gases from the zeolites and their average $N_2$ compositions.

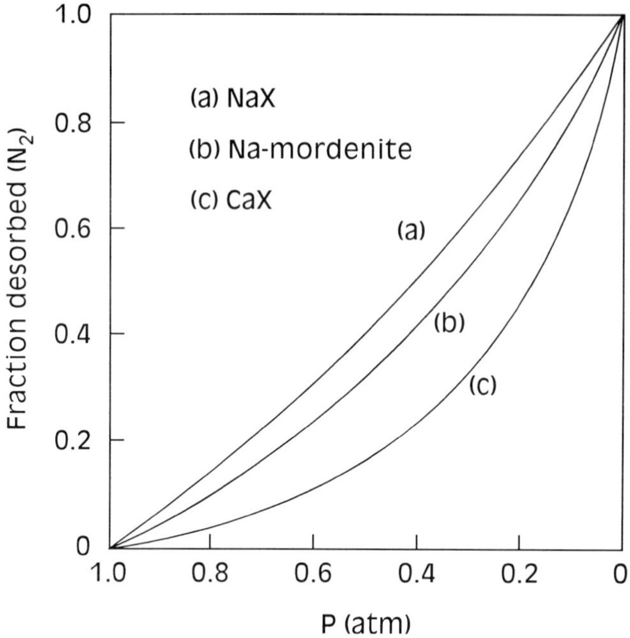


Figure 3. Fraction of $N_2$ desorbed by isothermal evacuation of a zeolite column saturated with air at 30°C as function of column pressure (P).

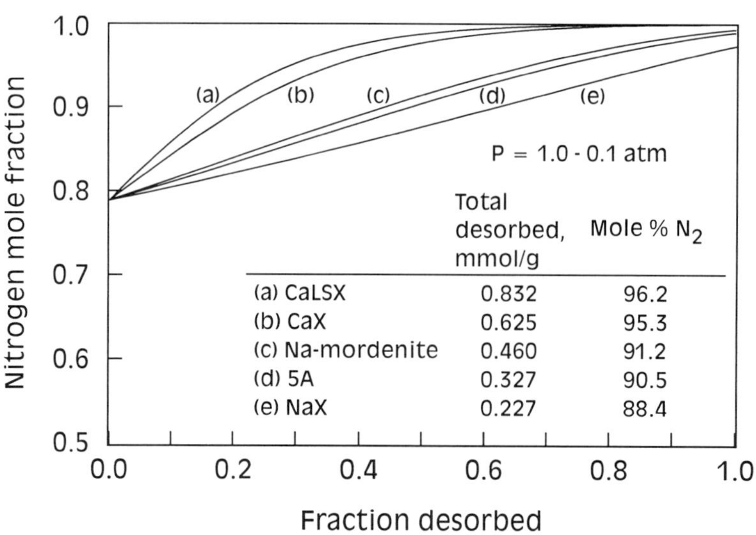


Figure 4. $N_2$ mole fraction of column effluent gas as function of total fractional amount desorbed for evacuation process of Figure 3.

The $N_2$ mole fraction of the effluent gas increases from its initial value of 0.79 as the column pressure is reduced and it eventually approaches unity. The zeolites with larger $N_2$ selectivities produce a substantial quantity of essentially pure $N_2$ at the lower pressure region of the evacuation process and the average $N_2$ composition of the desorbed gases are higher for these zeolites. However, it follows from Figures (3) and (4) that a relatively large amount of $N_2$ remains in the column packed with highly $N_2$ selective zeolites even after most of the $O_2$ molecules are evacuated out of the column, thereby reducing the $O_2$ recovery by a PSA process for production of $O_2$ enriched gas. This loss of $O_2$ per unit amount of $N_2$ desorbed increases as the selectivity of adsorption of $N_2$ by the zeolite increases.

Figure 5 shows the fraction of $N_2$ desorbed (f) from a column saturated with pure $N_2$ by purging with pure $O_2$ as a function of the specific amount of $O_2$ introduced (Q) into the column. It shows that the amount of $O_2$ purge gas needed to desorb a given fraction of $N_2$ from the column increases substantially when the $N_2$ is more selectively adsorbed over $O_2$. Most of the purge $O_2$ is lost with the desorbed $N_2$ during this process. Thus, like evacuation, the $O_2$ loss per unit amount of $N_2$ removed by purge also increases when the zeolite is more selective towards $N_2$.

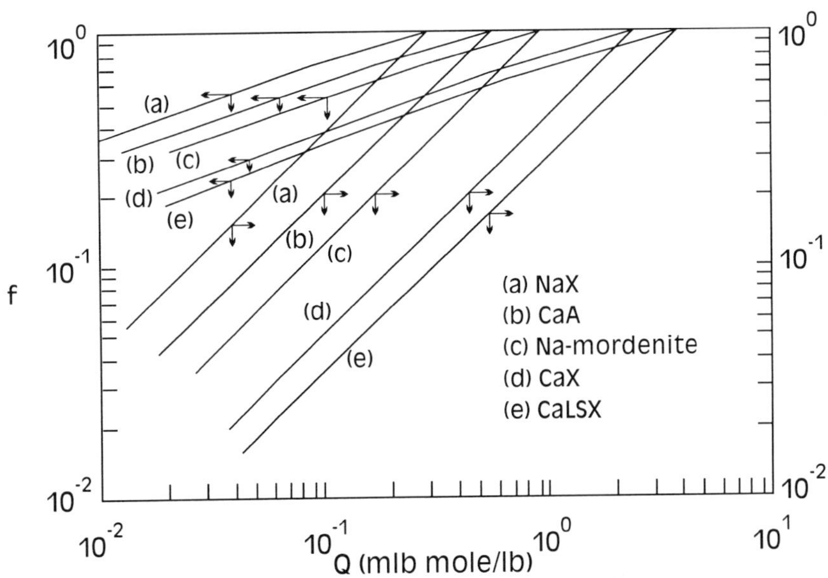


Figure 5. Fraction of $N_2$ desorbed by isothermal $O_2$ purge of zeolite columns saturated with $N_2$ at 30°C.

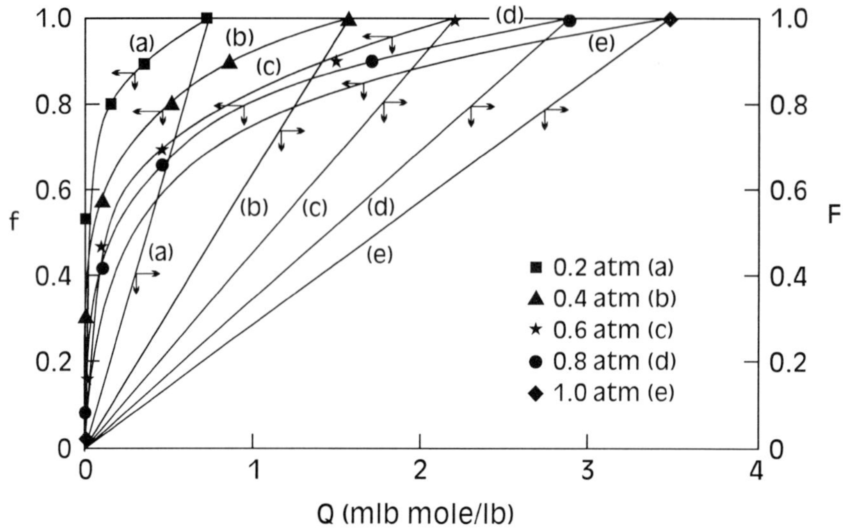


Figure 6. Fraction of $N_2$ desorbed by isothermal $O_2$ purge (under vacuum) of CaLSX zeolite columns saturated with $N_2$ at 30°C.

Figure 5 also shows the fractions of the column volume (F) at the purge $O_2$ inlet-end which is completely freed from $N_2$ during the purge process. The value of the variable F is much lower than that for variable f for a given quantity of $O_2$ purge and it decreases drastically as the $N_2$ selectivity of the zeolite increases. Thus, a larger quantity of $O_2$ purge (lower $O_2$ recovery) will be required to produce higher purity $O_2$ product using a zeolite with larger $N_2$ selectivity.

It is possible to decrease the purge $O_2$ quantity in order to obtain more efficient $N_2$ desorption (higher values of F and f) by carrying out the purge step at a lower column pressure. The column may be evacuated to a sub-atmospheric pressure level and then purged with $O_2$ at that pressure. Figure 6 shows such $N_2$ desorption characteristics [f and F as functions of Q] for CaLSX at different sub-atmospheric purge pressure levels. The initial column pressure is atmospheric and some $N_2$ is desorbed (finite f at Q=0) during the initial pressure reduction process before the $O_2$ purge begins. It may be seen by comparing Figures 5 and 6 that substantial reduction in specific $O_2$ purge gas quantity can be achieved by purging the column at a reduced pressure in order to obtain the same degrees of cleaning [f and F]. The quantity of purge $O_2$, however, still increases in the order CaLSX>CaX>Na-Mordenite>5A>NaX.

The above-described (idealized and simplified) effects of zeolite $N_2$ capacity and selectivity on the adsorption and desorption processes for air separation clearly demonstrate that the selection of the optimum zeolite for a given PSA process or designing an optimum process for a given zeolite can be fairly complex.

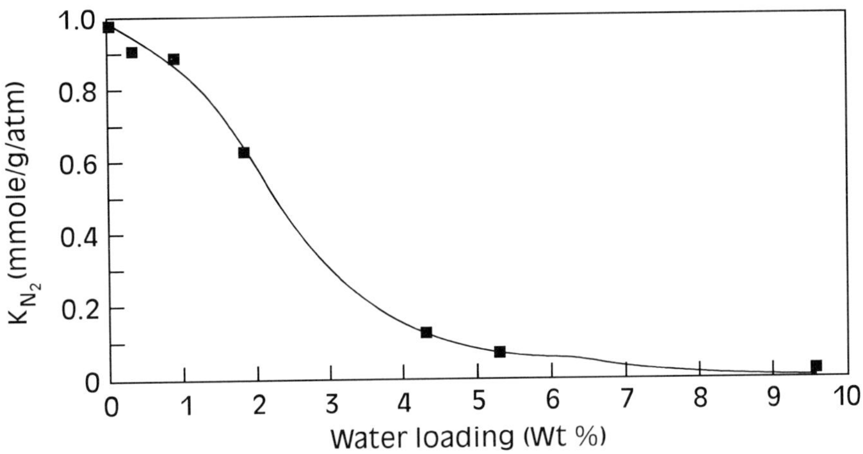


Figure 7. Henry's law constants for adsorption of pure $N_2$ on Na-mordenite at 30°C as functions of water loadings on the adsorbent.

**Adsorptive properties that help the adsorption step of the process inhibit the desorption step.** The real processes are further complicated by non-isothermal operation, non-isobaric process steps, adsorption kinetics, gas channeling and maldistribution, etc. **It is often necessary to experimentally evaluate the performance of the zeolite-process combination in a pilot scale unit before the optimum process design and adsorbent selection can be made.**

**It should be mentioned here that the zeolite regeneration is a critical issue in determining the process performance. A small amount of a polar adsorbate like water can significantly reduce its thermodynamic and kinetic adsorptive properties for air separation [7].** An example of this effect is depicted by Figure 7. It plots the Henry's law constants ($K_{N_2}$) for adsorption of pure $N_2$ at 30°C on Na-Mordenite as functions of water loadings (weight %) on the zeolite. $K_{N_2}$ (moles/g/atm) is the slope of the adsorption isotherm at the limit of zero pressure. The Figure shows that $K_{N_2}$ drastically decreases when the water loading on the adsorbent exceeds 1.0%. These data were measured in our laboratory. The effect is more pronounced when the cations in the zeolites have larger charge densities (Li~Ca>Na).

**Consequently, water must be removed from the ambient air before the zeolite can carry out the $N_2$-$O_2$ separation.** This is typically done by using a water selective desiccant like activated alumina or NaX zeolite for selectively removing the water from air. The desiccant generally forms an integral part of the zeolite PSA process for air separation (a layer of the desiccant followed by the zeolite layer in the same column or in two separate columns). The ratio of the amount of desiccant to zeolite used is typically between 0.2-0.5. Cyclic desorption of water from the desiccant is a key requirement for the practical PSA process. It is generally achieved by column pressure reduction and purging the desiccant with the desorbed gases from the zeolite layer for air separation. **The conditions of operation of the air separation PSA process may often be controlled by the desorption characteristics of water from the desiccant. This operational limitation is not always recognized.**

## 3. AIR SEPARATION PROCESSES USING ZEOLITES

We review several PSA processes for production of oxygen-enriched and nitrogen-enriched product gases using zeolites as adsorbents. Comparative process performances will be given when appropriate.

### 3.1. Production of $O_2$-enriched air

The zeolitic air separation processes for production of $O_2$-enriched air can be classified into two groups. They include (a) processes for production of medium purity (90-95%) $O_2$ and (b) processes for production of low purity (23-50%) $O_2$.

### 3.2. Medium purity $O_2$ product

**Process A:**

This process is patented by Union Carbide Corporation [8,9]. It is one of the earlier commercial processes for production of ~90% $O_2$. The adsorption step is carried out at a super-ambient pressure level (3-4 atmospheres) and the desorption steps are carried out at near-ambient pressure. There are six cyclic steps in the process as follows:

(a) Pressurization of the adsorbent column from near-ambient pressure level to an intermediate super-ambient pressure level ($P_I$) by simultaneously introducing compressed air through the feed end and $O_2$-enriched product gas through the product end.

(b) Further pressurization of the adsorber to the highest pressure level ($P_A$) of the cycle by introducing compressed air alone through the feed end.

(c) Depressurization of the adsorber by withdrawing an $O_2$-enriched gas through the product end (co-current). A part of this gas is used to pressurize another adsorber undergoing step (a) and the balance is withdrawn as product gas.

(d) Further co-current depressurization of the adsorber by withdrawing some more $O_2$-enriched gas through the product end. A part of this gas is withdrawn as the product and the balance is used as counter-current purge gas (Step f).

(e) Final depressurization of the adsorber to near-ambient pressure level by withdrawing $N_2$-enriched gas through the feed air end (countercurrent). This gas contains the desorbed $H_2O$, $CO_2$, $N_2$ and some $O_2$ and it is wasted.

(f) Counter-current purging of the adsorber using an $O_2$-enriched gas from step (d) to further desorb the $N_2$, $CO_2$ and $H_2O$. The effluent from this step is also wasted.

(g) Repeat cycle from step (a).

An oxygen-enriched product gas containing ~90% $O_2$ can be produced at near-ambient pressure by this process. Table 7 describes the performance of this process using 5A zeolite and an adsorption pressure (step b) of 3.0 atmosphere [9]. The waste gases (steps e and f) contain about 85.7% $N_2$. The process requires three adsorbers to obtain continuous flow of feed air and product gas and continuous operation of the compressor. Figure 8a shows a schematic diagram of the process flow sheet.

**Process B:**

This process is patented by Praxair Technology, Inc. [10]. It is one of the more recent commercial processes for production of ~90% $O_2$. The adsorption step is carried out at a super-ambient pressure level (~1.5 atmosphere) and the final desorption pressure is sub-atmospheric (~0.34 atmosphere). This process also has six cycle steps (as described below) but they are very different from those of Process A:

(a) The adsorber is pressurized to the highest pressure level ($P_A$) of the cycle from an intermediate sub-atmospheric pressure level ($P_I$) by introducing compressed air through the feed end.

(b) Continuing compressed air feed flow through the adsorber at the highest pressure level of the cycle while producing an $O_2$-enriched product gas at nearly feed gas pressure. A part of this gas is withdrawn as product, a part is used as the purge gas in step (e), and the balance is used as pressurization gas in step (f).

(c) Counter-current depressurization of the adsorber to near-ambient pressure level. The effluent desorbed gases are wasted.

(d) Counter-current evacuation of the adsorber to the lowest sub-atmospheric pressure level of the cycle ($P_D$). The effluent desorbed gases are wasted.

(e) Counter-current purging of the adsorber at the lowest sub-atmospheric pressure level by introducing a portion of the $O_2$-enriched product gas from step (b) through the product end. The effluent gas is wasted.

(f) Counter-current pressurization of the adsorber to an intermediate sub-atmospheric pressure level ($P_I$) by introducing a part of the $O_2$-enriched product gas from step (b).

(g) Repeat cycle from step (a).

Table 7
PSA process performance for production of medium purity oxygen enriched gas

| Process | Adsorbent | Operating Pressures: Adsorption (atm) | Operating Pressures: Desorption (atm) | Oxygen Product: Purity (Mole%) | Oxygen Product: Recovery (%) | Oxygen Product: Productivity (Milli Pound Moles/lb of zeolite/cycle) | Oxygen Product: Pressure (atm) |
|---|---|---|---|---|---|---|---|
| A | 5A | 3.00 | 1.00 | 90.0% | 38.0% | 0.018 | ~1-1.2 |
| B | LiX | 1.43 | 0.34 | 90.0% | 53.0% | 0.061 | ~1.4 |
| C | CaX | 3.00 | 0.10 | 95.5% | 34.0% | 0.094 | ~3.0 |

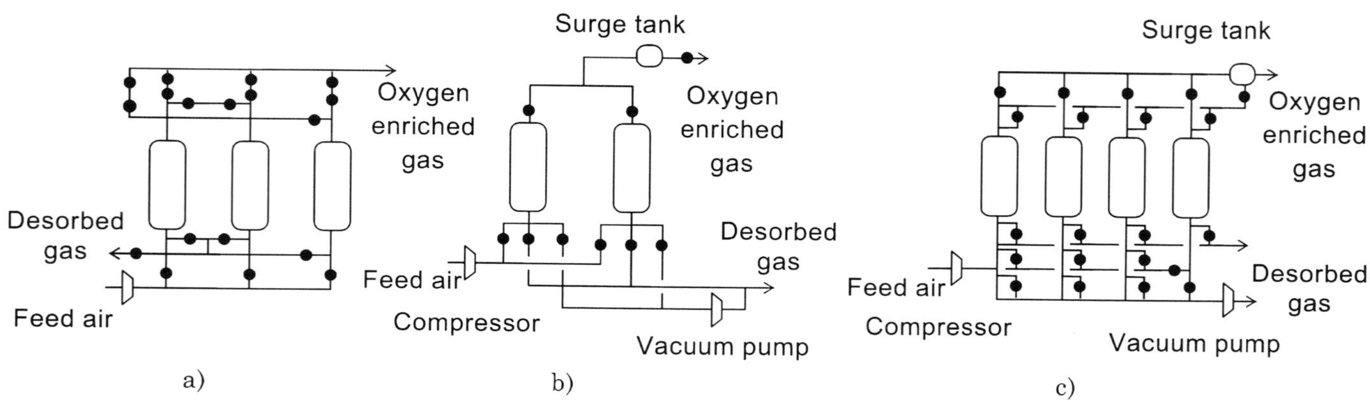


Figure 8. Schematic flow diagrams for PSA processes producing medium purity $O_2$ -enriched gas (a) process A, (b) process B, (c) process C.

The oxygen-enriched product gas (~90% $O_2$) is produced at the feed air pressure ($P_A$) by this process. Table 7 describes the performance of this process using LiX zeolite and specific pressure levels [$P_A$=1.43, $P_I$=0.61, $P_D$=0.34 atmospheres] of operation. The combined waste gases (steps c, d, and e) contain 88.7% $N_2$ along with the desorbed $CO_2$, $H_2O$ and some $O_2$. A two column embodiment of the process is described in Figure (8b). A surge tank is needed to smooth out product flow and composition. The operation of the vacuum pump is intermittent for this case.

**Process C:**

This process is patented by Air Products and Chemicals, Inc. [11]. It is designed to produce an essentially $N_2$-free, $O_2$-enriched product gas. $O_2$ and Ar cannot be separated by zeolites (based on thermodynamic selectivity) at normal conditions. Thus, the $O_2$: Ar molar ratio in the product gases by all PSA processes are approximately equal to that in the ambient air (21:1). Thus, the highest possible purity of $O_2$ product ($N_2$ free) by PSA is 95%.

The adsorption step in this process is carried out at a super-atmospheric pressure level ($P_A$<8.0 atmospheric) and the final desorption pressure ($P_D$) is sub-atmospheric. The process operates under an ad(de)sorption pressure ratio ($P_A/P_D$) of at least 20 and it consists of the following four cyclic steps:

(a) Compressed air is passed through the adsorber at the highest pressure level of the cycle ($P_A$) while producing an $O_2$-enriched gas at feed pressure. A part of this gas is withdrawn as the product gas and the balance is used as pressurization gas in step (d).

(b) Counter-current depressurization of the adsorber to near-ambient pressure level. The effluent is wasted.

(c) Counter-current evacuation of the adsorber to the lowest pressure level ($P_D$) of the cycle. The effluent is wasted.

(d) Counter-current pressurization of the adsorber from $P_D$ to $P_A$ using a part of the $O_2$-enriched product gas from step (a).

(e) Repeat cycle from step (a).

Table 7 describes the performance of this process using CaX zeolite and specific pressure levels ($P_A$=3.0, $P_D$=0.1 atmospheres) of operation [10]. The average $N_2$ concentration of the waste gases (steps b and c) is 86.0%. A schematic flow diagram for the process using four adsorbers (for continuous product withdrawal and operations of the air compressor and vacuum pump) is shown by Figure 8c.

All of the above described processes can be operated using fewer adsorbers and switch valves by incorporating gas storage tanks and allowing discontinuous operation of the rotating machines.

**These three examples demonstrate that a variety of PSA process cycles using different adsorbents and operating conditions can be designed to produce medium purity (90–95%) $O_2$-enriched product gas.** Numerous other PSA process cycles to achieve the same goal are reported in the patent literature. The process performance for production of $O_2$-enriched air by different combinations of process cycles, operating conditions and adsorbents, however, can be substantially different while delivering the same product purity. Table 7 shows that Process B produces nearly 3.5 times more $O_2$-enriched (90%) product gas per unit amount of the adsorbent per cycle of operation than Process A. It also provides larger $O_2$ recovery from feed air (moles of $O_2$ in product/moles of $O_2$ in feed air cycle). Process C delivers an essentially $N_2$-free $O_2$-enriched (95%) gas with a very high productivity. The actual $O_2$ production rate of these process will be determined by the cycle times used in the process design. A faster cycle time (number of cycles/day) will increase the daily production rate for a given $O_2$ productivity of Table 7. A typical total cycle time per adsorbent column for these processes vary between 50 and 120 seconds.

### 3.3. Low purity $O_2$ product

**Process D:**

This process (called oxy-rich process) is patented by Air Products and Chemicals [12]. It is designed to directly produce 23-40% $O_2$ enriched air from ambient air. It has only three cyclic steps as follows:

(a) Compressed air is introduced into the adsorber until the adsorber pressure reaches a specific super-atmospheric pressure level ($P_A$). The product end valve is kept closed during this period.

(b) Co-current depressurization of the adsorber to an intermediate pressure level ($P_I$) while withdrawing the $O_2$-enriched product gas through the product end

(c) Counter-current depressurization of the adsorber from $P_I$ to near-ambient pressure level to desorb the $H_2O$, $N_2$ and some $O_2$. The effluent gas is wasted.

(d) Repeat cycle from step (a).

The process was tested using NaX, 5A and Na-Mordenite as adsorbents [13]. Figure 9a and 9b, respectively, show the $O_2$ productivity (milli pound moles/lb of zeolite/cycle) and $O_2$ recovery (%) as functions of $O_2$ product purity by the oxy-rich process using an adsorption pressure ($P_A$) of 3.7 atmospheres. A very high $O_2$ productivity (>0.1 milli pound moles/lb/cycle) can be achieved at a high $O_2$ recovery (>80%) by the process when producing a low purity (23-25%) $O_2$ product. The figures show that both $O_2$ productivity and recovery for a given $O_2$ product purity increase in the order NaX>5A>Na-Mordenite. This is due to the fact that $N_2$ is most weakly adsorbed on NaX among the three zeolites and its desorption from the zeolites under the mild conditions of desorption employed by the process (no evacuation or back purging with $O_2$) controls the over-all process performance for production of low purity $O_2$. The process performs poorly when the product $O_2$ purity is above 30% and an $O_2$ purge step is required to improve its separation efficiency [12]. **This example demonstrates that zeolites having high $N_2$ capacity and selectivity may not be preferred for certain product specifications.**

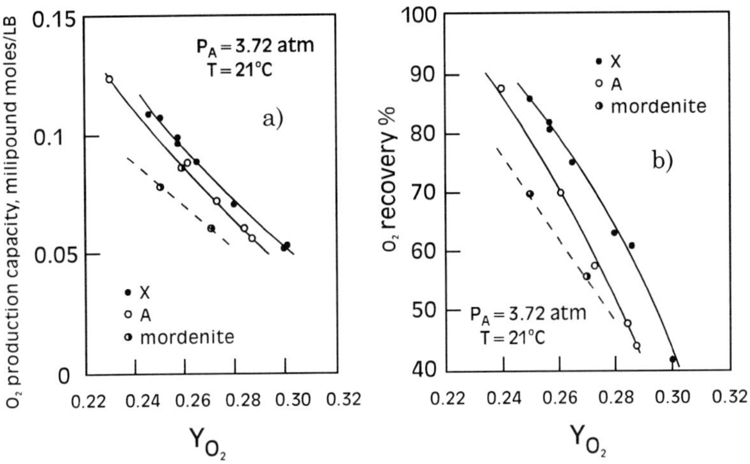


Figure 9. Performance of the oxy-rich process (D): (a) $O_2$ productivity as a function of $O_2$ product purity, (b) $O_2$ recovery as a function of $O_2$ product purity.

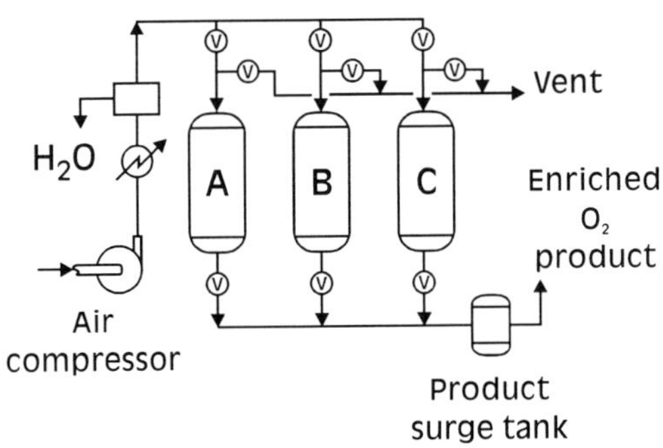


Figure 10. Schematic flow diagram for the oxy-rich process (D).

Figure 10 shows the flow sheet of a three adsorber embodiment for the Oxy-rich process. It needs a product surge tank to smooth out product flow and composition variations. An interesting feature of this process is that there is no interaction between the adsorbers during the process steps (unlike processes A, B and C). The total cycle time for each adsorber can vary between 30-60 seconds.

### 3.4. Rapid PSA

A new class of PSA processes called Rapid Pressure Swing Adsorption (RPSA) processes has emerged for production of low purity $O_2$-enriched air. They are designed to operate with a very rapid total cycle time (1-10 seconds) per adsorber. **The objective is to increase the specific rate of production of $O_2$-**

**enriched gas (volume/volume of adsorbent/hour) by more than an order of magnitude compared to the conventional PSA processes.** Three RPSA process schemes are described below:

**Process E:**

This RPSA process is patented by Union Carbide Corporation [14]. It uses a single adsorber vessel with the following three cyclic steps:

(a) Compressed air is introduced into the adsorber for a short period of time at a super-atmospheric pressure level ($P_A$).
(b) Stopping introduction of air for a short period of time.
(c) Counter-current depressurization of the adsorber to near-ambient pressure level.
(d) Repeat cycle from step (a).

A stream of $O_2$-enriched product gas is continuously withdrawn from the product end of the adsorber during steps (a)-(c) of the process. The product is obtained at near atmospheric pressure.

Figure 11a shows a schematic flow sheet for the process. The air compressor runs intermittently during this RPSA cycle. Very small adsorbent particles (20-120 mesh) are used to facilitate near-equilibrium operation (rapid kinetics) of the process. Extensive experimental evaluations of this process using 5A zeolites have been published [15, 16]. The process can produce a low purity (<30%) $O_2$-enriched product gas at a very high rate of production (>100 V/V/Hr) and at a decent $O_2$ recovery (>40%) but the $O_2$ recovery rapidly decreases at higher $O_2$ product purity. The total cycle time for the process can be between 3-6 seconds.

**Process F:**

This RPSA process is developed by the Institute of Industrial Science in Tokyo, Japan [17]. It uses an interesting apparatus as depicted by Figure 11b. The adsorbent chamber is mounted at the mouth of a cylinder fitted with a piston. The piston cylinder is fitted with two one-way valves (A, C) and the product end of the adsorbent chamber is fitted with a one-way valve (B). The cycle steps are as follows:

(a) Ambient air is drawn into the piston cylinder through valve A during the downward stroke of the piston.
(b) The air is compressed and forced to flow through the adsorbent chamber during the upward stroke of the piston producing an $O_2$ enriched product gas through valve B.
(c) The adsorbent is evacuated during the next downward stroke of the piston (valves A, B, C closed) which desorbs the $N_2$ from the zeolite.
(d) The desorbed gas is forced out of the piston cylinder through Valve C during the next upward stroke of the piston (valve B closed).
(e) The process is repeated from step (a).

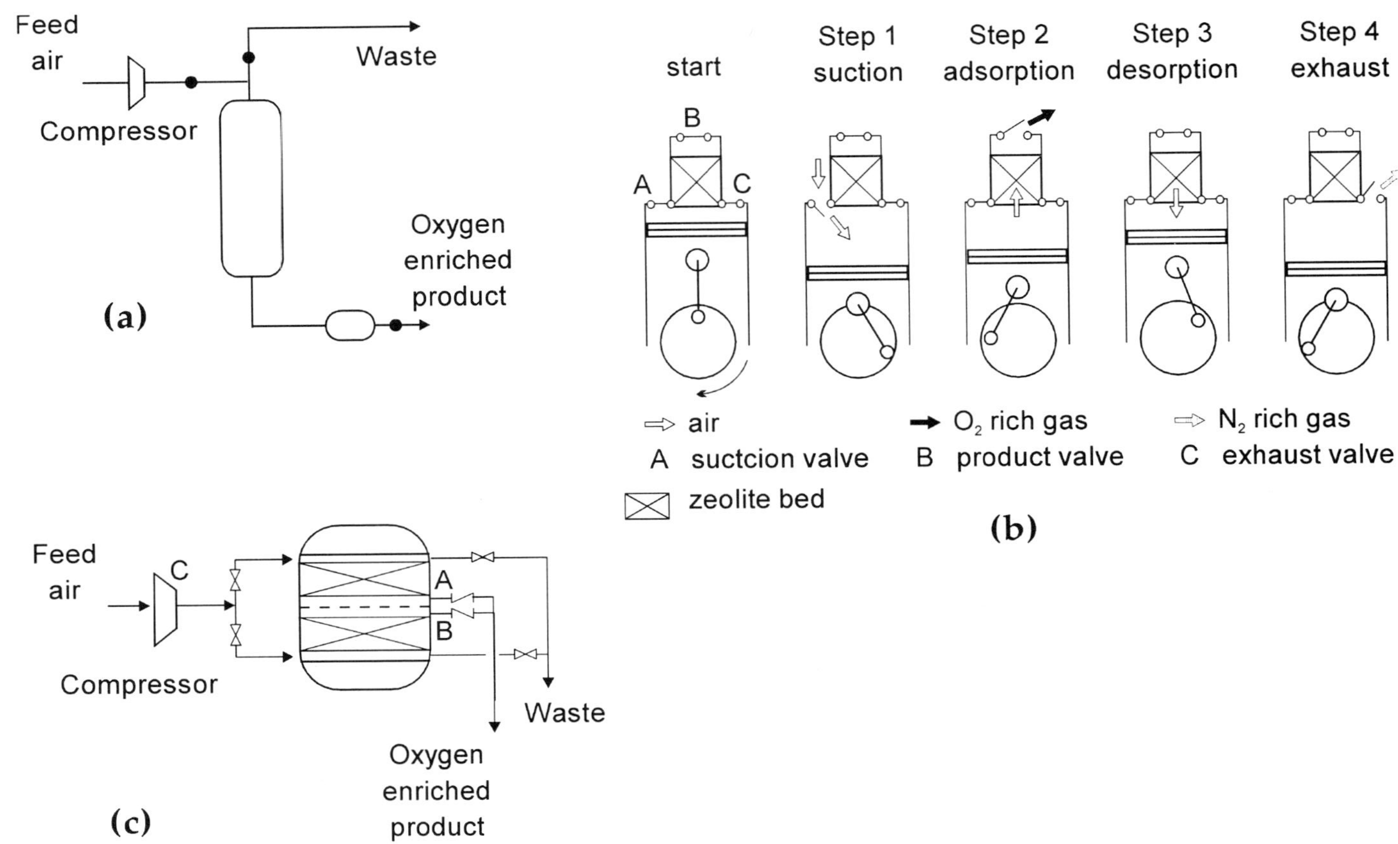


Figure 11. Schematic flow diagrams for RPSA processes producing low purity $O_2$-enriched gas: (a) process E, (b) process F, (c) process G.

The piston is driven by a motor which controls the frequency of its movement (cycle time). The $O_2$ production is intermittent and a gas storage tank is needed for continuous product flow. A very fast cycle time of less than one second could be used. Very small adsorbent particles (48-325 mesh) are used for fast adsorption kinetics. The process is capable of delivering large rate of production (>100V/V/Hr) of the $O_2$-enriched product gas using a 5A zeolite but the $O_2$ recovery is low (<5%).

**Process G:**

This RPSA process is patented by Air Products and Chemicals [18]. It uses a novel adsorber design consisting of two or more pairs of shallow adsorbent layers installed in a single vessel as shown by Figure 11c. The layers are separated by perforated metal plates for creation of flow resistance between the layers. The process cycle consists of the following two steps:

(a) Introduction of compressed air into one of the adsorbent layers for simultaneous pressurization of the layer from a near-ambient pressure level to the highest pressure level of the cycle ($P_A$) and adsorption of $N_2$ to produce an $O_2$-enriched gas. A part of this $O_2$-enriched gas is withdrawn as the product and the balance is passed through the companion adsorbent layer as purge gas after dropping its pressure through the perforated metal plate.

(b) Counter-current depressurization of the adsorbent layer from $P_A$ to near-ambient pressure level while simultaneously purging it counter-currently with a part of the $O_2$-enriched gas produced by the companion layer.

Relatively small adsorbent particles (30-50 mesh) are used in the process. The cycle times for each step (a and b) of the process are between 3-10 seconds. Figure 12 shows the performance of the process using NaX zeolite and final adsorption pressures of 2.1 and 3.1 atmospheres [19, 20]. The total cycle time for the data of Figure 12 is 12 seconds (6 seconds for each step). The process can deliver a very large rate of production of $O_2$ enriched gas (>200V/V/Hr) with a moderate $O_2$ recovery (>30%) when the $O_2$ purity is less then 50% and the adsorption pressure is ~3.0 atmosphere.

An alternative mode of operation of this process has also been developed where the air drying and air separation duties of the RPSA process are decoupled [20]. The RPSA process, in this case, is operated using a lower feed air pressure ($P_A$=2.1 atmosphere) and the air is pre-dried using a conventional thermal swing adsorption (TSA) unit. This option reduces the cost of the $O_2$ product.

Table 8 compares the performance of the three RPSA processes using 5A zeolite as the adsorbent. It shows that Process G performs much better (higher $O_2$ production rate and recovery) than the other two RPSA processes for production of low purity (27-45%) $O_2$-enriched gas even though the other processes are operated using a much faster cycle time than Process G. This may

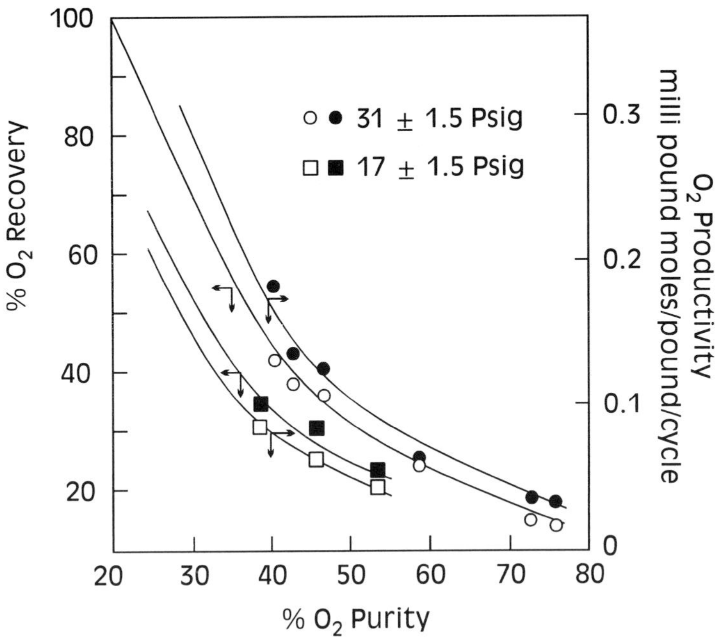


Figure 12. Air separation performance of process G.

be caused by the adsorption kinetic limitation for a RPSA process [19] as well as by the differences in the design of the process steps.

### 3.5. Production of $N_2$-enriched air

The actual average $N_2$ composition of the $N_2$-enriched desorbed gases from the PSA processes producing medium purity (90+%) $O_2$ product vary between 85-92%. The model calculations (Figure 4) of isothermal desorption from a zeolite column by evacuation, which is initially saturated with a binary gas mixture (79% $N_2$ + 21% $O_2$) at ambient pressure also show that the average $N_2$ mole fractions of total desorbed gas do not exceed 96% even for a highly $N_2$ selective zeolite. PSA processes have been developed to circumvent this problem and produce high purity (98+%) $N_2$ products. **These processes simultaneous produce a $N_2$-enriched product stream (98-99% $N_2$) and an oxygen-enriched product stream (90+% $O_2$).** Two of these processes are described below:

### Process H:

This commercial process is patented by Air Products and Chemicals [21]. It consists of the following four cyclic steps:

Table 8
Comparative performance of RPSA processes for production of oxygen enriched air using 5A zeolites

| Oxygen Purity (mole%) | Production Rate (V/V/Hr) | | | $O_2$ recovery (%) | | |
|---|---|---|---|---|---|---|
| | Process E | Process F | Process G | Process E | Process F | Process G |
| 27.2-27.8 | 173(a) | | 2290(d) | 44.2(a) | | 64.1(d) |
| 44.7 | 43(a) | 50-180(b) | 128(c) | 19.2(a) | ~2.0(b) | 22.1(c) |

(a) Cycle times for Process E (step a: 1 sec, step b: 0.5 Sec, step c=4 sec), Feed Air Pressure = 1.68atm
(b) Cycle times for Process F (0.5 sec), Feed Air Pressure = 2.0 atm
(c) Cycle times for Process G (step a: 10 sec, step b = 10 sec), Feed Air Pressure = 1.75 atm
(d) Cycle times for Process G (step a: 5 sec, step b=5 sec), Feed Air Pressure = 2.22 atm

(a) Ambient air is passed over a zeolitic adsorbent column at a slightly elevated pressure (~1.2 atmospheres) level ($P_A$) and a stream of $O_2$ enriched air (90%) is produced through the product end. A part of this gas is withdrawn as the $O_2$ product and the balanced is used as pressurization gas in step (d).

(b) The column is then co-currently rinsed with a part of the $N_2$ enriched product gas from step (c) in order to remove void and co-adsorbed $O_2$ from the column. The effluent gas is air like which is wasted or recycled as feed gas.

(c) Counter-current evacuation of the column to the lowest pressure level of the cycle ($P_D$) to produce a $N_2$ enriched (98+%) gas. A part of this gas is withdrawn as the $N_2$ product and the balance is used as rinse gas in step (b).

(d) The column is then pressurized from $P_D$ to $P_A$, counter-currently by introducing a part of the $O_2$ enriched gas from step (a).

(e) Repeat cycle from step (a).

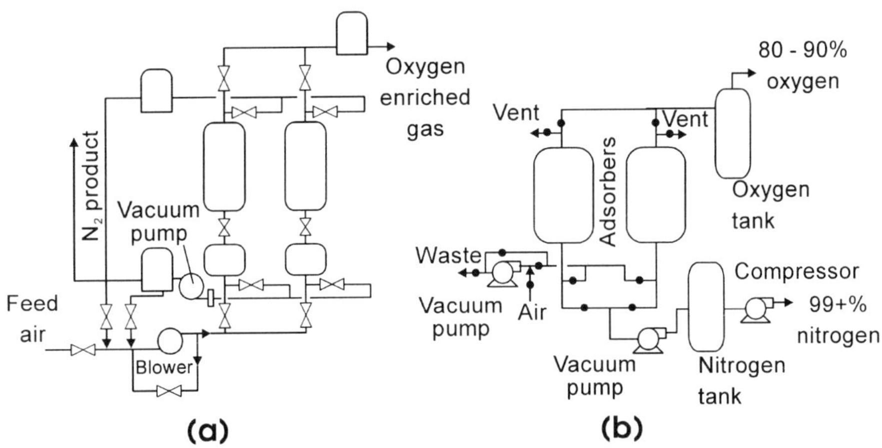


Figure 13. Schematic flow diagrams for VSA processes for simultaneous production of $N_2$ and $O_2$ enriched gases: (a) process H, (b) process I.

The process can be run with the desiccant and the zeolite layers packed in the same adsorber [21] or the two adsorbents can be placed in two adsorbers connected in series [22]. A flow diagram for the second option is shown in Figure 13a. In this case, the final evacuation levels for the two adsorbers can be different to facilitate $H_2O$ desorption.

**Process I:**

This process (called Fractionated Vacuum Swing Adsorption Process) is also patented by Air Products and Chemicals [23]. The concept of operation of this process is illustrated by Figure 4. It shows that the $N_2$ concentration of the evacuated gas from an air saturated column reaches a very high value (>98+%) at the lower pressure levels for zeolites with large $N_2$ selectivity. Thus, the desorbed gas from the zeolite can be fractionated to produce a high purity $N_2$ product without using the $N_2$ rinse step of process H. The process cycle, therefore consists of the following three steps:

(a) Ambient air is passed over a zeolite column (high $N_2$ selectivity) at a slightly elevated pressure (~1.2 atmospheres) level ($P_A$) and a stream of $O_2$-enriched air (90%) is produced.

(b) Counter-current evacuation of the adsorber to the lowest pressure level ($P_D$) of the cycle. The first part of the desorbed gas produced between $P_A$ and an intermediate pressure level ($P_I$) is rejected. The second part of the desorbed gas between pressure levels of $P_I$ and $P_D$ is collected as the $N_2$-enriched (98+%) product gas.

(c) Counter current pressurization of the adsorber from $P_D$ to $P_A$ using a part of the $O_2$-enriched gas produced by step (a).

(d) Repeat cycle from step (a).

A two column embodiment of the process I is described in Figure 13b. The design uses a single vacuum blower which also acts as air blower for step (a) of the process. The total cycle times for each adsorber of both processes H and I can be between 40-120 seconds. Both processes produce a wet $N_2$ enriched product gas at near-ambient pressure.

Table 9 compares the performance of process H and I for simultaneous production of $N_2$ and $O_2$ enriched product gases [24]. The $N_2$ purity of process H is higher because it was operated using a lower vacuum level. Both processes produce high purity $N_2$ product gas with comparable productivity in conjunction with a medium purity $O_2$ product gas. They, however, use two very different zeolites (Table 6). The zeolite for process H can have a moderate $N_2$ selectively (because of $N_2$ rinse step) while the zeolite for process I must have a large $N_2$ selectivity. **Thus, they represent good examples of tailoring the process steps of a PSA process to match the adsorptive properties of the zeolite for a given product specification.**

**It should be mentioned here that non-zeolitic carbon molecular sieves (CMS) are also commercially used for production of $N_2$-enriched (98+%) air from ambient air.** Numerous PSA cycles using CMS adsorbents have been developed [25, 26]. These processes do not produce a medium purity $O_2$-enriched gas as the by product. The air separation on the CMS is based on kinetic selectivity of $O_2$ over $N_2$. The smaller $O_2$ molecules diffuse into the pores of the CMS faster than the larger $N_2$ molecules creating an $O_2$-enriched adsorbed phase when air is contacted with the CMS for short time periods. The $N_2$-enriched gas is produced during the adsorption step and the product is dry.

Table 9
PSA process performance for simultaneous production of nitrogen and oxygen enriched gases

| | | Operating Pressures | | Nitrogen Product | | | Oxygen Product | | |
|---|---|---|---|---|---|---|---|---|---|
| Process | Adsorbent | $P_A$ (atm) | $P_D$ (atm) | Purity (%) | Recovery (%) | Productivity (millipound moles/lb of zeolite/cycle) | Purity (%) | Recovery (%) | Productivity (millipound moles/lb of zeolite/cycle) |
| H | Na-Mordenite | 1.05 | 0.07 | 99.9 | 53.2 | 0.21 | 90.0 | 58.8 | 0.067 |
| I | CaX | 1.10 | 0.09 | 99.0<br>98.0 | 30.0<br>42.6 | 0.17<br>0.25 | 90.0 | 24.2 | 0.040 |

The adsorption literature often classifies the adsorptive processes in terms of (a) Pressure Swing Adsorption (PSA) when the adsorption step is carried out at a super-atmospheric pressure level and the lowest desorption pressure is ambient (Processes A, D, E, F, G), (b) Vacuum Swing Adsorption (VSA) when the adsorption step is carried out at near-ambient pressure level and the lowest desorption pressure is sub-ambient (Processes H, I) and (c) Pressure Vacuum Swing Adsorption (PVSA) when the adsorption step is carried out at a super-atmospheric pressure level and the lowest desorption pressure is sub-atmospheric (Processes B, C).

## SUMMARY

Production of $O_2$-enriched (90+mole% $O_2$) and $N_2$-enriched (98+mole% $N_2$) air from ambient air by selective adsorption of $N_2$ on zeolites is an established chemical engineering operation. Many different zeolites having a large spectrum of properties like $N_2$ adsorption capacity from air, $N_2$ selectivity of adsorption over $O_2$, strength of $N_2$ adsorption, etc. are available for air separation. The $N_2$ adsorption and desorption characteristics of five commercial zeolites (NaX, 5A, Na-Mordenite, CaX and CaLSX) are described.

Many different PSA air separation processes can be designed using zeolitic adsorbents. They differ by the modes of operation of the adsorption, the desorption and the complementary steps used for improving the separation efficiency. The design and performance of nine different zeolitic PSA processes are reviewed and compared. Different zeolites can be used in conjunction with different PSA process cycle designs in order to satisfy a given product specifications. This process design and material flexibility is responsible for the large number of air separation process patents using zeolites.

## REFERENCES

1. S. Sircar and B. F. Hanley, Separation Sci. and Tech., 28 (1993) 2553.
2. R. M. Barrer and A. B. Robbins, Trans. Faraday Soc., 49 (1953) 807; 49 (1953) 929.
3. T. R. Gaffney, Porous Solids for Air Separation, in: Current Opinion in Solid State and Materials Science, 1 (1996) 64.
4. S. Sircar, Pressure Swing Adsorption Technology in: Adsorption Science and Technology, NATO ASI Series E, A.E. Rodrigues (ed.), Kluwer Academic Publishers, The Netherland, 158 (1989) 285.
5. T. C. Golden and S. Sircar, J. Colloid and Interface Sci., 162 (1994) 182.

6. S. Sircar and T. C. Golden, I&E.C. Research, 34 (1995) 2881.
7. D. W. Breck, Zeolite Molecular Sieves, R. E. Krieger Publishing Company, Malabu, Florida (1984).
8. L. K. Batta, US Patent No. 3636679 (1972).
9. J. J. Collins, US Patent No. 3973931 (1976).
10. F. W. Leavitt, US Patent No. 5415683 (1995).
11. S. Sircar, US Patent No. 4756723 (1988).
12. W. C. Kratz and S. Sircar, US Patent No. 4685939 (1987).
13. S. Sircar and W. C. Kratz, Separation Sci. Tech., 23 (1988) 437.
14. R. J. Jones, G. E. Keller and R. C. Wells, US Patent No. 4194892 (1980).
15. C. L. Pritchard and G. K. Simpson, Chem. Eng. Res. Des., 64 (1986) 467.
16. C. T. Chou and H. C. Wu, Zeolites and Related Microporous Materials: State of the Art, Studies in Surface Science and Catalysis, 84 (1994) 1255.
17. M. Suzuki, T. Suzuki, A. Sakoda and J. Izumi, Adsorption, 2 (1996) 111.
18. S. Sircar, US Patent No. 5071449 (1991).
19. S. Sircar and B. F. Hanley, Adsorption, 1 (1995) 313.
20. S. Sircar, Adsorption, 2 (1996) 323.
21. W. P. Schmidt, R. Kumar and A. D. Abel, US Patent No. 4813977 (1989).
22. S. Sircar and J. W. Zondlo, US Patent No. 4013429 (1977).
23. S. Sircar, US Patent No. 5084075 (1992).
24. S. Sircar, Separation Sci. Tech., 23 (1988) 2379.
25. K. Knoblauch, H. Heimbach and B. Harder, US Patent No. 4548799 (1985).
26. N. C. Lemcoff and R. C. Gmelin, US Patent No. 5176722 (1993).

Adsorption and its Applications in Industry and Environmental Protection
Studies in Surface Science and Catalysis, Vol. 120
A. Dabrowski (Editor)



# Production, characterization and applications of carbon molecular sieves from a high ash Greek lignite

P. Samaras, X. Dabou and G. P. Sakellaropoulos

Chemical Process Engineering Laboratory, Department of Chemical Engineering, Aristotle University of Thessaloniki and Chemical Process Engineering Research Institute, PO Box 1520, Thessaloniki 54006, Greece

## ABSTRACT

Activated carbons were produced from raw and demineralized Greek lignite under various experimental conditions. Demineralization was accomplished by acid washing of raw lignite, and pore structure of products was estimated by adsorption of $N_2$ at 77K and of $CO_2$ at 298K. Demineralization caused a decrease in coal reactivity but the activated products had a high surface area. Activated carbons produced from demineralized lignite at low burn-off were exclusively microporous, and had significant molecular sieving properties to gases, as derived from the uptake curves of $CO_2$, $CH_4$, $O_2$ and $N_2$ at room temperature. Carbon molecular sieves were also produced from activated carbons containing the ash minerals, by employing a method of propylene cracking for coke deposition and pore structure modification. High reaction temperatures resulted in a decrease of the mesopore surface area while the micropore surface area was unchanged. Samples prepared under these conditions had high $CO_2/CH_4$ selectivity ratios, exhibiting molecular sieving behavior for gas separations.

## 1. INTRODUCTION

Energy efficient gas separation processes, particularly in the chemical and petrochemical industry, have been a subject of increased attention in recent years. This is due to the high energy input, and thus high cost of industrial gas separations by conventional processes, such as cryogenic distillation and/or absorption. Adsorption processes, based on activated carbon or silica gel as adsorbents, have proven reliable, efficient and cost effective for a number of industrial gas separations. The search for a suitable adsorbent is the first step in this [1, 2].

The potential of adsorption as a basis of separation processes was greatly enhanced by the advent of the Pressure Swing Adsorption (PSA) process, and the development of new Carbon Molecular Sieve (CMS) adsorbents. The CMS are a specially prepared form of activated carbons having average pore dimensions similar to the critical dimensions of small molecules. The unique kinetic separation ability of CMS [3] made some new applications feasible, which otherwise required the use of other molecular sieving adsorbents like zeolites.

### 1.1. Properties

CMS have had a significant impact on separation processes and they have also found applications as catalysts. In recent years, most research and development efforts concerning CMS materials have been devoted to their preparation, property characterization and use in gas separations [4]. Less attention has been paid to their catalytic properties and their use as selective catalyst support, which is currently an active field of research [5, 6].

The literature on CMS is large and growing due to the advantages of CMS over zeolites, in being more stable at higher temperatures, less hydrophilic, and inert to acidic and basic media, except to oxidation. The properties of CMS are quite variable and depend upon initial manufacturing procedures and starting material properties. Some typical properties of CMS and activated carbons are given in Table 1.

Table 1
Some typical properties of CMS and ordinary activated carbons (AC)

| Property | CMS | AC |
|---|---|---|
| Average pore diameter (Å) | 3-12 | <20 |
| Pore volume ($cm^3/g$) | 0.5-0.5 | 0.5-0.15 |
| Surface area ($m^2/g$) | 300-1500 | 100-1000 |

CMS have substantially uniform micropores (3 to 12 Å) while ordinary activated carbons have a wide pore size distribution (2 to 2000 Å). Narrow pores of CMS favor the diffusion of small molecules, and inhibit penetration of large ones, thus providing a sharp separation as compared to ordinary activated carbons. Another important property of CMS is the slit shaped pore structure which makes them useful for kinetic separation and adsorption equilibrium studies.

Gas separation is achieved by two alternative mechanisms: kinetic separation and selective adsorption. Kinetic separation is based on the kinetically-controlled gas diffusion caused by the constrictions of the apertures of the pores. The diameters of the bottle-necked pores are in the same range as those of the adsorbed molecules. Thus, when a CMS is used in an air separation process, the oxygen molecules, which have a smaller diameter than the nitrogen molecules, can penetrate much quicker into the pores than the nitrogen molecules. Therefore, nitrogen is recovered to a high degree whilst almost all the oxygen is adsorbed. In the second separation mechanism, the pore system is sufficiently wide to enable fast diffusion; separation is caused by selective adsorption which depends upon the van der Waals forces between the carbonaceous substrate and the gas species [1].

CMS are amorphous materials. Their pore structure below 5 Å can not be studied by X-ray diffraction, in contrast to most mineral molecular sieves. Transmission electron microscopy has also not been found suitable for determining such small pore dimensions. The most effective method for characterization is the analysis of adsorption isotherms of small probe molecules with different critical dimensions, viz. $O_2$, $N_2$, $CO_2$, $CH_4$. These adsorption isotherms are useful in determining the pore size distribution, surface area, pore volumes and separation capacity of CMS. In addition, these isotherms give information on the potential industrial applications of these materials, e.g. for the separation of nitrogen from air or of carbon dioxide and methane from flue gases.

### 1.2. Applications

Carbon molecular sieves find today wide industrial applications in gas mixture separations, gas purification processes, and as catalysts and catalyst supports.

Gas mixture separation processes are based on the specific pore size distribution of CMS, which permits diffusion of different gasses at different rates. These processes aim to either recover and recycle valuable constituents from industrial waste gases, or to separate small gas molecules by preferential adsorption. The latter is at present the most important large scale application of CMS. Separations that have been accomplished include oxygen from nitrogen in air, carbon dioxide from methane in natural gas, ethylene from ethane, linear from branched hydrocarbons (such as n-butane from isobutane), and hydrogen from flue gases [6].

Production of low cost, high purity nitrogen (99.9%) from air is primarily based on the Pressure Swing Adsorption (PSA) technique, which exploits the micropore structure of CMS to achieve $N_2$-$O_2$ separation. The PSA-process utilizes the difference in loading of an adsorptive substance on CMS at high pressure (during adsorption), and at low pressure (during desorption). The available loading is related either to the equilibrium adsorption or to the diffusion rate of the gases. In the case of $O_2/N_2$ separation the kinetic separation mechanism is applicable due to the higher diffusivity of $O_2$ than $N_2$ molecules. Hence, in the PSA air separation process, the high pressure (rafinate) product is $N_2$ while $O_2$ is produced as the low pressure (extract) stream.

A typical PSA system generally consists of two beds filled with CMS, as shown in Figure 1, each being subjected to a series of four distinct processing steps in cyclic mode. In the first step, high pressure feed gas is introduced to one adsorption bed while cocurrently nitrogen gas is produced. The other bed undergoes desorption and regeneration, by venting to the atmosphere or by vacuum. In the next step of pressure equalization, the two beds are brought to an intermediate pressure, by fluid communication. In the third and fourth steps, the first and second steps are repeated with the two beds changing roles [7].

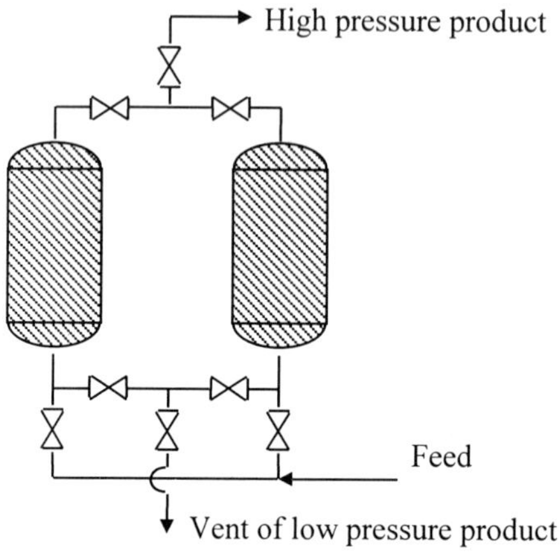


Figure 1. Typical PSA system with two adsorption beds.

In general, the main equipment of a PSA plant consists of the pressure vessels and the appropriate pipe connection network and automatic valves. The main advantages of the PSA process are the relatively low capital and maintenance costs, the automatic operation, the easy handling of devices during start up and shut down procedures, the system flexibility in the change of cycle time, the high purity of products, the long lifetime of vessels. PSA units for $N_2$ generation are now used in various applications such as the production of inert gas for the chemical industry and off-shore ships, metal heat treatment processes (steel annealing), tank or vessel purging or inertizing, food storage, pressure transfer, bottling of wine and beer, etc. [6,8-12].

In addition to air separation, the PSA technique is also used for hydrogen purification and/or separation from gas mixtures. Hydrogen is used in a variety of chemical and petrochemical processes. Hydrogen demand can be partially satisfied by recovering it from steam reforming gases, ethylene plant effluent gas, gases produced in coal utilization processes, or gases from ammonia synthesis plants. In general, gases produced in these processes contain about 65-90% by volume hydrogen, with the balance being carbon monoxide, carbon dioxide, methane, oxygen and nitrogen in various proportions, as well as small amounts of hydrocarbons and traces of tars and sulphur compounds. The success of a PSA hydrogen separation system is largely due to the high selectivities of the impurity gases over hydrogen. Best sorbents used in practice include a combination of carbon and zeolitic molecular sieves [8]. In this way, high purity hydrogen (over 99% by volume) can be produced, while product recoveries range from 75 to 85%.

An important application of CMS is related to the recovery of methane from bio-gases produced from municipal solid waste landfills. Landfill biogas contains substantial amounts of methane (40-60% v/v), carbon dioxide (30-50% v/v), oxygen (0-5% v/v), nitrogen (0-15% v/v), various undesirable contaminants such as aromatic and aliphatic hydrocarbons, halogenated hydrocarbons and sulphur compounds (mostly hydrogen sulfide). Methane is a valuable energy source and its recovery by utilizing the PSA technique can exceed 90% [3, 12]. Halocarbons and sulphur components contained in the landfill gas may cause considerable corrosion damage during its utilization, and PSA gas purification can assist in the removal of such compounds and in the protection of the plant.

Application of CMS for natural gas storage is related to their capacity for methane adsorption. Natural gas, being essentially methane, has a low molecular weight and it occupies a large volume per unit weight compared to liquid fuels. Thus, storage of natural gas at high weight/volume ratios is needed for its practical use as a fuel. Methods for doing this include storage at very high pressures (up to 200 atm), liquefaction and storage at low temperatures (-162°C), and adsorption on solids. Of these methods the latter is of particular interest, since compact storage of natural gas at moderate pressures may be achieved, while avoiding the cryogenic temperatures needed for liquid natural gas or the high pressures needed for storage as a gas in reasonable volumes. CMS packed in cylindrical vessels have been suggested for adsorption of natural gas at pressures of 14 to 45 atm, to deliver a gas volume at moderate pressures higher than 70 times the volume of the vessel [13-15].

Purification of gases can be also accomplished with CMS by tailoring their effective pore size distribution. CMS have been used in a variety of gas mixture separations such as: purification of chlorofluoracarbons produced from chlorinated hydrocarbons by complete removal of hydrochloric and hydrofluoric acid using a CMS with an average pore size of 3.5 Å [16]; removal of vinylidene chroride (from a concentration of 900 ppm to lower than 200 ppm), and of dichloroacetylene (from 4-20 ppm to lower than 2 ppm) from 1,1 dichloro-1-fluoroethane [17, 18]; adsorption and removal of volatile organic compounds, such as ketones and fluorocarbons, from vapor streams [19]; separation of a binary azeotrope of 3% HF/97% 1,1,1,2-tetrafluoroethane [20]; and drying and purification of liquids and gases [6].

The high surface area and structure of CMS has attracted attention for their possible exploitation as catalysts and catalyst supports. CMS have been found effective for the oxidative dehydrogenation and dehydration of a variety of substances like methanol, ethanol, 1- and 2-propanol and propanal, with activities superior to many inorganic oxide based systems. As catalysts, CMS function via a hydride or hydrogen atom abstraction mechanism, depending on the nature of the substrates, while as catalyst supports, a synergism has been demonstrated between the CMS support and the metal dopants, to yield a greater activity than that of either of the constituents alone [21]. The high surface area, the ability to stabilize and disperse metal clusters and the adsorption capabilities of CMS are the key contributors to their high activity. A catalyst consisting of 15% $MoO_3$/CMS gave significant activity towards methanol oxidation, where 70% of the feed was converted to methyl formate in a single pass with over 90% selectivity. Oxidative dehydrogenation of ethylbenzene to styrene, with an 80% conversion and a high product yield (selectivity 90%), was achieved over a commercial CMS at 350°C [22]. In this reaction 'active coke' is the active catalyst element. With metal oxide carriers alone as catalysts, higher temperatures (450 to 600°C) are usually necessary to form and maintain a certain stationary amount of 'active coke'.

CMS prepared from a coal, modified by pitch and phenol-formaldeyde resin, were used as catalyst supports for methanol decomposition at 350-450°C [5]. CMS supports with Ni, having a sharp pore size distribution of about 0.45 nm in diameter, gave a high selectivity in methanol decomposition to CO and $H_2$ (conversion up to 97%), with negligible by-products. CMS are also effective in the selective hydrogenation of linear over branched hydrocarbons due to the restricted diffusion of branched molecules through the CMS micropore structure [23]. During competitive hydrogenation over a CMS containing 1% platinum, conversions of 1-butene higher than 70% were achieved, compared to near zero 1-methyl-1-butene conversions, in contrast to results with commercial catalysts. In this case, branched molecules were unable to reach the active sites of platinum metals located within the pores of CMS. An improved composite material consisting of a CMS and a $SiO_2/Al_2O_3$ catalyst has been developed for the selective synthesis of monomethylamine and dimethylamine, at the expense of trimethylamine, from a starting feed of methanol and/or dimethylether and ammonia [24]. With this catalyst, selectivity was based on the high permeation of monomethylamine and dimethylamine and on the negligible permeation of trimethylamine into the CMS.

In recent years, there is a growing interest in the development of Carbon Molecular Sieve membranes for gas separations. The advantages of CMS membranes are their chemical inertness and their thermal stability up to more than 500°C [25]. CMS membranes can be derived by controlled pyrolysis of a thermosetting polymer precursor, usually polyimide [26,27]. Macroporous supports can be used to increase the mechanical strength of the membranes [28]. Selectivities obtained by carbonised membranes are generally higher than those of polymeric membranes. Selectivities for $H_2/CH_4$ of about 550 have been reported for carbon membranes produced from the pyrolysis of polyimide at 550°C [29]. The molecular sieving effect of a Kapton-type polyimide carbon membrane was confirmed by the high selectivity values of more than 1000, for $H_2/N_2$ mixtures [26]. Practical application of such membranes is currently under investigation due to problems concerning the preparation of membranes with reproducible properties in commercially available forms, such as asymmetric membranes in capillary or hollow fiber configuration, the lack of knowledge of the correlation between gas permeation properties and membrane structure, and the stability and durability of CMS membranes.

### 1.3. Preparation techniques

The earliest preparation of CMS was based on the decomposition of a Saran co-polymer (90/10 mixture of vinylidene chloride and vinyl chloride); today CMS with a wide range of physical properties are made from a variety of natural and synthetic precursors. These include coal, coconut shell, phenol-formaldehyde resin, polyfurfuryl-alcohol, polyacrylonitrile, polyvinyl-alcohol, and cellulose, with coal tar pitch used in most cases as a binder [30-33].

Several papers have been published on the preparation of CMS and on their sieving properties. Preparation methods can be grouped into the following four principle ones:

1. Carbonization of Saran or polyvinylidene chloride under controlled conditions [34]. In an earlier work, Walker et al., [35] have reported that a composite CMS can be produced by coating a thermosetting polymer on the surface of activated carbon which, upon carbonization, creates constrictions of a suitable size.
2. Partial activation of coal chars at low burn-off values for improvement of molecular sieving properties [36]. Anthracites, devolatilized under nitrogen at 950°C and slightly activated in air at 427°C up to 8% burn-off, presented molecular sieving behavior, as deduced from nitrogen and neopentane adsorption experiments. Verma and Walker [37], tried to alter the molecular sieving behavior of activated carbons and commercial molecular sieves by treatment at temperatures up to 1050°C in an inert and a reactive gas atmosphere, or under vacuum, and they underlined the significant role of the starting material. Selectivity for the separation of $O_2$ and Ar was enhanced by thermal treatment of a pelletized commercial CMS while no useful sieving properties were developed by the heating of the activated carbons.
3. Deposition of carbon inside the pores of activated carbons or various carbonized materials though the carbonization of additives, such as resins, coal tar pitch and sulfate pulp waste liquor [32,33,38,39], to reduce pore size. In these processes the starting material is ground and the powder is mixed with the appropriate binder and extruded in the form of pellets. Pellets are then heated to temperatures up to 1000°C. In this way, the binder penetrates the surface of the starting material and decomposes to form a carbon crystallite, growing on the inner surface, thus improving selectivity to gas mixtures.
4. Chemical deposition of the vapors of organic substances within the pores of carbonized materials for pore size adjustment. In this process, CMS are generally prepared by carbonization of the raw precursors, activation for the enlargement of pore structure, and finally chemical deposition for tailoring pore apertures. Suitable carbonaceous substances that can be

used in the treatment include methane, acetylene, propylene, benzene, toluene, etc. [32,37,40,41]. Upon pyrolysis, carbonaceous deposits are formed that modify the pore structure of the activated carbon, improving its molecular sieving capacity. Several modifications have been proposed in the preparation procedures in order to obtain products with desired properties. A two step hydrocarbon deposition method, involving pyrolysis of isobutylene in two different concentrations, was used by Cabrera et al., [30], to control the hydrocarbon cracking at the pore mouth.

The above methods intend to slightly modify the pore structure of a carbonaceous material, by employing a carbon deposition procedure, involving the formation of a gate-keeping layer onto the pores of the coal. This gate-keeping layer is a region near the pore opening that is narrowed sufficiently to allow molecules to pass through, in significantly different rates. Any molecule able to pass through this gate with a high diffusion rate, has access to a high internal pore volume of the particle. Preparation conditions should be carefully controlled so that carbon deposition would be restricted on the pore openings, and simultaneously excessive carbon deposition in the pore interior would be diminished.

The carbon deposition procedure is very much dependent on both the properties of the starting material and the preparation conditions. The primary properties of the starting material which affect CMS are the pore structure and the presence of catalytically active inorganic species. The effect of inorganic matter of raw materials on the characteristics of activated carbons, such as pore structure development and adsorption capacity, has been reported by several investigators [42-44]. While the important role of inherent inorganic matter is recognised in coal gasification reactions, its role in hydrocarbon cracking for the production of suitable CMS has attracted little attention. Few authors have studied the effect of mineral matter on hydrocarbon pyrolysis reactions for the production of CMS. Propylene pyrolysis over nickel - impregnated activated carbons showed the enhancement of carbon deposition at the nickel sites within the pores [41]. Vyas et al., [4], used a high ash bituminous coal (with 33% ash content) pre-treated in acid solutions to produce CMS. However, the effect of inorganic matter on carbon deposition could not be determined, since acid treatment resulted in the removal of mineral matter from the raw material. Samaras et al., [44] have shown that removal of catalytically active mineral matter from lignite caused a reduction in activation rate, but resulted in carbons exhibiting molecular sieving behavior, as deduced from differences in $N_2$ and $CO_2$ surface areas.

The objectives of this work were to study the production of carbon molecular sieves from a Greek lignite with a high mineral matter content, to investigate the influence of the starting material properties (inorganic matter and pore structure) on the produced CMS surface area and separation capacity for $CO_2/CH_4$ and $N_2/O_2$ gases, and to determine the optimum experimental conditions for the production of molecular sieves with desired properties.

## 2. EXPERIMENTAL

The coal used in this study was a Greek lignite from the Ptolemais seam. The proximate and ultimate analysis data are shown in Table 2. The coal was ground and sieved, and the 60-100 Tyler mesh fraction (particle size 150-250 ìm) was subjected to further treatment.

Table 2
Ultimate and proximate analysis of lignite (% wt dry basis)

| Ultimate Analysis | | Proximate Analysis | |
|---|---|---|---|
| C | 36.2 | Volatiles | 52.8 |
| N | 1.0 | Fixed carbon | 16.3 |
| H | 3.3 | Ash | 30.9 |
| S | 0.8 | | |
| O | 27.8 | | |
| Ash | 30.9 | | |

The raw lignite was demineralized in order to study the effect of mineral matter on the properties of the final product. Removal of the inorganic constituents of the raw coal was accomplished by acid washing of the initial material with HCl and HF acid solutions [45]. About 120 ml of 5N HCl were added to 18 g of coal in a glass beaker. The mixture was stirred for one hour at room temperature. The treated coal was then filtered and washed with distilled water. Subsequently, the coal was mixed with 120 ml of full strength (22N) HF in a plastic beaker. This mixture was also stirred for one hour at 25±2°C, then filtered off and washed. The coal residue was then mixed with 120 ml of full strength (12N) HCl for a third treatment, for one hour at the same temperature. Finally, the coal was filtered, washed, and dried at 373K in an inert atmosphere.

CMS production commenced with the production of activated carbons by thermal treatment of lignite. Activation was carried out in a fixed bed reactor made of 1-½ inch diameter 316 SS tube, with a porous disc of Hastelloy plate (pore diameter 50 ìm). The flow diagram of the reactor system is shown in Figure 2. The reactor was heated to the desired temperature by a 5 kW electric furnace. Gas flow and temperature were controlled by the ancillary instrumentation.

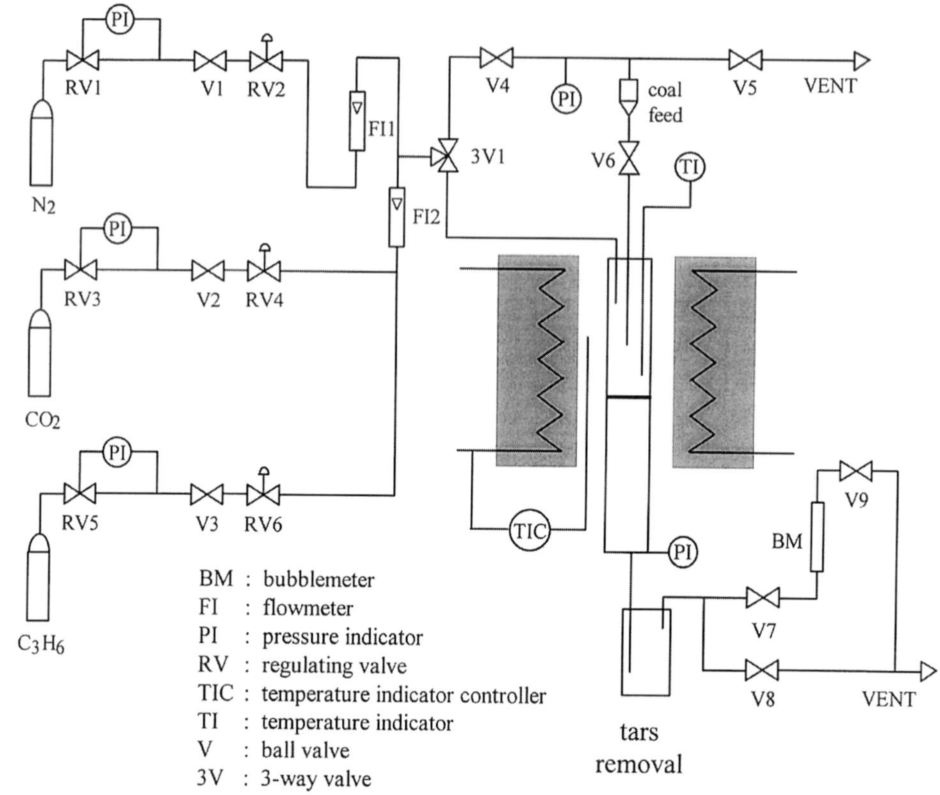


Figure 2. Pilot unit for activation of coal.

Activation was performed at temperatures of 800 and 900°C. The reactor was preheated to the desired temperature and purged with a continuous flow of 200 cc/min $N_2$. About 2-3 g of coal were flashed into the reactor and pyrolysed for 1 hour. The activation started when a mixture of $N_2$:$CO_2$=3:1 was introduced into the reactor. Char-coals were activated using different reaction times in order to prepare activated carbons with various burn-off levels. After a desired time elapsed, the reactor was rapidly cooled to room temperature while $N_2$ was passed through the reactor.

Carbon molecular sieves were prepared from activated carbons by coke deposition from the thermal cracking of propylene. The heat treatment was carried out in a smaller diameter U-shaped fixed bed reactor made of an ½ inch 316 SS tube. A 2 kW vertical electric furnace was used for heating the reactor, while the design of gas flow and temperature control instruments in the activation system permitted also their use in the smaller reactor system. The reaction cracking temperatures were 500 and 700°C. The reactor was preheated to the desired temperature under a continuous nitrogen flow, and 0.5 g of activated carbon were fed into the reactor. Coke deposition took place in flow, from a gas mixture containing 10% v/v propylene in nitrogen. CMS with different coke deposition levels were prepared by varying reaction times.

Activated carbons and CMS were characterized by measuring the adsorption isotherms of $N_2$ at 77 K and of $CO_2$ at 298 K. A low temperature, $N_2$ adsorption apparatus (Quantasorb, by Quantachrome) was used for the determination of $N_2$ adsorption isotherms. $CO_2$ adsorption data were obtained with a laboratory volumetric equipment. Surface areas were estimated from $N_2$ isotherms by using the BET multiple point equation and from $CO_2$ isotherms by using the Dubinin-Radushkevich equation. Prior to the measurement of $N_2$ adsorption, samples were outgassed at 383 K for 12 hours under helium flow. For $CO_2$ adsorption measurements, samples were first oven - dried at 383 K for 24 hours, and then outgassed overnight at 383 K, at a pressure of about 1 Pa.

Molecular sieving properties of produced CMS were evaluated by measuring the adsorption of $CO_2$, $CH_4$, $O_2$ and $N_2$ volumetrically, under ambient conditions. Samples were degassed at 383 K under vacuum for 1 hour prior to adsorption. Kinetic adsorption curves at 298 K were derived by calculating the amount of gas adsorbed with time. The selectivity or uptake ratio, expressed as the ratio of the adsorbed amount of $CO_2$ to $CH_4$ and of $O_2$ to $N_2$ at particular adsorption times, was also determined.

## 3. RESULTS AND DISCUSSION

Prior to coke deposition experiments, it was essential to establish the behavior of lignite during activation, to study the pore structure development of activated carbons, and to investigate the influence of mineral matter on the properties of the final product. Raw and acid treated lignites were used for the preparation of activated carbons. Pre-treatment of lignite with hydrochloric and hydrofluoric acid was effective in removing the inorganic constituents, the ash content being reduced from 31% to only 0.3%. Hydrochloric acid treatment of lignite resulted in the removal of HCl - soluble components of the mineral matter, such as carbonates, and in the extraction by ion exchange of some cations associated with the organic structure. Hydrofluoric acid treatment caused the removal of quartz and clay minerals [46].

*Activated carbon production*

Activated carbons were produced from raw and acid treated lignites by a two stage technique involving pyrolysis in an inert atmosphere for the removal of volatile matter followed by activation in carbon dioxide. The conversion of the overall reaction (carbonization and activation), generally known as burn-off, is defined as the fraction of the reacted organic matter, according to the following equation:

$$\text{Burn-off (\%)} = [(W_{in} - W_{out}) / W_{in}] \times 100 \qquad (1)$$

where $W_{in}$, the initial mass of lignite on a dry, ash free basis

$W_{out}$, the final mass of sample on a dry, ash free basis.

Pyrolysis of raw and acid lignite in a nitrogen flow took place at 800 and 900°C and pyrolysis time ranged from 15 min to 1 hour. Pyrolysis caused a reduction of carbon mass by 70 to 79% for raw lignite, the highest value been observed at the highest temperature. However, lower weight losses were obtained during demineralized lignite pyrolysis ranging from 60 to 65%. The lower weight loss values observed for the acid washed lignite could be attributed to the removal of a percentage of volatile matter during acid treatment, or to the removal of alkali and alkaline earth metals, whose presence can change the mechanisms of volatile gas release [47]. For both lignites, weight losses did not change significantly with reaction time. During pyrolysis of raw lignite at 900°C, weight loss ranged from 75% at 15 min to 79% at 1 hour. The corresponding weight losses for demineralized lignite were between 62 and 65%. It appears that a high proportion of volatile matter is released during the first minutes of pyrolysis, while no significant volatilization takes place at prolonged reaction times.

Activated carbons were produced from the chars obtained in the pyrolysis step, at temperatures of 800 and 900°C, using $CO_2$ as an activation medium. Figure 3 shows the overall burn-off as a function of activation time and temperature for both raw and demineralized lignite. Although reaction time did not play an

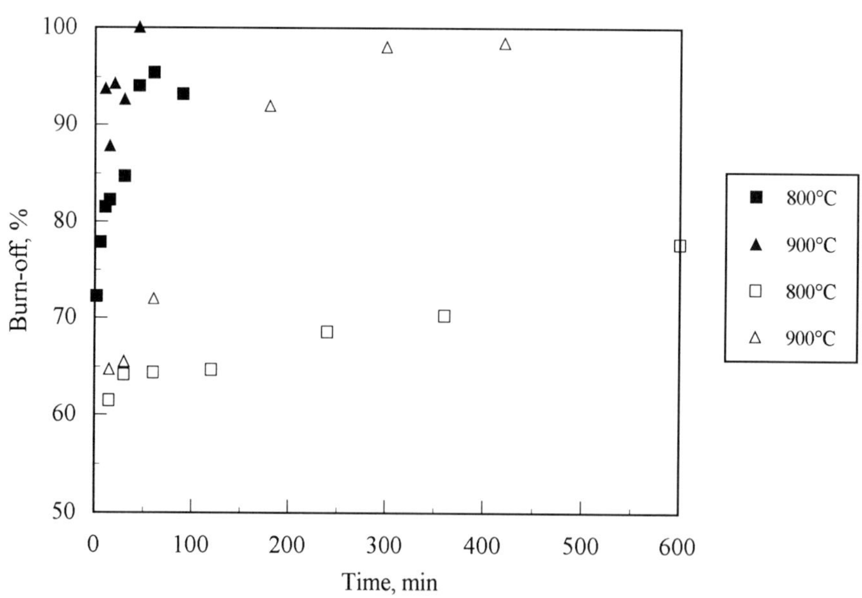


Figure 3. Burn-off of lignite as a function of reaction time (open symbols: demineralized lignite, closed symbols: raw lignite).

important role during carbonization, activation was strongly affected by this factor, as shown in this figure. Burn-off increased with time and temperature at a different rate for the two types of lignite. Thus, for carbons produced from raw lignite at 900°C, activation was complete after only 15 min of reaction in a flow of $N_2$:$CO_2$=3:1. At low temperatures (800°C), reaction rate decreased, and the burn-off was about 95% after 1.5 hours of reaction. Activation of acid washed lignite proceeded very slowly. At 800°C, burn-off was 61% after 30 min of activation, and increased to about 78% after 10 hours of reaction. However, at elevated temperatures of 900°C, gasification was more rapid and burn-off reached 98% after seven hours of gasification.

The differences in the activation rate between raw and acid treated lignite could be attributed to the removal of inorganic constituents from raw lignite. The mineral matter of raw lignite contains mainly Ca, K, Mg, Na and Fe. These inorganic species are excellent catalysts for char gasification in air, steam, $H_2$ and $CO_2$ [46, 48]. Among the metal cations, calcium and potassium have the highest effect on reactivity, while magnesium shows the lowest activity [49]. Pre-treatment of raw lignite in acid solutions resulted in the complete removal of catalytically active inorganic species, and thus in a significant change of the gasification rate.

The pore structure characterization of the activated carbons was studied by using $N_2$ and $CO_2$ as adsorbates. Figures 4 and 5 show the $N_2$ adsorption isotherms for carbons of various burn-off values, activated at 800 and 900°C,

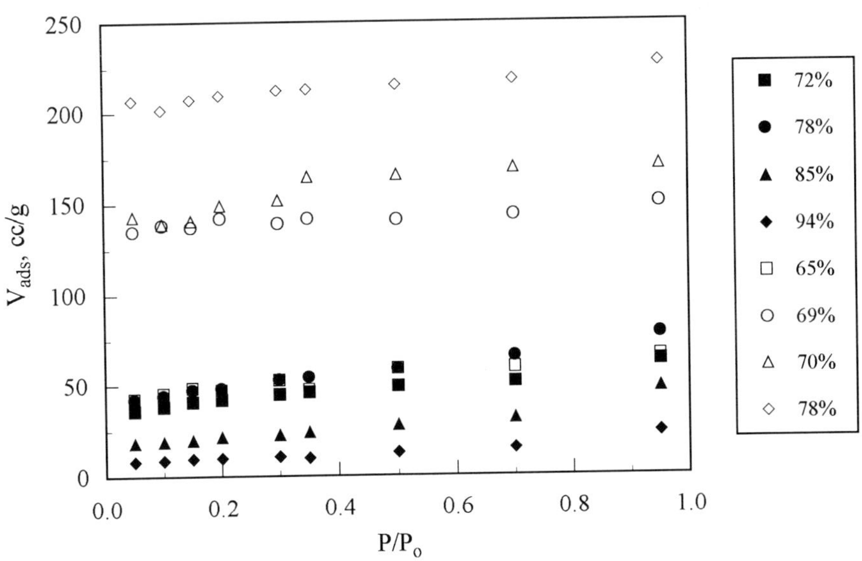


Figure 4. $N_2$ adsorption isotherms for carbons activated at 800°C (open symbols: demineralized lignite, closed symbols: raw lignite).

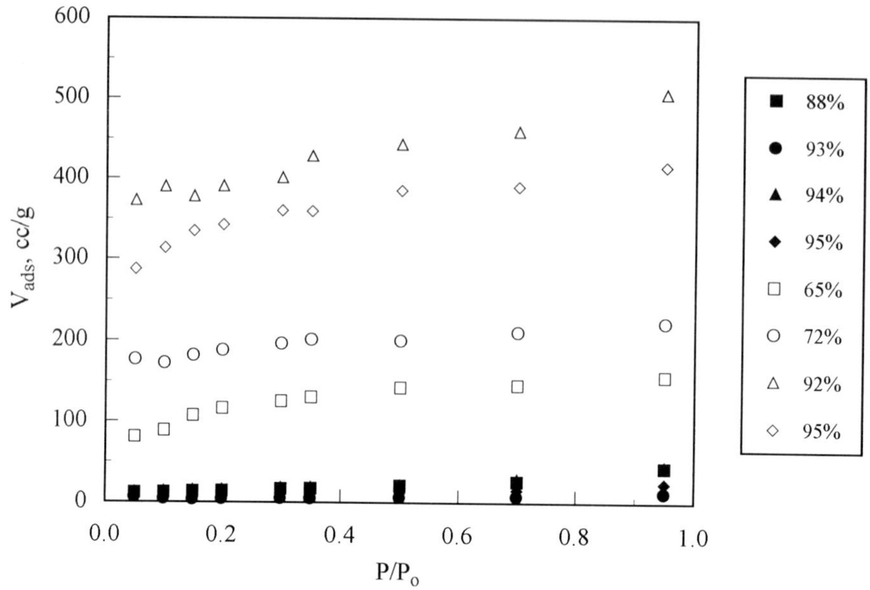


Figure 5. $N_2$ adsorption isotherms for carbons activated at 900°C (open symbols: demineralized lignite, closed symbols: raw lignite).

respectively. Since activated carbons are microporous materials, the $N_2$ adsorption isotherms are of type I, having a steep initial branch followed by a gradual approach to the linear branch at higher relative pressures. Once the micropores are filled with the adsorbate, the amount adsorbed did not increase significantly for higher relative pressures. About 50-75% of the pore volume was filled below $P/P_o$=0.10 indicating a narrow microporosity. However, differences were observed in the volume of gas adsorbed in samples from raw and acid treated lignite. Generally, samples prepared from acid treated lignite presented a higher adsorptive capacity than samples from raw lignite. This difference in adsorption capacity was more pronounced for high temperature (900°C) samples.

The BET surface area of all activated carbons produced is shown in Figure 6 as a function of overall burn-off, at temperatures 800 and 900°C. As anticipated from $N_2$ adsorption isotherms, there were clear differences in the surface areas of the two types of activated carbons. Activated carbons prepared from raw lignite had surface areas ranging from 7 to 153 $m^2/g$, while for activated carbons produced from acid-treated lignite the surface area reached about 1200 $m^2/g$.

Surface areas increased with burn-off up to a maximum value and then declined. At low burn-off, the two types of carbons exhibited almost comparable surface areas. However, as activation proceeded to higher burn-off, activated carbons from demineralized lignite had higher surface areas than carbons from

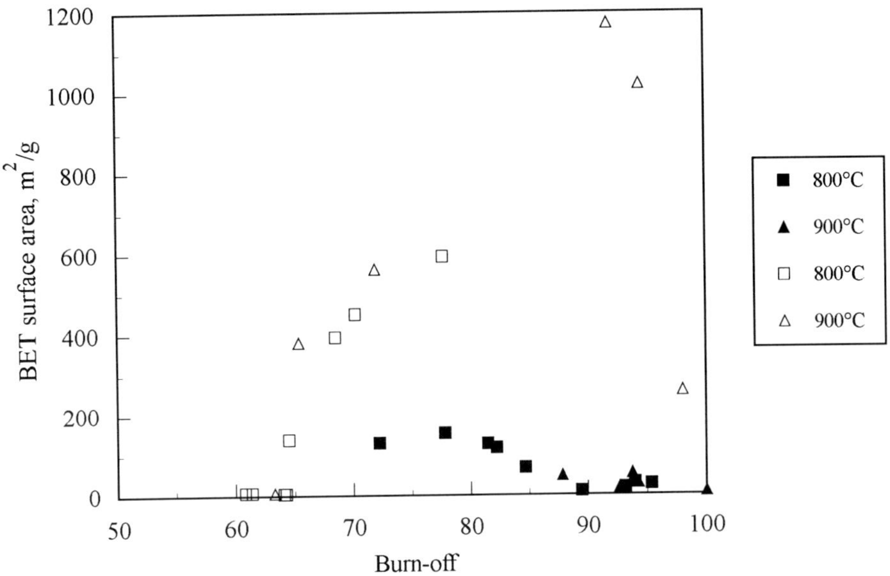


Figure 6. BET surface area of activated carbon as a function of burn-off (open symbols: demineralized lignite, closed symbols: raw lignite).

raw lignite. For the latter samples, the maximum surface area of 153 m$^2$/g was attained at a burn-off of about 78% at 800°C, while for activated carbons from acid treated lignite the highest surface area (1200 m$^2$/g) was measured for a sample activated at 900°C with 92% burn-off. This value is about eight times higher than the maximum surface area achieved for raw lignite.

In the case of microporous materials, surface areas estimated from $N_2$ adsorption usually suffer from activated diffusion of $N_2$, because of the low temperature of adsorption. Therefore, $CO_2$ adsorption at room temperature was also used for pore structure characterization of the activated carbons. Figures 7 and 8 show the $CO_2$ isotherms of samples with the same burn-off as those used for the $N_2$ adsorption isotherms (Figures 4 and 5). Adsorption isotherms show that the amount of $CO_2$ adsorbed, is affected by the degree of activation and increases considerably for carbons produced from demineralized lignite.

The $CO_2$ surface areas as a function of sample burn-off are shown in Figure 9, for the temperatures of 800 and 900°C. Surface areas were estimated by using the Dubinin-Radushkevich equation. Surface areas of carbons from raw lignite ranged between 20 and 205 m$^2$/g and were lower than those from acid treated lignite which ranged from 600 to 1200 m$^2$/g. For the latter samples, as is shown in Figures 6 and 9, BET areas are lower than $CO_2$ areas at low burn-off values, indicating their exclusively microporous structure.

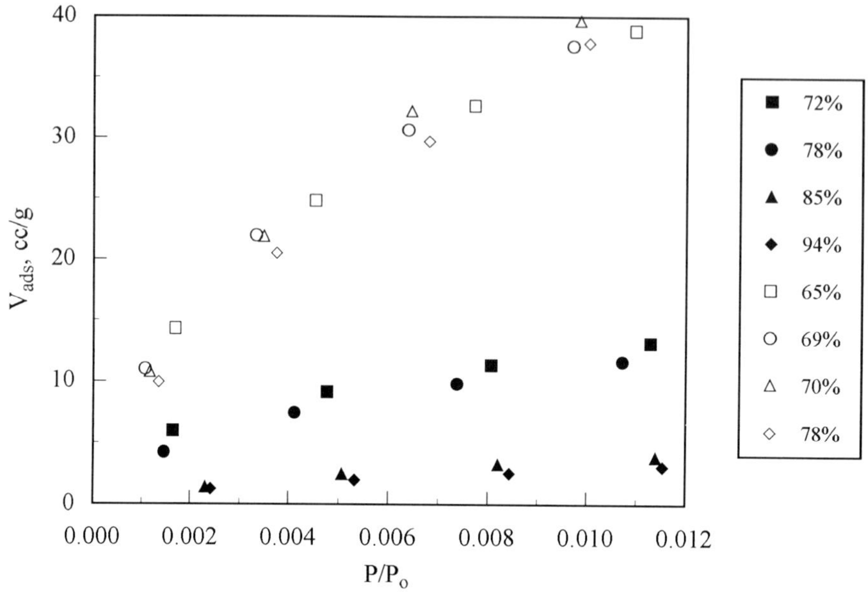


Figure 7. $CO_2$ adsorption isotherms for carbons activated at 800°C (open symbols: demineralized lignite, closed symbols: raw lignite).

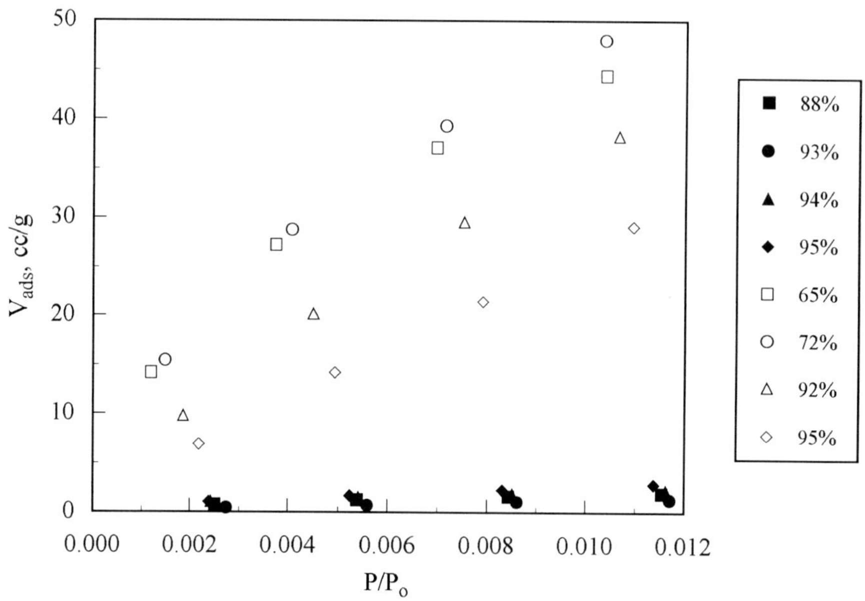


Figure 8. $CO_2$ adsorption isotherms for carbons activated at 900°C (open symbols: demineralized lignite, closed symbols: raw lignite).

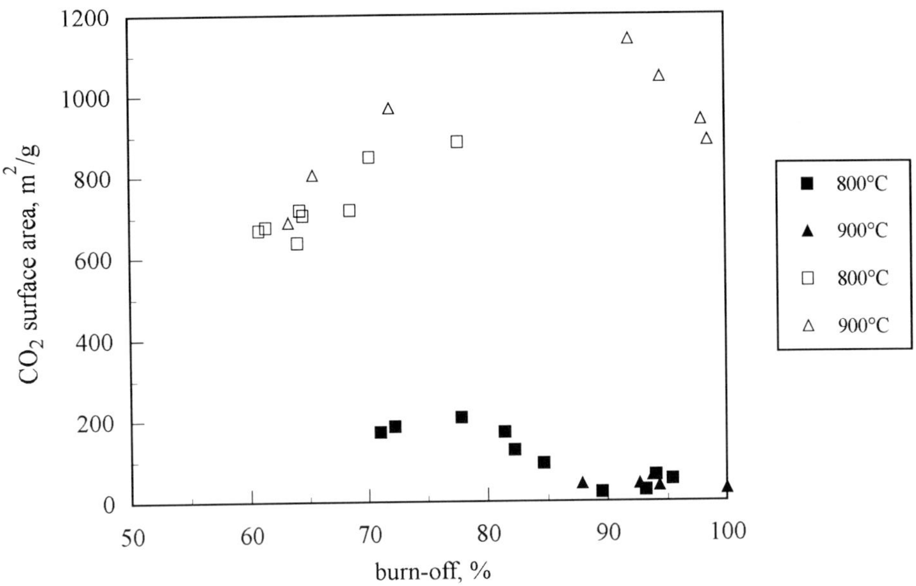


Figure 9. $CO_2$ surface area of activated carbon as a function of burn-off (open symbols: demineralized lignite, closed symbols: raw lignite).

These results show that pore structure development during activation is the result of a balance between two competing mechanisms: The first one is the formation of new pores and the opening of closed pores, while the second one is associated with micropore widening. During the initial stages of activation, the former phenomena appear to contribute to porosity development. As activation proceeds to higher values, the latter mechanism seems to dominate and results in a decrease of the surface areas. Furthermore, although temperature increased significantly the activation rate, it did not affect the pore structure development for both types of carbons, as samples activated at different temperatures with similar burn-offs presented similar surface areas.

Inorganic matter affects the mechanism of several reactions, such as the release and subsequent reaction of oxygen functional groups, and secondary hydrocarbon cracking, thereby changing the reaction rate and the pore structure of the final product. The presence of inorganic constituents seems to enhance the activation reaction locally, resulting in the development of mesopores and macropores, and in the decrease of micropores. However, activation of acid treated lignite proceeds uniformly, throughout the particle interior, with a very low rate, resulting in the formation of a highly microporous structure.

The molecular sieving behavior of produced activated carbons was examined by measuring the adsorption uptake curves of $CO_2$, $CH_4$, $O_2$ and $N_2$ at room temperature. These gases were selected due to the significance they present to industrial applications. $CO_2$ uptake curves for carbons activated at 800 and 900°C are shown in Figures 10 and 11, respectively, while the corresponding $CH_4$ curves for the same samples are shown in Figures 12 and 13. In all cases, carbon dioxide adsorption was higher than that of methane, indicating a molecular sieving property. Activated carbons from demineralized lignites gave higher adsorption capacities than carbons from raw lignite, due to the higher surface area of the former. The equilibrium adsorption capacity of demineralized samples, activated at 800°C, ranged between 22 and 30 cc/g for $CO_2$, while for activated carbons from raw lignite it reached up to 10 cc/g. These results are in accordance with the surface area trends of Figures 6 and 9. Samples activated at different temperatures, but with similar burn-off and surface areas, gave similar adsorption capacities, indicating that temperature did not affect the pore structure development.

One useful criterion for the evaluation of sieving quality achieved by a carbon is to compare the ratio of uptaken volumes of carbon dioxide and methane at a fixed period of time, say 1 min [37,40]. The estimated $CO_2/CH_4$ selectivity ratios for various activated samples are given in Table 3.

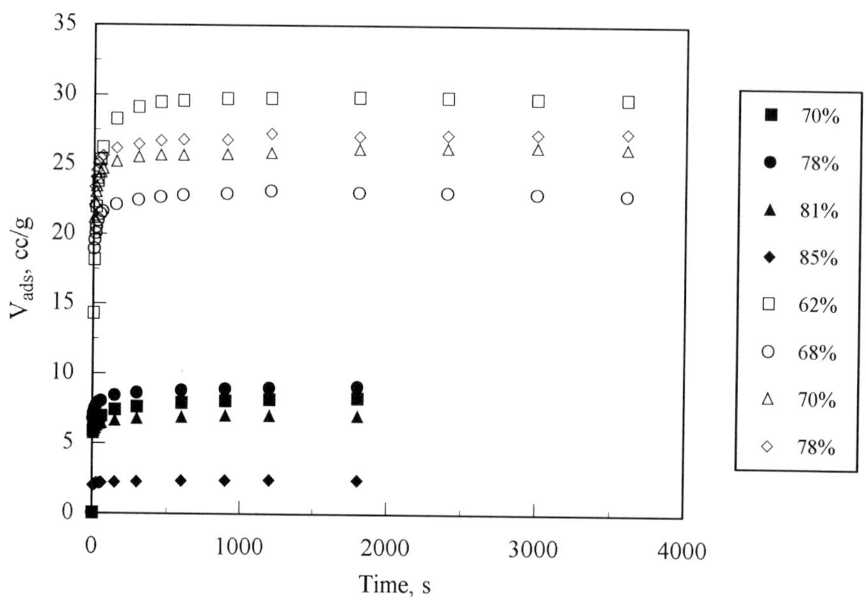


Figure 10. $CO_2$ uptake curves for carbons activated at 800°C (open symbols: demineralized lignite, closed symbols: raw lignite).

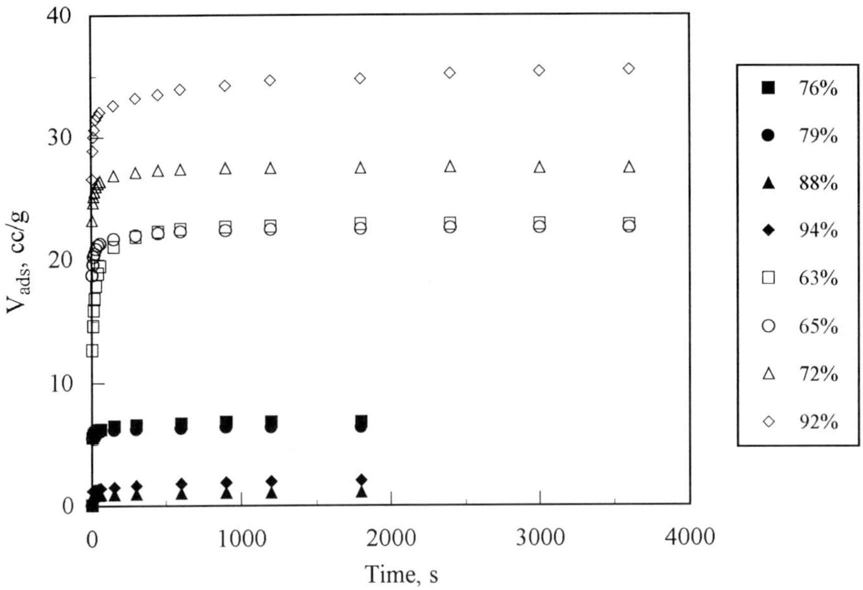


Figure 11. $CO_2$ uptake curves for carbons activated at 900°C (open symbols: demineralized lignite, closed symbols: raw lignite).

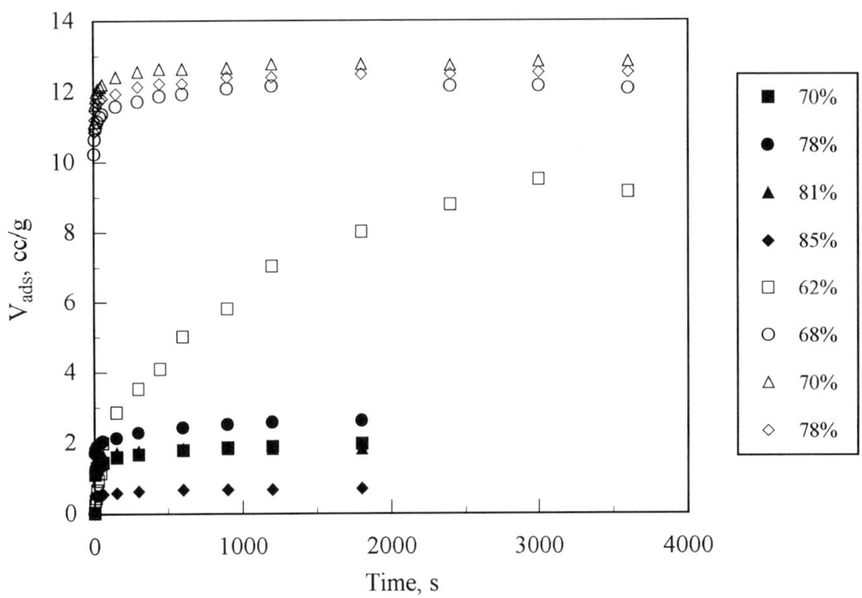


Figure 12. $CH_4$ uptake curves for carbons activated at 800°C (open symbols: demineralized lignite, closed symbols: raw lignite).

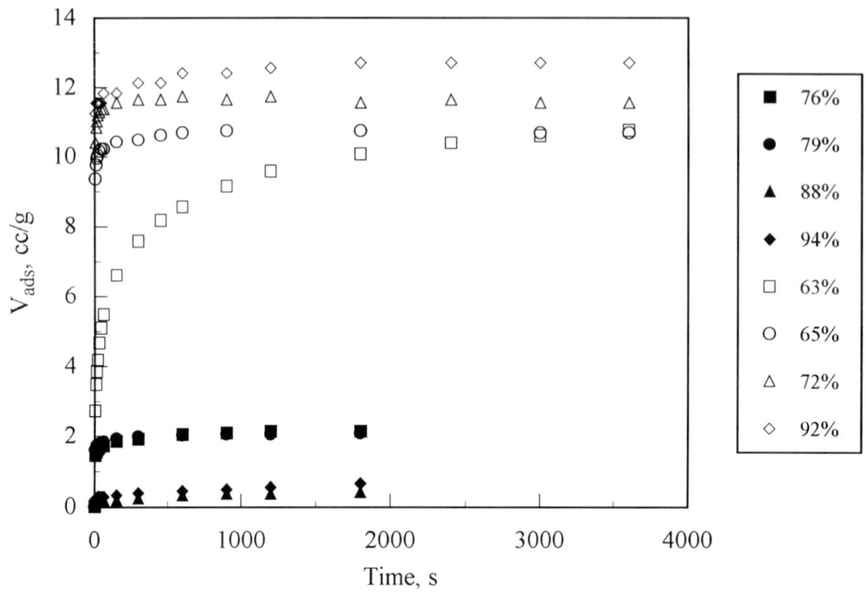


Figure 13. $CH_4$ uptake curves for carbons activated at 900°C (open symbols: demineralized lignite, closed symbols: raw lignite).

Activated carbons obtained from raw lignite gave comparable selectivities, ranging from 3.2 to 4.9, while demineralized samples exhibited enhanced selectivity ratios, up to 22.0. High selectivities were observed for demineralized samples of elevated $CO_2$ surface area (700 $m^2/g$) but low BET area (50 $m^2/g$), cf. Figures 6 and 9. Therefore, it seems that partial activation of demineralized lignite at short reaction times results in production of activated carbons with an increased microporous and a restricted mesoporous structure which exhibits some molecular sieving properties.

Similarly, the selectivity ratios of $O_2/N_2$, as obtained from their respective uptake curves (Figures 14 - 17), are shown in Table 3. The adsorption capacity for $O_2$ and $N_2$ is lower than for the carbon dioxide equilibrium capacity, with low selectivity ratios of 1.0 to 2.2. Selectivity improved slightly at low burn-off values but still the results indicated poor kinetic separation effects. The kinetic diameters of $CO_2$, $CH_4$, $O_2$ and $N_2$ are 3.3, 3.8, 3.46 and 3.64 Å respectively. Hence, the $CO_2$ molecule, being the smallest of all, could be quickly and preferentially adsorbed in the activated carbon pore structure, over the largest methane molecule. However the kinetic diameters of $O_2$ and $N_2$ differ only slightly, which makes their separation difficult in the existing activated carbons from raw lignite. Such separation would require the preparation of a product with a narrow range of micropores, which could impose greater resistance to $N_2$ adsorption compared to $O_2$.

Table 3
$CO_2/CH_4$ and $O_2/N_2$ selectivity ratios of activated carbons at 1 min

| Starting material | T (°C) | Burn-off % | $CO_2/CH_4$ | $N_2/O_2$ |
|---|---|---|---|---|
| Raw Lignite | 800 | 70 | 4.9 | 1.7 |
| | | 78 | 3.9 | 1.5 |
| | | 81 | 4.0 | 1.1 |
| | | 85 | 4.0 | 1.0 |
| | 900 | 75 | 3.2 | 1.5 |
| | | 76 | 3.6 | 1.0 |
| | | 79 | 3.3 | 1.5 |
| | | 88 | 3.7 | 1.2 |
| | | 94 | 3.5 | 1.0 |
| Acid-treated | 800 | 61 | 22.0 | 1.9 |
| Lignite | | 62 | 13.3 | 2.2 |
| | | 65 | 2.3 | 1.7 |
| | | 68 | 1.9 | 1.0 |
| | | 70 | 2.1 | 1.3 |
| | | 78 | 2.2 | 1.0 |
| | 900 | 63 | 3.6 | 1.7 |
| | | 64 | 12.0 | 2.0 |
| | | 72 | 2.3 | 1.0 |
| | | 92 | 2.7 | 1.0 |

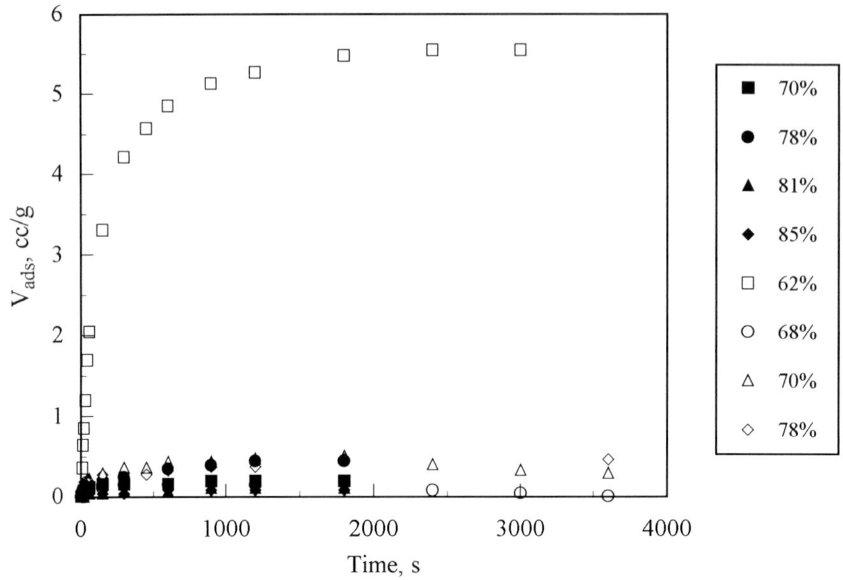


Figure 14. $O_2$ uptake curves for carbons activated at 800°C (open symbols: demineralized lignite, closed symbols: raw lignite).

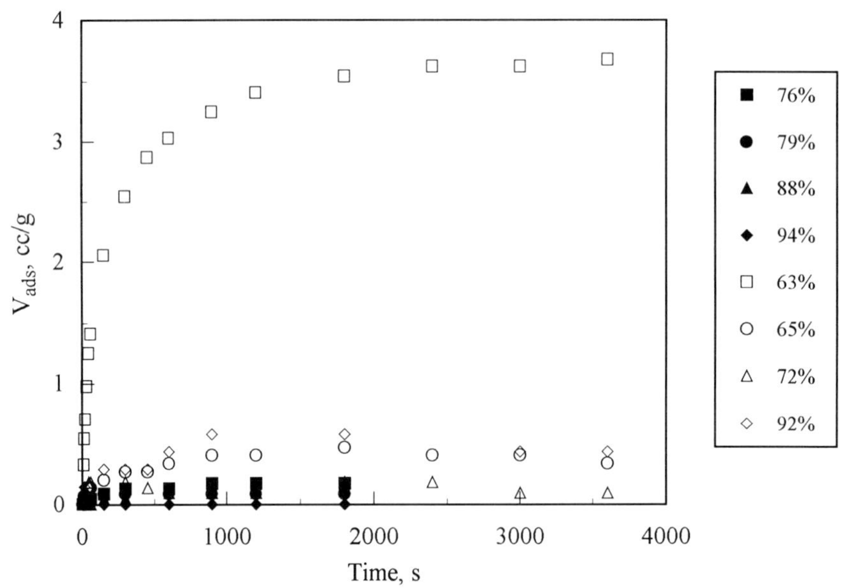


Figure 15. $O_2$ uptake curves for carbons activated at 900°C (open symbols: demineralized lignite, closed symbols: raw lignite).

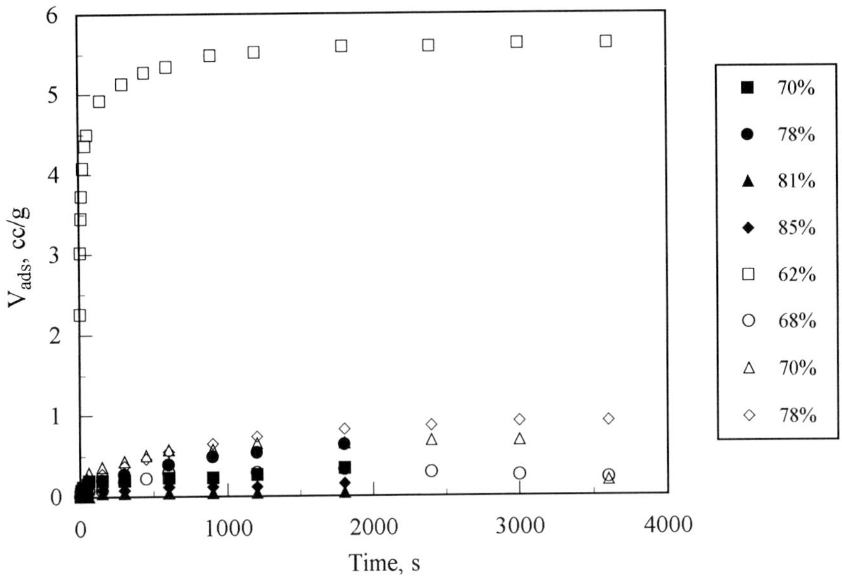


Figure 16. $N_2$ uptake curves for carbons activated at 800°C (open symbols: demineralized lignite, closed symbols: raw lignite).

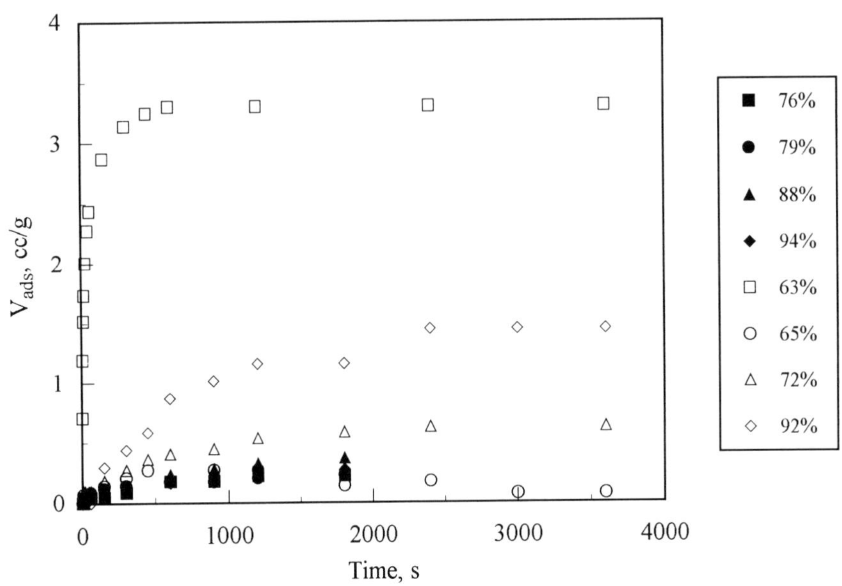


Figure 17. $N_2$ uptake curves for carbons activated at 900°C (open symbols: demineralized lignite, closed symbols: raw lignite).

*Propylene pyrolysis*

Partial gasification of demineralized lignite gave activated carbons with better molecular sieving properties than those produced from raw lignite. However, complete removal of inorganic matter by acid treatment caused a reduction in the activation and the hydrocarbon cracking reaction rates. Preliminary experiments with demineralized activated carbons, under pure propylene flow (150 cc/min) at 700°C, resulted in low propylene cracking rate, and low weight loading of the products. This may be attributed to the absence of any catalytic action by the removed inorganic matter. For this reason, further tests on hydrocarbon deposition were performed on activated carbon from raw lignite, where the presence of inorganic matter enhances the cracking reaction rate. The activated carbon selected as the starting material for the examination of hydrocarbon cracking was the sample with 78% burn-off. This sample has almost similar $CO_2/CH_4$ selectivity ratio (3.9) to the sample with 70% burn-off, but presents a higher $CO_2$ equilibrium capacity (8.0 cc/g compared to 6.9 cc/g), due to its higher $CO_2$ surface area (205 $m^2/g$ compared to 183 $m^2/g$).

The selected sample was exposed to a stream of 10 - 20% propylene in nitrogen, at two temperatures, 500 and 700°C, for a period of 5 to 60 min. Deposition results are depicted in Figure 18. At low propylene concentration

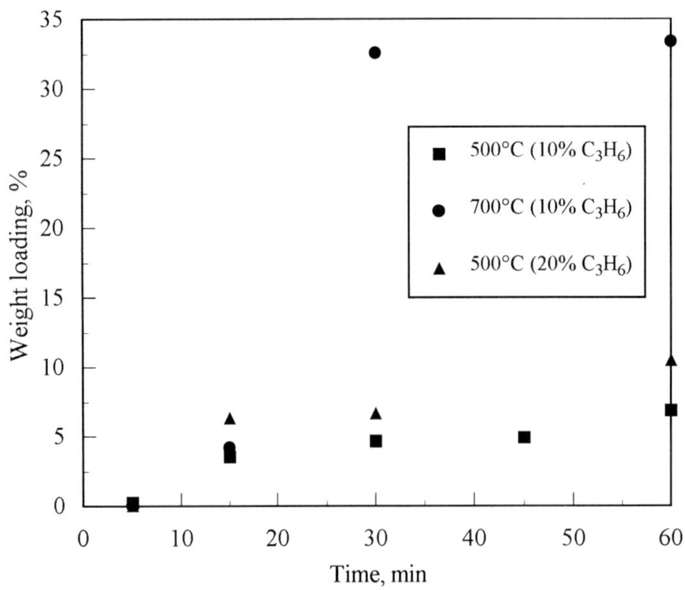


Figure 18. Coke deposition by propylene cracking on activated carbon, as a function of time and temperature.

(10%) and low temperature (500°C), the deposition rate was slow. About 5% carbon loading was achieved after 1 h of reaction. The deposition increased with temperature, although at 700°C a small carbon loading was observed during the first 20 minutes of reaction, compared to 500°C. However, upon further pyrolysis, a sharp increase in carbon deposition took place, resulting in about 33% loading after 30 min of reaction. Further reaction led to insignificant increase in carbon loading (34% after 1h), possibly due to the closure of pores accessible to propylene [40]. Weight loading slightly increased with propylene concentration in the gas stream. After 1 h of activated carbon exposure in a stream of 20% propylene at 500°C, carbon deposition was about 10%.

Results indicate that the rate and amount of propylene cracking over the microporous activated carbon depends upon cracking conditions such as temperature, time and propylene partial pressure. Intense propylene pyrolysis conditions, such as high temperature and propylene concentration, favor carbon deposition. Thus, selection of proper experimental conditions is important for successful conversion of microporous carbons to useful molecular sieves.

The pore structure of molecular sieves produced by propylene cracking, was evaluated by $N_2$ and $CO_2$ adsorption. $N_2$ adsorption isotherms, measured at 77K, and estimated BET surface areas, are shown in Figures 19 and 20 for samples prepared under various experimental conditions. These adsorption isotherms

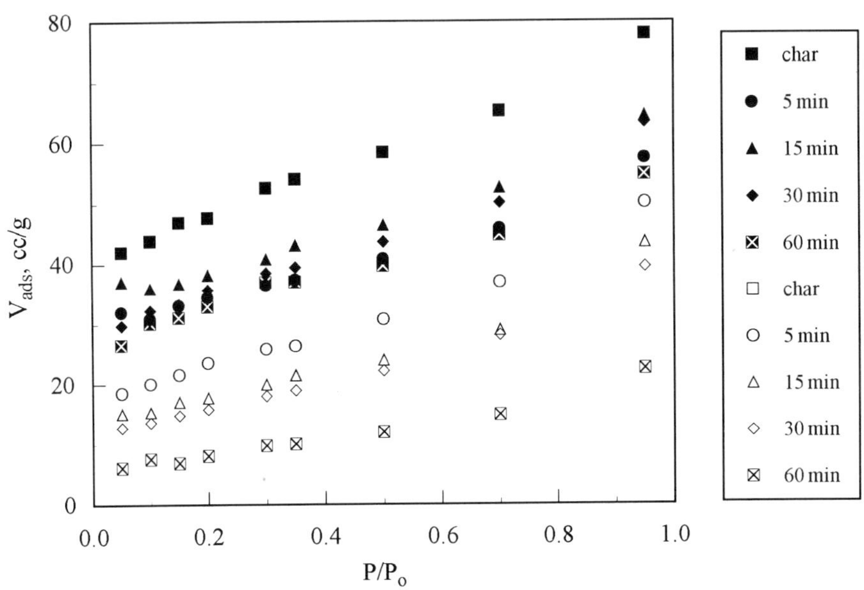


Figure 19. $N_2$ adsorption isotherms of CMS obtained at various $C_3H_6$ cracking times (open symbols: 700°C, closed symbols: 500°C).

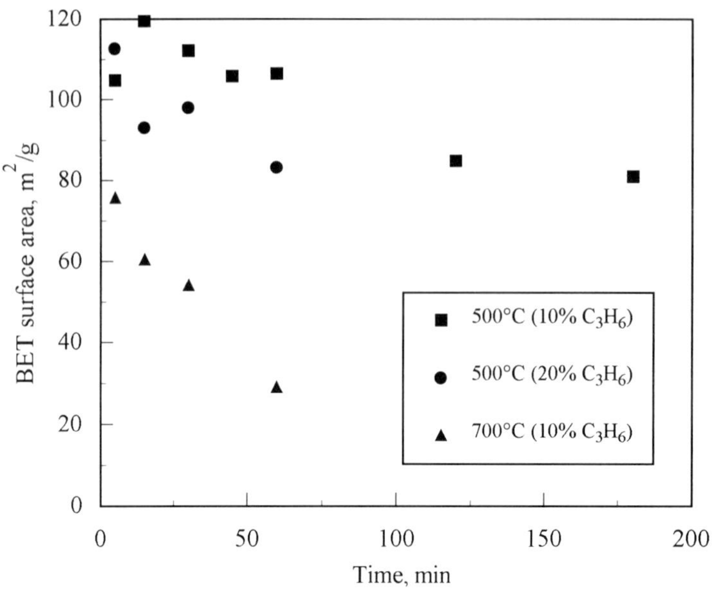


Figure 20. BET surface area of CMS as a function of reaction time and temperature.

indicate a material with a rather wide pore size distribution. The linear branch at relative pressures higher than 10%, was no more parallel to the horizontal axis indicating the filling of wide micropores and mesopores by adsorbate.

Samples prepared at 500°C in 10% propylene, had surface areas of 80 to 120 $m^2/g$, the lowest surface areas corresponding to extended coke deposition conditions and prolonged reaction times (2 and 3 hours). Similar surface areas were also observed for samples prepared in 20% propylene flow. Surface areas decline faster with cracking time at elevated temperatures (700°C), ranging between 75 and 25 $m^2/g$.

Carbon dioxide isotherms at room temperature, and the corresponding surface areas estimated by the Dubinin-Radushkevich equation, are shown in Figures 21 and 22 as a function of reaction time and temperature. $CO_2$ surface areas changed only slightly with $C_3H_6$ reaction time and ranged between 130 and 170 $m^2/g$ at 500°C, and 104 to 164 $m^2/g$ at 700°C. For all samples, no distinct trends are identified on the possible effect of temperature and cracking reaction time on $CO_2$ surface area development.

These results suggest that, under the experimental conditions used, propylene molecules concentrated mainly in the mesopore area, where the subsequent coke deposition resulted in the progressive closure of mesopores and the decrease in BET surface area. The higher the temperature and the reaction time, the higher the amount of coke deposited in the mesopores and the higher the decrease in

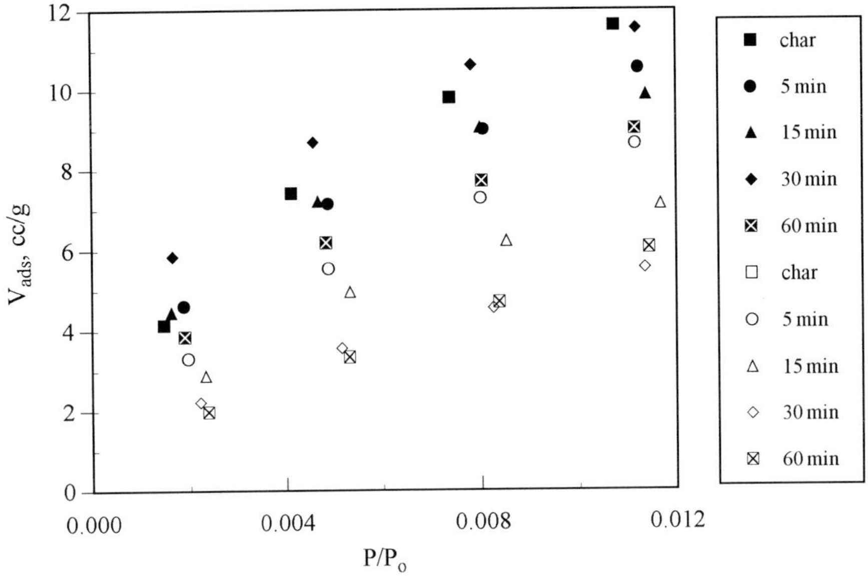


Figure 21. $CO_2$ adsorption isotherms of CMS obtained at various $C_3H_6$ cracking times(open symbols: 700°C, closed symbols: 500°C).

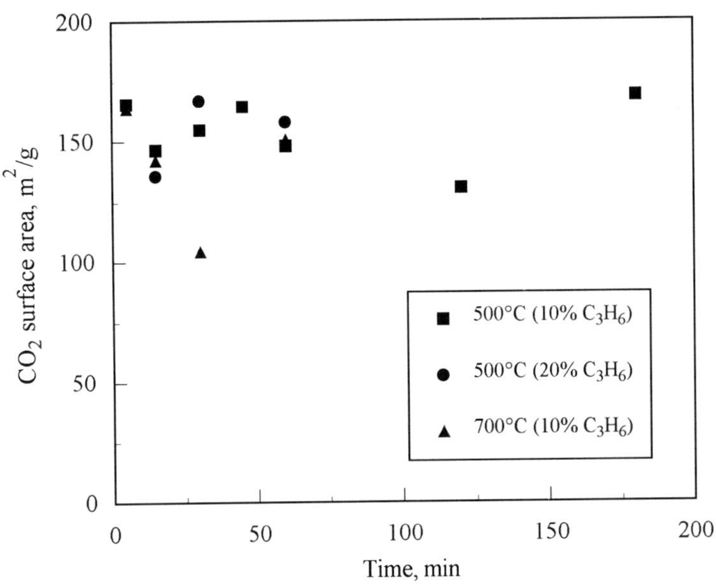


Figure 22. $CO_2$ surface area of CMS as a function of reaction time and temperature.

BET surface area. However, micropores were not affected significantly by propylene cracking, resulting in only minor changes in $CO_2$ surface area. Therefore, samples prepared by cracking at high temperature for prolonged periods, presented a higher micropore than mesopore surface area, and they may possess molecular sieving properties.

Kinetic adsorption curves of $CO_2$, $CH_4$, $N_2$ and $O_2$ (Figures 23 to 26), were used to estimate the molecular sieving behavior of samples obtained by $C_3H_6$ cracking for gas separation. The selectivity ratios of $CO_2/CH_4$ and $N_2/O_2$, estimated from the adsorption kinetic curves at 1 min, are given in Table 4.

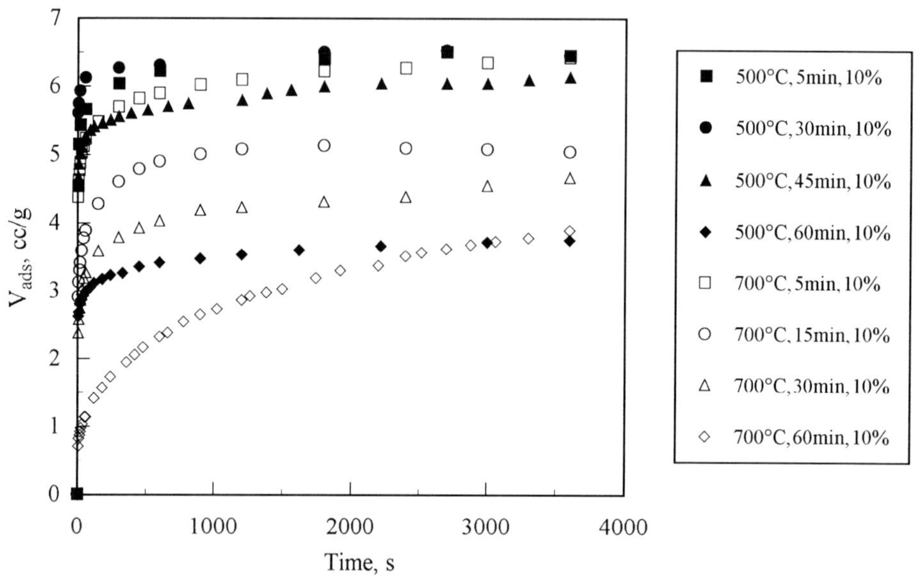


Figure 23. $CO_2$ uptake curves of CMS from raw lignite, after activation and $C_3H_6$ cracking.

Table 4
Selectivity ratios of activated carbons after propylene cracking, estimated from kinetic adsorption curves at 1 min

| Temperature (°C) | Cracking Time (min) | $CO_2/CH_4$ | $O_2/N_2$ |
|---|---|---|---|
| 500 | 5 | 3.5 | 2.0 |
| 500 | 30 | 2.9 | 2.3 |
| 500 | 45 | 3.5 | 2.4 |
| 500 | 60 | 4.1 | 2.8 |
| 500 | 180 | 3.7 | 2.2 |
| 700 | 5 | 5.3 | 2.5 |
| 700 | 15 | 6.7 | 1.0 |
| 700 | 30 | 7.4 | 3.0 |
| 700 | 60 | 10.4 | - |

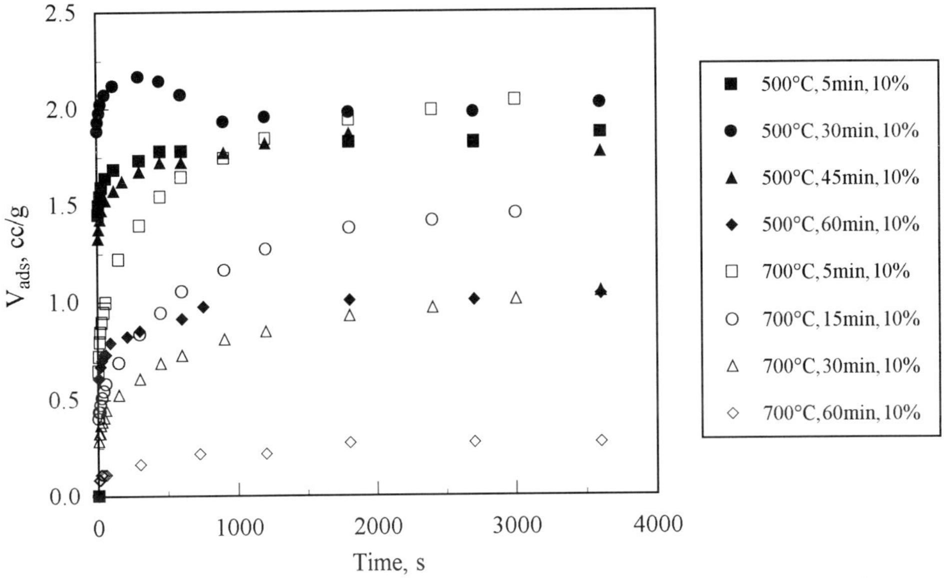


Figure 24. $CH_4$ uptake curves of CMS from raw lignite, after activation and $C_3H_6$ cracking.

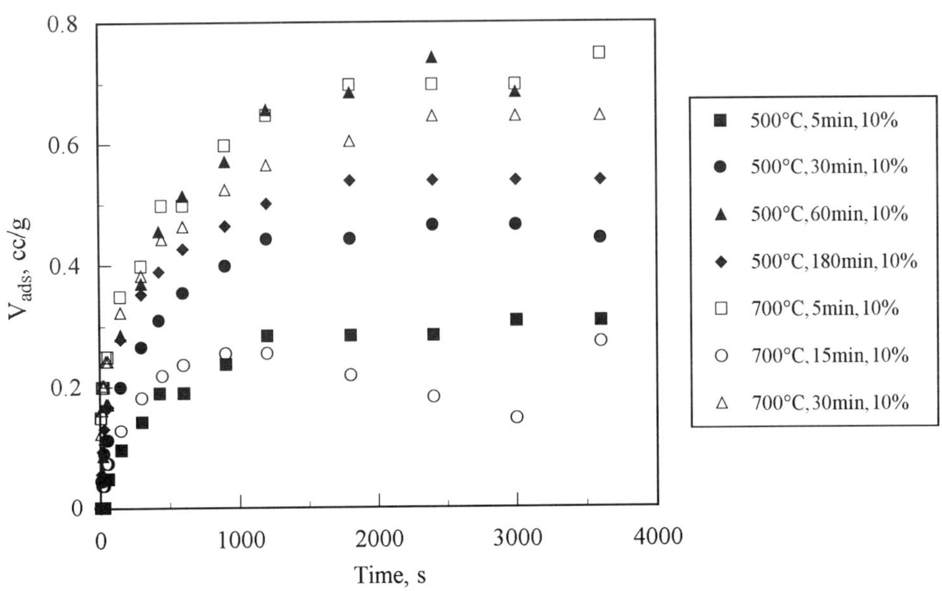


Figure 25. $O_2$ uptake curves of CMS from raw lignite, after activation and $C_3H_6$ cracking.

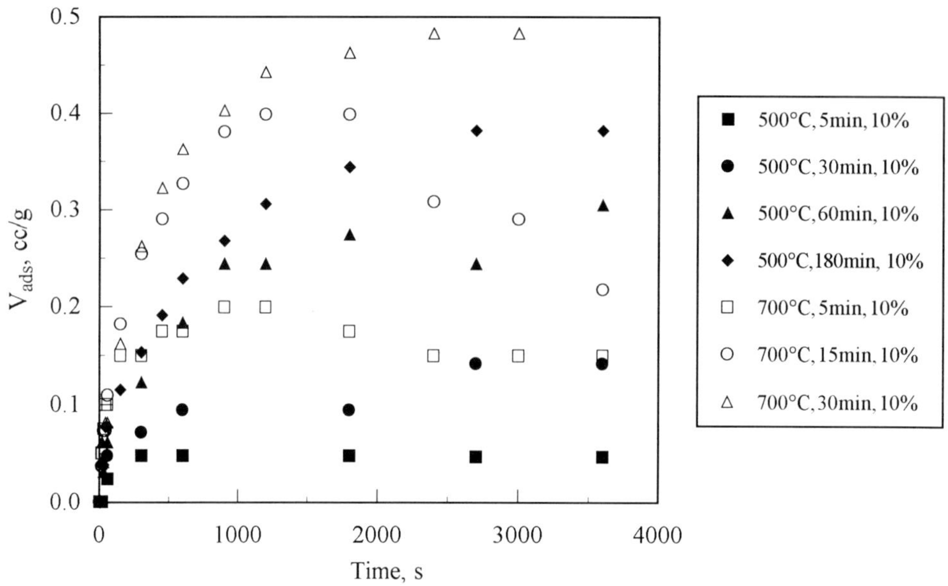


Figure 26. $N_2$ uptake curves of CMS from raw lignite, after activation and $C_3H_6$ cracking.

In all cases, $CO_2$ uptake was higher than that of $CH_4$. For samples prepared by $C_3H_6$ cracking at 500°C, the uptake curves were quite similar, and equilibrium uptake capacity decreased at long cracking times. A significant reduction in uptake capacity was observed for samples prepared at 700°C, especially for $CH_4$ and at long reaction periods. Selectivity ratios of samples prepared at 500°C varied between 2.9 and 4.1, while selectivities improved up to 10.4 for samples prepared at 700°C. The highest selectivity ratio was observed for CMS produced by propylene pyrolysis at 700°C for 60 min.

Similar results were obtained for oxygen and nitrogen uptakes, shown in Figures 25 and 26. For all samples oxygen uptakes were higher than the corresponding nitrogen ones. As shown in Table 4, rather low $O_2/N_2$ selectivity ratios were achieved, i.e. 2.0 to 2.8, for CMS prepared at 500°C, and 1.0 to 3.0 for CMS prepared at 700°C. These values are similar to the ones reported in the literature [33].

Results of this work indicate that propylene cracking takes place mainly in the mesopores, thus the micropore area remains nearly unchanged. At low temperatures and reaction times, coke deposition is restricted, and samples have comparable BET and $CO_2$ surface areas, exhibiting negligible molecular sieving properties and low $CO_2/CH_4$ and $O_2/N_2$ selectivity ratios. As the temperature increases, extended propylene cracking takes place, resulting in the decrease of

BET surface area, while the $CO_2$ surface area is only slightly affected. Samples prepared under these conditions present molecular sieving behavior, as ensured further from the selectivity ratio of $CO_2/CH_4$. Thus, selection of proper experimental conditions is important for successful conversion of microporous carbons to useful molecular sieves.

## 4. CONCLUSIONS

Greek lignite with a high content of inorganic matter was used for the production of activated carbons and carbon molecular sieves. High $CO_2$ activation rates were observed at high temperatures, possibly due to the catalytic action of inorganic constituents in the lignite ash. Removal of inorganic matter was accomplished by acid treatment of lignite and resulted in decreased reactivity of the products to $CO_2$ activation. Activated carbons from acid treated lignite had higher surface area than carbons from raw lignite. Partial activation of demineralized lignite at low burn-off values resulted in samples with a molecular sieving behavior to gases such as $CO_2$, $CH_4$, $O_2$ and $N_2$, which was attributed to the extended microporous and to the restricted mesoporous structure.

Carbon molecular sieves were produced from raw lignite activated carbons by employing propylene cracking as a method for coke deposition. The rate of coke deposition increased with temperature, reaction time and propylene concentration in the feed gas mixture. These factors also affected the pore structure development. Extended coke deposition was obtained at high cracking temperatures, resulting in closure of the mesopores, and in a sharp decrease of BET surface area, without affecting the micropore surface area. Kinetic adsorption curves of $CO_2$, $CH_4$, $O_2$ and $N_2$ on samples prepared at high temperatures, gave high selectivity ratio values of $CO_2/CH_4$ and $O_2/N_2$. Mild experimental conditions of low temperature and low propylene concentration resulted in restricted coke deposition into the particle interior. Samples prepared under such conditions, presented similar BET and $CO_2$ surface areas and negligible molecular sieving behavior, as indicated by low $CO_2/CH_4$ selectivities.

## ACKNOWLEDGEMENTS

The financial support of ECSC, Contract No. 7220-EC/766, is gratefully acknowledged.

## REFERENCES

1. H. Jüntgen, K. Knoblauch and K. Harder, Fuel, 60 (1981) 817.
2. S.K. Verma, Carbon, 29 (1991) 793.
3. A. Kapoor and R.T. Yang, Chem. Eng. Sci., 44 (1989) 1723.

4. S.N. Vyas, S.R. Patwardhan and H.B. Natraj, J. Chem. Soc. Faraday Trans., 86 (1990) 3455.
5. K. Miura, J. Hayashi, T. Kawaguchi and K. Hashimoto, Carbon, 31 (1993) 667.
6. M. Srivastava and H. Singh, Erdöl und Kohle-Erdgass-Petrochemie Vereinigt mit Brennstoff-Chemie, 47 (1994) 242.
7. A.I. Shirley and A.I. LaCava, AIChE J., 41 (1995) 1389.
8. R.T. Yang (ed.), Gas Separation by Adsorption Processes, Butterworths Publishers, Boston, 1987.
9. M. Suzuki (ed.), Adsorption Engineering, Chemical Engineering monographs, vol. 25, Elsevier Science Publishers, Amsterdam, 1989.
10. G.E. Keller II and R.T. Yang (eds.), New Directions in Sorption Technology, Butterworths Publishers, Boston, 1989.
11. H. Marsh, E.A. Heinz and F. Rodriguez-Reinoso (eds.), Introduction to Carbon Technology, University of Alicante, 1997.
12. J.W. Patrick (ed.), Porosity in Carbons: Characterization and Applications, Edward Arnold, 1995.
13. C.H. Chang, G.J. Seminara and A.E. VanTil, US Patent No. 5 461 023 (1995).
14. C.H. Chang, US Patent No. 5 308 821 (1994).
15. L.M. Aparicio, S.R. Keenan and L. Wang, US Patent No. 5 292 707 (1994).
16. C.H. Chang, US Patent No. 4 902 312 (1990).
17. S.F. Yates, US Patent No. 4 902 825 (1990).
18. S.F. Yates, US Patent No. 4 940 824 (1990).
19. C.H. Chang, G.J. Seminara, US Patent No. 4 820 318 (1989).
20. A. Hong, R.K. Mariwala, M.S. Kane and H.C. Foley, Ind. Eng. Chem. Res., 34 (1995) 992.
21. G.C. Grunewald and R.S. Drago, J. Am. Chem. Soc., 113 (1991) 1636.
22. F. Cavani and F. Trifiro, Applied Catalysis A: General, 133 (1995) 219.
23. J.L. Schmitt, Carbon, 29 (1991) 743.
24. H.C. Foley, G.C. Sonnichsen, L.D. Brake, R.K. Mariwala and D.S. Lafyatis, US Patent No. 5 482 909 (1996).
25. J.E. Koresh and A. Soffer, Sep. Sci. Techn., 22 (1987) 973.
26. K. Haraya, H. Suda, H. Yanagishita and M. Matsuda, J. Chem. Soc. Commun., 17 (1995) 1781.
27. C.W. Jones and W.J. Koros, Ind. Eng. Chem. Res., 34 (1995) 158.
28. M.B. Rao and S. Sircar, J. Membrane Sci., 85 (1993) 253.
29. C.W. Jones and W.J. Koros, Carbon, 30 (1995) 1419.
30. A.L. Cabrera, J.E. Zehner, C.G. Coe, T.R. Gaffney, T.S. Farris and J.N. Armor, Carbon, 31 (1993) 969.
31. R.K. Mariwala and H.C. Foley, Ind. Eng. Chem. Res., 33 (1994) 607.
32. S.N. Vyas, S.R. Patwarthan and B. Gangadhar, Carbon, 30 (1992) 605.
33. S.N. Vyas, S.R. Patwarthan, G. Vijayalakshmi and B. Gangadhar, Fuel, 72 (1993) 551.
34. T.G. Lamond, J.E. Metcalfe and P.L. Walker Jr., Carbon, 3 (1965) 59.

35. P.L. Walker Jr., T.G. Lamond and J.E. Metcalfe, Proc. 2nd Conf. Ind. Carbon and Graphite (1965) 7.
36. R.L. Patel, S.P. Nandi and P.L. Walker Jr., Fuel, 51 (1972) 47.
37. S.K. Verma and P.L. Walker Jr., Carbon, 28 (1990) 175.
38. T. Ohsaki and S. Abe, US Patent No. 4 458 022 (1984).
39. C. Marumo, E. Hayata and N. Shiomi, US Patent No. 4 933 314 (1990).
40. S.K. Verma and P.L. Walker Jr., Carbon, 30 (1992) 829.
41. S.K. Verma and P.L. Walker Jr., Carbon, 31 (1993) 1203.
42. O.P. Mahajan and P.L. Walker Jr., Fuel, 58 (1979) 333.
43. P.L. Walker Jr., Fuel, 60 (1981) 801.
44. P. Samaras, E. Diamadopoulos and G.P. Sakellaropoulos, Carbon, 32 (1994) 771.
45. L.R. Radovic, P.L. Walker Jr. and R.G. Jenkins, Fuel, 62 (1983) 209.
46. P. Samaras, E. Diamadopoulos and G.P. Sakellaropoulos, Fuel, 75 (1996) 1108.
47. H.N.S. Schafer, Fuel, 59 (1980) 295.
48. F. Kapteijn and J.A. Moulign, Fuel, 62 (1983) 221.
49. F. Kapteijn, O. Peer and J.A. Moulign, Fuel, 65 (1986) 1371.

Adsorption and its Applications in Industry and Environmental Protection
Studies in Surface Science and Catalysis, Vol. 120
A. Dabrowski (Editor)



# Development of carbon-based adsorbents for removal of mercury emissions from coal combustion flue gas

M. Rostam-Abadi[1,2], H-C Hsi[2], S. Chen[1], M. Rood[2], R. Chang[3], T. R. Carey[4], C. F. Richardson[4], B. Rosenhoover[5]

[1] Illinois State Geological Survey, 615 E. Peabody Dr., Champaign, IL 61820, USA

[2] University of Illinois, Environmental Engineering Program, 205 N. Mathews Ave., Urbana, IL 61801, USA

[3] Electric Power Research Institute, 3412 Hillview Ave., Palo Alto, CA 94403, USA

[4] Radian International, LLC, 8501 N. Mopac Blvd., Austin, TX 78759, USA

[5] CONSOL, 4000 Brownsville Rd., Library, PA 15129, USA

## 1. INTRODUCTION

The U.S. Clean Air Act Amendments of 1990 listed 189 substances as hazardous air pollutants, of which 37 substances have been detected in power plant emissions [1]. Of the 37 hazardous air pollutants, 11 are trace metals such as mercury. It is of greatest concern because of the perceived risks from its environmental release, it occurs mainly in the vapor form, and because is not captured effectively by existing particulate removal systems.

The total amount of mercury release to the atmosphere from anthropogenic and natural sources each year is estimated to be 5,500 tons [2]. Coal-fired power plants (CFPPs) are one the largest (about 60 tons per year in the U.S.) anthropogenic sources of vapor phase mercury emissions. The concentration of mercury in flue gases generated by CFPPs typically is 1 to 10 $\mu g/m^3$.

Speciation of mercury in CFPP flue gases is of interest because the molecular form of mercury influences the ability of air quality control devices to remove mercury from flue gas streams. The relative amounts of each mercury species strongly depends on the concentration of mercury in coal and conditions during combustion such as gas residence time, temperature, and gas composition [3]. For bituminous coal, mercury concentrations can be less than 0.01 ppmv up to 3.3 ppmv [4]. In the combustion zone (1200-1400°C), mercury is vaporized from the coal and exists as elemental mercury ($Hg^0$). As the flue gas temperature decreases, $Hg^0$ is partially oxidized to form $Hg^{2+}$ and partitions between gas, liquid or solid phases

[3,5]. $Hg^0$ is more difficult to remove from the flue gas streams due to its high equilibrium vapor pressure (0.25 Pa at 25°C) [6] and low solubility in water (60 μg/L at 25°C) [7]. However, $Hg^{2+}$ in the form of mercuric chloride ($HgCl_2$), is more water-soluble (69 g/L at 25°C) [8], and is more readily removed from flue gas streams by current air pollution control devices such as wet scrubbers [9].

Carbon-based processes (both direct injection and fixed-bed) have been developed for control of mercury emission from municipal- and hazardous-waste incinerators [10, 11]. Existing data from the incinerators provide some insight on mercury control, but these data cannot be used directly for coal-fired utilities because mercury concentrations, species, and process conditions differ greatly [1]. For example, municipal solid waste (MSW) mercury concentrations (200 to 1000 μg/m$^3$) are one to two orders of magnitude larger than for flue gases generated by coal combustion sources.

Injection of activated carbon upstream of a particulate control has the potential of providing a low-cost option for control of mercury emissions from utility flue gas [1]. In several bench [5,12-16] pilot and full-scale tests [17-19] of the method, the influence of carbon type [5,12,13,19], carbon structure [5,13] and carbon surface chemistry [5,12,14,16], injection methods (dry or wet) [19], amount of carbon injected [17-19], and flue gas temperatures [17-19] on mercury removal have been examined. The small concentrations of mercury in the flue gas, and limited exposure time (<3 seconds) of the sorbent, generally required large amounts of activated carbons in these sorbent injection tests. To achieve significant Hg removal (>90%), the required ratio of carbon to mercury (C/Hg) in the flue gas has generally (on weight basis) been found to be 10000-50000, depending on the process conditions. Carbon-to-mercury ratio in MSW incinerators is more than an order of magnitude less than that necessary to achieve similar mercury removal in coal combustors.

The large C/Hg ratio could be a result of either mass transfer limitation or a limited mercury capacity of carbon because of the extremely small concentration of mercury in the flue gas, or the limited reactivity of the carbon. To reduce the operating cost of the carbon injection process, either a more efficient sorbent that can operate at a smaller C/Hg ratio, or a lower-cost sorbent, or both is required. An understanding of physical and chemical processes that affects mercury removal from flue gas and a systematic sorbent development study would be required to develop an efficient, cost-effective carbon injection process for removal of mercury from coal-fired utility flue gas.

Sulfur-impregnated activated carbons has been shown to be an effective sorbent for the removal of vapor-phase $Hg^0$ from CFPP and MWC flue gases [3,12,14,20-26]. This material has a large mercury adsorption capacity, but the addition of sulfur to activated carbon requires additional production cost. To avoid such processing steps, it may be possible to produce activated carbons from a precursor that already contains sulfur.

The work presented here is part of an ongoing program to develop a low-cost, high-efficiency sorbent for removal of vapor phase $Hg^0$ and $HgCl_2$ from coal combustion flue gas. An initial objective of the program was to investigate Illinois high organic sulfur coal as a suitable precursor for producing activated carbon for mercury adsorption. It is important to determine whether a portion of the inherent sulfur which remains in the resulting activated carbon could make this material an effective mercury sorbent. In this paper, the results of a study to evaluate the influence of mass transfer on mercury removal in the carbon injection process, and a summary of bench- and pilot-scale carbon production and testing of several Illinois coal-based sorbents are presented.

## 2. MASS TRANSFER CALCULATIONS

### 2.1. Film mass transfer

When injected into flue gas, fine carbon particles will suspend and flow with the gas stream. In absence of internal (intraparticle) diffusion, the equation describing the transfer of mercury molecules from the bulk flue gas to the surface of the carbon particle per unit volume of duct is [19]:

$$N = k_g\left(\frac{a}{V}\right)\left(C_g - C^*\right) \tag{1}$$

where $N$ = mass flux (g/cm$^2$.s); $k_g$ = mass transfer coefficient (cm/s); $a$ = total interfacial area in the duct (cm$^2$/cm$^3$); $V$ = total volume of the duct (cm$^3$); and $C_g$ = mercury concentration (g/cm$^3$) in the bulk flue gas, and $C^*$ = mercury concentration in equilibrium with the adsorbed mercury on the carbon surface ($C_s$). $C_g$ -$C^*$ is considered as the driving force for mass transfer.

If no strong turbulence or back mixing occurs in the duct, the gas-solid phase can be modeled as a plug flow system. The mass balance equation for a plug flow system is:

$$k_g \frac{a}{V}\left(C_g - C^*\right)Sdz = -F_g dC_g \tag{2}$$

where $S$ is the cross section area of the duct (cm$^2$), $F_g$ is the flue gas flow rate (Nm$^3$/s), and $dz$ is the differential length of the duct (cm).

The velocity of the particles relative to the flue gas is practically zero. If the carbon particles are well dispersed and do not agglomerate during the process, the mass transfer coefficient at the gas-solid interface could be calculated as:

$$k_g = \frac{2D_{Hg}}{d_p} = \frac{2\cdot 0.261}{d_p} = \frac{0.522}{d_p}\,(\text{cm/s}) \tag{3}$$

where $d_p$ = particle size (cm) and $D_{Hg}$ = diffusivity of the mercury molecule in flue gas (cm$^2$/s), which is 0.261 (cm$^2$/s) for the diffusivity of mercury in air at 140°C.

Equation (3) shows that the mass transfer coefficient increases with decreasing carbon particle size. Any attempts to introduce turbulence to the flow may not have any significant effects on the mass transfer coefficient.

To examine the role of film mass transfer (the maximum mass transfer flux), assume $C^* << C_g$ at all positions in the duct ( this means that mercury adsorption capacity of the carbon and the carbon reactivity are not limiting the mass transfer rate). Equation (4) is obtained by integrating equation (2) using the following boundary conditions: 1) at z = 0 ( entrance), $C_g$=$C_0$; 2) at z = L (outlet), $C_g$=$C_g$.

$$\ln\left(\frac{C_0}{C_g}\right) = k_g \frac{a}{V}\frac{SL}{F_g} \tag{4}$$

where L = length (cm) of the duct and $SL/F_g$=t is the residence time of carbon particles in the duct. Eqs. (3) and (4) can be used to solve for the minimum activated carbon interfacial area required by mass transfer to remove 90% of the mercury, i.e. $C_g = 0.1\ C_0$, from one $Nm^3$ of utility flue gas:

$$\frac{a}{V} = \frac{2.303}{k_g}\frac{F_g}{SL} = \frac{2.303}{0.522}\frac{d_p}{t} = 4.412\frac{d_p}{t}\left(cm^2/cm^3\right) \tag{5}$$

Because only the external surface area of carbon particles serves as the gas-solid interfacial area, the minimum interfacial area needed for mass transfer implies that a certain minimum amount of carbon is required to achieve the desired mercury removal. For spherical particles the external surface area per gram of activated carbon is $6/d_p\ \rho_c$ where $\rho_c$ is the carbon particle density in g/cm³. The amount of carbon required for mercury removal from one $Nm^3$ of utility flue gas therefore is:

$$\left(4.412\frac{dp}{t}\right)\Big/\left(\frac{6}{d_p\rho_c}\right) = 0.7353\frac{d_p^2\rho_c}{t} \tag{6}$$

The carbon/mercury weight ratio for 90% mercury removal can be calculated from the following relationship:

$$R = \frac{\text{Carbon}}{\text{Mercury}} = 0.7353\frac{\rho_c d_p^2}{tC_0} \tag{7}$$

Equation (7) shows that the C/Hg ratio depends strongly on the particle size, residence time, and on the mercury concentration in the flue gas $C_0$ - $C_g$. Table 1 shows the C/Hg weight ratios required for 90% mercury removal from flue gas under mass transfer limited conditions, with activated carbon ranging in size from 1 to 20 μm, assuming the particle density of activated carbon of $\rho_c$= 0.5g/cm³, a contact time of 2 seconds, and an inlet mercury concentration of 10μg/Nm$^3$. The required C/Hg weight ratio for a 1 μm carbon particle is 184 but it increases to 73500 for a 20 μm carbon particle. This analysis indicates that a large C/Hg weight ratio is required when the carbon particle size is larger than 10 μm.

The required mercury capacities of sorbent as predicted by mass transfer analysis are presented in Table 1. Under a mass transfer limited process, a low capacity sorbent is required. For example for a 5 μm carbon particle size, the mass transfer capacity is only 217 μg/g carbon. When the mercury capacity of a sorbent is comparable to that of the mass transfer capacity, however the C/Hg ratio is determined by both mass transfer parameters and adsorbent capacity. Under some extreme conditions, the mercury capacity of the adsorbent could limit the removal efficiency, and the C/Hg ratio is determined by the sorbent capacity rather than the mass transfer capacities presented in Table 1.

Table 1
Mass transfer C/Hg weight ratios and mercury capacities for different carbon particle sizes

| Particle size (mm) | $k_g$ (cm/s) | Interfacial area (m$^2$/g) | C/Hg ratio (g/g) | Mercury Capacity (μg Hg/g C) |
|---|---|---|---|---|
| 20 | 261 | 0.6 | 73500 | 14 |
| 10 | 522 | 1.2 | 18375 | 54 |
| 5 | 1044 | 2.4 | 4593 | 217 |
| 1 | 5220 | 12.0 | 184 | 5435 |

The dependence of the minimum C/Hg weight ratio on the concentration of the mercury in the flue gas (for 90% removal and 2 seconds residence time) is presented in Figure 1. Equation (7) and Figure 1 can be used to compare the minimum C/Hg weight ratios required to remove 90% mercury from utility and incineration flue gases. The C/Hg ratio, according to Equation (7), for the MSW flue gas will be about two orders of magnitude less than that of the utility flue gas. For a 10 μm carbon particle (Fig. 1), the theoretical C/Hg ratio required for a MSW flue gas containing 600 μg/Nm$^3$ mercury is 306 which is about 60 times less than that for a utility flue gas containing 10 μg/Nm$^3$ mercury.

According to Equation (7), increasing residence time or mercury concentrations decreases the minimum C/Hg ratio. In practice, however, residence time and flue gas inlet mercury concentration can not be increased for an existing system. Only particle size of the injected carbon can be reduced. The relationship between mercury removal percentage and the C/Hg ratio under the typical process conditions can be obtained by rearranging Eqs. (4) and (5)

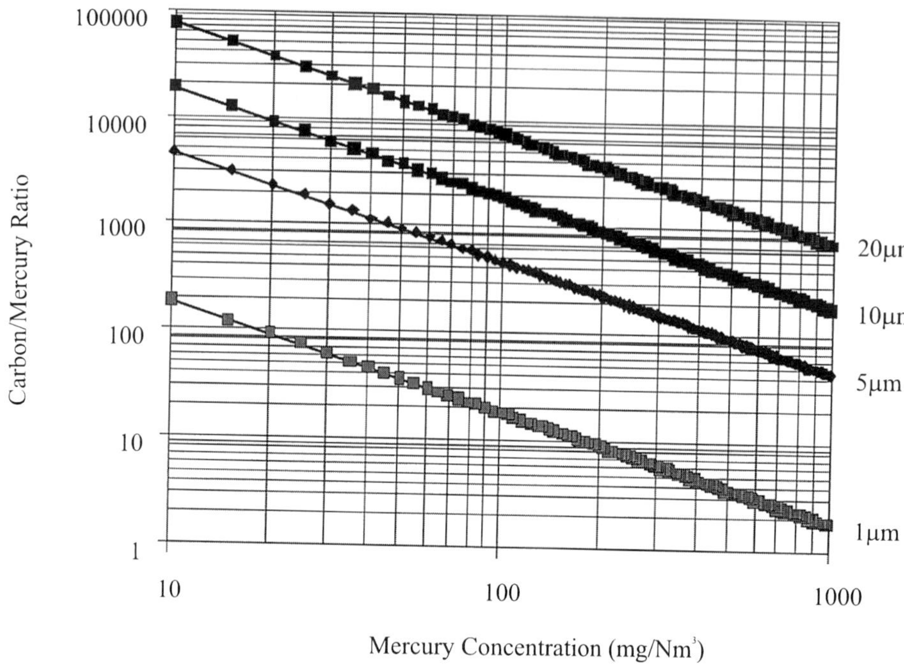


Figure 1. Dependence of C/Hg ratio on inlet mercury concentration in flue gas.

$$X = 1 - \frac{C_1}{C_0} = 1 - \exp\left[-12\frac{D_{Hg}m}{d_p^2\rho_c}t\right] = 1 - \exp\left[-12\frac{D_{Hg}\,C_{\in}R}{d_p^2\rho_c}t\right] \tag{8}$$

where X is the mercury removal percentage.

Equation (8) is useful for process design. It can be used to predict mercury removal performance of an injection process for a set of given process conditions or to predict the minimum C/Hg ratio for a required mercury removal percentage. For example, using the same process conditions employed to produce Table 1, Eq.(8) was used to generate plots showing in the dependence of mercury removal percentage on the minimum C/Hg ratio.

These calculated C/Hg weight ratios are the minimum needed and assume mass transfer limitations. At small C/Hg ratios (such as for the MSW flue gas at a C/Hg ratio of 200 to 300), it is possible that the carbon will become saturated. Also at higher temperatures (>160°C), most sorbents have very limited capacity for mercury. Under these conditions, much more sorbent will be needed than that predicted by mass transfer limitations.

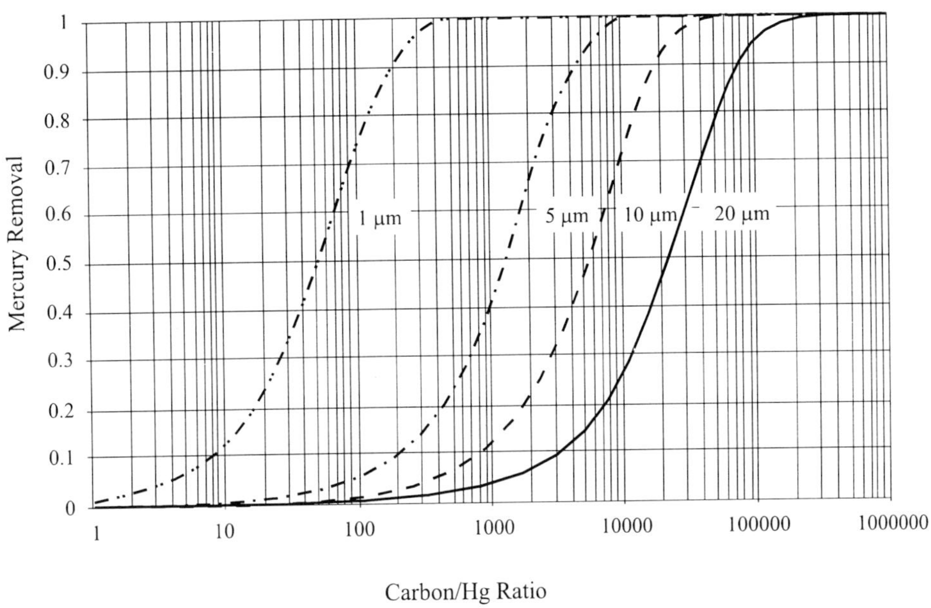


Figure 2. Dependence of mercury removal on C/Hg ratio.

## 2.2. Intraparticle diffusion

Because diffusivities in mircroporous materials vary considerably, depending on their pore structure and pore size, it is difficult to estimate diffusivity when the diffusion is in the configurational range. Configurational diffusion only occurs when the micropore size is comparable to the molecular size of the adsorbate. In this study, calculations were made for a single spherical carbon particle dispersed in a flue gas. The carbon particle is exposed to a step change in mercury concentration at its external surface at t = 0 (corresponding to the injection location). The diffusion of mercury molecules into the carbon, assuming a constant effective diffusivity, can be described by:

$$\frac{\partial q}{\partial t} = D_{Hg}\left(\frac{\partial^2 q}{\partial r^2} + \frac{2}{r}\frac{\partial q}{\partial r}\right) \tag{9}$$

where $D_{Hg}$ is the mercury effective diffusivity in carbon particle and q(r,t) is the adsorbed phase concentration of mercury at position r ( inside the carbon particle) and time t. If the mercury concentration at the external surface of the particle remains constant, the following initial and boundary conditions apply:

$$q\left(r_c = \frac{1}{2} d_{p,t}\right) = q_0 \tag{10}$$

$$q(r,0) = 0 \tag{11}$$

$$\left(\frac{\partial q}{\partial r}\right)_{r=0} = 0 \tag{12}$$

The solution to this problem is given by [27]:

$$\frac{\bar{q}}{q_0} = \frac{m_t}{m_\infty} = 1 - \frac{6}{\pi^2} \sum_{n=1}^{\infty} \frac{1}{n^2} \exp\left(-\frac{4n^2\pi^2 D_{Hg} t}{d_p^2}\right) \tag{13}$$

where $\bar{q}$ , the average concentration of mercury inside the particle, is defined by:

$$\bar{q} = \frac{3}{r_c^3} \int_0^{r_c} q r^2 \, dr \tag{14}$$

and $m_t$ and $m_\infty$ are the uptakes of mercury at time t and t=∞. When the fractional uptakes, $\frac{m_t}{m_\infty}$, are larger than 70%, the following simplified equation can be used:

$$\frac{m_t}{m_\infty} = 1 - \frac{6}{\pi^2} \exp\left(-\frac{4n^2\pi^2 D_{Hg} t}{d_p^2}\right) \tag{15}$$

For a particle size of $d_p$=10 μm, t=2s, and 90% mercury uptake ($m_t/m_\infty$=0.90), the mercury diffusivity in activated carbon can be calculated using Equation (15):

$$D_{Hg} = \frac{d_p^2}{4\pi^2 t} \ln\left[\frac{\pi^2}{6}\left(1 - \frac{m_t}{m_\infty}\right)\right] = \frac{\left(10 \cdot 10^{-4}\right)^2}{4\pi^2 \cdot 2} \ln\left(\frac{\pi^2}{6} \cdot 0.1\right) = 1.01 \cdot 10^{-8} \, (cm/s)$$

This value of diffusivity is in the range of configurational diffusion. The carbon particle was assumed to be exposed to a step change in mercury concentration at its external surface at t = 0 (corresponding to the injection location). The calculations indicate that with a 10 μm activated carbon particle, the intraparticle diffusion will be important only when the pore diameter is about 3 Å, i.e., the atomic diameter of mercury. Because the micropore size of the activated carbon is generally larger than 3 Å, it can be concluded that intraparticle diffusion is unlikely to be the controlling step in the carbon injection process.

## 2.3. Discussion of mass transfer study

The analysis presented above indicates that under certain carbon injection conditions, mercury removal from coal-fired flue gas is film mass transfer controlled. For example, Miller et al. [6] used a C/Hg ratio greater than 3000 for an activated carbon with a weight-averaged particle size of 5.5μm, to remove about 90% mercury from a flue gas. In the same study, for an iodine-impregnated activated carbon with a weight-averaged particle size of 3 μm, the C/Hg ratio was about 1000. Such C/Hg ratios are comparable to those shown in Figure 1. Full-scale tests of the carbon injection process in MSW incinerator flue gas also confirmed the results shown in Table 1. Licata et al.[28] reported that the equilibrium mercury capacity of an activated carbon (AC) used in their tests was about 0.33 gHg/gAC, which corresponds to a C/Hg ratio of 3 (temperature was not mentioned). In full-scale MSW tests with the same carbon, however a C/Hg ratio of more than 300 was used to reduce mercury concentration in the flue gas from 600 to 70 μg/Nm$^3$ at 135°C. This ratio corresponds to 3.3 mgHg/gAC, which is only 1% of the equilibrium capacity of the carbon. In another field test, White et al.[18] found that carbon injection methods (dry or wet) had a significant effect on mercury removal while the type and surface chemistry of the activated carbon had none. These data suggest that mass transfer was controlling the mercury removal. For conditions where mercury adsorption is mass transfer limited, measures should be taken to increase the mercury mass flux (from the bulk gas to the surface of carbon) rather than using a carbon with high adsorption capacity. To increase the mass transfer, either the mass transfer coefficient, $k_g$, or the interfacial area, a/V, should be increased. According to equation (3) the mass transfer coefficient increases with decreasing carbon particle size. Reducing carbon particle size also increases the interfacial area, without increasing carbon dosage. The most effective way to reduce the C/Hg ratio is therefore to decrease the carbon particle size.

Mass transfer limits only apply when the carbon has sufficiently large reactivity and capacity. Under certain injection conditions, the mercury capacity of the carbon may become limiting. To evaluate the effect of mercury capacity of carbon on the required C/Hg weight ratio, an adsorption isotherm which relates $C^*$ and $C_s$ is required. This isotherm can be used in conjunction with the mass balance Equation (2) to develop an expression which relates C/Hg weight ratio to mass transfer parameters, exposure time, and mercury capacity of the carbon. The details are given elsewhere [29].

The dependence of C/Hg weight ratio on the mercury capacity of carbon for several carbon particle sizes using a Henry's law isotherm (mercury capacity is represented by the Henry's law constant, H) and employing conditions used to generate Table 1. Three distinct regions were obtained (Fig. 3). In region I, the C/Hg weigh ratio is solely controlled by the mercury capacity of the carbon. An injection process could fall in this region when the mercury capacity or particle size of carbon is small. In region II, C/Hg weight ratio is determined by both the mass transfer parameters and mercury capacity of the carbon. An injection process could fall into this region when mercury capacity of the carbon is comparable to the mass transfer capacity. In region III, the C/Hg weight ratio is controlled only by the mass transfer parameters.

Figure 3 provides some guidance for designing a carbon injection process. For example, to remove 90% of the mercury from flue gas using a carbon with a Henry's law constant of 0.01 g Carbon/ $Nm^3$ (corresponding to a mercury capacity of 1000 μg/g Carbon at 10 μg/$Nm^3$ gas phase mercury concentration), the particle size of the carbon should be between 5 and 10 μm. If the particle size is larger than

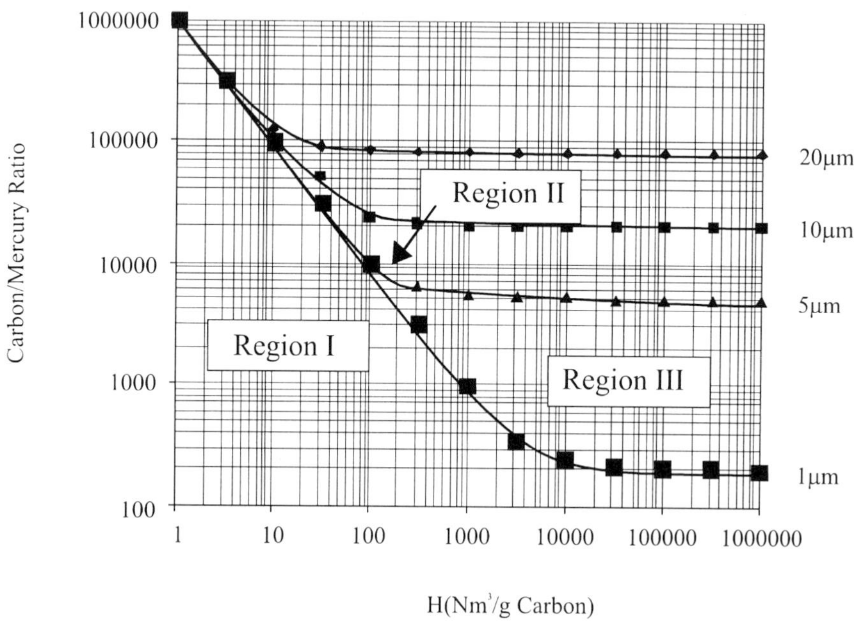


Figure 3. Dependence of C/Hg ratio on activated carbon mercury capacity (H $Nm^3$/g carbon) for 90% removal.

10μm, then mass transfer will limit the injection process and only a small fraction of the carbon capacity will be utilized. If the carbon particle size is less than 5 μm, then the mercury capacity of the carbon is smaller than the mass transfer capacity. Under this condition, the mercury removal efficiency can not be increased further by reducing the particle size of the sorbent.

## 3. CARBON DEVELOPMENT STUDIES

### 3.1. Characteristic of activated carbon for mercury vapor capture

To design and produce a suitable, low-cost sorbent for use in an injection process, the characteristics of the sorbent for optimum performance must be identified.

Because the typical concentrations of mercury in utility flue gas is about 1 to 10 $\mu g/Nm^3$ and sorbent exposure time in the duct is about less than 3 seconds, it is likely that only a small fraction of the carbon pore surface area is used in the mercury removal process. The large C/Hg weight ratios calculated from the mass transfer confirms this observation. It is, therefore, important to determine the carbon surface area required for monolayer mercury coverage. For example, if a large fraction of the pore surface area of carbon is utilized in the process, then increasing the pore surface area of the carbon would be desired (i.e., high capacity). In contrast, if a small fraction of the pore surface area is occupied by mercury, than a low surface area carbon which exhibits a large mercury adsorption rate and would be saturated during the limited exposure time to flue gas is desirable (i.e., large reactivity). In the extreme case, the surface area needed by the monolayer mercury coverage is less than the interfacial area needed for the mass transfer. In this situation, chemical properties of the carbon such as the type and amount of surface functional groups would be critical, and the internal structures of the carbon (pore size, pore volume and pore surface area) will not be critical.

Assuming the same conditions used in the mass transfer analysis and further assuming that each Hg molecule occupies 10 $Å^2$ of surface area and that the average carbon particle size of 10μm, then the pore surface area of the carbon required for removal (monolayer coverage) of 90% of the mercury molecules in one $Nm^3$ of utility flue gas is :

$$(10^{-1})(\mu g/Nm^3)/200(g)\times 6.02\times 10^{23}\times 10\times 10^{-16}(cm^2)= 27.1\ cm^2/Nm^3$$

The interfacial area required by film mass transfer is:

$$a/V=\frac{2.303}{t\,k_g}=\frac{2.303\,d_p}{2\,t\,D_{Hg}}=\frac{2.303\left(10\cdot 10^{-4}\right)}{2\cdot 2\cdot 0.522}=0.001103\left(cm^2/cm^3\right)=1103\left(cm^2/m^3\right)$$

Comparing this value with the surface area needed for a monolayer mercury coverage indicates that only 2.45% (27.1/1103) of the total carbon pore surface area is required for a monolayer mercury coverage. This analysis suggests that a carbon with low surface area (low capacity) and a fast adsorption rate is desirable for the carbon injection process.

### 3.2. Activated carbon production

A carbon development program was initiated at the Illinois State Geological Survey (ISGS) and the University of Illinois at Urbana-Champaign (UIUC) to investigate the effects of different carbon types, carbon structures, and carbon surface functional groups on the rate and extent of adsorption of vapor-phase mercury. The results from a study to prepare Illinois coal-based activated carbons are presented. Carbon products were made both in bench- and pilot-scale reactors.

### 3.3. Bench-scale carbon production

Bench-scale experiments were performed in a 5-cm ID fluidized-bed reactor (FBR). These experiments were intended to: 1) prepare carbon products with varying ash content, sulfur content, surface area, pore volume and production yield, 2) identify optimum conditions for producing activated carbon samples with desired properties for removal of various mercury species from utility flue gas, 3) evaluate the influence of inherent sulfur of the coal precursors on the mercury adsorption reactivity and capacity of the resultant activated carbon products, and 4) obtain scale-up date for producing larger quantities of carbon products in a pilot-scale FBR.

Two high-organic sulfur bituminous coals (IBC-107 and C-2) and one low organic sulfur bituminous coal (IBC-109) were selected as precursors to prepare the activated carbons. The IBC-107 and IBC-109 coals were obtained from Illinois Basin Coal Sample Program [30]. C-2 coal was directly mined from active Illinois coal basins. Samples were initially crushed to about 5 mm size particles and then pulverized (Holmes Model 500) and ground (Quaker City Model 4-E) to decrease the particle size to less than 1 mm. Samples were subsequently sieved with U.S. standard sieves (Tyler Standard Screen) to obtain -16 + 65 mesh (1.000-0.210 mm) samples.

Two series, AC and FBR, of Illinois coal-derived activated carbon (ICDAC ) were prepared ICDACs were produced according to a three-step process. Typically about 100 g samples of sieved coal was pre-oxidated (air, 225°C, 1-3 hr), pyrolyzed ($N_2$, 400°C, 1 hr) to form char, and then activated with steam (50%$H_2O$/50%$N_2$, 825°C, 1-7 hr) to produce ICDAC. AC-1 to AC-5 carbons were prepared using 3 hr of activation time. AC-63 and FBR-45 series carbons were prepared at activation times ranging from 1 to 7 hr. For example AC-63-3 refers to a AC-63 carbon prepared using 3 hr of activation time. For characterization and mercury tests, samples were ground 15-20 μm (Microtrac II analyzer).

### 3.4. Sulfur impregnation of carbon products

Sulfur-impregnated ICDACs were produced to obtain additional information related to the role of the carbon sulfur functional groups on mercury adsorption onto activated carbon. Sulfur-impregnated carbon were prepared by mechanically mixing AC-1 and AC-3 activated carbons with 20 wt.% elemental sulfur and heating the mixtures at 600°C for 8 hr in a tube furnace (Lindberg 59344). The tube furnace was purged with ultra-high pure nitrogen before heating the mixtures and during the heating and cooling periods.

### 3.5. Pilot-scale production

A batch, 48-cm ID pilot-scale FBR was used to produce more than 60 kg of activated carbon from C2 coal. Pilot tests were conducted at Svedala Inc., a chemical process equipment firm located in Oak Creek, Wisconsin. Five pilot production tests were made. The processing conditions employed in the pilot production runs were comparable to those identified during the bench-scale studies of coal C2 except that about 50 kg of <0.25 mm coal was typically used.

### 3.6. Sample characterization

Specific surface area, pore volume and pore size distribution of original coal samples and resulting ICDACs were determined with a Micromeritics ASAP2400 analyzer using $N_2$ adsorption at 77 K. Samples were degassed in a 10 to 20 μtorr vacuum at 150°C for 24 hr before adsorption measurements. Total surface area was calculated by BET (Brunauer, Emmett, and Teller) equation [31]. Micropore (pore size < 17 Å) surface area and volume were calculated from t-plot analysis using Jura-Harkins equation: $t = [13.99/(0.0340\text{-}\log(p/p_o)]^{0.5}$ [32]. The thickness t values used in the least-squares analysis were between 4.5 to 8.0 Å. Micropore size distribution was determined by MP (micropore analysis) method [33]. Mesopore and macropore size distribution was calculated by the Barret-Joyner-Halenda (BJH) method [34]. Ultimate and sulfur analyses of original coal samples and ICDAC were measured by LECO MAC-d SC-32 systems according to ASTM methods D5373-93 and D4239-94, respectively.

## 4. MERCURY REMOVAL TESTING

The mercury removal performance of the activated carbon samples produced in this study were measured both in a bench-scale test apparatus and in a sorbent injection pilot-plant.

### 4.1. Bench-scale, fixed-bed tests

Bench-scale mercury testing was conducted at Radian International's Austin, Texas laboratory. The screening tests evaluated mercury breakthrough characteristics of ICDAC samples and Darco FGD carbon, a lignite-based activated carbon manufactured by American Norit.

Two identical test units were constructed (Fig. 4). One unit was used to test elemental mercury adsorption while the other was used to measure mercuric chloride adsorption. In both systems, a simulated flue gas was prepared by mixing heated gas streams containing 1600 ppm sulfur dioxide ($SO_2$), 50 ppm hydrochloric acid (HCl), 12% carbon dioxide ($CO_2$), 7% water ($H_2O$), and 6% oxygen ($O_2$). Mercury was injected into the gas by contacting nitrogen carrier gas with either recrystallized mercuric chloride solids or with an elemental mercury diffusion tube (VICI Metronics) in a mercury saturation vessel. The mercury concentration (typically 40-60μg/Nm$^3$) is controlled by the temperature of the mercury saturator and the nitrogen flow rate through the saturator. All gas mixing, water saturation, and mercury injection occur within a closed, temperature-controlled box designed to prevent water condensation which could affect the behavior of mercury and the gas concentrations in the flow lines.

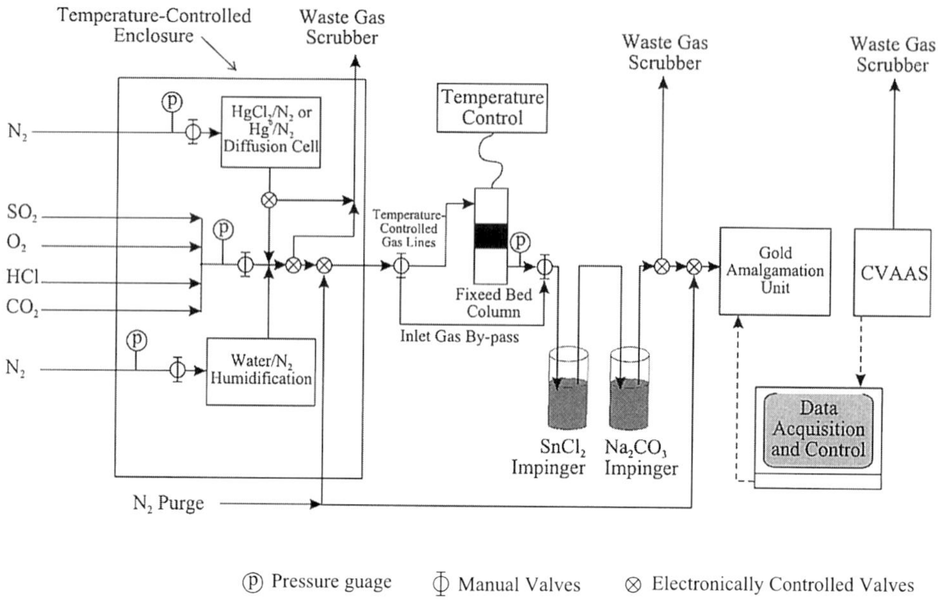


Figure 4. Bench-scale, fixed-bed mercury adsorption apparatus.

The reaction gas flows at about 1 L/min (at 24°C) through heated Teflon lines (107°C) to a temperature- controlled column (1.27 cm ID) containing the sorbent to be studied. The sorbent was mixed in a sand diluent prior to being packed in the reaction column. A ratio of 20 mg carbon to 10 g of sand was used in all tests reported in this paper. In the column, the gas was heated to the reaction temperature before contacting the sorbent by passing it across a bed of pyrex spheres designed to enhance heat exchange. The column temperature is controlled using an internally mounted thermocouple shielded from the gas with a glass sheath. The reaction gas flows downward through the column to minimize the chance of selective flow or channeling through the bed. The bed material was supported by a fritted glass disk and packed with quartz wool. The linear gas velocity through the empty column is approximately 18.2 cm/min at 135°C.

The effluent gas from the fixed-bed column flows through heated lines to an impinger containing $SnCl_2$ which reduces any oxidized mercury compounds to elemental mercury. The gas then flowed through a buffer solution ($Na_2CO_3$) to remove the $SO_2$ and HCl from the gas, thus protecting the downstream, analytical gold surface. Gas exiting the impinger solutions flowed through a gold amalgamation column housed in a tubular furnace where the mercury in the gas was adsorbed (<100°C). After adsorbing mercury onto the gold for a fixed period of time, the mercury concentrated on the gold was thermally desorbed (>750°C) and sent as a concentrated mercury stream to a cold-vapor atomic absorption (CVAA) spectrophotometer for analysis.

During each test, the sorbent/sand mixture is equilibrated at the desired adsorption temperature for at least one hour before introducing flue gas. During this time, the inlet gas bypassed the sorbent column and passed to the analytical system to determine the inlet mercury concentration. After the equilibration time was complete, the adsorption test was initiated by diverting the reaction gas through the sorbent column. The amount of mercury exiting the column was measured on a semi-continuous basis until 100% of the inlet mercury is detected at the outlet (100% breakthrough). The reaction gas continually passes through the sorbent column by sending the effluent gas to a waste scrubber during the analysis step. The typical sampling time was 6 minutes, followed by a 6-minute analytical period.

A typical adsorption curve is shown in Fig. 5. The percent breakthrough is determined as a function of time by normalizing the measured mercury concentration at the outlet of the sorbent bed to the inlet mercury concentration. The capacity of the sorbent to adsorb mercury (µg Hg/g carbon) was determined by summing the total mercury adsorbed to the sorbent through a given time. The „initial" breakthrough capacity was defined as the capacity of the sorbent at the time when mercury was first detected at the outlet. The 100% breakthrough „equilibrium" capacity was defined as the capacity at the time when the outlet mercury concentration was first equal to the inlet concentration. The time to initial breakthrough and the time to equilibrate have to be determined in order to calculate the respective capacities. These times were determined based on the slope of the breakthrough curve. A linear regression of the breakthrough curve is determined over the range of 20% to 80% breakthrough. The initial breakthrough and 100% breakthrough times were then based on this linear fit. To calculate the respective capacities, the amount of mercury adsorbed from the start of the test to the respective time was determined.

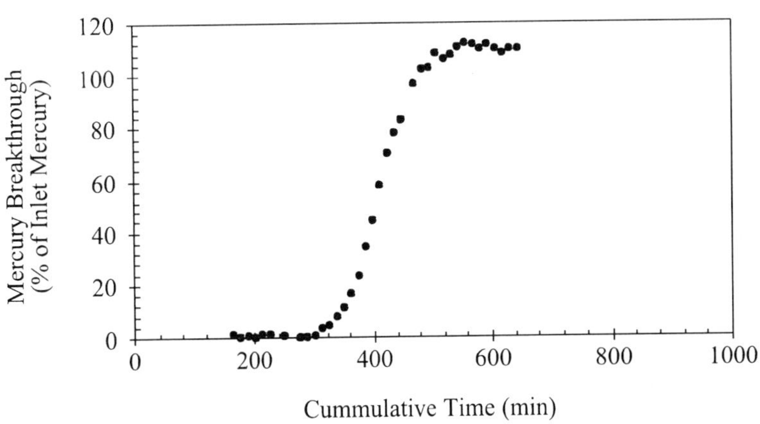


Figure 5. Typical adsorption curve.

### 4.2. Pilot-scale mercury tests

The mercury removal performance of pilot-scale ICDAC and of Norit's FGD carbon were determined in a 0.236 $m^3/s$ (0.25 MWe) pilot plant operated by CONSOL, Inc., Library, PA. The pilot plant can simulate flue gas conditions downstream of the air preheater in a coal fired utility power plant. The flue gas mercury concentration studied (10-15 $\mu g/m^3$) is typical of utility flue gas concentration. Mercury removals were evaluated in the flue gas duct, which provided a gas residence time of approximately 2 seconds, and in the baghouse, where the solids retention times can be as long as 30 min. Common test conditions were: flue gas flow, 0.165 $m^3/s$; flue gas wet bulb temperature, 50-53°C; flue gas composition, 1000 ppmv dry $SO_2$, 10 vol% dry $O_2$, and 10 vol% dry $CO_2$. All tests were conducted with a fly ash obtained from a coal-fired utility boiler firing an eastern bituminous coal. The fly ash feed rate was 4.5 kg/hr (solids loading of 90.6-104.7 gm/$dcm^3$). Mercury removal was determined from the mercury feed rate, the solids (carbon and fly ash) feed rate, and mercury analysis of the feed and recovered solids (by combustion followed by cold vapor atomic absorption spectroscopy). Except where noted, all mercury removal results discussed in this paper include mercury removal by the carbon sorbent and the fly ash. A more detailed description of the pilot test unit is given elsewhere [27].

## 5. CHARACTERISTICS OF COALS AND CARBON PRODUCTS

The total sulfur content of the high organic sulfur coals ranges from 3.7-4.1 wt.% with about 75% of the sulfur existing as organic sulfur (Table 2). The total sulfur content of the low organic sulfur coal was 1.2 wt.% and 60% of the sulfur was in organic form. The total sulfur contents of the ICDACs were between 0.85 to 2.6 wt.% (Table 3). Specific surface areas and total pore volumes ranged from 353 to 994 $m^2/g$ and 0.2 to 0.7 $cm^3/g$, respectively. Micropore contributes about 62 % to 87% of the total surface area and 44 % to 74 % of the pore volume for the ICDACs. The precursor coals had no microporosity and specific surface areas of less than 50 $m^2/g$.

For the AC-63 and the FBR-45 series carbons, increasing activation time from 3 to 6 hr resulted in an increase in ash content, a decrease in total sulfur content, an increase in surface area, and an increase in pore volume.

The pilot ICDACs were obtained by blending the activated carbon samples produced during two of the pilot production tests and ground to two particle size ranges (5 and 8μm). The finer particle size sample, AC-F, had a slightly greater ash content (24.7 wt.%) than did the coarser AC-C sample (18.8 wt.%). The surface areas of the pilot carbons (671 and 688 $m^2/g$) were comparable to the activated carbon produced from C2 coal in the bench FBR (FBR-39 series).

Table 2
Analyses of select Illinois basin coals (wt.%)

| Property | Coal Sample | | |
|---|---|---|---|
| | IBC-107 (wt. %) | C-2 (wt. %) | IBC-109 (wt. %) |
| Moisture | 9.3 | 13.7 | 9.2 |
| Ash[a, b] | 11.5 | 9.7 | 8.3 |
| Carbon[a] | 68.1 | 72.0 | 75.0 |
| Hydrogen[a] | 4.8 | 5.3 | 4.9 |
| Nitrogen[a] | 1.2 | 1.4 | 1.6 |
| Oxygen[a] | 10.5 | 7.7 | 8.4 |
| Total Sulfur[a] | 3.7 | 4.1 | 1.2 |
| Sulfatic Sulfur[a] | 0.2 | 0.0 | 0.0 |
| Pyritic Sulfur[a] | 0.4 | 1.1 | 0.4 |
| Organic Sulfur[a] | 2.9 | 3.0 | 0.7 |
| Total | 99.8 | 100.2 | 99.4 |

[a] moisture free values, [b] at 750°C with air.

Table 3
Physical characteristics of activated carbon samples[a]

| Sample Code | Precursor Coal | Ash Content (wt.%) | Sulfur Content (wt.%) | Production Yield (wt.%) | $N_2$-BET Total ($m^2/g$) | $N_2$-BET Micropore ($m^2/g$) | Total Pore Volume ($cm^3/g$) | Micropore Volume ($cm^3/g$) |
|---|---|---|---|---|---|---|---|---|
| AC-1 | IBC-107 | 27.0 | 1.60 | 30.0 | 787.3 | 578.3 | 0.504 | 0.265 |
| AC-2 | IBC-109 | 14.3 | 0.85 | 50.3 | 619.8 | 505.4 | 0.336 | 0.232 |
| AC-4 | IBC-107 | 20.0 | 1.83 | 48.8 | 556.4 | 427.6 | 0.328 | 0.197 |
| AC-63-3 | IBC-109 | 15.9 | 1.18 | N/A[b] | 748.7 | 633.3 | 0.418 | 0.293 |
| AC-63-4 | IBC-109 | 26.1 | 1.25 | N/A | 845.5 | 686.6 | 0.614 | 0.317 |
| AC-63-5 | IBC-109 | 21.8 | 1.13 | N/A | 884.3 | 676.2 | 0.565 | 0.312 |
| AC-63-6 | IBC-109 | 26.8 | 1.06 | N/A | 930.0 | 698.0 | 0.620 | 0.322 |
| AC-63-7 | IBC-109 | 43.4 | 0.90 | N/A | 858.4 | 655.9 | 0.657 | 0.303 |
| FBR-45-1 | C-2 | 14.0 | 2.52 | N/A | 538.1 | 467.0 | 0.294 | 0.216 |
| FBR-45-2 | C-2 | 16.4 | 2.20 | N/A | 717.6 | 559.1 | 0.421 | 0.259 |
| FBR-45-3 | C-2 | 19.0 | 1.76 | N/A | 928.7 | 674.5 | 0.592 | 0.312 |
| FBR-45-4 | C-2 | 25.5 | 1.46 | N/A | 875 | 599.2 | 0.607 | 0.278 |
| AC-C | C-2 | 18.8 | 2 | 27 | 688 | N/A | 0.382 | N/A |
| AC-F | C-2 | 24.7 | 2 | 27 | 671 | N/A | 0.423 | N/A |
| Norit FGD | N/A | 32.1 | 0.98 | N/A | 502.8 | 228.5 | 0.635 | 0.105 |

[a] All physical characteristics were obtained based on dry mass sample,[b] N/A: not available.

The specific surface area of Darco FGD carbon, manufactured by American Norit, was 502 $m^2/g$. FGD is a gas-phase carbon and has been used for removal of vapor-phase mercury from combustion flue gases. It had a mass median particle size of 15 μm. Compared with ICDACs, commercial Darco FGD carbon had a higher ash content (32.1 wt.%) and lower total sulfur content (0.98 wt.%). FGD carbon contained less microporosity, about 17 % of the total porosity than those of the ICDACs (44 to 74 %).

## 6. RESULTS FROM MERCURY TESTING

### 6.1. Bench-scale tests

The initial/equilibrium (I/E) $Hg^0$ and $HgCl_2$ adsorption capacities of AC-1 were 2243/ 2718 and 514/ 596 μg/g, respectively (Table 4). The I/E $Hg^0$ and $HgCl_2$ adsorption capacities of AC-2 were 0/1304 and 0/ 19 μg/g, respectively, which were much lower than those of the AC-1 carbon. AC-1 was derived from high organic sulfur IBC-107 coal which contained 3.7 wt.% total sulfur and 2.9 wt.% organic sulfur. AC-2 was derived from low organic sulfur IBC-109 coal with 1.2 wt.% total sulfur and 0.7 wt.% organic sulfur (Table 2). The total sulfur contents of AC-1 and AC-2 were 1.60 and 0.85 wt.%, respectively (Table 3). These data indicate that mercury adsorption reactivity and capacity of ICDAC from high organic sulfur coal were greater than those of the activated carbon derived from low organic sulfur coal.

Table 4
Mercury adsorption capacity of activated carbon samples

| Sample Code | Inlet $Hg^0$ Concentration (μg/Nm3) | Inlet $HgCl_2$ Concentration (μg/Nm3) | Initial $Hg^0$ Capacity (μg/g) | Equilibrium $Hg^0$ Capacity (μg/g) | Initial $HgCl_2$ Capacity (μg/g) | Equilibrium $HgCl_2$ Capacity (μg/g) |
|---|---|---|---|---|---|---|
| AC-1 | 54 | 61 | 2243 | 2718 | 514 | 596 |
| AC-2 | 73 | 46 | 0 | 1304 | 0 | 19 |
| AC-4 | 66 | N/A | 0 | 1508 | N/A | N/A |
| AC-63-3 | 50 | 46 | 636 | 674 | 334 | 407 |
| AC-63-4 | 45 | 37 | 508 | 551 | 342 | 360 |
| AC-63-5 | 53 | 40 | 705 | 843 | 302 | 330 |
| AC-63-6 | 53 | 43 | 0 | 866 | 407 | 420 |
| AC-63-7 | 52 | 42 | 588 | 617 | 475 | 542 |
| FBR-45-1 | 61 | 54 | 2072 | 2469 | 441 | 487 |
| FBR-45-2 | 67 | 59 | 0 | 1354 | 457 | 486 |
| FBR-45-3 | 62 | 62 | 1636 | 1670 | 466 | 507 |
| FBR-45-4 | 46 | 64 | 1419 | 1507 | 1108 | 1142 |
| AC-C | 76 | 42 | 1939 | 2188 | 431 | 450 |
| AC-F | 64 | 56 | 1721 | 1958 | 397 | 438 |
| FGD | 59 | 60 | 516 | 2566 | 1330 | 1570 |

Additional evidence showing the importance of the inherent presence of organic sulfur in the starting coal can be illustrated by comparing the results obtained for AC-1 and AC-63-3 samples, which had comparable surface area and micropore volume (Table 3). These samples were prepared from IBC-107 and IBC-109 coals, respectively. Both coals had equal amount of inorganic sulfur (0.4 wt.%), but the organic sulfur content of IBC-107 coal was about four times greater than that of IBC-109 coal (2.9 vs 0.7 wt.%) resulting in a greater fraction of organic sulfur in the activated carbon product (1.6 wt.% for AC-1 vs 1.2 wt.% for AC-63-3). This in turn resulted in higher adsorption of $Hg^0$ (2718 μg/g vs 674 μg/g). AC-63 series samples, prepared from low-organic sulfur IBC-109 coal, showed 500 – 860 μg/g $Hg^0$ adsorption capacities (except for AC-63-3 which showed no capacity) and 300-540 μg/g $HgCl_2$ adsorption capacities. The initial and equilibrium $Hg^0$ adsorption capacities of AC-63 samples were comparable to each other, but they were 3 to 4 times less than those of activated carbons derived from high-organic sulfur coals in the FBR series tests. The FBR-series samples exhibited $Hg^0$ adsorption capacities between 1300 to 2460 μg/g (except for FBR-45-2 which showed no initial mercury capacity).

The effect of inorganic sulfur (pyritic sulfur) in the precursor coal on the properties and mercury adsorption of the resultant activated carbon products can be illustrated by comparing AC-63-5 and FBR-45-4 samples. These samples were prepared under comparable processing conditions from IBC-107 and C-2 coals which had total sulfur contents of 1.2 and 4.1 wt.%, respectively. The inorganic sulfur contents of the coals were 0.4 and 1.1 wt.%, respectively. Carbon products AC-63-5 and FBR-54-4 had comparable physical properties as indicated by their surface area and pore volume data. The total sulfur content of the FBR-45-4 was 1.46 wt.% as compared with 1.13 wt.% for the AC-63-5 carbon. Data from an X-ray diffraction (XRD) test revealed that FBR-series samples contained magnetite, however, no iron sulfides (marcasite, pyrrhotite or pyrite) were presented in the samples. These results suggest that inorganic sulfur (pyrrhotite and pyrite) in the coal char was converted to magnetite during the steam activation step. Therefore, the difference in the sulfur content of the two carbons was because of the greater organic sulfur content of the FBR-45-4. The mercury adsorption capacity of this carbon was twice that of the AC-63-5 carbon (Table 4). These results indicate that the amount of organic sulfur, rather than either the total or inorganic sulfur content, is the important parameter for selecting a suitable coal for producing activated carbon for vapor-phase mercury adsorption.

There were no significant differences between the $HgCl_2$ capacities of IDCACs derived from high- and low-organic sulfur coals. For example, except for FBR-45-4, the equilibrium $HgCl_2$ capacities of all samples were within the range of 447 ± 79 μg/g. This implies that for the ICDACs, the adsorption of $HgCl_2$ was not as dependent on precursor coal and processing conditions as was observed for the adsorption of $Hg^0$.

I/E $Hg^0$ and $HgCl_2$ adsorption capacities of FGD carbon were 516/2566 μg/g and 1330/1570 μg/g (Table 4). The initial $Hg^0$ capacities of the ICDAC were generally greater than that of the FGD carbon. However, their saturation capacities were comparable. FGD carbon had much greater $HgCl_2$ capacity than the ICDAC samples. The average pore diameter of the FGD carbon was about 38.5 Å which was about twice that of the ICDAC samples. Pore size distribution data for the ICDAC samples indicated that almost all the pores were smaller than 100 Å, and a large fraction of the pore area was contained in pores with diameters smaller than 17 Å. In contrast, almost half the pore area of the FGD carbon was in pores between 100 and 1000 Å. The molecular diameter of mercuric chloride is about 5 Å as compared with about 3 Å for elemental mercury. Therefore it is possible that the larger mercuric chloride molecule was more accessible to the FGD carbon than the ICDAC samples.

The elemental mercury and mercuric chloride capacities of the pilot samples (AC-C and AC-F) were comparable to those prepared in the lab-scale FBR carbons (Table 4). There were no significant differences between the lab and pilot ICDACs. These data confirmed that scale up of the carbon production was successful and that the ICDAC could be manufactured on a commercial scale.

Physical and chemical characteristics including surface area, pore volume and sulfur content and equilibrium mercury capacities of ICDAC samples can be used to offer a general trend between carbon properties and mercury adsorption capacity. As mentioned previously, ICDACs derived from high-organic sulfur coal had greater $Hg^0$ adsorption capacities than those derived from low-organic sulfur coal. However, an examination of data presented in Tables 3 and 4 indicate that equilibrium $Hg^0$ capacities of ICDACs derived from high-organic sulfur coal decreased, in general, with increased total sulfur content. This implies that the presence of sulfur alone in carbon is not responsible for improved mercury adsorption capacity. Physical properties of carbon could also influence adsorption reactivity and capacity. $Hg^0$ adsorption capacity generally increased with increasing surface area and micropore volume. No significant change in $HgCl_2$ capacity was observed with increasing surface area and micropore volume. Overall, physical properties of activated carbon appeared to influence $Hg^0$ adsorption capacity while $HgCl_2$ did not demonstrate such dependence. It remains to be determined how the interactions between the physical (such as pore size distribution) and chemical properties (such as form and amount of different surface functional groups) affect vapor-phase mercury adsorption onto activated carbon.

### 6.2. Effects of sulfur impregnation

The sulfur contents of the ICDACs impregnated with 20 wt.% elemental sulfur at 600°C (AC1S-600 and AC2S-600) increased from 0.85 – 1.6 wt.% to about 12 wt.% (Table 5). Sulfur impregnation, however, did not affect the physical properties as the surface areas of the modified carbons decreased only about 15% when compared with those of the untreated carbons. Dissociation thermodynamics of elemental sulfur indicate that at 600°C a large portion of the vapor-phase sulfur molecules are in smaller $S_2$-$S_4$ forms [35]. Chemically and thermally stable organic-sulfur functional groups are formed when the elemental sulfur molecules react with unsaturated carbon sites and possibly other carbon functional groups [36].

Table 5
Physical characteristics of sulfur-treated ICDACs

| Sample Code | Precursor Coal/Activated Carbon | Ash Content (wt.%) | Sulfur Content (wt.%) | $N_2$-BET Total Surface Area ($m^2/g$) | Micropore Surface Area ($m^2/g$) | Total Pore Volume (cc/g) | Micropore Volume (cc/g) |
|---|---|---|---|---|---|---|---|
| AC-1 | IBC-107 | 27.0 | 1.60 | 787.3 | 587.3 | 0.504 | 0.265 |
| AC-2 | IBC-109 | 14.3 | 0.85 | 619.8 | 505.4 | 0.336 | 0.232 |
| AC1S-600 | AC-1 | 24.6 | 11.5 | 669.5 | 484.5 | 0.435 | 0.222 |
| AC2S-600 | AC-2 | 12.2 | 11.9 | 532.0 | 427.4 | 0.284 | 0.196 |

After sulfur impregnation, the initial/equilibrium $Hg^0$ adsorption capacities of the AC-1 carbon decreased slightly from 2243/2718 to 1648/2037 μg/g, but those of the AC-2 carbon increased from 0/1304 to 1646/2037 μg/g (Table 6). These results indicate that, for $Hg^0$ adsorption, activated carbon AC-1 derived from the high-organic sulfur coal (IBC-107) did not benefit from the sulfur impregnation process when compared to the activated carbon AC-2 prepared from the low-organic sulfur coal (IBC-109). In fact, the $Hg^0$ adsorption capacities of AC-1 carbon (1.6 wt.% S) were comparable to those of the sulfur impregnated AC-2 carbon (AC2S-600, 12 wt.% S). The $HgCl_2$ reactivity and capacity of both AC-1 and AC-2 carbons, however, improved substantially after sulfur impregnation, indicating that carbon-sulfur functional groups with high reactivity for $HgCl_2$ adsorption were formed in these samples.

Table 6
Mercury adsorption capacity of sulfur-treated ICDACs

| Sample Code | Inlet $Hg^0$ Concentration (μg/Nm³) | Inlet $HgCl_2$ Concentration (μg/Nm³) | Initial $Hg^0$ Capacity (μg/g) | Equilibrium $Hg^0$ Capacity (μg/g) | Initial $HgCl_2$ Capacity (μg/g) | Equilibrium $HgCl_2$ Capacity (μg/g) |
|---|---|---|---|---|---|---|
| AC-1 | 54 | 61 | 2243 | 2718 | 514 | 596 |
| AC-2 | 73 | 46 | 0 | 1304 | 0 | 19 |
| AC1S-600 | 62 | 89 | 1646 | 2037 | 1597 | 1668 |
| AC2S-600 | 64 | 80 | 1799 | 2051 | 886 | 1059 |

### 6.3. Pilot-scale tests

Under all test conditions studied, both the AC-C and AC-F samples were as effective as the FGD carbon, and in many cases were significantly more effective. At flue gas temperatures of 135 and 162.8°C and a C/Hg weight ratio of 10000, AC-C and AC-F achieved significantly greater mercury removals than did the FGD carbon (Figure 6). With AC-F, $Hg^0$ removal was 77% and $HgCl_2$ removals were 64 to 69%. With the FGD carbon, $Hg^0$ removals were 53% to 57% and $HgCl_2$ removals were 34 to 44%. $Hg^0$ and $HgCl_2$ removals at 135°C with AC-C were about 84%. The date shown in Figure 6 do not demonstrate a pronounced effect on $Hg^0$ removal as flue gas temperature was increased from 135 to 162.8°C. $HgCl_2$ removal with AC-F and FGD carbon decreased with increasing temperature. The absolute changes however, were small. At a C/Hg weight ratio of 3,000 and 107.2°C flue gas temperature, the system removals of $Hg^0$ and $HgCl_2$ with AC-F were 52 and 47%, respectively. Removals with the FGD carbon were 44% with both mercury species.

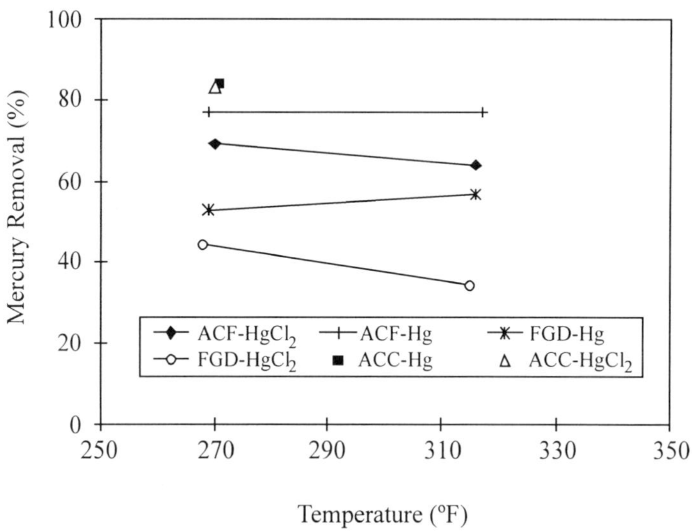


Figure 6. Effect of temperature on pilot system mercury removal.

Test results from injecting fly ash alone at 135°C showed $Hg^0$ and $HgCl_2$ removals of 9 ($Hg^0$) and 15 ($HgCl_2$) duct and 31 ($Hg^0$) and 23 ($HgCl_2$) system (duct+baghouse), and 8 ($Hg^0$) and 15 ($HgCl_2$) duct and 25 ($Hg^0$) and 33 ($HgCl_2$) system at 162.8°C. If the fly ash mercury removal is assumed to be constant, then the incremental removal resulting from the activated carbon can be estimated by subtracting the fly ash contributions from the combined carbon plus fly ash mercury removal tests. AC-F $Hg^0$ system removal at flue gas temperatures of 135 and 162.8°C, and at

10,000 C/Hg weight ratio were 46 and 52%, respectively, compared with 22 and 32% with the FGD carbon. AC-F $HgCl_2$ system removals at these conditions were 46 and 31%, compared with Norit FGD removals of 21 and 1%, respectively. AC-C $Hg^0$ and $HgCl_2$ system removals at 135°C were 53 and 60%, respectively. Duct mercury removals were small with both AC-F (7% $Hg^0$ and 2% $HgCl_2$) and FGD carbon (3% $Hg^0$ and 3% $HgCl_2$). AC-C duct removals were somewhat larger, 15% ($Hg^0$) to 21% ($HgCl_2$).

The pilot plant duct provided a 2 second residence time. For most tests, the duct mercury removal was between 9 and 19%, and was independent of carbon type, carbon feed rate, temperature, and mercury species. AC-C showed a somewhat larger in-duct $HgCl_2$ removal (36%). These results indicated that removal in the duct is limited by bulk gas mass transfer.

## 7. CONCLUSIONS

The results presented in this chapter demostrated that the minimum amount of carbon needed to achieve a specific mercury removal efficiency by sorbent injection into a flue gas stream can be predicted by assuming mass transfer limitations. Mercury removal effectiveness can be increased by decreasing the size of the carbon injected, increasing the residence time, or the amount of carbon injected. If mercury removal is limited by the reactivity and capacity of the carbon (i.e. not mass transfer limited), then significantly more carbon than the amount predicted by mass transfer limitations may be needed for effective mercury removal unless the reactivity and capacity of the carbon can be improved through structural and surface chemistry changes. Intraparticle diffusion is not important because of the small carbon sizes normally used for injection.

Illinois coal-derived activated carbon (ICDAC) samples with desired properties were prepared both in bench- and pilot-scale reactors. It was shown that activated carbons derived from high organic sulfur coal had greater mercury adsorption capacities than carbons derived from low organic sulfur coal. The equilibrium $Hg^0$ capacities of ICDACs derived from high-organic-sulfur coal, however, decreased with increasing in total sulfur content. This implies that the presence of sulfur alone in carbon is not responsible for improved mercury adsorption capacity. Physical properties of carbon could also influence adsorption reactivity and capacity. In fact, $Hg^0$ adsorption capacity generally increased with increasing surface area and micropore volume. No significant change in $HgCl_2$ capacity was observed with increasing surface area and micropore volume. Overall, physical properties of activated carbon appeared to influence $Hg^0$ adsorption capacity while $HgCl_2$ did not demonstrate such dependence.

The result also revealed that, for $Hg^0$ adsorption, an activated carbon derived from high-organic sulfur coal dod not benefit from a sulfur impregnation process at 600°C when compared with a sulfur-impregnated activated carbon prepared from low-organic sulfur coal. However, $HgCl_2$ reactivity and capacity of samples improved after sulfur impregnation.

The results from pilot-scale mercury tests showed ICDAC has mercury removal capacity comparable to or greater than a commonly used commercial product.

## ACKNOWLEDGEMENT

Supports from EPRI and the Illinois Clean Coal Institute made this research possible. Nitrogen adsorption surface analyses were carried out at the center for Microanalysis of Materials, University of Illinois at Urbana-Champaign, which is supported by the U.S. Department of Energy under grant DEFG)2-ER45439.

## REFERENCES

1. R.Chang and D. Owens, EPRI J., 46, July/August (1994).
2. US Environmental Protection Agency, Mercury Study Report to Congress, vol. 1: Executive Summary, Office of Air Quality Planning and Standards, December (1997).
3. K. C. Galbreath and C. J. Zygarlicke, Environ. Sci. Technio., 30 (1996) 2421.
4. C. D. Livengood, H. S. Huang and J. M. Wu, Proceedings of the 87th Annual Meeting of the Air and Waste Management Association, (1994), reprint, 14.
5. C. L. Senior, L.E. Bool III, G.P. Huffman, F.E. Huggins, N. Shah, A. Darofim, I. Olmez and T. Zeng, Presented at the 90th Annual Meeting and Exhibition of the Air and Waste Management Association, Toronto, Canada, June (1997).
6. W. H. Schroedor, G. Yarwood and H. Niki, Water, Air, Soil Pollut., 56 (1991) 653.
7. E. Schuster, Water, Air, Soil Pollut., 56 (1991) 667.
8. CRC Handbook of Chemistry and Physics, 63th ed., CRC Press, Inc., Boca Raton, FL, 1982.
9. C. Volland, presented at the 84th Annual Meeting and Exhibition of the Air and Waste Management Association, Vancouver, BC (1991) Paper 91-35.5.
10. S. Haythornthwaite, S. Sjostrom, T. Ebner, J. Ruhl, R. Slye, J. Smith, T. Hunt, R. Chang and T. D. Brown, EPRI-DOE-EPA Combined Utility Air Pollutant Control Symposium, Washington, DC (1997).
11. E. G. Waugh, presented at the EPRI-DOE-EPA Combined Utility Air Pollutant Control Symposium, Washington, DC (1997).
12. S. V. Krishnan, B. K. Gullett and W. Jozewicz, Environ. Sci. Technol., 28 (1994) 1506.
13. T. R. Carey, O. W. Hargrove Jr. and C. F. Richardson, 90th Annual Meeting and Exhibition of the Air & Waste Management Association, Toronto, Canada, June (1997).

14. R. D. Vidic and J.D. McLaughlin, J. Air Waste Manage. Assoc., 46 (1996) 241.
15. S. J. Miller, D. L. Laudal, R. Chang and P.D. Bergman, In Proc. of 87th Annual Meeting & Exhibition of the Air & Waste Management Association, Cincinnati, Ohio, June (1994).
16. S. Sjostrom, D. L. Roberts, G. Anderson, F. Sagan and J. Smith, 90th Annual Meeting & Exhibition of the Air & Waste Management Association, Toronto, Canada, June (1997).
17. K. Felsvang, R. Gleiser, G. Juip and K. K. Nielsen, Fuel Proc. Technol., 39 (1994) 417.
18. D. M. White, W. E. Kelly, M. J. Stucky, J. L. Swift and M. A. Palazzolo, US EPA, EPA/600/SR-93/181, January (1994).
19. S. Chen, M. Rostam-Abadi and R. Chang, Prep. Am. Chem. Soc., Div. Fuel, Chem., 41 (1996) 442.
20. M. Steijns, A. Peppelenbos and P. J. Mars, Colloid and Interface Sci., 57 (1976) 181.
21. H. Nishino, T. Aibe and K. Noguchi, US Patent No. 4 500 327 (1985).
22. Y. Otani, H. Emi, K. Chikao, I. Uchijima and H. Nishino, Environ. Sci. Technio., 22 (1988) 708.
23. J. Teller and J. M. Quimby, Proceeding of the 84th Annual Meeting of the Air & Waste Management Association (1991), reprint, 13.
24. J. A. Korpiel and R. D. Vidic, Environ. Sci. Technol., 31 (1997) 2319.
25. T. M. Matviya, R. S. Gebhard and M. Greenbank, US Patent No. 4 708 853 (1987).
26. R. L. Horton and S. Russel, US Patent No. 4 591 490 (1986).
27. M. R. Stouffer, W.A. Rosenhoover and F.P. Burke, A & WM 89th Annual Mtg, Nashville, TN, June (1996).
28. Licata, M. Babu and L. Nethe, Proceedings of the 1994 National Waste Processing Conference. Boston, Massachusetts, June (1994).
29. S. Chen, M. Rostam-Abadi and R. Chang, Mercury Removal from Flue Gas by Carbon Injection: Mass Transfer Analysis, in preparation.
30. R. D. Harvey and C. W. Kruse, J. Coal Qual., 7 (1988) 109.
31. S. Brunaer, P.H. Emmett and E. J. Teller, Am. Chem. Soc., 60 (1938) 309.
32. C. Lippens, B. G. Linsen and J. H. de Boer, J. Catal., 3 (1964) 32.
33. R. Sh. Mikhail, S. Brunauer and E. E. Bodor, J. Colloid Interface Sci., 26 (1968) 45.
34. S. J. Gregg and K. S. W. Sing, Adsroption, Surface Area and Porosity, Academic Press, New York, 1967.
35. J. Berkovitz, Elemental Sulfur, Chemistry and Physics, B. Meyer (ed.), Interscience, 1965, 125.
36. B. R. Puri and R. S. Hazra, Carbon, 9 (1971) 123.

Adsorption and its Applications in Industry and Environmental Protection
Studies in Surface Science and Catalysis, Vol. 120
A. Dabrowski (Editor)



# Sorption properties of gas/coal systems, degasification of coal seams

J. Tóth

Research Laboratory for Mining Chemistry
of the Hungarian Academy of Sciences
P.O. Box 2, H-3515 Miskolc-Egyetemváros, Hungary

## INTRODUCTION

It is well known that the coals, especially the bituminous high rank coals, are important sources of different energies, however, only the experts have knowledge about their special property: The coals are excellent sorbents both for gases and for liquids. For example, 1 metric ton high rank coal may contain 20-100 $m^3$ STP methane and 10-100 kg water.

It should be emphasized previously here that coals are sorbents and not adsorbents, that is the gases and liquids for the greater part are bound in the inside structure of the coals and only a smaller amount of gases are adsorbed on the surface (in the open pores) as it is the case of industrial adsorbents such are the activated carbons and silica gels. This sorption capacity and all its consequences will be discussed in this chapter, however, in advanced should be mentioned that this storage capacity of coals is quite different from that of the natural gas reservoirs.

The gas storage capacity of coals has a great disadvantage but, fortunately, it has an advantage as well, which can be utilized for power engineering. The disadvantage is the proneness of coalbed gases (methane and/or carbon dioxid) to very dangerous outbursts caused death of many miners. Over the past fifty years, according to careful appraisments, more than thousand miners died in world's coal mines on the occasion of gas-coal outbursts. Especially there were many victims of *sudden* outburst in spite of the up-to-date protective methods applied in the work places. The main cause for this tragic events is stimulating on further investigations: Although we have definite informations on the rock mechanical effects connected with the sudden outburst, so far we do not have exact physico chemical explanation for causes of absorption and sudden (or slower) liberation (desorption) of gases from coal seams. The advantage of gas capacity of coals is obvious: If we were able to bring under control the liberation (desorption) process of coalbed methane (especially the enormous great amount of gas desorbed during an outburst), then we should have an alternative energy source instead of coal. This controlled degasification of coal seams may have important role in energetics of countries where the traditional coal mine activities are already not profitable.

The relation of advantages to disadvantages of coalbed methane is similar to those discovered by the scientists in the past. For example, first was the A-bomb and *only* later were built the atomic power stations, thereafter came the H-bomb and we have to wait for economic fusion reactors. We hope that this chapter gives informations on "how to make" shorter the distance between the advantages and disadvantages of coalbed methane.

## 1. THE INSIDE STRUCTURE OF COALS

The first step to bring under control the liberation of coalbed gases is the knowledge of the inside structure of coals, that is, we have to know where is "place" for this great amount of gases, what kind of binding forces and, finally, what kind of processes take place before, during and after the desorption of gases. The molecular structure of coals is a very old problem for the coalchemists and this short chapter does not have the aim to summarize the history of investigations and models made to determine the coal structure. In this chapter the recent results achieved over the past 5-10 years are summarized only and an arbitrary classification is made among the results mentioned above.

In the first group are collected the investigations of the coalchemists and according to the second group can be formed opinion on the coal structure discussed and accepted by reservoir engineers.

### 1.1. The molecular structure of coals constructed by the coalchemists

In 1989 a very interesting debate was published in Fuel [1] where 14 world-wide known coalchemists discussed on this problem and the following model was accepted. The coal structure can be explained by a cross-linked, three-dimensional macromolecular model, that is, the coals (especially the bituminous high rank coals) have lattice structures. The modern physical measurements such are pyrolisis-field ionization mass spectrometry (py-f.i.m.s), N.m.r experiments have been proved that in the inside structure there are *low* molecular mass materials which are trapped in the macromolecular framework of coals. Also has been previously [2-4] proved that this low molecular mass materials (shortly: small molecules) are mobile and they can be taken out from the framework (by heating to 400 °C, by pyrolisis at 250 °C or by Soxhlet extraction at ambient temperature). Based on these facts Haenel [5] gave a very simple but very descriptive graph for the coal inside structure shown in Figure 1. Just in 1992 Nishioka [6] somewhat modified Haenel's model and this modified model is schematicly represented in Figure 2. The essence of this modification is that the small molecules present in the framework can physically associate with the immobile part of the structure, however, this association is a reversible process: By heating a dissociation can take place, however, by cooling the association process is dominant. So very shortly can be summarized: The up-to-date molecular structure model of high rank coals constructed by the coalchemists is a lattice structure with some elastic properties.

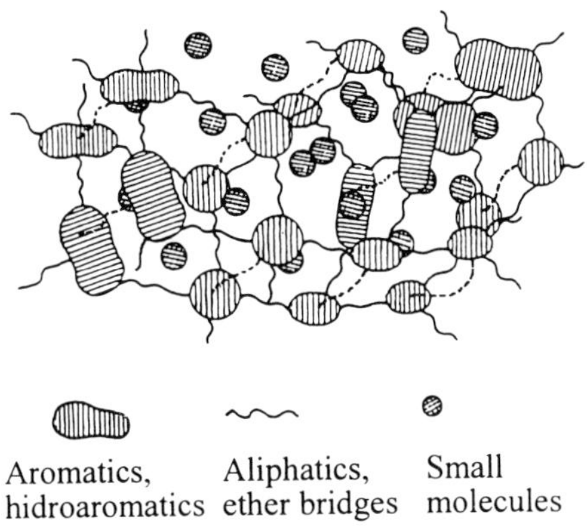


Figure 1. Macromolecular three-dimensional crosslinked network model of coal structure (After Haenel [5]).

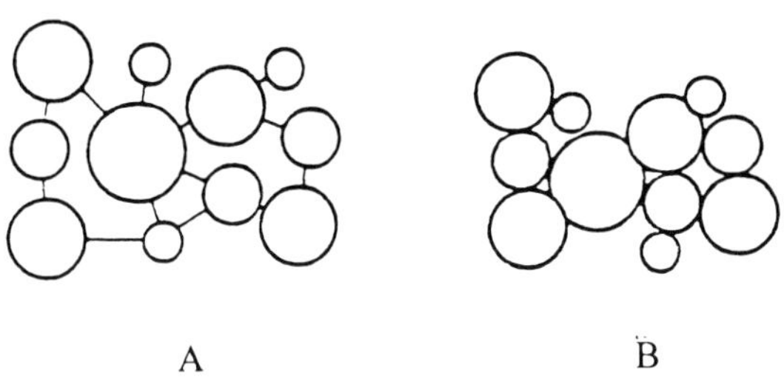


Figure 2. A: Covalently cross-linked and B: physically associated models of coal structure (After Nishioka [6]).

**1.2. The inside structure of coals constructed by the reservoir engineers**

The reservoir engineers have practical aims, therefore, they put the same practical questions directly connected with production of hydrocarbon gases from whatever reservoirs, let they are sandstones, limestomes or coal seams. The most important problems should be solved by the engineers are: How much is the gas content, what a level of recovery factor can be achieved, what is the prospective and specific time of the production ($m^3$ STP/day)? Furthermore, the most important properties of reservoirs are their permeability, the distribution of that in the whole production area, the average depht of the reservoir, and the pressure of gas to be produced. These parameters mostly influence the profitability of the gas production.

These are the reasons why the reservoir engineers are not interested in the detailed molecular structure of coals, their attention is rather concentrated on the problem: How can flow the methane in the inside structure of coals, that is, they think in micropores, cavities, fractures giving possibility to transport of methane through the coal. It is indifferent to them how these pores, cavities etc. taken shape from the concrete molecular structure. To demonstrate this statement only two examples, from the recent literate, are shown here. King and Ertekin [7] supposed a desorption process from "internal coal surfaces" then diffusion through matrix and micropores and finally a fluid flow in network (See Figure 3).

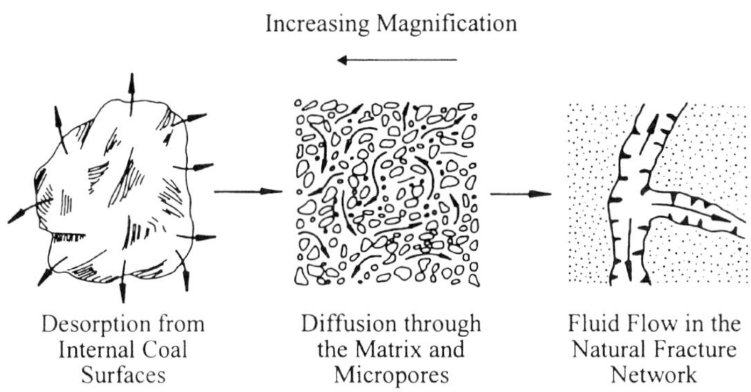


Figure 3. Transport model of methane through coal (After King and Ertekin [7]).

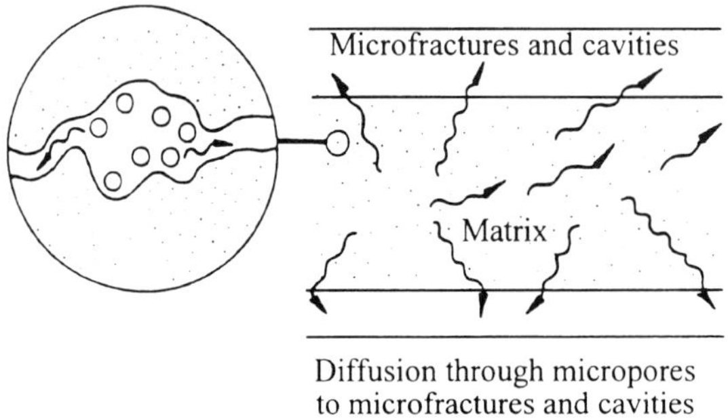


Figure 4. Model of methane flow through coal microstructure (After Gamson and his co-workers [8]).

Some years later Gamson and his co-workers [8] modified this model and they supposed such type of micropores wich are very similar to those present in activated carbons (See Figure 4).

This microstructure has been investigated by up-to-date methods especially the scanning electron microscope examination proved the presence of fractures, cavities and matrix porosities wich, in reservoir engineers' opinion, deciding determine the transmissibility of methane (the permeability of coal). It is true, that commercial gas production to date has been hampered by the low permeabilities of coal seams. For example the permeability of a "good" porous sandstone reservoir is between 0.1-0.5 $\mu m^2$ just then a coal seam has only 5-20.$10^{-3}$ $\mu m^2$ permeability but there are coals with permeability between $10^{-5}$-$10^{-7}$ $\mu m^2$. As it will be discussed in point 4 the low permeability is indeed the greatest problem of methane recovery from coal seams.

### 1.3. The inside structure of coals determined by sorption and other physico chemical methods

As we have interpreted above the coal structure were investigated by up-to-date physical measurements both by the coalchemist and by the reservoir engineers. However, in the literature can rarely be found papers dealing with *gas sorption* properties of coals [9-12], but up to the present have not been measured liquid mixture (excess) sorption isotherms combined with gas sorption data and other physical measurements. In the following parts these investigations are discussed separately but thereafter a synthesis of those are made.

#### 1.3.1. Gas sorption properties of high rank coals

It can be found some methane/coal and carbon dioxid/coal isotherms in the literature [9-12] but they were not entirely and together investigated from four standpoints. The first important task is to increase the domain of equilibrium pressure because in the massive coal seams may be much greater pore pressures than those in the work places and it is sure that during the coalification process the relatively great geostatic pressure determined the amount of methane sorbed in the coal structure. The second problem is to investigate the *desorption* processes in the same enlarged pressure range and to explain the hysteresis phenomena if they exist at all. The third task is to determine the isotherms at very different temperatures especially at low ones (under the critical temperatures) because in this domain energetic calculations can be made very easily. Finally, it should be proved which has only been stated above: The greatest amount of the gas is in sorbed form in the *inside* structure of the coal. There is in Hungary a very hazardous coal mine (Mecsek) where, in spite of the applied protective methods, relatively often occured sudden methane-coal outbursts. The same situation is in Poland where the carbon dioxid-coal outbursts are very dangereous [4].

This is the reason why over the past twenty years in both countries far-reaching investigations have been made and the results of them are very similar.

Nevertheless, the measurements made in Hungary are in detail discussed here because our aim was not only to elaborate a protective method against the unexpected outbursts but to bring under control the sudden desorption of methane and so get an alternative energy source from coal. It is also interesting that in Mecsek coal mines there are very hazardous work places (great proneness to sudden outburts) and are ones where never occured accidents, in spite of the fact, that the rank and age of coals (Jurassic) are equal. Therefore, the sorption and other properties of hazardous and non-hazardous coal samples were investigated separately. The data of samples are tabulated in Table 1.

Table 1
Characteristic data of coal samples used to gas and liquid sorption measurements

| Sample | Age | Ash | Volatile matter | C | H | N | Grain size (mm) |
|---|---|---|---|---|---|---|---|
| | | (W/W % dry bases) | | | | | |
| Hazardous Mecsek coal | Jurassic | 11.0 | 31.0 | 57.7 | 4.3 | 3.3 | 1.5-2.0 |
| Non-hazardous Mecsek coal | Jurassic | 9.1 | 32.8 | 61.6 | 4.6 | 3.2 | 1.5-2.0 |

The summarizing and characteristic results of gas sorption properties of coal samples are shown in Figures 5 and 6.

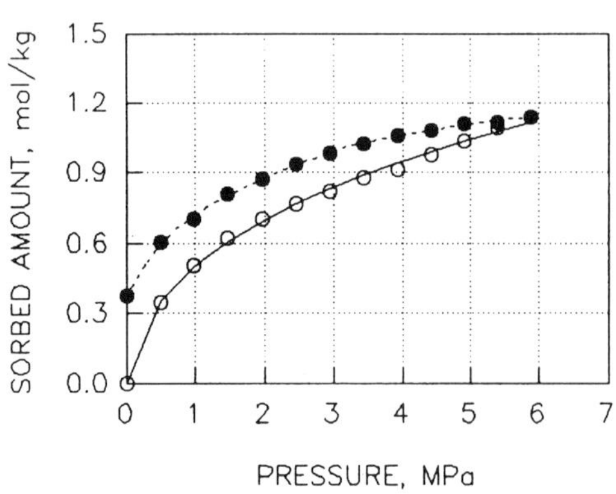


Figure 5. Adsorption hysteresis of methane measured on a Hungarian hazardous high rank (bituminous) coal at 25°C [13]. ○ : adsorption; ● : desorption.

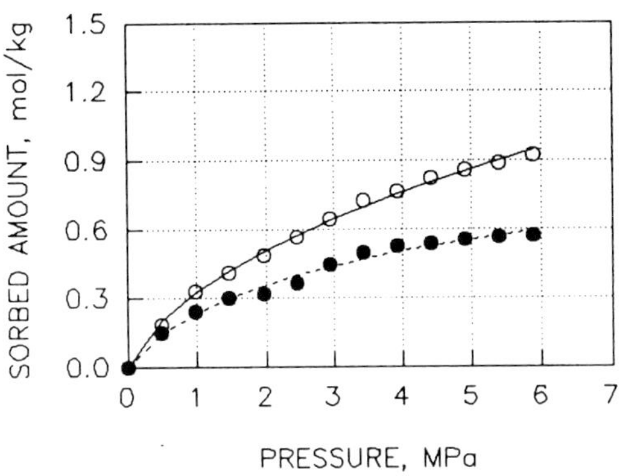


Figure 6. Adsorption and desorption isotherms of methane measured on Hungarian hazardous high rank (bituminous) coal at 25 °C [13]. ○: adsorption; ●: desorption.

In Figure 5 is represented the normal sorption hysteresis of the hazardous coal sample. It is an "open" hysteresis function measured by Sartorius 4112 Type micro balance applicable to 12 MPa pressure [13]. The "open" hysteresis indicates that at very low equilibrium pressure range a great amount of methane remains in the inside structure of coal. This phenomenon cannot only be explained by very slow kinetic process of desorption because the "remained" amount of methane is independent of time required to formation of desorption equilibrium pressure. The "remained" amount of methane can better be emphasized by data shown in Figure 6.

The empty circles indicate the normal sorption data, however, the full symbols represent non-equilibrum data measured by the following way [13]. After every sorption processes the methane was expanded to pressure 1 bar and was measured the desorbed amount of gas. It means, that the full circles in Figure 6 relate to the remained sorbed amount of methane when every current equilibrium pressure is reduced to 1 bar. It is very interesting that there are coals with the same age (Jurassic) which can absorb more methane (see Figure 7) than coals shown in Figures 5 and 6, however, these coals in Figure 7 have only smaller "open" hysteresis than that of the hazardous coals and in these coal entries never were observed any proneness to outburst.

The great "open" hysteresis, therefore, seems to be an indication for hazardous character of coals. The detailed explanation of this phenomenon, connected with the thermodynamic model of degasification process, is discussed in point 2.h.

The phenomenon of "open" hysteresis also proves that the coals and activated carbons have quite different inside and outside structures. The "open" hysteresis has never been observed on activated carbons although both sorbents are often

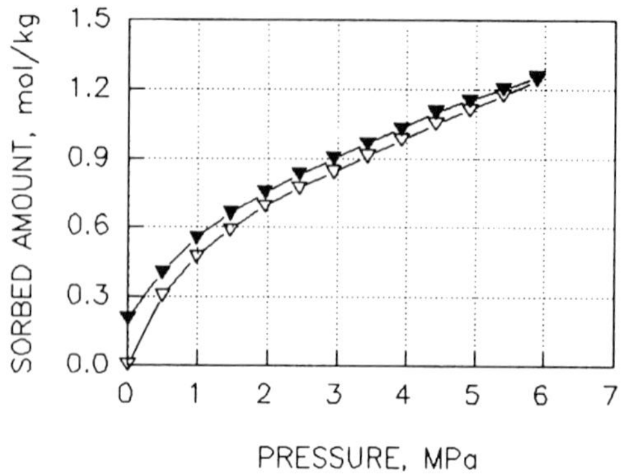


Figure 7. Adsorption hysteresis of methane measured on a Hungarian non-hazardous high rank (bitumenous) coal at 25 °C [13].▽: adsorption;▼: desorption.

called micro-porous ones. The measurements and investigations mentioned above prove that the micropores of coals and those of activated carbons are not similar.

This fact can be summarized in a simplified form: The activated carbons have surface - micropores, however, the bituminous coals have inside - micropores caused by the lattice structures discussed in point 1.1 (See Figures 1, 2 and 4). It is evident that these two types of micropores are connected with different sorption and desorption properties. These problems are discussed below.

### 1.3.1.1. Sorption isotherm, isobars and surface areas measured on high rank coals

We have mentioned in the previous part of this chapter that it is very important to determine the gas (especially the methane) isotherms at very wide temperature range including the critical temperatures as well. Such type of isotherm measurements have never been published in the literature, and we were convinced that these isotherms would give more informations on the energy structure of the sorption and desorption processes. We were not disappointed in our expectations.

In Figure 8 methane sorption isotherms measured on Mecsek high rank coal (Hungary) at two different temperatures [15] are shown. The important phenomenon is that the isotherms have inversion, i.e., the amounts of methane sorbed at lower temperature (120 K) are smaller than those measured at higher temperature (294 K). In order to better understand of this phenomenon many sorption isotherms have been measured between 120-294 K and the results as sorption isobars are represented in Figure 9 [15]. It is well-known that isobars of maximum type can only occur when much greater binding energies have also

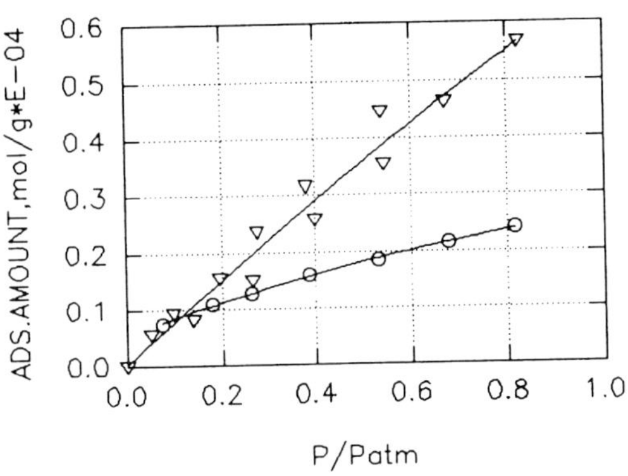


Figure 8. Sorption isotherms of methane measured on high rank coal (Mecsek, Hungary) at temperature 120 K and 294 K [15]. o: 120 K; ∇: 294 K.

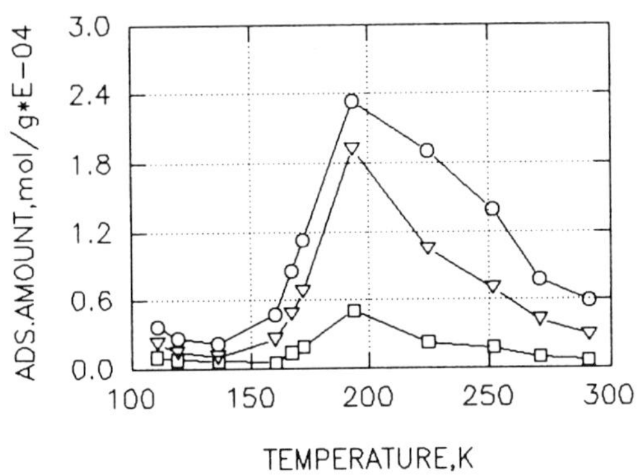


Figure 9. Sorption isobars calculated from coal-methane sorption isotherms measured between temperatures 120-294 K [15]. o: $P/P_{atm}$=0.7; ∇: 0.3; □: 0.05.

role in the uptake of the adsorptive than the van der Waals forces. It is also known that this type of sorption needs activation energy wich is assured by the kinetic thermal movement of the adsorptive molecules. This is the reason why is required 10-20 times longer time to achieve the equilibrium pressure at temperatures left from the maximum of isobars than at "right hand" temperatures.

Independent of the concrete mechanism of the sorption processes the maximum character of the isobars prove that the uptake of methane in coal takes place at two energy levels. The first is the normal physical adsorption in (on) the surface-(micro)pores and the second sorption takes place in the lattice structure (in the inside-micropores), where the sorbed molecules are bound with much greater energy than on the surface.

The penetration of gas molecules in the lattice (inside) structure of coal prove indirectly the fact that the monolayer equivalent surface are of coal depends on the temperature and on the molecular size of the sorptive gases [15]. In Figure 10 are shown these phenomena which can be explained by the inside structure of coal discussed in point 1.1.

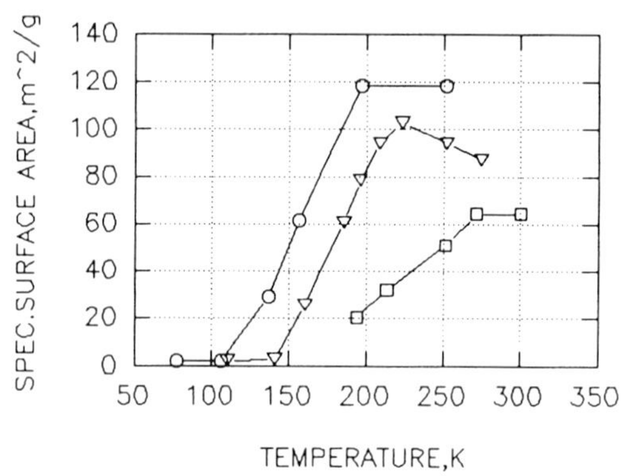


Figure 10. Monolayer equivalent surface area of coal calculated by BET and Dubinin method as functions of the temperature and the molecular size of the sorptives [15]. o: nitrogen; ∇: methane; □: ethane.

The temperature dependence of the measured surface area is evident when it is taken the Nishioka's [6] model (see Figure 2) into account. With increasing temperature the dissociation of mobile parts also increases, therefore, the gas molecules can better penetrate in the inside structure of coal. At a definite temperature, however, the distribution of the dissociation grade is heterogeneous, i.e., the coal sorbs gases similarly to an molecular sieve.

### 1.3.1.2. Calculation of amount of sorbed gas in coal at very high (geostatic) pressures

The maximum equilibrium pressures applied in laboratories in measuring the methane/coal sorption isotherm were about 6-10 MPa, however, the geostatic pressure, during and after the coalification process, may be much greater. This case especially may occur when the coal seam is a sunken lock and the methane was still present in the coal at the time of the tectonic movement. Therefore, it is required to describe not only the measured isotherm, but, as far as possible, to extrapolate the measured data to the supposed or calculated geostatic pressure. This calculation is especially important at new-discovered coal seams, because this is the only possibility to estimate the maximum sorbed amount. The experimental determination of this maximum amount is very difficult because during the drilling activities the coal loses its methane content up to 50-60 %. (The explanation see in point 3.)

Describing the measured data and extrapolation of those to the geostatic pressure is possible on condition that we suppose an equivalent adsorption, i.e., we suppose as if the *sorbed* amount were *adsorbed* on a heterogeneous surface. This is an unavoidable supposition because all known isotherm equations describe physical or chemical adsorption taking place on *surfaces* and does not do it for the bulk phases. One of the widely-used isotherm equation is the Tóth [16] equation also applied by Valanzuela and Myers in their handbook where are collected more than 100 simple gas isotherms [17]. The explicit form of this equation is the following:

$$n^s = \frac{n_\infty^s p}{\left[\left(K^{-1}\right) + p^m\right]^{1/m}} \tag{1}$$

where $n^s$ is the (ad)sorbed amount in mol/kg, p is the equilibrium pressure in MPa, $n_\infty^s$ is the adsorbed amount when p tends to infinity, m is a constant characterizing the heterogeneity of the surface, i.e., if m=1, then Eq. (1) transforms into the Langmuir equation, and in most cases it is valid that

$$0 < m < 1 \tag{2}$$

Finally, K is also a constant (it depends on the temperature only) and it is expressed in ($MPa^{-m}$). K is in direct connection with the adsorptive potentials of the heterogeneous surface [18]. The parameters of Eq. (1) applied to sorption isotherm shown in Figures 6 and 7 are tabulated in Table 2. From the extrapolated data of Eq. (1) to the estimated or sometimes measured geostatic equilibrium pressures can be calculated that coal seams beeing in a depth of 700-1000 m and having a mass about 40 million metric tons may contains 50-60 thousand million cubic meters methane wich is comparable with a "good" sandstone reservoir.

Table 2
Parameters of T equation applied to sorption and desorption isotherms shown in Figures 5, 6 and 7

| Isotherms | $n^s_\infty$ (mol/kg) | K ($MPa^{-m}$) | m |
|---|---|---|---|
| Isotherm in Figure 5<br>Symbol: o | 25.285 | 1.070 | 0.169 |
| Isotherm in Figure 6<br>Symbol: o | 14.433 | 0.515 | 0.285 |
| Isotherm in Figure 6<br>Symbol: • | 1.402 | 0.539 | 0.587 |
| Isotherm in Figure 7<br>Symbol: ∇ | 166.731 | 0.683 | 0.154 |

### 1.3.2. Liquid sorption properties of high rank coals

We hoped that it was possible to obtain more informations on the inside structure of coal by liquid mixture excess isotherms measured on coal. Our opinion was based on data published in the literature [19].

These measurements have proved that there are some interactions between high rank coals and methanol, however, it seems to necessary to investigate this phenomenon with exact liquid sorption measurements. The methanol-water excess isotherms measured on hazardous and non-hazardous high rank coal are shown in Figure 11 [20-21]. In this Figure can be seen that methanol

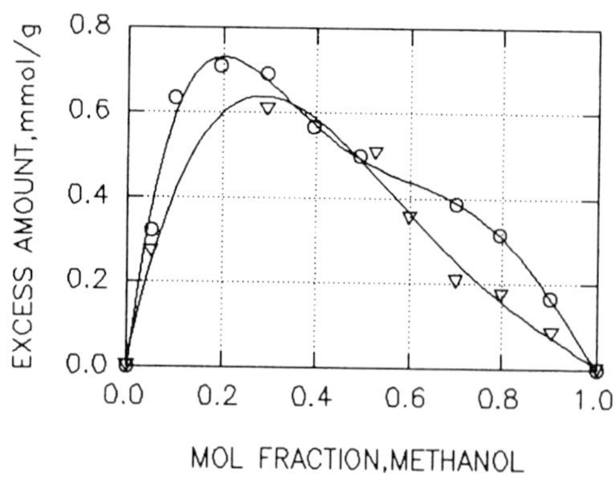


Figure 11. Methanol(1)-water(2) excess isotherms measured on hazardous and non-hazardous high rank coals at 25 °C [20]. o: hazardous; ∇: non-hazardous.

preferentially is sorbed in coal and because of the irregular form of the excess isotherms the Schay-Nagy method [22] for calculation of the equivalent surface area from the range of methanol molar fraction 0.8-1.0 can only be calculated. The so calculated surface area for the hazardous coal is 190 $m^2g^{-1}$ ± 10 % and for non-hazardous coal 91 $m^2g^{-1}$ ± 10 %, respectively. Further investigations have been made with methanol-benzene mixtures, the corresponding excess isotherms are shown in Figure 12.

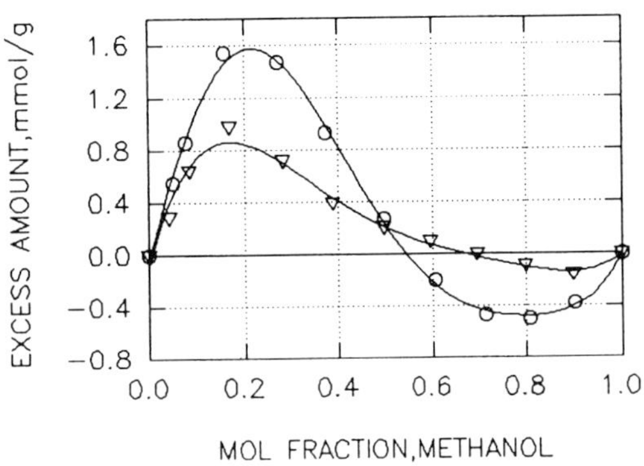


Figure 12. Methanol(1)-benzene(2) excess isotherms measured on hazardous and non-hazardous high rank coals at 20 °C [20-21]. o: hazardous; ∇: non-hazardous.

The results are surprising because not only the type of the isotherms has been changed but the calculable surface areas as well. These values are for hazardous coal 754 $m^2g^{-1}$ ± 10 % and for non-hazardous coal 600 $m^2g^{-1}$ ± 10 %, respectively. Comparing these results with the equivalent surface areas calculated from gas sorption isotherms (3-120 $m^2g^{-1}$) the following conclusion can be made. The methanol and especially the benzene penetrate in the inside structure of the coal, desaggregate the associated mobil small molecules (See Haenel and Nishioka's models in Figures 1 and 2), and, therefore, the equivalent or (apparent) surface areas increase [20]. This desaggregation may be promoted by the kaolinite wich is intercaleted in the inside structure of coals. This statement is supported by the X-ray diffraction analysis (CuKα-ray with Philips diffractometer) which proved the presence in all coal samples the basal planes of kaolinite with a distance of 7.14-7.15 [20]. Later the kaolinite was separated from the coal samples in form of a clay-suspension [20]. It was also a very interesting observation that from the coal samples saturated with methane, during the contact with methanol-benzene mixtures, the gas desorbed in form of visible bubbles.

The differences of surface areas calculated from gas sorption isotherms and from different excess isotherms measured on the same coal samples also prove the fundamental difference in structures of coals and in that of the microporous activated carbons. Namely, the equivalent surface area of an activated carbon is independent of the sorption system (gas or liquid excess isotherm) and of the composition of the liquid mixtures. This statement is valid in spite of the fact that the type of the excess isotherms (classified by Schay and Nagy) may depend on the composition of the liquid mixtures.

## 2. THERMODYNAMIC MODEL OF DEGASIFICATION AND OUTBURST OF COALS

Based on experimental results discussed in the previous parts of this chapter the following conclusions and statements can be made:

a) The gases (methane, carbon dioxid etc.) are bound in the coals in two forms. In open or surface-micropores the van der Waals forces and in the inside-micropores much greater forces (approx. 100-150 $kJmol^{-1}$) bind the gas molecules.

b) According to the two types of binding energies two equilibria should exist in an undisturbed coal/gas system: The equilibria between the bulk phase (inside-micropores) and the surface, and between the surface-pores and gas phase.

c) The inside structure of coals can be destroyed by liquids (especially methanol and benzene is applicable for this aim) and during this process the gases are desorbed at ambient pressure and temperature.

From these three facts a model-schema can be constructed which may be a basis for all industrial processes, both for protection against sudden gas/coal outburst and for degasification of coal seams. This model has already been published [23], however, some alterations were made in the past four years.

From existence of the two equilibria mentioned above it follows that, from the standpoint of thermodynamics, the thermodynamical potentials (free enthalpy or free energy) as a function of a variable-state must have two local minima at two levels of energy. This requirement is drafted in Figure 13 where the free energy (F), as a function of distance (r), is shown.

The starting point of the distance r is inside the structure of coal where the van der Waals forces acting on the surface pores do not have any effects. The distance is in direct relation to the gas-concentration in the coal (which is a state-variable), however, from a model's point of view, the concrete mathematical form of this relationship has no importance. The free energy is discussed here because the coal/gas system can be regarded as having a constant volume. It is known that the free energy is a part of the inside energy (U) which, during a process coming to an equilibrium, may transform to work or may release energy equal to this work. This is the reason why the starting point of the free energy in

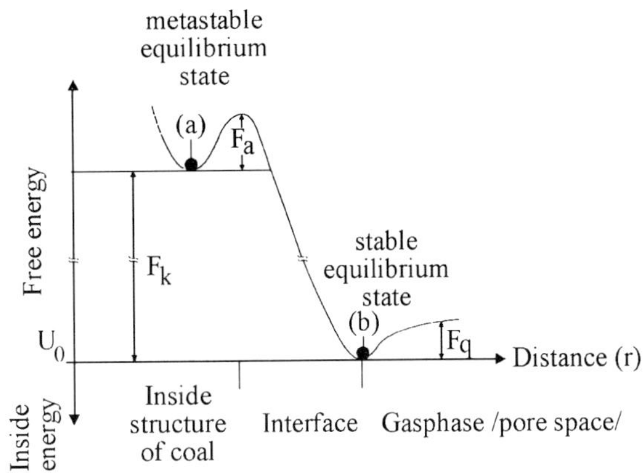


Figure 13. The free energy of a coal/gas system as a function of distance calculated from the inside structure to the gas phase [23].

Figure 13 is the inside energy $U_o$ which, in the process investigated, cannot already be transformed into work. The function of free energy in Figure 13 provides information on the direction of changes in energy. Namely, when a process goes towards the minimum (i.e. towards the equilibrium state) the system releases energy; in reverse, the same movement from the equilibrium state (from the minimum) needs energy (intake of energy).

It is evident that there is an energy-difference $F_k$ between the equilibria (a) and (b) and there exists an $F_a$ potential barrier between the inside and outside structure of the coal. The existence of such a potential barrier is a thermodynamic evidence in all cases when there are in a system two (or more) equilibria at different energy levels.

From this model it follows that the gas (methane) from equilibrium state (a) can change to equilibrium state (b) if the gas absorbs the energy $F_a$ and so crosses the potential barrier. When this happens the energy ($F_a$ +$F_k$) releases. The equilibrium (a) is called a metastable one, referring to the fact that in the system another equilibrium (b) exists at a lower energy level. This last equilibrium is a stable one because a third equilibrium does not exist where the gas can go by releasing some energy.

Based on this simple model of equilibria present in a coal/gas system the following phenomena can be explained and applied to plan a degasification technology:

a) The essential difference between the sandstone and coal seam reservoirs appears in the fact that the liberation of gas (gas production) from the coal is only possible if in the *whole* coal seam greater energy than $F_a$ is in some form transmitted (injected).

b) This transmission (or injection) of energy $F_a$ can be carried out by different energy impacts such are: hydraulic fractures, explosions or destroying the coal structures by liquids etc. These impacts should be touched the whole coal seam from that the gas is planed to produce.
c) In addition to this energy injection should be assured a zone or channel of much greater permeability than that of the other parts of the coal seam. This channel must assure the continuous flow of gas towards the gas production well(s).
d) The cause for the sudden (unexpected) gas/coal outbursts is that the energy barrier $F_a$, during the mining activities, is decreased to a critical level where a very little additional energy impact (for example a small knock) is enough to release a great energy of ($F_a$+ $F_k$). This situation is very similar to the state of an overheated liquid. The molecules cannot pass into the vapour phase because the surface tension is a potential barrier, however, a very small energy impact (a small vibration) is enough to bring the liquid, with the velocity of an explosion, into the vapour phase.
e) In this manner the essential difference between hazardous and non-hazardous coal seams can be explained. In undisturbed state the hazardous coal seams have great energy barrier $F_a$, however, this barrier, as an aftermath of the mining activities (shaft sinking, hard-works in the roadhead etc.) decreases to the critical level mentioned above. The non-hazardous coals have originally (i.e. before beginning the mining activities) small energy barrier $F_a$, therefore, the desorption of methane can take place easily. So in these coal mines the safety of the work places depends to a great extent on the good exhaust ventilation. In hazardous coal mines the safety is supported by destroying the barrier $F_a$ by different protective processes (for example provocative explosions).
f) Both in hazardous and in non-hazardous coal mines the safety partly depends on the permeability of the coal. In spite of the small energy barrier in non-hazardous coals may occur that because of the very low permeability the desorbed gases accumulate in the inside-pores, therefore, the pore pressure increases and it may cause a blow-out or outburst. However, in non-hazardous coal mines this phenomenon can always be avoided by protective methods mentioned above.
g) The cause of the "open" hysteresis is also the existence of the energy barrier $F_a$. The metastable gas molecules beeing "behind" the barrier cannot desorb from the inside structure or this process, as a result of thermal movement of gas molecules, needs very much time (more than one year) at ambient temperature.
h) The explanation of the phenomenon shown in Figure 7, based on the conclusions mentioned above, is very easy: The coals having very small "open" hysteresis also have small energy barrier $F_a$ before beginning of the mining activities, and the permeability of coals is enough great to avoid the accumulation of the desorbed gas in the inside pore structure.

## 3. LABORATORY MEASUREMENTS NEEDED TO PLAN A DEGASIFICATION PROCESS

From statements and conclusions mentioned in the previous parts of this chapter it follows that the basic aims of a degasification process should be the determination (or at least the comparative estimation) of the energy barrier $F_a$, the estimation of energy required to destroy this barrier and to know how much methane is desorbed after the elimination of the metastable state.

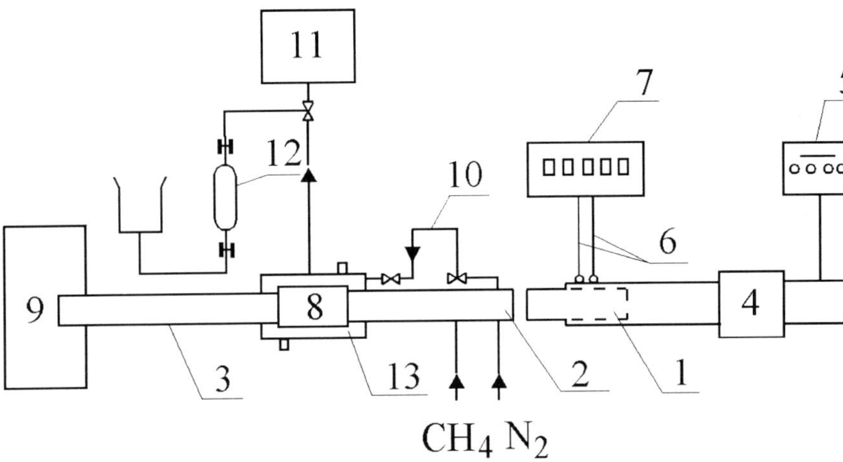


Figure 14. Schematic arrangement of an equipment applied to measure the desorbed gas before and after an energy impact [24].

To measure these parameters the equipment shown in Figure 14 has been applied [24]. The equipment comprises three bars, a striker bar 1, an incident bar 2 and a transmitter bar 3. All bars have a diameter of 20 mm and are made of low-impendance steel. Loading pulses are initiated by the impact of the striker bar against the incident bar. The striker bar is accelerated to a desired impact velocity with a pneumatic gun 4 which is operated by a launching controller 5. The impact velocity is measured with two photodiodes 6 producing sharp electrical signals during shadowing by the projectile face. The time counter 7 records the time interval between the electrical pulses. The measurement of the impact velocity is necessary in the calculation of the kinetic energy of the striker bar. During the impact, this kinetic energy excites wave energy in the incident bar. The pressure chamber 8 containing coal sample (13-15 g) is placed between bars 2 and 3. A liquid thermostat serves to maintain constant temperature in the chamber. The remained wave energy is absorbed in bar 9.

The *spontaneous* desorption of methane is measured as follows:

A fresh, air-dried and weighed coal sample with volume nearly equal to that of the chamber is maintained at a temperature of 30 °C. The coal sample is preflushed with pure nitrogen (flow rate 200 $cm^3h^{-1}$) through tubing 10 until methane can be detected with a HP 5750 gas chromatograph 11. After the pre-flushing (~ 30-90 min) the chamber is sealed for 24 h. The methane desorbed during this period is swept out with nitrogen in a 200 $cm^3$ gas-sampling bulb 12. This volume of nitrogen is always sufficient to deliver the desorbed gas into the bulb where 1.00 $cm^3$ gas taken with a Hamilton microsyringe for analysis. The quantity of the desorbed methane is determined with the gas chromatograph equipped with a flame ionization detector and packed a Porapak-Q column (length 1.5 m, diameter 4.5 mm). The oven temperature is maintained at 100 °C.

The *stimulated* desorption of methane is measured under conditions fairly similar to those used to measure the spontaneous desorption. The only difference is that before the impact, the desorbed methane is eluted with 200 $cm^3$ nitrogen into the gas bulk 12 in order to measure the rate of spontaneous desorption. Soon after, the shock wave is initiated by the impact of the striker bar against the incident bar. Immediately after the impact the desorbed gas is again flushed into the gas bulb.

This stimulated desorption always is associated with four phenomena:

a) The desorption of methane takes place instantly after the impact. This stimulated desorption is so quick that it is impossible to measure its rate.
b) The amount of methane desorbed is two-three order of magnitude greater than that desorbed spontaneously before the impact.
c) The higher hydrocarbons components ($C_2$-$C_4$) always appear in the desorbed gas (concentration max. 1 vol%).
d) The coal sample is crushed. The relative changes in the outside surface area is measured by sieve analysis, supposed that the crushed particles have spherical forms.

All these phenomena can be termed micro-degasification process, therefore, a great attention was directed towards determination of relationship between the kinetic energy of the striker bar (E) and the amount of methane desorbed immediately after the impact ($V_t$).

These functions can be seen in Figure 15. The proportionality is proved by good corelation coefficient (0.95-0.98) and there is a significant difference between the functions corresponding to hazardous and non-hazardous coals. In spite of the fact that the kinetic energy E can only be approximatively proportional to the excited wave energy passing through the coal sample the slope of the functions ($s_a$) in Figure 15 can be regarded as a comparative degree of the energy barrier $F_a$ expressed in volume of desorbed methane x mass of coal$^{-1}$ x energy$^{-1}$. Evidently, this value should be related to the same temperature and to the same rate of the spontaneous desorption before the impact.

For planning a degasification process it is also very important to answer the following questions: Does the amount of gas desorbed depend on the number of

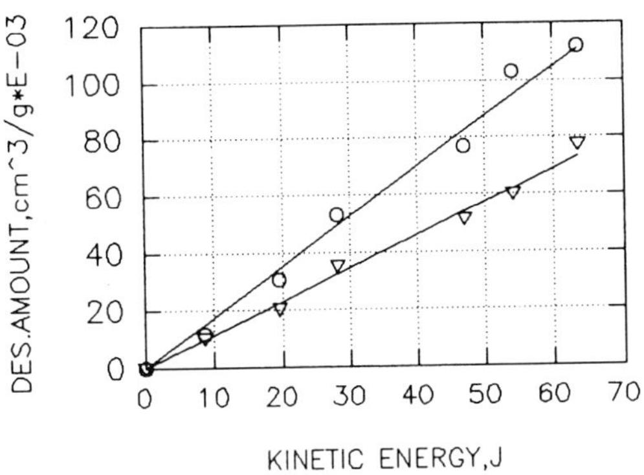


Figure 15. Amount of methane desorbed after the impact plotted as a function of the kinetic energy of the striker bar. Temperature 30 °C. Rate of the spontaneous desorption before the impact: 6.0 $10^{-3}$ $cm^3g^{-1}min^{-1}$. o: hazardous coal, $s_a$=1.75 $10^{-3}$ $cm^3g^{-1}J^{-1}$; ∇: non-hazardous coal, $s_a$=1.15 $10^{-3}$ $cm^3g^{-1}J^{-1}$.

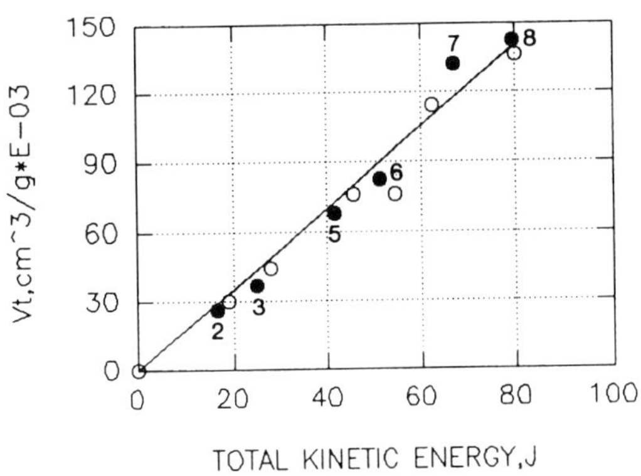


Figure 16. The total amount of methane desorbed as a function of repeated impacts is independent of the type of impacts. Temperature: 30 °C. Rate of the spontaneous desorption before the impacts: 6.0 $10^{-3}$ $cm^3g^{-1}min^{-1}$; o: single impacts; •: repeated impacts [24].

impacts, and which has a greater effect, one large shock, or several repeated small impacts? The answers to these questions may be found in Figure 16 where the total amount of methane desorbed is plotted against the total kinetic energy of the striker bar. The figures beside the symbols refer to the number of impacts required to achieve the given energy. The empty circles represent single impacts.

Figure 16 proves that the total amount of gas desorbed is independent of the type of impact. For example, it may be seen that within the limits of standard deviation, the same amount of methane is desorbed by 1x80 J of energy as by 8x10 J of energy. All results and consequences discussed in this point should be applied in planning and making degasification processes.

## 4. INDUSTRIAL DEGASIFICATION TECHNOLOGIES APPLIED IN COAL SEAMS

There are only three countries where since 10-15 years the degasifications of coal seams have been applied successfully: USA, Canada and Australia. In these three countries the following common features are characterizing the coalbed methane projects:

a) The average methane content of the coal seams are 20-50 STP $m^3$/metric ton and the whole amount of gas to be produced is more than 200 thousand million STP $m^3$.

b) The average permeability of coal seams are between $2.10^{-3}$ - $20.10^{-3}$ $\mu m^2$ (≈2-20 mD).

c) The degasification technologies are based on hydraulic fractures usually made with gel basis fracfluids, but several types of foam or water fracture treatments also have been tried successfully. During the fracture treatment a definite energy was injected in the coal seams for destroying the coal structure and the energy barrier $F_a$. The measure of energy injected depends on the volume of coal touching by the fracfluid and on the pressure of that because volume x pressure = energy.

d) The optimal well specing are determined both by theory and in the praxis. It means that in the higher-permeability areas ($15\text{-}20.10^{-3}$ $\mu m^2$) a well spacing of 30-60 ha is optimal and in the lower-permeability areas ($2\text{-}15.10^{-3}$ $\mu m^2$) the well spacing 16-30 ha is still acceptable [25]. In the past and recent literature cannot be found descriptions or case-histories on degasification technologies applied in very low permeability coal seams ($10^{-6}$-$10^{-7}$ $\mu m^2$), although, in many countries (also in Europe) there are many coal mines with this very low permeability but with very high methane content. This is the situation also in Hungary (Mecsek coal seams) where two years ago an unsuccessful hydraulic fracture with liquid $CO_2$ was carried out. The short case-history of this treatment is given here because it can provide more informations on "know-how" of fracturing in very low permeability coal seams.

The following process took place when 83 $m^3$ liquid $CO_2$ at temperature -20 °C and at pressure 20 MPa was injected in the coalbed with temperature 49 °C and with very low permeability mentioned above. The injection time was 15 minutes. During this time about 80 percentage of the injected $CO_2$ was evaporated and the temperature of the touched coal decreased to 31 °C. The common and final temperature would be 26 °C supposed that between temperature 31-26 °C did not take place significant condensation of the evaporated $CO_2$. The sorption of the evaporated (and partly of the liquid) $CO_2$ surely took place. The amount of coal, the activation energy and time needed to this sorption were available. Taking the sorption of $CO_2$ into account can be stated that the released heat of sorption was enough both for the evaporation of the total amount of the injected $CO_2$ and for increasing the temperature of coal from 26 °C to about the original value of 49 °C. In this sense the common effects of the thermodynamic and sorption processes were the followings: The injected $CO_2$ was totally evaporated and sorbed in the coal. The equilibrium temperature of the coal was approximatively equal to the original one, however, the most important conclusion of this unsuccessfull treatment is the following: Because of the rapid evaporation and sorption of the $CO_2$ an elastic connection became between the coal and the fracfluid, therefore, the volume of coal x pressure of the fracfluid = injected energy was not enough to destroy the energy barrier $F_a$, to destroy the inside structure of the coal and so to assure a minimum permeaility required to continuous flow of the desorbed methane.

The mathematical simulations also prove [25] that the hydraulic-fracture methods are not available to coalbeds with such a low permeability. The laboratory investigations and results described and discussed in this chapter suggest that in these cases a basically new technology is required for degasification of low permeability coal seams. The main principles of this project are the followings:

a) In low permeability coalbeds an optimal well-spacing is not enough for producing of methane, therefore, the new drilling technology should be applied.
b) Multilateral well drillings are needed in the coal seams.
c) In this non-horizontal holes repeated energy impacts are initiated. (For example, small explosions.) The sum of this energy impacts should be enough to destroy the energy barrier $F_a$ and to assure a permeability sufficient for continuous flow of methane.
d) The energy impacts can be combined with methanol-benzene base fracfluids destroying (desaggregating) the inside structure of coal. This fracfluids may be applied alone (without energy impacts) but the effectivness of this method is lower than that of the combined technology.
e) It may occur that the multilateral well-holes, after the energy impacts, get damaged. This damage does not influence the continuous methane production if two conditions has already been met: (i) the energy impacts destroid the barrier $F_a$, and (ii) the damaged hole has sufficient permeability (min $2.10^{-3}$ $\mu m^2$) to assure the flow of the desorbed methane.

This new degasification technology is being patented in Hungary and in other foreign countries.

## REFERENCES

1. F. Derbyshic, A. Marzec, H.R. Schulten, M.A. Wilson, A. Davis, P. Tekely, J.J.Delpuech, A. Jurkewicz, C.H. Bronnimann, R.A. Wind, G.E. Maciel, R. Narayan, K. Bartle and C. Snape, Fuel, 68 (1989) 1091.
2. A.R. Brown and P.L. Waters, Fuel, 45 (1966) 17.
3. P.M. Shaw, S.C. Brassal, O.J. Assinder and G. Eglington, Fuel, 67 (1988) 551.
4. K.D. Bartle, W.R. Ladner and T.G. Martin, Nature, 277 (1979) 284.
5. M.W. Haenel, Fuel, 71 (1992) 1211.
6. M. Nishioka, Fuel, 71 (1992) 941.
7. G.R. Kin and T.M. Ertekin, in: Proceedings of the 1989 Coalbed Methane Symposium, University of Alabama, 17-20 April, 1989.
8. P.D. Gamson, B.B. Beamish and D.P. Johnson, Fuel, 72 (1993) 87.
9. I.L. Ettinger and V.V. Serpinsky, Mining Sci. Techn., 13 (1991) 403.
10. E.Y. Alcin and S. Durucan, Mining Sci. Techn., 13 (1991) 215.
11. J. Jagiello, M. Lason and A. Nodzenski, Fuel, 71 (1992) 431.
12. J. Milewska-Duda, G. Ceglarska-Stefanska and J. Duda, Fuel, 73 (1994) 975.
13. L. Fejér and Zs. Radnai-Gyöngyös, Bányászat, 111 (1978) 550 (in Hungarian).
14. J. Litwiniszyn, Mining Sci. Techn., 3 (1986) 243.
15. J. Lakatos, J. Tóth and L. Bokányi, Proceedings of the Fifth Symposium on Mining Chemistry, pp. 163-172, Istanbul, The Turkey, 1995. Published by the Research Laboratory for Mining Chemistry of the Hungarian Academy of Sciences, Miskolc-Egyetemváros, P.O.Box 2. H-3515.
16. J. Tóth, Acta Chim. Acad. Sci. Hung., 69 (1971) 311.
17. D.P. Valenzuela and A. Myers, Adsorption Equilibrium Data Handbook, Prentice Hall, Englewood Cliffs, Nex Yersey, 1989.
18. J. Tóth, Acta Chim. Hung. Models in Chemistry, 129(1) (1992) 51.
19. M. Gumkowski, Q. Liu and E.M. Arnett, Energy and Fuels, 2 (1988) 295.
20. I. Dékány, J. Tóth and L. Bokányi, Magyar Kémiai Folyóirat, 101 (1995) 64 (in Hungarian).
21. J. Tóth, Proceedings of the Fifth Symposium on Mining Chemistry, pp. 183-192, Istanbul, The Turkey, 1995. Published by the Research Laboratory for Mining Chemistry of the Hungarian Academy of Sciences.
22. L.Gy. Nagy and G. Schay, Acta Chim. Acad. Sci. Hung., 39 (1963) 363.
23. J. Tóth and I. Lakatos, The Mining Engineer, 153 (1994) 359.
24. J. Tóth, J. Szabó-Lakatos and Z. Pindel, Mining Science and Technology, 10 (1990) 29.
25. D. Zuber, V.A. Kuuskraa and W.K. Sawyer, SPE Formation Evaluation, March 1990. Presented at the SPE Gas Technology Symposium, Dallas, 1989.

Adsorption and its Applications in Industry and Environmental Protection
Studies in Surface Science and Catalysis, Vol. 120
A. Dabrowski (Editor)



# The influence of properties within particles of active carbons on selected adsorption processes

B. Buczek*

Faculty of Fuels and Energy, University of Mining and Metallurgy, 30-059 Cracow, Poland

In spite of nearly ninety years of production and industrial application, carbonaceous adsorbents, and in particular active carbons, are still an interesting subject of studies, for both experimental researchers and theoreticians dealing with physical chemistry of interfacial phenomena [1].The increasing use of adsorption methods in separation of liquid and gaseous mixtures and in environmental protection leads to the raised demand for carbonaceous adsorbents. It is accompanied by the dynamic development of science concerning the production and properties of those materials. The knowledge of phenomena occurring on carbon surfaces is deepened [2].

Most activated carbons are manufactured by mild oxidation (using steam or carbon dioxide as activating agents) of chars obtained from various carbonaceous raw materials. Activation is the process through which an extended surface area and porous structure is developed in carbonised products. The activating agent basically burns away the more reactive portions of the carbon skeleton forming CO and $CO_2$ and the extent of burn-off depends on the nature of the employed gas and the temperature of activation. The burning out of the carbon skeleton also occurs at different rates and different parts of the exposed surface [3]. Moreover, the activating gas in the process is thought to penetrate the particle of the char as a result of diffusion accompanied by chemical reactions, the gas concentration decreasing with an increasing distance from the external surface. Usually the activation reactions are endothermic and this fact leads to a temperature decrease in the reaction zone while heat energy flows slowly to the deepest layers of the particle. Because of these factors, the burn-off of the carbonaceous substance is a function of the radial position within the carbon particle [4, 5].

* The author is grateful to the Polish State Committee of Scientific Research (KBN) for its partial financial support of this work (Project No. 8 T10B 054 12)

## 1. EVALUATION OF PROPERTIES WITHIN PARTICLES OF ACTIVE CARBONS

Several processes, both physical and chemical, are nowadays carried out in various type of fluidised and spouted bed apparatuses. The phenomena of abrasion and grindability, associated with the contact of phases are estimated in different ways, depending on the process in which they take place [6]. Usually it, is treated as a factor lowering the efficiency of the processes running in fluidised and spouted beds.

Recently new possibilities of application of the processing in the spouted bed to the basic studies have been outlined, particularly for the carbonaceous materials obtained as the result of processes such as carbonisation and activation. To carry out a quantitative evaluation of property change in active carbon, a method was used which successfully removes layers from a particle surface by abrasion in a spouted bed [5,7].

### 1.1. Active carbons

Investigations were carried out on commercially manufactured active carbons. Granular active carbons were obtained using following processes: (AG) from hard coal at 1173-1227K, (RN) the same method of preparation as carbon AG, except in this case of peat at 1227K.

### 1.2. Sampling of active carbon particles

A method of successive removal of the layers from the carbon particles was used. A diagram of the experimental equipment is shown in Figure 1.

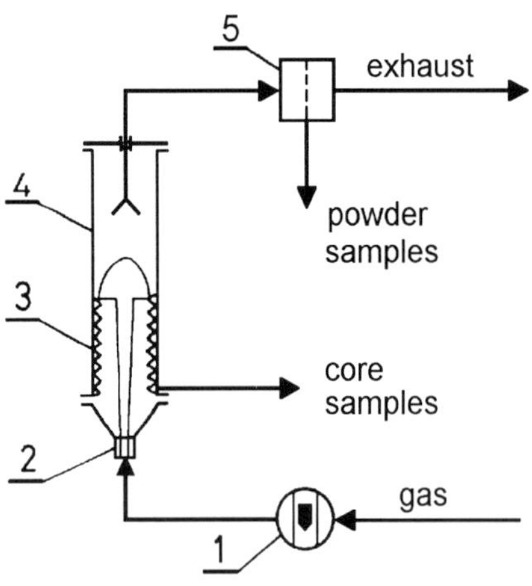


Figure 1. Experimental equipment used: 1, rotameter; 2, nozzle; 3, abrasive lining; 4, column; 5, filter.

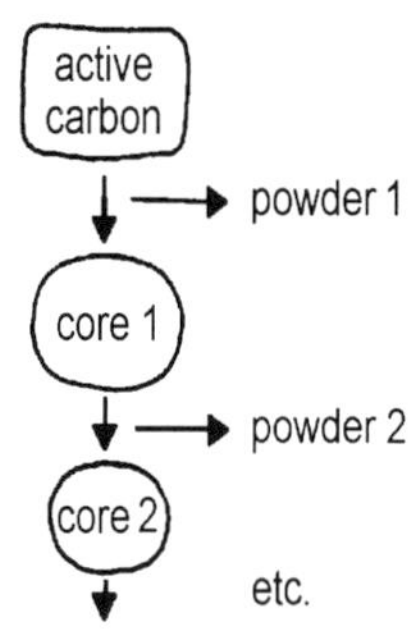


Figure 2. Change in the shape and dimensions of active carbon particle during abrasion process.

The active carbon particles were subjected to abrasion for a chosen length of time. The powdered samples were taken from positions more and more remote from the external surface of the particles. The remaining granular core samples were also obtained in such a way that external layers of different thickness were removed. In preliminary investigations, optimum conditions for the processing of the particles in the spouted bed were determined. This enabled the abrasion of the material from the particle surface to occur without crushing them. The abrasion for different batches under same conditions, show good reproducibility of the sampling.

The shape and dimensions of the active carbon particles change as a result of their processing in the spouted bed, as shown schematically for a particle with layers removed to different degrees in Figure 2.

The cores of different degree of surface layer removal and powders originating from different regions within the particles were investigated, paying special the radial attention to changes of properties of active carbons.

### 1.3. Burn-off, porous structure and surface character within particles of active carbons

Apparent density and ash content were determined for powder and core samples obtained from active carbons AG and RN. Both of these properties may be an indication of transformations proceeding in the carbon particles during steam activation. Using the results of ash content measurements, the burn-off was determined [5] and ascribed to a definite location within the particle of active carbon ($r/r_0$). Changes in burn-off (x) and in apparent density ($\rho_a$) in both studied adsorbents are presented in Figures 3a and 3b. Additionally, the values of the apparent density of the chars ($\rho_c$), from which the respective active carbons were obtained are drawn for comparison.

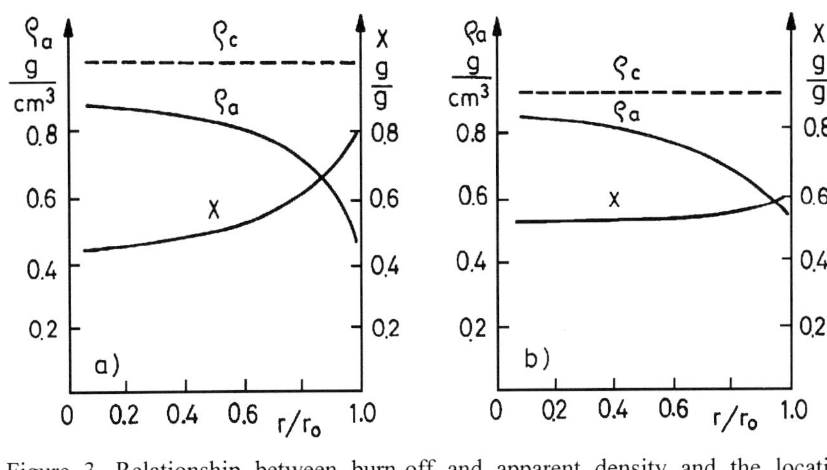


Figure 3. Relationship between burn-off and apparent density and the location within particles of active carbon: a) AG and b) RN.

The burn-off of the AG active carbon particle varied from 0.45 to 0.80, while the extent of changes was considerably lower for RN from 0.5 to 0.6. The apparent density diminishing radially from the particle core to its external surface, indicates that the changes in burn-off are accompanied by variations in the porous structure of the particles.

The analysis of the porous structure was carried out using densimetry and adsorption technique. True density of all investigated specimens was measured, and together with apparent density, used to estimate total volume of pores ($v_t$). Additionally, adsorption and desorption isotherms for benzene vapours were determined, which gave the volume of micropores ($v_{mi}$) and mesopores ($v_{me}$) [3,8]. Volumes of all type of pore throughout particles are shown in Figures 4a and 4b.

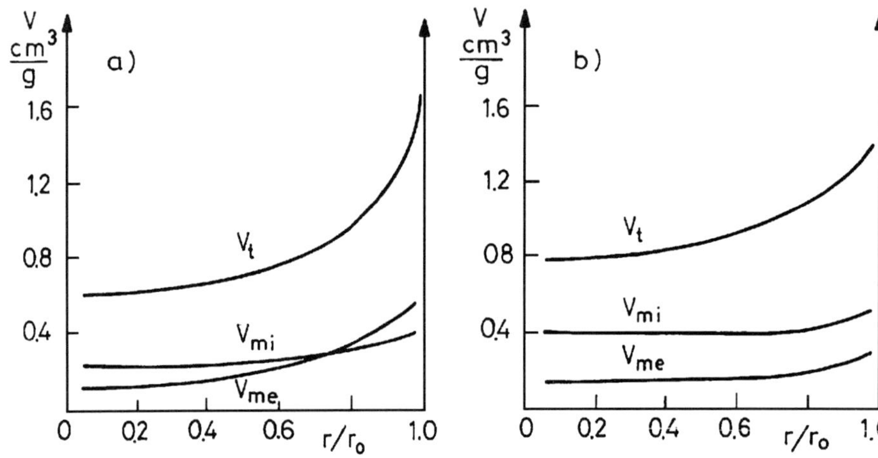


Figure 4. Textural changes inside particles of active carbon: a) AG and b) RN.

Both active carbons differ in the development of porous structure within particles, as well as in the contribution of various kinds of pores into their total porosity. AG carbon is a typical polydisperse adsorbent, of a comparable contribution of all kinds of pores. On the contrary, RN active carbon is a microporous adsorbent with dominant microporous structure and lower content of macro- and mesopores.

Properties of active carbons are determined not only by their porous structure, but also by chemical composition and structure of their surface layer [9]. The chemical composition of active carbons is determined primarily by the nature, number and way of binding of various heteroatoms, mainly oxygen ones. The surface character of the studied materials obtained from the particles of active carbons was evaluated basing on the water sorption and thermogravimetric analysis [7].

The number of adsorption sites in the form of oxygen groups, capable of binding water molecules was estimated from water sorption isotherms [3,10]. The results are given as the concentration of oxygen-containing groups per 1 $m^2$ of adsorbent surface area ($a_0$). The loss of mass ($\Delta m$) over the range of temperatures between 453 and 1273 K, related mainly to thermal decomposition of groups present on the surface of active carbons, was determined using the thermogravimetric data [11,12]. The changes in the nature of surface within the active carbon particles are presented in Figure 5a and 5b.

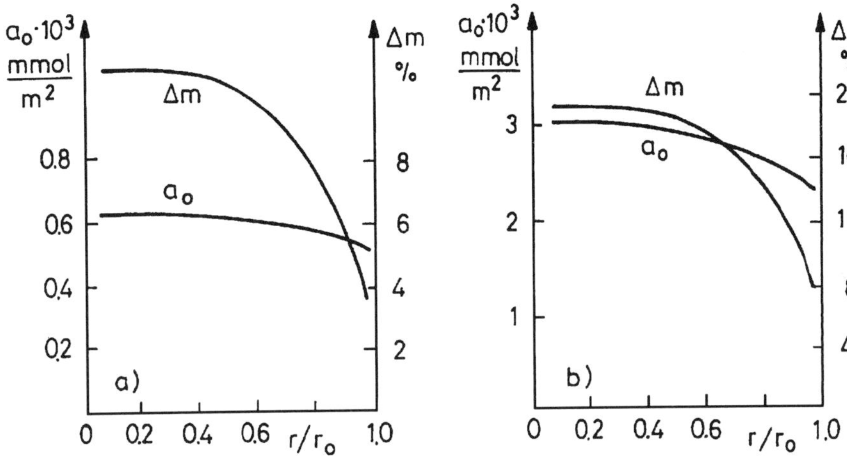


Figure 5. Surface character within particles of active carbon: a) AG and b) RN.

The presented values show that hydrophilic character of both carbons increases within particles towards the particle centre, i.e. with the decrease in burn-off of the carbon matter. RN active carbon had a considerably higher content of surface groups than AG active carbon which may be related to both its lower burn-off on one side and the type of raw material (peat) used for its production. Thus, using two independent experimental methods it has been shown that the character of surface within active carbon particles may be a function of its radial position.

## 2. ACTIVE CARBONS AND THEIR SAMPLES AS STUDIED BY ADSORPTION PROCESSES

Active carbons and samples obtained from these carbons, with different structural and adsorptive properties were used to investigate the effect of the properties of carbonaceous adsorbents on the efficiency selected adsorption processes. The choice of adsorptive processes was based on both theoretical and practical reasons.

Dynamic studies of the adsorption of methane, benzene and cyanogen chloride were stimulated in the fixed bed batch processes for several adsorbents to be used in various purification and separation techniques. The investigation of low temperature nitrogen adsorption from its mixtures with helium was carried out in order to elucidate the problem of the choice of adsorbents suitable for purification of noble gases. Finally, separation ability of methane and storage capacities of the samples obtained from the active carbons were estimated to find the prospective applications of the carbon porous materials for the process of enrichment of methane-containing gaseous mixtures and adsorptive storage of gaseous fuels.

Moreover, the performed studies should answer the questions if the abrasive treatment of active carbons may improve their properties by removal of the most external layers, thus affecting the efficiency and selectivity of separation and purification of mixtures of gases.

### 2.1. Recovery of solvents

Organic solvents are used in many branches of chemical industry. Many of them (petrols, benzene, toluene, xylenes, acetone, lower alkanols and n-alkanes, halogen derivatives of hydrocarbons, carbon disulphide) are used in several technological processes where the inherent operation accompanying them is solvent evaporation. Vapours are emitted together with air, and thus are usually highly diluted. Efficient solvent recovery and low losses during solvent recycling to the production are a necessary condition of cost improvement. The consumption of solvents is often very high and may cause substantial raise of production costs.

The best contemporary method of solvent recovery is its adsorption on active carbon [13,14]. In a properly designed equipment the cost of recovery should not exceed 5-20% of the solvent cost. The method is particularly suitable for mixtures with low concentration of a solvent. The operational costs depend on the solvent, process conditions, size and construction of the equipment and the degree of recovery [15, 16].

Designing a fixed bed adsorber for a particular solvent requires the knowledge of dynamic characteristics of active carbon used in the process. It is particularly important for active carbon, from which the external layers were abraded to a different degree. The mechanical abrasion and crushing of its particles during the process causes changes in the adsorptive capacity of the bed.

The dynamics of adsorption of benzene vapours, mixed with nitrogen at concentration $c_0$ = 10 g/cm$^3$ was investigated for active carbon AG and its samples, in the form of cores from which 40 and 80 wt.% external surface layers were removed (denoted as AG40 and AG80 respectively). The measurements were carried out in the apparatus presented in Figure 6, under the following conditions: 0.086 m/s flow through the bed of a constant height of 0.03 m, temperature of 298K.

During the experiments the curves of outflow, determining the course of the increase of the adsorbate concentration (c), behind the adsorbent bed in time (t) were determined. The results of measurements are presented in Figure 7 in the form of isoplanes in co-ordinate system of the relative concentration versus time.

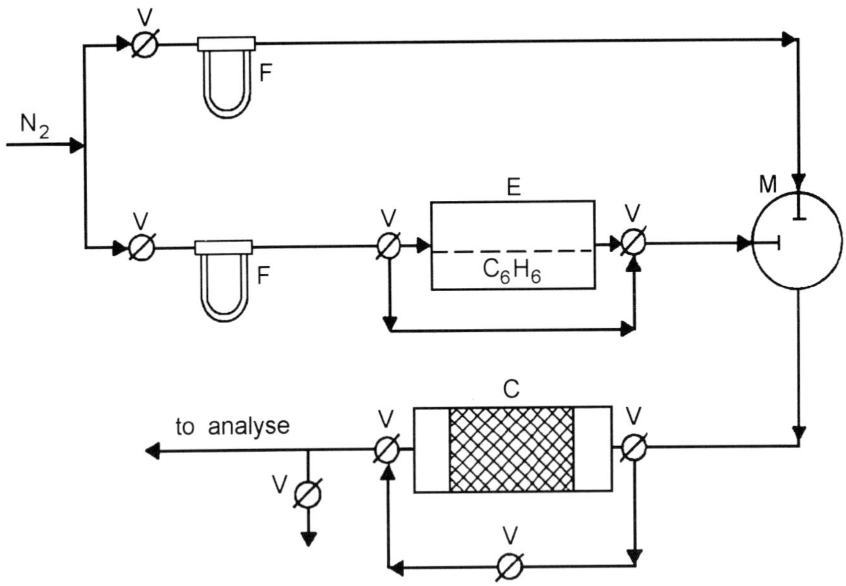


Figure 6. The experimental equipment to study the dynamics of adsorption of solvent vapours: C-column, E-evaporator, M-gas mixer, V-valve, F-flowmeter.

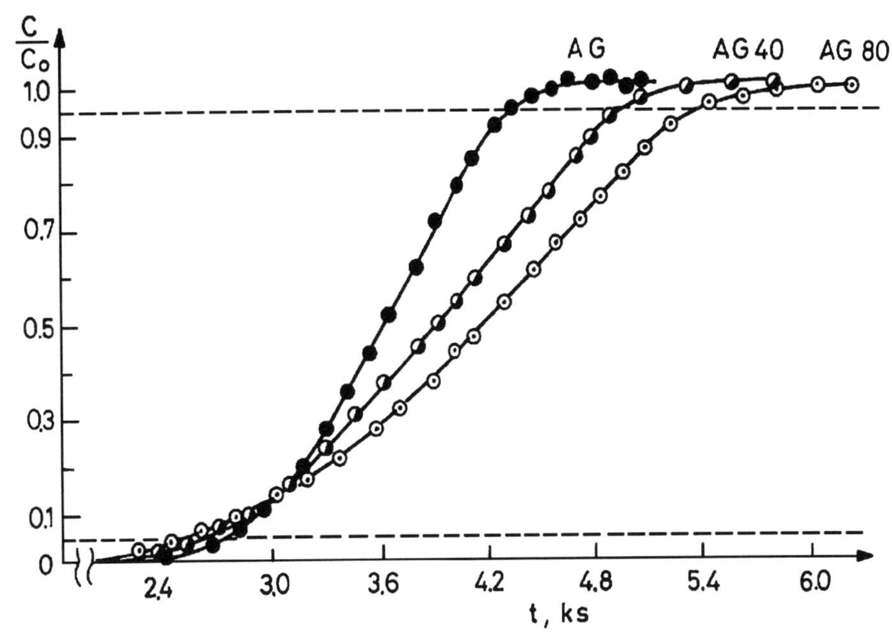


Figure 7. Breakthrough curve of benzene vapours behind the bed of AG, AG40 and AG80 carbons.

The slope of almost linear parts of the outflow curves decreases with the extent of AG carbon treatment by abrasion. It causes the shortening of adsorption time period needed for breakthrough of the adsorbent bed on one hand and the prolongation of time needed for total saturation of the bed on the other hand. These effects may be ascribed to the increase of resistance in mass transport within active carbon particles, which were deprived of external layers to different extent.

The quantitative results of dynamic investigations were analysed using the concept of the zone of mass transfer introduced by Michaels [17], and supplemented and developed by Campbell [18]. The height of the mass transfer zone ($h_a$) was calculated from the formula:

$$h_a = \frac{t_k - t_p}{t_k - (t_k - t_p)(1-f)} h \qquad (1)$$

where: h - height of the adsorbent bed,

$t_p$ - time of breakthrough of the bed,

$t_k$ - time of saturation of the bed,

f - coefficient of symmetry of the breakthrough curve.

The dimensionless coefficient f, which additionally characterises a degree of adsorbent saturation in the zone of mass transfer is given by the following formula:

$$f = \frac{\int_{t_p}^{t_k} (1 - \frac{c}{c_o}) dt}{t_k - t_p} \qquad (2)$$

and usually determined by graphical integration of the breakthrough curve, over the assumed time limits. The mass transfer zone, called also the adsorption zone is defined as this part of the bed, within which adsorbate concentration diminishes from 95 % to 5 % of its initial value under the fixed conditions. Using this definition, the height of the adsorption zone and its characteristic parameters, resulting from the assumed model of dynamics adsorption were calculated and presented in Table 1.

The height of the adsorption zone depends on several factors, such as: initial concentration, the rate of gas flow, the height of the adsorbent bed and the size of its particles. All above mentioned parameters were constant in the described experiments, except for the decreasing size of particles of the carbons AG, AG40 and AG80. It must be mentioned, however the height of the mass transfer zone undoubtedly increases, which leads to changes in parameters related to this value. When we take into account the fact that it corresponds to the decreasing

Table 1
Parameters characterising dynamics of adsorption in a fixed bed

| Active carbon | $h_a$ [m] | $t_p$ [ks] | $t_k$ [ks] | f [-] | $a_d$ [kg/m$^3$] |
|---|---|---|---|---|---|
| AG | 0.0133 | 2.736 | 4.326 | 0.539 | 104.9 |
| AG40 | 0.0180 | 2.622 | 4.956 | 0.541 | 120.0 |
| AG80 | 0.0207 | 2.520 | 5.346 | 0.559 | 111.7 |

porosity of particles, one may prove that the size of the adsorption zone is influenced by resistance in mass transfer of adsorbate inside the carbon particles.

Independently of the measurements of the outflow curves, the amount of benzene dynamically adsorbed ($a_d$) behind the adsorbent bed (after equilibration of concentration) was determined by weight. These values, expressed per unit of carbon volume are presented in Table 1. The maximum of benzene adsorption observed for AG40 carbon results both from the increase in the filling the carbon particles and the removal the layers with high burn-off from external surface.

The described results of dynamic experiments are not advantageous (irrespective of the increased benzene adsorption) from the practical point of view because the working time of adsorbent till the breakthrough is shortened. A similar phenomenon was observed in industrial installations but it is explained not only by mechanical erosion of adsorbent, but also by its ageing and poisoning during the multicycle exploitation.

## 2.2. Denitrification of helium concentrates

Natural gases are practically unique source of helium containing 0.1 to 0.6 % of this element. Total world production is based on the low temperature recovery of helium. The components accompanying helium are separated by fractional condensation and the process is carried out in two stages consisting of obtaining helium concentrate of up to 95 volume % helium and its purification from nitrogen, oxygen, argon and neon, of which nitrogen is the most difficult to remove.

Several methods of helium purification from nitrogen are used. The washing the concentrate with liquid hydrocarbons, particularly with propane, is based on abnormal helium solubility in liquid propane. There are also continued intensive studies and implementation works on helium purification using membranes [19,20]. The purification by washing does not, however give a product of a sufficient purity and efficiency of the diffusive separation on membranes is still too low. For these reasons, the basic process of helium denitrification is adsorption on solid adsorbents, mainly on active carbon.

As we prepared the series of samples differing in porous structure, obtained from the initial active carbons AG and RN we tried to estimate the effect of texture on nitrogen adsorption from the mixtures of compositions similar to those met for helium concentrates.

Nitrogen belongs to gases weakly adsorbing under ambient conditions. The adsorption isotherms are linear for the most of adsorbents and the Henry equation describes the process over a wide range of pressures. Only high pressures and low temperatures affect the equilibrium curves and may considerably increase the adsorbed amount of nitrogen. The determination of adsorption isotherms under such experimental conditions is much more difficult than under low pressures. The volumetric method of Antropoff-Vasiliev and its modifications are often used in high pressure investigations [21], but there are problems with the determination of the adsorbed amount of gas, as well as with the corrections for gas which was not adsorbed. The application of sensitive McBain and Bakr balances leads to the enormous complication of the equipment at pressures higher than 2 MPa, which influences the accuracy of the measurements [22].

We applied an original way of determination of the high pressure adsorption isotherms at low temperatures, using quartz resonators as a piezoelectric microbalance [23,24]. In order to compensate the influence of pressure and fluctuations of temperature we introduced a differential experimental system. For this reason, one of the resonators was covered with glue, then with 3-5 μm adsorbent layer (no more than 20 μg), while the reference resonator was covered only with the same amount of glue as the measuring one. The difference of vibration frequencies caused by the presence of adsorbent on the resonator ($\Delta f_c$), is given by the formula [25]:

$$\Delta f_c = -c_f \frac{\Delta m_c}{S} k_H \tag{3}$$

where: $c_f$ - a constant, resulting from the way of cutting the quartz crystal,
$\Delta m_c$ - mass of adsorbent,
S - surface of electrode,
$k_H$ - coefficient, taking into account the inhomogeneity of covering the electrodes with the studied material.

The change of frequency caused by adsorption is given by:

$$\Delta f_a = -c_f \frac{\Delta m_a}{S} k_H \tag{4}$$

where: $\Delta m_a$ - change of mass caused by adsorption.

The amount of adsorbed substance (a), is expressed by the ratio of $\Delta m_a$ to $\Delta m_c$. From equations (3) and (4) it may be derived that:

$$a = \frac{\Delta f_a}{\Delta f_c} \quad (5)$$

Therefore, when controlling the vibration frequency of the balance during covering the electrode with adsorbent, we are able to determine the value of adsorption (a) basing only on changes of vibrations ($\Delta f_a$) caused by adsorption.

The microbalance based on the presented relationships and the applied differential measuring system makes it possible to determine the adsorption equilibria over a wide range of temperature and pressure for mixtures of gases considerably differing in adsorptivity. Its block diagram is presented in Figure 8.

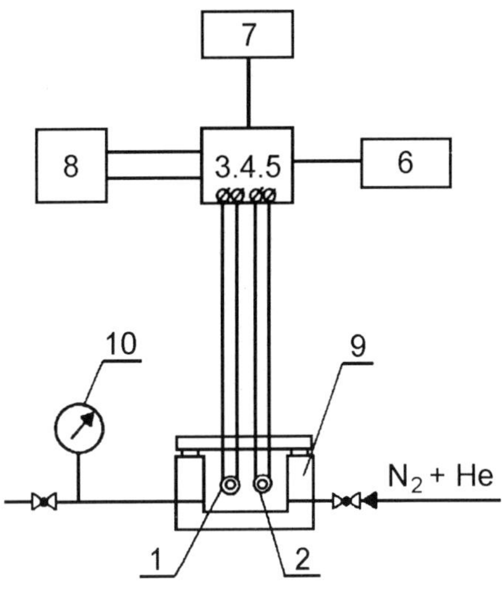


Figure 8. Diagram of the piezoelectric microbalance used for studies of high pressure adsorption: 1,2-quartz resonator, 3,4-oscillator, 5-mixer-amplifier, 6- frequency meter, 7-recorder, 8-power supplier, 9-high pressure vessel, 10-manometer.

Using this microbalance we studied changes in nitrogen adsorption on the samples of active carbons (cores after removal of 80 wt.% of surface layers AG80 and RN80 as well as on powders obtained from the most external layers of the initial active carbons after removal of 20 wt.% PAG2 and PRN2, respectively).

Before the experiment the resonators covered with appropriate sample were purified by a stream of spectrally pure helium, first hot (423 K) and then at a room temperature. The last operation was additionally repeated with the high pressure vessel immersed in a liquid nitrogen bath (77.5K). When the vibration frequency of the system of resonators stabilised the mixture of 3% $N_2$ and 97% He was introduced into the system, increasing the pressure by 0.2 MPa on average, within the range of pressures between 0.2 and 4 MPa. All readings were carried out after stabilisation of both pressure and the frequency. Results of experiments, in the form of the adsorbed amount as a function of pressure are presented in Figure 9.

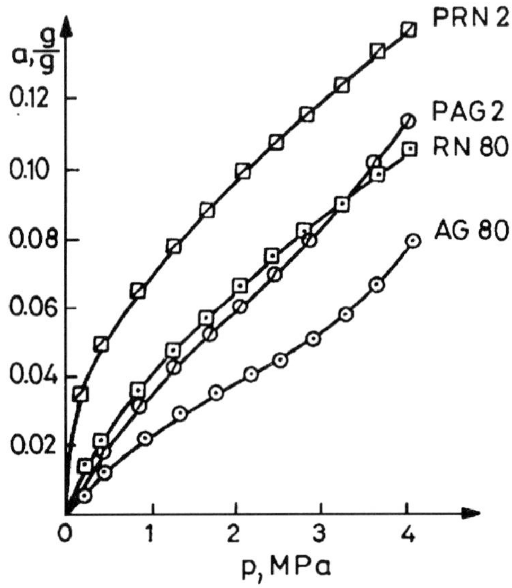


Figure 9. High pressure adsorption isotherms of nitrogen from $N_2$ + He mixture on active carbons samples.

The applied experimental method and insignificant changes of frequency after introduction of pure helium to the measuring system indicate that almost exclusively is adsorbed nitrogen from the mixture. It is confirmed in the work [26], where helium adsorption on 3 types of active carbon was studied, at the temperature of liquid nitrogen up to 3 MPa and very low helium adsorption, not exceeding 14 $Ncm^3/g$ (0.0024 g/g), corresponding to the presented results was found. The preferential adsorption of nitrogen results both from the dimensions of molecules (critical diameters are equal to 0.37 and 0.20 nm for $N_2$ and He respectively) and the state at which both gases are under experimental conditions, understood as the distance from the critical temperatures ($N_2$ = 126 K, He = 5.25 K). Electrostatic interactions of functional groups present on the surface of active carbon samples and nitrogen molecule may also play a role. Such possibility is indicated by comparison of the values of heats of adsorption for nitrogen and noble gases on graphite with and ionic solids. When nitrogen is adsorbed on ionic solids the heat is surprisingly high, which may be ascribed to large quadrupole moment of nitrogen molecule in comparison with noble gases [27, 28].

The porous structure exerts, however, a dominant effect on nitrogen adsorption. It may be seen from Figure 9, that powders collected from external layers, PAG2 and PRN2, adsorb more nitrogen that the inner parts of particles of both carbons, AG80 and RN80. Comparing the development of the porous structure within the particles of the active carbons (Figure 4a and 4b) a correlation may be found between the decreasing contribution of micropores and the reduction in nitrogen adsorption from the mixture with helium. A similar conclusion may also be drawn while comparing changes in total volume of pores, as determined by densimetry.

Considering the possibility of application of differently activated active carbons for the purification of helium concentrates basing on the obtained results we have also to take into account the conditions, under which the adsorbent will work. Deeply activated carbons, such as samples of external layers, have low mechanical strength, while those less activated higher one. Moreover, during purification both temperature and pressure change over wide range which increases the destruction of the adsorbent. Taking into consideration all those factors, we may conclude that choice of the carbonaceous adsorbent for helium concentrate denitrification is always a compromise between the optimum porous structure and sufficient mechanical strength.

### 2.3. Methane storage on active carbons

Development of efficient storage capabilities is an important factor in the utilisation of methane as an energy carrier. Methane is an alternative to liquid fuels in a variety of applications, ranging from metal welding, home heating to vehicular fuel. It has been shown that methane should be preferred as a fuel for internal combustion engines. The use of methane as a vehicular fuel, an application where storage volume is limited, has necessitated the use of high-pressure storage of the order of 20 MPa, to give an adequate if not entirely satisfactory driving range. The disadvantage of compressed methane is that its energy density is about 30 % of that of gasoline. Gas is stored in heavy steel cylinders and filling a tank requires an expensive multistage compression facilities. This disadvantage can be eliminated by using low-pressure storage systems containing adsorbents where high density of surface phases is exploited [29, 30].

The analysis of experimental data on changes in density and adsorptive properties of active carbon samples, from which external layers were removed to different extent showed the possibility of application of those materials in storage of gaseous hydrocarbons, particularly methane [2,31].

Results of studies on methane adsorption on active carbons indicate that the adsorbed amount increases considerably, particularly in the range of pressure up to about 4 MPa [32, 33], which would be particularly interesting because there is a possibility of filling the containers directly from the gas pipelines at such pressures. This creates optimum conditions for the effective application of adsorption system as the storage system.

The most important factors affecting the storage capacity in the presence of adsorbents are the development of their surface and densimetric properties. The maximum surface area of carbonaceous adsorbents is achieved by additional activation or by washing out the ballast substances from the porous structure. This leads, however, to the reduction in their bulk density and in results does not increase amount of methane in the storage system considerably [2, 34].

Because of the observed increase in apparent density of the particles of active carbons AG and RN (see Figure 3a and 3b) and variations of the specific surface area $S_{BET}$, especially for the samples of the carbon RN (Table 2), from which the most external layers were removed, it was decided to carry out investigations of their storage capacity. Total amount of gas contained in the adsorption system is used as the measure of storage capacity, and in the case of methane it includes adsorbed and non-adsorbed methane present both in adsorbent particles and in interparticle spaces.

Storage capacity for methane was determined on RN active carbon and its samples in the form of the cores, after the removal of 40 and 80 wt. % of the external layers, further denoted as respectively RN40 and RN80, by the volumetric method [31] using the equipment presented in Figure 10.

This equipment enables one to determine of adsorption isotherms and curves of storage capacity over the pressure range of 0.1 ÷ 10 MPa by measuring the volume of outflowing gas. The measurements for active carbons proceeded as follows: first the adsorbent container was filled with methane up to 6 MPa and

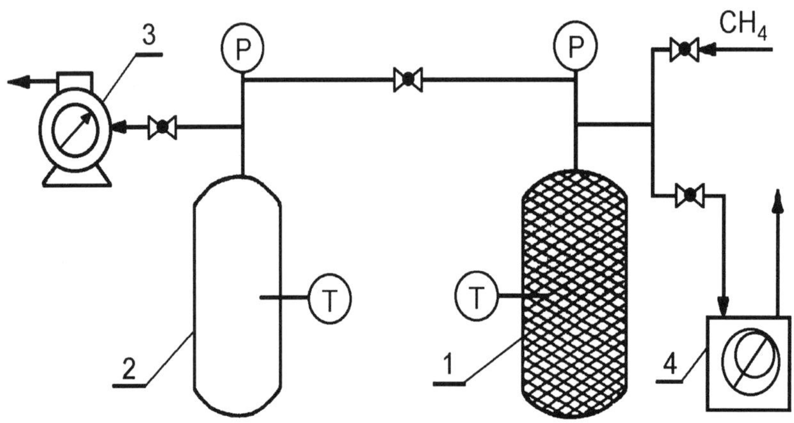


Figure 10. Experimental equipment to study high pressure storage of methane: 1- container with adsorbent, 2-methane container, 3 - gasmeter, 4-vacuum pump, P- pressure transducer, T-resistance thermometer.

after equilibration temperature and pressure was obtained, the storage capacity was determined by gradual outflow of methane up to various values of pressure at constant temperature of 298K. The volume of the outflowing gas was measured at each fixed pressure with the gasmeter. The storage capacity, calculated for volume unit of active carbon in the container is presented in Figure 11. Additionally, the storage capacity of methane in the same container in the absence of adsorbent was shown. In comparison to the empty container, the storage capacity of the systems containing any of the studied carbons is significantly higher over the whole range of investigated pressures. The most significant differences may be observed at low and medium pressures. They diminish with increasing pressure in the container. Moreover, the amount of the stored methane is the largest for RN40, lower for the initial RN active carbon and the smallest for RN80. The amount of methane outflowing from the system over the pressure range of 2.5 ÷ 0.1 MPa was calculated from the storage curves [31]. The limiting values of pressure were chosen for practical reasons. Firstly the costs of storage under high pressure which mainly comes from the energy of compression where use of multigrade compressors is required must be taken into account. Secondly, the pressure adsorption storage makes it possible to apply of simple two-stage compressors or the direct use of natural gas from the pressure gas pipelines. The obtained results show that the adsorption process under

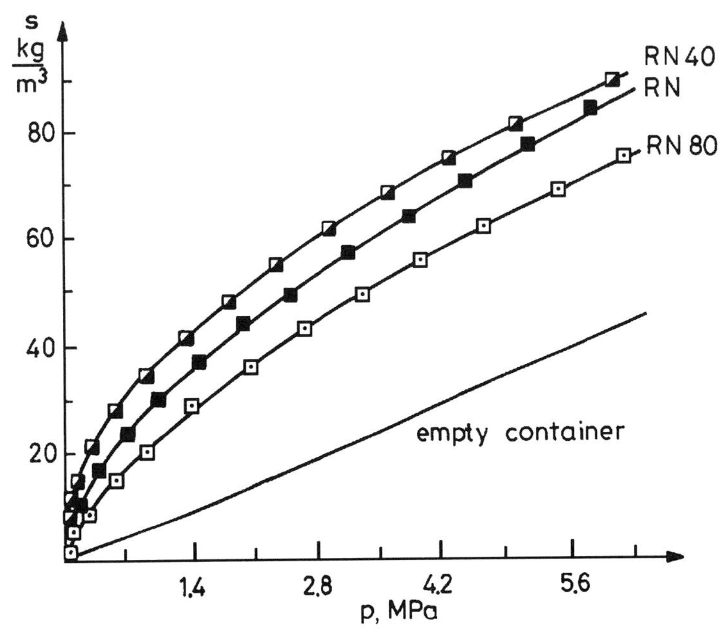


Figure 11. Storage capacity of the carbons RN, RN40 and RN80, in function of methane pressure.

deliberation enables one to store the same quantity of methane as a container of the same volume would contain under the pressure of 7 ÷ 8 MPa in the absence of active carbon. The values of the storage capacity (s) for the pressure of 2.5 MPa, the coefficient (f), defined as the excess of gas in respect to the container without carbon and methane adsorption (a) are listed in Table 2.

The product of the specific surface area ($S_{BET}$) and bulk density ($\rho_b$), given in same Table, was chosen as the parameter characterising the applicability of active carbon for methane storage. The values of this parameter correlate well with the quantity of methane stored in the container filled with active carbon. A similar correlation was found when $S_{BET}$ in this product was replaced with the volume of micropores.

Table 2
The characteristics of the storage system of active carbon-methane

| Active Carbon | $S_{BET}$ [g/cm³] | $\rho_b$ [g/cm³] | s [kg/m³] | f [-] | a [kg/kg] | $S_{BET}$*$\rho_b$* $10^{-8}$[m²/m³] |
|---|---|---|---|---|---|---|
| RN | 1280 | 0.413 | 50,62 | 2,99 | 0,088 | 5,59 |
| RN40 | 1298 | 0.458 | 57,68 | 3,41 | 0,089 | 5,84 |
| RN80 | 1133 | 0.479 | 41,62 | 2,46 | 0,076 | 5,40 |

In general, the obtained results indicate that the processing of active carbon obtained in the process of physical activation, by abrasion may lead to the increase in its storage capacity for about 12 %, with simultaneous enhancement of its mechanical strength. Main factor determining storage capacity of active carbons is the development of their structure, expressed by as specific surface area or volume of micropores per unit volume of the adsorbent. A further enhancement in storage capacity may be achieved by filling interparticle spaces with fine active carbon, which increases bulk density of the adsorbent and the contribution of micropores in the whole volume of container [35].

### 2.4. Enrichment of methane-containing gases

There has been growing interest lately in alternative raw materials containing methane which is an essential component of gaseous fuels as natural gas. They have not been utilised yet because of low content of combustible components, mainly methane. The main condition of their utilisation as fuels would be the enrichment of these gases in methane up to high concentrations in methane, which would results in the calorific value corresponding to that of natural gas.

Among potential sources of methane: biogases, nitrogenated natural gas and coal bed methane are most frequently described in the literature [36, 37].

One of the methods widely applied for separation of mixtures of poorly condensing gases is pressure swing adsorption (PSA) [38,39]. It is applied in industry for the production of: oxygen and nitrogen from air, hydrogen from gases after reforming process, as well as for separation of various hydrocarbons. Recently the attempts to use the PSA technique for obtaining carbon dioxide and methane has been discussed [40, 41].

In the case of mixtures such as nitrogen containing natural gas or mine gases, the enrichment consists practically of methane separation from nitrogen and oxygen. Previous investigations showed that the best separation of these gases may be obtained on carbonaceous adsorbents with an appropriate porous structure, making use of differences in adsorption isotherms of individual components of the mixtures [42, 43].

Methane is much better adsorbed on carbonaceous adsorbents than nitrogen. This difference in adsorption capacity determines the course of pressure swing separation. Thus, methane recovery proceeds in the stage of countercurrent desorption, while its enrichment is caused by nitrogen removal during cocurrent desorption. The amount of removed nitrogen, and thus gas phase enrichment in methane depend on adsorptive and separative properties of the adsorbent. In order to evaluate the suitability of adsorbents for enrichment of methane-containing gases the estimation method based on measurements under conditions simulating the basic stages of PSA process was developed [44, 45].

The fact that the microporous structure of both investigated adsorbents changes radially (compare Figure 4 a and b) led to the studies of their separation properties towards the nitrogen-methane mixture. Active carbons (AG, RN) and their samples in the form of internal cores devoid of 80 wt.% of the surface layers (AG80, RN80) were investigated under increasing pressure and during desorption. For comparison, commercial carbon molecular sieve (CMS), used in pilot plant for enrichment of mine gases [46] was also estimated.

The standard experiment consisted of the following operations: i) adsorbent degassing in a test column down to several mbar; ii) the increase of pressure of the column up to 0.3 MPa, with the supplying gas containing 50% methane and 50% nitrogen; iii) desorption of gas from the column up to 0.1 MPa with the simultaneous analysis of the composition of the outflowing gas. The diagram of the experimental equipment is presented in Figure 12.

The equipment enables one to determine gas phase content in the outlet part of the column as a function of pressure during either the cocurrent (through valve 9) or countercurrent (valve 9) gas desorption.

The column filled with an adsorbent was degassed, supplied with the mixture to be separated and desorption was started. The samples of desorbed gas were analysed, with simultaneous recording of the pressure in the column. The obtained results in the form of methane concentration in the outlet part of the column versus pressure during the desorption are shown in the Figure 13.

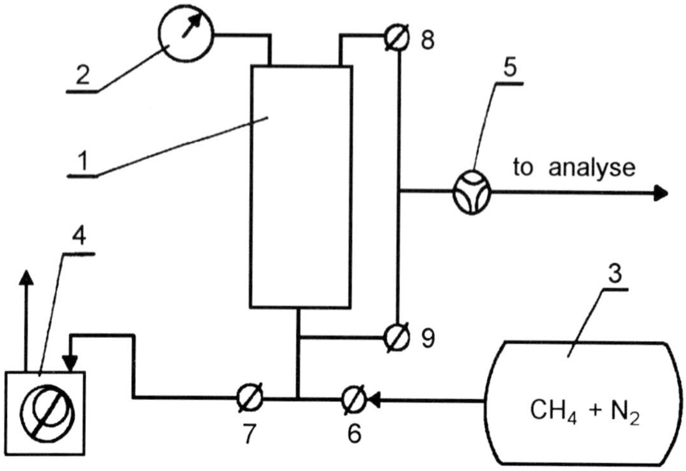


Figure 12. The experimental equipment for the estimation of the separation properties of the adsorbents under study: 1-adsorbent containing column; 2-manometer; 3-gas container; 4- vacuum pump; 5 six-way valve; 6-9 valves.

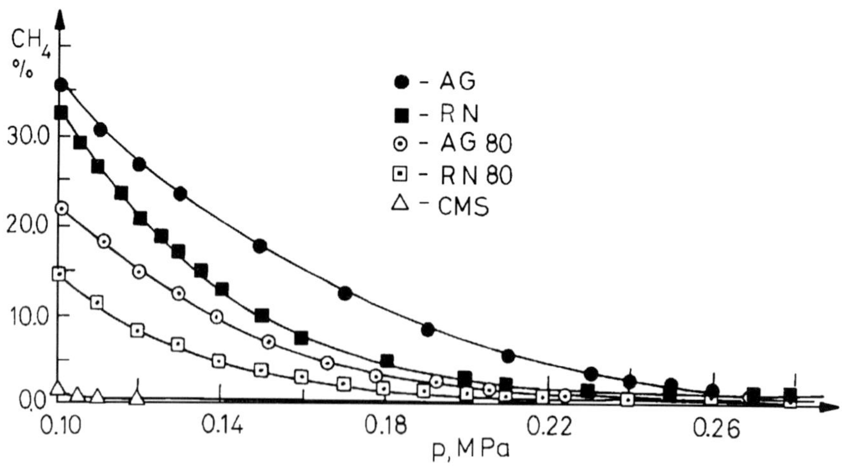


Figure 13. The composition of the desorbed gas as a function of the outlet pressure for AG, AG80, RN, RN80 and CMS adsorbents.

The experimental curves show a considerable differentiation in separation abilities of the studied adsorbents which forms a sequence: CMS > RN > AG. RN80 and AG80 samples have better separation properties than the initial active carbons. It indicates that the interior of active carbon particles has a porous structure similar to that of carbon molecular sieve.

Basing on both experimental and curve theoretically calculated from the equilibrium model of PSA process [47,48], the separation coefficient k, was evaluated for the studied adsorbents to be: 2.9; 3.5, for AG and AG80, 3.2; 3.7, for RN and RN80 and 5.1 for CMS, respectively.

The quantitative estimation of separation ability, based on the determination of the coefficients k, enables one to estimate of suitability of a particular adsorbent for methane enrichment in the swing pressure process. It may be seen from calculations that the adsorbents with k equal to about 5.0 it make possible obtain practically pure nitrogen during desorption stage. The investigated active carbons, despite of their well developed porous structure, show the inappropriate separation properties and practically are not suitable for the enrichment in methane, which concentration during desorption reaches 30 and more percent. One may, however, obtain adsorbents of middle separation coefficients (3.5 and 3.7) from the same active carbons, which may be used for methane enrichment up to about 70-80 % [44], if $CH_4$ concentration in the desorbed gas does not exceed 20 %. Similarly as in the case of methane storage (compare chapter 2.3), the treatment of active carbons by abrasion improves their adsorption properties, mainly where separation process are concerned, by removal of the over-activated layers from the surface of the particles.

### 2.5. Sorption of toxic gases on impregnated active carbons

Active carbons are used for removal of the majority of volatile vapours of organic compounds from the air (compare chapter 2.1). However, the removal by active carbon is not sufficient for all air contaminating gases and vapours are among which $SO_2$, $H_2S$, HCN, ClCN, $AsH_3$, CO, $NO_x$ and many other compounds, often highly toxic to live organisms, may be mentioned. The majority of vapours and gases that do not adsorb on active carbon show alkaline either or acidic character or the ability of forming complexes or oxidation in air to atoxic compounds. These properties create the basis for sorption of toxic vapours and gases on impregnated active carbons. The impregnating compounds are usually salts or metal oxides or their mixtures. Sometimes also elements are used, e.g. iodine or sulphur. In such system active carbon play role of carrier and impregnating substance - catalyst of physical and chemical transformations of toxic gases.

Active carbons impregnated with chromium, copper and silver salts [49-51] deserve special attention. Among the systems consisting of a carrier and an impregnating substance, playing an important role as catalysts of many chemical reactions in gaseous phase. They are commonly used in gas-masks and filtrating-venting safety systems for sorption of toxic gases.

Active carbons impregnated with salts of Cr, Cu and Ag have superior protective properties against many poorly adsorbing gases such as cyanogen chloride, hydrogen cyanide, dicyanogen, arsenous hydride, phosgene and others.

They are usually obtained by the impregnation with solutions of the above mentioned salts. The kind of active carbon and its properties (specific surface area, porous structure, chemical nature of its surface) influence the activity, selectivity and stability of the impregnated product [52-53].

The investigated active carbons have both the porous structure and the nature of surface varying within the particles (see Figures 4 and 5). The AG active carbon, of the greatest differentiation of textural and surface hydrophilic character was used for the estimation of the effect of radial heterogeneity on the properties of impregnated active carbon. For this reason, the initial carbon was impregnated and subsequently abraded in the spouted bed, so as to study the structural and protective properties of the obtained samples [54].

The impregnation of the active carbon AG was carried out by saturation them with aqueous-ammonia solutions of salts of chromium, copper and silver. In relation to 1 kg of carbon carrier the solutions contained: 46 g of chromic anhydride in 70 $cm^3$ distilled water and 25$cm^3$ of 25% aqueous ammonia, 760 $cm^3$ of aminocupric carbonate solution with density 1.26 g/$cm^3$, 2 g of silver nitrate in 5 $cm^3$ 25% aqueous ammonia. After saturation the active carbon was held at the temperature of 403K until the content of ammonia reached less than 0.6%.

The obtained impregnated active carbon (IAG) was abraded in the spouted bed in the equipment presented in Figure 1. The process was carried out in such a way that each abraded layer contained about 33 wt. % of the initial carbon. Two samples obtained in the form of cores devoid of either 34 or 67% of external layers are denoted as IAG34 and IAG 67, respectively.

Adsorption and desorption isotherms of benzene vapours were determined for the impregnated carbon and its samples and the volumes of micropores ($v_{mi}$) and mesopores ($v_{me}$) were calculated [8]. Moreover, the amount of the active phase present inside porous structure was determined. The results of calculations and measurements are given in Table 3.

It may be seen from the Table 3 that volume of micro- and mesopores diminishes as one proceed towards the interior of the impregnated carbon particle. The effect is caused by the diminishing of the burn-off within the

Table 3
The properties of impregnated active carbon and its core samples

| Carbon | $v_{mi}$ [$cm^3$/g] | $v_{me}$ [$cm^3$/g] | wt.cat. [%] | $t_p$ [min] | $a_d$ [mg/g] |
|---|---|---|---|---|---|
| IAG | 0.200 | 0.110 | 9.1 | 11 | 2.8 |
| IAG34 | 0.196 | 0.104 | 8.0 | 31 | 8.3 |
| IAG67 | 0.172 | 0.094 | 7.5 | 23 | 5.7 |

particle and the loading of the catalyst inside the pores (compare Figure 4 a; the volume of micropores in the whole AG active carbon particle is equal to about 0.305 $cm^3/g$). The results of this analysis and the material balance of the deposited active phase show that considerably greater amounts of the catalyst are deposited in the external layers of a particle than in its core one. The results are in good agreement with the data of Rossin [55] who compared the amount of deposited metal to carbon (M/C) using chemical analysis and XPS for active carbon impregnated with chromium, silver and copper compounds. He showed that the outer regions of granules are enriched in active phase and its distributions was not radially uniform.

The protective efficiency of carbons (IAG, IAG34 and IAG 67) was evaluated in the equipment working on the basis similar to that applied for adsorption of vapours of organic solvents (compare Figure 6). The breakthrough curves for cyanogen chloride were determined under the following conditions: ClCN concentration 2 mg/L; air flow 0.2 $L/cm^2min$; relative humidity 50 %; adsorbent bed depth 1.5 cm; temperature 293 K. The breakthrough time ($t_p$) and the amount of adsorbed cyanogen chloride ($a_d$) calculated from these curves are given in Table 3.

The breakthrough time of cyanogen chloride and adsorption capacity are not proportional to the amount of impregnant deposited inside porous structure of IAC samples which indicates that its properties depend on the degree of external layers removal from the particles. By the abrasion of external layers of carbon particles, it is possible to obtain core samples which have different physical and chemical properties in comparison to the starting impregnated active carbon. The removal outer layers leads to better adsorptive-catalytic properties of particles with the increased adsorption capacity of ClCN.

## 3. SUMMARY AND CONCLUSIONS

Active carbons have been commercially produced by steam reaction with carbonaceous materials. The activating gas in the process is thought to penetrate the particle as a result of diffusion accompanied by chemical reactions. The endothermic character of the reactions may lead to a temperature decrease in the reaction zone while heat flows slowly to the deepest layers of the particle. Due to these factors, the burn-off of the carbonaceous substance is a function of the position within the particle.

In order to carry out a quantitative evaluation of the change in the properties of active carbons, an original method was developed which successfully removes layers from a particle surface by abrasion in a spouted bed. This enables one to obtain samples from various parts of carbon particles. Detailed experimental investigations revealed the radial anisotropy of both the porous structure, the character of the surface and burn-off of the carbonaceous matter within steam-activated particles.

The abrasive treatment was applied to obtain samples of two commercial active carbons, originally used for adsorption of various vapours and gases from gaseous phase. The obtained samples of different physical and chemical properties, dependent on the location within the carbon particle were tested both in already established industrial and newly developed processes.

The following adsorption processes were studied: recovery of organic solvents, denitrification of helium concentrates, methane storage, enrichment of methane-containing gases and sorption of toxic gases. Basing on the results of experiments on benzene adsorption carried out under dynamic conditions we found enlargement of the zone of mass transfer, dependent on the extent of removal of external layers from the active carbon particles. This may affect the efficiency and working time of adsorption units. The investigations of samples originating from various places within the particles proved that there exists a relationship between the nature and contribution of narrow micropores and the suitability for both purification of helium concentrates and enrichment of methane-containing gases, using PSA method. The analysis of the adsorptive methane storage showed that a significant improvement of storage capacity may be achieved by the removal of the over-activated (external) layers from the surface of active carbon particles. In the case of abrasive treatment of the impregnated active carbon, the removal of the external layers results in the improvement of the adsorption capacity and protective efficiency, particularly for cyanogen chloride.

The results of the presented considerations contribute to the knowledge of the effect of texture and nature of the surface on adsorption processes taking place in the gas-solid systems. On the other hand, experimental proof of radial inhomogeneity make it possible to control the properties of carbonaceous adsorbents in manufacturing operations.

## REFERENCES


1. M. Smisek and S. Cerny, Active Carbon, Elsevier Publishing Company, Amsterdam, 1970.
2. J. W. Patrick (ed.), Porosity in Carbons, Edward Arnold, London, 1995.
3. R. C. Bansal, J. B. Donnet and F. Stoeckli, Active Carbon, Marcel Dekker, New York, 1988.
4. M. M. Dubinin, Izv. A.N. SSSR, Ser. Chim., 1 (1980) 18 (in Russian).
5. B. Buczek, in: Particle Size Analysis, N.G. Stanley-Wood and R.W. Lines (eds.), The Royal Society of Chemistry, Cambridge 1992, 33.
6. K. B. Mathur and N. Epstein, Spouted Beds, Academic Press, New York, 1974.
7. B. Buczek, Langmuir, 9 (1993) 2509.
8. M. M. Dubinin, in: Chemistry and Physics of Carbon, P. L. Walker Jr. (eds.), Marcel Dekker, New York 1966, vol. 2, 51.
9. T. van der Plas, in: Physical and Chemical Aspects of Adsorbents and Catalysis, B. G. Linsen (ed.), Academic Press, London 1970, 425.
10. M. M. Dubinin, Carbon, 18 (1980) 353.

11. J. Siedlewski and W. Smigiel, Chem. Stosow., 21 (1977) 175 (in Polish).
12. P. L. Walker, O. C. Cariaso and I. M. K. Ismail, Carbon, 18 (1980) 375.
13. T. S. Canson, Pollut. Eng. Technol., 2 (1977) 421.
14. G. E. Keller II, Chem. Eng. Progress, 10 (1995) 56.
15. H. L. Barnaby, J. Air Pollution Control Assoc., 15 (1965) 422.
16. M. H. Stenzel, Chem. Eng. Progress, 4 (1995) 36.
17. A. S. Michaels, Ind. Eng. Chem., 44 (1952) 1922.
18. J. M. Campbell, R. B. Ashford, R. B. Needham and L. S. Reid, Petrol. Ref., 42 (1963) 89.
19. M. Mulder, Basic Principles of Membranes Technology, Kluwer Academic, Boston, 1991.
20. W. J. Koros, Chem. Eng. Progress, 10 (1995) 68.
21. A. J. Sarachov, Viesy v fiziko-chimiczeskich issledovaniach, Nauka, Moscow, 1968.
22. S. J. Gregg and K.S.W. Sing, Adsorption, Surface Area and Porosity, Academic Press, New York, 1982.
23. J. Zelenka, Piezoelectric Resonators and Their Applications, Academia, Prague, 1986.
24. V. Bielkov and B. Buczek, Zurn. Fiz. Chim., 63 (1989) 2509 (in Russian).
25. G. Z. Sauerbrey, Z. Phys., 155 (1959) 206.
26. A. Czaplinski and E. Zielinski, Przem. Chem., 37 (1958) 640 (in Polish).
27. D. M.Young and A. D. Crowell, Physical Adsorption of Gases, Butterworths, London 1962.
28. B. G. Isirikijan and A. W. Kisielev, Zurn. Fiz. Chim., 37 (1963) 2520 (in Russian).
29. K. R. Matranga, A. L. Myers and E. D. Glandt, Chem. Eng. Sci., 47 (1992) 1569.
30. D. F. Quinn and J. A. MacDonald, Carbon, 30 (1992) 1097.
31. B. Buczek and L. Czepirski, Ads. Sci. Technol., 4 (1987) 217.
32. D. L. Peterson, F. Helffrich and R. J. Carr, AIChEJ., 12 (1966) 903.
33. R. Reich, W. T. Ziegler and K. A. Rogers, Ind. Engng. Chem. Process Des. Dev., 19 (1980) 336.
34. X. S. Chen, B. McEnaney, T. J. Mays, J. Alcaniz-Mange, D. Cazorla-Amos and A. Linares-Solano, Carbon, 9 (1997) 1251.
35. B. Buczek and L. Czepirski, Extended Abstracts and Program of CARBON'97, 23rd Conference on Carbon, 18-23 July 1997, The Pennsylvania State University, USA, vol. I, 338.
36. E. Richter, Erdol und Kohle-Erdgas-Petrochemie 40 (1987) 394.
37. S. Sircar, Sep. Sci. Technol., 23 (1988) 333.
38. R. T. Yang, Gas Separation by Adsorption Processes, Butterworths, Boston 1987.
39. M. Suzuki, Adsorption Engineering, Kodansha, Tokyo 1990.
40. H. Knoll, U. Harms, W. Hase and H. Furtig, Chem. Technol., 42 (1990) 260.

41. G. V. Baron, in Separation Technology, E. F. Vasant (ed.), Elsevier Science B. V. Amsterdam 1994, p. 201.
42. D. P. Valenzuela and A. L. Myers, Adsorption Equilibrium Data Handbook, Prentice-Hall, Englewood Cliffs, NJ 1989.
43. D. M. Ruthven, Chem. Eng. Sci., 47 (1992) 4305.
44. B. Buczek, in Mining Science and Technology, G. Yuguang, T. S. Golosinski, A. A. Balkema (eds.), Rotterdam, Brookfield 1996, p.19.
45. M. Balys, B. Buczek and J. Zietkiewicz, Inz. Chem. Proc., 2 (1997) 279 (in Polish).
46. E. Pilarczyk and K. Knoblauch, in Separation Technology, N. N. Li and H. Strathmann (eds.), Engineering Foundation, New York 1987, p. 522.
47. K. Knaebel, F. Hill, Chem. Eng. Sci., 40 (1985) 2351.
48. M. Balys, B. Buczek, L. Czepirski and J. Zietkiewicz, Inz. Chem. Proc., 4 (1993) 601 (in Polish).
49. C. Aharoni and Z. Barnir, Am. Ind. Hyg. Assoc. J., 39 (1974) 334.
50. M. M. Ross, R. J. Colton and V. R. Deitz, Carbon, 27 (1989) 492.
51. V. R. Deitz V and C. J. Karwacki, Carbon, 32 (1994) 703.
52. J. Kloubek, Carbon, 19 (1981) 303.
53. M. Molina-Sabino, V. Perez and F. Rodriquez-Reinoso, Carbon, 32 (1994) 1259.
54. B. Buczek, S. Zietek and A. Swiatkowski, Langmuir, 13 (1997) 1342.
55. J. A. Rossin, Carbon, 27, (1989) 611.

Adsorption and its Applications in Industry and Environmental Protection
Studies in Surface Science and Catalysis, Vol. 120
A. Dabrowski (Editor)



# Moisture sorption and transport processes in paper materials

B. V. Ramarao

Empire State Paper Research Institute SUNY College of Environmental Science and Forestry Syracuse, New York 13210 USA

**Abstract**

The moisture content of paper materials is an important variable which significantly effects their physical properties. At higher moisture contents, paper's stiffness, tensile strength and compression strength are all affected adversely. Similarly, the electrical properties also are strong functions of sheet moisture content.

In this chapter, some recent studies on moisture transport processes in paper materials are reviewed. The equilibrium aspect of moisture interaction with paper shows significant hysteresis which can be estimated by an application of Everett's theory of independent domain complexions. Thus, when a paper sheet is subjected to arbitrary cycles of humidity all the while allowed to reach equilibrium at each state, the sheet's moisture content evolution may be predicted by an analysis of the sorption isotherms and the interior of the sorption hysteresis loop. It is shown that the theoretical predictions of equilibrium moisture content are in good agreement with experimentally determined values.

Transient moisture sorption under ramp changes in external humidity is analyzed. A general model describing the dynamics of moisture sorption is derived. The paper sheet is considered as a composite structure of fibers and voids through which moisture is transported by diffusion. The mathematical description of moisture transport embodies two suitably averaged concentration fields, c and q. Two unsteady state diffusion equations describe the time and spatial evolution of these fields. The average moisture content of the sheet and the moisture flux at the surface are evaluated.

A set of limiting cases of transport is developed, comprising of situations where diffusion through the void space, fiber space and external boundary layers, each contribute significantly to transport. By means of a scaling analysis, the conditions under which each limiting case is valid are identified. Finally, a comparison of the model predictions with experimental data indicates that the model is capable of describing transient sorption dynamics quite well.

## 1. INTRODUCTION

The interaction of moisture with paper and board materials is an important process. The moisture content of paper materials effects their mechanical properties strongly. For example, there is drastic loss in the strength properties of paper when subjected to changes in the humidity of the environment resulting in higher moisture contents. Paper strength has been shown to decrease by 5 to 10% for each unit percentage change in the relative humidity [1]. Benson [2] showed that under increased humidity conditions (and thus at higher moisture contents) paper showed decreases in elastic modulus, yield stress and the ultimate tensile strength.

Besides the moisture content itself, moisture transients have also been known to adversely impact paper performance. The tensile stiffness of paper is known to be much lower when the sheets are subjected to varying relative humidity (denoted henceforth by RH) levels as compared to when the sheets were held at the corresponding equilibrium values [3]. The effects of transient variations in moisture content caused by cyclic humidity changes in RH are even more dramatic. The compressive creep deformation of linerboard (used for packaging boxes) has been shown to accelerate sharply with cyclic changes in RH [4-6]. This is particularly significant since the performance of boxes degrades significantly and collapse occurs even though they are designed to sustain the loads safely at the highest humidity [4-6]. Thus, shortened lifetime of boxes brought about by rapidly varying relative humidities is a major concern for the manufacturers of paper, containers and other board products.

Furthermore, the moisture content of paper impacts the dimensional stability of paper and board products. The physical dimensions of materials made of paper are sensitive to the moisture content as well as the history of its change. Most paper materials expand with moisture content due to the swelling of the fibers. This swelling is anisotropic and is predominant in the radial direction of the fibers. As a consequence of non-uniform fiber distributions inside them, paper sheets tend to curl and deform on a small scale locally. The local deformation is termed as cockle. Hence, knowledge of the interaction of moisture with paper will help in producing more dimensionally stable products [7].

This profound effect of moisture especially under dynamic humidity conditions has led to extensive investigations into the mechanical behavior of paper and board [8,9] under dynamic humidity conditions. However, investigation of the transport and sorption processes inside paper materials subjected to dynamic humidity conditions (e.g., step, ramp and sinusoidal changes) has received little attention.

In this chapter, we review some of our recent studies on moisture equilibrium and transport in paper materials. Equilibrium sorption is first studied with particular reference to sorption hysteresis. We found that the evolution of the equilibrium moisture content of a paper sheet subjected to arbitrary humidity changes can be predicted by applying the theory of independent domain complexions. In the second section, we develop a general moisture transport model incorporating diffusion and sorption effects. This model is used to study the effects of changing relative humidity on paper sheets.

## 2. EQUILIBRIUM MOISTURE SORPTION BEHAVIOR OF PAPER

By equilibrium moisture sorption behavior, we refer to the moisture content of paper sheets attained when the sheets reach a final thermodynamic equilibrium state with the environment around them. This represents the limiting moisture content under a particular relative humidity but also is a function of the past history to which the paper sheet was exposed.

Moisture sorption in paper is quite similar to that by other natural cellulosic materials such as wood and cotton fibers. Besides, the sorption by paper sheets is reflected by the sorption of the pulp fibers and other additives included in the paper sheet. Common to all these materials is the fact that moisture sorption isotherms are all of the high affinity type. The moisture content (defined as the ratio of moisture gained by the paper sheet to the sheet's dry weight) increases as a function of relative humidity at low RH values. It goes through a plateau around 40% but increases in humidity beyond this region show rapid increase in moisture gain. Venkateswaran [10] has provided a review of various theories applied to the sorption isotherms for wood and cellulose. The well known BET surface adsorption theory has not been successful in explaining the sorption isotherm especially above relative humidities of 60% [10-12]. Other theories, some of which are modifications of the BET theory have been more successful. The Hailwood-Horrobin theory for sorption models the water-cellulose system as a solid solution. Simpson [13] found that the Hailwood-Horrobin solid solution theory represents the sorption isotherms for paper reasonably well. For the case of food materials, van den Berg and Bruin [14] found that the Guggenheim-Anderson-deBoer equation was capable of representing the sorption isotherms better than the traditional BET model. The GAB equation is a three-parameter modification of the BET isotherm equation. Eagleton and Marcondes [15] have found that the GAB model fits the sorption data for fiberboard packaging materials. Chatterjee et al. [16] have also been able to fit the GAB model to the sorption data of Prahl [17] for kraft pulp fibers.

An important feature of moisture sorption in paper is the presence of hysteresis. Thus, the sorption isotherms obtained from samples when the humidity is reduced progressively in equilibrium steps from a high value to a low value are significantly higher than the corresponding adsorption isotherms. Similar hysteresis in the sorption by wood based materials has been well investigated by Barkas [18]. An extensive series of investigations by Seborg and co-workers [19-22] indicated that the sorption hysteresis for papers is independent of the pulping or bleaching actions. There was a mild dependence on beating though. Furthermore, the ratio of the adsorption moisture content to the desorption moisture content is relatively independent of the RH level, wood species, pulp digestion process and beating and bleaching processes.

Qualitative explanations for the hysteresis in sorption for cellulosic materials have been advanced. Urquhart [23] hypothesized that hysteresis could be caused by a differential availability of hydroxyl groups in cellulose during the adsorption and desorption branches of the RH cycle. A second explanation for hysteresis stems from the observed plasticity of the cellulose gels, which swells upon adsorption. Desorption results in some irreversible (plastic) deformation, which results in a higher moisture content of the structure as compared to the adsorption process. His theory postulated that the lost work during a stress strain cycle is equal to the lost work represented by the hysteresis loop between two relative humidity levels.

Hysteresis is found not only in sorption processes but also in the dewatering of porous media and the magnetization of ferromagnetic materials. A fundamental feature of hysteresis is that it is rate independent [24,30]. A traditional treatment of hysteresis by memory integrals [Volterra operators] is thus not possible. An elegant theory for equilibrium hysteresis has been put forward by Everett and coworkers [25-29]. This theory is based on the representation of hysteric phenomena by means of a large number of microdomains.

## 3. SORPTION ISOTHERMS AND HYSTERESIS

The sorption equilibrium of moisture in paper and board has been widely studied. However, only recently has it been recognized that a close examination of the interior of the hysteresis loop is necessary in order to understand the complete sorption behavior of paper materials in order to model the unsteady moisture sorption process under cyclic humidity variations [ 31]. Such an investigation of sorption hysteresis was reported by Chatterjee et al. [32]. In the following, we provide a description of the hysteresis and its representation using the theory of independent domains for the case of paper.

A representation of the typical moisture sorption isotherm for paper is shown in Figure 1. In this and the following Figures 1 through 5, the curves are to be understood as schematic representations and not the actual experimental data themselves. The lower curve is obtained when the material is subjected to progressively increasing humidity environments starting from a dry state whereas the upper curve is obtained when a completely saturated sample is exposed to progressively drier environments. These two curves are referred to as the boundary adsorption and desorption curves respectively. A sheet, which has been subjected to an arbitrary history of humidity cycling, will attain moisture content somewhere in between these two curves. It is our purpose to obtain an estimate of this moisture content when the humidity history of the sample is known.

Figure 2 shows a set of scanning curves and some sample spiral paths on the equilibrium diagram. These trajectories represent the evolution of the moisture

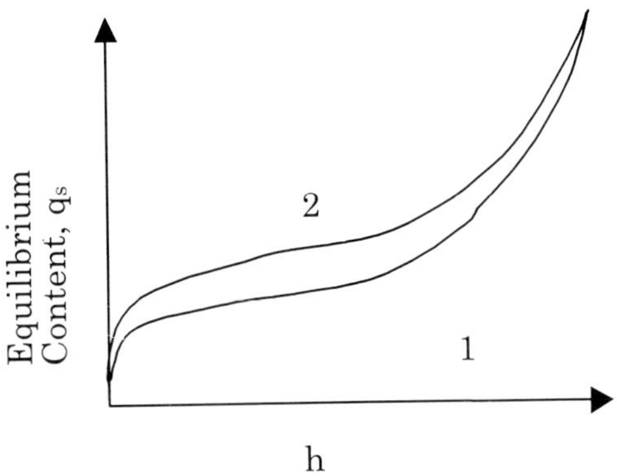


Figure 1. Typical equilibrium moisture sorption isotherms for paper. 1 - adsorption, 2 - desorption.

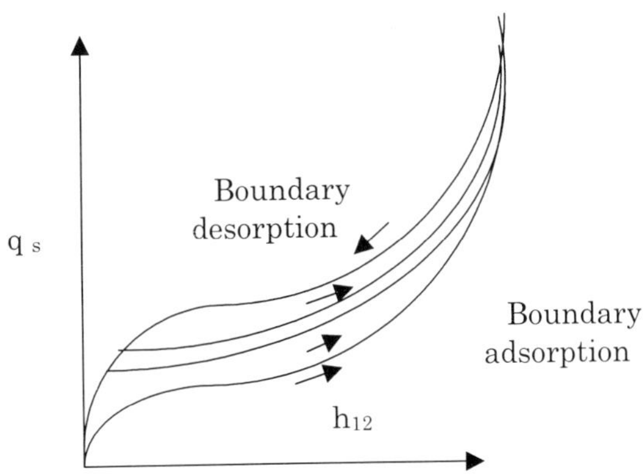


Figure 2a. Sorption equilibrium diagram showing adsorption scanning curves.

content of the paper sheets subjected to specific humidity histories. For instance, the adsorption scanning curves shown in (a) are obtained as follows. Starting with a sample at a high humidity state, the sample is exposed to progressively drier environments to follow the boundary desorption curve. Subsequently, the sample is exposed to more humid conditions in sequence to obtain the curves shown in Figure 2a. The turning point of the humidity i.e. the RH at which each of these scanning curves cross the boundary isotherm is a significant point and is labeled $h_{21}$. A complementary desorption scanning curves set is obtained by starting with a relatively drier environment which gets more humid and then switches to decreasing humidity levels. These curves are represented in (b). When the sheet is subjected to cyclic variation in humidity where the turning

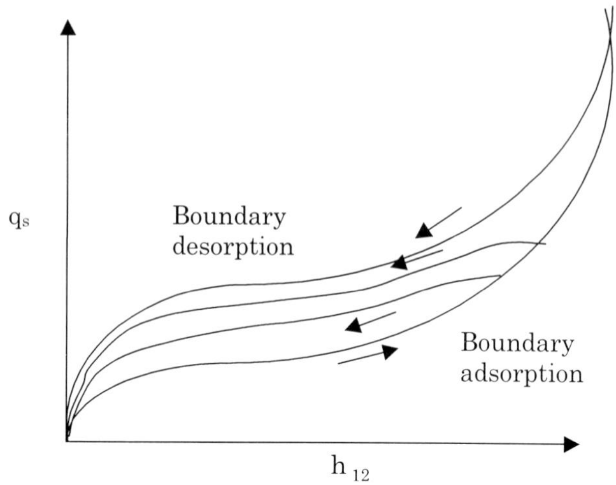


Figure 2b. Sorption equilibrium diagram showing desorption scanning curves.

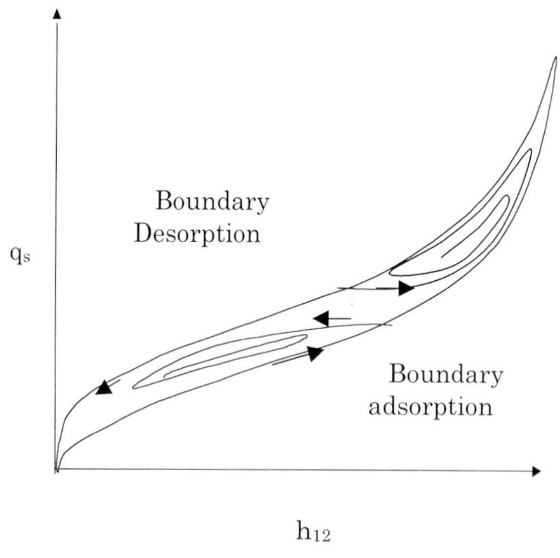


Figure 2c. Spiral trajectories inside the hysteresis loop starting from the boundary adsorption and desorption curves.

points of the RH progression are decreasing in order, a spiral curve as represented in (c) is obtained. Spirals could originate from the boundary desorption or the adsorption curves as shown.

Figure 3 shows the development of scanning surfaces as three-dimensional diagrams. When a set of scanning curves as shown in Figure 2 are redrawn with the turning point humidity (e.g. $h_{21}$) as the third axis, each scanning curve traces a different curve along the moisture surface as represented in this diagram. Obviously, there are two such surfaces that are complementary to each other and are obtained from the adsorption and desorption data. An infinite hierarchy of such surfaces will be obtained if scanning data are used for each level of scanning behavior (namely primary, secondary, tertiary, etc.).

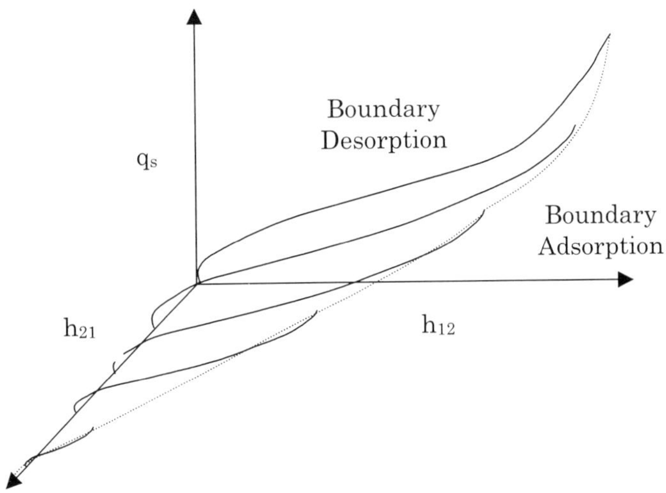


Figure 3. Three dimensional representation of the primary desorption surface.

Everett and Whitton [26] proposed a basic theory for sorption hysteresis idealizing the material to consist of a large number of independent domains. Each such domain can be thought to exist in one of two states (I and II) corresponding to being filled with condensed adsorbate (i.e. liquid water) or with vapor. A transition from a vacant state (I) to a filled state (II) occurs when the partial pressure of water or the RH is increased. The humidity level at which the transition (I → II) occurs is denoted by $h_{12}$ and the reverse transition (II → I) occurs at $h_{21}$. Since these two humidities are different from each other this indicates irreversibility in the filling and emptying behavior of the domains. Assuming that the paper possesses a large number of such micro-domains that empty and fill at different humidity levels, the resulting equilibrium curves will show hysteresis. This is illustrated in Figure 4a and Figure 4b.

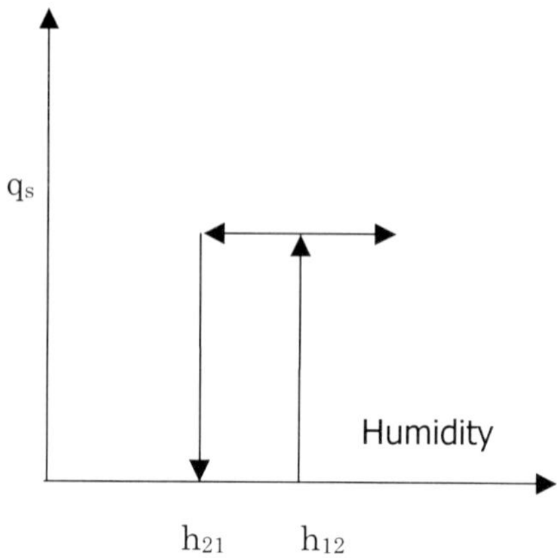


Figure 4a. Illustration of hysteresis effect due to the irreversible behavior during adsorption and desorption of a single microdomain. Domain fills at $h_{12}$ but empties at $h_{21}$.

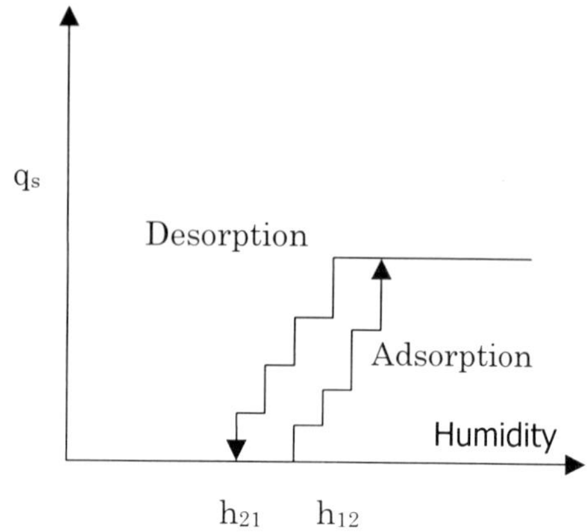


Figure 4b. Development of hysteresis due to the presence of a multitude of microdomains with different characteristics for filling and emptying adsorbate.

Everett and Smith [27] proposed a set of theorems governing the nature of the interior paths, loops and shapes of the scanning curves inside the hysteresis loop. A more formal treatment of hysteresis was developed by Everett [28,29] introducing the concept of a domain complexion function which serves as a memory of the system and tracks the evolution of the precise thermodynamic state of the material. The basis of the domain complexion theory for hysteresis is as follows. The population of domains which are filled with moisture i.e. those in state II is followed by means of a two dimensional diagram where the humidity level $h_{12}$ is graphed against the level $h_{21}$. Since for paper, the transition of II → I is always delayed as compared to the transition I → II, we expect $h_{12} >= h_{21}$ and the permissible states of the paper sheet will be represented by the right triangular region as shown in Figure 5a. Figure 5b shows a representative diagram of the equilibrium path when the system is taken through a series of equilibrium adsorption and desorption processes. One key assumption of this theory is that the distribution function of the domains is independent of the path itself. This is valid if the domains are independent of each other.

An example application of the theory of domain complexions to describe the moisture sorption equilibrium of a bleached kraft paperboard (240-gsm basis weight) is shown here. A detailed listing of the experimental set up and the experimental data can be obtained from [32,33,34]. Briefly, the experimental set up consists of a test chamber within which the humidity can be controlled to a

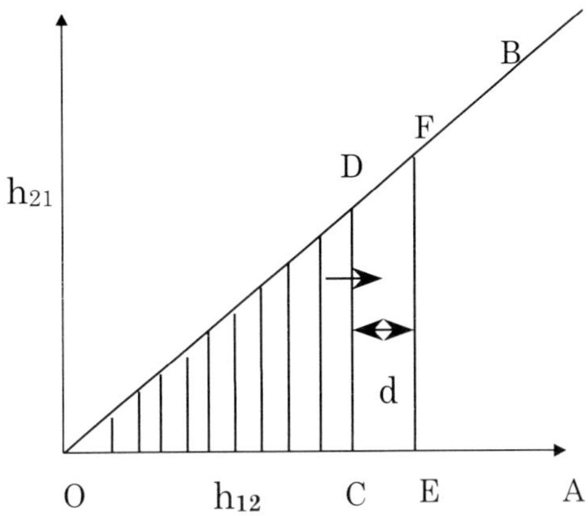


Figure 5a. Basic domain complexion diagram to track the filling up of microdomains during adsorption. Primary adsorption curve as in Figure 1.

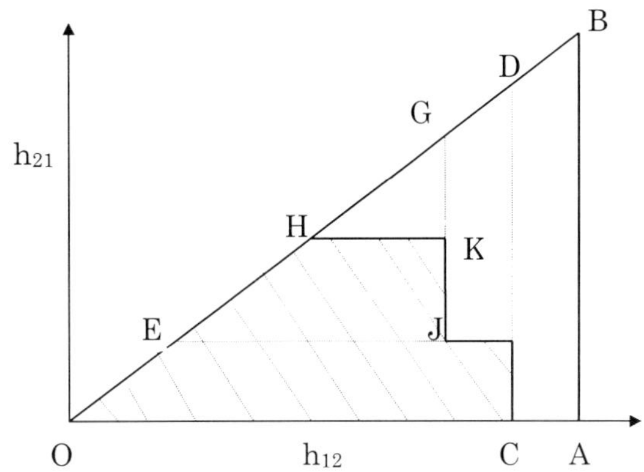


Figure 5b. Representative diagram of domain complexions when the system is taken through a spiral path as in Figure 2c.

prescribed variation. The test chamber is continuously swept with a mixture of dry and saturated air mixed in controlled proportions. Paper samples are hung from a recording balance and their weight is monitored continuously. The dry weight of the samples is determined after drying in an oven and the moisture content is obtained from the wet and dry sample weights.

Paper samples 10 cm by 10 cm and approximately 0.29 mm thick were used for the following experiments. The samples were preconditioned in order to remove the effect of their past history and bring all samples to a consistent level of behavior by exposing them to two successive humidity cycles between 90% and 15% RH. Figure 6 is a schematic of the experimental apparatus. Detailed experimental data and the complete calculations of secondary and higher order trajectories to check the validity of the independent domain theory has been presented elsewhere. Here only the essential features of the results will be presented.

The equilibrium sorption data for a bleached kraft paperboard are shown in Table 1 [32,33 ]. The sorption equilibrium is seen to increase sharply as the RH increases above 80%. Hysteresis between the boundary curves is observed to exist throughout the range as shown. The ratio of the adsorption moisture content to the desorption value is approximately 0.82 - a fact which has been reported previously [21]. Figure 7 shows the sorption isotherms and selected secondary desorption curves. The secondary curves extend all the way to the lower end of the hysteresis loop unlike the case of wood (Peralta). The boundary sorption

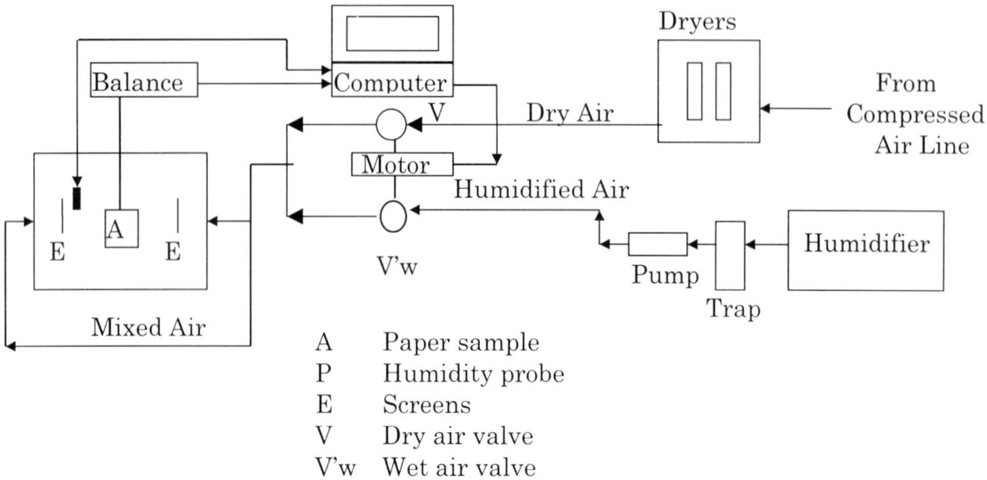


Figure 6. Schematic of experimental set up to determine equilibrium and transient moisture sorption behavior of paper materials. The experimental chamber is a constant humidity and temperature chamber in which the paper sample is hung from a balance.

Table 1
Measured equilibrium moisture content ($q_s$, EMC) data (weight fractions) for a bleached kraft paperboard sheet

| RH | Adsorption EMC | Desorption Equilibrium Moisture Content | | | | | | | |
|---|---|---|---|---|---|---|---|---|---|
| | | 100 | 90 | 80 | 75 | 60 | 45 | 30 | 15 |
| 15 | 0.0484 | 0.0504 | 0.0497 | 0.049 | 0.487 | 0.486 | 0.0485 | 0.0485 | 0.0484 |
| 30 | 0.0623 | 0.0689 | 0.0682 | 0.067 | 0.066 | 0.065 | 0.0633 | 0.0623 | |
| 45 | 0.0774 | 0.0879 | 0.0858 | 0.084 | 0.083 | 0.083 | 0.0774 | | |
| 60 | 0.0967 | 0.1120 | 0.1063 | 0.103 | 0.102 | 0.097 | | | |
| 75 | 0.1200 | 0.1480 | 0.1319 | 0.123 | 0.120 | | | | |
| 80 | 0.1290 | 0.1625 | 0.1410 | 0.129 | | | | | |
| 90 | 0.1723 | 0.2511 | 0.1723 | | | | | | |
| 100 | 0.2951 | 0.2951 | | | | | | | |

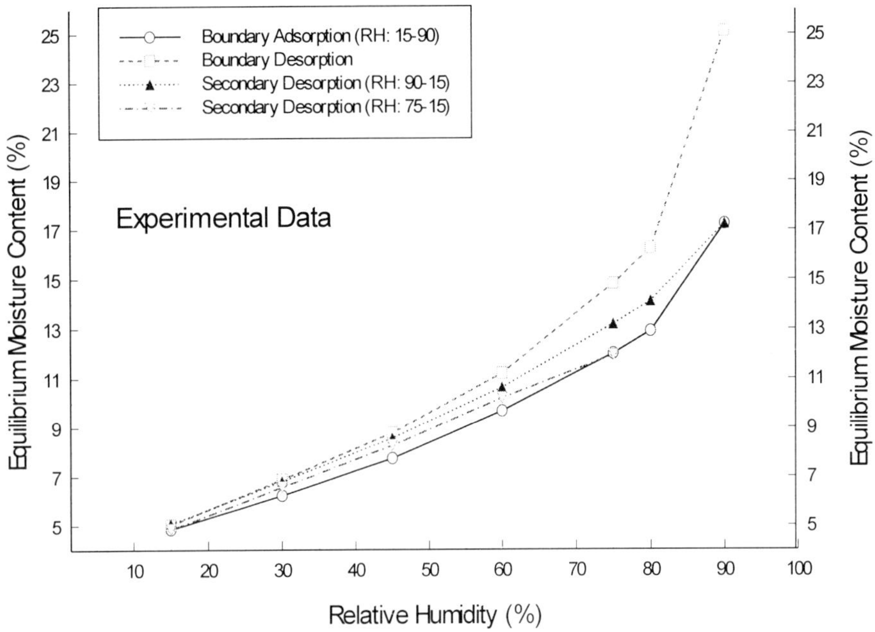


Figure 7. Sorption isotherms and secondary desorption isotherms representative of bleached kraft paperboard.

data were fitted to the Guggenheim-Anderson-deBoer (GAB) isotherm [16]. The GAB isotherm is given by

$$q^* = \frac{m_0 C_{GAB} K_{GAB} h}{(1 - K_{GAB} h)(1 - K_{GAB} h + C_{GAB} K_{GAB} h)} \qquad (1)$$

where the equilibrium moisture content is denoted as $q^*$, h is the relative humidity, $m_0$ , $K_{GAB}$ and $C_{GAB}$ are the GAB constants. Table 2 provides the GAB constants for the boundary sorption data given in Table 2.

The boundary desorption data presented here were obtained by equilibrating an initially waterlogged sample paper sheet to progressively drier environments. The first data point on this isotherm corresponded to 90% relative humidity. Experimental data at high humidities are particularly difficult to obtain since good control is difficult to achieve in this region. Another important issue is the actual equilibrium moisture content value of a paper sheet at saturation humidity (100%). By using a porous plate technique [35] or by using the dilution of a non-adsorbing high molecular weight polymer [36], the saturation water content of papermaking pulps has been determined. It has also been found that

Table 2
Constants in the GAB Isotherm representation of the boundary adsorption and desorption data of Table 1. Units correspond to those used for $q_s$ and h

| | Adsorption | Desorption |
|---|---|---|
| $m_0$ | 0.048 | 0.0562 |
| $K_{GAB}$ | 0.673 | 0.8487 |
| $C_{GAB}$ | 17.659 | 27.7379 |

the data for equilibrium moisture contents between 95% and 100% RH are in continuity with the saturation moisture content data obtained in these studies.

From the secondary desorption data presented in Table 1, a discrete approximation to the moisture density distribution function (defined below) can be determined.

$$m = \frac{\partial^2 M}{\partial h_{12} \partial h_{21}} \tag{2}$$

The values for m as a function of $h_{12}$ and $h_{21}$ are presented in Table 3.

Table 3
Discrete approximation to the moisture density function $[m(h_{12},h_{21})\Delta h_{12}\Delta h_{21}]$x10 for bleached kraft paper board

| Total | 0.484 | 0.139 | 0.151 | 0.193 | 0.233 | 0.09 | 0.433 | 1.23 | 2.951 |
|---|---|---|---|---|---|---|---|---|---|
| 100-90 | | | | | | | | 0.440 | 0.440 |
| 90-80 | | | | | | | 0.313 | 0.573 | 0.886 |
| 80-75 | | | | | | 0.055 | 0.036 | 0.054 | 0.145 |
| 75-60 | | | | | 0.180 | 0.021 | 0.055 | 0.104 | 0.360 |
| 60-45 | | | | 0.142 | 0.05 | 0.002 | 0.011 | 0.036 | 0.241 |
| 45-30 | | | 0.141 | 0.029 | 0 | 0.005 | 0.001 | 0.014 | 0.190 |
| 30-15 | | 0.138 | 0.097 | 0.021 | 0.002 | 0.004 | 0.01 | 0 | 0.185 |
| 15-0 | 0.484 | $7x10^{-4}$ | $3x10^{-4}$ | 0.001 | 0.001 | 0.003 | 0.007 | 0.007 | 0.504 |
| $h_{21}$ $h_{12}$ ↑ → | 0-15 | 15-30 | 30-45 | 45-60 | 60-75 | 75-80 | 80-90 | 90-100 | Total |

The detailed calculation from experimental data is presented in [32]. Table 4 shows various trajectories estimated by applying the theory of independent domain complexions and using the moisture density function generated in Table 3. Close agreement between experimental and theoretical predictions are found indicating that Everett's theory is a valid description of the hysteresis in paper materials.

An interesting point of comparison between the theoretical predictions and the experimental data may be made. For the case of trajectories starting from an initial wet state (T1, T2, and T3), the theory underestimates the actual moisture content. However, when the trajectories start from an initial dry state, the agreement between the theoretical prediction and experimental data is very good. Similar behavior has been found when more such trajectories were investigated [33]. The reason seems to be the fact that the moisture density function is estimated from approximate data for the desorption and adsorption isotherms between 90% and 100% RH. More accurate data in this region should provide better theoretical estimates. It appears that the theory of independent domain complexions advanced by Everett and coworkers is indeed applicable for paper materials showing moisture sorption hysteresis.

## 4. TRANSIENT MOISTURE SORPTION DYNAMICS IN PAPER SHEETS

The migration of moisture by diffusive mechanisms in paper is responsible for the dynamics of the paper sheet. Before these mechanisms can be identified, a few remarks regarding the structure of a paper sheet must be made. Paper sheets are composed of a number of intercrossing fibers that are laid on top of each other during the papermaking process. Paper shows a layered structure under examination by electron microscopes. The fibers are found to be oriented along planar directions within each layer and typical paper sheets are composed of between 10 to 50 such layers. The prominent moisture transport mechanisms in such sheets can thus be identified as: diffusion of water vapor through the inter-fiber void space (denoted hereafter as vapor diffusion), Knudsen diffusion (in pores of small dimensions of the order of 10 nm), surface diffusion, bulk solid diffusion (within fibers) and finally, at high humidities close to saturation, capillary transport. Thus, a number of alternative pathways to transport exist within a sheet and depending upon the external variables and the sheet structure, any combination of these may be dominant. In the following, we provide a generalized transport model based on two prominent mechanisms. The first of these is the diffusion of water vapor through the inter-fiber void space within the sheet. The second is the diffusion of water (within the fiber matrix in a bound state or in an adsorbed state at the fiber/void interface). This second mechanism is considered to be a composite solid diffusion mechanism.

Table 4
Comparison of predictions of equilibrium moisture content (in percent) for various trajectories using Everett's theory with experimental data. Trajectories are drawn schematically in Figure 5. RH is in percent

| T1 | | | T2 | | | T3 | | | T4 | | |
|---|---|---|---|---|---|---|---|---|---|---|---|
| RH | Expt. | Theory | RH | Expt. | Theory | RH | Expt. [1] | Theory | RH | Expt. | Theory |
| 45 | 8.72 | 8.79 | | | | 90 | 25.50 | 25.11 | 15 | 4.77 | 4.84 |
| 60 | 10.37 | 10.21 | 60 | 11.13 | 11.20 | 45 | 8.72 | 8.79 | 90 | 17.69 | 17.23 |
| 75 | 12.75 | 12.51 | 75 | 13.28 | 13.00 | 90 | 19.43 | 17.44 | 30 | 6.57 | 6.82 |
| 80 | 13.74 | 13.29 | 80 | 14.21 | 13.76 | 60 | 10.92 | 10.84 | 75 | 12.21 | 12.24 |
| 90 | 18.80 | 17.44 | 90 | 19.25 | 17.80 | 75 | 13.02 | 12.64 | 45 | 8.32 | 8.52 |
| | | | | | | 80 | 13.99 | 13.40 | 60 | 9.90 | 9.94 |
| | | | | | | 75 | 13.29 | 12.85 | | | |

[1]Average of two experimental measurements.

Figure 8 shows a schematic representation of a paper sheet for modeling moisture transport. As is conventional for paper sheets, we assume that the direction into the sheet is represented by the z co-ordinate whereas the two in-plane co-ordinates are represented by x and y. It is useful to identify two concentration fields, one for the moisture concentration within the void space, $c(\underline{x},t)$ and the moisture content within the fiber matrix, $q(\underline{x},t)$ where $\underline{x} = (x,y,z)$. Diffusive transport under transient conditions through this composite medium is described by the following equations.

$$\varepsilon \frac{\partial c}{\partial t} = D_p[\frac{\partial^2 c}{\partial x^2} + \frac{\partial^2 c}{\partial y^2} + \frac{\partial^2 c}{\partial z^2}] \tag{3}$$

$$\frac{\partial q}{\partial t} = \frac{\partial}{\partial x}[D_q \frac{\partial q}{\partial x}] + \frac{\partial}{\partial y}[D_q \frac{\partial q}{\partial y}] + \frac{\partial}{\partial z}[D_q \frac{\partial q}{\partial z}] \tag{4}$$

$\varepsilon$ is the porosity of the sheet which is assumed a constant. $D_p$ is the pore diffusivity and $D_q$ is the diffusivity of moisture in bound form inside the sheet. Bound water diffusion is idealized to occur within the fiber space and $D_q$ is generally dependent non-linearly on the local moisture content q.

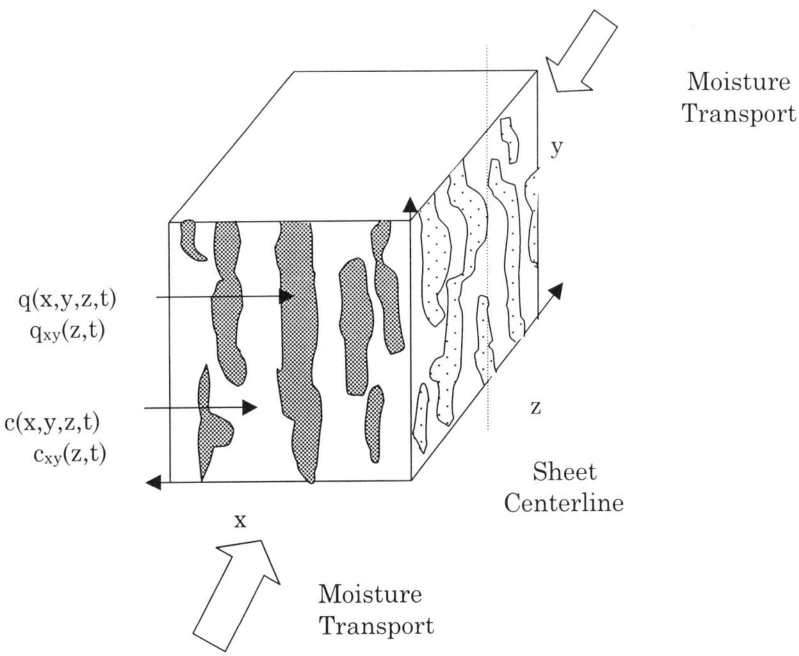


Figure 8. Schematic representation of structure of a paper sheet. z direction is the thickness. Concentrations in fiber and void spaces are shown as also the average concentrations in the x-y plane.

From the structural representation of the paper sheet as shown in Figure 8, the two concentration fields can be assumed to vary on two different length scales in the z direction. On the small scale, the concentrations vary within each fiber or void from location to location. However, a slower variation on the gross sheet thickness scale can be thought of as representing the overall transport.

We consider now two averaged concentration fields c and q as

$$\bar{c}(z) = \frac{1}{S_s} \int_{S_s} c\,dx\,dy \tag{5}$$

$$\bar{q}(z) = \frac{1}{S_s} \int_{S_s} q\,dx\,dy \tag{6}$$

where $S_s$ is a suitably large area in the x-y plane over which the fields c and q are averaged. Also note that the c field will be non-zero only in the void space and q is non-zero only in the fiber space. By averaging eqs. (3) and (4), we obtain

$$\varepsilon \frac{\partial \bar{c}}{\partial t} = D_p [\frac{\partial^2 \bar{c}}{\partial z^2}] + D_p \frac{1}{S_s} \int_{S_s} \nabla^2 c\,dS \tag{7}$$

$$\frac{\partial \bar{q}}{\partial t} = \frac{\partial}{\partial z} [D_q \frac{\partial \bar{q}}{\partial z}] + D_q \frac{1}{S_s} \int_{S_s} \nabla^2 q\,dS \tag{8}$$

The Laplacian operator is understood to be defined in the x-y plane in these equations. Each of the second terms on the right hand side of both these equations can be expressed as

$$\int_{S_s} \nabla^2 c\,dV = \int_{C_s} \nabla c.\underline{n}\,dC \tag{9}$$

where the curve $C_s$ is identified as the interface between the fiber matrix and the void space within $S_s$. The gradient and the normal vectors are in the x-y plane similar to above. Furthermore, we require the continuity of mass flux of water across this interface and hence,

$$-D_p \nabla c.\underline{n} = -D_q \rho_p \nabla q.\underline{n} \tag{10}$$

Note that q is defined as a moisture content making the appearance of the sheet density necessary for dimensional consistency in the above equation. Applying these equations in eqs.(7) and (8), we obtain

$$\varepsilon\frac{\partial\bar{c}}{\partial t}=D_p\frac{\partial^2 c}{\partial z^2}+D_q\rho_p\frac{1}{S_s}\int_{C_s}\nabla q.\underline{n}dC_s \tag{11}$$

$$\frac{\partial\bar{q}}{\partial t}=\frac{\partial}{\partial z}[D_q\frac{\partial\bar{q}}{\partial z}]+D_q\frac{1}{S_s}\int_{C_s}\nabla q.\underline{n}dC_s \tag{12}$$

In each of the above equations, the first term on the right hand side represents diffusive transport along the thickness direction whereas the second term represents diffusion in the x-y directions. The second term can be understood to represent the uptake of moisture by the individual fibers. As is common to analysis of adsorption phenomena in fixed beds [37], it is convenient to represent the moisture uptake by the fibers (i.e. the second term) by the following equation.

$$D_q\frac{1}{V_s}\int_{V_s}\nabla^2 c dV=k_i[q_s(\bar{c})-\bar{q}] \tag{13}$$

The coefficient $k_i$ may be assumed constant in which case the approximation is called as the Gluckauf approximation [37]. When diffusion along the z direction within the fibers is small, e.g. when the contact area between the fibers along the z-direction is small or when $D_q$ itself is very small, each fiber can be considered to take up moisture locally according to the following equation.

$$\frac{\partial\bar{q}}{\partial t}=k_i[q_s(\bar{c})-\bar{q}] \tag{14}$$

This equation is obtained by substituting eq.(13) in eq.(12) and neglecting the first term on the right hand side. Eq.(14) indicates that the rate of change of the local moisture content of the fibers, denoted by q(z) can be described as being proportional to the deviation of the moisture content from the equilibrium value at the boundary. Thus, the unsteady diffusion equation (12) is being effectively replaced by a kinetic equation. Thus, diffusion in the x-y plane is being viewed as a relaxation process towards thermodynamic equilibrium with spatial dependencies being lumped into a kinetic constant $k_i$. Substantial literature exists on the theoretical justification for this approximation [37]. For the case of spherical adsorbent particles, the value of $k_I$ is known to be a constant dependent upon the intra-particle diffusivity ($D_q$) and some geometrical parameters. In the case of paper, this parameter should be determined by comparing the theoretical predictions with experimental data.

At this point, the over-bars on the concentrations can be dropped with the understanding that all further concentrations are averages in the x-y plane of the paper. The unsteady diffusion equations, eqs.(11) and (12) can be written as

$$\varepsilon\frac{\partial c}{\partial t} = D_p[\frac{\partial^2 c}{\partial z^2}] - \rho_p k_i[q_s(c) - q] \tag{15}$$

$$\frac{\partial q}{\partial t} = \frac{\partial}{\partial z}[D_q \frac{\partial q}{\partial z}] + k_i[q_s(c) - q] \tag{16}$$

We consider a ramped change in humidity given as follows.

$$h_b(t) = h_0 + \frac{(h_1 - h_0)}{t_R}t \tag{17}$$

for $t < t_R$ and

$$h_b(t) = h_1 \tag{18}$$

for $t >= t_R$.

The humidities can be converted into water vapor concentrations according to

$$c = \frac{p_{sat}(T)h}{R_g T} \tag{19}$$

The initial condition can then be given as

$$c(t = 0, z) = c_0 \tag{20}$$

indicating that the void space concentration is the initial value inside the sheet. Further, equilibrium is assumed everywhere within the sheet. Hence, the local fiber moisture content q is given by

$$q(t = 0, z) = q_s(c_0) = q_0 \tag{21}$$

At the center $(z = L)$, the concentration profile is symmetric if the sheet structure is uniform and is also exposed to similar conditions on either side. Thus,

$$\frac{\partial c}{\partial z}(t, z = L) = 0 \tag{22}$$

$$\frac{\partial q}{\partial z}(t, z = L) = 0 \tag{23}$$

Finally, at the sheet's surface the water flux is assumed to first diffuse into the vapor space from where it is absorbed by the fibers and also diffuses into the interior. This means that the convective moisture flux at the surface balances the diffusive flux as given by the following equation.

$$-D_p \frac{\partial c}{\partial z}(t, z = 0) = k_f[c_b(t) - c(t, z = 0)] \tag{24}$$

This equation indicates an important approximation being made in this analysis. Since the sheet is exposed to the humid environment, the fibers within the x-y plane on the sheet's surface will receive moisture directly. Thus, the assumption that the fibers of the sheet take up moisture at a rate given by a first order relaxation process is violated. That is, application of eq. (13) at $z = 0$ is an approximation and must be recognized as such. In reality, we expect the fiber moisture content $q(t,z=0)$ to be that which is in equilibrium with $c(t,z=0)$ i.e.

$$q(t, z = 0) = q_s[c(t, z = 0)] \tag{25}$$

The following analysis is intended to provide a set of simplified models that can be used to analyze the sorption process under given conditions. We first begin by expressing the diffusion model equations (15) and (16) in dimensionless variables as defined in the nomenclature section. The fiber diffusivity, $D_q$ is assumed to be a constant for the purpose of this dimensional analysis. It can be easily modified to include a non-linear functional relationship with q.

$$\varepsilon \frac{\partial C}{\partial \tau} = \frac{1}{P} \frac{\partial^2 C}{\partial \zeta^2} - \alpha\beta[Q_s - Q] \tag{26}$$

$$\frac{\partial Q}{\partial \tau} = \frac{1}{P'} \frac{\partial^2 Q}{\partial \zeta^2} + \beta[Q_s - Q] \tag{27}$$

The initial and boundary conditions in dimensionless forms are

$$C(\tau = 0, \zeta) = Q(\tau = 0, \zeta) = 0 \tag{28}$$

$$\frac{\partial C}{\partial \zeta}(\tau, \zeta = 1) = \frac{\partial Q}{\partial \zeta}(\tau, \zeta = 1) = 0 \tag{29}$$

and

$$-\frac{\partial C}{\partial \zeta}(\tau, \zeta = 0) = \gamma P[C_b(\tau) - C] \tag{30}$$

The sorption equilibrium can be represented by the following relationship in dimensionless form.

$$Q_s(C) = F(C) \tag{31}$$

where F(C ) represents a non-linear function of C.

## 5. ANALYSIS OF LIMITING CASES

### 5.1. Case 1. Intra-fiber diffusion in thickness direction insignificant

The dimensionless parameter P' represents intra-fiber diffusion. When P' and βP' are large, the effect of intra-fiber diffusion is small and may be neglected in the analysis. This is tantamount to ignoring intra-fiber diffusion in the z-dimension. In terms of the system variables, these two conditions imply that

$$D_q << \frac{L^2}{t_R} \tag{32}$$

and

$$D_q << k_i L^2 \tag{33}$$

The coefficient $k_i$ is related to the intra-fiber diffusivity and the fiber geometry by the following approximate relationship.

$$k_i = G\frac{D_q}{R^2} \tag{34}$$

G is a constant and is equal to 15 for spherical geometry [37] and 6 for cylindrical geometry [39]. For the case of fibers, it is expected to be different yet of similar order of magnitude. Substituting in the inequality (33), we obtain

$$R^2 << GL^2 \tag{35}$$

as the second condition under which intra-fiber diffusion along the z-dimension in the sheet may be neglected. Thus, both inequalities (33) and (35) need to be satisfied for intra-fiber diffusion to be negligible. Under this condition, we can view the following model equations as the leading order simplifications to eqs.(26) and (27). Higher order approximations may be obtained and a perturbation expansion solution may be obtained easily.

$$\varepsilon\frac{\partial C}{\partial \tau}+\alpha\frac{\partial Q}{\partial \tau}=\frac{1}{P}\frac{\partial^2 C}{\partial \zeta^2} \tag{36}$$

$$\frac{\partial Q}{\partial \tau}=\beta[Q_s-Q] \tag{37}$$

The boundary condition at $\zeta = 0$ is

$$-\frac{\partial C}{\partial \zeta}=\gamma P[C_b-C] \tag{38}$$

For the case of a linear sorption isotherm, an analytical solution to the above equations, (36) and (37) subject to conditions (28), (29), (38) has been obtained using the integral transform technique by Ramarao and Chatterjee [34]. The solution is complex and will not be presented here. A numerical solution to the same equations when the isotherm is non-linear was also obtained by Bandyopadhyay et al. [40].

### 5.2. Case 2. Dominant diffusion in void space

When $\beta >> 1$, in addition to the above conditions, it implies that the pick up of moisture by the fibers is extremely rapid and the resistance to sorption resides primarily in the void spaces of the sheet. This situation may occur for the case of thick sheets whose diffusivity $D_p$ is low and yet the fibers may be very thin such that R is also low. The combined effect is that $\beta$ is high and P is not very large. The model reduces to a single partial differential equation which may in fact be simplified as

$$\frac{\partial C}{\partial \tau}=A(C)\frac{\partial^2 C}{\partial \zeta^2} \tag{39}$$

where A(C ) is a non-linear dimensionless diffusion coefficient. For the case of a linear isotherm this becomes a constant. The equation for A( C) is given as

$$A(C)=\frac{1}{P[\varepsilon+\alpha\frac{dQ}{dC}(C)]} \tag{40}$$

The initial and boundary conditions are still given by eqs.(28), (29) and (38).

### 5.3. Case 3. Dominant fiber diffusion and external resistance

When $\alpha P << 1$, C and Q are independent of $\zeta$. In this case, the resistance to transport in the z' dimension is negligible although there still may be significant resistance in the radial dimension of the fibers. This model is parallel to case 2 and can be obtained from case 1 equations. Integrating eq.(39) with respect to $\zeta$, and applying the boundary condition leads to

$$\varepsilon \frac{dC}{d\tau} + \alpha \frac{dQ}{d\tau} = \gamma[C_b - C] \tag{41}$$

$$\frac{dQ}{d\tau} = \beta[Q_s - Q] \tag{42}$$

The initial conditions are given at $\tau = 0$ as

$$C = Q = 0 \tag{43}$$

For the case of a non-linear isotherm, a numerical solution of the above equations is necessary. However, when the isotherm is linear, an analytical solution can be obtained.

When $\beta >> 1$, eq.(42) indicates that the moisture content within the sheet is at equilibrium with the local concentration. The model can be simplified as follows.

### 5.4. Case 4. Dominant external resistance

The accumulation of water vapor within the void spaces can be neglected in comparison to the accumulation within the fibers. Thus,

$$\varepsilon << \alpha K \tag{44}$$

Therefore eqs.(41) and (42) can be simplified as

$$\frac{dC}{d\tau} = \frac{\gamma}{\alpha K}[C_b(\tau) - C] \tag{45}$$

and

$$Q\ (\tau) = Q_s(C) \tag{46}$$

The initial condition is the same as the first part of eq.(43). When either $\gamma$ is very large or $\alpha$ or K are very small such that $\gamma >> \alpha K$, we can assume that $C(\tau)$ is equal to $C_b(\tau)$ and the entire sheet is at equilibrium. This gives us equilibrium behavior as follows.

### 5.5. Equilibrium behavior

$$C = C_b(\tau) \tag{47}$$

and

$$Q(\tau) = Q_s[C_b(\tau)] \tag{48}$$

## 6. SORPTION DYNAMICS AND COMPARISON OF TRANSPORT MECHANISMS FOR TYPICAL SITUATIONS

Let us choose a situation where a sample of a bleached kraft linerboard (240 gsm) is subjected to a ramp change in humidity from a base value of 15% to 45%. The parameters of the paper sheet are shown below in Table 5. Sorption dynamics for these sheets, the steady state diffusion coefficients for water vapor

Table 5
Parameters for experiments on sorption dynamics

| Parameter | Symbol & Units | Value or Range |
|---|---|---|
| Bound Water Diffusivity [2] | $D_q$, $m^2s^{-1}$ | Varies from $10^{-14}$ at low RH (below 75%) through $10^{-6}$ at higher values. |
| Water Vapor Diffusivity[3] | $D_p$, $m^2s^{-1}$ | $6.3x10^{-7}$ |
| Sheet Density | $\rho_p$, $kgm^{-3}$ | 839 |
| Porosity | $\varepsilon$ | 0.54 |
| Sheet Thickness | 2L, m | $0.325x10^{-3}$ |
| Fiber Radius | R, m | $10x10^{-6}$ |
| Mass transfer coefficient[4] | $k_f$, $ms^{-1}$ | 0.0025 |
| Internal diffusion parameter[5] | $k_i$, $s^{-1}$ | 0.003 |
| Temperature | T, C | 23.7 |
| Saturation Vapor Pressure | $p_{sat}$, kPa | 2.9 |
| Isotherm parameter | $q_0/c_0$, $m^3kg^{-1}$ | 15.24 |
| Slope of isotherm | $K_{iso}$, $m^3kg^{-1}$ | 4.570 |
| External humidity change | $H_1$-$H_0$ | 45%-15% |
| Ramp Time | $t_R$, s | 1800 |

[2] Estimated from steady state water flux through sheets at high RH (>75% avg).
[3] Calculated values from steady state water vapor flux through sheets at low RH (<75% avg.)
[4] Mass transfer coefficient determined from liquid water evaporation data in chamber.
[5] Obtained by comparing model predictions with experimental transient sorption data (for one ramp).

and the relevant mass transfer coefficients have been determined experimentally and reported in [38]. The dimensionless parameters corresponding to these values are given in Table 6.

Table 6
Dimensionless parameters at baseline values

| Dimensionless Parameter | Baseline Value |
|---|---|
| K | 0.3 |
| $\alpha$ | 12789 |
| $\gamma$ | 27692 |
| $\beta$ | 5.4 |
| P | $2.328 \times 10^{-5}$ |
| P' | $1.47 \times 10^{3}$ |
| Bi | 0.645 |

Figure 9 shows a plot of the dimensionless moisture content as a function of time calculated by solving case 1 numerically. Also shown in this figure are experimental data on the sorption dynamics as a function of time for a sample sheet consisting of bleached kraft linerboard. For an external ramp humidity change, we observe that the sorption dynamics is indeed well predicted by the model. In the baseline case, we observe that $P' >> 1$ and $\beta P' >> 1$ i.e. $D_q << L^2/t_R$ and $R << G^{1/2}L$ (assuming $G = 6$). Therefore, intra-fiber diffusion along the z-dimension is negligible under these conditions.

On the other hand, if the humidity enters the higher range (>75% say), the estimated value for $D_q$ increases to $10^{-6}$ $m^2s^{-1}$. Intra-fiber diffusion can no longer be ignored and the complete model equations must be solved numerically. For extremely thin sheets or slow ramps, such that inequalities (33) and (35) are violated, intra-fiber diffusion may be important. However, in such cases, equilibrium is likely to prevail throughout the sheets and so, these limits are not likely to be realized in practice.

We note that the model incorporating vapor diffusion should be able to describe the chosen experimental sorption data. However, since $\alpha P = 0.298$ which is not very small compared to unity, the effect of vapor diffusion in the baseline case is marginal but could become significant if any of the variables incorporated into this dimensionless grouping change suitably. For example, if the sheet thickness were to double, this parameter would quadruple. Vapor diffusion in the z-dimension of the sheet will be at least as important as the other transport mechanisms. Therefore, our baseline experiment corresponds to a situation where either case 3 or case 2 can be used but case 3 is expected to be a slightly more accurate representation.

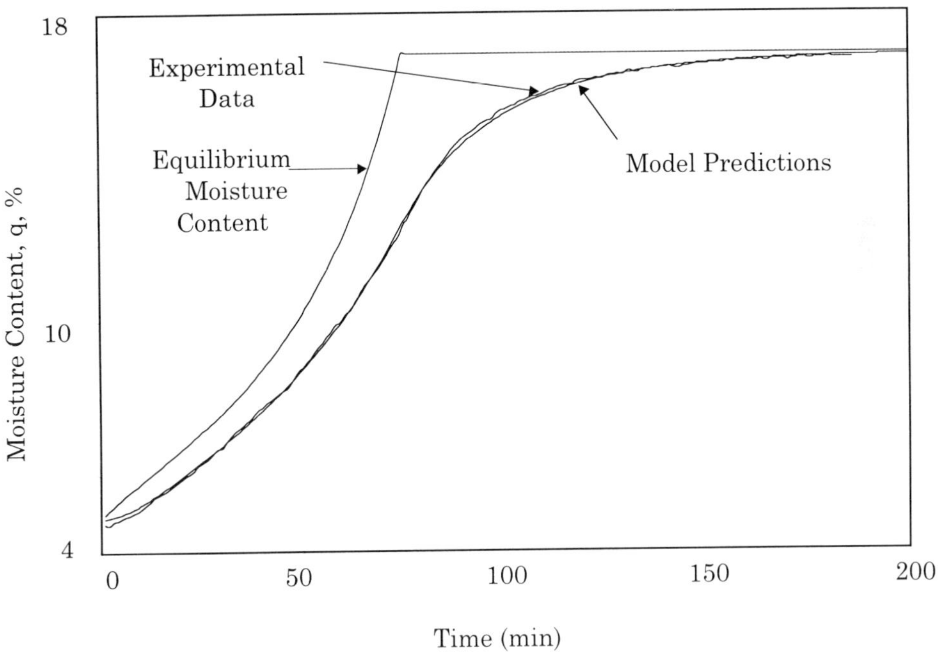


Figure 9. Transient moisture sorption predictions of model incorporating intra-fiber and inter-fiber diffusion [40]. Experimental data on transient moisture sorption of a paper sheet exposed to ramp change in humidity are shown as well as comparison with model predictions.

Another way in which vapor diffusion can become significant is if the ramp slope is increased i.e. the external humidity change occurs more rapidly. If sheets whose diffusivity is lower than the sheet for the baseline experiment were chosen, vapor diffusivity under identical conditions would be significant again as indicated by higher values of $\alpha P$.

When $\alpha P \ll 1$, case 3 reduces to case 2. In the baseline case, if the ramp slope were reduced further, or the thickness was lower, $\alpha P$ will be lower than 0.298. The model incorporating the mass transfer coefficient and the diffusion parameter $k_I$, case 3 can be applied with increased confidence. For the baseline case, we observe that $\beta^{-1}$ is 0.18. Since this is not much smaller than unity, the model incorporating external resistance alone cannot describe the sorption dynamics.

Finally, when $\alpha K/\gamma << 1$, the sheet will track the sorption equilibrium corresponding to the external humidity change. In the baseline case, this parameter group is equal to 0.14. Again, significant lag between the equilibrium values and the actual observed moisture content is seen. The condition will be satisfied better if sheet thickness is smaller or the mass transfer coefficient is enhanced or the ramp change occurs slower. Any of these will result in the sheet following equilibrium more closely. Table 7 shows example conditions under which each of these models is applicable.

Table 7
Example conditions under which each model is applicable

| | |
|---|---|
| All mechanisms are important. | |
| RH Change | 15% to 95%. |
| Time | 40 min |
| Diffusivity, $D_p$ | $6.3 \times 10^{-7} m^2 s^{-1}$ |
| $D_q$ | $18.0 \times 10^{-7} m^2 s^{-1}$ |
| Sheet thickness, 2L | $1.25 \times 10^{-3}$ m |
| All other parameters remain at their baseline values given in Table 5. | |
| Case 1. Intra-fiber diffusion along sheet z - dimension negligible in comparison to others. | |
| RH Change | 15% - 75%. |
| All other parameters as in Model 4 case above. | |
| Case 3. Intra-fiber and void space diffusion negligible. | |
| RH Change | 15% - 75%. |
| Sheet thickness, 2L | $2.5 \times 10^{-4}$ m |
| All other parameters remain at their baseline values. | |
| Case 4. Internal sheet transport resistances negligible compared to external resistances. | |
| RH Change | 15% - 75% |
| Time | 160 min |
| Sheet thickness, 2L | $1.25 \times 10^{-4}$ m |

## 7. CONCLUSIONS

The above cases and the models corresponding to them represent the isothermal sorption process in a complete sense. In any experimental situation, any subset of these models will be relevant in describing the dynamics. The effects of external and internal resistances can be identified by an appropriately designed 'critical experimental program' which takes advantage of the features of each of these transport mechanisms and their impact on sorption dynamics. If sorption lags the equilibrium moisture content, the sheet thickness and the external mass transfer coefficient should be varied in such a manner that their ratio is a constant. If the external resistance is the sole agent in controlling sorption dynamics, sorption will be dependent only on this ratio and independent of the individual values of L or $k_f$.

One may check if sheet's internal transport resistance can be attributed to diffusion within the fibers or the void space by changing sheet thickness while the external resistance is rendered insignificant. Under this condition, if the sheet's pore space diffusion resistance is negligible, dynamics would be independent of sheet thickness thus providing a 'critical test' of the hypothesis that it is only radial fiber diffusion that is important and not sheet transverse diffusion that is important.

Finally, when sheet internal resistance becomes important, the role of intra-fiber diffusion in the sheet transverse direction can be examined by steady state diffusion experiments under the same RH conditions. A strong dependence of the steady state moisture flux on the RH indicates intra-fiber diffusion providing yet another 'critical' test of the hypothesis.

The analysis described in this paper can be used to make judgements and estimates of the mechanical behavior of paper materials under transient humidity conditions. Changing moisture content results in altered physical and chemical properties of sheets. One example is the surface electrical resistivity of sheets which is known to be an exponential function of the local moisture content. An analysis of the type shown here allows us to predict the electrical property variation under transient humidity conditions. Another example of the applicability of this analysis is in the compressive strength of paper board. It is known that the compressive strength of paper board decreases sharply with changing moisture contents. An analysis based on the thermal and moisture transport effects can easily identify whether thermal and moisture transients are responsible for accelerated loss in compressive strengths of paper materials.

## REFERENCES

1. H. Markstrom, Lorentzen and Wettre, Stockholm, 1991.
2. R. E. Benson, Tappi J., 54, 4, (1971) 699.

3. E. L. Back, L. Salmen and G. Richardson. Svensk Papperstidning, 88, 6, (1983) 61.
4. V. L. Byrd, Tappi J., 55, 2, (1972) 247.
5. D. E. Gunderson, Tappi J., 64, 11, (1981) 67.
6. C. Soremark and C. Fellers. J. Pulp and Paper Sci., 19, 1, (1993) 19.
7. J. G. Anderson, Proc. IFAC Conference, 1, Brussels (1976) 195.
8. C. Fellers and T. L. Laufenberg, Proceedings of the conference on Moisture-Induced creep behavior of paper and board, STFI, Stockholm, Sweden (1994).
9. I. Chalmers, Proceedings of the third international symposium on Moisture-induced creep on paper, board and containers, Rotorua, NZ (1997).
10. A. Venkateswaran, Chem. Rev., 70, 6, (1970) 619.
11. J. D. Babbitt, Can. J. Res., NRC Can., 20 A, 9, (1942) 143.
12. H. M. Barnes, C. Skaar and P. Luner. J. Appl. Poly. Sci., 21 (1977) 1084.
13. W. T. Simpson, Wood and Fiber, 5, 1, (1973) 41.
14. C. van den Berg and S. Bruin, in: Water activity: Its influence on food quality, L. B. Rockland and G. F. Stewart (eds.), Academic Press, New York, (1981) 1.
15. D. G. Eagleton and J. A. Marcondes, Tappi J., 77, 7, (1994) 75.
16. S. G. Chatterjee, B. V. Ramarao and A. R. K. Eusufzai, Transient moisture transport in paper - A preliminary analysis of diffusion and adsorption, Empire State Paper Research Institute Report No. 102, Syracuse, New York (1995) 143.
17. J. M. Prahl, Thermodynamics of paper, fiber and water mixtures, Ph. D. Thesis, Harvard University, Cambridge MA (1968).
18. W. W. Barkas, The swelling of wood under stress: A discussion of its hygroscopic, elastic and plastic properties, His Majesty's Stationery Office, London, (1949).
19. C. O. Seborg and A. J. Stamm, Ind. Eng. Chem., (1931) 1271.
20. C. O. Seborg, F. A. Simmonds and P. K. Baird, Ind. Eng. Chem., 28, 11, (1936) 1245.
21. C. O. Seborg, Ind. Eng. Chem., 29, 2, (1937) 169.
22. C. O. Seborg, F. A. Simmonds and P. K. Baird, Paper Trade J., 107, (1938) 45.
23. A. R. Urquhart, in: Moisture in textiles, J. W. S. Hearle and R. H. Peters (eds.), Textile Book Publishers Inc., New York (1960).
24. A. Visintin, Differential models of hysteresis, Springer-Verlag, Berlin, (1991) 13.
25. D. H. Everett, in: The Solid-Gas Interface, E. A. Flood (ed.), Marcel-Dekker, New York NY, 2 (1967) 1055.
26. D. H. Everett and W. I. Whitton, Trans. Faraday Soc., 48 (1952) 749.
27. D. H. Everett and F. W. Smith, Trans. Faraday Soc., 50 (1953) 187.
28. D. H. Everett, Trans. Faraday Soc., 50 (1954) 1077.
29. D. H. Everett, Trans. Faraday Soc., 57-1557 (1955).

30. D. Pescetti, in: Models of hysteresis, A. Visintin (ed.), Pitman Research Notes in Mathematics Series, Longman Scientific & Technical, Harlow, Essex (UK), 286 (1993) 118.
31. B. V. Ramarao, S. G. Chatterjee, A. R. K. Eusufzai and C. Tien, Moisture transport and sorption by paper under cyclic variations in humidity. 195- 214 AMD-Vol. 209/MD-Vol. 60, Mechanics of Cellulosic Materials, ASME (1995).
32. S. G. Chatterjee, B. V. Ramarao and C. Tien, J. Pulp and Paper Sci., 23, 8, (1997) 366.
33. S. G. Chatterjee, and B. V. Ramarao, in: Proceedings of Third International Symposium on Moisture and creep effects on paper, board and containers, I. Chalmers (ed.), FRI, Rotorua NZ, (1997) 121.
34. B. V. Ramarao and S. G. Chatterjee, in: The Fundamentals of Papermaking Materials. Transactions of the Eleventh Fundamental Research Symposium, C. F. Baker (ed.), PIRA International , Leatherhead, England, (1997) 703.
35. G. N. Christensen and W. W. Barkas, Trans. Faraday Soc., (1955) 130.
36. J. E. Stone and A. M. Scallan, Tappi J., 50, 10, (1967) 496.
37. C. Tien, Adsorption calculations and modeling, Butterworth Publishers, Stoneham MA (1994).
38. K. Beck, H. Radhakrishnan, S. G. Chatterjee, B. V. Ramarao and H. Makkonen, Steady state diffusion of moisture through paper sheets, Empire State Paper Research Institute, Syracuse NY, Report 108, (1998) 65.
39. S. G. Chatterjee, Unpublished Results, (1998).
40. A. Bandyopadhyay, B. V. Ramarao and S. G. Chatterjee, Transient moisture sorption in paper materials, Submitted (1998).

## SYMBOLS

| | |
|---|---|
| A( C) | Dimensionless functional representation of diffusivity, given by eq.(40) |
| Bi | Biot number, = $k_f L/D_p$ |
| c | Water vapor concentration in void space inside sheet, $kgm^{-3}$ |
| $c_b$ | Water vapor concentration in bulk air, $kg/m^3$ |
| $c_0$ | Initial concentration of water vapor in air and void spaces in sheet |
| C | Dimensionless concentration, $C = (c-c_0)/c_0$ |
| $C_{GAB}$ | Constant in GAB isotherm, Values correspond to units for q and h |
| $D_{air}$ | Diffusivity of water vapor in air, $m^2/s$ |
| $D_p$ | Diffusivity of water vapor in void space of paper sheet, $m^2/s$ |
| $D_q$ | Diffusivity of bound water through fiber matrix, $m^2/s$ |
| F(C) | Dimensionless function representing isotherm |
| h | Relative humidity (fractional) |
| $h_0$ | Relative humidity of air in equilibrium with sheet initially |
| $h_b$ | Relative humidity of air during ramp change |
| $h_1$ | Relative humidity of air after change |
| $h_{12}$,$h_{21}$ | Relative humidity turning points on Figs. 2a through 2c. |

| | |
|---|---|
| $k_f$ | Mass transfer coefficient external to sheet, m/s |
| $k_i$ | Transport coefficient representing diffusion in fibers, 1/s |
| $K_{GAB}$ | Constant in GAB isotherm, value corresponds to units of q and h |
| L | Sheet thickness, m |
| m | Moisture density distribution function obtained from hysteresis loop |
| $m_0$ | Constant in GAB Isotherm |
| M | Equilibrium moisture content, $q_s$ |
| $M_{air}$ | Molecular weight of air, kg/kgmole |
| $\underline{n}$ | Unit vector |
| P | Dimensionless number $P = L^2/(t_R D_p)$ |
| P' | Dimensionless number $P' = L^2/(t_R D_q)$ |
| $p_{sat}$ | Saturation vapor pressure of water at temperature T, Pa |
| q | Average Moisture content of sheet at location z' (kg water/kg dry fiber) |
| $q_s$ | Equilibrium moisture content, fraction |
| $q^0$ | Reference moisture content for linearization of isotherm |
| Q | Normalized moisture content, $= (q - q_0)/q_0$ |
| $Q_s$ | Normalized equilibrium moisture content |
| R | Typical fiber radius, m |
| RH | Relative humidity, (fractional or percentage) |
| $R_g$ | Universal gas constant |
| t | Time, s |
| $t_R$ | Ramp time, s |
| $\underline{x}$ | Position vector |
| z | Sheet thickness dimension, m |
| $\alpha$ | $= \rho_p(dq/dc)$ |
| $\beta$ | $= k_i t_R$ |
| $\gamma$ | $= k_f t_R/L$ |
| $\varepsilon$ | Sheet void fraction |
| $\tau$ | Dimensionless time $= t/t_R$ |
| $\zeta$ | Dimensionless position $= z'/L$ |
| $\rho_p$ | Sheet density, Dry fiber mass/sheet volume $kg/m^3$ |

Adsorption and its Applications in Industry and Environmental Protection
Studies in Surface Science and Catalysis, Vol. 120
A. Dabrowski (Editor)



# Adsorption on aluminas – current applications

H. L. Fleming

Cochrane, Inc., 800 3rd. Avenue, King of Prussia, PA 19406, USA

Aluminas have been in use for many years as adsorbents. First introduced commercially in 1932 by Alcoa for water adsorption [1], activated aluminas have traditionally been known as desiccants for the chemical process industries. As early as 1901, references can be found for the use of synthetic aluminas in the chromatographic purification of biological compounds [2]. In recent years aluminas have found widespread usage in applications as diverse as municipal wastes, polymers, and pharmaceuticals. Novel design of these materials is extending their separations capability into even more nontraditional areas of adsorption technology.

This chapter surveys current applications of aluminas as adsorbents. Advances in materials design and adsorption process technologies are also discussed, relating recent trends in applications of adsorbent aluminas to developing specialty process requirements. The ability of these materials to selectively remove desired compounds from the remainder of the process stream has greatly accelerated their use in recent years.

## 1. HISTORY OF ADSORPTION WITH ALUMINAS

To place current events in their proper perspective, it is advantageous to briefly examine the chronological development of the use of aluminas in adsorption technology. The earliest known use of adsorbent alumina was in the chromatographic separation of liver extracts by Folkers and Shovel in 1901 [2]. Beginning in the 1930s, when laboratory-scale chromatography became a common analytical technique, aluminas were the packing material of choice [3,4]. Chromatography using aluminas continued to develop through the late 1950s until celluloses and dextran materials were synthesized and were perceived as having greater versatility in separations [5,6]. Continued research in alumina based media was virtually halted and this group of materials has been largely ignored since that time. There were many successful analytical separations with aluminas, however, which can be found in surveys on small-scale chromatography [6].

Table 1
Commercial alumina adsorption applications

| | Molecular Weight | Gas Phase Impurity | Gas Phase Product | Liquid Phase Impurity | Liquid Phase Product |
|---|---|---|---|---|---|
| I. Pre 1940 | Low | Water | Natural Gas | | |
| | | | Ammonia | | |
| II. 1960s | Low | Water | Natural Gas | Water | Polyethylene |
| | | | | | Aromatics |
| | | Isotopes | Isotopes | Halogens | |
| | | | | | Color bodies |
| | High | | | | Organometallics |
| III. 1980s | | Water | Natural Gas | Water | Aromatics |
| | | Isotopes | Isotopes | Halocarbons | |
| | Low | $SO_2$ | Flavors | Halogens | |
| | | $H_2S$ | Fragrances | | Phosphates |
| | | Acid gases | Xylenes | Inorganic acids | |
| | | Amines | Aromatics | Ketones | |
| | | | | | Carboxylic acids |
| | | | | | Mercaptans |
| | | | | | Alcohols |
| | | | | | Sulfides |
| | | | | | Metals Metals |
| | | | | | Color bodies |
| | High | | | | Organometallics |
| | | | | | Pharmaceutical |
| | | | | | $>C_5$ paraffins |
| | | | | Sugars | |
| | | | | Vitamins | |

As shown by Table 1, the only large-scale application of adsorbent aluminas prior to 1940 was as desiccants. Both air and natural gas dehydration were being routinely performed with waste-heat regeneration of the alumina columns. In later years, dehydration with aluminas was extended to cracked gas and heavier hydrocarbon streams, as well as those containing carbon dioxide and ammonia.

Liquid dehydration of aromatic and paraffinic hydrocarbons, halogenated hydrocarbons, and gasoline- among others- became common in the 1960s. Even today, drying applications are being extended to include cold bed adsorption, more complex liquid dehydration, and pressure swing operation [7].

Use of aluminas was extended in the 1950s and 1960s to isotopic separation of a number of the actinide series compounds, as well as some specialty organics. In the 1960s, additional uses were the impurity separation of halogenated compounds from aqueous streams and organic acids from hydrocarbons. Degenerative carboxylic acids from lubricating oils is an example of such a separation.

In recent years the variety and complexity of applications have become extensive. On a volume basis, desiccant usage now accounts for less than one-half of the total adsorbent alumina consumption each year in North America (Fig. 1). Section III of Table 1 is a representative listing of current large-scale generic applications. With the advent of more stringent conservation requirements in the early 1970s, removal of phosphate, mercaptan, and fluoride compounds in ground and drinking waters became successful applications of alumina adsorbents [8,9]. The passage of the Clean Air and Clean Water acts in the late 60s and early 70s led to several large-scale alumina adsorption processes in aqueous streams.

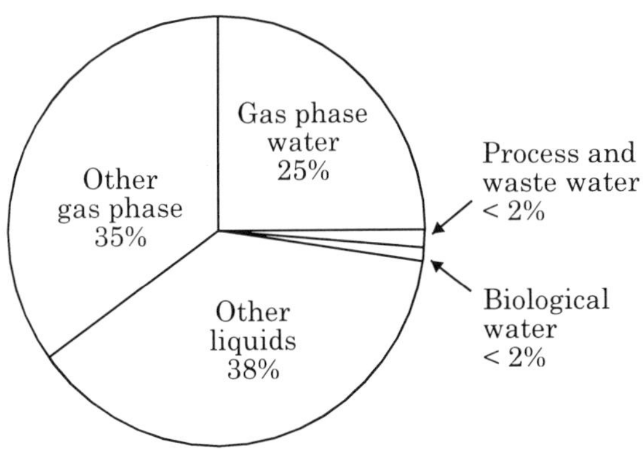


Figure 1. Commercial applications of adsorption with aluminas in North America (volume of aluminas).

Environmental concerns also established new applications for aluminas in gas treatment. Although not of the magnitude of air purification with activated carbon, adsorption of acid gases and sulfur species in petroleum-related industries constituted a significant portion of adsorbent alumina utilization in the 1970s. These areas have received even greater attention in the 1980s and are included in

Table 1. Unfortunately, it is a fact of almost every industry using aluminas that the application is proprietary. The examples presented here are a few representatives of the broad range of applications.

A second major driving force for the use of alumina-based chemicals as selective adsorbents has been the cost of energy. In the petrochemical industry, for example, processes which have traditionally used distillation are employing adsorption as the primary operation for separation and purification. Figure 2 illustrates this trend. The purification of ethylene dichloride, a commodity feed material for the production of polyvinyl chloride, is produced by a process utilizing a selective alumina-based adsorbent. Trace chloride and iron contaminants are effectively removed to undetectable limits from the liquid without the need to change the process temperature nor volatilize any of the stream to effect the separation.

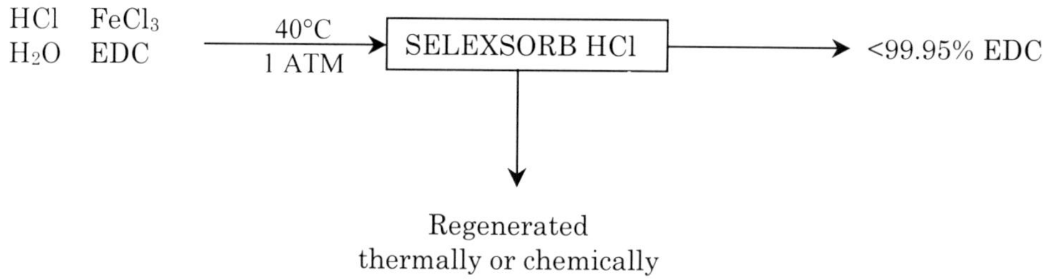


Figure 2. Ethylene dichloride (EDC) cleanup process.

## 2. ALUMINA STRUCTURE AND SURFACE CHEMISTRY

Considering the longevity of alumina usage in the catalysis and adsorption industries, it is surprising to note how many misconceptions still exist concerning its physical-chemical properties. Because the surface structure of an adsorbent essentially determines its adsorptive characteristics, an understanding of the surface chemistry of aluminum oxides is necessary to comprehend selective adsorption properties.

Alumina is a commonly used term for aluminum oxides and hydroxides, which exist as at least five thermodynamically stable phases and many more meta-stable transition forms. Other chapters include discussions of the physical forms of these materials and their methods of production. With the exception of some alpha aluminas, all forms possess surface hydroxyls which can have a certain degree of activity in adsorption. The treatise by Wefers and Bell is considered to be a good review of aluminum oxide phase chemistry [10].

The crystallographically stable oxides and hydroxides are not usually considered useful as adsorbents since they have little porosity and low surface area. Some of these aluminas are useful in water treatment and are reviewed in a later section. Because the primary mechanisms of separation are usually quite different, this discussion is limited to the use of aluminas as adsorbents in all applications other than water.

Transition forms of alumina constitute the largest group employed as adsorbents. Usually formed via thermal decomposition from the hydroxides, the loss of hydroxyls results in defect structures in the aluminum and oxygen lattice. Depending on the particular transition form, the $Al^{3+}$ and $O^{2-}$ ions can be either tetrahedrally or octahedrally coordinated in cubic or hexagonal close-packed systems. For example, gamma alumina, the best-known form, has a defect spinel structure in which the oxygens are in cubic close packing. The oxygen sub-lattice is fairly well ordered, although there is significant disorder in the tetrahedral aluminum sub-lattice.

As an example of the complexity of the alumina system, eta alumina, a transition form with a spinel structure similar to that of gamma alumina, has a strong one-dimensional disorder of the cubic close spaced lattice. This distortion results in a greater concentration of surface acid sites (and strengths) in eta than in gamma alumina. This is an important consideration in adsorption systems, such as water and amines, which require these functionalities.

Depending on the synthetic methodology, the initial form in producing transition aluminas can be amorphous (or at least indifferent, as determined by X-ray diffraction). In fact, a number of commercial products that are sold as "gamma alumina" are amorphous. It is also generally recognized that a large portion of the catalysis and adsorption literature using "gamma alumina" that was not well characterized is actually data on an amorphous material, or at least poorly formed gamma. As a rule, amorphous transition aluminas (as distinguished from amorphous hydroxide aluminas) have greater concentrations of defect structures, i.e., greater chemical activity than the more crystalline transition aluminas[11]. In the amorphous to crystalline transition, octahedral $Al^{3+}$ must reorder to tetrahedral with considerable rearrangement of the oxygen lattice. Large numbers of defect structures can be envisioned during this process, which result in active sites for adsorption. Most commercial transition materials fall somewhere between amorphous and more structured.

Gelatinous aluminas comprise an additional group of materials which can be important as adsorbents. As a group, these materials possess only short range lattice order and can be prepared in numerous ways, ranging from precipitation to hydrolysis reactions. Their extremely high surface areas and chemical activity make alumina gels of continued interest in catalysis and adsorption.

The poorly crystallized pseudoboehmite, for instance, exhibits good adsorptive properties for phosphates and arsenates in aqueous streams.

Another group of materials which exhibits adsorptive properties is the beta aluminas. These consist of alkali- substituted aluminates and their related compounds. The best known is beta sodium aluminate. Dawsonite ($Na_2O$-$Al_2O_3$-$2CO_2$-$2H_2O$) is a related compound which, with beta sodium aluminate, is the principal component of the so-called "alkalized alumina" process for desulfurizing flue gas[12]. Dawsonite can be synthesized as both crystalline and gelatinous and appears to have interesting chemical activity. Neither Dawsonite nor the beta aluminas have received much attention for their adsorptive properties.

To be complete, alpha alumina should be included in this list. Although not normally considered reactive because of almost total dehydroxylation, low surface area, and virtually no porosity, alpha alumina exhibits greater Lewis acidity per unit surface area than the transition aluminas and can be shown to have strong adsorptive properties through this acidity. For example, halides, water, and inorganic acids can be adsorbed well on alpha alumina [13].

The chemical nature of the active sites responsible for adsorption phenomena on aluminas is still not well understood. Defect structures formed via surface dehydroxylation result in localized regions of adsorptive and catalytic activity. The activity cannot be completely attributed to anion vacancies. Depending on the particular alumina, various concentrations of at least five distinct sites have been identified. They are sometimes listed as the A, B, E, I, and X sites, following the catalysis model [14], and roughly correspond to distinct surface hydroxyls. Depending on their synthesis, geometric orientation, and concentration, they can possess differing degrees of both acidic and basic character in adsorption. Although not complete, a good discussion of the hydroxyl properties can be found in the papers by Peri [15,16] and the review by Knozinger and Ratnasamy [14].

For adsorption, active aluminas can be considered as possessing both Lewis and Brønsted acidic and basic sites of various strengths and concentrations. Acidity is contributed by lattice $Al^{3+}$ ions, protonated hydroxyls, and some acidic hydroxyls. Basicity is a result of $O^{2-}$ anion vacancies and basic hydroxyls. A useful schematic representing the various surface functionalities is shown in Figure 3 [15].

Because of the variation of alumina surface properties with hydroxyl concentration and type, adsorption behavior can be modified in numerous ways on the materials. A relatively easy method, for example, is thermal dehydroxylation. As shown in Figure 4 for two levels of hydroxyl content, the acidic and basic site concentrations can be made quite different. This trend is summarized in Figure 5 for transition aluminas as a function of final activation temperature in the synthesis process. Higher temperatures tend to eliminate surface hydroxyls, which increase the number of Lewis acid sites, thus increasing total acid sites on a surface area basis.

A complicating factor in the design of alumina surface chemistry for adsorption is the tremendous complexity of the surface. A wide variety of distinct sites exist

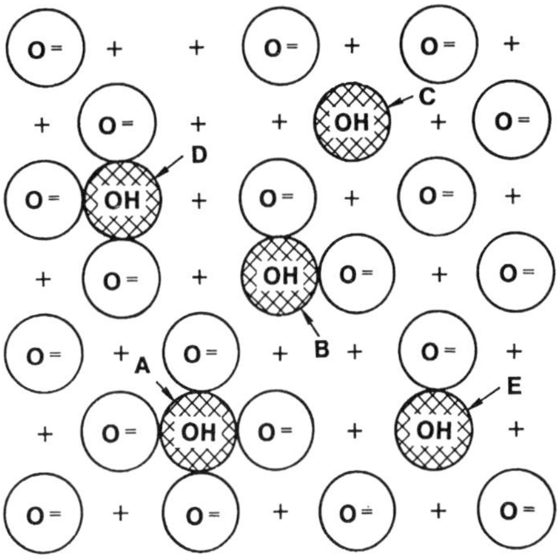


Figure 3. Transition aluminas surface functionalities (taken from Ref. [15] under permission).

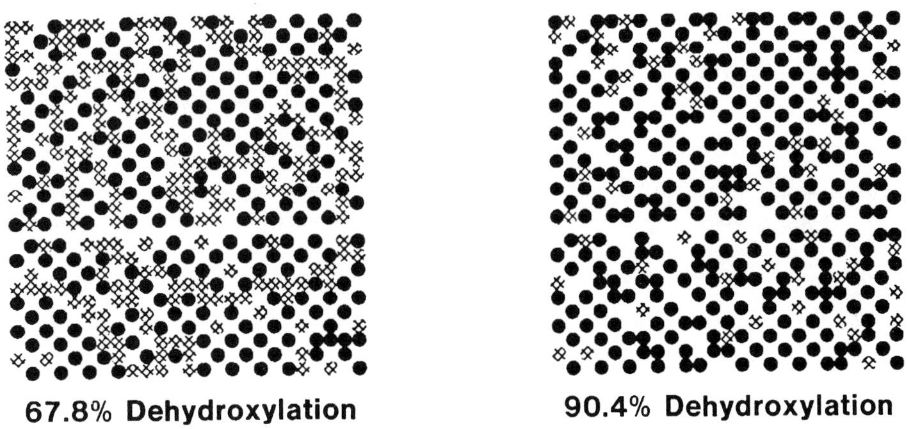


Figure 4. Transition alumina dehydroxylation (taken from Ref. [15] under permission).

simultaneously. Termed surface heterogeneity, this distribution of sites with different energies on a surface typically inhibits selectivity and makes specific design of a material for a particular application more difficult. There will always be, for example, some residual hydroxyls which act as basic sites, rather than the bulk of the hydroxyls which are acidic, ready to donate their protons in the attached hydrogen atoms. This phenomenon is demonstrated in Fig. 6, where the site-energy distributions of various aluminas for gas phase ammonia adsorption are given.

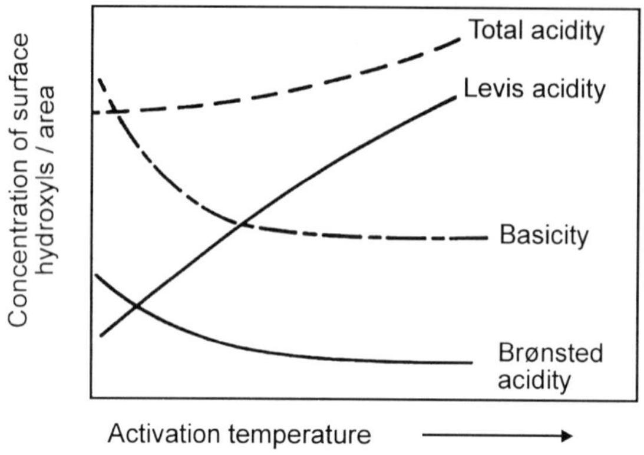


Figure 5. Transition aluminas (acid – based modifications).

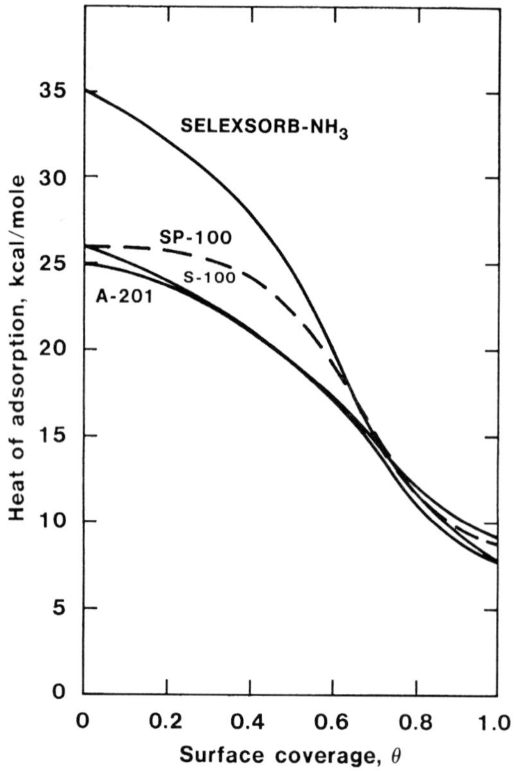


Figure 6. Site – energy distributions for $NH_3$ adsorbed on various commercial aluminas.

As measured by the heats of interaction, it is seen that the surface of each of these commercial alumina-based materials is quite different, and each is heterogeneous in its interaction with ammonia. There are some sites which bind very strongly (>20 kcal/mol), while others only physically interact with Van der Waals forces (<10 kcal/mol). Both SP-100 and SELEXSORB-$NH_3$ are aluminas whose surfaces have been chemically modified.

In addition to surface effects, there are other ramifications of changing activation temperature in design of the adsorbent. As seen in Fig. 7, total surface area is a sensitive function of temperature, as well as pore distribution and total pore volume. In gas phase separation of small molecules with fast pore transport, surface area is critical for good uptake capacities. Viscous liquids such as large hydrocarbons, on the other hand, tend to require more macroporosity, i.e., inherently lower surface areas, with well designed surface chemistries. Optimum selection of an alumina will depend on the relative importance of these factors to the end user.

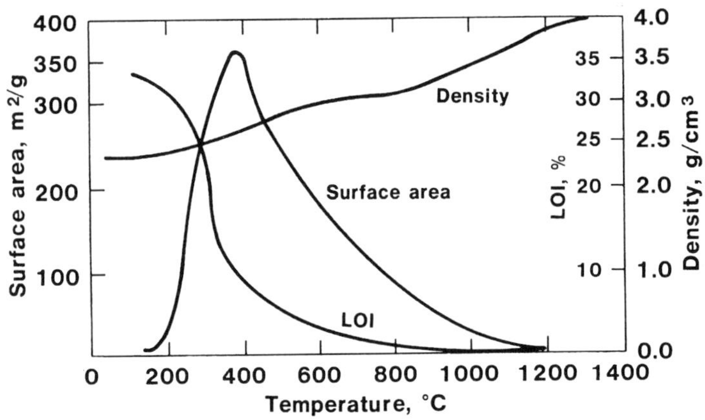


Figure 7. Physical properties during thermal activation (LOI = loss on ignition) (taken from Ref. [10] under permission).

Advances in pore structure control of the porous active aluminas have resulted in major improvements in commercial adsorbents. Zeolites have their pore structures determined simultaneously with the precipitation process and are constrained in size by the configuration of the sodalite cage. In contrast, active alumina porosity is relatively independent of the bulk phase formation process and is usually engineered following material synthesis. Microporosity is controlled via kinetics of the dehydroxylation process, whereas macroporosity is usually developed in the agglomeration process.

A pre-eminent theme in current adsorption requirements is the move toward greater specificity of separation with more complex streams. In the past there was a tendency to classify aluminas as nonspecific with limited adsorption capability. More recently, development efforts have begun to generate materials of a better characterized, highly specific nature. Several transition alumina forms are now available in commercial quantities for particular applications. At last count, they included gamma, chi, and eta, as well as amorphous. Alcoa's CP and CPN grades are examples of amorphous and transition forms, respectively. Alcoa CPN is a CP grade (activated powder) which has undergone an additional thermal activation. The result is a more crystalline structure, and a material that is much less likely to rehydrate in the presence of water (thus, the N in CPN represents Nonrehydratable).

Because of the influence of impurities on adsorption performance, greater attention is being given to chemical purity. Good examples are the ultrapure pseudoboehmites Catapal (Vista Chemical Company, Houston, TX) and Dispural (Condea, Martinswerk, Germany). Most manufacturers now offer varying grades of chemical purity for differing adsorption requirements. There are two reasons for this trend. First, more aluminas are being produced that are not products of a Bayer stream. Consequently, they are not constrained by the purity limitations of the Bayer liquor. Catapal and Dispural are prominent examples, as is fumed alumina. The second factor in improved chemical purity is the increased process control now being associated with Bayer precipitation. Greater care is being taken to achieve acceptable chemical purities for the technical aluminas.

Increased understanding of these phenomena, in conjunction with greater control of initial particle size, has allowed engineering of pore size distributions and degree of porosity. Representative pore distributions of various commercial transition aluminas are given in Figure 8. They cover the range of 3 to 100000 nm (30 to 1000000 Å) in bi-, tri-, and monomodal distributions. Total pore volumes ranging from 0.3 to 0.8 $cm^3/g$ are not uncommon, with some experimental materials having porosity up to 1 $cm^3/g$.

Currently, more manufacturers are tailoring pore structures to meet specific process requirements than was the case in the past, where only commodity materials were available. Within practical constraints, the technology is now in place for generation of both microporous and macroporous aluminas in any derived structural form. In addition, there is little question that the future in adsorbent alumina manufacture is in the direction of greater control of pore structures, particularly for liquid phase adsorption of deactivating species such as organometallics. Controlled pore structures can also be beneficial for more efficient regeneration of difficult adsorbates. A primary example is deposition of elemental sulfur, where the frequency of cyclic regeneration can be controlled by strong pore diffusion resistances.

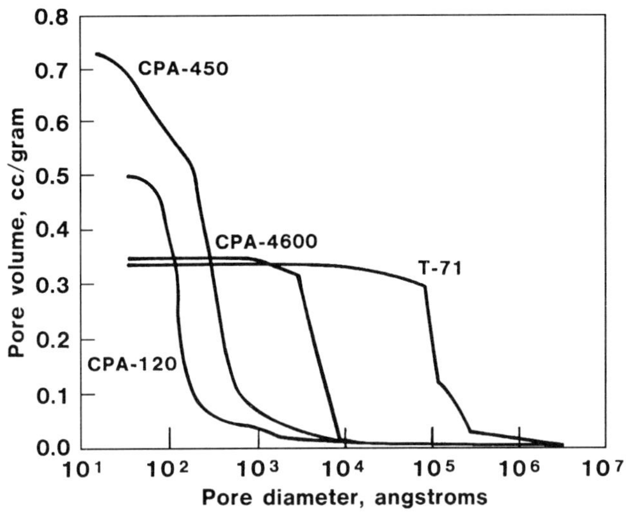


Figure 8. Hg pore distribution for various commercial aluminas.

Modification of alumina surface chemistry to enhance selective adsorption of particular compounds has become more prevalent in recent years. Active aluminas are multifunctional materials with ratios of surface sites. Engineering the alumina to contain advantageous surface functionalities while reducing undesired sites is fast becoming a science and is a powerful tool in the design of selective adsorption units.

The most common approach to surface modification with transition aluminas is through control of the thermal treatment. As shown in Figures 5 and 7, it is quite easy to alter the total surface Lewis and Brønsted acidity and basicity via control of the hydroxyl concentration. To some extent this technique has been used since the early 1950s, particularly in chromatographic aluminas. The so-called Brockmann scale for rating aluminas is based on thermal history, and commercial materials are available today in the various Brockmann grades. Only recently has the concept been applied to large scale adsorption systems.

A more complex approach to enhancement of surface properties is the addition of chemical modifiers which increase the concentration of the existing desired functionalities. Adsorption of acids or bases which alter the concentration and reactivity of alumina surface hydroxyls is a good example.

Although the basic technology has existed for over 20 years and is used routinely in catalysis, adsorbent aluminas have been noticeably neglected and are only today coming on stream in commercial systems. However, most of the applications are proprietary and the materials have been designed with the particular user in mind.

A simple example of chemical modification of transition aluminas is shown in Figure 9 for the addition of alkali cations. It is noted that total surface basicity is increased as alkali is impregnated on the alumina surface. It is also of interest to note that both Lewis and Brønsted acidity decrease, but only after an increase at low alkali concentration.

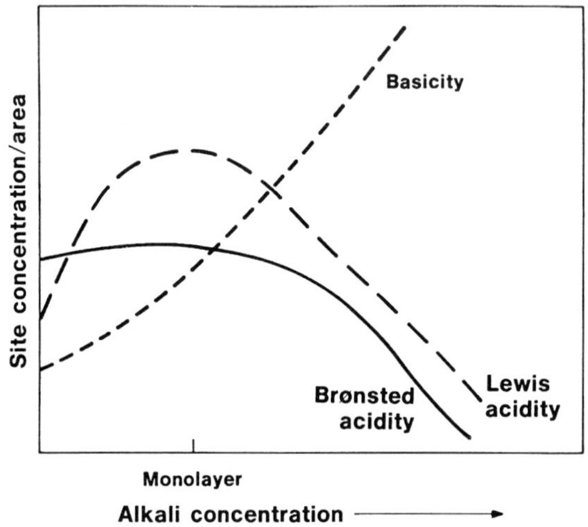


Figure 9. Modification of surface acidity/basicity by alkali addition.

Because the variety of chemical functional groups that can be adsorbed on aluminas is so diverse, a third group of surface modifications is becoming more popular. They involve chemical treatment of the alumina surface in order to introduce new functionalities which can enhance adsorption. An example of such a modification is given in Figure 10. A transition alumina surface is silanized through a condensation reaction with a commercial triethoxysilane. The synthesized surface exhibits quite different properties than the base alumina and can be used for applications such as reversed phase chromatography.

Although not modified to increased adsorption capacity, the first commercial example was the cobalt chloride-impregnated transition alumina marketed by Alcoa in the 1950s for water adsorption. Calcium, sodium, and potassium salts have also been used commercially for various reasons. The most recent published example is the modified alumina used to separate mercury from natural gas streams in the Arabian oil countries. A related example is the permanganate impregnation of a

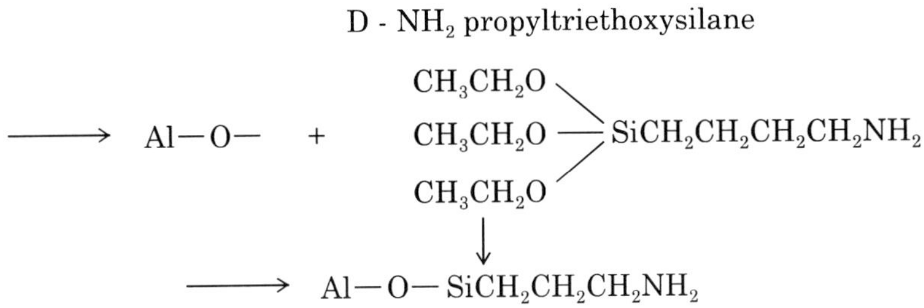


Figure 10. Silanization of alumina surface.

transition alumina for simultaneous adsorption and oxidation of selected air pollutants. Most of the material modifications have been done for specific proprietary applications and are not available for public disclosure. Aluminas tend to adsorb species via chemisorption, a form of chemical reaction. As such, the affinity is usually quite strong, with large heats of uptake, as well as desorption. Because of this property, aluminas have a distinct advantage over other materials, such as carbons and ion exchange resins, that operate by alternative mechanisms in certain applications. First, they are particularly good at removing trace quantities of substances from a bulk stream. This is illustrated in Figure 11. There is a very favorable isotherm, or affinity, for species that to adsorb at low concentrations.

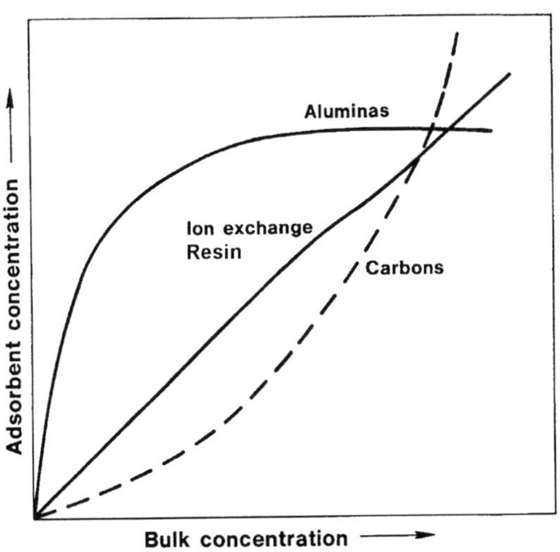


Figure 11. Adsorption isoterms on various materials.

Other materials become competitive at higher concentrations (typically >1000 ppm). Second, aluminas remove components of streams very selectively. When it is necessary to either purify or remove a product from a stream with multiple components, aluminas are generally applicable. A general outline of their affinity for species, as defined by functional group, is shown in Figure 12 for a number of common chemicals. Although difficult to generalize, this may represent a rough consensus of correct thought regarding chemisorption of these species.

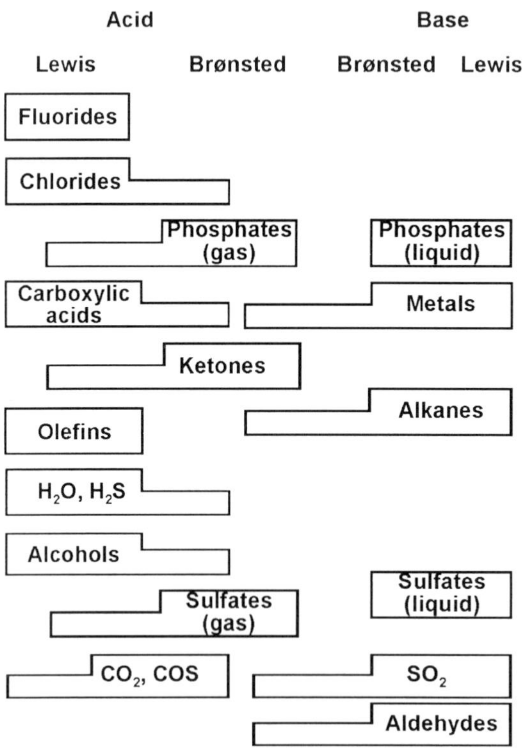


Figure 12. Adsorbate specifity on transition aluminas.

Although many methods of contacting the alumina adsorbent with the multicomponent stream of interest are possible, most processes today still use packed beds. A packed bed offers high efficiency of separation per volume, is easy to control, and has low capital and operating costs. Typical operation is outlined in Figure 13. A gas, liquid, or multiphase stream passes through the bed, with the component(s) of interest being removed (Y denoting gas phase concentration). A mass transfer zone of the adsorbate remaining in the fluid phase is established (theta denoting time on stream) which moves through the bed until it breaks through the end and is detected in the effluent. The bed is then considered spent and is either regenerated or discarded.

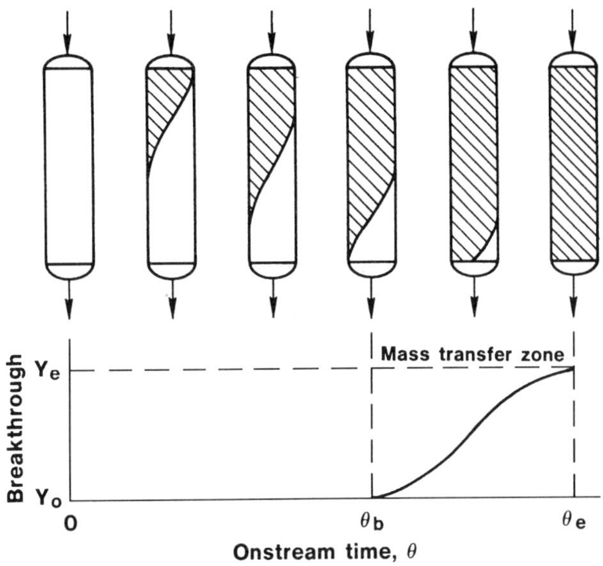


Figure 13. Adsorption dynamics.

Two general techniques are used for regenerating beds. When possible, thermal desorption via an inert gas at a high temperature is used to strip the adsorbate from the alumina for reuse. Typical temperatures for regeneration range from 150° to 400°C, depending on the affinity of the adsorbate for the alumina surface. The concentrated adsorbate can be either reused or vented. Chemical energy can be used instead of thermal energy in certain instances. For example, a liquid can be contacted with the bed which displaces the adsorbate and serves the same purpose as the hot gas in regenerating the bed. In either case, the alumina remains intact and is immediately reusable. These techniques are generally applicable in all applications, except aqueous streams.

## 3. CURRENT COMMERCIAL APPLICATIONS

Because of the unique versatility of aluminas, and a greater current understanding of their properties, these materials have moved into industries not formerly considered as applicable adsorption opportunities. In the past, petrochemical applications of alumina adsorption accounted for virtually all of the adsorbent alumina market. Today the list encompasses specialty chemicals, polymers, metals, waste management, and biological processing, among others.

An exhaustive compilation regarding the use of aluminas for chemical separations would be quite an undertaking. As shown in Figure 12, the broad range of surface affinities allows these materials to be used to remove a wide range of chemical functionalities. Adsorption of $C_1$ to $C_5$ alcohols on aluminas in gas and liquid phase is common [17,18]. Selective removal of alcohols and thiols and their interaction has been examined [19].

Sulfides and thiols can be separated [20,21] There are references to the removal of substituted phenols from various mediums and phases [22,23]. The selectivity of aluminas for different species from streams containing alcohols, carboxylic acids, and amines has been reported [24,25]. Selectivity has been shown for carboxylic acids from phenols, and vice versa [26,28].

Polyaromatic compounds, such as porphyrins and naphthalenes, are being removed from heavy oils [29]. Large molecule nitrogen-containing hydrocarbon species in oils are being considered in processes similar to that of the porphyrins [30]. Even small hydrocarbons, both saturated and olefinic, are being removed in both gas and liquid phases [31,32]. Acid and organic chlorides are routinely removed from streams as diverse as inert gases to heavy hydrocarbon liquids [33,34]. Species as diverse as cyanide [35] and hydrogen [36–38] can be adsorbed to some extent on aluminas.

Most applications of aluminas are proprietary and are generally not contained in the preceeding list. One get a sense of the trends in the use of these materials from a scattered sampling. A few selected cases are offered to illustrate current trends.

Removal of carbonyl sulfide from propylene, a key step in the production of polypropylene, can be accomplished via adsorption on aluminas. Figure 14 features some representative data on COS removal with three aluminas which are quite different, both chemically and physically. SELEXSORB-COS is a material produced by Alcoa with specific affinity to carbonyl sulfide.

Similar needs for purification are found in petroleum processing, where synthetic hydrocarbons are produced. Hydrochloric acid is a typical impurity which affects downstream catalytic processing. As shown in Figure 15, aluminas have the potential to handle this separation problem. SELEXSORB-HCl is a chemically modified alumina used commercially in these types of applications.

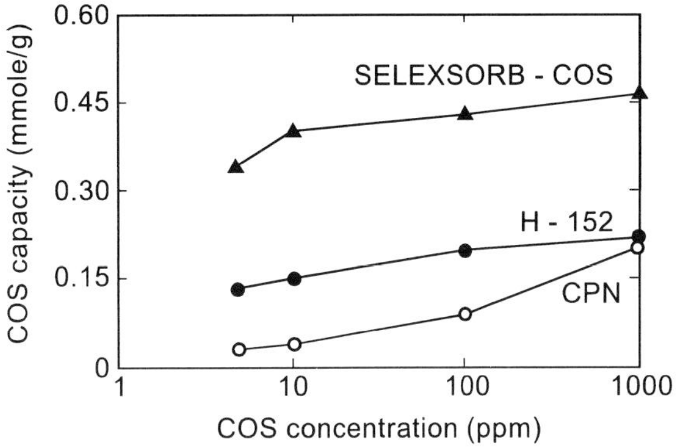


Figure 14. COS-adsorption isotherms for various alumina-based adsorbents (25°C, total pressure =20 PSIA).

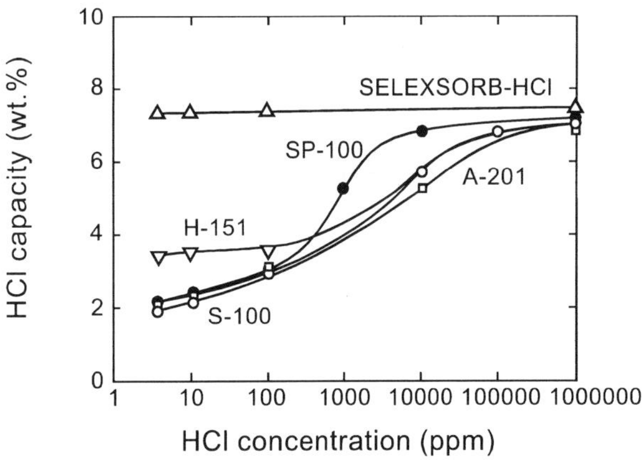


Figure 15. HCl adsorption isotherms on various alumina (25°C, 1 atm, dynamic excess flow).

Selective removal of carbon dioxide from water, as well as the converse, is typical in petrogas drying. These two species are mutually compatible in their adsorptive behavior, in that selectivity can be obtained by modification of surface Lewis acidity. This is represented in Figure 16 for a series of commercial aluminas and depicts the power of selectivity available to the design engineer.

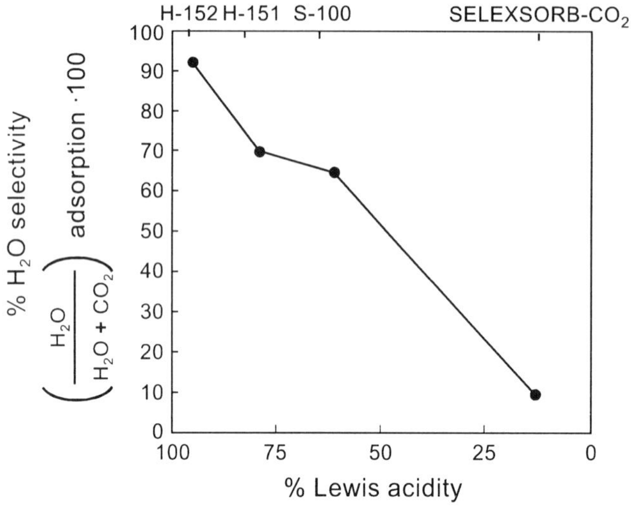


Figure 16. $CO_2/H_2O$ selective adsorption for series of commercial aluminas (25°C, 1 atm, dynamic flow, 50 ppm each).

An example of selectivity in the liquid phase is given in Figure 17. Silver and copper in water can be recovered individually by the choice of the correct alumina, as is commercially practiced in the catalyst manufacturing industry. Acidic and basic CPN simply refers to transition aluminas which have been chemically

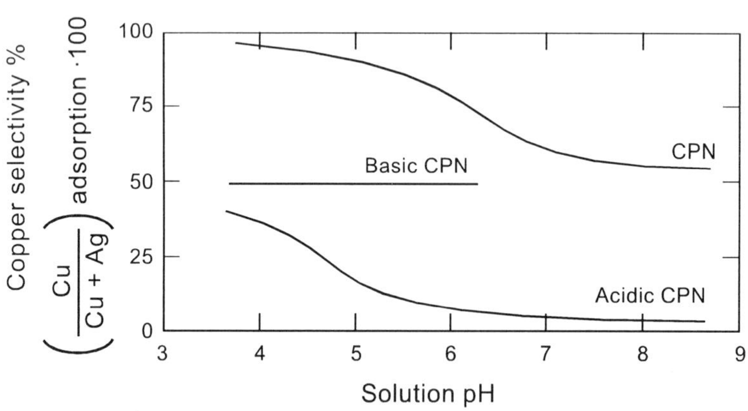


Figure 17. $Cu^{2+}/Ag^{+}$ adsorption selectivity.

modified so that their zero point of charge (ZPC) falls on positive and negative sides of the solution pH, respectively.

An additional field in which alumina adsorption technology has made inroads recently is biological processing. Because of the ability to modify surface characteristics, define pore structures, and exhibit good physical and chemical stability with consistency, aluminas are finding successful application in commercial processing of biological materials. A representative list of demonstrated bulk separations is given in Table 2. Although systems of pharmaceutical interest comprise the largest portion of the applications, a broad spectrum of biomolecules is included which encompasses foods, agriculture, medical, and specialty chemicals in addition to pharmaceuticals. Some of the larger volume commercial separations, all of which are chromatographic, include steroids, proteins and vitamins.

Table 2
Representative biological chromatographic purifications using aluminas

| Separation | Technique | Reference |
|---|---|---|
| Basic drugs | Ion Exchange Chromatography | 39, 40 |
| Hypnotic antiepileptic drugs | Liquid Chromatography | 41 |
| Antipyretic drugs/aspirin | Liquid Chromatography | 42 |
| Alkaloids | Thin-Layer Chromatography | 39, 43 |
| | High Performance Liquid Chromatography | |
| Proteins | High Performance Liquid Chromatography | 44 |
| | Reverse Phase Liquid Chromatography | |
| Catocholamines | Reverse Phase Liquid Chromatography | 44, 45 |
| | High Performance Liquid Chromatography | |
| Triglycerides/lipids/oils | Liquid Chromatography | 46, 47 |
| Estrogens | Liquid Chromatography | 48, 49 |
| Testosterones | Thin-Layer Chromatography | 32 |
| Steroids | High Performance Liquid Chromatography | 48, 50 |
| Cyclic nucleotides | Liquid Chromatography | 52 |
| Peptides | High Performance Liquid Chromatography | 49 |
| Mucopolysaccharides/ heparin | Liquid Chromatography | 53 |
| Phospholipids | High Performance Liquid Chromatography | 54 |
| Streptomycins | Liquid Chromatography | 55 |
| Glucoamylases | Adsorption Chromatography | 55 |
| Quinones | Liquid Chromatography | 56, 57 |
| Aminohydrolases | High Performance Liquid Chromatography | 58 |
| Fatty acid methyl esters | Thin-Layer Chromatography | 59 |
| Staphylocci | Liquid Chromatography | |

Aluminas have found applications in virtually every type of commercial chromatographic process. Adsorption chromatography in aqueous systems comprises a large portion of alumina usage because of the specific affinity of aluminas for a variety of chemical functionalities. Ion exchange chromatography is possible, particularly at high pH. The purification of proteins is an obvious example and is demonstrated in Figure 18.

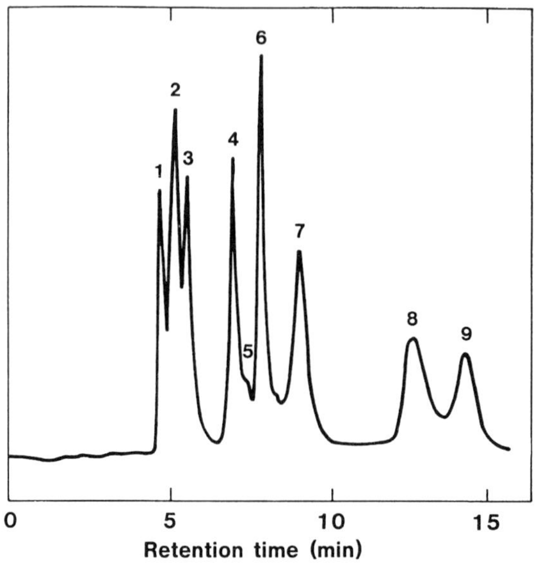


Figure 18. Chromatographic separation of standard proteins on Alcoa CG-20.
Mobile phase: 0.25 M $Na_2PO_4$ (pH 9).
Solutes: 1 and 2 – bovine albumin; 3 – ovalbumin; 4 – myoglobin; 5 – unknown; 6 – arginine vasopressin; 7 – trypsinogen; 8 – lysozyme; 9 – chymotrypsinogen.

Depolarization of active alumina surfaces and modification of the Brockmann number by many techniques may make their use possible in high performance liquid chromatography and reversed phase chromatography, respectively. Molecular exclusion chromatography has been demonstrated on aluminas with large concentrations of macropores. There is even some evidence to indicate that affinity chromatography for very specific attachment of biological functionalities may be successful via covalent ligand addition to particular alumina structures.

As the entire field of biotechnology continues to expand, with its associated requirements for greater separations processing, more specific aluminas with tailored properties are expected to be developed which meet specialty demands. It is interesting to note that the whole area of biological separations seems to have made a complete circle in its direction. For some of the same reasons that aluminas were first investigated and used for analytical purifications in the original work on aluminas over 40 years ago- namely physical stability and strong affinity for a wide variety of biomolecules- they are being rediscovered and further refined in the context of present day needs.

Chromatographic adsorption with aluminas is gaining acceptance as a viable kinetic separation alternative to molecular sieve zeolites and carbons. Although traditionally associated with biological systems, large scale chromatography is rapidly finding application in various industries, including nuclear and specialty chemicals. In fact, one of the largest chromatographic columns in commercial practice uses aluminas to produce various flavors and fragrances. In another industry, separations of isomers, such as meta- and ortho-xylene, which are impossible by distillation, are easily accomplished using high affinity chromatographic aluminas.

Aluminas have traditionally been considered to be limited to impurity separations of the minor stream components. With the advent of successful, continuous large scale chromatographic technology, numerous bulk separations have become practical which would not have been considered in the past. The same properties of commodity aluminas which were considered detrimental- e.g., strong affinity for a wide range of components- usually advantageous for alumina chromatographic processing. An example of the wide range of separation possibilities is included in Table 3. This compilation of the elutropic series of solvents on a transition alumina is illustrative of the considerable versatility potentially available for chromatographic processing.

As successful large scale chromatographic technology continues to be demonstrated, the trend toward use of aluminas in greater numbers of bulk, as well as impurity, separations is expected to increase. The simulated moving bid process developed by UOP (U.S.-based engineering firm), the U.S. Department of Energy-Oak Ridge annular rotating, multiple extraction system, as well as other designs for continuous chromatography, will provide additional options in commercial separations other than those that exist today. Through reviews on these and other large scale chromatographic technologies are given by Sussman [60] and Rendell [61].

Table 3
Elutropic series of common solvents on γ alumina (temperature = 25°C)

| Solvent | Snyder Elution Power Function E°($Al_2O_3$) | Viscosity (mPa·s) |
|---|---|---|
| Fluoroalkanes | -0.25 | |
| n-pentane | 0.00 | 0.23 |
| Hexane | | 0.00 |
| Iso-octane | 0.01 | |
| Petroleum ether, skellysolve B, etc. | 0.01 | 0.3L |
| n-decane | 0.04 | 0.92 |
| Cyclohexane | 0.04 | 1.00 |
| Cyclopentane | 0.05 | 0.47 |
| Di-isobutylene | 0.06 | |
| L-pentene | 0.08 | |
| Carbon disulfide | 0.15 | 0.37 |
| Carbon tetrachloride | 0.18 | 0.97 |
| Amyl chloride | 0.26 | 0.43 |
| Butyl chloride | 0.26 | |
| Xylene | 0.26 | 0.62 - 0.81 |
| L-propyl ether | 0.28 | 0.37 |
| L-propyl chloride | 0.29 | 0.33 |
| Toluene | 0.29 | 0.39 |
| n-propyl chloride | 0.30 | 0.35 |
| Chlorobenzene | 0.30 | 0.80 |
| Benzene | 0.32 | 0.65 |
| Ethyl bromide | 0.30 | |
| Ethyl ether | 0.38 | 0.23 |
| Ethyl sulfide | 0.38 | 0.45 |
| Chloroform | 0.40 | 0.57 |
| Methylene chloride | 0.42 | 0.44 |
| Methyl-l-butylketone | 0.43 | |
| Tetrahydrofurane | 0.45 | |
| Ethylene dichloride | 0.49 | 0.79 |
| Methyl ethyl ketone | 0.51 | |
| L-nitropropane | 0.53 | |
| Acetone | 0.56 | 0.32 |
| Dioxane | 0.56 | 1.54 |
| Ethyl acetate | 0.58 | 0.45 |
| Methyl acetate | 0.60 | 0.37 |

Table 3 (continued)
Elutropic series of common solvents on γ alumina (temperature = 25°C)

| Solvent | Snyder Elution Power Function E°($Al_2O_3$) | Viscosity (mPa·s) |
|---|---|---|
| Alyl alcohol | 0.61 | 4.1 |
| Dimethyl sulfoxide | 0.62 | 2.24 |
| Aniline | 0.62 | 4.4 |
| Diethyl amine | 0.63 | 0.38 |
| Nitromethane | 0.64 | 0.67 |
| Acetonitrile | 0.65 | 0.37 |
| Pyridine | 0.71 | 0.94 |
| Butyl cellusolve | 0.74 | |
| I-Propanol, n-propanol | 0.82 | 2.3 |
| Ethanol | 0.88 | 1.20 |
| Methanol | 0.95 | 0.60 |
| Ethylene glycol | 1.11 | 19.9 |
| Acetic acid | Large | 1.26 |
| Water | Larger | |
| Salts and buffers | Very Large | |

## 4. THE FUTURE IN SELECTIVE ADSORPTION

Virtually every trend present in separations technology today points toward selectivity as a key criterion. The demand for higher quality chemicals and feed streams, the energy driven need to remove trace contaminants from the desired bulk component, and growth and awareness of pollution abatement requirements, are all major driving forces for development of separations processes which are highly specific. Adsorption, particularly using aluminas, will be a front runner in this movement. As forming technology becomes more controllable, advances in understanding of the surface chemistry will lead to molecularly engineered surfaces which should considerably expand the horizons for aluminas in selective adsorption processes.

## REFERENCES

1. J.B. Barnitt, US Patent No. 1 868 869 (1932).
2. K. Folkers and J. Shovel, US Patent No. 2 573 702 (1901).

3. H. Brockmann, Disc. Faraday Soc., 7 (1949) 58.
4. P. Rcichstein and C. W. Shoppee, Disc. Faraday Soc., 7 (1949) 305.
5. P. Polesuk, J. Am. Lab., 27 (1970) 5.
6. J. Sherma, Handbook of Chromatography. CRC, Boca Raton, FL, 1971.
7. S. Joshi and J. Fair, Ind. Eng. Chem. Fund., 3rd Quarter (1986)
8. F. Rubel and R. D. Woosley, EPA Tech. Rep. 570/9-78-001 (1978).
9. W.C. Yee, JAWWA, 239 (1966) 2.
10. K. Wefers and G. Bell, Oxides and Hydroxides of Aluminum, Technical Paper 19, Aluminum Company of America, Pittsburgh, PA, 1972.
11. S.J. Teichner, Bull. Soc. Chim. Fr., 7-8 (1974) 1226.
12. D. Bienstock, USPM Report of Investigations 7021, Part 3 (7/67).
13. R.R. Bailey and J. Wightman, J. Colloid Interface Sci., 70 (1979) 112.
14. H. Knozinger and P. Ratnasamy, Catal. Rev.-Sci. Eng., 17 (1978) 31.
15. J.B. Peri, J. Phys. Chem., 69 (1965) 231.
16. J.B. Peri, J. Phys. Chem., 70 (1966) 3168.
17. H. Knozinger and B. Stubner, J. Phys. Chem., 82 (1978) 1526.
18. A.V. Deo, J. Phys. Chem., 75 (1971) 234.
19. J. Travert, Proc. Int. Conf. Vibrat. Surfaces, 333 (1982).
20. R.W. Glass and R. A. Ross, J. Phys. Chem., 77 (1973) 2576.
21. T.L. Slager and C. H. Amberg, Can. J. Chem., 50 (1972) 3416.
22. D.R. Taylor and K. H. Ludlum, J. Phys. Chem., 76 (1972) 2882.
23. G.L. Mundhara, J. Ind. Chem. Soc., 57 (1980) 306.
24. S.D. Williams and K. W. Hipps, J. Catal., 78 (1982) 96.
25. J. Koubek, J. Catal., 38 (1975) 285.
26. G.L. Mundhara and J.S. Tiwari, J. Ind. Chem. Soc., 56 (1979) 737.
27. L.I. Vladyko, Kinet.Catal, UDC 541.124.03+ 541.183, 1200 (1982).
28. E.A. Nechaev and G.V. Zvonareva, Kolloidnyi Zhurnal, 42 (1980) 511.
29. A. Chantong and F.E. Massoth, A.I.Ch.E. J., 29 (1983) 725.
30. C.D. Ford, Anal. Chem., 53 (1981) 831.
31. W.J. Nelson and D.G. Walmsely, Proc. 2nd Int. Conf. Birat. Surfaces, 471 (1982).
32. H. Tamon, J. Chem. Eng. Jpn., 14 (1981) 136.
33. M. Tanaka and S. Ogasawara, J. Catal., 16 (1970) 157.
34. E.B. Krasnyy and L.M. Iozefson, Trudy Kazanskogo Khimiko- Tekhno. Inst., imieni S.M. Kirova, 40 (1969) 273.
35. S.L. Regen, J. Am. Chem. Soc., 101 (1979) 7629.
36. Y.P. Borisevich, Zhurnal Fizicheskoi Khimii, 56 (1982) 1298.
37. Y. Amenomiya, J. Catal., 22 (1971) 109.
38. S.W. Weller and A.A. Montagna, J. Catal., 21 (1971) 303.
39. C. Laurent, Chromatographia, 17 (1983) 253.
40. C. Laurent, Chromatographia, 12 (1983) 395.
41. M. Sarsunova and T.H. Ba, Cesk. Farm., 15 (1966) 522.

42. S. Gocan, Stud. Univ. Babes-Balyeii, Ser. Chem., 22 (1977) 42.
43. L.N. Slepova, Tr. Leningred. Khim.-Farm. Inst., 28 (1969) 69.
44. B. Westerink, Clin. Chem., 28 (1982) 1745.
45. W. Drell, Anal. Biochem., 34 (1970) 142.
46. S.D. Gusekova, Khim. Prir. Soedin., 3 (1978) 310.
47. E. Pescucci and G. Ventura, Rass. Chim., 24 (1972) 124.
48. T. Ivanov, Nauch. Tri., 21 (1971) 55.
49. M.K. Pal and J. Neth, Anal. Biochem., 57 (1974) 395.
50. S. White, Methods Enzymol., 38 (1974) 41.
51. G. Fluoret and O. Hechter, Anal. Biochem., 58 (1975) 276.
52. I. Bartley, Biochem. J., 126 (1972) 251.
53. M. G. Luthra and A. Sheltavy, Biochem. J., 126 (1972) 251.
54. G.P. Mueller, J. Am. Ceram. Soc., 69 (1947) 195.
55. J. Cormon, US Patent No. 3 332 851 (1967).
56. O.P. Mitsan, Visn. L'viv. Politekh. Inst., 112 (1977) 69.
57. A. Steward, Disc. Faraday Soc., 7 (1949) 65.
58. C. Baron, Anal. Biochem., 124 (1982) 84.
59. C. Marutoiu, Romanian Patent No. R 071 380B (1980).
60. M.V. Sussman, Chemtech., 6 (1976) 260.
61. M. Rendell, Proc. Eng., 66 (1975) 4.
62. M. Yakovleva, Endokrinopatii Lech. Int. J. h. Gromonami, 5 (1970) 216.
63. E. Z. Naugol'nykh, Lab. Delo, 9 (1978) 539.
64. F. Orlandi, Biochim. Biol. Sper., 5 (1966) 281.

Adsorption and its Applications in Industry and Environmental Protection
Studies in Surface Science and Catalysis, Vol. 120
A. Dabrowski (Editor)



# Thermal stability of low-temperature alumina sorbents and catalysts

O.V.Andryushkova[a], O.A.Kirichenko[b] and V.A.Ushakov[c]

[a]Novosibirsk Medical Institute, Krasny prspekt, 52, Novosibirsk, 630091, Russia

[b]Institute of Solid State Chemistry, SB of RAS, Kutateladze, 18, Novosibirsk 630128, Russia

[c]Boreskov Institute of Catalysis, SB of RAS, Lavrentieva, 5, Novosibirsk 630090, Russia

The purification systems with low-temperature aluminas as adsorbents are widely used for protection of environment from hazardous organic compounds, especially, at their low content in wastes. The employment of alumina sorbents in adsorptive-contact drying is another way to improve safety of industrial processes. In both cases, as adsorptive capacity has been exhausted, the adsorbent must be regenerated by heat treatment. So the problem of thermal stability (i.e. stability of phase composition, high specific surface area and mechanical strength) of transition aluminas are of great importance during the development, production, employment and regeneration of sorbents and catalysts.

The mechanical properties of alumina sorbents and their changes under heating depend on the method of sorbent preparation and a modificator added. The most strong alumina granules and bodies are obtained if a production procedure includes the mechanical activation methods. Another profitable feature of the employment of mechanical activation is the design of the ecologically safe solution- and waste-free technologies. At the same time, intensive mechanical treatment in mills and activators can accelerate the phase transformations. From this point of view, it seems important to study the effect of the mechanical treatment conditions on the temperature ranges of existence of the low-temperature forms of alumina.

Usually, under heating the destruction of organic compounds to be evaporated occurs on the surface of active alumina sorbents. This phenomenon results in the formation of harmful volatile compounds and the coke that decreases adsorbent capacity. Therefore, it seems more preferable to oxidize organic compounds to carbon dioxide and water during adsorbent regeneration. The processes and purification systems that combine the adsorption and the catalytic combustion regeneration were developed. The transition aluminas doped with catalytic components are used as adsorbents in them. The similar adsorbents-catalysts were used for adsorption-contact drying of thermally unstable materials. The major problems arose for these systems are thermal and mechanical stability as well as thermal shock resistance of alumina adsorbents doped with catalytic components. The temperature and sequence of phase transformations in aluminas are extremely sensitive to the nature and loading of cations introduced in alumina as a catalytic component.

In the present work the effect of a wide range of the metal oxide additives and mechanical treatment conditions on the phase transformations and mechanical properties of transition aluminas was studied.

## 1. THE EFFECT OF MECHANICAL TREATMENT ON THE PHASE TRANSFORMATIONS IN LOW-TEMPERATURE ALUMINAS

The effect of mechanical treatment of aluminum hydroxides on the phase composition of aluminas formed after their heating was studied in detail previously [1,2]. The amorphous products obtained by prolong treatment of bayerite, boehmite and gibbsite in FRITSCH P-7 and EI-2x150 mills do not give any intermediate phase of $\delta$-, $\theta$-, $\kappa$-$Al_2O_3$ under annealing, the formation of $\alpha$-$Al_2O_3$ occurring at 1173 – 1373 K instead 1573 K which is characteristic for untreated ones. The only intermediate phase observed was classified as the alumina of $\eta$-like structure. In contradiction, the high energy ball-milling in a FRITSCH mill was found to result in the milling-induced transformation sequence "hydroxide $\rightarrow \chi$- $\rightarrow \kappa$- $\rightarrow \alpha$–$Al_2O_3$" irrespective of whether the starting powder was gibbsite or boehmite [3].

The thermal induced transformations in the mechanical treated aluminas are less studied. It is already known that mechanical treatment of $\gamma$-$Al_2O_3$ carried out with the compression of the material, or with the use of ball mills, accelerates the formation of the $\alpha$-$Al_2O_3$ phase without any changes of the temperature of this process [3]. More intensive affect with impact waves [5] or with heavy balls in planetary mills [6-9] reduces the temperature of the $\alpha$-$Al_2O_3$ formation, which was observed even directly after mechanical treatment.

The dependence of the phase transformation temperature and sequence on the mill type, milling media, time, acceleration is analyzed in this chapter for a few samples of transition aluminas and their mixtures. Mechanical treatment carried out in the high-powerful planetary ball mills developed at the Institute of Solid State Chemistry SB RAS. Since it is known [10] that mechanical activation with and without elimination of heat leads to essentially different processes in the powder under activation, experiments were carried out with the oxides under investigation using mills both with and without aquatic cooling.

### 1.1. The effect of the mechanical treatment conditions on the oxide microstructure features

In the initial samples of $\chi$-$Al_2O_3$ particles of irregular form with the dimensions near 0.3 - 0.5 μm are prevailing as the data of electron microscopy show (Fig. 1, A). The developed porous structure in the volume of the particles is seen on microphotographs. A dot character of the electron diffraction pattern suggests that a separate particle is a single crystal corresponding to $\chi$-$Al_2O_3$ by phase composition. After calcination at 1173 K the dimensions of particles increase by a factor of 2-3 as a result of sintering accompanied with annealing of micropores and the formation of mesoporous structure (Fig.1, B). The growth of the $\kappa$-$Al_2O_3$ fragments is observed as the result of volume recrystallization.

MA of $\chi$-$Al_2O_3$ for several minutes results in sharp decrease of the dimensions of oxide particles to 0.06 - 0.09 μm (Fig.2, A).

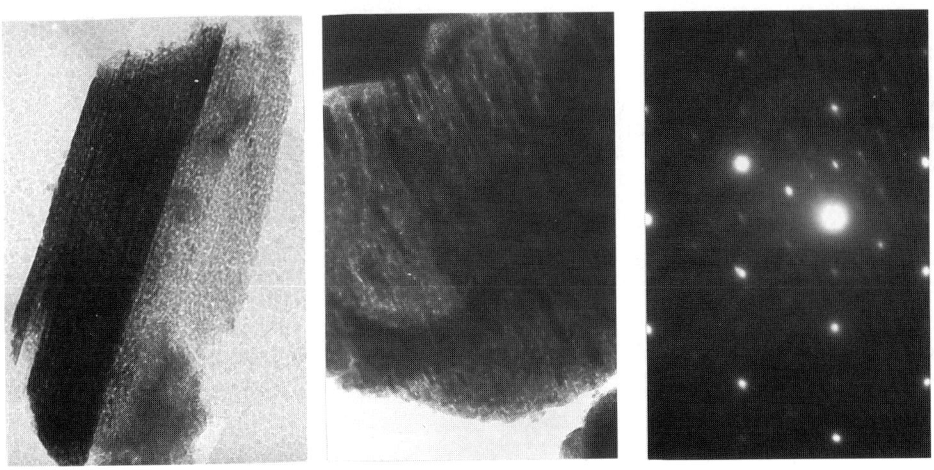

A B C

Figure 1. Microstructure of initial $\chi$-$Al_2O_3$ sample (A) and that of heated at 1173 K (B and C): A - ×120 000, B - × 80 000.

Phase composition of the sample is not changed, but due to the accumulation of microstresses and destruction of coarse particles the "blur" of diffraction maxima occurs (Fig.2, B). Microstructure of a sample differs gradually from that of initial sample. Thus, if a particle is a single crystal in the initial sample, it becomes a polycrystal after MA. This is a result of the decrease of micropore volume and the rupture of the regularity in docking crystal microblocks that compose the particle. A ring-type microdiffraction picture can be the evidence of this. It should be noted that the fragments of initial microcrystals sometimes remain unchanged, which is evident from the microphotograph (Fig.2 A). Heating at 1123 K of the sample activated at 25 W/g leads to the occurrence of single crystals of $\alpha$-$Al_2O_3$ which can be seen on microphotographs as smooth and non-porous particles (Fig.2, C). Along with this, the development of mesopore system is seen (Fig.2, C) in the volume of $\chi$-$Al_2O_3$ crystals which have not undergone the transformation.

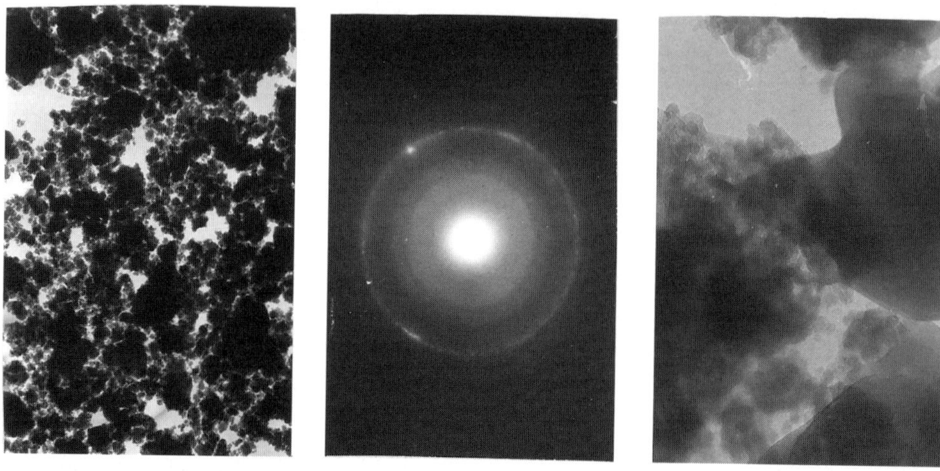

A B C

Figure 2. Microstructure of $\chi$-$Al_2O_3$ sample activated for 15 min at 25 W/g before (A and B) and after heating at 1123 K (C): A -×80 000 , C - ×200 000.

The morphology of the initial $\gamma$-$Al_2O_3$ particles (Fig.3, A) is characteristic of obtained from pseudoboehmit. This is a disperse material composed of thin long fibers which are sometimes linked up to form aggregates regular in form ("logs"). It can be seen from the electron diffraction patterns (Fig.3, B) that a separate fiber is a microcrystal. This is confirmed by the occurrence of dot reflexes on the ring-type microdiffraction picture.

Its ring-type character is explained by the chaotic (in most cases) orientation of fibers in the aggregates. After annealing at 1123 K, sintering of fibers is observed with the formation of particles typical for $\delta$-$Al_2O_3$ ("plates") with clearly defined cutting and the most well expressed (110) facet (Fig.3, C). The elevation of temperature to 1173 K leads to the further increase of the size of oxide particles and to the formation of the traces of $\alpha$-phase (single crystals with a great number of surface steps can be seen on microphotographs) (Fig.3, D). After annealing at 1223 K the number of single crystals of $\alpha$-oxide increases.

Reflexes and kikuchi-lines that appear on the electron diffraction patterns point out to the high degree of crystallization of the volume of these particles.

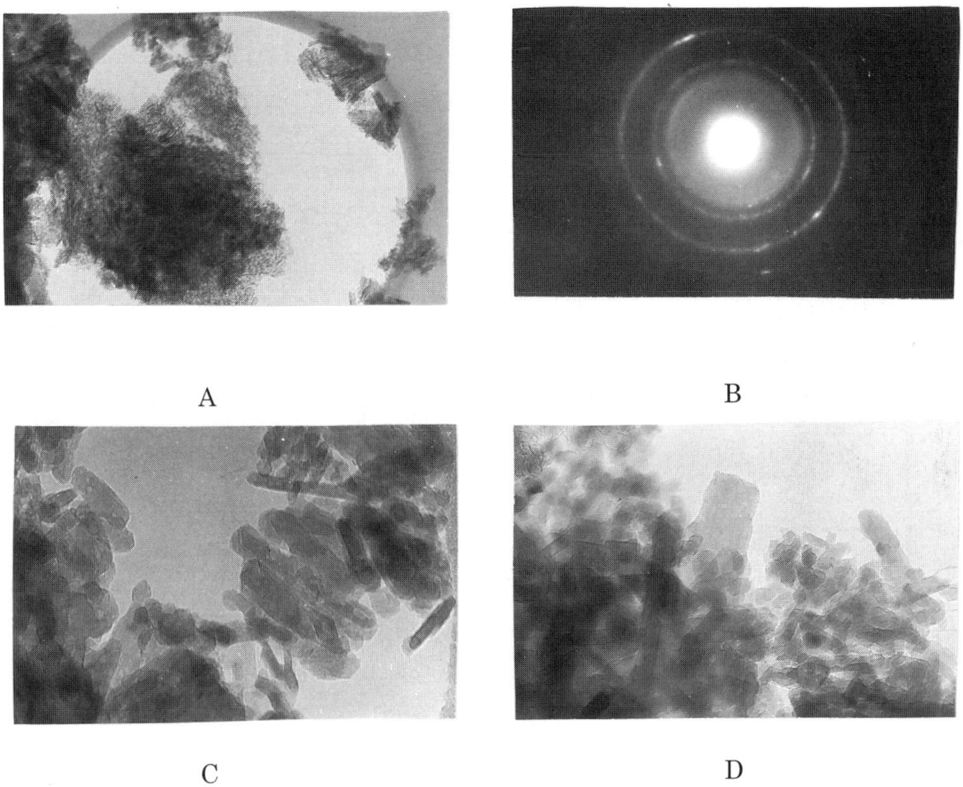


Figure 3. Microstructure of the initial $\gamma$-$Al_2O_3$ sample (A and B) and the sample heated at 1123 K (C), 1173 K (D): A - × 30 000, C - × 500000, D - × 300 000.

MA for 15 min at 25 W/g without elimination of heat causes a sharp change of microstructure of the sample (Fig.4, A). The aggregates corresponding to the composition of $\gamma$-$Al_2O_3$ are still present in the sample, but their structure is close to vitreous state ("polymerization"). Thermal treatment of this vitreous sample at 1123 K does not cause a sharp increase of the aggregate size, although the traces of $\alpha$-$Al_2O_3$ appear in the sample. The microphotographs of the $\gamma$-oxide calcined particles show that the formation of the nuclei of $\alpha$-$Al_2O_3$ takes place not in the whole volume of the particle as in activated $\chi$-oxide but only in some fragments (Fig.4, B). Further sintering and recrystallization cause the transformation of these fragments in coarse (larger than 1 μm) single crystals of $\alpha$-phase. Many surface steps can be seen on the microphotograph of the latter.

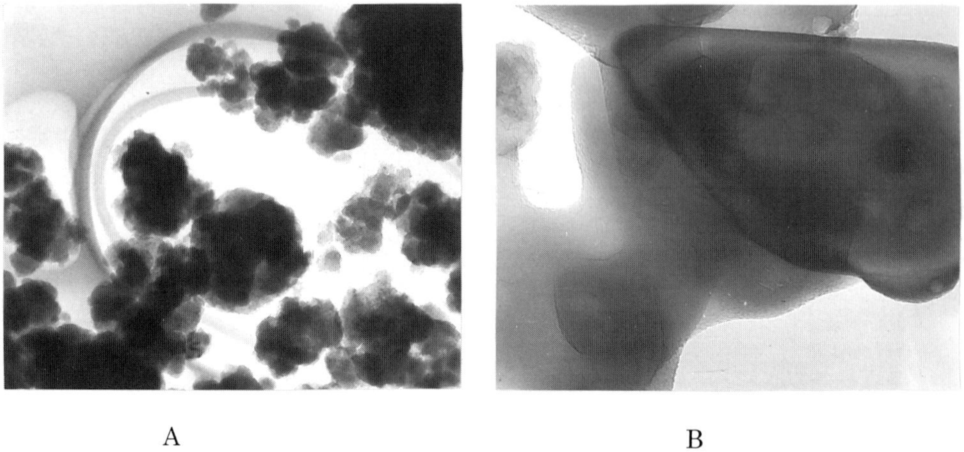


Figure 4. Microstructure of the $\gamma$-$Al_2O_3$ sample activated for 15 min at 25 W/g before (A) and after heating at 1123 K (B): A - × 120 000, B - × 200 000.

## 1.2. X-ray studies of mechanical activated aluminas and their thermal induced transformations

There are no any changes in X-ray diffraction pictures of the samples after mechanical treatment at 10 W/g. After MA at 25 W/g the intensity of the diffraction maxima in the region d/n = 2.12 Å ($2\Theta = 42.5^0$) was found to be increased, the position of maxima being the same (Fig.5, 6). It should be noted

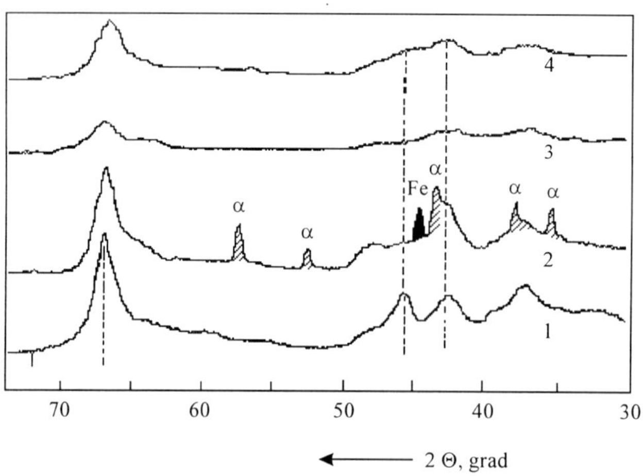


Figure 5. Diffraction pattern of initial $\chi$-$Al_2O_3$ (1), products of thermal decomposition (3) and mechanical treatment: 2 - 30 min, with heat elimination; 4 - 15 min, without heat elimination (steel balls ∅5mm).

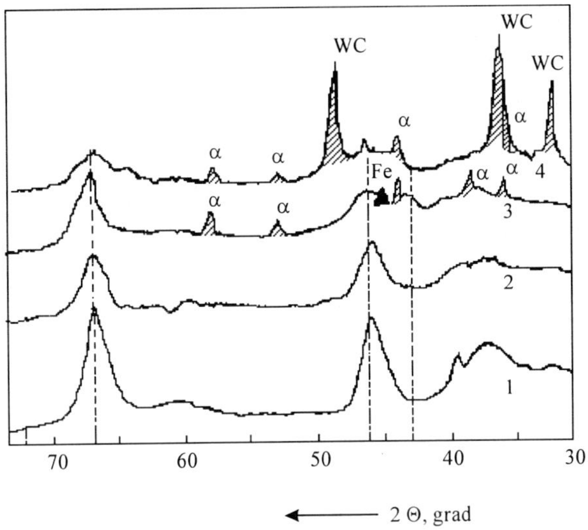


Figure 6. Diffraction pattern of initial $\gamma$-$Al_2O_3$ (1) and the products of its mechanochemical activation: 2 - 15 min without heat elimination; 3 - 30 min with heat elimination (steel balls ∅5mm), 4 - 4 min without heat elimination (WC balls ∅6mm).

that the line d/n = 2.12 Å is not characteristic of the initial $\gamma$-$Al_2O_3$ and appears only after MA of the samples.

The maxima in the range d/n = 1.95 Å ($2\Theta = 46^0$) and d/n = 1.40 Å ($2\Theta = 67^0$) were found to get flattened. Since the height of the background is also changed after MA, it is more correct to characterize the change of the line d/n = 1.40 Å with the relation $F = H_{67}/H_{72}$ involving the intensity (height) of the line in view of the background level to the background intensity. Its variation with increasing the time of MA correlates with the changes in surface area (Fig.7) pointing out the interrelation between the intensity of the line d/n = 1.40 Å and the deformation of initial particles in dense aggregates. After the activation of $\chi$-$Al_2O_3$ for 15 minutes without elimination of heat, the X-ray pattern in the region $2\Theta = 40$ - $50^0$ is similar to the diffraction picture characteristic of the product of thermal decomposition of gibbsite (Fig.5, curve 4) which corresponds most completely to amorphous aluminium oxide.

With increasing the time of MA and the ball mass, the X-ray pattern, besides the changes mentioned above, demonstrate several lines in the positions correlating with the most intensive lines of $\alpha$-$Al_2O_3$ (Fig.5, 2 and 6, 3). The sufficiently accurate determination of lattice parameters for $\alpha$-$Al_2O_3$ formed during mechanical treatment in hydromill is impossible because of small amounts of this form. This process is more clearly expressed for $\chi$-form. Similar

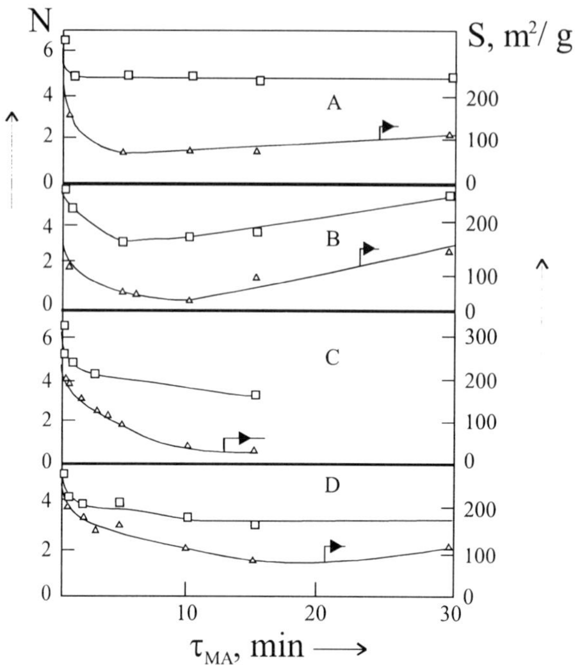


Figure 7. Relative diffraction line intensity F and specific surface area S *vs* MA duration for $\gamma$- (A,C) and $\chi$-$Al_2O_3$ (B,D). Activation with (A,B) and without (C,D) heat elimination.

changes were observed in X-ray patterns of $\gamma$-$Al_2O_3$ after the treatment with shock waves equivalent to the pressure of 40 kbar [5].

The sequence and temperature of thermal induced phase transformations in the activated samples are extremely sensitive to preliminary mechanical treatment duration (Fig.8). Mechanical treatment increases the upper temperature boundary of the $\gamma$- and $\chi$- $Al_2O_3$ existence up to coexistence with $\alpha$-$Al_2O_3$, i.e. direct formation of $\alpha$-$Al_2O_3$ proceeds. This stabilization of low-temperature aluminas never observed previously.

The content of intermediate $\kappa$–, $\delta$–, $\theta$–$Al_2O_3$ decreased with the duration and intensity of mechanical treatment to the point of disappearance. The duration of mechanical treatment and the temperature of thermal treatment determine the stability of initial low-temperature aluminas and their transformation to $\alpha$-$Al_2O_3$. In initial $\gamma$- and $\chi$- $Al_2O_3$ transformation to $\alpha$-$Al_2O_3$ observed starting from 1320 K with slow growth of its content: 3% and 7% in 30 and 60 minutes for $\gamma$-, 10% and 13% for $\chi$- respectively. Prolong annealing at 1070 – 1100 K of the samples after mechanical activation does not result in $\alpha$-form occurring even the surface area decreases to 40 – 45 $m^2/g$. The formation of $\alpha$-form is observed starting from

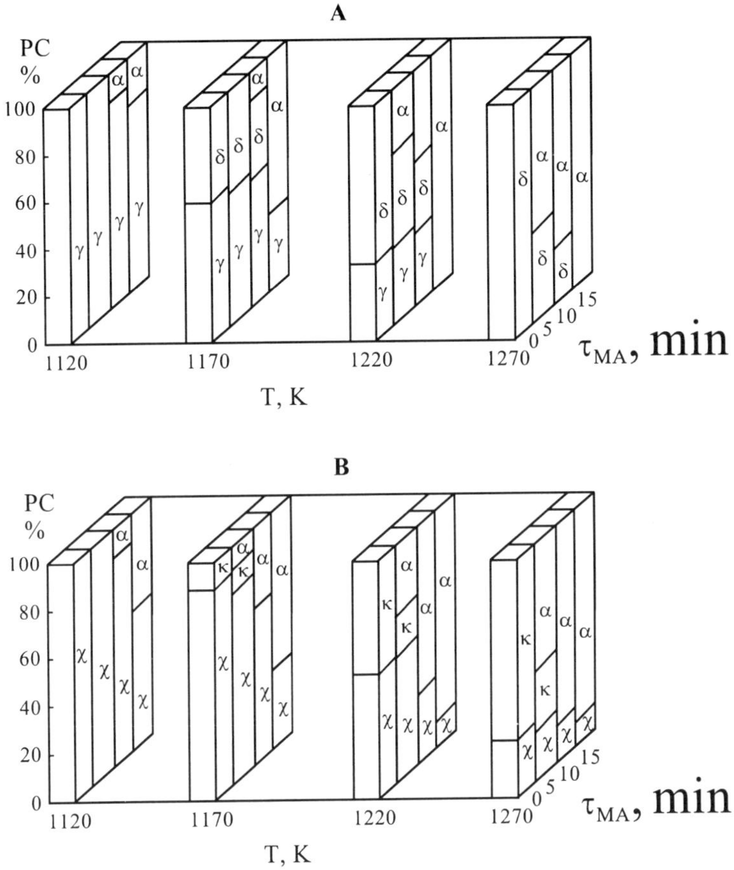


Figure 8. Dependence of phase content of $\gamma$- $Al_2O_3$ (A) and $\chi$- $Al_2O_3$ (B) vs MA duration at 25 W/g without heat elimination and temperature of the subsequent calcination.

1120 K. Lattice parameters of $\alpha$-$Al_2O_3$ thus formed coincide with standard parameters within the accuracy of ±0.003 Å.

The initial low-temperature alumina transition forms are ultradisperse and characterized by highly developed surface. Thermal induced transition to $\alpha$-oxide is accompanied by sintering and significant reduction of specific surface area. The specific surface area of the samples after MA is lower that of initial ones in the temperature region studied (Fig. 9, A and B) due to the formation of strong aggregates and enhanced sintering in them.

In the mixtures of initial aluminas except ($\theta$– + $\chi$–) the specific surface area decreases with increasing of $\chi$–, $\kappa$–$Al_2O_3$ content, which determines the formation of high-temperature phase as well (Table 1, 2).

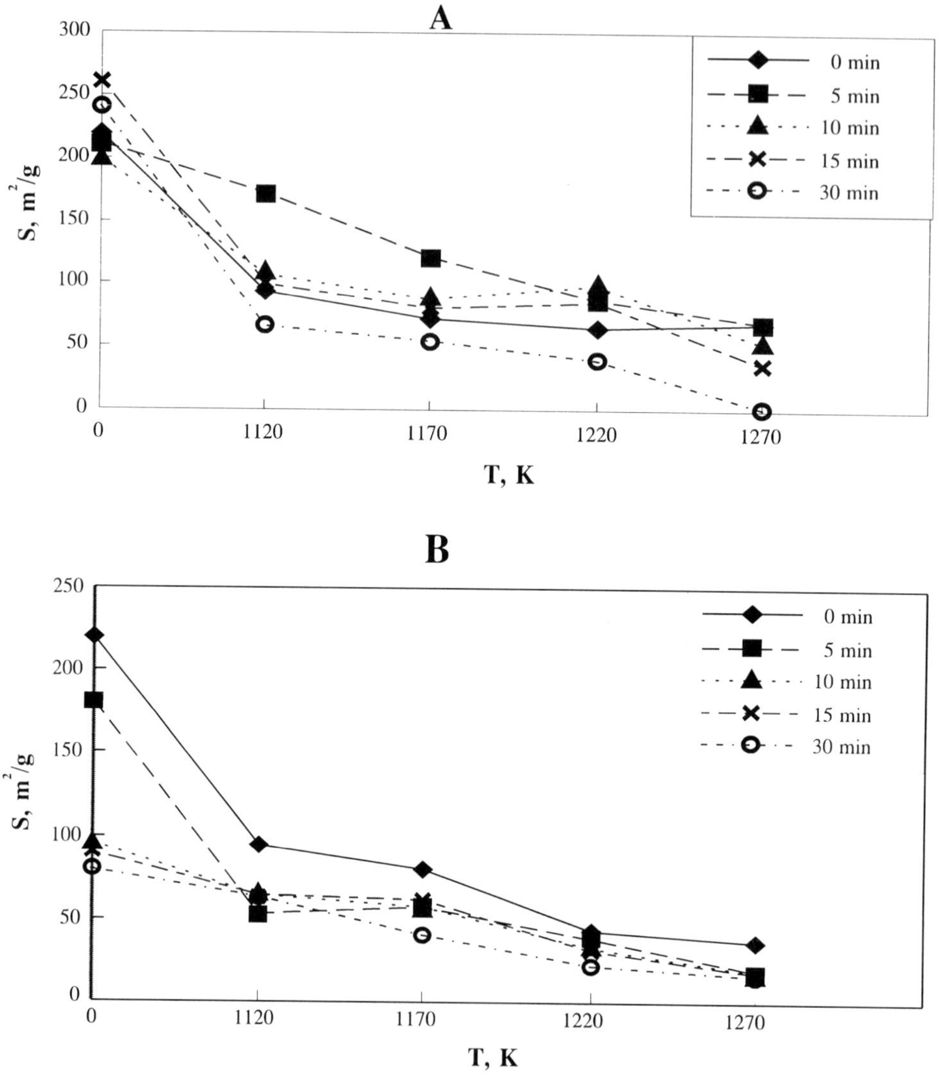


Figure 9. The effect of mechanical treatment time and annealing on specific surface area (S, $m^2/g$) for $\gamma$–(A) and $\chi$–$Al_2O_3$ (B).

The initial low-temperature alumina transition forms are ultradisperse and characterized by highly developed surface. Thermal induced transition to $\alpha$-oxide is accompanied by sintering and significant reduction of specific surface area.

Table 1
The effect of composition «$\gamma - \chi$» mixtures on specific surface area and $\alpha$-phase content (MA at 10 W/g)

| Sample | S, $m^2/g$ | | | $\alpha$–$Al_2O_3$, wt. % |
|---|---|---|---|---|
| | Prior annealing | 1123 K, 30 min | 1223 K, 15 min | 1223 K, 15 min |
| $\gamma$ | 140 | - | 64 | 7 |
| $\gamma(0.8) + \chi(0.2)$ | 140 | 67 | 50 | 0 |
| $\gamma(0.2) + \chi(0.8)$ | 110 | 56 | 37 | 59 |
| $\chi$ | 90 | 39 | 20 | 58 |

Table 2
The effect of mixture composition and MA conditions (1 - at 10 W/g; 2 – at 25 W/g) on the specific surface area and $\alpha$-phase content

| Sample | S, $m^2/g$ | | | | | $\alpha$–$Al_2O_3$, wt. % | |
|---|---|---|---|---|---|---|---|
| | 1 | | 2 | | | 2 | |
| | prior annealing | 1223 K, 15 min | prior annealing | 1123 K, 30 min | 1223 K, 15 min | 1123 K, 30 min | 1223 K, 15 min |
| $\kappa$ | 53 | 23 | 10 | 8.9 | 6.2 | 10 | 77 |
| $\kappa(0.2) + \theta(0.8)$ | 62 | 36 | 32 | 29 | 11 | 7 | 81 |
| $\theta(0.8) + \chi(0.2)$ | 84 | 48 | - | - | - | 0 | - |
| $\theta(0.8) + \chi^*(0.2)$ | 63 | 45 | 68 | 51 | 35 | 8 | 36 |
| $\chi$ | 90 | 20 | 85 | 20 | 15 | 40 | 62 |
| $\theta$ | 66 | 47 | 50 | 40 | 50 | 8 | 36 |

*the phase after preliminary MA

Typical "rows" of transition forms and their formation temperatures are given in the review [14]. The following forms and formation temperature values characterize oxides considered in the present paper:

$$\chi \xrightarrow{1173\,K} \kappa \xrightarrow{1473\,K} \alpha\text{-}Al_2O_3$$

$$\gamma \xrightarrow{1173\,K} \delta \xrightarrow{1273\,K} \theta + \alpha \xrightarrow{1473\,K} \alpha\text{–}Al_2O_3.$$

According to the present studies, mechanical treatment results in variation of the sequences and temperatures of the phase transitions in the aluminum oxides during heating:

a) mechanical treatment at 10 W/g for 30 min:

$$\chi \xrightarrow{1123\,K} \chi + \alpha \xrightarrow{1223\,K} \alpha\text{-}Al_2O_3$$

$$\gamma \xrightarrow{1123\,K} \gamma + \delta + \alpha \xrightarrow{1323\,K} \alpha\text{-}Al_2O_3.$$

b) mechanical treatment at 25 W/g for 15min:

$$\chi \xrightarrow{1123\,K} \chi + \alpha \xrightarrow{1223\,K} \alpha\text{-}Al_2O_3$$

$$\gamma \xrightarrow{1123\,K} \gamma + \alpha \xrightarrow{1223\,K} \alpha\text{-}Al_2O_3.$$

In order to interpret the reasons of the observed significant influence of MA on phase transitions of aluminium oxides, it is necessary to get understanding of the mechanism of these transitions and the structures of aluminas studied. The analysis of literature data reveals that the mechanisms of solid-phase transitions of low-temperature $\gamma$- and $\chi$-$Al_2O_3$ are poorly studied. The temperature range of $\delta$-$Al_2O_3$ strongly depends on boehmite crystallinity and varies from 600 to 1050°C, the lower temperature being attributed to more crystalline product [14]. It is a common supposition that the formation of $\alpha$-form proceeds via either two inseparable stages, namely, accompanied by synchroshift, or the diffusion mechanism of transformation including the stages of nucleation and growth of the particles of $\alpha$-form. Some authors [12,17,18] propose that the first mechanism is preferable, and synchroshift occurs when the sintered clusters (aggregates) achieve critical size.

Complete transformation of $\gamma$- and $\chi$-$Al_2O_3$ into $\alpha$-$Al_2O_3$ includes not only structural transformation but also the removal of OH-groups, which stabilize the lattice of low-temperature aluminium oxide. Detail analysis of this problem is depicted in the following chapter. The removal of OH-groups is accompanied by the redistribution of cations resulting in the formation of $\delta$-, $\theta$- and $\kappa$-$Al_2O_3$. It has been shown using X-ray phase analysis that the increase of MA duration results in the decrease of the amount of $\delta$- or $\kappa$-$Al_2O_3$, but for 15 min treatment, or longer, these transition phases do not appear. This fact, together with the decrease of bonded water content of activated samples, stated with the help of thermogravimetry, point to the decrease of the concentration of OH-groups. It has been marked earlier [10] that mechanical activation in energy-strained planetary mills causes the loss of hydroxyl ions in the form of water molecules from aluminium hydroxides. If low-temperature dehydration is supposed to occur under the action of MA, this would result in the formation of new $Al_2O_3$ clusters containing no OH-groups. These clusters can substantially differ in structure from δ-, θ-, κ-forms due to the low probability (especially at low temperature) of the diffusion of $Al^{3+}$ cations which is needed for the transitions in the chains γ –> δ –> θ and χ -> κ to take place. This may be the observed amorphous and vitreous phase (Fig.8). These clusters can serve as the nuclei of the direct $\alpha$-$Al_2O_3$ formation avoiding intermediate forms.

However, dehydration of γ- and χ-forms with simultaneous formation of amorphous aluminium oxide is not the only reason of $\alpha$-$Al_2O_3$ occurrence. Electron microscopic study reveals that aggregates of randomly directed particles are observed. Specific surface of the samples is decreased by several times which is the evidence of high density of these aggregates. Mechanical loading of porous materials similar in structure to the aggregates of γ- and $\chi$-$Al_2O_3$ causes simultaneous occurrence of compressing, stretching and shift tensions. Therefore, the most favorable conditions are created in the present case for the realization of diffusion creep and dislocation slipping. Moreover, the shift of $O^{2-}$ layers can occur if the applied shift tension reaches sufficiently high values. It can be easily shown that the shift of C and A layers in close-packed cubic lattice ABCAB leads to the formation of fragmentary hexagonal lattice packed as ABABC which contains four layers able to serve as nuclei of $\alpha$-$Al_2O_3$ formation. Compressing tension causes the growth of the nuclei and accelerates the formation of α-phase [18].

Evidently, there exist limits of the density of aggregates formed in MA and of the deformation of primary particles. When these limits are achieved, the relaxation of tensions occurs which proceeds via the violation of aggregate uniformity. The increase of the duration of pulse mechanical loading results in the formation and growth of the fatigue microcracks, followed by the destruction of the aggregates of treated material. An evidence for this fact is the increase of specific surface of samples treated for 30 min. Tension relaxation occurs in these samples, manifesting itself through the increase of the intensity of the line with d/n=1/40 Å. The fall of tensions decreases the rate of $\alpha$-$Al_2O_3$ nuclei formation, and this phase is observed at higher temperature and in essentially lower amounts, compared to the samples activated for 15 min. However, neither δ- nor $\kappa$-$Al_2O_3$ is formed in both cases, which indicates that the reason of such changes in the sequence of transitions is irreversible deep change of their structure during MA.

Thus, an intensive mechanical treatment of aluminum oxides during an adsorbent production is unfavorable, since it can lead to the decrease of thermal stability. Nevertheless, there are definite conditions of mechanical treatment after which the initial γ- and χ- aluminas became stable at higher temperatures.

## 2. CRYSTAL STRUCTURE AND COMPOSITION OF ALUMINA-BASED ADSORBENTS

Transition aluminas are commonly obtained by the dehydration of the aluminum trihydrates (gibbsite, bayerite, and nordstrandite), monohydrate (boehmite) or hydroxide gels (the most important one is pseudoboehmite) according to various decomposition sequences. The process coursed by heat treatment leads through different sorts of intermediate hydrates to anhydrous $\alpha$-$Al_2O_3$. More details can be found in the excellent reviews [14,19,20].

Schema displays the sequence of transition forms that occur during the thermal decomposition of trihydroxides and monohydroxides:

$$\text{Bayerite or Nordstrandite} \xrightarrow{523\,K} \eta \xrightarrow{1123\,K} \theta \xrightarrow{1473\,K} \alpha\text{-}Al_2O_3$$

$$\text{Pseudoboehmite} \xrightarrow{573\,K} \gamma \xrightarrow{1173\,K} \delta \xrightarrow{1273\,K} \theta + \alpha \xrightarrow{1473\,K} \alpha\text{-}Al_2O_3$$

$$\text{Boehmite} \xrightarrow{723\,K} \gamma \xrightarrow{1173\,K} \delta \xrightarrow{1323\,K} \theta + \alpha \xrightarrow{1473\,K} \alpha\text{-}Al_2O_3$$

$$\text{Gibbsite} \xrightarrow{523\,K} \chi \xrightarrow{1173\,K} \kappa \xrightarrow{1473\,K} \alpha\text{-}Al_2O_3$$

The transformation sequences of gibbsite and bayerite are affected with a particle size and partial pressure of water in the heating atmosphere. The transition forms $\chi$- and $\eta$-$Al_2O_3$ are usually developed from gibbsite and bayerite, respectively. In the coarse particles of these hydrates, from which the water of dehydroxylation cannot evaporate rapidly, boehmite is formed as a result of local hydrothermal conditions. If dehydroxylation of gibbsite takes place in vacuum or in rapidly moving air the amorphous form ($\rho$-$Al_2O_3$) is obtained at the temperature above 1000 K.

The transition aluminas contain varying amount of hydroxyl ions and are not well defined stoichiometrically; therefore, they cannot be considered as true polymorphs of alumina. The structures of the forms occurring at a given temperature range are determined by the structure of the starting materials, and they are different for gibbsite, bayerite, boehmite, or diaspore. The microcrystalline state of these substances and the poor character of their X-ray diffraction pattern preclude the application of refined methods of structure determination.

Among the wide variety of the phases formed by the thermal decomposition of aluminum hydroxides we shall consider only alumina forms which derive from pseudoboehmite, boehmite and bayerite. Information about these structures is rather poor. With the exception of the spinel form ($\eta$-, $\gamma$-), the species are mainly characterized by the dimensions of the unit cell and by some indications on their crystallographic system [20]. The structure of $\gamma$-$Al_2O_3$ was considered as a tetragonal distorted spinel lattice (a superstructure with one unit cell parameter tripled in the case of $\delta$-$Al_2O_3$) and is depicted by the formula $Al_{8/3}[\ ]_{1/3}O_4$, where [ ] is a cation vacancy. This formula is consistant with the constraints of stoichiometry and electrical neutrality. It is difficult to know the exact distribution of the cations in the octahedral or tetrahedral sites despite an abundance of work on this subject [21-23]. The $\theta$-phase, isomorphous with

$\beta$-$Ga_2O_3$, though indexed in the monoclinic system, has many similarities with the spinel structure with an approximately cubic close-packed oxygen lattice. From their X-ray powder-diffraction patterns the $\delta$- and $\theta$- phases appear to be more crystalline than the $\gamma$-phase. All these structural studies generally neglect the fact that the metastable aluminas are not pure oxides and contain water in the form of hydroxyl species [24,25].

Soled [26] proposed the description of $\gamma$-form, considering OH species as structural elements, and proposed formula with the stoichiometry - $Al_{2.5}[\ ]_{0.5}O_{3.5}(OH)_{0.5}$. The changes of $\gamma$-form to $\delta$- and $\theta$- by thermal treatment are than depicted as a particle growth which results from condensation of hydroxyl species and formation of water. This model accounts for a more dehydroxylated character of $\delta$- and $\theta$- aluminas compared to $\gamma$-. However, in this formulation the ratio between trivalent and divalent cationic sites does not meet the requirement of the spinel structure that must be equal to 2. To make it correct, Burtin *et al* [27] introduced into the formula the new structural element, which corresponds to oxygen vacancies, noted $< >$. Oxygen vacancies were believed to a result from removing of the water molecules. The common features of the of $\gamma$-, $\delta$- and $\theta$-forms of transition aluminas which were found in their structural and chemical properties allow to describe them by the general formula:

$$Al_2\,[\ ]\,O_{3-v/2}\,(OH)_v < >_{1-v/2},$$

where Al represents the aluminum ions in the trivalent sites, [ ] - the cationic vacancies in the divalent sites, OH - the hydroxyl groups substituting the oxygen in the normal positions of the anionic sublattice and $< >$ - the anionic vacancies in a spinel structure. This formula satisfies the conditions of electrical neutrality of the crystal, accounts for the required ratio of the number of sites in the structure and describes the range of stoichiometries from boehmite ($v = 2$) to $\alpha$-alumina ($v = 0$). According [27], the evolution of transition aluminas during thermal treatment consists of dehydroxylation by removal of water followed by the exothermic structural transformation to the $\alpha$- phase, which can be described in terms of an annihilation reaction between anionic and cationic vacancies.

The description proposed in [26, 27] is the empirical consideration of alumina structures. The evalution of the profile fitting and Rietveld refinement procedure allow new approaches to previously intractable patterns of structural disorder materials.

Spinels are compounds of structure space group Fd3m and of the general formula:

$$A^{tet}\,(II)\,B_2^{oct.}\,(III)\,O_4,$$

in which the formal valence and the cation coordination are indicated.

The unit cell of a «normal» spinel consists of 32 cubic-close-packed oxygen anions. In this unit cell, 8 of the 64 tetrahedral intersites (8a, 8b and 48f) are filled with divalent metal cations, and 16 of the 32 octahedral intersites (16d and 16c) are filled with trivalent metal cations. The space group Fd3m permits for spinels the existence of two pairs of cationic sites when the origin of coordinates is selected at the point 4(-) 3m: it is possible to select the coordinates A (8a) or A*(8b) for the eightfold site (tetrahedra) and B(16d) or B*(16c) for the sixteenfold site (octahedra). The use of only the coordinates A* and B* is equivalent to a shift of the origin of coordinates.

The refinement results [28] show that Al ions in η- alumina occupy only one octahedral site (16d), one tetrahedral site (48f), and one quasi-trihedral site (32e). However, in γ- alumina Al ions occupy only one octahedral site (16d), one tetrahedral site (8a), and one quasi-octahedral site (32e). The refinement results [29] show that the Al ions in η-alumina occupy 1.128 (16d) and 0.539 (16c) octahedral site, 0.503 (8a) and 0.497 (48f) tetrahedral site.

On the basis of the real diffraction pattern and using the method of the full-profile analysis, it was established [30] that the low-temperature forms γ- and η- cannot be described as spinel-type structures with a statistical or regular distribution of vacancies. If the space group Fd3m is adopted as a basis, then the above forms can be regarded as structures in which two allowed tetrahedral and two allowed octahedral sites are filled statistically. Each form has its own characteristic number of anions consisting of hydroxyl-groups or other residual groups of the type $NO^-_3$, $Cl^-$, etc. A model of the type

$$A^*_{x1}\, A_{x2}\, B^*_{y1}\, B_{y2}\, O_{32-z}\, (OH, ...)_z$$

was examined. It follows from steric considerations that the use of type A* coordinates is possible only when the four nearest octahedral cations type B are removed simultaneously and the disposition of the cations type B* requires in its turn the removal of two nearest cations type A. The low-temperature forms can be described as:

boehmite γ–:
$A^*_{1.75}\, A_{5.5}\, B^*_{2.0}\, B_{9.0}\, O_{22.5}\, (OH)_{9.25}$ or ($Al_2O_3 \cdot 0.5\, H_2O$),

pseudoboehmite γ–:

$A^*_{1.75}\, A_{6.75}\, B^*_{1.0}\, B_{9.0}\, O_{23.5}\, (OH)_{8.5}$ or ($Al_2O_3 \cdot 0.46\, H_2O$),

η–:
$A^*_{1.5}\, A_{6.75}\, B^*_{1.0}\, B_{10.0}\, O_{25.75}\, (OH)_{6.25}$ or ($Al_2O_3 \cdot 0.32\, H_2O$).

We consider the transition aluminas as thermodynamically unstable, but reproducible states of structural rearrangement. The starting hydroxides have only octahedral Al ions and quasi-cubic close-packed oxygen ions. Since the tetrahedral Al ions are the newcomers, from the dehydroxylation, they most reflect the diffusion induced disordering. Accordingly, protons still present as stabilizing agents of this protospinel structure.

Using the general Debye formula for the scattering intensity and comparing the synthesized and experimental diffraction patterns [31] confirmed the validity of the model of γ-alumina described above.

It has been shown by the RDF method [32] that, simultaneously with the loss of OH groups, dehydration involves a rearrangement of the structures associated with a decrease in the filling of the octahedral and an increase in the filling of tetrahedral cationic sites respectively. Under these conditions, new tetrahedral cationic sites (E is the site 48f in the space group Fd3m is adopted as a basis), uncharacteristic for the low-temperature forms, are filled. The high-temperature aluminum oxides differ in the filling of the cationic sites and are described by the general formula:

$E_R\, A^*_{x3}\, A_{x4}\, B^*_{y3}\, B_{y4}\, O_{32\text{-}f}\, (OH,\ )_f$,

where $R + x3 + x4 + y3 + y4 < 24$ and $f$ ($f \rightarrow 0$) is the number of structural OH and other groups. The formation of the δ- and θ–forms can be regarded as a result of the structural compensation for the partial removal of hydroxyl-groups and other anionic «impurities» during the complete dehydration of the hydroxides to anhydrous α-alumina. According [33] the conversion of γ- into δ- or of η- into θ-alumina lead to an increase in the tetrahedral aluminum contents but oxygen framework of the higher-temperature phase is better organized. The tetrahedral aluminum content of η- and θ-alumina are equal to 0.33 and 0.55 respectively. Solid state NMR spectra of alumina [34] show that the fraction of tetrahedral aluminum in δ-alumina is actually greater than that in γ-alumina.

Local overheating of the catalyst granules is known to occur on account of exothermal catalytic reactions under the catalyst-adsorbent regeneration. This leads to the transformation of active alumina into corundum and to the reduction of specific surface area and porosity. Thus, the improvement of the catalyst thermal stability depending on that of the support is of a key importance. γ-Alumina is the most thermally stable alumina since its transition to α- starts at the highest temperature, χ– alumina is the least stable. The temperature of the transformation is influenced by various factors such as crystalline form, particle size and morphology, the nature of the gaseous atmosphere, additives (or impurities) etc. [35].

## 3. THE EFFECT OF Me-OXIDE ADDITIVES ON PHASE COMPOSITION

The additives or impurities have a great influence on the kinetics of the transformation. Additives can be subdivided into mineralizers and stabilizers according to their effects on materials. Mineralizers are substances that accelerate the changes of the basic material. For γ– these substances accelerate sintering and polymorphous transformations during the calcination and decrease the temperature of the formation for the high temperature oxides. Stabilizers are substances that decelerate the changes of the basic material. Different authors have demonstrated the accelerating or inhibiting effects of a wide variety of added cations. The transition aluminas are found to be stabilized by $Sc^{3+}$, $Be^{2+}$, $Ce^{4+}$, $Sr^{2+}$, $Th^{4+}$, $Ca^{2+}$ [13] and $Cr^{6+}$ [12,36,37] and the transformation is facilitated by $Fe^{3+}$ [12,37], $Mn^{3+}$, $V^{5+}$, $Mo^{6+}$, $Co^{3+}$, $Zn^{2+}$, $Y^{3+}$ [13] and $Cr^{4+}$ [37]. Contradictory effects have been found with $Mg^{2+}$, $La^{3+}$ and $Zr^{4+}$ [13] and little or no effects have been found with $Ni^{2+}$ [12]. Some other additive effects have been reviewed in [38].

The effect of the additives on alumina has been widely studied with respect to the following parameters: stabilizing elements, concentrations (0,5 - 20% wt.), type of the precursor (salt, hydroxide, oxide), procedure of introduction, conditions of the subsequent thermal treatment. The controversy in the role and mechanism of various additives is caused by the fact that various researchers use different doping methods and different method producing alumina. Moreover, the term $\gamma$-alumina is frequently used to designate any transition forms which occur first in the decomposition sequence, i.e. $\chi$–, $\eta$–, $\rho$ – and $\gamma$–$Al_2O_3$.

The contradicting data on the efficiency of additives and on the influence of their quantity on the phase transformation in alumina do not allow describing in detail the mechanism of the additive stabilizing effect. There is common point of view that as the sintering of alumina occurs thermodynamically, the most convenient method of stabilization is the kinetic inhibition of sintering by additives. The additives occupy the surface sites or bulk sites to prevent diffusion of atoms (surface diffusion and bulk diffusion) which brings to the $\alpha$–phase transformation.

Commonly additives were introduced during the preparation of alumina or added by impregnation. During impregnation, drying and heat treatment a variety of interactions can occur between the phase precursor and the support.

The proposed mechanisms of additive action depend on the structural formula used for description of transition aluminas. According [27] the general formula for $\gamma$-alumina is $Al_2[\;]_m O_{3-v/2}(OH)_v<\;>_{1-v/2}$. The incorporation of a foreign cation M(z+), either in an Al(3+) site or in a divalent vacancy, leads to a modification of the stoichiometry of different structural elements as

$$(<Al^{3+}>_{3+}{}^{0})_{2(1-\beta N)}\;(<M^{z+}>_{3+}{}^{z-3})_{2\beta N}\;(<M^{z+}>_{2+}{}^{z-2})_{2\beta(1-N)}\;(<\;>_{2+}{}^{2-})_{1-2\beta(1-N)}$$

$$(<O^{2-}>_{2-}{}^{0})_{4-y-v}\;(<OH^{1-}>_{2-}{}^{1+})_{v}\;(<\;>_{2-}{}^{2+})_{y}$$

The condition of electroneutrality of the crystal leads to a relation between v, $\beta$, N, z and y as follows:

$$y = 1 + \beta\,(3N - z) - v/2.$$

It has been proposed that the evolution of transition aluminas (pure and doped) during the thermal treatment can be described as a single phase in which the stoichiometric concentrations of the quasi-chemical species vary. This evolution is consists of dehydroxylation by removal of water followed by the exothermic structural transformation to α-phase, which can be described in terms of an annihilation reaction between anionic and cationic vacancies:

$$( < >_{2+}{}^{2-} ) + ( < >_{2-}{}^{2+}) = 0.$$

The refinement of the structures of the low-temperature aluminum oxides [30] made it possible to put forward a hypothesis concerning the mechanism of the formation of the solid solutions and nonstoichiometric spinels obtained when cations and anions are introduced [32,39]. According to this model, γ- has a protospinel structure, slightly different from the spinel one, and is assigned to compounds of the common formula:

$$A^*_{x1}\, A_{x2}\, B^*_{y1}\, B_{y2}\, O_{32-z}\, (OH, ...)_{z},$$

where A and B are spinel tetrahedral and octahedral cation positions; A* and B* are «protospinel» cation positions considered in respect to spinel structure as defects. Low - temperature solid solutions formed in γ- have principally a similar structure, differing only in the values of $x$, $y$ and $z$ coefficients. The substitution of aluminum by a divalent cation in the protospinel structure, without a distinct preference for an octahedral or a tetrahedral environment, leads to the breakdown of the electroneutrality of the alumina molecule. The excess charge can then be compensated either by an increase in the fraction of hydroxyl-groups in the anionic sublattice (during formation of mixtures of hydroxides by coprecipitation) [40, 41] or by the partly removal of some anions (under the heat treatment). In both cases, a general disordering of the structure, in particular of the anionic sublattice, is observed, which is manifested by a frequently noted broadening of the diffraction maxima on the X-ray patterns. Simultaneously with the resulting removal of OH groups and the perfection of cubic anionic framework, the process of the compact filling of cationic positions takes place because in the protospinel the ratio of filled anionic and cationic positions is 32:18,5 instead of 32:24 for true spinel. This process can proceed in two principally different ways:

1. By the preferential filling of octahedral positions of Al at the expense of the packing of the defective (imperfect) positions. The final result of this process is the formation of stoichiometric spinel $AB_2O_4$.
2. By the preferential filling of tetrahedral positions of Al. In this case the new positions, untypical for spinel and low-temperature alumina, in practically cubic anionic framework are occupied. As the result of the structural rearrangement approximately 24 cationic sites filled forming θ-alumina structure. The final result of the sructural rearrangement is

the transition from practically cubic closely packed anionic framework to the hexagonal one with the formation of α-alumina or hexaluminate, as well as solid solutions on their basis.

Let us consider the examples of the effect of the additives on the alumina phase transformations.

**La.** $La^{3+}$ ions is most widely investigated as a stabilizer of transition alumina. Nevertheless, the exact nature of the interaction between La and alumina is not yet clearly understood. Stabilization of alumina appears to be a rather complicate process and depends upon several factors including La loading, calcination temperature and nature of alumina. There are following possible substances which formation is responsible for increased thermal stability:

- lanthanum oxide,
- perovskite-like lanthanum aluminate [56-58],
- lanthanum aluminate with the β–alumina structure [59],
- lanthanum ions in the surface defects of γ-alumina, or non-crystalline dispersed phase [12,60-62],
- surface compound, which does not correspond to known bulk compounds.

According [56] the influence of Ln elements on the texture of the adsorbents during the intermediate stage of the thermal treatment at $T < 1073$ K could be related to interaction between $Ln^{3+}$ ions and $OH^{-1}$ groups. As in the case with zeolite compounds, $La^{3+}$ ions are able to undergo a hydrolytic reaction with formation of species such as $La^{3+}$—$OH^{-1}$—$La^{3+}$ by electron withdrawing from the O-H bond leading to an increase in the acidity of the OH groups. These hydroxyl groups would be more difficult to remove from the surface of alumina.

Our results on the La cation effect on the thermal and mechanical stability of alumina are described in [48]. Even the small amount of La (up to 5wt.% by $La_2O_3$) causes deceleration of the α-alumina formation. At higher concentrations transition to α- is completely suppressed and alumina is stabilized either in θ-form or in a mixture of γ- and δ–forms. Observed results allow us to conclude that the La stabilizing effect occurs when the solid solution of La ions forms in the structure of the low temperature aluminas. La incorporation into the alumina lattice prevents the diffusion of Al ions and the rearrangements into the high temperature forms.

**Ce.** Ceria has been used to improve thermal stability of alumina. There are following possible Ce-containing phases in ceria-alumina system:

- $CeO_2$ [63, 64],
- solid solution between ceria and alumina as an intermediate for $CeAlO_3$ formation [65],
- Ce aluminate with the perovskite-like structure $CeAlO_3$ [66,67],
- Ce ions in the surface defects of □□alumina or non-crystalline dispersed phase.

According to [64] Ce occurs mainly as a surface disperse phase non-interacting with alumina. The fact that Ce has a weaker effect on alumina stability than La was explained by this phenomenon. The stabilizing effect is associated to the presence of superficial dispersed $Ce^{4+}$. However, $Ce^{4+}$ is unable to maintain the stabilizing effect over long periods, presumably because of a segregation process that leaves the alumina unprotected [63]. The results of the research [68] have shown that $Ce^{3+}$ species are important in the stabilization of alumina, even at very low concentration.

X-ray studies of two systems prepared with the same alumina samples by the same procedure and calcined at the same conditions allow us to conclude that alumina stabilization by Ce is weaker than by La [49]. The formation of the solid solution based on the γ-alumina structure by substitution of $Al^{3+}$ on $La^{3+}$ ions imposes limitations on the transitions to δ, θ, α–alumina. Alumina modified by Ce does not show essential stabilization of the structure of alumina. This is caused by the limitations on alumina interaction with Ce ions.

**Mg+La.** Simultaneous introduction of Mg and La ions into alumina stabilizes the solid solutions based on the alumina structure [52, 53]. The observed higher thermal stability of double modified alumina system MgO - $La_2O_3$ - $Al_2O_3$ in comparison with system $La_2O_3$ - $Al_2O_3$ appears to be adequately explained only by the existing of the both above mentioned, principally different, ways of structural rearrangement in $Al_2O_3$ (Fig. 10).

Coexistence in the same solid solution of Mg and La, different in the preference for filling the particular cationic positions, does not permit the realization of the solid solution structural rearrangements typical for each of the cations. Moreover, the stabilization of a «loose» structure of the low-temperature solution favors the admittance of lanthanum ions into the support structure. In contradiction to the system $La_2O_3$ - $Al_2O_3$, no formation of lanthanum aluminate is observed with

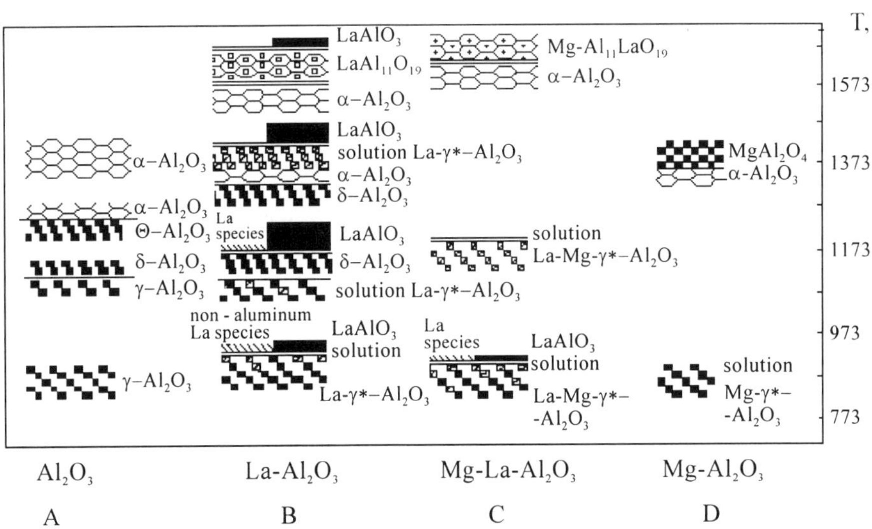


Figure 10. Scheme of the changes in the structure and phase composition of aluminas doped with magnesium (ca. 3 wt.%) and lanthanum (ca. 10 wt.%) with increased temperature.

temperature increase because in this case the preferential process is apparently the diffusion of lanthanum ions into the structure of the mixed solid solution. At temperatures above 1473 K two compounds with hexagonal packing of oxygen framework ($\alpha$-alumina and $MgLaAl_{11}O_{19}$ ) coexist in the system.

**Mg+Ce.** The simultaneous introduction of Mg and Ce ions into alumina did not increase its thermal stability in comparison to MgO - $La_2O_3$ - $Al_2O_3$. This effect is caused by the fact that only a small part of Ce introduced is able to interaction with the support to be stabilized as $Ce^{3+}$ and to improve slightly the thermal stability. The interaction product decomposes with the temperature increase, yielding a separate phase of $CeO_2$, $\alpha$-$Al_2O_3$ and stoichometric spinel. The major part of Ce introduced during impregnation transforms from $Ce^{3+}$ to $Ce^{4+}$ forming the $CeO_2$ phase, which provides no effect on the thermal stability of alumina.

**La-Si.** Silica is recognized stabilizer [69] but its use will introduce acidic properties which may be undesirable. Apparently, both La and Si, present in the system, decelerated the phase transitions in alumina with the temperature growth [55]. The $\alpha$–alumina appeared in $SiO_2$ - $La_2O_3$ - $Al_2O_3$ only at 1573 K. Note that the stabilization effect for $SiO_2$ - $La_2O_3$ - $Al_2O_3$ system was by far higher than that of MgO - $La_2O_3$ - $Al_2O_3$. In this case higher thermal stability also can be explained by existing of the two principally different ways of structural rearrangement of $Al_2O_3$.

**Cu.** It has been revealed that at T = 823-1473 K in inert media, as well as in air the solid solutions are formed based on alumina structures [42,45,46]. $Cu^{2+}$ ion incorporation stabilizes the $\chi$-alumina, however decreases the temperature of $\alpha$-alumina formation. Mineralizing action of Cu cations is suggested to be due to the decomposition of $Cu^{2+}$- containing aluminate to $Cu^{+}$ aluminate and $\alpha$-alumina. The mineralizing effect depends on the copper content, i.e. the higher the number cations introduced, the greater the mineralizing action. At low contents of copper oxide, when only a divalent copper aluminate is detected, no mineralizing effect was observed.

**Ni.** At high temperatures the samples with $Ni^{2+}$ ions, like those with $Cu^{2+}$, also contain solid solutions on the basis of $\gamma$–alumina and nickel aluminate, but in this case no changes in phase composition was observed, since, unlike the copper sample, nickel aluminate is stable at T < 1473 K.

**Fe, Cr, Me(Cu, Mg, Zn) + Cr on $\chi$-alumina.** At temperatures 973-1273 K solid solutions are formed and transformation to $\alpha$-alumina takes place differently in accordance with thermal stability of these solutions [46]. The stability of Me-Cr-Al-O systems decreases in the row Zn>Mg>Cu with Cu being mineralizer. The presence of the $\alpha$-phase structure nuclei introduced into the mixture with $\chi$- effects the formation of $\alpha$–$Al_2O_3$ more weakly than the decomposition of solid solutions based on the structures of $\chi$-$Al_2O_3$, $CuAl_2O_4$, $CuCr_2O_4$.

The mineralizing effect depends on the $Fe^{3+}$ loading, i.e. the higher the content, the stronger the mineralizing action, which enhances with temperature increase. The mineralizing effect depends on the nature of $Fe^{3+}$ precursor as well. Chromium stabilizes the $\chi$-alumina not effecting the $\alpha$– formation.

**Cu+La, Cr+La, Cu+Cr+La.** Lanthanum ions introduced into alumina structure, decrease the solubility of copper and chromium ions, thus providing the decrease of the mineralizing effect [50, 51] . Copper and chromium ions not involved in the support structure, yield an extra quantity of copper chromite, which enriches correspondingly the solid solution of $CuAl_2O_4$ and $CuCr_2O_4$ spinels, providing the observed changes in the lattice parameters. Based on these results it is concluded, that in order to suppress phase transformations in alumina, the introduced stabilizing ion must, first, restrict the solubility of the mineralizing catalytic components and, second, form a compound with the alumina stable at high temperatures (which does not decompose producing $\alpha$-alumina.)

Hence according to the results presented the structure of transition forms of alumina is protospinel with all three features (filled tetrahedral and octahedral positions, as well as OH-groups) equally determining the conditions and course of structural rearrangement under external action. Eventually these three structural elements define the condition of the transformation from cubic anion sublattice to hexagonal one. Introduction of foreign ions into alumina results to interaction of different extent depending on the precursor nature, content, presence of another foreign ions, calcination conditions etc. The major prerequisite for the effect of ions (cations and anions) on thermal stability of aluminas is that the transition aluminas are the substances which occurrence range is determined by the conditions stabilizing their structures. Formation of stable solid solutions based on the structures of low-temperature aluminas enhances the thermal stability of the transition aluminas.

## 4. THERMOMECHANICAL STABILITY

The purification system, which employs low-temperature alumina based adsorbents, involves the stage of adsorbent regeneration by heat treatment, with the temperature varying up to 1000K. If an adsorbent granule to be regenerated contains organic compounds inside its pores, the local internal temperature of pellet under heating can be higher than fixed one due to the combustion process on the internal pellet surface. The best purification systems operate uninterruptedly by means of the steady flow of adsorbent granules from adsorber to regenerator (or additional fluidized bed). Therefore, an adsorbent should possess a high mechanical strength. More over, the adsorbent pellets must be able to retain high mechanical strength after prolong treatment at temperatures about and above 1000 K. However, it is well known that the mechanical strength of low-temperature alumina adsorbents tends to decrease under these conditions

[70]. The simultaneous chemical, thermal and mechanical action on the adsorbent was shown to result in the pellet strength drop during the first operation hours in a fluidized bed and its slight decrease in the hours followed (up to 1000 hs) [71]. So the first problem arose was to predict the effect of prolonged heat treatment on adsorbent and catalyst wear resistance.

The mechanical strength of adsorbent and catalyst pellets is usually determined under static stress (crushing) and more rarely under the dynamic ones (impact testing) [72,73]. Pellets with crushing strength below 10 MPa were found to crush upon attrition in a PIG-2 apparatus and undergo rapid destruction in a fluidized bed. Meanwhile, the attrition rate of pellets with 10-36 MPa strength appeared to be constant [74]. This does not agree with the pilot testing data [71,75]. Impact testing of pellets is more sensitive to the physicochemical nature of aluminas [76]. For adsorbents and catalysts of a similar phase composition, the impact strength factor is proportional to the crushing strength, though various materials of the same crushing strength exhibit different impact strength values (Fig.11). The long-term strength appears to be most sensitive to structural and composition features of materials [77]. The catalyst long-term strength was determined under cyclic loading (fatigue strength) [78].

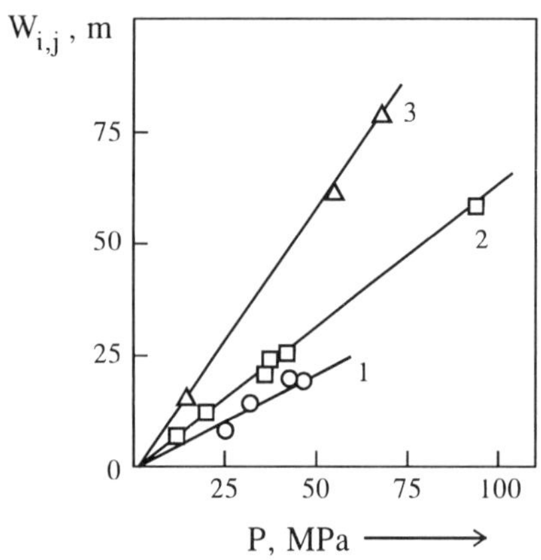


Figure 11. Correlation between impact ($W_i$) and crushin (P) strenght of CaO-doped aluminas: 1- CaO/$\gamma$-$Al_2O_3$; 2 - $\gamma$-$Al_2O_3$ – $CaO{\cdot}Al_2O_3$ – $CaO{\cdot}2Al_2O_3$; 3 – $CaO{\cdot}2Al_2O_3$ –$CaO{\cdot}6Al_2O_3$.

To a first approximation, thermomechanical stability of adsorbents and catalysts were considered to be the change in the pellet mechanical strength after thermal treatment under the temperatures higher that of adsorbent preparation. Adsorbent with a constant strength or the strength growing upon heating is the most desirable. Decrease of the mechanical strength upon thermal treatment indicates the restriction of the acceptable temperature region of regeneration.

The dependence of pellet crushing strength on the calcination temperature after short-time (2-4 hs) isothermal treatment was studied for a wide set of the alumina adsorbents produced by different companies and methods. While the initial mechanical strength values were high enough, for most samples the thermal treatment causes the decrease of mechanical strength, the values obtained depending strongly on the production procedure and initial phase composition of the adsorbents.

The additional doping of the alumina adsorbent with a catalitically active component often changes the temperature dependence of the mechanical strength. The crushing strength can either increase (Mg-Cr-O/$\gamma$-$Al_2O_3$), or decrease (Cu-Cr-O/$\gamma$-$Al_2O_3$, ($\gamma$+$\chi$)-$Al_2O_3$; Mg-Cr-O/($\gamma$+$\chi$)-$Al_2O_3$) or remain practically the same (Fe-O/$\gamma$-$Al_2O_3$) [79-81]. The highest thermomechanical stability was obtained for alumina adsorbents, which were produced via technologies developed at Boreskov Institute of Catalysis of SB RAS [82]. Modification of alumina with CaO by special procedure [83] enhances mechanical strength and thermomechanical stability. Variations in the crushing strength of pure and doped samples are depicted in Fig.12.

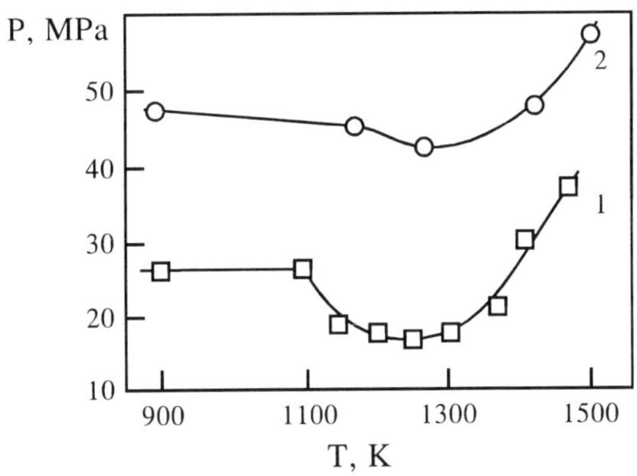


Figure 12. Crushing strength of pure (1) and CaO-doped (2) $\gamma$-$Al_2O_3$ *vs* temperature of calcination.

The above mentioned studies base on the short-time thermal treatment and give only the temperature tendency of the crushing strength, whereas adsorbent should endure the effect of prolong treatment. Therefore, the new laboratory tests were required to predict the long-term heat treatment. The concept of the sintering process as the governing factor for thermomechanical stability gives a line of attack on this problem.

To simulate a prolonged thermal treatment in the laboratory, one can rise the temperature, since both time and temperature enter the kinetic equations for sintering. Below 1250 K sintering kinetics of alumina-based adsorbents and catalysts can be described via the surface area reduction [79] by the equation proposed for the initial stage of sintering [84]. The temperature of thermal treatment for 4 hrs, which corresponds to a 1000 K temperature effect on the alumina adsorbents and various catalysts over half of a year period, was calculated from the kinetic data of [79]. Its value varies in the range 1140-1330 K. At T>1250K, the sintering mechanism of pure and doped $\gamma$- and $\chi$-$Al_2O_3$ changes [79], thus a short treatment at T>1250 does not correspond to prolong sintering at lower temperatures. Therefore, to predict the effect of prolonged treatment at temperatures about 1000 K, it was proposed to test the adsorbents and catalysts for crushing strength after their thermal treatment at 1250 K for 4 hrs. The minimal strength of individual pellets should exceed the value of 10 MPa ($P_c$).

If the crushing strength does not comply with this test requirement, the allowed regeneration temperature and adsorbent (catalyst) operating life can be estimated using the temperature dependence of crushing strength, surface area and the sintering kinetics data [81]. The time needed for the adsorbent strength to drop to the limit $P_c$ can be calculated from the equation obtained from its initial sintering (surface area reduction) kinetic equation [79-81]:

$$\tau = [(S_o - S_c)/S_c]^{\gamma} T / \beta \exp(-E_a/RT)$$

$$S = S_{BET}\, \rho$$

Here $S$ is the surface area of volume unit ($m^{-1}$); $S_{BET}$ is specific surface area; $\rho$ is the real density; $S_o$ is initial surface area of adsorbent; $S_c$ is surface area of the sample with the strength equal to $P_c$; $\gamma$, $\beta$, $E_a$ are parameters of sintering kinetic equation determined experimentally.

It should be mentioned that the fatigue strength of the aluminas doped with Fe-oxides were shown to be much more sensitive to the heat treatment than the crushing one (Fig.13).

Whereas, there were no changes in the crushing strength of the aluminas doped with Cr-Mg-oxides after calcination at 770-1270 K, the fatigue strength curves of the same samples changed, with the average number of cycles before granule disintegration (*K)* decreasing (Fig.14).

Thus this line of inquiry shows promise for better characterization of thermomechanical stability of the alumina adsorbents.

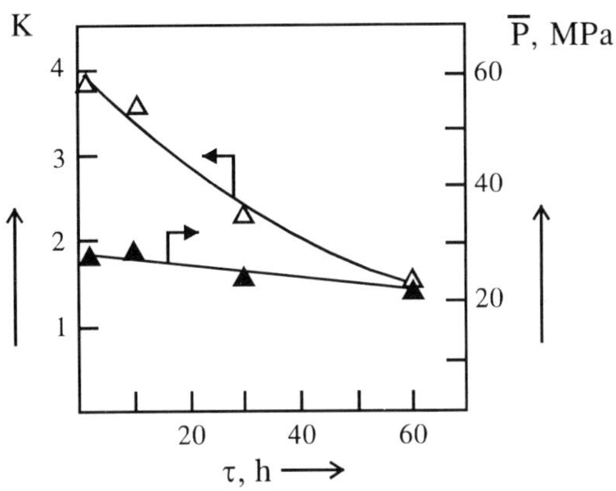


Figure13. Crushing (P) and fatique (K) strength of Fe-Al –O adsorbent *vs* time of annealing at 1070 K.

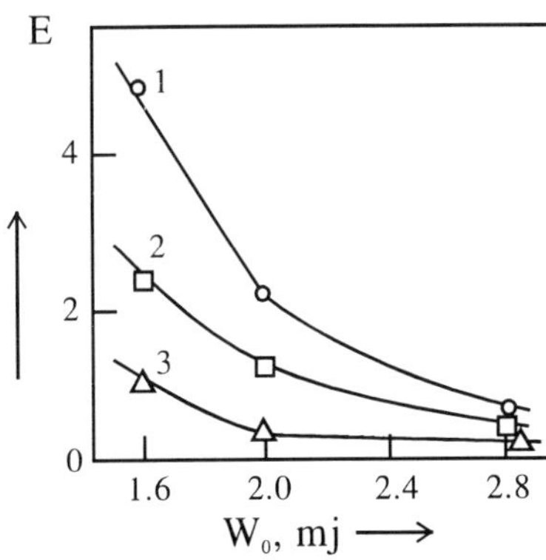


Figure14. The fatique curves of initial adsorbent (1), after calcination at 1170 K (2) and after thermal shock testing (3).

## 5. THERMAL SHOCK RESISTANCE

Adsorption and regeneration in a fluidized or moving bed of spherical pellets include contact with wet powders, and the rapid mass transport of liquids and their evaporation. Condensed moisture can affect the adsorbent and catalyst pellets, whose temperature is 500-1300 K in the bed, providing a relatively rapid cooling with a subsequent rapid heating. This effect can yield a decrease of pellet strength and even their disintegration. The capacity of adsorbents and catalysts to endure abrupt temperature differences is characterized by the «thermal shock resistance», commonly defined by pellet integrity preserved after repeated heating and cooling.

Catalyst pellets tested for thermal shock resistance are usually heated in flame burners or in furnaces to 1470 K or to their melting point and then cooled to ambient temperature in air. Various catalysts of methane conversion heated to 770 K were cooled via dipping into water [85]. Techniques assuming pellet heating to 1470 K are not applicable for testing of the alumina adsorbents, because of alumina phase transformations at this temperature. Spherical shape adsorbents and catalysts with crushing strength higher than 10 MPa reveal no pellet disintegration under testing via the procedure given in [85]. The alumina sorbents for adsorptive-contact drying processes were tested by means of multicycle thermal regeneration at 520 K after vapor saturation to water capacity values of 0.05-0.06 ml/g [86]. It was shown that the crushing and impact mechanical strength of the samples with micropore volume higher than $0.1 cm^3/g$ decreases significantly after 50 cycles independently of the alumina adsorbent nature. However, when used for other processes, adsorbent must be stable in more rigorous conditions [87,88].

Alumina supports and catalysts were treated under conditions simulating the real processes, and various strength factors were determined [89]. As a result, a technique was designed for evaluating the thermal shock resistance, which defines the pellet fatigue strength after a heating recycle (heating at a definite temperature in a furnace with subsequent rapid dipping into water and reheating of wet pellets). The catalyst fatigue strength was specified by the average number of cycles before disintegration at registered impact energy $W_o$ [73]:

$$K = \sum_{j=1}^{k} (k_j - 1)\, n_j / N$$

here $k_j$ is the number of impacts with energy $W_o$ until pellet disintegration and $n_j$ is the number of pellets failed after $k_j$ impacts; $N = 50$ is the total number of pellets tested. The fatigue strength of adsorbents and catalysts decreases with the heating-cooling cycle number, the largest drop taking place after the first cycle.

Three regions of the thermal shock resistance of modified aluminas and supported alumina-based catalysts can be recognized, depending on the temperature of preliminary heating of the pellet [90]. The region boundaries depend weakly on chemical and phase composition, as well as the adsorbent texture. In the first region (below 570 K), the fatigue strength does not vary significantly, though its value depends on the sample nature. Within 570 –670 K, the fatigue strength drops abruptly (Table 3 ). At higher temperatures the fatigue strength is weak and changes within the measuring accuracy. Only for the most strengthen Ca-modified adsorbents it does tend to decrease, though not so distinctly as in the second region.

Table 3

Fatigue strength of modified alumina sorbents under testing for thermal shock resistance

| $W_o$, mJ | $\gamma$-$Al_2O_3$ | | | Al-Mg-O | | | Al-Ca-O | | |
|---|---|---|---|---|---|---|---|---|---|
| | Initial | 620 K | 920 K | Initial | 620 K | 920 K | Initial | 620 K | 920 K |
| 0.85 | 16 | 5 | 0.2 | - | - | >65 | - | - | >100 |
| 1.6 | 2 | 1.5 | 0 | >100 | - | 5 | >100 | - | >18 |
| 4.0 | 0 | - | - | 13 | 1 | 0 | 22 | 10 | 0.1 |

Based on the results obtained one can conclude that the temperatures below 570 K are preferable for the alumina adsorbents. Thus the technique proposed makes it possible to characterize the thermal shock resistance of adsorbent and to define the limits of regeneration temperature from the viewpoint of mechanical strength features and operation durability.

## REFERENCES


1. T.Tsuchida and N.Ichikawa, Reactivity of Solids, 7 (1989) 207.
2. S.M.Paramzin, B.P.Zolotovsky and O.P.Krivoruchko, Izv. SO AN SSSR. Ser. Khim. Nauk, 1 (1989) 33.
3. A.Tonejc, A.M.Tonejc, D.Bagovic and C.Kosanovich, Materials Science and Engineering, A, 181, 112 (1994) 1227.
4. F.W. Dynys and J.W.Halloran, J. Am. Ceram. Soc., 65, 9 (1982) 442.
5. B.G. Adamenko, P.O. Pashkov and L.N. Tambovtseva, Poroshkovaya metallurgiya, 10 (1978) 93.
6. I.J.Lin, S.Navid and P.Bar-On, Thermochim. Acta, 148 (1989) 301.
7. P.A. Zielinski, R. Schulz, S. Kaliaguine and A. Van Neste, J. Mater. Res., 8, 11 (1993) 2985.
8. B. B. Bokhonov, I. G. Konstanchuk and V. V. Boldyrev, Mat. Res. Bull., 30, 10 (1995) 1227.
9. O.V.Andryushkova, V.A.Ushakov, G.N.Kryukova, O.A.Kirichenko and V.A.Poluboyarov, Chemistry for Sustainable Development, 4 (1996) 15.
10. L. T. Menzheres, V. P. Isupov and N. P. Kotsupalo, Izv. SO AN SSSR. Ser. Khim. Nauk, 3 (1988) 53.
11. V.A. Ushakov and E.M. Moroz, React. Kinet. Catal. Lett., 24, 1-2 (1984) 113.
12. G.C. Bye and G.T. Simpkin, J. Am. Ceram. Soc., 57, 8 (1974) 367.

13. V.I. Vereshchagin, V.Yu. Zelinsky and T.A. Khabas, ZhPKh, 55, 9 (1982) 1946.
14. B.G.Linsen (eds.), Physical and Chemical Aspects of Adsorbents and Catslysts, Academic Press, London and New York, 1970.
15. O.V.Andryushkova, O.A.Kirichenko, V.A.Ushakov and V.A.Poluboyarov, Solid State Ionics ( in print).
16. N.A. Toropov (ed.), Diagrammy sostoyaniya silikatnykh sistem: Spravochnik, Issue 2, Nauka, Leningrad, 1970.
17. V.A. Gagarina, V.N. Kuklina and L.G. Khomyakova, Kinet. Kataliz, 13 (1972) 174.
18. S.V. Raman, R.H. Doremus and R.M. German, Mat. Sci.. Res., Plenum Press, N.-Y., 16 (1984) 253.
19. R.K.Oberlander, in: Applications of Industrial Catalysts, B.E.Leach (eds.), V.3 Chapter 4, Academic Press, London, 1984.
20. Alumina Chemicals: Science and Technology Handbook, B.E.Leach (eds.), The American Ceramic Society Inc., Westerville, Ohio, 1990.
21. H.Knozinger and Ratnasamy, Catal. Rev. Sci. Eng., 17 (1978) 31.
22. S.J.Wilson, Proc. Brit.Ceram.Soc., 28 (1979) 281.
23. D.S.Tucker, J.Jenkins and J.J.Hren, J.Electr.Microsc.Techn., 2 (1985) 29.
24. L.Robert and Jr.Burwell, J.Catal., 86 (1984), 301.
25. C.J.H.Jacobsen, N.-Y.Topsoe, H.Topsoe, L.Kellberg and H.J.Jakobsen, J. Catal., 154 (1995) 65.
26. S.Soled, J.Catal., 81, (1983) 252.
27. P.Burtin, J.P.Brunelle, M.Pijolat and M.Soustelle, Appl.Catal., 34 (1987) 239.
28. R.L.Snyder, Mater. Science Forum , V.79 - 82 (1991) Pt. 2, 513.
29. K.Shirasuka, H. Yanagida and G. Yamaguchi , Yogyo-Kyokai-Shi, 84 (1976) 610.
30. V.A.Ushakov and E.M.Moroz, Kinet.Katal., 26 (1985) 968.
31. D.Espinat, F.Thevenot and K. Malkio, 12th European Crystallographic Meeting, Abstracts, Moscow, 3 (1989) 280.
32. V.A.Ushakov, E.M.Moroz and E.A.Levitskii, Kinet.Katal., 26 (1985) 1206.
33. A.J. Leonard, P.N.Semaille and J.J.Fripiat, Proceedings of the British Ceramic Society, 13 (1969) 103.
34. S. Srinivasan, Prepr. Symp. Am. Chem. Soc. Div. Fuel Chem., 36 (1991) 515.
35. D.L. Trimm, Studies in Surface Science and Catalysis, 68 (1991)38.
36. N.S. Kurkova, I.R. Katsobashvili and N.A. Akchurina, Zh. Prikl. Khim., 46 (1973) 1002.
37. T.Tsuchida, R. Furuichi, T.Ishii and K. Iton, Thermochim. Acta, 64 (1983) 337.
38. D. S. Tuker and J. J. Hren, Mat. Res. Soc. Symp. Proc., 31 (1984) 337.
39. E.M.Moroz, V.N.Kuklina and V.A.Ushakov, Kinet. Katal., 28 (1987) 699.
40. N.A. Koryabkina, Z.R. Ismagilov, R.A. Shkrabina, E.M. Moroz and V.A. Ushakov, Appl. Cat., 91 (1991) 63.

41. N.A. Koryabkina, N.A. Litvak, R.A. Shkrabina and Z.R. Ismagilov, Kinet. Katal., 34 (1993) 913.
42. E.M. Moroz, O.A.Kirichenko, V.A.Ushakov and E.A. Levitskii, React. Kinet. Catal.Lett., 28 (1985) 9.
43. S.V.Tsybulya, L.P.Solov'eva, G.N.Kryukova and E.M.Moroz, Zh. Strukt. Khim., 32 (1991) 18.
44. O.A.Kirichenko, M.P.Vorob'eva and V.A.Ushakov, International Congress on Catalysis, Proceeding, Pt. C, Budapest, 1989.
45. O.A.Kirichenko, V.A.Ushakov, E.M.Moroz and Z.R.Ismagilov, React. Kinet. Catal. Lett., 38, 2 (1989) 307.
46. O.A.Kirichenko, V.A.Ushakov, E.M.Moroz and M.P.Vorob'eva, Kinet. Katal., 34 (1993) 739.
47. O.A.Kirichenko, V.A.Ushakov and V.A.Poluboyarov, React. Kinet. Catal. Lett., 51 (1993) 167.
48. R.A.Shkrabina, N.A.Koryabkina, V.A.Ushakov, E.M.Moroz, M.Lausberg and Z.R.Ismagilov, Kinet. Katal., 37 (1996) 116.
49. N.A.Koryabkina, R.A.Shkrabina, V.A.Ushakov, E.M.Moroz, M.Lausberg and Z.R.Ismagilov, Kinet. Katal., 37 (1996) 124 .
50. V.A.Ushakov, O.A.Kirichenko, R.A.Shkrabina, N.A.Koryabkina, M.Lausberg, E.M.Moroz and Z.R.Ismagilov, Kinet. Katal., 37 (1996) 130.
51. R.A.Shkrabina, N.A.Koryabkina, O.A.Kirichenko, V.A.Ushakov, F.Kapteijn and Z.R.Ismagilov, Scientific Bases for Preparation of Heterogeneous Catalysts: 6th International Symposium: Prepr. V.3, Louvain la Neuve, 1994.
52. N.A.Koryabkina, R.A.Shkrabina, V.A.Ushakov and Z.R.Ismagilov, Catal. Today, 29 (1996) 427.
53. V.A.Ushakov R.A.Shkrabina, N.A.Koryabkina and Z.R.Ismagilov, Kinet. Katal., 38 (1997) 133.
54. N.A.Koryabkina, Z.R.Ismagilov, R.A.Shkrabina, E.M.Moroz and V.A.Ushakov, Kinet. Katal., 32 (1991) 1013.
55. R.A.Shkrabina, N.A.Koryabkina, Z.R.Ismagilov, V.A.Ushakov, L.T.Tsykoza and D.A.Arendarskii, Proc. 7th. Nordic Symp. Catalysis, Turku, Finland, 1996.
56. F.Oudet, P.Courtine and A.Vejux, J.Catal., 114 (1988) 112.
57. M. Bettman., R.E. Chase, K. Otto and W.H. Weber, J.Catal., 117 (1989) 447.
58. B.Beguin and E.Garbowski, M. Primet, Apll. Catal., 75 (1991) 119.
59. M.Machida, K.Eguchi and H.Arai, J.Catal., 103, (1987), 385.
60. J.S.Ledford, M.Houalla, A.Proctor and D.M.Hercules, J. Phys. Chem., 93 (1989) 6770.
61. M.F.L.Johnson, J. Catal., 123(1990) 245.
62. L.P.Haack, J.E. de Vries, K.Otto and M.S.Chattha, Appl. Catal., A: General, 82 (1992) 199.
63. J.S.Church, N.W.Cant and D.L.Trimm, Appl.Catal., A: General, 101 (1993) 105.

64. M.Ozawa, M.Kimura and A.Isogai, J.Less Common Met., 162 (1990) 297.
65. N. Kaufherr, I. Mendelovici and M. Steinberg , J. Less. Common. Metal., 107 (1984) 281.
66. M. Che, J.F.J.Kibblewhite and A.J. Tench, J.Chem.Soc.Faraday Trans., 69 (1973) 857
67. J.Z.Shyu, W.H.Weber and H.S.Gandhi, J.Phys.Chem., 92 (1988) 4964.
68. L.G.Appel, C.A.C.Perez, J.G.Eon and M.Schmal, in: Environmental Catalysis G.Centi, (ed.), Proceedings of the 1st World Congress -Environmental Catalysis -for a Better World and Life, Pisa, (1995). - SCI Pub: Rome (Italy) 391.
69. R.K.Iler, J.Amer.Ceram.Soc., 47 (1964) 339. Y.Bulent, J.Mater.Sci., 11 (1976) 465.
70. B.M. Fedorov, V.Ya. Danyushevskii, V.L. Balashov and A.S. Berenblyum, Kinet. Katal., 32 (1991) 447.
71. Z.R. Ismagilov, in: Problems of Catalysts Deactivation, R.A.Buyanov (ed.), Institute of Catalysis, Novosibirsk, 1985.
72. E.D. Shchukin, A.I. Bessonov and S.A. Paranskii, Mechanical Testing of Catalysts and Sorbents, Nauka, Moscow, 1971.
73. L.I. Titelman, Strength Properties of Catalysts and Sorbents. NIITEKHIM, Moscow, 1986.
74. L.F. Eliseeva, R.A.Shkrabina, M.N. Shepeleva, O.A. Kirichenko, V.A. Ushakov, Z.R. Ismagilov and A.S. Petrova, Proc. 1st All-Union Meeting on Catalyst Preparation. Novosibirsk, 1983.
75. L.F. Melgunova and A.A. Samakhov, Standardization of Instruments and Sets for Control of Commercial Catalysts, R.A. Buyanov and N.N. Bobrov (eds.), Institute of Catalysis, Novosibirsk, 1991.
76. R.A.Shkrabina, V.N. Kuklina, M.N. Shepeleva, O.A. Kirichenko, L.F. Eliseeva, G.B. Barannik, E.M. Moroz and Z.R. Ismagilov, Proc. 1st All-Union Meeting on Catalyst Deactivation, Ufa, 1985.
77. E.A. Stepanov, N.N. Peschanskaya and V.V. Shpeizman, Strength and Relaxation Phenomena in Solids, Nauka, Leningrad, 1984.
78. L.I. Titelman, Methods for Analysis and Control of Chemical Industry Commodities, vol.9, NIITEKHIM, Moscow, (1979) 43.
79. O.A. Kirichenko, M.P. Vorob'eva and V.A. Ushakov, Studies in Surface Science and Catalysis, vol.75C, Elsevier, (1993) 2035. Proc. 10th ICC, Budapest, 1992.
80. O.A. Kirichenko, Strength Problems of Granulated Supports and Catalysts, Z.R. Ismagilov (ed.), Institute of Catalysis, Novosibirsk, (1989) 76.
81. O.A. Kirichenko. React. Kinet. Catal. Lett., 51 (1993) 183.
82. M.N. Shepeleva, R.A. Shkrabina, Z.R. Ismagilov and V.B. Fenelonov, Preparation of Catalysts V, G.Poncelet (ed.), Elsevier, Amsterdam, (1991) 583.
83. O.A. Kirichenko, V.N. Kuklina, E.M. Moroz, N.N. Boldyreva, G.YU. Loskutova, L.F. Eliseeva and G.P. Vishnyakova, Method for

Preparation of Modified Granulated Alumina, Russia Patent No 1 135 053 (1984).
84. R.M. German and Z.A. Munir, J. Am. Ceram. Soc., 59 (1976) 379.
85. V.V. Veselov, N.P. Galenko and V.S. Kirpichov, Catalysis and Catalysts, Naukova Dumka, Kiev, 8 (1971) 68.
86. I.Zh. Zainieva, N.S. Kashun, R.A. Shkrabina and Z.R. Ismagilov, Strength Problems of Granulated Supports and Catalysts, Z.R. Ismagilov (ed.), Institute of Catalysis, Novosibirsk, 1989.
87. G.K. Boreskov, E.A. Levitskii and Z.R. Ismagilov, Zh. Vses. Khim. Ob-va Mendeleeva, 29 (1984) 19.
88. A.D. Simonov, Zh. Vses. Khim. Ob-va Mendeleeva, 35 (1990) 107.
89. O.A. Kirichenko, A.O. Bobyreva, R.A. Shkrabina, L.F. Melgunova and N.A. Koryabkina, Khim. Tekhnol., 4 (1991) 12.
90. O.A. Kirichenko, React. Kinet. Catal. Lett., 51 (1993) 197.

Adsorption and its Applications in Industry and Environmental Protection
Studies in Surface Science and Catalysis, Vol. 120
A. Dabrowski (Editor)



# Aluminium hydroxide as selective sorbent of lithium salts from brines and technical solutions

V.P.Isupov[a] , N.P.Kotsupalo[b] , A.P.Nemudry[a] , L.T.Menzeres[b]

[a]Institute of Solid State Chemistry and Mechanochemistry, Kutateladze-18, Novosibirsk, 630128, Russia

[b]Ekostar-Nautech Company, B. Chmelnitsky, 2, 630075 , Novosibirsk, Russia

The disordered aluminium hydroxide and the chloride of the double hydroxide of lithium and aluminium (LADH-Cl) synthesized on the basis of the former species are capable of extraction lithium selectively from complex salt chloride systems while the well crystalline compounds possess these properties to a lesser extent. The formation of LADH-Cl from aluminium hydroxide, desorption and sorption of lithium chloride are connected with the processes of its intercalation-deintercalation in the layered matrix. These processes can be described by the following scheme:

$$LiCl + 2Al(OH)_3 + mH_2O \longrightarrow LiCl \cdot 2Al(OH)_3 \cdot mH_2O;$$

$$LiCl \cdot 2Al(OH)_3 \cdot mH_2O + H_2O \longrightarrow x\,LiCl + (1\text{-}x)LiCl \cdot 2Al(OH)_3 \cdot qH_2O;$$

$$x\,LiCl(brines) + (1\text{-}x)LiCl \cdot 2Al(OH)_3 \cdot qH_2O \longrightarrow LiCl \cdot 2Al(OH)_3 \cdot mH_2O$$

Desorption occurs in the aqueous medium and sorption takes place during the treatment of the sorbent with lithium-containing brine. These sorbents are specific inorganic ones working in the chloride systems within the pH range $3 < pH < 8$. In alkali media at $pH > 8$ LADH- Cl is transformed into the double aluminium-lithium hydroxide LADH or its carbonate form LADH-$CO_3$. Both the latter forms do not possess any sorption properties. In an acid medium at $pH<3$ the sample is dissolved. The sorbent can be prepared in the granulated form which allows one to use the pulse Higgins columns. The obtained sorbent can be reused the many times , and it exhibits high degree of lithium recovery (93 - 96%) from the complex salt systems (natural brines, liquid wastes) and provides the possibility to obtain rather pure lithium chloride solution (5 - 7 g/l).

## 1. INTRODUCTION

Lithium and its compounds are widely used in modern industry to produce chemical current sources, light-weight alloys, special ceramics, etc. Traditional raw materials used to produce lithium are its aluminosilicates (spodumen, lepidolithe). However, in the recent decades, unusual sources of lithium-containing raw material have been developed extensively including naturally occurring lithium-containing mineralized water (continental liquors and solars, liquors of the marine type). One of the major sources of lithium are lithium-containing wastes (side petroleum water, water of diamond mines, etc.).

### 1.1. The types of lithium – containing brines and technical solutions. Industrial technologies of the lithium recovery from brines. The selective extraction methods of lithium from brines and liquid wastes

From the variety of the compositions of natural highly mineralized waters and liquors (see data given in Table 1 according to [1-7], it follows that the existing lithium-containing liquors always contain substantial amounts of chloride ions. In some cases, beside chloride they can contain sulphate and carbonate anions. According to the cation composition, the liquors can be divided into the following major groups: liquors with a single major cation (sodium, potassium, calcium, or magnesium); liquors containing several cations in comparable amounts at one and the same time. The same classification can also be used for liquid lithium-containing wastes (side petroleum water, water of diamond mines, etc.).

The choice of a method of lithium recovery is determined by many factors including concentrations of lithium and major macrocomponents ($NaCl$, $CaCl_2$, $MgCl_2$), general mineralization, technological possibility of performing the process and its economical expediency, as well as regional peculiarities. All the data available from literature concerning the existing and developing methods of lithium recovery from continental mineralized waters can be divided into two groups. The first one includes the methods involving the recovery of lithium in the form of its poorly soluble compounds ($Li_2CO_3$, $LiF$, etc.) after the liquors have been preliminarily repeatedly concentrated. These methods being profitable can be used to treat hydromineral raw material with high lithium content and low content of magnesium and alkaline earths metals in case if a deposit is situated in a region with arid climate, for example Silver Peak (Nevada) and Atakama (Chile).

When a deposit is situated in a region of moderate climate and liquor contains substantial amounts of alkaline earths, it is reasonable to recover lithium by using selective sorbents. The choice of these sorbents is limited by specific properties of lithium cation, namely, small radius (0.68 Å), high hydration energy ($\Delta G_{hydr}$=121 kcal/g·ion), and the complexity of the composition of natural liquors. This does not allow one to use the majority of organic cation exchangers. Among the known sorbents which are selective to lithium ion in liquors are cationites

Table 1
Composition of lithium-containing natural waters and brines of the world's major deposits

| Deposit country | Ion composition (g/l) | | | | | | | | | | Density (g/cm$^3$) | Total salt content (g/l) |
|---|---|---|---|---|---|---|---|---|---|---|---|---|
| | $Li^+$ | $Na^+$ | $K^+$ | $Mg^{2+}$ | $Ca^{2+}$ | $Sr^{2+}$ | $Cl^-$ | $Br^-$ | $SO_4^{2-}$ | $CO_3^{2-}$ | | |
| Lake S, USA (California) | 0.080 | 141.7 | 27.6 | traces | 0.02 | no data | 149.8 | no data | 59.1 | 43.4 | 1.30 | 413 |
| Lake Silver Peak, USA (Nevada) | 0.440 | 68.2 | 8.8 | 0.40 | 0.50 | " | 111.1 | " | 7.8 | — | 1.12 | 195 |
| Chile (Atakama) | 1.960 | 93.2 | 22.0 | 12.3 | 0.30 | " | 192.0 | " | 23.3 | — | 1.30 | 345 |
| Bolivia (de Yuni) | 0.45 | 99.1 | 6.8 | 6.2 | 0.66 | " | 180.0 | " | — | — | 1.18 | 293 |
| Great Salt Lake, USA (Utah) | 0.070 | 85.4 | 4.9 | 9.8 | 0.4 | " | 170.8 | " | 18.3 | — | 1.19 | 266 |
| Dead Sea Israel and Jordan | 0.020 | 37.2 | 6.8 | 37.5 | 15.0 | " | 197.1 | " | 0.7 | — | 1.22 | 298 |
| 111Kara-Bogaz-Gol Bay, Turkmenistan | 0.016 | 58.5 | 5.7 | 43.7 | — | " | 160.3 | " | 83.5 | — | 1.26 | 332 |
| North Caucasus, Russia (Dagestan) | 0.12 | 45.0 | 2.2 | 0.55 | 5.2 | 0.51 | 80.3 | 0.22 | — | — | 1.1 | 140 |
| Angara - Lena basin, Russia (Eastern Siberia) | 0.13 | 42.1 | 13.1 | 11.1 | 46.0 | 1.5 | 200.0 | 3.6 | — | — | 1.2 | 315 |

based on manganese, titanium and niobium oxides [8]; thorium arsenate [9,10], and sorbents based on aluminium hydroxide, crown – ethers.

Sorbents based on manganese and titanium have been tested for different types of liquors, in particular, for iodine and bromine containing waters and liquors of the chloride-sulphate type with high magnesium content (waters of the Kara–Bogaz–Gol Bay). However, despite high sorption capacity of these cationites ($\varepsilon_{Li} = 2 - 10$ mg·equiv./g) they are not widespread in industry that seems to be explained by the complicated procedure of their preparation and high cost. The use of crown-ethers in practice is limited by their high cost.

## 1.2. The application of aluminium hydroxide as selective sorbent of lithium salts from brines and technical solutions

For the liquors of multicomponent composition and of any mineralization, the most promising sorbents of lithium are the compounds based on aluminium hydroxide. The synthesis of aluminium compounds are rather simple and may be realized both in the reaction zone and out of it. The sorption capacity of aluminium hydroxides (~ 8 – 10 mg·equiv./g of $Al_2O_3$) is not worse than that of the cationites based on manganese and titanium. The degree of lithium recovery from liquors with aluminium hydroxide is also influenced by the method of aluminium hydroxide synthesis, molar ratio $Al_2O_3 : Li_2O$ in the reaction mixture, temperature and pH of the process, interaction time, macrocomponent composition of the liquor (concentrations of NaCl, $MgCl_2$, $CaCl_2$ and other electrolytes).

Many papers are devoted to the study of selective properties of freshly precipitated (amorphous) aluminium hydroxide $Al(OH)_n$ , $n \geq 3$ [5-7, 11-14]. The most widely used method of obtaining aluminium hydroxide is its precipitation from solutions. $Al(OH)_n$ can be obtained either by the interaction of soluble aluminium salts with the hydroxides of alkaline or alkaline-earth metals or by lowering pH of alkaline solutions of sodium or potassium aluminates. It is known that aluminium is present in aqueous solutions of acids (pH < 3) and alkalies (pH > 10) as an aquo ion $[Al(H_2O)_6]^{3+}$ and the anion $Al(OH)_4$, respectively. The change of pH of these solutions causes polymerization of ions and precipitation of amorphous aluminium hydroxide which is an efficient sorbent of lithium ions. The aging of these precipitates involves thickening and crystallization of gel followed by the decrease of sorptive capacity [13]. The use of this sorbent does not allow one to achieve high degree of lithium recovery from the liquors (90 - 98 %) in every case. Due to this circumstance, in the previous works freshly precipitated aluminium hydroxide obtained directly in the reaction mixture was used mainly to achieve high degree of lithium recovery from natural solutions and their synthetic analogs closely related by the composition to the former ones.

In many papers under consideration, the molar ratio between $Al_2O_3$ and $Li_2O$ is varied within the range 3.5 – 6.0. However, in some studies this ratio is much higher and reaches 40 [4,5,15]. Effect of sodium, potassium and calcium chlorides on lithium recovery from liquors is found to be not so significant [11,16], though according to [11] the increase of NaCl concentration from 0.5 to 4 M causes some decrease of the degree of lithium recovery. In a series of works magnesium ions were shown to have a negative effect on lithium recovery. Many authors [6,14,16–18] believe this is connected with the possibility of the formation of difficultly soluble double compounds of magnesium and aluminium those are described in literature in detail. In has been shown in [11] that the degree of lithium recovery in the system $MgCl_2(2M)\cdot LiCl(0.2M)\cdot Al(OH)_n\cdot H_2O$ is also substantially dependent on the temperature. At 298K lithium does not interact with aluminium hydroxide for a long time, while at a temperature elevated up to 363K lithium is practically completely precipitated. It should be noted that in the major works mentioned above data on chemical composition of the precipitates are absent. The precipitates formed at lithium recovery with freshly deposited aluminium hydroxide gel are gel-like, poorly sedimented and filtered, with water content contain up to 80-95 %. To improve technological properties of the sediments, amorphous aluminium hydroxide can be deposited onto solid inert supports, e.g., activated carbon.

Data on the mechanism of lithium sorption from liquors with freshly precipitated aluminium hydroxide are limited and contradictory. For example, Goodenough reports on the isolation of "lithium-aluminate" complex [17]. Some authors [11] suggest that lithium is adsorbed on the surface of aluminium hydroxide, and high adsorption coefficient is provided by the presence of salt background which prevents aluminium hydroxide from crystallization when the temperature of the process is raising. Other authors assume that lithium coprecipitation with aluminium hydroxide leads to intercalation of lithium into the structure of the latter and to the formation of lithium-containing aluminium hydroxide characterized by pseudo-boehmite structure [12]. The authors of [13] conclude that the overall interaction of lithium with amorphous aluminium hydroxide can be represented by the following scheme:

$$Al_2O_3\cdot nH_2O + OH^- + Li^+ + (5-n)H_2O \longrightarrow LiH(AlO_2)_2\cdot 5H_2O \quad (1)$$

Thus, there is no doubt that the process of lithium recovery leads finally to the formation of anionic form of double hydroxide of lithium and aluminium (LADH-X).

Double hydroxide of lithium and aluminium, $LiOH\cdot 2Al(OH)_3\cdot pH_2O$ (LADH), was obtained from lithium-containing solutions of aluminates prepared by dissolving metal aluminium in the solution of lithium hydroxide for the first time in 1900 by Allen and Rodgers[19]. Lithium-containing aluminate solutions can be obtained by other methods, e.g. by the interaction of aluminium hydroxide with the solution of lithium hydroxide, or by adding lithium hydroxide to the solutions of sodium aluminate. LADH is formed in all these cases. Acidification of lithium-containing aluminate solutions and alkalization of aqueous solutions of lithium salts lead to the formation of anionic form of double hydroxide of lithium and aluminium (LADH-X).

These data suggest that more detailed studies of the structure of compounds formed and mechanisms of their formation are needed to achieve better understanding of the nature of selective sorption of lithium with aluminium hydroxide.

## 2. THE INVESTIGATION OF SORPTION MECHANISM OF LITHIUM SALTS BY ALUMINIUM HYDROXIDE

### 2.1. Study of LADH-X structure

In literature, there are two alternative models of LADH-X structure. One of the main reasons of this circumstance is that the knowledge of LADH-X structure is based on the studies of powdered samples of these compounds. The first model is based on Feitknecht's idea that double hydroxides consist of alternating layers of the metals hydroxides [20]. Ideas of Feitknecht were developed in [21] with respect to the structure of LADH and its anionic forms. It has been assumed that LADH, LADH-Cl and LADH-$SO_4$ are composed of alternating layers of aluminium hydroxide and lithium salts. Double hydroxide of lithium and aluminium, its sulphate and chloride forms were believed to have a hexagonal cell with the following parameters: LADH - a=8.75 Å, c=22.76 Å; LADH-Cl - a=8.75 Å, c=22.76 Å; LADH-$SO_4$ - a=5.84 Å, c=26.60 Å. It should be noted that the reliability of identification carried out in this work is not so strong due to the poor crystallization of the synthesized compounds and small number of reflexes observed in their X-ray diffraction patterns. An assumption on the possible localization of lithium ions between aluminium-hydroxide layers has been made in [22,23] on the basis of the studies of products formed as a result of the interaction of crystal aluminium hydroxide (gibbsite) with lithium salts. On the basis of the crystal chemistry considerations it is assumed that the basic layer in LADH and LADH-X (X is $CO_3$, $SO_4$, $PO_4$) is similar to that of bayerite[24,25]. Lithium together with water molecules and anions forms another layer. Lithium cations are localized in the trigonal holes located close to vacant octahedral voids of bayerite layer. The same conclusion on the location of lithium in $LiOH{\cdot}2Al(OH)_3{\cdot}2H_2O$ and other anion forms of double hydroxide was made later in [26,27] on the basis of the analysis of crystal optical data.

The second model of the structure is based on the works of Allman and Tailor who identified structures of the minerals (hydrotalcite and manasseite) [28,29] and showed that magnesium and aluminium ions are located in octahedral voids of hydroxide layer. Anions X and water molecules form another layer. Using this approach, Serna assumed on the basis of electron microdiffraction data and X-ray diffraction patterns of powders that the structure of LADH-$CO_3$ obtained by the hydrolysis of aluminium-tri-(sec-butoxide) in lithium carbonate is based on closely packing two-dimensional layer [30 ]. Two thirds of the octahedral voids in this layer are occupied by aluminium cations located in a manner similar to that of gibbsite. The remaining one third of voids is occupied by lithium cations. On the basis of X-ray data this structure is described as having the hexagonal cell with a=5.32 Å and c=15.24 Å. A similar model of the basic layer was proposed by using X-ray and neutron diffraction analysis of $LiOH{\cdot}2Al(OH)_3{\cdot}2H_2O$ obtained by keeping the mixture of bayerite and solid lithium hydroxide over water vapour [31], as well as for $LiX{\cdot}2Al(OH)_3$ (X is Cl, Br, or $NO_3$) obtained by intercalation of lithium salts into gibbsite [32].

In order to study the structure of anionic forms in more detail (including lithium localization) we carried out NMR investigations of $LiCl{\cdot}2Al(OH)_3{\cdot}1.5H_2O$ obtained by the interaction of lithium chloride with gibbsite, as well as the product of its dehydration $LiCl{\cdot}2Al(OH)_3$. The best description of the experimental peak shape in $^{27}$Al NMR spectrum of gibbsite $Al(OH)_3$ (with the mean square difference between the experimental and calculated spectra of approx. 5%) was obtained with the use of superposition of two lines of identical intensity described by the parameters:

$K_1 = 4{,}4 \pm 0{,}1$ MHz, $\eta_1 = 0{,}4$ , $\beta = 3{,}1 \pm 0{,}1$ kHz (Al 1);
$K_2 = 1{,}5 \pm 0{,}1$ MHz, $\eta_2 = 0{,}9$, $\beta = 3{,}1 \pm 0{,}1$ kHz (Al 2). (Fig.1A).

The obtained expansion of $^{27}$Al NMR spectrum in two lines is in agreement with the crystallographic data providing evidence for the presence of two nonequivalent positions of aluminium nuclei in the structure of gibbsite [33]. The shape of $^{27}$Al NMR spectrum of the sample $LiCl{\cdot}2Al(OH)_3$ corresponds to a single position of $^{27}$Al nuclei characterized by quadrupole splitting parameters:

$K = 1{,}9 \pm 0{,}1$ MHz; $\eta \approx 0{,}3$; $\beta = 1{,}9 \pm 0{,}1$ kHz (Fig.1B).

Thus, gibbsite intercalation is accompanied by substantial changes of the gradient of electric field in the sites of the localization of aluminium nuclei, compared to the initial $Al(OH)_3$. Since the main contribution into these gradients

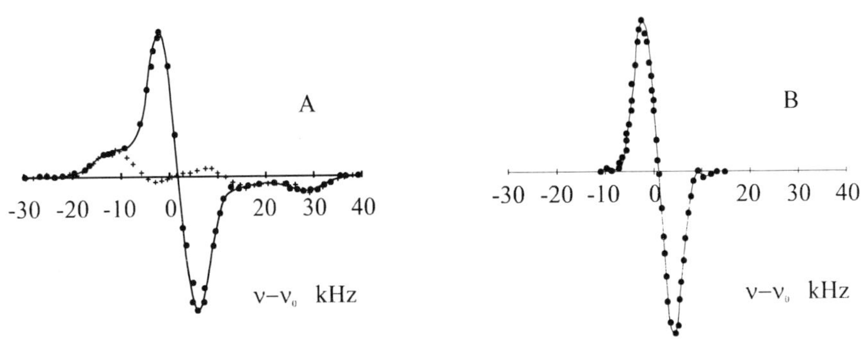


Figure 1. NMR ( $^{27}$Al) spectra of $Al(OH)_3$ (gibbsite) (A) and $LiCl{\cdot}2Al(OH)_3$ (B).

is that from Al interaction with the nearest O and H atoms, then intercalation of gibbsite by lithium salts is accompanied by the change of the first coordination sphere of aluminium ions in gibbsite lattice. The obtained data fit well with the results of the authors [32] who showed, on the basis of X-ray and neutron diffraction studies, that both $LiCl \cdot 2Al(OH)_3 \cdot 1.5H_2O$ and $LiCl \cdot 2Al(OH)_3$ contain one structurally equivalent position of aluminium, variation of aluminium-oxygen distance being in both cases much less than that of gibbsite. $^7Li$ NMR spectra of $LiCl \cdot 2Al(OH)_3$ sample (Fig.2A) are characterized by a fine splitting typical for crystal lithium salts. Fine splitting of the peakes of absorption spectra is described by first-order quadrupole effects. The shape of spectrum was found to be sensitive to the change of radio frequency (RF) field amplitude. In particular, the increase of RF field amplitude (~0.2 G) by a factor of 6 caused a 1.4-fold amplitude decrease of the central part of the spectrum that corresponded to the $-1/2 \leftrightarrow 1/2$ change. Such saturation of the central change and the absence of side line saturation is connected with the existence of an additional mechanism of relaxation at the changes $3/2 \leftrightarrow 1/2$ and $-1/2 \leftrightarrow -3/2$. One of the possible mechanisms can be the influence of changes of the type $3/2 \leftrightarrow -1/2$ with $|\Delta m| = 2$ induced by Raman spin-photon interaction [34]. However, for a sufficiently high efficiency of this process, it is necessary that the dynamics of $Li^+$ ions should be essentially connected with the acoustic branches of the vibrational spectrum of the system. This condition can be met only when lithium ions are incorporated into aluminium-hydroxide layers of the structure. Spectral parameters were determined from the quantitative analysis of the shape of spectrum obtained at the minimum level of RF field generation:

$K = 55{,}2 \pm 0{,}4$ kHz; $\eta \approx 0$; $\beta = 4{,}3 \pm 0{,}2$ kHz .

These parameters are close in the order of magnitude to the quadrupole splitting constants of $^7Li$ in $LiNbO_3$ (K=46 kHz), in lithium nitrate $LiNO_3$ (K=39 kHz) [35]. Therefore, it can be assumed that the structure of the first coordination sphere of

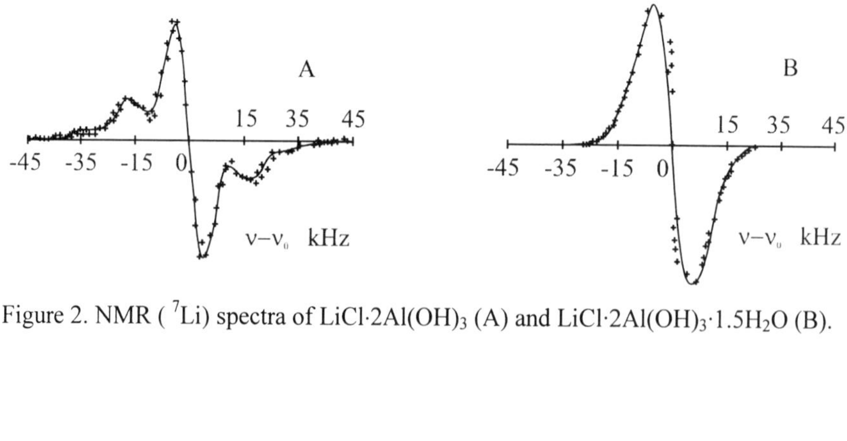


Figure 2. NMR ( $^7Li$) spectra of $LiCl \cdot 2Al(OH)_3$ (A) and $LiCl \cdot 2Al(OH)_3 \cdot 1.5H_2O$ (B).

Li ions in $LiCl \cdot 2Al(OH)_3$ is approximately the same as that of the listed compounds, namely a distorted octahedron. The obtained data on the presence of lithium in octahedral void of aluminium-hydroxide layer are in agreement with the results of [32]. $^7Li$ NMR spectra of hydrated $LiCl \cdot 2Al(OH)_3 \cdot 1.5H_2O$ samples (Fig.2B) strongly differ from those of anhydrous samples recorded under similar conditions. Their width is approximately two times less, fine splitting is not observed. The application of the above algorithm of the analysis leads to the following parameters: K=27,0 ± 0,7kHz; β= 4,3 ± 0,3 kHz. Sharp decrease of the quadrupole splitting constant of lithium nuclei during hydration of the intercalate points out to a substantial influence of water molecules on the gradients of electric field in the sites where Li is located. One can assume that the observed high sensitivity of $^7Li$ NMR to hydration of the system, at a simultaneous insensitivity of $^{27}Al$ NMR, indicate on the localization of water molecules close to lithium ions in the interlayer space. It should also be noted that a narrow signal of low intensity presents in the centre of $^7Li$ NMR spectrum of the hydrated sample. This signal can be assumed to correspond to lithium ions which leave their localization sites in gibbsite layers and diffuse over the interlayer space of intercalate.

NMR spectra of protons of the initial gibbsite sample are wide, without clearly expressed fine structure. Their shape is close to the Gauss one. The mean value of the second moment measured for these spectra is $M_2$=235±10 $kHz^2$. The obtained experimental value is in agreement with that calculated for gibbsite structure. $^1H$ NMR spectra of gibbsite samples intercalated with lithium salts preserve the initial shape, but their widths decrease. $M_2$ values were equal to 112±4 and 156±4 $kHz^2$ for hydrated intercalate and for anhydrous one, respectively. A sharp decrease of $M_2$ in anhydrous intercalate in comparison with the initial value for gibbsite points out to a sharp change of the mode of hydrogen atom localisation into the intercalate and to the increase of mean H-H distances which is in agreement with the above data of $^{27}Al$ NMR and results of neutron diffraction study carried out in [32]. Since $M_2$ of hydrated intercalate is essentially lower than the expected value for hydrates with rigidly bonded water (~360 $kHz^2$), one can assume that the translation diffusion of water molecules is observed in hydrated intercalate. This conclusion is in agreement with the disordering of chlorine ions and water molecules that was observed in $LiCl \cdot 2Al(OH)_3 \cdot 1.5H_2O$ by means of neutron diffraction analysis and X-ray diffraction patterns [32].

Thus, the presented data indicate that lithium cations in anhydrous intercalation compounds synthesized by the interaction of gibbsite with lithium salts are most probably localized in octahedral voids of aluminium-hydroxide layers. It is likely that addition of water causes partial release of lithium into the interlayer space.

## 2.2. Investigation of gibbsite interaction with lithium salts

The interaction of gibbsite with aqueous solutions of lithium salts is described by the equation (2) [36]:

$$Li_nX + 2nAl(OH)_3 + aq = Li_nX{\cdot}2nAl(OH)_3{\cdot}pH_2O. \quad (2)$$

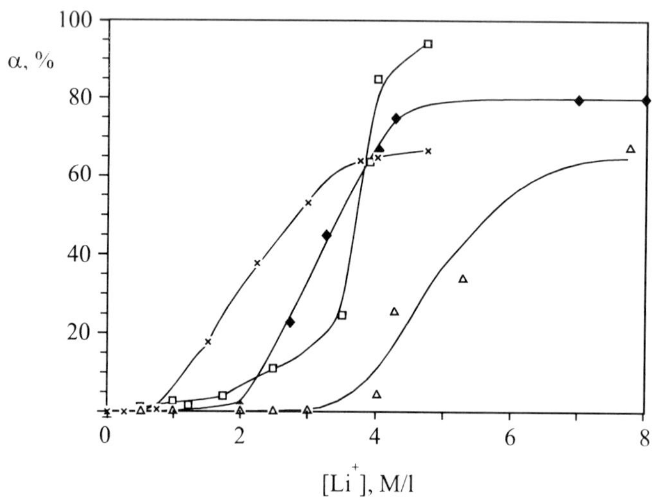


Figure 3. Dependence of the transformation degree (α) of $Al(OH)_3$ into LADH-X on concentration of lithium salt solution (M/l Li). – ◆ –LiCl, – ❍ – $LiNO_3$, —×—$Li_2SO_4$, —Δ— LiBr. Time of treatment: 1hour for LiCl, LiBr and $LiNO_3$; 2 hours for $Li_2SO_4$. T-363K.

The dependence of the conversation degree (α) of gibbsite versus $Li_nX$ concentration is shown in Fig.3. One can see that the conversation degree is non-linearly dependent on lithium salt concentration. The synthesis of LADH-X is essentially dependent on the temperature at which the process is performed. The effect of temperature has been studied in detail for the interaction of lithium chloride with gibbsite fraction 10 - 20 μm as an example. Kinetic curves for different temperatures are shown in Fig.4. Taking into account sharply anisotropic character of gibbsite, kinetic curves were treated in the coordinates of the equations describing the reaction limited by two-dimensional diffusion, as well as the processes at the boundary. Neither of these equations describes intercalation kinetics in the whole range of the observed values of conversation degree. Above 75-80% conversation degree, experimental data are observed to deviate from linearization curves. Activation energy for the initial region of

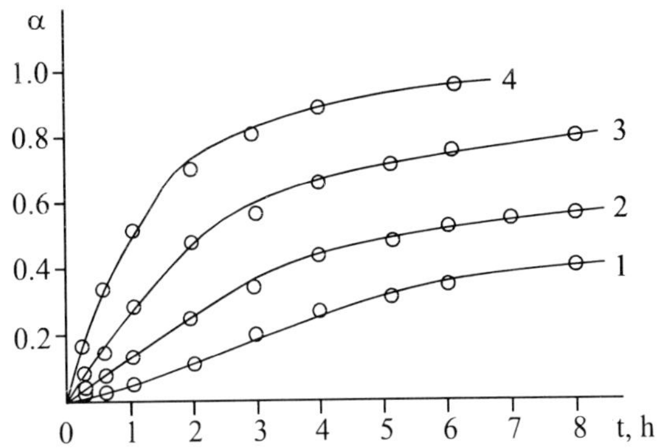


Figure 4. Temperature dependence of the transformation degree of gibbsite into LADH–Cl: 1–323K; 2–333K; 3–348K; 4–363K. [LiCl] – 190 g/l.

kinetic curves is 54±5 kJ/mole for the diffusion equation and 55±5 kJ/mole for the reaction controlled by the processes at the interface. In order to reveal the mechanism of the reaction, X-ray studies of the products of gibbsite single crystal interaction with LiCl, LiBr, $Li_2SO_4$, LiJ were carried out [22,23,36]. Interaction of gibbsite single crystal with lithium salts leads to the formation of highly textured pseudomorphous. It has been shown according to the Laue method and electron microscopy that the product is an aggregate of crystallites with high degree of orientation with respect to each other. To determine orientation relations between the initial gibbsite and the reaction product, the method of crystal rotation was used. The following orientation relations were obtained:

$$[100]_G \text{ II } [100]_{LADH\text{-}X}, (a_G \approx a_{LADH\text{-}X}); [010]_G \text{ II } [010]_{LADH\text{-}X}, (b_G \approx b_{LADH\text{-}X}) \quad (3)$$

It follows from these relations that (100) planes of LADH-X are parallel to the planes (100) of gibbsite. Microscope observations of gibbsite single crystal interaction with an aqueous solution of lithium salts showed a sharp anisotropy of gibbsite-LDAH-X interface movement, namely, its preferred direction is that parallel to basal planes. The anisotropy of the boundary movement is confirmed by the measurements of the profile of chlorine concentration at the boundary [37]. Intercalation is accompanied by fracturing of gibbsite single crystal along the cleavage planes. In order to understand the process of gibbsite interaction with lithium salts, let us turn to gibbsite structure. Gibbsite is a layered mineral crystalline in monoclinic system, lattice parameters being a=8,684A°, b=5,078A°, c=9,736A°, β =94,54°, space group $P2_1/n$. Its structure was determined for the first time by Megaw [38] and defined more accurately by Saalfeld [34]. Gibbsite consists of double layers of $OH^-$ anions in which the $Al^{3+}$ cations occupy two thirds of the octahedral holes within layers. Due to alternating filled and vacant octahedral positions, octahedrons are somewhat distorted. As a result of this deformation, vacant octahedrons increase in size while the filled ones decrease. The layers are linked with each other by hydrogen bonds. The sequence of layers is AB-BA-AB-... Thus, the hydroxide anions of adjacent layers are situated directly opposite each other. Two types of hydrogen bonds can be observed. One of them is connected with the interaction of the hydroxide anions of adjacent layers. Another type of hydrogen bonds is conditioned by the interaction of hydroxide ions within $Al(OH)_3$ layers. These hydrogen bonds are longer.

During the reaction, lithium cations, anions and water molecules attack gibbsite surface, penetrate between the layers and diffuse from the periphery to the crystal centre. One can state that it is rather easy for lithium cation to penetrate the space between aluminium-hydroxide layers. The comparison of lithium cation radius (R=0.68 Å) with the radius of gaps existing between aluminium-hydroxide layers confirms this statement. It is clear from this comparison that lithium cation is to lose of the hydration sheath. Intercalation of lithium cation can lead to the weakening of hydrogen bonds due to electrostatic repulsion between lithium and protons bonding these layers to each other. Weakening of hydrogen bonds provides possibility for anions and water molecules to intercalate between the layers. At the same time, lithium cations seem likely to move from interlayer space inside octahedral holes of the aluminium-hydroxide layer. Fixing of lithium cations inside octahedral holes must lead to the change of the positions of hydroxide anions that form these holes. Firstly, the distance between hydroxide ions and lithium cation should be changed. Secondly, hydroxide ions should turn in a manner providing the decrease of repulsion between lithium cation and protons of hydroxide ions (Fig.5). Penetration of anions and water molecules results the increase of the distance between aluminium-hydroxide layers by 2.8 Å and more. This considerable increase of interlayer distance at the conservation of integrity of aluminium-hydroxide layers

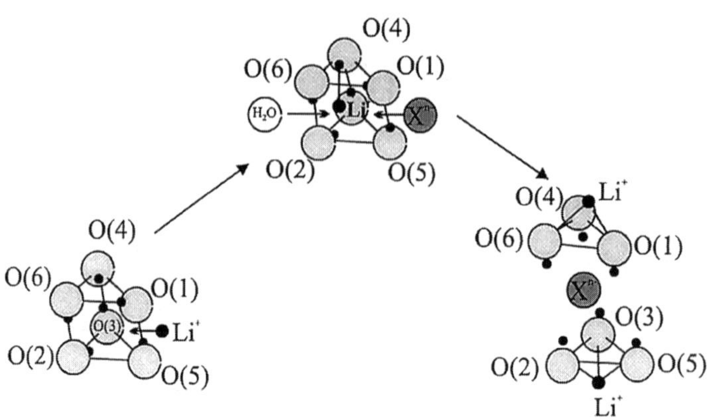


Figure 5. Scheme of $Li_nX$ intercalation into gibbsite.

leads to the appearance of substantial elastic stress at the interface gibbsite–intercalation compound . Relaxation of elastic stress at the boundary leads both to fracturing of the initial gibbsite and to the formation of the zone of plastic deformations in the initial gibbsite in front of the formed LADH-X. For layered compounds, plastic deformation usually occurs as a result of the formation and propagation of Shockly partial dislocations [39]. Partial dislocations lead to a random shift of layers in dislocation ribbon by a distance defined by their geometry, character and localization of interlayered bonds. Intercalation of the lithium salts into plastically deformed gibbsite leads to the fact that the product inherits the disorder in the stacking of Al-OH layers. This formation of stacking faults are observed in block single crystals formed in the interaction of gibbsite single crystals with lithium salts.

This scheme of interaction allows us to explain the dependence of the rate intercalation on lithium salt concentration, type of solvent, as well as on the state of solid phase. In our opinion, non-linear dependence of conversation degree on concentration can be explained as follows. Intercalation process is connected with partial desolvation of lithium cations and anions. The increase of lithium salt concentration in the solution must cause the decrease of desolvation energy which increases the rate of synthesis. Desolvation energy is solvent-dependent. In particular, the change from water to alcohols leads to the decrease of this energy. So, change of solvent should lead to change of the rate of gibbsite interaction with lithium salts which is really observed in the experiment [40].

The proposed scheme of interaction allows one to understand unusual dependence of lithium salt (LiCl, LiBr, $LiNO_3$, $Li_2SO_4$) intercalation degree on the size of gibbsite particles. The decrease of particle size leads to the decrease of the degree of gibbsite conversation into the intercalation compound [41]. A similar anomalous influence of the particle size of solids on intercalation was observed in [42] for kaolinite interaction with organic reagents. The authors believe that elastic stresses arising during the reaction as a result of intercalation into one part of crystal can stop intercalation process at the opposite part of the crystal. This leads to an unusual dependence of the rate intercalation on the particle size. Since the stress should exist at the interface of gibbsite with the reaction product, it can be assumed that one of the possible explanations of the unusual dependence of intercalation rate on particle size is similar to that presented above.

Taking into account that intercalation causes change of hydrogen and van der Waals bonds between aluminium-hydroxide layers and deformation of layers themselves, we can state that intercalation rate can depend on the degree of order of the solid phase.

## 2.3. The methods of the preparation of disordered aluminium hydroxide

The formation of defects in the initial aluminium hydroxide can be initiated by different means: by mechanical activation of aluminium hydroxide, by its thermochemical treatment. Mechanical activation of aluminium hydroxide in high-energy activator mills have been studied in detail in [43-45]. The treatment leads not only to fracturing of the material but also to substantial changes of the compound structure. At the initial stages of mechanical treatment, aluminium-hydroxide layers are shifted with respect to each other in the **ab** plane. Further activation causes destruction of aluminium-hydroxide layers with the formation of X-ray amorphous phase. Amorphous phase is characterized by close packed of oxygen atoms, contains six-coordinated aluminium and substantial amount of water formed during partial dehydroxylation of hydroxide ions. The formed X-ray amorphous phase is composed of porous aggregates with a size of 1-2 μm - several tens of μm. The size of particles comprising the aggregates can be determined from the specific surface area. It is approximately 0.03 - 0.04 μm.

Another known method of preparing disordered aluminium hydroxide is pulse thermal heating of aluminium hydroxide at relatively low temperatures [46]. Pulse heating can be carried out by different means: in the flow of hot gas, with electron beam . Pulse heating in the flow of hot gas leads to a partial dehydration of the initial gibbsite. The degree of dehydration depends on the conditions of thermal treatment. In particular, under certain conditions the formation of completely X-ray amorphous product is possible, its composition being close to that of boehmite (AlOOH).

### 2.4. Intercalation of lithium salts into disordered forms of aluminium hydroxide

The dependence of the degree of gibbsite conversation into LADH-Cl on the time of mechanical treatment is shown in Fig.6 . It can be seen that preliminary mechanical activation results in an essential increase of lithium chloride intercalation degree, compared to the initial crystalline aluminium hydroxide. The conversation degree is also increased for other lithium salts that were studied (bromide, sulphate, nitrate). A continuous increase of the conversation degree with the time of mechanical treatment is observed up to 10 min of activation while X-ray amorphous product is formed. Further growth of the activation time causes the decrease of the conversation degree. This seems to be due to the partial rearrangement, as a result of activation, of the ol-bonds (Al - OH - Al) into oxo-bonds (Al - O - Al) which are characteristic of boehmite, the latter being unreactive towards lithium salts. The interaction of the X-ray amorphous phase with lithium salts leads to the appearance of broadened reflections in the X-ray diffraction patterns which is characteristic of intercalation compounds LADH-X. Thus, the process of lithium salt intercalation into X-ray amorphous aluminium hydroxide is connected with some ordering of its structure. Since the mechanical activation of aluminium hydroxide leads not only to structural distortions in the solid, but is also accompanied by fracturing, a comparison of the total kinetics of lithium chloride intercalation into X-ray amorphous and crystalline aluminium hydroxide with identical specific surface area has been carried out to reveal the contribution of structural distortions into the increase of intercalation rate. The reaction rate was found to be substantially

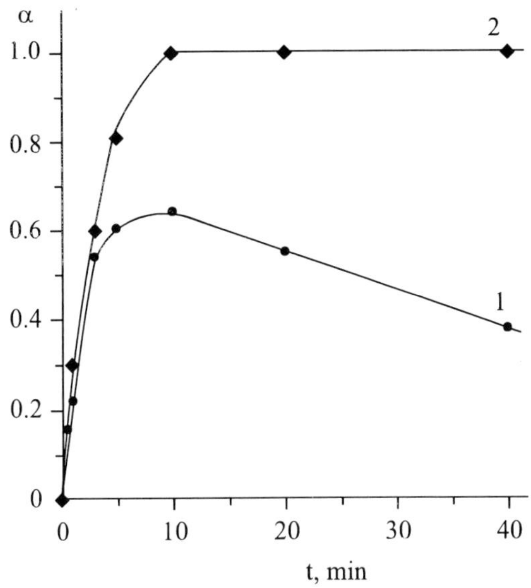


Figure 6. Influence of the time of preliminary mechanical activation of gibbsite on: 1 – the degree of aluminiium hydroxide transformation into LADH–Cl; 2 – X-ray amorphous phase content. Interaction condition: T-363K, time-3h, [LiCl]-60g/l.

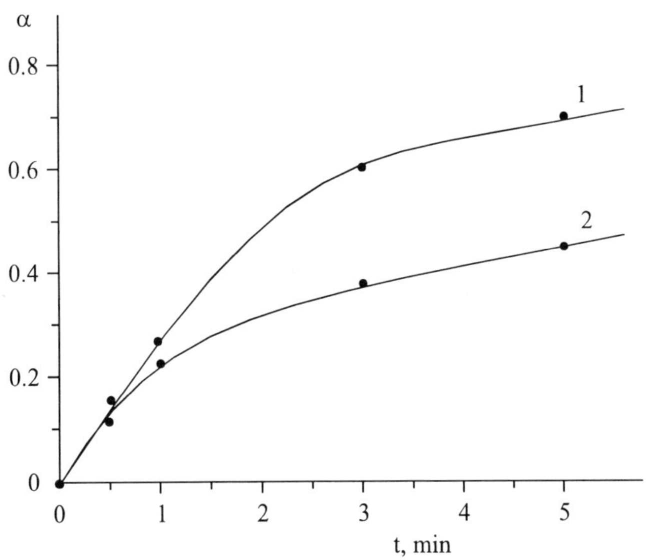


Figure 7. Dependence of the degree of gibbsite transformation into LADH - Cl on the activation time for different fractions of aluminium hydroxide: 1 - 1-3 μm; 2 - >30 μm. T - 363 K, time - 3 h, [LiCl] - 60 g/l.

higher for X-ray amorphous aluminium hydroxide than for crystalline one. This means that a decisive contribution to the growth of chemical reactivity of aluminium hydroxide is determined not by the growth of surface but the change of the degree of order of the solid phase. As we have shown earlier [43], mechanically activated aluminium hydroxide consists of the set of porous aggregates of a micron size which are composed of more fine particles. In order to check the possibility of the influence of diffusion processes inside a particle on the process kinetics, we have separated two fractions, one being 1 - 3 μm in size,and another larger than 30 μm . Fig.7. represents the dependence of the conversation degree of $Al(OH)_3$ in LADH-Cl for the two fractions obtained at different time of mechanical activation. It can be seen that for larger aggregates the conversation degree of aluminium hydroxide is lower. The ratio of the conversation degree to the specific surface area is insignificantly dependent on the size of the fractions for 3 and 5 min of activation. This allows us to assume that the rate of intercalation is not limited by the diffusion of LiCl through the pores of the aggregates but is determined by the processes taking place in the solid.

Preliminary pulse thermal treatment of gibbsite also leads to the increase of the rate of lithium salt intercalation in comparison with crystal gibbsite. It should be noted that the maximum value of atomic ratio of lithium to aluminium is much less than 0.5. This is likely to be due to the presence of inactive, X-ray amorphous boehmite in the products of thermal treatment.

### 2.5. The investigation of lithium chloride deintercalation from well crystalline LADH-Cl

The process of lithium salt intercalation into aluminium hydroxide is reversible. The treatment of intercalation compound with water at $T>70^oC$ and ratio liquid/solid=100 leads to practically complete deintercalation of lithium chloride from the solid phase and to the formation of aluminium hydroxide [47]. Since this value of liq/sol ratio is too high from the technological point of view, it is necessary to study deintercalation at lower liq/sol values. The changes of solid phase composition with the interaction time for lower liq/sol ratio are presented at Fig.8. One can see that the overall rate exhibits its maximum at the beginning of the process and then decreases, and for 0.5 h the kinetic curve reaches its plateau. The similar behaviour is observed for other LADH-X. The decrease of rate can be connected either with the accumulation of lithium salts in the liquid phase or with the inhibiting effect of the solid product formed during intercalation. The influence of lithium chloride concentration in the solution on the rate of deintercalation from LADH-Cl has been investigated (Fig.9). The change from pure water to the solution of lithium chloride with the concentration 5 - 6 g/l causes the decrease of the rate of the process by an order of magnitude. The concentration of the formed lithium chloride at ratio liq/sol=10 is close to 4.5 g/l. Therefore, the decrease of deintercalation rate can be caused by the accumulation of lithium chloride in the solution. In order to estimate the inhibiting effect of the formed solid product (aluminium hydroxide), we treated

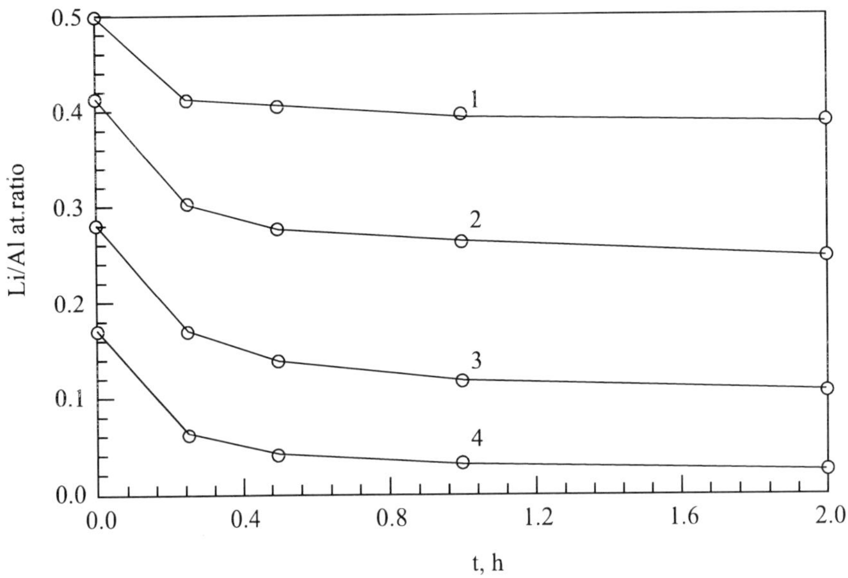


Figure 8. Dependence of composition of solid phase on time of water treatment of LADH-Cl. 1, 2, 3, 4 - number of repeated treatment. T - 348K, liq/solid ratio - 10.

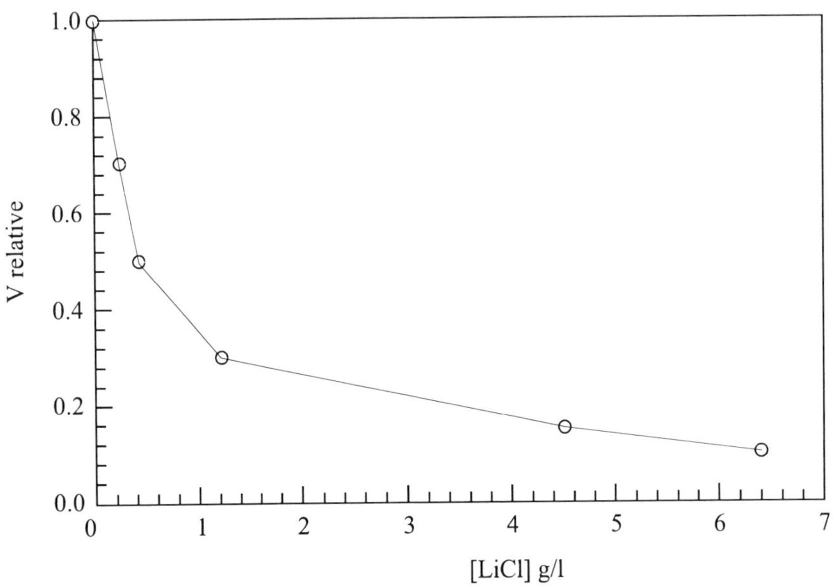


Figure 9. Dependence of the relative rate of lithium chloride deintercalation from LADH–Cl on the initial lithium chloride concentration in the solution. T=348K.

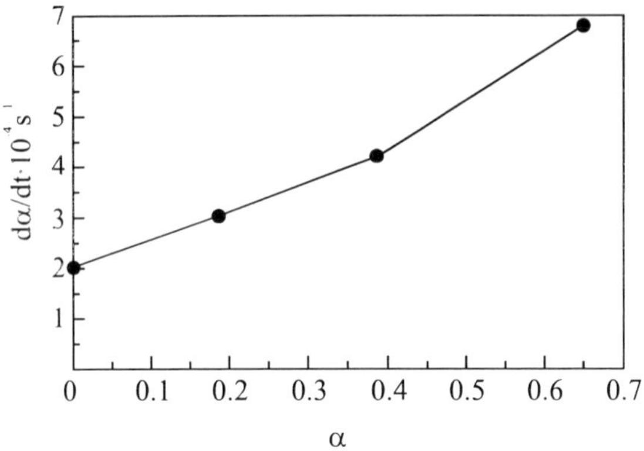


Figure 10. Dependence of the initial rate of LiCl deintercalation from LADH–Cl on deintercalation degree. T=348K.

the solid phase repeatedly with fresh portions of water. It can be seen that for the repeated treatment, deintercalation rate is the highest at the beginning of the process. Then it decreases and the curves reach a plateau. On the basis of kinetic curves at 348K we have evaluated the dependence of the initial deintercalation rate on the degree of deintercalation (Fig.10). Overall rates are seen not to decrease but increase after the repeated treatment. This points out to the fact that the solid product of the reaction (aluminium hydroxide) does not act as an inhibitor up to rather high degree of deintercalation.

### 2.6. Physical-chemical characteristics of solid phases obtained by lithium chloride deintercalation from well crystalline LADH-Cl

The initial gibbsite is composed of spherulite-like splices of prismatic crystals. Aluminium hydroxide formed after the intercalation-deintercalation cycle is composed of particles formed by thin plates 0.1 - 0.3 μm thick. Numerous cracks are often observed at the cleavage face {001}. The diameter of the particles varies within a wide range (1 - 30 μm) (Fig.11). The largest changes in the dispersity of the initial aluminium hydroxide occur during intercalation. Deintercalation causes smaller changes in the dispersity of the solid phase. The formed aluminium hydroxide differs from the initial gibbsite by physicochemical characteristics. Specific surface area of the former (6-7 $m^2/g$) is much higher than that of the initial gibbsite (0.2 $m^2/g$). The position of some reflection peaks originating from the hydroxide formed is close to that of gibbsite reflexes (002, 112, 330, 211, etc.). However, the former reflexes are broadened. Besides, some gibbsite reflexes (200, 202, $\underline{2}02$, $\underline{3}16$, $\underline{4}24$) are observed to disappear from the X-ray diffraction patterns (Fig.12). The presence of some reflexes characteristic of bayerite allows one to assume that this phase is present in small amounts in the deintercalation products. The broadening reflexes of the

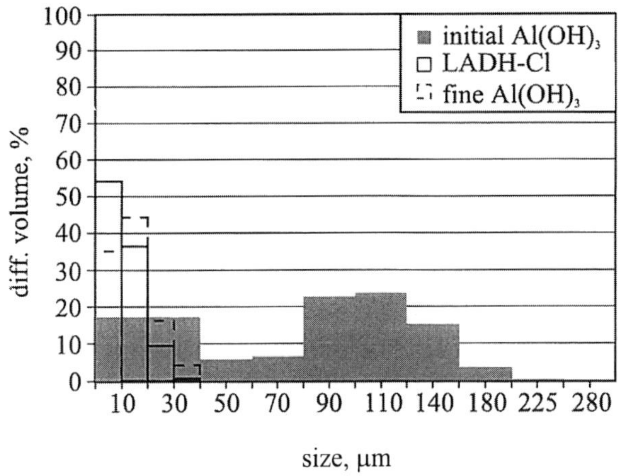


Figure 11. Distribution particle size of initial crystal gibbsite, LADH–Cl and fine $Al(OH)_3$.

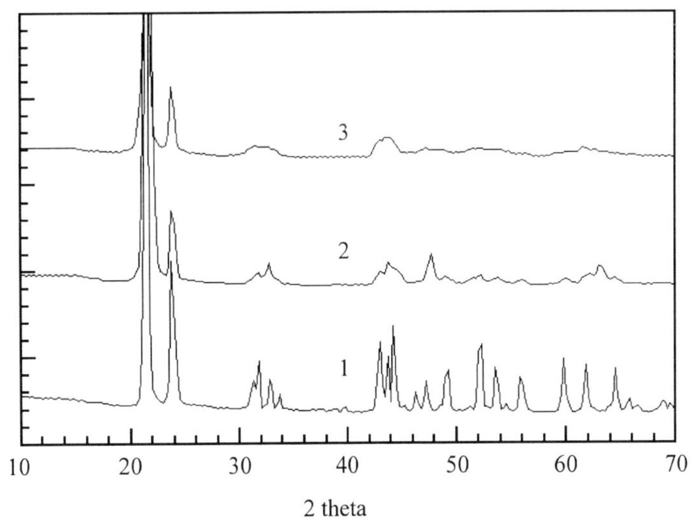


Figure 12. XRD patterns of initial gibbsite (1) and aluminium hydroxide obtained by deintercalation of: LiCl from LADH–Cl (2); Li–benzoate from LADH–benzoate (3).

aluminium hydroxide that is being formed can be due either to small size of the formed particles or to the presence of defects in the solid. According to the Sherer equation, taking into account the dispersity of the particles, we estimated microdeformation ($\Delta d/d$) for the reflex (002) for aluminium hydroxide samples obtained at the complete and partial deintercalation of lithium chloride. The mean value of microdeformation (0.0051) in the hydroxide formed by partial deintercalation is larger than the corresponding value for the complete deintercalation (0.0028). Taking into account the layered structure of aluminium hydroxide, one can assume that the presence of microdeformations in the formed hydroxide is due to shift of aluminium-hydroxide layers with respect to each other in the ab plane. The formation of microdeformations in a solid is likely to occur not only at the stage of lithium intercalation into gibbsite which has been mentioned above but also during deintercalation. In view of the layered structure of LADH-Cl, its interaction with water should be initiated by the intercalation of water molecules in the layer containing anions. Then the lithium chloride molecules should leave the periphery of the LADH-Cl crystal. The removal of lithium chloride will lead to the formation of the aluminium hydroxide phase at the periphery of the crystal. In course of deintercalation, the zone of aluminium hydroxide should move towards the crystal centre. Lithium chloride deintercalation is accompanied by a substantial decrease in the molar volume of the substance (by about 30 %). This circumstance, as well as the fact that the initial substance and the reaction product incorporate common aluminium-hydroxide layers explain why the elastic stress arises at the interphase between aluminium hydroxide and the intercalation compound. Relaxation of the elastic stress can lead either to fracturing of the solid or to the shift of aluminium-hydroxide layers with respect to each other in the ab plane, i.e. to microdeformations. The existence of the elastic stress zone in the product of partial deintercalation near the interphase boundary is likely to increase the deficiency of the aluminium hydroxide formed. The elastic stress at the interphase boundary should increase when the distance between aluminium-hydroxide layers increases. Because of this, we can assume that larger anion X would lead to higher degree of deficiency in aluminium hydroxide formed during deintercalation. Really, X-ray reflection peaks of aluminium hydroxide obtained by lithium benzoate deintercalation from LADH-$C_6H_5COO$ are wider than the reflexes of the hydroxide synthesized from LADH-Cl (Fig.12).

## 2.7. The synthesis of disordered LADH-Cl. Lithium chloride deintercalation from disordered LADH-Cl

The synthesis of disordered LADH-Cl was performed according to different procedures: (1) the interaction of mechanically activated $Al(OH)_3$ with the aqueous solution of lithium chloride [48]; (2) the precipitation from the solutions of lithium chloride and aluminium by adding an alkali [17,18]; (3) the anode dissolution of metal aluminium in the solution of LiCl [49]; (4) the treatment of the mixture composed of crystalline $Al(OH)_3$ and LiCl ($LiCl{\cdot}H_2O$) in the mill-activators [50]. One can see (Table 2) that all the samples of the disordered LADH-Cl (No. 1 - 4) obtained under different synthesis conditions are close to each other in chemical composition and are distinguished by the nonstoichiometry of lithium, compared to the well crystalline sample (5), the

Table 2
Compositions of the synthesized samples of disordered LADH-Cl (1-4) and well crystalline LADH-Cl(5)

| Nr | Chemical composition, % mass , % | | | Molar ratio | | S $m^2g^{-1}$ | Formula |
|---|---|---|---|---|---|---|---|
| | LiCl | $Al(OH)_3$ | $H_2O$ | $LiCl/Al(OH)_3$ | $H_2O/LiCl$ | | |
| 1 | 16.1 | 65.6 | 18.3 | 0.45 | 2.7 | 5.5 | $LiCl{\cdot}2.2Al(OH)_3{\cdot}2.7H_2O$ |
| 2 | 11.0 | 58.1 | 28.9 | 0.35 | 6.2 | 5.8 | $LiCl{\cdot}3.1Al(OH)_3{\cdot}6.2H_2O$ |
| 3 | 13.8 | 66.7 | 19.5 | 0.38 | 3.3 | 6.2 | $LiCl{\cdot}2.6Al(OH)_3{\cdot}3.3H_2O$ |
| 4 | 13.3 | 66.8 | 19.9 | 0.36 | 3.5 | 4.5 | $LiCl{\cdot}2.7Al(OH)_3{\cdot}3.5H_2O$ |
| 5 | 18.3 | 70.5 | 11.2 | 0.48 | 1.4 | 3.0 | $LiCl{\cdot}2.1Al(OH)_3{\cdot}1.4H_2O$ |

composition of the latter being close to the ideal $LiCl{\cdot}2Al(OH)_3{\cdot}1.5H_2O$. An indication of this fact is the decrease of the molar ratio $LiCl{:}Al(OH)_3$ from 0.5 to 0.35 - 0.45. The X-ray diffraction patterns of all the samples of the disordered LADH-Cl are characterised by the close location of the reflexes and their broadening, compared to the well crystalline LADH-Cl. The broadening of the reflexes is evidence of both the decrease of the coherent length and the appearance of disordering in the packing of aluminium-hydroxide layers. In spite of lithium nonstoichiometry, X-ray diffraction patterns do not exhibit reflexes characteristic of aluminium hydroxides (gibbsite, bayerite). The treatment of disordered LADH-Cl with water leads to deintercalation of lithium chloride from them. However, unlike well crystalline LADH-Cl, deintercalation leads finally not to the disordered gibbsite with bayerite as an admixture but to the disordered bayerite. As in the case of well crystalline LADH-Cl, temperature elevation leads to the increase in deintercalation rate while the increase in lithium concentration in the liquid phase causes the decrease of deintercalation rate. It is interesting to note that a partial (<35 %) deintercalation of lithium chloride from the disordered LADH-Cl goes on with the conservation of the structure of the initial compound. The increase of deintercalation degree above this value causes the appearance of bayerite rerflections in the X-ray diffraction patterns. It can be assumed that in case of disordered LADH-Cl, the formation of solid solutions on the basis of the initial compounds is possible.

## 2.8. The study of the influence of conditions of the disordered aluminium hydroxide preparation and other factors (temperature, time, type of brine ) on selective lithium sorption

A substantial increase in the rate of interaction between the disordered aluminium hydroxide and lithium chloride, compared to the rate observed for the initial gibbsite, allows to use the disordered forms for the selective sorption of lithium from brines and technological solutions. Mechanically activated samples of aluminium hydroxide were obtained by the activation of commercially available gibbsite samples in a planetary activator AGO-2. The weight ratio between the sample of $Al(OH)_3$ and steel balls was 1:20. Acceleration created in the mill drums was ca. 60 g. Mass of $Al(OH)_3$ to be treated- 10 g; the diameter of steel balls-5 mm. Time of activation was varied from 0.5 to 15 min. The treatment of the initial aluminium hydroxide and the products of mechanical activation with water solutions of the lithium salt and with the solutions imitating natural brine was carried out in hermetically sealed teflon vessels equipped with agitators and hydrolocks. After the experiment, the solid phase was separated from the liquid one by filtering. Lithium was determined in the liquid and the solid by means of flame photometry. The degree of conversation of aluminium hydroxide into LADH-Cl was estimated from the atomic ratio of lithium to aluminium in the obtained samples. The degree of lithium recovery from brines was estimated from the changes of lithium content of the brines. The initial crystalline gibbsite does not sorb lithium from a model solution (Table 3). The activation of the initial gibbsite only for 3 min causes a substantial (66 %) recovery of lithium from the solution. The increase of the activation time up to 10 min causes practically complete sorption of lithium from liquid phase to the solid. Because of this, the time of preliminary mechanical treatment was chosen to be 10 min for further experiments. In order to test the influence of the molar ratio Al/Li, experiments were carried out in which this ratio was changed from 2.6 to 10.8. As Table 3 suggests, practically complete lithium recovery is observed for the ratio value within the range 9 - 10.8. Temperature dependence of the recovery degree presented in the same Table shows that heating up to 333K or even higher is needed to achieve a high degree of lithium recovery from the model solution. Half an hour is sufficient for a complete lithium recovery. The substitution of calcium chloride by sodium chloride in the initial solution does not introduce any substantial changes into the sorption activity. The use of magnesium chloride as a salt background causes a sharp decrease in the lithium recovery degree to 32 % which is due to the competitive effect of magnesium ions. The possibility of magnesium chloride to form a compound with the X-ray amorphous aluminium hydroxide is confirmed by the chemical analysis of the solid phase obtained in the experiment 22 (atomic ratio Li/Mg = 0.66). In order to check the influence of the salt background on lithium chloride sorption, we studied the interaction of pure sodium, calcium and magnesium chloride solutions with the X-ray amorphous aluminium hydroxide. In the case of calcium and sodium chlorides, the interaction does not lead to the formation of

Table 3
The influence of some parameters on the degree of lithium recovery from model solutions (experiments 1 - 22) and natural brine of North Yakutia (23)

| No. exp. | Activation time, min. | Molar ratio Al/Li | Sorption temperature K | Sorption time, h | The degree of recovery, α % |
|---|---|---|---|---|---|
| 1 | 0 | 10.9 | 343 | 2 | 0 |
| 2 | 3 | – | – | – | 66 |
| 3 | 5 | – | – | – | 85 |
| 4 | 10 | – | – | – | 98 |
| 5 | – | 2.6 | – | – | 37 |
| 6 | – | 3.9 | – | – | 50 |
| 7 | – | 5.1 | – | – | 64 |
| 8 | – | 6.6 | – | – | 77 |
| 9 | – | 7.7 | – | – | 88 |
| 10 | – | 9.2 | – | – | 94 |
| 11 | – | 10.8 | – | – | 98 |
| 12 | – | – | 293 | – | 85 |
| 13 | – | – | 313 | – | 95 |
| 14 | – | – | 323 | – | 95 |
| 15 | – | – | 333 | – | 98 |
| 16 | – | – | 343 | – | 97 |
| 17 | – | – | 353 | – | 96 |
| 18 | – | – | 333 | 0.5 | 98 |
| 19 | – | – | – | 1.0 | 95 |
| 20 | – | – | – | 2.0 | 98 |
| 21 | – | – | – | – | 95 |
| 22 | – | – | – | – | 32 |
| 23 | – | 12.0 | 342 | – | 96 |

In exp.1-20 $CaCl_2$- 117 g/l. In exp. 21 NaCl- 117 g/l. In exp. 22 $MgCl_2$- 95 g/l.
In exp.23 $CaCl_2$ -180 g/l, $MgCl_2$- 52 g/l, NaCl- 76 g/l. In exp. 1-23 LiCl- 0.96 g/l.

intercalation compounds. The interaction of magnesium chloride leads to the appearance of substantial amounts of magnesium in the solid phase. The X-ray diffraction patterns of the solid phase contain broadened reflexes, their positions being close to those of LADH-Cl. It should be noted that the X-ray diffraction patterns of the products obtained in the interaction of the X-ray amorphous aluminium hydroxide with the model solutions also contain (besides the LADH-Cl reflexes) the low-intensity reflexes characteristic of bayerite. So, the interaction is accompanied not only by lithium chloride intercalation into aluminium hydroxide but also by partial crystallization of the X-ray amorphous aluminium hydroxide. This circumstance is one of the reasons explaining the necessity of using a substantial excess of the sorbent compared to the stoichiometric amount. Experiments carried out using the brines of North Yakutia show the possibility of practically complete lithium recovery from them (Table 3). To obtain X-ray amorphous aluminium hydroxide, other energy-tense activators of the planetary and other types can be used.

### 2.9. The application of well crystalline and disordered LADH-Cl as a reversible selective sorbent for the extraction of lithium from brines and technological wastes

The studies into the possibility of using intercalation compounds of aluminium hydroxide as reversible selective sorbents of lithium from brines are of special interest. The works in this direction were described for the first time by Lee and Bauman [51-53]. These works involved the synthesis of intercalation compounds inside the pores of macroporous anion exchange resin with the inner surface not less than 10 $m^2/g$ and the porosity $\geq 15$ % (pore size was ~ 20 - 200 nm). The content of the sorbent in the composite material was not more than 50 %. The presence of the anion exchange resin complicates the investigation of the intercalation compound during the sorption-desorption cycles. Because of this, at the first stage of investigations it was reasonable to study sorption-desorption of lithium chloride using well crystalline LADH-Cl.

### The study of well crystalline LADH-Cl

We have studied the interaction of the aluminium hydroxide (obtained by lithium chloride deintercalation from LADH-Cl) with the aqueous solution of lithium chloride (Fig.13). Kinetic curves are composed of the two regions with sharply different rates. The reaction rate slows down when the composition of the solid phase corresponds to one mole of LiCl per 2.5 moles of $Al(OH)_3$. The increase in the temperature leads to the rate increase in the first region. At the same time, the change of temperature has only a weak effect on the rate in the second region. The rate of the lithium chloride intercalation into the disordered hydroxide is higher than that observed for well crystalline gibbsite with the same specific surface area. These data allow us to assume that the intercalation rate is correlated with the degree of order of the solid phase. The initial intercalation rate depends on the degree of lithium chloride deintercalation (Fig.14). When the intercalation degree is higher than 0.50 - 0.60 the intercalation rate decreases sharply. This fact allows one to reuse the well crystalline LADH-Cl in the repeated intercalation-deintercalation cycles in case of the partial

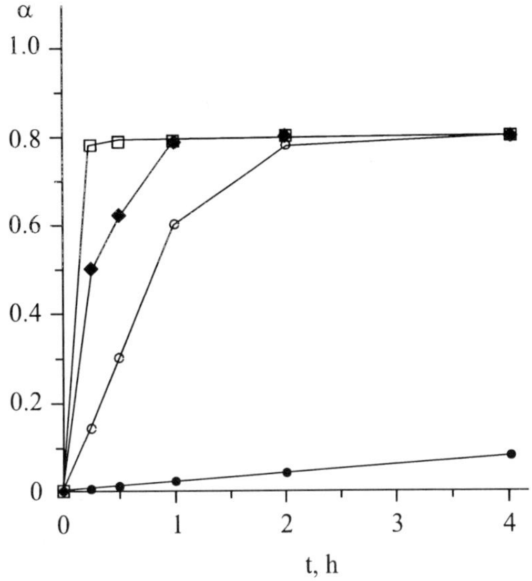


Figure 13. Temperature dependence of the transformation degree of: (□,◆,○)–$Al(OH)_3$ obtained by deintercalation lithium chloride from well crystalline LADH–Cl; ● – fine well crystalline gibbsite. ○ – 303K; ◆ – 308K; □,● – 313K. [LiCl]–190g/l.

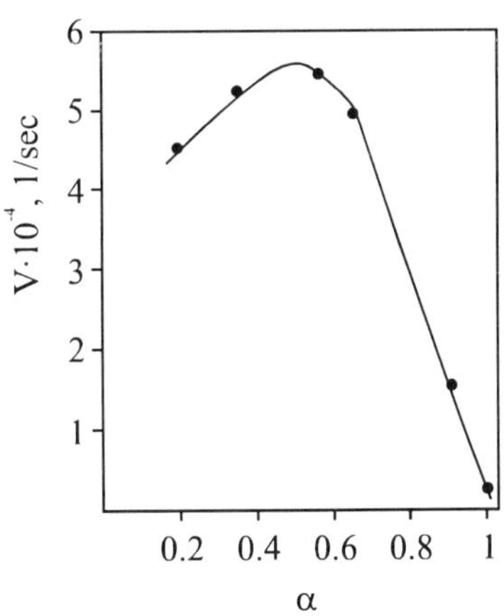


Figure 14. Dependence of the relative rate of LiCl intercalation into $Al(OH)_3$ on deintercalation degree (α). T=360K. [LiCl]=120g/l.

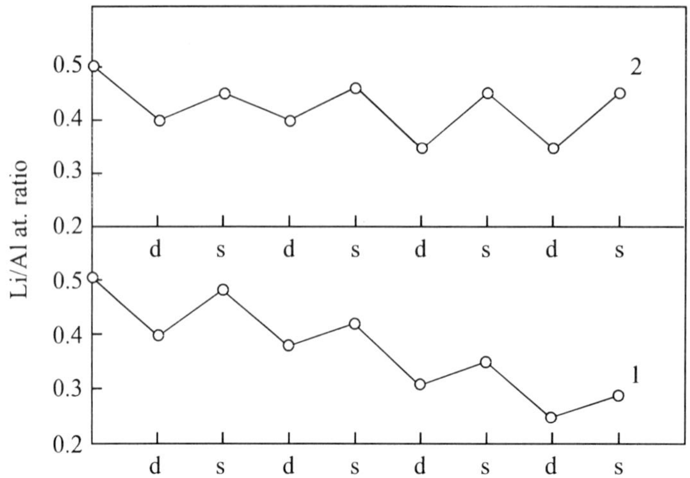


Figure 15. Variation of the chemical composition of well crystalline LADH–Cl during desorption (d) and sorption (s) processes. Desorption conditions: T–348K; t–0.5h; liq/sol. ratio–10; [LiCl]–0g/l (1), 0.25g/l (2). Sorption conditions: T–363K; t–1h; [LiCl]–60g/l.

deintercalation. The intercalation degree was found to be not higher than 20 - 30 % (Fig.15). Besides, a stability of the sorbent requires that desorption should be carried out using not pure water but a diluted lithium chloride solution.

**The study of disordered LADH-Cl**

In order to study the disordered forms of the intercalation compound as the reversible sorbents we synthesized the samples based on mechanically activated aluminium hydroxide. The reversibility of the sorbent can be possible in case if the intercalation degree is not higher than a definite value (Fig.16). When the intercalation degree exceeds this value, the sorbent degrades. The investigation of the phase composition shows that crystal bayerite is gradually accumulated in the sorbent. Similar results were obtained when we used the disordered LADH-Cl obtained by other methods. The studies of sorption from the brines shows that the presence of the salt background helps recovering the initial composition of the sorbent at smaller lithium chloride concentration. The use of the defect forms of LADH-Cl for sorption from the brines allows to increase the sorbent capacity in comparison with well-crystalline LADH-Cl (Fig.17) and to conduct the sorption process at a room temperature. It is necessary for the conservation of the sorbent efficiency that the desorption degree should not exceed 40 %. The data obtained are in agreement with the results reported by Lee who found that up to 50 % of lithium chloride could be removed from the sorbent based on macroporous anionites at the conservation of the sorbent efficiency.

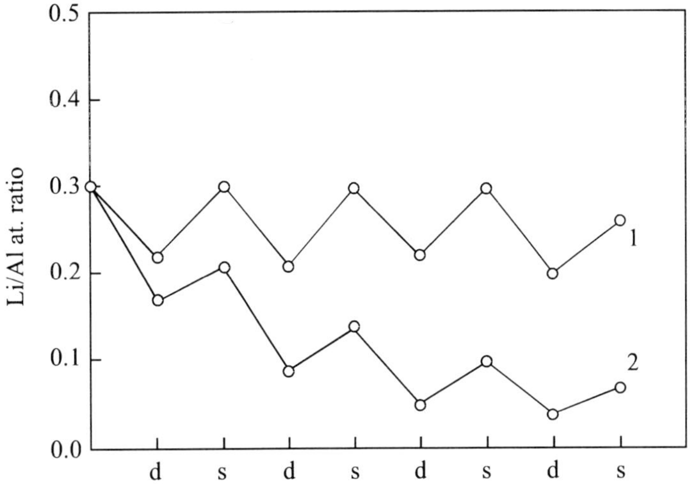


Figure 16. Variation of the chemical composition of disordered LADH–Cl during desorption (d) and sorption (s) processes. Desorption conditions: T–348K; t–0.5h; liq/sol. ratio–10; [LiCl]–0.25g/l (1), 0g/l (2). Sorption conditions: T–323K; t–1h; [LiCl]–60g/l.

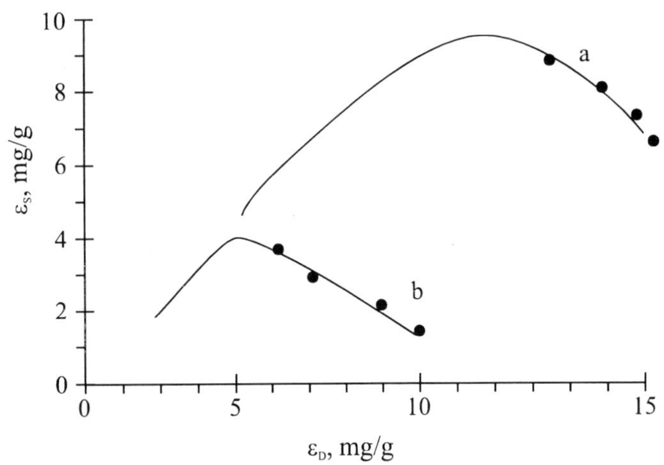


Figure 17. Dependence of the sorptive capacity ($\varepsilon_S$) on the value of lithium desorbed from LADH–Cl ($\varepsilon_D$). A–disordered LADH–Cl, B–well crystalline LADH–Cl. ● – formation of gibbsite (a), bayerite (b).

## 2.10. The technology of lithium extraction process

Since the disordered LADH-Cl is a finely dispersed powder, it is necessary to granulate it when using in columns. In spite of the synthesis of many inorganic ion exchange materials during the last few decades, there are only few methods of their granulation. The authors of [54,55] pressed the fine sorbent as the composite fibers, the maximum sorbent content being 50-80 %. Polyacrylic fiber was used for this purpose, dimethylformamide being the solvent [54]. Complicated composites based on fluorocarbon plastics, ethylene glycole and surfactants were used. Dimethylformamide or acetone was used as a solvent [55]. The authors of [56, 57] considered the possibility of preparing granules or uniform porous blocks composed of ionite and a polymer binder. Polytetrafluoroethylene [56] or polyurethane-based plastics [57] were used as a binder. The content of the ionite powder with a particle size 30 μm and smaller did not exceed 50 % of the composite mass. The methods considered in these works involve either a complex procedure of the composite preparation or by low content of the sorbent in the composite.

Acetone-soluble fluorocarbon plastics were selected for use in the granulating procedure. These plastics form a film (when dried) which is stable against the corrosive medium, e.g. lithium chloride or brines. Tetrafluoroethylene-vinylidene fluoride copolymers (grades F-42L, F-32L, F-23L) and vinylidene-hexafluoropropylene copolymers (grades F-26, F-26L) were used [58]. Mechanical strength of the granules increases with increasing the amount of fluorocarbon plastics added into the mixture; however, the increase of the binder above 8% is accompanied by decreasing of the kinetic properties of the granulated material. Because of this, the following mixture composition was considered as an optima one (mass. %):

disordered LADH-Cl 67.0-55.8
fluorocarbon plastic F-26 4.4- 5.9
acetone 28.6- 38.6

As calculated for a dried solution-free product, LADH-Cl accounted for 91 - 93 % and fluorocarbon plastic for 7 - 9 %. The results of the studies show that the sorption kinetics depends on the granule size. When increasing the granule size from 2 - 4 mm to 5 - 9 mm, the sorption capacity of the sample decreases by about 1.5 times. Because of this, all the further studies have been conducted using the granules with a size of 2 - 4 mm containing 7.5 - 8 % of the binder. The use of fluorocarbon plastic as a binder allows to obtain mechanically strong granules. However this leads to some decreasing of the sorption-desorption kinetics. In case of multiple sorption-desorption cycles the kinetic characteristics of the sorbent are improved. Under the static conditions when the equilibrium is achieved, sorption and desorption isotherms (Fig.18) show that the reversible overall exchange capacity is ~ 7 mg of lithium per 1g of the granulated sorbent. One can see from the behaviour of the desorption isotherm that the concentration of LiCl in the eluates cannot exceed 10 - 12 g/l. The sorption capacity of the granulated sorbent under the dynamic conditions was tested using a laboratory-type column in the filtering regime. 14 cycles of lithium sorption and desorption were carried

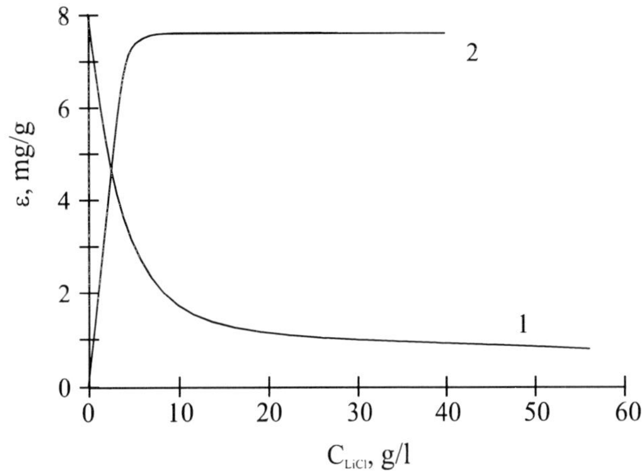


Figure 18. 1 – isotherm of lithium desorption. 2 – isotherm of lithium sorption from brine. $T_{exp}$ – 293K.

out. The volume of the brine (with the mineralization of about 330 g/l and LiCl content ~ 1 g/l) that passed through the column was 270 l. We obtained 90 l of the eluate with the mean LiCl concentration ~ 3 g/l. LiCl concentration reached 5 - 7 g/l in separate portions of the eluate. The rate of the brine flow was varied within 0.4 - 1.5 l/h (1.8 - 6.7 l/h per 1 kg of the sorbent). The total dynamic exchange capacity (TDEC) was ~ 6 mg/g, while the working exchange capacity (WEC) was ~ 40 % of the TDEC value. Lithium recovery prior to the achievement of WEC was 93 - 97 %. Unlike organic ionites, the inorganic sorbent LADH-Cl swells only slightly in solutions. During prolonged tests (~6 months) in the filtering regime we did not observe any changes in the granulometric composition. We also did not observe the appearance of any fractions with the particle size < 0.25 mm. The industrial tests of the granulated sorbent using a Higgins-type column filled with ~ 0.4 t of the granulated sorbent [59,60] under the conditions involving step-counterflow hydrodynamic regime of the sorbent movement (with the pulse shift of the sorbent 4 times an hour) confirmed the data obtained under laboratory conditions. Granule abrasion in this regime was ~0.5 % per day.


## ACKNOWLEDGEMENTS

We thank Prof. S.P.Gabuda and Dr. S.G.Kozlova for NMR investigations of intercalation compounds.


## REFERENCES


1. E.V. Pinneker, Rassoly Angaro - Lenskogo artezianskogo basseyna, Nauka, Moscow, 1966.

2. B.I. Kogan, Redkie metally, Nauka, Moscow, 1979.
3. M. Coad, Industrial Minerals No. 205 (1984) 27.
4. P.A. Sturm, Bull. Utah Geol. and Miner. Surv., 116 (1980) 147.
5. D. Kaplan, Israel J. Chem., 1 (1963) 115.
6. I. Pelly, J.Appl. Chem. Biotechnol., 28 (1978) 469.
7. J.A. Epstein, E.M. Feist, I. Zmora and Y. Marcus , Hydrometallurgy, 6 (1981) 269.
8. V.V. Volkhin, G.V. Leontyev and S.A. Onorin, in: Khimiya i tekhnologiya neorganicheskikh sorbentov , Izd. PPI, Perm, 1980, 67.
9. V. Vasely and V. Pekarek, Talanta, 19 (1972) 219.
10. G. Alberti and C. Palocco, Method of purifying and concentrating lithium ions, US Patent No. 3 851 040 ( 1975).
11. Kang Ding-Xue, Shen Xiang-Mu and Wan Xue-Yuan, J. Chem. Tech. Biotechnol., 31 (1981) 687.
12. M. Frenkel, A. Glasner and S. Sarig, J.Phys.Chem., 84 (1980) 507.
13. H. Wada, N. Kitamura, A. Fujii and S. Katon, The Chemical Society of Japan, 7 (1982) 1152.
14. V.V. Volchin, E.I. Ponomarev and V.L. Zolotavin, Izv. AN SSSR, Neorgan. materialy, 1 (1965) 1573.
15. R.L. Retallack, Electrolytic recovery of lithium from brines, US Patent No. 4 251 338 (1979).
16. A. Ramazanov, Khimiya i tekhnologiya vody, 13 (1991) 140.
17. R.D. Goodenough , Recovery of lithium , US Patent No. 2 964 4381 (1960).
18. N.P. Neipert and C.L. Bon, Improved method for recovering lithium from brines, US Patent No. 3 306 700 (1966).
19. E.F. Allen and H.F. Rodgers, Amer. Chem. J., 24 (1900) 304.
20. W. Faitknecht, Fortschritte der Chemischen Forschung, 2 (1953) 670.
21. V.P. Danilov, I.N. Lepeshkov and L.T. Kotova, in: Redkie schelochnie elementy, Izd. PPI, Perm, 1969, 65.
22. A.P. Nemudry, V.P. Isupov, N.P. Kotsupalo and V.V. Boldyrev, Izv. SO AN SSSR, Ser. Khim. Nauk, 5 (1984) 47.
23. A.P. Nemudry, V.P. Isupov, N.P. Kotsupalo and V.V. Boldyrev, Reactivity of solids, 1 (1986) 221.
24. E.T.Devyatkina, N.P. Kotsupalo, N.P. Tomilov and A.S. Berger, Zhurnal Neorgan. Khimii, 28 (1983) 1420.
25. E.T. Devyatkina, N.P. Tomilov and A.S. Berger, Zhurnal Neorgan. Khimii, 30 (1985) 86.
26. I. A. Poroshina and N.P. Kotsupalo, Zhurnal Strukturnoy Khimii, 31 (1990) 74.
27. I.A. Poroshina, N.P. Kotsupalo, L.T. Menzheres and V.P. Isupov, Zhurnal Strukturnoy Khimii, 35 (1994) .
28. R. Allman and H.P. Jepsen, N.Jhb.Miner.Mh., 12 (1969) 544.
29. H.F.W. Taylor, Miner.Mag., 39 (1973) 377.
30. C.J. Serna, J.L. Rendon and J.F. Iglesias, Clays and Clay Minerals, 30 (1982) 180.

31. J.P. Thiel, C.K. Chiang and K.R. Poeppelmeier, Chem. Mater., 5 (1993) 297.
32. A.V. Besserguenuv, A.M. Fogg, R.J. Francis, S.J. Price, D. O'Hare, V.P. Isupov and B.P.Tolochko, Chem.Mat., 9 (1997) 241.
33. H. Saalfeld and M. Wedde, Z.Kristallogr., 139 (1974) 129.
34. A. Abragam, Yadernij Magnetism, IL, Moscow, 1963.
35. G.K. Semin, T.A. Babushkina and G.G. Jakobson, Primenenie Yadernogo Kwadrupolnogo Resonamsa v Khimii, Khimija, Leningrad, 1972.
36. A.P. Nemudry, I.A. Poroshina, V.P. Isupov, N.P. Kotsupalo and V.V. Boldyrev, Izv. SO AN SSSR, Ser. Khim. Nauk, 2 (1987) 48.
37. A.P. Nemudry, I.A. Poroshina, G.N. Goldenberg, V.P. Isupov, N.P. Kotsupalo and V.V. Boldyrev, Izv. SO AN SSSR, Ser. Khim. Nauk, 2 (1988) 58.
38. H.D. Megaw , Z.Kristallogr., 87 (1934) 185.
39. S.Amelinckx, The Direct Observation of Dislocations, Academic Press, New York, London, 1964.
40. A.Yu. Jagodin and A.P. Chupakhin, Izv. SO AN SSSR, Ser. Khim. Nauk, 2 (1988) 63.
41. V.P. Isupov, Jurnal Prikladnoj Khimii, 69 (1996) 12.
42. A. Weiss, H.O. Becker, H. Orth, G. Mai, H. Lechner and K.J. Range, in: Proc. Int. Clay Conf., Jerusalem, 1970, 2, 180.
43. L.T. Menzheres, V.P. Isupov and N.P. Kotsupalo, Izv. SO AN SSSR, Ser. Khim. Nauk, 5 (1988) 53.
44. V.P. Isupov, L.T. Menzheres, M.I. Tatarintseva, E.L. Goldberg, A.F. Eremin and N.P.Kotsupalo, Izv. SO AN SSSR, Ser. Khim. Nauk, 6 (1988) 99.
45. S.M. Paramzin, Yu.D.Pankratiev, E.A.Paukshtis, O.P. Krivoruchko, B.P. Zolotovsky and R.A. Bujanov, Izv. SO AN SSSR, Ser. Khim. Nauk, 4 (1984) 33.
46. R. Poisson, J.P. Brunelle and P. Nortier, in: Nositeli i Nanesennie Katalizatori, A.B. Stiles (ed.), Khimija, Moscow, 1991, 31.
47. V.A.Pushnjakova, V.D.Belykh, V.P.Isupov, A.P.Nemudry and N.P.Kotsupalo, Izv. SO AN SSSR, Ser. Khim. Nauk, 6 (1984) 57.
48. V.P. Isupov, V.D. Belykh, L.T. Menzheres and N.P. Kotsupalo, Sposob poluchenija sorbenta dlja izvlechenija litija iz rassolov, RF Patent No. 1 665 581 (1989).
49. N.P. Kotsupalo, L.L. Sitnikova and L.T. Menzheres, Sposob poluchenija sorbenta dlja izvlechenija litija iz rassolov, RF Patent No. 2 028 385 (1992).
50. N.P. Kotsupalo, L.T. Menzheres and A.D. Ryabtsev, in : Mechanochemistry and Mechanical Activation, N.Z.Lyakhov (ed.), Izd. SO RAN, Novosibirsk, 1997, 175.
51. J.M.Lee and W.C. Bauman, Recovery of lithium from brines, US Patent No. 4 116 856 (1976).
52. J.M.Lee and W.C. Bauman, Recovery of lithium from brines, US Patent No. 4 159 311 (1979).
53. J.M.Lee and W.C. Bauman, Recovery of lithium from brines, US Patent No. 4 221 767 (1980).

54. M.N. Mikheeva, B.F. Myasoedov, Yu.P. Novikov, V.M. Komarevsky, M.S. Mezhirov, R.K. Idiatulov and A.N. Fedorowa, Radiokhimiya, 25 (1987) 638.
55. Yu.G. Kryazhev, N.V. Grishina , Z.A. Zazylina, E.V. Dobizha, Yu.P. Truhin and R.A. Shuvalov, Sposob poluchenija sorbenta na osnove dioksida margantsa dlja izvlechenija litija iz rassolov, RF Patent No. 1 494 970 (1987).
56. L.N. Moskvin, V.S. Miroshnikov, A.A. Melnikov and G.S. Slytskiy, Radiokhimiya, 16 (1974) 50.
57. L.N. Moskvin, A.A. Melnikov, A.A. Besedin, V.N. Epimahov, E.N. Krigin, E.V. Lebedeva and V.D. Golubeva, Zhurnal prikl. Khimii, 56 (1983) 516.
58. Yu.A. Panshin, S.T. Malkevich and Y.S. Dynaevskaya, Ftoroplasty, Khimiya, Liningrad, 1978.
59. L.T.Menzheres, N.P.Kotsupalo, L.B.Orlova and V.P.Isupov, Sposob poluchenija granulirovannogo sorbenta dlja izvlechenija litija iz rassolov, RF Patent No. 2 009 714 (1992).
60. A.D. Ryabtsev, L.T. Menzheres and N.P. Kotsupalo, Sposob i ustrojstvo dlja selektivnogo sorbtsionnogo izvlechenija litija iz rassolov, RF Patent No. 2 020 333 (1993).

Adsorption and its Applications in Industry and Environmental Protection
Studies in Surface Science and Catalysis, Vol. 120
A. Dabrowski (Editor)



# Sol-gel preparation of nanostructured adsorbents

Y.S. Lin[a] and S.G. Deng[b]

[a] Department of Chemical Engineering, University of Cincinnati, Cincinnati, Ohio 45221

[b] BOC Group Technical Center, 100 Mountain Avenue, Murray Hill, New Jersey 07974

This paper reports sol-gel synthesis and properties of several major ceramic adsorbents. Nanostructured γ-alumina, zirconia and titania adsorbents with uniform pore size distribution and an average pore diameter of about 3 nm are prepared by hydrolysis and condensation of corresponding metalorganic precursors. The as-synthesized adsorbents are in their meta-stable phases, and transform to their stable phases upon heat-treatment. The phase transformation accompanies a decrease in surface area and increase in pore size. Microporous (< 2 nm) silica adsorbents are prepared by the sol-gel method following a polymeric sol route. Doping alumina in silica adsorbents improves their thermal and hydrothermal stability. A sol-gel granulation process based on the oil-drop principle is described for preparation of γ-alumina granular particles. A solution-sol mixing coating method is used to coat active species on the pore surface of the granular γ-alumina particles. The sol-gel derived granular adsorbents exhibit high crush strength, good attrition resistance and desired pore structure. Experimental results show that the sol-gel derived CuO/γ-alumina and CuCl/γ-alumina granular adsorbents have better properties for sorption of $SO_2$ and ethylene than similar adsorbents prepared by the conventional methods.

## 1. INTRODUCTION

Sol-gel processing refers to the process of fabrication of ceramic materials by preparation of a sol, gelation of the sol, and removal of the solvent [1]. Sols are dispersion of colloidal particles in a liquid solvent. A gel is a solid matrix encapsulating a solvent. In a sol-gel process the sol can be formed from a solution of colloidal powders or hydrolysis and condensation of alkoxides or salt precursors. In the latter approach which is much more popular, primary particles of uniform size are formed and grow in a sol and connect to each other to form aggregates during gelation. These aggregates forming the network of gel are broken apart into the primary particles in the drying step. Upon calcination and sintering, these primary particles are bound together strongly to form a very stiff solid network, and large interparticle space with uniform nanoscale pores is formed.

A typical sol-gel process using an alkoxide precursor involves the following steps: (1) formation of stable sols; (2) casting or shape formation; (3) gelation of the sols; (4) aging of the gel; (5) drying of the gel; (6) calcination; (7) sintering if necessary. It should be pointed out that sol-gel processing can also be done with a wide variety of precursors in addition to alkoxides. With alkoxide ($M(OR)_n$) as a precursor, sol-gel chemistry can be described in terms of two classes of reactions.

Hydrolysis: $-MOR + H_2O \Leftrightarrow -MOH + ROH$

Condensation: $-MOH + ROM \Leftrightarrow -MOM- + ROH$

or $-MOH + HOM- \Leftrightarrow -MOM- + H_2O$

The above simple description of sol-gel chemistry identifies two key ideas. First, a gel forms because of the condensation of partially hydrolyzed species into a three-dimensional polymeric network. With time the colloidal particles and condensed species link together to become a 3-D network. The physical characteristics of the gel network depend greatly upon the size of the particles and extent of cross-linking prior to gelation. At gelation, viscosity increases sharply, and solid object results in the shape of the mold.

Second, any factors that affect either or both of these reactions are likely to impact on the properties of the gel. In fact it is the control of many of these factors, generally referred to as sol-gel parameters, that separates sol-gel preparation from other methods. A representative but not exhaustive list of these parameters includes type of precursors, type of solvent, water content, acid or base content, precursor concentration, and temperature. These parameters affect the structure of the initial gel and, in turn, the properties of the material at all subsequent processing steps.

The time between the formation of a gel and its drying, known as aging, is also an important parameter. Aging of the gel, also called syneresis, involves maintaining the cast object or wet gel granules for a period of time, hours to days, completely immersed in liquid solvent. During aging, polycondensation continues along with localized solution and reprecipitation of the gel network, which increases the thickness of interparticle necks and decreases the porosity. The strength of the gel thereby increases with aging. An aged gel must develop sufficient strength to resist cracking during drying.

Another parameter that affects a sol-gel product is the drying condition. During drying the liquid is removed from the interconnected pore network, and large capillary stresses can develop during drying when the pores are small (<20 nm). These stresses will cause the gels to crack catastrophically unless the drying process is controlled by decreasing the liquid surface energy by addition of surfactants or elimination of very small pores, by supercritical evaporation, which avoids the solid-liquid interface, or by obtaining monodisperse pore sizes by controlling the rates of hydrolysis and condensation. Xerogel is obtained under conventional evaporative drying conditions. The resultant materials of supercritical drying of wet gels, known as aerogels, have high surface area, porous structure, and low density. In a way drying can be viewed as part of the overall aging process because the material can, and often does, undergo physical and chemical changes during this stage.

Calcination of gels at temperatures higher than drying temperatures causes further removal of solvent and dehydration of the gels. Those physically adsorbed water molecules are removed at initial stage of calcination, and thermal dehydroxylation occurs subsequently. Porosity develops when, due to additional cross-linking or neck formation, the gel network becomes sufficiently strengthened to resist the compressive forces of the surface tension created during the calcination stage. Moderate densification of the gel skeleton also occurs due to the condensation reactions and structural relaxation. Sintering of the calcined gels at even higher temperatures will further consolidate the gel skeleton by eliminating the majority of the porosity of the gels. The primary particles in sintered gels are cross-linked together so close that they generate the strongest binding for the material.

Ceramic thin films, sensors, nanoscale materials, multi-functional ceramic composites, optical fibers, ceramic membranes and many other products can be manufactured by the sol-gel process [1-3]. The major applications of sol-gel processing are in ceramic industry for fabrication of oxide ceramics and glasses. Several studies have been reported on the preparation of supported catalysts and zeolite granular particles using the sol-gel technique [4-10]. Sol-gel derived inorganic thin films and membranes have recently attracted attentions from both academia and industry [11-13]. Only limited studies have been carried out on the sol-gel fabrication of adsorbents for industrial separation or purification purposes.

A few studies on the preparation of adsorbents by the sol-gel method for the waste water treatment in nuclear power plant were conducted in the former USSR. They reported briefly on the sol-gel derived adsorbents in the form of inorganic metal hydroxide, phosphate, ferrocyanide and/or composite from the salt of zirconium, titanium, tin, manganese, and niobium[14-17]. Sol-gel prepared alumina adsorbents have also been used as flue-gas cleaning adsorbents and proved to be superior to the conventional adsorbents with respect to their chemical and mechanical stability [7, 18]. Hydrophobic alumina and silica aerogels with large specific surface area (~600 $m^2/g$) and large pore volume (~ 17 $cm^3/g$) were synthesized by supercritical drying [6, 19, 20], but no report on their adsorption property was found. Most recently, aerogel-like mesoporous siliceous molecular sieve material with large specific surface area (1075 $m^2/g$), large pore volume (1.36 $cm^3/g$) and promising adsorption properties was also successfully synthesized by sol-gel method at conventional drying conditions [21, 22].

Modified silica or diatomaceous earth oxides with impregnated active species have been applied as packings or supports in gas or liquid chromatography [23, 24]. These materials generally have much lower surface area and were not studied for large scale separation. Zirconia and titania were prepared by the sol-gel method to replace silica as packing materials for high pressure liquid chromatography (HPLC) [25-28]. These adsorbents were used to separate structural isomers, amino acids and proteins by HPLC [29, 30] and to separate peridines by ligand-exchange GC [31].

As a result of the unique characteristics of the sol-gel processing, the sol-gel derived single component materials have high purity because of the quality of the available precursors. Furthermore, the textural properties of the product, most notably surface area and pore size distribution, can be tailored. These materials usually have large surface area, uniform pore size distribution, controlled average pore size, and good mechanical strength, which are particularly important to the materials for separation applications. Besides, the microstructure of the sol-gel derived material can be controlled and tailored together with its bulk form (i.e., spherical particles, fibers, and thin films).

However, it is believed that the area in which sol-gel preparation is going to make the most significant impact is multicomponent systems because of the following specific advantages of the sol-gel approach: (i) the ability to control structure and composition at a molecular level; (ii) the ability to introduce several components in a single step; (iii) the ability to impose kinetic constrains on a system and thereby stabilize metastable phases; and (iv) the ability to fine-tune the activation behavior of a sample and thereby trace the genesis of active species [32].

Our research conducted at Cincinnati have been primarily focused on sol-gel synthesis of alumina, zirconia, titania and silica. These metal oxides not only are commonly used as adsorbent or catalyst support but also have recently emerged as excellent materials for ceramic membranes. The objective of this article is to report synthesis and properties of these sol-gel derived adsorbent materials with emphasis on development of a sol-gel granulation method and the properties of the sol-gel derived granular adsorbents.

## 2. SOL-GEL DERIVED ADSORBENT MATERIALS

The adsorptive properties of an adsorbent depends mainly on the pore structure (surface area, pore size and pore volume) and surface chemical properties of the adsorbent. The sol-gel derived ceramic adsorbents possess unique pore structure defined by the microstructure of the sol-gel derived materials. This section describes the synthesis and microstructure of several crystalline and amorphous adsorbents prepared by the sol-gel method in our laboratory.

### 2.1. Crystalline materials

Common crystalline adsorbent materials include $\gamma$-alumina, zirconia and titania. These porous ceramic bodies consist of small crystallites of alumina, zirconia or titania. The alumina, zirconia or titania sols could be directly prepared by dispersing the fine solid particles of these oxides in aqueous solution. They could also be prepared from the inorganic or metalorganic precursors of these oxides.

In our laboratory, alumina sol was prepared by the Yoldas process [33]. Typically, stable 1M alumina (boehmite) sol was synthesized by dissolving 260 ml of aluminum tri-sec-butoxide in 1 liter of water at 70-90°C. The boehmite precipitate formed from the hydrolysis and condensation was peptized by adding 70 ml 1M $HNO_3$ solution at 90-100°C under refluxing condition. Stable 0.25M titania sol was prepared by dissolving 74 ml titanium tetra-isopropoxide (with 500 ml isopropanol) in 450 ml water in a nitrogen box. The titania precipitate was washed with water to remove alcohol and diluted with 1 liter of water. The product was then peptized by adding 72 ml of 1M $HNO_3$ at 75°C under refluxing condition.

Stable 0.25 M zirconia sol was synthesized by modifying a procedure reported earlier by our laboratory [34]. It was prepared by hydrolysis and condensation of 0.25 mole zirconium n-propoxide in a water (900 ml)/isopropanol (500 ml) solution. The white zirconia precipitates were filtered with vacuum suction and washed in water several times to remove the isopropanol. In this process, a small amount of water was added to the zirconia precipitates to help filtering, and the washed water was filtered again to prevent the loss of zirconia precipitates. The filtered zirconia cake was diluted in 1 liter of water and peptized with 125 ml of 1 M $HNO_3$ solution at 90-100°C overnight.

Xerogels of alumina, titania or zirconia in the form of thin sheet (20-200 μm thick) were prepared by pouring respective sols in given quantities in petri dishes. The sols in petri dishes were dried at 40°C and 40-50% relative humidity. Gelation occurred after a sufficient amount of solvent (water) had evaporated. The xerogel samples were calcined in a temperature programmable box furnace at 450°C for 3 hours, with carefully controlled heating and cooling rates.

XRD data reveal that alumina particles in the sol are of boehmite crystalline structure and the particles in zircornia and titania sols are of amorphous structure [34]. The alumina, titania and zirconia samples obtained from the sols after gelation and calcination at 450°C are respectively in the phases of γ-alumina, tetragonal zirconia and anatase. These are thermodynamically meta-stable phases, and may transform to the thermodynamically stable phases, which are α-alumina, monoclinic zirconia and rutile. The crystallite structure and lattice parameters of these phases are listed in Table 1.

Fig.1 shows the pore size distributions of γ-alumina, titania (anatase) and tetragonal zirconia (after calcination at 450°C for 3 hours). The pore structure data of these three adsorbent samples are compared in Table 2. The pore structure data were obtained from nitrogen adsorption isotherms measured by Micromeritics ASAP-2000 adsorption porosimeter. As shown in Fig.1, the pore size distributions of these materials are rather narrow, with an average pore diameter of about 3 nm. Such narrow size distribution and nanoscale average pore size are determined by the primary crystallite particles. The particles of the

Table 1
Characteristics of phase transformation of sol-gel derived alumina, zirconia and titania

| Material | Initial phase and lattice parameter (A) | Final phase and lattice parameter (A) |
|---|---|---|
| Alumina | γ-alumina (cubic)<br>a=7.8 | α-alumina (hexagonal)<br>a=4.8, b=13.0 |
| Zirconia | Tetragonal<br>a=b=5.1, c=5.3 | Monoclinic<br>a=5.14, b=5.2, c=5.3, β=99.2° |
| Titania | Anatase (tetragonal)<br>a=b=3.8, c=9.5 | Rutile (tetragonal)<br>a=b=4.9, c=3.0 |

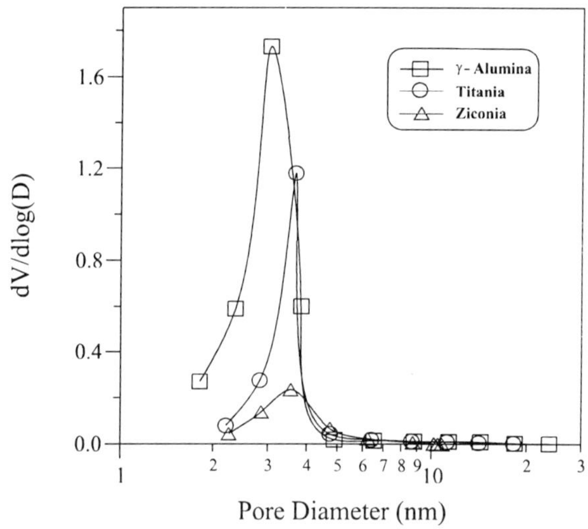


Figure 1. Pore size distribution of sol-gel derived alumina, zirconia and titania.

sol-gel derived alumina, titania and zirconia, due to the Ostwald ripening mechanism, are usually in nanoscale size, with a uniform particle size distribution [1]. γ-alumina crystallites are of plate-shape [35, 36] with a size in the range from about 5 to 20 nm. The sol-gel derived γ-alumina consists of such plate-shaped crystallite particles, which give rise to a relatively large surface area. Crystallites of tetragonal zirconia and rutile are of more spherical shape, with a crystallite size of about 15 nm and 11 nm, respectively [34].

Table 2
Pore structure of $\gamma$-$Al_2O_3$, titania and zirconia (calcined at 450°C for 3h).

| Materials | Average pore size (nm) | Pore volume (ml/g) | Surface area ($m^2/g$) |
|---|---|---|---|
| $\gamma$-$Al_2O_3$ | 2.8 | 0.33 | 373 |
| $TiO_2$ | 3.4 | 0.21 | 147 |
| $ZrO_2$ | 3.8 | 0.11 | 57.2 |

Alumina, titania and zirconia in their metastable phases will transform to their stable phases. Such phase transformation usually occurs via a nucleation and crystal growth process [34, 37]. Kinetically, however, the phase transformation will not occur or be observed at low temperatures. For alumina, the $\gamma$-$Al_2O_3$ to $\alpha$-$Al_2O_3$ (via $\delta$- and $\theta$-aluminas) phase transformation is found to start at temperature around 900°C. This is accompanied with a sharp decrease in the surface area and increase in the pore size of the alumina adsorbent. The surface area of $\gamma$-$Al_2O_3$ adsorbent will decrease with time at temperatures lower than 900°C due primarily to sintering [34].

From the metastable tetragonal phase, zirconia starts transforming to the more stable monoclinic phase around 700°C. At temperatures lower than 700°C, sintering causes change of the pore structure of zirconia (decrease in surface area and increase in pore size). At temperatures higher than 700°C, phase transformation plays a major role in affecting the pore structure of the zirconia. Phase transformation also results in a decrease in surface area and an increase in pore size [34, 38]. For titania the phase transformation from anatase to rutile, occurring around 450°C, also accompanies a substantial change in the pore structure of titania [39].

The activation energy for phase transformation is respectively about 600, 570 and 213 kJ/mol for the sol-gel derived alumina, zirconia and titania [34]. These data were obtained when these materials were exposed to air. Presence of steam in the atmosphere appears to reduce the activation energy for phase transformation, thus enhancing the rate of the pore structure change of the adsorbent materials at a given temperature. For practical application, the pore structure of the materials can be kinetically stabilized by heat-treating the materials at a temperature a few hundred degrees higher than the application temperature.

### 2.2. Noncrystalline silica

Although silica has several crystalline forms, only amorphous silica finds uses as adsorbents for separation and purification applications. Both particulate silica sol (with dense particle larger than 1 nm) and polymeric silica sol (without dense particle larger than 1 nm) could be prepared, depending on the precursor, solvent and, most importantly, the catalyst used. Silica adsorbents prepared from particulate silica sol have mesopore size with surface area in the range comparable to $\gamma$-alumina. Microporous silica adsorbents with pore diameter smaller than 2 nm could be prepared from polymeric silica sol. These microporous silica has a surface area much larger than the crystalline adsorbents discussed above.

Silica sol is prepared by hydrolysis of a metal alkoxide and subsequent or simultaneous polycondensation (polymerization), which can be either acid or base catalyzed. The acid catalyzed hydrolysis proceeds through an electrophilic attack of the $H^+$ ions. This means that the reactivity decreases as the number of OR groups decreases with the progression of hydrolysis [40-42]. The probability of formation of fully hydrolyzed silicon, $Si(OH)_4$, is thus very small. Since the condensation reaction starts before the silicon alkoxide is completely hydrolyzed, silicon alkoxide molecules will polymerize with non-hydrolyzed alkoxyl groups, in which the degree of crosslinking is low. The gyration radius of these small molecules is typically in the order of 1.5 - 1.7 nm [43].

The base catalyzed hydrolysis reaction proceeds through the nucleophilic substitution of $OH^-$ ions. The reactivity increases as the number of OR groups decreases. The silicon alkoxide will tend to be completely hydrolyzed. The condensation reaction is the rate determining step and the more crosslinked species will grow at the expense of the smaller one [43]. Therefore the base catalyzed hydrolysis condensation reactions yield large, highly crosslinked polymers. Gels formed from these large polymers normally contain large pore [44]. Therefore, the base catalyzed synthesis will not result in microporous materials due to the large highly crosslinked particles. Only the acid catalyzed reactions could result in microporous materials.

In our laboratory, microporous silica sol was experimentally prepared by the nitric acid catalyzed hydrolysis and condensation of tetraethoxysilane (TEOS). Typical sol composition (in molar ratio) was 3.8 for ethanol/TEOS, 6.5 for deionized water/TEOS and 0.09 for nitric acid/TEOS. This composition was adopted from Uhlhorn et al [45]. Solutions of TEOS/ethanol and nitric acid/water were prepared separately and then mixed in a spherical flask equipped with a reflux condenser. The mixture was rapidly heated to and stayed at 90°C for 3 hours with vigorously stirring. Finally, the resulted sol was cooled naturally to room temperature.

1.5 mol% $Al_2O_3$ doped $SiO_2$ adsorbents were also prepared by the sol gel method. In this case, solution of 10% wt/vol $Al(NO_3)_3{\cdot}9H_2O$ in ethanol was prepared by heating $Al(NO_3)_3$ at moderate temperature for 15 minutes to ensure complete dissolution of the salt in ethanol. Ethanol was chosen as the solvent for the salt because it was also used in preparing the silica sol. Doping the silica sol with alumina was performed by mixing the sol with a controlled amount of the ethanol solution of aluminum nitrate.

Xerogel samples of silica or alumina doped silica were obtained by drying the corresponding silica sols in petri-dishes at 40°C and 60% relative humidity for one to two days. Silica adsorbent (in the form of thin sheet) was obtained by calcining the xerogel samples at 400°C for 3 hours with heating and cooling rate of 250°C/hour. Pore structure and surface properties of microporous silica were analyzed by nitrogen adsorption using adsorption porosimeter (Micromeritics ASAP 2000 with micropore capability). The sample was first degassed at 300°C under vacuum for more than 10 hours until the sample passed the degassing check up test. The sample was then weighted and transferred to the measurement port. The nitrogen adsorption isotherm was measured at liquid nitrogen temperature (78 K) and nitrogen pressure ranging from $10^{-7}$ to 1.0 $P/P_o$ (relative pressure of nitrogen). The Horvath-Kawazoe model [46], which was pre-installed in the micropore analysis program of the ASAP 2000 adsorption porosimeter, was applied to calculate the pore size distribution and micropore volume of the silica samples.

Nitrogen adsorption isotherm on a pure silica sample after drying and calcination is shown in Fig.2. It is of type I, characteristics of microporous materials [47]. The corresponding pore size distribution of this sample is given in Fig.3. As shown, the sol-gel derived silica sample contains microporous (< 2 nm) pores, with an average pore diameter of about 0.6 nm. The pore structure of the pure and alumina doped silica samples is summarized in Table 3. The surface area of sol-gel derived silica is much larger than the sol-gel derived crystalline samples, described in Section 2.1.

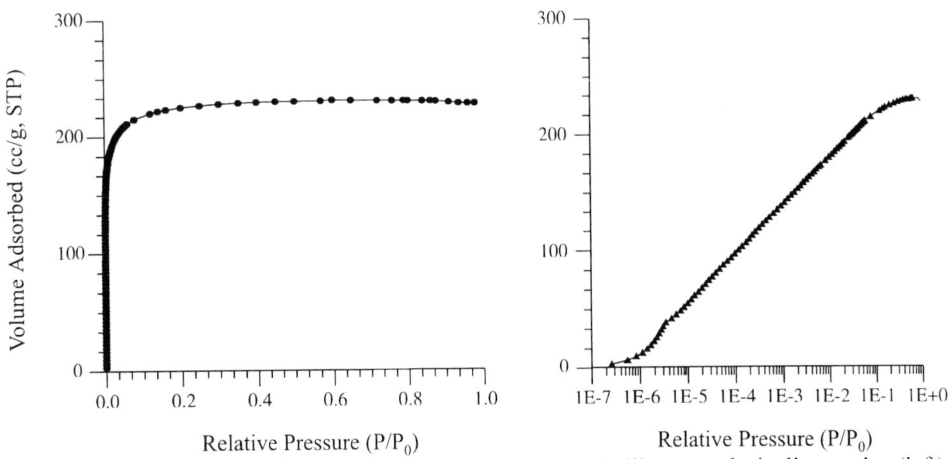


Figure 2. Nitrogen adsorption isotherm on sol-gel derived silica sample in linear plot (left) and in semi-logarithmic plot (right).

Table 3
Pore structure of pure microporous silica and alumina doped silica

| Materials | BET Surface Area ($m^2/g$) | Average Pore Diameter (Å) | Pore Volume ($cm^3/g$) |
|---|---|---|---|
| Pure $SiO_2$ | 588 | 6.4 | 0.24 |
| 1.5%$Al_2O_3$-$SiO_2$ | 660 | 6.5 | 0.26 |

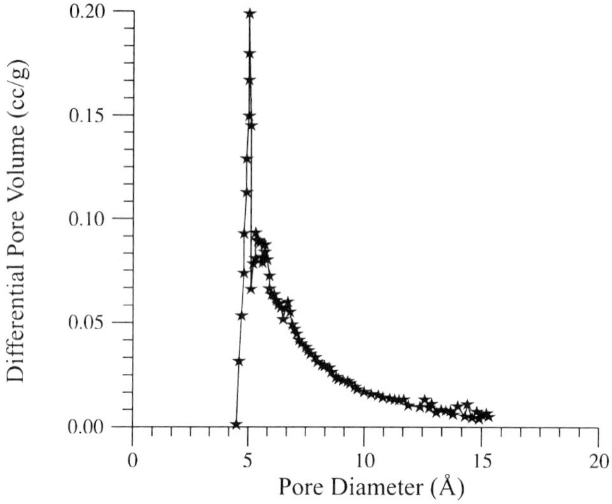


Figure 3. Pore size distribution of the silica sample shown in Fig.2.

As compared with silica gels prepared by other methods, the sol-gel derived microporous silica is not thermally stable at high temperatures, especially under humid atmosphere. Table 4 shows the pore structure data of both silica samples after heat-treatment at 600°C for 30 hours. Fig.4 compares the pore size distributions of a silica sample before and after the heat-treatment. For pure silica, the heat treatment results in an 89% reduction in the surface area and a loss of 87% micropore volume. These results agree with the previous study which indicated similar effects of steam and heat on the stability of silica [48]. Such a structural change of silica at high temperature is believed due to continuous condensation of surface silano groups and sintering.

Table 4
Pore structure of pure microporous silica and alumina doped silica after heat-treatment at 600°C for 30 hours (data in parenthesis are the changes with respect to data in Table 3)

| Materials | BET Surface Area ($m^2/g$) | Average Pore Diameter (Å) | Pore Volume ($cm^3/g$) |
|---|---|---|---|
| Pure $SiO_2$ | 64.4 (-89%) | 8.4 (31%) | 0.03 (-87%) |
| 1.5%$Al_2O_3$- $SiO_2$ | 447.1 (-32%) | 6.9 (6%) | 0.18 (-30%) |

As to alumina-doped silica sample, the reduction of surface area and pore volume after the heat-treatment was approximately 32% and 30%, and the increase of average pore diameter was just 6%. This improvement can be more clearly illustrated by comparing the pore size distribution of the alumina-doped silica sample before and after the heat treatment, as shown in Fig.5. The alumina doped silica clearly has a better thermal stability than pure silica adsorbent.

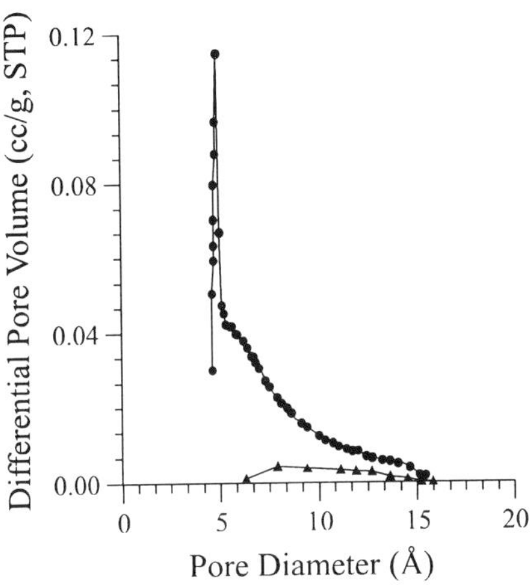


Figure 4. Comparison of the pore size distribution of a silica sample before (●) and after (▲) heat-treatment at 600°C for 30 hours.

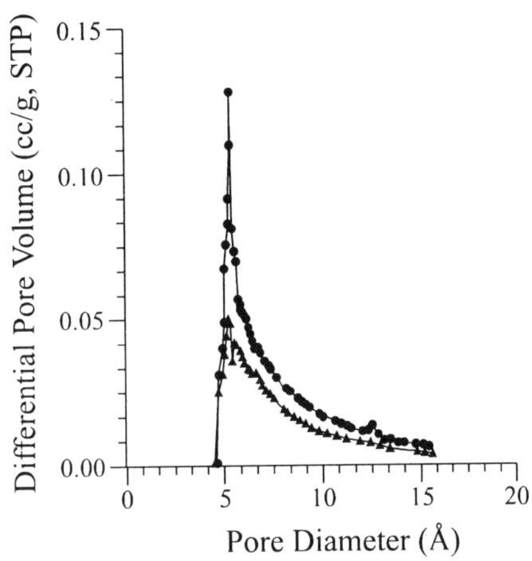


Figure 5. Comparison of the pore size distribution of 1.5% alumina-doped silica sample before (●)and after(▲) heat-treatment at 600°C for 30 hours.

## 3. SOL-GEL GRANULATION PROCESS

Adsorbents are practically used in the granular form. The performance of an adsorption process critically depends on the size, shape, pore texture and mechanical strength of the support body used. Sol-gel process has commonly been employed for preparation of ceramic films, fibers and powders. This section reports our work on synthesis and properties of the sol-gel derived alumina granular particles. The principle of the sol-gel derived granulation process should be applicable to preparation of granular particles of other ceramic materials discussed in the preceding section.

### 3.1. Granulation processes

Conventional granulation processes can be classified as agitation methods (tumbling agglomeration and mixer agglomeration), pressure methods (compaction, extrusion, and rolling), thermal methods (sintering), spray and dispersion methods and agglomeration from liquids [49, 50]. With these established granulation methods, relatively large particles are usually form from small particles through different intraparticle forces of cohesion. After granulation processes the microstructures of granules generally do not change significantly, but the bulk properties of the granules improve substantially. Granules can be prepared to have useful structural forms and shapes, better definition of quantity units, larger bulk density, less dusting losses, easy to handle, better heat transfer and controllable porosity which are very important for adsorbents and catalysts.

The process to make spherical granular supports, catalysts and adsorbents is mainly based on the methods which rely upon the build-up of the smaller particles into spheres by means of a rolling, or "snow-ball" technique. The production costs for this type of processes are low, but the spheres are irregular and the particle size are randomly distributed. A few studies were reported on the sol-gel routes for preparing spherical gel particles of various sizes ranging from 0.5 μm to 2 mm for different applications [25, 49-53]. Large spherical granules (typically in the range of 100 μm to 5 mm) are generally prepared by gelation of individual droplets generated from the starting sol by the so called "oil-drop" method. The spheres manufactured by individual shaping are more expensive. However their shape and calibration are quite regular.

Among these reported research works on sol-gel granulation or "oil-drop" process, most were on the synthesis of spherical silica ($SiO_2$) supports or zeolite granules [8-10, 54, 55]. For example, Spek and van Beem [9] reported a sol-gel granulation process for fabrication of spherical silica supports. In their process, an aqueous sodium silicate (waterglass) solution comprising 12 wt.% of $SiO_2$ was mixed continuously in a mixing chamber with another aqueous 1.2 N sulfuric acid solution in a volume ratio (acid/waterglass) of 0.65. The hydrosol formed in the mixing chamber was converted into droplets which were allowed to fall through a vertical disposed cylindrical tube with a length of 1.8 m filled with paraffinic hydrocarbon oil at 25°C. Spherical hydrogels particles formed were separated, washed with water and dried. The spherical silica particles can be used directly as catalyst supports. Li et al. [8] used aqueous solution of sulfuric acid, ammonium sulfate and waterglass as the starting materials to fabricate spherical $SiO_2$ particles, and synthesized binderless zeolite type A from $SiO_2$ in an aqueous solution of alkaline sodium aluminate at a given condition. Binderless faujasite zeolites were also synthesized this way by Mirsky et al. [10].

Granular alumina supports were also fabricated by the "oil-drop" sol-gel method [7, 56-59]. All these reported "oil-drop" processes for preparation of granular alumina supports start with a boehmite sol using pseudo-boehmite powder as precursor, followed by introduction of the sol dropwise through an orifice into a column of heated paraffin oil. Each drop of oil-insoluble mixture forms a sphere and converts to spherical gel in the hot oil. The microstructure, especially the pore texture, of the granular alumina supports prepared by these methods are basically determined by the properties of the primary particles of the starting pseudo-boehmite powder. It is difficult to control and tailor the microstructure of the granular alumina supports to meet various needs by these "oil-drop" processes since the properties of the starting pseudo-boehmite powder are source dependent and generally not controllable in the granulation process.

Sol-gel granulation processes were also used to fabricate high density microspheres as nuclear fuels [60-62]. In the sol-gel process of fabrication of $ThO_2$-2% $UO_2$ fuel pellets, the starting sols were prepared by passing a controlled amount of ammonia gas through the thorium and uranyl nitrate solution. Droplets of the sol were then introduced through an electromechanical vibrator with horizontal jetting nozzle inside a containment box that houses two horizontal ammonia gas pipes and the gelation bath. Thus, the droplets passed through a curtain of ammonia gas and quickly coated themselves with a gel skin before falling into the gelation bath filled with 1% $NH_4OH$ and 4M $NH_4NO_3$ with pH about 8. The suitable $NH_4^+/NH_3$ buffer system established the conditions essential to achieve the desirable gelation rate. The gels were then washed, dried and calcined before they were pressed into pellets and sintered at high temperature. It was reported that this sol-gel approach could achieve the pellet density of higher than 94% of the theoretical density, and a uniform dispersion of the uranium in the fuel, because the sol-gel process has a much better control of the properties of the primary particles and microspheres [61].

### 3.2. Sol-gel preparation of alumina granular particles

γ-alumina is perhaps the most common crystalline material used as support for catalysts or adsorbents. Preparation of porous γ-alumna granules with excellent mechanical properties and desirable pore structure is of great importance to development of novel catalysts and adsorbents for various applications. The superior mechanical properties can be derived from the unique microstructure of the granule which is defined by compacting small γ-$Al_2O_3$ crystallite particles bound together by the bridges of the same material formed through coarsening or sintering. To obtain high surface area of the granules the γ-$Al_2O_3$ crystallites should be preferably within a few nanometer. Such nanostructured γ-$Al_2O_3$ can be prepared by the Yoldas process, as described earlier. In what follows we report preparation of sol-gel derived nanostructured γ-$Al_2O_3$ granules based on a granulation process that combines the Yoldas process and the "oil-drop" method.

A flow diagram of the formation of spherical supports and adsorbents by the sol-gel granulation method is shown in Fig. 6. The process includes generating sol droplets, shaping and partially gelating the droplets into spherical wet-gel granules, and gelating and aging the wet-gel granules into solid gel granules. The wet-gel granules were then washed with water, dried and calcined. In our laboratory, the starting material for preparing the γ-$Al_2O_3$ granules was high concentration (>1.0M) boehmite sol. Two methods were used to prepare the high concentration boehmite sol. The first method is based on the exact Yoldas process, as described earlier. The resulting sol at 1M aluminum concentration was concentrated by evaporation on a hot plate. The second method, which avoids the evaporation step and is referred to here as the modified Yoldas process, was direct preparation of the high concentration boehmite sol. This method is described in more detail next using preparation of 2M boehmite sol as an example.

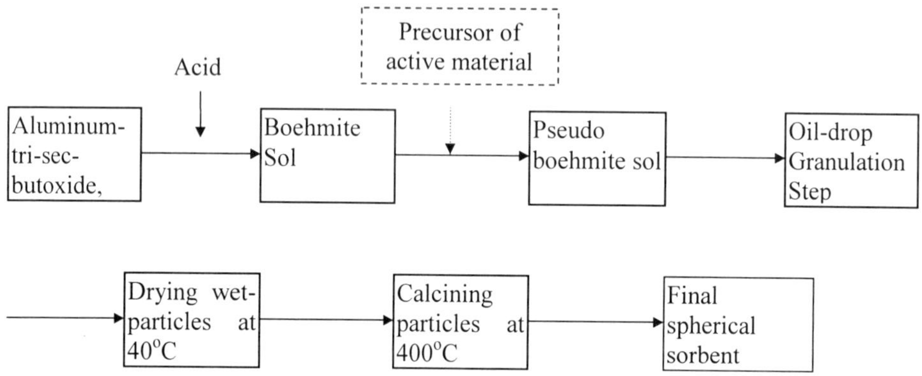


Figure 6. Schematic diagram of the sol-gel granulation process.

The synthesis process started with slow addition of a total of 520 ml of aluminum tri-sec-butoxide (ALTSB) in 1000 ml distilled water which was well stirred and controlled at 80°C. Large irregularly-shaped clusters started to appear, and gradually precipitated in the flask during the hydrolysis and condensation process when the amount of ALTSB added exceeded 400 ml (equivalent to about 1.56 M). The slurry was then dispersed with addition of 140 ml of 1 M $HNO_3$ (that gave the $H^+/Al^{3+}$ ratio of 0.07) and/or dilution with about 200 ml deionized water. Continuous addition of 520 ml of ALTSB in water took about 5 hours. After that, the ALTSB/water mixture was vigorously stirred for half an hour and refluxed overnight at 90-100°C.

The 2M boehmite sol was first mixed with a small amount of 1M $HNO_3$ (acid/sol volume ratio of 1:5). The modified sol was stirred at 70-80°C with a magnetic stirrer. A dramatic decrease in the pH with the addition of $HNO_3$ accelerated the gelation process of the sol. After aging for 1/2 hour at 60-70°C, the 2 M boehmite became so viscous that it could not be stirred with the magnetic stirrer set at the maximum power. It was then transferred to droppers as the starting material for the granulation process.

The granulation process included generating sol droplets by the droppers, shaping and partially gelating the droplets into spherical wet-gel granules in a paraffin oil layer, and consolidating the structure of the wet-gel granules in a 8 wt.% ammonia solution layer. The device used to prepare the gel particles is schematically shown in Fig.7. The temperature of the paraffin oil layer (white color, density: 0.7864 g/cm$^3$, kinematic viscosity: 34.5 centistokes at 40°C, from Fisher Scientific) varied from 25 to 100°C, and the ammonia solution layer was kept at room temperature. The interface between the oil and ammonia-solution was slowly stirred with a stirrer (at 18-50 rpm) to facilitate transport of the wet-granules across the interface. After aging in the ammonia solution for at least 45 minutes, the spherical wet-gel particles were removed from the ammonia solution, and carefully washed sequentially with water and alcohol solution, dried at 40°C for 48 hours, and finally calcined in air at 450°C for about 4 hours.

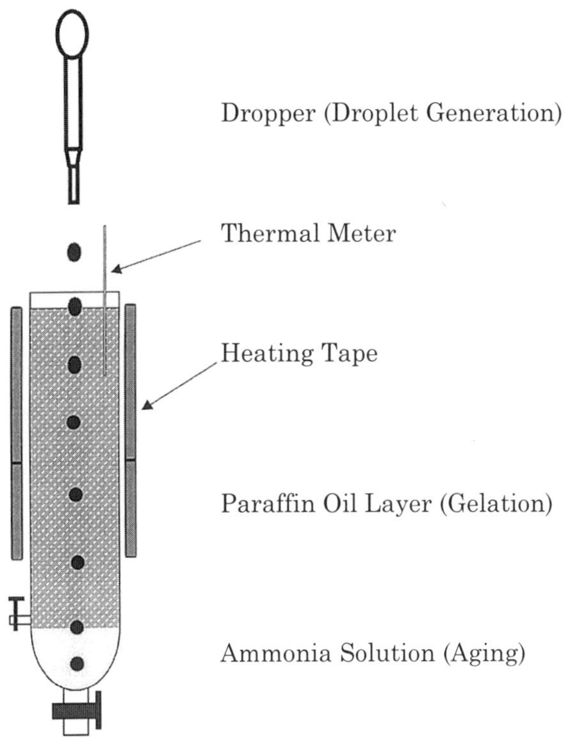


Fig.7 Sol-drop generation device.

One of the key steps in this sol-gel granulation process is the formation of the sol droplets. There are basically two ways for droplet formation: gravity-force assistant method [63] and shear-force assistant method [64]. In both cases, the orifice of the dropper is immersed in the paraffin oil. In the second case, the shear-force is created by either moving the paraffin oil (orifice is fixed) or moving the orifice in the paraffin oil. When the sol is continuously pushed out through the orifice, the sol in the oil experiences the gravity force or drag-force (in the shear-force assistant method) that acts to separate the drop from the orifice.

When the gravity force (minus buoyancy force) or the drag-force exceeds the forces acting to keep the drop in the orifice (e.g., interfacial tension), the drop breaks away from the orifice. For the gravity-force assistant method, balancing the two forces acting to separate and hold the drop gives the following equation for the volume of a droplet at low flow rate of sol [53]:

$$V = \frac{F\pi\gamma D}{g\Delta\rho}$$

where $\gamma$ is the interfacial tension (between the sol and orifice material), D is the diameter of the orifice, $\Delta\rho$ is the density difference between the sol and oil, g is the acceleration due to the gravity, and F is a correction factor.

This equation can be used to select preparation conditions for controlling the diameter of the spheres. In general, larger spheres are obtained by increasing diameter of orifice. The sol concentration or density and interfacial tension are other important parameters. Thus, increases in the interfacial tension ($\gamma$) yield larger droplet and therefore larger spheres. If the kinetic and drag forces are considered, it can be shown that the drop size increases with increasing flow rate of the sol through the orifice and increasing viscosity of the continuous phase (oil phase).

Continuous generation of the droplets by the gravity-force assistant method may be facilitated by the use of a drop-generation device that generates sol droplets by intermittently forcing the sol through the orifice [63]. The orifice of the dropper could be placed in the air above the oil layer. In this case the droplets experience more gravity because the buoyancy force in minimized. The shear-force assistant method [64] appears to be more convenient for continuous generation of the sol droplets. In this method the size of the droplets is determined by the diameter of orifice, relative velocity of the orifice with respect to the oil phase, and flow rate of the sol being pushed out of the orifice.

The rheological properties (viscosity and viscoelasticity) of the sol flowing through the orifice are very important in determining the size, uniformity and break-up of the droplet. Such properties are generally complex, since concentrated sol and liquid containing dissolved polymers, which are frequently employed in these processes, exhibit non-Newtonian behavior. With boehmite sol, the important parameters that control the rheological properties of the sol are the solid concentration and pH, and aging temperature and time of the sol before the droplet-generation step. These effects have been discussed in detail elsewhere [65, 66].

The droplets generated from the dropper usually were not perfectly spherical, but they would gradually become good spheres while falling through the hot oil layer. This is basically because the immiscibility between the boehmite sol and the paraffin oil makes the spherical shape thermodynamically most stable. At the meantime when the droplets changed to spheres, partial gelation of the thin outer layer of the droplets occurred due to the increase of temperature. The droplets would become rigid enough to keep the spherical shape in the oil layer and then fell into the ammonia solution layer for further aging and conditioning.

During aging in the ammonia layer, ammonia would penetrate through the oil film to neutralize the acid in the partially gelled sol in the wet-gel granules. Thus, the wet-gel granules were further gelated, aged and became solid granules in the ammonia solution layer. After aged in the ammonia solution and separated from the liquid media the alumina wet-gel granules became rigid enough to be transported, washed and dried. The washing step may be required to remove those impurities such as ammonium nitrate, nitric acid or ammonia and hydrocarbons in the wet-gel granules before the drying process. In this case, a careful washing procedure should be followed in order to avoid breakage of the gel particles [65, 66]. Washing could also result in a loss of about 10-15% of alumina into the washing solvents. Thus, if possibly, the washing step should be skipped. In these cases, the impurities in the wet-gel granules can be removed by directly drying and calcining the wet-gel granules. In the drying step, the aggregates in the gels broke apart into primary particles of pseudo-boehmite or boehmite, which would bind together to form the skeleton of the final granules. Pores and porosity gradually developed as calcination step progressed.

### 3.3. Physical properties

A microscopic view of the sol-gel derived alumina granules is shown in Fig. 8. The multiple granules prepared under the same experimental conditions are non-aggregated spherical particles without lumps. The particle sizes of these granules are uniform. This smooth spherical shape and narrow particle size distribution provide the necessary homogeneity of γ-alumina support for catalysts or adsorbents for use in either fixed-bed, moving-bed or fluidized-bed reactors.

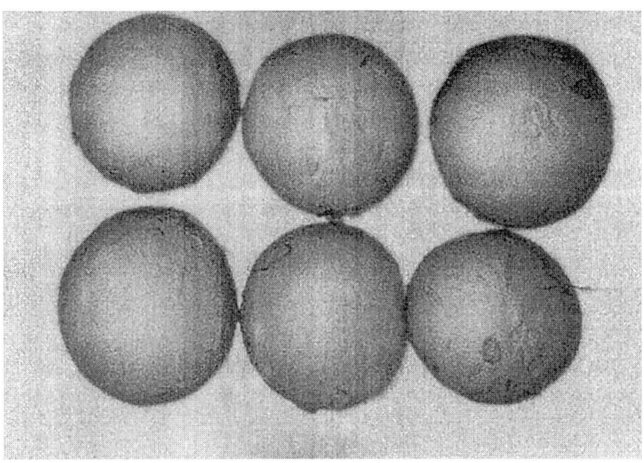

Figure 8. Appearance of the sol-gel derived γ-alumina granules (particle diameter 2mm).

Experiments performed in our laboratory showed that it was easy to make good spherical and smooth γ-alumina granules with diameters smaller than 3 mm using the sol-gel granulation method. Bigger particles had the tendency to deform during the washing and drying stages.

Nitrogen adsorption and desorption isotherms and the pore size distribution (calculated from the desorption isotherm) of the granule sample are shown in Fig. 9. Similar to the nongranular sample, the alumina granules have a narrow pore size distribution (20~60 Å) with an average pore diameter of around 35 Å. The uniform pore size distribution indicates presence of monodisperse pores in the granular particles defined by the intercrystalline space of the primary particles. The BET surface area and pore volume of the sol-gel derived alumina granules are about 390 $m^2/g$ and 0.5 cc/g, larger than the nongranular alumina samples (see Table 2). It should be noted that preparation procedure and purity level of the aluminum butoxide precursor may affect somewhat the pore structure of the sol-gel derived alumina.

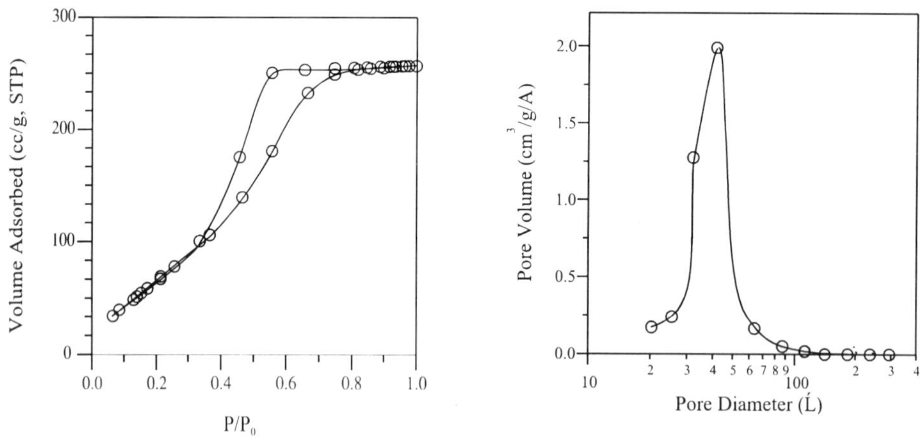


Figure 9. Nitrogen adsorption/desorption isotherms (left) and pore size distribution (right) of a sol-gel derived alumina granular sample.

The mechanical properties such as crush strength and attrition resistance are important properties for industrial adsorbents, especially for uses in moving or fluidized bed processes. Adsorbents with poor mechanical properties are usually not appropriate for industrial applications in fixed-bed processes due to the problems of dust formation in the adsorbent vessels, which will change the particle size distribution of the adsorbents and increase in the pressure drop of the adsorption system. Furthermore, the loss of the adsorbent weight due to dust formation will also deteriorate the performance of the adsorption system.

The crush strength of individual granular γ-alumina supports and adsorbents prepared by the sol-gel method and some commercial supports and adsorbents were experimentally determined by a universal testing instrument. During these experiments, a single granule was placed between two smooth and parallel compression surfaces made of steel. One of the flat surfaces was mounted on the base of loading frame of the instrument and the other flat plate attached to a crosshead moving towards the granule at a controlled speed of 2 mm/min. A compression load cell (force sensor, 0-400 1b) mounted on the moving crosshead was used to measure the force acting on the particle. Both the load force and displacement of the sphere were recorded by a computer. At the end of test, the granule crushed or collapsed when the force applied on the granule was large enough, and an abrupt decrease of the force signal was detected. The maximum force load applied to break the granule was taken as the side crush strength.

Fig. 10 shows the typical load-displacement curves obtained in the crushing strength test for a sol-gel derived spherical γ-alumina granule and a commercial alumina granule sample. The alumina particles are elastic and the maximum displacements before crushed are about 40~75% of the diameter of the spheres. The experimental side crushing strength of the sol-gel derived alumina granules are compared with several commercial granular samples in Table 5. The crush strength of the sol-gel derived γ-alumina granules (diameter = ~2 mm) prepared in this work is 160 N, several times larger than the commercial alumina and zeolite granular samples.

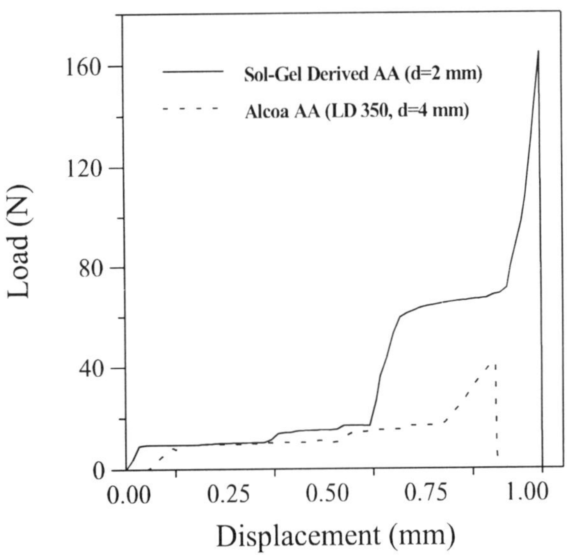


Figure 10. A comparison of load-displacement curves for the sol-gel derived granular alumina particle and a commercial alumina particle (from Alcoa).

Table 5
Comparison of mechanical properties of sol-gel derived alumina granules with commercial adsorbents

| Adsorbents | Particle Diameter (mm) | Crush Strength (N) | Attrition Rate (wt.%/h) |
|---|---|---|---|
| Sol-Gel Alumina | 2.0-2.5 | 160 | 0.033 |
| Sol-Gel Alumina | 2.6-2.8 | 190 | - |
| Alcoa Alumina (LD-350) | 4.0-4.6 | 42 | 0.177 |
| UOP Silicalite | 1.4-1.6 | 16 | 0.575 |
| Degussa DAY Zeolite | 3.5-3.7 | 40 | 0.073 |

Two types of particle degradation are usually observed in the mechanical attrition of solid particles: "fracture" or deep disintegration of the adsorbent or catalyst particles; and "grinding" or the abrasive removal of the particle skin. Particle fracture yields an assortment of fragments from small size to large size [67]. The grinding mechanism leaves behind a particle reduced in size and quantity of fine particles. The forces to which the adsorbent or catalyst particles are exposed in the moving bed or fluidized bed unit involve mechanical motion between any given particle-particle or particle-wall interactions and high speed-velocity impingement during movement of particles from vessel to vessel. These forces will result in both particle fracture and grinding in the moving beds and fluidized beds.

Fig. 11 shows a Peter Spence attrition set-up [67] used in this work to measure the attrition rate of the alumina adsorbents. It consists of a geared motor and a stainless steel attrition testing cylinder (315 mm long, 10 mm in inter diameter) mounted at a point about 80 mm from the center. The testing cylinder rotated about an axis normal to the length at a controlled speed ranging from 20 to 200 rpm. In the attrition test performed in this work, about 5 g of granules were placed in the testing cylinder and tested at a rotation speed of 60 rpm for 24 h. Thereafter the powder material generated during the attrition experiments was sieved off over a No. 35 sieve with mesh opening of 500 μm. The weight loss percentage rate (wt.%/h) calculated by the following equation was used as attrition index:

Attrition Index = (Initial Weight - Remaining Weight) / Initial Weight / Time x 100%

In attrition experiments, each granule sample after attrition run was examined for its breakage and formation of particle fragments and fine powder. No breakage and particle fragments were found in all samples listed in Table 5 except for silicalite. Most of the silicalite extrudates were found broken into small particles and a certain amount of fine powder was generated after the attrition test. This may indicate that the attrition of silicalite pellets followed both particle fracture and abrasion mechanisms, and the attrition of other granules follows the

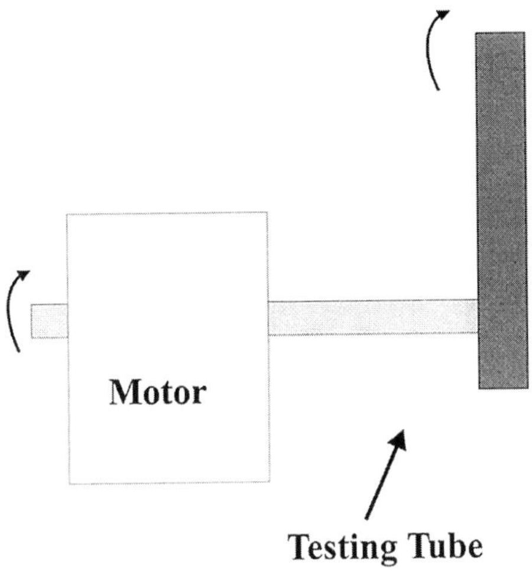


Figure 11. Schematic of Peter Spence method for testing attrition rate.

abrasion mechanism only. The attrition index of the sol-gel derived spherical γ-alumina granules along with the commercial supports and adsorbents determined by the Peter Spence method are summarized in Table 5. Among all samples listed in Table 5, the sol-gel derived spherical γ-alumina granules prepared in this work have the smallest physical attrition index: 0.033 wt.%/h, about 1/5 of that of the commercial alumina granule (LD-350 of Alcoa). This clearly demonstrates that the sol-gel derived spherical γ-alumina granules have exceptional attrition resistance. These results also indicate that the attrition of the sol-gel derived γ-alumina granules prepared in this work follows the "fracture" mechanism. In this case the attrition of the sol-gel derived alumina granules is extremely low if the mechanical stress is not large enough to break the particles in the attrition test.

The excellent mechanical strength, high attrition resistance, and desirable pore structure of the spherical γ-alumina granules prepared in this work are unique properties of sol-gel derived ceramics. In synthesizing these granules, plate-shaped boehmite primary particles were formed and grew via the hydrolysis and condensation of aluminum alkoxide in aqueous solution [35]. The particles reached a final uniform nanoscale size (about 5-10 nm) due to the Ostwald ripening mechanism [1]. The boehmite particles were connected during the partial gelation process while the sol droplets were falling through the oil layer. Complete gelation occurred when aged in the ammonia solution. After removal of solvent, the structure of the solid boehmite gel granules was consolidated by calcination in air at 450°C. The boehmite particles also transformed to γ-alumina crystallites during the calcination step. The final granules consist of nanoscale primary $\gamma$-alumina particles of rather uniform size. These uniform particles are strongly bound together by the alumina bridges formed in the calcination step, and, as a result, the granules become mechanically very strong.

## 4. SOL-GEL DERIVED GRANULAR ADSORBENTS

The sol-gel derived ceramic granular particles described above may be directly used as the adsorbents for certain applications. More often, however, these granular particles are used as the support body on which an active species is coated. The active species coated on the internal pore surface of the support body provides active sites for preferential adsorption of selected gas and liquid species. Because of the unique coating process and microstructure of the sol-gel derived support body, the sol-gel derived supported adsorbents are expected to exhibit better properties than similar adsorbents prepared by the conventional methods. This section describes synthesis and adsorption properties of the sol-gel derived granular adsorbents.

### 4.1. Synthesis of adsorbents

Coating a metal or oxide species on the surface of adsorbent supports have been traditionally accomplished by the solid dispersion method [68] and wet-impregnation method [65, 69]. The first method is done by mixing physically the support body and solid active specie followed by heat-treatment to disperse the active species on the internal pore surface of the support. In the second method, the support body is brought in contact with a liquid solution containing the precursor of the active species. The precursor is transported into the pores of the support by capillary force. Solvent is removed by drying. The adsorbent is then calcined under specific atmosphere to convert the precursor to the desired active species.

The major disadvantage of the solid-dispersion coating method is its difficulty to disperse oxide precursor with a high melting point on the internal pore surface of the support body. With the wet-impregnation method there is a problem of controlling the amount of the active species coated on the support body. In the case of the sol-gel derived support body, the wet-impregnation method requires another drying and calcination steps in addition to those used in preparing the support body.

In our laboratory, we have developed a so-called solution-sol mixing method to prepare supported adsorbents by the sol-gel method [66]. In this method, the stable particulate sols of the alumina, titania and zirconia are first prepared from corresponding metalorganic precursors. A aqueous solution containing appropriate precursor (usually a salt) is mixed with the stable sol, as shown in Fig.6. After a controlled aging period, the salt doped sol is used as the starting material for preparing the coated granular adsorbents. In this method, the precursor is coated on the surface of primary particles in the sol, either during the mixing or subsequent drying step. This method allows coating an active species in a precisely controlled amount on the support. Other advantages include: (1) the active precursor can be homogeneously coated on the grain surface of the support as a result of the homogeneous mixing taking place in the liquid phase; (2) the method does not require additional drying/calcination steps in comparison with the wet-impregnation coating procedure.

Table 6 summarizes the pore structure of sol-gel derived alumina coated with various active species by the solution-sol mixing method. Synthesis of the boehmite sol was described above. Nitric acid was used to peptize the particulate sol. The sol was mixed with the aqueous solution of corresponding salt solution in a predetermined ratio to give the loading listed on the second column in Table 6. The coated samples were dried and calcined in air at 450°C for 3 hours.

Table 6
Pore structure of alumina coated with an active species by solution-sol mixing method using the boehmite sol as the starting material

| Doping Salt | Loading (M/Al) | $S_{BET}$ ($m^2/g$) | Vp (ml/g) | $d_{av}$ (nm) |
|---|---|---|---|---|
| Pure | 0 | 323 | 0.36 | 3.1 |
| $Ca(NO_3)_2$ | 0.20 | 298 | 0.25 | 2.9 |
| $Cu(NO_3)_2$ | 0.14 | 302 | 0.28 | 2.9 |
| $AgNO_3$ | 0.10 | 262 | 0.25 | 2.8 |
| $La(NO_3)_3$ | 0.03 | 290 | 0.28 | 3.1 |
| $CuCl_2$ | 0.20 | 207 | 0.18 | 2.9* |

* *Sol becomes unstable after mixing, with large aggregates precipitated.*

For most cases shown in Table 6, the sols after doping still remained stable. The pore structures of the coated samples after calcination do not differ that much from the corresponding uncoated samples, as shown in Table 6. However, mixing a salt containing an anion different from the anion of the acid used in pepetization could destabilize the sols. The pore structure of the samples derived from the unstable sols is very different from those derived from the stable sols. Similar results were observed for zirconia and titania sols [38, 39].

The pore structure of the samples after calcination is determined by the shape and size of the primary particles, and, to a less extent, by the manner these particles are packed. The slight differences in the pore structure between the uncoated and coated samples are due to the effects of coating material on the size and shape of the primary particles in the sol. The particulate sols contain fine particles of zirconia, boehmite or titania. Mixing a salt solution in the sol might change slightly the cation charge distribution, affecting the size and shape of the primary particles. The following were found to be critical to ensuring the stability of a sol after doping: (1) the precursor is preferred to be the salt containing the same anion as that of the acid used in pepetization of the sol; (2) the pH of the doped sol should remain essentially the same as the undoped sol and (3) the sol should be rigorously mixed while the precursor solution is added into the sol.

The exact location of the dopant ions in the doped sol is not clear. This requires further fundamental study on the solution-sol mixing method. However, it can be assumed that the doped salt is coated on the grain surface of the primary particles in the drying process. After calcination at 450°C, the coated salt is converted to corresponding oxide which may still remain on the grain surface of the primary particles.

A more detailed discussion on the solution-sol mixing coating method is given next with preparation of CuO coated $\gamma$-$Al_2O_3$ granular particles as an example. The adsorbent was prepared from the 2 M boehmite sol described earlier. In experiments, a given amount of $Cu(NO_3)_2$ solution (with concentration varied from 0.52 M to 9 M, depending on the required amount of coating) was mixed with the 2 M boehmite sol in the volume ratio of 1:6. After aging for 1/2 hour at 60-70°C, the Cu doped 2 M boehmite was transferred to droppers as the starting material for the granulation process. Following the same granulation process for preparation of the pure $\gamma$-$Al_2O_3$ granular particles, the Cu doped alumina particles were aged in ammonia solution, washed sequentially with water and alcohol solution, dried at 40°C for 48 hours, and finally calcined in air at 450°C for about 4 hours.

CuO coated $\gamma$-$Al_2O_3$ granular adsorbents were also prepared by the wet-impregnation method for comparison. A given amount of the sol-gel derived $\gamma$-$Al_2O_3$ granular particles was brought in contact with $Cu(NO_3)_2$ solution for over 16 hours, followed by the same drying and calcination procedures described above. The amount of active species coated was estimated from the weights of support and $Cu(NO_3)_2$ that were brought in contact with each other (assuming that all $Cu(NO_3)_2$ was coated on the support). The CuO/$\gamma$-$Al_2O_3$ samples, prepared either by the solution-sol mixing method or the wet-impregnation method were further calcined in air at 550°C for 6 hours to convert $Cu(NO_3)_2$ to CuO.

Table 7 compares the pore structure of pure $\gamma$-$Al_2O_3$ granular particles and CuO/ $\gamma$-$Al_2O_3$ granular particles prepared by the wet-impregnation method and by the solution-sol mixing method. For the wet-impregnated CuO/$\gamma$-$Al_2O_3$ granular particles with CuO loading smaller than about 10 wt.%, the particles exhibit gray color with smooth surface. The particles of the samples with CuO loading exceeding 10 wt.%, however, show gray color with a rough surface. XRD peaks of CuO crystallites appear in the samples with CuO loading of about 10 wt.%. According to the monolayer loading theory [68], this result indicates that only limited amount of CuO (<10 wt.%) could be well dispersed on the internal pore surface of the alumina support by the wet-impregnation method used here. Excessive CuO may crystallize and is present on the external surface of the particles. For the CuO/$\gamma$-$Al_2O_3$ granular particles, the color of the wet samples (before drying) was blue. The color became darker as CuO loading increased. The color of the final calcined samples was green. XRD data indicate that CuO in a loading up to at least 18 wt.% could be dispersed on the grain surface of these particles by the solution-sol mixing coating method.

Table 7
Pore structure of pure and CuO coated $\gamma$-$Al_2O_3$ granular particles (particle diameter 2 mm) prepared by different methods

| Coating Method | CuO wt.% | Surface Area ($m^2/g$) | Average Pore diameter (nm) | Pore Volume (ml/g) |
|---|---|---|---|---|
| Pure Alumina | 0 | 325 | 6.2 | 0.5 |
| WI Method | 9.3 | 239 | 8.0 | 0.48 |
| WI Method | 13.4 | 237 | 8.3 | 0.45 |
| SSM Method | 9.0 | 244 | 7.5 | 0.45 |
| SSM Method | 13.5 | 243 | 7.4 | 0.46 |

It should be noted that in a previous study [65] CuO at a much higher loading (>15 wt.%) could be coated on the sol-gel derived $\gamma$-$Al_2O_3$ granular particles by the wet-impregnation method. The only difference between these two studies is that the concentration of the $Cu(NO_3)_2$ solution used in the study described in the above paragraph is about 5 times that of the solution used in the earlier study [65] for wet-impregnation. Apparently, it is more difficult to disperse the precursor on the support surface with impregnation solution of higher concentration. These results indicate the difficulty in controlling the coating process in the wet-impregnation method.

The surface area of the CuO coated $\gamma$-$Al_2O_3$ samples prepared by both coating methods is about 25% smaller than the uncoated alumina samples. Coating CuO on $\gamma$-$Al_2O_3$ results in about 25% increase in the average pore size. The pore volume of the coated samples is the same as the uncoated ones. The pore size distribution data for the granular samples of pure $\gamma$-$Al_2O_3$, CuO/$\gamma$-$Al_2O_3$ prepared by the wet-impregnation, and CuO/$\gamma$-$Al_2O_3$ alumina by the solution-sol mixing method, are shown in Fig.12. The two coated samples have similar pore size distribution. Compared with the pure alumina sample, the pore size distribution data suggest that the coated CuO fills the smaller pores of $\gamma$-$Al_2O_3$, resulting in an appreciable decrease in the number of smaller pores, and, consequently, an increase in the average pore size. The results in Table 7 and Fig.12 also show that CuO coated $\gamma$-$Al_2O_3$ samples prepared by the two coating methods have essentially same pore structure.

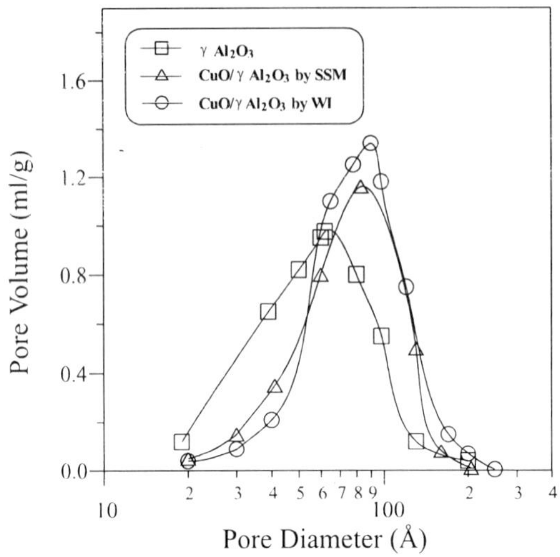


Figure 12. Pore size distribution of pure $\gamma$-$Al_2O_3$ granule, 9.3 wt% CuO coated $\gamma$-$Al_2O_3$ granule prepared by wet-impregnation method, and 9.0 wt.% CuO coated $\gamma$-$Al_2O_3$ granule prepared by the solution-sol mixing method.

## 4.2. Adsorption properties

Adsorption of $SO_2$ on a 30 wt.% CuO/$\gamma$-$Al_2O_3$ granular samples prepared by the sol-gel method was measured by using a Cahn electronic microbalance, as described in detail in reference [70]. The pore structure of the sorbent (Sample 1) is given in Table 8. The experimental results on the sulfation, regeneration and oxidation of the sorbent at 500°C and total pressure of 1 atm are presented in Fig. 13. Curve A is the weight gain of the sorbent when exposed to air stream containing 0.8 % $SO_2$. In this step $SO_2$ and $O_2$ react with CuO to form $CuSO_4$. The sulfation results in about 30% weight gain of the sorbent, equivalent to 3.75 mmol/g sorption capacity of $SO_2$. This sulfation capacity is much larger than similar sorbents prepared by the conventional method [71].

Table 8
Pore structure of sol-gel derived adsorbents used in adsorption study

| Sample # | Method | Active Species | Loading | $S_{BET}$($m^2$/g) | $V_p$(ml/g) | $d_p$(nm) |
|---|---|---|---|---|---|---|
| 1 | Sol-gel | CuO | 30 wt% | 228 | 0.5 | 5.4 |
| 2 | Sol-gel | CuCl | 30 wt% | 283 | 0.43 | 4.9 |
| 3 | Sol-gel | CuO | 9 wt.% | 243 | 0.45 | 7.3 |
| 4 | From UOP | CuO | 7 wt.% | 158 | 0.60 | 15.2 |

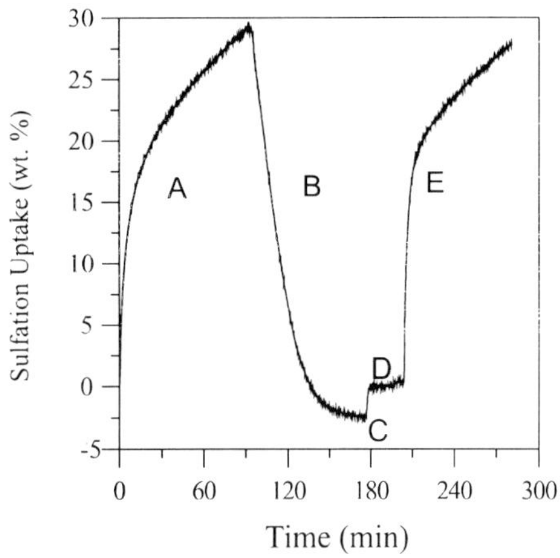


Figure 13. Sulfation, regeneration and oxidation curves of CuO/ γ-alumina granular sorbent.

The sulfated sorbent can be completely regenerated using 10 v.% methane in nitrogen, as shown in Curve B. Curve C-D shows weight gain of the sorbent when exposed to air due to oxidation of copper formed in the regeneration step. Curve E corresponds to weight gain in the second sulfation step, which is essentially identical to curve A. This indicates that the sorbent is highly recyclable.

Ethylene adsorption was measured on a CuCl/ γ-alumina granular particles prepared by sol-gel/wet-impregnation method. A cuprous solution was synthesized by mixing cuprous chloride, ammonia hydroxide, ammonia citrate and deionized water with the composition of 1.98g: 8.8ml : 0.4g : 20ml. About 180 mg of alumina sample was used in each adsorption measurement run by Cahn balance. Before each run, the sample was activated at 250°C for 4 hours in ethylene atmosphere to ensure that the $Cu^{2+}$ ions, if any, were reduced to $Cu^{+}$ ions. Following that, the sample was exposed to nitrogen at 250°C for 2 more hours to completely remove the residual ethylene adsorbed on the sample. Finally, the sample was cooled down under vacuum (1.0 Pa) to the room temperature for adsorption measurement.

Fig. 14 shows adsorption isotherm of ethylene on the sol-gel derived CuCl/ $\gamma$-$Al_2O_3$ sorbent (Sample # 2 in Table 8). The isotherm is of type I. The equilibrium adsorption data of $C_2H_4$ at 1 atm pressure and 14-18°C on CuCl/ $\gamma$-$Al_2O_3$ (prepared by the solid state mixing method) was previously reported in the literature [68]. The amount of ethylene adsorbed on CuCl/ $\gamma$-$Al_2O_3$ was only about 0.29 mmol/g. Yang and Kikkindes [72] reported very good data of adsorption of 0.73 mmol/g of ethylene on CuCl/ $\gamma$-$Al_2O_3$ adsorbent with support surface area of 340 $m^2$/g. The result of our sorbent is similar to the best result reported.

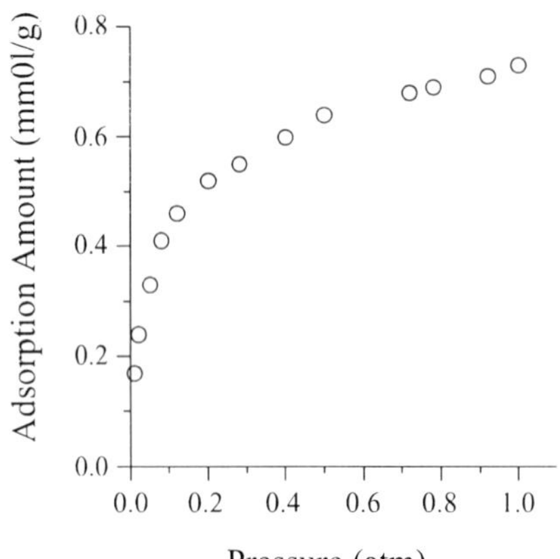


Figure 14. Ethylene adsorption isotherm on sol-gel derived CuCl/ $\gamma$-$Al_2O_3$ granular sorbent.

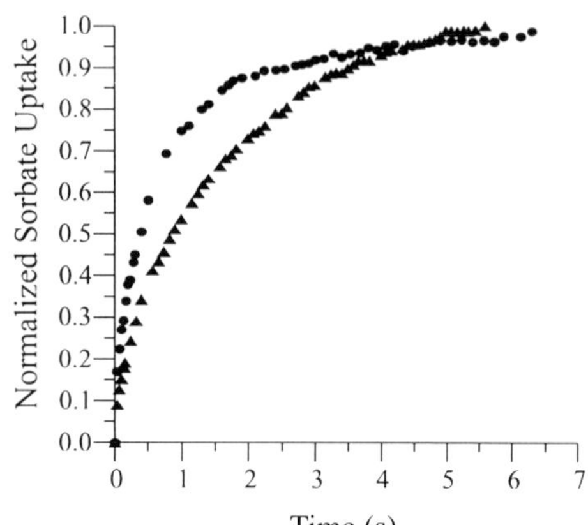


Figure 15. Ethylene adsorption uptakes in CuCl/ $\gamma$-$Al_2O_3$ granule (2 mm) (▲) and thin sheet (0.1 mm) (●).

The availability of the sol-gel derived CuCl/ $\gamma$-$Al_2O_3$ samples in the granular form (particle diameter 2 mm) and in thin sheet form (0.1 mm thickness) allows us to examine the rate-limiting step of ethylene adsorption on the adsorbents [73]. As shown in Fig.15, we found that the time for reaching 95% uptake for both adsorbents subjected to an increase in the ethylene pressure (corresponding to q change from 0 to 0.2 mmol/g) is essentially the same (about 5 seconds) although they are different in the physical dimension. This indicates that adsorption of ethylene on the internal pore surface of adsorbent is the rate-limiting step.

$SO_2$ removal capacity of the sol-gel derived sorbents was also studied in a fixed-bed adsorber system, which includes an adsorber column packed with sorbent pellets, a feed flow control system, and a sulfur analyzer (HORIBA, PIR-2000). A computer data acquisition system was incorporated into this adsorber system in order to obtain continuous breakthrough curves of $SO_2$. The fixed-bed was made from dense $\gamma$-$Al_2O_3$ tube of 6 mm ID and 8 mm OD. The central portion of the bed was packed with the sorbent, with both ends of the bed filled with quartz particles (0.5 mm in diameter). The feed was air containing 2000 ppm $SO_2$, and the fixed-bed temperature was 400°C.

Fig. 16 shows the $SO_2$ breakthrough curve from the fixed-bed packed with 0.5 g of one of the sol-gel derived $\gamma$-$Al_2O_3$/CuO sorbent prepared by the solution-sol mixing method (Sample #3 in Table 8). The feed flow rate is 8.7 ml/min. The temperature for the experiment is 400°C. As shown, the concentration of sulfur dioxide in the effluent stream is zero in the first 3 hours. After 3 hours, the concentration of $SO_2$ in the effluent stream begins to increase, but it takes more than 50 hours to reach the input concentration. These results indicate that the sorbent used in the experiment has high capacity for $SO_2$.

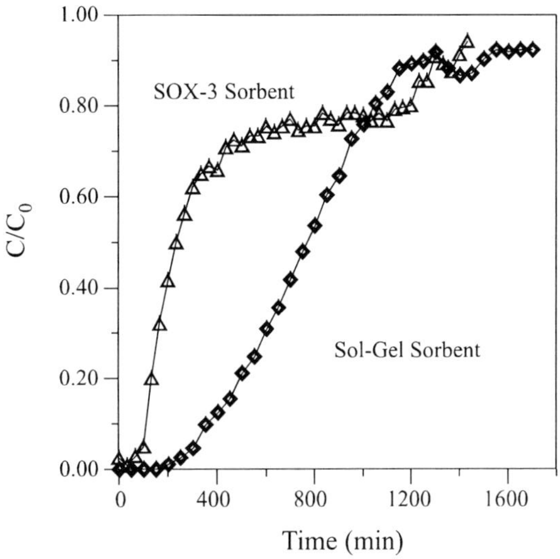


Figure 16. Comparison of $SO_2$ breakthrough curves at 400°C from fixed-bed packed with CuO/ γ-alumina sorbents prepared by solution sol mixing sol-gel method and conventional method.

$SO_2$ breakthrough curves from fixed-bed packed with CuO/ γ-alumina prepared by the conventional process (UOP SOX-3 sorbent [74,75], 7 wt.% CuO on γ-alumina, Sample #4 in Table 8) were also measured for comparison purpose under the same conditions as for the sol-gel derived sorbent. The result is compared in Fig.16 with that for the sol-gel derived sorbent (both at 400°C). At 100% $SO_2$ removal efficiency the sol-gel derived sorbent could treat about 1800 ml flue gas (corresponding to 200 min breakthrough point), about 3 times that of the conventional sorbent (80 min). Considering the total area between C/Co axis and the breakthrough curve, the total sorption capacity of the sol-gel derived sorbent is also about 3 times that of the conventional sorbent. Obviously, the sol-gel derived sorbent has much higher sulfation capacity than the conventional sorbent. The slightly higher amount of CuO coated on the sol-gel derived sorbent alone would not contribute to the significantly improved $SO_2$ removal capacity of sol-gel derived sorbent. The excellent sulfation reactivity of the sol-gel derived sorbents is a result of more uniform dispersion of CuO on the support surface.

## 5. CONCLUSIONS

γ-alumina, zirconia and titania adsorbents with uniform pore size distribution and an average pore diameter of about 3 nm are prepared by hydrolysis and condensation of their metalorganic precursors. These materials consist of nanoscale (5-20 nm) crystallites of which the intercrystalline space gives rise to the pore space. The as-synthesized adsorbents are in their meta-stable phases, and will transform to their stable phases upon heat-treatment at high temperatures. The phase transformation accompanies a decrease in surface area and increase in pore size. The sol-gel derived γ-alumina, zirconia and titania adsorbents should be stable below 400, 300, 200°C, respectively.

Microporous (< 2 nm) silica adsorbents are prepared by the sol-gel method following a polymeric sol route. The adsorbents have a large surface area (600-900 $m^2/g$). Doping alumina in silica adsorbents improves their thermal and hydrothermal stability. These silica adsorbents can be safely used at temperatures lower than 200°C.

A sol-gel granulation process based on the oil-drop principle is described for preparation of γ-alumina granular particles. The sol-gel derived granular sorbents exhibit high crush strength, good attrition resistance and desired pore structure. A solution-sol mixing coating method is used to coat active species on the pore surface of the granular γ-alumina particles. This coating method is particularly useful in preparation of the sol-gel derived supported adsorbents. The method avoids additional drying and calcination steps as otherwise required in the wet-impregnation method. The supported adsorbents prepared by this method exhibit better dispersion of the active species on the support surface. Experimental results show that the sol-gel derived CuO/γ-alumina and CuCl/γ-alumina granular adsorbents have better properties for sorption of $SO_2$ or ethylene than similar adsorbents prepared by the conventional methods.

## ACKNOWLEDGEMENT

We are grateful to the Ohio Department of Development, the National Science Foundation, Amoco Co. and the University of Cincinnati for support on this project. We would like to thank the following members of our laboratory at Cincinnati: C.-H. Chang, K. Boyd, R. Gopalan, Y. Wang, M. Chandak, Z. Wang and G. Buelna, for their contributions to the work summarized in this article.

## REFERENCES

1. C.J. Brinker and G.W. Scherer, Sol-Gel Science: The Physics and Chemistry of Sol-Gel Processing, Academic Press, San Diego, CA, 1990.

2. L.C. Klein (ed.), Sol-Gel Technology for Thin Films, Fibers, Performs, Electronics and Specialty Shapes, Noyes Publications, Park Ridge, NJ, 1988.
3. Materials Research Society Symposium Proceedings: Better Ceramics Through Chemistry, I (vol. 32, 1984); II (vol. 73, 1986); III (vol. 121, 1988); IV (vol. 180, 1990); V (vol.271, 1992) and VI (vol.346, 1994); Materials Research Society, Pittsburg, PA.
4. C.J.G. van der Grift, A. Mulder and J. W. Geus, Colloids and Surfaces 53 (1991) 223.
5. E.H.P. Wolff, A.W. Gerritsen and P.J.T. Verheijen, Powder Technol., 76 (1993) 47.
6. A.J. Fanelli, S. Verma, T. Engelmann and J.V. Burlew, Ind. Eng. Chem. Res., 39 (1991) 126.
7. A.E. Duisterwinkel and G. Frens, in: Preparation of Catalysts VI, G. Poncelet et al. (eds.), Elsevier, Amsterdam, (1995) 1051.
8. S.A. Li, Y. Ke, C. Tang and E. Zhao, The Preparation of Zeolite Molecular Sieve Type A in the Form of Bideless Spherical Granules, Chinese Patent No. CN 1010301B (1990).
9. T.G. Spek and M.J.L. van Beem, Silica Particles and Method for Their Preparation, European Patent EP 0067459 A1 (1982).
10. Y.V. Mirsky, A.Z. Dorogochinskyy, N.F. Meged and A. P. Kosolapova, Process for the Production of Zeolites in the Form of Binderless Spherical Granules, US Patent 4113843 (1978).
11. R.R. Bhave, Inorganic Membranes: Synthesis, Characterization and Applications, Van Nostrand Reinhold, New York, 1991.
12. H.P. Hsieh, Inorganic Membranes for Separation and Reaction, Elsevier, Amsterdam, 1996.
13. A.J. Burggraaf and L. Cot (eds.), Fundamentals of Inorganic Membrane Science and Technology, Elsevier, Amsterdam, 1996.
14. V.E. Moiseev, L.M. Sharygin, V.P. Pyshkin, R.V. Kuzmina, T.G. Malykh and V.M. Galkin, J. Appl. Chem. Russ., 61 (1988) 891.
15. V.N. Krylov, S.M. Borisov and A.G. Burov, Soviet Radiochemistry, 30 (1988) 615.
16. T.N. Perekhozeva, L.M. Sharygin and G.P. Albantova, Inorg. Mater., 25 (1989) 1295.
17. V.M. Galkin, L. M. Sharygin, V.E. Moiseev, T.N. Perekhonzhevea and E.V. Loguntesv, J. Appl. Phys. Chem. Russ., 62 (1989) 2051.
18. G.C.M. Hakvoort, J.C. van der Bleek, J.C. Schouten and P.J.M.Valkenburg, Thermochim. Acta, 114 (1987) 103.
19. J.N. Armor and E.J. Carlson, J. Mater. Sci., 22 (1987) 2549.
20. S.J. Teichner, G.A. Nicolaon, M.A. Vicarini and G.E.E. Grades, Adv. Colloid Interface Sci., 5 (1976) 245.
21. C.J. Guo, in: Zeolites: A Refined Tool for Designing Catalytic Sites, L. Bonneviot and S. Kaliaguine (eds.), Elsevier, Amsterdam, (1995) 165.

22. C.J. Guo, C.W. Fairbridge and J.P. Charland, Synthesis of Mesoprous Catalytic Materials, US Patent No. 5538710 (1996).
23. G. Grassini-Strazza, V. Carunchio and A.M. Girelli, J. Chromatogr., 466 (1989) 1.
24. J.C. Kohli and K.K. Badaisha, J. Chromatogr., 320 (1985) 455.
25. U. Trüdinger, G. Müller and K. K. Unger, J. Chromatogr., 535 (1990) 111.
26. M.P. Rigney, T.P. Werber and P.W. Carr, J. Chromatogr., 484 (1989) 273.
27. T.P. Werber, P.W. Carr and E. F. Funkenbusch, J. Chromatogr., 519 (1990) 31.
28. J.A. Blackwell and P.W. Carr, J. Chromatogr., 549 (1991) 59.
29. T.P. Weber and P.W. Carr, Anal. Chem., 62 (1990) 2620.
30. J.A. Blackwell and P. W. Carr, J. Liquid Chromatogr., 15 (1992) 1487.
31. K. Fujimara and T. Ando, Anal. Chem., 49 (1977) 1179.
32. D.A. Ward and E.I. Ko, Ind. Eng. Chem. Res., 34 (1995) 421.
33. B.E. Yoldas, Amer. Ceram. Soc. Bull., 54 (1975) 289.
34. C.-H. Chang, R. Gopalan and Y. S. Lin, J. Membr. Sci., 91 (1994) 27.
35. A.F.M. Leenaars and A. J. Burggraaf, J. Mater. Sci., 19 (1984) 1077.
36. Y.S. Lin, K.J. de Veries and A.J. Burggraaf, J. Mater. Sci., 26 (1991) 715.
37. J.W. Christan, The Theory of Transformation in Metals and Alloys, 2nd Ed., Pergamon Press, New York, Chap.10, (1975) 12.
38. R. Gopalan and Y.S. Lin, Ind. Eng. Chem. Res., 34 (1995) 1189.
39. R. Gopalan, C.H. Chang and Y.S. Lin, J. Mater. Sci., 30 (1995) 3075.
40. C.J. Brinker, K.D. Keefer, D.W. Schaeffer, T.A. Assink, B.D. Kay and C.S. Ashley, J. Non-Cryst. Solids, 63 (1984) 45.
41. K.D. Keefer, in: Better Ceramic through Chemistry, C.J. Brinker, D. E. Clark and D. E. Ulrich (eds.), Wiley, New York, III (vol. 121), (1988) 15.
42. S. Sakka, H. Kozuka and S.-H. Kim, in: Ultrastructure Processing of Advanced Ceramics, J.D. Mckenzie and D.R. Ulrich (eds.), Wiley, New York, (1988) 159.
43. C.J. Brinker, K. D. Keefer, D. W. Schaeffer and C. S. Ashley, J. Non-Cryst. Solids, 48 (1982) 47.
44. R.K. Iler, The Chemistry of Silica, Wiley, New York, NY, 1979.
45. R.J.R Uhlhorn, K. Keizer and A. J. Burggraaf, J. Membrane Sci., 66(2/3) (1992) 271.
46. C. Horvath and K. Kawazoe, J. Chem. Eng. Japan, 16 (1983) 470.
47. P.A. Webb and C. Orr, Analytical Methods in Fine Particle Technology, Micromeritics, Norcross, GA, Chap.3, 1997.
48. G.P. Fotou, Y.S. Lin and S.E. Pratsinis, J. Mater. Sci., 30 (1995) 2803.
49. C. E. Capes and A.E. Fouda, in: Handbook of Powder Science and Technology, M.F. Fayed and L. Otten (eds.), Van Nostrand Reinhold, New York, (1984) 332.

50. C.E. Capes, in: Particle Size Enlargment in Handbook of Powder Technology, J C. Williams and T. Allen (eds.), vol. 1, Elsevier, Amsterdam, 1986.
51. S. Komarneni and R. Roy, Mater. Lett., 3 (1985)165.
52. M. Marella, M. Tomaselli, L. Meregalli, M. Battagliarin, P. Geronopoulosa, F. Pinna, M. Signoretto and G. Strukul, in: Preparation of Catalysts VI, G. Poncelet et al. (eds.), Elsevier, Amsterdam, (1995) 327.
53. J.D. F. Ramsay, in: Controlled Particle, Droplet and Bubble Formation, D. J. Wedlock (ed.), Butterworth Heinemann, Oxford, (1994) 1.
54. B. Barten, F. Janssen, F. V. D. Kerkhof, R. Leferink, E. T. C. Vogt, A. J. van Dillen and J. W. Geus, in: Preparation of Catalysts IV, B. Delmon, P. Grange, P. A. Jacobs and G. Poncelet (eds.), Elsevier, Amsterdam, (1987) 103.
55. J.H.A. Kiel, W. Prins and W. P. M. van Swaaij, Appl. Catal. B. Environ., 1 (1992) 13.
56. R.M. Cahen, J.M. Ander and H.R. Debus, in: Preparation of Catalysts II, B. Delmon, P. Grange, P. A. Jacobs and G. Poncelet (eds.), Elsevier, Amsterdam, (1979) 585.
57. A. Meyer and K. Noweck, Verfahren zur Herstllung von kugelförmiger Tonerde, Europäische Patentanmeldung, No. 0 090 994 A2 (1982).
58. M.N. Shepeleva, R.A. Shkrabina, Z.R. Ismagilov and V.B. Fenelonov, in: Preparation of Catalysts V, G. Poncelet, P. A. Jacobs, P. Grange and B. Delmon (eds.), Elsevier, Amsterdam, (1991) 583.
59. K. Svoboda, W. Lin, J. Hannes, R. Korbee and C. M. van den Bleek, Fuel, 73 (1994) 1144.
60. P.A. Haas and S.D. Clinton, Ind. Eng. Chem. Product Res. Dev., 5 (1966) 236.
61. C. Ganguly, H. Langen, E. Zimmer and E.R. Merz, Nuclear Technology, 73 (1985).
62. C. Ganguly, P.V. Hegde and G.C. Jain, Nuclear Technology, 105 (1994) 346.
63. H. Barten, F. Janssen, F.V.D.Kerkhof, R. Leferink, E.T.C. Vogt, A.J. van Dillen and J.W. Gels, in: Preparation of Catalysts, VI, B. Delmon et al, (eds.), Elesvier, Amsterdam, 1987.
64. M.Z.-C. Hu and M. Reeves, Biotechnol. Prog., 13 (1997)60.
65. S.G. Deng and Y.S. Lin, AIChE J., 43 (1997) 505.
66. Z. Wang and Y.S. Lin, J. Catalysis, 174 (1998) 43.
67. P.K. Doolin, D.M. Gainer and J.F. Hoffman, J. Testing & Evaluation, 21 (1993) 481.
68. Y.-C. Xie and Y.Q. Tang, Adv. Catal., 37 (1990) 1.
69. S.G. Deng and Y.S. Lin, AIChE J., 41 (1995) 559.
70. S.G. Deng and Y.S. Lin, Ind. Eng. Chem. Res., 35 (1996) 1429.
71. G. Centi, A. Riva, N. Passarini, G. Brambilla, B.K. Hodentt, B. Delmon and M. Ruwet, Chem. Eng. Sci., 45 (1990) 2769.
72. R.T. Yang and E.S. Kikkinides, AIChE J., 41 (1995) 509.

73. Y. Wang and Y.S. Lin, J. Sol-Gel Sci. Tech., in press (1998).
74. P. Harriott and J.M. Markussen, Ind. Eng. Chem. Res., 31 (1992) 373.
75. J.T. Yeh, R.J. Demski, J.P. Strakey and J.I. Joubert, Environ. Prog., 4 (1985) 232.

Adsorption and its Applications in Industry and Environmental Protection
Studies in Surface Science and Catalysis, Vol. 120
A. Dabrowski (Editor)



# Ceramic membranes - characterization and applications

E.S. Kikkinides[a], A.K. Stubos[b], K.P. Tzevelekos[a], A.Ch. Mitropoulos[a] and N. Kanellopoulos[a]

[a] Institute of Physical Chemistry, NCSR Demokritos, 15310 Ag. Paraskevi Attikis, Greece

[b] Institute of Nuclear Technology and Protection, NCSR Demokritos, 15310 Ag. Paraskevi Attikis, Greece

## ABSTRACT

A combination of characterization techniques for the pore structure of meso- and microporous membranes is presented. Equilibrium (sorption and Small Angle Neutron Scattering) and dynamic (gas relative permeability through membranes partially blocked by a sorbed vapor) methods have been employed. Capillary network and EMA models combined with aspects from percolation theory can be employed to obtain structural information on the porous network topology as well as on the pore shape. Model membranes with well defined structure formed by compaction of non-porous spherical particles, have been employed for testing the different characterization techniques. Attention is drawn to the need for further development of more advanced sphere-pack models for the elucidation of dynamic relative permeability data and of Monte-Carlo Simulation for the analysis of equilibrium sorption data from microporous membranes.

The application of ceramic membranes in separations of condensable from non-condensable vapors is explored both theoretically and experimentally. Capillary condensation greatly enhances the permeability of the condensable vapor through the mesoporous membrane resulting in large selectivities over non-condensable vapors. Attention is drawn to the fact that the dynamic membrane properties depend on a wider range of microstructural characteristics, relevant to the separation efficiency of the membrane. Therefore, measurements of the dynamic relative permeability and of condensable vapor permeability provide significant saving in effort for the determination of the optimum pressure and temperature operation conditions and for the development of a model predicting the membrane performance.

# 1. CHARACTERIZATION OF CERAMIC MEMBRANES

The evaluation of the commercial potential of ceramic porous membranes requires improved characterization of the membrane microstructure and a better understanding of the relationship between the microstructural characteristics of the membranes and the mechanisms of separation. To this end, a combination of characterization techniques should be used to obtain the best possible assessment of the pore structure and provide an input for the development of reliable models predicting the optimum conditions for maximum permeability and selectivity. The most established methods of obtaining structural information are based on the interaction of the porous material with fluids, in the static mode (vapor sorption, mercury penetration) or the dynamic mode (fluid flow measurements through the porous membrane).

## 1.1. Equilibrium methods

### 1.1.1. Intrusive methods - sorption measurements

While a number of established characterization methods exist for mesopores and macropores, the assessment of microporosity is much less advanced, due to experimental difficulties and the lack of a suitable model for the interpretation of the isotherm data. Obtaining accurate experimental isotherms is hampered by the long equilibration times required at the low liquid nitrogen temperatures. In order to overcome this limitation the micropore structure evaluation can be based on isotherms of carbon dioxide or other vapors obtained at higher temperatures, provided that a suitable equilibrium model for the sorption of non spherical molecules is available.

The Grand Canonical Monte Carlo (GCMC) method is ideally suited to adsorption problems because the chemical potential of each adsorbed species is specified in advance [1,2]. At equilibrium, this chemical potential can then be related to the external pressure making use of an equation of state. Consequently, the independent variables in the GCMC simulations are the temperature, the pressure and the micropore volume, which is a convenient set, since temperature and pressure are the adsorption isotherm independent variables. Therefore, the adsorption isotherm for a given pore can be obtained directly from the simulation by evaluating the ensemble average of the number of adsorbate molecules whose chemical potential equals that of the bulk gas at a given temperature and pressure.

Three types of trials have been used, i.e. attempts to move (translate or reorient) particles, attempts to delete particles and attempts to create particles in the simulation box. A decision is made on whether to accept each trial or to return to the old configuration based on a probability which in the case of an attempted move takes the form:

$$p_{move} = \min\left[\exp\left(-\frac{\Delta U}{kT}\right)\right] \tag{1}$$

where $\Delta U = U_{new} - U_{old}$ is the difference in the potential energies of the new and old configurations. A detailed presentation of the method is given in [1-4].

The realistic character of simulations and the accuracy of the results depend largely upon the potential energy model used. Here, the carbon dioxide molecule is modeled as Lennard-Jones interaction sites on the atoms plus point charges to account for the quadrupole (three center LJ model). The interactions are cut (but not shifted) at 2.0 nm. Because of this relatively large cutoff and the confinement of the molecules in the micropores, no long range corrections have been employed.

The graphitic surface is treated as stacked planes of Lennard-Jones atoms. The interaction energy between a fluid particle and a single graphitic surface is given by the 10-4-3 potential of Steele as:

$$u_{sf}(z) = 2\pi d_s \varepsilon_{sf} \sigma_{sf}^2 \Delta \left\{ \left( \frac{2}{5} \frac{\sigma_{sf}}{z} \right)^{10} - \left( \frac{2}{5} \frac{\sigma_{sf}}{z} \right)^{4} - \frac{\sigma_{sf}}{3\Delta(0.61\Delta + z)^3} \right\} \tag{2}$$

where $\Delta$ is the separation between graphite layers and $d_s$ is the number of carbon atoms per unit volume in the graphite layer. The values used for $\Delta$ and $d_s$ are 0.335 nm and 114 $nm^{-3}$ respectively. The solid-fluid Lennard-Jones parameters $\sigma_{sf}$ and $\varepsilon_{sf}$ are calculated by combining the graphite parameters with the appropriate fluid parameters according to the Lorentz-Berthelot rules. The external field, $u^{(1)}$ for a single Lennard-Jones site in a slit pore of width H is the sum of the interactions with both graphitic surfaces and can be expressed as:

$$u^{(1)} = u_{sf}(z) + u_{sf}(H-z) \tag{3}$$

where H is the C center - C center separation across the pore. This equation ignores the surface corrugation, which is unlikely to significantly affect the results at high enough temperatures. The effective pore width H′ (which is determined by the experiments) is in general given by:

$$H' = H - \Delta \tag{4}$$

Alternatively, it has been proposed to use an estimation of the effective pore width based on the reduction in accessible by the adsorbate pore volume, caused by the physical size of the carbon atoms. Taking as the distance of closest possible approach of adsorbate to adsorbent the position $z_o$, where the potential function for a plane wall passes through zero, results in:

$$H' = H - 2z_o + \sigma_g \tag{5}$$

where $\sigma_g$ is the hard sphere diameter of an adsorbate atom.

Periodic boundary conditions have been applied in the directions other than the width of the slit. For a given simulation, the size of the box (i.e. the two dimensions other than H) was varied in order to ensure that sufficient particles

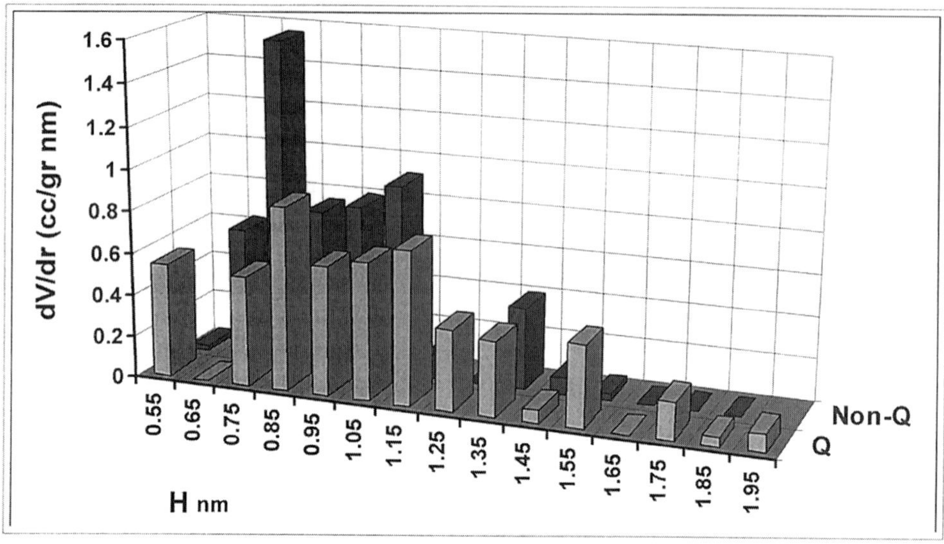


Figure 1a. Calculated PSD's from isotherm data and GCMC simulations.

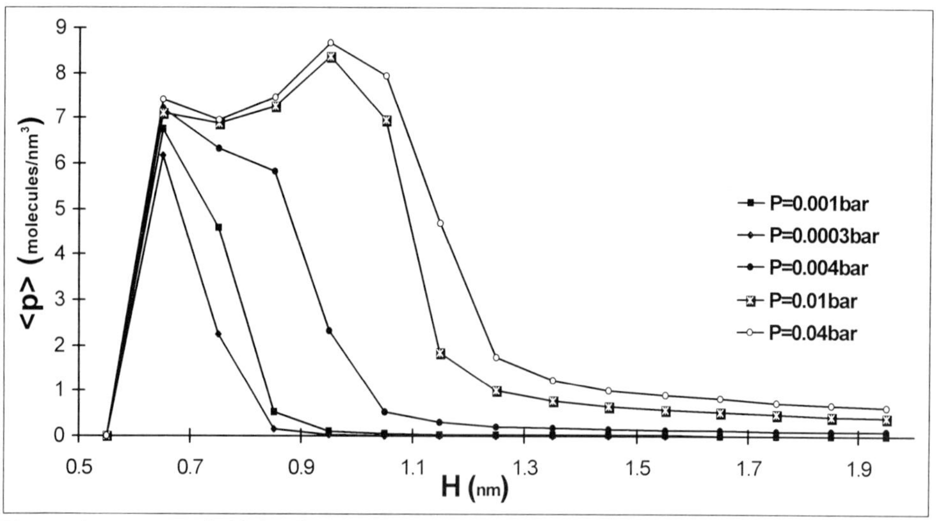


Figure 1b. Average fluid density vs. pore size for different bulk pressures (T=195.5 K).

(ca. 200-450) remained in the simulation at each pressure and was always greater than twice the cutoff distance. Statistics were not collected over the first $2{\cdot}10^6$ configurations to assure adequate convergence of the simulation. The uncertainty on the final results (ensemble averages of the number of adsorbate molecules in the box and the total potential energy) is estimated to be less than

4%. Typical calculations for a single point require between 1.7 and 4.2 hours of CPU time on a Convex 3820.

A method for the determination of the micropore size distribution based on Monte Carlo simulations has been proposed [3]. In this work, the method is used in its grand ensemble variant in combination with experimentally derived isotherm data to characterize microporous structures and obtain the corresponding pore size distribution (PSD). Specifically, the mean $CO_2$ density inside a single slit shaped graphitic pore of given width, is found on the basis of Grand Canonical Monte Carlo simulations for a pre-defined temperature (195.5 K) and different relative pressures. Starting from an initial PSD guess, it is then possible to produce a computed $CO_2$ sorption isotherm and compare it to the measured one. After a few iterations, the procedure results in a PSD which, if desired, can be further refined at the cost of additional computational effort. Pore size distributions of activated carbon membranes obtained by a Nitrogen porosimeter (with Krypton upgrade) via the conventional (DR) approach has been employed for the sake of comparison with the method of [3]. The effects of molecular parameters (i.e. quadrupole moment) are taken in account in the GCMC simulations.

The selection of the fluid and the low temperature is based on practical considerations regarding the relative ease of obtaining experimental isotherms at dry ice temperature with a molecule that is known for its ability to enter into the narrow microporosity and the realistic equilibration times required. In addition, the temperature (dry ice) chosen is conveniently low to avoid large effects from excess quantities calculations and, at the same time, high enough to ignore corrugation of the solid surface. Presently, the method is further validated by extending it to considerably higher temperatures (298 K and 308 K, i.e. slightly below and above the $CO_2$ critical temperature, respectively) and comparing the resulting PSD with that obtained at dry ice conditions. The latter is shown in fig. 1a, where the effect of omitting quadrupole moment is also included. In addition, further insight to the structure of $CO_2$ molecules packing in the individual micropores at dry ice temperature has been obtained through the simulations. Figure 1b presents the average (excess) density of the fluid under varying bulk pressure as function of the micropore size.

### 1.1.2. Non intrusive static methods - Small Angle Scattering of partially blocked membranes by sorbed vapors

According to Small Angle scattering (SAS) theory, the intensity I(h) (h is the scattering vector) scattered by a two phase system is related to the electron (SAXS) or scattering length (SANS) densities $\rho_1$ and $\rho_2$ of the phases in terms of the expression :

$$I(h) \approx (\rho_1 - \rho_2)^2 \tag{6}$$

In the case of adsorption of a vapor by a porous material, a three phase system in terms of SAS is produced: pore/adsorbed film or capillary condensed vapor/ solid. Since the Scattering Length Densities (s.l.d.) of $H_2O$ and $D_2O$ are known, and the s.l.d. of pores equals to zero, Contrast Matching conditions are achieved if an appropriate mixture of $H_2O/D_2O$ that has the same s.l.d. as the solid is used as the adsorbate. In this case the adsorbed film as well as the condensed cluster of pores will cease to act as scatterers, and only the remaining empty pores will produce measurable scattering. In terms of SANS, Contrast Matching reduces the solid/film/pore system to a binary one. In this way, an interface of adsorption and SANS can be achieved. By determining a number of scattering curves corresponding to the same porous sample equilibrated at various relative pressures, for both the adsorption and desorption branches of the adsorption isotherm, a correlation of the two methods could be possible. If the predictions of the Kelvin equation are in accordance with the SAS analysis, a reconstruction of the adsorption isotherm can be obtained from the SAS data [5,6].

SAS can detect the presence of non-accessible to vapor pores, which cannot be detected by the intrusive methods, such as sorption. If no such pores exist, a good agreement is expected between the intrusive sorption and the non-intrusive SANS, as illustrated in fig.2. In this figure, an experimental water adsorption isotherm on a mesoporous alumina membrane is compared to the corresponding one reconstructed from the SANS data. An appropriate $H_2O/D_2O$ mixture was used as the adsorbate for the SANS experiments. It should be noted that at

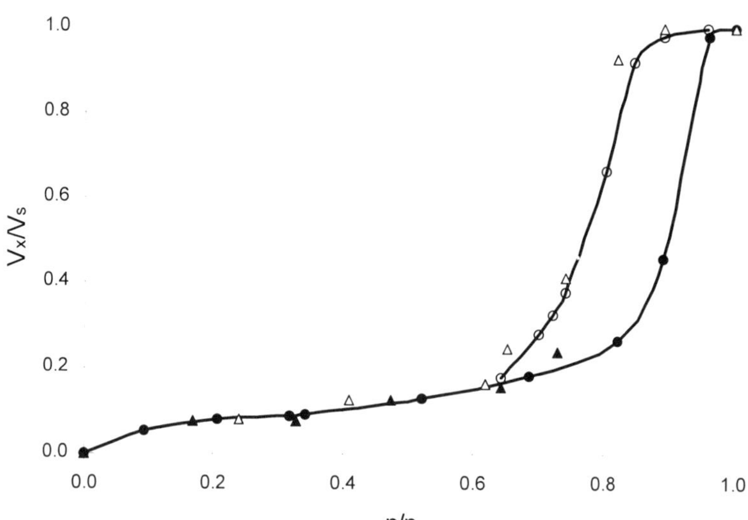


Figure 2. Comparative presentation of experimental $H_2O$ adsorption isotherm (filled and empty circles for adsorption and desorption respectively) and the reconstructed from the SANS data isotherm (filled and empty triangles).

relative pressures close to unity, surprisingly high scattering intensity is observed. This can be attributed to the existence of large voids, which are formed due to the agglomeration of the primary particles.

### 1.2. Dynamic methods

#### 1.2.1. Membrane modeling

Dynamic methods rely on the study of fluid flow properties of porous membranes, which are extremely sensitive functions of the pore size distribution, f(r), and of additional pore structural characteristics, such as the pore connectivity, z. The dynamic data, if analyzed in combination with other measurements obtained by equilibrium methods, can offer important structural information, relevant to the evaluation of performance of membranes, provided that an appropriate pore structural model is used for the data interpretation.

The most widely used way of representing a porous structure is as a bundle of tortuous capillaries with radii obeying the pore size distribution f(r) and effective length $L_{eff}$, along the axis of the flow. The model is completely defined by f(r) and the tortuosity factor, $\tau=(L_{eff}/L)^2$, where L is the straight distance in the direction of the flow.

Alternatively, the capillary network model constitutes a significant improvement over the aforementioned mentioned tortuosity model, since it can provide realistic modeling, especially for systems involving membranes partially blocked by condensed vapors. In this model the degree of connectivity of the pores, z, is replacing the less tangible tortuosity factor $\tau$. The estimation of the z can be based on gas and condensed vapor relative permeability measurements, presented in this section.

#### 1.2.2. Gas relative permeability

The gas relative permeability, $P_R$, is defined as the permeability of a fluid through a porous medium partially blocked by a second fluid, normalized by the permeability through the same porous solid, when the pore space is free of this second fluid. In most cases, the gas relative permeability diminishes at the "percolation threshold", at which a significant portion of the pores are still conducting but they don't form a continuous path through the membrane along the direction of flow. The tortuous capillary model fails to predict this; the percolation threshold arises only when all pores are blocked by capillary condensation. In comparison, the network model can provide a satisfactory analysis of the percolation threshold problem, without, as noted earlier, increasing the number of the model parameters.

Nicholson et al. [7] introduced a simple network model, and applied it on gas relative permeability [8, 9]. For the gas relative permeability, an explicit approximate analytical relation between the relative permeability and the two network parameters, namely the z and the first four moments of the pore size distribution, f(r), has been developed, based on the Effective Medium Approximation (EMA) [10, 11]. Bethe trees, which are lattices that do not admit reconnections, can also be considered since they give simplified expressions for several properties of the porous medium, while at the same time retain most features of percolation theory.

If a porous solid is in equilibrium with bulk vapor at a vapor relative pressure $P/P_0$ the adsorbate consists of a capillary condensed liquid filling the pores with radii smaller than the Kelvin radius, $r_K$ (subcritical pores, $r \leq r_K$) and an adsorbed layer of thickness t covering the walls of the supercritical pores ($r > r_K$). For the classic case of $N_2$ sorption on a mesoporous medium at 77 K, the following expressions have been employed [12-14]

$$t = \left( \frac{c_2}{\ln\left( P/P_0 \right)} \right)^{\frac{1}{3}} \tag{7}$$

$$r_k = \frac{c_1}{\ln\left( P/P_0 \right)} + t \tag{8}$$

where $c_1$, $c_2$ are constants, characteristic of the particular adsorbate (for the case of $N_2$ adsorption at 77 K, $c_1$=-0.477 nm and $c_2$=-0.218 $nm^3$ ). The flux expression for an open cylindrical pore (bond) of the network connecting two nodes (sites), i and j, in the Knudsen regime can be written as follows (assuming long capillaries,[15])

$$J_{ij} = \frac{2\pi x_{ij}^3}{3} \left( \frac{8 R_g T}{\pi M} \right)^{1/2} \frac{(P_{gi} - P_{gj})}{l} \tag{9}$$

where $x_{ij}=r_{ij}-t$ is the open core radius of a capillary partly filled with adsorbate of thickness t, $R_g$ is the universal gas constant, T is the ambient temperature, M is the molecular weight of the non-adsorbed gas (e.g. He) and l is the length of the capillary pore (assumed to be the same for all network pores). Writing the material balance equation at each pore junction results in a set of linear algebraic equations which can be solved for the nodal pressures using successive over-relaxation methods. The network permeability is then determined from the total flux, J, obtained for a given pressure drop across the network. (Standard resistor network analysis; see, for example [9,16]). The above computation scheme is repeated for a range of values of $P/P_0$ between zero and unity and the relative permeability $P_R = \frac{J\left(P/P_0\right)}{J(0)}$ is determined as a function of the relative pressure $P/P_0$ or the normalized adsorbed volume $V_S$. This volume is calculated by the following expression:

$$V_S = 1 - \frac{\sum\sum x_{ij}^2}{\sum\sum r_{ij}^2} \tag{10}$$

Application of EMA for the determination of $P_R$ gives [14]:

$$\frac{1-f_b}{z/2-1} + \int_{r_a}^{r_b} \frac{\left(Pe_M - x^3\right)\cdot f(x)}{x^3 + (z/2-1)\cdot Pe_M} dx = 0 \tag{11}$$

where $Pe_M$ is the dimensionless effective medium permeability which is determined by the solution of the integral equation, x=r-t and $f_b$ is the number fraction of the open (supercritical) pores at a certain value of $P/P_0$. For a normalized distribution function it follows that:

$$f_b = \int_{x_{ac}}^{x_b} f(x)\cdot dx \tag{12}$$

where $x_{ac}=\max\{x_a, x_K\}$ and f(x) differs in shape from f(r) because the subcritical tail of f(r) is now missing [14]. On the other hand, $V_S$ is given by:

$$V_S = 1 - \frac{\int f(x)x^2 dx}{\int f(r) r^2 dr} \tag{13}$$

$V_S$ is related to $f_b$ through the following expression:

$$V_S = 1 - f_b \cdot \Phi \tag{14}$$

where $\Phi(P/P_0)$ is a complicated function of $P/P_0$ and contains first and second moments of the distribution functions f(r) and f(x)=f(r-t). The exact relation between $f_b$ and $V_S$ can be found elsewhere [11,14]. Thus for a certain value of $P/P_0$, the quantities $V_s$, $f_b$ and $P_R = Pe_M(P/P_0;V_S,f_b)/Pe_M(0;0,1)$ are determined from the above equations. It is important to note that when $f_b$=2/z then $V_S = V_{SC}$ and $P_R$=0 according to EMA. This result, although generally accurate in two dimensional networks [17,18], totally breaks down for the case of three dimensional networks and other methods should be employed in such a case [19,20].

For the case of a Bethe network with coordination number z,. it has been shown [21-23] that the effective conductivity can be calculated by solving an integral equation for the effective conductivity, $g_M$, of the network. By expanding the aforementioned integral equation in n inverse powers of z-1, $g_M$ can be determined to the order of $(z-1)^{-n}$ by solving the following integral equation [21,24]:

$$\int_{r_a}^{r_b}[\tau(r)+g_M/z]\,f(r)dr+\sum_{m=2}^{\infty}b_m I_{m10}=0 \qquad (15)$$

where

$$\tau(r)=\frac{r^3(1-z)g_M}{zr^3-(1-z)g_M} \qquad (16)$$

and $b_m$, $I_{m10}$ are implicit functions of the unknown $g_M$ in complicated integral forms (see also [21,24]).

The infinite series in eq. (14) can be treated in the same way as in Stinchombe, 1974, keeping terms up to the order of $(z-1)^{-4}$, and resulting in an integral equation which can be solved to determine $g_M$ with an error of $(z-1)^{-5}$ (see also [24]). An alternative approach that is valid for $f_b$ near $f_{bc}$ ($f_b - f_{bc} \leq 1/z$) has been derived both by Stinchombe [21], and Heinrichs and Kumar [22] according to which $g_M$ is given by:

$$g_M=-\frac{f_{-3}\cdot(1.522z)}{z-2}\cdot\left(\frac{1-f_b}{f_{bc}}\right)^2 \qquad (17)$$

where

$$f_{-3}=\int h(g)\cdot g^{-1}dg=\int f(r)\cdot r^{-3}dr \qquad (18)$$

assuming Knudsen type of flow ($g \propto r^3$). Using the above equations, $g_M$ can be determined both away and near the percolation threshold of the Bethe tree.

Once $g_M$ is determined, $P_R$ can be calculated from $P_R = g_M(P/P_0;V_S,f_b)/g_M(0;0,1)$, where again $V_S$ and $f_b$ are calculated by eqs (12) and (13) for all cases.

In fig. 3a relative permeability curves computed by the network model are plotted for z=4,6 and 8. EMA results are also shown in this figure, for comparison purposes. It appears that as *z* increases the $P_R$ curve becomes broader as it approaches the percolation threshold, $V_{SC}$. In all cases EMA is in very good agreement with the network solution, except in the neighborhood of $V_{SC}$.

In that region, the EMA predicted $P_R$ curve decreases linearly with $V_S$, while the network solution results in a non-linear behavior and reaches a higher percolation threshold, $V_{SC}$. This is because $V_{SC}$ predicted by the network model corresponds to the theoretical $f_{bc}$ predicted by percolation theory ($f_{bc} \sim 1.5/z$, [25]), while $V_{SC}$ found by EMA corresponds to $f_{bc}=2/z$ [17].

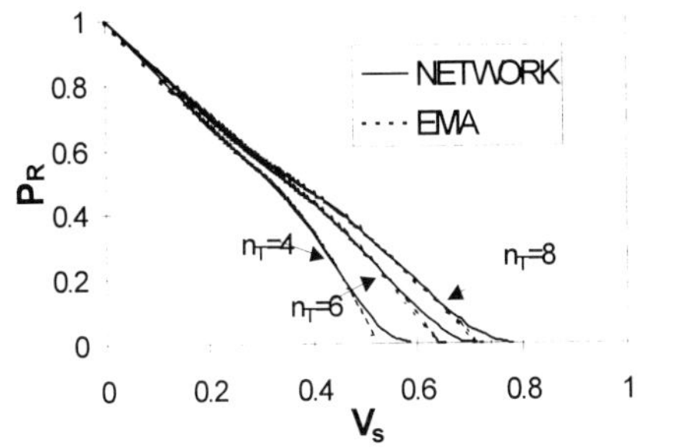


Figure 3a. Relative permeability against volume of adsorbate, during adsorption for 3D network and EMA models at different connectivities.

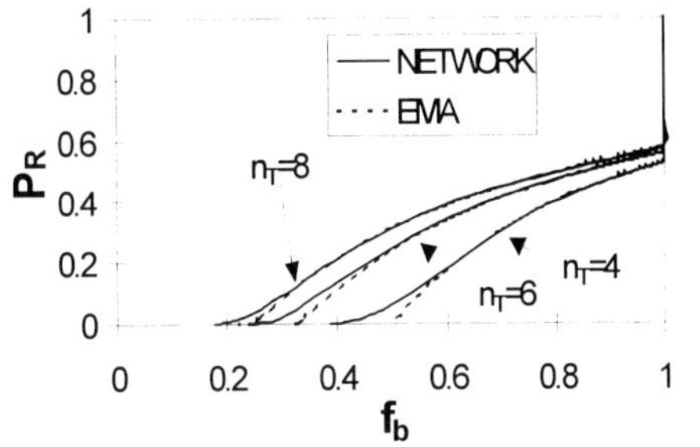


Figure 3b. Relative permeability against fraction of open pores, during adsorption for 3D network and EMA models at different connectivities.

A similar picture is obtained in figure 3b, where $P_R$ is plotted as a function of the fraction of the open pores, $f_b$. It can be seen that for all z, near the percolation threshold, EMA shows a linear decrease of $P_R$ with $f_b$ . On the other hand, network results indicate that, in the same region, $P_R$ decreases with $f_b$ according to a power law. For an infinitely large network percolation theory states:

$$P_R \propto (f_b - f_{bc})^t \tag{19}$$

where t is a universal critical exponent that depends only on the dimensionality of the network [26]. For a three dimensional lattice t≅1.87. For a network of size L×L×L finite size scaling effects are expected to influence the above behavior [27]. However, a standard approach is employed for the case of finite size networks in order to apply percolation theory and to determine critical exponents [26]. According to this, eq. (19) is replaced by the following:

$$P_R \sim L^{-t/\nu} g\left[(f_b - f_{bc})L^{1/\nu}\right] \tag{20}$$

where ν is another universal critical exponent, with ν =0.88 for three dimensional lattices [23] and g(x) is a scaling function, which varies from lattice to lattice but has a common asymptotic behavior:

$$g(x) \rightarrow \begin{cases} 1; & x << 1 \\ x^t; & x >> 1 \end{cases} \tag{21}$$

Thus at the percolation threshold, $P_R$ is size dependent:

$$P_R \sim L^{-t/\nu} \tag{22}$$

Figure 4 is a plot of $P_R$ against $(f_b\text{-}f_{bc})L^{1/\nu}$ with L=20, ν=0.88. It can be seen that at the percolation threshold, $P_R$ is not exactly zero but reaches an asymptotic value of around $1.6\text{-}1.8\cdot10^{-3}$, which from eq. (22) results in t ~1.9. In addition, in the other asymptotic region where $(f_b\text{-}f_{bc})L^{1/\nu}>>1$, there is a power law behavior, again with t~1.9. These results are the same for all three different values of z=4, 6, 8 for which $f_{bc}$ is, 0.39, 0.25 and 0.18 respectively for the bond percolation problem in three dimensions.

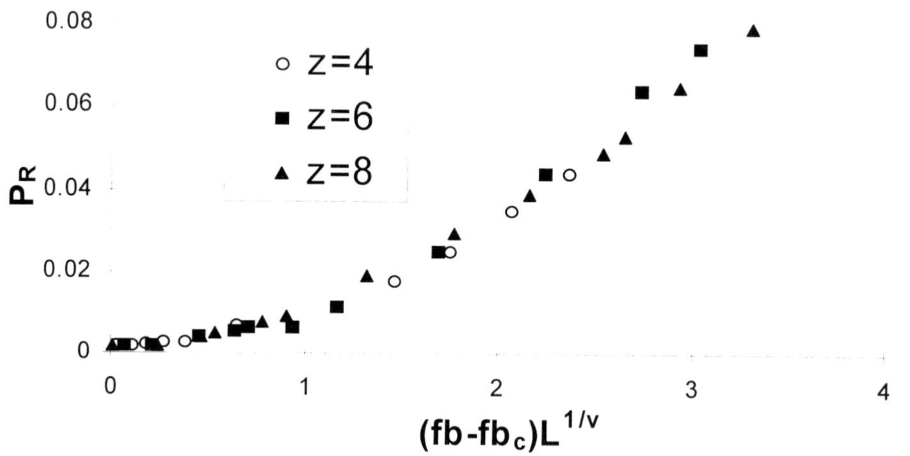


Figure 4. Generalized relative permeability curves for 3D networks of different connectivities.

Thus it appears that relative permeability curves follow percolation theory, since they satisfy both the theoretical percolation threshold and the scaling law for three dimensional networks [25]. More importantly, relative permeability curves of different connectivity exhibit the same behavior with $f_b\text{-}f_{bc}$ as $f_{bc}$ is approached. The same conclusion is valid for different pore size distribution functions f(r) provided that $f_{-3}< \infty$ [26].

Since eq. (17) is similar to the scaling law for three dimensional networks, it is evident that a Bethe tree can in principle provide a prediction for relative permeability which will be close to the one found from regular networks. However, the percolation threshold of a Bethe network is different from the one of a three dimensional regular lattice, for the same network connectivity, z. Therefore, when comparing Bethe trees with regular lattices, the comparison should be made on the basis of the same percolation threshold, $f_{bc}$, and not on the basis of the same network connectivity, z [23,28].

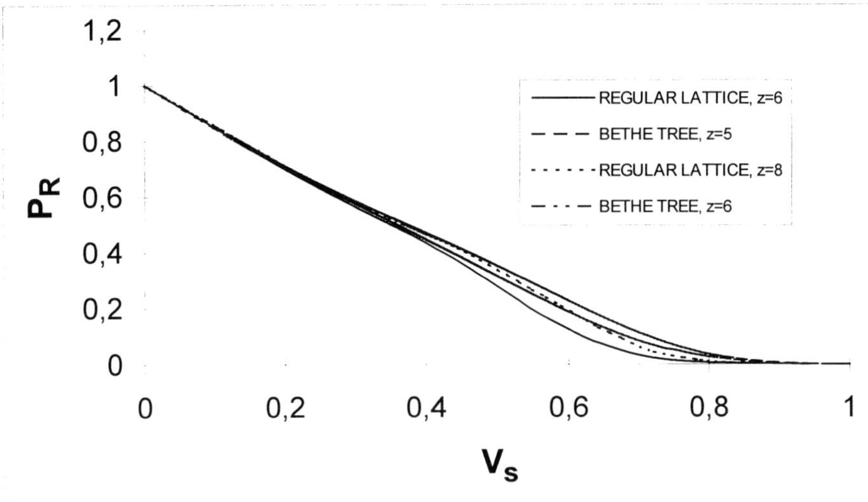


Figure 5. Relative permeability against volume of adsorbate during adsorption for 3D networks and Bethe trees at different connectivities.

For a simple cubic lattice, (z=6), $f_{bc}$=0.25. Thus the corresponding Bethe tree should have z=5 in order to match the value of $f_{bc}$. Equivalently, for a bcc lattice (z=8) with $f_{bc}$=0.18, the corresponding Bethe tree should have a value of z~6. Results in terms of relative permeability against $V_S$ are shown in Figure 5. In this figure, Bethe network curves are obtained using eq.(16), keeping terms up to the order $(z-1)^{-4}$, in the whole $V_S$ range except close to $V_{SC}$, where this equation is slowly convergent [24]. In the neighborhood of $V_{SC}$, eq. (17) was used. It appears that $P_R$ from a Bethe network for z=5 and 6 is in very good agreement with the network solution for the corresponding three dimensional lattices (z=6 and 8, respectively), even in the region close to the percolation threshold. This is expected since the critical exponents are very close for both types of lattices (2 and ~1.9 respectively). A similar agreement between Bethe and three dimensional lattices, has been observed by Sahimi [28], for the case of diffusion-controlled reactions, and by Sahimi [23], for the case of power law transport, in disordered porous media. Thus, Bethe trees can be employed to give a very good approximation of the relative permeability curve of three dimensional regular networks, provided that the connectivity, z, of the Bethe lattice is properly adjusted so that the same percolation threshold is obtained with the corresponding regular network.

## 2. APPLICATIONS OF CERAMIC MEMBRANES IN GAS SEPARATIONS

Pores, and especially mesopores (with sizes between 2 and 50 nm) and micropores (with sizes less than 2 nm), play an essential role in physical and chemical properties of industrially important materials like adsorbents, membranes, catalysts etc. In addition to pore structural characterization described above, the description of transport phenomena in porous materials has received attention due to its importance in many applications such as drying, moisture transport in building materials, filtration etc. Although widely different, these applications present many similarities since they all depend on the same type of transport phenomena occurring in a porous media environment. In particular, transport in mesoporous media and the associated phenomena of multilayer adsorption and capillary condensation have been investigated as a separation mechanism for gas mixtures [29].

A few publications have reported the permeation of capillary condensate in mesoporous materials. Carman and Raal [30], measured permeability of $CF_2Cl_2$ in Linde silica porous plugs at 240 and 251.5 K. Lee and Hwang [31], measured freon and water vapor permeabilities on vycor membranes. These permeabilities were found to exhibit maxima at relative pressures around 0.6-0.8, with values 20-50 times the Knudsen permeability. Ulhorn et al. [32], reported a similar behavior for propylene at 263K in $\gamma$-alumina membranes. Sperry et al. [33] demonstrated the ability of mesoporous $\gamma$-alumina membranes in methanol separation at 473 K, provided the applied pressure of methanol is increased at a partial pressure of 23 bar, which corresponds to a relative pressure of 0.65 at that temperature.

More recently, novel developments in manufacturing and controlling thin film structures, which can grow inside or at the surface of mesoporous substrates, have boosted industrial interest in the study of the underlying mechanisms for the case of micropores too [34]. However, the inability of continuum models in this regions necessitates the use of computer intensive stochastic models to study both equilibrium (see also section 1.1.1 above) and dynamic properties .

In this section, reference is made to discrete approaches for the modeling of gas/condensate flow through mesoporous structures. Capillary network models are presented along with typical predictions and sensitivity studies. Comparison is made with results from the literature. Finally, experimental results obtained in our laboratory are presented for the system helium-water vapor on two mesoporous membranes, made by compaction of alumina microspheres, with porosities 0.41 and 0.48, respectively.

### 2.1. Condensable vapor permeability

The flow of a condensable vapor through a mesoporous membrane is a phenomenon of great complexity [29,30]. Despite the early experimental findings and the immediate impact in gas separations using membrane technology [32] the above physical process has been only recently simulated using two-dimensional network models [35], Bethe networks [36] and effective medium theory [37].

As the membrane is exposed to a certain vapor pressure gradient, adsorption, capillary condensation and surface flow phenomena occur at the same time, during the initial stages of the experiment [31]. As the system reaches a steady state, a film of adsorbate has been formed on the pore walls, while at the same time capillary condensation occurs in the subcritical pores (pores with radius smaller than the Kelvin radius) according to the modified for the case of adsorption, Kelvin equation (eq. 8).

It is clear that the three phases of the penetrating fluid coexisting in the porous matrix, contribute independently to the overall permeability.

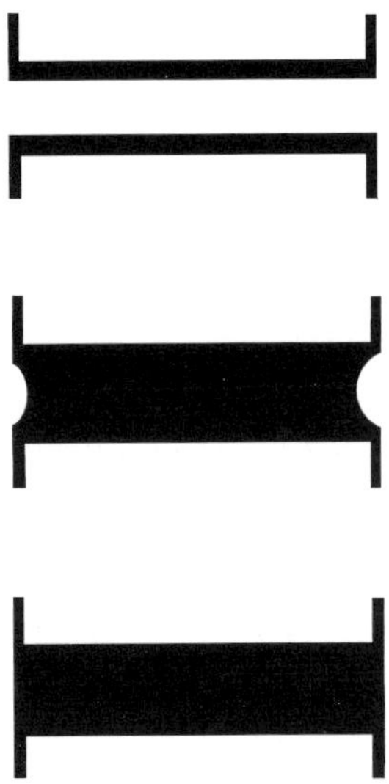

Gas + Surface Flux ($r > r_k$)

$$J_g = \left(\frac{32 \cdot \pi \cdot M}{9 \cdot R \cdot T}\right)^{\frac{1}{2}} \cdot (r-t)^3 \cdot \frac{\Delta P_g}{l} \qquad J_s = \frac{2 \cdot \pi \cdot r \cdot \chi^2 \cdot R \cdot T}{P_m \cdot C_R \cdot S^2} \cdot \frac{\Delta P_g}{l}$$

Capillary Enhanced Flux ($r_k \geq r > t$)

$$J_c = \frac{\pi \cdot r^4 \cdot \rho_l}{8 n_l} \left[1 + \frac{(r-t)^2}{r^2} \cdot \frac{\rho_l \cdot R \cdot T}{M \cdot P_m}\right] \cdot \frac{\Delta P_g}{l}$$

Liquid Poiseuille Flux ($r \leq t$)

$$J_l = \frac{\pi \cdot r^4 \cdot \rho_l}{8 n_l} \cdot \frac{\Delta P_g}{l}$$

Figure 6. Different flow regimes developed inside the individual pores, depending on the specified pressure gradient across the network.

Depending on the specified pressure gradient across the network, different flow regimes may develop inside the individual pores (Figure 6). At low enough mean pressures, the observed mass flux is considered to be made up of non-adsorbed molecules moving in the free pore space, (gas phase component, $J_g$) and of adsorbed molecules moving along the pore wall surface (surface flow component, $J_s$). The mechanism of the gas-phase varies from diffusive to viscous depending on the gas concentration (or equivalently pressure). Following multilayer adsorption on the pore wall, capillary condensation occurs at high enough pressures, as indicated by the Kelvin equation [12].

The steady state viscous condensate flux, $J_c$, is assumed to obey Poiseuille's formulation and is given by the following equation:

$$J_c = \frac{\pi r^4 \rho_l}{8 n_l} \cdot \frac{\Delta P_l}{l} \tag{23}$$

where $\rho_l$ is the liquid density and $n_l$ is the viscosity of the fluid. As has been already pointed out in [35], equation (23) may be recast in terms of gas pressure, and transformed to:

$$J_c = \frac{\pi r^4 \rho_l}{8 n_l} \left[ 1 + \frac{(r-t)^2}{r^2} \cdot \frac{\rho_l \cdot R \cdot T}{M \cdot P_m} \right] \cdot \frac{\Delta P_g}{l} \tag{24}$$

Equation (24), when compared to Poiseuille's law (eq. (23)), is characterized by an enhancement factor ($\rho_l RT/MP_m$), which is physically attributed to capillary pressure gradients [31]. Indeed, an additional driving force occurs due to the difference in the curvatures of the menisci that are formed between nodes and bonds filled with condensate. This capillary action is gradually diminishing as the mean pressure increases for a given bond. The reason is that the menisci begin to flatten as the pressure is raised above Kelvin equilibrium conditions. This effect is taken into account in eq. (24) by multiplying the enhancement factor with the term $(r-t)^2/r^2$, as suggested by Lee and Hwang [31]. It is clear that the effect of the enhancement factor decreases as t increases, and is totally eliminated at the relative pressure where the condition r=t is fulfilled. At this particular relative pressure the liquid surfaces at the end of the bonds are planar [12,38]. From this relative pressure on and up to saturation conditions, the flow obeys the liquid Poiseuille formulation, given by equation (23).

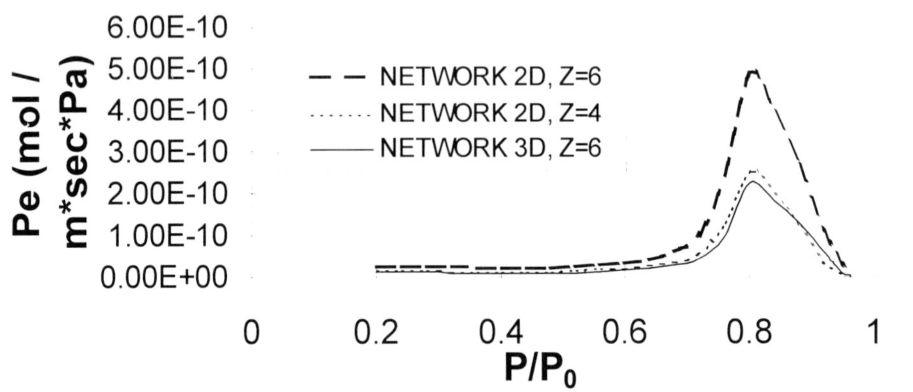


Figure 7. Freon permeability on Vycor glass (Uniform p.s.d. with Rm=30Å and σ=15Å) for various network dimensionalities and connectivities.

Presently, two and three-dimensional networks with pore connectivity of z=4, 6 and 8, are considered. The theoretical case study involves the flow of freon 113 on Vycor glass at 314.5K; surface flow has been neglected for the shake of simplicity. The results are presented in figure 7. It is evident that as the pore connectivity increases the maximum permeability value also increases. In addition, the higher the network connectivity, the lower the relative pressure, at which the capillary enhancement effects start becoming significant. This behavior can be better viewed in figures 8a and 8b where snapshots for two-dimensional square (z=4) and triangular (z=6) networks, at $P/P_0$=0.653 are presented. This relative pressure corresponds to a fraction of capillary condensed pores, $f_c$, of 0.348 (note that $f_c$=1- $f_b$). From these snapshots one can observe that a percolating cluster of capillary condensed pores is formed only for case of the triangular network (z=6), whereas the capillary condensed pores for the case of the square lattice (z=4) do not form such a cluster. This result is in accordance with bond percolation theory where the value of the percolation threshold, $f_{cc}$, for a regular lattice with z=6 is 0.3473 while the threshold value for a lattice with z=4 is 0.5 [25]. Thus, as shown from fig. 7, the permeability curve for z=6 has started its rapid increase at a lower relative pressure compared to the permeability curve for z=4.

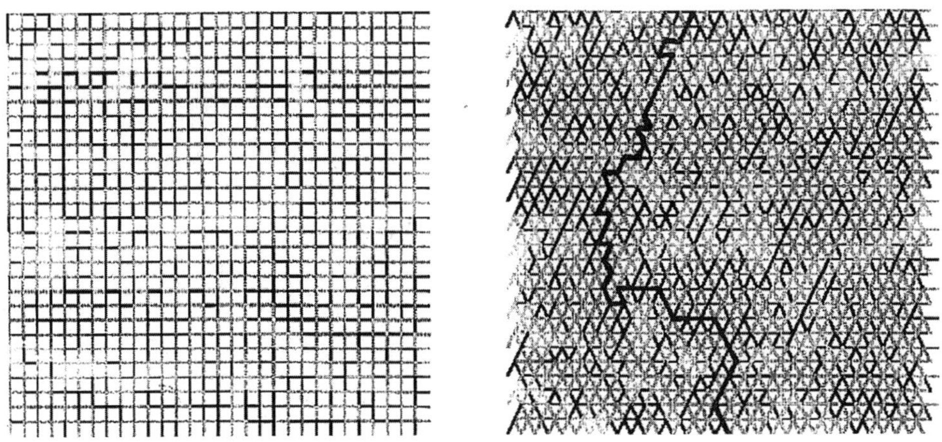

(a) (b)

Figure 8. Network snapshots for connectivity values a) z=4 and b) z=6 at P/Po=0.653. Gray pores follow gas + surface flow, while black pores follow capillary enhanced flow.

In the same figure (fig. 7) the permeability curve for the case of a three dimensional regular network with z=6 is also shown in comparison with the curves for the two dimensional networks. When looking at the two curves with z=6, it appears that the 3D case shows a considerably lower maximum permeability value compared to the 2D case. This result should be expected because for the 3D case although the connectivity is 6, there are only 4 bonds in each plane in the direction of flow. Thus the permeability curve for z=6 in three dimensions looks closer to the permeability curve for z=4 in two dimensions. In order to visualize the percolating behavior of the capillary enhanced "super-conducting" pores two 3D snapshots for a 10×10×10 network at $P/P_0=0.55$ and at $P/P_0=0.622$ are presented in figures 9a and 9b. The first snapshot corresponds to a fraction, $f_c$, of only ~10% of capillary condensed pores where no percolating cluster is formed across the network. In the second snapshot where ~25% of capillary condensed pores appear, a percolating cluster spans across the network. This result is again in accordance with ordinary percolation theory, according to which the percolation threshold, $f_{cc}$, for the case of simple cubic lattice (z=6) is ~0.25[25].

Experimental data from the literature [31] concerning freon 113 permeability on a vycor glass membrane were simulated by the 3D network model. An average effective length of each pore was selected in a way that the (non-condensing) helium permeability predicted by the network matches the experimental values, and at the same time gives a porosity and surface area close to the experimental ones for this material. Subsequently, the pore size distribution obtained from porosimetry and the effective pore length were used for the simulation of the condensable vapor permeability.

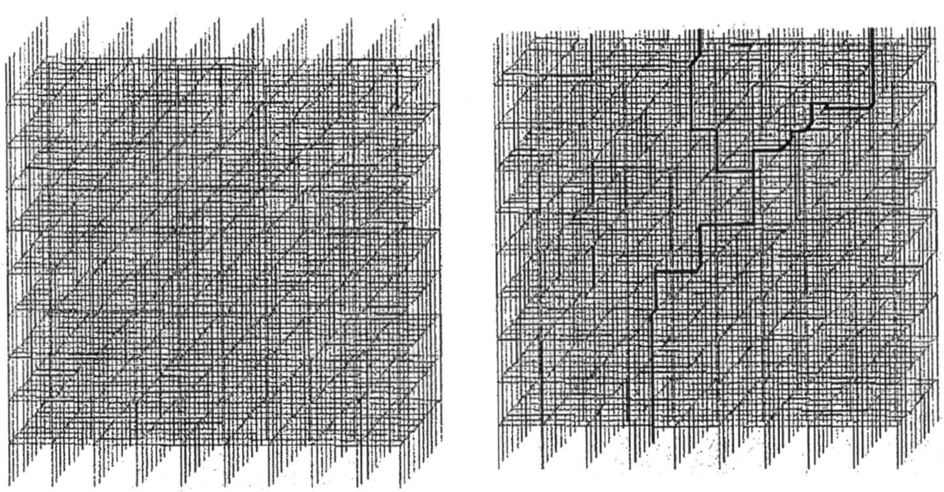

(a) (b)

Figure 9. Network snapshots for regular cubic lattice (z=6) at a) P/Po=0.55 and b)P/Po=0.622. Gray pores follow gas + surface flow, and black pores follow capillary enhanced flow.

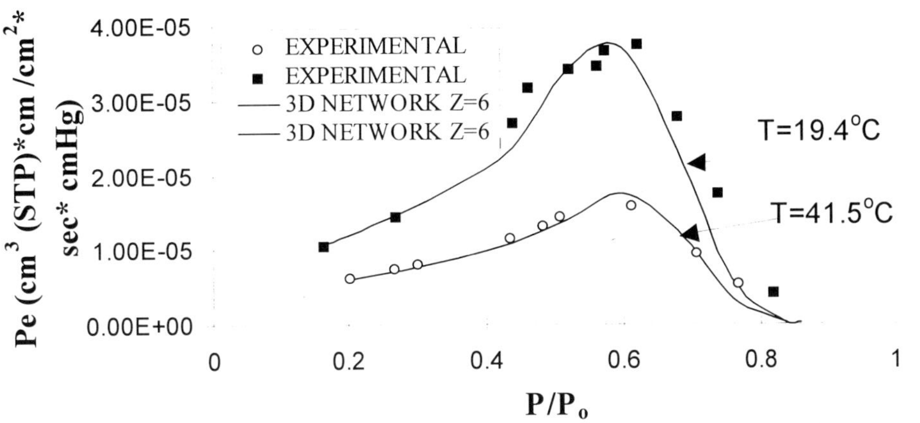


Figure 10. Comparison of network and experimental results [35] of Freon 113 permeability on Vycor glass membrane.

The agreement between the experimental points and the theoretical results is excellent, for two different temperatures, as can be observed from Figure 10. This agreement is attributed mainly to the narrow pore size distribution of the vycor membrane, as well as to the shape of the pores for vycor which appear to be well represented by cylinders.

## 2.2. Experimental measurements

Professor R. M. Barrer and his group, were the first to measure gas relative permeability $P_R$ of He and $H_2$ through membranes formed by compaction of carbolac blocked by sorbed $SO_2$ and $NH_3$ [39-41].

Very few measurements have been reported thereafter due to experimental complications [42]. In order to overcome these difficulties, a new simplified relative permeability technique has been demonstrated recently by Steriotis et al. [10]. A similar technique, known as permporometry, has been developed [43-45], based on counter diffusion measurement through the partially blocked membrane.

The membrane characterization data reported in this section have been obtained by means of the apparatus shown in fig. 11. It is made of stainless steel and can operate from high vacuum up to 70 bars. In addition, it is characterized by the unique capability of performing a broad range of porous membrane characterization and evaluation measurements, namely: equilibrium isotherms, absolute (integral and differential) and relative gas and condensed vapor permeabilities and selectivities. Each of the two membranes, accommodated by the facility, is connected to the high and low pressure sections through four valves MiH, MiL, MiHF, MiLF (i=1, 2, 3).

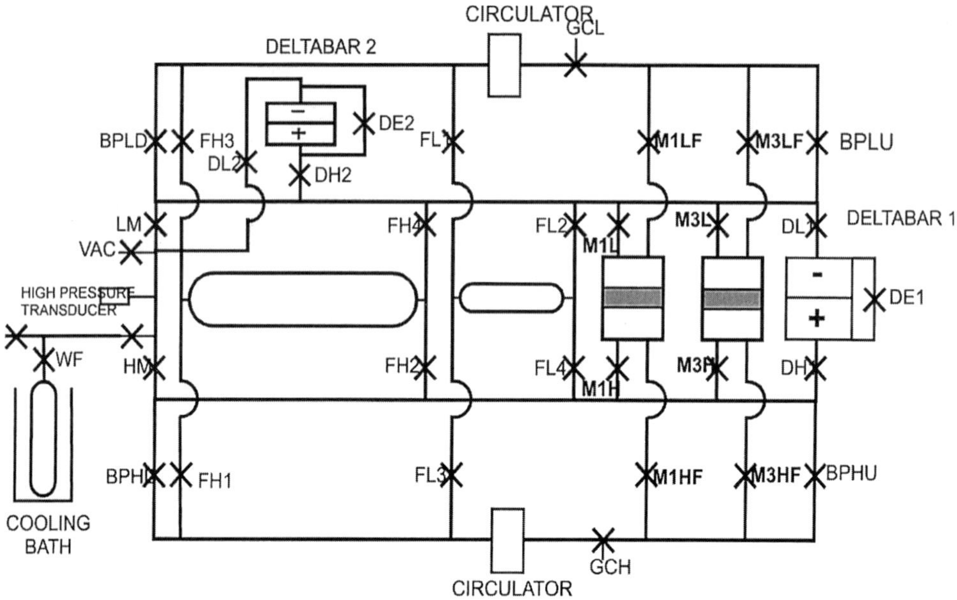


Figure 11. Experimental apparatus.

Both sections are connected to the high vacuum line (HV, LV), the gas line (LG, HG) and the Gas Chromatographer loop (GCHi, GCLi, i=1, 2). The two flasks (5 l of volume connected to the high pressure section and 0.5 l volume connected to the low pressure section) are used to minimize the high pressure section pressure drop and the low pressure section pressure uptake during permeation measurements respectively [46].

For the case of gas relative permeability and selectivity experiments, the required mixing of gases is achieved by means of two gas circulators (CH, CL). The analysis of gas mixtures for such experiments is performed by means of a Gas Chromatographer. The high pressure side of a differential manometer (4...20 mA, 0...1250 mbar) is used for measuring the pressure at the entrance of the chromatographer. Another similar differential manometer is used to monitor the pressure difference between high and low pressure sections, while the absolute high and low pressure sections values, are continuously monitored by means of pressure transducers.

In order to minimize the temperature variation effects on the differential manometer pressure measurements, the whole apparatus is immersed in an oil bath maintained in constant temperature, by means of a PID temperature controller. The bath temperature variations are minimized by insulating the space over the bath and keeping its temperature constant to ± 0.1 °C by means of a PD temperature controller and an air circulation fan. In addition, the room temperature over the insulating cover is kept constant to ± 1 °C.

Two pellets, of porosities 0.41 and 0.48 respectively, were made by means of coaxial compaction of Alumina powder consisting of non porous spherical particles of size ca. 200Å in diameter. Each pellet consists of 11 sections, and the compaction pressures of those sections were selected in such a way that no macroscopic porosity inhomogeneities would be present on the final pellet. The BET specific surface area of the pellets was calculated $100 \pm 15$ $m^2/gr$.

Experimental results of water vapor adsorption, Helium relative permeability, $P_R$, and water vapor permeability, Pe, for the two alumina pellets are presented in figures 12a and 12b, for water relative pressures up to unity. It may be noticed that, as the amount of water adsorbed starts to rapidly increase with $P/P_o$ , due to capillary condensation, a significant increase of its permeability may also be observed due to the resulting capillary enhancement of flow. At a certain value of $P/P_o$ where $V_s$ is close to unity (saturation), all pores of the membrane are in the capillary condensation regime and thus follow the capillary enhanced type of flux. At this point water vapor permeability reaches its maximum value. In comparison, helium relative permeability decreases rapidly and falls to zero well below the point of saturation. This may be attributed, according to percolation theory, to the fact that in a simple cubic lattice, if ~75% of the pores are blocked by capillary condensate the system has reached its percolation threshold and helium can no longer percolate through the membrane. The point where helium ceases to percolate through the membrane is closely related to the point where water permeability starts becoming "super-conducting" due to capillary enhancement effects. The maxima regions in the condensed phase permeability plots, are very important in assessing the conditions for maximum permeability and selectivity of the membrane.

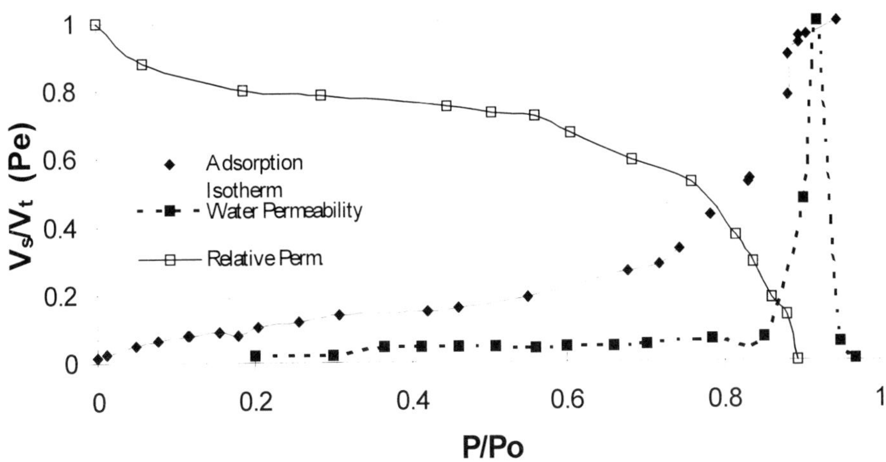


Figure 12a. Experimental results for Alumina membrane ε=41%.

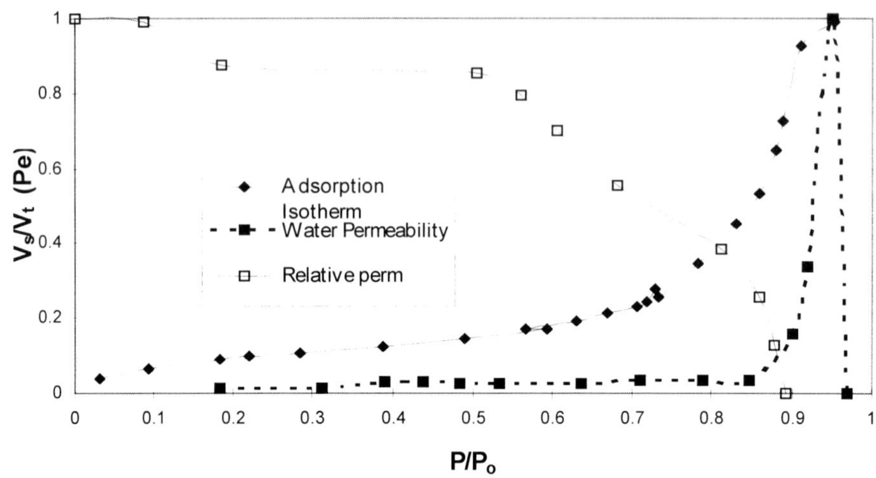


Figure 12b. Experimental results for Alumina membrane ε=48%.

Since the dominant feature of packings of spheroidal particles is the constrictions between the tetrahedral cavities formed by the Alumina microspheres, a more realistic model is required, based on the random sphere packing models. Such models are obviously more complex. Conversely, they permit a more realistic representation of the pore space among the spheroidal particles. A preliminary model has been reported for sorption [47] and relative permeability $P_R$ [48]. A more realistic model of the sorption and permeability at different porosities is under development, aiming at predicting the sorption and the condensed phase permeabilities, given the porosity and the spheroidal particle size distribution.

## 3. CONCLUSIONS

The microstructure characterization of porous membranes is of great importance for understanding the separation process mechanism. Due to the complexity of the structural effects on the membrane performance, a combination of independent equilibrium and dynamic methods is required, applied on the same model membrane systems.

For the case of equilibrium methods, for mesoporous materials gas porosimetry is complemented by Small Angle Neutron Scattering to obtain information on pore size distribution. For microporous membranes the extraction of structural information from the equilibrium sorption measurements can be based on techniques like Grand Canonical Monte Carlo Simulation.

Dynamic methods such as gas relative permeability combined with aspects from percolation theory can be employed to obtain structural information on the porous network topology (dimensionality and connectivity) as well as on the pore shape. Model membranes with well defined structure formed by compaction of non-porous spherical particles, are ideal for testing the different characterization techniques. One should bear in mind though, that dynamic methods should be used as a complement of, rather than as an alternative to, the equilibrium methods. Furthermore, capillary network models are not always appropriate for the derivation of structural information, and for membranes made by compaction of spheres the use of advanced pore modeling tools based on random sphere packing geometry is required.

Finally, the application of ceramic membranes in separations of condensable from non-condensable vapors is explored both theoretically and experimentally. Capillary condensation greatly enhances the permeability of the condensable vapor through the mesoporous mebrane resulting in large selectivities over non-condensable vapors. Attention is drawn to the fact that the dynamic membrane properties depend on a wider range of microstructural characteristics, relevant to the separation efficiency of the membrane. Therefore, measurements of the dynamic relative permeability and of condensable vapor permeability provide significant saving in effort (which is normally required by "blind" selectivity experiments) for the determination of the optimum pressure and temperature operation conditions and for the development of a model predicting the membrane performance.

## ACKNOWLEDGEMENTS

This research was partially supported by the EPET II program of the Greek General Secretariat of Research and Technology (contract number: 724)

### List of symbols

| | |
|---|---|
| $d_s$ | number of carbon atoms per unit volume in the graphite layer |
| $f(r)$ | distribution of pore radii |
| $f_b$ | fraction of open pores |
| $f_c$ | fraction of pores perfoming capillary condensation |
| $f_{bc}$ | percolation threshold of open pores |
| $f_{cc}$ | percolation threshold of capillary condenced pores |
| G | conductance of a branch of a Bethe network |
| g | bond conductance |

| | |
|---|---|
| $g_{av}$ | effective conductivity predicted by EMA |
| $g_M$ | effective conductivity of a Bethe network |
| H | slit pore width |
| I | Scattering intensity |
| J | flux of gas flowing through an open pore |
| L | network size in each direction |
| l | bond length |
| M | molecular weight |
| $R_g$ | universal gas constant |
| $R_m$ | mean pore radius |
| $P_g$ | gas pressure |
| P | adsorbate pressure |
| $P_0$ | adsorbate saturation pressure |
| $V_{sA}$ | fraction of total pore volume occupied by adsorbate during adsorption |
| $V_{sD}$ | fraction of total pore volume occupied by adsorbate during desorption |
| Pe | absolute gas permeability |
| $P_R$ | gas relative permeability |
| $r_K$ | Kelvin radius |
| T | fluid temperature |
| t | thickness of adsorbed layer; critical exponent |
| u | potential energy |
| $x=r-t$ | open core radius of a pore partly filled with adsorbate |
| z | pore connectivity |

**Greek letters**

| | |
|---|---|
| $\Delta$ | separation between graphite layers |
| $\varepsilon$ | Lennard - Jones parameter |
| $\eta$ | viscosity |
| $\nu$ | critical exponent in eq.(23) |

| | |
|---|---|
| $\rho$ | density |
| $\sigma$ | standard deviation in pore size distribution function |
| $\sigma_g$ | hard sphere diameter of an adsorbate atom |
| $\sigma_{sf}$ | Lennard Jones parameter |
| $\chi$ | amount adsorbed |

**Subscripts**

| | |
|---|---|
| C | threshold |
| G | gas phase |
| K | Kelvin |
| L | liquid phase |

## REFERENCES

1. D. Nicholson and N. G. Parsonage (eds.), Academic Press, Computer Simulation and the Statistical Mechanics of Adsorption, New York, 1982.
2. R. F. Cracknell, D. Nicholson and N. Quirke, Mol. Phys., 80 (1993) 885.
3. S. Samios, A. K. Stubos, N. K. Kanellopoulos, R. F. Cracknell, G. K. Papadopoulos and D. Nicholson, Langmuir, 13 (1997) 2795.
4. R. F. Cracknell, D. Nicholson and N. Quirke, Mol. Sim., 13 (1994) 161 .
5. T. A. Steriotis, A. Ch. Mitropoulos, N. K. Kanellopoulos, U. Keiderling and A. Wiedenmann, Physica B, 234-236 (1997) 1016.
6. A. Ch. Mitropoulos, T. A. Steriotis, F. K. Katsaros, K. P. Tzevelekos, N. K. Kanellopoulos, U. Keiderling, S. Sturm and A. Wiedenmann, J. Membrane Sci., 129 (1997) 289.
7. D. Nicholson, J. K Petrou and J. H. Petropoulos, Chem. Eng. Sci., 43 (1988) 1385 .
8. D. Nicholson and J. H. Petropoulos, J. Chem. Soc. Faraday Trans., 80 (1988) 1069.
9. N. K. Kanellopoulos and J. K. Petrou, J. Membrane Sci., 37 (1988) 1.
10. T. A. Steriotis, F. K. Katsaros, A. Ch. Mitropoulos, A. K. Stubos and N. K. Kanellopoulos, J. Porous Mat., 2, (1995) 73.
11. J.H.Petropoulos, J.K. Petrou and N.K. Kanellopoulos, Chem Eng. Sci., 44 (1989) 2967.
12. S.L.Gregg and K.S.W. Sing (eds.), Adsorption, Surface Area and Pororsity, Academic Press, New York, 1982.
13. D. Nicholson and J.H. Petropoulos, J. Chem Soc. Faraday Trans. I, 80 (1984)1069.

14. J.H Petropoulos, J.K.Petrou and N.K.Kanellopoulos (eds.), Characterization of Porous Solids, Elsevier, Amsterdam, 1988.
15. J. Karger and D.M. Ruthven (eds.), Diffusion in Zeolites and Other Microporous Solids, Wiley and Sons, New York, 1991.
16. V.N.Burganos and S.V.Sotirchos, AIChE J., 33 (1987) 1678 .
17. S. Kirkpatric, Rev. Mod. Phys., 45 (1973) 574 .
18. J. Koplic, J. Phys. C, 14 (1981) 4821.
19. G. Ahmed and J.A.Blackman, J. Phys. C, 12 (1979) 873.
20. M. Sahimi, B.D. Hughes, L.E. Scriven and H.T.Davis, Phys. Rev. B, 28 (1983) 307.
21. R.B.Stinchcombe, J. Phys. C: Solid state phys., 7 (1974) 179.
22. J. Heinrichs and N.Kumar, J. Phys. C: Solid state phys., 8 (1975) L510 .
23. M. Sahimi, AIChE J., 39(1993) 369.
24. S.Reyes and K.Jensen, Chem. Eng. Sci., 40 (1987) 1723.
25. V.K.S. Shante and S. Kirkpatrick, Adv. Phys. B, 28 (1971) 307.
26. M. Sahimi, B.D. Hughes, L.E. Scriven and H.T.Davis, J. Chem. Phys., 78 (1983) 6849 .
27. D. Stauffer, Taylor and Francis (eds.), Introduction to Percolation Theory, London, 1985.
28. M. Sahimi, Chem. Eng. Sci., 43(1988) 2981.
29. P.B. Weisz, Bunsegres Phys. Chem., 79 (1975) 798 .
30. P.C.Carman and F.A.Raal, Proc. Roy. Soc., A209 (1951) 38 .
31. K.H. Lee and S.T. Hwang, J. Colloid Interface Sci., 110 (1986) 554 .
32. R.J.R. Uhlhorn, K. Keizer and A.J. Burgraaf, J. Membrane Sci., 66 (1992) 259.
33. D.P. Sperry, J.L. Falconer and R.D. Noble, J. Membrane Sci., 60 (1991) 185.
34. M-D. Jia, B. Chen, R.D. Noble and J.L. Falconer, J. Membrane Sci., 90 (1994) 1.
35. M. E. Kainourgiakis, A. K. Stubos, N. D. Konstantinou, N. K. Kanellopoulos and V. Millicic, J. Membrane Sci., 114 (1996) 215.
36. P.Rajniac and R.T. Yang, AIChE J., 42 (1996) 319.
37. E. Kikkinides, K. P. Tzevelekos, A. K. Stubos, M. E. Kainourgiakis and N. K. Kanellopoulos, Chem. Eng. Sci., 52, (1997) 2837.
38. D.E. Everett and J.M. Haynes, J. Colloid Interface Sci.,38 (1972) 125.
39. R. Ash, R.M. Barrer and C.G. Pope, Proc. Roy. Soc., London, 19 (1963) A271.
40. R. Ash, R. M. Barrer and R. T. Lowson, J.Chem. Soc. Faraday Trans. 1, (1973)2166
41. R. Ash, R. M. Barrer and R. Sharma, J. Membrane Sci., 17 (1976) 1.
42. N. K. Kanellopoulos and J. H. Petropoulos, J. Chem. Soc. Faraday Trans.1, 79 (1983) 517.
43. C. Eyraud, E. Dridi and M. Nakagaki (eds), Proc. Eur.-JPN Cong. Memb. Processes, 1984, 629.
44. M. Katz and G. Barush, Desalination, 58 (1986) 199.

45. F.P. Cuperus, D. Bargeman and C.A. Smolders, J. Membrane Sci., 71 (1992) 57.
46. T.A.Steriotis, F.K. Katsaros, A.Ch. Mitropoulos, A.K.Stubos, P.Galiatsatou, N.Zouridakis and N.K. Kanellopoulos, Rev. Sci. Instr., 67 (1996) 2545.
47. N. K. Kanellopoulos, J. K. Petrou, and J. H. Petropoulos, J. Colloid Interface Sci., 96 (1983) 90.
48. N. K. Kanellopoulos, J. K. Petrou, and J. H. Petropoulos, J. Colloid Interface Sci., 96 (1983) 101.

Adsorption and its Applications in Industry and Environmental Protection
Studies in Surface Science and Catalysis, Vol. 120
A. Dabrowski (Editor)



# Adsorption controlled preparation of heterogeneous catalysts

S. Haukka[a], E.-L. Lakomaa[b] and T. Suntola[a]

[a]Microchemistry Ltd., P.O. Box 132, FIN-02631 Espoo, Finland

[b]Neste Oy, Technology Center, P.O. Box 310, FIN-06101 Porvoo, Finland

## 1. INTRODUCTION

The trend in materials processing today is toward structures with atomic scale dimensions. This is apparent in the increasing use in the literature of the term nanotechnology, an as yet poorly defined term for all processing technologies aiming at atomically controlled structures. One important application of atomically controlled structures is in industrial processes relying on heterogeneous catalysis. Here, even a minor improvement in selectivity—through better process control or through the development of more highly tailored catalysts—could lead to a huge decrease in undesired side reactions. A method of catalyst preparation capable of producing precise surface structures would thus be an important step on the way toward more economical and environmentally friendly catalytic processes [1].

The most common catalyst supports are high surface area oxides such as $SiO_2$ and $Al_2O_3$. Not only are these materials low cost and harmless, they are also easy to produce in large quantities. Nevertheless, their highly porous structures make it difficult to prepare well-defined catalytically active species by conventional impregnation. Typically many different species exist on the surface, and the behavior of these in catalytic reactions is difficult to predict. Current research on well-defined, single crystal surfaces is attempting to overcome this problem, and indeed much new knowledge of the relationship between surface structure and catalytic performance has been gained [2]. The gap between commercial heterogeneous catalysts and model catalysts is still wide, however. More controlled means for the preparation of catalysts are needed for further catalyst development.

One way to prepare precise surface structures on catalyst supports is to utilize adsorption controlled methods. Adsorption covers a variety of interactions from physisorption (weak interaction) to chemisorption (strong interaction). The term control, however, restricts the adsorption to strong interactions. The requirements for adsorption controlled catalyst processing have been shown to be met in both liquid and gas phase. In this paper we focus on the use of a gas phase method, called Atomic Layer Epitaxy (ALE), in the preparation of heterogeneous catalysts. The method was first developed for thin film growth in the early 70s [3-6] and was extended to catalyst preparation in the 90s [7-9]. The characteristic feature of ALE is systematic utilization of surface saturation through chemisorption thus providing the means for adsorption control in catalyst processing.

In section two below, we briefly describe the degree to which adsorption control is achieved in various methods of catalyst preparation. Sections three and four describe the practical means for successful catalyst preparation by ALE, including the basis for adsorption control and the means for regulating the metal loading. Section five offers a more detailed account of the surface structures processed by ALE.

## 2. CATALYST PREPARATION METHODS AND ADSORPTION CONTROL

In classifying the methods of catalyst preparation we follow the recommendation of the International Union of Pure and Applied Chemistry (IUPAC) that they be divided into the following groups: precipitation, deposition, encapsulation, and selective removal [10]. The deposition methods, which are dealt with below, comprise impregnation, ion exchange, grafting (anchoring), and chemical vapor deposition (CVD) (Figure 1). Atomic layer epitaxy (ALE), which is based on saturating gas–solid reactions, is a deposition method closely related to grafting and CVD.

In the deposition methods, adsorption control is determined by the strength of the interaction between the precursor and the support. The primary factors influencing this strength are the reaction medium (water solution, organic

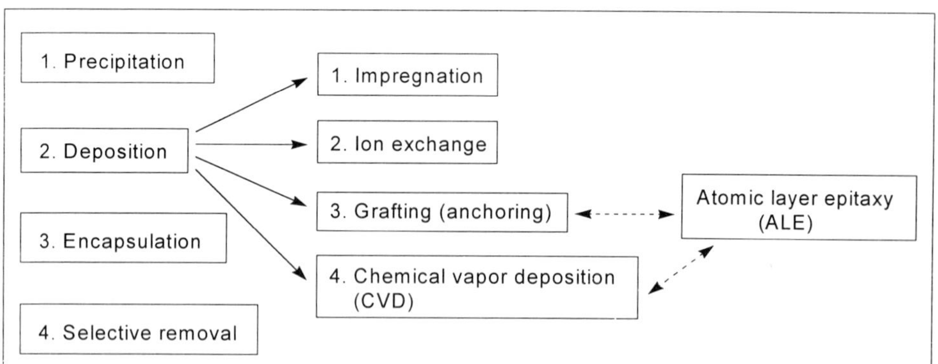


Figure 1. The Atomic Layer Epitaxy (ALE) method in relation to other methods of processing catalysts. The other methods are classified according to the International Union of Pure and Applied Chemistry (IUPAC) [10].

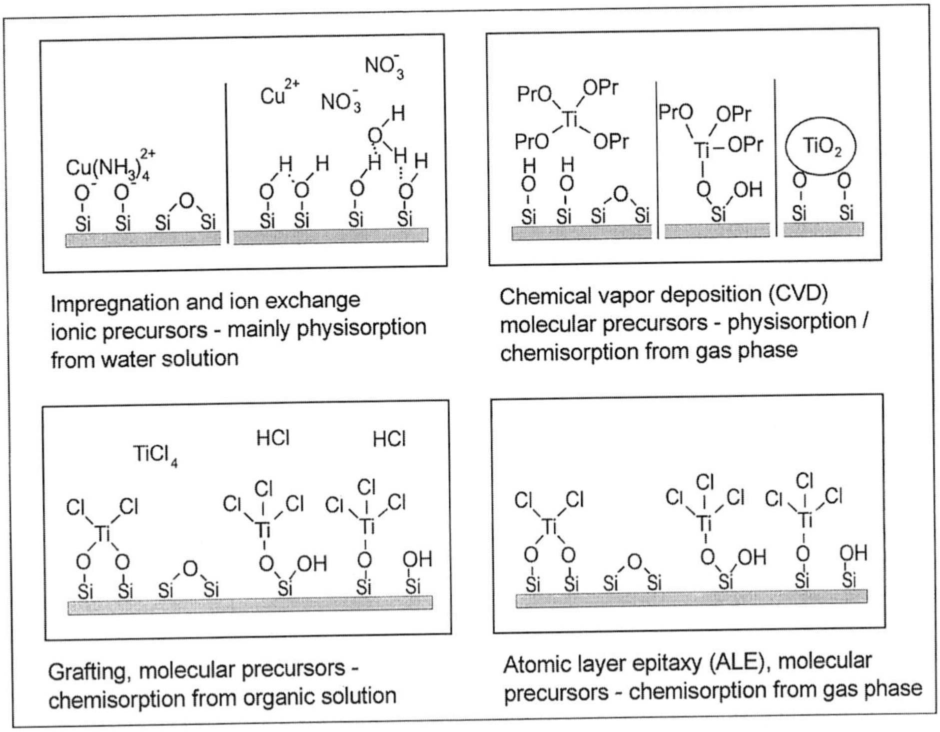


Figure 2. The strength of interaction between precursor and support in deposition methods of catalyst preparation.

solution, or gas phase) and the chemical form of the precursor (ionic or molecular), as shown in Figure 2. In general, the basic difference between the catalyst preparation from water solution with ionic precursors and that from organic solution and gas phase with molecular precursors is the following: in the preparation from water solution, drying or more usually, further calcination, at elevated temperatures is required to create a covalent bond between the support and the precursor, whereas from organic solution and gas phase the covalent bond between precursor and support can be formed directly.

The strength of the interaction is responsible for essential catalyst properties such as the dispersion and distribution of metal species through the catalyst particles, and also for practical matters such as the means of control of the metal loading. For instance, weak interaction leads to the formation of crystallites rather than monolayers on the surface; the formation of monolayers requires very strong interaction.

In the following, the adsorption control in ALE is briefly compared with that in the other deposition methods of processing catalysts. For more information about the specific details and fine tuning of the other methods, the reader is referred to the literature [10,11].

### 2.1. Preparation from water solution

Incipient wetness impregnation carried out in aqueous solution is the most common method of catalyst preparation and at the same time the method least controlled by adsorption. Metal salts used as catalyst precursors are simply dissolved in the impregnating solution, the volume of which is made to match the pore volume of the support. The metal loading is thus controlled by the concentration of metal ions in solution, which means that the support surface does not play an important role, but merely acts as a physical support.

To take an example, to achieve adsorption control by using $Cu^{2+}$ ions to modify the silica support with copper oxide requires the utilization of specific adsorption sites such as $Si\text{-}O^-$ groups. In larger quantities, however, these surface groups are formed only in solution of high pH, where the copper ions will precipitate as copper hydroxide. If copper hydroxide is not preferred for some reason, the impregnation must necessarily be carried out at lower pH with fewer number of $Si\text{-}O^-$ surface groups available. In such a situation the interaction between the support and most $Cu^{2+}$ ions used as precursors will be relatively weak making adsorption control impossible. The weak interaction, furthermore, means that, during removal of the solution and in further calcination, there will be a strong tendency to formation of crystallites.

Wet impregnation is somewhat more adsorption controlled, since the solution volume used is greater than the pore volume and only those ions interacting specifically with the surface groups of the support are left on the surface after removal of the water solution by filtration or centrifuging. Unfortunately, only a small number of metal ions such as $Cu^{2+}$, are selectively adsorbed, for instance, on silica.

Zeolites have very specific ion exchange sites and ion exchange is the most common method used to introduce metal ions into zeolite cages. Ion exchange on zeolites can be highly adsorption controlled and pure metal ions can be used as precursors.

In the selective ion exchange on silica and alumina, in turn, particular metal complex ions are required to achieve adsorption control. Ion exchange on silica, for instance with $Cu^{2+}$, can be carried out by using $NH_3$ to form at the same time $Si\text{-}O^-$ adsorption sites and $Cu(NH_3)_4^{2+}$ complex ions, the formation of the latter then prevents the precipitation of copper hydroxide. $Cu(NH_3)_4^{2+}$ ions will interact very strongly with the $Si\text{-}O^-$ groups on the silica surface, which will suppress crystallization during calcination. In addition, the amount of adsorbed copper ions can be very precisely controlled by the pH of the solution. The drawback is that the pH does not stay constant during drying and the adsorptive character of the surface changes. This is a problem in all catalyst preparations carried out in water solution. In addition, the support may start dissolving at certain pH. Silica will totally dissolve in very basic conditions and alumina begins to dissolve in strongly acidic medium.

### 2.2. Preparation from organic solution

Grafting carried out in non-polar organic solvents differs from impregnation and ion exchange, in that a covalent bond between the molecular precursors and the functional groups present on the surface can be formed in the reaction medium, without the need for after treatment at elevated temperatures [12,13]. Grafting requires unperturbed adsorption sites, which means that the removal of physisorbed water is essential. This is most readily done through heat treatment. Other prerequisites for the formation of covalently bonded metal species are that the molecular precursors must be soluble in the organic solvent and be reactive with the adsorption sites. Some metal beta-diketonates rather than forming covalent bonds, specifically interact with the adsorption sites through the ligands providing another form of adsorption control [13-15].

### 2.3. Preparation from gas phase

In CVD, the molecular precursors are brought into contact with the support surface in a vapor phase, and depending on the reaction temperature, will either interact weakly or form a chemical bond with the support. Thus the adsorption control ranges from that in incipient wetness impregnation to that in grafting. In addition, the precursor molecules can be intentionally decomposed either by use of a reaction temperature high enough to decompose them or by simultaneous introduction of oxygen. The intentional decomposition makes the catalyst processing difficult to control and in the worst case may lead to blocking of the support pores.

Condensation and decomposition of the vaporized molecular precursors are prohibited in ALE, which means that the reaction temperature must be carefully chosen to ensure that neither process takes place. In other words, a chemical bond must be formed between the support surface and the precursor, leading to chemisorption. The prerequisite in ALE, as in grafting, is that adsorption sites are free from physisorbed water. Any water molecules present on the surface would fatally interfere with the adsorption of precursors. One further requirement is that the reaction temperature used in ALE processing should never exceed the heat treatment temperature of the support. If it did, the release of water during the reaction from OH groups serving as adsorption sites could lead to undesirable secondary reactions with the precursors. In addition, in the preparation of oxides the feeding of the precursors - the metal compound and the oxidizing agent - is fully separated with an inert gas purge to avoid unwanted gas phase reactions.

ALE is closely related to both grafting and CVD, and gas–solid reactions like those carried out in ALE are usually referred to as grafting in the literature [12]. In ALE *the surface saturation is systematically utilized* and the processing is truly adsorption controlled. The use of a distinct term is accordingly warranted.

## 3. ADSORPTION CONTROL IN CATALYST PREPARATION BY ALE

Adsorption control in catalyst preparation by the atomic layer epitaxy (ALE) method is based on the simultaneous fulfillment of the following conditions:

- a reactive volatile metal compound used as precursor.
- a support with well-characterized adsorption sites.
- a reaction temperature, at which the precursor is thermally stable and forms a covalent bond to the support (*chemisorption*).
- a precursor dose high enough to *saturate the support surface.*

When all these conditions are satisfied, ALE provides the means not only for excellent dispersion and distribution of metal species and molecular structures but also for a highly reproducible process which is easy to scale-up. In the following we examine the various requirements for a successful adsorption controlled preparation of catalysts by ALE and describe the actual ALE processing and scale-up of the process.

### 3.1. Volatile metal compounds

The volatile metal or semimetal compounds used as precursors in ALE can be selected from a wide range of compounds including metal chlorides, β-diketonates, alkyls, alkoxides, and metallocenes etc. [16,17]. The precursor can be liquid, solid, or gas at room temperature and must be stable enough to withstand heating. It should also be free of impurities that volatilize below or at the temperatures used in the processing. Often moisture is the most serious impurity, and the compound should be dried before use. Any solvent coordinated to metal compounds such as β-diketonates needs to be removed.

Metal alkyls, being highly reactive with moisture in the air, must be handled in inert atmosphere. Special care should also be taken in studying new reactants not earlier used in ALE so that toxic compounds are not accidentally released in the vaporization or reaction stage.

### 3.2. Adsorption sites

All oxide surfaces are terminated by OH groups and oxygen bridges, which act as adsorption sites. In adsorption controlled processing it is important to know the number and character of these different adsorption sites, and also the heat treatment conditions needed to stabilize them to a selected level. For purposes of ALE processing, the OH groups have been studied qualitatively by FTIR and quantitatively by $^1$H MAS NMR [18,19].

On silica surface the OH groups are the primary adsorption sites for most of the precursors utilized in ALE, while siloxane bridges may serve for dissociative adsorption of some highly reactive precursors. OH groups may appear in the form of isolated single/geminal groups or weakly/strongly H-bonded groups. The weakly H-bonded OH groups, or inaccessible OH groups as they are called, are not involved in the reactions. The most reactive siloxane bridges are only formed at heat treatment temperatures above 450°C, where the isolated OH groups start condensing. The proportions and quantities of the different OH groups determined by $^1H$ MAS NMR depend on the heat treatment of the silica surface as shown in Figure 3 [9,19].

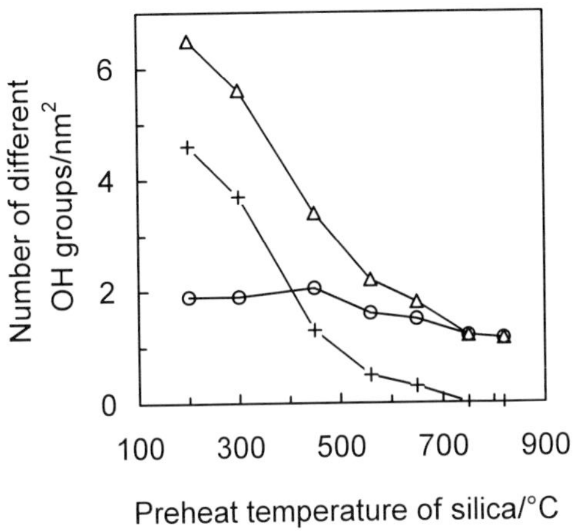


Figure 3. Number of surface species/nm$^2$ on silica samples preheated at different temperatures: (Δ) total OH groups, (+) H-bonded groups, and (o) isolated OH groups (surface area 300 $m^2/g$) [9,19].

As shown by $^{29}Si$ CPMAS NMR analysis, the ratio of geminal to single OH groups on the silica surface can be changed, and triple OH groups can be formed through saturating the surface with trimethyl silyl groups and then removing the groups by subsequent air treatment [20]. Interestingly, although the ratio is changed and the triple OH groups are formed the total number of OH groups stays at the same level as on pure silica heat-treated at the same temperature.

If alumina is heat-treated at elevated temperatures the surface structure is complex consisting not only of basic, neutral, and acidic OH groups but also of coordinatively unsaturated aluminum and oxygen (c.u.s.) sites [21,22]. All these sites can be involved in the reactions, though the basic OH groups are the primary reactive sites for certain metal beta-diketonates. DRIFT spectra reveal qualitatively the presence of the different OH groups [18]. As shown by $^1H$ MAS NMR, the total number of OH groups on γ-alumina (200 $m^2/g$) decreases from 8.3 OH/nm$^2$ at a preheat temperature of 200 °C to 4.1 OH/nm$^2$ at 400 °C and 2.0 OH/nm$^2$ at 600 °C [18]. Furthermore, the combination of $^1H$ MAS NMR analysis and HCl reaction has revealed the number of basic OH groups on alumina heat treated at 600 °C to be 0.8 OH/nm$^2$ [18].

Knowledge of the number and nature of the adsorption sites needs to be extended to include the new OH groups formed in the modification of the support surfaces with various oxides (see section 5). However, the presence of these OH groups is sometimes difficult to verify, because the peaks in the IR and $^1$H NMR spectra due to the new OH groups often tend to appear at almost the same positions as those of alumina and silica [23].

### 3.3. Chemisorption

Chemical bonding to the support surface is essential to achieve adsorption control in ALE on porous materials. Suntola [5] in his review of the ALE process, describes in a general way the effect of activation energy on bond formation in ALE reactions, and he considers the desorption of precursors from the surface. However, in most of the reactions of metal compounds with porous supports that have been studied the reaction is directed towards chemisorption through ligand exchange. An equilibrium condition does not therefore exist where adsorption and desorption take place equally. Desorption is more likely to occur when elemental precursors are used, such as metallic zinc in the preparation of Zn/zeolite catalysts [24].

#### 3.3.1. Verification of chemisorption

To verify the covalent bond formation (chemisorption), the results from several different analyses (element determinations, spectroscopic analysis) must be combined. Element determinations can provide an indication of ligand exchange reactions by showing whether the number of ligands in the surface complex is fewer than in the original metal complex, in the following way:

$$/\!\!/\!\!/|\text{-OH} + ML_4 \rightarrow /\!\!/\!\!/|\text{-O-}ML_3 + HL\uparrow$$

On alumina, despite covalent bond formation, the ligand/metal ratio on the surface may be close to that in the original metal complex. This is because it is possible for the precursor or the ligand released in the ligand exchange reaction to dissociate on the surface [25,26]. In metal chloride reactions carried out at elevated temperatures it has also been observed that a highly reasonable ligand/metal ratio can be obtained on the silica surface, even though agglomerated metal oxides are formed [19].

In general IR spectroscopy can reveal the disappearance or decrease in the number of OH groups through the decrease in intensity of their absorptions in the high-frequency region. In the case of silica a specific absorption in the low-frequency region due to Si-O-M bond formation can also appear. This is clearly indicated by the DRIFT spectra in Figure 4, where, after silica heat-treated at 200°C was reacted with $TiCl_4$ at 175 °C, the absorption at 3740 cm$^{-1}$ due to the isolated OH groups decreased in intensity simultaneously with the appearance of the absorption at 926 cm$^{-1}$ due to Si-O-Ti bond formation [27].

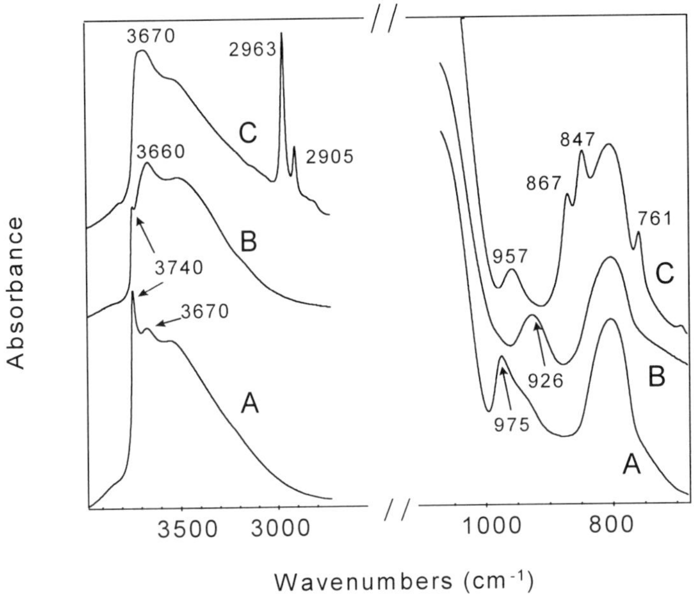


Figure 4. DRIFT spectra of silica preheated at 200°C: (A) original spectrum, (B) after treatment with $TiCl_4$ at 175°C, and (C) after treatment with hexamethyldisilazane (HMDS) [27].

As shown by our group $^1H$ MAS NMR can give basically the same information about the reactions as the high-frequency IR spectra [19,28,29]. The great advantage of $^1H$ MAS NMR is that the number of OH groups present on the surface after the reaction can be determined. This advantage has been exploited to confirm the unreactivity of $TiCl_4$ towards siloxane bridges [19].

One cannot rely exclusively on IR and $^1H$ MAS NMR spectra to provide information about OH groups on alumina and silica, however. Organic ligands in the bound metal surface complex, in particular, may interact through H-bonding with the unreacted OH groups causing severe bordening and even shifting of the peaks - and wrong conclusions about the extent of the removal of the adsorption sites in the reactions.

When the unreacted isolated OH groups on the silica surface are perturbed by H-bonding, their absorptions in the high-frequency IR spectra (originally at 3745-40 $cm^{-1}$) are shifted towards lower wavenumbers and become hidden under the absorptions of the weakly H-bonded OH groups at 3660-3670 $cm^{-1}$ [20,27]. This is shown in the spectrum C in Figure 4, which was recorded after the reaction of hexamethyldisilazane (HMDS) with silica heat treated at 200°C (see also Figure 18). That shift in the high-frequency region really does occur for the unreacted isolated OH groups is verified in the low-frequency region by the presence of a peak at 957 cm$^{-1}$. Without perturbation, the peak due to the unreacted isolated OH groups would be visible in the high and low frequency region at 3940 and 975 cm$^{-1}$, respectively. It must be added that absorptions of the weakly H-bonded OH groups at 957 cm$^{-1}$ can be excluded because no peak is visible at 957 cm$^{-1}$ in the spectrum measured from $TiCl_x$/silica sample even though the high frequency spectrum verifies their presence.

FTIR and NMR spectroscopy can also be used to confirm whether the organic ligands stay intact in the surface complex. For instance owing to the ring structure, the beta-diketonate ligands have very specific absorptions in the IR spectra at 1500-1600 cm$^{-1}$. In the case of decomposition these characteristic absorptions of the ligands are lost as shown in Figure 6.

### 3.3.2. Results from chemisorption studies

In the following, studies at temperatures where the precursor is thermally stable and forms a covalent bond to silica or alumina are presented for three precursors, that differ in reactivity and stability: namely $La(thd)_3$, $Cr(acac)_3$, and $Ti(OPr^i)_4$ (see Figure 15 for the structure of the acac and thd ligands). Of the three, $Cr(acac)_3$ is the least inclined to the ligand exchange reaction. In contrast to silica, which is a relatively inert material, alumina catalyzes reactions itself. As is clearly shown by the reactions of $Ti(OPr^i)_4$, great care must therefore be taken in selecting the processing temperature

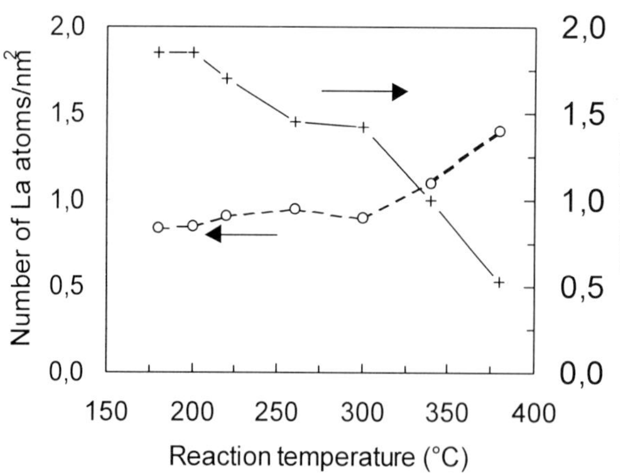


Figure 5. $La(thd)_3$ reactions on silica heat-treated at 600 °C: (o) number of La at/nm$^2$ and (+) thd/La ratio [9].

Figure 5 [9] and the DRIFT spectra in Figure 6 reveal that the ligand exchange reaction of $La(thd)_3$ with OH groups on silica without any decomposition takes place at 180-300 °C. The spectra suggest that the ring structure with its characteristic absorptions at 1500-1600 $cm^{-1}$, stays intact up to 300 °C, and the absence of these bands at 340 °C indicates the decomposition. At lower temperatures, primarily one thd ligand is removed as Hthd, indicating the formation of singly bonded species. At 250-300 °C the possibility increases for simultaneous ligand exchange with two OH groups. In addition to the DRIFT spectra, the decomposition of $La(thd)_3$ above 300 °C is reflected in Figure 5 by the gradual increase in the number of La atoms along with the decrease in thd/La ratio of the bound surface complex. Furthermore, the color changes from white at 180 - 300 °C to brown at 380 °C. It must be pointed out here that the reaction of $La(thd)_3$ probably would take place at even lower temperatures than 180 °C, but the low vapor pressure of $La(thd)_3$ below 200 °C leads to unrealistic catalyst processing times.

The interaction mechanism of $Cr(acac)_3$ with silica differs from that of $La(thd)_3$. Whereas $La(thd)_3$ is prone to ligand exchange reaction, forming Si-O-$La(thd)_2$ species below 200°C, a reaction temperature of 200°C and above is needed to form chemically bonded Si-O-$Cr(acac)_2$ species [30,31]. Below 200 °C the interaction of $Cr(acac)_3$ with silica occurs selectively through the acac ligands

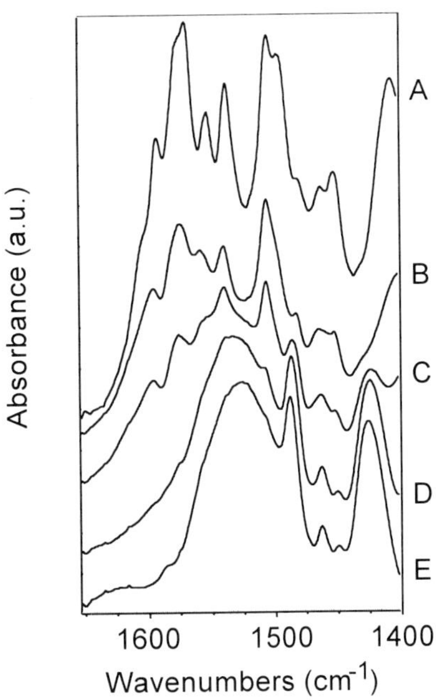


Figure 6. DRIFT spectra of (A) pure $La(thd)_3$ and (B-E) of the surface metal complexes formed in the reactions of $La(thd)_3$ with 600°C silica at 180°C (B), 300°C (C), 340°C (D), and 380°C (E). All samples were diluted in KBr.

[31,32]. As shown by Hakuli and Kytökivi [31], the change in the interaction mechanism is indicated by the acac/Cr ratio of the surface complex, which decreases with increasing temperature from 3 at 150 °C to 2 at 200 °C and above. In the DRIFT spectra the interaction between the ligands and the isolated OH groups is seen as a broad absorption in the high-frequency region [31]. The intensity of this absorption diminishes with increasing reaction temperature and is absent after reaction of $Cr(acac)_3$ with silica at 200 °C, where according to the acac/Cr ratio ligand exchange takes place. The decomposition of $Cr(acac)_3$ occurs above 280 °C restricting the ALE processing temperature range to 200-280 °C on silica [30].

The DRIFT spectra at 1500-1600 $cm^{-1}$ suggest that the ligands of the bound chromium complex stay complete on silica [30,31]. In the light of the gas FTIR spectroscopic analysis [33], however, partial opening of the ring structure also seems to be possible, because besides Hacac a small amount acetone without any acetate, being another decomposition product of acac ligand, evolves during the reaction. Hence, the possible formation of surface chromium species with both acac and acetate ligands must be taken into account. The combination of the information from the DRIFT and $^{13}C$ CPMAS NMR spectra verifies that, unlike $Cr(acac)_3$, the ring opening does not take place with $La(thd)_3$ [34].

The most feasible temperature range for the $La(thd)_3$ reaction on alumina ranges from 220 to 260°C, some decomposition taking place at 260°C, however, as indicated by the coke formation observed from the $^{13}C$ CP MAS NMR spectrum [35]. Most of the ligands remain intact at 260 °C. The reaction of $Cr(acac)_3$ with alumina has been studied at 200 °C [36]. At this temperature the ligand exchange reaction with OH groups of alumina takes place and one acac ligand is released as indicated by the calculated acac/Cr ratio of two.

Figure 7 shows the difference in the Ti isopropoxide reactions on silica and

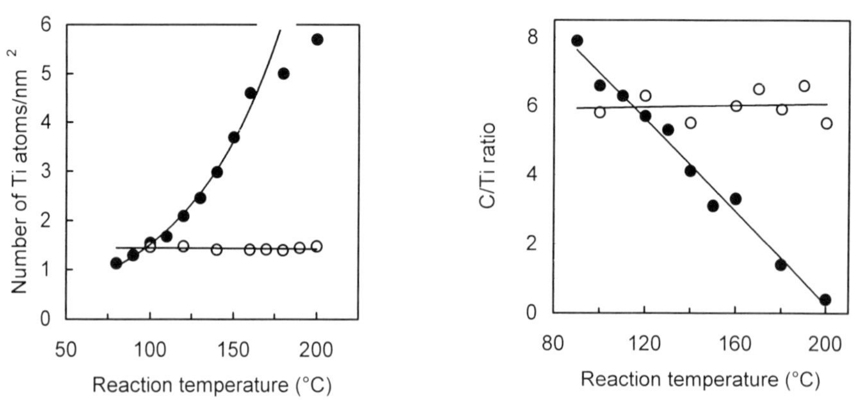


Figure 7. Effect of reaction temperature on the number of Ti atoms/nm$^2$ and the C/Ti ratio, when $Ti(OPr^i)_4$ is bound to alumina (●) and silica (O) [37].

alumina [37]. On silica, reaction takes place over a wide reaction temperature range (100-200 °C) leading by means of ligand exchange to titanium surface species with two intact isopropoxide ligands (C/Ti ratio close to 6). On alumina the processing temperature is restricted to a narrow temperature range of 100-110 °C. This is because on the catalytically active alumina surface Ti isopropoxide starts decomposing to $TiO_2$ at low temperatures. As shown by the carbon content, the complete decomposition takes place at 200 °C. Additional information about the decomposition has been obtained from the disappearance of the absorptions due to isopropoxide ligands from the IR spectra [8]. The decomposition of Ti isopropoxide starts at much lower temperature than that of $La(thd)_3$, indicating that not only the alumina but also the type of compound is responsible for the organic ligand removal. Specifically the strength of the interaction between the ligand and the metal, seems to influence the decomposition temperature. On silica the observed wide temperature range for Ti isopropoxide can be taken as granted only for the first reaction sequence. This is because the silica becomes a catalytically active material after it is modified with $TiO_2$.

### 3.4. Verification of surface saturation

The power of the ALE method lies in the systematic utilization of surface saturation. Under saturating reaction conditions, the surface species adopts the most favorable position with respect to the surface and to the chemical and physical characteristics of the precursor. This means that all accessible adsorption sites on a surface are filled, while those beyond reach, for instance, those sites shaded by bulky organic ligands, are left unoccupied. No new bonding sites for the precursor are created during the reaction.

To illustrate how surface saturation can be verified, Figure 8 shows a

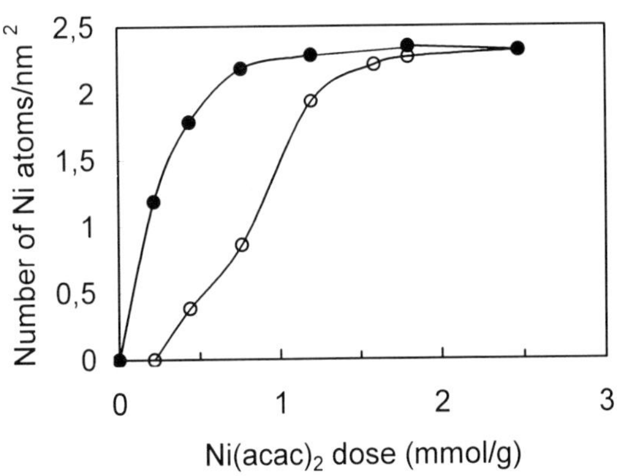


Figure 8. Number of Ni atoms/$nm^2$ in samples taken from (•) the top and (o) bottom parts of a fixed alumina bed as a function of $Ni(acac)_2$ dose [25].

simple procedure for determining the dose of a saturating precursor in catalyst processing under appropriately selected reaction conditions. The example is from a study of the reactions of $Ni(acac)_2$ on alumina [25]. A similar saturation curve for silica surface can be found in ref. 8. The vaporized $Ni(acac)_2$ was fed with the aid of nitrogen stream from the top to the bottom of a support bed supported on sinter. As can be seen from Figure 8, after a certain precursor dose level the surface densities are equal in the top and bottom parts of the support bed and the surface density cannot be increased with a further increase in the precursor dose. This ability of the surface to saturate itself is of great importance for process control, for the growing surface itself takes over the external process control.

Determination of the saturation curve is not always required, since one can also use information about nature and number of the adsorption sites obtained from study of the support surfaces. This information allows a rough but adequate estimate.

Under saturating reaction conditions the growth is highly reproducible and the samples are homogeneous. This is clearly indicated by Table 1, which shows that differences between the saturation densities achieved for repeated runs are small and close to the accuracy of the determination [8]. Checking of the saturation densities in the top and bottom parts of the bed gives only a macroscopic picture, however, and does not rule out the existence of a concentration gradient within the particles. The uniformity throughout the catalyst particles was therefore

Table 1
Reproducibility of the saturation density of different metals on alumina and silica supports. Samples taken from the top and bottom of the support bed indicate the uniformity of saturation [8]

| Saturated gas-solid reactions | | | Saturation density (atoms/nm $^2$) | |
|---|---|---|---|---|
| Precursor / support pair | Preheating / reaction temperature °C | Number of process runs | top | bottom |
| $ZrCl_4$ / $SiO_2$ | 300 / 300 | 5 | 1.4 ± 0.1 | 1.4 ± 0.1 |
| $WOCl_4$ / $Al_2O_3$ | 200 / 200 | 10 | 1.8 ± 0.1 | 1.8 ± 0.1 |
| $Ni(acac)_2$ / $Al_2O_3$ | 200 / 200 | 5 | 2.5 ± 0.3 | 2.3 ± 0.1 |
| $Cr(acac)_3$ / $SiO_2$ | 820 / 200 | 5 | 0.32 ±0.01 | 0.30±0.00 |
| $Mg(thd)_2$ / $SiO_2$ | 600 / 200 | 6 | 1.0 ± 0.1 | 0.8 ± 0.1 |

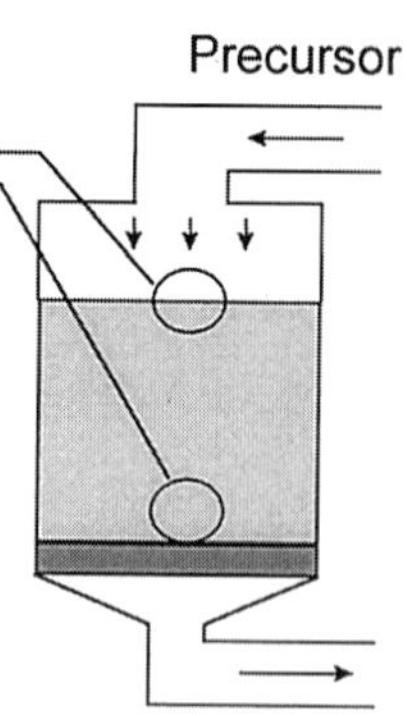

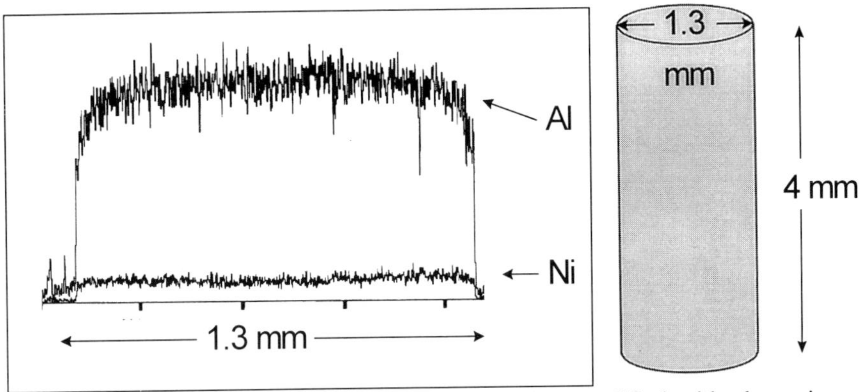


Figure 9. EDS line-scan over a cut alumina extrudate modified with alternating saturating reactions of $Ni(acac)_2$ and air (4 reaction cycles) producing a NiO/alumina catalyst with 10 wt-% of Ni.

further verified by embedding the particles in epoxy resin, cutting cross-sections of particles with a microtone, and analyzing by SEM-EDS. The EDS line-scan over a cut alumina extrudate modified with alternating saturating reactions of $Ni(acac)_2$ and air (4 reaction cycles) producing a NiO/alumina catalyst with 10 wt-% of Ni, is shown in Figure 9. As can be seen from the figure, Ni is evenly distributed through the extrudate with a diameter of 1.3 mm and length of 4 mm. The amount of extrudates in the processing was 100 g.

The homogeneity of the distribution of metal species has also been tested by binding $Cr(acac)_3$ on a large-scale monolith coated with an alumina-based mixed oxide (see Figure 10) [37]. Samples, six in total, were taken at three different radial positions and two different axial positions of the monolith, and analysis gave an average of 1.0 ± 0.1 wt-% chromium.

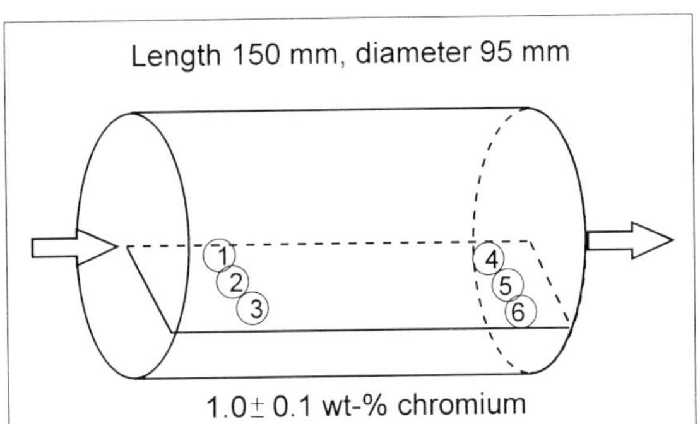


Figure 10. Reaction of $Cr(acac)_3$ on a large-scale monolith coated with an alumina-based mixed oxide. Cr was determined in samples taken from six different positions on the monolith [37].

The excellent run-to-run reproducibility allows a straightforward scale-up of the ALE process. Figure 11 presents the scale-up of the $ZrCl_4$ reaction on silica from 10 g to 1 kg [8].

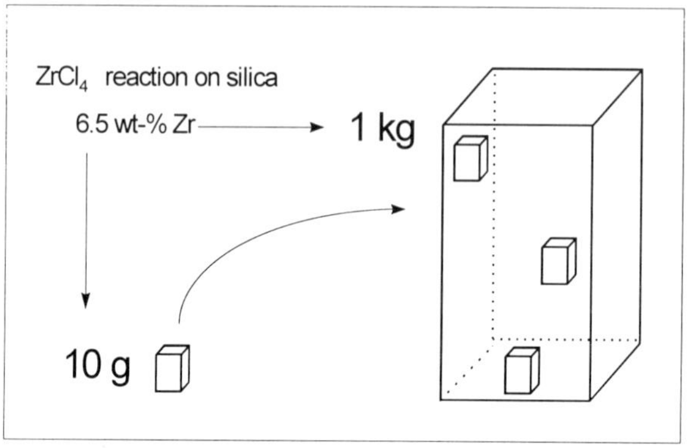


Figure 11. $ZrCl_4$ reactions at 300 °C on 10-g and 1-kg batches of silica. Both reactions gave a Zr concentration of 6.5 wt-% [8].

### 3.5. ALE processing

ALE processing can be carried out in a fluidized bed or solid bed reactor, or in any reactor designed to bring the precursor in vapor phase into contact with the support. Once the ALE conditions are met the mode of processing does not play any difference for the final result. In a fluidized bed, some information on the saturation and distribution of the metal inside the particles may be lost, for all particles simultaneously get the same precursor dose, not necessarily always a saturating dose. This means that the penetration of the precursor to the center of the particle must be studied separately in order to assure saturation. In a solid bed the decomposition of the precursor, as well as an inadequate dose, is most simply noticed by checking the color of the sample (white samples constitute an exception): under non-saturating conditions, the sample will be only partly colored, the bottom part being colorless.

Figure 12 presents a schematic diagram of a solid bed reactor where the vaporized precursors are fed to meet the solid support in an inert gas (usually nitrogen), while Figure 13 presents the ALE procedure. As can be seen from Figure 13 the precursors are fed one at a time, an inert gas purge in between. This purging prevents gas phase reactions, which could lead to uncontrolled material growth.

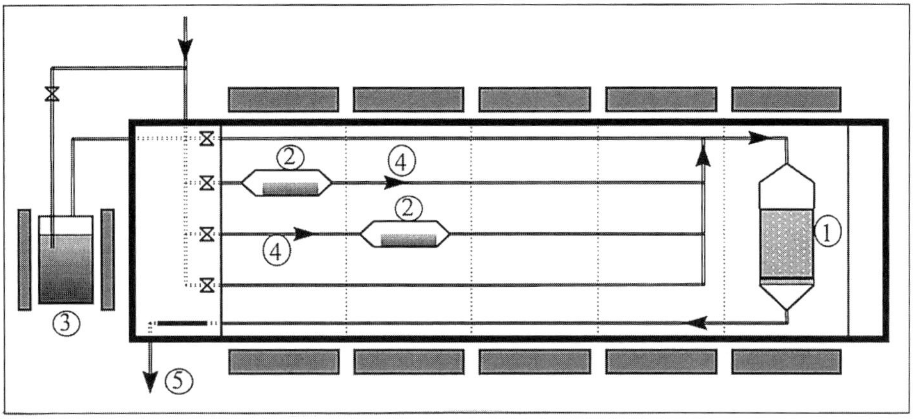


Figure 12. A schematic diagram of the flow-type ALE reactor. The reaction chamber (1) consists of a quartz cup which can hold up to 20 g of porous material. The precursor vessels are for solid (2) and liquid or gaseous precursors (3). The reaction chamber, the precursor vessels, and the tubes leading the precursor vapors into the chamber are resistively heated. The heating and the flow of the precursor and carrier gas (4) are computer controlled. The reaction chamber is kept at 6-10 kPa, and the pumping takes place from the bottom of the fixed bed of the porous material (5).

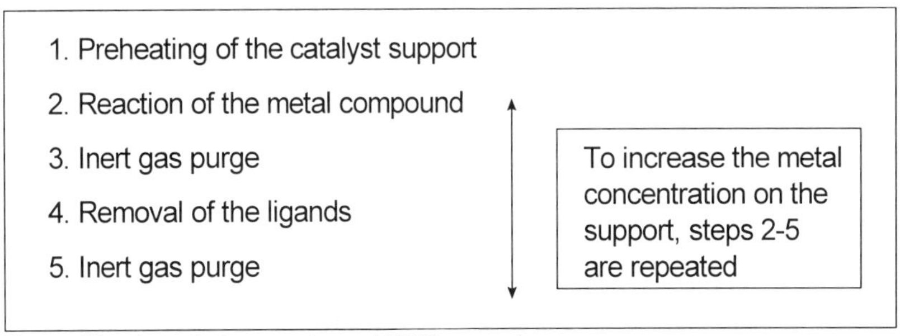


Figure 13. Atomic layer epitaxy (ALE) procedure.

## 4. CONTROLLING THE METAL CONCENTRATION ON THE SUPPORT

Controlling the metal concentration or the density of modifying species on the support surface is essential. The metal concentration in catalysts is most commonly expressed as weight percentage. For the ALE method, however, it is more informative to use the expression atoms/nm$^2$, which can be directly related to the number of possible adsorption sites. In addition, for very light elements the use of weight percentage may give a misleading impression of the actual amount. For instance, 1 wt-% of Mg and 5 wt-% of La on a silica surface of a surface area 300 m$^2$/g corresponds to 0.8 and 0.7 atoms/nm$^2$, respectively.

There are many ways to control the metal concentration in catalysts prepared by ALE. It must be emphasized once more that, in ALE, the precursor brought into contact with the support, however, always is in excess relative to the amount capable of chemisorbing on the adsorption sites. Thus the metal concentration is *not controlled* by the dose.

The primary means to control the metal concentration are the following:
- control by the number of bonding sites
- control by the size of the metal compound
- control by the number of reaction cycles

The control can be achieved by using just one of these means or by combining two or more of them, providing thereby a wide variety of possibilities to achieve the metal concentration needed for a specific catalytic reaction or surface modification. These means are not restricted to one metal but can be used to control the concentration of two or several metals in the catalyst. The more complex the multimetal structures we want to prepare the clearer the advantage of using ALE becomes, as demonstrated in section 5 below.

### 4.1. Control by bonding sites

Heat treatment of the support to control the number of adsorption sites is the easiest means to control the metal concentration (Figure 14). The use of a particular temperature during a particular time gives a support with a reproducible number of OH groups. Independent of the precursor the number of metal atoms bonded to silica follows the number of these OH groups, as shown in Figure 14. A similar tendency has also been observed on alumina.

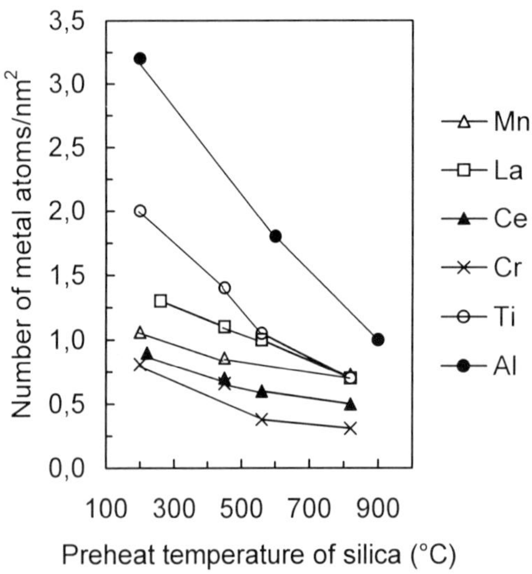


Figure 14. The effect of pretreatment temperature of silica on the number of Mn, La, Ce, Cr, Al, and Ti atoms/$nm^2$ bound from $Mn(thd)_3$ at 180 °C [35], $La(thd)_3$ at 260 °C [35], $Ce(thd)_4$ at 220 °C [9], $Cr(acac)_3$ at 200 °C [31], $Al(CH_3)_3$ at 200 °C [38], and $TiCl_4$ at 175°C [19].

### 4.2. Control by molecular precursor size

Another means to control the density of the metal atoms to be bound is by the size of the ligand in the precursor. The bulkier the ligand the lower is the metal density achieved on the surface. A good example of this is the reactions of acetylacetonate (acac) and 2,2,6,6,-tetramethyl-3,5-heptanedionate (thd) compounds of the same metal, in which a higher metal density is achieved with the acac than the thd compound. This is well illustrated in Figure 15, which shows the results for the reactions of $Ni(acac)_2$ and $Ni(thd)_2$ on alumina [40].

The effect of precursor size not only applies to the same metal but also to different metals. This has been clearly shown by the reactions of metal beta-diketonates on alumina [41]: the smallest density was obtained with the largest precursor, namely $Ce(thd)_4$ (see Table 2).

| | | |
|---|---|---|
| precursor | $Ni(C_5H_7O_2)_2$ acac | $Ni(C_{11}H_{19}O_2)_2$ thd |
| surface structure | 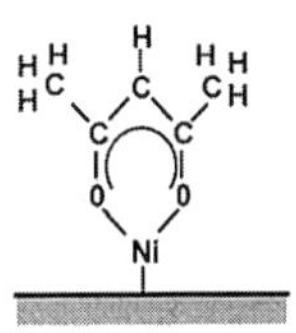 | 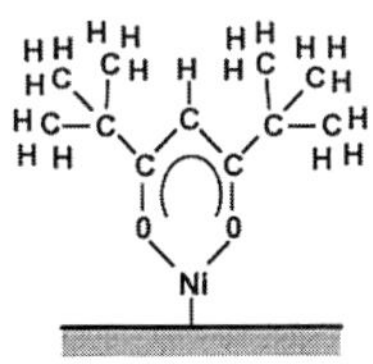 |
| calculated filling area | A= 0,47 $nm^2$ | A= 0,87 $nm^2$ |
| calculated max density | 2,1 Ni / $nm^2$ | 1,1 Ni / $nm^2$ |
| observed saturation density | 2,5 Ni / $nm^2$ | 0,92 Ni / $nm^2$ |

Figure 15. Effect of precursor size on the surface density of Ni bound on alumina [40].

The effect of the size of the metal beta-diketonate on the surface density is not as clear on silica as it is on alumina [9,35]. For instance, the number of metal atoms bound on 820 °C silica from $Cu(thd)_2$, $La(thd)_3$ and $Ce(thd)_4$ is 0.5, 0.7 and 0.5/$nm^2$, respectively. The difference in the number of La and Ce atoms can be explained by the size of the surface complexes to which two and three intact thd ligands are bound, respectively. In the case of copper the number of atoms would have been close to 1/$nm^2$ if the size of the surface complex having one intact ligand had been the dominating factor. The much smaller value suggests that the activation energy needed for the reaction to occur with all sterically available bonding sites was not exceeded at the reaction temperature of 170 °C used. Since

Table 2
Metal beta-diketonate reactions on alumina heat-treated at 600 °C [41]. See Figure 15 for the structure of the thd and acac ligands

| Metal beta-Diketonate | Reaction temperature (°C) | Metal atoms/nm$^2$ | Carbon atoms/nm$^2$ | Ligand*/metal ratio |
|---|---|---|---|---|
| $Ce(thd)_4$ | 220 | 0.5 | 14 | 2.5 |
| $Mn(thd)_3$ | 180 | 0.6 | 13 | 2.0 |
| $La(thd)_3$ | 260 | 0.8 | 12 | 1.4 |
| $Cu(thd)_2$ | 170 | 0.7 | 12 | 1.6 |
| $Mg(thd)_2$ | 180 | 0.7 | 13 | 1.7 |
| $Co(acac)_3$ | 180 | 1.5 | 13 | 1.7 |

* thd or acac

decomposition of $Cu(thd)_2$ starts at 190 °C, the saturation density cannot be increased to 1 atom/nm$^2$ by using a higher reaction temperature.

### 4.3. Control by number of reaction cycles

The original objective in ALE studies was to grow thin films with several thousand atomic layers on flat substrates [5]. The same procedure can be applied on porous surfaces, but for catalytic purposes there is no need to grow thousands of layers; usually one to ten layers suffices. Growing new layers is a simple way to increase the metal concentration of the catalyst.

The layer-by-layer growth is not as feasibly utilized in the other catalyst preparation methods. For example, growing layers from organic liquids requires removal of the liquid and drying of the sample before ligand removal, which is carried out separately through calcination, and after that the sample must be reinserted in the organic liquid for a new reaction. In addition, in the case of metal chlorides, water cannot be used to remove the ligands because the bound metal species would be released into the water solution.

The growth of different binary oxides on alumina and silica is presented in Figure 16 [9]. Each reaction cycle in the figure comprises the reaction of precursor and ligand removal agent. As can be seen from the figure the growth proceeds more or less evenly and the saturation density increases by about the same amount in each reaction cycle giving a highly effective and reproducible way to obtain a desired metal concentration on the support surface. The effect of the size of the metal beta-diketonate on the metal concentration is also clearly seen in Figure 16. For instance, the growth rate of cerium oxide from $Ce(thd)_4$ on alumina is much lower than that of NiO from $Ni(acac)_2$. In addition to the oxides presented in Figure 16, others successfully processed include $SiO_x$/alumina from HMDS and air [42], CoOx/silica from $Co(acac)_3$ and air [43], and $Al_2O_3$/silica from $Al(CH_3)_3$ and $H_2O$ [38]. The number of reaction cycles was 6, 5, and 5, respectively.

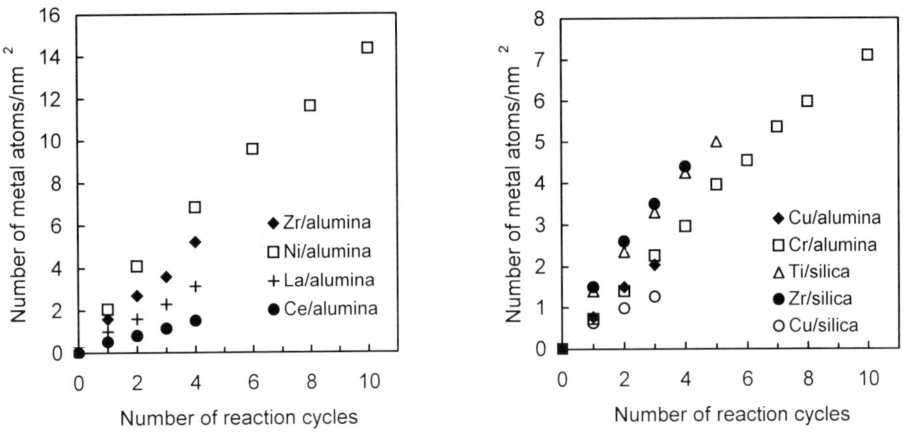


Figure 16. Controlling of metal concentration on alumina and silica by the number of reaction cycles. One growth cycle comprises the reaction of precursor and ligand removal agent (reaction temperature in parenthesis). The following precursors were used: $ZrCl_4$ (300°C) [29], $Ni(acac)_2$ (200°C) [25], $La(thd)_3$ (240°C) [9], $Ce(thd)_4$ (220°C) [9], $Cu(thd)_2$ (170°C) [9], $Cr(acac)_3$ (200°C) [36], and $Ti(OPr^i)_4$ (160°C) [37]. The organic ligands were removed by air treatment at 450-600 °C and the chloride ligands by water vapor treatment at 450 °C.

## 5. SURFACE STRUCTURES PROCESSED BY ALE

The number of surface structures (species) that can be processed by ALE is enormous. This includes the singly, doubly or even triply bonded species in the very first reaction prepared particularly on silica by varying the heat treatment temperature and the binary and mixed oxides prepared by carrying out repeated reaction cycles. In this section we describe various surface structures processed up to the present along with the reaction mechanisms that lead to these structures.

### 5.1. First reaction cycle

#### 5.1.1. Surface structures controlled by heat-treatment temperature of support

Heat-treatment temperature of the support determines the number of OH groups and thereby also the distance between these groups, which has an effect on the bonding mode. This is clearly seen in Figure 17 where results are shown for the reactions of $Ce(thd)_4$ and $TiCl_4$ on silica [9,19]. Irrespective of the precursor, a similar trend is seen with increasing heat-treatment temperature (200-800 °C) of silica, in the number of metal atoms bound and in the bonding mode the latter being indicated by the ligand/metal ratio of the bound surface complexes. For both $Ce(thd)_4$ and $TiCl_4$ the number of metal atoms bound decreases with the decrease in the number of OH groups on the silica surface, and the bonding mode, deduced from the ligand/metal ratio, changes from the triply and doubly bonded species on 200-220 °C silica to the predominantly singly bonded species on 820 °C silica. Use of the thd/Ce ratio as calculated from the carbon content is made possible by the fact that the ring structure of the ligands stays intact in the bound surface complexes. This has been verified by both $^{13}C$ CPMAS NMR and FTIR [34].

The effect of heat-treatment temperature is different on alumina than on silica. As noted already above, whether the species on alumina are singly or doubly bonded is difficult to determine because the precursor and the ligand released in the ligand exchange reaction may dissociate when heat-treatment is at elevated temperatures. The dissociation is most probably brought about by coordinatively unsaturated aluminum and oxygen (c.u.s) sites, whose amount is increased with increase in the heat-treatment temperature.

The dissociation of the precursor was clearly demonstrated in a study of the reaction of HMDS ($(CH_3)_3Si\text{-}NH\text{-}Si(CH_3)_3$) with 600 °C alumina by $^{29}Si$ CPMAS NMR [42]. After the reaction, NMR peaks due to both $Al\text{-}O\text{-}Si(CH_3)_3$ and $Al\text{-}NH\text{-}Si(CH_3)_3$ surface species were discernible. At lower heat-treatment temperatures, where the number of c.u.s sites is smaller, the intensity of the peak due to $Al\text{-}NH\text{-}Si(CH_3)_3$ was noticeably weaker.

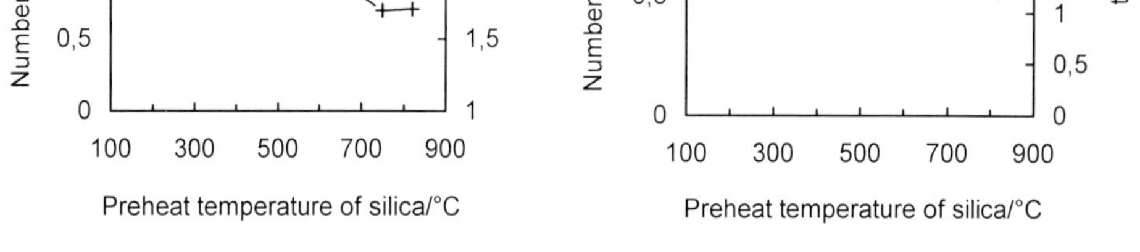


Figure 17. $TiCl_4$ reactions at 175 °C and $Ce(thd)_4$ reactions at 220 °C with 200-820 °C silica: (+) metal atoms/$nm^2$ and (□) ligand/metal ratio in the surface complex [9,19].

**5.1.2. Surface structures controlled by blocking**

As mentioned above, the reaction of HMDS with silica heat treated at 200 °C can be used to decrease the amount of titanium species bonded from $TiCl_4$. At the same time, the decrease in Ti concentration leads to the exclusive formation of doubly bonded Ti species. The presence of these species was verified by element determinations [19] and was further supported by the low-frequency DRIFT spectrum [27]. This specific blocking cannot only be performed from gas phase but, as shown by Blitz [44], the same surface species can be achieved by carrying out the HMDS and $TiCl_4$ reactions in organic liquid. A scheme for the blocking is shown in Figure 18, which also makes clear the reactivity of HMDS towards the terminal H-bonded OH groups [39].

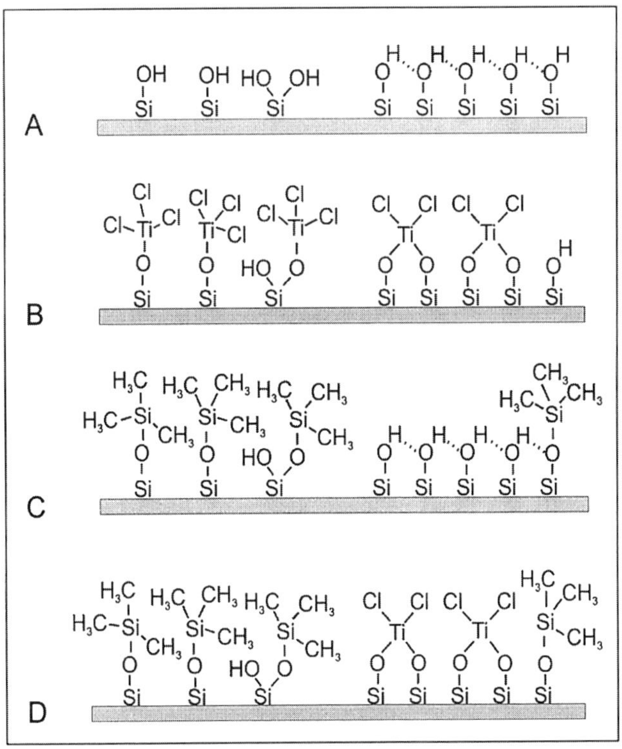


Figure 18. Surface structures controlled by blocking [39]: surface species (A) on silica heat treated at 200 °C and (B) on 200 °C silica after $TiCl_4$ reaction at 175 °C; (C) on 200 °C after silylation with HMDS and (D) on 200 °C silica after $TiCl_4$ reaction at 175 °C. The siloxane bridges are omitted for clarity.

### 5.1.3. Surface structures influenced by difference in precursor reactivity

Reaction mode in a gas–solid interaction varies with the precursor. Some precursors form covalent bonds vigorously, whereas others are less reactive. Surfaces with different metal density but also with surface species bound to different bonding sites can be achieved by selecting metal alkyls instead of metal β-diketonates. An example of this is shown in Figure 19 where aluminum is bound on the silica surface from two different compounds, namely aluminum acetylacetonate ($Al(acac)_3$) and aluminum alkyl ($Al(CH_3)_3$, TMA). TMA reacts with all isolated OH groups [38], whereas $Al(acac)_3$ is too large a molecule to engage in ligand exchange with all. Besides occupying all the OH groups TMA also reacts with siloxane bridges. Like all other metal beta-diketonates studied, $Al(acac)_3$, in turn, exhibits reactivity solely towards to the OH groups.

Although TMA gives the highest Al concentration it also forms a highly stable $Si\text{-}CH_3$ species as verified by $^{29}Si$ CPMAS NMR [38]. This species is extremely difficult to eliminate below air treatment temperatures of 450 °C. In addition to this species, $\text{-}Si(CH_3)_3$ has been observed after reaction at elevated temperatures. Its presence indicates a simultaneous breakage of several Si-O-Si bridges and means that, at the same time as the methyl groups are transferred to silicon atoms, the oxygen atoms are bonded to aluminum leading to agglomerated aluminum oxide species.

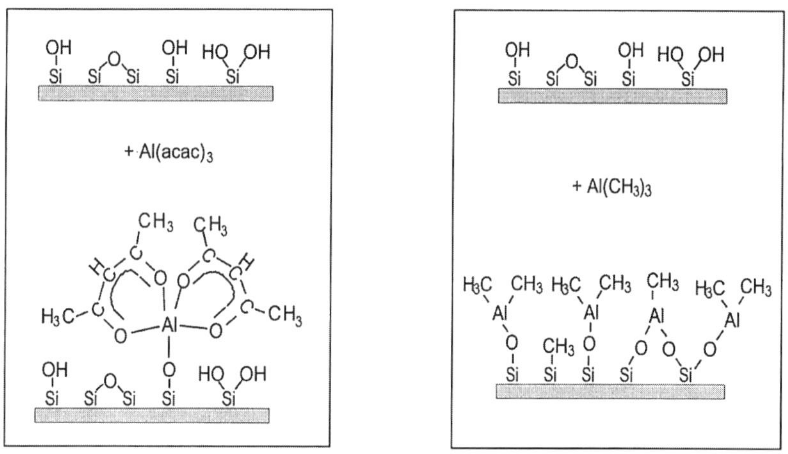


Figure 19. Difference in reactivity of $Al(acac)_3$ and $Al(CH_3)_3$ towards silica.

### 5.1.4. Surface structures controlled by different reaction temperatures

Depending on the reaction temperature, the reactions of $ZrCl_4$ and $TiCl_4$ on silica have been found to lead to amorphous surface species directly bonded to OH groups and to agglomerated $ZrO_2$ and $TiO_2$ species interacting with siloxane bridges (for $TiCl_4$, see Figure 20) [19,27,28,45,46]. The agglomeration takes place without any water treatment by means of direct chlorination. Most probably volatile $M(OH)_xCl_y$ species are formed, which decomposing on the surface gives rise to the corresponding oxides together with HCl. Despite agglomeration the reactions are highly surface controlled, i.e. controlled by the number of OH groups. The agglomeration is not only typical for silica but also for alumina as shown by the $ZrCl_4$ reaction [28].

Water treatment of Cl-terminated surfaces leads to the reappearance of the OH groups of silica and also the formation of OH groups on titanium and zirconium as shown by both the DRIFT and $^1H$ MAS NMR spectra [19,27,28]. The number of OH groups on the metals is higher in the amorphous samples.

The bonding mode of $TiCl_4$ on 200 °C silica can be directed to doubly bonded species through HDMS blocking of the isolated OH groups leaving the the strongly H-bonded OH groups unoccupied. At higher preheat temperatures of silica the strongly H-bonded OH are not present. However, in the absence of these reactive sites one can obtain doubly bonded species by applying a higher reaction temperature [45]. Evidently, just before agglomeration, $TiCl_4$ forms primarily doubly bonded surface species with the isolated OH groups.

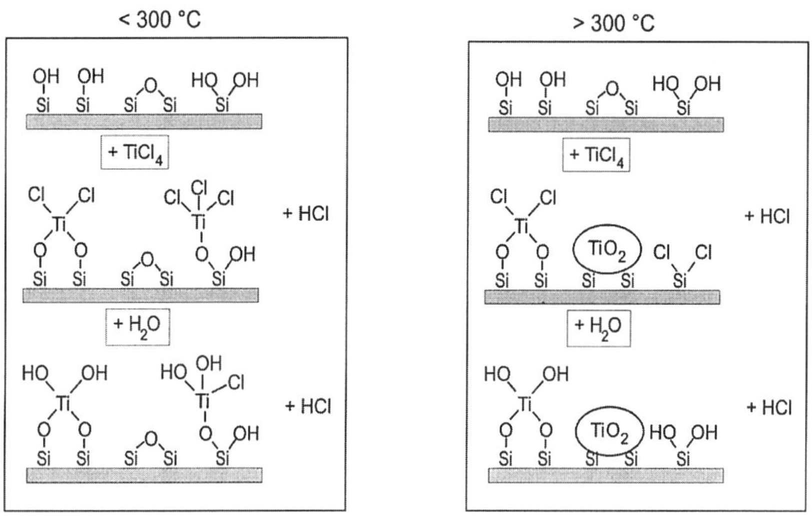


Figure 20. The effect of reaction temperature of $TiCl_4$ on the titanium surface species formed on silica [46]. Both reaction modes are surface controlled, i.e controlled by the number of OH groups on the silica surface.

### 5.1.5. Surface structures produced through silylation

Many alkoxysilanes containing a variety of functional groups are available and can be applied in the gas phase modification of a silica surface. The advantage of processing from the gas phase is that many of the tedious operations involved in the liquid phase processing can be eliminated. This has been clearly shown by Iiskola et al. [47,48], who obtained a highly reproducible organic cyclopentadienyl surface by means of saturating gas–solid reaction of $Cp(CH_2)_3\text{-}Si(OCH_2CH_3)_3$ with silica. The preparation of this organic surface was one step in the successful heterogenization of a homogeneous metallocene catalyst.

The gas–solid reaction of another alkoxysilane, namely $NH_2(CH_2)_3\text{-}Si(OCH_2CH_3)_3$, results in a highly reproducible number of amine groups on the surface [49]. This amine-terminated surface has been used to increase in a controlled manner the number of palladium atoms bound to the silica surface from $Pd(thd)_2$. The $Pd(thd)_2$ reaction was also carried out from gas phase.

## 5.2. Repeated reaction cycles

In section 4.3 we presented results obtained in growing various binary oxides on silica and alumina as a means to control the metal concentration. In this section we consider the development of the various surface structures during growth.

### 5.2.1. Surface structures produced in the growth of metal oxides from metal chlorides

In the absence of water treatment the first reaction cycle in the growth from metal chlorides at elevated growth temperatures leads to metal oxide agglomerates. Depending on the growth temperature anatase or rutile crystalline structures of $TiO_2$ are produced [45,46,50]. On 560 °C silica the reaction temperature 350 °C leads exclusively to anatase, 450°C to both anatase and rutile, and 550 °C exclusively to rutile. The anatase/rutile transformation temperature is somewhat lower than that for bulk $TiO_2$ oxide, and perhaps because of the small particle size.

After $TiCl_4$ reactions on silica the presence of crystalline $TiO_2$ was revealed by XRD diffraction measurements, together with etching with sulfuric acid which removes only the amorphous surface species [45], whereas after $ZrCl_4$ reaction on silica the presence of agglomerated $ZrO_2$ could only be revealed by etching [28]. The XRD peaks due to crystalline, tetragonal $ZrO_2$ appeared after two reaction cycles of $ZrCl_4$ and $H_2O$ as shown by Kytökivi et al. (see Figure 21) [29]. It has been suggested that the number of $ZrO_2$ crystallites is too small after the first reaction cycle for detection by XRD but during the second reaction cycle the number is increased to a detectable amount. This explanation is based on their finding that up to the third reaction cycle the domain size stays constant, at 100 Å, and after the fifth reaction cycle it increases only up to 120 Å. It is stated in the literature [50] that at room temperature only $ZrO_2$ crystallites of diameter smaller than 100 Å should be tetragonal, and the values presented by Kytökivi et al. can be considered within this range.

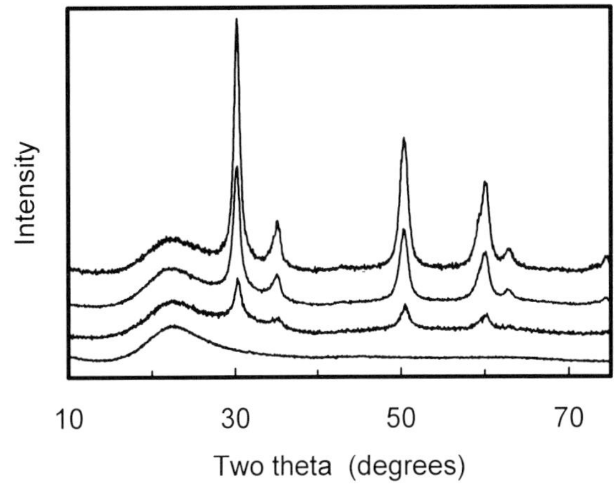


Figure 21. X-ray diffractograms of silica samples after (from bottom to top) one, two, three, and five reaction cycles of $ZrCl_4$ and $H_2O$ [29]. The reaction of $ZrCl_4$ and the water vapor treatment were carried out at 300 °C.

As revealed by transmission (TEM) and scanning electron microscopy (SEM) measurements, the domain size of $TiO_2$ on silica varied from 15 to 30 nm and the particle size from 1-2 μm to 10-20 nm in the same sample after the first reaction cycle [45]. The particle size then gradually increased with each of four reaction cycles [51], and the increase in the particle size of the biggest particles could even be observed by light microscopy [52]. This was not, however, the case with $ZrO_2$ particles, which were not seen by SEM even after five reaction cycles. The larger size of the $TiO_2$ particles than the $ZrO_2$ particles is assumed to be due to the difference in the lifetime of the proposed volatile $M(OH)_xCl_y$ species: the much longer lifetime for titanium leads to larger $TiO_2$ particles.

To estimate the surface coverage, the water-treated $ZrO_2$/silica samples were characterized by LEIS (low energy ion scattering), which is a surface- sensitive technique probing only the topmost atomic layer [29,53]. As determined by LEIS about 50% of the silica surface was covered with $ZrO_2$ during the five reaction cycles. It is extremely difficult to cover the silica surface completely, because of the presence of siloxane bridges, which are reactive towards very few precursors. One rare example is TMA, as discussed above. On the other hand, because of the agglomeration through direct chlorination any interaction of $ZrO_2$ oxide with siloxanes can assist in achieving a more complete coverage.

### 5.2.2. Surface structures produced by $Ti(OPr^i)_4$

In studies on the growth of $TiO_2$ on silica and alumina from titanium isopropoxide and air, Lindbald et al. [37] found the growth rates to be similar on the two supports. The reaction temperatures were 160 °C on silica and 100 °C on alumina and the air treatment was carried out at 450 °C. After five reaction cycles on alumina, no $TiO_2$ crystallites were observed by XRD, and the amorphous surface species were thermally stable up to 800 °C, suggesting strong interaction with the support. On silica, the growth led after four reaction cycles to small anatase crystallites. Si-O-Ti bonds are known to be highly susceptible to water [27], and during burning off of the organic ligands water vapor is formed, which can cause cleavage of the bonds and allow movement of the titanium species. On the other hand, breakage of the Si-O-Ti bonds was not observed when a lower water treatment temperature of 200 °C was used for Si-O-$TiCl_x$ surface species [27]. A similar difference in the strength of the interaction of surface species and the support has been observed with the growth of CuO on alumina and silica, as will be shown below.

It is worth adding here that the growth rate of $ZrO_2$ on silica from $ZrCl_4$ and $H_2O$ corresponds to that of $TiO_2$ from $Ti(OPr^i)_4$ and air indicating that, irrespective of the partial cleavage of the bonds, the growth is controlled by the OH groups on the silica surface (see Figure 16).

### 5.2.3. Surface structures produced through lateral and vertical growth

The growth of nickel oxide [54] and chromium oxide [36] on alumina has been studied by low energy ion scattering (LEIS). These studies show that the growth of Cr oxide proceeds laterally and that of NiO both laterally and vertically, which leads to different types of surface structures. The double growth mode has been shown to be the most common in ALE; the full monolayer coverage leading only to vertical growth is seldom achieved. This means that after removal of the ligands the reactive sites of the support shaded by the ligands are exposed and available for reaction.

The LEIS technique makes it possible to study the growth after each reaction cycle. From LEIS study of the growth of chromium oxide from $Cr(acac)_3$ and air on alumina, it was concluded that during six reaction cycles, corresponding to deposition of 7.4 wt-% of Cr (4.5 atoms/$nm^2$), the growth proceeds laterally [36]. In other words, $CrO_x$ is present on the surface as a monolayer. Only during the seventh reaction cycle does multilayer growth begin.

A possible explanation for the growth mechanism of Cr oxide is that the oxidation state of chromium changes from +3 to +6 during air treatment, and primarily chromate structures are formed. These cannot serve as reactive sites for the next $Cr(acac)_3$ reaction and $Cr(acac)_3$ prefers to react with the bonding sites on alumina. In the case of NiO the multilayer growth is already observed during the second reaction cycle. OH groups are formed on Ni as shown below and these can then act as reaction sites, too, facilitating the multilayer growth. The difference in the growth modes of NiO and $CrO_x$ is illustrated in Figure 22.

Both lateral and vertical growth thus take places in the growth of NiO on alumina leading to small NiO particles. The reduced NiO/alumina catalyst has been tested in hydrogenation of toluene, and an optimum particle size for the hydrogenation was formed after four reaction cycles, corresponding to 10 wt-% Ni [25,54].

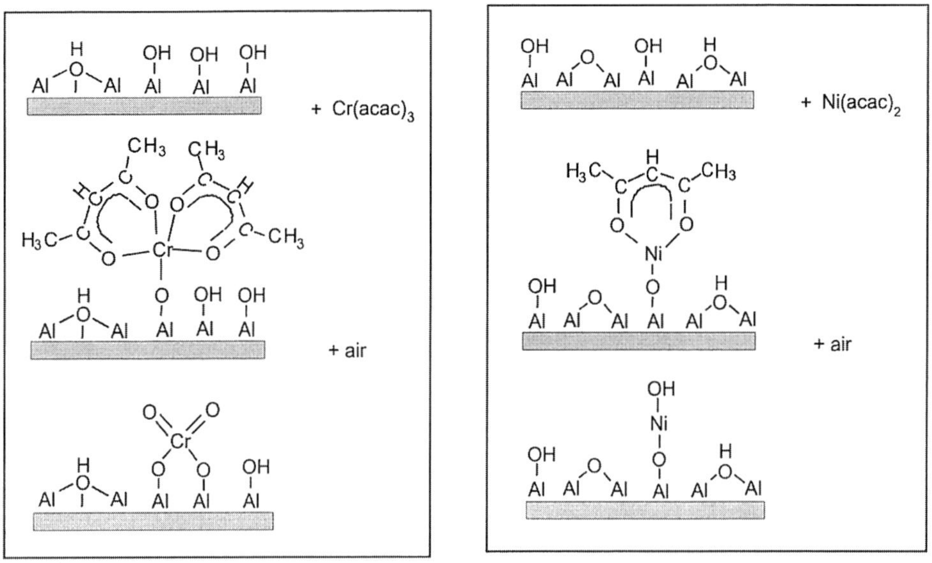


Figure 22. A simplified illustration of the different growth modes of $CrO_x$ and NiO on alumina, the former being an example of monolayer growth and the latter of simultaneous vertical and lateral growth [36,54]. After the air treatment $Cr(acac)_3$ is directed to react with the reactive sites on alumina, whereas $Ni(acac)_2$ finds both the reactive sites of alumina and the OH groups of Ni available.

The simultaneous lateral and vertical build-up of surface species, the most common form of growth in ALE, also takes place during the growth of silica on alumina [42]. Six reaction cycles of HMDS and air were carried out, and the growth was followed by DRIFT and $^{29}Si$ CPMAS NMR. During the first reaction cycle the air treatment of the silylated alumina surface leads to the appearance of Si-OH groups. Besides these new groups, also the reactive sites on alumina are still available for the HMDS to react with during the second and third reaction cycle as shown by the formation of Al-O-Si$(CH_3)_3$ species. The number of these species gradually decrease during the fourth and fifth reaction cycles and only Si-O-Si$(CH_3)_3$ species are formed in the sixth reaction cycle. This indicates that, in the sixth reaction cycle only Si-OH groups had served as reactive sites for HMDS. Evidently the alumina surface became totally covered with silica during five reaction cycles.

### 5.2.4. Study of new reactive sites during surface structure build-up

DRIFT spectra were recorded to study the development of new adsorption sites during the growth of NiO, CuO, and $CeO_2$ on alumina from $Ni(acac)_2$, $Cu(thd)_2$, and $Ce(thd)_4$ and air (Figure 23) [9]. The spectra shown were recorded after the

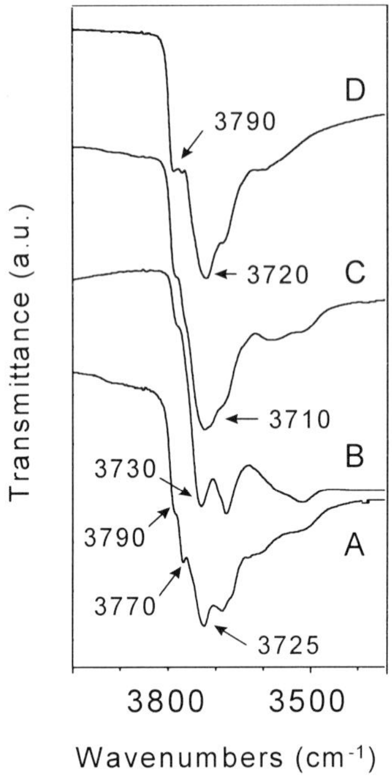


Figure 23. Formation of new adsorption sites (OH groups) in the growth of NiO, CuO, and $CeO_2$ on alumina. DRIFT spectra of (A) 600 °C alumina, (B) NiO/alumina (four reaction cycles), (C) CuO/alumina (three reaction cycles), and (D) $CeO_2$/alumina (four reaction cycles). The growth of NiO, CuO and $CeO_x$ was carried out by using the corresponding beta-diketonates, $Ni(acac)_2$, $Cu(thd)_2$, and $Ce(thd)_4$ as precursors and air as ligand removal agent.

fourth, third, and fourth reaction cycle, respectively. Judging from the disappearance of the peak at 3770 $cm^{-1}$ and the decrease in the intensity of the peak at 3790 $cm^{-1}$ in the IR spectra, $Cu(thd)_2$ and $Ni(acac)_2$ consume most of the basic OH groups on alumina during the growth of the oxides. In the reactions of $Ce(thd)_4$, in turn, only the basic OH groups at 3770 $cm^{-1}$ are consumed, and partially, while the ones at 3790 $cm^{-1}$ are left unoccupied. Thus only the intensity of the peak at 3770 is decreased and the peak at 3790 $cm^{-1}$ is now more clearly discerned.

New OH groups are formed during air treatment. Unfortunately the absorptions of Al-OH groups can partly overlap those of the possible new OH groups, making interpretation difficult. The spectrum of pure alumina shows a peak centered at 3725 $cm^{-1}$, whereas the growth of NiO, $CeO_2$, and CuO gives rise to peaks at 3732, 3720, and 3710 $cm^{-1}$ and the peak at 3725 $cm^{-1}$ partly disappears. The new peaks can be attributed to Ni-OH, Ce-OH, and Cu-OH, respectively. According to the literature the absorbance due to Ni-OH of bulk NiO appears at 3735 $cm^{-1}$ and also at 3690 $cm^{-1}$ [23,55] and that due to Ce-OH of bulk $CeO_2$ at 3710 and 3640 $cm^{-1}$ [23]. The formation of new OH groups is not as clear for $CeO_x$ as it is for NiO, because the number of Ce atoms is much lower (see Figure 16). The shift in the peak position to 3720 $cm^{-1}$ can, however, be tentatively attributed to the formation of Ce-OH. To our knowledge no absorbance values have been reported for Cu-OH of bulk CuO. Nevertheless, the spectrum recorded after the growth of CuO on alumina is different from that of pure alumina and also from that of alumina modified with NiO and $CeO_2$. The absorbance at 3710 $cm^{-1}$ is probably due to Cu-OH therefore.

Both $^1$H MAS NMR and DRIFT spectra have been recorded as a means of following the development of Zr-OH during the growth of $ZrO_2$ on silica and alumina from $ZrCl_4$ and $H_2O$ [29]. Five reaction cycles were carried out. The spectra revealed the formation of Zr-OH groups and also the reappearance of part of the Si-OH and Al-OH groups during water treatment. The formation of OH groups on $TiO_2$ processed on silica from five reaction cycles of titanium isopropoxide and air has also been confirmed [56]. Clear IR peaks due to $TiO_2$ were detected at both 3670 and 3720 $cm^{-1}$, these peaks being typical for anatase.

### 5.2.5. Surface structures produced in mixed oxide growth

The ALE method allows the preparation of very different types of mixed oxides. The reproducible growth of LaCeCuO, MgMnCoO [41] and LaSiO on alumina has already been demonstrated. Two combinations of the first two oxides were processed by carrying out five to nine reaction cycles. The corresponding metal beta-diketonates were used as precursors. As in the processing of binary oxides, the metal surface densities determined stepwise after each reaction give valuable information about the growth mechanism. As examples, the growth of LaSiO on alumina is shown in Figure 24 and that of LaCeCuO in

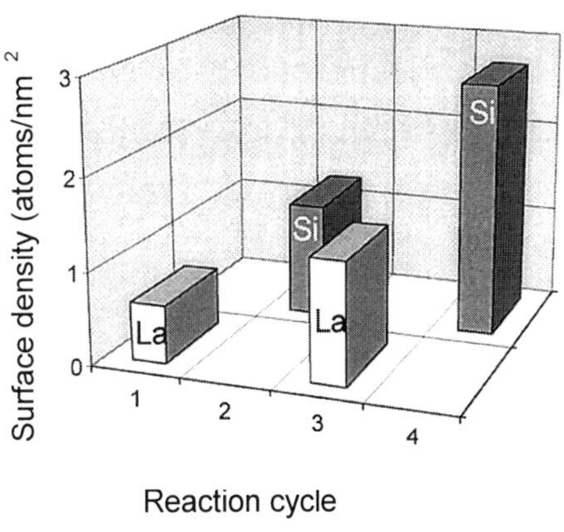


Figure 24. Surface densities of Si and La after each reaction cycle in the preparation of LaSiO/alumina samples. HMDS, $La(thd)_3$, and air were used as precursors.

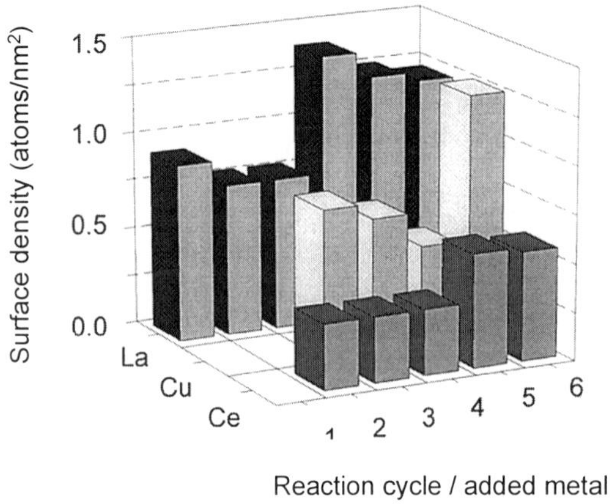


Figure 25. Surface densities of metals determined after each reaction cycle of metal beta-diketonate and air in the preparation of LaCeCuO/alumina. $La(thd)_3$, $Ce(thd)_4$ and $Cu(thd)_2$ were used as precursors.

Figure 25. Etching played an important role in the growth when only metal beta-diketonates were used. Despite this the growth was found to be reproducible and thus surface controlled. The etching did not occur when LaSiO/alumina samples were prepared from alternating reactions of $La(thd)_3$, HMDS, and air. This is because the thd released in the reaction of $La(thd)_3$ cannot release silicon into the vapor phase, and similarly the $NH_3$ formed in the HMDS reaction cannot release lanthanum.

DRIFT spectra were measured to follow the development of the reactive sites in the growth of mixed oxides. The spectra showed that, during the first reaction cycle, $La(thd)_3$ removes all the basic OH groups from the alumina surface [41]. This was also revealed by the number of La atoms, which corresponded to the number of basic OH groups determined, 0.8 La atoms/nm$^2$. When the Al-O-La(thd)$_x$ surface was then reacted with air, new OH groups most probably on La appeared. The presence of reactive sites on La has been indirectly verified by reacting the lanthana-modified alumina with HMDS in the preparation of silica-lanthana mixed oxide. The new surface species were characterized by $^{29}$Si CPMAS NMR, which revealed the presence of Al-O-La-O-Si($CH_3$)$_3$ groups. In addition, the $^1$H MAS NMR spectrum suggested that La-OH groups had been formed. The OH groups of lanthanum in the mixed oxide growth were then further consumed in the reactions of $Ce(thd)_4$ and $Cu(thd)_2$. Unfortunately, as in the growth of binary oxides, the possible new OH groups that are formed are difficult to distinguish. In future LEIS study on the growth of mixed oxides would be useful in providing information about the growth by revealing, for instance, whether a new metal atom is sitting on the original surface or on a metal atom bonded to the surface in earlier reaction cycles.

### 5.2.6. Surface structure size

After the first reaction on the support, before ligand removal, the surface species are atomically dispersed. An exception is the metal chloride reactions on silica and alumina carried out at elevated temperatures, where metal oxide crystallites, big enough to be seen by XRD, are formed. In general, when the surface structures are produced during several reaction cycles the ALE growth on alumina leads to XRD-amorphous samples, as in the case of the $TiO_2$/alumina prepared from five reaction cycles of $Ti(OPr^i)_4$ and air. Likewise, the NiO/alumina and $CrO_x$/alumina samples prepared from $Ni(acac)_2$ and $Cr(acac)_3$ using air as a ligand removal agent were still XRD-amorphous after six and eight reaction cycles, respectively. The interaction is weaker on silica than on alumina and small anatase crystallites were present on the silica after just four reaction cycles of $Ti(OPr^i)_4$ and air.

Owing to the excellent dispersion attainable in ALE, other means than XRD are needed to evaluate the size of the surface structures formed. In the following we review a TEM and EXAFS (extended X-ray absorption fine structure) study of CuO processed on silica and alumina from three reaction cycles of $Cu(thd)_2$ and air [57]. The Cu concentration was 4.1 wt-% on alumina and 4.3 wt-% on silica corresponding to 2.0 and 1.4 Cu atoms/$nm^2$, respectively. From the TEM pictures of the supported CuO samples it was apparent that the particles must be very small, because they could not be seen as separate particles. This was further supported by the EXAFS results for the samples indicating that the CuO particles are extremely small, probably less than 5 Å. The CuO particles on silica seemed to be somewhat larger than those on alumina. This also indicated that the interaction of the precursors is stronger with alumina than with silica. The small particle size determined by EXAFS further confirmed the excellent dispersion achievable by the ALE method when proper reaction conditions are applied.

## 6. CONCLUSIONS

Adsorption control in catalyst preparation can be achieved from both liquid and gas phase once the necessary conditions for the strong interaction between precursor and support have been created. This review has focused on the atomic layer epitaxy (ALE) method where the gas–solid reactions of precursors are directed to the strong interaction of covalent bond formation. In ALE, surface saturation is systematically utilized, providing the means for precise control of metal density and rendering the method truly adsorption controlled.

The simultaneous presence of covalent bond formation and surface saturation has been shown to provide good homogeneity of metal concentration throughout the particles, excellent reproducibility of the process as evaluated on both macroscopic and atomic scale, and feasible scale-up. The nanotechnological approach, which is dominating materials science today, is demonstrated in the ALE method, which can produce atomically controlled structures even on high surface areas. The build-up of structures in nano-scale, not only with one component but with multicomponents, is already a reality.

## ACKNOWLEDGEMENTS

The other members of the catalyst group are thanked for their significant contribution to this paper.

## REFERENCES

1. F. H. Ribeiro and G. A. Somorjai, Recl. Trav. Chim. Pays-Bas, 113 (1994) 419.
2. P. S. Cremer and G. A. Somorjai, J. Chem. Soc. Faraday Trans., 91 (1995) 3671.
3. T. Suntola, J. Antson, A. Pakkala and S. Lindfors, SID 80 Digest (1980) 108.
4. T. Suntola and J. Antson, US Patent No. 4 058 430 (1977).
5. T. Suntola, Mat. Sci. Rep. 4, (1989) 261.
6. T. Suntola, Atomic Layer Epitaxy, in: Handbook of Crystal Growth 3, Thin Films and Epitaxy, Part B: Growth Mechanisms and Dynamics, Chapter 14, Elsevier, 1994.
7. E.-L. Lakomaa, Appl. Surf. Sci., 75 (1994) 185.
8. S. Haukka, A. Kytökivi, E.-L. Lakomaa, U. Lehtovirta, M. Lindblad, V. Lujala and T. Suntola, Stud. Surf. Sci. Catal., 91 (1995) 957.
9. S. Haukka and T. Suntola, Interface Sci., 5 (1997) 119.
10. J. Haber, Pure & Appl. Chem., 63 (1991) 1227.
11. R. L. Augustine, Heterogeneous Catalysis for the Synthetic Chemist, Marcel Dekker, Inc., New York, 1996.
12. C. Louis and M. Che, in: Handbook of Heterogeneous Catalysis, Vol. 2, G. Ertl, H. Knözinger and J. Weitkamp (eds.), VCH, Weinheim, 1997, 207.
13. J. C. Kenvin, M. G. White and M. B. Mitchell, Langmuir, 7 (1991) 1198.
14. P. Van Der Voort, K. Possemiers and E. F. Vansant, J. Chem. Soc. Faraday Trans., 92 (1996) 842.
15. P. Van Der Voort, I. V. Babich, P. J. Grobet, A. A. Verberckmoes and E. F. Vansant, J. Chem. Soc. Faraday Trans., 92 (1996) 3635.
16. M. Leskelä and L. Niinistö, in: Atomic Layer Epitaxy, T. Suntola and M. Simpson (eds.), Blackie, London, 1990, 1.
17. M. Tiitta and L. Niinistö, Chem. Vap. Deposition, 3 (1997) 167.

18. A. Kytökivi, M. Lindblad and A. Root, J. Chem. Soc. Faraday Trans., 91 (1995) 941.
19. S. Haukka, E-L. Lakomaa and A. Root, J. Phys. Chem., 97 (1993) 5085.
20. S. Haukka and A. Root, J. Phys. Chem., 98 (1994) 1695.
21. H. Knözinger and P. Ratnasamy, Catal. Rev.-Sci. Eng., 17 (1978) 31.
22. A. A. Tsyganenko and P. P. Mardilovich, J. Chem. Faraday Trans., 92 (1996) 4843.
23. H.-P. Boehm and H. Knözinger, in: Catalysis - Science and Technology, J. R. Anderson and M. Boudart (eds.), Springer, Berlin, 1983, Vol. 4, 39.
24. E.-L. Lakomaa, M. Lindbald and T. Suntola, Stud. Surf. Sci. Catal., 92 (1994) 363.
25. M. Lindblad, L.P. Lindfors and T. Suntola, Catal. Lett., 27 (1994) 323.
26. A. Kytökivi, A. Rautiainen and A. Root, J. Chem. Soc. Faraday Trans., 93 (1997) 4079.
27. A. Kytökivi and S. Haukka, J. Phys. Chem. B, 101 (1997) 10365.
28. A. Kytökivi, E.-L. Lakomaa and A. Root, Langmuir, 12 (1996) 4395.
29. A. Kytökivi, E.-L. Lakomaa, A. Root, H. Österholm, J.-P. Jacobs and H.H. Brongersma, Langmuir, 13 (1997) 2717.
30. S. Haukka, E.-L. Lakomaa and T. Suntola, Appl. Surf. Sci., 75 (1994) 220.
31. A. Hakuli and A. Kytökivi, unpublished results.
32. I. V. Babich, Yu. V. Plyuto, P. Van Der Voort and E. F. Vansant, J. Colloid Interface Sci., 189 (1997) 144.
33. A. Hakuli, A. Kytökivi, E.-L. Lakomaa and A. O. I. Krause, Anal. Chem., 67 (1995) 1881.
34. S. Haukka, A. Kytökivi and A. Rautiainen, submitted to Langmuir.
35. S. Haukka, A. Kytökivi and A. Rautiainen, unpublished results.
36. A. Kytökivi, J.-P. Jacobs, A. Hakuli, J. Meriläinen and H. H. Brongersma, J. Catal., 162 (1996) 190.
37. M. Lindblad, S. Haukka, A. Kytökivi, E.-L. Lakomaa, A. Rautiainen and T. Suntola, Appl. Surf. Sci., 121/122 (1997) 286.
38. E.-L. Lakomaa, A. Root and T. Suntola, Appl. Surf. Sci., 107 (1996) 107.
39. S. Haukka, E.-L. Lakomaa and T. Suntola, Appl. Surf. Sci., 82/83 (1994) 548.
40. T. Suntola, Appl. Surf. Sci., 100/101 (1996) 391.
41. S. Haukka, M. Lindblad and T. Suntola, Appl. Surf. Sci., 112 (1997) 23.
42. M. Lindblad and A. Root, submitted to Stud. Surf. Sci. Catal.
43. L. Backman, A. Rautiainen, A. O. I. Krause and M. Lindblad, Catal. Today, in press.
44. J.P. Blitz, Colloids Surf., 63 (1992) 11.
45. S. Haukka, E.-L. Lakomaa, O. Jylhä, J. Vilhunen and S. Hornytzkyj, Langmuir, 9 (1993) 3497.
46. S. Haukka, E.-L. Lakomaa and T. Suntola, Thin Solid Films, 225 (1993) 280.
47. E. I. Iiskola, S. Timonen, T. Pakkanen, O. Härkki, P. Lehmus and J. V. Seppälä, Macromolecules, 30 (1997) 2853.

48. E. I. Iiskola, S. Timonen, T. Pakkanen, O. Härkki, and J. V. Seppälä, Appl. Surf. Sci., 121/122 (1997) 372.
49. S. Haukka, A. Rautiainen and H. Siro, submitted to Stud. Surf. Sci. Catal.
50. R. C. Garvie and M. F. Goss, J. Mater. Sci., 21 (1986) 1253.
51. E.-L. Lakomaa, S. Haukka and T. Suntola, Appl. Surf. Sci., 60/61 (1992) 742.
52. S. Hornytzkyj and S. Haukka, unpublished results.
53. H. H. Brongersma, P. A. C. Groenen and J.-P. Jacobs, in: Science of Ceramic Interfaces II, J. Nowotny (ed.), Vol. 81, 113. Mat. Sci. Monographs, Elsevier Science, Amsterdam, 1994.
54. J.-P. Jacobs, J. G. H. Reintjes, H. H. Brongersma, L. P. Lindfors and J. Jylhä, Catal. Lett., 25 (1994) 315.
55. A. A. Tsyganenko and V. N. Filimonov, Spectrosc. Lett., 5 (1972) 477.
56. M. Lindblad, personal communication.
57. A. M. Molenbroek. S. Haukka, B. S. Clausen and J. K. Norskov, to be published.

Adsorption and its Applications in Industry and Environmental Protection
Studies in Surface Science and Catalysis, Vol. 120
A. Dabrowski (Editor)




# The use of activated carbon as catalyst and catalyst carrier in industrial applications

W.M.T.M. Reimerink

NORIT Nederland B.V., P.O Box 105, 3800 AC Amersfoort, The Netherlands

## 1. INTRODUCTION

Activated carbon is applied in the first place as adsorbent for water purification and for gas/air purification. Production streams in the chemical, pharmaceutical and food industry are purified for decolorization and deodorization etc. Recovery of the adsorbates by desorption brings a number of other applications. Known examples are the recovery of solvents in solvent recovery installations and the recovery of gold in mines. Activated carbon is not only applied as an adsorbent. It is also used as a catalyst for a limited number of reactions and as a catalyst carrier. As catalyst carrier activated carbon is among others applied as support for precious metals. These catalysts are mainly applied in the fine chemical and in the pharmaceutical industry. Nevertheless a limited number of precious metal catalyst applications for the bulk chemical industry is known. Not only precious metal catalysts based on activated carbons are produced. A limited number of other catalytic active compounds are supported by activated carbon.

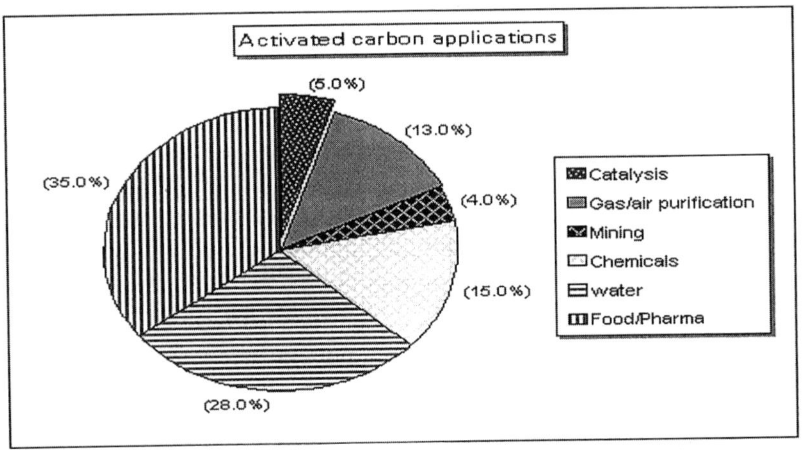


Figure 1. Application of activated carbon (NORIT).

In Figure 1 the importance of activated carbon in catalysis is compared to other applications for NORIT, one of the largest activated carbon producers in the world.

In spite of the fact that activated carbon as a catalyst carrier is not a high volume application it is one of the most interesting applications because more specialized products are required in this market. New developments in the field of catalysis are expected for the future. These developments offer also possibilities for new applications with activated carbons. Anticipating a large number of studies concerning activated carbon as catalyst and as carrier are carried out and published by university groups all over the world.

This review however deals in the first place with industrial relevant applications.

## 2. ACTIVATED CARBON

Activated carbons are a carbonaceous material of which the structure and the properties depending on it are more or less similar to graphite, although the structure is less perfectly ordered. On the basis of X-ray analysis Riley [1] proposed two types of structure. The first type of structure consists of elementary crystallites. The structure differs from graphite in that the parallel planes are not perfectly oriented with respect to their common perpendicular axis. The angular displacement of one layer with respect to another is random and the layer overlap each other irregularly. Hetero atoms as oxygen, hydrogen, nitrogen etc are bound to the carbon atoms, which form the edges and the corners of these elementary graphitic crystallites [2]. Riley [1] describes the second type of structure as a disordered, cross linked space lattice of the carbon hexagons. The carbon atoms in the disordered structure are more reactive. These energetically richer carbon atoms saturate their residual valencies forming bonds with neighbouring elementary crystallites and they tend to form bonds with oxygen, hydrogen and other elements [2]. They bind products of thermal decomposition during the process of carbonisation and they form centres where impurities concentrate by diffusion, when the carbonaceous material is subjected to high temperatures. Activated carbons are dominating micro porous. Micro and meso pores are present caused by selective oxidation of the graphite layers in the elementary crystallites, forming slit-like and V-shaped pores. Meso pores can also be present in the disordered structure between the elementary crystallites.

Activated carbons available in the market differ in pore distribution, in form and in chemical composition. The differences between activated carbon types are a consequence of the choice of activation process, the choice of activation conditions and to some extend of the choice of the raw material. Activated carbons are produced using raw materials as peat, wood, lignite, anthracite, fruit, kernels and shells. The raw materials are converted in activated carbon by steam activation or by chemical activation.

With steam activation the raw material is carbonised and/or oxidized depending on the degree of carbonisation. Activation takes place above 900°C with steam. Process variations as residence time in the kiln, the activation temperature, the type of kiln and other conditions, allow carbonised materials to develop small micro pores which are enlarged up to large micro pores or small meso pores [2]. Low activated carbons have a large volume of small micro pores, High activated carbons have more larger micro pores and small meso pores. By longer activation in a rotary kiln or by activation in a multiple head furnace a heterogeneous product is always formed existing of more and less activated particles. Thus an activation degree is always the mean value of the activation degrees of the independent particles. After activation, during cooling and by exposing to air at ambient temperature oxygen reacts with still unsaturated carbon atoms, forming various functional groups, which make the activated carbon shows basic properties.

With chemical activation an activating agent, normally phosphoric acid, is mixed with young carbonaceous vegetable material, which is carbonised at about 500°C. Carbonisation is followed by recovery of the activation agent by washing. In this method the phosphoric acid is not volatilised to any degree at the carbonisation temperature, it alters the mechanisms of charring resulting in higher yields and supports the evolving carbon matrix during carbonisation and coincident volatile removal. The resulting product contains a well developed meso porosity which is attributed to the voids separating the carbonised micro fibrils of the cellulose which make the majority of the plant tissue [3]. Chemical activated carbons are less graphitised than steam activated products. Chemical activated carbons possess by virtue of the production process a much larger number of varying functional groups, which give the carbon an acid character. Chemical activated carbons are more homogeneous activated than steam activated carbons.

For steam activated carbons as well as chemical activated carbons the kind and the number of the functional groups can be changed by oxidation, by reduction with different reactants or by heat treatment in inert atmosphere under various conditions.

Besides functional groups containing oxygen, hydrogen, nitrogen etc activated carbons contain ash compounds. The ash content is determined by the raw materials and can be diminished by washing. Most times washing of steam activated carbons is carried out with HCl. In that case small amounts of chloride can stay behind. Washing with hydrogen nitrate is more complicated but also possible. Chemical activated carbons contain by virtue of the production process less mineral components.

Activated carbons are produced in different forms:

- powders
- granular carbons
- extrudated carbons.

All three forms are used for activated carbons as catalyst and catalyst carrier. By special milling techniques and a separation step the particle size distribution of powders can be adjusted. Also extrudates and granular carbons are produced in various sizes.

## 3. THE PRODUCTION OF CATALYSTS WITH ACTIVATED CARBON AS SUPPORT

In industrial applications activated carbons are used as supports for precious metal and metal oxide catalysts. These catalysts can be prepared by adsorption from solution, impregnation, precipitation and other techniques as gas phase deposition [4]. The most common way of preparation is adsorption and impregnation by bringing the activated carbon in contact with a solution of the desired metal compound or with a solution of a metal precursor, in most times a complex salt. Then the impregnated activated carbon is dried. Reduction is carried out when the metal precursor has to be transformed in the metal. Important quality criteria are:
- the distribution of the catalytic active metal(compound) throughout the activated carbon particle
- the particle size distribution of the catalytic active metal (compound)
- the metal content.

The distribution of the active phase through the activated carbon particles and the dimensions of metal(compounds) are determined by activated carbon properties in relation to the chosen preparation method. Important activated carbon quality criteria are:
- surface area and pore size distribution
- wettability and hydrophobic/hydrophilic character
- the charge of the external and internal carbon surface
- the number of anchoring sites
- reducing properties
- activated carbon particle size
- the purity of the carrier
- homogeneity within a lot
- reproducibility of lots.

Important choices for the production of the catalyst are:
- impregnation method (excess of solvent, incipient wetness or dry impregnation )
- the concentration of the metal compound or precursor in the solution
- the charge of the metal compound or precursor in the solution
- the polarity of the solution.

For the preparation of catalysts the activated carbon must have pores which are large enough to deposit metal particles, which can be a few nm or more in size. So the pores in the activated carbon must be large enough that the active phase can deposit and small enough that a large internal surface for deposition is available. For activated carbons as catalyst carrier, mostly high activated carbons with many large micro pores and a well developed meso pores structure are applied. To prevent transport limitations during the preparation of the catalyst and during reactions also macro pores are desirable especially in extrudates and granular carbons.

Boehm [5] shows that activated carbons contain a large number of various functional groups. Some of the groups are neutral other groups contribute to the basic character and still others to the acid character of the activated carbon. In Table 1 a number of the various groups is given.

Table 1
Examples of functional groups contributing to the acid/basic character of activated carbon

| | |
|---|---|
| Carboxyl, lactone, phenol, carbonyl | Acid |
| Pyrone, ether, chromene | Basic |

The easiest way to characterize the oxygen containing functional groups is to measure the amount of CO and $CO_2$ which is released as a function of the temperature in inert atmosphere.

The number and the kind of functional groups determine the wettability and the hydrophobic/hydrophilic character of activated carbons. Steam activated carbons are hydrophobic caused by their graphitic structure. However drying experiments carried out by L.M. Knijff [6] show that only the surface of the micro pores and part of the meso pores is hydrophobic. The macro pore surface and a part of the meso pore surfaces is more hydrophilic. Chemical activated carbons contain by their way of production much more oxygen groups and are more hydrophilic. Rodriguez-Reinoso [4] shows that $CO_2$ groups are related with the wettability and the hydrophilic character. Wettability and hydrophilic/hydrophobic character of the activated carbon in relation to the polarity of the solution determine among others how fast the solution with the metal compounds or precursors penetrate in the carbon particles. Apolar solutions of metal compounds or precursors will penetrate much faster in a hydrophobic activated carbon than polar solutions of the metal compounds or precursors.

Depending of the solvent-support interaction more or less competition with the metal compound or precursor takes place. In a relatively strong adsorbing solution the metal precursor will penetrate deeper in the particles. With a relatively moderate adsorbing solution strong adsorption of the precursor at the outside of the particles is possible. By varying the solution-carrier interaction or by the addition of strong adsorbing additives to the solution the distribution of the active phase can be influenced.

The penetration rate as well as the solvent-support interaction determine if so called egg-shell or homogeneous divided catalysts are formed. Depending of the application homogeneous or egg-shell catalysts are applied.

To adsorb metal compounds or precursors on the surface of activated carbon the charge of the activated carbon must be opposite to the charge of the metal compound or precursor. The charge of the activated carbon surface will be determined by the pH of the solution of the precursor. By measuring the zero point of charge (ZPC) the charge of the activated carbon as function of the pH can be determined. By measuring the iso electric point (IEP) the charge at the external surface is determined. The charge of a surface is the sum of positive and negative charged groups. So a positive charged surface means that more positive charged groups are present than negative charged groups.

The functional groups doesn't only determine the carbon charge. The groups act also as anchoring sites for the metal oxides and the metal precursors. Rodriguez-Reinoso [4] shows that the CO groups are related with the number of exchanging anchoring places. Not only the functional groups but also the graphite layers with their electron donating qualities contribute to the anchoring of metal compounds or precursors [7]. For creating a fine dispersed precious metal catalyst the growing rate of the metal particles must be low in relation to the anchoring rate.

Anchoring is important to keep the metal divided during reduction. Hoogenraad [8] shows that carbon fibrils containing by nature much less oxygen groups than activated carbons have to be oxidised to create anchoring sites for the desired dispersion. In activated carbons these groups are already present.

After impregnation with a (precious) metal precursor the catalysts have to be reduced. The graphite layers have reducing qualities. The precious metal precursor can (partly) be reduced by the activated carbon itself [7].

Activated carbons are relatively inert. However catalysts with a chemical activated carbon as carrier can be bi-functional. The carrier can be used as acid co-catalyst [9].

Next to the solid-solute interaction also the particle size is important for achieving a good distribution of the metal through the carbon particles. The smaller the particle the more homogeneous the metal division. The particle size is also important for the reaction rate during the application of the catalysts. Smaller particles give higher activity and as a rule less undesired side reactions.

To prevent side reactions by impurities and to prevent catalyst poising activated carbons applied as catalyst carrier have to be pure. Especially transition metals as iron can give these side reactions. But also small contents of sulphur, silicium and chloride can be undesired. Next to the above mentioned metals, functional groups can give undesired side reactions.

Next to chemical and physical characteristics of the activated carbons also hardness, attrition, filterability and pressure drop are important. To prevent metal losses the activated carbon must be hard and give a low attrition. This is particularly important for precious metal catalysts. However a higher degree of activation gives a lower hardness and more attrition. Thus a compromise can be necessary. Also oxidation, so a better wettability for water and more anchoring places lowers the hardness and increase the attrition. A fast filterability is important to reduce the duration of production. A small particle size distribution of relative large particles give a better filterability than a broad particle size distribution of smaller particles. Larger particles can influence the metal distribution on a negative way and this can be to the expense of activity and selectivity of reactions. So also here a compromise is necessary.

## 4. ACTIVATED CARBON PROFILE AND CHARACTERIZATION

As catalysts and catalyst carriers mostly high activated carbons are applied as mentioned in § 3. Micro, meso pores distributions and internal surface are characterized by gas phase isotherms. Mercury porometry can be used for measuring macro (and apart of the meso-)pores inside extrudates and granular carbons. With the help of the Kelvin equation the pores distribution can be calculated. In Figure 2 a range of steam and chemical activated carbon powders as produced by NORIT and their pores distribution is given. In Figure 3 a range of steam and

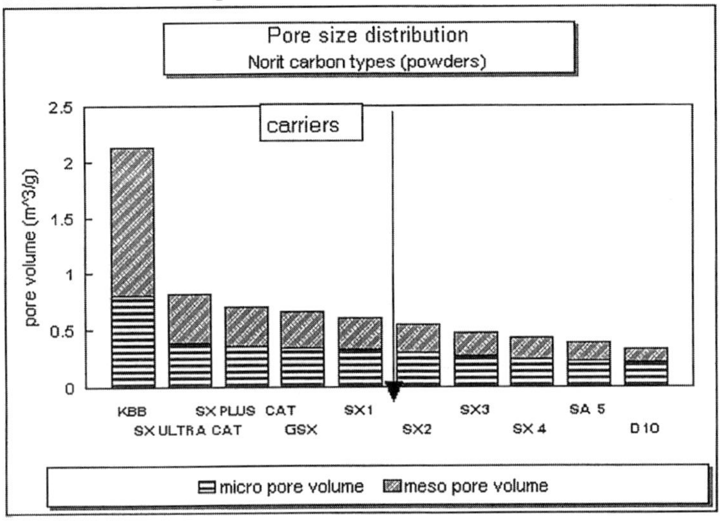


Figure 2. Pore size distribution of activated carbon powders applied as catalyst carriers.

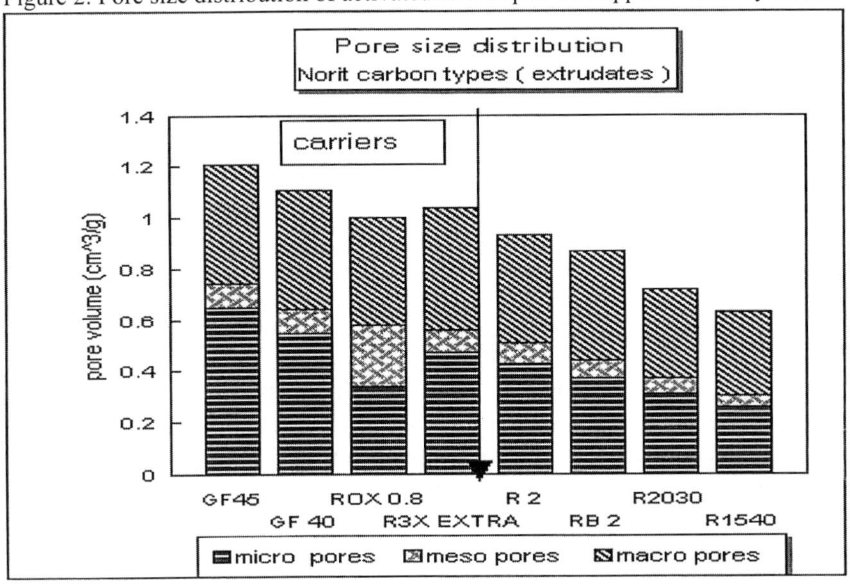


Figure 3. Pore size distribution of activated carbon extrudates applied as catalyst and catalyst carrier.

chemical activated extrudates is given. In both graphs the activated carbon types used as catalyst and catalyst carrier have been indicated. For catalysts and catalyst carriers the activation degree is often given as BET surface. However for activated carbon this adsorption theory is not applicable.

The CO and $CO_2$ contents for a steam and a chemical activated carbon is given in Table 2. The data in Table 2 show that chemical activated carbons contain more CO as well as $CO_2$ groups. The $CO_2$ groups give the chemical activated DARCO KBB carbon a more hydrophilic character. The DARCO KBB has also more anchoring places. By oxidation CO as well as $CO_2$ groups are formed. Oxidation can influence the $CO/CO_2$ ratio. New methods have been developed to measure the hydrophilic/hydrophobic character. Hoogenraad [8] measures the division of powdered activated particles in a two phase water-nitro toluene solution. Van Zon [10] measures the adhesion in bubble pick-up experiments.

Table 2
The chemical characterization of a steam activated and a chemical activated carbon

| | NORIT SX 1G<br>Steam activated washed powder | DARCO KBB<br>Chemical activated powder |
|---|---|---|
| CO (%) | 0.92 | 6.43 |
| $CO_2$ (%) | 0.31 | 2.40 |
| $CO/CO_2$ | 3 | 2.7 |
| $CO + CO_2$ | 1.23 | 8.83 |
| ZPC | 9.9 | 4.7 |

The charge of an activated carbon can be determined by measuring the ZPC with the help of a mass titration curve. In a mass titration curve the pH is measured as a function of the carbon addition to an acid and a basic solution. The results are given in Figure 4. The results in Figure 4 show the more basic character of a washed steam activated carbon and the more acid character of a chemical activated carbon. By oxidizing a steam activated carbon the ZPC moves away in the direction of a chemical activated carbon.

The particle size of powders can be measured with laser diffraction. In Table 3 the particle size distribution is given for 2 steam activated carbons and a chemical activated carbon. The steam activated carbons differ in particle size distribution. The SX 1G contains larger particles and a smaller particle size distribution than the SX 1 caused by the absence of very small and very large particles. The DARCO KBB contains many small particles. The particle size distribution determines the filterability. Filterability can be measured in tests where in the filtration time at constant pressure is measured in a slurry of activated carbon and water. The results are also given in Table 3. The results in Table 3 show the much better filterability of the NORIT SX 1 G type. However producers of precious metal catalysts use also the NORIT SX 1 because the presence of the small particles which contribute substantially to the performance of the catalyst.

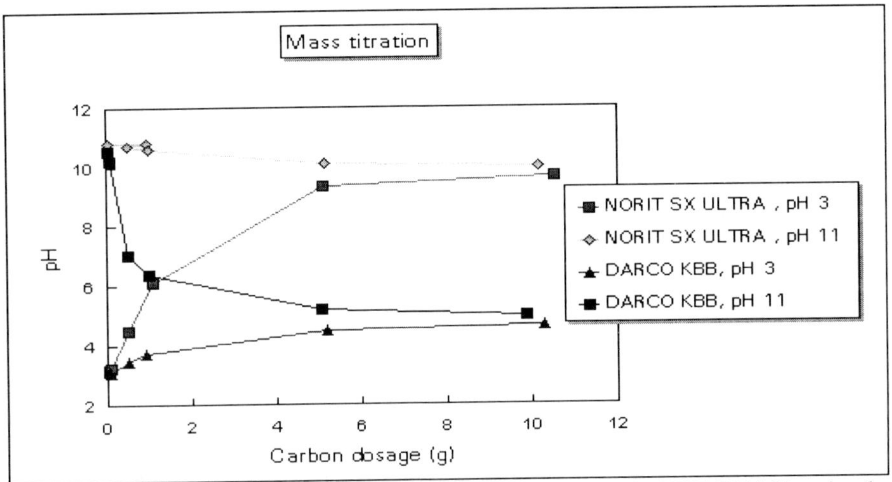

Figure 4. The mass titration curve of a steam activated carbon (NORIT SX 1G) and a chemical activated carbon (DARCO KBB).

Table 3
Particle size distribution and filtration time for various activated carbon powders

| | NORIT SX 1 G | NORIT SX 1 | DARCO KBB |
|---|---|---|---|
| Particle size ($D_{10}$) (μm) | 9.4 | 5.2 | 4.6 |
| Particle size ($D_{50}$) (μm) | 37.2 | 29.1 | 33.9 |
| Particle size ($D_{90}$) (μm) | 97.3 | 111.0 | 91 |
| Filtration time (min) | 5.46 | 27.21 | 46.50 |

Hardness and attrition of particles and extrudates can be measured according to the ASTM hardness test or by measuring of the abrasion index. For activated carbons as catalyst carrier bulk and particle crushing strength tests have been developed. In Table 4 the hardness, the abrasion index and crushing strength are given. Extrudated carbons are extremely hard.

Abrasion and coagulation of powdered carbons can be measured by shaking activated carbons according to a standard procedure and measuring the changes in particle size distribution. Chemical activated carbons are relatively soft.

Table 4
Crushing strength, hardness and abrasion index for various activated carbon extrudates/granular

| | NORIT R 3 EXTRA steam activated 3 mm extrudate | NORIT GF 45 chemical activated 2.2 mm extrudate | NORIT ROX 0.8 steam activated 0.8 mm extrudate |
|---|---|---|---|
| particle crushing strength (lb/1/8") | 16.8 | 3.2 | 3.8 |
| STD-crushing strength | 5.7 | 1.1 | 1.4 |
| hardness ( w/w% ) | 99.9 | 89 | 97 |
| abrasion index (mg/minute ) | 3.6 | 43 | |

As a rule activated carbons applied as catalyst carriers have to be very pure. Chemical activated carbons are pure because of the choice of raw material and the production process. Steam activated carbons are mostly washed. In Table 5 the ash content and the Fe, Ca and Mg contents of the various carbon types are given. To show the influence of washing also an unwashed steam activated carbon with the same activation degree is given. The results show that washing decreases the mineral content.

Tabel 5
The ash composition of various types of activated carbon

| | NORIT SX 1 G Steam activated HCl washed | NORIT SA 1 Steam activated unwashed NORIT SX 1G | DARCO KBB Chemical activated |
|---|---|---|---|
| Ash content (w/w%) | 6 | 8.4 | 2.5 |
| Acid soluble ash (w/w%) | 0.5 | 5 | 1.7 |
| Calcium (acid exr.) (mg/Kg) | 150 | 8000 | 25 |
| Magnesium (acid exr.) (mg/Kg) | 100 | 8000 | <10 |
| Iron (acid exr.) (mg/Kg) | 100 | 3000 | 100 |
| Chloride (acid exr.) (w/w%)) | 0.2 | 0.01-0.1 | <0.005 |

# 5. ACTIVATED CARBONS COMPARED TO OTHER CARRIERS AND CATALYSTS

Activated carbons have a number of advantages and disadvantages compared to other carriers. Some of the characteristics are an advantage in one application and a disadvantage in another application.

Inertness. Activated carbon is inert. The interaction between carrier and active phase (most times noble metals) is small. Thus the qualities of the metals are less influenced by the activated carbon as carrier than by other carriers. Because of the inertness activated carbon hardly shows activity as acid catalyst and does not decompose hydrocarbons resulting in carbon deposition on the metals and causing poisoning of the catalyst. Steam activated carbons are more inert than chemical activated carbons .

Stability In contrast to oxide carriers activated carbons are resistant against attacking in acid and basic solutions. Because of this activated carbons are often applied in reactions in aggressive liquids. Activated carbons can be applied at high temperature in an inert atmosphere because they do not deform even at high temperatures. Steam activated carbon remains stable up to far above the activation temperature of about 900°C. Alumina goes through changes of crystallisation form starting with $\gamma$ alumina at about 600°C up to $\gamma$ alumina at 1200°C. Also silica's change at higher temperatures because of decomposition of the hydroxyl groups.

Surface area Activated carbons contain a relatively large internal surface for deposition of catalytic active compounds compared to oxide carriers.

Hardness and attrition Steam activated extrudated carbons in particular are very hard also compared to oxide carriers. Granular chemical activated carbons are much less hard.

Ignition Activated carbons ignite at lower temperature than oxide carriers. Activated carbons show an ignition temperature of about 200°C (chemical activated carbons) up to 500°C (steam activated carbons). This characteristics is used advantageously by burning the carrier to reclaim the precious metal.

Hydrophobicity Activated carbon is hydrophobic compared to oxide carriers. A hydrophobic character can be profitable in reactions in which water disturbs the activity and selectivity. Hydrophobicity can also be an advantage for the production of the catalysts.

Resistance Activated carbons show a faster heat transport and a better electric conductivity than the oxide carriers. Through this the risk for hot spots is lower.

Cost Activated carbon is compared to other carriers relatively cheap.

# 6. PRECIOUS METAL CATALYST

## 6.1. The production of precious metal catalysts

Precious metal catalysts are prepared by wet impregnation or precipitation with a solution of the metal compound or precursor. Salts used are $H_2\,PtCl_6$, precious metal amines, precious metal chlorides as $PdCl_2$, $CuCl_3$, $RhCl_3$, $IrCl_4$, alkali metal chloropalladites and similarly salts of Ru, Rh, Ir and Os, precious metal nitrates etc. To produce a Pt catalyst for instance the compounds or precursors used are $H_2PtCl_6$ or $Cl_2Pt(NH_3)_4$. A solution of $H_2PtCl_6$ in water gives negative charged $PtCl_6^{2-}$ ions. A solution of $H_2Pt(NH_3)_4$ in water gives positive charged $Pt(NH_3)_4^{2+}$.

## 6.2. Applications of precious metal catalysts [11]

Precious metal catalysts with activated carbons as carrier are applied in the fine chemical industry and at the production of pharmaceuticals. This means that the catalysts are applied in a large variety of different processes. The reactions in the fine chemical industry have the following common characteristics [12].

- The reactions are carried out in liquids.
- The reactions are carried out in 3 (4) phase systems. Gas as reactant/product, liquid as reactant/product or solvent and catalysts as the solid phase. The liquid phase can consist of two not mixing liquids.
- During the reaction water can be formed not mixing with the reactants and/or the solvent.

The most important type of reactions are [11]:

- Hydrogenation
  - Pd (Rh) is used for the hydrogenation of acetylenic compounds to olefinic compound.
  - Ru is used for the hydrogenation of aliphatic aldehydes and ketones
  - Pd is used for the hydrogenation of aromatic aldehydes and ketones
  - Re, Os (Pt, Ru) are used for the hydrogenation of unsaturated aldehydes to unsaturated alcohols
  - Pt, Pd,Ru en Rh are used for the hydrogenation of the aromatic nucleus
  - Pd, Pt and Rh are used for the hydrogenation of nitriles, resulting in primary, secondary and tertiary amines
- Hydrogenolyses
  - Re and Ru are used for the hydrogenolyses of carboxylic acids to alcohols
  - Pd and Pt for the hydrogenolyses of aromatic nitro groups (Pd gives aniline, Pt gives hydroxylamine)
  - Pt is used for the hydrogenolyse of nitro haloaromatics to the corresponding halomine
  - Pd is used for the hydrogenolyse of aromatic alcohols to aromatic hydrocarbons
- Dehalogenation
  - Pd is used for the selective dehalogenation of a dihalide to a monohalide
- Dehydrogenation
  - Pd is used for the dehydrogenation of alicyclic hydrocarbons to fully or partically aromatic hydrocarbons
- Coupling reactions

  Pt on an acid activated carbon can be used for the coupling reaction of nitrobenzene to different products.

# 7. OTHER APPLICATIONS OF ACTIVATED CARBON AS SUPPORT

## 7.1. The production of vinyl chloride monomer

One of the oldest applications of activated carbon as catalyst carrier is at the production of a $HgCl_2$ supported catalyst. The $HgCl_2$ content is about 10 %. This catalyst is used for the production of vinyl chloride monomer from HCl (in the vapour phase) and acetylene

$$C_2H_2 + 2\ HCl \rightarrow C_2H_2Cl_2 + H_2$$

The process is carried out at high pressure and at temperatures between 150 and 180°C.The reaction is exothermic, so heat control in the production process is important.

Vinylchloride monomer is used as intermediate for the production of polymers in a large number of production processes [13].

Activated carbons used in this application are high steam activated and acid washed extrudates like the NORIT RX 3 EXTRA.

## 7.2. The production of vinyl acetate

A zinc acetate supported catalyst is used for the production of vinylacetate from acetylene and acetic acid [14]

$$C_2H_2 + CH_3CO_2H \rightarrow CH_2CHOCOCH_3$$

The process is carried out in the gas phase at a temperatures above 160°C. The reaction is exothermic, so heat control in the production process is important.

Vinylacetate is used as intermediate in a large number of production processes.

The catalysts can be produced by impregnation of an activated carbon with zinc acetate, followed by drying. The zinc concentration is in the order of 11 up to 13%. The optimal activated carbon support is a high steam activated and acid washed 3 or 4 mm extrudate like the NORIT RX 3 EXTRA or NORIT RX 4 EXTRA.

Important characteristics of the impregnated carbon are:

- a homogeneous Zn distribution through the individual support particles
- a high residual internal surface
- low contents of other minerals as Fe, Cu and Cl.

## 7.3. Treatment of mercaptan-rich hydrocarbon streams

Mercaptans are converted in disulfides by oxidative dehydrogenation. The reaction is carried out for sweetening hydrocarbons, such as jet fuel gasoline kerosine and gas oil. The reaction is catalysed by a mixture of transitional metals as Cu, Mn, Fe and Co with activated carbon as catalyst carrier. The process is developed by UOP under the trade name MEROX [15-17]

$$2\ RSH + O_2 \rightarrow RSSR + H_2O_2$$

$4\ RSH + O_2 \rightarrow 2\ RSSR + 2\ H_2O$

The process is carried out at a temperature of 40-50°C and at a pressure of 500 KPa. The reaction is exothermic.

By sweetening the offensive mercaptan odour is eliminated, however the sulphur remains in the oil. The activated carbon used in this process must have a granular form with a particle size distribution of 12*30 mesh and an optimal pores distribution for adsorption of methyl mercaptans. The NORIT PKDA 12*30 mesh fulfills these requirements. This carbon is produced in the same way as the powdered carbons shown in Figure 2 but not milled. The activation degree agrees with the SX 2/SX 3 type. The carbon is not washed.

### 7.4. The decomposition of CFC's [18-20]

CFC's is supposed to deplete the ozone layer. This has led to a stepwise phase out of CFC's and other ozone depleting substances. Last years a number of destruction technologies have been proposed. For example CFC reacts with alcohol to yield carbon monoxide, carbon dioxide, methyl chloride and hydrogen fluoride. As catalyst a metal halide supported on activated carbon [18] is applied. In this way CFC-113 decomposes by reacting with methyl alcohol

$$CF_2ClCFCl_2 + 3CH_3OH \rightarrow CO + CO_2 + 3CH_3Cl + 3\ HF$$

The catalyst functions as follows:

$$2FeCl_3 + 6\ CH_3OH = Fe_2O_3 + 3\ CH_3OCH_3 + 6\ HCl$$

$$Fe_2O_3 + 3\ CuCl_2 = 2\ FeCl_3 + 3CuO$$

$$CuO + 2HCl = CuCl_2 + H_2O$$

The main catalyst $FeCl_3$ reacts with the alcohol forming $Fe_2O_3$, which lowers the activity. $CuCl_2$ is used as co-catalyst and regenerates the main catalyst. HF regenerates the co-catalyst. For this application relatively low steam activated extrudates are applied as support. The NORIT R 2 or RB 2 as mentioned in Figure 3 have the desired porosity.

Other new developments are the conversion of waste CFC-12 ($CCl_2F_2$) into high added value product $CH_2F_2$ (HFC-32) [19,20]. Also other CFC's can be converted according to the same reaction mechanism For this conversion a Pd supported catalyst is used. As support a very pure steam activated carbon with a relatively low activation degree is selected. Washed NORIT RB 2 fulfill the requirements.

# 8. ACTIVATED CARBON AS A CATALYST

On industrial scale activated carbon is applied in a limited number of processes as catalysts. The largest applications are:
- the production of cyanurylchloride and sulfurylhalides
- the production of phosgene
- the production of polysulphides
- the production of isopropylamine salt of N-(phosphonomethyl)glycine
- the production of hydrogen sulphate acid
- the decomposition of phosgene
- the decomposition of hydrogen peroxide.

In most of reactions activated carbons act as halogenation catalyst or as oxidation catalyst.

## 8.1. The production of cyanurylchloride

Cyanurylchlorid is used as intermediate for the production of certain pesticides (wheat protection), as optical brighteners and as colouring agent [13], Cyanurylchloride is produced from cyanogenchloride

$$3\ ClCN \rightarrow (ClCN)_3$$

In most processes the starting material is hydrogen chloride. The hydrogen chloride is chlorinated in-situ, forming cyanogen chloride. The reaction temperature is 270-450°C. The process is a 2 phase fixed bed gas phase system. For this process the following activated carbon qualities are important:
- degree of activation. The activated carbon must be a high activated carbon with large micro pores, small meso pores as produced by steam activation [21]. Also a good developed macro porosity is likely to be important [22]
- purity. The activated carbon must be produced from very pure raw material or must be acid washed [23].
- for use the carbon must be dry.

The NORIT RX 3 EXTRA fulfills these requirements.

## 8.2. The production of sulfurylhalides

Sulfurylchloride is used as a chemical for selective chlorination of side chains of aromatics. Sulfurylchloride also sulfonates and dehydrates organic compounds. In the industry it is used as intermediate for the production of organics for colouring agent, for pesticides, for disinfection matters and for pharmaceutical products [13]. Sulfurylchloride is also used at the production of sulfonic acid and for the sulfochlorination of polyethylene. Sulfurylchloride is produced from sulfur dioxide and chloride

$$SO_2 + Cl_2 \rightarrow SO_2Cl_2$$

The reaction is exotherm, thus heat control is important. The process takes place at low temperature < 0°C [24-27]. Both reactants are in the gas phase. The sulfurylchloride product a liquid.

Next to sulfurylchloride also sulfurylfluoride is produced on a small scale. Sulfurylfluoride is also applied as pesticide

$SO_2 + Cl_2 + HF \rightarrow SO_2F_2 + 2\ HCl$

The process is carried out at temperatures above 230°C [28,29]. $SO_2F_2$ is a gas. Thus the process is carried out in the gas phase.

For the production of sulfurylchloride and sulfurylfluoride the activated carbon catalyst has to fulfill to the same requirements as mentioned for the production of cyanurylchloride.

### 8.3. The production and decomposition of phosgene

Phosgene is an intermediate in the chemical industry for the production of di-iso cyanates, polycarbonate plastics, chloroformic esters, carbonate esters, dyes [13] etc. Phosgene is produced from carbon monoxide and chlorine

$CO + Cl_2 \rightarrow COCl_2$

The process is a fixed bed gas phase system. The reaction is carried out at temperatures up to 400°C. The reaction is exotherm, so cooling is important. To prevent emission, waste steams of phosgene originating in the production plant are lead over an activated carbon to decompose [30,31]. For both reactions (production and decompostion) a relatively low steam activated, unwashed carbon is used, in spite of the fact that for the production a higher activated carbon gives a better performance. However the type of activated carbon in this application is not very critical. An important characteristics of the activated carbon is the water content, which must be very low.

### 8.4. Oxidizing $Na_2S$ in polysulfidide for increasing wood pulp yield

Chiyoda Corporation and Mitsubishi Paper Mills [32-34] developed a system for in situ oxidation of sodium sulfide in white liquor to polysulfide increasing the pulp yield in the pulping process.

Sodium sulfide is converted to polysulfide as an intermediate compound, then it is oxidized to thiosulfate

$2Na_2S \rightarrow Na_2S_2 \rightarrow Na_2S_2O_3$

Activated carbon catalyses the conversion of Sodium sulfide in combination with a high selectivity for polysulfide. The result of catalysts screening shows that activated carbons with a high pores volume of pores with over 10 nm in diameter give a high conversion. Also hardness and attrition behaviour of activated carbon is important.

The ROX 0.8, as shown in Figure 3, fulfills these requirements.

### 8.5. The production of the isopropylamine salt of N-(phosphonomethyl)glycine

N-(phosphonomethyl)glycine can be oxidised using oxygen, optionally in the presence of primary or secondary amine [35-37]. The catalyst used in this process is a steam activated carbon. The activation degree is lower than in the most catalytic applications, indicated by the arrow in Figure 2. The activated carbon has to be washed to decrease the mineral substances. The activated carbon acts as a oxidation catalyst. The oxidation activity can be increased by decreasing the number of oxides on the surface. The isopropyl amine salt of N-(phosphonomethyl)glycine is applied as herbicide [13].

### 8.6. The decomposition of hydrogen peroxide

For disinfection hydrogen peroxide is used. To decompose the hydrogen peroxide activated carbon is used as a catalyst. In this application the choice of the activated carbon type is not very critical. Dependent of the fact whether some leaching of mineral is allowed a washed or unwashed type is applied.

### 8.7. The production of hydrogen sulphate

Hydrogen sulphate can be produced from $SO_2$ containing waste streams by oxidation in the presence of water [38].

$$SO_2 + O_2 + H_2O \rightarrow H_2SO_4$$

The reaction is carried out at temperatures of 45 up to 70°C.

## REFERENCES


1. H.L. Riley, Quart. Revs., 1 (1947) 59.
2. T. Wigmans, Fundamentals and practical implications of activated carbon production by partial gasification of carbonaceous materials, NORIT. NATO ASI Ser.,Ser.E (1986) 105 Carbon, Coal Gasification, 559.
3. J.D. MacDowall, P.D.A. McCrae and W.M.T.M. Reimerink, The development of an effective activated carbon for evaporative emission control. Paper presented 25th ISATA Silver Jubilee Meeting, Florence, 1-5th June (1992).
4. J.W. Patric (ed.), Porosity in carbons: Characterization and applications. Edward Arnold, Great Britain, 1995.
5. C. Morterra, A. Zecchina and G. Costa (eds.), Structure and reactivity of surfaces, Elsevier, Amsterdam, The Netherlands, (1989).
6. L.M. Knijff, The application of active components into catalyst support bodies, Dissertation, Univ. Utrecht, (1993).

7. H.E. van Dam, Carbon supported noble metal catalysts in the oxidation of glucose 1-phosphate and related alcohols, Dissertation, Univ. Delft, (1989).
8. M.S. Hoogenraad, Growth and utilization of carbon fibrils, Dissertation, Univ. Utrecht, (1995).
9. D.S. Cameron, S.J. Cooper, I.L. Dodgson, B. Harrison and J.W. Jenkins, Catalysis Today, No. 7 (1990) 113.
10. M. van der Zon, Publication in preparation, IOP catalysis project, The Netherlands.
11. A.B. Stiles, Catalyst supports and supported catalysts, Butterworths, Stoneham, USA (1987).
12. Innovation oriented research programmes (IOP) in Catalysis, The Netherlands, (1994).
13. Ullmann's encyclopedia of industrial chemistry, VCH Verlagsgesellschaft MbH, Weinheim, (1985-1996).
14. T. Kawaguchi, J. Nakagawa and T. Wakasugi, Applied Catalysis, 32 (1987) 23.
15. C.A. Embry, A.W. Tindle and J.F. Wood, Hydrocarbon Processing, February, (1972).
16. T.A. Verachtert, G.F. Asselin and B.E. Staehle, Merox treatment of kerosine, UOP Process Division Technology Conference, September, (1982).
17. K.M. Brown, Hydrocarbon Processing, 52 (1973) 69.
18. Komatsu, Publication in preparation.
19. A. Wiersma, E.J.A.X. van de Sandt, H. van Beckum, M. Makkee and J.A. Moulijn, Difluormethaanbereiding, NL octrooi A9401574 (1994).
20. V.N. Rao, Catalytic hydrogenolysis, US Patent No. 5136113 (1992).
21. H. Huemer and H. Schultz, Verfahren zur Herstellung von Cyanuridchlorid, DE Patent No. 842067 (1949).
22. Y.S. Suryanarayana and L.J. Reid, Process for the production of cyanuric chloride, US Patent No. 3867382 (1975).
23. J. Riethmann, Process for the preparation of cyanuric chloride, US Patent No. 3312697 (1967).
24. R.H. McKee and C.M. Salls, Ind Eng. Chem., 16 (1924) 279.
25. H. Danneel, Z. Angew. Chemie, 39 (1926) 1553.
26. Verfahren zur von Herstellung Sulfurylchlorid aus schwefliger Saure und Chlor in Gegenwart von Kohle, DE Patent No. 364519 (1919/1922).
27. Herstellung von Sulfurylchlorid, DE Patent No. 419521 (1925).

28. R.P. Ruh, R.A. Davis and K.A. llswede, Process for the production of sulphurfluoride, US Patent No. 3092458 (1963).
29. A.G. Bisignani and C. del R.H. Edgecomb, Production of sulphurfluoride, US Patent No. 3320030 (1967).
30. H. Wollthan and W. Groves, Verfahren zur Verbichtung von Phosgen, DE Patent No. 961681 (1957).
31. H. Wollthan, Process for the decomposition of phosgene, US Patent No. 2832670 (1957).
32. Y. Ohguchi and S. Hara, Kami Pa Gikyoshi, 42(1) (1988) 46 (Japanese).
33. M. Suzuki and S. Hara, Method of oxidizing sulfide containing liquor, US Patent No. 4855123 (1989).
34. H. Nishijima and R. Inaba, Corporate Person Petrochemical Society, No.18-6 (1995).
35. C.S. King, Improved oxidation catalyst, US Patent No. 4696772 (1985).
36. A. Hershman and J.W. Gambell, Oxidation of phosphomethylamines, US Patent No. 4072706 (1978).
37. W.J. Berg and H.G. Schwabe , Verfahren zur Herstellung von N-Phosphomethylglycin, DE Patent No. 2519388 (1975).
38. Process development for sulfur dioxide-containing waste gas with sulfuric acid as the end product, Dechema, Monogr., 80 (1976) 1639.
39. E. Peter, Chem. Eng., 76, 13 (1969) 62.

Adsorption and its Applications in Industry and Environmental Protection
Studies in Surface Science and Catalysis, Vol. 120
A. Dabrowski (Editor)



# Adsorption of proteins and its role in modern technology and environmental protection

V.N. Izmailova and G.P.Yampolskaya

Chemical Department, Lomonosov Moscow State University
119899, Moscow, B-234, Leninskie Gory, Russia

## 1. INTRODUCTION

Proteins are natural high molecular mass surface active substances. Interfacial behavior of proteins determines peculiarities of natural and technological processes. Model investigations of the protein adsorption are of great interest for biology. This is true not only because a number of biochemical processes proceeds at interfaces with participation of adsorbed proteins (some of them, been water-soluble proteins, to be activated, must adsorb at an interface), but and following that, that physical principles of protein adsorption discover new approaches to development of insights of the life origin as well as evolutionary pathways. Colloid chemistry of proteins is the source of current knowledges about adsorption processes proceeding at different interfaces such as "mobil" surfaces created by a contact of protein solutions with air or another immiscible liquid and with solid surfaces of various types.

Researchs in this area form the basis for industrial applications of adsorbed proteins as well as for environmental protection from waste-proteins. It is accepted to consider proteins as biodegradable substances, but due to their great affinity to interfaces, proteins rapidly adsorb on them imparting a high stability to colloids. Such stabilized dispersions together with gas bubbles tend to accumulate and coagualate on surfaces of water reservoirs resulting in a drastic disturbance of eco-systems. Partial proteolysis of adsorbed proteins does not reduce essentially the stability of surface structures. In such cases waste-proteins act as pollutants. On the other hand, surface activity of proteins allows to use the stage of their adsorption at any interfaces for waste purification from proteins as well as from fine dispersed contaminations of various origins: solids, organics, ions of heavy metals, etc. Current experience in this area shows significant effiency of adsorption purification technologies and their large operating potential. This offers new opportinities in separation science and environmental science for waste minimizations. It is hard to overestimate the positive role of the protein adsorption for industrial activities. Adsorption of proteins long long ago is used in food productions. New applications of adsorbed proteins push forward successful development of biotechnology, pharmacology and medicine, determining the usefulness of new drugs and the control of drug delivery. Regulation of the preferential adsorption on solid surfaces from mixtures of proteins is main problem of biocompatibility of synthetic materials for medicine purposes. Applications of immobilized / or adsorbed enzymes as a specific catalyst open new routes in moderm applied chemistry. Number of such relevances of the protein adsorption for industrial and environmental protections are considered in this paper.

The formation of adsorption layers of proteins at different interfaces is not clear completely in this time and many questions, such as kinetics, reversibility of adsorption, dependence of adsorption on conformation stability of the protein and others must be elucidated additionally. The question of most importance is to understand what really proceeds in an adsorption layer after an attachment event. This stage usually is marked as “arrangement”. It is not clear also how large the contribution of conformational changes of a protein molecule is in an arrangement and final state of the adsorbed protein and adsorption layer. At present many efforts are directed to find the correlation between surface properties of the protein and its surface behavior. An influence of surface properties of an adsorbent on the surface protein behavior remains so far yet obscure. Most researchers limited their studies by estimations of dependences of the protein adsorption on total hydrophobicity or hydrophilicity of surfaces or on a sign of the surface charge.To predict the surface behavior of proteins one uses the value of their surface hydrophobicity, paying no attention to that, that this value must be determined under specific conditions accounting the nature of the second phase (air, liquid or solid). For example, to characterize the protein adsorption at solid surfaces it is possible to use data obtained by means of hydrophobic chromatography and it becomes incorrect, when such data are used to compare adsorption of proteins at boundaries of their solutions with vapour or with other liquid phase.

It is impossible to say, in which systems the adsorption of proteins does not take place and rather it can suppose that, if an interface between two phases is created in the presence of a protein solution, protein will accumulate on it, resulting in essential modification of the surface, that is the basis of practical applications of this phenomenon in different area of human activities.

For successful discussion of results of investigations of the protein adsorption it is useful to characterize in short properties of protein molecules and protein behavior in aqueous solutions.

Spatial structure of protein molecules governs their essential properties, including surface activity. Native proteins (further, proteins) have the unique structure of molecules, ideal energetically balanced and created due to principle of structures hierarchy with optimal using of hydrogen bonds, dispersion, electrostatic, as well as hydrophobic interactions to stabilize the compact form of polypeptide chains. All structure elements of the protein molecule, including ordered ones (secondary structures), domains and globules, as well as protein structures of higher level are specified by aminoacid sequences of polypeptide chains. Macromolecular chains of protein are much different from those of typical synthetic polymers because they are invariably represented by heteropolymers with nonrepeating sequences providing the optimal folding of macromolecular chains and compact forms of protein molecules.

Proteins carry functional groups dissociating in water, but properties of proteins are not equivalent to typical polyelectrolytes because of the conformation pH - stability of native protein with a part of functional groups located inside a globule remains undissociated. For a number of proteins boundaries of pH and temperature stability of the native conformation are well investigated. Number of proteins in some extent (up to 1-20 tens w.%) are soluble in water resulting in the real molecular solution. However, molecule of a protein is a compact particle of the size that is essentially larger than molecules of water. In case of proteins macroscopic concepts can be extended to the molecular level, that allows to consider area (surface), volume, protein-solvent interface, etc. The mean packing density of interior protein atoms is essentially identical with that found in crystals of small organic molecules, and the interior of protein molecule is not an oil drop but resemble rather molecular crystal. The surface of protein has atomic level of the roughness with coefficient of about $1.7 \pm 0.2$, which is not expected to vary much from one protein to other. Compact molecules of globular proteins have a specific surface topography with asymmetrically located polar and nonpolar atoms groups. These surface features determine amphiphilic character and, consequently, surface activity of native proteins. About half the surface of average protein is nonpolar yet in contact with water. Surface of membrane proteins (water insoluble) is somewhat more nonpolar (of about 70-80% of total surface), than surface of water soluble proteins. These peculiarities of protein surface topography,determining protein surface activity, seem to have functional meaning in biological reactions as well as in the formation of protein structures of higher level (aggregates, dimers and polymers and other condensed forms and crystals). Relative stability of the spatial structure and surface topography of proteins allow to consider native protein as a surface active substance. The macroscopic approach allows to consider protein solutions as thermodynamically stable dispersions and respectively protein molecules of typical sizes (1-10 nm) as colloid particles. In this case protein solutions gain some general properties with micellar solutions of common (low molecular mass) surfactants. Both types of systems are treated often "complex fluids". Of course, unlike micelles of surfactants, proteins behave in solutions as much more hard core particles. All these properties of protein molecules must be accounted discussing the protein adsorption.

## 2. ADSORPTION OF PROTEINS AT THE SURFACE WATER -AIR

The surface active behavior of proteins is frequently employed in medicine and industry, for example, in the development of new biomaterials, drug delivery, and stabilization of foams. Foaming is the technological stage commonly used in the food production. The adsorption of proteins at an interface is accompanied by changes in the physical properties of the interface, particulary, adsorption at the surface water - air (a/w) decreases the surface tension. At equilibrium the surface tension is controlled with gradients of the energy density. The adsorption process on the "mobil" surface, such as water-air or liquid-liquid differs in many aspects from that at water-solid surfaces. Adsorption sites are energetically similar and can additionally arise due to a surface expansion; it is anticipated, that lateral processes in the adsorption layer will be essentially enchanced.

### 2.1. Surface tension isotherms of protein solutions

Proteins form stable monolayers at water-air surface and adsorb from solution at the surface solution-vapour. Monolayers of proteins are well investigated [1-14] and Langmuir-Blodgett technique is wide applied to obtain samples of important practical uses [15]. In [1] two-dimensional isotherms more than 20 proteins and polyaminoacids have been analysed taking into account for theories of liquid-crystal transitions in monolayers [16-21] and conclusion was made, that all peculiarities of these isotherms can be explained by orientational ordering of different elements of protein molecules in the plane of monolayer. α-Helices, asymmetric structures of higher levels and native globules of a protein can be as the element of ordering. β-Structure also can play the role of the ordering element, but monolayers of β-structured proteins are poorly investigated. Orientation phenomena, evidently, in some extent take place within adsorption layers of proteins in the buildup process and under shear conditions,especially, for mobil surfaces.

Direct quantitative measurements of the protein adsorption at the surface a/w are very difficult to be performed experimentally. Decreases in the surface tension of protein solutions testify adsorption of the protein at the a/w surface. Possibility to evaluate the protein adsorption using the Gibbs adsorption equation is doubtful for several reasons. First, the chemical potential (the state) of adsorbed proteins is poorly defined, relative to their solvated native states. Second, the Gibbs equation is valid only for reversible, equilibrium adsorptive processes, while proteins, as it frequently regarded, adsorb at the a/w surface in an irreversible manner. Third, the Gibbs equation is valid only for the adsorption event with attachment of an adsorbat molecule at the surface and change in its orientation with an increase in the surface concentration and is not applicable for any structure formation with participation of adsorbed molecules. Adsorption of proteins at the a/w surface is accompanied with a jumping in rheological behavior of adsorption layer, that often can be considered as two-dimension phase transition [22]. The problem of applicability of the Gibbs equation for proteins as surface active substance was considered many times [23-28]. In all cases dynamic and equilibrium surface tension of aqueous protein solutions are not directly proportional to the adsorption value, at all possible values of adsorption. However, conditions when the adsorption rate is equal to the rate of diffusion mass transfer of the protein towards the solution-vapour surface, i.e. absence of any adsorption barrier or simultaneous processes of an adsorption layer arrangement, allow to produce some adsorption characteristics using the Gibbs equation. Dependences of surface tension on time are studied systematically and they reflect the relaxation character of the surface layer buildup [22]. Minimal values of the surface tension are attained during several hours. However, when surface tension is changed during a longer time, side processes, such as assosiation, formation of structures of higher levels, conformational shifts, hydrolysis of peptide bonds and others, can take place and have to be carefully studied, with accounting for a possibility of their proceeding in the protein solution as well as in the protein adsorption layer. In general, kinetics of surface tension decreases can be presented by equation:

$$(\sigma_t - \sigma_\infty) = (\sigma_0 - \sigma_\infty)\exp(-t/\tau) \quad (1)$$

where $\sigma_0$, $\sigma_t$ are the initial surface tension and the same value at the time t, respectively; $\sigma_\infty$ is the minimal surface tension, achieved for protein solution of a given concentration, $\tau$ is the effective relaxation time, that can be represented as a spectrum of magnitudes in accordance with the time scaling of each kinetic stage contributing in the surface layer formation. Equilibrium or quasi-equilibrium value of the surface tension is measured by static or quasi-static methods, which allow usually to record changes in the surface tension during time interval from a fraction of seconds to several hours and obtain characreristics of slow processes in the surface layer formation. Note, that time scale of chemical reactions consists of $1\text{-}10^{-11}$ s. Sometimes [23] kinetics the surface tension decrease is described by equation:

$$\sigma_t = \sigma_\infty + \frac{\sigma_0 - \sigma_\infty}{1 + \left(t/t^*\right)^m} \quad (2)$$

where t* is the time constant, which is equal to the half time of attaining the initial meso-equilibrium surface tension and m is a dimensionless exponent. Kinetic parameter, the rate of surface tension decay at t*, can be computed from formula:

$$d\sigma/dt = (\sigma_0 - \sigma_\infty)/2t^* \quad (3)$$

Diffusional contributions in protein adsorption are described by equation of non-steady diffusion:

$$\Gamma \cong 2C_0\sqrt{D_0 t/\pi} \quad (4)$$

where $\Gamma$ is the value of adsorption, $C_0$ is initial concentration of the protein solution, $D_0$ is diffusion coefficient of the protein. At very low concentrations the equation (5) is valid:

$$\Delta\sigma \approx 2C_0RT\sqrt{D_0t/\pi} \tag{5}$$

Thus, kinetics of serum albumin adsorption, according to data given in Table 1, is determined by diffusion mass transfer to the surface. Times of the decrease in the surface tension to the value of $\Delta\sigma = 0.1$ mN/m at different concentrations of serum albumin were measured experimentally and calculated using equation (5).

Table 1
Diffusion contribution in the bovine serum albumin adsorption [22]

| $C_0 \cdot 10^8$, M | t, s | | $C_0 \cdot 10^7$, M | t, s | |
|---|---|---|---|---|---|
| | experiment | eq. 5 | | experiment | eq. 5 |
| 1.4 | 45 ± 5 | 107 | 1.4 | 55 | 64 |
| 2.8 | 19 ± 1 | 27 | 2.8 | 15±2 | 16 |
| 4.2 | 9 ± 0.2 | 11.9 | 4.2 | 6±1 | 7 |

Deviations from equation (5) become pronounced at larger decrease in the surface tension, for example, the time of achievment the value $\Delta\sigma$=2 mN/m measured experimentally is equal to 120 - 135 s at protein concentration within 2.8 - $3.2 \cdot 10^{-7}$ M and calculated values are one order of magnitude lower.Increase in the adsorption time shows the barrier of adsorption. Reasons of this barrier is poorly so far understood. It can be supposed, in general, that including of a new molecule in the surface layer requirs the creation of the additional free surface. Causes of adsorption barrier are of interest to be considered on the basis of results revealing a long time kinetics of adsorption.

In the work [23] the pendant drop technique was used to measure the dynamic surface tension of eight globular proteins (hen egg lysozyme, bovine ribonuclease A, equine myoglobine, tuna heart cytochrome C, bovine etythrocyte superoxide dismutase, bovine milk β-casein, recombinate human growth hormone and bovine serum albumin) at air/water interface as a function of protein bulk concentration (0.01, 0.1 and 1 mg/ml). All eight proteins achieved mesoequilibrium surface tension within 15 hours at the highest concentration, its value does not vary much among the eight proteins, with an average value of $47 \pm 6$ mN/m. Measurements were performed at pH 7.0 using phosphate buffer for the solution preparing. Ultimate decrease in surface tension values is lower for aqueous protein solutions with no added electrolytes. Data of dynamic surface tension were numerically modeled using Hua-Rosen equation (2) and analyzed with respect to protein surface hydrophobicity, conformational stability, bulk depletion effects, and the apparent vs theoretical diffusion-limited rate of adsorption. Authors conclude, that conformational stability (hardness) and surface hydrophobicity have a large influence on the adsorption rate of globular proteins at the a/w surface, with hardeness is more important variable. However, discussion the proteins adsorption at a/w surface requires another characteristic parameters to comparing surface tension isotherms for different proteins. For example, the data on hydrophobicity of the protein surface obtained using hydrophobic chromatography are not ideally applicable for the surface a/w and rather available for events of the protein adsorption at water-solid surfaces. To characterize the behavior of proteins at a/w surface it is useful to determine the surface activity (G) for each protein:

$$G = -(d\sigma/dC)_{C\to 0} \tag{6}$$

It is true, that the determination of the surface tension value is hard to be performed at very low protein concentrations. G is rigorously thermodynamic magnitude, that has been induced by Rehbinder for surface active substances of any nature [29,30]. For practical uses, when adsorption at a/w surface has to be compared for a series of proteins, the value $[(d\sigma/dt)_{init}/C]$ can be recommended as a measure of hydrophobicity of the protein surface, which better characterizes the protein affinity to the a/w surface. As for adsorption kinetics, it is strongly depended on the initial protein concentration. Diffusion-induced adsorption is significant only at initial concentrations at/or below 0.01 mg/ml, that coincides with data of Table 1. At high concentration diffusion mass-tranfer rates do not appear to be limiting factor of the adsorption rate, with exception of β-casein. As for other seven proteins, kinetics of the surface tension decrease is not diffusion-limited, but is much slower, due to an energy or probability barrier to adsorption. This barrier can be connected, in according to our data [1,22] with the slow process of the surface orientation and lateral association of native globules of the protein resulting in an ordering of adsorption layer and slow appearance of additional surface for the adsorption of new protein molecules.

Determinations of the adsorption are facilitated by enhancing of the surface area following the higher sensitivity of methods to measure the equilibrium protein concentration due to the larger depletion of the solution. In the case of adsorption at a/w surface this task can be solved by foaming of protein solutions with the determination of the solution depletion after the complete foam drainage. This method have obvious limitations and can not be used in a wide region of protein concentrations, electrolyte additives, and pH of aqueous phase, because foams have to be stable to be dried and allow to performe a positive dispersion analysis. Such approach was used for studying α-chymotrypsin adsorption at the a/w surface. Measurements were successeful in the range of protein concentration $1\cdot10^{-6}$ - $1.6\cdot10^{-5}$ M. At low concentration the value of adsorption is equal to $6\cdot10^{-8}$ mol/m$^2$ and area occupied by one molecule of the protein is ≈ 28 nm$^2$, that respects to the tightly packed monolayer with the native-like protein, adsorbed side-on. The molucule of this protein is an spheroid with half-axises 1.6×1.6×3.2 nm. At higher concentrations of the protein recorded values of adsorption allow to assume the polymolecular character of adsorption. Analogeous conclusion was made, when adsorption of the protein was measured at the plane a/w surface by means of radioactive indicators method, using the tritium-labeled protein. Isotherms of the α-chymotrypsin adsorption Γ(C) at the plane a/w surface were obtained in a wide interval of the protein concentrations. Adsorption is formally described by modified Langmuir equation:

$$\Gamma = \frac{l\Gamma_{max}KC}{1+KC} \qquad (7)$$

where $\Gamma_{max}$ is ultimate adsorption with the ultimate surface packing by non-changed molecules of the protein, l is the number of such layers in vicinity of the surface. Experimental values of adsorption respecting to the adsorption curve calculated using equation (7) using the value of K equal to $5\cdot10^5$ l/mol. Depending on the type of adopted orientation of globules of the protein in the plane of surface, (adsorbed as side-on or end-on) l is varied for α-chymotrypsin adsorption from 1.5 to 3.0.

In [31] kinetics of the surface tension decrease was described using the model accounting for diffusion-controlled adsorption of protein molecule and for conformational changes of adsorbed molecule. The model corresponds to one proposed by Serrien [32] and describes diffusion toward a/w surface and subsequent reorientation and other changes in adsorption layer, which usually one gives a sence of conformational changes the adsorbed protein. The model yields the diffusion relaxation time (τ) and ($k_c$) - the rate constant of conformational changes.

$$\sigma(t) = \sigma_\infty + \left\{\alpha \exp\left(-\sqrt{4t/\pi\tau}\right) + \beta\right\}\exp(-k_c t) \qquad (8)$$

In this equation (8) the term with α determines the diffusion transfer of proteins and with β corresponds to an arrangement process. Independently on the physical sence of β the character of dependences α + β and β on protein concentration is very complex and accounting for value of β was calculated as a difference between α + β and α, it is difficult to evaluate the reliability of obtained results. The conclusion, that decrease in the surface tension is determined not only by diffusional mass transfer of the protein toward the surface, to say by adsorption, but in some extent and at any initaial protein concentrations by an arrangement the surface layer with participation of already adsorbed molecules of the ptotein, seems to be of interest to understand in detail the formation of protein adsorption layer. Particulary, it is clear, that an arrangement of the surface layer leads side by side with adsorption to a decrease in the surface tension. This conclusion has a fundamental significance. In the work [33] the new method of the adsorption measurement at interfaces was elaborated, including the transfer of adsorption layers from a/w surface into solid surface of an element for ATR spectrum registration in IR region. Time of the gelatin adsorption and run of dependences $\Gamma(C)$ is affected by the nature of the second phase: adsorption has been measured at a/w surface and at interfaces aqueous gelatin solution with benzene or $CCl_4$. The highest plateau adsorption is observed for interface gelatin solution - $CCl_4$. Constants of the gelatin adsorption are 230 ml/g, 560 ml/g and 1900 ml/g for indicated surfaces, respectively [34]. According to results of the adsorption determination at different temperatures, the adsorption is the function of gelatin conformation in a solution. It is of interest to note, that near the solution concentration of about 1 g/100 ml of gelatin in aqueous phase gelation takes place and the adsorption determination becomes difficult. In the work [35] isoterms of the surface tension were obtained for the systems with gelatin concentration near the threshold of gelation. In accordance with theory of fractal clustering scale parameter of length δ determines the minimal surface occupied by one molecule in the adsorption layer $S_0$. This value is connected with hydrodynamic volume of a macromolecular particle with the formula

$$S_0 \sim V_0^{D/3} \tag{9}$$

where D is fractal dimension of a cluster. For the ideal case the surface tension of a solution is:

$$\sigma = \sigma_b + \frac{kT}{\omega \ln\left[\varphi_b^{(1)} / \varphi_b\right]} \tag{10}$$

$\sigma_b$ is the surface tension of a polymer, ω has the sence of a surface area occupied by one unit of the lattice, $\varphi_b^{(1)}$ is volume fracture of macromolecular component in the surface layer, $\varphi_b$ is the same for the bulk. The surface tension decrease is proportional to $V_0^{D/3}$, or surface tension of a solution is:

$$\sigma = \sigma_b - kV_0^{(-D/3)} \tag{11}$$

Approaching to the critical concentration of gelation the surface tension asymptotically trends to the constant value, which does not depend on concentration of the gelatin. This minimal surface tension depends on $V_0$. Applicability of scaling pricipale connecting an area occupied by one molecule $S_0$ with scaling length δ it can obtain:

$$\sigma = \sigma_a - k_0 S_0^{(-d/2)} \tag{12}$$

where $\sigma_a$ is the surface tension of solvent, d is fractal dimensions of a percolation cluster originated on the surface, with $d = 1.75 \pm 0.10$ for the surface gelatin solution-air and $k_0$ is constant. The obtained value d coincides with known value of fractal dimension of a percolation cluster on the plane lattice. Thus, the surface tension value in vicinity of gelation of gelatin solutions is a function of hydrodynamic volume of macromolecules and fractal dimensions of the inner percolation cluster on the cubic lattice, or a function of surface areas with fractal dimensions on flat lattice. This shows the fractal mechanism of the formation of gelatin adsorption layer, probably, it acts, when adsorption layer forms and at lower gelatin concentration because of powerful increase in two-dimensional concentration of the protein.

Rheological properties of formed surface layers can be considered as the strong demonstration of protein adsorption at a/w surface. These properties show the formation of a surface film of definite structure, that can be regarded as a thin layer of a third phase [36].

Table 2 represents dependence of critical shear stress (Prs) of adsorption layers of lysozyme formed at the surface lysozyme solution-air on pH values of aqueous phase.

Table 2
Dependence of critical strength (critical shear stress) Prs (mN/m) of surface layer of lysozyme on pH of aqueous protein solution

| pH | Prs | pH | Prs | pH | Prs | pH | Prs |
|---|---|---|---|---|---|---|---|
| 2.0 | 0.8 | 4.0 | 1.5 | 4.8 | 1.4 | 6.2 | 0.9 |
| 2.8 | 0.9 | 4.3 | 1.7 | 5.1 | 1.1 | 7.0 | 0.9 |
| 3.7 | 1.2 | 4.6 | 1.7 | 5.7 | 1.0 | 9.0 | 0.9 |

Concentration of lysozyme solution is $6.9 \cdot 10^{-4}$ M. T = 293 K.

The set of rheological parameters, characterizing properties of adsorption layers of α-chymotrypsin at a/w surface is listed in Table 3.

Table 3
Rheological characteristics of adsorption layers (a/w surface) of α-chymotrypsin in dependence on the concentration of the protein solution

| $C \cdot 10^{-6}$, M | $E_{1s}$ | $E_{ss}$ | $E_{2s}$ | $\lambda$ | $\eta_s$ | $\eta^*$ | $\tau$ | $P_{k1}$ | $P_{k2}$ |
|---|---|---|---|---|---|---|---|---|---|
| 1 | 5.6 | 1.4 | 1.9 | 67 | 517 | 0.06 | 260 | 0.009 | 0.22 |
| 6 | 19.4 | 7.7 | 15.4 | 51 | 2840 | 0.12 | 180 | 0.012 | 0.66 |
| 16 | 28.6 | 12.0 | 19.0 | 81 | 4890 | 0.02 | 250 | 0.030 | 0.78 |

$E_{1s}$, $E_{2s}$ and $E_{ss}$ are modules of elasticity, respectively, of fast, slow and equilibrium elastic deformations (mN/m); $\lambda$ is elasticity of adsorption layers (%); $\eta_s$ and $\eta^*$ are shear surface viscosities (mN·s/m), with $\eta^*$ represents Bingam viscosity of partially destroyed structure; $P_{k1}$ and $P_{k2}$ are Shvedov and Bingam critical shear stresses for flow of adsorption layers, respectively; $\tau$ is the period of relaxation. Experiments have been performed at pH 7.8.

Other examples of rheological behavior of adsorption layers of protein at the a/w surface one can find in the reference book [37]. In accordance with indicated above adsorption data for α-chymotrypsin it can be seen, that rheological parameters become stronger with a gain in adsorption values. Formation of a surface layer of some structure already near tightly packing of monolayer or even at lower surface packing with the protein shows that an arrangement of the surface layer can be mainly attributed with this phenomenon. Kinetics of the rheological parameters increase, studied in detail for different systems [22,36,37] gives the values of period relaxation close coinciding with ones evaluated from kinetics of surface tension decreases. Surface layer buildup is accompanied with a slow decrease in $\sigma$ and reasons for surface tension change consist in adsorption events and separation of thin layer of a surface phase with specific rheological properties. In the same time, a surface structure with a mechanical strength can give some uncertainty in experimental determinations of surface tension values. This problem is poorly discussed literary. In according to all indicated data, measurements of surface tension of protein solutions are of great significance for the evaluation of surface activity of each protein, because this value is determined at protein concentrations tending to zero, decrease surface tension is very small and adsorption of protein, it is anticipated, is determined by diffusion of the protein to the surface. This parameter is of importance as the characteristic of amphiphilicity of the protein surface. For an average globular protein with molecular mass 20-50 kDa the adsorption corresponding to the monolayer with ultimate packing with native protein molecules, as a rule, is observed at equilibrium bulk concentrations of about $10^{-7}$ M and the determination of the protein surface activity has be performed at much lower concentrations of solutions. Surface activity of proteins, expressed in the units mN·l/mol, is a function or their molecular mass and proportional to the size of the protein. From theories of protein folding [38] it is known, that the fraction of the total hydrophobic area on protein surfaces increases with the protein size. Surface activity of the protein determines its properties in solution (for example, tending to dimerization, association) and the behavior in adsorption state. Surface protein activity is a function of total surface charge and double electric layer arround the surface of molecule, consequently, it is correct to suppose, that each pH-conformers of the protein will be characterized with own surface activity. Existing scanty data of the surface activity measurements allow to treat this characteristic of the protein as the lowest one at pI, and at small deviation from pI, no more, than ± 2-3 units of pH, the surface activity of the protein globule can become 1.5-2 times larger, apparently, following to increase of amphiphility of the surface of the protein molecule. Induction periods in kinetics of surface tension decrease, observed in [23] at lowest studied concentration (0.01 mg/ml) can be regarded as a characteristic of protein surface activity. This parameter well correlate with data of foam stability for all proteins, indicating, that details of inner molecular structure of globular proteins do not affect so much the protein adsorption at a/w surface. Under conditions resulting in protein denaturation the sence of the surface activity for protein becomes uncertain. Adsorption at pI results in more energeticaly ideal balanced surface structure of adsorption layers, which can form in the absence of lateral repulsions between adsorbed molecules in the plane of adsorption layer, and both electrostatic and hydrophobic interations act as an attraction. Observed tending to multilayers formation due to adsorption events at the a/w surface may have, as it is clear at present, two origins. First consists in that, that the following layer forms attaching to the first adsorption layer with approximately tightly packing by the protein molecules. Electrostatic and hydrophobic attraction interactions betweeen protein molecules are as driving forces in multilayer formation, because the surface of the protein molecule represents a mosaic of hydrophilic and hydrophobic areas. The second way to form the multilayers includes the stacking up the surface with high content of protein molecules, this process is specific for rigid adsorption layers.

## 2.2. Foam stability and its practical relevance

Surface adsorption layers of proteins are of a large importance, because their rheological properties supply a high stability and industrial uses of foams in food production and waste purification. To better understand principles of the foams stabilization with proteins, thin free films of proteins have been investigated thoroughly [22,39-45]. Such systems model properties of macroscopic dipersions, i.e. foams. Lysozyme, α-chymotrypsin and serum albumins were used as a stabilizers of free (foam) films. In all cases conditions of the stable black (limit thin) films formation and their thermodynamic parameters were determined. It can deduce some general rules, which correspond to well stabilization of foams with proteins. Concentration of the protein solution used for foaming has to be lightly higher, than one providing the adsorption respecting to the tightly packed monolayer. This value has been determined as Cs, which is the concenration of the formation of the black films [39-40]. In Table 4 optimal conditions of the free black film formation and their thermodynamic parameters are given.

Table 4
Thermodynamic characteristics and conditions of the formation of free black films, stabilized with proteins

| Protein | pH | Cs, M | θ′ | $\sigma \cdot 10^3$ | $\Delta \cdot 10^3$ N/m | $\Delta^* \cdot 10^3$ |
|---|---|---|---|---|---|---|
| Lysozyme | 11.8 | $3.3 \cdot 10^{-5}$ | 110 | 54.7 | 109.3 | 5.5 |
| α-chymotrypsin | 7.8 | $6.0 \cdot 10^{-6}$ | 63 | 58.1 | 116.2 | 2.3 |
| Human serum albumin | 4.9 | $7.7 \cdot 10^{-5}$ | 66 | 56.3 | 112.6 | 2.1 |

θ is contact angle between the film and meniscus of the solution (min), from which the film is obtained, this parameter characterizes the thermodynamic state of a thin film. Thermodynamic relationships have been received in [46-50] and relate tension of film Δ, dissjoining pressure, surface tension of the film $\sigma_f$, surface tension of the solution σ, which is in equilibrium with foam film, thickness of film (h). Value Δ* is free energy of the black film and determined as:

$$\Delta^* = 2\sigma - \Delta = 2\sigma(1 - \cos\theta) \tag{13}$$

Contact angles of foam black films of proteins coincide with ones of non-ionic surfactants.

Formation of the stable black films is observed at pH respecting to the isoelectric point of the protein. Addition of electrolytes leads to decrease in the stability of foam films. However, it was found, that free films of α-chymotrypsin remain stable, when NaCl is added in the range of $10^{-6}$-$10^{-4}$ M. Under this condition the protein does not preferentially link ions and the charge of the protein remains close to zero. It is clear, that the black films, stabilized with proteins shows a non-DLVO behavior. Preliminary evaluation of the thickness of films showed its approximate equality to the double linear size of the protein globule, and accounting for specific rheological properties of these films, as well as adsorption layers, such films can be considered as Newtonian ones. Such films are formed due to close contact of adsorption layers. Essential stabilization of black films of proteins is connected with elastic repulsion of adsorption layer at the low distance between them, when black film forms [22]. In all cases, investigations of the thinning of protein films using solutions with higher protein concentrations show the formation of stable films of larger thicknesses [39-45]. Theory of stability of thick films is not elaborated in present, but rheological properties of respective adsorption layers allow to suppose the same (elastic) nature of repulsion of thick adsorption layers. Methods of $\theta$ measurements are given in [22].

For practical uses it is of important to determine stability of foam films of proteins in the presence of lipids. Stabilization of real commercial products occurs when proteins and lipid are in mixtures. In [51] the influence of cardiolipine on the stability and thermodynamics of free films of human serum albumin was investigated. Addition of the lipid leads to a decrease in the film stability in parallel to a decrease in rheological parameters of the surface layers. However, the molar ratio of components (protein/lipid = 4.9) was found, when black foam films form and reveal a well stabilty. Salt additives ($7.8 \cdot 10^{-2}$ M) promote the stable films formation. These results indicate the necessity to vary the surface active components ratio to control the stability and properties of foam products of practical application. It is useful to remember, that stability of foam films, as well as foams, stabilized by natural surface active substances is determined by elasticity of black films, that is determined by elasticity of surface adsorption layers. These data are of interest for wastes purification from proteins and acompanying components using foam separation i.e. foam chromatography. Optimization ways for this chromatography technique for wide practical applications are based on the current knowlege of free films stability, rheology of surface layers and thin films and surface activity of protein. Let us consider the application of foam chromatography for goals of the wastes purification. Sources of wastes containing proteins are various, including food industry (milk, meet, vegetables remaking and others).

As an example, it is useful to show the possibility of wheys purification from milk proteins using foam chromatography. Milk processing for cruds, cheeses, or technical products results in the large amount of a whey. In dependence on the technology the whey composition varies too. On average one kilogram of a final product gives 30 kilograms of whey with the content of high nutritional compounds [52]: β-lactalbumin (β-Lg)-3.3, α-lactalbumin (α-La)-1.2, serum albumin (Sa)-0.3, immunoglobulins-0.5, proteoglycans-0.3, peptons-0.3 and others (concentration of the protein is given in g/l). Waste whey is a large source of the environment contaminations and disposal of untreated whey leads to loss of valuable biological components. Dairy farms, which are not equiped with purifying works, inevitablly contaminate water, soil and air and increase in biological and chemical consumption of oxygen.

For goals of whey purification different technology can be used. One of them, membrane separation, gets a wide spreading [53-54] because of many advantages, first of all, the possibility to separate from aqueous phase dispersed particles of different sizes (bacteria, lipid drops, solid particles using microfiltration, or cold sterilization; proteins, using ultrafiltration; salts and lactose are removed by hyperfiltration). Reverse osmosys allows the one- set removing of all dry substances from a whey. However membrane processes sometimes make difficult to operate revealing phenomena of fouling and membrane breaking and others [54]. High performance liquid chromatography was used as an analytical method to estimate the protein concentraton in the whey before and after foam separation. For model experiments the sub-crud whey was used. According to high performance liquid chromatography data the peak with the retention time of 15.4 min respects to α-La (m.m.14000), the fraction with 13.5-14.2 min respects to monomer, dimer and oligomer forms of β-Lg (18000-36000), the fraction with 18.8 min consists of serum albumin. The foaming process can be divided into several stages, which take place during the bubbling of column filled with a whey: adsorption layer formation on the bubble surface, formation of foam with black stable films, drainage and others. It is clear, that the floating up bubble has to be in a protein solution during the time, that will provide the protein adsorption with a approximately densely packed surface. The rate of a bubble moving through aqueous phase depends on its radius, and increases, for example, from 1.1 to 12.5 cm/s with increase in bubbles radius from 20 to 530 mkm. In according to the time of the formation of adsorption layers with available rheological properties and, sequently, with high probability of the formation of stable foam film, the time of bubble floating up must be no less, than half one hour. This condition will be performed at high dispersivity of bubbles and increasing of height of a column.

Application of foam chromatography will be efficient at concentrations of proteins to be isolated from a whey close to Cs and pH of aqueous medium in vicinity of isoelectric point. These conditions are the basis for a selective separation of proteins. The stage of the formation of cell-like, polyhonal foams is monitored with following parameters: well stabilization of bubbles with the protein, stability and sizes distribution constancy. The advantage application of foaming for separation of surface active substances, including proteins, is determined with the concentration coefficient and coefficient of foam filtration. Whey proteins were concentrated by foaming of liquid phase by means of bubbling using two columns with diameters of 2.0 and 2.5 cm and with height of 5 and 55 cm. In preliminary experiments it was found, that the whey revealed the well foaming ability at 40°C. This is, probably, connected with decrease in the content of association forms of β-Lg at higher temperature, this affects the foaming process because this protein presents in the whey in highest concentraction. Further, all whey treatments were performed under this temperature. Coefficient of concentration $R_f$ is $R_f = C_f/C_o$, where $C_f$ is protein concentration in the foam and $C_o$ is the same in the initial whey. Efficiency of whey purification can be characterized in two manners, as a purification coefficient β or extraction level α:

$$\beta = C_0/C_f \tag{14}$$

and

$$\alpha = (C_0 - C_f)/C_0 \tag{15}$$

Using foam separation method protein components with molecular mass larger, than $1\cdot10^5$ was separated from whey on 50-70%, BSA on 64-65%, α-lactalbumin was separated on 70%. The poorest extent of separation was found for β-lactglobulin (35%) and peptones, as products of degradation of proteins, are separated on 30%. The initial protein mixtures consisted of 50% β-Lg, 20% α-La, 5% BSA, 8% IgGs, 5% proteoglycans and others substances. The dry mixture obtained by foam chromatography procedure reveals the following content: 50% β-Lg, 35% α-La, 8% BSA and 4% of products of hydrolysis of polypeptide bond in proteins. Thus large amount of β-Lg is transfered in surface adsorption layers following the high concentation of this protein in the whey, but the content of dry mixture of separated proteins after one foaming stage is not enriched with this protein. The content of α-La essentially increases up to 15%. This protein tends to be completely isolated, that is in a coincidance with data indicating specific higher surface activity of this protein. The reason of some increase in the BSA content in the separated mixture is clear: this protein, been the most high molecular mass among proteins of whey, reveal a higher surface activity. This rule is operated largely for the adsorption of proteins at boundary a/w. Number parameters, including a height of the foam column, pH of protein solutions, ionic strength and others, must be adjusted to reach the complete separation of each type of protein from whey and thus to control the effiency of this technology for its applications for environmental protection. The similar method has been employed for extraction of potato proteins, which are enriched with irreplaceable aminoacids, from washing water on vegetable processing factories [55] and for other environmental protection goals [56,57].

Extraction of gelatin from aqueous solution is of great practical importance. When individual free film of gelatin were investigated, it was established, that gelatin forms stable thick films in accordance with non-globular character of folding of its polypeptide chains. Tacking into account, that gelatin gives adsorption layers at the surface solution-vapour with suitable rhelogical parameters and free stable films, foaming of gelatin solution will results in well stable foams formation, which can be used for the foam separation leading to gelatin exraction. Maximal coefficient of concentration $R_f$ = 10-20 was found under condition of isoelecric state of gelatin in the solution. In the case of gelatin it was shown, that addition into its solutions of low-molecular mass surfactants gives the possibility to control surface activity of this macromolecular component in parallel with rheological properties of adsorption layers and free film stability, as well as foam stability. Earlier it was shown [58], that mixtures of gelatin solutions with high concentrations (up to CCM) of surfactants exibit synergism in the decrease surface tension, revealing the high surface activity of complexes formed in the bulk at some their compositions. This allows to authors [58] to consider such complexes as a new surfactant. Free foam films stabilized with such complexes were investigated in the work [59] and regions of foam films stability and conditions of the formation of limit thin films and conditions of foam drying were determined. It was established, that addition of a surfactant can be used for enhanced gelatin extraction. At the ratio gelatin/surfactant in aqueous phase of 1.74 (optimal complex formation) gelatin extraction of a large level is attained, depending on technology of foaming. Thus under periodical foaming (in stationary foam) $R_f$ consists of 100%, when foaming was performed in continuous regime the value of $R_f$ consists of 50% [60].

Industrial applications of foam chromatography for many reasons are determined by the problem of the foam drying, because foams are able to retain continuous medium in large extent [61-62]. Dry foams are produced following the drainage process. Foam beds have been extensively investigated both as gas-liquid contactors and as separation devices [63]. The performance of these devices depends strongly on the liquid holdup with the foam. The liquid holdup varies with a height of foam column and with time as well for batch foams. A number of attemps exists in the literature for predicting the holdup values for both types of foams. The earlier drainage expressions were emperical, later workers have tryed to develop models taking into accout the actual structure of the foam. For low holdups, typical of foams, the gas bubbles are distored and are polyhedral in shape. This divides the total liquid of a foam into to types: entrapped in thin films and Plateau borders, revealing the necessity of separate investigation of drainage from two different systems [64]. The flow through the films due to gravity is negligible and they drain into adjacent Plateau borders due to capillary suction. The Plateau borders form a network, through which the liquid flows down due to gravity. It is useful to study the variation of bubble distributions in foams with time, this information allows to evaluate contribution of different processes, such, as drainage, coalescence, or Ostwald ripening [61]. In [63] realistic tetrakaidecahedral structure of foams bubble and model of statistic foam drainage have been considered and well coincidence between calculated rates of drainage and experimental results were obtained. In the discussed works [22,39-45] the drainage of individual foam films stabilized with protein adsorption layers were investigated in details. The low rate of films thinning and specificity of initial thinning at the border of the circular film, as well as hysteresis of contact angles, phenomenon of the film dimpling [65] allow to conclude that all dynamic and thermodynamic properties of foam films of proteins are determined by rheology of adsorption layers, that in its turn, results in the elasticity of the free black film.The rate of drainage in such cases is described with Reynolds equation:

$$d\left(1/h^2\right)/dt = 4P_\sigma/3\eta r_f^2 \quad (16)$$

where h is thickness of foam film, $P_\sigma$ is capillar pressure ($P_\sigma = 2\sigma/r_f$), $r_f$ is radius of film and $\eta$ is viscosity of liquid inside of the foam film.

It is known, that the thinning rate of a individual circular film is equal to the thinning rate of polygonal films in foams, when both films have the same thickness and area.

## 3. ADSORPTION OF PROTEINS AT LIQUID - SOLID SURFACE

The surface of proteins is heterogeneous, and normally exposes positive and negative charges groups with hydrogen bonding abilities, as well as non-polar regions. A consequence of this complexity of the protein surface is, that each type of protein can interact with other molecules and surfaces in a great number of ways. There are possibilities for ionic interactions (both repulsion and attractive), hydrogen bonding, hydrophobic interaction, hydration forces, acid-base interactions, and, it is evedently, the van der Waals force. The most important driving forces for protein adsorption are hydrophobic and ionic interactions. The contribution of these interactions during an adsorption event will affect the orientation of the adsorbed protein. In this moment it is difficult to generalize the behavior of different proteins at surfaces. Nevertheless, the main problem in this scientific area is to elaborate the possibility to predict the behavior of as yet not studied systems, which are very quickly incorporeted in human activities as well as to use adsorption characteristics to ctreate the reliable technological stages for their fast industrial adoptations. Adsorption at solid/liquid interfaces has some peculiarities as compared with liquid/fluid interfaces. The chemical nature of the solid surface and its properties (charge, hydrophobicity, etc.) determine the mode and strength of binding, as well as, in many cases, the conformational changes in adsorbed protein molecules. Usually, in contrast to fluid surfaces, solid surfaces are nor chemically or energetically uniform, and their heterogeneity may result in nonuniform of protein adsorption layers. The solid surface can be easily modified and turned out for specific types of interactions. The current knowledge on the protein adsorption on solids is based on results obtained for very well-characterized proteins. Investigations of adsorption of genetic variats or site-directed mutants of a single protein promise new developments in this field. Exposure of an aqueous protein solution to a solid surface results in an accumulation of protein molecules at the solid/liquid interface. The tendency of proteins to spontaneously adsorb is of fundamental importance for natural processes, including the genesis of terrestrial life. During the last thirty years considerable interest for protein interactions with materials used in medicine devices can be noted, including contact lens industry in a connection with interaction of tear proteins with contact lenses or implanted polymers with blood components.

The physical adsorption of immunoglobulins (IgGs) at solid surfaces is a process which is commonly applied in biomedicine and food industry where adsorbed IgGs molecules are used in detection systems such as an immunological tests and biosensors. The desire to control and manupulate protein adsorption in these applications requires a detailed understanding of the adsorption process. In many cases spontaneous adsorption of proteins leads to undesired consequences such as thrombosis on synthetic cardiovascular implants, fouling of hemodialysis membranes, contact lenses and bioprocessing equipments, as well as plaque formation on theeth and dental restoration. In other cases the protein adsorption allows to use controlled drug delivery and drug release systems, diagnostic tests, biosensors and protein purification techniques, including protein chromatography and membrane devices. Adsorbed enzymes give practical uses of special interest. Such kind of proteins is applied for different industrial goals for many centures, for example as leather tanning, cheese making and the leavering of bread. The commercial application of enzymes started at the end of the 19th century. Nowdays the interest in the industrial application of enzymes is growing and connects with adsorbed or immobilized enzymes as catalysts with very high specificity, renewable and biodegradable. Among a number of adsorbed enzymes working in a heterogeneous state lipases are of most interest. A wide application of protein adsorption phenomena is connected with stabilizing effect of interface layers of proteins in different dispersion. Milk serum proteins are commonly used as emulsifiers and foam stabilizers for various food products. The *in vitro* adsorption behaviour of fibrinigen and platelet adhesion on artificial materials may be related to the *in vivo* antithrombogenicity of various materials.

### 3.1. Methods to measure the protein adsorption

Study of a protein adsorption from solution at a solid surface is executed using two approaches. One of them is basesd on measurements of the solution depletion in consequence of the protein adsorption. Using this approach isotherms of "excess" adsorption of protein can be obtained. The second approach allows to measure the protein concentration directly at the surface. Different methods used for the determination of protein adsorption are given in short further. Depletion of a protein solution is measured by spectrophotometric determination of protein concentration at the characteristic wave length, or using colorimetric Lowry method. In such experiments the adsorbed amount is calculated from mass balance. Seldom the determination of a biological activity of proteins is used for this goal. The last method is applied for enzymes and based on kinetic measurements of the rate of reactions (V) catalysed by enzyme in according to Mechaelis-Menten kinetic equation:

$$V = V_{max}/K_m + [S] \qquad (17)$$

where $V_{max}$ is the maximal rate of the reaction, $V_{max}= k_{kat}$ [E], $k_{kat}$ is the catalytic constant, and [E] is the enzyme concentration.

Isotherms of "absolute" adsorption for the case of the monomolecular adsorption are determined from conditions:

$$\Gamma_i(x) = (1 - x)\Gamma_1^{(u)} - x_2\Gamma_2^{(u)} \qquad (18)$$

and

$$S_1\Gamma_1^{(u)} + S_2\Gamma_2^{(u)} = 1 \qquad (19)$$

where $\Gamma_1^{(x)}$ is excess adsorption of protein, $\Gamma_1^{(u)}$ and $\Gamma_2^{(u)}$ are absolute adsorption of protein and solvent (water), respectively; $x_1$ and $x_2$ are fractions of moles of protein and water in equilibrium solution; $S_1$ and $S_2$ are areas occupied by one mole of protein and water in adsorption layer. This leads to the equations:

$$\Gamma_1^{(u)} = S_2\Gamma_1(x) + x_1/S_2x_1 + S_2(1 - x_1) \tag{20}$$

and

$$\Gamma_2^{(u)} = S_1\Gamma_2^{(x)} + (1 - x_1)/S_1x_1 + S_1(1 - x_1) \tag{21}$$

Thus, when $x_1$ is much lower than 1, $\Gamma_1^{(x)} \cong \Gamma_1^{(u)}$

Exact determination of the absolute amount of adsorbed protein at the solid/water surface will require measurement of both residual protein and water after adsorption equilibrium. A positive excess adsorption thus does not mean that only protein and no water is adsorbed. It rather means that protein is adsorbed in excess (in term of the proportion in which protein and water exist in solution) of any water that is adsorbed. Negative adsorption indicates higher preference of the surface for water than for protein molecules. In [66], assuming the constant amount of water (b) bound to the adsorbent surface with zero absolute adsorption of protein the equations are given to discuss unusual behavior of protein adsorption with increase in initial protein concentrations:

$$\Gamma = -\,b/W_aC_e \tag{22}$$

where $\Gamma$ is the protein adsorption, $W_a$ is the amount of adsorbent, $C_e$ is equilibrium concentration of protein solution. When P' is absolute amount of adsorbed protein and b is a function of the initial protein concentration, at $P'V_p > P \cdot b$, adsorption of protein is positive and if $P'V_p < P \cdot b$, adsorption is negative. Note, that P is amount of adsorbed protein and $V_p$ is the volume of the protein solution.

When a flat solid surface is exposed to the protein solution, the adsorbed amount may be measured using piezo-electric weighing [67].

Ellipsometry is one of widely spread method for direct measurements of protein adsorption. Prior to adsorption ellipsometry measurements require determination of the optical properties of the solid surface [68-69]. For obtaining the amount of adsorbed protein the method described in [70] is applied using the values of the refractive index increments of the protein. The optical technique reflectometry is a simplified version of ellipsometry [71]. Reflectometry is especially suited to study adsorption kinetics because it can rapidly monitor the rates of adsorption using relatively simple and cheap instruments. During the experiment the protein layer is adsorbed on top of the sorbent surface and the problem left for determining the Q-factor to find the expressions for refractive index and for the thickness of protein film; for inhomogeneous layer with adsorbed amount $\Gamma$ of average thickness $d_p$ is given by:

$$n_p = n_s + (dn/dc)\Gamma/d_p \qquad (23)$$

where $n_s$ is the refractive index of the buffer solution and dn/dc is the refractive index increment of adsorbed protein layer.The factor Q equals to the change in the adsorbed amount per relative change in the signal, i.e. it is defined as

$$Q = d\Gamma/d(R_p/R_s)/(R_pR_s)_0 \qquad (24)$$

where $R_p$ and $R_s$ are parallel and perpendicular reflectivity coefficients, respectively. $(R_p/R_s)_o$ is reflectivity of the bare surface, and the magnitude with no index is related to $d\Gamma$. Another method for adsorption measuring in situ is described in [72-73]. It is based on surface plasmon oscillations (SPO) and involves the interaction of an evanescent photon field with boundary condition found at the interface between a thin metal film and a dielectric material. The plane-polarized beam penetrates the optical prism coated with thin layer of gold and reflects from the metal/dielectric interface. When the wave vector of the incident light matches with the wave vector of the surface plasmons, a resonance occurs and the reflected beam is strongly attenuated. In practice it needs to achieve a minimum in reflected signal ($\theta_{spo}$). If a thin film forms on the surface, there will be a shift in $\theta_{spo}$ proportional to the mass adsorbed per unit area, $\Gamma$. Methods of Fourier transform infrared [74-75], total internal reflection fluorescence [76-77], $\gamma$-photon spectroscopy [78] and streaming potential measurements [79] also wide used for protein adsorption studies. Since last five years new methods for the protein adsorption monitoring are used, including different surface forces apparatus (SFA) [80]. The technique involves the direct measurements of the intermolecular forces between two separated layers of a protein adsorbed on solid surface. The forces are obtained as a function of separation distance, which usualy is measured with an interferometry technique. This approach has been used for investigations of adsorption of surfactant, polymers, lipid mono- and bilayers, proteins and glycolipids on the mica. The bimoraph surface force apparatus (MASIF) [80] is a variant of surface force measurements, which are carried out by bringing the surfaces together, that allows to determine the forces as a function of separation from the hard wall contact with a high precision of about 0.1-0.2 nm, but this method does not allow to determine the thickness of firmly adsorbed layer. Adsorption of a number proteins has been investigated using total internal reflection intrinsic fluorescence spectroscopy (FTIR), especially for contact lens industry. The goal of studies in this case is to control lysozyme adsorption and its minimization to prevent contact lens deposits.

To characterize the ability of the protein to adsorb on the surface solution/solid investigations are performed using agreed-upon programe, including kinetics of adsorption, measurement of adsorption as a function of the protein concentration in solutions, the same at different pH values with paying attention to isoelectric point of the protein, and studying the adsorption on the solid surfaces of different hydrophobicity. This is a set of typical measurements, which are performed for number of most investigated proteins. In some extent this approach has been elaborated in our works [22]. Adsorption data are often presented as adsorption isotherms, where at constant T, the amount adsorbed protein $\Gamma$ is plotted against the concentration of sorbate $C_e$ in solution.

### 3.2. Driving forces and the problem of (ir)reversibility of the protein adsorption

Now much reviews have been published on the protein adsorption at solid/liquid surfaces [80], and a great activity in this scientific field of W.Norde with co-authors must be pointed out [81-86]. Detailed adsorption data are combined with thermodynamic arguments in an effort to determine whether protein adsorption to solids is a reversible or irreversible process. According to thermodynamics in a reversible adsorption process, the ascending (increasing concentration in the bulk) and descending (decreasing concentration in the bulk) branches of the adsorption isotherm must overlap at all protein concentrations. It is clear, that only for such reversible process the adsorption isotherm can be used to determine the equilibrium adsorption constant K. This is not correct for high-affinity isotherms, because determinations of reversibilty are hard to be performed and require of special methods to measure the protein concentration for very dilute solutions. However, low-affinity isotherms of protein adsorption are observed often, but dilution of the system does not lead to detectable desorption of the adsorbat from the surface, when shear is excluded. Hysteresis indicates, that the system has equilibrium and meta-stable states with the transition from meta-stable state requires the work against energy barrier. This fact shows, that during the transition from the adsorption state to desorption one a physical change in adsorption layer takes place. Authors of [81-87] mean the main reason for protein adsorption irreversibility due to conformation changes of adsorbed molecules resulting in an entropic gain. In the same time, when attraction interactions between adsorbed protein molecules are strong and exceed repulsion interactions and energy of interproteins attraction is the same in magnitude with energy of interaction with surface, the conditions turn out for a favourable change in Gibbs energy [88] for an orientation long-range ordering with attributes of a new phase, which is meta-stable with respect to dilution. However, conditions for the structured adsorption layers formation at solid surfaces are, rather, different, than at fluid surfaces. It is well-known the ability of proteins to form protective envelopes round various particles, resulting in encapsulation of dispersed phases. In such cases partially denaturation of a protein must be induced, for example, due to limited hydrolysis of peptide bonds.

Norde et al. consider protein adsorption as a nonequilibrium irreversible process, whose description should be, generally based on the lows of irreversible thermodynamics. But in this moment similar treatments of the protein adsorption are not published. Number of data, which show reversibility of protein adsorption at different solid surfaces as well as its irreversibility, is approximately the same [89]. Furthermore, the adsorption event of the protein is not as a irreversible one, but an arrangement of adsorption layer taking place during a long period of time is ended by attributes of irreversibility. Frequently, the adsorbed protein can be redealy desorbed from solids by changing the pH or ionic strength, or using surfactants. It needs to remember, that in case of truly irrevesible adsorption the adsorption isotherms would rise instantly to the plateau value or in very diluted systems complete depletion would be observed, but this behavior is not characteristic to the protein adsorption. In [90] reversibility of savinase adsorption on latex A were investigated. The savinase (m.m. 27000) is the extracellular enzyme, which is a serine endopeptidase of the subtulisin family from *B. lentus*. Latex A consists of polysterene particles with diameter of 495 nm and negatively charged surface carrying $OSO_3^-$ groups. In the former series of experiments savinase adsorption was achieved according to preliminary measured isotherm of adsorption, than a new portion of the protein was added. Measurements of adsorption shown, that the final adsorption value of the protein corresponds to one, recorded at the total protein concentration in independent measurements of isotherm adsorption, and no difference among one step or two step adsorption is observed. The reverse experiment is the determination of the desorption on dilution of a latex-savinase solution after equilibrium was reached. In this case it was established, that adsorbed savinase is not desplaced by solvent molecules. The second type of experiment consisted of changing the pH from 8 to 11 and from 11 to 8. This execution showed the complete reversibility of the protein adsorption. In all cases the adsorption corresponded to measured isotherms under constant experimental conditions. The last type of reversibilty that was measured by authors, was the replacement of adsorbed savinase by protein molecules with identical adsorption characteristics. Quantitative exchange took place within 30 min, which prove a dynamic equilibrium of savinase adsorption. No significant loss of enzymatic activity due to adsorption was measured. These data are presented because they show a typical behavior of adsorbed proteins. It seems to be correct the Magdassi assumption [89], that the adsorption of proteins (especially of those with the rigid globular structure of molecules) is usually inherently reversible, and irreversibility is only apparent and has kinetic origins. This provides the possibilty to estimate from experimental isotherms characteristics of the protein adsorption process, such as adsorption equilibrium constant K and corresponding standard free energy (or its lower limits, if irreversible entropy and enthalpy changes cannot be completely ignored).

Protein adsorption is complicated process determined by a delicate balance between several attractive and repulsion interactions. It is clear, that the relative importance of various driving forces is different for various types of proteins. Conclusions of recent reviews [91-97] allow to divide interactions determining the protein adsorption into four categories. Electrostatic interactions (attractive or repulsive) origin due to net charge on the solid surface and on the surface of the protein. Mosaic charge distribition on the surface of protein results in dipole moment and aslo contribute to the electrostatic interactions. The electrostatic interactions are further complicated by the incorporation of ions into the adsorption layer. Hydrophobic interactions also contribute to the adsorption process and can take place between protein molecules and between the protein molecule and surface of the solid. In this moment these interactions are considered as a dehydration of interfaces on the protein surface and on solid surfaces. Such interactions promote protein adsorption on hydrophobic surfaces and act oppositevely when adsorption takes place on hydrophilic solid surface. But this type of interactions can be realized between protein molecules during the adsorption at hydrophilic surface. As protein molecules have larger radii (2-4 nm) as compared with that of water molecule (0.14 nm), adsorption has lead to release of a large number of water molecules. The contribution of these hydrophobic interactions often overrule any other driving forces and are the sources of enthropy gain. Defined contributions in the protein adsorption layer formation can be possible due to conformational changes and/or lateral protein-protein interactions. But, such contributions are poorely estimated in this moment.

### 3.3. Kinetics of the protein adsorption and conformational state of the adsorbed protein

As a rule kinetic dependences of the protein adsorption are analysed using equation (4). Accumulation of the protein at the surface often shows much slower rates of this process, than that can be calculated according to eq.4. This slowing-down is, according to generally accepted point of view, not diffusion limited and connected with an arrangement of the adsorption layer. It seems, that the problem is at a loss. In [72] another approach was offered, that consists in the determination of characteristic time, $t_c$,which is proportional to $C^{-2}$, indicating the satisfactory coincidance of kinetics with the diffusion-limited model. It is of interest to consider in short principal postulates of this work. Adsorbance values are asymptotic ($t \to \infty$) and this emphsizes a deviation on Langmuir type adsorption. When binding energy is very strong, every molecule that makes contact with the surface is adsorbed. The rate of adsorption is limited by the rate at wich molecules diffuse from the bulk to the surface. Because the solution adjacent to the surface becomes depleted, the rate of adsorption is reduced because additional molecules must arrive from deeper within the bulk. The adsorption rate is described by

$$d\Gamma/dt = C(D_0/\pi t)^{1/2}\Phi \qquad (25)$$

where $\Phi$ is the fraction of surface available for adsorption and may have a complicated dependence on $\Gamma$. This value was approximated as $\Phi(\Gamma) \approx 1 - \Gamma/\Gamma_{max}$ and after integration eq.24 gives the adsorbance as a function of time:

$$\Gamma(t) \approx \Gamma_{max}\left\{1 - \exp\left[-(t/\tau)^{1/2}\right]\right\} \quad (26)$$

where $\tau = \tau_D \equiv \pi\Gamma_{max}^2/4DC^2$ ; $\Gamma_{max}$ is limit value of adsorption, that realizes at the given protein concentration, this value accounts for different manners of the orientation of protein molecule at the surface (side-on and end-on) and reflects concentration-dependent changes of the orientation.

The $t_c$ value is significantly greater than predicted from diffusion-limited model. This result authors attributed to a requirement that proteins have to be correctly oriented if they are to be adsorbed. For experimental studies rabbit anti-BSA immunoglobulins were used and their adsorption on antigen-coated (BSA-coated) surface was studied by means of surface plasmon oscillation method. Kinetics of the protein adsorption accompanied with specific interactions shows that for high protein concentrations the theory of random sequential adsorption has be applied in addition, because the adsorbed molecules can exclude more surface area than they physically occupy. The extreme slowness of the adsorption was only partly explained by addition of an orientational degree of freedom. In random seqential adsorption particles are placed at random on a lattice. Once placed, they are unable to desorb or diffuse along the surface. The adsorption proceeds until a shape-dependent "jumping limit" is reached. Started point for the protein adsorption is described by Langmuir equation.

Studying the protein adsorption it is of important to remember, that as a model of protein molecules a fluid particle with enchanced spreading ability cannot be used, because almost all globular proteins have to be regarded as a compact rigid particles. Lysozyme is one among a number of proteins with well known structure, adsorption of which was investigated in details. In particular, the absence of conformational changes following adsorption was proved for this protein many times, including our investigations of hen lysozyme adsorbed on the quartz surface [98]. The effect of structural stability of T4 lysozyme was monitoring at hydrophobic and hydrophilic surfaces with *in situ* ellipsometry [99]. Three mutants of this protein were obtained from *Escherichia coli* strains with different structural stability. Comparative studies with site-directed mutants of single protein demonstrate the importance of structural stability in the protein adsorption behavior.

Adsorbed mass of wild type of T4 lysozyme was registrated on hydrophilic and hydrophobic surfaces after 30 min and, respectively, consists of 2.84 mg/m$^2$ and 3.28 mg/m$^2$. For Ile3→Cys (S-S), Ile3→Trp mutants these values were much reduced, in the same time the mutant Ile3→Ser shows a much greater adsorption, respectively, 4.29 and 3.80 mg/m$^2$. The end-on T4 lysozyme adsorption corresponds to ultimate value of 3.96 mg/m$^2$, and side-on one to

2.05 mg/m$^2$. All mutant variants of the protein show the slower adsorption and lower conformational stability. Using FTIR spectroscopy changes in secondary structure of two variants of IgG in adsorbed state were studied [100]. IgGs are proteins with high content of β-structure (76%). IgGs are monoclonal immunoglobulins IgG1B and IgG2A, both mouse anti-hCG (human chronic gonadotropin). These proteins differ one from other with isoelectric points. For both proteins the several reduce in β-structure content is observed (from 76% up to approximately 60%), when proteins adsorb at the solid surface of both types with adsorption-induced reduction in the β-sheets content is larger at hydrophobic surfaces, but some increase in α-helices content is noted. Essential reduction in content of β-structure is observed, when the adsorption time was higher than 3 hours. Investigations of individual domains of IgGs (Fab and Fc) show the surface affinity of such fragments of IgGs and their excellent structure stability in adsorbed state. Authors concluded, that IgGs adsorb onto surface with Fc fragment, which has a stronger affinity for sorbent surfaces because it gains more conformational entropy upon adsorption. But entropic origin of IgGs adsorption is not clear, because α-helices are very rigid elements of structure of protein molecules. As for unperturbed Fab parts, they expose toward the solution. This is reasonable, because it is known, that IgGs react with antigen molecules with Fab-s fragments.

Data on catalytic activity of adsorbed proteins give the information of great importance about conformational state in adsorbed states. Proteolytic enzymes show the ability to autholysis, that can proceed in the solution and in adsorption layers. We observed inhibition of autholysis of protein molecules in adsorption layers [22]. Sometimes the addition of suspensions with hydrophibic or hydrophilic particles leads to change in the rate of autholysis [101]. For savinase the half-life of enzymatic activity in solution is 3.5 hours, and this period is strongly reduced in the presence of particles with hydrophobic surfaces. On the contrary, hydrophilic silica particles stabilize the adsorbed enzyme against autholysis.

New data were obtained using surface force techniques. The proof was given, that for the small globular no large-scale structural changes occur upon adsorption and the layer thikcness is consistent with the dimension of the protein. This is in line with measurements by CD-spectroscopy, that show only minor loss of secondary structure in such proteins upon adsorption [102]. Surface forces technique is very sensitive to a fraction of protein molecules adsorbed end-on, and to the presence of some molecules in outer layer. This is different compared to methods like ellipsometry and FTIR, which primarily measure the adsorbed amount. In all cases a steep short-range force is present between surfaces coated with globular proteins. This force, observed in the range of 1-2 nm, has been attributed to a compression of the adsorption layers and to dehydration of the proteins. Multilayer adsorption is observed under conditions, when the short-range protein-protein interaction is attractive. This type adsorption was shown for lysozyme, cyrochrome C, insulin and RNase A [80].

### 3.4. Typical adsorption behavior of globular proteins

In our work [103] the adsorption of α-chymotrypsin and bovine serum albumin at the hydrophobic surface of fluoroplast and at the hydrophilic silica has been investigated. In a tube 0.3 ml the adsorbent in an amount of 75-80 mg was induced (as a suspension in the case of hydrophobic adsorbent), than 5 ml of the protein solution of definite concentration was induced in a tube, carefully stirred during 2 hours and than centrifugated. The range of concentrations varying was 0.1-0.9 mg/ml for α-chymotrypsin and 0.2-3 mg/ml for BSA. Results are given in Table 5.

Table 5
Adsorption of BSA and α-chymotrypsin on hydrophilic silica and hydrophobic fluoroplast

| $C_e$, mg/ml | BSA Γ mg/g | α-chymotrypsin Γ mg/g | | α-chymotrypsin-sodium oleate complex | |
|---|---|---|---|---|---|
| | silica | fluoroplast | silica | fluoroplast | fluoroplast |
| 0.2 | 7.0 | 2.5 | 4.5 | 4.0 | 5.0 |
| 0.5 | 20.0 | 6.0 | 11.0 | 8.0 | 15.0 |
| 0.8 | 23.0 | 9.0 | 15.0 | 8.0 | 20.0 |
| 1.0 | 28.0 | 9.6 | 15.0 | 8.0 | 22.0 |
| 1.5 | 38.0 | 12.0 | - | - | 24.0 |
| 2.0 | 45.0 | 15.0 | - | - | 24.0 |
| 3.0 | 48.0 | 15.0 | - | - | - |

The ultimate adsorption of both proteins on the hydrophilic adsorbent is of two-three times higher, than adsorption on hydrophobic one. This fact indicates the contribution of hydrophobic intermolecular interactions in adsorption layer buildup. This difference can be in some extent attributed to peculiarities of the conformation of adsorbed proteins. The conclusion was made, that conformational shifts are larger in the case of hydrophobic surfaces. This can be connected with different extent of the surface packing, which is larger for hydrophilic silicagel. The experimental values are essentially lower than these respect to the ultimate adsorption according to Langmuir equation; maximal surface coaverage achieved for hydrophobic adsorbent equals to 0.3 and for hydrophilic one - 0.7. α-Chymotrypsin is lightly and completely displaced from hydrophilic surface with addition of electrolyte, like as KCl.

To examine the properties of the adsorbed protein on hydrophobic surface the enzyme activity of α-chymotrypsin was determined at the different extent of the surface packing. The rate of enzymatic hydrolysis of N-acetyl-l-tyrosine ethyl ether was measured using pH-state at pH 7.4. It was established, that the relative specific activity of adsorbed enzyme (with respect to enzyme activity at the same concentration in aqueous solution) lowers in great extent, but increases with increase in amount of adsorbed enzyme. When the surface coverage is low, for example, 0.09, the enzyme activity consists with 10% of the enzyme activity in the solution at the same amount of the α-chymotrypsin, and respectively grows to 16% and 24% at the surface packing of 0.20 and 0.30. It is difficult to distinguish the contributions of conformational shifts, hindered diffusion for enzyme-substrate formation due to blocking of the active centre with the surface or two-dimension assosiation of the enzyme. The theoretical and experimental evaluations of the contibution of the last process has been considered in [104]. Thus, according to this work, assuming a negligeable contribution of conformation changes, the increase in enzyme activity in parallel with the increase in adsorbed amount of the enzyme can be connected with a change of orientations of the protein molecules resulting in a decrease of blocking effects of active sites of the enzyme due to, for example, dimers formation. But this conclusion is poorly consistent with observed irreversibility of the α-chymotrypsin adsorption on the hydrophobic surface of solids.

To control the protein adsorption on hydrophobic surface the adsorption of α-chymotrypsin on such surfaces was studied under conditions of a complex formation between this protein and surfactant in the bulk. As a surfactant sodium oleate was used. Results are given in Table 5 and show, that amount of the adsorbed protein is increased in large extent. As for catalytic activity of the adsorbed complexes, it is much larger as compared with the activity of enzyme adsorbed with no added surfactants. Dependence of catalytic activity on the surface packing extent is reverse to one observed in the surfactant free system. The magnitude of the ultimate adsorption gives the area occupied by the protein molecule, that equals to 23 $nm^2$ and is very close to the area of native globule of this enzyme (28 $nm^2$) adsorbed side-on. Probably, a surfactant addition (if it does not lead to an inhibition of the enzymatic activity in the bulk) can also modify the surface of hydrophobic solids resulting in its hydrophilization. These results show, that fluoroplast is a very complex adsorbent for model investigations of protein adsorption. Futher a comparative analysis of adsorption isotherms was performed using pancreas lipase, α-chymotrypsin, and BSA [105]. As an adsorbent methylated silica beads were used with specific area 152 $cm^2/g$ (Serva). Sizes of adsorbent particles were additionally examined using the device "Particle Measurement Coumputer, Millipore". All proteins adsorb at hydrophobic surface and essential differences in the kinetics of adsorption did not established with the adsorption time of about 20-30 s. The isotherms of lipase and BSA adsorption formally are in accordance with equation of Langmuir type. Thus, using the criterion of protein dimerization in adsorbed state, according to [104]:

$$\theta_s/[C(1-\theta_s)] = K + 2\chi KC(1-\theta_s) \qquad (27)$$

where K is the constant of the adsorption equilibrium characterising the protein-surface interaction; $\chi$ is the constant of the twin interaction of the protein molecules at the surface; $\theta_s$ is the extent of the surface packing.

The value of $\chi$ that is distinctive from zero was found only for α-chymotrypsin ($\chi$ = 2.5) indicating the role of dimers formation accompanying the adsorption of this protein at hydrophobic surface. The more homogeneous surface of methylated silica as compared with the surface of fluoroplast allows to measure the higher values of the protein adsorption and areas occupied by the protein molecule at the ultimate adsorption are very close to the surface area of the native protein. Ultimate values of the protein adsorption are $0.8 \cdot 10^{-7}$ mol/m$^2$ for BSA and $2.3 \cdot 10^{-7}$ and $3.4 \cdot 10^{-7}$ mol/m$^2$ for lipase and α-chymotrpysin, respectively. These values were calculated in according to Langmuir equation with adsorption constants K for the proteins in the same row: $6.8 \cdot 10^6$, $2.3 \cdot 10^7$ and $1.6 \cdot 10^6$ l/mol. The surface affinity increases in the series α-chymotrypsin < BSA < lipase. Despite of anticipating of an increase in the surface activity with increase of molecular mass of the protein, it can be seen, that this behavior is not fully satisfied the protein adsorption. Because of this fact it is impossible to consider molecules of different proteins as a homologeous substances. It is better to compare surface activity of proteins with data of their surface hydrophobicity. For the estimation of the contribution of hydrophobic forces in protein adsorption it is useful to determine the surface hydrophobicity of the protein using one of several elaborated methods. None of these approaches can be accepted as an universal method, however, the hydrophobic chromatography may be regarded as a satisfactory way, when further data are used for prediction of the protein adsorption on the solid surface The surface affinities of α-chymotrypsin and BSA are in accordance with surface hydrophobicities of these proteins determined in the work [106], respectively, 81.3 and 68.4 min. Accounting this correlation it becomes clear that even formal description of isotherms of protein adsorption using available equations does not lose a sence and is of importance for deep insight into this complex phenomena.

Sometimes isotherms of protein adsorption show points of inflection, that idicates a possibility of change the energy of adsorption. Similar surface behavior is observed for α-chymotrypsin at silica [103]. Heteroenergetics of BSA adsorption at different solid surfaces was found using FTIR, when BSA adsorbed from a thermodynamically good solvent commonly used as a buffer for the protein crystallization [107]. Quarz, $SnO_2$ and alkyl-modified silica surfaces demostrate broad continuous distribution of energies of adsorption activation and protein surface binding. For heteroenergetic adsorption the type of the isotherm and kinetics depends on the type of energy distribution functions and the type of correlation between energies. Hydrophobic interactions seem to be the principal driving force in determination the behavior of BSA at the solid-liquid interface and likely play an important role in the protein crystallogenesis. Authors concluded, that heteroenergetic adsorption is possibly a general property of protein interfacial behavior. A well correlation of adsorption plateau values and the protein surface hydrophobicity was shown by Haynes and Norde [81]. From reviews [108-109] it becomes clear, that the surface hydrophobicity of the protein strong affects its surface affinity. The other essential type of driving forces promoting the protein adsorption is electrostatic interaction resulting in higher adsorption at pI of the protein and under condition of opposite charges of the solid surface and protein [22,81].

Influence of the prtotein self-association in the solution on adsorption layer buildup at hydrophilic and methylated silica surfaces has been investigated using in suti ellipsometry method for β-lactoglobulin A and B [110]. These proteins are genetic variants of cow β-lactoglobulin. Polypeptide chain of the proteins consists of 162 aminoacids and A and B variants differ only at two position, 64 and 118, which are Asp and Val for A variant and Gly and Ala for B. The proteins do not show any significant differences in the structure molecules. The adsorption isotherms all have a similar shape and are essentially only shifted by the concentration. The isoterms have two distinct levels of adsorbed amout at 1 and 2.5 mg/m$^2$, respectively, this indicates the possibility of different modes of binding at high and low concentration. Since it is well-known, that the protein under these physical conditions, exists as monomers, dimers and it seems likely, that the two levels of adsorption are related to this equlibrium. Adsorbed amount is proportional to the fraction of dimers. Accounting the constants of dissociation of a dimer for each type of the ptotein ($6.3 \cdot 10^{-5}$ and $0.8 \cdot 10^{-5}$ M, respectively for A and B variants) a nearly complete overlap of isotherms was obtained. Preferential adsorption of larger oligomers has been observed many times, including lysozyme, human serum albumin and insulin. Isotherms of adsorption on hydrophilic silica are similar to the isotherms recorded for hydrophobic surface, with only difference, that it was impossible to measure adsorption at low protein concentration, when the dimer fraction tends to be zero. Low rate of the adsorption at hydrophilic surface is explaned by preferential dimer adsorption with displacement of adsorbed protein monomers. The influence of pH on adsorption also is related to the variation in self-association within the pH interval (2-7).

Adsorption isotherms of β-lactoglobulun and acethylated β-lactoglobulin at the alumina/water surface were obtained at different pH of solutions. Both proteins showed the highest adsorption at pH 5.1. Above pH 5.1 adsorption decreases to almost zero for both proteins. These results were related to the aggregation/dissociation property of the proteins in the entire pH range. With this respect the data on immunoglobulins (IgGs) adsorption may be discussed. One important aspect of adsorption is the change in the environment of the protein which may result in conformational changes and, in turn, may induce an alteration in the biological activity. In [71] the adsorption of two monoclonal IgGs which differ in their isoelectric point and their fragments F(ab)2 and Fc has been compared. The proteins are adsorbed on hydrophilic silica and on hydrophobic methylated surfaces at different pH and ionic strength. The results show that for IgG the adsorbed amounts decrease with an increasing net charge density on the protein. At higher ionic strength the influence of the pH-dependent electrostatic interaction is suppressed. Fc part of IgGs has a stronger affinity for sorbent surfaces and apparantly, that molecules of IgGs preferentially adsorb with their Fc parts attached on the sorbent surface and their, relatively unperturbed Fab parts pointing toward the solution.

### 3.5. Dependence of the efficiency of membrane technologies on the protein adsorption process

In this modern society great amounts of food, water and fuel flow into human dwellings from outer environment and waste is discarded to the outer environment again. Among these metabolic substances for cities water is consumed in the large amount. The modern period based upon the rapid volume transportation with large scale simple production way is about to be ended by environmental restrictions. It can be concluded, that all water systems are in crisis. To purify industrial water and to preserve water sources membrane technology obtained a wide spreading [54].

Adsorption of proteins at water-polymer surface in accounting its importance for many applications gives rise many questions to be investigated. Current knowledges in this area have to be discussed. Many years ago development of technical membrane processes are originated, among them microfiltration has been used in Germany (1920) for bacteria separation, since 1950 year (Netherlands) hemodialysis is used as an artificial kidney and since 1960 (USA) ultrafiltration is used to concentrate macromolecules, including protein. Other membrane applications fall out our consideration. The ultrafiltration is carried out at low applied pressure in the range of 1-10 bar. In this case membranes have asymmetric porous structure (pore sizes are equal to 1-100 nm) with the thickness of actual separating layer of 0.1-1 mkm and separation is based on particle sizes. As a membrane material polysolfone, polyacylonitrile and ceramics (oxides of zirconium and aliminium) are used. Main industrial applications, connected with the discussed problem of protein adsorption, are dayry (milk, whey, cheese making), food (potato starch and proteins) and pharmaceutial (enzymes, antibiotics, pyrogens). In an industrial scale electrodialysisis applied for desalination and aminoacids separation in food and pharmaceutical productions. The membrane performance changes very much with time resulting in a considerable decrease the flux through the membrane. Causes of this phenomenon are pore-blocking, adsorption on solid walls of a membrane, formation of a gel layer and concentration polarization. The larger decline effect is known as membrane fouling, which can be defined as the (ir)reversible deposition (adsorption) of retained particles, colloids, macromolecules, etc. on or in membrane. To help understand the effects of protein adsorption on membrane filtration performance in [111] the molecular interactions between cellulose acetate films and proteins using a surface force apparatus have been investigated. Contributions of several phenomena, such as protein adsorption, aggregation and gel formation, due to the effect of pH and ionic strength on the solute-solute and solute-membrane interactions and of reduced driving force caused by effect of the osmotic pressure due to the buildup of proteins at the solution-membrane surface, were discussed. It was known on the example of BSA, that the membrane fouling decreases, when the solution pH is away from pI of the protein, when little salt is present and when the membrane surface was hydrophilic. The main conclusion of the work [111] is 1) at low pH values HSA and RNase A are denaturated and would induce membrane fouling adsorbing at its surface; 2) at higher pH values these proteins remain as a native and their ability to membrane foul is low; 3) influence of pH is larger in the case of HSA, but interaction of HSA with the surface of the membrane in much stronger. These data are in accordance with data of HSA adsorption on polysterene latex and well known data on conformational transitions of serum albumins at low pH values. This is the first prove of enhanced membrane fouling, when conformationally modified protein adsorb on membrane. Most fouling studies have been performed using well defined protein solution models such as BSA solutions. In [112] fouling was studied using a fermentator protein mixture of bacteria, dominated by the methanothropic *Methylococeus Capsulatus*. Fouling phenomena was attributed to solute adsorption on the membrane surface and pores plugging with pore blocking type membrane fouling is the most dominant mechanism. Back flushing can be a good approach to prevent pore blocking. In addition to this method, the membrane surface purification from protein layers causing the fouling enzyme solutions can be used. As cleaners enzymes with function to cut molecules at specific aminoacid sequences are most useful for removal of proteinaceous foulants. Using the same method, diffusion and adsorption of proteins in a model pore have been studied in [113]. Transport paths, i.e. pores are often of molecular dimensions, this results in significant wall effects with the mobility of molecules being affected by hydrodynamic, steric and energetic interactions between protein molecules and the wall. Wall effects become especially pronounced when energetic interations lead to protein adsorption on the pore walls. For such studies experimental methods used traditionnaly do not allow to obtain reliable conclusions, and techniques must be applied that can directly probe events occuring inside pores. Among these that have been reported are holographic relaxation spectroscopy, microscopy interferometry and pulsed field gradient NMR spectroscopy [114]. When the energetic interactions between proteins and pore surfaces are favorable, both hindered diffusion and adsorption govern the overall transport rates in fine pores. Results obtained in [113] show the effect of ionic strength on the transport rates for lysozyme diffusing inside mica pore and influence of electrolytes on the extent of the protein adsorption. Adsorption on opposite charged surface leads to pore blockage resulting in the pore size decreases to 2-3 times the molecular diameter. These effects were observed, when lysozyme adsorption on pore wall respects to 2 $mg/m^2$, consistent with a monolayer of its molecules and do not take place, when other protein such as pepsin was treated with the same experimental technique. Pepsin has a high negative charge around neutral pH and does not adsorb on the mica surface. Thus electrostatic effects are central to the adsorption process. The protein adsorption on the membrane surface and its pores determines the fouling. In the same time it is clear, that protein-protein interactions whithin adsorption layer give the essential contribution to fouling process. Connecting with this, properties of the cake formed on the retentive membranes in dead-end ultrafiltration of protein solutions have been investigated using mixtures of BSA and lysozyme in solutions. Ultrafiltration membrane which is retentive for BSA and permeable for lysozyme was used. In the pH range, where both protein molecules have opposite electrical charges, the cake becomes compact and dry due to attractive forces. This leads to decrease in filtration ability. Porosity of the cake and average specific filtration rate depend on pH and ratio of concentration of BSA and lysozyme. The structure of cake has a major impact on the flux and the rejection in protein ultrafiltration [115]. The performance of a hemodialyzer during clinical treatments is lower, than that obtained from in vitro measurements. The reduction in diffusive permeability of dialysis membrane caused by protein adsorption is one of factors reducing the performance. In [116] was shown, that the protein adsorption varies with membrane structure and the asymmetrical membrane has an advantage of less permeability reduction for low molecular weight solutes after protein adsorption.

However, adsorption of the protein on the membrane allows to its separation from water and can be used for other useful goals. In [117] adsorption of BSA on hollow fibers was used to separation of optical isomers of phenylalanine. BSA links the L-isomer with the binding constant of $1.8 \cdot 10^6$ $M^{-1}$, that is one order higher than for D-isomer. Separation factor two-three times is higher for immobilized BSA, demonstrating channel proteins for optical resolution. Many other examples of useful applications of systems consisting of proteins adsorbed on ultrafiltration membrane can be expected.

Membrane technology has been used widely for separations and purifications of bio-products in dairy industry. In order to concentrate protein in cheese whey, the ultrafiltration and reverse osmosis methods have been operated commercially. Durig these processes of cheese whey filtration, the fouling of membranes and concentration polarization cause some serious problems which are reduction of efficiency due to decrease in permeate flux and shortening membrane life. These phenomena are due to adsorption of protein and precipitation of salt at the membrane, as it was considered above. Adsorption of proteins is accompanied by precipitation of calcium phosphate resulting in large permeate resistance. Membrane operations are satisfied only in narrow range of whey pH of about 5.8 [118]: at pH $<$ 5.8 proteins stong adsorb on membrane, at pH $>$ 5.8 calcium phosphate precipitate on the membrane surface.

### 3.6. Adsorption of proteins and the problem of biocompatibility of materials

Modern medicine extensively uses synthetic materials, polymers or ceramics, in substitution surgeon, when surgical intervention requires to replace different organs, vessels, bones and tissues by synthetic ones. This human activity brings to light the problem of biocompatibility, and bloodcompatibility is its particular case. Current knowledges show, that the problem is connected with the extent of a surface energy decrease due to contact of material with aqueous medium [119]. Biocompatible materials equilibrated with liquid phase have to possess the optimal surface energy of about several units of $mJ/m^2$, that allows to prevent the adsorption on the surface of a potentially compatible material of proteins, which promote intensively the adhesion of platelets resulting in the thrombosis. In the same time the surface energy value cannot be too low to prevent spontaneous dispersing and thus destruction of a material [119]. This task must be investigated using blood-material contacting systems. But such studies are difficult to be performed and controlled. The problem is solved now executing the blood protein adsorption on synthetic materials. It becomes clear at present, that preferential adsorption of some proteins, for example, serum albumins, results in higher reliability in bloodcompatibilty of potentially suitable materials *in vivo*. On the contrary, preferential adsorption of fibrinogen and other protein factors of blood coagulation decreases in the serviceability these materials. Thus, the adsorption of proteins at biomedical surfaces, such as implants, cathethers and insulin pumps, is the first step in a complex series of biophysical/biochemical processes, which determine the biological response to the foreign material.

Interactions of contact lens with tear proteins are the most simple example of the biocompatibility importance. Thus, peculiarities of the lysozyme adsorption on the surface of contact lens can determine the quality and applicability of contact lens. Lysozyme is the main protein of tear. Using FTIR method, the comparable study of human and hen lysozyme was performed and the surface hydrophobic/hydrophilic characteristics of the two lysozymes were evaluated [120]. Human lysozyme reveals a higher adsorption affinity to both hydrophilic and hydrophobic surfaces as compared with hen lysozyme. This proteins behaviour lightly can be explained with structural difference between their molecules. Polypeptide chains differ one from other in 40% aminoacids positions, with all these varying aminoacids are located on the surface of the protein. Proteins are differed in a number of disulfide bonds (4 in hen and 3 in human). The proteins have the similar secondary structure (CD, far UV), but different Trp and Tyr environments (CD, near UV). Human lysozyme reveals the lower termal stability. The check of antibody reactivity shown different surface epitopes. Crystal group also shows different surface groups and mode of intermolecular association. Data of self-association studies lead to conclusion, that hen lysozyme dimerizes and oligomirizes at $pH > 5$ and high protein concentrations (5-10 mg/ml). The similar kind of self-association might be going on at the surface, where the adsorption process inherently concentrates the proteins. Besides, the adsorption of monomer, dimer and oligomer forms from the bulk, that occurs at different protein concentrations must be accompanied with changes adsorption characteristics. Adsorption of both lysozymes is higher on hydrophobic surfaces with participation of hydrophobic area located at the back side of both proteins. It seems that peculiarity of the charge distribution influence adsorption of these proteins. Both lysozymes have long positive side chains extending out into the solvent while the surface negative charges are not nearly accessible. This point further suggests a preferential adsorption of lysozymes onto negative surfaces. The hen lysozyme showed non-saturated adsorption behavior on charged surfaces. This indicates that charged surfaces promote the formation of multiple protein layers. When adsorption of hen lysozyme on mica is measured by means of surface forces technique the thickness of adsorption layer increases with concentration and consists with 9-10 nm. At low concntration of lysozyme (0.02 mg/ml, 1 mM NaCl, pH 5.6) steric repulsion due to adsorption layer is present below about 6 nm. This corresponds to the side-on monolayer of lysozyme on each surface. It is of interest to note, that adsorption at low concentration leads to strong reducing the surface potential (from 84 mV to 14 mV) and the repulsion due to double-layer forces. Comparison of in vivo spoilage with in vitro adsorption of tear proteins to optimize properties of materials for soft contact lens is carried out in [121].

It seems, that lipids have a profound role towards the thrombosis and hemostasis mechanism, but lipids action is not clear in the thromboembolic phenonema that accur at the blood-foreign material interface. Contribution of platelets may have an important role in the increased tendency to thrombosis in hyperlipidemia. It has been reported that in hyperlipoproteinemic humans the platelet survival time is shortened and that in patients with myocardial infarction platelets seem to be activated [122]. Lipids cause coagulation via the generation of thrombin by phospholipids and in the activation of platelets. Adsorption of lipids causes the cracking of silicone rubber prosthetic hand joints due to concentration of the surface stress crack tendency of these implants. It is also known, that vitamin C influences the surface behavior of proteins, leading to increase of the protein adsorption and reducing the platelet adhesion toward an artificial surface. The adsorption of three lipids, namely, cholesterol, cephalin and sphingosine from their individual solutions to polycarbonate surface and their interaction with proteins and vitamin C has been studied using contact angle measurements and labeled proteins. Lipids adsorbed at the surface interact with proteins and platelets. Contact angles and surface energies of polycarbonate exposed to buffer are given in Table 6. Data allow to evaluate the modification extent of the polycarbonate surface due to lipid or lipid-protein adsorption.

Table 6 shows that the surface of polycarbonate with adsorbed serum albumin is the most suitable one to be used in implant devices. The behavior of all lipids toward blood-polymer interaction is not similar and may change depending on the nature of lipid, net charge of the lipid-adsorbed surface and the lipid-protein/lipid-platelet interaction at the interface. Under conditions of high cholesterol concentrations addition of vitamin C leads to suitable surface characteristics of polycarbonate. The question is how to garantee the preferential the albumin adsorption on an implant surface? In works of Malmsten and Lassen [123] competitive adsorption at hydrophobic surfaces from binary protein solutions was investigated. Methylated silica or silica modified by plasma deposition of hexamethyldisiloxane were used as hydrophobic adsorbent, *in situ* ellipsometry and FTIR were as methods of the protein adsorption measurements. Preferential adsorption of IgG and fibrinogen from mixtures with HSA was established. However, in sequential adsorption experiments, where HSA was first allowed to adsorb, followed by rinsing and addition of either IgG or fibrinogen, the additional adsorption is quite limited and adsorbed HSA is not removed by either IgG or fibrinogen to any larger extent.

Table 6
Modification of the polycarbonate surface due to cholesterol, protein or cholesterol-protein adsorption

| Surface | $\sigma_{bl}$, mJ/m$^2$ | $\theta^o$ | Platelet adhesion |
|---|---|---|---|
| No modification | 6.42 | 70.0 ± 2.0 | 6.9 ± 1.5 |
| Cholesterol adsorption | 20.66 | 77.0 ± 2.0 | 10.8 ± 2.0 |
| Cholesterol-albumin adsorption | 11.86 | 62.0 ± 2.5 | 4.0 ± 1.5 |
| Cholesterol-fibrinogen adsorption | 18.22 | 73.0 ± 1.5 | 14.0 ± 2.5 |
| Albumin adsorption | 10.58 | 59.5 ± 2.5 | 3.0 ± 1.0 |
| Fibrinogen adsorption | 17.91 | 72.5 ± 2.0 | 14.5 ± 1.5 |

$\sigma_{bl}$ is the surface energy of polycarbonate equilibrated with buffer, $\theta$ is contact angle of water drop on the surface of polycarbonate.

The role of the surface charge of blood-contacting implants on their blood compatibility is still controversial. Although most solids with positive zeta potential are thrombogenic, the negative surface potential is not a guarantee of nonthrombogenicity. In [124] adsorption of proteins was studied onto low-temperature isotropic carbon (LTIC), which are reputed as blood compatible material. Results of many studies on protein adsprption on the solid surface with different electric surface potentials allow to conclude, that in solution of normal ionic stregth, electrostatic interactions do not play a major role in protein adsorption. In [124] impedance measurements, which allow to change the potential of electrodes and to record variations of double-layer capacitance, from which one can acquire information on protein adsorption kinetics, surface coverage, and isotherms. Under neutral and acidic conditions the more negatively charged LTIC adsorbs more protein regardless of the nature of the proteins. Hydrophobic interactions are stronger than the electrostatic ones. The former can be altered through a change in the structure of interfacial water by the imposition of charge on the LTIC surface and adsorption of a protein is due to competition between proteins and water for the solid surface. The state of interfacial water and its interactions with the surface play an integral role in the protein adsorption. Thus, it is probaly the surface charge cannot be the sole factor in determining the trombogeniety of blood-contacting material.

Successful solid phase diagnostics depends on specific interactions between a protein, which must be detected, and the surface. For this purpose specific IgGs can be preadsorbed or covalent immobilized on the surface of adsorbent. Thus regenerated cellulose was activated by tresyl or tosyl groups, which provide a easy direct coupling of a protein. When BSA-antibody was adsorbed on activated surfaces the 30-fold increase in BSA bonding was attained [125]. In this case fraction of nonspecific adsorption does not exceeds 5%. Dependence of controlled drug delivery and drug targeting on kinetics of plasma protein adsorption is considered in [126].

### 3.7. Catalytic activity of adsorbed lipases

Lipases occupy a prominent place among biocatalysts and have a wide spectrum of biotechnological applications. Now it can be said, that a new field of applied chemistry - catalysis by adsorbed lipases formed.

Lipases are enzymes with the natural function to hydrolyze triglycerides to glycerol, mono- and diglycerides and fatty acids. They are found in the pancreas, blood plasma, saliva, pancreatic juice, milk, in number of triglyceride-producing plants, in molds and bacteria. Lipolysis occurs at the lipid/water interface and can thus be enhanced by increasing the interface area. Because of this considerable rates of lipolysis are measured when substrates are used as emulsions, micelles or, espesially for water soluble triglycerides, in the presence of hydrophobic suspensions. It is of interest, that lipases can reverse the reaction direction in non-aqueous media. In all cases the stage of important for liposysis is the lipase adsorption at an interface. The stability of these enzymes at water-nonpolar surfaces have pushed them into the frontier area of organic synthesis leading to the designing of novel drugs, surfactants, bioactives compounds and oleochems. In addition, lipases catalyze trans-esterification and inter-esterification reactions and can be exploatated in the fat industry [127-135]. Additions of a lipase into washing compounds essentially increase their washing abilility. Lipases are widely used in textile industry to enhance wettability and absolency of products. Looking into the wide scenario of lipases applications, commercialization of lipases production is a prime are of interest for micro-biologists, process engineers and biochemists. Researchs carried out in this field have revealed that microbs, especially fungi and bacteria, are the tools for the choice of the lipase for commercial production. The structure determination of a few microbial lipases has widened our knoweledge about unique mechanism of catalysis of these enzymes. The problem of the industrial application of lipases is only to optimize conditions (the source of lipase, polarity of the reaction medium, temperature, pH and dispersed form of a substrate, as well as ways to immobilize or absorb the enzyme at solid surfaces) for the task to be solved. This become possible as the result of long time investigations of the lipases activity and their adsorption at interfaces of different nature. Lipases from different sources differ one from others by molecular mass and details of secondary structures show identical mechanism of the surface catalysis. Mechanism of lipolysis is investigated in detail using high purified pancreas lipases.

Pancreas lipase consists of 435 aminoacid resideaus, has 2- SH-groups and is protein of one polypeptide chain. M.M.-48000, pI -5. The contain of element of secondary structure is following: α-helices 19%, β-sheets 34% and irregular structure 47%. The drastic increase in catalytic activity of lipases is observed, when a substrate concentration exceeds the concentration of aqueous solution saturation and the interface between the enzyme solution and substate is appeared. This effect is determined not only by simple increase the surface area lipid-solution, that would be analogeous to increase in substrate concentration. When the scheme of lipase activity is represented as

$$E + S = [ES] \rightarrow EA + P_1 \quad (28)$$

and

$$EA \rightarrow E + P_2 \quad (28a)$$

where E is enzyme, S is substrate, ES enzyme-substrate complex, EA is acylated form of enzyme and $P_1$ and $P_2$ are products of lipolysis, it is of interest to present [S] as surface concentration. In this case Km is such a surface concentration of substrate, when $V_o = V_{max}/2$.

However, Km is depended on interfacial tension as well as surface potential and cannot be used as a measure of the formation of bimolecular enzyme-substrate complex. The apparent Km value of pancreatic lipase for triglycerides strongly depends on physical properties of the interface, and, respectively on adsorbed state of the lipase. Sometimes Km is determined as a reverse value of the constant of adsorption equilibrium $K_m = 1/K$.

High purified pancreatic lipase is obtained using labour-consuming method with lots of stages with enormeous volumes of organic solvents [136], that is the reason of its limitation for wide and industrial applications. The more simple methods of purification include adsorption of a lipase on hydrophobic surfaces, that allows to consider an especial type of chromatography as "hydrophobic" one. Sorbents for this chromatography are natural and synthetic polymers with hydroxylic groups esterified by aliphatic acids or modified by aliphatic amines with long hydrophobic chains. It is possible to eliminate stages of a sorbent synthesis. Silicagels with applied on their surface trioleine (olive oil) [137], or solid tripalmitine [138] were treated as specific adsorbents for lipases purification. Aqueous solutions of salts or surfactants can be used as an eluent for chromatography with solid sorbents. Table 7 shows the well correlation of lipase adsorption and catalytic activity.

Table 7
Correlation between the amount of adsorbed lipase and activity

| Amount of adsorbed lipase, % from total lipase in the system | Activity, referred to activity in homogeneous medium |
|---|---|
| 10 | 1.8 |
| 20 | 2.5 |
| 40 | 4.0 |
| 50 | 4.5 |
| 60 | 6.0 |
| 70 | 7.0 |

This is hydrolysis of water soluble tripropionine [So]=60 mg/ml, bulk enzyme concentration is $7.9 \cdot 10^{-8}$ M, pH = 8. In the reaction medium hydrophobizied glass beads were added, than specific activity and adsorption were recorded using the equation (29).

$$A[E] = [E_s]A_{rel}A_1 + ([E]-[E_s])A_1 \tag{29}$$

where [E] is the total enzyme concentration, $[E_s]$ is the quantity of lipase adsorbed on the solid surface, $A_1$ is specific activity of lipase in homogeneous phase, $A_{rel}$ is the relative activity of adsorbed lipase, A is the total enzymatic activity recorded for the system. Ten-fold increase in activity was recorded for adsorbed lipase at the hydrophobic surface, that is coated by lipase molecules within 0.3-1. The preliminary adsorption of the lipase at the same solid surface and inducing of thus immobilized enzyme in reaction medium with water-soluble substrate leads to the close value of activity of adsorbed lipase [139]. Thus immobilized enzyme does not lose the activity, when kept under dry conditions during a long time. Enhanced interface between aqueous solution and decane affects on lipolysis in an analogeous manner [140]. Other result, that can be deduced from these investigations is the essential dependence of activity of adsorbed lipase on the degree of the surface packing with a higher activity of isolated molecule occupying infinite surface. This result indicate the role of water molecules in adsorption layer in the process of lypolysis. Participation of water molecules in lypolisis leads to accelaration of the deacylation stage (eq.28a) for enzyme adsorbed on hydrophobic surfaces. The same conclusion was obtained studying the lipases from Candida rugosa and Humicola lanuginosa on cellulose. Number of facts shows, that lipase activity increases by addition of hydrophobic indifferent carriers, and does not change in the presence of hydrophilic ones. In the case of emulsified substrates indifferent surfaces added in reaction media lead to decrease in lipase activity due to competition in adsorption of lipase between two kinds of the surfaces. Bisides of this it has been noted, that organic solvents influence the catalytic ability of lipases [141,142]. Table 8 shows the dependence of adsorption of pancreatic lipase on hydrophobizied and hydrophilic glass beads on its concentration in aqueous solution, according to adsorption isotherm measurements. Adsorption of the lipase calculated according to Langmuir monolayer adsorption equation with $\Gamma_{max}$ $4.8 \cdot 10^{-8}$ mol/m$^2$ and $7.0 \cdot 10^{-8}$ mol/m$^2$ and adsorption constants $2.6 \cdot 10^7$ and $1.3 \cdot 10^7$ $M^{-1}$. Adsorption on hydrophobic solid surface is well described by Langmuir equation, with $\Gamma_{max} = 4.8 \cdot 10^{-8}$ mol/m$^2$ corresponding to real area ($s = 1/\Gamma_{max} \cdot N$) that can be occupied by native, non-deformed globule of the lipase. Data allow to conclude, that lipase adsorption layers at different surfaces (hydrophilic and hydrophobic) are quite different due to peculiarities of an orientation of the protein at interface, intermolecular interactions, induced conformational shifts or others resulting in the formation of an adsorption layer with different pattern.

Limiting stage of adsorption also depends on the adsorbent nature. Lipase cannot be adsorbed at an interface with low interfacial tension, except cases when colipase is in the content of the adsorption layer and specific enzyme-co-enzyme interactions of high energy promote the lipase adsorption. This was established using different interfaces, including holic salt additives[143]. Reasons of interfacial activation of lipase: orientation due to hydrophobic binding area on the lipase surface (hydrophobic head and hydrophilic tail), increase in concentration, a role of the surface diffusion of substrate molecules, an activation conformation change due to adsorption, contribution of water structure. The last question was considered in [144].

We have investigated substrates obtained by adsorption of olive oil on solids: silicagel, silicagel modified by lethitin and fluoroplast 3M with specific surfaces 0.1-0.4 m$^2$/g, as well as tripalmitin, siliconised glass beads (Serva Feinbiochemie Heidelberg).

In case of proteins as well as common surface active substances part of an interface is occupyed by molecule of adsorbed substance and another one by water molecules. The last part of a surface can be regarded as “free” surface, that allows to use for adsorption evaluation the theory of adsorption of gases.

Adsorption of lipase at these adsorbents seems to be as Langmuir isotherms and data are well linearised in corresponding coordinates. In spite of that, it was of interest to evaluate the contribution of protein-protein interaction at the surface in adsorption of different proteins, including lipase. Only lipase adsorption on the surface of hydrophilic glass beads revealed the value of $\chi$ distinguished from 0, in this case $\chi$=23.

Table 8
Adsorption of lipase on different solids

| Solids | Specyfic area of adsorbent (cm²/g) | Experiment | | | Langmuir Eq. | | | Equation (27) | |
|---|---|---|---|---|---|---|---|---|---|
| | | $\Gamma_{max}$ mg/m² | S nm² | time s | $\Gamma_{max}$ mg/m² | S nm² | K l/mol | K l/mol | $\chi$ |
| Glass beads | 158 | 2.5 | 32 | 5400 | 3.3 | 24 | $1\cdot10^7$ | $2\cdot10^6$ | 23 |
| Hydrophobizied glass beads | 152 | 1.8 | 43 | 15-20 | 2.3 | 35 | $2\cdot10^7$ | $2\cdot10^7$ | 0 |
| Fluoroplast | 7800 | 2.6 | 31 | 30-40 | 2.8 | 28 | $4\cdot10^6$ | $4\cdot10^6$ | 0 |

According to results of adsorption and activity of adsorbed enzyme investigations a fast method of lipase determination in any liquid has been elaborated, that can be use in sensors [145]. A drop of a liquid with supposed lipase activity is located on the glass plate coated with olive oil. For recording one can use high specific spreading out of the aqueous drop through the solid surface. The rate of sreading is proportional to lipase activity. Plates coated with the substate for lipase were prepared by following: a suitable quatity of solution of olive oil in ethyl ether is spreaded as a monolayer on the surface of glass. After evaporation of the solvent plates are ready to be applied. pH of an aqueous drop must be 8. Speading of the drop can be recorded using optical microscope using measurements of diameter of the drop during several time. The specific lipase activity is expressed as the ratio of spreading rate to the total dissolved protein. The limit of the lipase determination by means of this method is $1\cdot10^{-9}$ M. This method can be used for the determination of a lipase contain in foods, microbiological lipase sources, in drugs and for clinic uses.

Lipases show broad substrate specificity with high selectivity and operate in both aqueuos and anhydrous organic solutions. In the last case enzyme must be surrounded by tnin film of water and a system behaves as two phastic with liquid/liquid interfaces. In this area microemulsions with lipase located at the interface are of most interest. Reverse micelles (or "microemulsions" at higher water content) offer, in pricipale, a unique possibility to overcome problems caused by medium heterogeneity. Such systems spontaneously are formed in an organic solvent (such as alkanes, benzene or chloroform) when certain surfactant molecules and small amount of water are added. The resulting optically transporant, highly dynamic and thermodynamically stable system is build up of small water droplets, which are surrounded by a layer of surfactant molecules which form the interphase between dispersed water droplets and the organic phase. Radius of water droplets mainly depends on the ratio of molarity of water to the molarity of surfactant. Enzymes are solubilized in such water pool with remaining of their catalytic activity. Different studies on lipases in reverse micelles have been carried out so far have recently summarized. When lipase from Pceudomonas cepacia was studied in two microemulsions using lecithin or AOT as a surfactant, the preferential localization near the surfactant membrane was recorded for enzyme. The interfaces of microemulsion dropds on the basis of lecithin were less penetrable [146]. Published data on lipase catalysis in microemulsions are emormeous and by several reasons fall out from topic of this part.

Immobilization of lipases due to adsorption on membrane surfaces allows to control specificity and influence the direction of reversible enzymatic reations [147] and to realize processes using biphastic (water/organic) media.

How surfactants influence on lipase activity? As a rule an action of such molecules is connected with the "surface inhibition", but at high concentration of surfactant conformational change of lipase has been recorded.

Ellipsometry was used to study the adsorption/desorption of highly purified lipase from Humicola lanuginosa in mixtures with surfactants, at the solid/liquid interface [148]. The effect of surfactant was studied both when it was allowed to adsorb in mixture with lipase and when added after lipase adsorption. Silica surfaces, totally or partially methylated, were used and surfactants were SDS (anionic), $C_{12}E_5$ and a commercial alcohol ethoxylate (AE) (both nonionic). The system models laundary process (pH 9). Lipase did not adsorb until the system was diluted, that is during the rinsing period. This effect was found for all lipase-surfactant mixtures. The amount of adsorbed lipase was larger after rinsing in mixture with SDS, compared with $C_{12}E_5$. Results from addition of surfactant after lipase adsorption indicate that lipase was replaced by surfactant. This was found for SDS and $C_{12}E_5$ at both the hydrophobic surface and the surface with intermediate hydrophobicity.

A quite new field of application for lipase is as a component in commercially available detergent formulations to aid in the removal of fat containing substances. Laundary detergents are composed of a range of substances, e.g. enzymes (proteases, lipases), chelating agents and surfactants. In such complex system it is difficult to extract and define the mode of action of lipase specifically. Study of lipase adsorption from mixtures with some surfactants may be use to deep insight and to understand what happens at the interface, for example to what extent and at which stage lipase adsorbs.

Method of ellipsometry allows to observe in situ adsorption of proteins or other surface active substances at interfaces, it was applied for studies cleaning processes. The main observation is the increase of protein adsorption from a mixture with surfactant with dilution of solvent due to exchange of surfactant for protein when the concentration of surfactant decreses below a critical concentration in the range of its CCM.

Interactions of surfactants and proteins at air/water and oil/water interface have been discussed by Dickenson and Woskett [149]. They pointed out several parallels between these interactions and the behavior at a hydrophobic solid surface. At concentrations around CMC and above, the surfactants are most likely to dominate at the interface, because they have a higher ability to decrease the surface tension. The complex formation between the lipase and surfactant may influence the adsorption behavior. Adsorption of lipase on hydrophobized silica is maximal at pH 5 and is independent of pH in the range 7-10.6. The laundary processes are performed usually at pH 9.

Geluk [150] studied the adsorption of lipase from *Candida rugosa* on cellulose. Adsorption does not exeeds the monolayer and the lower was the degree of surface coating, the higher was the specific lipase activity. And for the lipase from *Humicola lanuginosa* it was found that the amount of adsorbed lipase decreased, activity increased with increment of wettability of the surface. In situ ellipsometry was used to study the adsorption/desorption of highly purified lipase from *Humicolla lanuginosa* in mixtures with surfactants at solid/liquid interface. Such studies are connected with a quite new field of applications for lipases in commercially available detergent formulations to aid in the removal of fat containing substances. Laundary detergents are composed of a range of substances, e.g. enzymes (proteases, lipases), chelating agents and surfactants. In such a complex system it is difficult to define the specyfic mode of the lipase actions. A study of the adsorption of lipase together with some relevant surfactants might shed some light on the interfacial processes, particularly, on the peculiarities of lipase adsorption. The effect of surfactants on the lipase adsorption was studied both, when enzyme was in mixtures with surfactants or surfactants were added after lipase adsorption. Solid surfaces were represented by hydrophobizied in different extent silica, and surfactants of different classes were used. Results indicate that lipases might be less effective in the laundary process when surfactant is present, but they may become active when the surfactant concentration is lowered due to rinsing. In all cases surfactants replace lipases from the solid surfaces. In our works the interation of lipase and other globular proteins (BSA and α-chymotrypsin) has been investigated in details. All proteins bond surfactant molecules in aqueous solutions with amount of bonded molecules depends on the protein-surfactant ratio. At low these values formation of such complexes does not lead to large perturbutions of the protein molecule. Inhibition of the lipase activity by sodium octhylbenzosulfonate was measured, when surfactant was added in the solution of lipase, or the surface of substrate was modified by preliminary adsorption of the surfactant on its surface. At the surfactant concentration of about $10^{-2}$ M the activity of lipase consists of only half of the lipase activity, measured in the absence of the surfactant. The same amount of the surfactant corresponds to adsorbed one at the surface coverage of substrate of about 0.1 with the similar inhibition of the lipase activity. The more the surfactant adsorption on the substrate surface the more inhibition extent is observed. Below a list of new practical applications of lipases adsorbed or immobilized on different carriers is given. Development of new lipases for removal of strains formed by greasy food is reviewed in [151]. Lipase from *Pseudomonas cepacia* rapidly and completely adsorbed on the surface of microporous polypropylen with high activity with respect to water-soluble substrate and very low activity to insoluble one. In the last case the low activity is due to a slow rate of substrate diffision inside the porous support. The same diffision phenomena can explain the complete change of fatty acid specificity of adsorbed lipase: lipase become mainly specific for short-chain acid esters, whereas the free enzyme is specific for long-chain ester.

## 4. ADSORPTION OF PROTEINS AT LIQUID/LIQUID INTERFACES.

Water and oil do not mix with each other because of high energetic cost associated with replacing water-water and oil-oil contacts by water-oil ones. This energetic cost is responsible for high interfacial tension between bulk oil and water (30-50 mN/m) and for sharpness of the interfacial region. The interfacial tension between immiscible liquids is determined by contributions of dispersion interactions and non-dispersion ones.

All peculiarity of the protein adsorption at liquid/liquid interfaces were recently reviewed in our paper [152].

When isotherms of the surface tension are compared with isotherms of the interfacial tension [22,152] the larger interfacial activity of proteins was established. This fact is connected with a series of phenomena. In the presence of proteins an increase in the hydrocarbon content in the aqueous phase comparing with two-phastic equilibrium water-hydrocarbon is observed. This is so called solubilization, which is well familiar for micellar systems of low molecular mass surfactants. Apparantly, hydrocarbons are bound to the protein non-polar sites resulting in a change of the hydrophilic-hydrophobic balance of the protein and a growing ability to act as a surfactant. The ultimate decrease in the interfacial tension correlates with the value of hydrocarbon solubilisation in aqueous protein solution and both parameters deminute with increasing of the size of hyrdocarbon molecule. In parallel with increase in hydrocarbon solubilization the formation of associated particles of protein and hydrocarbon in the aqueous phase with association number larger for small hydrocarbons. Often it can be observed, that in parallel to the solubilization decrease, adsorption of a protein at the interface protein solution-hydrocarbon is reduced approaching a value slightly higher, than that at a/w surface. Thus proteins reduce the excess interfacial energy and adsorb at interface differently, depending on the nature of equilibrium organic phase, the protein itself and the ability to solubilze small non-polar molecules.

For direct measurements of the protein adsorption at an interface the method of radioactive indicators was elaborated using tritium labeled proteins [153]. The method allows to measure the adsorption of proteins in a wide range of concentration down to very low values of $10^{-8}$-$10^{-5}$ M. Labeled proteins are obtained by means of tritium thermal activation. This operation replaces hydrogen atoms in C-H bonds by tritium, which also partially penetrates into labile positions of O-H, N-H and S-H bonds. Protein is purified from labile tritium by dyalisis. Measurements were performed using MARK-2 (USA) liqiud-scintillation counter in a standart dioxane scintillator.

In [154] the influence of added electrolytes was investigated on the system of two immissible liquids with proteins as the surface active substance. The necessity of such investigations is caused by the fact that addition of electrolyte affects the phase behavior of water-hydrocarbon-surfactant systems and often used to reach ultra low interface tensions. In the same time this problem is not investigated systematically for systems containing proteins. Three regions corresponding to various interfacial coverage are distinguished: dependences lgΓ(lgCe) and lg Csph (lgCe) have two linear regions and intermediate region. The first one corresponds to low interface packing with the protein, i.e at $\theta_s < 1$; the second one corresponds to the model of densely packed monolayer, and the third region corresponds to the formation of the third phase between two liquids with this phase may be visually observed. This phase is a thin layer of a middle phase. It should be pointed out that the amount of the protein, determined in the phase of hydrocarbon is large enough, and the transfer of protein is promoted by the contact of two liquid phases, this leads to the protein adsorption at interface and hydrocarbon solubilization in aqueous phase. After addition of salts the interface adsorption layer, that can be modelled as ultimate packed with protein monolayer becomes stable in the wider range of equilibrium concentration of the protein in aqueous phase and a transition to multilayers is observed only at larger protein concentrations. When multilayers are formed at the interface drastic reduction in the rheological parameters is observed. The addition of a salt decreases the flow limits (LiCl < KCl < CsCl < NaCl), the interface layer of the protein loses the solid-like properties and behaves as a non-newtonian (anomalous) liquids. The dependences of effective viscosities on the shear stress and differences between viscosities of multilayer with non-destroyed and destroyed structure show a real phase transition under salt addition.

To some extent the effect of the salt addition is similar to that observed for analogeous systems with low molecular mass surfactants: surfactant is concentrated at the interface resulting in the formation of a middle phases.

This method allowed to establish, that besides adsoption layer buildup at the interface the protein solution-organic phase, the distribution of the protein between liquid phases takes place. BSA and α-chymotrypsin at aqueous the protein solution-toluene were investigated in detail [155,156]. Transfer of the protein from the solution into interface layer as well as into organic phase was shown. In [157] the average hydrodynamic radius (r) of associates of BSA with non-polar molecules in the adjacent phases of aqueous solution and hydrocarbon were determined by the quasi-elastic light scattering method. Results are given in Table 9.

Associates of BSA with lipids were recorded after mutual saturation of the contacting aqueous and benzene phases. The saturation process is accompanied by transfer of protein and water from aqueous solution into the benzene phase and simultaneous transfer of lipid and benzene into the aqueous phases. While

Table 9
The influence of the nature and the concentration of a lipid and molarity of an acetic buffer on the diffusion coefficient ($D_o$) and the mean hydrodynamic radius (r) of scattering particles in aqueous and benzene phases: pH 6.7, T=298 K, $C_{BSA}=1.2\cdot10^{-5}$ M

| The initial composition of the nonpolar phase | Water phase | | | Benzene phase | | |
|---|---|---|---|---|---|---|
| | $D_o\times10^7$ cm²/s | r, nm | Measurement error,% | $D_o\times10^7$ cm²/s | r, nm | Measurement error,% |
| Buffer 0.01 M with: | | | | | | |
| Benzene | 3.27 | 7.9 | 1.5 | 0.34 | 107.0 | 0.5 |
| Lecithin solution in benzene | | | | | | |
| $1.3\cdot10^{-4}$M | 1.52 | 17.0 | 0.5 | 0.28 | 127.0 | 0.7 |
| $1.3\cdot10^{-3}$M | 1.43 | 18.0 | 0.6 | 5.43 | 6.6 | 1.9 |
| Cholesterol solution in benzene | | | | | | |
| $2.6\cdot10^{-4}$M | 1.61 | 16.0 | 1.0 | 0.40 | 90.0 | 1.3 |
| $2.6\cdot10^{-3}$M | 1.53 | 17.0 | 0.7 | 0.47 | 76.0 | 1.3 |
| Buffer 0.1 M with: | | | | | | |
| Benzene | 4.09 | 6.3 | 1.3 | 0.58 | 61.0 | 1.7 |
| Lecithin solution in benzene | | | | | | |
| $1.3\cdot10^{-4}$M | 1.84 | 14.0 | 1.4 | 1.29 | 28.5 | 1.2 |
| $1.3\cdot10^{-3}$M | 2.13 | 12.1 | 0.9 | 5.52 | 6.5 | 0.7 |
| Cholesterol solution in benzene | | | | | | |
| $2.6\cdot10^{-4}$M | 1.94 | 13.3 | 0.8 | 0.72 | 50.6 | 1.1 |
| $2.6\cdot10^{-3}$M | 2.15 | 12.0 | 1.7 | 2.07 | 17.3 | 0.9 |
| Buffer 1 M with: | | | | | | |
| Benzene | 4.58 | 5.9 | 0.8 | 0.74 | 48.3 | 1.2 |
| Lecithin solution in benzene | | | | | | |
| $1.3\cdot10^{-4}$M | 1.84 | 14.0 | 0.7 | 1.89 | 18.9 | 1.3 |
| $1.3\cdot10^{-3}$M | 2.48 | 10.4 | 1.5 | 5.87 | 6.1 | 0.6 |
| Cholesterol solution in benzene | | | | | | |
| $2.6\cdot10^{-4}$M | 2.21 | 11.7 | 1.2 | 0.81 | 44.1 | 0.8 |
| $2.6\cdot10^{-3}$M | | 11.2 | 1.0 | 2.32 | 15.4 | 1.2 |

considering two flows with opposite directions, it is necessary to take into account the process of the mass condensation of protein at interface. Note that the size of scattering particles in the aqueous phase in equilibrium with the hydrocarbon are essentially larger than the size of BSA moleculees in the aqueous solution. The size of BSA molecule in aqueous solutions of different concentration were measured using SANS [158]. In low concentrated solutions the diameter of hydrated BSA is 6.1 nm and increases slowly with concentration. The size of associates of protein-hydrocarbon reduces with increasing buffer molarity. Therefore large scattering particles were recorded in the organic phase. All processes lead to the equilibrium in the system with distribution of the protein between liquid phases and adsorption layer. At any rate the achievement of such equilibrium was proven for the both proteins. The last in this case, apparantly, consists of interacting particles, which formed by all components present in the system with two immiscible liquids. It is of interest to understand the reason for the large differences in size of the associates in water and in non-polar phases.

When the number of components in the system is increased, for example, by addition of lipids or salts, associates of another sizes are appeared. When lipid concentration is increased, but the protein concentration is kept constant, essential lyophilization of the system is observed. The amount of the protein in the interface layer increases, the interfacial tension is significantly lower, the rheological parameters of the protein interface layers are simultaneously reduced, revealing a phase transition in these layers. Besides of these changes the difference of the particles sizes present in aqueous and organic phases becomes negligeable. Such particles can be considered as a droplets of microemulsions. Their composition is determined by properties of the phases, in which they are found and by composition of the contacting liquid phases. But it is evident, that the structure of a particle has to satisfy to the requirement of integrating with the corresponding continuous liquid phase, i.e. reveal attributes of a direct or reverse dispersion.

All the processes taking place in parallel or sequentally in the system of two immissible liquids with proteins and other components may be divided into several groups: 1) formation of associates, 2) adsorption and fixation of the protein or its associate with other components at liquid-liquid interface, 3) phase transition from an intermediate interface layer into equilibrium one, 4) protein equilibrium distribution between interface and liquid macrophases. Apparentlly all components of the system have to be disrtibuted between liquid phases and interface layer in a determined manner. As for hydrocarbons, its solubilization in aqueous solutions of proteins has been investigated in detail, recently new data were obtained by NMR method, that reveal a specific state of benzene molecules in the gelatin interface layer differed from liquid benzene [159] and allow to conclude including of molecules of organic phase in the contain of interface layer. The distribution of water in such system is not sufficiently investigated. Peculiarity of the protein adsorption at liquid/liquid interfaces are of interest to stabilize emulsions of great practical significance. A number of questions on emulsion properties and production was the object of many investigations during a long period of time, and emulsion stability is the most general one among them. Current knowledge indicates that the problem of regulation of the emulsion stability is quite far from completion. At present a number of interesting new developments as well as new problems can be noted in emulsion chemistry and technology: the making of emulsions of small droplets (< 1 μm), involvement of the phase behaviour of two immiscible liquids in the presence of surface active components (thermodynamic factor) into the emulsion stability, destruction and phase inversion, influence of critical phenomena on the process of coalescence, multiple emulsion. It is also clear, that the description of concentrated emulsions (they are of interest for many practical cases) call for new kinetic and rheological approaches. Success in this scientific area determines the existence of many biotechnological processes, food productions and others. The study of concentrated emulsions can be regarded now as a special part of colloid science. Such systems can be obtained in different way. Although they are thermodynamically unstable, they can be perfectly well stabilised and show a long life-time and are often called "stable emulsions". This means, that investigations of rheological properties of interfacial layers, rheology (mechanics and flow) of emulsions and high molecular mass emulsifiers gain special significance. Many efforts have been undertaken to study the dependence of the coalescence process on rheological properties of protein interface layers in works of Izmailova and co-workers [22,152]. In these studies different proteins (human serum albumin (HSA), chymotrypsin, casein, gelatin) were used as emulsion stabilisers. The role of a structure-mechanical barrier in the stabilisation of concentrated emulsions is discussed in a book [36]. Stable emulsion drops behave as hard spheres [160], they are stable to the aggregation and reveal hindered sedimentation. The rate of this process is much lower than the rate of sedimentation of Stokes particles and decreases with increasing volume fraction of the emulsion ($\varphi$) [161]. At $\varphi \cong 0.4$ the sedimentation rate can be only 0.1 of the Stokes velocity. The deviation from the Stokes equation depends on the volume fraction, the width of the droplet size distribution, the Peclet number and viscosity of the continuous phase, which can even become non-Newtonian. The increase in concentration of the stabiliser leads to a lower degree of packing. Systems reveal a quasi-equilibrium state and tixotropic behaviour. The coalescence process, taking place in three-dimensional aggregated structures, is apparently the main reason for instability of emulsions stabilized by protein interfacial layers. The determining role of structure-rheological properties of interfacial layers formed at interfaces was confirmed investigating elements of emulsions, thin emulsion films [22,152].

## 5. CONCLUSIONS

Adsorption of a protein from an aqueous solution on interfaces is one of numerous natural properties of such substances and investigations of physicochemical principles of adsorption phenomena in this case are of important for biology, modern technology and environmental protection.

Ability of a native globular (folded) protein to be adsorbed is determined by its unique energetically ideally balanced molecular structure, with the surface of such molecules represented by hydrophobic and hydrophilic regions imparting to these components surface activity. Unfolding of protein molecules following adsorption is not obligatory attribute of adsorption process, because unfolding does not always result in energy gain of adsorption.

Protein adsorption is characterised by long time dependences (especially for adsorption at interfaces solution-air, or -organic phase), which are not described by simple diffusion models. Adsorption and desorption processes proceed with activation barriers, slowness of such processes reflects a number of processes parallel or successive to stohastic event of adsorption, resulting in free energy minimization. Several symptoms of adsorption irreversibilty led to insight on arrangement of protein adsorption layers after adsorption, reasons of which have to be investigated additionally by means of new methods and new approaches.

The main driving forces of the protein adsorption are hydrophobic and elecrostatic interactions. Nevertheless, insight in detail of protein adsorption in dependence on specific properties of solid surfaces can be regarded as the problem of important significance for current investigations.

New possibilities of gen engineering alow to control modification the surface properties of proteins and application of such proteins to investigate protein adsorption opens new horizonts in development of insight in this area. Change only one the "surface" amino acid in polypeptide chain of a protein does not affect conformation of protein molecule and its stability, but influences the hydrophobicity of the protein surface resulting in a modification of all properties, depending on its, including adsorption.

Essential influence of protein dimerization in bulk and at surfaces on adsorption can be noted because high values of dimerization constants, which are of about only one order of magnitude lower than known average constants of protein adsorption. "Natural" dimers of protein form due to specific intermolecular interactions as a rule have a higher surface affinity.

Adsorption of protein at "mobil" surfaces as compared with adsorption of proteins from solutions on solid adsorbents reveals additional phenomena: adsorption layers aquire specific rheological properties indicating formation of a new middle phase; adsorption can be accompanied by an extension and or stacking of rigid monolayer with nearly ultimate packing by protein and in the case of two liquid immiscible phases by a protein distribution between these phases, evidently in a form of associates of protein with other components of the system. These phenomena remain poorely investigated. Adsorption of a protein on mobil interfaces results in the formation adsorption layer with strong ability to stabilize foams or emulsions, respectively, and their elements- free and emulsion films, which can be as ultimately thin, black (thickness of such films is of about 10nm). Thermodynamics of black films determines the stability of dispersions and conditions of the protein stable film formation allow to optimize foam chromatography processes for well protein separation from different wastes. The example of the milk proteins separation from sub-crud whey is considered according protein adsorption at the surface solution-air, black films formation and peculiarities of thinning of foam films. Other possibilities to optimize technology of waste purification, such as parameters of a foam column and rate of whey bubbling also have been considered.

Negative results of protein adsorption, such as membrane fouling, thromboembolitic effects of implants, deffects of contact lens and others have been considered and some ways to reduce undesirable processes are given. To remove these difficulties it is useful to use low energetic surfaces of optimal surface energy with well water saturated surfaces, to avoid essential contributions of electrostatic interactions between the surfaces of protein and adsorbent, as well as conditions inducing denaturation of proteins in a solution. Materials with pre-adsorbed HSA, or a mixture of HSA and vitamin C are useful especially for biocompatibility improving.

Catalysis by adsorbed enzymes opens new outlooks for near-term chemical applications. Development of this area has been considered using lipases as an example.

## 6. NOTATION

Latin symbols

| | |
|---|---|
| $A$ | total enzyme activity |
| $A_l$ | enzyme activity in solution |
| $A_{rel}$ | activity of adsorbed enzyme relative to activity in solution |
| b | amount of water molecules bound to the surface of an adsorbent |
| C | concentration |
| $C_e$ | equilibrium concentration after adsorption |
| $C_f$ | concentration of a component in foam |
| $C_o$ | initial concentration |
| $C_s$ | concentration of protein solution forming black foam film |
| $C_{sph}$ | concentration of protein in organic phase in equilibrium with aqueous protein solution |
| D | fractal dimension of a cluster |
| Do | diffusion coefficient |
| d | fractal dimension of a surface cluster |
| $d_p$ | average thickness of adsorption layer |
| $E_{1S}$ | modulus of fast elasticity |
| $E_{2S}$ | modulus of slow elasticity |
| $E_{SS}$ | modulus of equilibrium elasticity |
| G | surface activity |
| h | film thickness |
| K | adsorption constant |

$K_m$ Michaelis constant
$k_{kat}$ catalytic constant of enzymatic reation
$k$ Boltzman's constant
$k_c$ rate constant
$l$ number of Langmuir monolayer
$m$ dimensionless exponent
$n$ refractive index
$Pi$ product of enzymatic reaction
$P_{K1}$ limit shear stress of Shvedov's creep
$P_{K2}$ limit shear stress of Bingham's flow
$P_{RS}$ critical shear stress
$P_\sigma$ capillar pressure
$Q$ adsorption function of reflectivity
$R$ universal gase constant
$R_f$ coefficient of concentrating
$r$ hydrodynamic radius
$r_f$ radius of thin film
$S$ surface occupied by adsorbed molecule in adsorption layer
$S_o$ minimal surface occupied by adsorbed molecule
$S_i$ surface fraction of a component
$T$ temperature
$t$ time
$V$ reaction rate
$V_o$ hydrodynamic volume of macromolecule
$V_{max}$ maximal rate of enzymatic reaction
$V_p$ volume of protein solution
$x_i$ mole fraction of a component
$W_a$ amount of adsorbent

Greek symbols
$\alpha$ extraction level
$\alpha_D$ constant of diffusion mass transfer
$\beta$ purification coefficient
$\beta_L$ adsorption layer arrangement coefficient
$\Gamma$ adsorption
$\Gamma_{max}$ ultimate adsorption
$\Gamma_1$ excess adsorption
$\Gamma_2^{(x)}$ excess adsorption of solvent
$\Gamma_1^{(u)}$ absolute adsorption of a component
$\Gamma_2^{(u)}$ absolute adsorption of a solvent
$\Delta$ tension of film
$\Delta^*$ free energy of a black film

$\delta$ length scale parameter
$\eta$ viscosity
$\eta_s$ surface viscosity
$\eta^*$ surface Bingham's viscosity
$\theta$ contact angle
$\theta_s$ surface packing extent
$\theta_{spo}$ reflection signal
$\chi$ constant of protein-protein interaction in adsorption layer
$\lambda$ elasticity of adsorption layer
$\pi$ constant = 3.14
$\sigma$ surface tension
$\sigma_a$ surface tension of solvent
$\sigma_b$ surface energy of a polymer
$\sigma_f$ surface tension of a black film
$\sigma_o$ initial surface tension
$\sigma_t$ surface tension at time t
$\sigma_\infty$ mesoequilibrium surface tension
$\Phi$ fraction of free surface
$\varphi$ emulsion volume fraction
$\varphi_b$ volume fraction of a component in bulk
$\varphi_b^{(1)}$ volume fraction of adsorbed component
$\omega$ surface of lattice unit

Abbreviations
AE etoxylate alcohol
Asp aspartic acid
ATR attenuated total reflection
BSA bovine serum albumin
CCM critical concentration of micellization
CD circular dichroism
Cys cystine
DLVO Deryaguin-Landay-Vervej-Overbeek theory
E enzyme
EA acylenzyme
Fab domain of IgG molecule
Fc domain of IgG molecule
FTIR fluorescence total internal reflection
Gly glycine
HSA human serum albumin
IgG immunoglobulin
Ile isoleucine
IR infrared spectroscopy
$\alpha$-La $\alpha$-lactalbumin
$\beta$-Lg $\beta$-lactglobulin

LTIC low temperature isotropic carbon
MASIF surface forces adhesion method
NMR nuclear magnetic resonance
RNase ribonuclease
S substrate
SA serum albumin
SANS small angle neutron scattering
SDS sodium dodecyl sulphate
SPA surface forces apparatus
SPO surface plasmon oscillation
Trp tryptophan
Val valine
UV ultraviolet spectroscopy

## REFERENCES

1. G.P.Yampolskaya, V.N.Izmailova, G.Z.Raznikova and P.V.Nuss, Izvestia Akademii Nauk, ser. phisicheskaia, 59, 3 (1995) 109.
2. T.Yamashita and H.B. Bull, J. Colloid. Interface Sci., 24 (1967) 310.
3. G.Z.Raznikova, G.P.Yampolskaya, P.V.Nuss and V.N.Izmailova, Kolloidn. Zh., 55, 6 (1993) 110.
4. M. Said, D.Rosen and J.B.Hasted, J.Chem.Soc.Faraday Trans.1, 85, 1 (1989) 99.
5. W. Nitsch, R. Maksimov and M.Erdmann, J.Colloid Interface Sci., 141 (1991) 322.
6. T.Nakagawa, M. Kakimoto, T. Miwa and M. Airawa, Thin Solid Films, 202 (1991) 151.
7. M.Demeny, S. Kochwa and H. Sobotka, J.Colloid.Interface Sci., 22 (1966) 144.
8. F.J.Holly, K. Dolow and K.M. Yamada, J. Collod Interface Sci., 100 (1984) 210.
9. V.V.Erohin, Thesis, Moscow, Institut Crystallographii Akad.Nauk, Russia, 1991.
10. W.Nitsch and R. Maksimov, Colloid and Polym. Sci., 268 (1990) 452.
11. N.F. Zhou and B.A. Pethica, Langmuir, 2 (1986) 4750.
12. F. McRitchie, Adv. Colloid and Interface Sci., 25 (1986) 341.
13. L.TerMinnassian Saraga, Pure Appl. Chem., 57 (1985) 621.
14. G.P.Yampolskaya, P.V.Nuss, G.Z.Rasnikova and V.N.Izmailova, Uspehi kolloidnoi khimii, Leningrad, Khimia (1991) 292.
15. Yu.M.Lvov and G.B.Suhorukov, Biological membranes, 14, 3 (1997) 229.
16. P.G.de Gennes, Preprint Faraday Symposium on molecular motion, 1972.
17. F. Matsubara and Y.Kurokawa, Colloid and Surfaces, 29 (1987) 209.
18. D. Frenkel and K. Eppenga, Phys. Rev., 31 (1985) 1776.
19. D.A.Ward and F.Lado, Mol. Phys., 63 (1988) 623.

20. L. Onsager, Ann. N.Y. Acad. Sci., 52 (1949) 627.
21. J.Wesemann, L. Qin and P. Siders, Langmuir, 5 (1989) 1358.
22. V.N.Izmailova, G.P.Yampolskaya and B.D.Summ, Surface phenomena in protein systems, (ed.) Khimia, Moscow, Russia, 1988.
23. C.Tripp, J.J. Magda and J.D. Andrade, J. Colloid. Interface Sci., 173 (1995) 16.
24. J. Benjamins, J.A. de Feijter, M.T.A. Evans, D.E. Graham and M.C.Phillips, Faraday Discuss. Chem.Soc., 59 (1975) 218.
25. J.Benjamins, J.A. de Feijter, M.T.A. Evans, D.E.Graham and M.C. Phillips, Faraday Discuss. Chem. Soc., 59 (1975) 254.
26. D.E.Graham and M.C. Phillips, J.Colloid Interface Sci., 70 (1979) 415.
27. M.C.Phillips and C.E.Sparks, Ann. N.Y. Acad.Sci., 348 (1980) 122.
28. J.M.G.Lankveld and J. Lyklema, J.Colloid Interface Sci., 41 (1972) 454.
29. P.Rehbinder, Z.Phys. Chem., 111 (1924) 447.
30. P.A.Rehbinder, ZhRPhChO ser. phys., 56 (1924) 530.
31. Weisje van der Vegt, W. Norde, H.C. van der Mei and H.J. Busscher, J.Colloid Interface Sci., 179 (1996) 57.
32. G.Serrien, G. Geeraerts, G. Ghosh and P. Joos, Colloids Surf., 68 (1992) 219.
33. V.N.Izmailova, B.N.Tarasevich, T.F.Busol and G.M.Pismennaya, Kolloidn. Zh., 39 (1977) 958.
34. T.F.Busol, G.M.Pismennaya, B.N.Tarasevich and V.N.Izmailova, Kolloidn. Zh., 41 (1979) 1055.
35. A.N. Krasovskii and I.A. Belousova, Zhurn. prikl. khimii, 69, 1 (1996) 142.
36. V.N.Izmailova and P.A.Rehbinder, Structure formation in protein systems, (ed.) Nauka, Moscow, Russia, 1974.
37. V.N.Izmailova, G.P.Yampolskaya, L.E.Bobrova and Z.D.Tulovskaya, in: Surface phenomena and surface active substances, A.A. Abramson and E.D.Shchukin (eds.), Khimia, Leningrad, Russia, 1984.
38. F.M.Richards, Ann.Rev. Biophys. Bioeng., 6 (1977) 151.
39. G.P.Yampolskaya, N.Rangelova, L.E.Bobrova, D.Platikanov and V.N.Izmailova, Biofisika, 5 (1977) 939.
40. D.Platikanov, G.P.Yampolskaya, N. Rangelova, Zh. Angarska, L.E.Bobrova and V.N.Izmailova, Kolloidnyi Zh., 6 (1980) 893.
41. Zh. Angarska, G.P.Yampolskaya, L.E.Bobrova and V.N.Izmailova, Kolloidnyi Zh., 3 (1980) 424.
42. D.Platikanov, G.P.Yampolskaya, N. Rangelova and V.N.Izmailova, Kolloidnyi Zh., 1 (1981) 177.
43. N.Rangelova, V.N.Izmailova, D.Platikanov, G.P.Yampolskaya and Z.D.Tulovskaya, Kolloidnyi Zh., 3 (1990) 515.
44. B. Dimitrova, Zh. Angarska, L.E.Bobrova, G.P.Yampolskaya and V.N.Izmailova, Vestnik Moskow Univ., ser. Khimia, 5 (1982) 476.
45. V.N.Izmailova, Zh. Angarska, G.P.Yampolskaya and Z.D.Tulovskaya, Kolloidnyi Zh., 1 (1981) 155.
46. B.V.Deryagin, G.A Martynov and Yu. V. Gutop, Kolloidnyi Zh., 3 (1965) 357.

47. A.I.Rusanov, Phase equilibrium and surface phenomena, Khimia, Leningrad, Russia, 1967.
48. I.A. deFeijter and A.Vrij, J.Colloid Interface Sci., 2 (1978) 258.
49. B.Toshev and I. Ivanov, Colloid Polymer Sci., 7 (1975) 593.
50. A.Sheludko, B.Radoev and T.Kolarov, Trans. Faraday Soc., 8 (1968) 2213.
51. N.Rangelova, V.N.Izmailova, D.Platikanov, G.P.Yamploskaya, Z.D.Tulovskaya and S.M.Levachev, Kolloidnyi Zh., 3 (1992) 133.
52. J.N. de Wit and G. Klakenbeek, J. Dairy Sci., (1984) 2701.
53. V.A.Pavlov, New methods for whey processing, Russia, Stavropol, (1990).
54. M.Mulder, Basic Principles of Membrane Technology, Kluwer Academic Publishers, 1991.
55. K.Ostermaier and B. Dobias, Colloid Surf., 14 (1985) 199.
56. S.F.Kuzkin, Flotation of Ions and Molecules, Moscow, Nedra, 1971.
57. I.L.Marhasin, V.N.Izmailova, L.K.Utyasheva and V.D.Nasarov, Itogi nauki et techn., ser. Okhrana Prirody Vosproisv. Prirodn. Resursov, Moscow, 20,1988.
58. V.N.Izmailova, S.R.Derkach, K.V.Zotova and R.G.Danilova, Kolloidn.Zh., 55, 3 (1993) 54.
59. S.R.Derkach, V.N.Izmailova and K.V.Zotova, Kolloidn.Zh., 53, 6 (1991) 1030.
60. V.N.Izmailova, S.R.Derkach, G.P.Yampolskaya, K.V.Zotova and V.N.Chernin, Colloid Journal, 59, 5 (1997) 654.
61. A.I.Rusanov, O.A.Levichev and V.T.Zharov, Surface separation of substances. Theory and Methods, Khimia, Leningrad, Russia, 1981.
62. P.M.Kruglyakov and D.Ekserova, Foam and Foam Films, Khimia, Moscow, Russia, 1990.
63. M.Gururaj, R.Kumar and K.S. Gandhi, Langmuir, 4 (1995) 1381.
64. P.A.Haas and H.F. Johnson, Ind.Eng.Chem.Fundam., 6 (1967) 225.
65. Yu.A.Ivashchuk, V.N.Izmailova and G.P.Yampolskaya, Kolloidn.Zh., 48, 4 (1986) 556.
66. A. Bhaduri and K.P. Das, J.Disper.Sci., 15, 2. (1994) 165.
67. L.M.Malinovskaya, Yu.V.Goryunov, V.N.Izmailova and B.D.Summ, Vysokomol. soed. B., 26, 2 (1984) 88.
68. M.Landgren and B.Jonsson, J.Phys.Chem., 97 (1993) 1656.
69. B.Lassen and M. Malmsten, J.Colloid Interface Sci., 179 (1996) 470.
70. J.A.de Feijter, J.Benjamnis and F.A. Veer, Biopolymers, 17 (1978) 1759.
71. J.Buijs, P.A.W. van der Berg, J. W.Th.Lichtenbelt, W.Norde and J.Lyklema, J.Colloid Interface Sci., 178 (1996) 594.
72. J.R.Rahn and R.B. Hallok, Langmuir, 11, 2. (1995) 650.
73. C.E.Jordan, B.L.Frey, S.Kornguth and R.M.Corn, Langmuir, 10, 10 (1994) 3642.
74. V.Straaten and N.A.Peppas, J.Biomater. Polym., (ed.) 2 (1991).
75. F-N. Fu and B.R. Singh, Appl.Spectroscopy, 47 (1993) 98.

76. C-G. Golander, Y-S.Lin, V.Hlady and J.D. Andrade, Colloids Surf., 49 (1990) 289.
77. D.Kim, W.Cha and R.L.Beissinger, J.Colloid Interface Sci., 159, 1 (1993) 1.
78. V.Dulm and W.Norde, J.Colloid Interface Sci., 91, 2 (1983) 248.
79. W.Norde and E.Rouwendal, J.Colloid Interface Sci., 139, 1 (1990) 169
80. P.M.Claesson, E.Blomberg, J.C. Froberg, T.Nylander and T.Arnebrant, Adv.Colloid Interface Sci., 57 (1995) 161.
81. C.A.Haynes and W.Norde, Colloid and Surf.B: Biointerfaces, 2 (1994) 517.
82. M. Kleijn and W.Norde, Heterogeneous Chem.Rev., 2 (1995) 157.
83. W.Norde, Cells and Materials, 5, 1 (1995) 97.
84. C.A.Haynes and W. Norde, J.Colloid Interface Sci., 169 (1995) 313.
85. W.Norde, IOP-Industrial Proteins, 2 (1995) 34.
86. W.Norde, Macromol.Symp., 103 (1996) 5.
87. T.A. Horbett and J.L.Brash (eds.), Proteins at Interfaces, ACS Symp.Ser., 1995.
88. A.Sheer and A.Smolders, J.Colloid Interface Sci., 63, 1 (1978) 7.
89. S.Magdassi (ed.), Surface Activity of Proteins, Marcel Dekker, Inc., N.Y., Basel, Hong Kong (1995).
90. S.Duinhoven, R.Poort, G. Van der Voet, W.G.M.Agterof, W.Norde and J.Lyklema, J.Colloid Interface Sci., 170 (1995) 340.
91. W.Norde, Adv.Colloid Interface Sci., 25 (1986) 267.
92. S.M.A.Cohen, M.A.G.J. Fleer, J.Lyklema, W.Norde and J.M.H.M. Scheutjens, Adv.Colloid Interface Sci., 34 (1991) 477.
93. W.Norde and J.Lylkema, J.Biomater.Sci.Polym., (ed.) 2 (1991) 183.
94. H.P.Jenninssen, Macromol. Chem. Macromolec. Symp., 17 (1988) 111.
95. B.Ivarsson and I. Lundsrom, CRC Crit. Rev. Biocompat., 2 (1986) 1.
96. J.D.Andrade, in: Surface and Interfacial Aspects of Biomedical Polymers 2, Plenum Press, N.Y., 1985.
97. W.Norde, Dispersion Sci.Technol., 13 (1992) 363.
98. V.N.Izmailova, G.P.Yampolskaya, P.Lapina and M.M.Sorokin, Kolloidnyi Zh., 44, 2 (1982) 217.
99. J. McGuire, M.C. Wahlgren and T.Arnebrant, J.Colloid Interface Sci., 170, 1 (1995) 182.
100. J. Buijs, W.Norde and J.W.Th Lichtenbelt, Langmuir, 12, 6 (1996) 1605.
101. M.C.L.Maste, H.A.Rinia, C.M.J.Brands, M.R.Egmond and W.Norde, Biochim. Biophys. Acta, 1252 (1995) 261.
102. A.Kondo, S.Oku and K. Higashitani, J.Colloid Interface Sci., 143 (1991) 214.
103. G.P.Yampolskaya, V.K.Maso, V.N.Izmailova and M.V.Citkovskii, Vestnik Moskovskogo Univ., 2 (1974) 208.
104. O.M.Poltorak and E.S.Chuhrai, Physical and chemical basis of enzymatic catalysis, Wysshaya shkola (ed.), Moscow, Russia, 1971.
105. A.V.Elenskii, G.P.Yampolskaya and V.N.Izmailova, Kolloidn. Zh., 1 (1987) 143.
106. Y.Kata and T.Hasimoto, J.Chromatography, 266 (1983) 49.

107. A.Asapov, L.Deluck, P.Oldman and W.Wilson, J.Colloid Interface Sci., 191, 1 (1997) 222.
108. F.E.Reiner, Science, 328 (1984) 321.
109. A.A.Gorbunov, J.Chromatogr., 365 (1986) 205.
110. U.M.Elofsson,M.A.Paulsson and Th.Arnebrant, Langmuir, 13 (1997) 1695.
111. F.Pincet, E. Perez and G. Belfort, Langmuir, 11 (1995) 1229.
112. I.G.Wenton and G. Jonsson, Proceedings ICOM'96 (1996) 216.
113. S.A. Palkar and A.M.Lenhoff, Colloids and Surfaces, A. 110, 1 (1996) 119.
114. C.S.Chern, C.K.Lee, C.Y.Chen and M.J.Yeh, Colloids and Surfaces, B. 6, 1 (1996) 27.
115. E.Iritani, Y. Mukai and T. Murase, Proceedings ICOM'96 (1996) 222.
116. K. Kokubo, T.Sunohara and K.Sakai, Proceedings ICOM'96 (1996) 98.
117. T. Hashimoto, M. Yonehara and A. Higuchi, Proceedings ICOM'96 (1996) 262.
118. H.Tsuge, M. Hiruma, M.Tsuzuki and N. Kamiyama, Proceedings ICOM'96 (1996) 854.
119. E.Ruckenstein and Gorisiankar, J. Colloid Interface Sci., 101, 2 (1984) 436.
120. D.Hardey, J.Herron and J.D.Andrade, in: T.Horbott and J.Brash (eds.), Proteins at Interfaces, ACS symp.ser.A, (1987) 290.
121. E.J. Castillo, J.L.Koenig and J.M.Anderson, Biomaterials, 7, 2 (1986) 89.
122. T.Chandy and C.P.Sharma, J.Colloid Interface Sci., 108, 1 (1985) 104.
123. M.Malmsten and Bo Lassen, J.Colloid Interface Sci., 166, (1994) 490.
124. L.Feng and J.D.Andrade, J.Colloid Interface Sci., 166, 2 (1994) 419.
125. M.Harvey and R.Mc. Donogh, Proceedings ICOM'96 (1996) 770.
126. T.Blunk, T. Lueck, A.Calvoez, D.F.Hochstrasser, J.C.Sanchez, B.W.Mueller and H.Rainez, Eur. J.Pharm. Biopharm., 42, 4 (1996) 262.
127. N.Lloyd, F. Thomas and M. William, J. Amer. Oil Chem. Soc., 73, 9 (1996) 1191.
128. M.D.Jonzo, A.Hiol and D. Druet, J.Chem.Technol. Biotechnol., 69, 4 463.
129. A.Tuli and A.Telefocu, Clin.Acta Turc., 24, 2 (1996) 145.
130. S.Matsumura, H.Beppa, K.Tsukada and K.Toshima, Biotechnol.Lett., 18, 2 (1996) 1041.
131. T.Shau-Wei and D. S.Jonathan, Biotechnol.Bioeng., 52, 2 (1996) 296.
132. P.K.Closh, R.K.Saxema, G.Rani, R.P.Yadov and D.Sheba, Sci.Progr., 79, 2 (1996) 119.
133. M.Wataru, S.Hirota, M.Nobukazu and H.Toshuki, Japan Patent No. 08 214 890 (1996).
134. H.de Castro, E.B. Pereiza and W.A. Anderson, J.Braz. Chem. Soc., 7, 4 (1996) 219.
135. A. Makasci, K.Arisay and A.Telefocu, Turk. J. Chem., 20, 3 (1996) 258.
136. R.Verger, G.H.de Haas, L.Sarda and P.Desnuelle, Biocem.Biophys.Acta, 188 (1969) 272.
137. G.P.Yampolskaya, A.V.Elenski, V.K.Maso, V.N.Izmailova and V.A.Shaternokov, SU Patent No. 5 608 88, M.Kl C 07 G 7/026 (1977).

138. A.V.Elenski, V.K.Maso, G.P.Yampolskaya and E.Yu.Zlotnikova, Vestnik Moskovskogo Univ., Ser. Biologie, 2 (1977) 87.
139. A.V.Elenski, G.P.Yampolskaya and V.N.Izmailova, Kolloidnyi Zh., 1 (1987) 143.
140. B. Estressangles and P. Denuelle, Biochim. Biophys.Acta, 341, 1 (1974) 437.
141. H.Stamatis, A.Xenakis and M.Provelegiou, Biotechnol. Bioeng., 42, 1 (1993) 103.
142. I.Uemasu and W.L.Hinze, Chirality, 6, 3 (1994) 649.
143. G.P.Yampolskaya, V.N.Izmailova and Z.D.Tulovskaya, Itogi nauki i techniki, Biotechnologia, 4 (1987) 199.
144. A.V.Elenski, G.P.Yampolskaya, V.N.Izmailova and I.D.Parfenova, Vestnik Moskovskokgo Univ., Ser.Khimia, 5 (1986) 531.
145. V.N.Izmailova, B.D.Summ, G.P.Yampolskaya, K.D.Higginson, A.V.Elenski and M.V.Kozhinskaya, SU Patent No.1 341 175 (1986).
146. G. Hedstrom, S. Bacjlund and F.Eriksson, Colloid. Polymer Sci., 2 (1997) 146.
147. K.Belafi-Bako, E.Csanyi and D.E. Nagy, Proceedings of ICOM'96 (1996) 812.
148. K.Wannerberger, M.Wahlgren and T. Arnebrant, Colloids Surfaces, B, 6, 1 (1996) 27.
149. E.Dickenson and C.M.Wosklett, in: R.D. Bee (ed.) Food and Colloids, London, 1989.
150. M.A.Geluk, W.Norde, H.K.A. Van Kalsbeek and K. Van't Riet, Enzyme Microbiol.Technol., 14, 3 (1992) 748.
151. O.Masset, Surfactant Sci.Ser., (Enzyme in Detergency) 69 (1997) 107.
152. V.N.Izmailova and G.P.Yampolskaya, in: Studies of Interface Sci., D.Mobius and R.Miller (eds.) 7 (1998) 103.
153. A.Yu.Alentiev and E.S.Filatov, Radiokhimia, 33 (1991) 80.
154. V.V.Pelekh, A.Yu.Alentiev, G.P.Yampolskaya and V.N.Izmailova, Kolloidnyi Zh., 56, 1 (1994) 73.
155. G.P.Yampolskaya, E.N.Bogacheva and V.N.Izmailova, Kolloidnyi Zh., 44 (1982) 1151.
156. A.Yu.Alentiev, V.N.Izmailova and G.P.Yampolskaya, Kolloidnyi Zh., 54, 1 (1992) 14.
157. S.M.Levachev and V.N.Izmailova, Kolloidnyi Zh., 56, 1 (1994) 193.
158. R.Nossal, C.J.Glinka and S-H.Chen, Biopolymers, 25 (1986) 1157.
159. V.V.Rodin and V.N.Izmailova, Colloids Surfaces, A, 106, 1 (1996) 95.
160. T.G.Mason, J.Bibette and D.A. Weitz, J. Colloid Interface Sci., 179, 2 (1996) 439.
161. P.Walstra, in: Encyclopedia of emulsion technology, P.Becher (ed.), Marcel Dekker, Inc., N.Y., Basel, Hong Kong, 4 (1994) 1.

Adsorption and its Applications in Industry and Environmental Protection
Studies in Surface Science and Catalysis, Vol. 120
A. Dabrowski (Editor)



# Protein adsorption dissociation constants in various types of biochromatography

D.C. Patel[1] and Robert G. Luo

Department of Chemical Engineering, Chemistry and Environmental Science
New Jersey Institute of Technology, Newark, New Jersey 07102, USA

The equilibrium kinetics, adsorption and desorption of various proteins on chromatographic adsorbents have been widely studied and published in the literature. What is lacking is a comprehensive review of the available information. In this article the equilibrium data has been summarized and presented in the form of dissociation constants. This summary leads to a chart of relative protein adsorption strength of various types of chromatography in terms of dissociation constants. Some of the factors affecting dissociation constants such as ligand concentration, pH, ionic strength and temperature have also been discussed. The results of this study can be used as important information for design, development, scale up and optimization of chromatographic processes for protein separations.

## 1. INTRODUCTION

Usually over half of the production costs of biotechnology products occur in downstream processing, which often involves centrifugation, filtration, extraction, adsorption and chromatography steps. Liquid chromatographic techniques play a very important role in meeting high purity demands of these products. Separations of proteins and other bioproducts require extensive usage of chromatography. In recent years significant part of chromatographic literature has been devoted to improving chromatographic techniques associated with protein separation. These studies focus on broad range of parameters related to unique design of mobile and stationary phases, optimum elution conditions, adsorption kinetics etc. Each of these parameters also has a bewildering amount of published literature. In this study, we have sorted through this entanglement, summarized and classified protein adsorption data based on one such parameter, dissociation constant, $K_d$, for various types of

---

[1] D.C. Patel wishes to thank the Society of Plastic Engineers, Newark Section for scholarship support.

chromatography. It is our hope that the results of this study will provide an individual working on protein separation with necessary information about design and development of chromatographic processes.

In chromatographic process the solute or in our case the protein is dissolved in the mobile phase and then fed at the top of the chromatography column which is packed with the stationary phase. The dissolved protein is then transported convectively through the length of the column where it comes in to intimate contact with stationary phase. During this contact protein, designated by P, binds to different extent to an available active site, designated by S. This interaction may be described by a second order reversible adsorption process where the protein is assumed to interact with the ligand by a monovalent interaction,

$$P + S \underset{k_2}{\overset{k_1}{\Leftrightarrow}} PS \tag{1}$$

where PS is the protein-ligand complex. The parameters $k_1$ and $k_2$ are forward and backward rate constants for the adsorption process respectively. The rate of adsorption for this type of interaction is given by

$$r_q = k_1 c(q_m - q) - k_2 q \tag{2}$$

where c and q are the protein concentrations in the liquid and the adsorbent, respectively, and $q_m$ represents the maximum adsorption capacity of the adsorbent. At equilibrium, Eq. (2) reduces to the well known Langmuir isotherm,

$$q^* = \frac{(q_m c^*)}{(K_d + c^*)} \tag{3}$$

where $c^*$ is protein concentration in mobile phase at equilibrium, $q^*$ is protein concentration in stationary phase at equilibrium, and $K_d$ is dissociation constant given by

$$K_d = \frac{k_2}{k_1} \tag{4}$$

In Eq. (3), $K_d$ is a measure of the protein-adsorbent binding strength. The lower the $K_d$, the stronger the binding. When $c^*$ becomes sufficiently large, $q^*$ approaches $q_m$. The other limiting case occurs when $c^*$ is "small" relative to $K_d$, the isotherm then becomes linear [1].

Some dissociation constants listed in this article were directly taken from the literature, while others were calculated by the authors based on the protein adsorption isotherms reported in the literature. In order to calculate $K_d$, Eq. (3) was rearranged into the following form,

$$\frac{c^*}{q^*} = \frac{c^*}{q_m} + \frac{K_d}{q_m} \tag{5}$$

This equation represents a linear form in a plot with $\frac{c^*}{q^*}$ vs. $c^*$. $K_d$ and $q_m$ can then be calculated by curve fitting the experimental isotherm data. The dissociation constants given by literature and calculated by the above described method are grouped according to the types of chromatographic techniques: affinity, hydrophobic, inorganic, ion exchange and reversed phase.

## 2. AFFINITY CHROMATOGRAPHY

Affinity chromatography is a method widely used in separation and purification of proteins. This process takes advantage of protein's natural "affinities" to certain ligand sites. The stationary phase of an affinity chromatography is comparised of a porous organic polymer or a silica gel support which is bonded via a reactive spacer arm to a ligand site. Affinity chromatography can be subdivided into biospecific, dye-ligand and metal chelate.

### 2.1. Biospecific chromatography

There are two forms of biospecific chromatography. In one form of biospecific chromatography, ligands such as enzymes and co-factors affixed to stationary supports bind to proteins. Smith and Sefton [2] investigated the equilibrium behavior of protein bovine thrombin binding to one such ligand. They reported a dissociation constant of around $10^{-6}$ M for bovine thrombin binding to a heparin on polyvinyl alcohol at 22°C in polyethylene glycol. Dalmia and Nikolov [3] found the dissociation constant of glucoamylase I adsorbing to raw starch cellulose varies with different pH values as shown in Figure 1. Figure 1 shows that the maximum binding of glucoamylase I occurring at pH 4.5.

Another form of biospecific chromatography employs the unique properties of monoclonal or polyclonal antibodies and their antigens. The $K_d$ of immobilized antibodies for the target protein is usually quite low as shown in Table 1. Low $K_d$ is an indication of very strong binding. Liapis et al. [4] examined the equilibrium behavior of hen egg lysozyme in a 50 mM TRIS-HCl buffer (pH 7.4) and 0.2 M NaCl at ambient temperature. $K_d$ values of $4.7 \cdot 10^{-10}$ M and $1.4 \cdot 10^{-8}$ M were reported for lysozyme binding to the antibody supported by silica and Sepharose respectively. The range of $K_d$ for biospecific chromatography is $10^{-4}$ to $10^{-10}$ M as shown in Table 1.

Table 1
Protein dissociation constant, $K_d$, in various types of biochromatography

| Column | Surface | Protein | Kd (M) | pH | Ref. |
|---|---|---|---|---|---|
| Affinity Adsorption | | | | | |
| *Biospecific* | | | $10^{-4} – 10^{-10}$ | | |
| | anti-β-galactosidase Sepharose 4B | β-galactosidase | $1.4 \cdot 10^{-8}$ | 7.2 | [28] |
| | silica anti- β -galactidose | β -galactosidase | $4.7 \cdot 10^{-10}$ | 7.4 | [28] |
| | silica monoclonal antibody | lysozyme | $3.5 \cdot 10^{-6}$ | 7.4 | [4] |
| | polyvinyl alcohol - heparin | bovine thrombin | $2.5 \cdot 10^{-6}$ | 7.2 | [2] |
| | p-aminobenzamidine-aminopropyl silica | trypsin | $1.32 \cdot 10^{-4}$ | | [27] |
| | raw starch | glucoamylase I | $2.0\text{-}5.0 \cdot 10^{-7}$ | | [3] |
| *Dye Ligand* | | | $10^{-5} – 10^{-8}$ | | |
| | Cibacron Blue Sepharose CL-6B | alcohol dehydrogenase | $1.5\text{-}2.6 \cdot 10^{-6}$ | 7.0 | [1] |
| | Cibacron Blue Sepharose CL-6B | bovine serum albumin | $2.5\text{-}0.6 \cdot 10^{-6}$ | 7.5 | [1] |
| | Cibacron Blue Sepharose CL-6B | lysozyme | $0.1\text{-}1.0 \cdot 10^{-7}$ | 7.5 | [1] |
| | Cibacron Blue Sepharose CL-4B | bovine serum albumin | $1.1\text{-}1.3 \cdot 10^{-6}$ | | [7] |
| | Cibacron Blue Sepharose CL-6B | lysozyme | $1.7 \cdot 10^{-6}$ | 7.2 | [28] |
| | Cibacron Blue Sepharose CL-6B | bovine serum albumin | $8.5 \cdot 10^{-6}$ | 7.2 | [28] |
| | Cibacron Blue 3GA | formate dehydrogenase | $5.0 \cdot 10^{-8}$ | | [36] |
| | Cibacron Blue 3GA | L-lactate dehydrogenase | $1.6 \cdot 10^{-7}$ | | [36] |
| | Cibacron Blue 3GA | alcohol dehydrogenase | $2.6 \cdot 10^{-6}$ | | [36] |
| | Cibacron Blue-Dextran conjugates (20D) | lysozyme | $9.7 \cdot 10^{-7}$ - $3.7 \cdot 10^{-5}$ | | [38] |
| | Cibacron Blue-Dextran conjugates (5D) | lysozyme | $5.8 \cdot 10^{-7}$ - $2.39 \cdot 10^{-5}$ | | [38] |
| *Metal Affinity* | | | $10^{-4} – 10^{-7}$ | | |
| | IDA $Cu^{2+}$ -modified zirconia (3M) | myoglobin | $2.83 \cdot 10^{-7}$ | | [9] |
| | IDA $Cu^{2+}$ -modified zirconia (PDZ) | myoglobin | $1.76 \cdot 10^{-7}$ | | [9] |
| | Con A modified zirconia | horseradish peroxidase | $1.45 \cdot 10^{-6}$ | | [9] |
| | $Cu^{2+}$ -chelating Sepharose | lysozyme | $3.7 \cdot 10^{-5}$ | | [10] |
| | $Cu^{2+}$ TSK gel. chelate-5PW | Lysozyme | $7.7 \cdot 10^{-5}$ | | [10] |

Table 1
Protein dissociation constant, $K_d$, in various types of biochromatography

| Column | Surface | Protein | Kd (M) | pH | Ref. |
|---|---|---|---|---|---|
| | $Cu^{2+}$-chelating Sepharose | ovalbumin | $6.8\cdot10^{-6}$ | | [10] |
| | $Cu^{2+}$ TSK gel chelate-5PW | ovalbumin | $6.8\cdot10^{-6}$ | | [10] |
| | $Cu^{2+}$ -chelating Sepharose | bovine serum albumin | $4.2\cdot10^{-6}$ | | [10] |
| | $Cu^{2+}$TSK gel chelate-5PW | bovine serum albumin | $9.6\cdot10^{-6}$ | | [10] |
| | $Cu^{2+}$ -chelating Sepharose | pig serum albumin | $1.0\cdot10^{-6}$ | | [10] |
| | $Cu^{2+}$ TSK gel chelate-5PW | pig serum albumin | $2.7\cdot10^{-6}$ | | [10] |
| | TSK Guardgel Chelate-5PW (12.0 Cu mM/ml gel) | horse heart cytochrome c | $1.88\cdot10^{-4}$ | | [11] |
| | TSK Guardgel Chelate-5PW (10.5 Cu mM/ml gel) | horse heart cytochrome c | $2.381\cdot10^{-4}$. | | [11] |
| | TSK Guardgel Chelate-5PW (9.5 Cu mM/ml gel) | horse heart cytochrome c | $2.5\cdot10^{-4}$ | | [11] |
| | TSK Guardgel Chelate-5PW (7.6 Cu mM/ml gel) | horse heart cytochrome c | $2.32\cdot10^{-4}$ | | [11] |
| | $Cu^{2+}$ metal chelate Superose | Myoglobin | $6.56\cdot10^{-4}$ | | [33] |
| | $Cu^{2+}$ metal chelate Superose | lysozyme | $5.61\cdot10^{-5}$. | | [32] |
| | $Cu^{2+}$ metal chelate Superose | RNase A | $4.93\cdot10^{-5}$ | | [32] |
| | $Cu^{2+}$ metal chelate Superose | Lactoferrin | $2.77\cdot10^{-6}$ | | [32] |
| | $Cu^{2+}$ metal chelate Superose | α-chrymotrypsin | $2.116\cdot10^{-4}$ | | [32] |
| | IDA $Cu^{2+}$ Sepharose 6B | Carboxypeptidase | $5.14\cdot10^{-5}$ | | [35] |
| Hydrophobic Interaction | | | $10^{-4} - 10^{-6}$ | | |
| | Hydrophobic Silica | Modified forms of ovalbumin | $1.3\text{-}1.9\cdot10^{-5}$ | 10.0 | [13] |
| | SynChroprop Propyl HIC @ 15°C | α-chymotrypsinogen A | $3.70\cdot10^{-6}$ | 5.0 | [14] |
| | SynChroprop Propyl HIC @ 25°C | α-chymotrypsinogen A | $1.49\cdot10^{-6}$ | 5.0 | [14] |
| | SynChroprop Propyl HIC @ 35°C | α-chymotrypsinogen A | $1.86\cdot10^{-6}$ | 5.0 | [14] |
| | Vydac Silica Gel | β-lactoglobulin A in 0.85 $M(NH_4)_2SO_4$ | $1.11\cdot10^{-4}$ | 4.5 | [26] |
| | Vydac Silica Gel | β-lactoglobulin A in 1.0 $M(NH_4)_2SO_4$ | $3.7\cdot10^{-5}$ | 4.5 | [26] |
| | Butyl Toyopearl 650S | ovalbumin | $4.41\cdot10^{-6}$ | | [39] |
| | TSK-Butyl polystyrene | lysozyme | $6.6\cdot10^{-6}$ | | [25] |

Table 1
Protein dissociation constant, $K_d$, in various types of biochromatography

| Column | Surface | Protein | Kd (M) | pH | Ref. |
|---|---|---|---|---|---|
| | TosoHass 650M Phenyl | lysozyme | $9.12 \cdot 10^{-4}$ | | [34] |
| Inorganic Adsorption | | | $10^{-5} - 10^{-8}$ | | |
| | Human Enamel Beads (Calcium Hydroxyapatites) | human serum albumin | $1.2\text{-}3.3 \cdot 10^{-7}$ | | [15] |
| | Hematite Spheres (α-Fe203) | ovalbumin | $1.2 \cdot 10^{-7}$ | 5.0 | [16] |
| | Hematite Spheres (α-Fe203) | lysozyme | $1.09 \cdot 10^{-6}$ | 6.0 | [16] |
| | Chromium hydroxide ($Cr(OH)_3$) | γ-globulin | $7.12 \cdot 10^{-8}$ | 6.0 | [16] |
| | Hydroxyapatite | lysozyme | $2.76 \cdot 10^{-5}$ | | [30] |
| | Hydroxyapatite | bovine serum albumin | $3.38 \cdot 10^{-6}$ | | [30] |
| Ion Exchange | | | $10^{-5} - 10^{-7}$ | | |
| | "Protion" DEAE-Sephadex | bovine serum albumin | $3.73 \cdot 10^{-7}$ | 6.5 | [29] |
| | Sephadex A-50 | bovine serum albumin | $6.6 \cdot 10^{-6}$ | 6.9 | [20] |
| | Amberlite CG-50 | RNase A | $2.0 \cdot 10^{-7}$ | 6.2 | [19] |
| | DEAE Sepharose CL-6B | β-lactoglobulin A | $1.34 \cdot 10^{-5}$ | 7.9 | [39] |
| | DEAE Sepharose CL-6B | β-lactoglobulin B | $2.25 \cdot 10^{-5}$ | 7.9 | [39] |
| | DEAE Sepharose CL-6B | bovine serum albumin | $3.6 \cdot 10^{-6}$ | 7.9 | [39] |
| | DEAE Trisacryl M | human serum albumin | $2.63 \cdot 10^{-7}$ | 6.0 | [31] |
| | DEAE Sepharose FF | human serum albumin | $1.25 \cdot 10^{-6}$ | 6.0 | [31] |
| | DEAE FractogeI 650 m | human serum albumin | $3.7 \cdot 10^{-7}$ | 6.0 | [31] |
| | DEAE Trisacryl M | carbonic anhydrase | $2.38 \cdot 10^{-7}$ | 11.5 | [31] |
| | Chitopearl 3510 | bovine serum albumin | $2.23 \cdot 10^{-5}$ | 6.9 | [40] |
| | Chitopearl 3510 | bovine serum albumin | $1.03 \cdot 10^{-6}$ | 5.1 | [40] |
| | Chitopearl 3510 | bovine serum albumin | $7.43 \cdot 10^{-5}$ | 8.8 | [40] |
| | DEAE Sephadex A-50 | bovine serum albumin | $4.56 \cdot 10^{-7}$ | 6.9 | [40] |
| | DEAE Sephadex A-50 | bovine serum albumin | $6.6 \cdot 10^{-6}$ | 6.5 | [40] |
| | Chitosan ion exchanger - Ch-2503 | bovine serum albumin | $1.617 \cdot 10^{-6}$ | | [41] |
| | LiChrospher Silica | β-lactoglobulin B | $2.7 \cdot 10^{-6}$ | | [37] |

Table 1
Protein dissociation constant, $K_d$, in various types of biochromatography

| Column | Surface | Protein | Kd (M) | pH | Ref. |
|---|---|---|---|---|---|
| Reversed Phase | | | $10^{-3}$ – $10^{-5}$ | | |
| | Ocytl-and Octadecyllkylsilicas* | BSA (0.05 M Phosphate) | $7.5 \cdot 10^{-5}$ | 3.2 | [24] |
| | Ocytl-and Octadecyllkylsilicas | BSA (30% 2-Propanol**) | $1.01 \cdot 10^{-3}$ | 3.2 | [24] |
| | Ocytl-and Octadecyllkylsilicas | BSA (40% 2-Propanol**) | $1.94 \cdot 10^{-3}$ | 1.9 | [24] |
| | Ocytl-and Octadecyllkylsilicas | BSA (0..01% Neodol**) | $1.64 \cdot 10^{-4}$ | 2.6 | [24] |
| | Ocytl-and Octadecyllkylsilicas | BSA (1.0% Neodol**) | $1.64 \cdot 10^{-3}$ | 2.6 | [24] |
| | Ocytl-and Octadecyllkylsilicas | BSA (0.05 M Phosphate) | $4.48 \cdot 10^{-4}$ | 1.3 | [24] |
| | Ocytl-and Octadecyllkylsilicas | BSA (0.01% Neodol**) | $5.97 \cdot 10^{-4}$ | 1.2 | [24] |
| | Ocytl-and Octadecyllkylsilicas | BSA (1.0 Neodol**) | $4.48 \cdot 10^{-3}$ | 1.5 | [24] |

*$C_8$ and $C_{18}$ alkylsilicas. **in 0.05 M Phosphate.

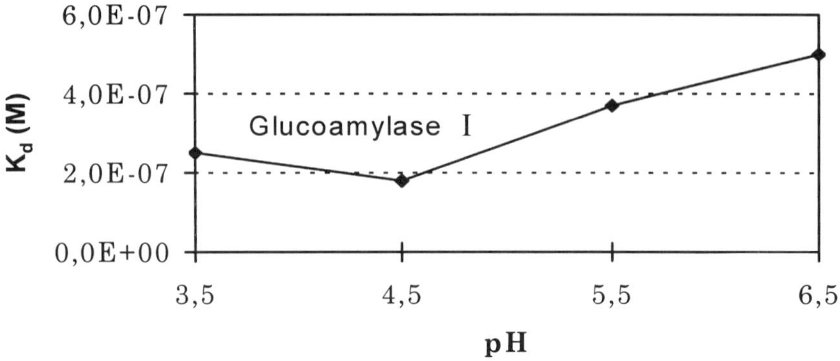


Figure 1. Effect of pH on dissociation constant, $K_d$, for glucoamylase I on adsorbent raw starch [3].

## 2.2. Dye ligand chromatography

The next form of affinity chromatography uses immobilized, reactive dyes as ligands (for example, Cibacron Blue and Procion Red [5]). The ligands are linked to polymeric supports via spacer arms. This type of affinity chromatography is also called dye-ligand chromatography. Dye-ligands are more general adsorbents but they do show their preferences for certain classes of proteins. Proteins and enzymes interact with the spacer arm, the matrix, and the dye itself. The nature of these interactions can be quite complicated (there are specific and nonspecific hydrophobic, electrostatic and other interactions that come into play).

In an effort to better understand the specific protein adsorption mechanisms in dye ligand chromatography and to improve the process of selecting buffers, Boyer [6] investigated the adsorption of lysozyme and bovine serum albumin to Cibacron Blue F-3GA on Sepharose CL-6B. In both cases, the author varied pH and ionic strength of buffers MES-OH and TRIS-HCl. Ionic strength was varied from 0.02 to 1.00 M and pH from 5.5 to 8.5. Boyer then fit the experimental data to Langmuir isotherm. Langmuir isotherm revealed that the dissociation constant values for lysozyme increased with increase in pH and ionic strength as shown in Figures 2 and 3, respectively. $K_d$ values for lysozyme ranged from $1.74 \cdot 10^{-6}$ M (pH of 5.5 and ionic strength of 0.02 M in a MES-OH buffer) to $2.14 \cdot 10^{-6}$ M (pH of 8.5 and ionic strength of 1.00 M in a TRIS-HCl buffer) as shown in Table 2. The author noted that the relative insensitivity to variations in these two parameters suggested that the forces in the protein adsorption were more of the hydrophobic, Van der Waals, or charge transfer type rather than cation exchange. $K_d$ of bovine serum albumin also increased with increase in pH and ionic strength as shown in Figures 2 and 3, respectively. $K_d$ values for bovine serum albumin ranged from $1.3 \cdot 10^{-7}$ M (pH 5.5 and ionic strength 0.02 M in a MES-OH buffer) to $1.02 \cdot 10^{-4}$ M (pH 6.0 and ionic strength of 0.5 M in a MES-OH buffer) as shown in Table 3.

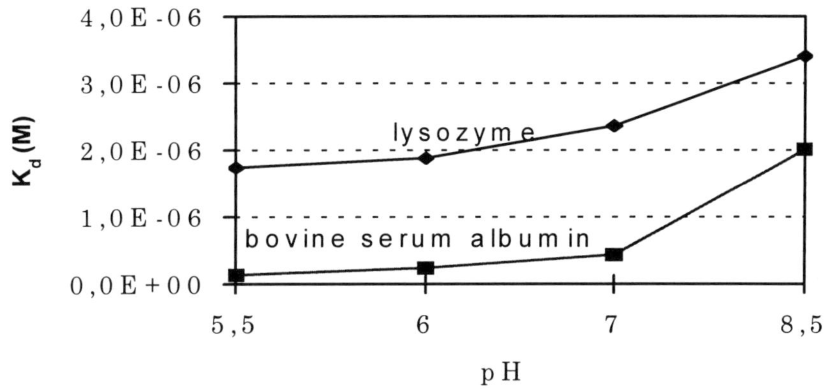


Figure 2. Effect of pH on dissociation constant, $K_d$, for lysozyme and bovine serum albumin on Cibacron Blue Sepharose CL-4B [6].

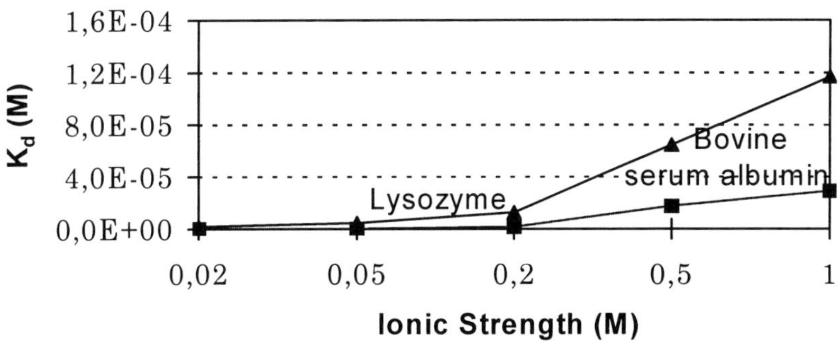


Figure 3. Effect of ionic strength on dissociation constant, $K_d$, for lysozyme and bovine serum albumin on Cibacron Blue Sepharose CL-4B [6].

Table 2
Variation of $K_d$ of lysozyme on Cibacron Blue F-3 GA with buffer pH and ionic strength [6]

| Ionic Stength [M] | $K_d$ | | | | | |
|---|---|---|---|---|---|---|
| | pH 5.5* | 6.0* | 7.0* | 7.0** | 7.5** | 8.5** |
| 0.02 | $1.74 \cdot 10^{-6}$ | $1.88 \cdot 10^{-6}$ | $2.36 \cdot 10^{-6}$ | $2.36 \cdot 10^{-6}$ | $2.01 \cdot 10^{-6}$ | $3.4 \cdot 10^{-6}$ |
| 0.05 | $4.58 \cdot 10^{-6}$ | $5.83 \cdot 10^{-6}$ | $3.68 \cdot 10^{-6}$ | $3.2 \cdot 10^{-6}$ | $1.88 \cdot 10^{-6}$ | $5.42 \cdot 10^{-6}$ |
| 0.20 | $1.28 \cdot 10^{-5}$ | $1.22 \cdot 10^{-5}$ | $1.28 \cdot 10^{-5}$ | $1.35 \cdot 10^{-5}$ | $5.3 \cdot 10^{-6}$ | $1.86 \cdot 10^{-5}$ |
| 0.50 | $6.46 \cdot 10^{-5}$ | $4.92 \cdot 10^{-5}$ | $4.43 \cdot 10^{-6}$ | $4.72 \cdot 10^{-5}$ | $5.86 \cdot 10^{-5}$ | $4.09 \cdot 10^{-5}$ |
| 1.00 | $1.17 \cdot 10^{-4}$ | $1.32 \cdot 10^{-4}$ | $1.47 \cdot 10^{-4}$ | $1.49 \cdot 10^{-4}$ | $9.51 \cdot 10^{-5}$ | $2.14 \cdot 10^{-4}$ |

*MES-OH buffer **TRIS-HCl buffer

Table 3
Variation of $K_d$ of bovine serum albumin on Cibacron Blue F-3 GA with buffer pH and ionic strength[6]

| Ionic Strength [M] | $K_d$ (M) | | | | | |
|---|---|---|---|---|---|---|
| | pH 5.5* | 6.0* | 7.0* | 7.0* | 7.5** | 8.5** |
| 0.02 | $1.34 \cdot 10^{-7}$ | $2.39 \cdot 10^{-7}$ | $4.33 \cdot 10^{-7}$ | $5.07 \cdot 10^{-7}$ | $5.07 \cdot 10^{-7}$ | $2.01 \cdot 10^{-6}$ |
| 0.05 | $2.39 \cdot 10^{-7}$ | $3.28 \cdot 10^{-7}$ | $7.46 \cdot 10^{-6}$ | $7.16 \cdot 10^{-6}$ | $9.7 \cdot 10^{-7}$ | $6.27 \cdot 10^{-7}$ |
| 0.20 | $1.49 \cdot 10^{-6}$ | $6.12 \cdot 10^{-5}$ | $1.87 \cdot 10^{-5}$ | $3.22 \cdot 10^{-5}$ | $2.89 \cdot 10^{-6}$ | $1.21 \cdot 10^{-5}$ |
| 0.50 | $1.58 \cdot 10^{-5}$ | $1.02 \cdot 10^{-5}$ | $1.78 \cdot 10^{-6}$ | $1.63 \cdot 10^{-5}$ | $3.93 \cdot 10^{-5}$ | $3.08 \cdot 10^{-5}$ |
| 1.00 | $2.91 \cdot 10^{-5}$ | $1.69 \cdot 10^{-4}$ | $5.07 \cdot 10^{-6}$ | $2.61 \cdot 10^{-5}$ | $4.46 \cdot 10^{-5}$ | $2.94 \cdot 10^{-5}$ |

*MES-OH buffer **TRIS-HCl buffer

Boyer and Hsu [1] studied the equilibrium adsorption of alcohol dehydrogenase onto Cibacron Blue F-3GA affixed to Sepharose CL-6B in a potassium phosphate, pH 7.0 mobile phase at 4°C and found that depending on the concentration of the immobilized dye, $K_d$ fell in the range of 1.0 - $2.6 \cdot 10^{-6}$ M. The relative effect of ligand concentration on $K_d$ for various proteins based on the results of Boyer and Hsu are shown in Figure 4. Arnold and Blanch [7] have found that the $K_d$ values fall within the range 1.1 - $1.3 \cdot 10^{-6}$ M for bovine serum albumin adsorbing to Cibacron Blue Sepharose CL-4B. They also reported the variation of $K_d$ for bovine serum albumin is due to temperature, as shown in Figure 5. The range of $K_d$ for dye ligand chromatography is $10^{-5}$ to $10^{-8}$ M as shown in Table 1.

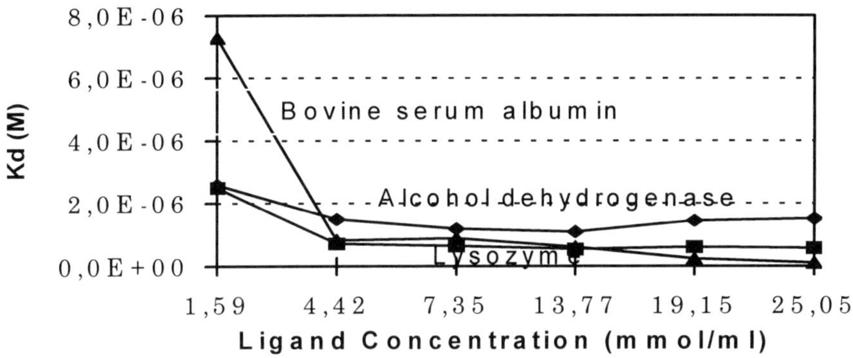


Figure 4. Effect of ligand concentration on dissociation constant, $K_d$, for alcohol dehydrogenase, bovine serum albumin, and lysozyme. Alcohol dehydrogenase experiments were carried out in 10 mM K-Pi, pH 7.0; bovine serum albumin and lysozyme experiments in 50 mM Tris-HCl, pH 7.5 [1].

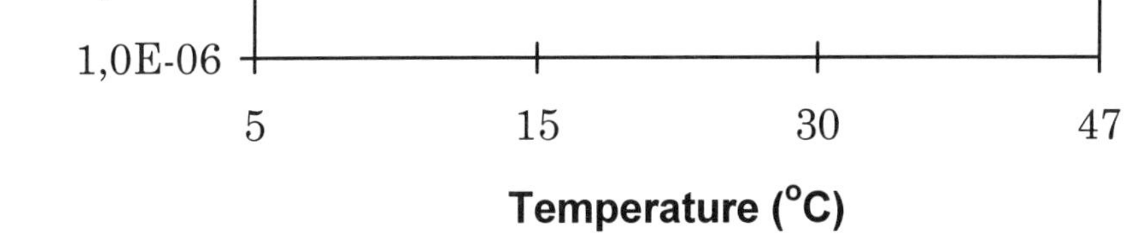


Figure 5. Effect of temperature on dissociation constant, $K_d$, for bovine serum albumin on Cibacron Blue Sepharose CL-4B [7].

## 2.3. Immobilized metal affinity chromatography

Another form of affinity chromatography that is applied in separation of protein is immobilized metal affinity chromatography (IMAC), also known as metal chelate affinity chromatography (MCAC). In IMAC, metal ions are affixed to adsorbents by chelating agents such as iminodiacetic acid (IDA), and tris (carboxymethyl)-ethylenediamine (TED) via a long hydrophilic spacer arm. The principle behind IMAC is that many transition metal ions like zinc and copper can coordinate to amino acids such as histidine, cysteine and tryptophan via electron donor groups on the amino acid side chain [8].

Wirth and Hearn [9] examined the equilibrium behavior of myoglobin adsorbing to IDA $Cu^{2+}$ immobilized on zirconia. They found $K_d$ for myoglobin to be around $10^{-7}$ M. Belew et al. [10] investigated the equilibrium behavior of several proteins adsorbing to $Cu^{2+}$ affixed to TSK gel chelate-5PW and chelating Sepharose. Their $K_d$ values are in the range of $10^{-5}$ to $10^{-6}$ as shown in Table 1. Todd et al. [11] reported the effect of $Cu^{2+}$ concentration on dissociation constant of horse heart cytochrome c. Variation of $K_d$ of horse heart cytochrome c with different $Cu^{2+}$ concentration is shown in Figure 6. It can be seen that the maximum binding of the horse heart cytochrome c to $Cu^{2+}$ occuring at $Cu^{2+}$ concentration of $9.50 \cdot 10^{-4}$ M. The range of $K_d$ for IMAC chromatography is $10^{-4}$ to $10^{-7}$ as shown in Table 1.

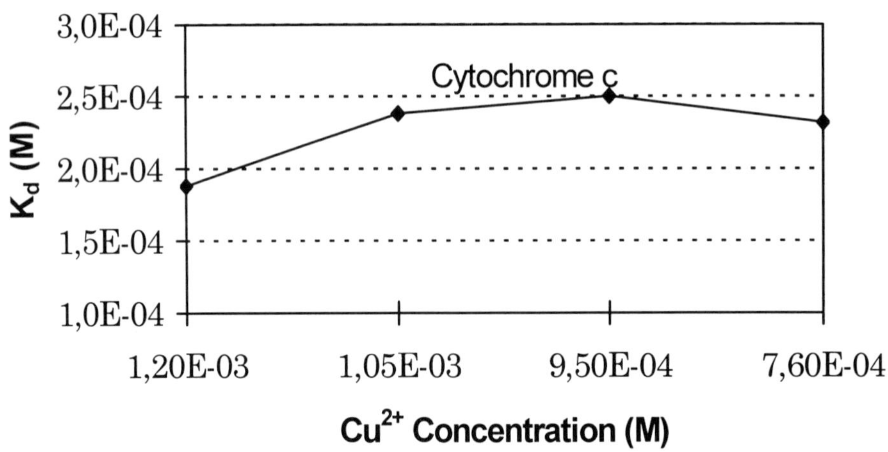


Figure 6. Effect of copper concentration on dissociation constant, $K_d$, for cytochrome c on TSK-Guardgel Chelate 5PW [11].

## 3. HYDROPHOBIC INTERACTION CHROMATOGRAPHY

Hydrophobic interaction chromatography (HIC) works on the principle that proteins and enzymes have hydrophobic regions that bind to aliphatic chains, phenyl groups, butyl groups and other hydrophobic ligands affixed to silica and polymeric matrices. Sample molecules containing hydrophobic characteristics are usually applied to a HIC column in a high ionic strength (high salt concentration) buffer [12].

Magdassi et al.[13] have reported $K_d$ values in the range of 1.3 - $1.9 \cdot 10^{-5}$ M for ovalbumin in sodium carbonate buffered solution interacting with hydrophobic silica. Tongta et al. [14] have examined the effect of temperature on dissociation constant of α-chymotrypsinogen A adsorbing to SynChroprep Propyl HIC. This study showed that maximum and minimum binding of α-chymotrypsinogen to SynChroprep Propyl HIC occurs at 25°C and 15°C respectively. The range of $K_d$ for hydrophobic chromatography is $10^{-4}$ to $10^{-6}$ as shown in Table 1.

## 4. INORGANIC ADSORPTION CHROMATOGRAPHY

In this type of adsorption, buffered proteins bind to inorganic surfaces through polar and dipole-dipole forces. Vogel et al. [15] examined the equilibruim behavior of human serum albumin at 20-27°C range in a 0.15 M NaCl aqueous solution binding to human enamel beads (calcium hydroxyapatite). Their study revealed a $K_d$ of $10^{-7}$ M. Other examples of inorganic adsorbents are hematite and chromium hydroxide. The experimental data from Johnson and Matijevic [16] were curve fitted by us to yield dissociation constant values of $1.2 \cdot 10^{-7}$, $1.09 \cdot 10^{-6}$, and $7.12 \cdot 10^{-8}$ M for ovalbumin, lysozyme and γ-globulin respectively. The range of $K_d$ for inorganic chromatography is $10^{-5}$ to $10^{-8}$ M as shown in Table 1.

## 5. ION EXCHANGE CHROMATOGRAPHY

Ion exchange resins are used as the stationary phase in ion exchange chromatography, a separation procedure in which ions of like charges are separated by elution from a column packed with finely divided resin. Ion exchange chromatography is very effective in separation of proteins since proteins have $NH_2$, and COOH functional groups that readily ionizes to $NH_3^+$ and $COO^-$ respectively under appropriate pH of the solution. The charged protein in this type of solution or the buffer diffuses through the pores of the packing, displaces the same charged counter ion and then binds to the opposite charged resin, thus ions are exchanged on the resin. Separations occur since each protein binds to different extent depending on their isoelectic points [17].

Synthetic ion exchange resins are high molecular weight polymeric materials containing many ionic functional groups per molecule and associated counter ions bonded to the inside surface layer of their pores. Cation exchange resins can be either a strongly acidic SP (sulphopropyl) group or a weakly acidic CM (carboxy methyl) group. Anion exchange resins can be either a strongly basic QAE (quarternary amino ethyl) group or a weakly basic DEAE (diethylamino ethyl) group [18].

Aranyi and Boross [19] examined the adsorption equilibria between the weak cation exchanger Amberlite CG-50 and various enzymes including bovine pancreatic RNase A. They obtained a Langmuir isotherm for RNase A in a 0.1 M phosphate buffered solution at pH 5.9. These data were than curve fitted by us to yield a dissociation constant, $K_d$, of $2.0 \cdot 10^{-7}$ M. Tou and Graham [20] experimented on an anionic exchanger DEAE Sephadex A-50 with protein bovine serum albumin (BSA) in 1.02% NaCl and 0.005 M phosphate buffer system at pH 6.9. The equilibrium data for this experiment also fit the Langmuir isotherm with a $K_d$ of $6.6 \cdot 10^{-6}$ M. Yamamoto et al. [21] used DEAE Sepharose CL-6B to study the adsorption of various proteins in a 0.11 ionic strength potassium phosphate pH 7.9 aqueous solution. Our curve fittings for these data revealed $K_d$ of $6.3 \cdot 10^{-6}$ M, $7.8 \cdot 10^{-6}$ M, and $2.3 \cdot 10^{-6}$ M for BSA, $\beta$-lactoglobulin A, and $\beta$-lactoglobulin B respectively. The range of $K_d$ for ion exchange chromatography is $10^{-5}$ to $10^{-7}$ M as shown in Table 1.

## 6. REVERSED PHASE CHROMATOGRAPHY

In reversed phase liquid chromatography (RPLC), one uses hydrophobic interactions between the biological molecules and the ligands on the chromatographic support to obtain seperation. RPLC stationary phases differ from HIC packings by a higher density of hydrophobic ligands [22]. RPLC also has a non-polar stationary phase and a polar mobile phase (as opposed to "normal" phase where the opposite is applied). In reversed phase the order of elution of components is determined by their polarity, i.e., the most polar elutes first [17].

The support for this type of chromatography is usually silica or polymer matrices and the ligands are usually aliphatic chains and other non-polar groups. Although reversed phase chromatography sometimes disrupts protein tertiary structure because of the high ligand density of the adsorbent medium and the organic solvents that are used, many proteins regain activity when re-suspended in aqueous buffer. Trypsin, poliovirus proteins and HIV protease are three cases where full activity is restored after RPLC purification [23]. Hancock [24] reported $K_d$ values of $10^{-3}$-$10^{-5}$ M for the adsorption of bovine serum albumin with various mobile phases at 23°C on octyl and octadecyllyl bounded silicas as shown in Table 1.

## 7. CONCLUSIONS

In this article we have summarized the dissociation constants for various types of chromatography frequently used in separation of proteins. The complete summary of all dissociation constants discussed in this paper is given in Table 1. Some of the factors that influence dissociation constants are also presented. From this summary, a chart of relative protein adsorption strength was developed based on the values of dissociation constants, as shown in Figure 7. This chart describes the range of protein dissociation constant in the term of Log $K_d$ in different types of biochromatography and compares the relative protein adsorption strength among various chromatographic processes. Notice that Figure 7 was developed based on the data collected from the currently available literature, and should be updated when new data and information are published. Also notice that the $K_d$ for biospecific chromatography covers a wider range while the $K_d$ for others such as HIC covers relatively narrower range in Figure 7.

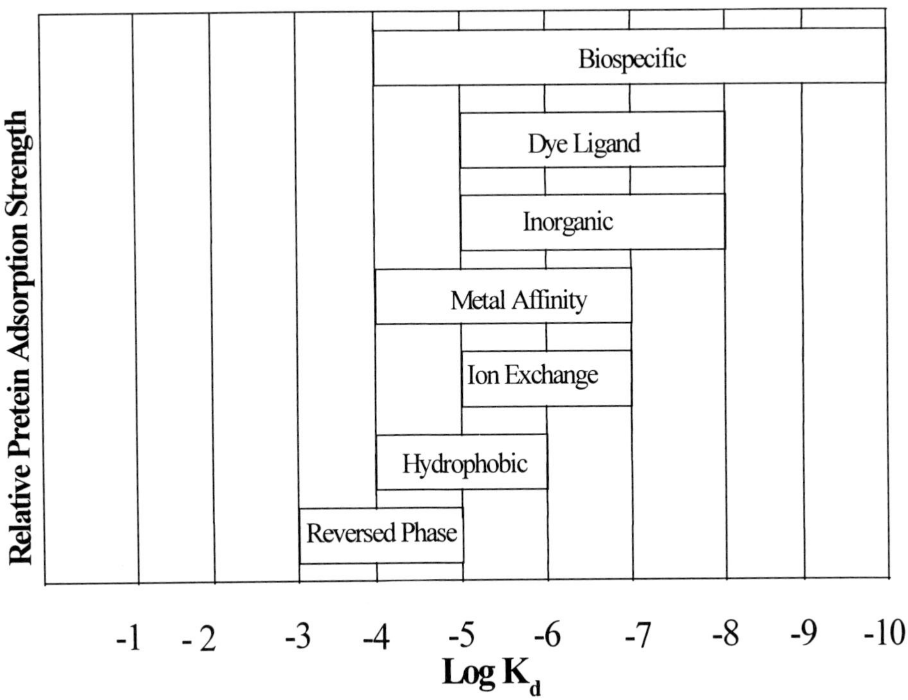


Figure 7. Relative protein adsorption strength of various types of biochromatography for protein seperation based on dissociation constants.

Figure 7 together with Table 1 can be used to help the selection of an appropriate form of chromatography for protein separation and purification. If a strong and specific protein adsorption is desired in a process, an affinity chromatography such as biospecific with a low $K_d$ can be chosen, though affinity resins are usually much more expensive comparing to ion exchange or HIC resins. Therefore, a compromise may need to be made between strong and specific protein adsorption and low price of the resin when choosing a chromatographic process. In this article we also discussed the effects of mobile phase conditions such as pH and ionic strength on the protein dissociation constant. Tables and figures are shown to illustrate these effects, which can be employed as useful references for chromatographic process optimization.

## NOTATION

$k_1$ = forward rate constant
$k_2$ = backward rate constant
$r_q$ = rate of adsorption
c = protein concentration in mobile phase, M
q = protein concentration in stationary phase, M
$q_m$ = maximum adsorption capacity, M
$q^*$ = protein concentration in stationary phase at equilibrium, M
$c^*$ = protein concentration in mobile phase at equilibrium, M
$K_d$ = dissociation constant, M

## REFERENCES


1. P.M. Boyer and J.T. Hsu, Chem. Eng. Sci., 47 (1992) 241.
2. B. Smith and M. Sefton, J. Biomed. Mat. Res., 36 (1992) 947.
3. B.K. Dalmia and Z.L. Nikolov, Enzyme and Microbiol. Tech., 13 (1991) 982.
4. A.I. Liapis, B. Anspach, M.E. Findley, J. Davies, M.T. Hearn and K.K. Unger, Biotech.and Bioeng., 34 (1989) 467.
5. Pharmacia, Affinity Chromatography; Principles and Methods, p.71, Snits & Design, Sweden, 1993.
6. P.M. Boyer, Engineering Analysis of Protein Adsorption by Dye-ligand Adsorbents, Ph. D. Dissertation, Lehigh University, 108 (1991).
7. F.H. Arnold and H.W. Blanch, J. Chromatogr., 355 (1986) 13.
8. S. Ostrove and S. Weiss, Affinity Chromatography Specialized Techniques, in: Guide to Protein Purification, Methods in Enzymology; M. Deutscher (ed.), Academic Press, San Diego, 1990.
9. H.T. Wirth and M.T.W. Hearn, J. Chromatogr., 646 (1993) 143.

10. M. Belew, T.T. Yip, L. Anderson and J. Porath, J. Chromatogr., 403 (1987) 197.
11. R.J. Todd, R.D. Johnson and F.H. Arnold, J. Chromatogr., 662 (1994) 13.
12. Bio-Rad Laboratories, Life Science Research Products Cataloge, p. 107, Hercules, CA, 1995.
13. S. Magdassi, D. Leibler and S. Braun, Langmuir, 6 (1990) 376.
14. A. Tongta, A.I. Liapis and D.J. Siehr, J. Chromatogr., 686 (1994) 21.
15. J.C. Voegel, S. Behr, M.J. Mura, J.D. Aptel, A. Schmitt and E.F. Bres, Colloids and Surfaces, 40 (1989) 307.
16. J. Johnson and E. Matijevic, Colloid and Polymer Science, 270 (1992) 353.
17. D.A. Skoog, D.M. West and J. Holler, Fundamentals of Analytical Chemistry, 6th. Edition, Saunders College Publishing, New York, 1992.
18. Pharmacia, Ion Exchange Chromatography; Principles and Methods. p.10, Uppsala, Sweden, 1996.
19. P. Aranyi and L. Boross, J. Chromatogr.,88 (1974) 239.
20. H.S. Tsou and E.E. Graham, AIChE J., 31 (1985) 1959.
21. S. Yamamoto, K. Nakanishi, R. Matsuno and T. Kamikubo, Biotech. and Bioeng., 25 (1983) 1373
22. Toso Haas, The Seperation Catalogue, p. 6, Montgomerville, PA, (1996).
23. A. DePalma, Genetic Engineering News, 6, June 1, 1996.
24. W.S. Hancock, High Performance Liquid Chromatography in Biotechnology, New York, John Wiley & Sons, 1990.
25. F.D. Antia, I. Fellegvari and C. Horvath, Ind. Eng. Chem. Res., 34 (1995) 2796.
26. R. Blanco, A. Arai, N. Grinberg, D.M. Yarmush and B.L. Karger, J. Chromatogr., 482 (1989) 1.
27. R.M. Chang and W.C. Lee, J. Chem. Tech. Biotechnol., 59 (1994) 133.
28. H.A. Chase, J. Chromatogr., 297 (1984) 179.
29. Chen, F. Fook and E. Graham, AIChE J., 28 (1982) 245.
30. R. Freitag and J. Breier, J. Chromatogr., 691 (1995) 101.
31. A. Johnston and M.T. Hearns, J. Chromatogr., 557 (1991) 335.
32. Y.J. Kim and S.M. Cramer, J. Chromatogr., 549 (1991) 89.
33. Y.J. Kim and S.M. Cramer, J. Chromatogr., 686 (1994) 193.
34. R.J. Klimchak and S. Wang, Biotechnology Techniques, 9 (1995) 731.
35. Krishnan, Sivadasan and M.A. Vijayalakshmi, J. Chromatogr., 397 (1987) 339.
36. N.E. Labrou, A. Karagouni and Y.D. Clonis, Biotech. and Bioeng., 48 (1995) 278.
37. R. Lemque, A. Jsulmes, B. Sebille and C. Vidal-Madjar, J. Chromatogr., 599 (1992) 255.
38. A.G. Mayes and R. Eisenthal, Biotech. and Bioeng., 40 (1992) 1263.
39. S. Yamamoto, M. Nomura and Y. Sano, Chem. Eng. Sci., 47 (1992) 185.
40. H. Yoshida and T. Kataoka, The Chem. Eng. J., 41 (1989) B11.
41. H. Yoshida and M. Yoshikawa, AIChE J., 40 (1994) 2034.

Adsorption and its Applications in Industry and Environmental Protection
Studies in Surface Science and Catalysis, Vol. 120
A. Dabrowski (Editor)



# Interactions of solutes with monodispersed colloids – practical aspects*

Egon Matijević

Center for Advanced Materials Processing
Clarkson University, Potsdam, NY 13699-5814, USA

## 1. INTRODUCTION

Adsorption of complex inorganic and organic solutes on particles in aqueous solutions can greatly alter the charge of the solids, their dispersion stability, solubility, adhesion on solid surfaces, and some other physical and chemical properties. In order to develop an understanding of interactions leading to adsorption, it is necessary to know the composition of the solutes and the surface characteristics of the solid adsorbents. In the latter case, it is essential to use as model systems dispersions consisting of particles uniform in size and shape (preferably spheres). For this reason, polymer latexes have been frequently employed in such studies. However, many monodispersed inorganic colloids of simple and mixed compositions, in a variety of shapes and modal sizes, are now available [1,2], a few of which are illustrated in Figure 1. These dispersions offer opportunities to investigate systems of direct interest in a variety of applications, such as in corrosion of metals, formation of catalysts and pigments, medical diagnostics, etc.

The present review considers adsorption phenomena of aqueous solutes on uniform dispersed solids, based on the bond formation between the interacting species. Examples are offered on the effects of simple and composite inorganic ions and of organic molecules of increasing complexity and chelating properties. The roles of specific solute species, the bond strength, and surface composition of adsorbates are dealt with in some detail. Since inorganic dispersions are available with particles of various morphologies, it is possible to investigate effects related to shapes other than spherical.

Various changes of properties of inorganic particles of different compositions, due to chemisorption of complex solutes are discussed. The consequences of these modifications in numerous applications are illustrated.

It should be noted that the accumulation of charged complexes at the adsorbent/solution interface, i.e. in the double (triple) layer, is not discussed. Furthermore, different models developed to quantitatively describe such systems,

*Supported by the NSF Grant 9423163

especially the metal oxides dispersed in aqueous media, are not included in this review.

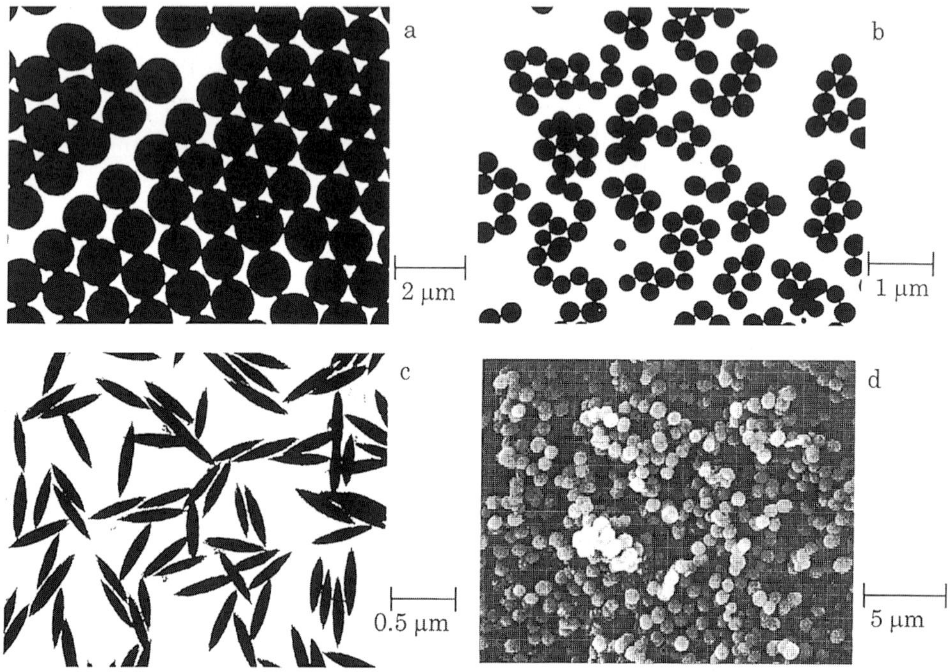


Figure 1. Transmission electron micrographs (TEM) of (a) silica particles obtained by aging for 1 h at 40°C a solution containing 0.30 mol $dm^{-3}$ tetraethoxysilane (TEOS), 1.13 mol $dm^{-3}$ $NH_3$, 6.3 mol $dm^{-3}$ $H_2O$, and 390 $cm^3$ isopropanol (for a total volume of 500 $cm^3$) [3]. (b) Chromium hydroxide particles obtained by aging for 26 h at 75°C a $4\cdot10^{-4}$ mol $dm^{-3}$ $KCr(SO_4)_2$ solution [4]. (c) Hematite ($\alpha$-$Fe_2O_3$) particles obtained by aging for 2 days at 100°C a solution containing $2.0\cdot10^{-3}$ mol $dm^{-3}$ $FeCl_3$ and $4.5\cdot10^{-4}$ mol $dm^{-3}$ $NaH_2PO_4$ [5]. (d) Scanning electron micrograph (SEM) of magnetite ($Fe_3O_4$) particles obtained by aging for 4 h at 90°C a $2.5\cdot10^{-2}$ mol $dm^{-3}$ $Fe(OH)_2$ gel containing $2.5\cdot10^{-2}$ mol $dm^{-3}$ $KNO_3$ and $5.0\cdot10^{-3}$ mol $dm^{-3}$ excess $FeSO_4$ [6].

## 2. ADSORPTION OF METAL IONS

### 2.1. Simple cations

Interactions of metal ions with solid particles suspended in aqueous solutions have been a subject of countless studies. Intuitively, one may expect cations to adsorb on negatively charged surfaces, and the efficiency of the uptake to increase with higher charge of the solute species. However, this assumption is not supported by experimental evidence; instead, the adsorption depends on the specific interactions of the cationic complexes with the surface sites of the adsorbent, i.e., chemisorption.

As an example, the system consisting of metal oxide particles in solutions of $Co^{2+}$ ions is described here [7]. Figure 2 gives the adsorption densities of cobalt(II) ions on monodispersed magnetite ($Fe_3O_4$) particles as a function of the pH, while Figure 3 shows the temperature effect on the same system. The latter clearly indicates the chemisorption nature of the processes involved.

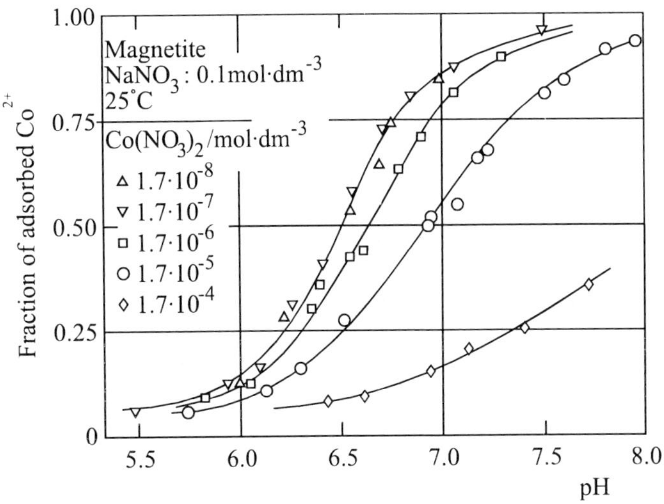


Figure 2. The fraction of adsorbed $Co^{2+}$ on 0.2 g of spherical magnetite particles (1 μm in diameter) in 20 $cm^3$ of a suspension as a function of the pH at a constant ionic strength (0.1 mol $dm^{-3}$) at 25°C [7].

In separate experiments it was established that under the described experimental conditions the ≡FeOH site represented the majority of surface groups on magnetite. The above results, and additional ones, in which the influence of the ionic strength was investigated, could be interpreted by assuming the following equilibria:

$$\equiv FeOH + Co^{2+} \overset{K_1}{\rightleftarrows} \equiv FeOCo^{+} + H^{+} \tag{1}$$

$$2 \equiv FeOH + Co^{2+} \overset{\beta_2}{\rightleftarrows} \equiv (FeO)_2Co + 2H^{+} \tag{2}$$

resulting in the relationship:

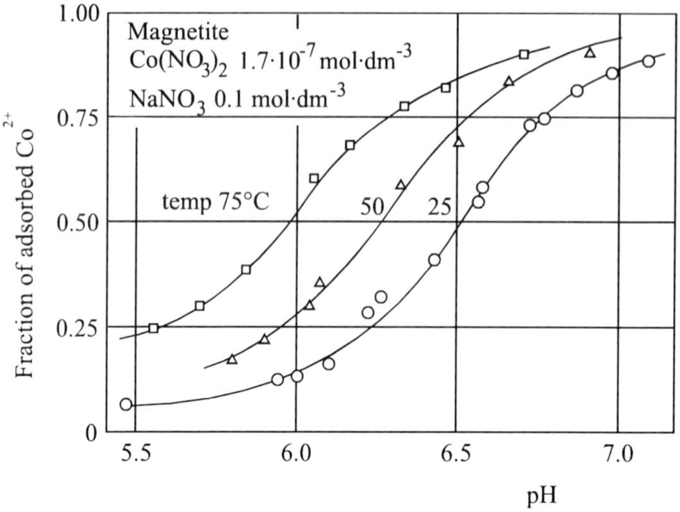


Figure 3. The fraction of adsorbed $Co^{2+}$ as a function of the pH (25°C) at a constant concentration of $Co(NO_3)_2$ ($1.7 \cdot 10^{-7}$ mol dm$^{-3}$) and ionic strength (0.1 mol dm$^{-3}$) for the system described in Figure 2 at three different temperatures [7].

$$\left[H^+\right]\Gamma/C = 5.4 \cdot 10^{-6}\gamma\left(K_1 + 5.6 \cdot 10^{-6}\beta_2/\left[H^+\right]\right) \tag{3}$$

where $C$ is the equilibrium concentration of $Co^{2+}$, $\Gamma$ is the amount of $Co^{2+}$ adsorbed per unit surface area, and $\gamma$ is the Debye-Hückel parameter:

$$\gamma = 10^{-2\sqrt{I}/\left(1+\sqrt{I}\right)} \tag{4}$$

The plot in Figure 4 indicates that the experimental results can be indeed described by the proposed interfacial reactions (1) and (2). A linear relationship between $\Gamma/C$ and $\gamma$ further supported the described mechanism.

Analogous results were obtained with $Co^{2+}$ ions in contact with colloidal hematite ($\alpha$-$Fe_2O_3$) [8,9], and the same effects can be expected with other unhydrolyzed cations that would react with surface ≡MOH sites, as long as strong bonds are formed by the substitution with protons.

Silica is used as another example to illustrate interactions of simple metal ions with inorganic particles. The silanolic groups on the surface of such particles are acidic and, thus, react with bases, but the proton at the ≡SiOH site is also exchangeable for cations, as shown in Figure 5, which is a mass action law presentation for Ludox HS silica in the presence of counterions of different charges

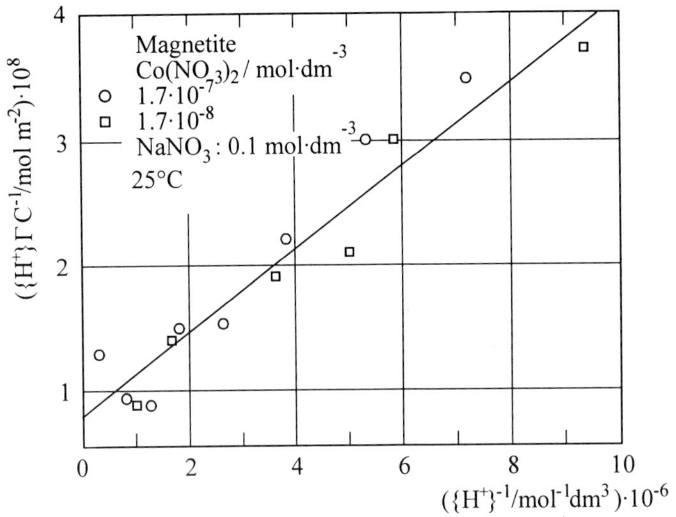


Figure 4. The plot of $\{H^+\}\Gamma/C$ as a function of $1/\{H^+\}$ for two concentrations of $Co(NO_3)_2$ ($1.7\cdot10^{-8}$ and $1.7\cdot10^{-7}$ mol dm$^{-3}$) for the system described in Figure 2, according to eq. (3) [7].

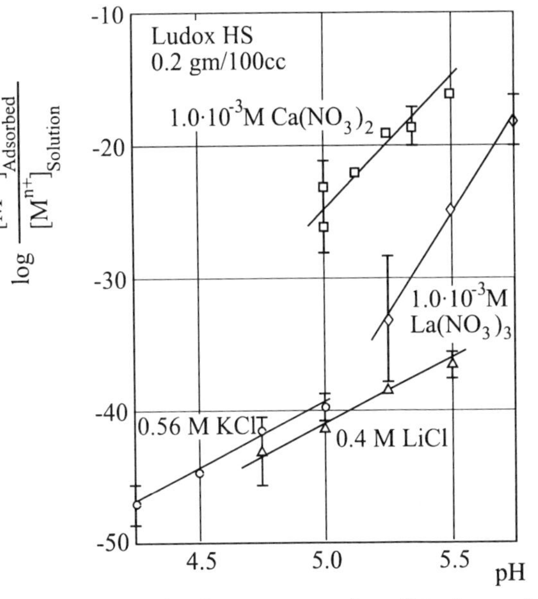


Figure 5. Mass action law representation of the ion exchange for Ludox HS at low pH in the presence of $1.0\cdot10^{-3}$ mol dm$^{-3}$ $La(NO_3)_3$ (◇), $1.0\cdot10^{-3}$ mol dm$^{-3}$ $Ca(NO_3)_2$ (□), 0.56 mol dm$^{-3}$ KCl (○), and 0.40 mol dm$^{-3}$ LiCl (Δ). The maximum possible error is represented by the vertical bars. The solid lines drawn through the data for the salts with + 1. +2. and +3 charged cations are the best linear fits with slopes +1, +2 and +3, respectively [10].

at low pH values [10]. Under these conditions only the equilibrium

$$M^{n+} + x(\equiv SiOH) \rightleftharpoons M(\equiv SiO)_x^{(n-x)+} + H^+ \tag{5}$$

needs to be taken into consideration, which leads to the expression

$$\log \frac{\left[M(\equiv SiO)_x^{(n-x)+}\right]}{\left[M^{n+}\right]} = \log \frac{\left[M^{n+}\right]_{adsorbed}}{\left[M^{n+}\right]_{solution}} = Z + x\,pH \tag{6}$$

where $Z = \log K_e + x \log[\equiv SiOH]$, $K_e$ being the equilibrium constant for the reaction (5).

The ion exchange capacity and the strong hydration control many properties of nanosized silica. For example, very high concentrations (up to 40% by wt) of stable

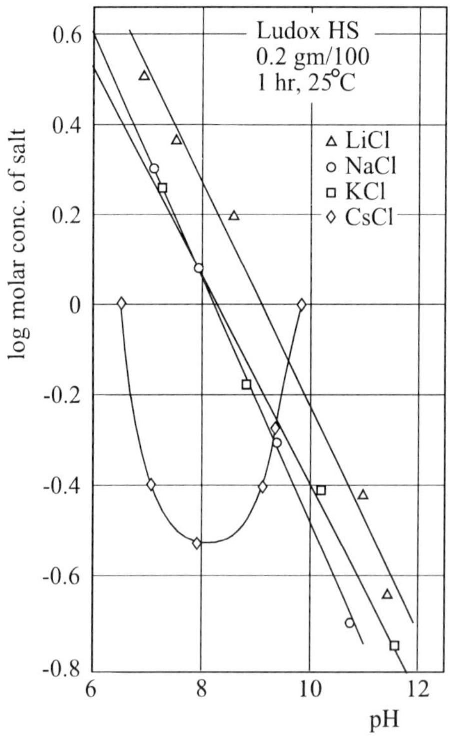


Figure 6. Critical coagulation concentrations of LiCl, NaCl, KCl, and CsCl for Ludox HS as a function of the pH, 1 h after the addition of salts. The sols are coagulated above the solid lines [10].

dispersions can be obtained in acidic aqueous media, under which these solids are weakly charged and highly hydrated. Under these conditions the solvation layer on the particles keeps them too far apart for the attractive van der Waals force to be operational. The stability towards electrolytes decreases with increasing pH values, i.e. with higher particle charge, due to the deprotonation of silanolic groups by acid-base and ion exchange reactions, resulting in partial dehydration of the particles [10]. These effects are displayed in Figure 6, with a Ludox HS sample in the presence of several singly charged counterions [10]. The somewhat unusual behavior of the potassium ion is explained in the same publication.

## 2.2. Adsorption of hydrolyzed cations

It is well known, that the vast majority of metal ions hydrolyze in aqueous solutions, yielding a variety of solutes, the complexity of which increases with the charge of the cation. It has been established with numerous adsorbents in aqueous solutions of metal salts, that the adsorptivity of cations increases dramatically, as a result of their hydrolysis [11]. Indeed, the ions which do not specifically interact with solid surfaces, do so once they form hydrolyzed complexes. The enhanced uptake has been documented by direct adsorption measurements and indirectly by determining electrokinetic mobilities (electrokinetic potentials) as a function of the pH. The latter experiments invariably show the formation of charged sites on neutral surfaces or charge reversal on negatively charged surfaces, due to the chemisorption of hydrolyzed cationic solutes.

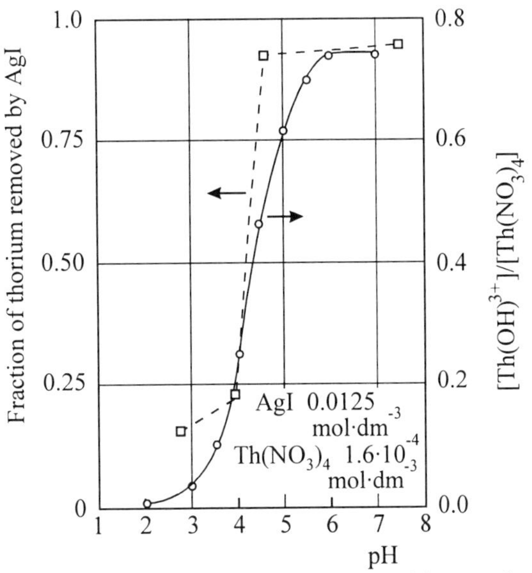


Figure 7. Fraction of thorium removed by a AgI sol ($1.25 \cdot 10^{-2}$ mol dm$^{-3}$) from a $Th(NO_3)_4$ solution ($1 \cdot 10^{-4}$ mol dm$^{-3}$) as a function of the pH (solid line) [11].

The tetravalent thorium ion was known to reverse the charge of hydrophobic colloid particles from negative to positive. It would be reasonable to assume, that the very small $Th^{4+}$ cation should be strongly attracted by a negatively charged surface and, consequently, alter the adsorbent's charge. Figure 7 shows the results on the uptake of thorium species on negatively charged silver iodide colloids as a function of the pH, by using a radioactive Th isotope. It is quite apparent that the adsorption of the unhydrolyzed $Th^{4+}$ ion is insignificant, but increases sharply at the pH which causes the onset of the hydrolysis of this cation [11,12]. Significantly, the charge of the adsorbing complex, $Th(OH)^{3+}$, is lower than of the hydrated cation, $Th^{4+}$. The ratio of the solutes, as calculated assuming the first hydrolysis step,

$$Th^{4+} + H_2O \rightleftharpoons ThOH^{3+} + H^+ \tag{7}$$

using the equilibrium constant from the literature [13], follows closely the adsorption trend. The fit could be made even better, if additional equilibria were taken into consideration.

As mentioned above, the adsorption of hydrolyzed species can alter the sign of the electrokinetic potential as illustrated in Figure 8 on the example of styrene-

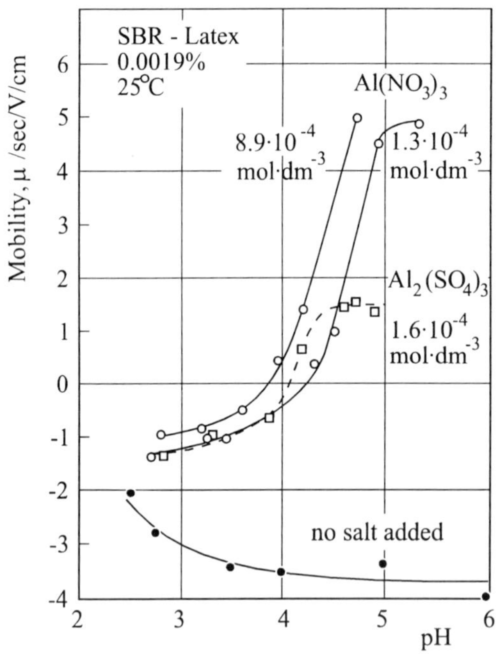


Figure 8. Mobility as a function of the pH of a styrene-butadiene rubber (SBR) latex in the presence of constant concentrations of $Al(NO_3)_3$ (open circles), $Al_2(SO_4)_3$ (squares), and without added salt (blackened circles). Concentrations: SBR latex: 0.0019% by wt; $Al(NO_3)_3$: $8.9 \cdot 10^{-4}$ mol dm$^{-3}$, and $Al_2(SO_4)_3$: $1.6 \cdot 10^{-4}$ mol dm$^{-3}$ [14].

butadiene rubber (SBR) latex. The mobility of the negatively charged particles decreases somewhat with the lowering of the pH, mostly due to the higher ionic strength. Obviously, just the accumulation of counterions at the interface cannot cause the charge reversal. In the presence of aluminum ions the particles are still negatively charged at pH <4, but become positively charged above this pH value, at which aluminum ion undergoes hydrolysis [14].

Analogous results have been obtained with a large number of colloid dispersions and a variety of hydrolyzable ions. Since the hydrolysis products vary with the metal ion and the pH, their adsorption and charge reversal phenomena will differ in each case, as shown with ferric [15], chromium [16], zinc [17], and other ions.

It is important to recognize that adsorbed hydrolyzed ions may be detached as long as the hydrolysis can be reversed by acidification of the dispersions; i.e., as long as the adsorbed complexes are in the olated state. Figure 9 illustrates desorption of Al-complexes from the SBR latex [18].

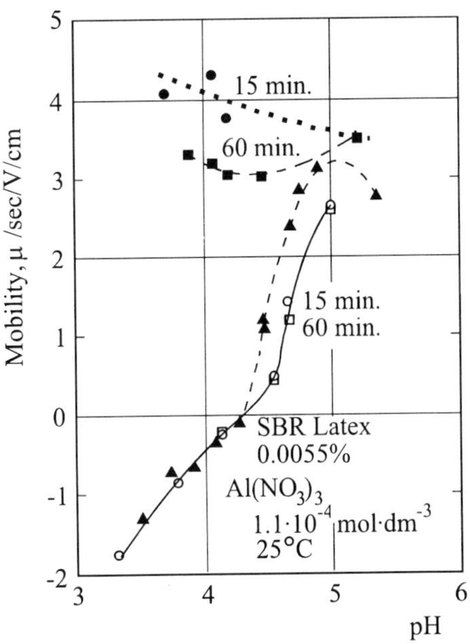


Figure 9. Electrophoretic mobilities of a SBR latex (0.0055% solids by wt) in the presence of $1.1 \cdot 10^{-4}$ mol $dm^{-3}$ $Al(NO_3)_3$ as a function of the pH. Solid line gives the mobilities 15 min (○) and 60 min (□) after adding the salt and adjusting the pH of the latex. The other curves are for the latex, the charge of which was first reversed by $Al(NO_3)_3$ at the highest pH on the plot and subsequently acidified to various lower pH values. The measurements were carried out at 15 min (●), 60 min (■), and 26 h (▲) after acidification [18].

On acidification of the recharged latex, the mobility of the particles at first becomes more positive, due to interactions with adsorbed protons (the dotted curve). The dehydrolyzation of attached Al-complexes and the diffusion of the resulting aluminum ion is obviously a slow process, as it takes hours to reach essentially the original charge conditions as a function of the pH (dashed line).

The detachment of hydrolyzed Al ions can be accelerated by the addition of $F^-$ ions, because the complexes of the $AlF_n^{(3-n)+}$ type (with $n = 1 - 6$) are quite strong, but do not adsorb, yet they promote the dehydroxylation process [18].

The effects of the hydrolysis of metal ions, which cause the change of the charge of the resulting solute complexes, and their enhanced absorbability have numerous repercussions in various applications. For example, the uptake of different cations in soils and their propagation are greatly affected by the natures of the complexes formed at different pH values. The uses of metal salts as coagulants, such as in the water purification, are also governed by the hydrolysis reactions, especially since the most common inorganic electrolytes in these applications contain multivalent cations, e.g. aluminum or iron(III). Indeed the optimum flocculation effects can be achieved either by neutralization of the negative charge of the suspended particles, caused by the adsorption of complex cations, or by incorporation of such particles in the precipitated metal hydroxide flocks, which represent a late stage of the hydrolysis process [15].

The ability of hydrolyzed cations to reverse the charge of particles from negative to positive is also employed in controlling their interactions with other solids, such as in the paper industry where pulp (which is as a rule negatively charged) is mixed with different dispersed solids to achieve a specific color, opacity, strength, weight, and other characteristics of the final products.

Furthermore, the particle adhesion from their aqueous dispersions on different substrates is greatly affected by the sign of the charge on interacting surfaces. In general, the attachment takes place on oppositely charged surfaces. In the absence of any bond formation, the removal of adhered particles can be achieved, if both surfaces bear the same sign of charge (as long as the ionic strength of the solution part is sufficiently low). Since the charge on the solids can be altered with the adsorption of hydrolyzed cations, adhesion phenomena may be manipulated as needed [19,20].

Finally, the surface charge modifications by hydrolyzed ions can be used to control the adsorption behavior of a solid. For example, if the sign is converted from negative to positive, anionic species will be attracted and vice versa. In addition, the adsorbed solutes may then act as coupling agents for other adsorbates, such as dyes, amino acids, etc. [21,22].

### 2.3. Adsorption of organic acids

The carboxyl group, characteristic of organic acids, is very reactive, because it can be neutralized with bases forming either metal carboxylates or strong ion pairs with cations, and - in many cases - yielding sparingly soluble salts. All these reactions enhance the adsorbability of organic acids, especially on inorganic solids, such as metal (hydrous) oxides. Furthermore, the carboxyl group can react with different basic groups forming covalent bonds, analogously to the formation of peptides on condensation with amino groups. Several examples of interactions of a variety of organic acids with monodispersed colloids are described below.

Figure 10 compares the adsorption densities of dipicolinic (2,6-pyridine dicarboxylic) acid on monodispersed spherical crystalline hematite ($\alpha$-$Fe_2O_3$) particles of ~100 nm in diameter and on amorphous chromium hydroxide spheres of ~ 350 nm in diameter, as a function of the pH [23]. Note the different scale on the ordinates referring to the two adsorbents. The maximum in this plot is expected, due to the effect of the pH on the dissociation of the acid and on the charge on the adsorbent particles [23].

It should be mentioned, that the adsorption isotherm of this acid on hematite is of the Langmuir type and on chromium hydroxide it is of the Freundlich type. On crystalline $\alpha$-$Fe_2O_3$ the saturation is achieved in less than four hours, and at the maximum uptake essentially a monolayer of this acid is formed on the adsorbent.

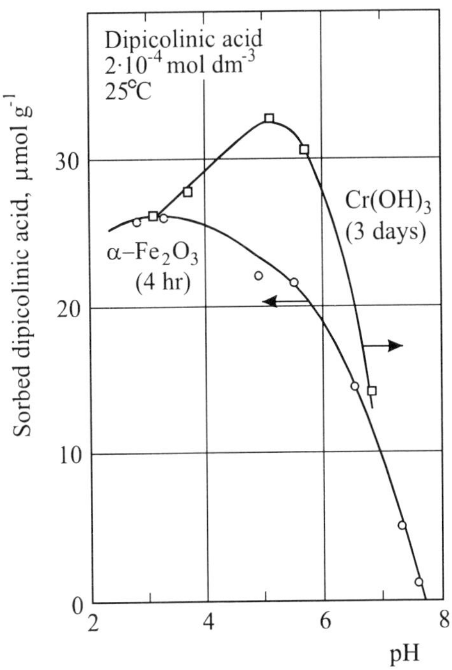


Figure 10. A comparison of dipicolinic acid adsorption on spherical ($\alpha$-$Fe_2O_3$) (○, at 4 hr) and $Cr(OH)_3$ (□, at 3 days) particles as a function of the pH at 25 °C and ionic strength of $5 \cdot 10^{-2}$ mol dm$^{-3}$. The initial dipicolinic acid concentration is $2 \cdot 10^{-4}$ mol dm$^{-3}$ [23].

In contrast, 25 times as much dipicolinic acid is adsorbed per unit weight of chromium hydroxide, despite the larger particle size of the latter as compared to hematite. Since it is quite improbable that this acid would form multilayers, it must be concluded that its molecules can actually diffuse into the hydrated amorphous particles of chromium hydroxide, i.e. in this case absorption is taking place. The crystalline character of hematite allows only for surface interactions, resulting - at best - in a monolayer.

Analogous effects were observed with aspartic acid in contact with chromium hydroxide particles [24]. However, only negligible amounts of tryptophan adsorbed on the same solid. The isoelectric point of this amino acid is much higher than of the aspartic acid, and it is close to that of chromium hydroxide. Furthermore, the larger aromatic (hydrophobic) tryptophan molecules may not be able to penetrate into the hydrophilic adsorbent.

A comparison of the adsorption effects of oxalic and citric acids on hematite leads to some rather useful conclusions. Electrophoretic mobilities as a function of the pH show a substantial shift in the isoelectric point of the iron oxide on addition of these acids to the dispersion, even at concentrations as low as $10^{-6}$ mol dm$^{-3}$ (Figure 11) [25]. The latter is a clear indication of the strong bond between the solutes and the

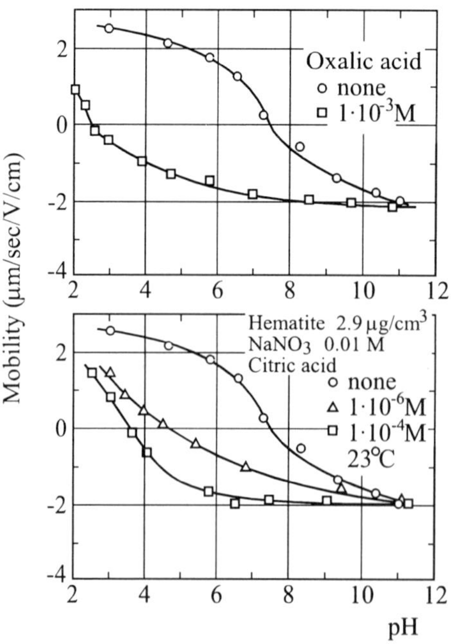


Figure 11. Electrophoretic mobilities of hematite ($\alpha$-$Fe_2O_3$) particles (0.09 μm in diameter) in the absence and in the presence of oxalic acid (upper) and citric acid (lower), as a function of the pH [25].

adsorbent. This conclusion is further substantiated by the adsorption measurements, which show that the uptake of oxalic acid on hematite increases with rising temperatures [25].

In view of the fact that both these acids are polyprotic, it is of interest to elucidate which solute species adsorb, i.e. the neutral molecules, $XH_2$, and $XH_3$, respectively, or their fully or partially deprotonated anions. In order to answer this question, the problem was approached by comparing the measured electrokinetic potential of hematite in the presence of these acids with the calculated adsorption potentials, assuming all possible solute species and their combinations as a function of the pH [26]. The results of such an analysis for the α-$Fe_2O_3$/oxalic acid system is given in Figure 12.

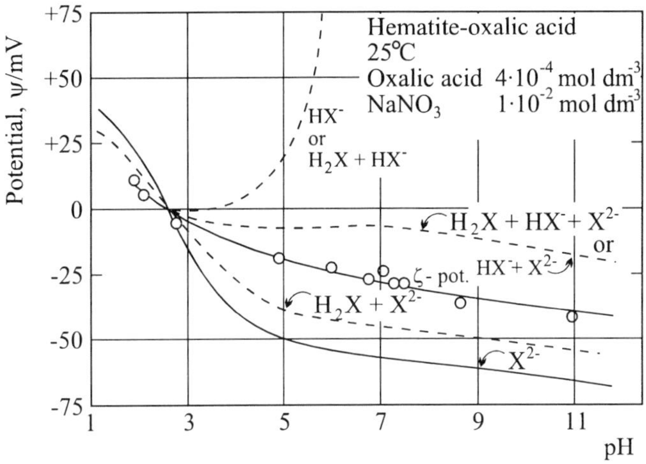


Figure 12. Plots of adsorption potentials, ψ, as a function of the pH calculated from adsorption data of oxalic acid on hematite [25], assuming different combinations of adsorbed solute species. Circles give the values of ζ-potentials of hematite particles in the presence of oxalic acid ($4 \cdot 10^{-3}$ mol dm$^{-3}$) and $NaNO_3$ ($1 \cdot 10^{-2}$ mol dm$^{-3}$) [26].

Since the ζ-potential must be lower than the surface potential, only species $X^{2-}$ and the combination $XH_2+X^{2-}$ should be considered. However, the last two solutes cannot coexist alone in the solution, leading to the necessary conclusion that only the doubly charged oxalate ion, $X^{2-}$, is interacting with hematite. Analogous treatment of the adsorption data for citric acid in contact with α-$Fe_2O_3$ yielded $HX^{2-}$ and $X^{3-}$ as adsorbates. The latter conclusion helped explain a rather unusual adsorption density profile of this system as a function of the pH, as given by circles in the upper part of Figure 13 [26]. In the lower part of this figure is given the experimentally determined speciation diagram for citric acid as reported in the literature. Taking into consideration the above established adsorbing solutes and their relative concentrations at different pH values, the established surface concentrations (circles) are very well correlated by the calculated solid line (upper part of Figure 13).

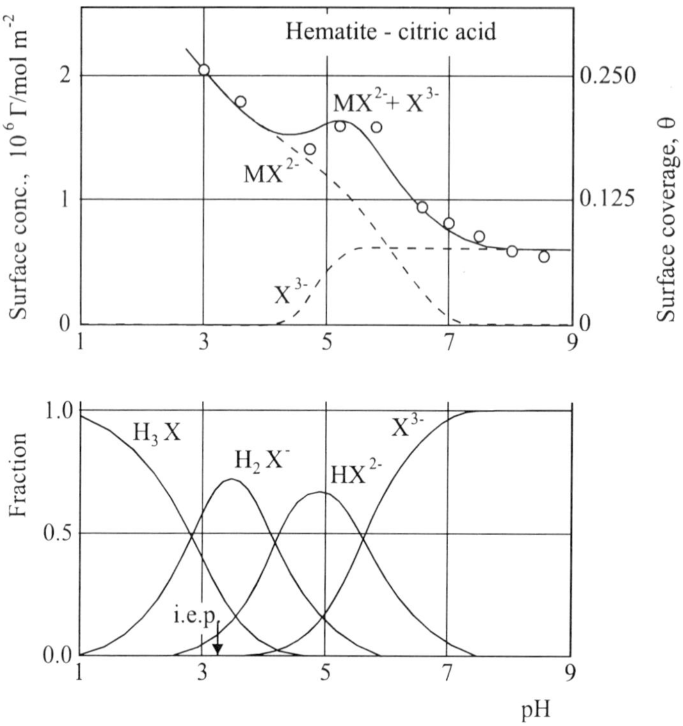


Figure 13. Lower: Species distribution in $2{\cdot}10^{-3}$ mol dm$^{-3}$ citric acid solution as calculated by using $K_1 = 1.2{\cdot}10^{-3}$, $K_2 = 4.9{\cdot}10^{-5}$, and $K_3 = 2.5{\cdot}10^{-6}$ in the presence of $1{\cdot}10^{-2}$ mol dm$^{-3}$ $NaNO_3$. Upper: Surface concentration (or coverage, θ) of citric acid on hematite as a function of the pH, calculated for the adsorption of its $HX^{2-} + X^{3-}$ anions (solid line). The circles represent the experimental data taken from ref. [25]. Dashed lines give the calculated contributions of individual ions to the total adsorption density [26].

The information described above is useful in various applications. One specific example is offered here. The corrosion products in nuclear power plants consist mostly of metal oxides (e.g. magnetite, hematite, etc.) and ferrites (e.g. of cobalt), which must be removed after a certain time of operation. The common approach to do so is by the so-called chemical decontamination, which involves dissolving the "rust" by means of various complexing agents, including organic acids. Obviously, it is of interest to find the most efficient additives to complete this process. Since the dissolution involves leaching out metal complexes, formed between the corrosion products and the reacting solutes, one of the primary conditions is the ability of the organics to interact with iron oxides (and other metals in ferrites) and form soluble complexes. Both oxalic and citric acids can be used as such agents and, in view of its stronger complexation with iron, one would expect citrate to be a more efficient additive. However, Figure 14 clearly demonstrates that the oxalic acid is by orders of magnitude a better dissolution agent for hematite than citric acid. This finding can be explained by considering the adsorption data presented above, showing much higher uptake of oxalate than citrate anions [25]. Indeed, at saturation oxalic acid occupies 8.3 Å$^2$ per molecule, and citric acid 33 Å$^2$ [25]. Since the rate determining step in the release of iron is breaking the bonds between constituent ions of the crystal lattice, the dissolution efficiency must be related to the adsorption density of the complexing species. It is also obvious that the citric acid covers a larger area requiring additional energy for breaking more bonds. Finally, the pH affects both the active sites on the surface and the speciation of anionic species in solution.

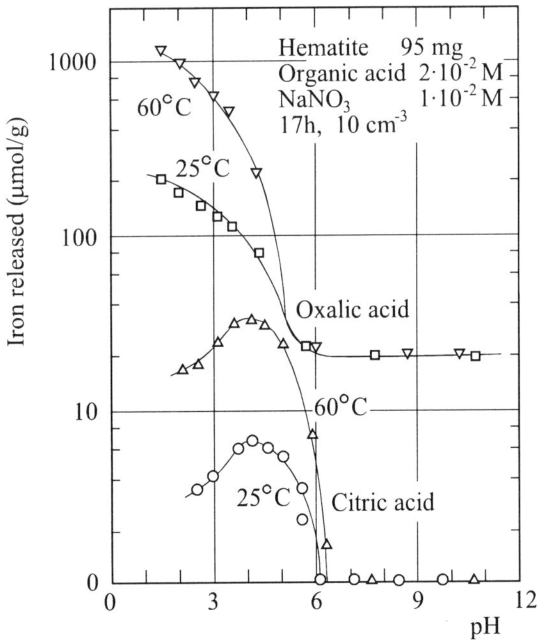


Figure 14. Specific amount of iron(III) released from hematite suspended in $2 \cdot 10^{-2}$ mol dm$^{-3}$ oxalic or citric acid solutions after 17 h of aging at 25 and 60°C, respectively, as a function of the pH [25].

One can expect the same kinds of effects to be responsible in other corrosion processes, based on the interactions of complexing acids with different metals.

Another area of practical importance is in controlling (or altering) the stability of colloid dispersions by organic acids. In analogy to hydrolyzed cations, polyprotic acids that react with the particles may either promote coagulation or inhibit it. The effect that prevails depends on the charge and the adsorptivity of anions in question. As example, the stability domain for finely dispersed hematite as a function of the citric acid concentration and the pH is given in Figure 15 [27], in which the shaded area indicates coagulated systems. The stable dispersions are either due to the original positive charge on iron oxide particles, or to the reversal of charge to negative (c.s.c.) (as determined by electrokinetic measurements in Figure 11), caused by the adsorption of citrate ions. The destabilization of hematite in the area between boundaries depicted with circles and triangles is caused by decreasing repulsion among particles, because their charge has been reduced by the uptake of the acid. The shape of these boundaries depends on the speciation of citric acid as a function of the pH. Finally, the destabilization region above the boundary designated by squares is caused by monovalent cations in the system.

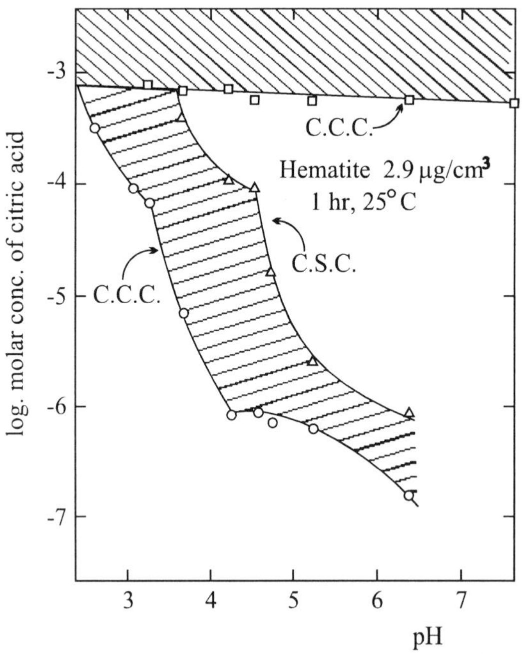


Figure 15. Stability domain of a hematite hydrosol (2.9 μg $cm^{-3}$) in the presence of citric acid as a function of the pH, 1 h after mixing the reacting components at 25°C. The critical coagulation concentrations (c.c.c.) are indicated by ○, □ and critical restabilization concentrations (c.s.c.) by Δ. Shaded areas designate coagulated dispersions [27].

## 2.4. Adsorption of chelating agents

It is to be expected that any solute, capable of forming a chelated bond with a surface group, should strongly adsorb on such a solid. Specific interactions in each case will be responsible for the adsorbate uptake, controlled by the stability constant of the pair in question. However, in addition to forming a complex, the adsorption density will depend on the population of the surface active sites and on the surface conformation of the chelating molecules. Several examples will illustrate these effects and some consequences of the relevant solid/solute interactions.

It was shown that the stability of negatively charged colloidal silver halides in aqueous solutions can be greatly affected by the addition of chelating molecules, such as ethylenediamine (en), 2,2'-dipyridyl (dipy), or 1,10-phenanthroline (phen). These complexing agents can coagulate such dispersions, but depending on the pH, the same compounds can restabilize them. The latter effect was explained by the charge reversal from negative to positive of silver bromide particles as demonstrated in Figure 16 [28].

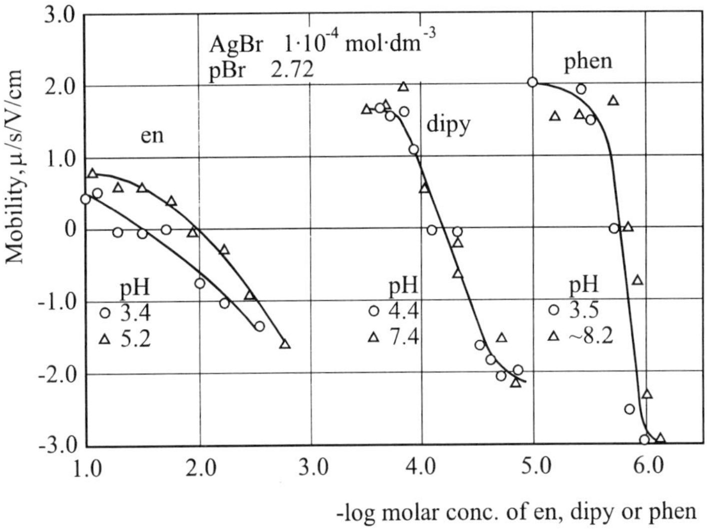


Figure 16. Electrophoretic mobilities of a silver bromide sol *in statu nascendi* (AgBr, $1.0 \cdot 10^{-4}$; excess $Br^-$, $1.9 \cdot 10^{-3}$ mol $dm^{-3}$) as a function of varying concentrations of ethylenediamine (en), 2,2'-dipyridyl (dipy), and 1,10-phenanthroline (phen) at pH values as indicated in the diagram [28].

The solid lines in Figure 17 give the boundaries (critical coagulation concentrations, c.c.c.) as a function of the pH above which the originally negatively charged colloidal AgBr is destabilized in the presence of the three chelating agents, while dashed lines (critical stabilization concentrations, c.s.c.) represent conditions above which the dispersions are stable again, due to the charge reversal. Consequently, between these two boundaries, the dispersions are unstable [28,29].

It is noteworthy that the c.c.c. for dipy and phen are the lowest at the higher pH values, at which these chelating agents are *neutral*, but they are less efficient once these molecules become protonated. Finally, the charge reversal effect is essentially eliminated in sufficiently acidic systems. The necessary conclusion is that the neutral chelating agents replace excess halide (bromide) ions and, thus, neutralize the surface charge causing the destabilization of the colloid. Further exchange of halide ions results in a surplus of silver cations in the form of Ag(dipy)$^+$ or Ag(phen)$^+$ complexes, rendering the particles positively charged and stable. The presence of such chelated species under the same conditions was confirmed by the IR spectroscopy.

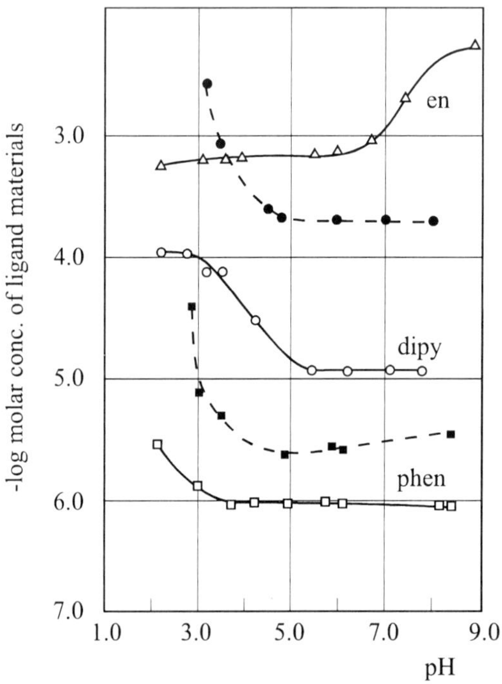


Figure 17. Plot of the critical coagulation concentrations (ccc, solid lines) and of the critical stabilization concentrations (csc, dashed lines) of a silver bromide sol *in statu nascendi* as a function of the pH for ethlenediamine (en, Δ), 2,2"-dipyridyl (dipy, ○, ●) and 1, 10-phenanthroline (phen, □, ■). AgBr, $1.0 \cdot 10^{-4}$; excess Br-, $1.9 \cdot 10^{-3}$ mol dm$^3$ [28].

Taking into consideration the known equilibrium constant for the protonation of dipy, it was possible to construct the diagram shown in Figure 18, which clearly correlates specific solute species with the coagulation (c.c.c.) and restabilization (c.s.c.) effects [29].

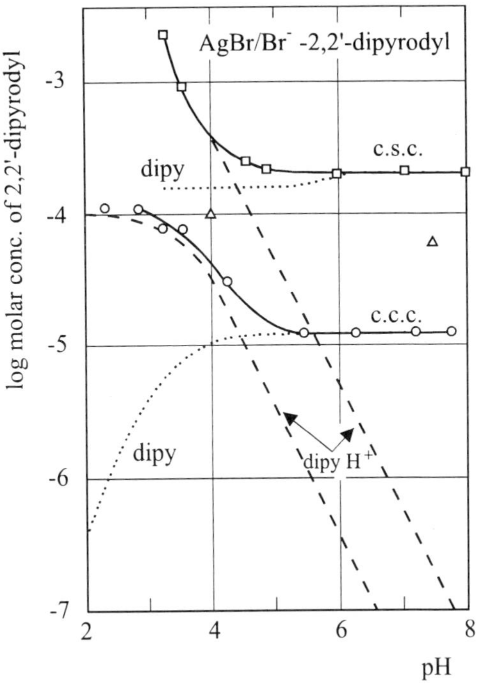


Figure 18. Experimentally determined critical coagulation concentrations (c.c.c., ○) and critical stabilization concentrations (c.s.c.) due to charge reversal (□) of 2,2"-dipyridyl (dipy) as a function of the pH for a negatively charged silver bromide sol. The molar concentrations of dipy (.....) and dipyH$^+$ (----) in the system are calculated for the corresponding c.c.c. and c.s.c. (symbol Δ indicates the isoelectric point) [29].

The adsorption data of trimeric hydroxamic acid (the methanesulfonate salt of desferrioxamine-B, Desferioxamine mesylate, Desferal, DFOB, from Ciba Pharmaceutical Co.) on monodispersed spherical hematite particles show that the uptake of this complex chelating agent increases with the rising pH (Figure 19). This compound was also used to evaluate the dissolution of the same iron oxide and the pH effect was opposite; i.e. the leaching of iron was most efficient at the lowest pH value of the medium (Figure 20), which can be explained by considering the shape and the charge characteristics of this chelating agent [30].

DFOB contains four acidic protons, one of which is attached to the peripheral ammonium group:

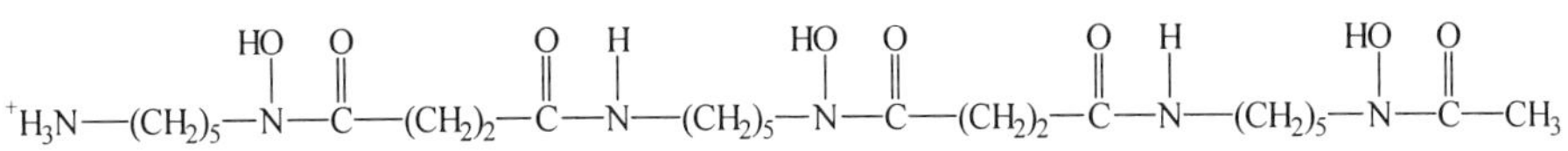

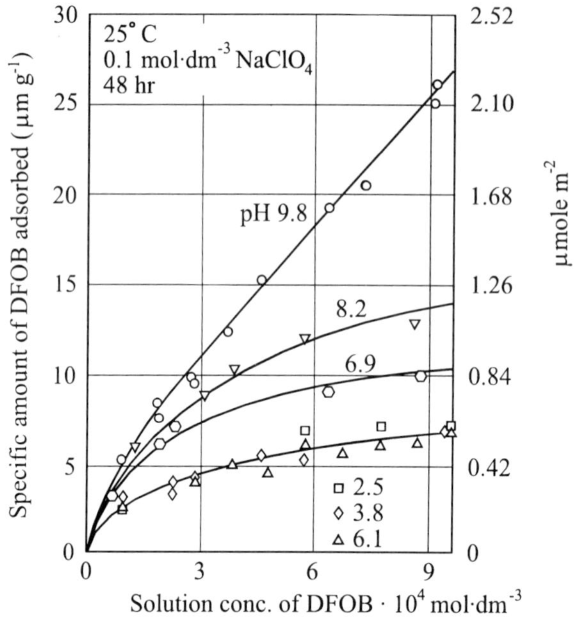


Figure 19. The specific amount of desferrioxamine (DFOB) adsorbed on the spherical hematite particles after 48 hr of equilibration as a function of the solution concentration of DFOB ($[DFOB]_{soln} = [DFOB]_{free} + [DFOB]_{comp}$) for different pH values at 25°C and 0.1 mol dm$^{-3}$ [30].

The corresponding p$K$ values were determined to be [31]:

$$H_4\text{-}DFOB \rightleftharpoons H^+ + H_3\text{-}DFOB \qquad pK - 8.39 \tag{8}$$

$$H_3 - DFOB \rightleftharpoons H^+ + H_2 - DFOB^- \qquad pK = 9.03 \tag{9}$$

$$H_2 - DFOB^- \rightleftharpoons H^+ + H - DFOB^{2-} \qquad pK = 9.70 \tag{10}$$

$$H - DFOB^{2-} \rightleftharpoons H^+ + H - DFOB^{3-} \qquad pK > 11 \tag{11}$$

Based on these equilibria, one can schematically visualize the surface orientation of DFOB species at pH values below and above the isoelectric point of hematite (pH ~ 8), by taking into consideration the charge on the adsorbate species, as illustrated in Fig. 21 [30]. At the low pH value the adsorption density of DFOB is low, due to its flat configuration on the solid surface, which also provides for

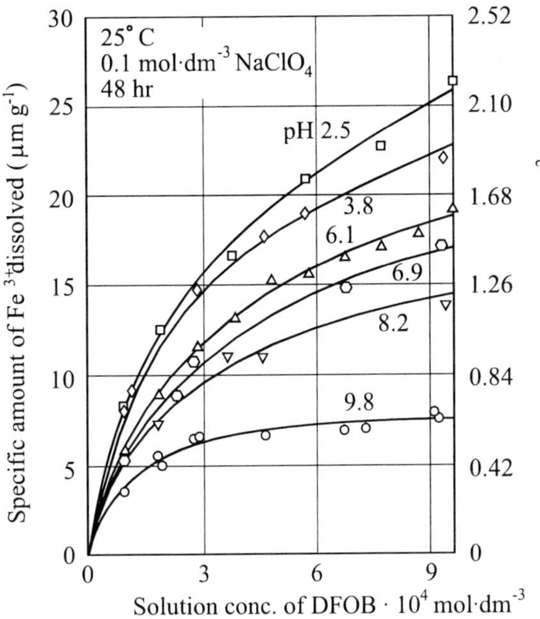


Figure 20. The specific amount of iron (III) dissolved from hematite particles after 48 h of equilibration as a function of the initial concentration of DFOB ($[DFOB]_{tot}$) for different pH values at 25°C and 0.1 mol $dm^{-3}$ $NaClO_4$ [30].

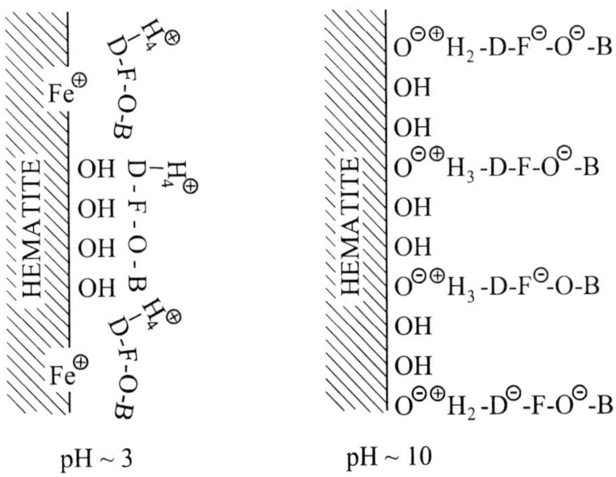


Figure 21. Schematic representation of the proposed models for the configuration of the adsorbed DFOB molecules at the hematite/solution interface at pH ~3 (a), and pH ~10 (b) [30].

interaction of chelating groups with the constituent $Fe^{3+}$ ions. The resulting complex is then leached into the solution. At the high pH values the DFOB chain is negatively charged and only the protonated end amino group is likely to be attached to hematite. Under such conformation only a few chelate complexes will form with iron and the dissolution is negligible.

A hydroxamic acid (HA) polymer with 11 atomic distances between two following HA groups in the side chains of the composition

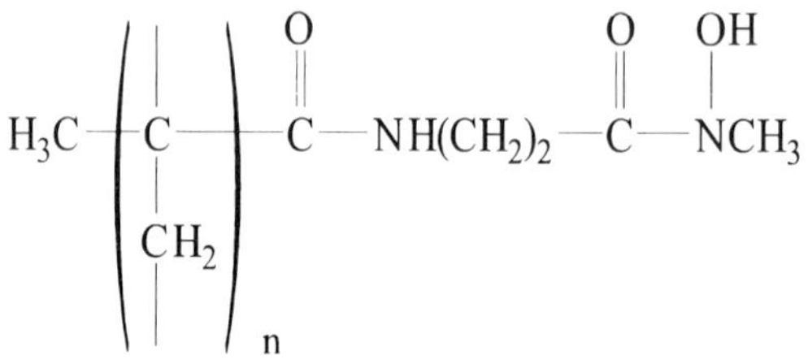


was synthesized [32] and interacted with the same iron oxide particle. The adsorption isotherms showed a great affinity of this macromolecule for the solid, yielding at the saturation an adsorbed amount of $1.25 \cdot 10^{-2}$ mol $m^{-2}$ at low pH, which is a rather high value for a polymer of the low molecular weight. This property lends itself for the use of such polyelectrolytes in corrosion protection. Indeed, the adsorption of the same compound on iron beads prevented their dissolution in HCl solutions until the chelated layer was decomposed by the acid.

The most commonly employed chelating agents are probably the ethylenediamine tetraacetic acid (EDTA) and related compounds. Among other applications, they are of special interest in corrosion, such as in chemical decontamination of water-cooled nuclear power plants. These compounds are especially reactive in contact with metal oxides or other solid metal compounds, because of their ability to form strong surface complexes between the adsorbate and the adsorbent.

Figure 22 gives the electrokinetic mobilities of hematite particles as a function of the pH in the presence of different concentrations of EDTA. An addition of only one micromole of the latter causes a shift in the i.e.p. of more than three pH units, as compared to the same iron oxide in the absence of this agent [33]. The adsorption densities of EDTA and several related amino carboxylic acids, listed in Table 1, are shown in Figure 23 [33]. It is noteworthy, that the trend for all studied chemically related solutes is similar. It is even more important that these complexing agents adsorb at pH values higher then the i.e.p. of hematite; i.e. on negatively charged substrates, despite the fact that their carboxylic acid groups are essentially fully dissociated under these conditions. Plotting the adsorption densities of all compounds listed in Table 1 at a given pH ($\Gamma_T$) as a function of their maximum density ($\Gamma_{max}$) yields one single curve (Figure 24). Up to the isoelectric point, this trend could be interpreted by taking into account the electrostatic and chemisorption potentials. However, to reproduce the experimental trend at pH>i.e.p., it was necessary to consider a configurational change of the chelating molecules on the

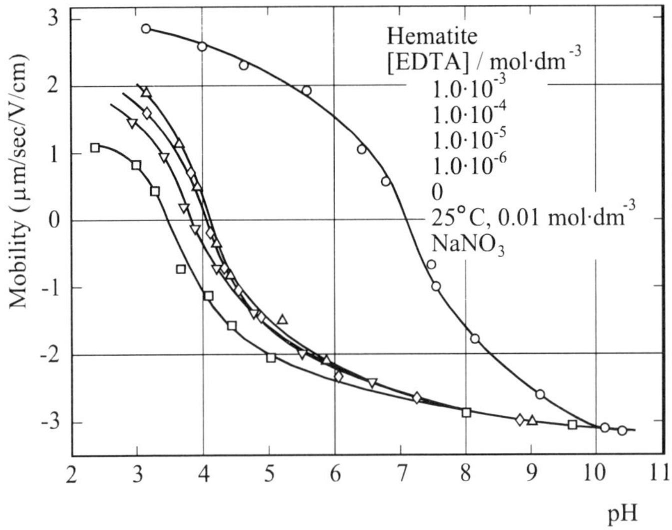


Figure 22. Electrophoretic mobilities of spherical $\alpha$-$Fe_2O_3$ (hematite) particles (1.1 μm in diameter) in the absence and in the presence of various concentrations of ethylene diamine tetraacetic acid (EDTA) as a function of the pH at 25 °C. Ionic strength = 0.01 mol dm$^{-3}$ ($NaNO_3$) [33].

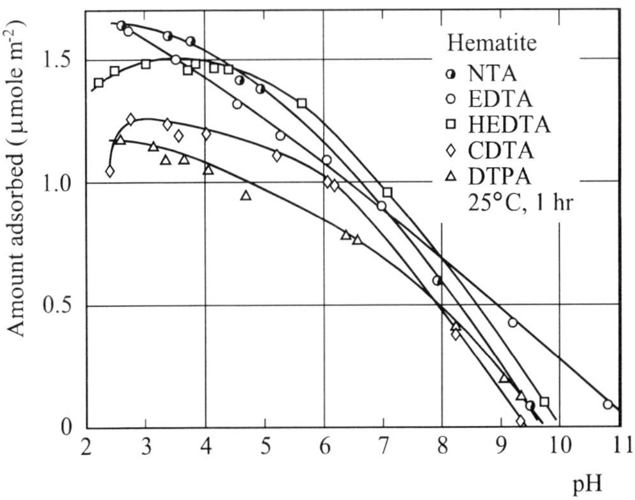


Figure 23. Specific amounts of five aminocarboxylic compounds adsorbed on $\alpha$-$Fe_2O_3$ particles after 1 h of equilibration at 25°C as a function of pH (adsorption envelope) [33].

Table 1

| Chelating agent | log. acid constants | | | |
|---|---|---|---|---|
| | $K_{HL}$ | $K_{H_2L}$ | $K_{H_3L}$ | $K_{H_4L}$ |
| $HO_2CCH_2—N(CH_2CO_2H)_2$ (NTA) | 9.81 | 2.57 | 1.97 | - |
| $(HO_2CCH_2)_2NCH_2CH_2N(CH_2CO_2H)_2$ (EDTA) | 10.34 | 6.24 | 2.75 | 2.07 |
| $(HO_2CCH_2)_2NCH_2CH_2N(CH_2CH_2H)(CH_2CO_2H)$ (HEDTA) | 9.81 | 5.41 | 2.72 | - |
| 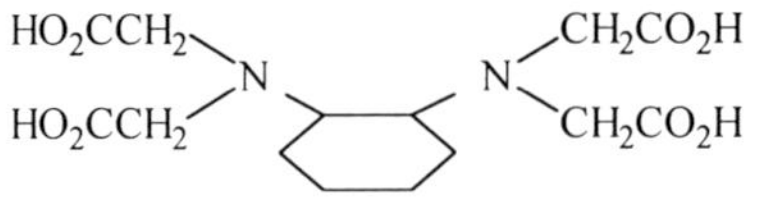 (CDTA) | 11.78 | 6.20 | 3.60 | 2.51 |
| $(HO_2CCH_2)_2NCH_2CH_2N(CH_2CO_2H)CH_2CH_2N(CH_2CO_2H)_2$ (DTPA) | 10.56 | 8.69 | 4.37 | 2.87 |
| $(HO_2CCH_2)_2N(CH_2)_2O(CH_2)_2O(CH_2)_2N(CH_2CO_2H)_2$ (EGTA) | 9.54 | 8.93 | 2.73 | 2.08 |
| $HO_2CCH_2—NCH_2CH_2N—CH_2CO_2H$ (EDDA) | 9.59 | 6.53 | - | - |

hematite surface. In the case of EDTA and related solutes, it is reasonable to assume that the nitrogen dentates are oriented away from the surface at pH<i.e.p., and that the molecule "flips over" with protonated amino groups facing the surface at pH>i.e.p. For this reason, an additional energy term was added to account for the adsorption in the latter case (Figure. 24). In doing so the experimental data and the calculated curve were brought into a reasonable agreement [33].

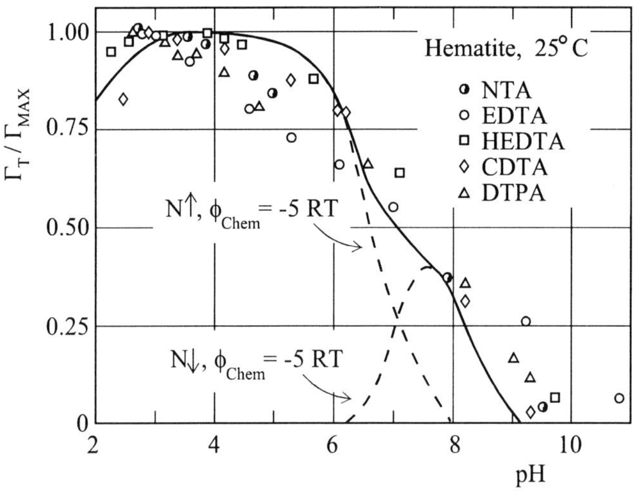


Figure 24. Data in Fig. 23 plotted as the fraction of the chelating agents adsorbed relative to the maximum adsorption density ($\Gamma_T/\Gamma_{max}$) for each chelating agent (open symbols). The solid line is calculated taking into consideration the electrokinetic and chemical interactions and a term to account for the configuration change of reacting molecules when adsorbed above the isoelectric point of hematite [33].

The adsorption of EDTA has a major effect on the dissolution of hematite, which is illustrated in Figure 25 [34]. The most efficient leaching of iron is observed at low pH values, at which the adsorption density is the highest. The dissolution increases dramatically over a rather narrow temperature range, i.e. from 100 to 120°C. Since the ferric-EDTA complex is released into the solution, a critical temperature is needed to break the lattice bonds in order for leaching of the chelated iron compound to take place.

Under the same conditions EDTA dissolves magnetite ($Fe_3O_4$) much more efficiently, which is known to be due to the presence of ferrous ions in the crystal lattice of this solid [35, 36]. Indeed, the dissolution of ferric oxides can be greatly enhanced by the addition of a reducing agent, such as hydrazine ($N_2H_4$), capable of converting $Fe^{3+}$ constituent ions in these solids to $Fe^{2+}$, as demonstrated in Figure 26 [36].

The influence of hydrazine can be interpreted in terms of a direct attack on the surface Fe(III)-EDTA complexes and the reduction of the same chelates in the dissolved state, i.e.:

$$Fe(III) - EDTA + N_2H_4 \qquad Fe(II) - EDTA + N_2H_3^{\bullet} + H^+ \tag{12}$$

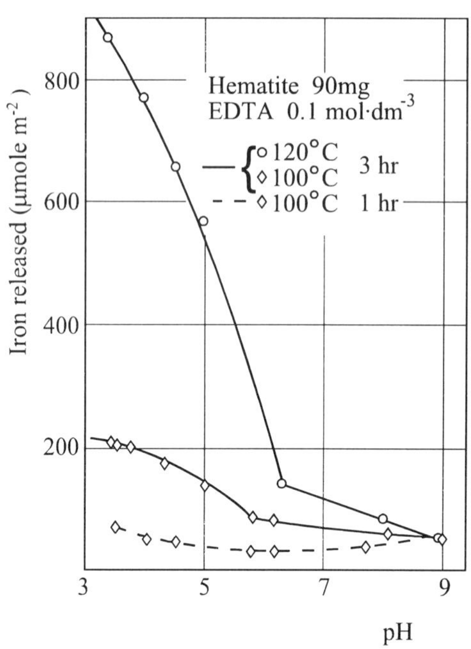


Figure 25. The amount of iron released from hematite in the presence of EDTA as a function of the pH at two different temperatures and reaction times [34].

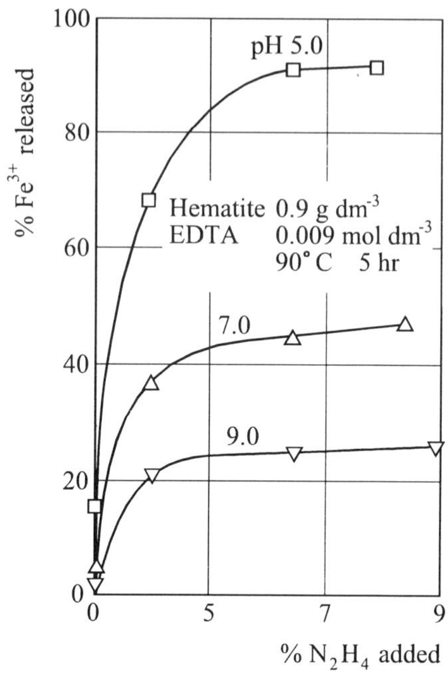


Figure 26. The influence of hydrazine ($N_2H_4$) concentration on the amount of iron released from hematite (0.9 g $dm^{-3}$) into $9 \cdot 10^{-3}$ mol $dm^{-3}$ EDTA solution after 5 h at 90°C at different pH values: 5.0 (□), 7.0 (Δ), and 9.0 (∇) (36).

The hydrazine radicals, $N_2H_3^{\bullet}$, further react with the ferric complexes, eventually consuming 4 moles of Fe(III) per mole of $N_2H_4$, yielding $N_2$ as the final product of the oxidation of hydrazine.

It should be mentioned that EDTA decomposes at elevated temperatures and it was of interest to establish, if this process affects the interactions with and dissolution of iron oxides. The resulting fragments of EDTA at moderately elevated temperatures were identified as iminodiacetic (IDA) and N-hydroxyethyl-iminodiacetic (HIDA) acids. (At higher temperatures the decomposition proceeds to still smaller subunits). By using IDA and HIDA it was found that analogous dissolution results with hematite were obtained as with EDTA, although considerably less efficiently [37,38].

The principles established in the studies of adsorption of chelating agents on iron oxides and the dissolution of the latter are applicable to other simple or composite metal oxides or metal compounds, providing soluble surface complexes can be formed. However, the rate of the dissolution will differ depending on a number of parameters, specific for each solid/solute system. To illustrate such a case, the dissolution of a nickel ferrite of the composition $Ni_{0.53}Fe_{2.47}O_4$ in the presence of EDTA is shown in Figure 27 [39]. It is noteworthy that the amount of nickel in the solution is consistently higher than expected, assuming a uniform composition of ferrite particles throughout the solid phase. However, electrokinetic mobilities of aqueous dispersions of nickel ferrites of varying nickel fractions as a function of the pH yielded the same i.e.p., which would not be expected, if the surface and the bulk of these particles had the same composition in terms of the two constituent metals [40]. This finding helps the understanding of the dissolution behavior of such solids. The inhomogeneity in the composition of particles consisting of more than one metal compound, prepared by precipitation from homogenous solutions, is a rather common occurrence [41].

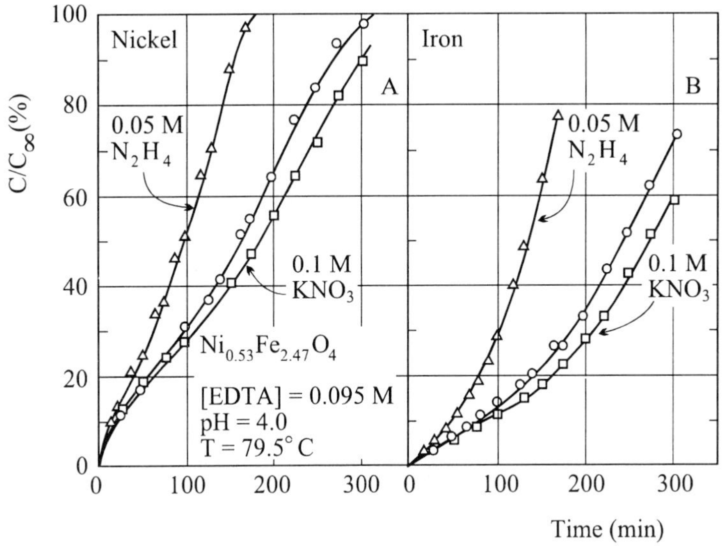


Figure 27. Plot of the ratio of the concentration of the respective metals in solution (C) to the total concentration, ($C_0$), that would result on dissolving the entire amount of the nickel ferrite, $Ni_{0.53}Fe_{2.47}O_4$, in the same volume as a function of the reaction time in the presence of EDTA with or without added hydrazine. (a) Refers to nickel and (b) to iron [39]. Circles give curves with no additive.

## 2.5. Adsorption of dyes

As a special case, the interactions of soluble dyes with uniform inorganic particles is considered here, with emphasis on the preparation of well defined pigments.

It is well known that optical properties of a given pigment, in particular the color and its purity, depend strongly on its particle size and shape, in addition to the chemical composition (i.e., the refractive index). Obviously, in order to ensure reproducibility of such colored materials, it is essential to synthesize them as uniform powders as possible. Spherical pigment particles have a significant advantage, because the optical properties of such dispersions can be exactly calculated and, consequently, the production of pigments for specific applications can be designed [42].

There are only a few colorants that meet the above described constraints; these are essentially some inorganic compounds such as iron oxides [43] or cadmium sulfide. Most commercial pigments, and especially the commonly used metal dye "lakes", are of irregular particle shapes and polydisperse. Because these solids are mostly insoluble in any solvent and decompose on melting, it is not feasible to prepare them as uniform particles.

The objective to prepare reproducible pigments of predictable optical properties, consisting of uniform spheres can be, however, met by interacting soluble dyes with uniform spherical colloidal particles [21,44,45]. By selecting the cores of a given refractive index and diameter, it is possible to obtain colorants of different opacities and chromaticities and, thus, design them for specific applications, such as printing inks, color filters, cosmetics, etc.

Again, the natures of the cores and the chemical composition of dye molecules will be the controlling factors in the adsorption effects. It is to be expected, that basic dyes would react with acidic solids, such as silica, or acid dyes with basic solids, e.g. aluminum hydroxide. It is further possible to modify inorganic particles in order to promote an interaction; thus, by coating silica with aluminum hydroxide to bind acid dyes, or by attaching aminosilane groups on silica particles to initiate specific reactions with certain dyes [21].

These procedures were proven useful in practice as described in the literature [44,45]. Figure 28 illustrates such a pigment prepared by reacting silica particles, modified with aminosilane, with Acid Red 183 dye [21]. Reflectance spectra of the same silica with this and several other dyes, described in the legend, are shown in Figure 29 [21].

Recently, nanometer size cores were reacted with dyes and incorporated into polymer films suitable for optical devices [46]. In view of the very small pigment size, 100% transparency of such colored films was achieved even with 55% by wt solid loadings, using layers as thick as 8 μm (Figure 30). Since the powder was perfectly dispersible in the polymer, the contrast of such color filters was ~150% higher than that of the best commercial products.

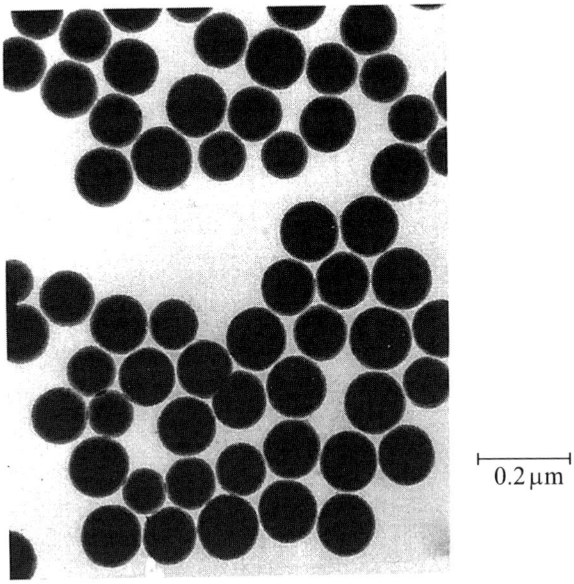


Figure 28. TEM of pigment particles obtained by interacting Acid Red 183 dye with colloidal silica modified with an aminosilane [21].

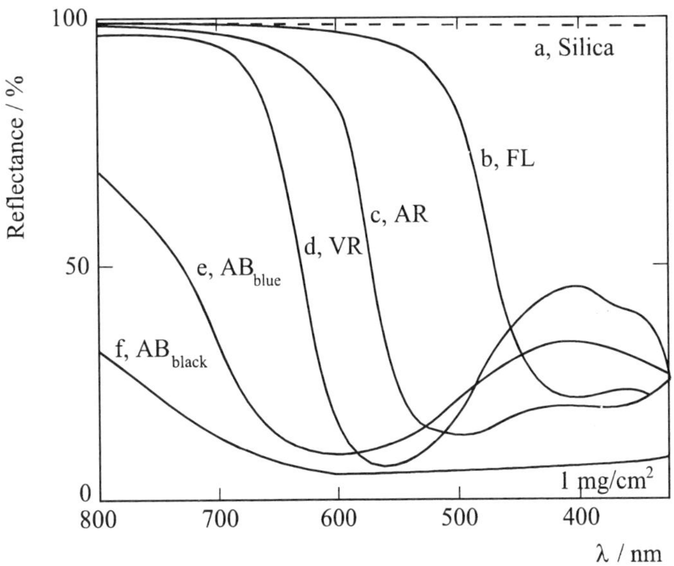


Figure 29. Reflectance spectra of powder samples of amino-modified silica particles (a), and of the same cores with incorporated dyes: (b) Flavazin L; (c) C.I. Acid Red 183; (d) Violamin R; and (e) and (f) C.I. Acid Blue 45. The spectra refer to a constant amount of 1 mg sample per $cm^2$ [21].

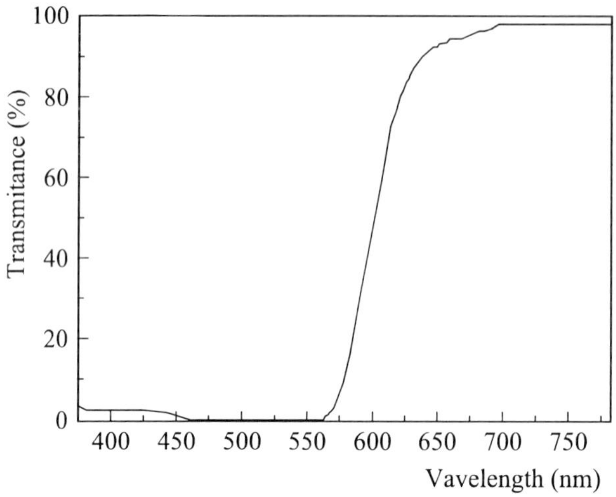


Figure 30. The transmittance spectrum of a polymer film contraining 55% by wt of a nanosize pigment prepared by interacting Red Dye # 6 with titania particles of 10-15 nm in diameter [46].

## 3. CONCLUDING REMARKS

This brief overview of the adsorption of various solutes on uniform adsorbents has emphasized the essential role which chemical interactions play in these processes. By choosing examples of metal ions and organic solutes of increasing complexity, it was demonstrated that specific bond formations between sites on the solid surfaces and active groups on the adsorbates control the uptake of the latter. As a consequence, various properties of the collectors are altered, including their surface charge, dispersion stability, solubility, adhesion, color, etc. These changes can be manipulated by the choice of solutes in contact with a given solid and then employed in numerous applications, including corrosion, catalysis, medical diagnostics, cosmetics, and pigment preparations, to mention a few.

## REFERENCES

1. E. Matijević, Langmuir, 10 (1994) 8.
2. E. Matijević, Chem. Mater., 5 (1993) 412.
3. W.P. Hsu, R. Yu and E. Matijević, J. Colloid Interface Sci., 156 (1993) 56.
4. R. Demchak and E. Matijević, J. Colloid Interface Sci., 31 (1969) 257.
5. M. Ozaki, S. Kratohvil and E. Matijević, J. Colloid Interface Sci., 102 (1984) 146.
6. T. Sugimoto and E. Matijević, J. Colloid Interface Sci., 74 (1980) 227.

7. H. Tamura, E. Matijević and L. Meites, J. Colloid Interface Sci., 92 (1983) 303.
8. I. Kobal, P. Hesleitner and E. Matijević, Colloids Surf., 33 (1988) 167.
9. N. Ogrinc, I. Kobal, E. Matijević and N. Kallay, Fine Fine Particles Science and Technology, E. Pelizzetti (ed.), NATO ASI., Series 3, Kluwer Acad. Publ., Vol. 12 (1996) 85.
10. H. Allen and E. Matijević, J. Colloid Interface Sci., 33 (1970) 420.
11. E. Matijević, in: Applications of Water Chemistry, Faust and Hunter (eds.), Wiley, New York, 1967, 328.
12. E. Matijević, M.B. Abramson, R.H. Ottewill, K.F. Schultz and M. Kerker, J. Phys. Chem., 65 (1961) 1724.
13. K.A. Kraus and R.W. Holmberg, J. Phys. Chem., 64 (1960) 1157.
14. E. Matijević and C.G. Force, Kolloid Z.Z. Polym., 225 (1968) 33.
15. E. Matijević and G.E. Janauer, J. Colloid Interface Sci., 21 (1966) 197.
16. S. Kratohvil and E. Matijević, J. Colloid Interface Sci., 24 (1967) 47.
17. E. Matijević, J.P. Couch and M. Kerker, J. Phys. Chem., 66 (1962) 111.
18. L. Eriksson, E. Matijević and S. Friberg, J. Colloid Interface Sci., 43 (1973) 591.
19. E. Matijević and N. Ryde, J. Adhesion, 51 (1995) 1.
20. N. Ryde and E. Matijević, J. Colloid Interface Sci., 169 (1995) 468.
21. H. Giesche and E. Matijević, Dyes and Pigments, 17 (1991) 323.
22. E. Matijević, J. Colloid Interface Sci., 58 (1977) 374; Colloid and Interface Science, M.Kerker, A.C. Zettlemoyer and R.L. Rowell (eds.), Academic Press, New York, Vol. I (1977) 397.
23. C.G. Pope, E. Matijević and R.C. Patel, J. Colloid Interface Sci., 80 (1981) 74.
24. H. Kumanomido, R.C. Patel and E. Matijević, J. Colloid Interface Sci., 66 (1978) 183.
25. Y. Zhang, N. Kallay and E. Matijević, Langmuir, 1 (1985) 201.
26. N. Kallay and E. Matijević, Langmuir, 1 (1985) 195.
27. Y. Zhang and E. Matijević, Colloid Polymer Sci., 262 (1984) 723.
28. E. Matijević, N. Kolak and D.L. Catone, J. Phys. Chem., 73 (1969) 355.
29. E. Matijević, J. Colloid Interface Sci., 43 (1973) 217.
30. J. Eisenlauer and E. Matijević, J. Colloid Interface Sci., 75 (1980) 199.
31. G. Schwarzenbach and K. Schwarzenbach, Helv. Chim. Acta, 46 (1963) 1391.
32. A. Winston and D. Kitchener, Macromolecules, 11 (1978) 597.
33. H.-C. Chang, T.W. Healy and E. Matijević, J. Colloid Interface Sci., 92 (1983) 469.
34. H.-C. Chang and E. Matijević, J. Colloid Interface Sci., 92 (1983) 479.
35. H.-C. Chang and E. Matijević, Finnish Chem. Lett., (1982) 90.
36. R. Torres, M.A. Blesa and E. Matijević, J. Colloid Interface Sci., 134 (1990) 475.
37. R. Torres, N. Kallay and E. Matijević, Langmuir, 4 (1988) 706.
38. R. Torres, M.A. Blesa and E. Matijević, J. Colloid Interface Sci., 131 (1989) 567.
39. A.E. Regazzoni and E. Matijević, Corrosion, 40 (1984) 257.
40. A.E. Regazzoni and E. Matijević, Corrosion, 38 (1982) 212.
41. E. Matijević, in Ultrastructure Processing of Advanced Materials, D.R. Uhlmann and D.R. Ulrich (eds.), Wiley, New York, 1992, 257.

42. M. Kerker, The Scattering of Light and Other Electromagnetic Radiations, Academic Press, New York, 1969.
43. M. Kerker, P. Scheiner, D.D. Cooke and J.P. Kratohvil, J. Colloid Interface Sci., 71 (1979) 176.
44. A. Tentorio, E. Matijević and J.P. Kratohvil, J. Colloid Interface Sci., 77 (1980) 418.
45. W.P. Hsu, R. Yu and E. Matijević, Dyes and Pigments, 19 (1992) 179.
46. G. Carotenuto, Y.-S. Her and E. Matijević, Ind. Eng. Chem. Res., 35 (1996) 2929.

Adsorption and its Applications in Industry and Environmental Protection
Studies in Surface Science and Catalysis, Vol. 120
A. Dabrowski (Editor)



# Solid desiccant dehumidification systems

T. K. Ghosh and A. L. Hines

Particulate Systems Research Center and Nuclear Engineering Program
University of Missouri-Columbia, Columbia, Missouri 65211, USA

## 1. INTRODUCTION

Desiccant dehumidification assisted air conditioning systems are finding increasing applications for humidity control in commercial and institutional buildings, such as supermarkets, schools, ice arenas, cold warehouses, hotels, theaters, and hospitals. The increasing use of desiccant systems particularly in supermarkets, hotels, and hospitals is due mainly to their better handling of the latent heat load compared to conventional vapor compression refrigeration systems. Over the past several years these systems and desiccant materials themselves have been going through significant improvements in their design to more cost effectively control humidity in buildings with higher moisture and ventilation loads from moist ambient air [1]. Desiccant systems can provide a number of other benefits to both the occupants and building owners.

Hotels are using desiccant systems to supply dry makeup air so that the mold and mildews growth on the furnishings can be reduced. This is turning out to be a major saving for hotels since these furnishings now have to be changed less frequently.

In hospitals, surgeons prefer a lower temperature in operating rooms (around 289 to 291 K) while performing lengthy procedures. A desiccant system is best suited for providing a comfortable atmosphere in such cases. According to Harriman [1], hospital operating rooms that use a desiccant system do not always save money or energy compared to low temperature cooling reheat systems. However, hospital administrators noted that surgeons can heavily influence the revenue generation. If the doctors are not satisfied with the facility, they tend to refer the patients to other hospitals. As a consequence, hospitals are using desiccant systems to make operating rooms more comfortable. Therefore, a desiccant system can indirectly increase the revenue for the hospitals. Also, the dry air can reduce the post operative fungal infection of the patients reducing the time a patient spends in the hospital, which is also beneficial to the hospitals. In some special applications, such as in supermarkets and icerinks, desiccant systems can save both money and energy compared to conventional vapor compression systems.

In the USA, the ANSI/ASHRAE Standard 62-1989, Ventilation for Acceptable Indoor Air Quality [2], is becoming more and more a part of the building codes in designing commercial buildings. As a consequence, a twofold to fourfold increase in outside air requirements for many commercial buildings was experienced. This has resulted in a significant increase in the latent cooling load (moisture removal) compared to sensible heat load (reduction in temperature). As shown in Figure 1, the latent cooling load can be substantial in humid climate [3]. A desiccant cooling system can reduce the cooling loads and at the same time it can improve the indoor air quality.

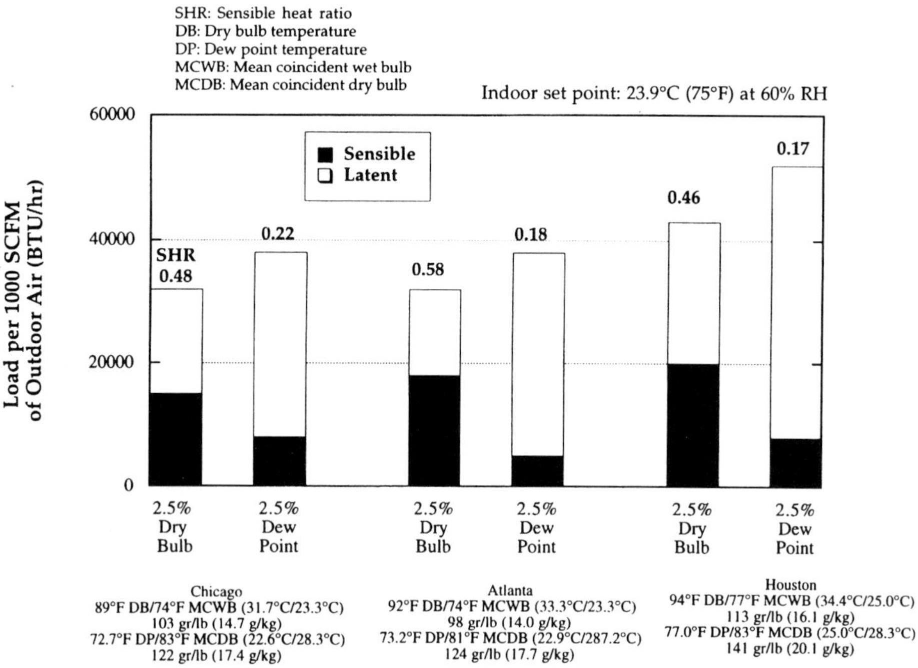


Figure 1. Outside air loads at three different cities (adapted from [3]).

The growing public awareness of the relationship between poor indoor air quality and the adverse health effect has drawn the attention of the building designers and owners to mitigate this problem. From a literature review, Sterling et al. [4] noted that by maintaining the humidity within a certain range risks to human health can be minimized. Bacterial populations were found to be at a minimum when the humidity was between 30% and 60%. Viruses were found to be more susceptible to death in the humidity range of 50% to 70%, while the maximum growth of fungi occurred above 95% relative humidity. The growth of fungi was almost totally inhibited when the humidity was below 80%. The mite population indoors appeared to increase when the relative humidity was above 60%. Based on the findings from the literature, Sterling et al. prepared a nomograph for optimum humidity ranges for good health. The nomograph is shown in Figure 2. They concluded that a humidity range between 40% and 60% at normal room temperature would minimize risks to human health. However, they were concerned that in cold climates relative humidities in that range may be high enough for moisture to condense on surfaces and may lead to fungal growth.

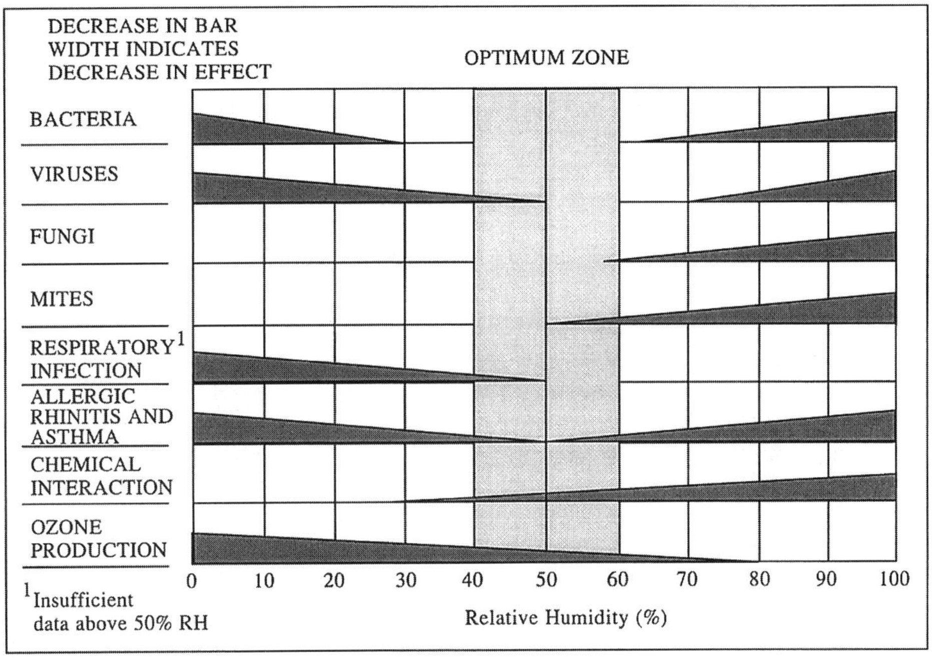


Figure 2. Optimum relative humidity range (adapted from [4]).

A wide variety of systems are available for controlling humidity including enhanced reheat, enthalpy wheel, and desiccant systems. As can be seen from Figure 3, desiccant dehumidification based systems may be most cost effective compared to other systems in handling the latent cooling loads. In spite of having various advantages, as noted by Collier [5], the desiccant cooling technology must attain high levels of thermal performance to become competitive with other systems for a wide range of applications. Gauger et al. [6] assessed various air-conditioning technologies that are alternative to vapor compression refrigeration for use in domestic air conditioning, commercial air conditioning,

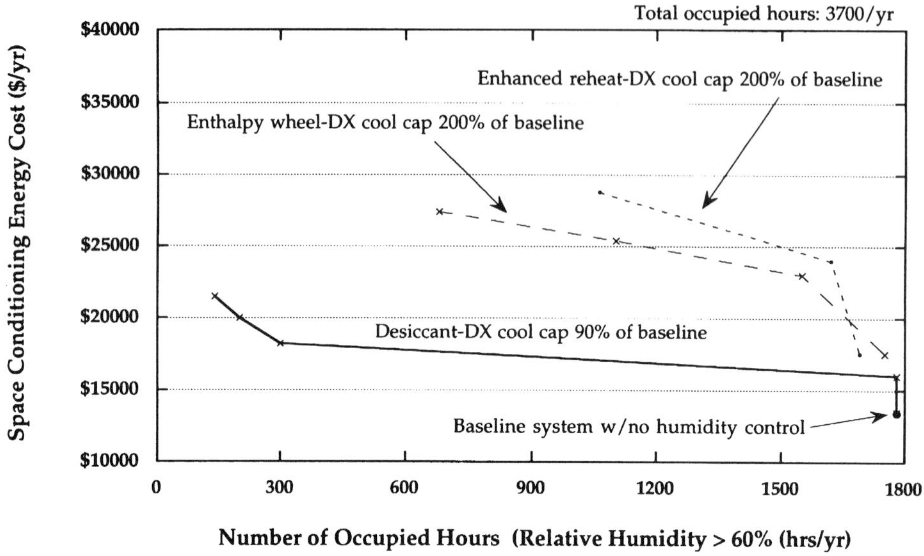


Figure 3. Costs for space conditioning by various humidity controlling systems (adapted from [3]).

mobile air conditioning, domestic refrigeration, and commercial refrigeration. They agreed with the earlier assessment of Collier regarding the desiccant cooling technology. Gauger et al. compared the alternative technologies based on the following criteria: state-of-the-art, complexity, size and weight, maintenance, useful life, and efficiency. They concluded that solid sorption is the most promising alternative technologies to vapor compression in terms of technical feasibility, particularly for air conditioning and refrigeration, where batch processes can be used.

Solid desiccant cooling systems can be broadly divided into two categories: a) Closed-Cycle and b) Open-Cycle. In this chapter, only open-cycle solid desiccant cooling technology has been discussed. However, for interested readers, closed-cycle adsorption technology is briefly described below. The working principle of a closed-cycle solid desiccant based cooling system is similar to that of a vapor compression refrigeration system. The adsorbent, maintained at a low temperature, adsorbs low pressure refrigerant vapor. When the adsorbent is heated at a high temperature, refrigerant vapor is desorbed at a high pressure. This adsorption/desorption process is equivalent to a mechanical compressor in a conventional vapor-compression system. Therefore, it is thermally driven by a heat source instead of electric power. A variety of working fluid-pairs, including zeolite and water, silica gel and water, and alcohol and activated carbons have been investigated. The performance analysis for closed-cycle systems has been carried out by a number of researchers and a variety of methods have been proposed to improve its performance [7-10]. However, most of the analyses showed that the coefficient of performance (COP) is less than unity. Techernev and Clinch [11] developed a prototype closed-cycle system that had an additional regenerative heat exchanger with a cooling COP of around 1.2 using zeolite-water as the working fluid. However a commercial application of this system has not been reported.

## 2. WORKING MECHANISMS OF SOLID DESICCANT COOLING SYSTEMS

Open-cycle desiccant cooling systems operate based on the Pennington cycle (see Figure 4). Moist and warm indoor air after mixing with the make-up outdoor air stream (which is called the process air stream) is filtered and is passed through a desiccant wheel to dehumidify the air stream. Since adsorption is an exothermic process, heat is released as the desiccant adsorbs moisture increasing the temperature of the process air stream. The hot, dry air next flows through a rotating thermal wheel (sensible heat exchanger) to cool the stream, and, thereby, to recover the heat from the process air stream. The relative humidity and temperature of this dry, cool process air stream may be further adjusted to a desired level by passing it through an evaporative cooler where water is added back into the air.

The regeneration of the desiccant material is carried out by using a second air stream, called reactivation or regeneration air. Either the building exhaust or the outdoor air may be used for this purpose. The regeneration air stream, after filtration, may be further cooled in an evaporative cooler to facilitate heat exchange between the process air and the regeneration air in the heat exchanger. The main purpose of this heat exchanger is to cool the process air as much as possible. Various type of heat exchangers including rotating heat wheel, also called enthalpy wheel, heat pipe, and plate-type may be used. The air stream from the heat exchanger is then heated in a direct-fired gas heater before sending it through the desiccant wheel. Pennington's cycle is now routinely referred to as the ventilation cycle. Basic components of a commercial desiccant assisted air-conditioning system are shown in a photograph (Figure 5). A schematic diagram of the system is given in Figure 6. A commercial solid desiccant dehumidification unit is shown in Figure 7.

The details of a desiccant wheel are shown in Figure 8. Generally, the desiccant material is coated onto a ceramic honeycomb structure formed into the shape of a wheel. The wheel constantly rotates between the process air and reactivation or regeneration air streams. Generally, a complete cycle (rotation) of the wheel takes about 6 minutes. However, it may vary between 10 to 20 rph. A section of the wheel is always in the adsorption mode where the moist air is dried, while the rest of the wheel is in the regeneration mode. A cycle may include 270 degrees during which adsorption occurs, and 70 degrees for regeneration. Often the remaining wheel section is used for purging. These correspond to: adsorption - 270 seconds, regeneration - 70 seconds, and purging - 20 seconds.

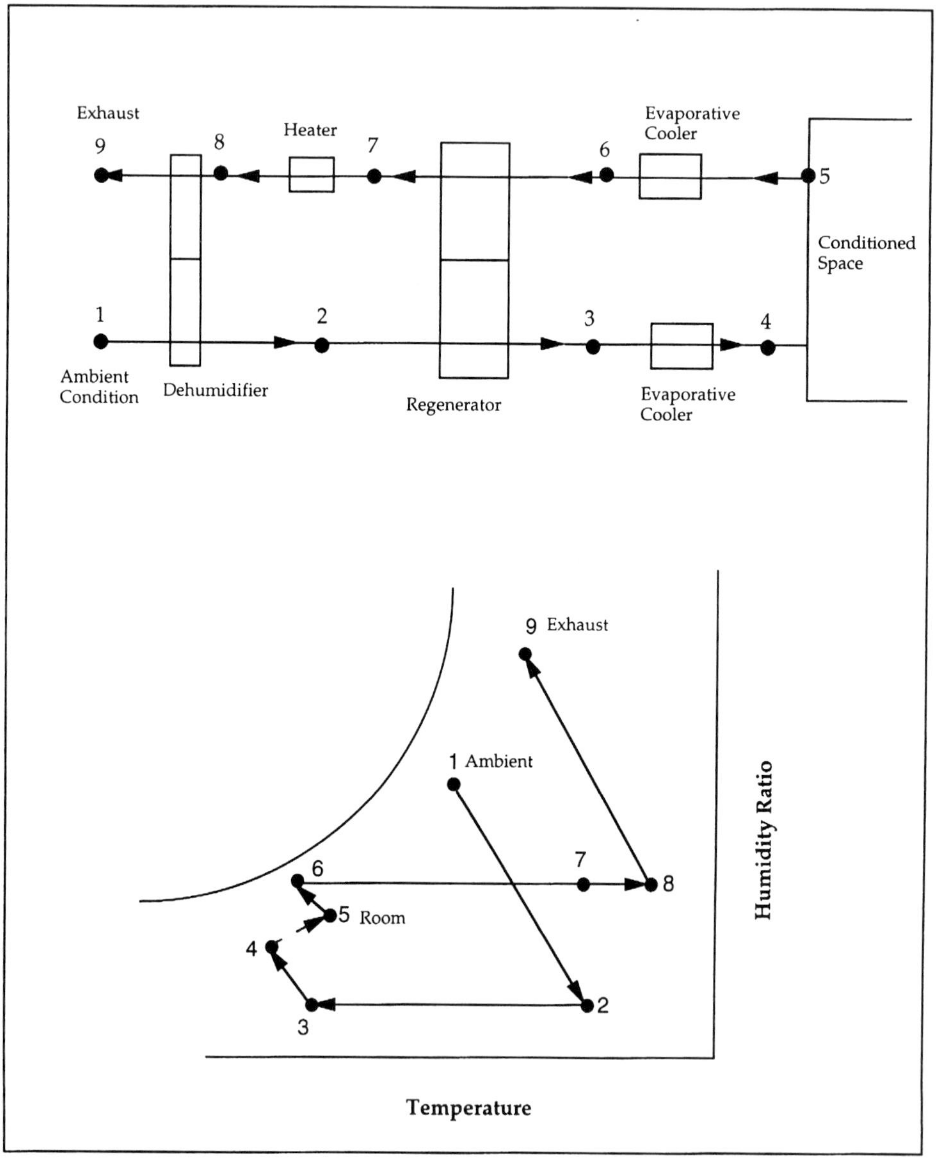


Figure 4. Pennington cycle for desiccant cooling system (adapted from [12]).

Figure 5. Various components of a commercial size solid desiccant dehumidification system (photo courtesy of Semco Inc., Columbia. MO).

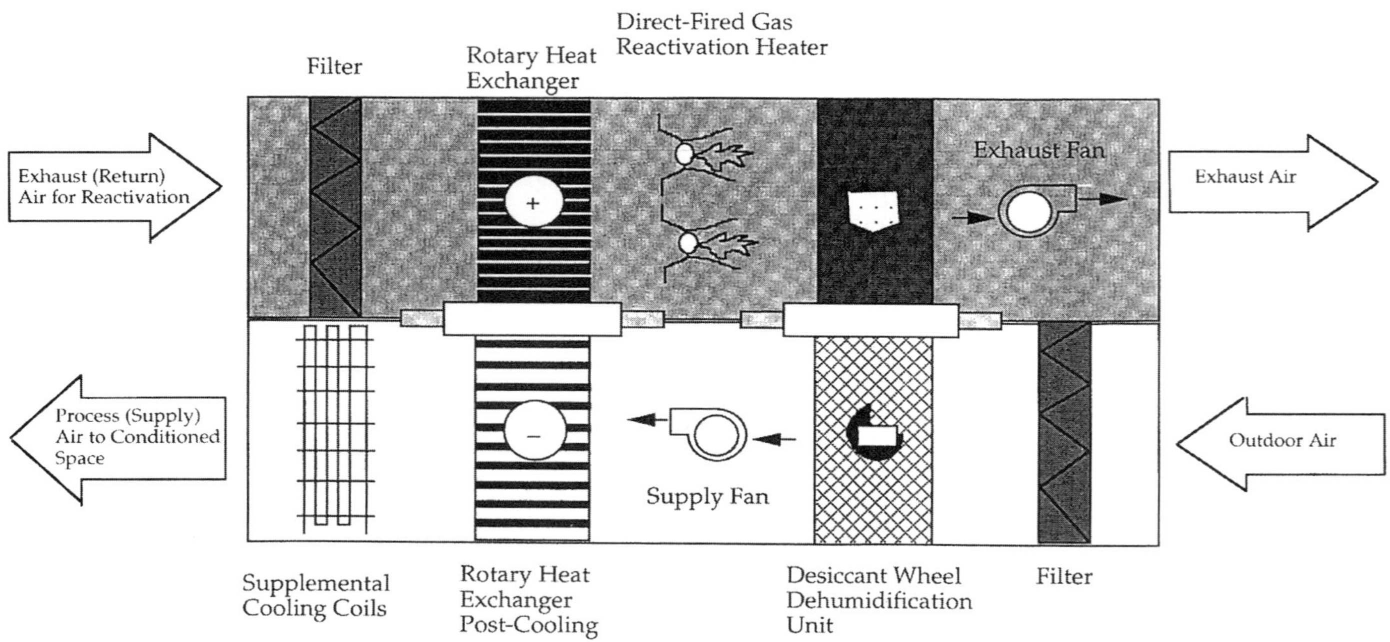


Figure 6. Schematic of the Desiccant Cooling System (adapted from [13]).

Figure 7. A packaged commercial solid desiccant cooling system.

Figure 8. Details of a desiccant wheel assembly.

The functions of various units of a desiccant assisted air-conditioning system are further explained in Figure 9 (a, b) using a psychrometric chart. In Figure 9(a), conditions of process air at various units are explained. Figure 9(b) shows the conditions of regeneration air.

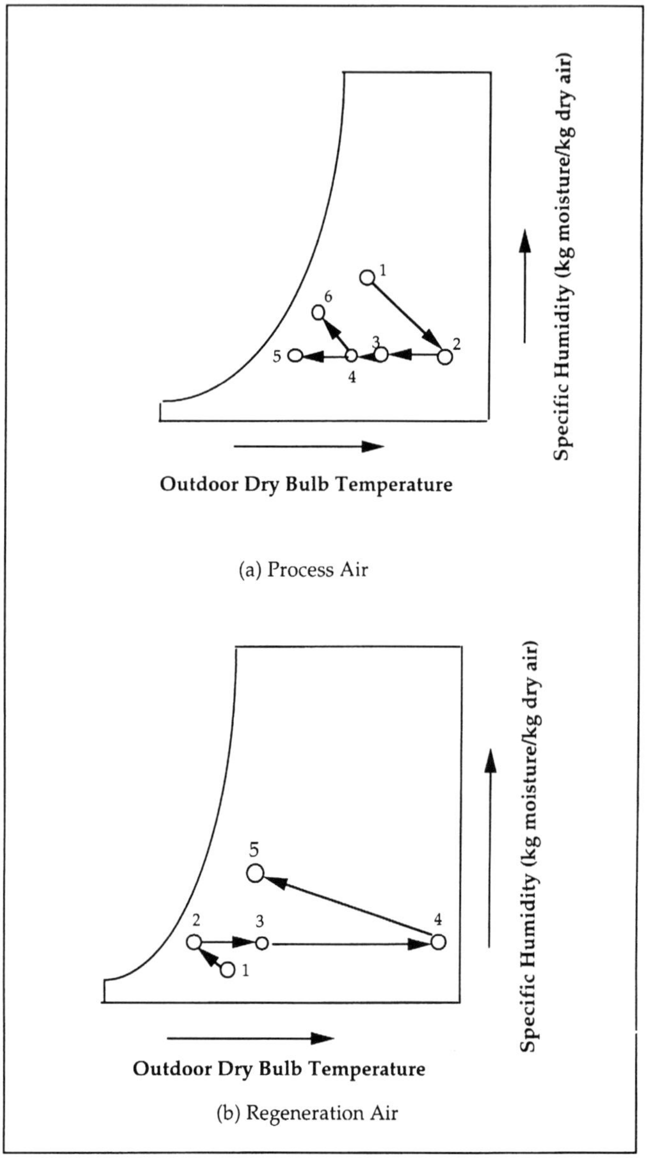


Figure 9. State points in a psychrometric chart within a desiccant system.

Process air

1. Process air is generally hot (305 K) and humid prior to entering the desiccant wheel.
2. Air leaving the desiccant wheel is relatively hotter (325 K) than point 1, but it is much drier than when it entered the wheel.
3. The air is cooled (to about 310 K) by lowering its sensible temperature by exchanging heat with the regeneration air.
4. The temperature of the air may be further reduced (300 K) using a gas cooling or conventional cooling system.
5. The air may be further cooled (288 K) by a vapor compression system depending on the application.
6. If the relative humidity of the air needs to be further adjusted, a direct evaporative cooling system may be used (air temperature at this point is about 297 K). However, this option is not desirable if the process air is to be used in human-occupant environments. The process air can become saturated with moisture at this point and the final humidity of the air may not be controlled adequately.

Regeneration air

1. Hot, moist outside air (305 K) is generally used for regeneration of the desiccant wheel.
2. The outside air stream may be further cooled (302 K) to facilitate heat exchange in the heat exchanger.
3. A second heat exchanger may be used to recover the waste heat from the process air. This increases the temperature of the regeneration air (325 K).
4. A direct fired gas heater is generally used to increase the temperature to about 408 K before sending through the wheel.
5. Following regeneration, the temperature of the stream drops to about 325 K, at which point it may be exhausted to the atmosphere, or the waste heat may be recovered if it becomes economical.

A desiccant cooling system can be operated in four different modes: recirculation, ventilation, make-up air, and mixed mode [13]. The system configurations for these four modes of operation are shown in Figure 10. The distinction in the operation is made mainly based on the sources of process and regeneration air streams. Outdoor air or a mixture of outdoor and exhaust air may be used for regeneration, and similarly, outdoor air or a mixture of outdoor and return air may be used for the process air stream. In the recirculation mode, the source of process air is return air from the conditioned space, and outdoor air is used for regeneration. In the ventilation mode, the pure outdoor air is used as the process air stream and exhaust air is used for regeneration. If outdoor air is used as a source for both process and regeneration air, it is called make-up mode

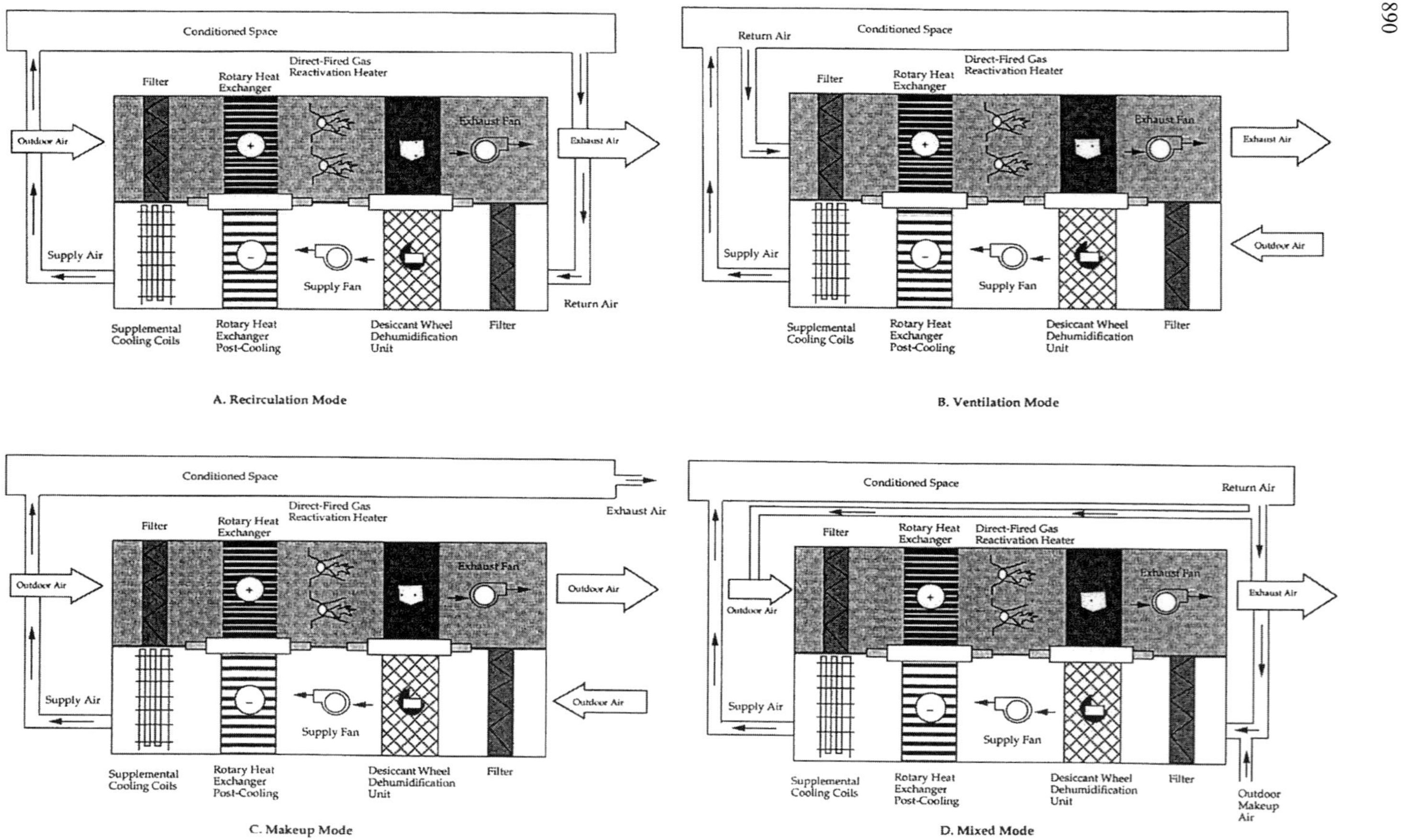


Figure 10. Various modes of operation of desiccant cooling system (adapted from [13]).

of operation. In the mixed mode, the source for process and regeneration air can be a combination of outdoor air and return air, outdoor air and exhaust air, and other combinations as discussed above. The choice of a particular mode of operation depends on the application requirement, and on the economic viability of the system.

## 3. DESICCANT MATERIALS AND THEIR RELATIONSHIP TO SYSTEM PERFORMANCE

A number of inorganic porous solids have strong affinity for adsorption of water vapor. For desiccant cooling applications, a number of factors including the water vapor isotherm shape, adsorption capacity, heats of adsorption, adsorption kinetices, desorption characteristics (hysteresis), and chemical and physical stability of the material from cycle to cycle should be considered in selecting a desiccant material. In the desiccant dehumidifier, the desiccant material is cyclically exposed between ambient temperature, 25°C, during the adsorption cycle and a regeneration temperature, which may be in the range of 90°C to 160°C.

Among various desiccant materials, silica gel is most frequently used in commercial units because of its thermal stability and higher water vapor adsorption capacity than other desiccants. Also it is relatively easy to regenerate. Although molecular sieves have water vapor adsorption capacity and thermal stability comparable to silica gel, a relatively higher temperature and a longer regeneration time are required, which make them unfavorable for desiccant cooling applications. Other desiccant, such as activated alumina, have not been seriously studied as potential materials because silica gel has better adsorption properties. The thermal coefficient of performance (COP)* of a desiccant cooling system using silica gel is usually in the range of 0.85 to 1.05 depending on the regeneration temperature and the efficiency of the sensible heat exchanger. The cooling capacity of such a system is generally in the range of 12 to 22 kJ/kg (if operated in a ventilation cycle). An increase in the COP value and a reduction in kJ/kg, which can reduce the physical size of desiccant cooling systems, could make desiccant cooling systems cost competitive with other air-conditioning systems.

The regeneration characteristics of the desiccant materials play a key role in determining the system performance. An analysis of desiccant cooling cycles by Collier [14] and Jurinak [15] showed that the adsorption and desorption characteristics of desiccant materials can have large influence on the cooling capacity and thermal coefficient of performance of the cycle.

---

* The performance of desiccant cooling systems is generelly assessed in terms of cooling capacity (CC) and thermal coefficient of performance (COP). The cooling capacity is defined as the amount of cooling (usually expressed in terms of kW or tons of cooling) delivered by the system divided by the amount of air passing through the system (in kg/s or cfm). The thermal COP is obtained as the amount of cooling delivered by the system divided by the thermal energy input for regeneration of the desiccant dehumidifier.

Various system configurations have been proposed for improving the COP of desiccant cooling systems. These methods are: (1) the increase of the regeneration air temperature and flow rate, (2) the addition of inert heat capacity to the desiccant matrix, and (3) staging of the regeneration air stream [16].

Several investigators [15, 17-19] studied the effect of balanced and unbalanced flows in both the heat exchangers and in the desiccant wheel. In the balanced flow mode of the operation, entire regeneration flow is heated to a maximum regeneration temperature. When an unbalanced flow is used, a portion of the preheated regeneration air stream from the sensible heat exchanger is vented to the atmosphere. An increase of 10 to 15 percent in the COP value was reported when unbalanced flow was used compared to a balanced flow system.

A large gain in the thermal COP, in the range of 50 to 150 percent, was predicted by several researchers if an inert heat capacity was added to the desiccant wheel [20-23]. Schlepp and his coworkers [21, 22] also calculated the amount of inert materials in a composite, parallel-passage desiccant wheel design. Collier and Cohen [16] examined the COP gain in a staged regeneration desiccant system. In the staged regeneration system, the regeneration of the desiccant was carried out in two stages. In the first stage, the desiccant was heated by air leaving the sensible heat exchanger. No additional heat was added in this stage. In the second stage, only a fraction of the regeneration air was heated to a maximum regeneration temperature before flowing through the desiccant wheel. Results form their computer simulation (see Table 1) showed that the staging the regeneration air stream can improve the system performance, while minimizing the amount of inert material in the desiccant wheel.

Collier [14] noted that the shape of the water adsorption isotherm could significantly affect the system performance. The performance of various materials having different shapes of the water adsorption isotherm was evaluated using a computer model. Results showed that certain desiccant materials can enhance the dehumidification performance, but require a greater amount of energy for regeneration, thus reducing the regeneration efficiency and vice versa. Therefore, the best desiccant material would be a compromise between these two factors. Collier also assumed various shapes of the water isotherm based on a constant separation factor [24] and performed the same analysis. The constant separation factor is defined as follow

$$R = \frac{FC \cdot (1 - SC)}{SC \cdot (1 - FC)} \tag{1}$$

where,

SC = relative solid concentration of water

FC = relative air concentration of water

R = separation factor

Table 1
Desiccant cooling system performance with staging regeneration

| | Low Heat Capacity | | Moderate Heat Capacity | | High Heat Capacity | |
|---|---|---|---|---|---|---|
| | W/O | With | W/O | With | W/O | With |
| COP | 0.76 | 1.02 | 0.78 | 0.92 | 0.75 | 0.84 |
| Specific Cooling Capacity (kJ/kg) | 20.9 | 18.3 | 19.0 | 18.0 | 14.5 | 14.2 |
| Regeneration Fraction (%) | 100 | 50 | 100 | 65 | 100 | 70 |

*Low Heat Capacity*: A desiccant bed consisting only of the desiccant material (silica gel)
*Moderate Heat Capacity*: A desiccant bed consisting of 50% desiccant and 50% inert material
*High Heat Capacity*: A desiccant bed consisting of 20% desiccant and 80% inert material

Other Parameters:

NTU for desiccant wheel: 28.5
Sensible heat exchanger effectiveness: 93%
Evaporative cooler effectiveness: 95%
Regeneration temperature: 368 K
Desiccant material: silica gel
Mode of operation: ventilation
ARI standard air conditions: 300 K (80°F), 0.011 kg/kg indoors; 308 K (95°F), 0.014 kg/kg outdoors.

(Adapted from reference [16])

Collier et al. [25] noted that a material that has a constant separation factor of 0.1 is best suited for the desiccant-dehumidification applications. They designated this material as type 1M material and its water adsorption isotherm can be described by Brunauer type 1 isotherm. The theoretical shape of the water adsorption isotherm on 1M material having a separation factor of 0.1 is compared with that on other materials, such as silica gel, molecular sieve 13X, and hydrophobic molecular sieve in Figure 11. Their analyses showed (see Figure 12) that with 1M desiccant the operating costs would be at least 20% lower compared to conventional materials. The use of 1M desiccant also lowers the initial cost of the desiccant cooling system due to the reduction in physical size. Interestingly, the same simulation study showed that the effects of heat and mass transfer resistances on desiccant cooling system performance were independent of desiccant type for ventilation cycle operation. Worek et al. [26] noted that the operation of a desiccant system on ventilation mode over the recirculation mode may be preferable due to a small increase in COP and a decrease in cooling capacity (Table 2). A number of researchers have computer

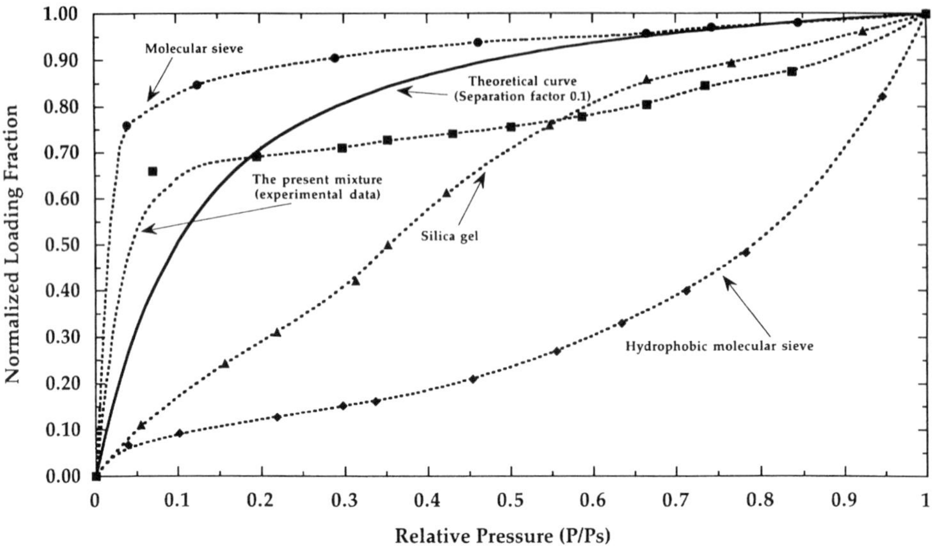


Figure 11. Comparison of shapes of water isotherms on various materials at 288 K with that of a theoretical material which has a constant separation factor of 0.1.
* Normalized Loading Fraction = amount of water adsorbed at a particular relative pressure/the amount of water adsorbed at saturation pressure.

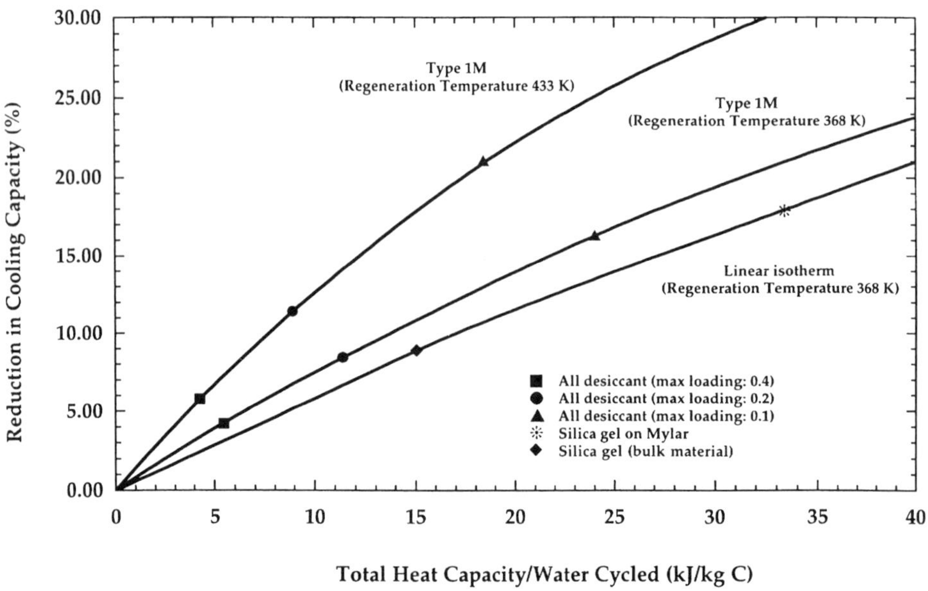


Figure 12. Reduction in cooling capacity when using 1 M Type desiccant (adapted from [25]).

Table 2
A comparison between recirculation and ventilation mode of operation of a desiccant cooling system

Recirculation mode of operation; R = 0.1, and $f^*W_{max} = 0.2$

| | $\eta_{HX}$=0.87 | | | | $\eta_{HX}$ = 0.93 | | | |
|---|---|---|---|---|---|---|---|---|
| $T_g$ (°C) | $COP_{opt}$ | $CC_{opt}$ kJ/kg | $PERC_{opt}$ | $TF_{opt}$ | $COP_{opt}$ | $CC_{opt}$ kJ/kg | $PERC_{opt}$ | $TF_{opt}$ |
| 90 | 0.79 | 14.8 | 0.37 | 314 | 0.91 | 16.1 | 0.36 | 299 |
| 125 | 0.84 | 18.4 | 0.27 | 310 | 0.96 | 20.0 | 0.26 | 298 |
| 160 | 0.83 | 18.6 | 0.19 | 308 | 0.96 | 20.7 | 0.18 | 297 |

Recirculation mode of operation; R = 0.1, and $f^*W_{max} = 0.2$

| | $\eta_{HX}$=0.87 | | | | $\eta_{HX}$ = 0.93 | | | |
|---|---|---|---|---|---|---|---|---|
| $T_g$ (°C) | $COP_{opt}$ | $CC_{opt}$ kJ/kg | $PERC_{opt}$ | $TF_{opt}$ | $COP_{opt}$ | $CC_{opt}$ kJ/kg | $PERC_{opt}$ | $TF_{opt}$ |
| 90 | 0.81 | 13.9 | 0.43 | 244 | 1.09 | 14.3 | 0.33 | 219 |
| 125 | 0.85 | 17.6 | 0.28 | 254 | 1.12 | 17.7 | 0.22 | 233 |
| 160 | 0.85 | 18.9 | 0.21 | 261 | 1.12 | 18.8 | 0.16 | 241 |

R: Desiccant isotherm separation factor; f: Mass fraction of desiccant in the dehumidifier; W: Equilibrium moisture content at Psat.; Tg: Regeneration temperature; COP: Coefficient of performance; CC: Cooling capacity; PERC: Fraction of the regeneration stream that is heated; TF: Nondimensional sorption time; Psat: Saturation pressure of water at system temperature., opt: Optimum condition.

Other input variables for numerical calculations

Desiccant fraction, f: 0.04
Inlet temperature of the air stream: 298 K
Inlet humidity ratio: 0.01225 kg/kg
Desiccant weight: 34.26 g

(Adapted from reference [23])

simulated the desiccant cooling system performance under various conditions and have calculated the system COP. Pesaran et al. [12] have summarized these findings as shown in Table 3.

Table 3
A summary of COP values obtained by various researchers from the computer simulation of the desiccant cooling system at ARI standard conditions

| Cycle | Thermal COP | Comments | References |
|---|---|---|---|
| Pennington | 1.2 | $\eta_{HX} = 0.93$, $\eta_{EC} = 0.95$, high dehumidifier performance | Schlepp and Barlow, 1984 [22] |
| Pennington | 0.53-1.02 | $\eta_{HX} = 0.85$, $\eta_{EC} = 0.90$ | Jurinak et al., 1984 [27] |
| Pennington | 1.4 | $\eta_{HX} = 0.95$, $\eta_{EC} = 0.95$, high heat capacity | Grolmes and Epstein, 1982 [28] |
| Pennington | 1.1 | $\eta_{HX} = 0.90$, $\eta_{EC} = 0.90$, staged regeneration | Husky et al., 1982 [29] |
| Pennington | 0.5 | Low dehumidifier performance | Majumdar et al., 1982 [30] |
| Recirculation | 0.7 | Normal heat exchanger effectiveness | Majumdar et al., 1982 [30] |
| Cooled Bed | 0.6 | Cooled bed configuration | Lavan et al., 1981 [31] |
| Recirculation | 0.50-0.78 | $\eta_{HX} = 0.85$, $\eta_{EC} = 0.90$ | Jurinak et al., 1984 [27] |
| Pennington | 1.1 | $\eta_{HX} = 0.93$, $\eta_{EC} = 0.90$, type 1M desiccant and staged regeneration | Collier et al., 1990 [14] |
| REVERS | 1.25 | Ideal components | Kang, 1985 [32] |
| SENS | 2.58 | Ideal components | Kang, 1985 [32] |
| DINC | 1.10-1.93 | $\eta_{HX} = 0.95$, $\eta_{EC} = 0.95$, high dehumidifier performance | Waugaman and Kettleborough, 1987 [33] |

$\eta_{HX}$ : Heat exchanger efficiency; $\eta_{EC}$ : Evaporative cooler efficiency

(Adapted from reference [12]).

## 4. MODELING OF DESICCANT DEHUMIDIFICATION SYSTEMS

The key unit of the desiccant cooling system is the dehumidifier. The potential for further increase in the system performance lies in the design of this dehumidifier. Therefore, much of the attention was in modeling and optimizing the performance of the dehumidifier. The simultaneous heat and mass transfer in air-water-desiccant systems has been modeled by a number of investigators [34-41]. These models have been used to optimize the performance in terms of flow rates, rotation speed, desiccant properties, and desiccant-matrix composition. The moisture balance equation can be written as [42]:

$$G_a = \left(\frac{\partial W}{\partial y}\right)_\theta + \rho_a F\left(\frac{\partial W}{\partial \theta}\right)_y = -\rho_A\left(\frac{\partial \omega}{\partial \theta}\right)_y \tag{2}$$

The local rate of adsorption is assumed to be proportional to the difference between the humidity ratio W of the moist air and the humidity ratio $W_e$ of moist air in equilibrium with the adsorbent, which is at the system temperature and exposed to local moisture concentration. Therefore,

$$\rho_A\left(\frac{\partial \omega}{\partial \theta}\right)_y = h_D A_V (W - W_e) \tag{3}$$

The energy balance equation yields

$$G_a = \left(\frac{\partial h}{\partial y}\right)_\theta + \rho_a F\left(\frac{\partial h}{\partial \theta}\right)_y = -\rho_A\left(\frac{\partial h_A}{\partial \theta}\right)_y \tag{4}$$

The local rate of energy transfer may be expressed as

$$\rho_A\left(\frac{\partial h_A}{\partial \theta}\right)_y = h_c A_v (T - T_e) + h_D A_V (W - W_e) h_W \tag{5}$$

Where,

| | | |
|---|---|---|
| $A_V$ | = | Adsorbent surface area, $cm^2/cm^3$ |
| F | = | Fractional volume occupied by voids between adsorbent particles |
| $G_a$ | = | Dry air mass velocity, $kg/s{\cdot}cm^2$ |
| h | = | Enthalpy of moist air, kJ/kg dry air |
| $h_A$ | = | Enthalpy of moist adsorbent, kJ/kg of adsorbent |
| $h_c$ | = | Convective heat transfer coefficient, $kJ/s{\cdot}cm^2{\cdot}K$ |
| $h_D$ | = | Convective mass transfer coefficient, kg water/$s{\cdot}cm^2{\cdot}$(kg water/kg) |
| $h_w$ | = | Enthalpy of water vapor, kJ/kg |
| T | = | Moist air dry bulb temperature, K |
| $T_e$ | = | Average temperature of adsorbent particle, K |
| W | = | Humidity ratio of moist air, kg water vapor/kg of dry air |
| $W_e$ | = | Humidity ratio of moist air in equilibrium with local adsorbent, kg water vapor/kg of dry desiccant |
| y | = | Adsorbent bed depth, cm |
| $\omega$ | = | Average moisture content of adsorbent particle, kg water vapor/kg adsorbent |
| $\theta$ | = | Time, s |
| $\rho_A$ | = | Density of adsorbent particle, $kg/cm^3$ |
| $\rho_a$ | = | Density of moist air, kg air/$cm^3$ dry air |

For calculating heat and mass transfer resistances during desiccant dehumidification, both the resistance in the gas phase and the resistance within a desiccant particle should be considered. The gas side heat and mass transfer characteristics have been studied extensively, and a wealth of information is available in the literature to analyze them for desiccant dehumidification processes. The flow through the individual channels of the desiccant matrix is usually laminar. The Nusselt number can adequately describe the gas side heat transfer, whereas the Sherwood number can be used for mass transfer. Diffusion processes within a desiccant particle are generally complex. However, conductive heat transfer within a particle has been studied extensively for desiccant particles. Most of the complexities are related to the mass transfer process. The nonhomogeneity within the particle, which is due mainly to the presence of pores, increases the complexity in the calculation. The size and distribution of pores vary throughout the particle and are difficult to characterize. Also, the concentration gradient within the particle changes continuously making the mass diffusion dependent upon the degree of uptake. Therefore, a rigorous method for solving the mass diffusion problem is to set-up the differential equations and then to solve them using appropriate boundary conditions, which are dependent on the particle geometry. The desiccant particles usually have irregular shape making the task even more challenging. However, such an approach requires greater computational effort because of the solution of an additional diffusional equation with the space variable in the radial direction. Kerestecioglu [36] using a finite-element method and Ko [37] using a finite-difference method solved these equations. However, the predicted values showed significant deviations from the experimental data. The lack of experimental data for diffusion within the particles limited the accuracy of the results.

A simple approach to predict the mass transfer within a desiccant particle is known as "Lewis number" analogy. The Lewis number may be defined as the ratio of the mass diffusivity divided by the thermal diffusivity in the same medium. For moisture transport in air, the Lewis number is approximately equal to unity. The gas side mass transfer, therefore, can be predicted from the knowledge of heat transfer using correlations developed for gas side heat transfers. The Lewis number analogy has been extended by many researchers for predicting mass transport phenomenon within desiccant particles by defining an "effective Lewis number". The sum of resistances to moisture transport within the desiccant particle and the resistance through gas phase is divided by the gas side thermal resistance to obtain the effective Lewis number. The objective is to assume that the mass transfer is controlled by this pseudo gas side resistance. Therefore, its value may be assumed to be larger than the gas side resistance when considered alone. This will simplify the solution of the problem. Although the researchers have noted that the Lewis number required for the adsorption process was different than the Lewis number required for regeneration step, the Lewis number did not have to be varied during the individual process. It was also noted that for silica gel particle diameter of less than 3 mm, the internal resistance to mass transfer was negligible. In desiccant dehumidification applications, the particle diameters are usually less than 3 mm, therefore, the use of Lewis number equal to one yielded a good agreement with experimental data. However, it should be noted that the use of effective Lewis number may lead to different conclusion regarding the mass transfer mechanisms. Within a particle, the driving force for mass transfer is the concentration gradient that exists within the particle and not the concentration gradient that exists between the surface of the particle and the bulk fluid stream. The effective Lewis number cannot provide information regarding true internal resistance.

Besides heat and mass transfer coefficients, a number of other factors also affect the moisture transport between the desiccant particle and the air stream including desiccant matrix channel size and geometry, and the flow rates through the channels. Therefore, another approach to analyze the desiccant systems with respect to heat and mass transfer is to use a dimensionless number called Number of Transfer Unit (NTU). The NTU may be viewed as the number of stages required to approach the equilibrium by the desiccant system. Although different NTU for heat transfer and mass transfer may exist, if the effective Lewis number is assumed to be unity then the number of heat transfer units become equal to the number of mass transfer. This further simplifies the calculation procedure. For example, equations (2) and (4) can be simplified by assuming negligible changes in the fluid mass and heat storage terms $\partial W/\partial q$ and $\partial h/\partial q$, and the simplified equations are given by:

$$G_a\left(\frac{\partial W}{\partial y}\right)_\theta + \rho_A\left(\frac{\partial \omega}{\partial \theta}\right)_y = 0 \tag{6}$$

$$G_a\left(\frac{\partial h}{\partial y}\right)_\theta + \rho_A\left(\frac{\partial h_A}{\partial \theta}\right)_y = 0 \tag{7}$$

By introducing NTU and Lewis number (Le), Equations (3), (5), (6) and (7) can be rewritten as:

$$\frac{\partial \omega}{\partial \tau} = \frac{h_D A_V L}{G_a}\frac{G_a \theta_c}{L\rho_A}(W - W_e) = -NTU\beta(W_e - W) \tag{8}$$

$$\frac{\partial W}{\partial z} = NTU(W_e - W) \tag{9}$$

$$\frac{\partial h_A}{\partial \tau} = -NTU\beta\left[Le\,c_{p,a}(T_e - T) + h_w(W_e - W)\right] \tag{10}$$

$$\frac{\partial h}{\partial z} = NTU\left[Le\,c_{p,a}(T_e - T) + h_w(W_e - W)\right] \tag{11}$$

where,

$$z = \frac{y}{L}, \ \tau = \frac{\theta}{\theta_c}, \ Le = \frac{h_c}{h_D c_{p,a}}, \ NTU = \frac{h_D A_V L}{G_a}, \ \text{and} \ \beta = \frac{G_a \theta_c}{L \rho_A}$$

Finite difference techniques have been used by a number of researchers to solve these equations. Maclaine-cross and Banks [38, 39] argued that these approximation may not be valid for rotating desiccant wheels and proposed two alternative approximation methods. Heat and moisture are transferred periodically from one fluid stream to the desiccant matrix and then from the desiccant to the other fluid streams. They used a single film transfer coefficient to describe both heat transfer and mass transfer. Mass transfer Equations (2) and (3) were transformed into a form analogous to heat transfer Equations (4) and (5) by introducing new variables, which they called characteristic potentials. The lack of experimental data prevented detailed evaluation of their models. However, a number of investigators adopted their finite difference scheme and the analogy between heat and mass transfers to develop similar models and studied the effects of other parameters on the system performance. Jurinak and Mitchell [18] studied the performance of a dehumidifier considering that a portion of the processed air stream was purged from the bulk flow and was recirculated with the regeneration stream. A symmetric and balanced counterflow silica gel dehumidifier was used. The analysis showed a small increase in the performance that ranged from 4% to 8% depending on the thermal capacitance of the dehumidifier.

Barlow [40] analyzed the moisture adsorption by a packed bed of silica gel using a pseudo-steady state model for coupled heat and mass transfers. The computer program developed by Barlow (called DESSIM) is a well documented program that is easy to follow and modify to suite a variety of desiccant materials, geometrics, and flow patterns. Schultz and Mitchell [29] compared the results from the DESSIM program with that developed by Maclaine-cross [39] and found that the difference in the predicted values by these two programs was only 5%. Because of its simplicity, the DESSIM program became an attractive tool for system performance analysis and design. Collier and Cohen [16] modified the DESSIM program for use in parallel flow operations rather than original counter flow, and for vapor pressure driving potential in the mass transfer rather than concentration driving potential. They also changed the structure of the program to make it more stable numerically and for modeling solid rotary dehumidifiers. A number of researchers [43-45] have verified these codes with experimental data with a varying degree of accuracy. However, all these codes agree well with each other when only gas-side mass transfer resistance was considered. When the internal mass transfer can no longer be ignored, these models required some manipulation to match experimental data.

The rotational speed of the wheel plays an important role in determining the design and optimum performance of the desiccant wheel as it indirectly influences the moisture uptake and heat and mass transfer rates [46]. For a desiccant wheel, there exists a rotational speed that can provide the lowest process air outlet humidity resulting in an optimal performance. The simulation study showed that a desiccant wheel utilizing a material with most favorable isotherm shape, excellent moisture uptake property, and a higher NTU values would not provide optimal performance if the rotational speed of the wheel is not optimized.

Zheng et al. [46] noted that the three key parameters that affect the dehumidifier performance are: (1) desiccant isotherm shape, (2) moisture uptake by desiccant matrix, and (3) heat and mass transfer characteristics of the desiccant matrix. Their computer simulation showed that for the optimum performance the isotherm shape of the material should have a separation factor of 0.07. Although the performance increased with the increase of moisture uptake by the desiccant matrix, no significant increase in the performance was noted if the moisture uptake was larger than 25% of the weight of the desiccant matrix. Similarly, for NTU larger than 12, the performance did not change significantly.

## 5. COMPONENTS OF DESICCANT SYSTEMS

Major components of a desiccant cooling system are a) desiccant wheel assemblies, b) reactivation heaters, c) heat exchangers for post cooling, d) indirect evaporative post coolers, e) filtration systems, and f) control systems. An excellent description of these units is given by Harimann [47].

A desiccant wheel assembly consists of: i) a core that contains the desiccant materials, ii) a core/wheel support structure, iii) a drive system to rotate the wheel at a very low speed, and iv) a set of air seals to separate the process air stream from the regeneration air stream. The core has a monolith honeycomb type structure. Both sinusoidal and hexagonal channels are used in commercial equipment. Various types of materials have been used for support matrix that includes ceramic, glass fibers, and corrugated aluminum sheets. The desiccant materials may be washcoated, impregnated, or formed in situ on the support matrix.

The support structure of the wheel depends on its size and diameter. For a smaller wheel, diameter typically below 150 cm, the core material itself is formed in the shape of a wheel and a round casing at its perimeter protects its edges. For a larger diameter wheel, the core material is manufactured into sections and is assembled into a spoke-and-rim structure, supported in a frame by a hub and bearing assembly in the middle (see Figure 8).

The wheel rotates at speeds in the range of 6 rph to 20 rph depending on the desiccant material, regeneration air temperature, and a host of other factors described earlier. Generally, a belt or chain attached to the perimeter of the wheel is driven by a motor. This results in a low power consumption compared to if rotated from the center hub.

The air seals at the edge of the desiccant wheel are used to separate the process air from the regeneration air. The seals are subjected to a high temperature on the regeneration side and a relatively cold process air on the other side. The system performance depends to some extent on how effectively the seals can prevent air leaks from either sides. Generally, two types of seals are used by the manufacturers. In the first type, flexible rubber strips are attached on the wheel face and wipe the surface of the wheel lightly. Air pressure from fans keeps these strips tight against the wheel. Although these type of seals are cheap and have long life, significant leakage of air has been experienced with these seals. Compressible bulb seals pressing against the face of the wheel are most effective in preventing air leakage across the seal. However, they are more expensive than wiper seals and also requires frequent maintenance. The rotor casing which provides support for the seals and the wheel is also an integral part of preventing the leakage. The casing must have smooth outer surface for free rotation and proper sealing of the surface by the seals.

The design of the reactivation heater depends on the heat sources, which may be electric resistance heaters, solar hot water coils, hot water coils, steam coils from the steam boilers, and natural gas heaters. A reactivation heater generally recovers about 20% heat from the process air and the rest is obtained from the above auxiliary sources. Among these heat sources, natural gas heaters are becoming popular because of the ease of the availability of the natural gas and its cost. Natural gas heaters may save between 50% and 75% of the cost of reactivating the desiccant compared to other means.

Both direct-fired and indirect-fired natural gas heaters are used by the manufacturers. In a direct fired heater, natural gas is burned directly into the reactivation air stream and the exhaust gas is used for regeneration. Although 90% to 95% heating efficiency can be achieved, there is always a danger of local hot spots that can scorch the desiccant and other components as the temperature can be as high as 977 K. These types of problems can be avoided in indirect-fired heaters, but the heating efficiency for indirect-fired heaters is in the range of 50% to 70%.

Heat exchangers commonly used for transfer of sensible heat from the process air stream to the regeneration stream include plate-type, heat wheels, and heat pipes. Among these heat exchangers, heat wheels are most frequently used because of their efficiency which is in the range of 80% to 95%. A heat wheel is similar to a desiccant wheel in design, except the composition of the core material. It also rotates at 10 to 20 rpm, which can provide significant flexibility in the design and in the operation. The efficiency of the heat wheel depends on the wheel rotation speed, since the cooling effect can be changed by varying its speed.

Often an evaporative cooler is used in conjunction with the desiccant wheel to further adjust the humidity and temperature of the process air. The process air is flowed through a wetted medium. Since it is an adiabatic process and the air is dry, only a small amount of energy is needed to circulate water through the wetted medium. However, the water reservoir needs to be cleaned and treated periodically to prevent microbial growth. In high-rise buildings and supermarkets where a water cooling tower already exists, the use of an evaporative cooling pad is even more economical.

Other common components required for desiccant systems include filtration systems and control modules. Unless desiccant systems are used in hospitals or sensitive industrial processes, the purpose of using filters is to remove large particles such as insects, grass clippings, etc. Like any other systems, common control modules such as that for direct or indirect fired burners, motors, etc., are used.

## 6. SYSTEM MODELS

Several system models are available for calculating cooling loads of a building. These models have combined together various individual component models. The program TRaNsient System Simulation (TRNSYS) developed by Klien et al. [48] considered models for solar collection and storage systems, building cooling loads, and weather data. The simulation can provide hour-by-hour control strategies based on the changing conditions and the total power consumption. Other researchers have combined the TRNSYS program with the desiccant dehumidifier model to study the cooling loads. Howe [49], who developed a model for desiccant/vapor compression hybrid system for a small office building, combined his model with the TRNSYS to calculate the building cooling loads.

Another system program has been developed by SEI Associates, Boulder, CO, incorporating a desiccant dehumidification system model into the Trane Company's energy and economic simulation program TRACE (Trane Air Conditioning Economics). This program has not been used widely by the researchers and other HVAC users.

The US Department of Energy has developed one of the most sophisticated building energy simulation programs (DOE-2) that is used most frequently by both practitioners and researchers throughout the world [50]. The DOE-2 program basically contains five subprograms and needs to be executed in sequence, since the output of the previous subprogram becomes input to the next subprogram. The Building Description Language (BDL) program requires the description of the building and its HVAC systems and equipment as the input. The LOADS subprogram calculates the sensible and latent heat loads on hourly basis for each space in the building. The SYSTEMS subprogram determines the hourly demand for hot water, chilled water, and other utilities for the HVAC distribution system. The cooling module is incorporated in the SYSTEMS subprogram. The remaining two subprograms are called PLANT and ECONOMICS. The PLANT program simulates the operation of various other components such as boilers, chilers, cooling towers, solar collectors, etc. It can also determine the monthly cooling and heating loads. The ECONOMICS package can provide both monthly and annual costs of the energy and thereby provides a means for comparing the three desiccant modules. The DOE-2 program has been further upgraded to meet the changing environment. Several researchers have combined various desiccant cooling system models with the computer model of US Department of Energy (called DOE 2.1) building load and HVAC simulation program. This simulation program has been further modified (DOE-2.1E) for comparing three competing systems; 1) packaged desiccant cooling systems, 2) solid desiccant dehumidifier module, and 3) liquid desiccant dehumidifier module. Florida Solar Energy Center has also developed a similar building energy simulation program (FSEC 2.3) that incorporated thermal and moisture transport and storage in typical building materials and furnishings, moisture infiltration loads when the fan systems are off, and moisture evaporation from the cooling coil during part-load conditions [51].

## 7. APPLICATION CONSIDERATIONS

Although a desiccant cooling systems can be used in any air-conditioning operation as it can provide precise control of temperature and humidity of the space, the high initial costs of such systems typically limit their use. Harriman [47] developed a chart (Figure 13) for initial screening and or evaluation of the applicability of a desiccant cooling system. Before final selection, a detailed cost-calculation needs to be performed that should include maximum allowable moisture level in the conditioned space, ratio of the latent cooling load to the sensible cooling load, amount of fresh air required, costs of various utilities and regeneration heat, and indoor air quality requirements. A number of case studies are available in the literature that can provide further guidance regarding the selection of desiccant systems for an application [52-55].

## 8. INDOOR AIR QUALITY ENHANCEMENT CAPABILITY

As mentioned earlier, during design of a HVAC system of any space, increasing attention is being given for improvement of the air quality. One of the strategies is to increase the ventilation or air exchange rate. Solid desiccant systems offer the added advantage of enhancing indoor air quality by co-adsorbing various air pollutants without any further increase of ventilation rate. However, limited data on the co-adsorption of pollutants in the presence of water vapor are available to access the full capability of solid desiccants.

The Gas Research Institute (Chicago, IL) is one of the leaders in developing solid desiccant based systems for commercial use, such as in homes, hotels, and supermarkets. One of their objectives is to enhance the pollutant removal capabilities of such systems. A number of projects have been initiated by Gas Research Institute to study these systems for their capability to remove a variety of pollutants. Relwani and Moschandreas [56] evaluated the performance of a

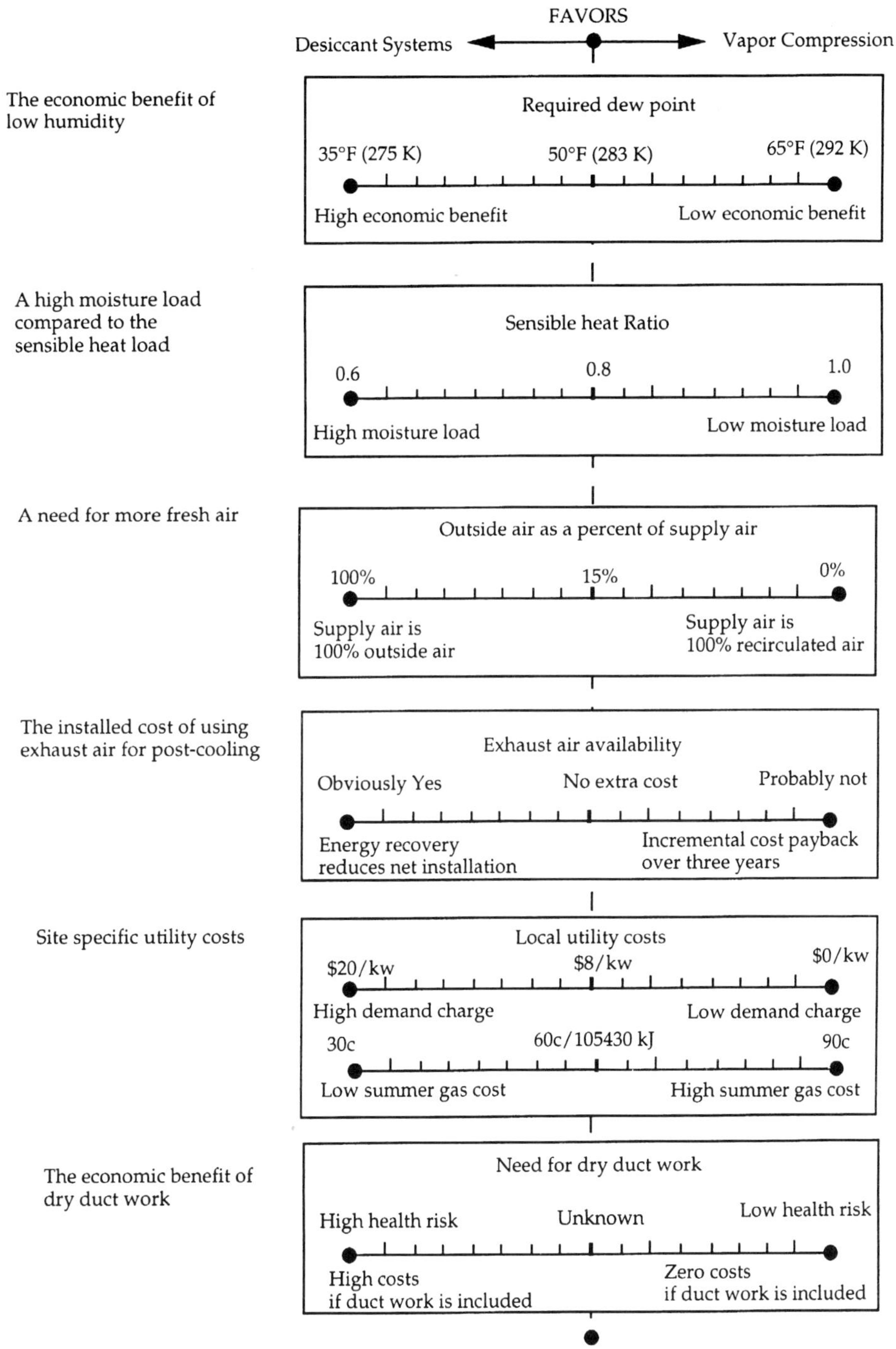


Figure 13. A guidance chart for desiccant applications (adapted from [47]).

commercially available rotary desiccant unit that employed either silica gel or a molecular sieve by determining its capabilities for removing CO, $NO_2$, and $SO_2$, along with water vapor. Later, Relwani et al. [57] also conducted a similar test with a bench scale apparatus as well as with a rotary desiccant unit that included some unspecified hydrocarbons along with the above contaminants. Experiments were conducted at various inlet contaminant concentrations, inlet humidities of air, and temperature. Factors that impact the regeneration of desiccants also were studied in their experiments. The concentration levels of pollutants in an air stream were 0.3 to 0.5 ppm for $NO_2$, 2 to 3 ppm for CO, 20 to 50 ppb for $SO_2$, and 8 to 12 ppm for hydrocarbons. The relative humidity of the air stream was varied from 10% to 25%. The molecular sieve exhibited higher adsorption capacities for $NO_2$, CO, and $SO_2$ than did silica gel both in the bench and pilot scale experiments. Silica gel demonstrated a higher adsorption capacity for hydrocarbons than the molecular sieve in the bench scale experiments. Silica gel was also easier to regenerate.

Hines and his coworkers [58-60] conducted laboratory scale experiments with a number of pollutants (representing various pollutant groups) and various adsorbents, including silica gel, molecular sieve 13X, and activated carbon. Pollutants included in their study were $CO_2$, toluene, trichloroethane, formaldehyde, tobacco smoke, and various bacteria and fungi. The above research produced many interesting results. Desiccants used in desiccant cooling systems are found to be capable of dehumidifying indoor air while enhancing its quality. As can be seen from Table 4, both silica gel and molecular sieve removed

Table 4
Pollutants removal capability of silica gel and molecular sieve 13X at various relative humidity of air and 298 K

| RH (%) | Percent Reduction in Adsorption Capacities | | | | | | | | |
|---|---|---|---|---|---|---|---|---|---|
| | Silica gel | | | | | Molecular Sieve 13X# | | | |
| | Water | $CO_2$ | Water | Toluene | 1,1,1 TCE* | Water | $CO_2$ | Water | Toluene |
| 18 | 9.1 | 50 | 46.4 | 48.9 | 58.2 | 27.3 | 75 | 19.9 | 98.7 |
| 45.3 | 0.0 | 60 | 19.8 | 67.2 | 87.5 | 10.6 | 76.9 | 12.9 | 98.9 |
| 64.4 | 0.6 | 75 | 19.8 | 77.2 | 94.6 | 1.3 | 87.8 | 9.1 | 99.7 |

RH: Relative humidity of inlet air stream
1,1,1 TCE: 1,1,1-Trichloroethane

* Water data was measured using a humidity probe from Viasala Inc. Water adsorption data were not obtained because of interference in the measurement by the humidity probe from 1,1,1-Trichloroethane.

\# Adsorption data for 1,1,1-Trichloroethane were not reported because of the catalytic activity of molecular sieve 13X towards 1,1,1-Trichloroethane.

a number of hydrocarbon pollutants, however, the degree of removal depended on the relative humidity of the air. Molecular sieve 13X adsorbed almost 100 times more $CO_2$ than either silica gel or activated carbon but its water adsorption capacity decreased in the presence of $CO_2$. The effect was negligible for silica gel and activated carbon, since both had an extremely small adsorption capacity for $CO_2$. However, the adsorbed $CO_2$ was easily displaced from the molecular sieve pores by water vapor as the adsorption progressed. Higher temperatures and times were required to regenerate a molecular sieve bed loaded with water vapor than those required for a silica gel bed. Although significant amounts of toluene and trichloroethane could be adsorbed on silica gel and molecular sieve in the presence of water vapor, they were displaced by water vapor from the pores as the adsorption continued [58]. Also, as shown in Figure 14, the water vapor and toluene adsorption capacities of both silica gel and molecular sieve 13X decreased with the increase of relative humidity of process air stream. Therefore, the length of adsorption-regeneration cycles should be chosen carefully. The opposite phenomenon - displacement of water vapor by toluene and trichloroethane - was observed in activated carbon beds. Molecular sieve was found to act as a catalyst for trichloroethane, converting it to more toxic vinylidene chloride. The reaction rate increased with an increase in temperature and all of the adsorbed trichloroethane was converted to vinylidene chloride during regeneration in the temperature range of 100 to 200°C. The reaction also proceeded in the presence of water vapor, but both trichloroethane and vinylidene chloride were displaced from the pores as the adsorption continued. Because of the catalytic effects of molecular sieve, any new desiccant materials that include molecular sieve, natural zeolites, or aluminosilicates must be evaluated to obtain a complete characterization of their performance.

Bayer and Downing [61] field tested a 1M desiccant wheel for its capability for removal of various indoor pollutants that included $CO_2$, $SO_2$, hexane, ozone, formaldehyde, and toluene. Their findings are given in Table 5. As can be seen from this table, 1M desiccant wheel coadsorbed significant amount of these pollutants in the presence of moisture. Popescu [62] also evaluated pollutant removal capability of 1M desiccant in a laboratory using experimental conditions that were comparable with that of Bayer and Downing. The pollutants used by Popescu included 1,1,1 trichloroethane, formaldehyde, toluene, and carbon dioxide. A gas mixture containing all these pollutants along with water vapor was used in the experiment. The concentrations of the individual pollutants at the bed outlet at different time intervals are shown in Figure 15. As can be seen from this figure, significant amounts of all the pollutants could be removed at the beginning of the cycle, within one minute. The bed became saturated with these pollutants rather quickly. Also, the displacement of weakly adsorbed pollutants by water or by another strongly adsorbed pollutant was not observed in this study. However, a comparison between the removal efficiencies for these pollutants by 1M desiccant obtained by Popescu and those obtained by Bayer and Downing [61] is presented in Table 6. As can be seen, a close agreement between these two studies was observed for most of the compounds.

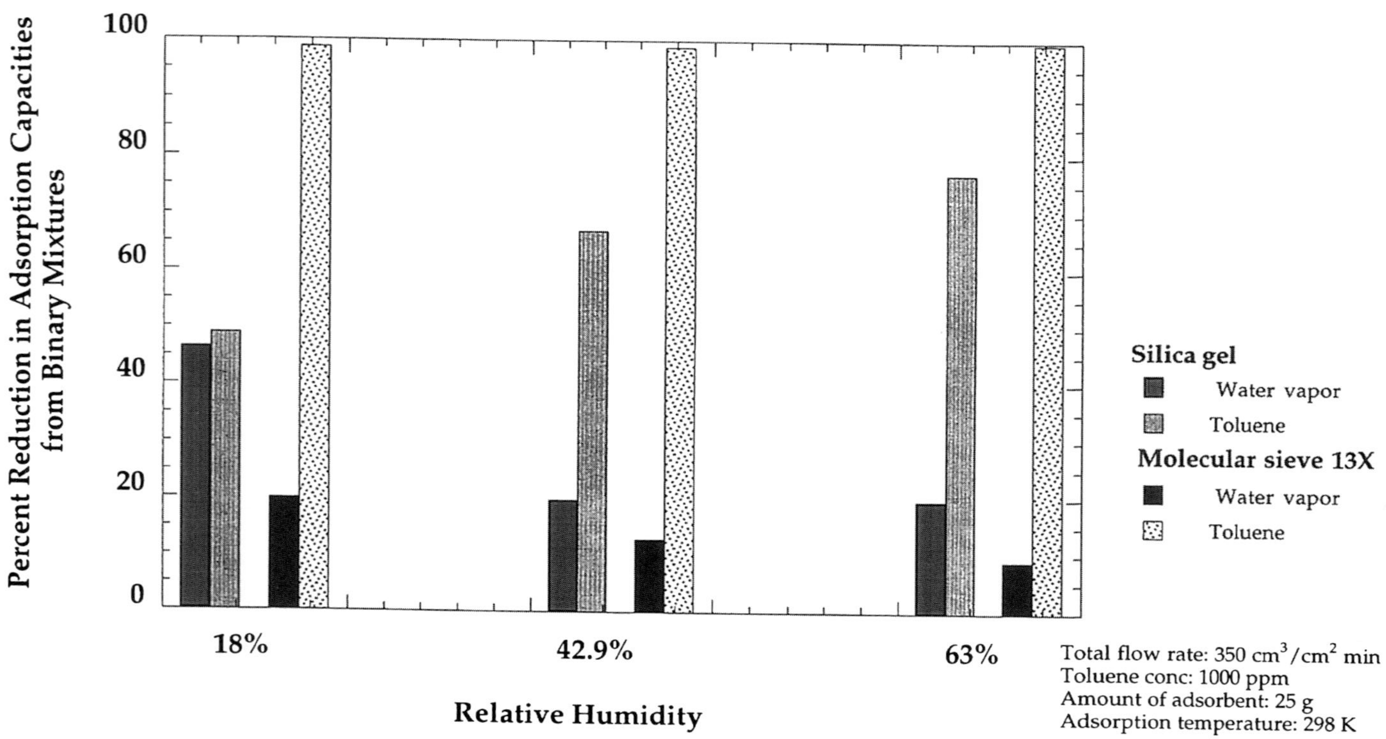


Figure 14. Reduction in adsorption capacities for water vapor and toluene from their pure component capacity when present in binary mixtures from various desiccant beds.

Table 5
Percent removal of some selected pollutants by an 1M desiccant wheel obtained from a field test

| Pollutants | Inlet Concentration | Outlet Concentration | Percent Removal (%) |
|---|---|---|---|
| $SO_2$ | 53 ppb | 4 ppb | 96 |
| Hexane (Low RH) | 1.72 ppm | 0.18 ppm | 92 |
| Hexane (High RH) | 1.72 ppm | 1.1 ppm | 36 |
| Ozone | 60 ppm | 12 ppm | 80 |
| Formaldehyde | 1.2 ppm | 0.3 ppm | 75 |
| $CO_2$ | 1325 ppm | 425 ppm | 68 |
| Toluene (High RH) | 5.75 ppm | 2.4 ppm | 61 |

High RH: 118 gr/lb (16.8 g/kg)
Low RH: 80 gr/lb (11.4 g/kg)

(Adapted from reference [61]).

Table 6
Comparison of removal efficiencies between field test data and the data obtained in a laboratory test for some seclected pollutants for 1M desiccant system after 1 minute of adsorption

| Pollutants | Inlet Concentration (ppm) | Laboratory Test Data* (% Removal) | Field Test Data# (% Removal) |
|---|---|---|---|
| Water vapor | 7220 | 52 | 58 |
| Toluene | 62 | 74 | 61 |
| Formaldehyde | 4.5 | 80 | 75 |
| Carbon Dioxide | 830 | 42 | 68 |
| 1,1,1-Trichloretane | 218 | 66 | - |

* Reference [62].
\# Reference [61].

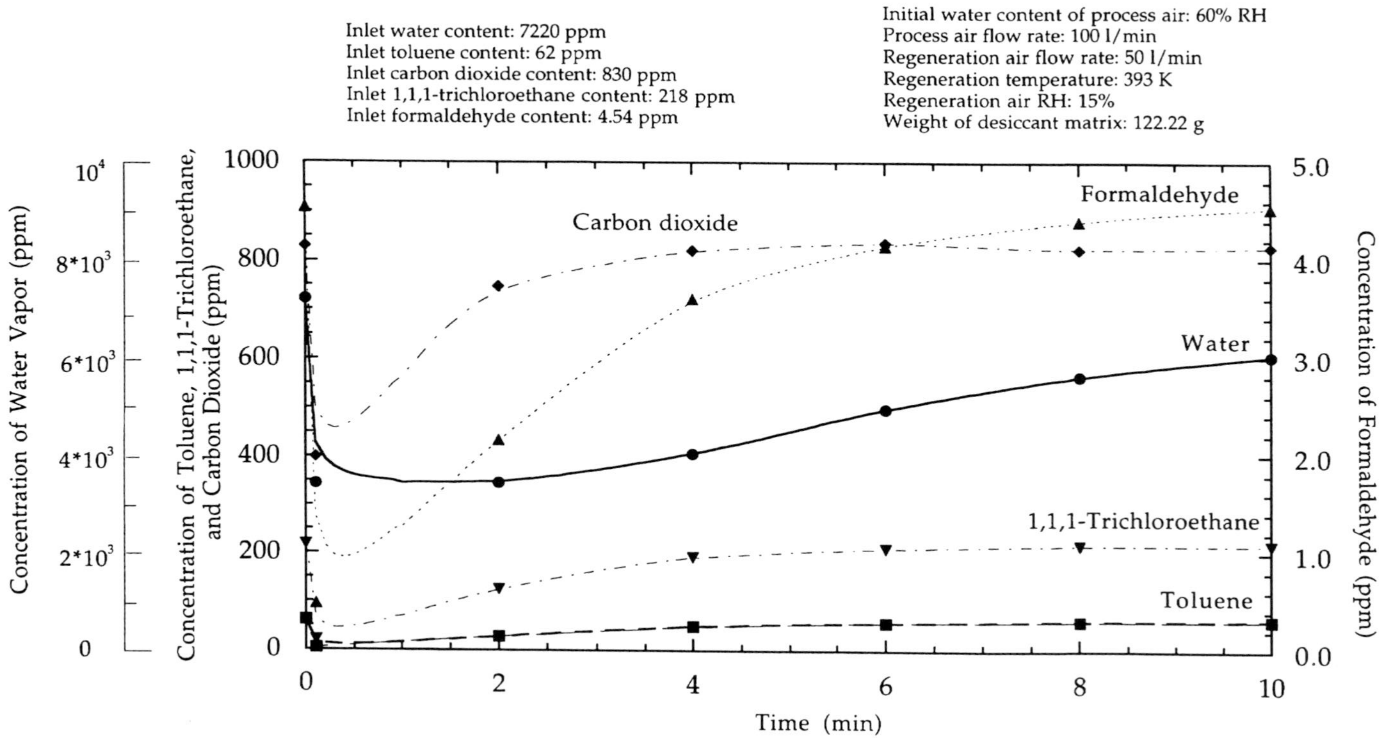


Figure 15. Breakthrough characteristics of water and various pollutants from an 1 M desiccant bed when all the compounds were present together in the air at 298 K.

The removal efficiencies for pollutants for two cases, one from their binary mixtures with water, and other when all of them were present together along with water vapor are compared in Table 7. As expected, the water adsorption capacity of 1M desiccant decreased when other pollutants were present. However, the removal efficiency for most of the pollutants except for formaldehyde remained almost the same. In the case of formaldehyde, the discrepancy may be due to the fluctuation in the formaldehyde concentration in the inlet air stream. It appears that the competition for adsorption sites on the desiccant was mainly between water and pollutants and not among the pollutants.

Table 7
Removal efficiencies of pollutants by 1M desiccant from their binary mixtures with water vapor and when present together with water vapor

| Pollutants | Removal Efficiency (%) | | | | |
|---|---|---|---|---|---|
| | Water vapor | Toluene | Carbon dioxide | 111-TCE | Formaldehyde |
| Binary mixtures with water vapor | 66 | 70 | 29 | 54 | 35 |
| Pollutants are present together along with water vapor | 52 | 66 | 32 | 66 | 70 |

Song [62] also studied the removal of the same pollutants, which were also used by Popescu, by 1M desiccant under a more controlled environment to gain a better understanding of the adsorption mechanisms of these pollutants in the presence of water vapor. As can be seen from Figure 16, water displaced both toluene and 1,1,1 trichloroethane from the pores. Once water broke through the column, its outlet concentration rose sharply to about 65% of its inlet concentration within 20 minutes, but then it took about 160 minutes to saturate the bed. It may be noted that the adsorption cycle in a desiccant wheel lasts 4 to 6 minutes. Therefore, only a fraction of the total adsorption capacity of the desiccant material is utilized and this should be taken into consideration in designing of the wheel. The adsorption capacity of 1M desiccant decreased for both water vapor and a pollutant as the relative humidity of the air stream was increased (see Figure 17). Again, this needs to be taken into consideration in the design of the desiccant cooling system if used for simultaneous dehumidification and removal of pollutants.

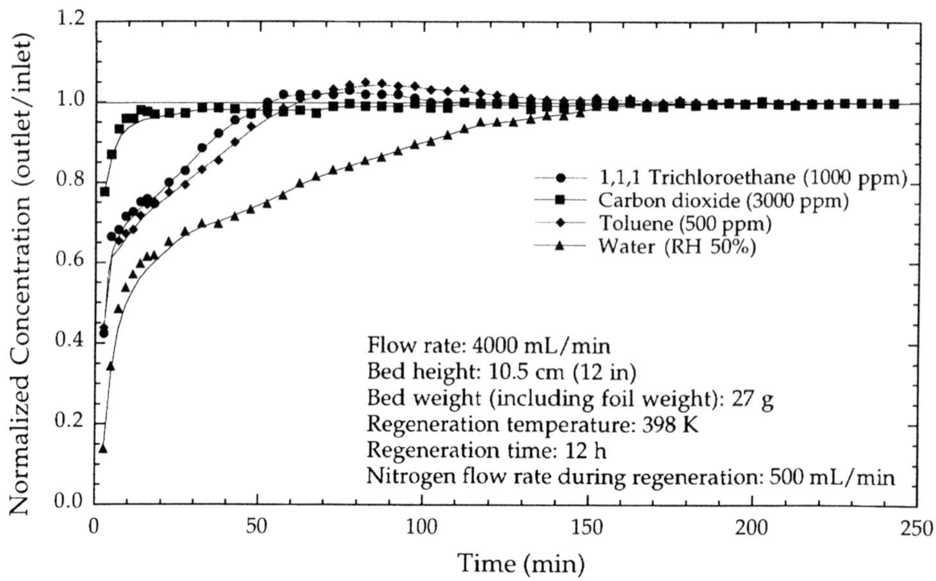


Figure 16. Breakthrough curves for various pollutants from a 1M desiccant bed in the presence of water at 298K.

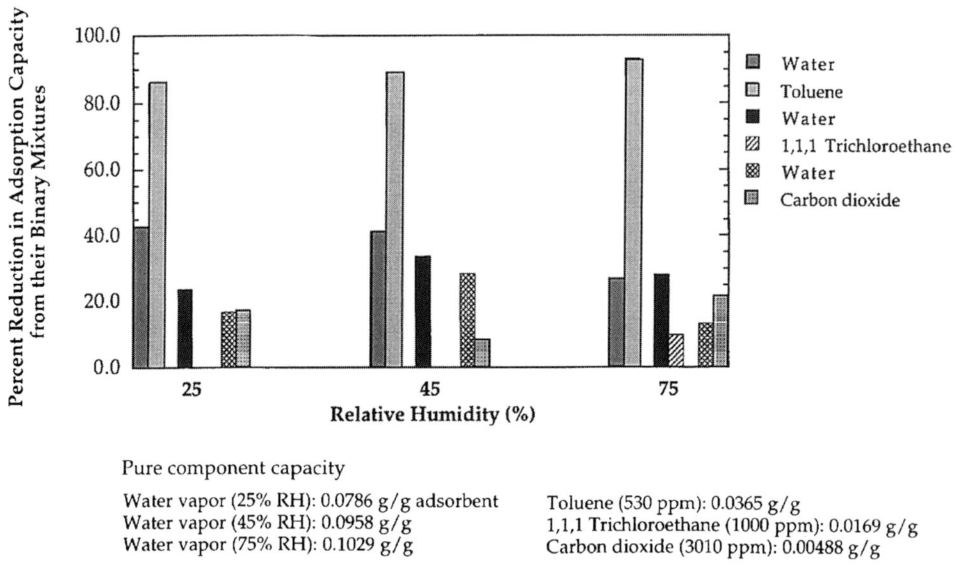


Figure 17. Reduction in adsorption capacities for water vapor and some selected indoor pollutants from their binary mixtures compared to their pure component capacity in a 1M desiccant bed at 298 K.

## 9. SUMMARY

The desiccant dehumidification technology is emerging as an alternative to conventional vapor compression systems for cooling and conditioning air for a space. Desiccant based systems are cost effective because of their use of low-grade thermal energy for removal of moisture from the air, particularly where the thermal energy required for regenerating the desiccant is readily available, the price for electricity is high, and the latent load fraction is greater than 25%. Desiccant based systems can improve indoor air quality both directly and indirectly. A number of indoor air pollutants are removed directly due to coadsorption by the desiccant material. Also, a number of microorganisms are removed or killed by the desiccant. In humid climates, desiccant systems can prevent microbial growth in the ducts and condensate drain pans, because basically no water is present on a post desiccant cooling coils, in the drain pans, and in the air distribution ducts.

## REFERENCES


1. L.G. Harriman, Heating/Piping/Air Conditioning, July (1994) 77.
2. Ventilation for Acceptable Indoor Air Quality, ASHRAE Standard 62-1989, American Society of Heating, refrigerating, and Air-Conditioning Engineers, Inc., Atlanta, GA.
3. D.R. Kosar, M.J. Witte, D.B. Shirey and R. L. Hedrick, ASHRAE J., 40(3), (1998) 71.
4. E.M. Sterling, A. Arundel and T.D. Sterling, ASHRAE Trans., 91(1B), (1985) 611.
5. R. K. Collier, Advanced desiccant materials assessment, Phase 2, Report No. GRI-88/0125, Gas Research Institute, Chicago, IL, 1988.
6. D.C. Gauger, H.N. Shapiro and M. B. Pate, Alternative technologies for refrigeration and Air-Conditioning applications, Report No. EPA/600/SR-95/066, US EPA, Air and Energy Engineering Research Laboratory, Research Triangle Park, NC, May 1995.
7. A. Hajji, W.M. Worek and Z. Laven, Numerical simulation of a regenerative closed-cycle adsorption cooling/heating system, 11th Annual American Society of Mechanical Engineers Solar Energy Conference, Solar Energy-1989, San Diego, CA, April, 1989, 339.
8. W. Zheng, W.M. Worek and G. Nowakowski, J. Solar Energy Eng., 117 (1995) 181.
9. D. Miles and S.V. Shelton, Heat Pump Design, Analysis and Application, AES- 26 (1991) 33.
10. L. Turner, Improvement of Activated Charcoal Ammonia Adsorption Heat Pumping/Refrigeration Cycles-Investigation of Porosity and Heat/Mass

Transfer Characteristics, Ph.D. Thesis, University of Warwick, Coventry, UK, 1992.
11. D. I. Tchernev and J.M. Clinch, Closed Cycle Zeolite Regenerative Heat Pump, 11th Annual American Society of Mechanical Engineers Solar Energy Conference, Solar Energy-1989, San Diego, CA, April, 1989, 347.
12. A.A. Pesaran, T.R. Penney and AI. W. Czanderna, Desiccant Cooling State-of-the-Art Assessment, National Renewable Energy Laboratory, Golden, CO, Report No. NREL/TP-254-4147, 1992.
13. US Department of Energy, Two-Wheel Desiccant Dehumidification System, Federal Technology Alerts.
14. R.K. Collier, D. Novosel and W.M. Worek, ASHRAE Trans., 96(1), (1990) 1262.
15. J.J. Jurinak, Open Cycle Solid Desiccant Cooling-Component Models and System Simulation, Ph.D. Thesis, University of Wisconsin, Madison, WI, 1982.
16. R.K. Collier and B.M. Cohen, J. Solar Energy Eng., Trans. ASME, 113 (1991) 157.
17. J.S. Nelson, W.A. Beckman, J.W. Mitchell and D.J. Close, Solar Energy (USA), 21(4), (1978) 273.
18. J.J. Jurinak and J.W. Mitchell, J. Heat Transfer, 106(3), (1984) 638.
19. M. Epstein, K. Davidson, M. Grolmes and D. Kosar, J. Solar Energy Eng. (USA), 107(2), (1985) 21.
20. B. Husky, Advanced solar/gas desiccant cooling system, Gas Research Institute, Chicago, IL Contract 5081-343-0477, 1982.
21. D.R. Schlepp, A High Performance Dehumidifier for Solar Desiccant Cooling Systems, Solar Energy Research Institute, Golden, CO, Report No. SERI/TP-252-1979, 1983.
22. D.R. Schlepp and R. Barlow, Performance of the SERI parallel-passage dehumidifier, NTIS SERI/tr-252-1951, Solar Energy Research Institute, Golden, CO, 1984
23. M.A. Grolmes and M. Epstein, Property and Composition Effects on the Cyclic Performance of Fixed Bed Desiccant Air Dehumidification, ASME/AIChE National Heat Transfer Conference, Denver, CO, August, 1985.
24. R.H. Perry, D. W. Green and J. O. Maloney, Chemical Engineering Handbook, McGraw Hill, New York, 6th ed., 1984.
25. R.K. Collier, ASHRAE Trans, 95(1), (1989) 823.
26. W.M. Worek, W. Zheng, W.A. Belding, D. Novosel and W.D. Holeman, ASHRAE Trans., 97(2), (1991) 609.
27. J.J. Jurinak, W.A. Beckman and J.W. Mitchell, J. Solar Energy Eng., 106(3), (1984) 252.
28. M. Epstein and M.A. Grolmes, Desiccant Cooling System Performance: A Simple Approach, Gas Research Institute, Chicago, IL, Report No. GRI-81/0089, 1982.

29. B. Husky, J. Sharp, A. Venero and H. Yen, Advanced Solar/Gas Desiccant Cooling System, Gas Research Institute, Chicago, IL, Report No. GRI-81/0064, 1982.
30. P. Majumdar, W.M. Worek and Z. Lavan, Simulation of a Cooled Bed Solar Powered Desiccant Air-Conditioning System, Progress in Solar Energy: The Renewable Challenge, International Solar Energy Society Conference, Houston, TX, 1982, 627.
31. Z. Lavan, W.M. Worek and J. Monniėr, Cooled Bed Solar-Powered Desiccant Air-Conditioning, IECEC Conference Proc. Intersoc. Energy Convers. Eng. Conf., Atlanta, GA, 1981, 1654.
32. T.S. Kang, Adiabatic Desiccant Open Cooling Cycles, M.S. Thesis, The University of New South Wales, Australia, 1985.
33. D.G. Waugaman and C.F. Kettleborough, Combining Direct and Indirect Evaporative Cooling with a Rotating Desiccant Wheel in Residential Applications, ASME/JSME/ISES Solar Energy Conf., Honolulu, HI, 1987, 848.
34. W. Zheng and W.M. Worek, Numerical Heat Transfer, Part A, 23 (1993) 211.
35. V.C. Mei, Heat and Mass Transfer in Air Cooled Desiccant Dehumidifiers, Ph.D. Thesis, Illinois Institute of Technology, Chicago, IL, 1979.
36. A.A. Kerestecioglu, The Detailed Mathematical Prediction of Simultaneous Heat and Mass Transfer in Cavities, Ph.D. Thesis, Florida Institute of Technology, Florida, 1986.
37. Y.J. Ko, Advanced Desiccant Cooling System with Staged Regeneration Including Solid Side Resistance, Ph.D. Thesis, Illinois Institute of Technology, Chicago, IL, 1989.
38. I.L. Maclaine-Cross and P.J. Banks, Int. J. Heat and Mass Transfer, 15 (1972) 1225.
39. I.L. Maclaine-Cross, A Theory of Combined Heat and Mass Transfer in Regenerators, Ph.D. Thesis, Monash University, Australia, 1974.
40. R. S. Barlow, Analysis of the Adsorption Process and Desiccant Systems-A Pseudo Steady State Model for Coupled Heat and Mass Transfer, Solar Energy Research Institute, Golden, CO, Report No. SERI/TR-631-1330, 1982.
41. K.J. Schultz and J.W. Mitchell, J. Solar Energy Eng., 111(4), (1989) 286.
42. C.E. Bullock and J.L. Threlked, ASHRAE Trans, 72 (1966) 301.
43. K.J. Schultz, Rotary Solid Desiccant Dehumidifiers: Analysis of Models and Experimental Investigation, Ph.D. Thesis, University of Wisconsin, Madison, WI., 1987.
44. T.S. Kang and I.L. Maclaine-Cross, J. Solar Energy Eng., 111(2), (1989) 176.
45. P.J. Banks, J. Heat Transfer, 107(1), (1985) 230.
46. W. Zheng, W.M. Worek and D. Novosel, J. Solar Energy Eng., 117 (1995) 40.
47. L. Harriman, Applications Engineering Manual for Desiccant Systems, American Gas Cooling Center, Arlington, VA, 1996.

48. S. A. Klein, TRNSYS-A Transient Simulation Program, Engineering Experiment Station Report No. 38-12, Version 12.1, University of Wisconsin, Madison, WI., 1983.
49. R.R. Howe, Model and Performance Characteristics of a Commercially-Sized Hybrid Air Conditioning System Which Utilizes a Rotary Desiccant Dehumidifier. M.S. Thesis, University of Wisconsin, Madison, WI, 1983.
50. DOE-2 Version 2.1E, Lawrence Berkeley Laboratory, Berkeley, CA, 1991.
51. A.A. Kerestecioglu, M.V. Swami, A.A. Kamel, ASHRAE Trans., 96(1), (1990) 447.
52. P.R. Burns, J.W. Mitchell and W.A. Beckman, ASHRAE Trans., 91(1B), (1985) 457.
53. B.M. Cohen and R.B. Slosberg, , ASHRAE Trans., 94(1), (1988) 525.
54. T.J. Marciniak, R.N. Koopman, D.R. Kosar, , ASHRAE Trans., 97(1), (1991) 657.
55. V.C. Mei, F.C. Chen, Z. Lavan, R.K. Collier and G. Meckler, An Assessment of Desiccant Cooling and Dehumidification Technology, Oak Ridge National Laboratory, Report No. ORNL/Con-309, 1982.
56. S.M. Relwani and D.J. Moschandreas, Indoor Pollution Control Capabilities of a Desiccant Dehumidifier System, Gas Research Institute, Chicago, IL., Report No. GRI 86/0200, 1986.
57. S.M. Relwani, D.J. Moschandreas and I. H. Billick, Indoor Air Quality Control Capabilities of Desiccant Materials, Proceedings of the 4th International Conference on the Indoor Air Quality and Climate, West Berlin, Germany, 3 (1987) 236.
58. A.L. Hines and T.K. Ghosh, Water Vapor Uptake and Removal of Chemical Pollutants by Solid Adsorbents, Gas Research Institute, Chicago, IL, Report No. GRI- 92/0157.2, pp. 234, 1992, NTIS No. PB95-104691.
59. A.L. Hines and T.K. Ghosh, Co-Adsorption of Indoor Air Contaminants by Adsorbents, Ch. 10 in Improving Indoor Air Quality Through Design, Operation & Maintenance, M. Meckler (ed.), The Fairmont Press, 1996.
60. S-Y. Lee and T.K. Ghosh, Removal of Selected Air Pollutants by A Mixed Adsorbent, Topical Conference on Separation Science and Technologies, AIChE Annual Meeting, Los Angeles, CA, November, 1997, 1220.
61. C. W. Bayer and C.C. Downing, Cosorption of Airborne Pollutants by SEMCO 1M Dehumidification Wheel. Phase 1: Test Results, Georgia Tech. Research Institute, 1993.
62. M. Popescu, Simultaneous Dehumidification and Removal of Air Pollutants from a Working Environment, M.S. Thesis, University of Missouri-Columbia, Columbia, MO., 1995.
63. S-Y. Lee, Adsorption of Moisture and Indoor Pollutants on a Mixed Adsorbent, Ph.D. Thesis, University of Missouri-Columbia, Columbia, MO., 1997.

Adsorption and its Applications in Industry and Environmental Protection
Studies in Surface Science and Catalysis, Vol. 120
A. Dabrowski (Editor)



# Applications of Self-Assembled Monolayers (SAMs) of Alkanethiolates on Gold

S.-W. Tam-Chang and I. Iverson

Department of Chemistry/216
University of Nevada-Reno, Reno, NV 89557, USA

## 1. INTRODUCTION

Extensive research has been performed on self-assembled monolayers (SAMs) formed by spontaneous adsorption of long-chain, terminally functionalized alkanethiols ($HS(CH_2)_nX$) on gold films. These molecules form stable monolayers on gold with the sulfur atom attached as a gold (I) thiolate; the long-chain becomes oriented away from the surface and is presented to the air or liquid interface. The interfacial properties are dependent on the nature of the terminal functional group. SAMs have demonstrated stability in air and under solvents for periods of months–although they desorb under extreme thermal or oxidative conditions. These interactions were first described in 1983 by Nuzzo and Allara [1]; today these systems continue to be studied as fundamental models of organic surfaces as well as being applied to many technological problems.

The exquisite control of surface parameters, ease of preparation, and ability to couple all this with the use of electrodes under the monolayer have spurred many investigations into applying SAMs on gold to a wide variety of scientific and technological areas; these include corrosion protection, surface wetting, construction of bio- and analytical sensors, colloidal nanostructure construction (linked cluster networks), biological immobilization of DNA, protein, and cells, fabrication of nanodevices, and controlling the alignment and orientation of liquid crystals, growth and deposition of polymer films to name some of the enormous volume of applied reports. SAMs have been functionalized with a wide range of novel surface moieties including fullerenes [2], hexasaccharides [3], cyclodextrins [4,5], and calixarenes [6,7]. Other materials besides gold have been found to adsorb thiol containing compounds; among these are other metals such as silver, copper, and even liquid mercury [8], or semiconductors such as GaAs(100) [9,10]. SAMs on gold have been characterized by a variety of techniques. These include the use of X-ray photoelectron spectroscopy (XPS), grazing angle FT-IR, surface plasmon resonance

(SPR) [11,12], surface-enhanced Raman spectroscopy, quartz-crystal microbalance (QCM) [13], ellipsometry, cyclic voltammetry, and imaging techniques such as ultrahigh-vacuum scanning tunneling microscopy (STM), atomic force microscopy (AFM) [14], and micromechanical analysis using gold coated scanning probe cantilevers [15].

Investigations concerning preparation, mechanism [16], characterization, and properties of SAMs have been reviewed before [9,17-19]. We will focus on the applications of SAMs of alkanethiolates on gold, emphasizing the recent developments of the last 2-3 years. Some of these advances are extensions of older work that is reviewed elsewhere. This chapter summarizes recent progress in the following areas:

(1) The methods and applications of patterning SAMs including microcontact printing, photolithography, and various other techniques such as scanning probe lithography using AFM tips.
(2) The immobilization of interesting biosystems such as DNA, protein, enzymes, and cells on SAM modified gold.
(3) The use of SAMs for functionalizing, stabilizing, and immobilizing colloidal nanoparticles on the surface.
(4) The immobilization of polymeric systems on a surface.
(5) The role of SAMs in newly developed electrochemical or optical devices and sensors.

## 2. PATTERNING OF SELF-ASSEMBLED MONOLAYERS (SAMS) ON GOLD

Many applications of surface chemistry require surfaces to be patterned in the plane of the surfaces. These applications include the fabrication of microelectronics, analytical and biosensors, and optical devices such as waveguides or displays. Biological attachment to patterned SAMs on gold has led to the fabrication of novel microengineered cellular structures. Several of these applications demand that these patterns be made as small as possible to accommodate the miniaturization of the products of these applications. The patterning of SAMs on gold has been expounded upon greatly in recent years-achieving detail resolution ranging from centimeters down to the 2 nm regime [20]. The interest in these systems is not a result of their technological promise alone, it is also due to their ability to act as model organic ultrathin films.

Often a need for patterning organic films exists that does not allow for a metal substrate, however, patterning systems found useful for thiolates on gold can be carried over to other substrates such as $Si/SiO_2$. The ease of preparation as well as the depth of understanding of these fundamental systems already firmly embedded in the literature have caused this extensive study of the patterning and microfabrication of these systems. Along with older methods such as photolithography [21-28], electron beam (or ion) writing [29,30], and micromachining [31,32] newer methods for patterning SAMs on gold are being seriously considered, such as: 1) Microcontact printing [33-53], 2) Microwriting [40,54,55], 3) Scanning probe lithography [14,20,56,57], 4) Atomic lithography [58,59], and 5) Thermodynamic electrochemical control of SAM formation on laser ablated gold arrays [60]. These different methods break down into two basic categories. Either the self-assembled monolayers can be patterned *during composition*, or they can be patterned by *decomposition* methods. The decomposition methods have been more prominent in the literature until recently. In the last several years however, a new compositional patterning method (micro-contact printing) has been developed primarily by the Whitesides group at Harvard, and now seems poised to become the dominant patterning method for many applications.

### 2.1. Microcontact printing

This method is accessible to a large body of scientists since it does not require complex equipment and clean rooms and thus is commonly used now. The fundamental principle behind microcontact printing (μCP) is the alkanethiol's ability to form a SAM upon "dry" contact with a gold surface. The traditional method for creating self-assembled monolayers of alkanethiolates on gold is immersion of the gold substrate into a dilute solution of alkanethiol. However, if the solution is evaporated [41,61] onto the elastomeric stamp and is physically placed in contact with bare gold a monolayer is also formed.

The development of the procedure for μCP has involved the use of an elastomeric "stamp" with relief features comprising the pattern to be created on the gold. This stamp is usually composed of PDMS polymer-poly(dimethylsiloxane)-cast on a master prepared from photolithography (Figure 1). The stamp is then "inked" with alkanethiol solution, allowed to dry, and pressed (often simply by hand) against the gold substrate creating a SAM with the pattern of the stamp (Figure 2). These SAMs are indistinguishable in comparison to SAMs created by immersion when the concentration of the "inking" solution is sufficiently high (>10 mM thiol) [35,41]. In fact, simply varying the concentration of the initial alkanethiol solution (<10 mM thiol) allows reproducible control over the wettability of the SAM formed from μCP, as well as the distribution of domains within the monolayer, their size and organizational state, and the pattern of depressions on the surface [41].

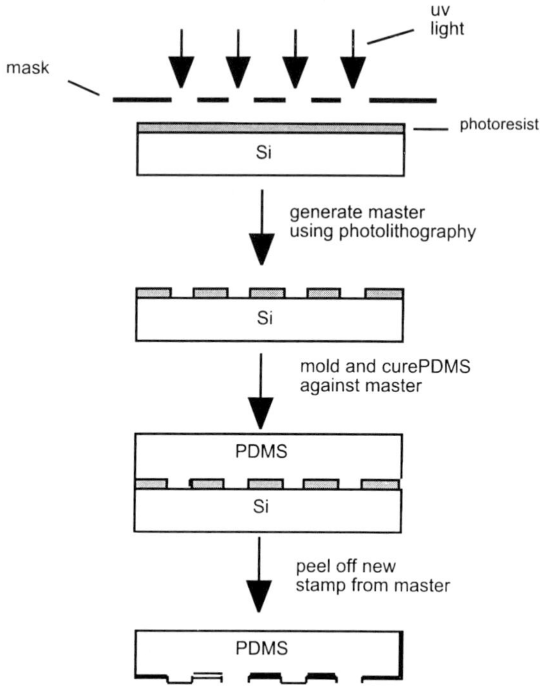


Figure 1. Formation of the elastomeric stamp for μCP.

Recently, several refinements have also been reported by Xia and coworkers [50,51,53] utilizing the elastomeric nature of the PDMS stamps (Figure 2). After the stamps are molded over the master with relief features, the stamp's features can be reduced in size by mechanically deforming the stamp using lateral compression or uniaxial stretching in the plane of the stamp's features (keeping pressure perpendicular to that plane). However, it is hard to reproduce such deformations uniformly. This problem can be addressed by using the elastomeric stamp while deformed by compression or bending as a master for another molding process [50]. Using a UV or thermally curable polymer (such as polyurethane) cast on the deformed elastomeric master surface, relief structures are replicated. These simple techniques offer ways to fabricate patterned features on even non-planar surfaces with flexibility since the master itself is flexible.

Once the patterned SAM is on the surface, the organic monolayer is often used as a resist for reactive chemical etching (RIE) [47]. The densely packed long chain alkyl groups do not allow the gold they are covering to be reached by the etching solution.

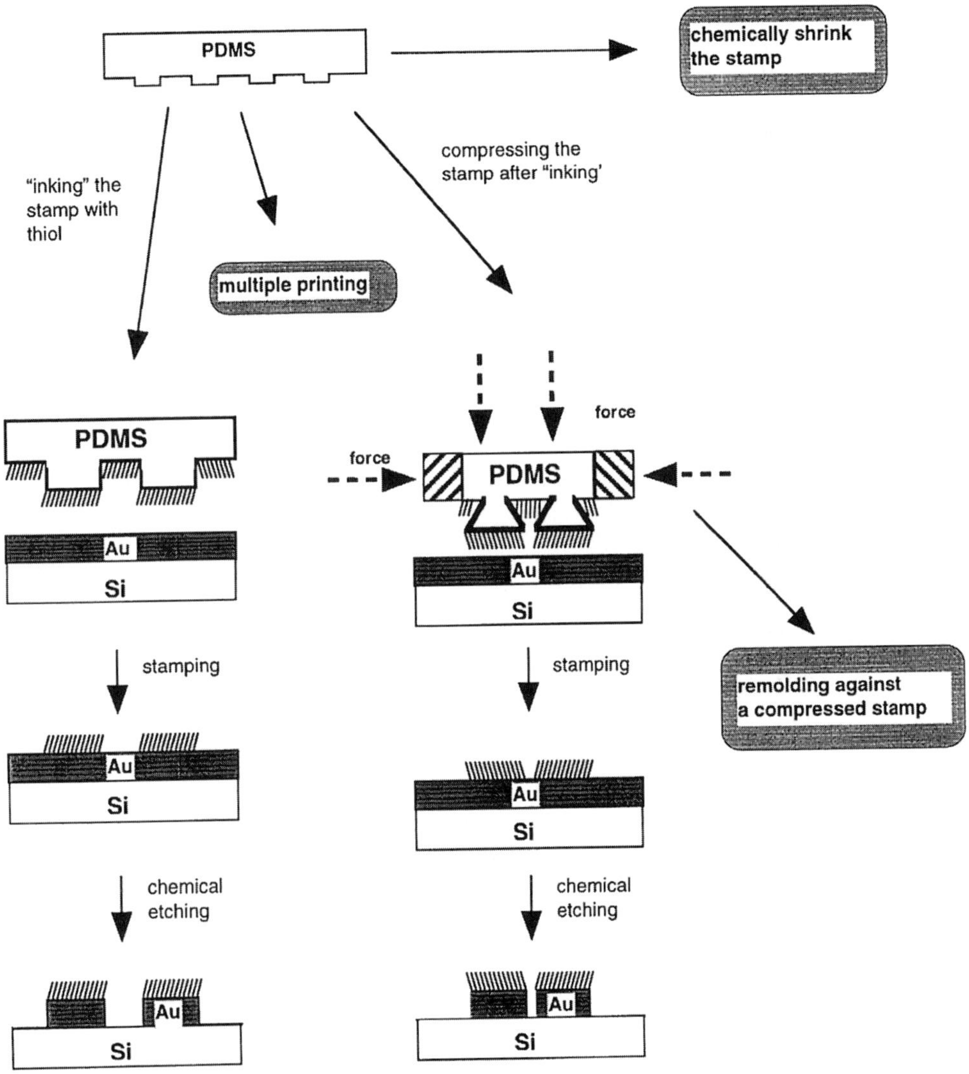


Figure 2. Using the same PDMS stamp for the various models of microcontact printing followed by gold microfeatures created by cyanide etching.

These methods allow the pattern of the SAMs created by μCP to be transferred to the substrate below [47,49]. If the substrate is silicon the features of the patterned SAM can be transferred directly to the silicon, by etching the substrate after the gold in that region is removed. The surface can then be cleaned of all organic and metallic material to reveal a patterned silicon surface. This pattern transfer technology could prove useful in production of microelectronics. Microcontact printing has also been shown to work with alkanethiols on silver [46] and copper [46,52] as well as with siloxanes on $Si/SiO_2$ [37,38,46]. The advantages [48] to this method of patterning SAMs include the ability to pattern nonplanar (especially curved) [37,43,48] surfaces as efficiently as planar surfaces, low cost accompanied by the lack of routine photolithographic work, and the formation of patterns with resolution down to 0.1 μm [40].

## 2.2. Microwriting

Along with μCP the only other method of patterning SAMs that uses controlled composition instead of decomposition of the monolayer is microwriting. This is a simple technique for producing patterned SAMs on gold with larger lateral resolution (>10 μm) [40,55]. Neat alkanethiol is dispensed from the tip of a very fine capillary or "pen" onto the surface of the gold. Line width can be controlled by varying the distance between tip and surface. The "autophobicity" of SAMs of thiols on gold forces the ordering of the SAM to occur. The main disadvantages of this system are its inefficient serial patterning method and its inability to create dimensions smaller than 10 microns.

A very different method that technically falls under the category of microwriting uses the tip of an atomic force microscope (AFM) that has been dipped in a solution of thiol and allowed to dry [62]. Using the precision of the AFM tip, SAMs can be fabricated as the tip grazes over the surface. This technique has much in common with some of those mentioned below in the micromachining category.

There are many more examples in the literature of patterning SAMs using the decomposition methods described below.

## 2.3. Photolithography

A widely used method of creating micropatterns or microfeatures in industrial applications of many varieties (including the manufacturing of microprocessors) is photolithography. It has therefore found application in patterning SAMs on gold both directly or indirectly.

The direct patterning applications utilize selective photooxidation of alkanethiolates to alkanesulfoxides [25,26] with UV light and $O_2$ that causes a patterned SAM [63,64] with a chrome mask usually protecting the undamaged alkanethiolates. Tam-Chang et al. have reported [23] that the rate of this photo-oxidation of the SAM depends on the structure of the thiol used which in turn can also be manipulated to add complexity to the lithographic pattern produced. Alkanesulfoxides are easily desorbed from the surface; a different functionalized alkanethiol can then be added to create patterned mixed SAMs, or chemical etching of the newly scrubbed surface can then transfer the photolithographic pattern to the substrate underneath using the remaining SAM as a chemical resist. The use of a Schwarzschild microscope to focus the light has been incorporated by Behm and coworkers [22] into projection photolithography, patterning SAM surfaces and readily achieving resolutions down to 100 nm. Recently, an interesting method combining photolithography and components of microcontact printing has been described by the Whitesides group [21,28] using an elastomeric stamp as a phase mask creating sub-100 nm (~90 nm) detail in the optical near field.

Indirectly, photolithography is used to pattern SAMs extensively. For example, the masters for the molding of stamps in μCP are generally created by photolithography. Photolithography has been an excellent tool that is well capable of generating submicron features. However, photolithography has several drawbacks: it is expensive with specialized equipment and clean room conditions required, shallow depth of field requires that the substrates be planar, and other problems arise from the optics involved such as standing wave effects [48]. The ultimate resolution possible with photolithography is of course dependent on the magnitude of the light's own wavelength. UV light has wavelengths of several hundred nanometers, therefore the lateral resolution of SAMs using conventional photolithography is submicrometer and wavelength-limited. In spite of these drawbacks, photolithographic techniques have been a staple of many industrial processes creating small patterned devices.

### 2.4. Beam lithography

A similar decomposition patterning methodology uses other types of beams instead of light waves to damage the SAM. These damaging agents include neutral atomic, ionic, or electronic beams that can be patterned with masks. Neutral atomic beams [58,59] seem to promise the best resolution since there are no charged particles (such as in electron and ionic beams) that limit focusing or flux density due to electrostatic interactions, and no large wavelengths to consider (such as in photolithographic beams)-their deBroglie wavelength is on the order of 0.01 nm. Neutral cesium beams [59] have been used by Younkin et al. forming nanometer scale (<100 nm) features on gold and silicon substrates. Berggren [58] and coworkers have also demonstrated nm scale resolution of features in SAMs using beams of neutral argon in a metastable excited state. Different damaging agents have varied effects on substituted thiols allowing tailoring of the technique to the individual system.

## 2.5. Micromachining

Micromachining simply consists of forming the monolayer and then physically scratching the surface to remove the thiolates in serial fashion [31,32]. After removal, etching can ensue or a second alkanethiol solution can be introduced generating a mixed SAM. Many tools have been used to create these scratches; among these are scalpel blades, wires, carbon fibers, and scanning probe tips. The use of AFM and STM probe tips in patterning SAMs has been examined recently and more investigators are studying *scanning probe lithography* [56,57,65-67].

The monolayer does not prevent electron tunneling from the STM tip [56]. However, recent reports [56,57] indicate that patterning of the SAM does not occur until biases between +2.3 and +3.0V are sustained. AFM tips have also been used by the Lui group [14,20] to create nanostructures with dimensions down to a few nanometers in monolayers on gold. *Edge resolutions better than 2 nm are easily achieved using these methods.* This process referred to as nanoshaving can also be coupled with nanografting, a process for the *in situ* replacement of the shaved monolayer with different alkanethiolates (Figure 3) [20]. The AFM tip plows through the SAM removing those thiolates and replacing them with thiol molecules present in solution at the time of the machining. Even though these methods rely on serial patterning, small scale applications for nanometer-scale electronic devices are possible.

## 2.6. Other techniques

These methods include modifying SAMs with electrochemical methods, and combined methodologies such as the method put forth by the Lopez group [60]. Individually addressable gold microelectrodes are created from continuous gold films by laser ablation followed by selective electrochemical desorption of the SAMs at certain microelectrodes producing arrays potentially useful as chemical or biosensors. Microscopic electrodes (individually electrically-addressable) are easily fabricated on thin gold films using commercially available microscope adaptations focusing the laser directly through the objective lens. The self-assembly process can then be blocked at individual electrodes by application of a voltage while the solution of alkanethiol is present. In this manner thermodynamic control of SAM formation is possible at each microelectrode in the array. SAMs from an individual microelectrode can also be selectively desorbed by cycling a voltage between -1.0 and -1.5 V for one minute. These electrodes can have edge definitions at or below 1 μm. The dimensions of these arrays are limited only by the method of fabricating the microelectrodes. The control over each individual component of the array is a major asset to this system with significant technological potential.

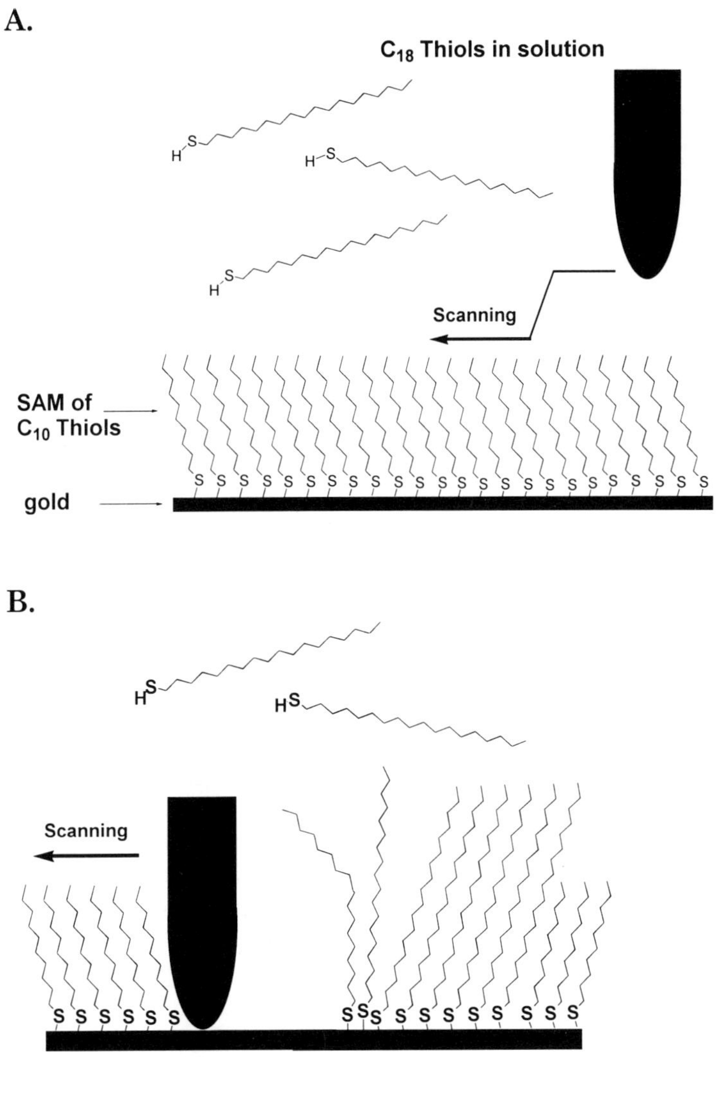


Figure 3. Representations of nanografting (not to scale), A) The AFM tip above the ordered $C_{10}$ monolayer in $C_{18}$ solution, B) The monolayer is displaced during the scan allowing a $C_{18}$ monolayer to form in its wake.

### 2.7. Limitations of patterned SAMs

Even though self-assembled monolayers of alkanethiolates on gold are the best developed examples of non-biological, self-assembling systems currently available, they have limitations that have yet to be overcome and have kept them from being used extensively in fields like the one of the largest patterning applications-microelectronic device fabrication. These limitations include the following [48]: 1) the most developed SAMs use a *gold* surface with a *titanium* adhesion layer on $Si/SiO_2$, the existence of the substantial amounts of heavy metals is not attractive, 2) there is a relatively high level of defects (especially when etching the SAMs) for consideration in high-resolution electronic applications, and 3) multilevel patterned fabrication still needs further development.

Despite present limitations in the applications of patterned SAMs of thiols on gold, the ease of preparation and exquisite control of surface structure continue to catalyze investigation into the industrial application of patterned SAMs.

## 3. BIOLOGICAL ADSORPTION

Recently many reports in the literature have demonstrated the usefulness of self-assembled monolayers of alkanethiolates on gold with respect to biologically important bound substrates or sensors. These findings promise great application in the medical field as biosensors for specific DNA sequences, in biochemistry as tools to understand cell adhesion, and in molecular biology screening for antibodies. Since molecules tethered to a self-assembled monolayer can be controlled easily (not able to diffuse away) and formation of these systems occurs under conditions that allow biomolecules to maintain their activity, these systems seem like ideal candidates for all kinds of biological investigation. In the last two years there have been numerous studies of SAMs on gold with biological implications done in areas such as: 1) DNA immobilization [2,68-73], 2) Protein attachment [44,74-86], 3) Cell, lipid bilayer and vesicle immobilization [43,44,80,87-91], 4) Scanning probe microscopy of biomolecules on SAMs [70,92,93], and 5) Biosensor applications [42,69,79,94-96]. There have also been examples of biopolymers adsorbed to the surface using SAMs [97].

### 3.1. DNA immobilization

Many reports have detailed the study of DNA bound to surfaces-among them has been thiol derivatized DNA on gold as well as DNA on existing SAMs. It is extremely desirable to create surface bound DNA probe arrays due to their potential usefulness in gene mapping and DNA sequencing technology. One successful approach to surface-confined DNA probe arrays consists of immobilizing single strand DNA of known sequence to a known position on the surface. The probe is then subjected to an analyte containing fluorescently or radioactively labeled complementary DNA. Very large DNA probe arrays on glass (not involving the sulfur-gold interaction) have been reported [68]. However, little is known concerning how surface immobilized DNA hybridization compares to solution hybridization. Therefore, since SAMs offer appealing surface immobilization control and order, systems for creating DNA probe arrays on a gold surface using SAMs are actively being explored.

Herne and coworkers have rigorously examined hybridization efficiency of immobilized single strand DNA (HS-ssDNA) [71], determining that hybridization on the surface is an efficient process which reaches a maximum at greatest surface coverage of the SAM. This can be optimized by creating a mixed monolayer consisting of a majority of underivatized alkanethiols and the thiol modified single strand DNA. These mixed monolayers of alkanethiols and HS-(ssDNA) have been characterized by X-ray photoelectron spectroscopy (XPS), ellipsometry, surface plasmon resonance spectroscopy, reflective-UV spectroscopy, dynamic scanning force microscopy and $^{31}P$ radiolabeling experiments. Ihara et al. have also recently demonstrated [72] a gene sensing system (DNA sequence probe) using immobilized single strand DNA on the gold coupled with complimentary analyte DNA tagged with a redox center (ferrocene). The successful hybridization on the surface is detected by electron transfer from the now immobilized ferrocene redox center to the gold electrode under the SAM.

Higashi and coworkers have shown that SAMs with a terminal cationic amino group can immobilize underivatized DNA duplexes through electrostatic attraction by the anionic phosphonate groups of the DNA without destroying its native structure [2]. Elghanian and coworkers have reported [69] thiol derivatized single strand DNA bound to gold *nanoparticles* (as substrates) that act as DNA sequence probes. The gold particles have complementary half-strands of the DNA being detected; as hybridization occurs the different Au nanoparticles are brought close together causing the solution's color to change from red to blue due to the surface plasmon resonance of the gold. This method differs significantly from other DNA surface bound probes due to the fact that the signal transduction mechanism is based on the substrate itself (and not a surface tag) serving as the reporter group.

### 3.2. Interaction of proteins with SAMs

Understanding protein adsorption and the factors that control it are important goals involved in understanding biological systems and it is being applied in fields like biomaterials chemistry. Protein adsorption often mediates cell adhesion having great impact on clinical practice of foreign implants in the body [76,98]. SAMs that adsorb protein are therefore being studied as model systems in which the surface properties are well-defined. Protein adsorption on SAMs is also being used as a vehicle to study bioprocesses such as biospecific recognition and enzyme catalysis.

The nonspecific adsorption of proteins on a SAM surface with different terminal groups has been found to occur in a noncovalent manner [44,76,80,83,99-101]. Perhaps just as importantly, certain end-groups have been shown to resist (nearly completely) the nonspecific adsorption of proteins on the surface [75,80,99]. This is important in that patterned SAM surfaces with certain regions of desired protein attachment are differentiated from other regions of the SAM. Resisting nonspecific adsorption is also critical in biomaterials science creating biologically inert materials (such as contact lenses and artificial hearts). The most commonly used end-group for this purpose has been poly(ethylene glycol) ($Eg_n$ n=2-7), however tripropylene sulfoxide can also be used. This resistance is purposed by De Gennes, Andrade, and coworkers [102,103] to be a result of the entropic penalty associated with compressing the disordered glycol chains and desolvating them upon protein adsorption; it is not certain that this hypothesis applies to oligoethylene glycols on SAMs. The nonspecific method of protein interaction on SAMs is simple but not easily controlled.

Recently, there have been several investigations into specific protein adsorption on SAMs [74,77,82,84-86,104-106]. This specific adsorption can be accomplished either directly by thiol derivatized proteins attaching to the surface or indirectly either by covalently linking the protein to the terminal group on the SAM or by having specific receptors immobilized at the gold surface by a thiol linkage that will recognize an analyte protein in solution above the surface. Often when biomolecules come into *direct* contact with heavy metals (including gold) they are denatured and lose their bioactivity [107,108]. This fact limits the direct method of immobilizing proteins on the surface. There needs to be insulation between the gold surface and the end-group (the protein-monolayer interface); this has been successfully accomplished using mixed monolayers to prevent direct interaction between the active biomolecular moiety and the surface [82].

Indirect specific protein immobilization can be affected by general covalent attachment to the SAM. A recent example of this put forth by Delamarche and coworkers [74] involves immobilization of antibodies using a photoactivated self-assembled monolayer on gold. The long-chain thiol is functionalized with benzophenone which is activated by near-UV radiation (>350 nm). The oxygen radical generated abstracts a proton from an $\alpha$ carbon in the protein followed by carbon-carbon bond formation upon radical recombination; thus a covalent bond is formed between the monolayer and the immuno-globulins which then specifically recognize alkaline phosphatase-linked antibodies. Since the immobilization occurs primarily in areas that have been photoactivated, patterning can easily be added to this strategy using photolithography. Another example [77] involves alkanoic acid functionalized SAMs that are capable of immobilizing cytochrome c oxidase on the surface without loss of its electroactivity and exhibiting quasi-reversible electron transfer at the gold electrode. Cytochrome C in solution above the surface can be both oxidized and reduced by the immobilized enzyme. Patel and coworkers [81] have demonstrated covalent attachment of a protein to existing SAMs by creating an *N*-hydroxysuccinimide ester (NHS) from an immobilized long chain alkanoic acid. The protein is then added creating an amide link between the immobilized long chain and the protein. Protein covalent attachment could also be achieved using the terminal amide formation strategy [85] of Yan et al. SAMs of long chain alkanoic acids are exposed to trifluoroacetic anhydride forming a surface of terminal, interchain carboxylic anhydrides-addition of a molecule with an amino group forms mixed SAMs of acids and newly formed amides.

A second general manner of indirect specific protein adsorption using biospecific recognition of the protein to the surface has been reported. The Whitesides group has used mixed monolayers consisting of mostly ethylene glycol end-groups and thiol derivatized "ligands" for protein recognition and adsorption displaying successful specific adsorption of proteins to self-assembled monolayers [80,105]. Carbonic anhydrase has been shown to adsorb to SAMs that present benzenesulfonamide groups (as the "ligands"). Sigal and coworkers [84] have also reported the generation of SAMs that *selectively* bind proteins whose primary sequence terminates with a histidine tag [78,109]. Mixed monolayers of nitrilotriacetic acid (NTA) and ethylene glycol terminated alkanethiols were prepared. The histidine interacting with the two vacant sites on a Ni(II) center chelated by the NTA terminal group. The attachment of His-tagged proteins to the SAM was observed by surface plasmon resonance (SPR). Several reports [104,106] have also utilized the specific molecular recognition of certain proteins such as streptavidin for biotin (vitamin H) bound to a SAM terminus. Surface plasmon resonance allows the *in situ* monitoring of this binding. Besides SPR, proteins adsorbed on SAM surfaces have also been characterized by FT-IR [78], acoustic plate mode (APM) devices [83,110], and quartz crystal microbalances (QCM); if these methods are coupled with a specific molecular recognition event between the immobilized protein and an analyte in solution, a biosensor can be fabricated. Madoz and co-workers [79] have utilized a mixed monolayer containing a terminal choline moiety to recognize a receptor protein fused to an enzyme (Figure 4). In this study, β-galactosidase from *E. coli* fused to the choline-binding domain of (acetylmuramoyl)-alanine amidase from *Streptococcus pneumoniae* is selectively immobilized on a SAM surface using thiol chains functionalized with choline. The neighboring amides in the mixed monolayer form a network of hydrogen bonds (Figure 4). This amide network greatly stabilizes the immobilization of the large macromolecules not only isolating the bioactive enzyme from the gold surface but also intrinsically stabilizing the monolayer itself-a phenomenon reported by Tam-Chang [23] *et al.* The bound enzyme maintained its bioactivity catalyzing the formation of p-aminophenol from (p-aminophenyl)-β-D-galactopyranoside (detected electrochemically using the gold surface as a working electrode).

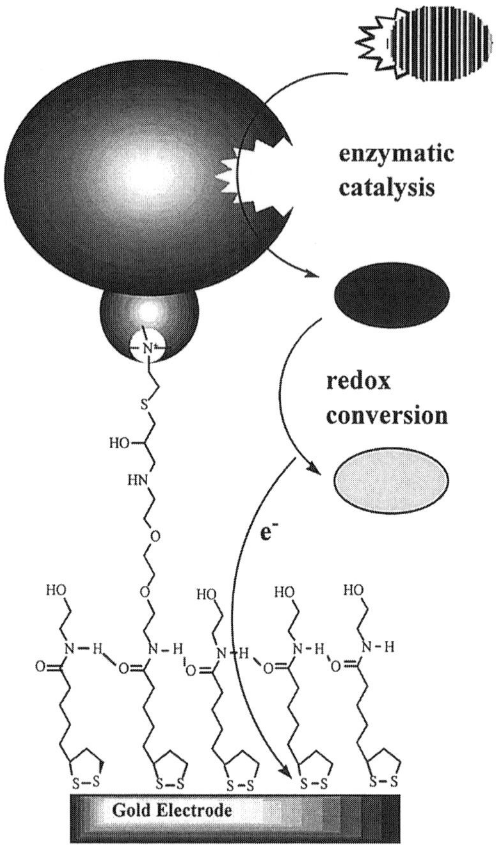


Figure 4. The protein-fused enzyme adsorption onto a monolayer that can catalyze a reaction in which the product is detected electrochemically with the gold electrode.

### 3.3. Cell attachment

Significant interest has been generated in the analysis of individual cultured cells. This analysis is important in drug screening, *in vitro* toxicology, and genetic engineering technologies. Factors essential for these studies are the ability to reproducibly isolate large numbers of individual cells, control their shape, and define their distribution [45]. Cell adhesion on SAMs is being studied as a model of these processes as well as a technologically viable tool for direct application. Coupling the controlled cell adhesion on SAMs with patterning methodologies allows the culturing of cells at high density while also holding cells in designated locations and keeping them isolated from their neighbors [45]. Microassays on individually indexed arrays of cultured cells using patterned SAMs can be envisioned.

Much has been learned about cell attachment to surfaces through studying cell adsorption onto self-assembled monolayers on gold. A common strategy for attaching anchorage-dependent cells to the surface relies on the adsorption of extracellular matrix (ECM) proteins onto the surface first [80]. The attachment between these cells and ECM proteins-fibronectin, laminin, vitronectin, heparin, collagens, etc.-is controlled by specific interactions between the integrin receptors of the cell membranes and peptide sequences of the ECM proteins [42]. Once the protein or an appropriate sequence of amino-acids [111] (mimicking a natural ECM protein) is presented to the surface (as described above), cells can adhere and grow on the surface. Mrksich and coworkers have demonstrated controlled cell attachment [43,44] with cell adhesion in methyl-terminated regions of the SAM and utter attachment resistance in ethylene glycol terminated regions of the SAM. For example, fibronectin-coated methyl-terminated regions of patterned SAMs anchor bovine capillary endothelial cells. The Allara group has reported [89] that cell growth can be augmented by increasing the binding tightness of the ECM protein, which in turn is dependent significantly on *surface functionality*. Cell growth has been shown to increase in the following order of these end-groups: $-CH_2OH < -CO_2CH_3 < -CH_3 << -CO_2H$.

Self-assembled monolayers of alkanethiolates on gold are the best system available to study fundamental aspects of biointerfacial science. If the gold layers are thin enough (producing a transparent substrate), these systems can be studied by inverted microscopy and fluorescence microscopy [43]. The SAMs are also easily patterned controlling the size and shape of cells and cell growth. In fact, recently Hutt et al. have demonstrated [87] control over cell morphology of cells immobilized on patterned SAMs *directly* (without ECM proteins)-inducing morphologies that are not normally observed for spreading cells-creating "cellular wires."

### 3.4. Bilayers and vesicle immobilization

Phospholipid vesicles and bilayers have also been immobilized on gold. One approach to vesicle adsorption uses mixed SAMs with short chain ethyleneoxythiols and cholesterol derivatized alkanethiols acting as anchoring units which become integrated into the lower leaflet of the phospholipid bilayer [91]. Another method of anchoring a phospholipid bilayer to a monolayer is simple fusion. Pierrat and coworkers [90] allowed very long chain alkanethiolates (octadecylmercaptan) to be ordered into a SAM, after which by immersing the surface in a vesicle solution phospholipids were fused into the SAM. A biomimetic assembly is then formed as pyruvate oxidase of *E. coli* is then incorporated into the phosphonate head of the

SAM supported (half) bilayer. Upon incorporation of the enzyme, its activity remains intact.

### 3.5. The role of SAMs in imaging biomolecules

Another biological application of self-assembled monolayers of alkanethiolates on gold is their role in scanning probe microscopy (SPM) of biological molecules. In the early stages of the use of scanning tunneling microscopy (STM) for studying biological molecules, sample preparation consisted of allowing a drop of the analyte solution to evaporate onto a graphite substrate [93]. This caused most of the analyte to form dense aggregates at the edge of the evaporated drop and was consequently no good for observing molecular scale structure of the individual biomolecules; the molecules that were not aggregated were spread out so far apart that it was very difficult to locate them. These problems were intensified by the fact that the scanning tip imposes a repulsive force on the biomolecules with sufficient force to move the analyte out of the scanned region [112]. These problems have been solved by immobilizing biomolecules *covalently* on a gold surface utilizing the stability of a self assembled monolayer of thiolates on gold. Individual biomolecules are clearly observed on the monolayer without being swept away by the STM tip [70,92,93], although the monolayers can be destroyed if the tunneling gap is decreased too much (increasing the interaction force).

### 3.6. Biosensors

The development of biosensors-molecular sensors with a biological recognition mechanism and a signal transduction device-is an extremely attractive target for biomedical technology. Since it is possible to selectively immobilize proteins, cells, or other biomolecules (or aggregates) using self-assembled monolayers of alkanethiolates on gold and control the details of these systems through patterning, SAM technology offers exciting possibilities for production of elegant, efficient biosensors. Surfaces can be tailored with respect to their molecular recognition properties and of course the conductive nature of the gold is a major asset to the system. The above mentioned report [79] by Madoz et al. shows potential as a biosensing system containing both the biorecognition and signal transduction mechanisms. DNA sequence probes with radiolabeled analytes (mentioned above) are also another example of the potential use of SAMs in biosensing systems.

A glucose biosensor incorporating SAMs on gold was described by Rubin and coworkers [96] using mixed SAMs of derivatized alkanethiols as an electronic relay device between an immobilized glucose oxidase enzyme and the gold electrode. The relay molecules of the mixed SAM are independent components mediating between the redox site in the immobilized enzyme and the electrode surface. Domains of short aminoethanethiol in the mixed SAM bind the enzyme while long chain ferrocene derivatized thiols bound near the enzyme mediate electron transfer between the enzyme and the surface allowing the catalyzed oxidation of glucose in solution-an example of using SAMs to effect a multi-step catalysis relying on successive electron transfers. The presence of glucose in solution is signaled by the catalytic reaction occurring that causes electron transfer.

Ding et al. have reported [95] a novel potassium ion sensor based on valinomycin-incorporated bilayers supported on a gold electrode modified by a hexadecylmercaptan monolayer. The ion selectivity for $K^+$ is 400 times greater than for $Na^+$. The system is stable for two months at -10 °C which is much more practical than the reported stability of many cationic bilayer membranes with longevities of less than 120 hrs [113]. The lipid membrane is prepared by the painted-frozen method [114] in which the SAM modified electrode is coated with the lipid solution and allowed to freeze (removing the solvent); the supported bilayers are then immersed into a solution of valinomycin to create the $K^+$ sensor.

Another significant report [94] of a biosensor system fabricated to take advantage of the properties of SAMs has been described by Cornell et al. The signal transduction method depends on the conductance of a population of molecular ion channels that is switched by the molecular recognition event. A mixed monolayer is formed with long anchors for lipid bilayer attachment and short stabilizing thiolated benzyl groups. A lipid bilayer containing "floating" gramicidin ion channels half as long as the bilayer is anchored to the monolayer forming an ionic reservoir between the SAM and the bilayer. When a potential is applied through the gold electrode, the gramicidin ion channels dimerize forming a complete channel to the bulk solution and ions flow between the reservoir and the outer solution. The channels are switched off when a recognition event occurs on the outer face of the bilayer thus cross-linking the half-channels and disallowing them to float into alignment for dimerization. This complex biosensing technique can be adapted to many types of receptor recognition scenarios and could be used in applications involving the detection of proteins, viruses, antibodies, DNA, drugs, pesticides, or other small molecules of interest. With the increasing understanding that has been accumulating concerning SAMs of alkanethiolates on gold it is only a matter of time until these systems become integrated into other elaborate biodetection strategies.

## 4. COLLOIDS AND NANOPARTICLE APPLICATIONS

The study of colloidal systems is a large field with many facets; applications of these systems include optoelectronics, thin film growth, and catalysis. This is due to their "exotic" physiochemical properties lending credibility to the claim that these systems are an "intermediate state of matter" [115]. Colloidal metal clusters have also been examined as materials suitable for quantum confinement and quantum dots which may serve as models for studying single electron tunneling (SET) and may lead to applications in nanoelectric digital circuits [116]. Self-assembled monolayers of alkanethiolates on gold have recently been used in the colloids/nanoparticle field in three different ways: 1) Thiolates have been found to form self-assembled monolayers on colloidal gold producing a stabilizing effect on the colloids [117-123], 2) SAMs have been used to order and confine 2 and 3 dimensional arrays [124-127] of clusters and nanoparticles in a variety of ways, and 3) SAMs have been used for the immobilization and growth of nanoparticles [128-132] on surfaces.

Thiolate monolayers on small gold particles similar to SAMs of alkanethiolates on planar gold substrates have been reported. Several reports have compared the two types of monolayers [121,123] and the basic nature of these "three dimensional SAMs" continues to be actively examined. The scarcity of analytical methods available for the study of two dimensional SAMs of thiolates on gold make this comparison interesting. Gold cluster monolayers are 3D-SAMs-they can be investigated in bulk samples (solid or solution) and are confined to molecular sized spatial dimensions [118,119,121]. Their properties have been characterized by NMR [118-120,133], FT-IR [119,121], differential scanning calorimetry (DSC) [119,120], UV-Vis spectroscopy [122,123,127], surface enhanced Raman spectroscopy [127], X-ray photoelectron spectroscopy (XPS) [122], and transmission electron microscopy (TEM) [118,120,124-126,128]. IR data suggests that the number of surface gauche defects for 3D-SAMs is similar to those in 2D-SAMs [121] (or only slightly higher). It has also been found that SAMs of colloidal gold help increase the stability of the particles [122,125]; the insulating alkyl chains keep unwanted crystal growth and precipitation from occurring.

Self-assembly processes involving gold and thiols have been used creatively to produce ordered colloidal arrays or cluster networks in either two or three dimensions. One approach involves using ($\alpha,\omega$)-dithiols and thus interconnecting gold particles [124,128]. An analogous method uses patterned mixed SAMs of dithiols on a planar gold surface to immobilize gold particles into patterned arrays [129,132]. The dithiol spacer molecule acts as an insulator causing the array of gold nanoparticles to take on conducting properties that mimick semiconductor activity [128]. Another approach [124] has been to spin-cast SAM coated gold clusters onto a planar substrate causing an unstable close-packed cluster monolayer (2D network). Then a solution of aryl dithiols is introduced to the substrate, competitively replacing the alkanethiols to form a *linked* cluster network (LCN).

Several groups have utilized LCNs [124-128,134], in some cases grafting biological recognition mechanisms [125,126] to nanoparticles creating rational order. Mirkin et al. have recently demonstrated [125] a method for rationally assembling nanoparticles into macroscopic materials using the specificity of DNA base pairing interactions. Their method uses two batches of gold particles each derivatized by a different non-complementary single strand (thiolated) DNA sequence. The batches are mixed without any interaction taking place since the DNA hybridization is very precise. Upon addition of an oligonucleotide duplex with single strand ends that are each complementary to one of the batches of derivatized gold clusters, self-assembly of the nanoparticles into aggregates occurs. The process is reversible upon thermal denaturing and can then be cycled over and over. Alivisatos and coworkers have devised another method [126] of rationally ordering gold clusters (Figure 5), also using the specificity of DNA base pairing interactions; this system differs in the fact that *discrete* numbers of gold clusters are ordered spatially creating nanostructures with well defined intercluster distances.

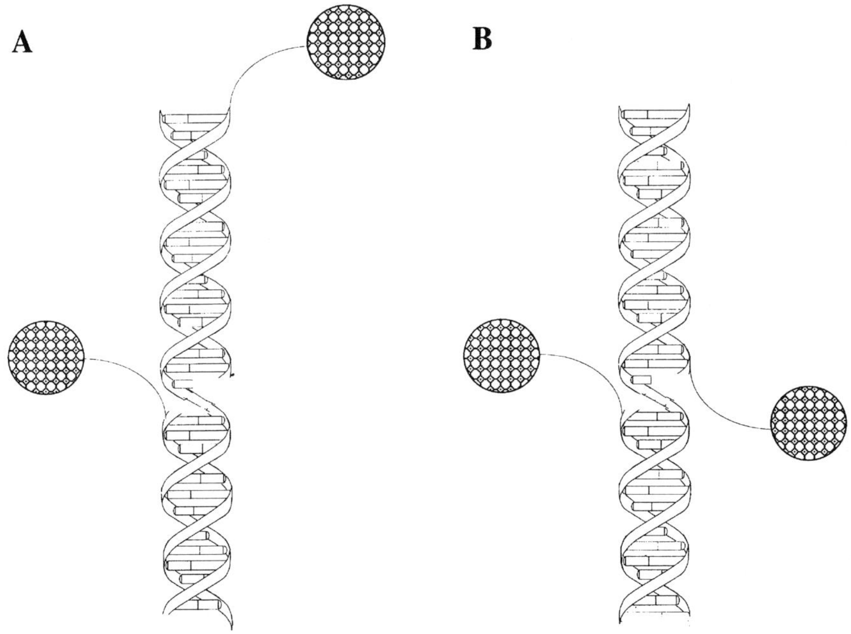


Figure 5. Rationally controlling the distance between two gold nanoclusters using a single strand DNA template hybridized with complementary half-strands derivatized with gold nanoclusters at the 3' or 5' ends.

Gold nanoparticles have been deposited on SAM surfaces using long chain ($\alpha,\omega$)–dithiols. Andres et al. have used (Au cluster - dithiol - Au substrate) systems to reproducibly demonstrate single electron transfer processes that are due to "coulomb staircase" phenomena [129]. Scanning tunneling microscopy was used to measure the current-voltage responses. Nanoparticles have also been "grown" on the surface of SAMs on gold. Grummt and coworkers [131] have chemically deposited silver nanoclusters on mixed monolayers of *n*-dodecane thiol and a terminally functionalized alkanethiol presenting a four-electron reducing agent to the surface. Silver (I) in solution above the surface is reduced on the SAM forming silver clusters, although the cluster grows larger than 4 atoms. Deposition of iron oxide on a SAM of mercaptohexadecanoic acid on gold has also been recorded [130]. Surface plasmon spectroscopy (SRS) was used *in situ* to monitor the iron oxide nanoparticle growth on the monolayer.

In the future it is foreseeable that the number of studies on "three-dimensional SAMs" will continue to grow and become more complex as the applications of colloids and clusters become manifest.

## 5. POLYMERIC IMMOBILIZATION

Self-assembled monolayers of thiols on gold have been used to rationally attach polymeric structures to the surface. Using conventional methods, it has not been an easy task to study the electrochemical properties of a polymeric film because of the difficulty in fabricating uniform films on metal electrodes. Once formed, these films also had poor surface adhesion and easily peeled off [135]. SAMs of gold bring order and uniformity as well as stability to this arena and are thus ideal candidate systems for studying polymer films. Reports of immobilizing polymeric structures onto SAMs have begun to grow more frequent. Our increasing understanding of SAMs allows the control of the supramolecular architecture of ultrathin polymer films and enables the exploration of their potential in electro-optical devices. These films often have properties such as nonlinear optical (NLO) activity, and the ability to conduct [63,64] and luminesce.

In the literature there are two basic motifs of polymers immobilized on SAMs: 1) The polymer is *attached to* the SAM surface at the surface/air interface [33,34,135-142], or 2) The polymer is *within* the SAM and helping to comprise it [7,143-145].

The former category can further be broken down into polymers *grown* on the SAM [135-137] (either epitaxially or in a protruding manner), or polymers already preformed *deposited* onto the SAM [33,34,97,139-141]. Niwa and coworkers have used photoinitiation [136] to grow polymers of methacrylic acid from aqueous solution on xanthate terminating long chain disulfides in the mixed monolayer. This forms poly(methacrylic acid) brushes that interact with cytochrome c through electrostatic interactions. These interactions and the growth of the polymer itself are monitored *in situ* using a gold-coated quartz crystal microbalance (QCM).

Hyperbranched poly(acrylic acid) films have been grown on SAMs of alkanoic acid [137]. These films, rich in carboxylic acid groups could be utilized to selectively bind metal ions or serve as sites for further derivatization. Wurm et al. have studied electrochemically induced epitaxial polymerization [135] of *N*-alkylpyrrole in solution onto a SAM composed of ((*N*-pyrrolyl)-*n*-undecyl)disulfide. These optically smooth films of long chain poly(*N*-alkylpyrrole) have excellent stability in air. The Collard group [142] has studied the kinetics of electro-oxidatively polymerized polyaniline and polypyrrole on surfaces modified with SAMs; formation of conducting films is initially blocked by the monolayer but nucleation at SAM defect sites leads to eventual deposition of rough polymer films.

Formation of ultrathin non-conducting polymer films has also been reported by Hammond et al. using ionic attraction on the surface of the carboxylate terminated SAM allowing layer-by-layer deposition of ionic multilayers [33,34,139-141]. These systems have been combined with patterning techniques and have been examined with respect to their optically responsive properties. Viologen-based polyelectrolytes have also been shown to adsorb onto a carboxylate SAM surface, forming an electroactive thin film [146].

The latter category can be differentiated into the lateral polymerization of the components of the pre-existing SAM [7,145,147,148], or the absorption of a preformed polymeric sulfur containing structure onto the gold surface [143,144] thus forming an internally polymerized SAM on gold. The Crooks group [7,145,147,148] has pioneered a successful method of photopolymerizing long chain thiol containing compounds that are already part of a monolayer. SAMs containing a diacetylene group can be polymerized with UV light at 254 nm effectively cross-linking the SAM and increasing the stability of the monolayer-creating the ability to withstand repeated electrochemical cycling to extreme potentials and thermostability to 200°C. These sytems have been combined with photolithography creating patterned polymer surfaces [149] on the gold (Figure 6). The *stabilized* monolayers can then incorporate a host of different functional groups at the terminus [7].

Sulfur containing polymers have also been reported to adsorb directly to a gold surface. Poly(3-octylthiophene) (POT), with a mass-average molar mass ranging from 80,000 to 100,000 g/mol, formed a stable self-assembled film on a gold electrode and was characterized by electrochemical methods [143]. The insulating POT layer can be converted to a conducting film upon doping with chemical oxidants such as iodine.

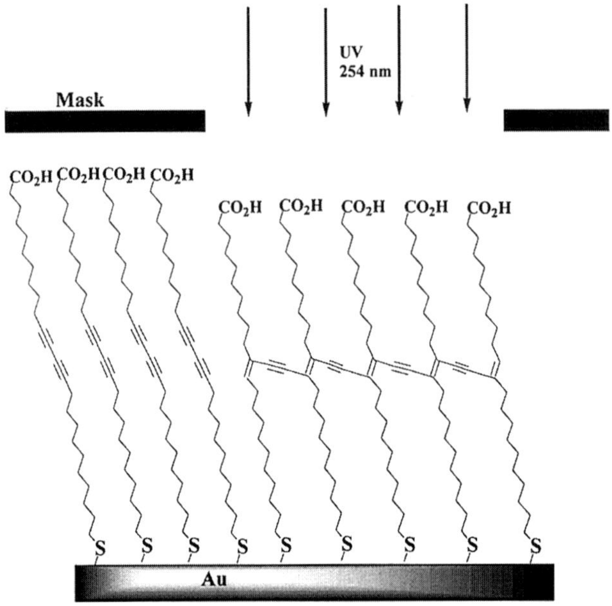


Figure 6. Patterned photo-initiated internal polymerization of diacetylene containing long chain thiolates.

## 6. OPTICAL DEVICES

Organic materials have long been looked to with the hope of applying them as components in electro-optical devices such as flat-panel displays (FPDs). In many of these devices it is essential to control the order and alignment of these materials. Organic liquid crystalline materials have been used extensively in this role, however controlling the alignment of these materials at a surface interface has proven difficult and much effort has gone into this end with the hope of creating better devices.

Of course, many investigators have realized the utility of employing SAMs on gold to help control and define the alignment and position of liquid crystals (LCs) on a surface. Several reports have demonstrated anchoring LCs on SAMs or including LC molecules into the SAM itself [36,150-156]. The ability to pattern SAMs and therefore LCs on nonplanar (curved) surfaces should allow development of new types of tunable electrooptical devices including diffractive optical phase gratings and light valves for orthogonal polarizers [157]. Abbott's group has recently achieved several types of molecularly designed surfaces using patterned SAMs on gold that anchor nematic LCs including LC diffraction gratings with μm features and patterned alignment on curved surfaces [36]. The authors were able to control both the polar and azimuthal orientations of the anchored LCs. The LCs orient differently on SAMs of alkanethiolates than on mixed SAMs that incorporate both long and short chain alkanethiols together. The mixed SAMs cause the LC to orient homeotropically (perpendicular to the surface) while the uniform alkanethiolates cause planar LC orientation [151]. It is also intriguing to note that the azimuthal orientation-the direction of orientation of the planar LC with respect to the direction of gold evaporation-can be controlled by using *an odd or even* alkanethiol in the constitution of the SAM [153]. This phenomenon is only observed on obliquely evaporated gold (gold evaporated with the source and the substrate at a fixed angle and direction of incidence). These characteristics of LCs on monolayers are now being studied as potential molecular probes in the characterization of SAMs themselves [152]. Smectic liquid crystals have been found to be a sensitive probe of molecular defects in the SAMs resulting from differences in orientations of the alkanethiol terminal groups. The immobilization of LCs on a surface directly using a sulfur containing liquid crystalline prone system would have enormous technological potential; active investigation in this area is in progress [156].

Other studies utilizing SAMs in systems involving the release of, or response to, light have been detailed with possible application in solar energy conversion processes [158], information storage devices, optical signal transduction [159-161], optical waveguides [39], and nonlinear optical materials [162]. Research into photoisomerization and photodimerization on the surface has been done in the Fox group [163-165]. Good photochemical reversibility and regioselectivity for [2+2] photocycloaddition reactions have been observed in SAMs of 7-(10-thiodecoxy)coumarin. These systems could offer information storage potential including write-erase capability since the processes have been shown to be reversible.

Kim et al. have described [39] the use of SAMs in the process of fabricating arrays of channel waveguides [166,167] which are optically useful as components in optical couplers, splitters, and interferometers. SAMs of methyl-terminated alkanethiols are patterned as arrays of closely spaced thin channels 45μm wide. In a second step, mercaptoalkanoic acid is allowed to self-assemble everywhere on the gold surface except the channel area covered with the methyl-terminated SAMs. Drops of a prepolymer of polyurethane are then placed on this surface; excess prepolymer can be drained away leaving a selective wetting of the acid terminated surface. The prepolymer is then cured with UV light leaving the channels as inset grooves in the surface. These processes are flexible and can incorporate non-planar substrate surfaces.

Non-linear optical (NLO) materials contain a lack of a center of inversion, therefore, thin films with NLO properties must be noncentrosymmetric. SAMs have been used as the mediator between a surface (gold) and a multilayer noncentrosymmetric film immobilized on the surface [162,163]. This avoids having to use poling with strong electric fields to create the noncentrosymmetric film. Instead a multilayer film is generated one monolayer at a time starting with a SAM on gold. Multilayer films involving SAMs have also been studied as photoluminescent materials [168]. An entire field of self-assembled multilayer films orignating on thiolates on gold or siloxanes on hydroxylated surfaces [169,170] has been elaborated on significantly and is not within the scope of this review.

## 7. ELECTROCHEMICAL DEVICES

The fundamental processes in electrochemical devices/systems are electron transfer and electrocatalysis [171-173]. Self-assembled monolayers of thiols on gold are model systems for the study of interfacial, long-range electron transfer [174-178] because the interfacial microenvironments of the electron transfer on the surface can be exquisitely controlled. One of these variables that is crucial to control is the distance between the redox center and the electrode. Many investigators have recognized that SAMs offer great potential to control the distance of the long-range electron transfer. This is coupled with the other advantages of a built in electrode, ease of preparation and functionalization-spurring many reports and generating many complex and creative SAM-Au incorporated systems. The general electrochemical properties of SAMs on gold have been previously reviewed [9,17-19].

SAMs have been used to study long range electron transfer with many non-biological redox centers such as ruthenium (II/III), azobenzene, ferrocene, conjugated aryl oligomers, surface bound metal clusters, and viologen. Typically, a redox center is attached to the terminal functional group of a thiol or disulfide and the electrochemical properties of the system can then be studied by cyclic voltammetry (CV), ac impedance spectroscopy (ACIS), chronoamperometry (CA), or electrochemical quartz crystal microbalance (EQCM) techniques. Creager and coworkers [179] have studied and tuned the microenvironment of immobilized ferrocenyl-alkanethiolates on gold concluding that the ferrocene moiety is "nestled" into the alkane chains of the surrounding *methyl* terminated mixed monolayer, while the ferrocene group sits on top of a partially solvated *hydroxyl* terminated mixed monolayer. Campbell et al. have also made important contributions [180] to the understanding of the microenvironment of an immobilized redox center within the film; they have shown that azobenzene-terminated absorbate molecules form densely packed SAMs on gold causing electrochemical inaccessibility due to the prohibiting of incorporation of charge-compensating cations upon reduction (double-layer effect). However, this system does allow electron transfer under conditions expanding the free volume of the monolayer (by co-adsorbing short alkanethiols) or by decreasing the size of the cation (using $H^+$, for example). This *ion-gating* of the electrochemical activity of the monolayer modified electrode is an important example of mimicking natural biological self-assembled systems.

SAMs on gold electrodes are also being used to study important electron transfer processes in biologically relevant systems. Martin and co-workers have demonstrated [181] promoted electron transfer (ET) through carboxylate terminated SAMs between horse spleen ferritin and a gold electrode while Clark et al. have studied [182] ET processes of cytochrome c adsorbed on SAMs of alkanoic acids on gold electrodes. Microperoxidase-11 (MP-11), a heme peptide digestive product of cytochrome c which is electrochemically inactive at the naked gold electrode can be induced to engage in electron transfer on surfaces modified with 4,4'-dithiodipyridine. Unexpectedly, even surfaces modified with octadecanethiol induced ET in MP-11, although only SAM based microelectrode assemblies of octadecanethiol (partially coated gold surfaces) exhibited this behavior. This is probably due to a favorable orientation of MP-11 inside the hydrophobic microdefects of the monolayer.

SAMs on gold have also made entry into other applied electrochemical devices such as their incorporation onto electrode interfaces for potential use in organic light-emitting diodes [183], modification of gold electrodes to function as detection probes for pH determination [184] as well as the selective detection of chromium (VI) [185], and even their integral role in a recently reported artificial photosynthesis mimic system [158]. Uosaki et al. have demonstrated [158] a SAM containing a porphyrin moiety as a photoactive electron donor, utilizing ferrocene as an electron relay between the porphyrin and the gold electrode. They have confirmed the largest quantum efficiency yet reported for any system with such lofty goals.

## 8. OTHER APPLICATIONS

SAMs on gold have been used as key components of studying "molecular wires"-conjugated molecules that form one-dimensional conductors [186-188]. Mixed SAMs of "non-conducting" dodecanethiol and conjugated molecules were probed by STM [188]; the conjugated rod-like molecules are isolated and diluted by the dodecanethiol allowing the observance of a single conjugated molecule conducting between the probe tip and the surface (Figure 7). The observance of such single "molecular wires" could have significant potential in proposed devices like single electron transistors or molecular switches [189].

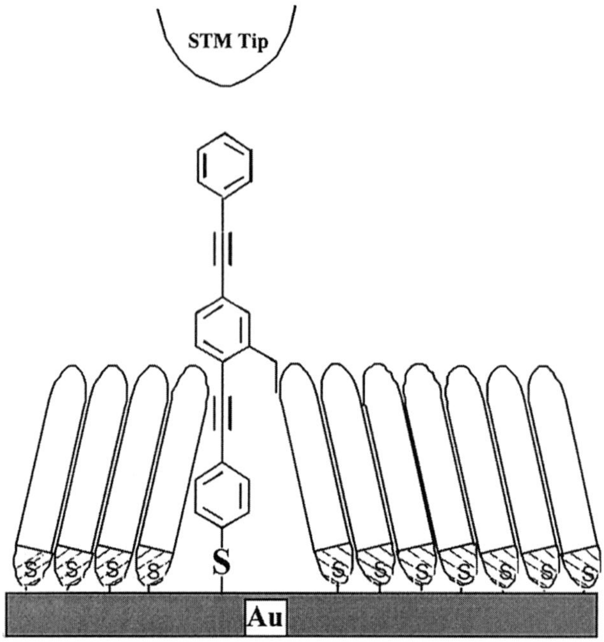


Figure 7. Single conjugated rod-like molecule isolated in a alkanethiolate SAM imaged by conductance between the STM tip and the Au through the conjugated molecule.

Synthetically prepared molecular receptors have also been immobilized on SAMs [4-6,190-192]. These systems allow the probing of molecular recognition events at organic interfaces, modeling biological membrane molecular recognition while controlling individual parameters of the surface [190]. Calixarenes and cyclodextrins have been immobilized on the surface followed by the monitoring of surface inclusion complexes [4-6]. Yan and coworkers demonstrated the first example [4] of SAM formation of alkanethiols prepared from *aqueous* solution by using cyclodextrins to form inclusion complexes with the hydrophobic alkyl chain. These water-soluble inclusion complexes organize spontaneously into SAMs in aqueous solution. SAMs of thiol-derivatized cyclodextrins on gold have also been prepared by Henke et al. [5]-thus covalently immobilizing the cyclodextrin to the surface. These systems could offer new avenues for desirable applications such as biosensors for small molecules in which host-guest interactions could be monitored by impedance analysis.

## 9. CONCLUDING REMARKS

Self-assembled monolayers (SAMs) of alkanethiolates on gold have been studied more and more extensively over the last fifteen years with tremendous potential as model systems for all kinds of complex interfaces and as important and practical tools for novel applications such as nanofabrication. SAMs on gold have become foundational systems from which an enormous amount of study has developed concerning a plethora of new, well defined, self-assembled surface interfaces.

## ACKNOWLEDGEMENTS

We would like to thank Jeffrey Stehouwer for his helpful corrections to the manuscript. We are greatful for support from the National Science Foundation (grant No. CHE-9510344) and under NSF EPSCOR cooperative agreement OSR-9553227 (WISE Program) and OSR-9553369 (TREC Program).

S.-W. Tam-Chang acknowledges the Junior Faculty Research Award granted by the University of Nevada, Reno.

## REFERENCES

1. R.G. Nuzzo and D.L. Allara, J. Am. Chem. Soc., 105 (1983) 448.
2. N. Higashi, T. Inoue and M. Niwa, Chem. Commun., (1997) 1507.
3. M.C. Fritz, G. Hahner and N.D. Spencer, Langmuir, 12 (1996) 6074.
4. J. Yan and S. Dong, Langmuir, 13 (1997) 3251.
5. C. Henke, C. Steinem, A. Janshoff, G. Steffan, H. Luftmann, M. Sieber and H.-J. Galla, Anal. Chem., 68 (1996) 3158.
6. J.H. Huisman, D.M. Rudkevich, H.C.J. M. VanVeggel and D.N. Reinhoudt, J. Am. Chem. Soc., 118 (1996) 3523.
7. D.L. Dermody, R.M. Crooks and T.J. Kim, Am. Chem. Soc., 118 (1996) 11912.
8. O.M. Magnussen, B.M. Ocko, M. Deutsch, M.J. Regan, P. S. Pershan, D. Abernathy, G. Grubel and J.-F. Legrand, Nature, 384 (1996) 250.
9. D.L. Allara, Nanoscale Structures Engineered by Molecular Self-Assembly of Functionalized Monolayers, in: Nanofabr. Biosyst., H.C. Hoch, L.W. Jelinski and H.G. Craighead (ed.), Cambridge University Press: Cambridge, UK, (1996) 180.
10. C.W.J. Sheen, A.N. Parikh, J. Martensson and D.L. Allara, J. Am. Chem. Soc., 114 (1992) 1514.
11. T.T. Ehler, N. Malmberg and L.J. Noe, J. Phys. Chem. B, 101 (1997) 1268.
12. G.B. Sigal, M. Mrksich and G.M. Whitesides, Langmuir, 13 (1997) 2749.

13. Y. Okahata, K. Matsuura, K. Ito and Y. Ebara, Langmuir, 12 (1996) 1023.
14. G.-y. Liu and S. Xu, Cruchon-Dupeyrat, S. Atomic Force Microscopy Studies of Self-Assembled Monolayers of Thiols, in: Thin Films: Self-Assembled Monolayers of Thiols., A. Ulman (ed.), Academic Press: New York, 1998, in press.
15. R. Berger, E. Delamarche, H.P. Lang, C. Gerber, J.K. Gimzewski, E. Meyer and H.-J. Gutherodt, Science, 276 (1997) 2021.
16. G.E. Poirier and E.D. Pylant, Science, 272 (1996) 1145.
17. L.H. Dubois, R.G. Nuzzo, Annu. Rev. Phys. Chem., 43 (1992) 437.
18. A. Ulman, An Introduction to Ultrathin Organic Films - From Langmuir-Blodgett to Self-Assembly; Academic Press, Inc.: San Diego, 1991.
19. G.M. Whitesides and C.B. Gorman, Self-Assembled Monolayers: Models for Organic Surface Chemistry, in: The Handbook of Surface Imaging and Visualization, A.T. Hubbard (ed.), CRC Press: Boca Raton, 1995, 713.
20. S. Xu and G.-y. Liu, Langmuir, 13 (1997) 127.
21. J.A. Rogers, K.E. Paul, R.J. Jackman and G.M. Whitesides, in press.
22. J.M. Behm, K.R. Lykke, M.J. Pellin and J.C. Hemminger, Langmuir, 12 (1996) 2121.
23. S.-W. Tam-Chang, H.A. Biebuyck, G.M. Whitesides, N. Jeon and R.G. Nuzzo, Langmuir, 11 (1995) 4371.
24. J.M. Calvert, M.-S. Chen, C.S. Dulcey, J.H. Georger, M.C. Peckerar, J.M. Schnur and P.E. Schoen, J. Vac. Sci. Technol. B, 9 (1991) 3447.
25. J. Huang and J.C. Hemminger, J. Am. Chem. Soc., 115 (1993) 3342.
26. M.J. Tarlov, R. F. Donald, J. Burgess and G. Gillen, J. Am. Chem. Soc., 115 (1993) 5305.
27. C.D. Frisbie, E.W. Wollman and M.S. Wrighton, Langmuir, 11 (1995) 2563.
28. J.A. Rogers, K.E. Paul, R.J. Jackman and G.M. Whitesides, in press.
29. N. Mino, S. Ozaki, K. Ogawa and M. Hatada, Thin Solid Films, 243 (1994) 374.
30. D.L. Allara, Appl. Phys. Lett., 62 (1993) 476.
31. N.L. Abbott, G.M. Whitesides, L.M. Racz and J. Szekely, J. Am. Chem. Soc., 116 (1994) 290.
32. N.L. Abbott, J.P. Folkers and G.M. Whitesides, Science, 257 (1992) 1380.
33. S.L. Clark, M.M. Montague and P.T. Hammond, Polym. Mater. Sci. Eng., 77 (1997) 400.
34. S.L. Clark, M. Montague and P.T. Hammond, Supramol. Sci., 4 (1997) 141.
35. D. Fischer, A. Marti and G. Hahner, J. Vac. Sci. Technol. A, 15 (1997) 2173.
36. V.K. Gupta and N.L. Abbott, Science, 276 (1997) 1533.
37. N.L. Jeon, R.G. Nuzzo, Y. Xia, M. Mrksich and G.M. Whitesides, Langmuir, 11 (1995) 3024.
38. N.L. Jeon, P. Clem, D.Y. Jung, W. Lin, K. Finnie, M. Erhardt, G.S. Girolami, D.A. Payne and R.G. Nuzzo, Polym. Mater. Sci. Eng., 77 (1997) 398.

39. E. Kim, G.M. Whitesides, L.K. Lee, S.P. Smith and M. Prentiss, Adv. Mater., 8 (1996) 139.
40. A. Kumar, K.L. Abbott, E. Kim, H.A. Biebuyck and G.M. Whitesides, Acc. Chem. Res., 28 (1995) 219.
41. N.B. Larsen, H. Biebuyck, E. Delamarche and B. Michel, J. Am. Chem. Soc., 119 (1997) 3017.
42. M. Mrksich and G.M. Whitesides, Trends in Biotechnology, 13 (1995) 228.
43. M. Mrksich, C.S. Chen, Y. Xia, L.E. Dike, D.E. Ingber and G.M. Whitesides, Proc. Natl. Acad. Sci. USA, 93 (1996) 10775.
44. M. Mrksich, L.E. Dike, J. Tien, D.E. Ingber and G.M. Whitesides, Exp. Cell Res., 235 (1997) 305.
45. R. Singhvi, A. Kumar, G.P. Lopez, G.N. Stephanopoulos, D.I.C. Wang, G.M. Whitesides and D.E. Ingber, Science, 264 (1994) 696.
46. S. Palacin, P.C. Hidber, J.-P. Bourgoin, C. Miramond, C. Fermon and G.M. Whitesides, Chem. Mater., 8 (1996) 1316.
47. T.K. Whidden, D.K. Ferry, M.N. Kozicki, E. Kim, A. Kumar, J. Wilbur and G.M. Whitesides, Nanotechnology, 7 (1996) 447.
48. G.M. Whitesides, A.J. Black, P.F. Nealy and J.L. Wilbur, Proceedings of the Robert A. Welch Foundation 39th Conference on Chemical Research/Nanophase Chemistry, 25 (1995) 109.
49. J.L. Wilbur, A. Kumar, H.A. Biebuyck, E. Kim and G.M. Whitesides, Nanotechnology, 7 (1996) 452.
50. Y. Xia, E. Kim, X.-M. Zhao, J.A. Rogers, M. Prentiss and G.M. Whitesides, Science, 273 (1996) 347.
51. Y. Xia, J. Tien, D. Qin and G.M. Whitesides, Langmuir, 12 (1996) 4033.
52. Y. Xia, E. Kim, M. Mrksich and G.M. Whitesides, Chem. Mater., 8 (1996) 601.
53. Y. Xia and G.M. Whitesides, Langmuir, 13 (1997) 2059.
54. N.L. Abbott, A. Kumar and G.M. Whitesides, Chem. Mater., 6 (1994) 596.
55. A. Kumar, H.A. Biebuyck, N.L. Abbott and G.M. Whitesides, J. Am. Chem. Soc., 114 (1992) 9188.
56. J.K. Schoer and R.M. Crooks, Langmuir, 13 (1997) 2323.
57. J.K. Schoer, F.P. Zamborini and R.M. Crooks, J. Phys. Chem., 100 (1996) 11086.
58. K.K. Berggren, A. Bard, J.L. Wilbur, J.D. Gillaspy, A.G. Helg, J.J. McClelland, S.L. Rolston, W.D. Phillips, M. Prentiss and G.M. Whitesides, Science, 269 (1995) 1255.
59. R. Younkin, K.K. Berggren, E. Cheung, K.S. Johnson, M. Prentiss, A.J. Black, G.M. Whitesides, D.C. Ralph, C.T. Black and M. Tinkham, SPIE 2995 (1997) 109.
60. L.M. Tender, R.L. Worley, H. Fan and G.P. Lopez, Langmuir, 12 (1996) 5515.
61. A. Kumar and G.M. Whitesides, Appl. Phys. Lett., 63 (1993) 2002.
62. M. Jaschke and H.-J. Butt, Langmuir, 11 (1995) 1061.

63. L.F. Rozsnyai and M.S. Wrighton, Langmuir, 11 (1995) 3913.
64. L.F. Rozsnyai and M.S. Wrighton, Chem. Mater., 8 (1996) 309.
65. T.S. Corbitt, R.M. Crooks, C.B. Ross, M.J. Hampden-Smith and J.K. Schoer, Adv. Mater., 5 (1993) 935.
66. K.C. Chan, T. Kim, J.K. Schoer and R.M. Crooks, J. Am. Chem. Soc., 117 (1995) 5875.
67. J.K. Schoer, C.B. Ross, R.M. Crooks, T.S. Corbitt and M.J. Hampden-Smith, Langmuir, 10 (1994) 615.
68. M. Chee, R. Yang, E. Hubbell, A. Berno, X.C. Huang, D. Stern, J. Winkler, D.J. Lockhart, M.S. Morris and S.P.A. Fodor, Science, 274 (1996) 610.
69. R. Elghanian, J.J. Storhoff, R.C. Mucic, R.L. Letsinger and C.A. Mirkin, Science, 277 (1997) 1078.
70. M. Hegner, M. Dreier, P. Wagner, G. Semenza and H.J. Guntherodt, J. Vac. Sci. Technol. B, 14 (1996) 1418.
71. T.M. Herne and M.J. Tarlov, J. Am. Chem. Soc., 119 (1997) 8916.
72. T. Ihara, M. Nakayama, M. Murata, K. Nakano and M. Maeda, Chem. Commun., (1997) 1609.
73. K.A. Peterlinz, R.M. Georgiadis, T.M. Herne and M.J. Tarlov, J. Am. Chem. Soc., 119 (1997) 3401.
74. E. Delamarche, G. Sundarababu, H. Biebuyck, B. Michel, C. Gerber, H. Sigrist, H. Wolf, H. Ringsdorf, N. Xanthopoulos and H.J. Mathieu, Langmuir, 12 (1996) 1997.
75. L. Deng, M. Mrksich and G. Whitesides, J. Am. Chem. Soc., 118 (1996) 5136.
76. M. Lestelius, B. Liedberg and P. Tengvall, Langmuir, 13 (1997) 5900.
77. J. Li, G. Cheng and S. Dong, J. Electroanal. Chem., 416 (1996) 97.
78. M. Liley, T.A. Keller, C. Duschl and H. Vogel, Langmuir, 13 (1997) 4190.
79. J. Madoz, B.A. Kuznetzov, F.J. Medrano, J.L. Garcia and V.M. Fernandez, J. Am. Chem. Soc., 119 (1997) 1043.
80. M. Mrksich and G.M. Whitesides, Annu. Rev. Biophys. Biomol. Struct., 25 (1996) 55.
81. N. Patel, M.C. Davies, M. Hartshorne, R.J. Heaton, C.J. Roberts, S.J.B. Tendler and P.M. Williams, Langmuir, 13 (1997) 6485.
82. D.D. Schlereth, J. Electroanal. Chem., 425 (1997) 77.
83. R.R. Seigel, P. Harder, R. Dahint, M. Grunze, F. Josse, M. Mrksich and G.M. Whitesides, Anal. Chem., 69 (1997) 3321.
84. G.B. Sigal, C. Bamdad, A. Barberis, J. Strominger and G.M. Whitesides, Anal. Chem., 68 (1996) 490.
85. L. Yan, C. Marzolin, A. Tefort and G.M. Whitesides, Langmuir, 13 (1997) 6704.
86. L. Yan, J. Rao and G.M. Whitesides, in press .
87. D.A. Hutt, E. Cooper, L. Parker, G.J. Leggett and T.L. Parker, Langmuir, 12 (1996) 5494.

88. L.K. Ista, H. Fan, O. Baca and G.P. Lopez, FEMS Microbiol. Lett., 142 (1996) 59.
89. C.D. Tidwell, S.I. Ertel, B.D. Ratner, B.J. Tarasevich, S. Atre and D.L. Allara, Langmuir, 13 (1997) 3404.
90. O. Pierrat, N. Lechat, C. Bourdillon and J.-M. Laval, Langmuir, 13 (1997) 4112.
91. L.M. Williams, S.D. Evans, T.M. Flynn, A. Marsh, P.F. Knowles, R.J. Bushby and N. Boden, Langmuir, 13 (1997) 751.
92. P. Wagner, F. Zaugg, P. Kernen, M. Hegner and G. Semenza, J. Vac. Sci. Technol. B, 14 (1996) 1466.
93. G.J. Leggett, M.C. Davies, D.E. Jackson and S.J.B. Tendler, Journal of Electron Spectroscopy and Related Phenomena, 81 (1996) 249.
94. B.A. Cornell, V.L.B. Braach-Maksvytis, L.G. King, P.D.J. Osman, B. Raguse, L. Wieczorek and R.J. Pace, Nature, 387 (1997) 580.
95. L. Ding, J. Li, E. Wang and S. Dong, Thin Solid Films, 293 (1997) 153.
96. S. Rubin, J.T. Chow, J.P. Ferraris, A. Thomas and J. Zawodzinski, Langmuir, 12 (1996) 363.
97. C.E. Jordan and R.M. Corn, Anal. Chem., 69 (1997) 1449.
98. B.D. Ratner, J. Biomed. Mater. Res., 27 (1993) 837.
99. K.L. Prime and G.M. Whitesides, Science, 252 (1991) 1164.
100. M. Mrksich, G.B. Sigal and G.M. Whitesides, Langmuir, 11 (1995) 4383.
101. K.L. Prime and G.M. Whitesides, J. Am. Chem. Soc., 115 (1993) 10714.
102. S.I. Jeon, J.H. Lee, J.D. Andrade and P.G. DeGennes, J. Colloid Interface Sci., 142 (1991) 149.
103. S.I. Jeon and J.D. Andrade, J. Colloid Interface Sci., 142 (1991) 159.
104. D. Piscevic, W. Knoll and M.J. Tarlov, Supramolecular Sci., 2 (1995) 99.
105. M. Mrksich, J.R. Grunwell and G.M. Whitesides, J. Am. Chem. Soc., 117 (1995) 12009.
106. J. Spinke, M. Liley, F.-J. Schmitt, H.-J. Guder, L. Angermaier and W. Knoll, J. Chem. Phys., 99 (1993) 7012.
107. F.A. Armstrong, H.A.O. Hill and J.N. Walton, Q. Rev. Biophys., 18 (1986) 261.
108. A. Sadana, Chem. Rev., 92 (1992) 1799.
109. R. Gentz, C.-H. Chen and C.A. Rosen, Proc. Natl. Acad. Sci. USA, 86 (1989) 821.
110. R. Dahint, M. Grunze, F. Josse and J. Renken, Anal. Chem., 66 (1994) 2888.
111. S.P. Massia and J.A. Hubbell, Anal. Biochem., 187 (1990) 292.
112. C.J. Roberts, M. Sekowski, M.C. Davies, D.E. Jackson, M.R. Price and S.J.B. Tendler, Biochem. J., 283 (1992) 181.
113. M. Rehak, M. Snejdarkova and M. Otto, Electroanalysis, 5 (1993) 691.
114. L. Ding, J. Li, S. Dong and E. Wang, J. Electroanal. Chem., 416 (1996) 105.
115. A. Henglein, Top. Curr. Chem., 143 (1988) 113.

116. D. Averin and K.K. Likharev, Single-Charge Tunneling; Plenum: New York, 1992.
117. M. Brust, M. Walker, D. Bethell, D.J. Schiffrin and R. Whyman, J. Chem. Soc. Chem. Commun., (1994) 801.
118. A. Badia, W. Gao, S. Singh, L. Demers, L. Cuccia and L. Reven, Langmuir, 12 (1996) 1262.
119. A. Badia, L. Cuccia, L. Demers, F. Mortin and R.B. Lennox, J. Am. Chem. Soc., 119 (1997) 2682.
120. A. Badia, S. Singh, L. Demers, L. Cuccia, G.R. Brown and R.B. Lennox, Chem. Eur. J., 2 (1996) 359.
121. M.J. Hostetler, J.J. Stokes and R.W. Murray, Langmuir, 12 (1996) 3604.
122. C.S. Weisbecker, M.V. Merritt and G. Whitesides, Langmuir, 12 (1996) 3763.
123. K.S. Mayya, V. Patil and M. Sastry, Langmuir, 13 (1997) 3944.
124. R.P. Andres, J.D. Bielefeld, J.I. Henderson, D.B. Janes, V.R. Kolagunta, C.P. Kubiak, W.J. Mahoney and R.G. Osifchin, Science, 273 (1996) 1690.
125. C.A. Mirkin, R.L. Letsinger, R.C. Mucic and J.J. Storhoff, Nature, 382 (1996) 607.
126. A.P. Alivisatos, K.P. Johnsson, X. Peng, T.E. Wilson, C.J. Loweth, P. Marcel, J. Brunchez and P.G. Schultz, Nature, 382 (1996) 609.
127. K.C. Grabar, K.J. Allison, B.E. Baker, R.M. Bright, K.R. Brown, R.G. Freeman, A.P. Fox, C.D. Keating, M.D. Musick and M.J. Natan, Langmuir, 12 (1996) 2353.
128. D. Bethell, M. Brust, D.J. Schiffrin and C. Kiely, J. Electroanal. Chem., 409 (1996) 137.
129. R.P. Andres, T. Bein, M. Dorogi, S. Feng, J.I. Henderson, C.P. Kubiak, W. Mahoney, R.G. Osifchin and R. Reifenberger, Science, 272 (1996) 1323.
130. P. Stroeve, M. Nagtegaal, W. Tremel and W. Knoll, Polym. Prep., 38 (1997) 965.
131. U.-W. Grummt, M. Geissler and T. Schmitz-Huebsch, Chem. Phys. Lett., 263 (1996) 581.
132. J.I. Henderson, S. Feng, G.M. Ferrence, T. Bein and C.P. Kubiak, Inorganica Chemica Acta, 242 (1996) 115.
133. W. Gao, L. Dickinson, C. Grozinger, F.G. Morin and L. Reven, Langmuir, 13 (1997) 115.
134. M. Brust, D. Bethell, D.J. Schiffrin and C.J. Kiely, Adv. Mater., 7 (1995) 795.
135. D.B. Wurm, S.T. Brittain and Y.-T. Kim, Langmuir, 12 (1996) 3756.
136. M. Niwa, M. Date and N. Higashi, Macromolecules, 29 (1996) 3681.
137. Y. Zhou, M.L. Bruening, D.E. Bergbreiter, R.M. Crooks and M. Wells, J. Am. Chem. Soc., 118 (1996) 3773.
138. V.G. Gregoriou, R. Hapanowicz, S.L. Clark and P.T. Hammond, Appl. Spectrosc., 51 (1997) 470.

139. S.L. Clark, M.F. Montague and P.T. Hammond, Polym. Prepr., 38 (1997) 967.
140. P.T. Hammond, Polym. Mater. Sci. Eng., 73 (1995) 308.
141. S.L. Clark and P.T. Hammond, Polym. Mater. Sci. Eng., 75 (1996) 446.
142. C.N. Sayre and D.M. Collard, Langmuir, 13 (1997) 714.
143. Z. Gao and K.S. Siow, J. Electroanal. Chem., 412 (1996) 179.
144. U.B. Steiner, W.R. Caseri and U.W. Suter, Langmuir, 10 (1994) 1164.
145. T. Kim, K.C. Chan and R.M. Crooks, J. Am. Chem. Soc., 119 (1997) 189.
146. L.A. Godinez, R. Castro and A.E. Kaifer, Langmuir, 12 (1996) 5087.
147. T. Kim and R.M. Crooks, Tetrahedron Lett., 35 (1994) 9501.
148. K.C. Chan, T. Kim, J.K. Schoer and R.M. Crooks, Langmuir, 12 (1996) 6065.
149. K.C. Chan, T. Kim, J.K. Schoer and R.M. Crooks, J. Am. Chem. Soc., 117 (1995) 5875.
150. N.L. Abbott, Current Opinion in Colloid and Interface Science, 2 (1997) 76.
151. N.L. Abbott, V.K. Gupta, W.J. Miller and R.R. Shah, Polym. Prepr., 38 (1997) 941.
152. V.K. Gupta, W.J. Miller, C.L. Pike and N.L. Abbott, Chem. Mater., 8 (1996) 1366.
153. V.K. Gupta and N.L. Abbott, Phys. Rev. E, 54 (1996) R4540.
154. V.K. Gupta and N.L. Abbott, Langmuir, 12 (1996) 2587.
155. S.D. Evans, H. Allinson, N. Boden, T.M. Flynn and J.R. Henderson, J. Phys. Chem. B, 101 (1997) 2143.
156. H. Schonherr, F.J.B. Kremer, S. Kumar, J.A. Rego, H. Wolf, H. Ringsdorf, M. Jaschke, H.-J. Butt and E. Bamberg, J. Am. Chem. Soc., 118 (1996) 13051.
157. P.J. Shannon, W.M. Gibbons and S.T. Sun, Nature, 368 (1994) 532.
158. K. Uosaki, T. Kondo, X.-Q. Zhang and M. Yanagida, J. Am. Chem. Soc., 119 (1997) 8367.
159. E. Katz, M. Lion-Dagan and I. Willner, J. Electroanal. Chem., 382 (1995) 25.
160. I. Willner, R. Blonder and A. Dagan, J. Am. Chem. Soc., 116 (1994) 9365.
161. M. Lion-Dagan, E. Katz and I. Willner, J. Am. Chem. Soc., 116 (1994) 7913.
162. D.G. Hanken, R.R. Naujok, J.M. Gray and R.M. Corn, Anal. Chem., 69 (1997) 240.
163. M.O. Wolf and M.A. Fox, Langmuir, 12 (1996) 955.
164. W. Li, V. Lynch, H. Thompson and M.A. Fox, J. Am. Chem. Soc., 119 (1997) 7211.
165. M.O. Wolf and M.A. Fox, J. Am. Chem. Soc., 117 (1995) 1845.
166. O. Parriaux, in: Fiber Optic Chemical Sensors and Biosensors, O.S. Wolfbeis (ed.), CRC Press: Boca Raton, Fl, 1991, 111.
167. E. VanTomme, P.P. VanDaele, R.G. Baets and P.E. Lagasse, IEEE J. Quantum Electron., 27 (1991) 778.
168. M. Brust, P.M. Blass and A.J. Bard, Langmuir, 13 (1997) 5602.

169. S.B. Roscoe, S. Yitzchaik, A.K. Kakkar and T.J. Marks, Langmuir, 12 (1996) 5338.
170. S.B. Roscoe, A.K. Kakkar and T.J. Marks, Langmuir, 12 (1996) 4218.
171. J.P. Collman, M.S. Ennis, D.A. Offord, L.L. Chng and J.H. Griffin, Inorg. Chem., 35 (1996) 1751.
172. J. Zak, H. Yuan, M. Ho, L.K. Woo and M.D. Porter, Langmuir, 9 (1993) 2772.
173. J.E. Hutchinson, T.A. Postlethwaite and R.W. Murray, Langmuir, 9 (1993) 3277.
174. C.E.D. Chidsey, Science, 251 (1991) 919.
175. C.E.D. Chidsey and D.N. Loiacono, Langmuir, 6 (1990) 682.
176. M.D. Porter, T.B. Bright, D.L. Allara and C.E.D. Chidsey, J. Am. Chem. Soc., 109 (1987) 3559.
177. E.E. Polymeropoulos and J. Sagiv, J. Chem. Phys., 69 (1978) 1836.
178. C. Miller, P. Cuendet and M. Gratzel, J. Phys. Chem., 95 (1991) 877.
179. S.E. Creager and G.K. Rowe, J. Electroanal. Chem., 420 (1997) 291.
180. D.J. Campbell, B.R. Herr, J.C. Hulteen, R.P.V. Duyne and C.A. Mirkin, J. Am. Chem. Soc., 118 (1996) 10211.
181. T.D. Martin, S.A. Monheit, R.J. Niichel, S.C. Peterson, C.H. Campbell and D.C. Zapien, J. Electroanal. Chem., 420 (1997) 279.
182. R.A. Clark and E.F. Bowden, Langmuir, 13 (1997) 559.
183. W. Chen, S. Burnham, C.E.D. Chidsey and J.C. Scott, Polym. Prepr., 38 (1997) 936.
184. J.C. Mason, S.-W. Tam-Chang, K. Leonard and K. Hwang, manuscript in preparation .
185. I. Turyan and D. Mandler, Anal. Chem., 69 (1997) 894.
186. J.M. Tour, L.J. II, D.L. Pearson, J.J.S. Lamba, T.P. Burgin, G.M. Whitesides, D.L. Allara, A.N. Parikh and S.V. Atre, J. Am. Chem. Soc., 117 (1995) 9529.
187. L.I. Jones, D.L. Pearson, J.J.S. Lamba, J.M. Tour, G.M. Whitesides, C.J. Muller and M.A. Reed, Polym. Prepr., 36 (1995) 564.
188. L.A. Bumm, J.J. Arnold, M.T. Cygan, T.D. Dunbar, T.P. Burgin, L. Jones, D.L. Allara, J.M. Tour and P.S. Weiss, Science, 271 (1996) 1705.
189. J.S. Miller, Adv. Mater., 2 (1990) 378.
190. K. Motesharei and D.C. Myles, J. Am. Chem. Soc., 116 (1994) 7413.
191. G. Nelles, M. Weisser, R. Back, P. Wohlfart, G. Wenz and S. Mittler-Neher, J. Am. Chem. Soc., 118 (1996) 5039.
192. F. Davis and C.J.M. Stirling, J. Am. Chem. Soc., 117 (1995) 10385.

Adsorption and its Applications in Industry and Environmental Protection
Studies in Surface Science and Catalysis, Vol. 120
A. Dabrowski (Editor)



# Adsorption methods in technology of chemooptical interface for fiber optic chemical sensors

A.Dybko and W.Wróblewski

Warsaw University of Technology, Department of Analytical Chemistry, Noakowskiego 3, 00-664 Warsaw, Poland

## 1. INTRODUCTION

Fiber optic sensors and systems are finding increasing number of applications in industry, environmental monitoring, medicine and chemical analysis. Fiber optic sensors can measure physical or chemical quantities. Their development has been stimulated by advances in optoelectronic technology mainly for applications in telecommunication. Light sources (light emitting diodes, laser diodes, lamps etc.), photodetectors (photodiodes, photomultiplier tubes etc.), connectors and optical fibers can be easily adapted for sensing purposes.

Fiber optic chemical sensors (FOCS) can be utilized in harsh and hardly accessible places. Many of the advantages of FOCS are due to the characteristics of fiber optics. In comparison with the electrochemical sensors they do not require any reference electrode or cell. They can be designed in a form of very small and flexible device. There is no possibility to generate an electric spark or electric shock (this is very important particularly in medical applications). Transmitted optical signal is not subject to electric or magnetic interferences. The measurements can be done in real time with extremely low consumption of measured species. Notwithstanding, there are several limitations and disadvantages of FOCS. One of the major limitations is an ambient light. The system should be preserved from the possibility of interfering by the laboratory light or the light source used should be modulated. Long term stability of the sensor is limited by photodecomposition of the transducer material, aging and washing out processes. More details on FOCS can be found in some reviews published over the last decade [1-4].

The schematic configuration of a FOCS is presented in figure 1.

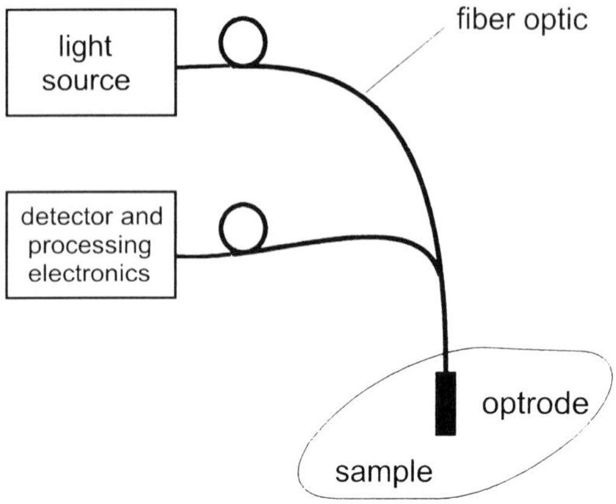


Figure 1. A schematic configuration of fiber optic chemical sensor.

A light beam from an appropriate source (e.g. a light emitting diode) is launched into the fiber optic and guided to the sensor (so-called optrode), where it interacts with the sensing layer. This interaction depends on the changes in optical properties of sensing layer and results in a light modulation. The modulated optical signal bearing chemical information is guided to the detector (e.g. a photodiode) by the same or another fiber optic. The scheme shown in figure 1 seems to be too simple for a contemporary sensor. It has been based on the function parts of the sensor and details on construction are neglected. In fact, both the light source and the detector parts are quite sophisticated including optical power stabilization or synchronous detection of low level optical signal.

From the point of view of signal conversion the FOCS comprises several interfaces (fig.2).

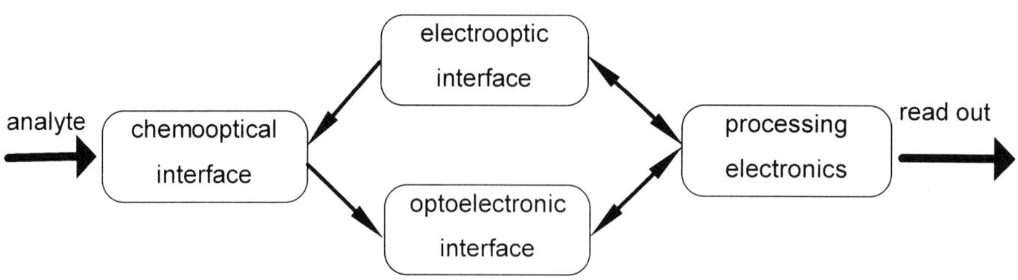


Figure 2. Signal conversion in fiber optic chemical sensor.

Since the number of analytes exhibiting their own spectral changes is very limited, it is necessary to match an appropriate reagent. The reagent should recognize the analyte and convert information on its concentration into changes of optical properties (e.g. absorbance, fluorescence, reflectance). The reagent is immobilized in the chemooptical interface, which is in contact with the sample under test. The principle of operation of an optical sensor depends on the chemooptical interface. The chemooptical interface can be designed as a separate layer - membrane, which contains reagent molecules, held against the end of a single fiber optic or bundle (so-called extrinsic sensors) [1]. In the second group the reagent phase is directly incorporated into the structure of fiber optic e.g. it replaces the removed cladding (so-called intrinsic sensors) [1] or it is immobilized in the porous sol-gel cladding [5]. The exciting beam of light is delivered by the electrooptic interface. Its main part is an appropriate light source, which is matched to the maximum of absorbance or excitation of the reagent. After having interacted with the chemooptical interface, this light is converted into an electrical signal by the optoelectronic interface, which contains a photodetector (e.g. photodiode) and an amplifier. Both optoelectronic and electrooptic interfaces can be driven by the processing electronics. Contemporary measuring systems are based on the use of microprocessors (microcomputers) and they give a direct read out after applying a calibration of the sensor.

## 2. DESIGN OF THE CHEMOOPTICAL INTERFACE

The chemical sensing and the signal processing of the chemooptical interface are achieved by the incorporation of a reagent. Selective complex-forming indicator dyes (chromoionophores) or acid-base indicators are frequently used. The reagent molecules include an ion-recognizing complexing center coupled with a chromophoric group. The chemical interactions between the reagent and the analyte are based on an acid-base, complexation or redox reaction [1,2].

The first step in the design of the chemooptical interface is the chemical matching of the reagent molecules to the analyte molecules. The recognition reaction must be followed by the change in optical properties of the reagent depending on the concentration of the analyte. The performances of the sensor (sensitivity, selectivity, dynamic range, response time etc.) are affected by the reagent properties. The selectivity is governed by the recognition ability of the reagent towards the analyte. The reagent used should exhibit large changes in spectral properties in dependence on small changes in the concentration of the analyte. In this way wide dynamic range and sensitivity can be obtained. Up to a point, the dynamic range depends on the concentration of the reagent in the chemooptical interface. Usually wider dynamic range is obtained for larger amounts of the immobilized reagent. On the other hand, a membrane containing excessive amounts of the reagent (a so-called high loaded membrane) cannot be used for the measurements of samples with low concentration of the analyte. In this case one of the fundamental requirements of the sensor i.e. the low analyte consumption is not fulfilled.

The following step includes the attachment of the reagent phase to the fiber optic. The chemooptical interface may consist of a reagent solution separated from the sample by a semipermeable membrane. The membrane retains the reagent phase and, in addition, its permeability can exclude interfering substances, thus improving the selectivity of the sensor. Another solution is based on incorporation of the reagent molecules on a support membrane. The reagent can be immobilized either directly on the surface or in the bulk of the membrane.

### 2.1. Immobilization of the reagent molecules

The incorporation of the reagent molecules involves their physical or chemical immobilization in the chemooptical interface [6]. Durability and efficient reagent immobilization in the chemooptical interface seem to be a serious problem that influences strongly further performances of the optical sensor. The technique of immobilization should meet several requirements:

- effective anchoring of the reagent molecules,
- active chromophoric system after immobilization,
- easy access of the analyte to the immobilized reagent.

Many types of chemooptical interfaces for FOCS have been developed [1,2,6]. Immobilization of the reagent leads to two types of the optrode sensing layers: surface and bulk ones [7]. The membrane can be prepared as a polymeric foil and then the reagent is immobilized on its surface, so the active site is located directly in the sample solution. It is the so-called surface membrane. Physical and chemical procedures of anchoring can be utilized to manufacture such a membrane. The surface membranes tend to suffer from a loss of reagent due to its washing out to the sample tested. Another disadvantage is the not-well defined mechanism (specific and non-specific influences of the sample matrix) and the influence of the ionic strength (as well as the surface charge density) in the sensor response.

Another procedure is based on physical entrapment of the reagent molecules in the bulk of the solution separated from the sample by a semipermeable membrane [8-10] or, more frequently, in the bulk of the polymer membrane. The reagent should be highly lipophilic to minimize the leaching out. In such a membrane a reversible mass transfer of the analyte from the sample into the bulk of the membrane is observed. The response time of the membrane depends on the mass transfer equilibrium and on the diffusion coefficients of the analyte in the bulk of the membrane. Therefore, very thin sensing layers are recommended for decreasing the response time. An advantage of the bulk membranes is the possibility of the immobilization of greater amounts of a reagent as comparing with the surface membranes.

The physical immobilization of a reagent is based on its surface adsorption [7] or bulk encapsulation [7] (e.g. sol-gel entrapment). A polymeric support matrix such as cellulose [11, 12], PVC [13], PVA [14], PTFE [15], polyacrylamide [16], Nafion [17,18], Amberlite polymer resins [19,20] and sol-gel silica [21] are the most frequently used materials for physical immobilization. Electrostatic interactions can be utilized to immobilize a charged reagent on an ion-exchange resin [22-24]. A limited life-time due to the leaching out of the reagent is the main disadvantage of the chemooptical interface based on the physical immobilization. Thus the design of durable optical sensors requires the transduction of the chemical bonds between the reagent and membrane matrix. In this case covalent bonds allow for efficient anchoring of the reagent on hydrophilic supports such as cellulose [25], glass [26, 27], silica [28, 29] or the following polymers: PTFE [30], polyHEMA [31, 32], PVA [33], polyacrylamide [34] and imidazole derivatives [35]. Other sophisticated methods involve the electropolymerization of the reagent phase on a support [36] or the formation of the chemooptical interface using the Langmuir-Blodgett technique [37]. The reagent molecules can also be directly chemically bonded to the silica surface of the fiber optic.

## 3. CHEMOOPTICAL INTERFACE BASED ON THE ADSORPTION

The adsorption involves a selective deposition of the external phase molecules on the adsorbent surface [38]. The concentration of molecules, if it occurs spontaneously, should lead to a state of higher stability, thus providing a driving force for adsorption. The state of higher stability and thus of lower energy is created by the release of the free energy of adsorption. The concentration phenomena occur due to the intermolecular attractive interactions between chemical species. The adsorption process may be reversible or irreversible depending on the type of bonds (physical or chemical) involved in the dye-substrate interaction. The identification of the kind of interactions is rather difficult as they can act simultaneously. One can expect that the free energy of chemical adsorption is greater than the physical adsorption. In addition, the chemisorption and the physical adsorption result in different thicknesses of obtained layers: considerably thicker one in the first case.

Reversible binding forces arise between dipoles or between ionized charged centers. The first process gives physical adsorption described more precisely by weak van der Waals dye-adsorbent interactions similar to the intermolecular forces in non-ideal gases. The forces exist between molecules having a permanent dipole moment as well as between molecules with no such internal charge separation e.g. gases. The van der Waals interactions are observed between dipoles of a permanent or induced nature. Another consideration introduces the perturbational mechanism of binding by van der Waals forces common in the hydrogen bond. It consists of interactions between a permanent dipole and an induced dipole molecule approaching one another.

The ion-sensing chemooptical interfaces of FOCS are often based on physical adsorption using non-polar polymer adsorbents. The Amberlite resins: XAD-2, XAD-4 and XAD-7 seem to be the most frequently used support type for physical immobilization of reagent. XAD-2 and XAD-4 - non-polar cross-linked copolymers of styrene and divinylbenzene are chosen as the immobilizing matrix because of their hydrophobic nature, which gives an ability to interact with the non-polar phenol derivative molecules of the indicator. They have to be hydrated before being used in aqueous solutions, which can be achieved by soaking the polymer first in methanol and then in water [39]. These adsorbents (especially XAD-2) may be considered as solid solvents for organic reagents containing the benzene ring and should affect the speed of formation of the dye-analyte complex [19, 40]. On the other hand, the chromogenic reagent is adsorbed on the surface of the beads of the solid support, which allows an easy access of the analyte to the active sites of the chemooptical interface [20]. In this case, the response time of the sensor, shorter than in the case of the PVC bulk membrane [41], depends only on the mass transfer to the optomembrane layer and the kinetics of complexing. This is the main advantage of physical adsorption of the reagent on the resin surface over the entrapment in the bulk of optomembrane. Amberlite XAD-4 was used for the immobilization of rhodamine derivative and this chemooptical interface had a longer fluorescence lifetime than the one adsorbed on XAD-2 resin [42]. Another polymer adsorbent Amberlite XAD-7, weakly polar cross-linked polymer of methacrylate, is suitable for the physical immobilization of metal complexes of phenanthroline derivative used as the reagent [43].

Teflon tape can also be used as the immobilization support due to its structural difference from the XAD resin. It is expected that the reagent molecules are layered on the PTFE tape in contrast to that encountered with the resin beads [44]. This structure affects the access of the analyte molecules to the obtained Teflon based chemooptical interfaces and thus their performance (response time, detection limit), which are considerably more favorable than for resin based layers. Microporous Teflon membrane used as a support for physical adsorption of reagent allows the immobilization of quite a large amount of the reagent. A simple and effective procedure for physical dye immobilization of hydrophilic pH-indicators on the surface of microporous PTFE matrix was elaborated [15]. The porous structure suitable for the physical adsorption of the dye can also be achieved in chemical way by the hydrolysis of cellulose acetate film. The obtained porous structure of this polymer minimizes the barriers to mass transfer between the analyte and the active layer, which enhances its response time [45]. The chemooptical interface based on physical adsorption can be obtained using the Langmuir-Blodgett thin-film technique. The obtained well-defined and ordered lipid thin layer containing the reagent is formed for example on silylated glass support [46]. Silica gel, the widely used adsorbent in column chromatography, and other porous organic polymers are also used as supports for the physical adsorption of reagent [43, 47].

The binding forces of physical adsorption may arise between two charged chemical species. In this case, there is a so-called electrostatic sorption, where the attractive forces are related to the coulombic interactions, which are inversely related to the distance between the charged centers. The electrostatic sorption may occur on the surface of a non-charged support where the process of adsorption gives a surface charge and thus the surface potential is governed by the adsorbed reagent ions. In other cases, the presence of electric charge on an adsorbing surface forms an electrical potential, which attracts oppositely charged counter-ions.

Synthetic ion-exchangers are applied as optomembrane supports giving the possibility of electrostatic immobilization of reagent molecules. They consist of a synthetic polymeric resin containing chemically anchored functional groups with an ionic bond. The functional groups dissociate in water solution giving charged active centers of the polymer resin and mobile counter-ions, which can be exchanged to another ions of the same charge. Depending on the functional group involved cation-exchanger (sulphonyl acid or carboxylic functionalities) and anion-exchanger (quaternary ammonium or tertiary amine functionality) are known. The selectivity of the electrostatic adsorption depends on the acidity or basicity of the functional groups. Amberlite, Dowex and Wofatit are ion-exchange resins frequently applied as a matrix for chemooptical interfaces [22-24, 48]. An important feature of this resin includes its porous microstructure, which provides a short equilibrium time and thus a rapid response time upon changes in analyte concentration. The procedure of electrostatic immobilization is quite simple i.e. in many cases it may be accomplished by immersing the polymeric support into a reagent solution. The amount of the dye attracted to the resin can be controlled by varying the concentration of the solution. Generally, the amount of the reagent in the prepared solution is lower than the capacity of the ion-exchange resin. The binding properties of the immobilized dyes remain practically unchanged compared with the reagent in solution because its functional charged groups e.g. sulphonyl functionalities are not involved in the complex formation with the analyte.

An efficient electrostatic adsorption may be achieved if the reagent molecules are weakly acidic or basic e.g. sulfonate- or carboxylate-functionalized. In this manner the dye molecules may be protonated or deprotonated in aqueous solution (or an another solvent if the reagent is not soluble in water) giving charged species capable of being electrostatically attracted to a polymeric resin. The pH of the dye bath should be matched to the $pK_a$ of the reagent to allow its complete dissociation. The reagent can be electrostatically trapped on an another commercially available ion-exchange perfluorinated polysulfate polymer Nafion [18]. Physical immobilization on Nafion is attractive because the reagent is both electrostatically and hydrophobically bound to the polyanion resin in aqueous solution. The anionic character of the polymer gives also very small anionic interferences.

The application of ion-exchangers to immobilize the reagent in the chemooptical interface can also solve the problem of slow response because the dye is located on the surface of the support. However, such materials are charged even after the dye immobilization, which results in some disadvantages [49]. Due to these limitations neutral copolymers such as styrene-divinylbenzene have been used but electrostatic binding forces are too weak for long-term applications.

Chemooptical interfaces based on physical immobilization are easy to obtain but often the observed leaching out of the reagent reduces their life-time. For this reason the design of durable optomembranes requires the presence of the chemisorption i.e. chemical interactions between the reagent and membrane matrix. The irreversible anchoring of the dye molecules is achieved by the formation of covalent bonds.

Cellulose esters are the most frequently used support for chemisorption of the reagents. Cellulose acetate, a hydrogel type polymeric material, is especially suitable as an optomembrane matrix because of its high permeability for ions and for water molecules. Additionally, it has been recognized that hydrophilic materials such as cellulose lead to chemooptical interfaces characterized by good reversibility and stability in contrast to optomembranes made of dyes immobilized on hydrophobic support such as Amberlite resin (pH hysteresis and long-term drift). The process of the chemisorption on the cellulose based-polymers can be very simple when unmodified direct dyes are applied [45]. It arises from the ability of direct dyes to be adsorbed on cellulose by immersing the support into a hot solution of a dye with a high ionic strength. The cellulose acetate film is first hydrolyzed in order to be converted to a highly porous material. It is expected that the chemical interactions of both the amine and the azo groups of the dye with the hydroxy groups of cellulosic support will increase the strength of chemisorption. The covalent bonds can also be achieved by the activation of the hydrolyzed cellulose polymer by means of urea and formaldehyde [50, 51]. In this way the activated matrix contains hydroxymethyl groups, which become chemical sites able to react with the amino groups or with the free ortho position of the unmodified indicators.

More complicated chemisorption methods involve the chemical modification i.e. the functionalization of the reagent in order to allow the chemical anchoring to the cellulose support [52, 53]. The reagent is functionalized with hydroxyethylsulfonyl groups. These groups are converted into sulfo esters and finally into vinylsulfonyl group. This group reacts with hydrolyzed cellulose giving dye/cellulose conjugate, which does not alter the optical properties of the reagent. A perfluorinated alkyl chain may be substituted to the nitrogen atom of fluoroionophore, which is efficiently embedded in a PTFE membrane [30]. Another procedure proposed the functionalization of the reagent, which allows its chemical immobilization on the surface of silica gel [28]. The covalent anchoring was achieved by formation of alkyl siloxane linkage to silica, thus giving a chemically stable surface bond, which does not appreciably affect the quantum yield of fluorescence.

It is also possible to obtain a reagent with a photoreactive group and to immobilize it on polyHEMA film [54]. The polyHEMA membrane was also applied to bind covalently a reagent containing an amino group. In this case HEMA was photocopolymerized with acrylic acid to give a membrane support containing carboxylic groups capable of dye binding [55]. The photochemical immobilization of the sensitive component in a hydrophilic polymer on the distal end of silanized optical fiber included also other materials such as polyacrylamide [34]. The reagent can also be immobilized on cellulose or glass support using EDC [26, 56] or cyanuric chloride [25, 57] as coupling agents. These procedures require also the modification of the support prior to the binding reaction i.e. hydrolysis of cellulose esters or silanization of glass surface.

## 4. FIBER OPTIC CHEMICAL SENSORS WITH ADSORPTION BASED OPTOMEMBRANES

Every optical sensor requires the use of an appropriate light source. There are many types of light sources, e.g. halogen lamps, xenon lamps, lasers, laser diodes, light emitting diodes (LED). Among them the semiconductor light sources offer several advantages over the others. The laser diodes and LEDs can be modulated electronically and their optical power can be stabilized. These features are very important in design of fiber optic sensors, because most of them are based on measurement of the light intensity [2, 4].

The basic version of the experimental set-up used in our experiments is shown in figure 3.

Light from an LED is modulated to a square wave by a generator (GEN). One arm of the fiber optic bundle is connected to the LED. The light emitted by the LED is matched to the maximum of the molar absorbance of the indicator

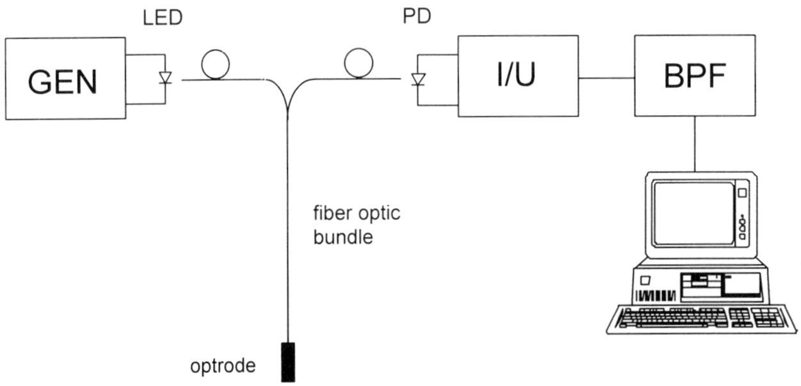


Figure 3. A set-up for measurements of fiber optic sensor.

used. Fiber optics guide the light to the optrode with an optomembrane where it is reflected in relation to the variations of the measuring parameter and then it is transmitted to a photodiode (PD) through the second arm of the bundle. The photodiode is connected to a transimpedance amplifier (I/U) and band-pass active filter (BPF). The active filter filters out the frequency of modulated light. Often a synchronous detection is applied. The electrical signal obtained is acquired and processed by a pc-lab card with 12-bit A/D converter.

The measuring properties of the sensing system can be enhanced by introducing the second wavelength, the so-called reference (which is not shown in the figure). The reference wavelength is not absorbed by the indicator immobilized in the membrane (usually it is infrared radiation). By combining together reference and analytical signals it is possible to compensate the changes in the read-outs, which are not caused by measured quantity.

The optrode was manufactured on the surface of the common end of the bundle, which is shown in figure 4.

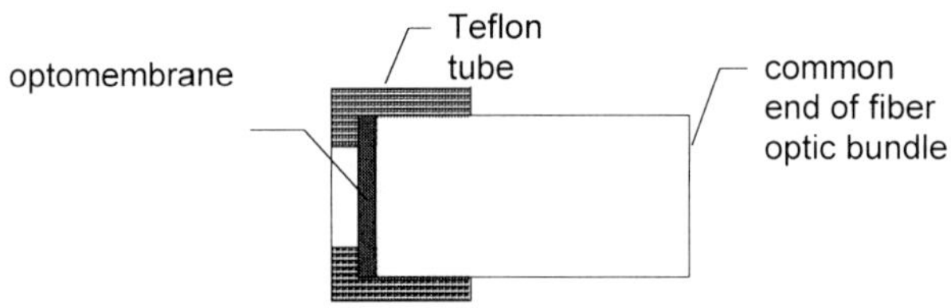


Figure 4. Schematic cross-section of an optrode.

The optomembrane obtained by the immobilization procedure is held against the optical fiber bundle by a removable Teflon tube. Such an arrangement allows for easy access of measured ions to the membrane and the membrane can be quickly exchanged.

Common practice in investigation of FOCS is the use of arbitrary units for data presentation. The measured data are normalized against their maximum value and then plotted. In this way, it is possible to compare curves obtained for various membranes.

### 4.1. Fiber optic pH sensor

Determination of pH is one of the most frequently applied measurements in analytical chemistry. Depending on the required accuracy a litmus paper or electrochemical sensors (e.g. glass electrode) are used. Application of such sensors in medicine or environmental monitoring is quite limited. Electrochemical sensors are subject to electromagnetic interferences, their input impedance is very high and they require a reference electrode. This causes serious problems in the design of measuring system or network of the sensors. The advantages of fiber optic sensors are particularly important in medical applications. The sensor can be very small, it does not require a reference electrode and it is safe in use. The first fiber optic chemical sensor for pH monitoring utilized phenol red immobilized on polymer beads covered with cellulose membrane [58]. Fiber optic pH sensors can also be used in medicine, environmental monitoring, biotechnology etc. [59, 60].

We have investigated the immobilizing properties of the Amberlite resins and of cellulose acetate as solid supports for fiber optic pH sensors. The assumption was to design a sensor with the working range pH=6.5 - 8.5 for environmental applications.

#### 4.1.1. pH sensor based on Amberlite resins

Immobillization procedure

The Amberlite resins were used as a support polymer for reagent immobilization. Two types of Amberlite beads were chosen: IRA 401 (strongly basic anion-exchanger) and XAD 1168 (non-ionic polymeric adsorbent). Since Amberlite beads cannot be directly attached to the fiber optic, poly(vinyl chloride) (PVC) was used as a matrix. PVC was dissolved in tetrahydrofurane (THF) and the solution was poured into a glass ring of i.d. 20 mm. At the end of slow evaporation of THF the Amberlite beads were put onto the polymer membrane. In this way, beads were imbedded into PVC and their top part might be used for reagent immobilization. The obtained membrane was dried and then immersed into aqueous - methanolic solution of pH indicator: bromothymol blue (BTB). After 15 minute adsorption the membrane was washed with distilled water and left overnight in water. During that time the not immobilized part of the indicator was washed out from the membrane.

Results

Bromothymol blue is an acidic pH chromoionophore where acid to base change corresponds to its neutral to anionic form, respectively. For this reason it can be electrostatically immobilized on the surface on anion-exchanger resin like Amberlite IRA 401.

Solutions of different concentration of the indicator were used. The time of the adsorption and temperature of the solutions were maintained in all cases. Calibration curves of the obtained sensors based on bromothymol blue attracted to Amberlite IRA 401 ion-exchanger are shown in figure 5.

It was found that the working range of BTB was shifted in comparison to the bulk solution [61], which can be attributed to change in the dissociation constant of the dye during the immobilization procedure. The linear range of the sensor depends on the concentration of the immobilized dye. The narrowest linear range was obtained for the membrane with the largest amount of the indicator, which was also accompanied by the widest dynamic range of the sensor. When the concentration of the indicator decreased the linear range broadened and was shifted to the base range. The dynamic range was also reduced. The figure 5 shows a possibility of the optimization of the sensor by changing the amount of the immobilized indicator. The other way is to change the nature of indicator applied.

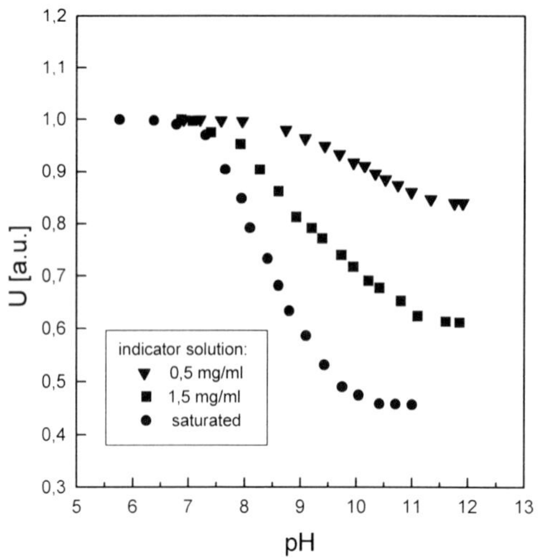


Figure 5. Calibration curves of pH sensors based on Amberlite IRA 401 related to immobilized dye concentration.

Figure 6 shows the response time characteristic of a pH sensor containing the widest dynamic range membrane.

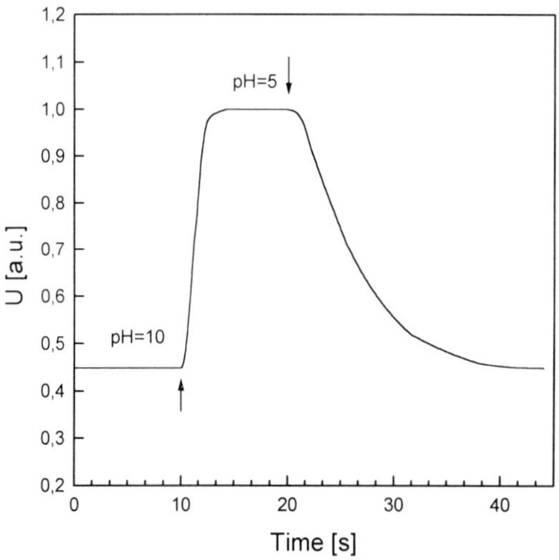


Figure 6. Response time of the pH sensor based on Amberlite IRA 401 (arrows mark points where acid or base were added).

The response of the sensor was rapid (less than 30 s) because the hydrogen ions interacted directly with the indicator molecules immobilized on the surface of the Amberlite beads. There was no additional membrane, which could cause a delay in the response. Typical differences in response time for changes from basic to acidic solutions and vice-versa were observed.

We have chosen another type of Amberlite resin, polymeric adsorbent XAD 1168, as a support in order to investigate its influence on dissociation constant of the dye. An identical, as previously described, procedure of immobilization was applied and a saturated solution of BTB was used. In this case the dye-support binding forces were mainly based on physical sorption interactions. Figure 7 compares calibration curves of the pH sensors based on the membranes prepared from different Amberlite resins.

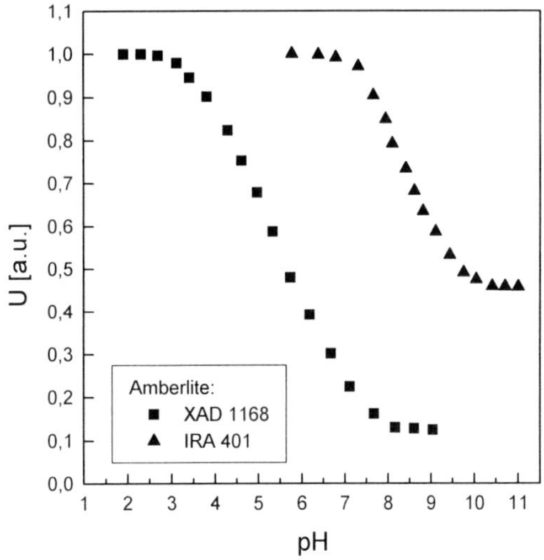


Figure 7. Calibration curves of pH sensors based on different types of Amberlite resins.

The linear range of the pH sensor was shifted in comparison with the previous one due to the different mechanism of the dye immobilization. Although the indicator was adsorbed from identical solution and under comparable conditions (time, temperature) it was found that the sensor based on XAD 1168 exhibited a wider dynamic range. This can be attributed to higher adsorptive properties of this kind of Amberlite, which leads to a greater amount of immobilized reagent.

### 4.1.2. pH sensor based on cellulose membrane

Cellulose acetate was chosen because of its high water permeability, dye-retaining capability and possibility to increase porosity by the hydrolysis of surface. A cellulose acetate membrane (Micro Filtration Systems - Cole Parmer) with porosity of 75% and micropore diameter of 0.8 μm was used in experiments.

Microporous membranes have larger specific surface than plain (non porous) membrane. Assuming that micropores are cylindrical it is possible to calculate a rough approximation of the gain in the active surface. In fact, the total surface is greater than calculated because the walls of the pores are not smooth. The total surface depends on the number of pores per square centimeter, the pores diameter and the thickness of the membrane. It can be expressed as:

$$S_T = S_{PL} - S_H + S_C$$

where: $S_T$ - the total active surface of a membrane,
$S_{PL}$ - the surface of a plain membrane,
$S_H$ - the surface of the holes,
$S_C$ - the lateral surface of cylindrical pores.

It is easy to find that the total active surface of microporous membrane is greater than that of a plain membrane if the thickness of the membrane is greater than pore radius. The calculated ratio $S_T/S_{PL}$ for our membrane was 235. The larger active surface allows to immobilize larger amount of the indicator, which leads to the wider dynamic range of changes in the output signal.

Two pH indicators with amino functional groups (Congo red, neutral red) were chosen for immobilization [62]. Prior to the immobilization procedure the foil was conditioned overnight in the 0.1M NaOH solution in order to hydrolyze its surface and increase its porosity. Then it was immersed for 1 minute into an aqueous-methanolic saturated solution of the indicator. Afterwards the membrane was washed with redistilled water, dried and conditioned alternately in basic and acid solutions. The procedure of soaking with the reagent was repeated several times to increase the amount of the immobilized reagent. Before the measurements the membranes were kept in the redistilled water at least one day.

A strong immobilization of the indicator was observed. It can be attributed to the chemical interactions of functional groups of the dyes with the hydroxy groups of cellulosic support, which govern the chemisorption effect. Calibration curves of two pH optrodes are shown in figure 8.

The linear range of the sensor based on neutral red was within pH=3.8-8.2 and for the optrode with immobilized Congo red at pH=4.1-6.2. The working ranges of these indicators were shifted in comparison to their aqueous solutions [61]. The optrode with neutral red exhibits a wider working range of pH especially suitable for environmental monitoring. On the other hand, the sensor based on Congo red exhibits a wider dynamic range.

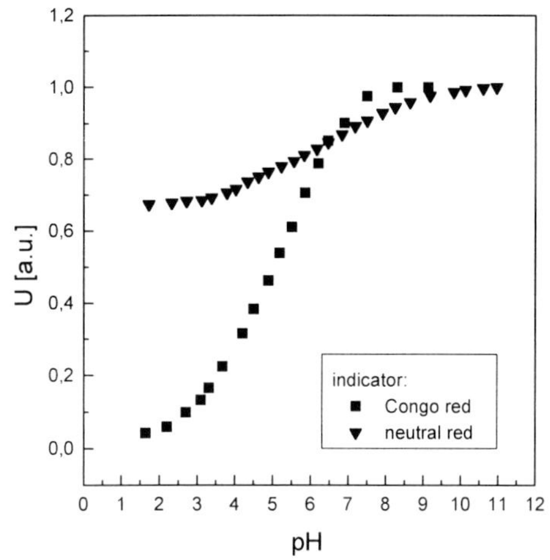


Figure 8. Calibration curves of pH sensors involving different indicators.

The response time of optomembranes made of Congo red and of neutral red was determined by changes of pH from 10 to 2. Similar results were observed in both cases so only the response time of the optrode with immobilized neutral red is shown in figure 9. The membrane was tested 10 times and reproducible steady state signals were achieved in time less than 30 s.

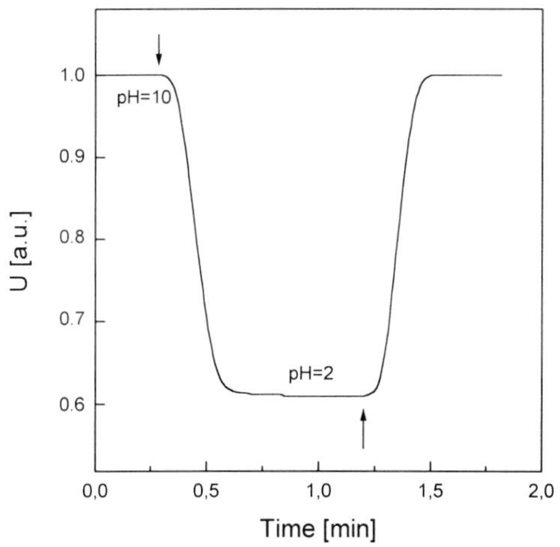


Figure 9. Response time of a pH sensor based on neutral red chemisorbed on cellulose (Reprinted from Sensors and Actuators B, 38-39, 207, 1997, with permission from Elsevier Science, The Boulevard, Langford Lane, Kidlington 0X5 1GB, UK)

A fiber optic pH-meter was set-up in order to study the properties of optomembranes. The results of pH evaluation were compared to the pH values obtained by means of a laboratory pH-meter with a glass electrode. A good correlation between these results was observed. Figure 10 shows the comparison of the two methods of pH determination using the optrode based on neutral red. A good agreement of results was observed i.e. the linear regression line had a standard deviation of 0.07 pH. Similar results were obtained for the fiber optic pH sensor based on chemically immobilized Congo red.

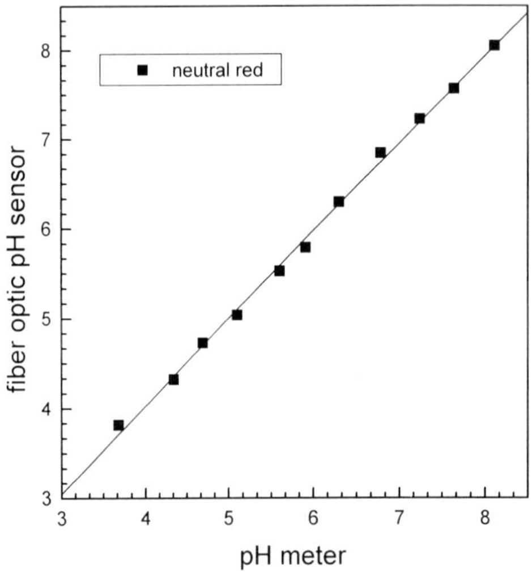


Figure 10. Comparison of the pH measurements carried out by pH-meter and fiber optic pH sensor (Reprinted from Sensors and Actuators B, 38-39, 207, 1997, with permission from Elsevier Science, The Boulevard, Langford Lane, Kidlington 0X5 1GB, UK).

### 4.2. Fiber optic redox titrator

Photometric redox titrations can be carried out in a so-called self-indicator system i.e. the color of a solution changes during the titrations. The analytical wavelength should be matched to the pair of the oxidant and the reductant, which is the main disadvantage of this method. Another method is to use an indicator, the oxidized form of which has a different color than the reduced one. This can be utilized in the fiber optic sensor. One indicator can be used for the titrations with various oxidants and reductants.

As an example of redox reaction we chose the titration of cerium (IV) with tin (II). The titration is based on the following reaction:

$$2Ce^{4+} + Sn^{2+} \leftrightarrow 2Ce^{3+} + Sn^{4+}$$

Standard potential $E^o$ of the redox indicator should lie between the potentials of the ions involved: $E^o_{Ce4+/Ce3+}$=+1.36 V and $E^o_{Sn4+/Sn2+}$=+0.15 V. We used N,N'-diphenylbenzidine (DPB) with the redox potential of +0.76 V. The reduced form of DPB is colorless and the oxidized one is violet. Details on the redox experiments can be found in [63].

#### 4.2.1. Redox titrator based on polymer track membrane

Immobilization procedure

We have used a polymer track membrane (PTM) [64, 65] as a support for reagent immobilization. Such membranes are used for microfiltration in medicine, electronics, biotechnology etc. PTM is made by irradiation of a polymer with high energy ions. The irradiated film is etched and in this way the tracks after passing ion beams are removed. Various polymers can be used as a base film for fabrication of PTM (e.g. polypropylene, polyamide, polycarbonate etc.). The diameter of the pores obtained depends on the type of ions used and on their energy.

A microporous polyethyleneterephthalate (PETP) membrane (30 μm thick, 0.2 μm micropores) was used in the experiments. The membrane was immersed into tetrahydrofurane (THF) and the pressure in the beaker was reduced in order to remove air from the micropores and to allow their filling with the solvent. The membrane was then dipped into a saturated solution of DPB in THF containing different amounts of poly(vinyl chloride) (PVC), (high molecular weight). After 2 hour soaking the membrane was dried at room temperature. Before the use, the membrane was kept overnight in distilled water.

Results

Optimization of time characteristics for PETP membrane

It was checked that PVC itself does not retain the indicator. The indicator was readily washed out from the PVC membrane. We have checked that also sole PETP cannot immobilize the reagent effectively. The reagent was quickly washed out from such a membrane (the changes in the signal obtained from the PETP membrane were similar to the dotted curve presented in figure 11).

From these experiments a conclusion can be drawn that the redox indicator is physically adsorbed on the surface as well as inside pores of the PETP microporous foil but the binding forces are not strong enough. The additional PVC layer should enhance the immobilization efficiency of the dye by its encapsulation. We have investigated the influence of the amount of PVC added to the bath dye solution, which governs the thickness of the covering layer. Figure 11 presents time characteristics for PETP membrane in function of PVC addition.

When Ce(IV) ions were added the reflectance signal decreased because the membrane changed from colorless to violet. Then Sn(II) ions restored the redox potential and the signal increased. The two-step response after addition of Ce(IV) ions occurred for the membrane prepared from the solution containing 10 mg of PVC. The membrane exhibited wide dynamic range but the envelop function of the curve decreased due to the washing out of the indicator. Such an amount of the

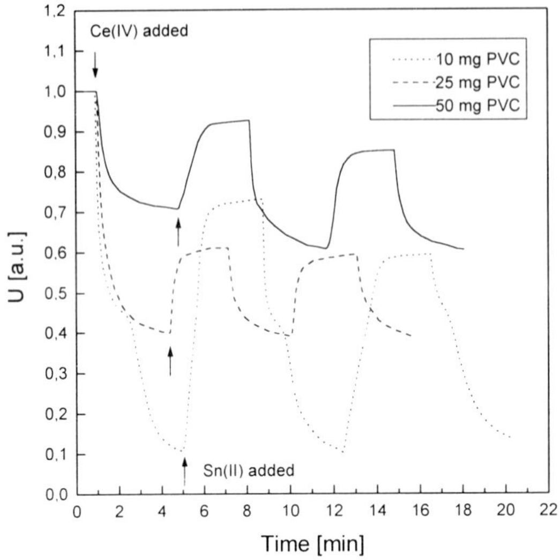


Figure 11. Response time of the redox optomembrane in function of PVC (arrows mark the addition of oxidant and reductant solutions).

PVC is insufficient for the proper encapsulation of the adsorbed dye and the membrane could not be used in redox titrations.

Increase of the amount of PVC up to 25 mg caused the decrease of the relative change in the signal. However, the optrode had stable signal levels in the oxidizing and the reducing environments and the smooth curve was observed. The response time of the optrode is of the order of several minutes. An addition of 50 mg of PVC has no further influence on the shape of the signal. The relative change of reflectance is similar to that of a membrane with 25 mg of PVC but there are different levels of the signal for oxidized and reduced forms of DPB. It can be related to the thicker PVC layer, which leads to the larger reflectance. Moreover, for the membrane with 50 mg of PVC a slight drift is observed. So the optimal membrane is the one with 25 mg of PVC. Such a membrane has good response and can be repeatedly used in titrations.

<u>Titration with solutions of different concentrations</u>

The redox optrode was tested in titrations with solutions of different concentrations. The curves obtained are shown in figure 12.

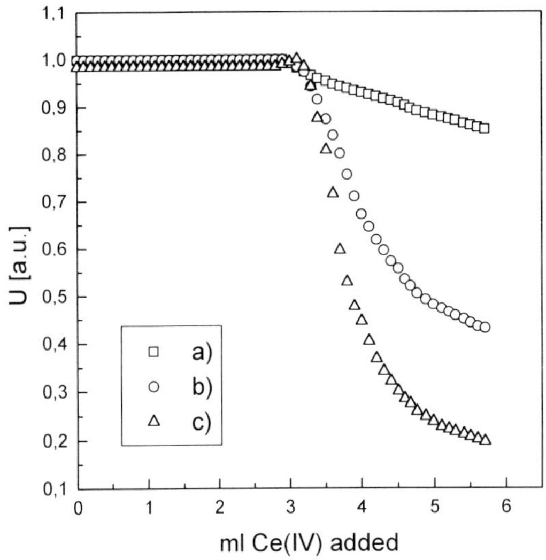


Figure 12. Titration curves for different concentrations of the solutions: a) 0.0006 mol/dm$^3$ $SnCl_2$ with 0.001 mol/dm$^3$ $Ce(SO_4)_2$, b) 0.006 mol/dm$^3$ $SnCl_2$ with 0.01 mol/dm$^3$ $Ce(SO_4)_2$ and c) 0.06 mol/dm$^3$ $SnCl_2$ with 0.1 mol/dm$^3$ $Ce(SO_4)_2$ (Reprinted from Sensors and Actuators B, 29, 374, 1995, with permission from Elsevier Science, The Boulevard, Langford Lane, Kidlington 0X5 1GB, UK).

The same optrode was used for the titrations. The titrations were carried out from low to high concentration of the solution. Between the measurements the redox potential was restored by means of Sn(II) ions. The shape of the titration curves changed in dependence on concentration. The end-point of the titration with 0.001 mol/dm$^3$ solution of Ce(IV) cannot be precisely determined since the signal decreases almost linearly. Moreover, a small relative change was observed. The relative changes were greater for solutions of the Ce(IV) concentrations: 0.01 mol/dm$^3$ and 0.1 mol/dm$^3$, and the end-point of the titrations can be determined.

### 4.2.2. Redox titrator based on cellulose acetate

Immobilization procedure

We have prepared optomembranes containing diphenylamine (DPA) as a redox indicator, which is irreversibly oxidized in the membrane to the DPB:

$$2DPA \rightarrow DPB + 2H^+ + 2e^-$$

The immobilization procedure was similar to that one applied for pH sensitive optomembranes. The membrane was conditioned alternately in the reductant and the oxidant solutions. Then the procedure of soaking with the reagent was repeated several times to increase the amount of the reagent immobilized. Before the measurements the membranes were kept in redistilled water at least for one day.

Results

The optrode exhibits relatively great changes in the measured signal due to the large amount of the indicator immobilized. In addition, high porosity of optomembranes reduces the barrier of mass transport, thus enhancing its response time presented in figure 13. Unfortunately, the envelop function of the signal decreases during the experiments because the non immobilized indicator is washed out from the pores. After a few changes in the redox potential the signal reaches reproducible values, but it is not shown in the figure.

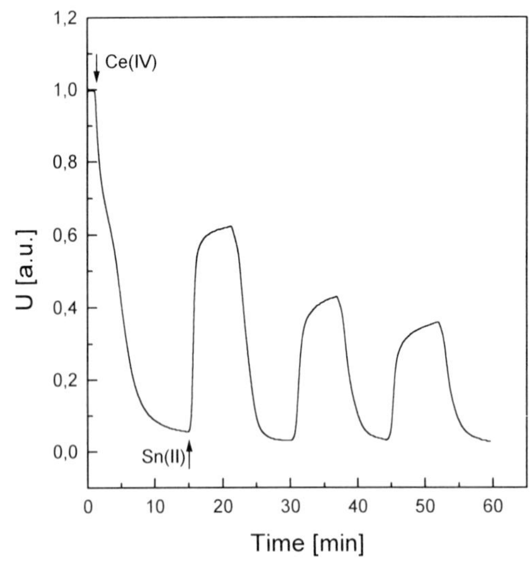


Figure 13. Response time of the redox optomembrane based on cellulose membrane containing DPA (arrows mark the addition of oxidant and reductant solutions) (Reprinted from Sensors and Actuators B, 38-39, 207, 1997, with permission from Elsevier Science, The Boulevard, Langford Lane, Kidlington 0X5 1GB, UK).

### 4.3. Fiber optic calcium ion sensor

Optomembranes for $Ca^{2+}$ measurements can be based on absorption or fluorescent reagents. Absorption indicators form stable complexes with the calcium ions at basic pH values. This obviously requires a preparation of the sample, which is not a common practice in measurements by sensors. On the other hand, fluorescent optical sensors utilize more complex measuring system.

Our task was to monitor the hardness of drinking water without any modification of the sample. We used a $Ca^{2+}$ - sensitive chromoionophore Chlorophosphonazo III which can work in neutral pH [66].

Immobilization procedure

The $Ca^{2+}$-sensitive chromoionophore Chlorophosphonazo III was physically immobilized on the PTFE polymer track membrane with micropore diameter of 0.2 μm. The PTFE foil was dipped into satured methanolic solution of the reagent and after a few seconds of soaking the foil was washed with distilled water to remove this part of the reagent, which has not been immobilized in the sensing layer. It was found that to immobilize a greater amount of the indicator the foil should be conditioned in an acid solution before treatment with the reagent.

Results

The calcium sensor was tested on a laboratory prepared sample containing calcium ions. An EDTA solution was used to achieve a base signal of the optrode before the interaction with the analyte. This regenerating step was necessary for the proper work of the sensor and was caused by a very slow process of decomplexation of $Ca^{2+}$ ions. Figure 14 shows the changes in the signal of the sensor. The concentration of calcium cations in the prepared sample solution was similar to the typical hardness of drinking water. The steady state of the signal after the addition of calcium ions was achieved after approx. 5 min. This relatively slow response time was due to the kinetic properties of the reaction between chromoionophore and the analyte, and the mass transport of the hydrophilic cations to the lipophilic membrane phase.

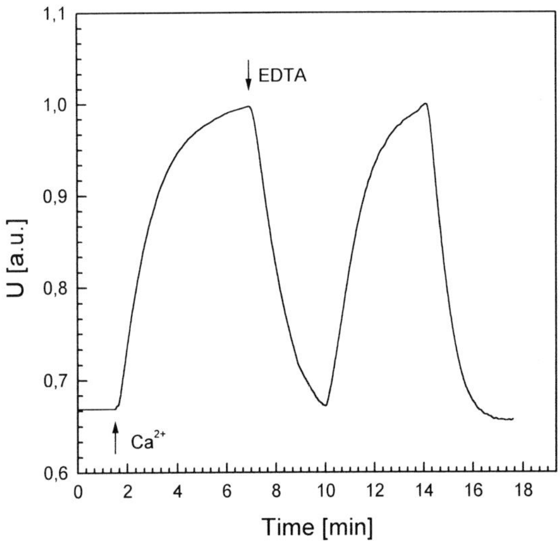


Figure 14. Time response of the calcium sensor (arrows mark the appropriate addition of calcium salt and EDTA 0.01 mol/dm$^3$ solution) (Reprinted from [66], with permission from SPIE).

The calibration curve of the calcium sensor is presented in figure 15. The lower limit of detection was found at about $10^{-4}$ mol/dm$^3$. The membrane exhibited linear dependence of the signal in function of the calcium ions concentration. However, at high concentrations (about $10^{-2}$ mol/dm$^3$) the indicator started to be washed out from the membrane because of weak adsorption forces.

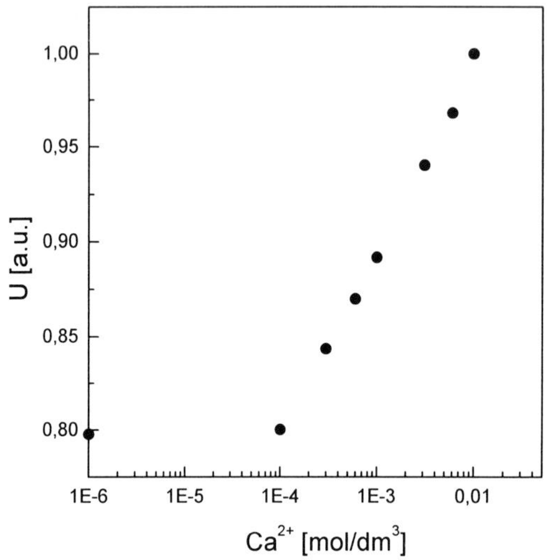


Figure 15. Calibration curve of the calcium sensitive fiber optic sensor.

## 5. CONCLUSIONS

Successful development of fiber optic chemical sensors requires co-operation of many specialists in various fields of science. Scientists in analytical chemistry, polymer science, material science, optoelectronics and electronics etc. can be involved in this multidisciplinary task. In dependence on the application of the sensor biologists, medical doctors or environmentalists can be incorporated to the working group. Although, the contribution of all specialists cannot be classified by the importance, analytical chemistry and material science seem to be the key to the success.

Appropriate matching of the indicator to the analyte and the design of its durable and efficient immobilization procedure governs the performances of the sensor. Adsorption is one of the technique, which can be utilized for immobilization of the reagent. The chapter described a few possibilities of preparation of the optomembrane for FOCS. We have investigated immobilization properties of Amberlite resins, microporous membranes made of

PETP, cellulose acetate and Teflon. Durable and reproducible immobilization procedures have been elaborated. The membranes prepared have been used over months and no significant leaching out of the reagent was observed. Quite promising materials are microfiltration membranes (e.g. cellulose acetate), which are commercially available. They have large active surface for reagent immobilization and the response time of the sensor is short due to the easy access of the analyte to the reagent molecules.

Adsorption immobilization of the reagent in chemooptical interface was utilized in fiber optic chemical sensors i.e. pH sensor, redox titrator and calcium sensitive optrode. The choice of the indicator as well as the support and the immobilization procedure governs the working and dynamic ranges of the pH sensor. The optrode based on optimized redox optomembrane (consisting of PETP additionally covered with PVC) may be used in various redox titrations e.g. Fe(II) with Mn(VII) or As(III) with Cr(VI). Hardness of water can be measured by the fiber optic calcium sensor directly in the studied medium without any sample preparation.

The sensors described in this chapter concern ion detection in liquid samples. They can be used in analytical, industrial, biomedical and environmental application. Based on designed chemooptical interfaces another sensors can be constructed e.g. for ammonia or carbon dioxide involving the pH transduction. Combining individual sensors a multiparameter probe can be built up for continuous monitoring of chemical quantities.

## REFERENCES


1. W.R. Seitz, Crit. Rev. in Anal. Chem., 19 (1988) 135.
2. O.S. Wolfbeis (ed.), Fiber optic chemical sensors and biosensors, Vol. 1 and 2, CRC, Boca Raton, FL, 1991.
3. D. Wise and L.B.Wingard, Biosensors with fiberoptics, Humana Press, Clifton, New Jersey, 1991.
4. G. Boisde and A. Harmer, Chemical and biochemical sensing with optical fibers and waveguides, Artech House, Boston - London, 1996.
5. C.A. Browne, D.H. Tarrant, M.S. Olteanu, J.W. Mullens and E.L. Chronister, Anal. Chem., 68 (1996) 2289.
6. R. Narayanaswamy and F. Sevilla, J. Phys. E.: Sci. Instrum., 21 (1988) 10.
7. K. Seiler and W. Simon, Sensors and Actuat. B, 6 (1992) 195.
8. Z. Zhujun, J.L. Mullin and W.R. Seitz, Anal. Chim. Acta, 184 (1986) 251.
9. T.L. Blair, S-T. Yang, T. Smith-Palmer and L.G. Bachas, Anal. Chem., 66 (1994) 300.
10. X. Zhou and M.A. Arnold, Anal. Chem., 68 (1996) 1748.
11. T.J. Cardwell, R.W. Cattrall, L.W. Deady, M. Dorkos and G.R. O'Connell, Talanta, 40 (1993) 765.

12. X-M Chen, K. Itoh, M. Murabayashi and C. Igarashi, Chem. Letters, 103 (1996).
13. K. Seiler, Ion-selective optode membranes, Fluka Chemie AG, Buchs, 1993.
14. C. Preininger, G.J. Mohr, I. Klimant and O.S. Wolfbeis, Anal. Chim. Acta, 334 (1996) 113.
15. W. Sellien, R. Czolk, J. Reichert and H.J. Ache, Anal. Chim. Acta, 269 (1992) 83.
16. A. Song, S. Parus and R. Kopelman, Anal. Chem., 69 (1997) 863.
17. J.E. Madden, T.J. Cardwell, R.W. Cattrall and L.W. Deady, Anal. Chim. Acta, 319 (1996) 129.
18. F.V. Bright, G.E. Poirier and G.M. Hieftje, Talanta, 35 (1988) 113.
19. R. Narayanaswamy, D.A. Russell and F. Sevilla, Talanta, 35 (1988) 83.
20. M.E. Diaz-Garcia, F. Alava-Moreno and A. Sand-Medel, Mikrochim. Acta, 113 (1994) 211.
21. A.K. McEvoy, C.M. McDonagh and B.D. MacCraith, Analyst, 121 (1996) 785.
22. I. Klimant and M. Otto, Mikrochim. Acta., 108 (1992) 11.
23. Z. Zhujun and W.R. Seitz, Anal. Chim. Acta, 171 (1985) 251.
24. L.K. Chau and M.D.Porter, Anal. Chem., 62 (1990) 1964.
25. L.A. Saari and W.R. Seitz, Anal. Chem., 56 (1984) 810.
26. E. Urbano, H. Offenbacher and O.S. Wolfbeis, Anal. Chem., 56 (1984) 427.
27. H. Offenbacher, O.S. Wolfbeis and E. Furlinger, Sensors and Actuators B, 9 (1986) 73.
28. M.R. Weaver and J.M. Harris, Anal. Chem., 61 (1989) 1001.
29. M. Ayadim, J.L.H. Jiwan, A.P. Desilva and J.P. Soumillion, Tetrahedron Lett., 37 (1996) 7039.
30. T.L. Blair, T. Cynkowski and L.G. Bachas, Anal. Chem., 65 (1993) 945.
31. C. Munkolm, D.R. Walt and F.P. Milanovich, Talanta, 35 (1988) 109.
32. B.G. Healey and D.R. Walt, Anal. Chem., 67 (1995) 4471.
33. R. Czolk, J. Reichert and H.J. Ache, Sensors and Actuators B, 7 (1992) 540.
34. M. Shortreed, R. Kopelman, M. Kuhn and B. Hoyland, Anal. Chem., 68 (1996) 1414.
35. S. Roosli, E. Pretsch, W.E. Morf, E. Tsuchida and H. Nishide, Anal. Chim. Acta, 338 (1997) 119.
36. S-T. Yang and L.G. Bachas, Talanta, 41 (1994) 963.
37. O.S. Wolfbeis and B.P.H. Schaffar, Anal. Chim. Acta, 198 (1987) 1.
38. I.D. Rattee and M.M. Breuer, The Physical Chemistry of Dye Adsorption, Academic Press, London and New York, 1974.
39. D.C. Ashworth, H.P. Huang and R. Narayanaswamy, Anal. Chim. Acta, 21 (1988) 251.
40. M. Bacci, F. Baldini, F. Cosi, G. Conforti and A.M. Scheggi, Proc. SPIE, Vol. 1014 (1988) 73.
41. F. Alava-Moreno, R. Pereiro-Garcia, M.E. Diaz-Garcia and A. Sand-Medel, Sensors and Actuators B, 11 (1993) 413.

42. W.A. Wyatt, G.E. Poirier, F.V. Bright and G.M. Hieftje, Anal. Chem., 59 (1987) 572.
43. R. Narayanaswamy and F. Sevilla, Analyst, 111 (1986) 1085.
44. W.A. Wyatt, F.V. Bright and G.M. Hieftje, Anal. Chem., 59 (1987) 2272.
45. T.P. Jones and M.D. Porter, Anal. Chem., 60 (1988) 404.
46. B.P.H. Schaffar and O.S. Wolfbeis, Mikrochim. Acta, III (1989) 109.
47. Z. Gong and Z. Zhang, Anal. Chim. Acta, 325 (1996) 201.
48. F. Lazaro, M.D. Luque de Castro and M. Valcarcel, Anal. Chim. Acta, 214 (1988) 217.
49. M. Bacci, F. Baldini and A.M. Scheggi, Anal. Chim. Acta, 207 (1988) 343.
50. Y. Kostov, S. Tzonkov, L. Yotova and M. Krysteva, Anal. Chim. Acta, 280 (1993) 15.
51. A. Safavi and M. Pakniat, Anal. Chim. Acta, 335 (1996) 227.
52. T. Werner and O.S. Wolfbeis, Fres. J. Anal. Chem., 346 (1993) 564.
53. G.J. Mohr and O.S. Wolfbeis, Anal. Chim. Acta, 292 (1994) 41.
54. K.S. Bronk and D.R.Walt, Anal. Chem., 66 (1994) 3519.
55. H. Hisamoto, M. Tsubuku, T. Enomoto, K. Watanabe, H. Kawaguchi, Y. Koike and K. Suzuki, Anal. Chem., 68 (1996) 3871.
56. O.S. Wolfbeis, N.V. Rodriguez and T. Werner, Mikrochim. Acta, 108 (1992) 133.
57. L.A. Saari and W.R. Seitz, Anal. Chem., 55 (1983) 667.
58. J.I. Peterson, S.R. Goldstein and R.V. Fitzgerald, Anal. Chem., 52 (1980) 864.
59. A.M. Scheggi and F. Baldini, Optica Acta, 33 (1986) 1587.
60. F. Baldini, S. Bracci and F. Cosi, Sensors and Actuators B, 37-38 (1993) 180.
61. E. Bishop, Indicators, Pergamon Press, Oxford, 1972.
62. A. Dybko, W. Wróblewski, J. Maciejewski, R. Romaniuk and Z. Brzózka, Sensors and Actuators B, 38-39 (1997) 207.
63. A. Dybko, J. Maciejewski, Z. Brzózka, W. Wróblewski and R. Romaniuk, Sensors and Actuators B, 29 (1995) 374.
64. A. Dybko, W. Wróblewski, J. Maciejewski, R. Romaniuk, Z. Brzózka and J. Kiełkiewicz, J. Applied Polym. Sci., 59 (1996) 719.
65. A. Dybko, W. Wróblewski, J. Maciejewski, Z. Brzózka and R. Romaniuk, Proc. SPIE, Vol. 2508 (1995) 351.
66. A. Dybko, W. Wróblewski, J. Maciejewski, R. Romaniuk and Z. Brzózka, Proc. SPIE, Vol. 3105 (1997) 361.

Adsorption and its Applications in Industry and Environmental Protection
Studies in Surface Science and Catalysis, Vol. 120
A. Dabrowski (Editor)



# Adsorption principles, design data and adsorbent materials for industrial applications: A Bibliography (1967-1997)

M. S. Ray

School of Chemical Engineering, Curtin University of Technology,
GPO Box U1987, Perth 6845, WESTERN AUSTRALIA

**Abstract** This chapter provides a bibliographic listing of published journal papers from 1967 to 1997 concerned with adsorbent materials, adsorption principles (theory and models), and design data which are used in industrial situations. The bibliography provides a quick and easy, but comprehensive, reference source. The references are taken from the fifty most important chemical engineering journals, but do not include papers from the chemistry journals, or books, or conference/symposium series. A listing of the journals surveyed is included at the end of this chapter. The references are arranged chronologically (and then alphabetically by first author surname) within the following subject groups:

*Fundamental Principles of Adsorption, Theory and Models*
*Adsorption Design Methods and Data*
*Adsorbent Materials*

The following topics, having an emphasis towards environmental protection, are included in a separate bibliography in Volume 2: *PSA and Cyclic Systems, and Applications; Liquid-Phase Adsorption; Ion Exchange, Chromatography, and Related Separations.*

A general bibliography of the chemical engineering journal literature from 1967-1988 has been published by the author [1], and can provide access to a wider range of topics. An earlier bibliography [2] provides access to the literature prior to 1967. A complete bibliographic listing of the chemical engineering journal literature from 1989 to 1997 (with subsequent six-monthly updates) is available on a CD-ROM database [3].

**Keywords:** Adsorption theories; isotherm data; adsorbents; zeolites; activated carbons; industrial applications.

References

1. M.S.Ray, Chemical Engineering Bibliography (1967-1988), Noyes Publications, New Jersey, USA (1990).
2. K.Bourton, Chemical and Process Engineering Unit Operations: A Bibliographical Guide; MacDonald and Co., London (1967).
3. Engineering and Applied Science Database on CD-ROM (includes CHERUB chemical engineering database, 1989-1997), published by Royal Melbourne Institute of Technology, Australia. Full details are available from the author.

## FUNDAMENTAL PRINCIPLES OF ADSORPTION, THEORY AND MODELS


### 1967

Hoory, S.E., and Prausnitz, J.M., Monolayer adsorption of gas mixtures on homogeneous and heterogeneous solids, Chem. Eng. Sci., 22(7), 1025-1034 (1967).

Lee, J.H., Statistical mechanical theory of adsorption life, Int. Chem. Eng., 7(1), 63-72 (1967).

Meyer, O.A, and Weber, T.W., Nonisothermal adsorption in fixed beds, AIChE J., 13(3), 457-465 (1967).

Niac, C., Kinetic method for determination of adsorption isotherms, Int. Chem. Eng., 7(4), 666-672 (1967).

Pan, C.Y., and Basmadjian, D., Constant-pattern adiabatic fixed-bed adsorption, Chem. Eng. Sci., 22(3), 285-298 (1967).

Satterfield, C.N., and Frabetti, A.J., Sorption and diffusion of gaseous hydrocarbons in synthetic mordenite, AIChE J., 13(4), 731-738 (1967).

Stuart, F.X., and Camp, D.T., Comparison of kinetic and diffusional models for packed bed adsorption, Ind. Eng. Chem. Fund., 6(1), 156-158 (1967).

### 1968

Carter, J.W., Developments in adsorption processes, Brit. Chem. Eng., 13(2), 229-234 (1968).

Carter, J.W., Isothermal and adiabatic adsorption in fixed beds, Trans. IChemE, 46, T213-T222 (1968).

Payne, H.K.; Sturdevant, G.A., and Leland, T.W., Improved two-dimensional equation of state to predict adsorption of pure and mixed hydrocarbons, Ind. Eng. Chem. Fund., 7(3), 363-374 (1968).

Punwani, D.; Chi, C.W., and Wasan, D.T., Dynamic sorption by hygroscopic salts, Ind. Eng. Chem. Process Des. Dev., 7(3), 410-415 (1968).

Schneider, P., and Smith, J.M., Adsorption rate constants from chromatography, AIChE J., 14(5), 762-771 (1968).

### 1969

Cerro, R.L., and Smith, J.M., Effects of heat release and nonlinear equilibrium on transient adsorption, Ind. Eng. Chem. Fund., 8(4), 796-802 (1969).

Eberly, P.E., Diffusion studies in zeolites and related solids by gas chromatography, Ind. Eng. Chem. Fund., 8(1), 25-30 (1969).

Eteson, D.C., and Zwiebel, I., Hybrid computer solution of the simple fixed bed adsorption model, AIChE J., 15(1), 124-126 (1969).

Kurbanaliev, T.G.; Shabataev, S.A., and Rasulov, A.M., Determination of activity of adsorbent in short-cycle adsorption systems, Int. Chem. Eng., 9(3), 451-453 (1969).

Pigford, R.L.; Baker, B., and Blum, D.E., An equilibrium theory of the parametric pump, Ind. Eng. Chem. Fund., 8(1), 144-149 (1969).

Van Ness, H.C., Adsorption of gases on solids, Ind. Eng. Chem. Fund., 8(3), 464-473 (1969).

### 1970

Cooper, R.S., and Liberman, D.A., Fixed-bed adsorption kinetics with pore diffusion control, Ind. Eng. Chem. Fund., 9(4), 620-623 (1970).

Gonzalez, A.J., and Holland, C.D., Adsorption of multicomponent mixtures by solid adsorbents, AIChE J., 16(5), 718-724 (1970).

Nemeth, E.J., and Stuart, E.B., Pore-diffusion mechanisms during vapor-phase adsorption, AIChE J., 16(6), 999-1004 (1970).

Pan, C.Y., and Basmadjian, D., Analysis of adiabatic sorption of single solutes in fixed beds: Pure thermal wave formation and its practical implications, Chem. Eng. Sci., 25(11), 1653-1664 (1970).

### 1971

Clough, P.S.; Dollimore, D., and Nicklin, T., Comparison of isothermal adsorption data by technique of absolute adsorption, J. Applied Chem. Biotechnol., 21, 137-138 (1971).

Forrester, S.D., and Giles, C.H., Gas-solid adsorption isotherms: A historical survey to 1918, Chem. Ind. (London), 24 July, 831-839 (1971).

Gupta, R., and Sweed, N.H., Equilibrium theory of cycling-zone adsorption, Ind. Eng. Chem. Fund., 10(2), 280-283 (1971).

Pan, C.Y., and Basmadjian, D., Equilibrium theory analysis of adiabatic sorption of single solutes in fixed beds, Chem. Eng. Sci., 26(1), 45-58 (1971).

Safanov, M.S., Criterion for formation of steady-state sorption front, Sep. Sci., 6(1), 35-42 (1971).

Yoon, S.M., and Kunii, D., Physical adsorption in a moving bed of fine adsorbents, Ind. Eng. Chem. Process Des. Dev., 10(1), 64-70 (1971).

**1972**

Brunet, J., Three-dimensional treatment of adsorption, Chem. Eng. Sci., 27(4), 685-694 (1972).

Cooney, D.O., and Strusi, F.P., Analytical description of fixed-bed sorption of two Langmuir solutes under nonequilibrium conditions, Ind. Eng. Chem. Fund., 11(1), 123-126 (1972).

Garg, D.R., and Ruthven, D.M., Effect of concentration dependence of diffusivity on zeolitic sorption curves, Chem. Eng. Sci., 27(2), 417-424 (1972).

Rhee, H.K.; Heerdt, E.D., and Amundson, N.R., Analysis of adiabatic adsorption columns, Chem. Eng. J., 1(2), 241; 1(3), 279 (1970); 3(1), 22-34; 3(2), 121-135 (1972).

**1973**

Blaisdell, C.T., and Kammermeyer, K., Countercurrent and cocurrent gas separation, Chem. Eng. Sci., 28(6), 1249-1256 (1973).

Chen, H.T.; Reiss, E.H.; Stokes, J.D., and Hill, F.B., Separations via semicontinuous parametric pumping, AIChE J., 19(3), 589-595 (1973).

Danner, R.P., and Wenzel, L.A., Gas-mixture adsorption on molecular sieves, AIChE J., 19(4), 870 (1973).

Fleck, R.D.; Kirwan, D.J., and Hall, K.R., Mixed-resistance diffusion kinetics in fixed-bed adsorption under constant pattern conditions, Ind. Eng. Chem. Fund., 12(1), 95-99 (1973).

Furusawa, T., and Smith, J.M., Diffusivities from dynamic adsorption data, AIChE J., 19(2), 401-403 (1973).

Furusawa, T., and Smith, J.M., Dynamics of packed-bed adsorbers using the cell model, Ind. Eng. Chem. Fund., 12(3), 388-390 (1973).

Garg, D.R., and Ruthven, D.M., Fixed-bed sorption behavior of gases with nonlinear equilibria, AIChE J., 19(4), 852-853 (1973).

Hashimoto, N., and Smith, J.M., Macropore diffusion in molecular sieve pellets by chromatography, Ind. Eng. Chem. Fund., 12(3), 353-359 (1973).

Hori, Y., and Kobayashi, R., Thermodynamic properties of adsorbate for high-pressure multilayer adsorption, Ind. Eng. Chem. Fund., 12(1), 26-30 (1973).

Kocirik, M.; Zikanova, A., and Dubsky, J., Numerical solution of the adsorption kinetics with a nonlinear isotherm, Ind. Eng. Chem. Fund., 12(4), 440-443 (1973).

Kyte, W.S., Freundlich isotherm for nonlinear adsorption in fixed beds, Chem. Eng. Sci., 28(10), 1853-1856 (1973).

Myers, A.L., Adsorption of gas mixtures on molecular sieves, AIChE J., 19(3), 666-667 (1973).

Ruthven, D.M.; Loughlin, K.F., and Holborow, K.A., Multicomponent sorption equilibrium in molecular sieve zeolites, Chem. Eng. Sci., 28(3), 701-710 (1973).

Sircar, S., and Myers, A.L., Surface potential theory of multilayer adsorption from gas mixtures, Chem. Eng. Sci., 28(2), 489-500 (1973).

**1974**

Bowen, J.H., Gas-solid, non-catalytic reaction models, Trans. IChemE, 52, T282-T284 (1974).

Cooney, D.O., Numerical investigation of adiabatic fixed-bed adsorption, Ind. Eng. Chem. Process Des. Dev., 13(4), 368-373 (1974).

Gilliland, E.R.; Baddour, R.F.; Perkinson, G.P., and Sladek, K.J., Diffusion on surfaces, Ind. Eng. Chem. Fund., 13(2), 95-105 (1974).

Gregory, R.A., Comparison of parametric pumping with conventional adsorption, AIChE J., 20(2), 294-300 (1974).

Kocirik, M., and Zikanova, A., Analysis of adsorption kinetics in materials with polydisperse pore structure, Ind. Eng. Chem. Fund., 13(4), 347-350 (1974).

Weber, T.W., and Chakravorti, R.K., Pore and solid diffusion models for fixed-bed adsorbers, AIChE J., 20(2), 228-238 (1974).

Zwiebel, I.; Gariepy, R.L., and Schnitzer, J.J., Fixed-bed desorption behavior of gases with nonlinear systems, AIChE J., 18(6), 1139-1147 (1972); 20(5), 915-923 (1974).

**1975**

Basmadjian, D.; Ha, K.D., and Pan, C.Y., Nonisothermal desorption by gas purge of single solutes in fixed-bed adsorbers, Ind. Eng. Chem. Process Des. Dev., 14(3), 328-347 (1975).

Choi, P.S.K.; Fab, L.T., and Hsu, H.H., Modeling and simulation of adiabatic adsorber, Sep. Sci., 10(6), 701-722 (1975).

Garg, D.R., and Ruthven, D.M., Linear driving force approximations for diffusion controlled adsorption in molecular sieve columns, AIChE J., 21(1), 200-202 (1975).

Greco, G.; Iorio, G.; Tola, G., and Waldram, S.P., Unsteady-state diffusion in porous solids, Trans. IChemE, 53, T55-T58 (1975).

Karger, J., and Bulow, M., Theoretical prediction of uptake behaviour in adsorption kinetics of binary gas mixtures using irreversible thermodynamics, Chem. Eng. Sci., 30(8), 893-896 (1975).

Lunde, P.J., and Kester, F.L., Chemical and physical gas adsorption in finite multimolecular layers, Chem. Eng. Sci., 30(12), 1497-1506 (1975).

**1976**

Chihara, K.; Suzuki, M., and Kawazoe, K., Effect of heat generation on measurement of adsorption rate by gravimetric method, Chem. Eng. Sci., 31(6), 505-507 (1976).

England, R., and Thomas, W.J., The significance and measurement of surface diffusion coefficients in catalysis, Trans. IChemE, 54, T115-T118 (1976).

Ferrell, J.K.; Rousseau, R.W., and Branscome, M.R., Development and testing of mathematical model for complex adsorption beds, Ind. Eng. Chem. Process Des. Dev., 15(1), 114-122 (1976).

Gidaspow, D.; Dharia, D., and Leung, L., Gas purification by porous solids with structural changes, Chem. Eng. Sci., 31(5), 337-344 (1976).

Hsu, C.C.; Rudzinski, W., and Wojciechowski, B.W., Three-dimensional mobile adsorption: Evaluation of density distribution in adsorbed phases, Chem. Eng. Sci., 31(12), 1123-1130 (1976).

Nagy, L.G., Study of porous adsorbents by isotopic molecular exchange method, Period. Polytech. Chem. Eng., 20, 25-36 (1976).

Ruthven, D.M., Sorption and diffusion in molecular sieve zeolites, Sep. Purif. Methods, 5(2), 189-246 (1976).

Szirmay, L., Relative diffusivities from breakthrough curves through exchange adsorption, Sep. Sci., 11(2), 159-170 (1976).

Tien, C.; Hsieh, J.S.C, and Turian, R.M., Application of h-transformation for solution of multicomponent adsorption in fixed bed, AIChE J., 22(3), 498-505 (1976).

Waksmundzki, A., et al., Application of gas-adsorption chromatography data to investigation of adsorptive properties of adsorbents, Sep. Sci., 11(1), 29-38; 11(4), 411-416 (1976).

**1977**

Jaroniec, M.; Borowko, M., and Rudzinski, W., Gonzalez-Holland model for adsorption of gas mixtures, AIChE J., 23(4), 605-607 (1977).

Klaus, R.; Aiken, R.C., and Rippin, D.W.T., Simulated binary isothermal adsorption on activated carbon in periodic countercurrent column operation, AIChE J., 23(4), 579-586 (1977).

Ma, Y.H., and Lee, T.Y., Diffusion of binary gas mixtures in zeolite-X pellets, Ind. Eng. Chem. Fund., 16(1), 44-48 (1977).

Ozil, P., and Bonnetain, L., Dynamical adsorption in fixed bed, Chem. Eng. Sci., 32(3), 303-310 (1977).

Shah, D.B., and Ruthven, D.M., Measurement of zeolitic diffusivities and equilibrium isotherms by chromatography, AIChE J., 23(6), 804-809 (1977).

Various, Adsorption and ion exchange (topic issue), Chem. Eng. Prog., 73(10), 44-64 (1977).

von Rosenberg, D.U.; Chambers, R.P., and Swan, G.A., Numerical solution of surface-controlled fixed-bed adsorption, Ind. Eng. Chem. Fund., 16(1), 154-157 (1977).

**1978**

Carleton, F.B.; Kershenbaum, L.S., and Wakeham, W.A., Adsorption in non-isobaric fixed beds, Chem. Eng. Sci., 33(9), 1239-1246 (1978).

Chihara, K.; Suzuki, M., and Kawazoe, K., Adsorption rate on molecular sieving carbon by chromatography, AIChE J., 24(2), 237-246 (1978).

Chung, I.J., and Hsu, H.W., Analysis of packed bed adsorption, Chem. Eng. Sci., 33(3), 399-403 (1978).

DiGiano, F.A.; Baldauf, G.; Frick, B., and Sontheimer, H., Simplified competitive equilibrium adsorption model, Chem. Eng. Sci., 33(12), 1667-1674 (1978).

Dyer, A., Separation of closely related systems by molecular sieve zeolites, Sep. Sci. Technol., 13(6), 501-516 (1978).

Glandt, E.D.; Myers, A.L., and Fitts, D.D., Physical adsorption of gases on graphetized carbon black, Chem. Eng. Sci., 33(12), 1659-1666 (1978).

Liapis, A.I., and Rippin, D.W.T., Simulation of binary adsorption in activated carbon columns, Chem. Eng. Sci., 33(5), 593-600 (1978).

Ozil, P., and Bonnetain, L., Theoretical prediction of temperature profile in adsorbent fixed-bed, Chem. Eng. Sci., 33(9), 1233-1238 (1978).

Razavi, M.S.; McCoy, B.J., and Carbonell, R.G., Moment theory of breakthrough curves for fixed-bed adsorbers and reactors, Chem. Eng. J., 16(3), 211-222 (1978).

Szirmay, L., Dynamic behaviour and relative mass transfer coefficients of a porous adsorbent through exchange adsorption, Trans. IChemE, 56, T101-T106 (1978).

Weber, T.W., Batch adsorption for pore diffusion with film resistance and irreversible isotherm, Can. J. Chem. Eng., 56, 187-197 (1978).

Wiedemann, K.; Roethe, A.; Radeke, K.H., and Gelbin, D., Modelling of adsorption-desorption breakthrough curves using statistical moments, Chem. Eng. J., 16(1), 19-26 (1978).

**1979**

Gelbin, D., and Bunke, G., Equilibrium adsorption dynamics in the cyclic steady state with concave isotherms, Chem. Eng. J., 17(3), 191-200 (1979).

Karger, K.J., and Zikanova, A., Influence of heat generated on adsorption in bidisperse adsorbents, J. Chem. Technol. Biotechnol., 29, 339-345 (1979).

Kiselev, A.V., Molecular statistical study of thermodynamic characteristics of hydrocarbon adsorption on zeolites, J. Chem. Technol. Biotechnol., 29, 673-685 (1979).

Lee, L.K.; Yucel, H., and Ruthven, D.M., Kinetics of adsorption in bi-porous molecular sieves, Can. J. Chem. Eng., 57, 65-80 (1979).

Liapis, A.I., and Rippin, D.W.T., Simulation of binary adsorption in continuous countercurrent operation and comparison with other operating modes, AIChE J., 25(3), 455-460 (1979).

Liaw, C.H., et al., New solution to kinetics of fixed-bed adsorption, AIChE J., 25(2), 376-381 (1979).

**1980**

Basmadjian, D., Rapid procedures for prediction of fixed-bed adsorber behavior, Ind. Eng. Chem. Process Des. Dev., 19(1), 129-144 (1980).

Danner, R.P.; Nicoletti, M.P., and Al-Ameeri, R.S., Determination of gas mixture adsorption equilibria by the tracer-pulse technique, Chem. Eng. Sci., 35(10), 2129-2134 (1980).

Gelbin, D., and Fiedler, K., Concentration dependence of diffusion coefficients in zeolites, AIChE J., 26(3), 510-513 (1980).

Harwell, J.H.; Liapis, A.I.; Litchfield, R., and Hanson, D.T., Non-equilibrium model for fixed-bed multicomponent adiabatic adsorption, Chem. Eng. Sci., 35(11), 2287-2296 (1980).

Hayhurst, D.T., Gas adsorption by some natural zeolites, Chem. Eng. Commun., 4(6), 729-736 (1980).

Kelly, J.F., and Fuller, O.M., An evaluation of a method for investigating sorption and diffusion in porous solids, Ind. Eng. Chem. Fund., 19(1), 11-17 (1980).

Liapis. A.I., and Litchfield, R.J., Ternary adsorption in columns, Chem. Eng. Sci., 35(11), 2366-2370 (1980).

Moharir, A.S.; Kunzru, D., and Saraf, D.N., Theoretical prediction of sorption curves for molecular sieves, Chem. Eng. Sci., 35(6), 1435-1442 (1980).

Mor, L.; Mor, L.A.,; Sideman, S., and Brandes, J.M., Time dependent packed bed adsorption of a chemically-bound adsorbate, Chem. Eng. Sci., 35(3), 725-736 (1980).

Oberoi, A.S.; Fuller, O.M., and Kelly, J.F., Methods of interpreting transient response curves from dynamic sorption experiments, Ind. Eng. Chem. Fund., 19(1), 17-21 (1980).

Ranade, M.G., and Evans, J.W., Reaction between gas and solid in nonisothermal packed bed: Simulation and experiments, Ind. Eng. Chem. Process Des. Dev., 19(1), 118-123 (1980).

Ruthven, D.M., and Kumar, R., An experimental study of single-component and binary adsorption equilibria by a chromatographic method, Ind. Eng. Chem. Fund., 19(1), 27-32 (1980).

Ruthven, D.M.; Lee, K.L., and Yucel, H., Kinetics of non-isothermal sorption in molecular sieve crystals, AIChE J., 26(1), 16-23 (1980).

Schweich, D.; Villermaux, J., and Sardin, M., Introduction to nonlinear theory of adsorptive reactors, AIChE J., 26(3), 477-486 (1980).

Suwanayuen, S., and Danner, R.P., Gas adsorption isotherm equation based on vacancy solution theory, AIChE J., 26(1), 68-83 (1980).

Tan, H.K.S., Kinetics of fixed-bed sorption processes, Chem. Eng. (N.Y.), 24 March, 117-119 (1980).

Wang, M.L.; Liou, C.T., and Chang, R.Y., Numerical technique for solving partial differential equations with applications to adsorption process, Comput. Chem. Eng., 4(2), 85-92 (1980).

Wilson, D.J. and Clarke, A.N., Theory of adsorption by activated carbon, Sep. Sci. Technol., 14(3), 227-242; 14(5), 415-430 (1979); 15(1), 1-22 (1980).

**1981**

Belfort, G., Similarity of adsorbed solution and potential theories for adsorption from a bulk phase onto a solid surface, AIChE J., 27(6), 1021-1022 (1981).

Jacob, P., and Tondeur, D., Non-isothermal gas adsorption in fixed beds, Chem. Eng. J., 22(3), 187-202 (1981).

Knox, D., and Dadyburjor, D.B., Bounds for acceptable values of adsorption entropy, Chem. Eng. Commun., 11(1), 99-112 (1981).

Liapis, A.I., and Litchfield, R.J., Off-diagonal terms of the effective pore diffusivity matrix, Trans. IChemE, 59, T122-T124 (1981).

Okazaki, M.; Tamon, H., and Toei, R., Adsorbed gas flow through porous media, AIChE J., 27(2), 262-277 (1981).

Ortlieb, H.J.; Bunke, G., and Gelbin, D., Separation efficiency in the cyclic steady-state for periodic countercurrent adsorption, Chem. Eng. Sci., 36(6), 1009-1016 (1981).

Paderewski, M.; Majkut, A., and Jedrzejak, A., Simplified model of adiabatic fixed-bed adsorption, Int. Chem. Eng., 21(1), 129-135 (1981).

Peel, R.G., and Benedek, A., Simplified driving-force model for activated carbon adsorption, Can. J. Chem. Eng., 59, 688-692 (1981).

Peel, R.G.; Benedek, A., and Crowe, C.M., Branched-pore kinetic model for activated carbon adsorption, AIChE J., 27(1), 26-32 (1981).

Radeke, K.H.; Ortlieb, H.J., and Gelbin, D., Evaluating breakthrough curves with the method of moments for systems obeying the Langmuir isotherm, Chem. Eng. Sci., 36(1), 11-18 (1981).

Rasmuson, A., Exact solution of model for diffusion and transient adsorption in particles and longitudinal dispersion in packed beds, AIChE J., 27(6), 1032-1035 (1981).

Rice, R.G., Adsorptive distillation, Chem. Eng. Commun., 10(1), 111-126 (1981).

Ruthven, D.M., and Lee, L.K., Kinetics of nonisothermal sorption, AIChE J., 27(4), 654-663 (1981).

Sircar, S., and Gupta, R., Semi-empirical adsorption equation for single component gas-solid equilibria, AIChE J., 27(5), 806-812 (1981).

Vanderschuren, J., Plate efficiency of multistage fluidized-bed adsorbers, Chem. Eng. J., 21(1), 1-10 (1981).

Yang, R.T., and Wong, C., Role of surface diffusion in Langmuir-Hinshelwood mechanism, Chem. Eng. Commun., 11(4), 317-326 (1981).

**1982**

Calligaris, M.B., and Tien, C., Species grouping in multicomponent adsorption calculations, Can. J. Chem. Eng., 60, 772-780 (1982).

Friday, D.K., and LeVan, M.D., Solute condensation in adsorption beds during thermal regeneration, AIChE J., 28(1), 86-91 (1982).

Karger, J., et al., Importance of dimension variation in determining the limiting steps in adsorption kinetics, J. Chem. Technol. Biotechnol., 32, 376-381 (1982).

Knopf, F.C., and Rice, R.G., Adsorptive distillation: Optimum solids profiles, Chem. Eng. Commun., 15(1), 109-124 (1982).

Linek, F., and Dudukovic, M.P., Representation of breakthrough curves for fixed-bed adsorbers and reactors using moments of the impulse response, Chem. Eng. J., 23(1), 31-36 (1982).

Mathews, A.P., Analytical solution for fixed bed adsorption with variable input concentration, Chem. Eng. Commun., 15(5), 313-322 (1982).

Moharir, A.S.; Kunzru, D., and Saraf, D.N., Sorption of non-uniform zeolite crystals for various bulk concentration profiles, Chem. Eng. Commun., 18(1), 15-28 (1982).

Morbidelli, M.; Servida, A.; Storti, G., and Carra, S., Simulation of multicomponent adsorption beds: Model analysis and numerical solution, Ind. Eng. Chem. Fund., 21(2), 123-131 (1982).

Rice, R.G., Approximate solutions for batch, packed tube and radial flow adsorbers: Comparison with experiment, Chem. Eng. Sci., 37(1), 83-92 (1982).

Rousar, I., and Ditl, P., Kinetic characteristics of batch adsorber or ion exchange device operated under nonisothermal conditions, Chem. Eng. Commun., 18(5), 341-354 (1982).

Sacco, A.; Chung, B., and Aksoy, Y., Nondestructive method to measure residual adsorption capacity of charcoal filters, Chem. Eng. Commun., 17(1), 43-56 (1982).

Urano, K.; Yamamoto, E., and Takeda, H., Regeneration rates of granular activated carbons containing adsorbed organic matter, Ind. Eng. Chem. Process Des. Dev., 21(1), 180-185 (1982).

Wilson, D.J., Theory of adsorption by activated carbon, Sep. Sci. Technol., 17(11), 1281-1292 (1982).

**1983**

Aris, R., Interpretation of sorption and diffusion data in porous solids, Ind. Eng. Chem. Fund., 22(1), 150-151 (1983).

Cooney, D.O., and Hines, A.L., Extractive purification on activated carbon, Ind. Eng. Chem. Process Des. Dev., 22(2), 208-211 (1983).

Coppola, A.P., and LeVan, M.D., Adsorption with axial diffusion in shallow beds, Chem. Eng. Sci., 38(7), 991-998 (1983).

Do, D.D., Adsorption in porous solids having bimodal pore size distribution, Chem. Eng. Commun., 23(1), 27-56 (1983).

Garza, G., and Rosales, M.A., Adsorption and diffusion rate parameters from dynamic gravimetric techniques, Ind. Eng. Chem. Process Des. Dev., 22(1), 168-169 (1983).

Gelbin, D.; Bunke, G.; Wolff, H.J., and Neinass, J., Adsorption separation efficiency in the cyclic steady state, Chem. Eng. Sci., 38(12), 1993-2002 (1983).

Knaebel, K.S., and Pigford, R.L., Equilibrium and dissipative effects in cycling zone adsorption, Ind. Eng. Chem. Fund., 22(3), 336-346 (1983).

Raghavan, N.S., and Ruthven, D.M., Numerical simulation of a fixed-bed adsorption column by orthogonal collocation method, AIChE J., 29(6), 922-925 (1983).

Ruthven, D.M., Axial-dispersed plug-flow model for continuous countercurrent adsorbers, Can. J. Chem. Eng., 61(6), 881-883 (1983).

Sheindorf, C.; Rebhun, M., and Sheintuch, M., Prediction of breakthrough curves from fixed-bed adsorbers with Freundlich-type multisolute isotherm, Chem. Eng. Sci., 38(2), 335-342 (1983).

Sircar, S., and Kumar, R., Adiabatic adsorption of bulk binary gas mixtures: Analysis by constant pattern model, Ind. Eng. Chem. Process Des. Dev., 22(2), 271-280 (1983).

Sircar, S., and Kumar, R., Adsorption of dilute adsorbate: Effects of small changes in column temperature, Ind. Eng. Chem. Process Des. Dev., 22(2), 280-287 (1983).

Sircar, S.; Kumar, R., and Anselmo, K.J., Effects of column nonisothermality or nonadiabaticity on the adsorption breakthrough curves, Ind. Eng. Chem. Process Des. Dev., 22(1), 10-15 (1983).

Viswanathan, K.; Khakhar, D., and Rao, D.S., Fluidized-bed adsorber modelling and experimental study, Chem. Eng. Commun., 20(1), 235-252 (1983).

Yoshida, H., and Ruthven, D.M., Dynamic behaviour of an adiabatic adsorption column, Chem. Eng. Sci., 38(6), 877-884 (1983).

**1984**

Aharoni, C., Review of kinetics of adsorption: The s-shaped z-t plot, Adsorpt. Sci. Technol., 1(1), 1-29 (1984).

Al-Ameeri, R.S., and Danner, R.P., Improved tracer-pulse method for measurement of gas adsorption equilibria, Chem. Eng. Commun., 26(1), 11-24 (1984).

Berzins, A.R., et al., Isothermal chemisorption upon oxide-supported platinum, Adsorpt. Sci. Technol., 1(1), 51-76 (1984).

Birnholtz, H.; Nir, A.; Lotan, N., and Aharoni, C., Surface diffusion as rate determining step in activated chemisorption, Can. J. Chem. Eng., 62(2), 233-240 (1984).

Carbonell, R.G., and Whitaker, S., Adsorption and reaction at a catalytic surface, Chem. Eng. Sci., 39(7), 1319-1321 (1984).

Friday, D.K., and LeVan, M.D., Thermal regeneration of adsorption beds: Equilibrium theory for solute condensation, AIChE J., 30(4), 679-682 (1984).

Friedrich, S., and Gelbin, D., Model of constant-volume adsorption kinetics allowing for two internal resistances, Chem. Eng. Sci., 39(5), 912-915 (1984).

Hills, J.H., and Pirzada, I.M., Examination of the accuracy of an approximate solution to isothermal packed bed adsorption with a linear isotherm, Chem. Eng. Sci., 39(5), 919-923 (1984).

Kadlec, O., Mechanisms of volume filling of micropores during adsorption of vapours, Adsorpt. Sci. Technol., 1(2), 133-150 (1984).

Kaul, B.K., Correlation and prediction of adsorption isotherm data for pure and mixed gases, Ind. Eng. Chem. Process Des. Dev., 23(4), 711-716 (1984).

Kumar, R., and Sircar, S., Adiabatic sorption of bulk or dilute single adsorbate from an inert gas: Effect of gas-solid mass and heat transfer coefficients, Chem. Eng. Commun., 26(4), 339-354 (1984).

Lee, T.V.; Huang, J.C., and Madey, R., Separation-factor method for analysis of ideal binary mixtures in gas-solid adsorption, Sep. Sci. Technol., 19(2), 157-172 (1984).

Linares-Solano, A., et al., The n-nonane preadsorption method applied to activated carbons, Adsorpt. Sci. Technol., 1(2), 123-132 (1984).

Mansour, A.R., et al., Numerical solution of general nonequilibrium multicomponent adsorption model, Sep. Sci. Technol., 19(8), 479-496 (1984).

Mansour, A.R.; Shahalam, A.B., and Darwish, N., Comprehensive study of parameters influencing performance of multicomponent adsorption in fixed beds, Sep. Sci. Technol., 19(13), 1087-1112 (1984).

Mehrotra, A.K., and Tien, C., Further work on species grouping in multicomponent adsorption calculation, Can. J. Chem. Eng., 62(5), 632-643 (1984).

Radcliffe, D.F., Lumped parameter models of adsorption kinetics in fixed beds, Chem. Eng. Commun., 25(1), 183-192 (1984).

Raghavan, N.S., and Ruthven, D.M., Dynamic behaviour of an adiabatic adsorption column, Chem. Eng. Sci., 39(7), 1201-1212 (1984).

Srivastava, R.K., and Joseph, B., Simulation of packed-bed separation processes using orthogonal collocation, Comput. Chem. Eng., 8(1), 43-50 (1984).

Valenzuela, D., and Myers, A.L., Gas adsorption equilibria, Sep. Purif. Methods, 13(2), 153-183 (1984).

Wankat, P.C., New adsorption methods, Chem. Eng. Educ., 18(1), 20-25, 44-48 (1984).

**1985**

Basmadjian, D., and Karayannopoulos, C., Rapid procedures for prediction of fixed-bed adsorber behavior, Ind. Eng. Chem. Process Des. Dev., 24(1), 140-149 (1985).

Blasinski, H., and Krauze, S.M., Method of moments for analyzing and predicting outlet curve for adsorption process, Int. Chem. Eng., 25(1), 182-191 (1985).

Bobyleva, M.S., et al., Relation between structure of molecules of nitrogen-containing heterocyclic compounds and their adsorption on graphitized carbon black, Adsorpt. Sci. Technol., 2(3), 165-176 (1985).

Cagliostro, D.E.; Changtai, W., and Smith, J.M., Gas adsorption on carbon-containing fabrics, Ind. Eng. Chem. Process Des. Dev., 24(2), 377-381 (1985).

Cochran, T.W.; Kabel, R.L., and Danner, R.P., Vacancy solution theory of adsorption using Flory-Huggins activity coefficient equations, AIChE J., 31(2), 268-277 (1985).

Cochran, T.W.; Kabel, R.L., and Danner, R.P., Vacancy solution model of adsorption, AIChE J., 31(12), 2075-2081 (1985).

Do, D.D., Discrete cell model of fixed-bed adsorbers with rectangular adsorption isotherms, AIChE J., 31(8), 1329-1337 (1985).

Evteeva, V.A., et al., Adsorption from binary gas mixtures and its theoretical description, Adsorpt. Sci. Technol., 2(3), 153-164 (1985).

Friedrich, S., and Gelbin, D., Simplified mathematical model describing activation of carbon, Chem. Eng. Process., 19(3), 143-150 (1985).

Ghosh, A.K., and Sridhar, T., Non-dissociative adsorption on activated carbon: A statistical rate theory of interfacial turbulence, Can. J. Chem. Eng., 63(5), 784-788 (1985).

Hyun, S.H., and Danner, R.P., Gas adsorption isotherms by use of perturbation chromatography, Ind. Eng. Chem. Fund., 24(1), 95-101 (1985).

Hyun, S.H., and Danner, R.P., Adsorption equilibrium constants and intraparticle diffusivities in molecular sieves by tracer-pulse chromatography, AIChE J., 31(7), 1077-1085 (1985).

Jacob, P., and Tondeur, D., Nonisothermal gas adsorption in fixed beds, Chem. Eng. J., 22(3), 187-201 (1981); 26(1), 41-58 (1983); 31(1), 23-38 (1985).

Knaff, G., and Schlunder, E.U., Experimental confirmation of Graham's law of diffusion up to pore diameters of two microns, Chem. Eng. Process., 19(3), 167-174 (1985).

Liang, S., and Weber, W.J., Parameter evaluation for modeling multicomponent mass transfer in fixed-bed adsorbers, Chem. Eng. Commun., 35(1), 49-65 (1985).

McKay, G., and Bino, M.J., Application of two-resistance mass transfer model to adsorption systems, Chem. Eng. Res. Des., 63, 168-174 (1985).

Mehta, S.D., and Danner, R.P., An improved potential theory method for predicting gas-mixture adsorption equilibria, Ind. Eng. Chem. Fund., 24(3), 325-330 (1985).

Nemeth, J.; Vasanits, E.V., and Virag, T., Mathematical modelling of countercurrent adsorption, Int. J. Heat Mass Transfer, 28(4), 859-866 (1985).

O'Brien, J.A., and Myers, A.L., Rapid calculations of multicomponent adsorption equilibria from pure isotherm data, Ind. Eng. Chem. Process Des. Dev., 24(4), 1188-1191 (1985).

Palancz, B., Modelling and simulation of heat and mass transfer in a packed bed of solid particles having high diffusion resistance, Comput. Chem. Eng., 9(6), 567-582 (1985).

Pfeifer, H., et al., Concentration dependence of intracrystalline self-diffusion in zeolites, Adsorpt. Sci. Technol., 2(4), 229-240 (1985).

Ruthven, D.M., and Wong, F., Generalized statistical model for the prediction of binary adsorption equilibria in zeolites, Ind. Eng. Chem. Fund., 24(1), 27-32 (1985).

Sircar, S., New isotherm for multilayer adsorption of vapours on non-porous adsorbents, Adsorpt. Sci. Technol., 2(1), 23-30 (1985).

Sircar, S., and Kumar, R., Equilibrium theory for adiabatic desorption of bulk binary gas mixtures by purge, Ind. Eng. Chem. Process Des. Dev., 24(2), 358-364 (1985).

Smith, D.M., and Keller, J.F., Nonlinear sorption effects on the determination of diffusion/sorption parameters, Ind. Eng. Chem. Fund., 24(4), 497-499 (1985).

Wojciechowski, B.W.; Hsu, C.C., and Rudzinski, W., Adsorption from multicomponent gas mixtures on the heterogeneous surfaces of solid catalysts, Can. J. Chem. Eng., 63(5), 789-794 (1985).

Wojsz, R., and Rozwadowski, M., Thermodynamical analysis of adsorption isotherms measured for microporous adsorbents, Chem. Eng. Sci., 40(1), 105-110 (1985).

**1986**

Adschiri, T., and Furusawa, T., Relation between carbon dioxide reactivity of coal char and BET surface area, Fuel, 65(7), 927-931 (1986).

Biswas, J.; Do, D.D.; Greenfield, P.F., and Smith, J.M., Importance of finite adsorption rate in the evaluation of adsorption and diffusion parameters in porous catalysts, AIChE J., 32(3), 493-496 (1986).

Blasinski, H., and Krauze, S.M., Method of moments for analyzing and predicting outlet curve for adsorption process: Estimation of outlet time and bed height, Int. Chem. Eng., 26(2), 340-348 (1986).

Cen, P.L., and Yang, R.T., Analytic solution for adsorber breakthrough curves with bidisperse sorbents (zeolites), AIChE J., 32(10), 1635-1641 (1986).

Ching, C.B., and Ruthven, D.M., Experimental study of a simulated countercurrent adsorption system, Chem. Eng. Sci., 41(12), 3063-3072 (1986).

Chitra, S.P., and Govind, R., Application of a group contribution method for predicting adsorbability on activated carbon, AIChE J., 32(1), 167-169 (1986).

Do, D.D., Analysis of a batch adsorber with rectangular adsorption isotherms, Ind. Eng. Chem. Fund., 25(3), 321-326 (1986).

Do, D.D., and Rice, R.G., Validity of the parabolic velocity profile assumption in adsorption studies, AIChE J., 32(1), 149-154 (1986).

Fairbridge, C.; Ng, S.H., and Palmer, A.D., Fractal analysis of gas adsorption on syncrude coke, Fuel, 65(12), 1759-1762 (1986).

Fiedler, K., and Grauert, B., Monte Carlo simulation of thermodynamic function in zeolites, Adsorpt. Sci. Technol., 3(3), 181-188 (1986).

Gelbin, D., et al., Breakthrough curves for single solutes in beds of activated carbon with a broad pore-size distribution, Chem. Eng. Sci., 41(3), 541-554 (1986).

Hidajat, K.; Ching, C.B., and Ruthven, D.M., Simulated countercurrent adsorption processes: A theoretical analysis of the effect of subdividing the adsorbent bed, Chem. Eng. Sci., 41(11), 2953-2956 (1986).

High, M.S., and Danner, R.P., Treatment of gas-solid adsorption data by the error-in-variables method, AIChE J., 32(7), 1138-1145 (1986).

Kumar, R., Column dynamics for multicomponent adsorption: Constant pattern formation, Sep. Sci. Technol., 21(10), 1039-1046 (1986).

Lee, C.S., and O'Connell, J.P., Statistical mechanical model for adsorption and flow of pure and mixed gases in porous media with homogeneous surfaces, AIChE J., 32(1), 96-122 (1986).

Lee, T.V.; Rothstein, D., and Madey, R., Moment analysis of time-dependent transmission of step-function input of radioactive gas through adsorber bed, Sep. Sci. Technol., 21(6), 689-700 (1986).

Matteson, M.J., and Schirmer, W., Adsorption of gases at evaporating and condensing water surfaces, Chem. Eng. J., 33(1), 27-38 (1986).
Moon, H., and Lee, W.K., A lumped model for multicomponent adsorptions in fixed beds, Chem. Eng. Sci., 41(8), 1995-2004 (1986).
Nakahara, T., Calculation of adsorption equilibria for the binary gaseous mixtures on a heterogeneous surface, Chem. Eng. Sci., 41(8), 2093-2098 (1986).
Palekar, M.G., and Rajadhyaksha, R.A., Sorption in zeolites, Chem. Eng. Sci., 40(7), 1085-1092 (1985); 41(3), 463-468 (1986).
Palekar, M.G., and Rajadhyaksha, R.A., Sorption accompanied by chemical reaction on zeolites, Catal. Rev. Sci. Eng., 28(4), 371-429 (1986).
Patwardhan, V.S., and Tien, C., Effect of particle stratification on the performance of fluidized adsorption beds, AIChE J., 32(2), 321-324 (1986).
Rousar, I., and Ditl, P., Numerical simulation of multicomponent isobaric adsorption in fixed-bed columns, Adsorpt. Sci. Technol., 3(2), 49-60 (1986).
Sircar, S., and Kumar, R., Column dynamics for adsorption of bulk binary gas mixtures on activated carbon, Sep. Sci. Technol., 21(9), 919-940 (1986).
Sircar, S., and Myers, A.L., Characteristic adsorption isotherm for adsorption of vapors on heterogeneous adsorbents, AIChE J., 32(4), 650-656 (1986).
Talu, O., and Zwiebel, I., Multicomponent adsorption equilibria of nonideal mixtures, AIChE J., 32(8), 1263-1276 (1986).
Tan, C.S., Pseudo-steady-state approximation for a packed-bed adsorber, Chem. Eng. Sci., 41(11), 2956-2958 (1986).
Tien, C., Incorporation of IAS theory in multicomponent adsorption calculations, Chem. Eng. Commun., 40, 265-280 (1986).
Utrilla, J.R., and Garcia, M.F., Effect of carbon-oxygen and carbon-nitrogen surface complexes on adsorption of cations by activated carbons, Adsorpt. Sci. Technol., 3(4), 293-302 (1986).
Wittkopf, H., and Brauer, P., Thermodynamic formulations of excess and absolute values of adsorption on solid surfaces: One and two phase approaches, Adsorpt. Sci. Technol., 3(4), 271-292 (1986).

**1987**

Aharoni, C., Adsorption by nonhomogeneous porous solids: Effect of adsorption energy gradient on surface flow, AIChE J., 33(2), 303-306 (1987).
Allen, S.J., Equilibrium adsorption isotherms for peat, Fuel, 66(9), 1171-1175 (1987).
Altshuller, D., et al., Analysis of countercurrent adsorber, Chem. Eng. Commun., 52, 311-330 (1987).
Aracil, J., Use of factorial design of experiments in the determination of adsorption equilibrium constants (methyl iodide on charcoals), J. Chem. Technol. Biotechnol., 38(3), 143-152 (1987).
Burganos, V.N., and Sotirchos, S.V., Diffusion in pore networks: Effective medium theory and smooth field approximation, AIChE J., 33(10), 1678-1689 (1987).
Chen, T.L., and Hsu, J.T., Prediction of breakthrough curves by fast Fourier transform method, AIChE J., 33(8), 1387-1390 (1987).
Davis, M.M., and LeVan, M.D., Equilibrium theory for complete adiabatic adsorption cycles, AIChE J., 33(3), 470-479 (1987).
Jaroniec, M., and Madey, R., Gas adsorption on structurally heterogeneous microporous solids, Sep. Sci. Technol., 22(12), 2367-2380 (1987).
Kapoor, A., and Yang, R.T., Roll-up in fixed-bed multicomponent adsorption under pore-diffusion limitation, AIChE J., 33(7), 1215-1217 (1987).
Mansour, A.R., Comparison of equilibrium and nonequilibrium models in simulation of multicomponent sorption processes, Sep. Sci. Technol., 22(4), 1219-1234 (1987).
Marutovsky, R.M., and Bulow, M., Sorption kinetics of multicomponent gaseous and liquid mixtures on porous sorbents (review paper), Gas Sep. Purif., 1(2), 66-76 (1987).
Talu, O., and Kabel, R.L., Isosteric heat of adsorption and vacancy solution model, AIChE J., 33(3), 510-514 (1987).
Tine, C.B.D., Single pellet model application to a two component fixed bed adsorber, Chem. Eng. Res. Des., 65(2), 199-206 (1987).
Wankat, P.C., Intensification of sorption processes, Ind. Eng. Chem. Res., 26(8), 1579-1586 (1987).

Wittkopf, H., and Brauer, P., Statistical thermodynamic calculations for adsorption data of gases: Two-phase approach, Adsorpt. Sci. Technol., 4(4), 251-274 (1987).

**1988**

Annesini, M.C.; Gironi, F., and Marrelli, L., Multicomponent adsorption of continuous mixtures, Ind. Eng. Chem. Res., 27(7), 1212-1217 (1988).

Bhatia, S.K., Combined surface and pore volume diffusion in porous media, AIChE J., 34(7), 1094-1105 (1988).

Biyani, P., and Goochee, C.F., Nonlinear fixed-bed sorption when mass transfer and sorption are controlling, AIChE J., 34(10), 1747-1751 (1988).

Buso, A.; Paratella, A., and Trotta, A., Solution of dynamic adsorption beds using the finite element method, Comput. Chem. Eng., 12(2/3), 247-252 (1988).

Crittenden, B., Selective adsorption, Chem. Eng. (Rugby, Engl.), September, 21-24 (1988).

Do, D.D., Asymmetry of adsorption and desorption in microporous solids, Chem. Eng. Commun., 74, 123-136 (1988).

Do, D.D., and Nguyen, T.S., A power law adsorption model and its significance, Chem. Eng. Commun., 72, 171-186 (1988).

Doong, S.J., and Yang, R.T., A simple potential-theory model for predicting mixed-gas adsorption, Ind. Eng. Chem. Res., 27(4), 630-635 (1988).

Filippov, L.K., Theoretical basis of adsorption processes for separation of multicomponent mixtures, Gas Sep. Purif., 2(3), 138-143 (1988).

Haas, O.W.; Kapoor, A., and Yang, R.T., Confirmation of heavy-component rollup in diffusion-limited fixed-bed adsorption, AIChE J., 34(11), 1913-1916 (1988).

Huang, C.C., and Fair, J.R., Adsorption and desorption of multiple adsorbates in a fixed bed, AIChE J., 34(11), 1861-1877 (1988).

Jasra, R.V., and Bhat, S.G.T., Adsorptive bulk separations by zeolite molecular sieves, Sep. Sci. Technol., 23(10), 945-990 (1988).

Karger, J., et al., NMR study of mass transfer in granulated molecular sieves, AIChE J., 34(7), 1185-1189 (1988).

Klotz, W.L., and Rousseau, R.W., Anomalous mass transfer for vapor adsorption on activated carbon, AIChE J., 34(8), 1403-1406 (1988).

Kluge, G., and Nagel, G., Modelling of non-isothermal multicomponent adsorption in adiabatic fixed beds, Chem. Eng. Sci., 43(10), 2885-2889 (1988).

LeVan, M.D., et al., Fixed-bed adsorption of gases: Effect of velocity variations on transition types, AIChE J., 34(6), 996-1005 (1988).

Moon, H., and Tien, C., Adsorption of gas mixtures on adsorbents with hetergeneous surfaces, Chem. Eng. Sci., 43(11), 2967-2980 (1988).

O'Brien, J.A., and Myers, A.L., Comprehensive technique for equilibrium calculations in adsorbed mixtures: Generalized Fast-IAS method, Ind. Eng. Chem. Res., 27(11), 2085-2091 (1988).

Rota, R., et al., Generalized statistical model for multicomponent adsorption equilibria on zeolites, Ind. Eng. Chem. Res., 27(5), 845-851 (1988).

Rudisill, E.N., and LeVan, M.D., Multicomponent adsorption equilibrium: Henry's law limit for pore-filling models, AIChE J., 34(12), 2080-2082 (1988).

Schlunder, E.U., et al., Desorption of a binary mixture from activated carbon and the resulting separation effect, Chem. Eng. Sci., 43(9), 2391-2398 (1988).

Smith, D.M.; Ross, S.B., and Ciftcioglu, M., Calculation of continuous pore size distributions from adsorption isotherms, Powder Technol., 55(3), 225-228 (1988).

Sotirchos, S.V., and Burganos, V.N., Analysis of multicomponent diffusion in pore networks, AIChE J., 34(7), 1106-1118 (1988).

Storti, G., et al., Adsorption separation processes: Countercurrent and simulated countercurrent operations, Comput. Chem. Eng., 12(5), 475-482 (1988).

Talu, O., and Myers, A.L., Rigorous thermodynamic treatment of gas adsorption, AIChE J., 34(11), 1887-1893 (1988).

Valenzuela, D.P.; Myers, A.L.; Talu, O., and Zwiebel, I., Adsorption of gas mixtures: Effect of energetic heterogeneity, AIChE J., 34(3), 397-402 (1988).

Whitaker, S., Diffusion in packed beds of porous particles, AIChE J., 34(4), 679-683 (1988).

Wotzak, G.P.; Kim, H.I., and Koronich, E., Transient adsorption by use of dynamic balance method, Chem. Eng. Commun., 69, 53-64 (1988).
Yoshida, H.; Kataoka, T., and Ruthven, D.M., Dynamic behaviour of an adiabatic adsorption column, Chem. Eng. Sci., 43(7), 1647-1656 (1988).
Young, B.D., and van Vliet, B.M., Effect of surface roughness on fluid-to-particle mass transfer in a packed adsorber bed, Int. J. Heat Mass Transfer, 31(1), 27-34 (1988).

**1989**

Bhaskar, G.V., and Do, D.D., Simple solution for nonisothermal adsorption in single particle, Chem. Eng. Sci., 44(5), 1215-1220 (1989).
Burganos, V.N., and Sotirchos, S.V., Knudson diffusion in parallel, multidimensional or randomly oriented capillary structures, Chem. Eng. Sci., 44(11), 2451-2462 (1989).
Burganos, V.N., and Sotirchos, S.V., Effective diffusivities in cylindrical capillary-spherical-cavity pore structures, Chem. Eng. Sci., 44(11), 2629-2638 (1989).
Buzanowski, M.A.; Yang, R.T., and Hass, O.W., Direct observation of effects of bed pressure drop on adsorption and desorption dynamics, Chem. Eng. Sci., 44(10), 2392-2394 (1989).
Chatterjee, S.G., and Tien, C., Adsorption with chemical reaction in single modified-carbon pellet, Chem. Eng. Sci., 44(10), 2283-2294 (1989).
Chen, T.L., and Hsu, J.T., Application of fast Fourier transform to nonlinear fixed-bed adsorption problems, AIChE J., 35(2), 332-334 (1989).
Chiang, A.S.T., and Hwang, G.W., Simulation of breakthrough curves by moving-zone collocation method, Comput. Chem. Eng., 13(3), 281-290 (1989).
Czepirski, L., and Jagiello, J., Virial-type thermal equation of gas-solid adsorption, Chem. Eng. Sci., 44(4), 797-802 (1989).
Do, D.D., Sorption rate of bimodal microporous solids with an irreversible isotherm, Chem. Eng. Sci., 44(8), 1707-1714 (1989).
Do, D.D., Effect of surface barrier on sorption of gases in microporous solids, AIChE J., 35(4), 649-655 (1989).
Do, D.D., Role of temperature and length scales on sorption of gases in microporous solids, Chem. Eng. Commun., 77, 229-246 (1989).
Dudziak, W., and Kowalski, S.J., Theory of thermodiffusion for solids, Int. J. Heat Mass Transfer, 32(11), 2005-2014 (1989).
Filippov, L.K., Dynamics of physical adsorption for porous grains with a complex internal structure, Gas Sep. Purif., 3(1), 29-36 (1989).
Filippov, L.K., Modelling of adsorption separation of gas mixtures with allowance made for thermal effects accompanying adsorption, Gas Sep. Purif., 3(1), 37-45; 3(2), 84-101 (1989).
Filippov, L.K., Theoretical basis of separation processes and adsorption dynamics of multicomponent mixtures, Chem. Eng. Sci., 44(3), 575-582 (1989).
Gamba, G., et al., Adsorbed solution theory models for multicomponent adsorption equilibria, AIChE J., 35(6), 959-966 (1989).
Graham, P.; Hughes, A.D., and Rees, L.V.C., Sorption of binary gas mixtures in zeolites, Gas Sep. Purif., 3(2), 56-64 (1989).
Gray, P.G., and Do, D.D., Adsorption and desorption of gaseous sorbates on a bidispersed particle with Freundlich isotherm: Theoretical analysis, Gas Sep. Purif., 3(4), 193-200 (1989).
Haselden, G.G., Gas separation fundamentals, Gas Sep. Purif., 3(4), 209-215 (1989).
Holgado, M.J., and Rives, V., A comparative study of the suitability of Sircar's and BET isotherms to study the adsorption of vapours on non-porous adsorbents, Adsorpt. Sci. Technol., 6(3), 110-118 (1989).
Huang, T.C., and Cho, L.T., Relationships between constants of Freundlich equation and temperature for gaseous adsorption, Chem. Eng. Commun., 75, 181-194 (1989).
Jaroniec, M., et al., Use of a polynomial equation for analyzing low-concentration adsorption measurements of ethane on activated carbons, Sep. Sci. Technol., 24(15), 1355-1362 (1989).
Jedrzejak, A.; Gorius, A., and Tondeur, D., Consistency problems arising in description of equilibrium adsorption of mixtures, Chem. Eng. Sci., 44(6), 1315-1328 (1989).
Kaguei, S.; Ono, N., and Wakao, N., Parameter estimation in batch adsorption with a linear isotherm, Chem. Eng. Sci., 44(11), 2565-2570 (1989).
Kaguei, S.; Shemilt, L.W., and Wakao, N., Models and experiments on adsorption columns with constant wall temperature: Radially varying and radially lumped models, Chem. Eng. Sci., 44(3), 483-492 (1989).

Kapoor, A.; Yang, R.T., and Wong, C., Surface diffusion, Catal. Rev. Sci. Eng., 31(1), 129-214 (1989).
Kast, W., and Otten, W., Breakthrough in fixed bed adsorbers: Methods of calculation and effects of process parameters, Int. Chem. Eng., 29(2), 197-212 (1989).
Kawai, Y., et al., Control of axial temperature distribution in moving-bed adsorber, Int. Chem. Eng., 29(1), 152-158 (1989).
Kuznetsov, B.V.; Tuan, N.A., and Rachmanova, T.A., Calorimetric and isosteric heats of adsorption of substances with molecules of different electronic structure, Adsorpt. Sci. Technol., 6(1), 27-34 (1989).
LaCava, A.I.; Koss, V.A., and Wickens, D., Non-Fickian adsorption rate behavior of some carbon molecular sieves: Slit-potential rate model, Gas Sep. Purif., 3(4), 180-186 (1989).
Limbach, K.W.; Nitsche, J.M., and Wei, J., Partitioning of non-spherical molecules between bulk solution and porous solids, AIChE J., 35(1), 42-52 (1989).
Mansour, A.R., Simulation of multicomponent sorption processes with axial diffusion, Sep. Sci. Technol., 24(12), 1047-1058 (1989).
Mansour, A.R., et al., Computer prediction of multicomponent sorption with variable initial concentrations using a complex model, Sep. Sci. Technol., 24(9), 697-708 (1989).
McKay, G., and Al Duri, B., Prediction of multicomponent adsorption equilibrium data using empirical correlations, Chem. Eng. J., 41(1), 9-24 (1989).
Mees, P.A.J.; Gerritsen, A.W., and Verheijen, P.J.T., Fast calculation of breakthrough curves in nonisothermal fixed-bed adsorbers, AIChE J., 35(8), 1380-1385 (1989).
Nagel, G., and Kluge, G., Non-isothermal multicomponent adsorption processes and their numerical treatment by means of integro-differential equations, Comput. Chem. Eng., 13(9), 1025-1030 (1989).
Okubo, T., and Inoue, H., Analysis of surface diffusion of adsorptive gases through porous glass by square-pulse method, Int. Chem. Eng., 29(3), 539-547 (1989).
Patwardhan, V.S., Diffusion and sorption in zeolites: A Markov process formulation, Chem. Eng. Sci., 44(11), 2619-2628 (1989).
Payne, G.F., et al., Adsorption of nonpolar solutes onto neutral polymeric sorbents, Sep. Sci. Technol., 24(5), 457-466 (1989).
Richter, E.; Schutz, W., and Myers, A.L., Effect of adsorption equation on prediction of multicomponent adsorption equilibria by ideal adsorbed solution theory, Chem. Eng. Sci., 44(8), 1609-1616 (1989).
Riekert, L., Transport of mass and heat in porous adsorbents, Chem. Eng. Process., 26(1), 59-62 (1989).
Ritter, J.A., and Yang, R.T., Thermodynamic analysis for rapid measurements of equilibrium adsorption from binary gas mixtures, Ind. Eng. Chem. Res., 28(5), 599-609 (1989).
Rudzinski, W., and Jagiello, J., A simple approach to the 2D mobile adsorption of gases on heterogeneous solid surfaces exhibiting random surface topography, Adsorpt. Sci. Technol., 6(1), 35-51 (1989).
Schwarzbach, J., Diffusion in porous materials partially wetted with a binary mixture, Chem. Eng. Process., 26(1), 35-44 (1989).
Seidel, A., and Carl, P.S., Concentration dependence of surface diffusion for adsorption on energetically heterogeneous adsorbents, Chem. Eng. Sci., 44(1), 189-194 (1989).
Sotirchos, S.V., Multicomponent diffusion and convection in capillary structures, AIChE J., 35(12), 1953-1961 (1989).
Tan, H.K.S., General solutions of second-order kinetics for fixed-bed sorption processes, Chem. Eng. Sci., 44(11), 2756-2760 (1989).
Tarasevich, Y.I., et al., Non-specific and specific interactions of hydrocarbons with the surface of a non-swelling organo-substituted layer silicate, Adsorpt. Sci. Technol., 6(3), 147-154 (1989).
Vafai, K., and Tien, H.C., Numerical investigation of phase change effects in porous materials, Int. J. Heat Mass Transfer, 32(7), 1261-1278 (1989).
Valsaraj, K.T., Partitioning of hydrophobic nonpolar volatile organics between the aqueous and surfactant aggregate phases on alumina, Sep. Sci. Technol., 24(14), 1191-1206 (1989).
Van Loon, W.K.P.; Van Haneghem, I.A., and Schenk, J., New model for non-steady-state probe method to measure thermal properties of porous media, Int. J. Heat Mass Transfer, 32(8), 1473-1482 (1989).
Vrettos, N.A.; Imakoma, H., and Okazaki, M., Transport properties of porous media from the microgeometry of a 3-D Voronoi network, Chem. Eng. Process., 26(3), 237-247 (1989).
Wong, T., and Frey, D.D., Matrix calculation of multicomponent transient diffusion in porous sorbents, Int. J. Heat Mass Transfer, 32(11), 2179-2188 (1989).

Yang, B.L.; Goto, M., and Goto, S., Comparison of operating time between batchwise and columnwise adsorption, Sep. Sci. Technol., 24(9), 741-754 (1989).
Yao, C., and Tien, C., Wetting of polymeric adsorbents and its effect on adsorption, AIChE J., 35(9), 1559-1564 (1989).
Yeh, Y.T., and Yang, R.T., Diffusion in zeolites containing mixed cations, AIChE J., 35(10), 1659-1666 (1989).
Yu, Q., and Wang, N.H.L., Computer simulations of dynamics of multicomponent ion exchange and adsorption in fixed beds: Gradient-directed moving finite element method, Comput. Chem. Eng., 13(8), 915-926 (1989).

**1990**

Al-Duri, B., and McKay, G., Comparison in theory and application of several mathematical models to predict kinetics of single component batch adsorption systems, Process Safety Environ. Prot., 68(4), 254-268 (1990).
Chen, Y.D.; Ritter, J.A., and Yang, R.T., Nonideal adsorption from multicomponent gas mixtures at elevated pressures on a 5A molecular sieve, Chem. Eng. Sci., 45(9), 2877-2894 (1990).
Cho, B.K., Determination of coverage-dependent heat of adsorption from transient pulse experiments, Chem. Eng. Sci., 45(5), 1422-1425 (1990).
Cooney, D.O., External film and particle phase control of adsorber breakthrough behavior, AIChE J., 36(9), 1430-1432 (1990).
Cooney, D.O., Rapid approximate solutions for adsorption bed concentration profile and breakthrough curve behavior: Favourable isotherms and both phase resistances important, Chem. Eng. Commun., 91, 1-10 (1990).
Cooney, D.O., On the basis for the Freundlich adsorption isotherm, Chem. Eng. Commun., 94, 27-34 (1990).
Do, D.D., Hierarchy of rate models for adsorption and desorption in bidispersed structured sorbents, Chem. Eng. Sci., 45(5), 1373-1382 (1990).
Do, D.D., and Rice, R.G., Applicability of the external-diffusion model in adsorption studies, Chem. Eng. Sci., 45(5), 1419-1421 (1990).
Do, D.D.; Pham, T.V., and Do, H.D., The role of temperature and length scales on the sorption of gases in microporous solids: Nonlinear isotherm, Chem. Eng. Commun., 89, 1-24 (1990).
Filippov, L.K., and Filippova, I.V., Isothermal and non-isothermal multicomponent adsorption kinetics. Chem. Eng. Commun., 93, 147-166 (1990).
Filippov, L.K.; Bulow, M., and Filippova, I.V., Determination of concentration dependent diffusion coefficient matrix inside porous sorbent particles, Gas Sep. Purif., 4(1), 41-49 (1990).
Filippov, L.K.; Filippova, I.V., and Czepirski, L., Technological computation of frontal modes in adsorption separation of multicomponent mixtures, Chem. Eng. Sci., 45(4), 927-934 (1990).
Frey, D.D., The entropy condition for the dynamics of nonlinear multicomponent sorption in porous media, Chem. Eng. Sci., 45(1), 131-142 (1990).
Gamba, G., et al., Adsorption equilibria of nonideal multicomponent systems at saturation, AIChE J., 36(11), 1736-1742 (1990).
Gray, P.G., and Do, D.D., Adsorption and desorption dynamics of sulphur dioxide on a single large activated carbon particle, Chem. Eng. Commun., 96, 141-154 (1990).
Gray, P.G., and Do, D.D., Adsorption and desorption of gaseous sorbates on a bidispersed particle with Freundlich isotherm: Contribution of surface diffusion to sorption dynamics of sulphur dioxide on activated carbon, Gas Sep. Purif., 4(3), 149-157 (1990).
Gubbins, K.E., Molecular adsorption in micropores, Chem. Eng. Prog., 86(8), 42-44 (1990).
Hidajat, K., and Ching, C.B., Simulation of the performance of a continuous counter-current adsorption system by the method of orthogonal collocation with non-linear and interacting adsorption isotherms, Chem. Eng. Res. Des., 68(1), 104-108 (1990).
Hines, A.L.; Kuo, S.L., and Dural, N.H., A new analytical isotherm equation for adsorption on heterogeneous adsorbents, Sep. Sci. Technol., 25(7), 869-888 (1990).
Kapoor, A., and Yang, R.T., Surface diffusion on energetically heterogeneous surfaces: An effective medium approximation approach, Chem. Eng. Sci., 45(11), 3261-3270 (1990).
Karger, J., et al., Influence of molecular shape on probing mass transfer resistances on zeolites, AIChE J., 36(10), 1500-1504 (1990).
Kim, D.H., Single effective diffusivities for dynamic adsorption in bidisperse adsorbents, AIChE J., 36(2), 302-306 (1990).
Knapp, H., Knowledge of thermodynamic properties for natural gas processing, Gas Sep. Purif., 4(3), 123-136 (1990).

Krishna, R., Multicomponent surface diffusion of adsorbed species: A description based on the generalized Maxwell-Stefan equations, Chem. Eng. Sci., 45(7), 1779-1792 (1990).
Limbach, K.W., and Wei, J., Restricted diffusion through granular materials, AIChE J., 36(2), 242-248 (1990).
Lin, S.Y.; McKeigue, and Maldarelli, C., Diffusion-controlled surfactant adsorption studied by pendant drop digitization, AIChE J., 36(12), 1785-1795 (1990).
Lipscomb, G.G., Unified thermodynamic analysis of sorption in rubbery and glass materials, AIChE J., 36(10), 1505-1516 (1990).
MacPherson, S.R.; Maneval, J.E., and McCoy, B.J., Diffusion with nonlinear adsorption: An eigenfunction expansion solution, Ind. Eng. Chem. Res., 29(7), 1571-1574 (1990).
Marczewski, A.W.; Derylo-Marczewska, A., and Jaroniec, M., A simple method for describing multi-solute adsorption equilibria on activated carbons, Chem. Eng. Sci., 45(1), 143-150 (1990).
McKay, G., and Al-Duri, B., Study of the mechanism of pore diffusion in batch adsorption systems, J. Chem. Technol. Biotechnol., 48(3), 269-286 (1990).
Micke, A., and Bulow, M., Application of Volterra integral equations to the modelling of the sorption kinetics of multicomponent mixtures in porous media: Fundamentals and elimination of apparatus effects, Gas Sep. Purif., 4(3), 158-170 (1990).
O'Shea, J.D., and Liapis, A.I., Evaluation of simple and complex models for mass transfer in the non-isothermal gas adsorption of multiple adsorbates in a single adsorbent particle, Chem. Eng. Res. Des., 68(3), 242-250 (1990).
Onyegbado, C.O., and Susu, A.A., Theoretical study of nonlinear adsorption isotherm by chromatography: Application to methyl linoleate and methyl oleate adsorption on supported copper catalyst, Chem. Eng. Commun., 91, 79-90 (1990).
Payne, G.F., and Ninomiya, Y., Selective adsorption of solutes based on hydrogen bonding, Sep. Sci. Technol., 25(11), 1117-1130 (1990).
Rajadhyaksha, R.A.; Pitale, K.K., and Tambe, S.S., Correlation effects in counterdiffusion in zeolites, Chem. Eng. Sci., 45(7), 1935-1939 (1990).
Rho, S.; Park, J.K., and Kim, J.D., Adsorption isotherms and heats of immersion in the adsorption of binary mixtures on activated carbon, Chem. Eng. Commun., 88, 1-10 (1990).
Ruthven, D.M., and Rojo, J.C., A simple model for a non-isothermal adsorption column, Can. J. Chem. Eng., 68(5), 795-798 (1990).
Sircar, S., and Rao, M.B., Estimation of surface diffusion through porous media, AIChE J., 36(8), 1249-1254 (1990).
Takao, M., A comment on two modified numerical methods for solving steady, one-dimensional diffusion, Int. Chem. Eng., 30(2), 297-299 (1990).
Tong, C.H., and Lund, D.B., Effective moisture diffusivity in porous materials as a function of temperature and moisture content, Biotechnol. Prog., 6(1), 67-75 (1990).
Tsotsas, E., and Schluender, E.U., Numerical calculation of the thermal conductivity of two regular bidispersed beds of spherical particles, Comput. Chem. Eng., 14(9), 1031-1038 (1990).
Yao, C., Comments and reply on "The relative importance of pore and surface diffusion in non-equilibrium adsorption rate processes", Chem. Eng. Sci., 45(10), 3199-3201 (1990).

**1991**

Astakhov, A.V., and Shirochin, D.L., Capillary-like condensation of sorbed gases in coals, Fuel, 70(1), 51-56 (1991).
Aviles, B.E., and LeVan, M.D., Network models for nonuniform flow and adsorption in fixed beds, Chem. Eng. Sci., 46(8), 1935-1944 (1991).
Baik, M.H., and Lee, K.J., Analysis of nonlinear diffusion phenomena in the compacted bentonite backfill material, Chem. Eng. Sci., 46(12), 3296-3299 (1991).
Bardakci, T., Temperature effect on diffusion of carbon dioxide through upper layers of Yucca Mountain, Gas Sep. Purif., 5(1), 11-15 (1991).
Berninger, J.A., et al., A versatile model for simulation of reaction and nonequilibrium dynamics in multicomponent fixed-bed adsorption processes, Comput. Chem. Eng., 15(11), 749-768 (1991).
Bhatia, S.K.; Gray, P.G., and Do, D.D., Modelling of sorption of gaseous sorbates in bidispersed structured solids: The concept of heterogeneity of the microphase, Gas Sep. Purif., 5(1), 49-55 (1991).
Buzanowski, M.A., and Yang, R.T., Approximations for intraparticle diffusion rates in cyclic adsorption and desorption, Chem. Eng. Sci., 46(10), 2589-2598 (1991).

Buzanowski, M.A., and Yang, R.T., A nondriving-force approximation for intraparticle diffusion in cyclic adsorption and desorption with short cycle times, Chem. Eng. Commun., 104, 257-266 (1991).
Chen, Y.D., and Yang, R.T., Concentration dependence of surface diffusion and zeolitic diffusion, AIChE J., 37(10), 1579-1582 (1991).
Davis, B.H., BET: The scientists behind surface science, Chemtech, January, 18-25 (1991).
Essenhigh, R.H., Rate equations for the carbon-oxygen reaction: An evaluation of the Langmuir adsorption isotherm at atmospheric pressure, Energy Fuels, 5(1), 41-46 (1991).
Fan, L.T., et al., The master equation for linear adsorption and desorption of gases on solid surfaces, Chem. Eng. Commun., 108, 127-146 (1991).
Farooq, S., and Ruthven, D.M., Dynamics of kinetically controlled binary adsorption in a fixed bed, AIChE J., 37(2), 299-301 (1991).
Filippov, L., A combined method for measurement of isotherms of adsorption of vapors and gases, Chem. Eng. Sci., 46(9), 2323-2330 (1991).
Filippov, L.K., Theoretical models for adsorption separation processes in non-isothermal cases, Chem. Eng. Commun., 103, 11-26 (1991).
Floess, J.K., et al., A method for decreasing baseline noise in thermogravimetric measurements, Energy Fuels, 5(1), 138-140 (1991).
Gorius, A.; Bailly, M., and Tondeur, D., Perturbative solutions for non-linear fixed-bed adsorption: Approximate analytical solutions for asymptotic fronts, Chem. Eng. Sci., 46(2), 677-684 (1991).
Gorius, A.; Bailly, M., and Tondeur, D., Perturbative solutions for non-linear fixed-bed adsorption: Approximate analytical expressions for the dynamics of micro-columns, Chem. Eng. Sci., 46(2), 685-692 (1991).
Gu, T.; Tsai, G.J., and Tsao, G.T., Multicomponent adsorption and chromatography with uneven saturation capacities, AIChE J., 37(9), 1333-1340 (1991).
Guo, C.J., A new approach for measuring multicomponent gas-solid sorption, Chem. Eng. Commun., 105, 3-10 (1991).
Hajji, A., and Worek, W.M., Thermodynamics of constant-pressure and constant-volume sorption processes in closed-cycle solid desiccant systems, Chem. Eng. Commun., 104, 21-40 (1991).
Hills, J.H., Non-isothermal adsorption in a pellet, Chem. Eng. Sci., 46(1), 69-74 (1991).
Jaroniec, M.; Choma, J., and Lu, X., An improved method for evaluating the micropore-size distribution from adsorption isotherm, Chem. Eng. Sci., 46(12), 3299-3301 (1991).
Jelinek, L.; Dong, P., and Kovats, E., The influence of the calibration procedure of a volumetric adsorption apparatus on the form of the isotherm and its BET parameters, Adsorpt. Sci. Technol., 7(3), 140-162 (1991).
Kapoor, A., and Yang, R.T., Contribution of concentration-dependent surface diffusion to rate of adsorption, Chem. Eng. Sci., 46(8), 1995-2002 (1991).
Keefer, B.G., and Ruthven, D.M., Constant pattern profiles for linear bulk adsorption systems, Chem. Eng. Sci., 46(10), 2756-2761 (1991).
Kluge, G., et al., Estimation of component loadings in fixed-bed adsorption from breakthrough curves of binary gas mixtures in nontrace systems, Chem. Eng. Sci., 46(1), 368-371 (1991).
Lai, C.C., and Tan, C.S., Approximate models for nonlinear adsorption in a packed-bed adsorber, AIChE J., 37(3), 461-465 (1991).
Lane, A.M., Single-pore model: A simplified diffusion simulation, AIChE J., 37(8), 1245-1248 (1991).
Larin, A.V., Theory of the adsorption dynamics of vapours on microporous adsorbents: Introduction into the model of a layer of equilibrium adsorption, Adsorpt. Sci. Technol., 7(2), 65-74 (1991).
Maity, N.; Payne, G.F., and Chipchosky, J.L., Adsorptive separations based on the differences in solute-sorbent hydrogen-bonding strengths, Ind. Eng. Chem. Res., 30(11), 2456-2463 (1991).
McKay, G., and Al-Duri, B., Extended empirical Freundlich isotherm for binary systems: A modified procedure to obtain the correlative constants, Chem. Eng. Process., 29(3), 133-138 (1991).
Nitsche, J.M., and Wei, J., Window effects in zeolite diffusion and Brownian motion over potential barriers, AIChE J., 37(5), 661-670 (1991).
Pan, D., and Mersmann, A.B., Methods for incorporating adsorbent activity into zeolitic adsorption isotherms, Gas Sep. Purif., 5(4), 210-214 (1991).
Petersen, E.E., Adsorption in bidisperse-pore systems, AIChE J., 37(5), 671-678 (1991).
Petersen, F.W., and Van Deventer, J.S.J., Inhibition of mass transfer to porous adsorbents by fine particles and organics, Chem. Eng. Commun., 99, 55-76 (1991).

Petropoulos, J.H.; Petrou, J.K., and Liapis, A.I., Network model investigation of gas transport in bidisperse porous adsorbents, Ind. Eng. Chem. Res., 30(6), 1281-1289 (1991).
Razmus, D.M., and Hall, C.K., Prediction of gas adsorption in 5A zeolites using Monte Carlo simulation, AIChE J., 37(5), 769-780 (1991).
Ritter, J.A., and Yang, R.T., Equilibrium theory for hysteresis-dependent fixed-bed desorption, Chem. Eng. Sci., 46(2), 563-574 (1991).
Rodrigues, A.E.; Zuping, L., and Loureiro, J.M., Residence time distribution of inert and linearly adsorbed species in a fixed bed containing "large-pore" supports: Applications in separation engineering, Chem. Eng. Sci., 46(11), 2765-2774 (1991).
Rudisill, E.N., and LeVan, M.D., Constant pattern behavior for adsorption processes with nonplug flow, Ind. Eng. Chem. Res., 30(5), 1054-1060 (1991).
Saito, A., and Foley, H.C., Curvature and parametric sensitivity in models for adsorption in micropores, AIChE J., 37(3), 429-436 (1991).
Schmidt-Naake, G., and Pippel, W., A Markov chain approach to complex diffusion processes superimposed by adsorption or chemical reaction, Chem. Eng. Sci., 46(7), 1685-1696 (1991).
Scholl, S.E., and Mersmann, A.B., On intraparticle total pressure change during gas phase adsorption, Gas Sep. Purif., 5(2), 77-82 (1991).
Schweighart, P.J., and Nicholson, D., Two-phase flow in a two-dimensional bidisperse anisotropic pore network model, Adsorpt. Sci. Technol., 7(2), 75-86 (1991).
Schweighart, P.J., and Nicholson, D., Adsorption and relative permeability in a two-dimensional bidisperse anisotropic pore network model, Adsorpt. Sci. Technol., 7(2), 97-106 (1991).
Sircar, S., Role of adsorbent heterogeneity on mixed gas adsorption, Ind. Eng. Chem. Res., 30(5), 1032-1039 (1991).
Sun, L.M., and Meunier, F., An improved finite difference method for fixed-bed multicomponent sorption, AIChE J., 37(2), 244-254 (1991).
Wittwer, A., and Lavanchy, A., X-ray computer tomography: A non-destructive method for investigating dynamic sorption processes on activated carbon, Adsorpt. Sci. Technol., 7(2), 87-96 (1991).
Yang, R.T.; Chen, Y.D., and Yeh, Y.T., Prediction of cross-term coefficients in binary diffusion: Diffusion in zeolite, Chem. Eng. Sci., 46(12), 3089-3100 (1991).
Yu, Q., Improving the computational efficiency of step inputs in adsorption processes, Comput. Chem. Eng., 15(9), 675-677 (1991).

**1992**

Al-Duri, B., and McKay, G., Pore diffusion dependence of the effective diffusivity on the initial sorbate concentration in single and multisolute batch adsorption systems, J. Chem. Technol. Biotechnol., 55(3), 245-250 (1992).
Al-Duri, B.; Khader, Y., and McKay, G., Prediction of binary component isotherms for adsorption on heterogeneous surfaces, J. Chem. Technol. Biotechnol., 53(4), 345-352 (1992).
Bingkui, W., and Shulin, L., Investigation of polymer adsorption on to a metal surface by an optical method, Adsorpt. Sci. Technol., 8(3), 134-141 (1992).
Ching, C.B., et al., Comparative study of flow schemes for a simulated countercurrent adsorption separation process, AIChE J., 38(11), 1744-1750 (1992).
Deak, P., and Solyom, A., Theoretical study of adsorption on a silica surface, Period. Polytech. Chem. Eng., 34(1), 121-126 (1992).
Douss, N., An experimental study of solid adsorption heat-pump cycles, Int. Chem. Eng., 32(4), 718-725 (1992).
El-Halwagi, M.M., and Srinivas, B.K., Synthesis of reactive mass-exchange networks, Chem. Eng. Sci., 47(8), 2113-2119 (1992).
Filippov, L.K., Adiabatic adsorption in fixed beds for quasi-linear isotherms, Chem. Eng. Commun., 111, 177-195 (1992).
Filippov, L.K., Coherent and incoherent frontal patterns for multi-component dynamics of adsorption: Analysis of frontal-pattern isothermal dynamics for convex mixture isotherms of adsorption, Chem. Eng. Sci., 47(5), 1199-1210 (1992).
Filippov, L.K., Coherent and incoherent frontal patterns for multi-component adsorption dynamics: Non-uniqueness of frontal patterns in the isothermal case for convex isotherms of two-component mixtures, Chem. Eng. Sci., 47(5), 1211-1218 (1992).

Frey, D.D., Local equilibrium theory for the dynamics of non-isothermal mixed-gas adsorption of high-concentration components, Chem. Eng. Commun., 117, 143-162 (1992).

Frey, D.D., Mixed-gas adsorption dynamics of high-concentration components in a particulate bed, AIChE J., 38(10), 1649-1656 (1992).

Hassan, N.M., et al., An isotherm model for adsorption of gases and vapours on heterogeneous adsorbents, Gas Sep. Purif., 6(4), 229-234 (1992).

Helfferich, F.G., Langmuir's isotherm: Kinetics or thermodynamics? Chem. Eng. Educ., 26(1), 23, 51 (1992).

Kaminsky, R.D., and Monson, P.A., An analysis of the statistical model adsorption isotherm, AIChE J., 38(12), 1979-1989 (1992).

Karavias, F., and Myers, A.L., Molecular thermodynamics of adsorption from gas mixtures: Composition of adsorbed phase from gravimetric data, Chem. Eng. Sci., 47(6), 1441-1452 (1992).

Koopman, D.C.; Cole, J.V., and Lee, H.H., Assumption of local equilibrium in adsorption processes, AIChE J., 38(4), 623-625 (1992).

Liu, H.; Zhang, L., and Seaton, N.A., Determination of the connectivity of porous solids from nitrogen sorption measurements: Generalisation, Chem. Eng. Sci., 47(17), 4393-4404 (1992).

Lu, Z.P., et al., Effect of intraparticle forced convection on gas desorption from fixed beds containing "large-pore" adsorbents, Ind. Eng. Chem. Res., 31(6), 1530-1540 (1992).

Lu, Z.P., et al., Dynamics of pressurization and blowdown of an adiabatic adsorption bed: Intraparticle diffusion/convection models, Gas Sep. Purif., 6(2), 89-100 (1992).

Matranga, K.R., et al., Molecular simulation of adsorbed natural gas, Sep. Sci. Technol., 27(14), 1825-1836 (1992).

Mayagoitia, V., et al., Simulation of heterogeneous surfaces of adsorbents by Monte Carlo methods, Gas Sep. Purif., 6(1), 35-42 (1992).

Myers, A.L.; Belfort, G., and Suzuki, M., Prediction of multicomponent adsorption-equilibrium: Discussions on energy of adsorption in relation to adsorption equilibrium, Int. Chem. Eng., 32(3), 585-590 (1992).

Park, I., and Knaebel, K.S., Adsorption breakthrough behavior: Unusual effects and possible causes, AIChE J., 38(5), 660-670 (1992).

Rudisill, E.N., and LeVan, M.D., Standard states for the adsorbed-solution theory, Chem. Eng. Sci., 47(5), 1239-1246 (1992).

Sun, W., and Costa, C.A.V., Fast method for solving nonequilibrium fixed-bed adsorption models with variable velocity and linear isotherm, Comput. Chem. Eng., 16(6), 535-544 (1992).

Tsibranska, I.; Penchev, I., and Assenov, A., The effect of microparticle-size distribution on the adsorption with biporous zeolite pellets, Chem. Eng. Sci., 47(2), 479-483 (1992).

Yao, C., and Tien, C., Approximation of intraparticle mass transfer in adsorption processes: Linear and non-linear systems, Chem. Eng. Sci., 47(2), 457-474 (1992).

**1993**

Carta, G., Exact solution and linear driving force approximation for cyclic mass transfer in a bidisperse sorbent, Chem. Eng. Sci., 48(9), 1613-1618 (1993).

Chen, Y.D., and Yang, R.T., Surface diffusion of multilayer adsorbed species, AIChE J., 39(4), 599-606 (1993).

Chen, Y.D.; Yang, R.T., and Sun, L.M., Further work on predicting multicomponent diffusivities from pure-component diffusivities for surface diffusion and diffusion in zeolites, Chem. Eng. Sci., 48(15), 2815-2816 (1993).

Ching, C.B., et al., Experimental study of a simulated counter-current adsorption system: Effects of non-linear and interacting isotherms, Chem. Eng. Sci., 48(7), 1343-1351 (1993).

Cooney, D.O., Comparison of simple adsorber breakthrough curve method with exact solution, AIChE J., 39(2), 355-358 (1993).

Davis, A.M.J., and Ethier, C.R., Transport through materials bounded by porous surfaces, Chem. Eng. Sci., 48(9), 1655-1664 (1993).

Do, D.D., and Hu, X., An energy-distributed model for adsorption kinetics in large heterogeneous microporous particles, Chem. Eng. Sci., 48(11), 2119-2128 (1993).

Filippov, L.K., Study of adsorption on polymer by ellipsometry, Chem. Eng. Sci., 48(19), 3391-3398 (1993).

Gilbert, S.W., Linear algebraic solution for multicomponent adsorption on porous media, AIChE J., 39(3), 518-520 (1993).

Golden, T.C., and Kumar, R., Adsorption equilibrium and kinetics for multiple trace impurities in various gas streams on activated carbon, Ind. Eng. Chem. Res., 32(1), 159-165 (1993).

Goto, M., and Hirose, T., Approximate rate equation for intraparticle diffusion with or without reaction, Chem. Eng. Sci., 48(10), 1912-1915 (1993).
Gutsche, R., Concentration-dependent micropore diffusion analysed by measuring laboratory adsorber dynamics: Study of the adsorber flow behaviour and micropore diffussion mechanism, Chem. Eng. Sci., 48(21), 3723-3742 (1993).
Harriott, G.M., Memory-integral mass-transfer models for adsorption process simulation (see errata: 40(6), 956, 1994), AIChE J., 39(3), 422-433 (1993).
Hu, X., and Do, D.D., Role of energy distribution in multicomponent sorption kinetics in bidispersed solids, AIChE J., 39(10), 1628-1640 (1993).
Hu, X.; Rao, G.N., and Do, D.D., Effect of energy distribution on sorption kinetics in bidispersed particles, AIChE J., 39(2), 249-261 (1993).
Jordi, R.G., and Do, D.D., Analysis of the frequency response method for sorption kinetics in bidispersed structured sorbents, Chem. Eng. Sci., 48(6), 1103-1130 (1993).
Kaguei, S., and Wakao, N., Diffusion of adsorbate through slab adsorbents: Investigation of diffusion modes, Chem. Eng. Sci., 48(7), 1353-1356 (1993).
Kikkinides, E.S., and Yang, R.T., Further work on approximations for intraparticle diffusion rates in cyclic adsorption and desorption, Chem. Eng. Sci., 48(6), 1169-1173 (1993).
Kikkinides, E.S., and Yang, R.T, Effects of bed pressure drop on isothermal and adiabatic adsorber dynamics, Chem. Eng. Sci., 48(9), 1545-1556 (1993).
Krishna, R., A unified approach to the modelling of intraparticle diffusion in adsorption processes, Gas Sep. Purif., 7(2), 91-104 (1993).
Lu, Z., and Rodrigues, A.E., Intensification of sorption processes using "large-pore" materials, Ind. Eng. Chem. Res., 32(1), 230-235 (1993).
Masthan, S.K., et al., Derivation of the expanded form of the BJH equation and its application to the pore structure analysis of mesoporous adsorbents, Adsorpt. Sci. Technol., 9(4), 212-230 (1993).
McKay, G.; Al-Duri, B., and McKee, S., Development of solutions to two-resistance mass transport models based on external and pore diffusion: Theoretical development and experimental results, Dev. Chem. Eng. Mineral Process., 1(2), 129-157 (1993).
Moon, H., and Tien, C., Further work on the prediction of gas-mixture adsorption equilibrium using the potential theory, Sep. Technol., 3(3), 161-167 (1993).
Muralidhar, R., and Talbot, J., Polydisperse mixture adsorption kinetics, AIChE J., 39(8), 1322-1329 (1993).
Northrup, M.A., et al., Direct measurement of interstitial velocity field variations in a porous medium using fluorescent-particle image velocimetry, Chem. Eng. Sci., 48(1), 13-22 (1993).
Ochoa-Tapia, J.A.; del Rio, J.A., and Whitaker, S., Bulk and surface diffusion in porous media: An application of the surface-averaging theorem, Chem. Eng. Sci., 48(11), 2061-2082 (1993).
Rota, R.; Gamba, G., and Morbidelli, M., On the use of the adsorbed solution theory for designing adsorption separation units, Sep. Technol., 3(4), 230-237 (1993).
Ruthven, D.M., and Stapleton, P., Measurement of liquid phase counter-diffusion in zeolite crystals by the ZLC method, Chem. Eng. Sci., 48(1), 89-98 (1993).
Scholl, S.; Kajszika, H., and Mersmann, A., Adsorption and desorption kinetics in activated carbon, Gas Sep. Purif., 7(4), 207-212 (1993).
Sereno, C., and Rodrigues, A., Can steady-state momentum equations be used in modelling pressurization of adsorption beds? Gas Sep. Purif., 7(3), 167-174 (1993).
Sotirchos, S.V., and Zarkanitis, S., A distributed pore size and length model for porous media reacting with diminishing porosity, Chem. Eng. Sci., 48(8), 1487-1502 (1993).
Storti, G., et al., Robust design of binary countercurrent adsorption separation processes, AIChE J., 39(3), 471-492 (1993).
Sun, L.M.; Meunier, F., and Karger, J., On the heat effect in measurements of sorption kinetics by the frequency response method, Chem. Eng. Sci., 48(4), 715-722 (1993).
Tien, C., Multicomponent adsorption calculations, Sep. Technol., 3(1), 32-45 (1993).
von Gemmingen, U., A new approach to adsorption isotherms (see corrigendum: 8(4), 272, 1994), Gas Sep. Purif., 7(3), 175-182 (1993).
Watson, J.S., The effects of flow through highly porous adsorbents on adsorption rates with favorable (nonlinear) isotherms, Sep. Sci. Technol., 28(1), 519-532 (1993).

Xiu, G.H., and Wakao, N., Batch adsorption: Intraparticle adsorbate concentration profile models, AIChE J., 39(12), 2042-2044 (1993).

Yao, C., and Tien, C., Approximations of uptake rate of spherical adsorbent pellets and their application to batch adsorption calculations, Chem. Eng. Sci., 48(1), 187-198 (1993).

Zhong, G.M.; Sun, L.M., and Meunier, F., Interference theory: Moment solution for linear isothermal multicomponent adsorption in a pellet, Chem. Eng. Sci., 48(24), 4115-4118 (1993).

**1994**

Akman, U., and Sunol, A.K., Equilibrium theory for exsorption: A gas-liquid-adsorbent mass-transfer operation, Chem. Eng. Sci., 49(21), 3555-3564 (1994).

Anand, N.; Manoja, B.G.R., and Gupta, A.K., Kinetics of adsorption on biporous solids, for a system with rectangular equilibrium, reanalysed, Chem. Eng. Sci., 49(19), 3277-3290 (1994).

Annesini, M.C.; Giona, M., and Gironi, F., Continuous model for complex mixture adsorption, Ind. Eng. Chem. Res., 33(11), 2764-2770 (1994).

Bakaev, V.A., and Steele, W.A., The characteristic curve in physical adsorption, Adsorpt. Sci. Technol., 10, 123-136 (1994).

Carrott, P.J.M., On the Dubinin-Serpinski equation, Adsorpt. Sci. Technol., 10, 63-74 (1994).

Chen, A., and Hirtzel, C.S., Monte Carlo simulation of multicomponent system by Marco State Markov Chain Model, Sep. Technol., 4(3), 167-173 (1994).

Croft, D.T., and LeVan, M.D., Periodic states of adsorption cycles: Direct determination and stability, Chem. Eng. Sci., 49(11), 1821-1830 (1994).

Croft, D.T., and LeVan, M.D., Periodic states of adsorption cycles: Solution spaces and multiplicity, Chem. Eng. Sci., 49(11), 1831-1842 (1994).

Dunne, J., and Myers, A.L., Adsorption of gas mixtures in micropores: Effect of difference in size of adsorbate molecules, Chem. Eng. Sci., 49(17), 2941-2952 (1994).

Filippov, L.K., Multicomponent non-isothermal adsorption dynamics, Chem. Eng. Commun., 127, 75-108 (1994).

Filippov, L.K., Study of adsorption of polymers on particulate surfaces by impregnation, Chem. Eng. Commun., 129, 1-18 (1994).

Filippov, L.K., Coherent and incoherent frontal patterns of multicomponent adsorption dynamics for variable linear fluid velocity in the fixed bed: Frontal patterns for linear adsorption isotherms, Chem. Eng. Sci., 49(20), 3499-3510 (1994).

Gomes, V.G., and Fuller, O.M., Fixed-bed adsorber dynamics in binary physi-sorption diffusion, Can. J. Chem. Eng., 72(4), 622-630 (1994).

Hassan, M.M.; Rahman, A.K.M.S., and Loughlin, K.F., Numerical simulation of unsteady continuous countercurrent adsorption system with nonlinear adsorption isotherm, Sep. Technol., 4(1), 15-26 (1994).

Jin, X.; Talbot, J., and Wang, N.H.L., Analysis of steric hindrance effects on adsorption kinetics and equilibria, AIChE J., 40(10), 1685-1696 (1994).

Jordi, R.G., and Do, D.D., Analysis of the frequency response method applied to non-isothermal sorption studies, Chem. Eng. Sci., 49(7), 957-980 (1994).

Kaminsky, R.D., and Monson, P.A., A simple mean field theory of adsorption in disordered porous materials, Chem. Eng. Sci., 49(17), 2967-2978 (1994).

Mazzotti, M.; Storti, G., and Morbidelli, M., Shock layer analysis in multicomponent chromatography and countercurrent adsorption, Chem. Eng. Sci., 49(9), 1337-1356 (1994).

Mazzotti, M.; Storti, G., and Morbidelli, M., Robust design of countercurrent adsorption separation processes: Multicomponent systems, AIChE J., 40(11), 1825-1842 (1994).

Mendes, A.M.M.; Costa, C.A.V., and Rodrigues, A.E., Linear driving force approximation for diffusion in spherical adsorbents with binary non-linear adsorption, Gas Sep. Purif., 8(4), 229-236 (1994).

Miyahara, M.; Kato, M., and Okazaki, M., Liquid-phase capillary condensation and adsorption isotherm, AIChE J., 40(9), 1549-1557 (1994).

Rozwadowski, M., et al., Equation of the sorption isotherm on zeolite-Y including heterogeneity of the sorbent: A novel modification, Adsorpt. Sci. Technol., 10, 241-253 (1994).

Scholl, S., and Mersmann, A., Binary adsorption on single adsorbent particles, Int. Chem. Eng., 34(3), 288-291 (1994).

Serpinski, V.V., and Jakubov, T.S., Dubinin-Radushkevich equation as the equation for the excess adsorption isotherm, Adsorpt. Sci. Technol., 10, 85-92 (1994).

Shah, D.B., A course on the fundamentals of adsorption, Chem. Eng. Educ., 28(4), 250-253 (1994).
Shen, C.M., and Worek, W.M., Cosorption characteristics of solid adsorbents, Int. J. Heat Mass Transfer, 37(14), 2123-2130 (1994).
Sircar, S., Non-isothermal differential adsorption kinetics for binary gas mixture, Ind. Eng. Chem. Res., 33(6), 1585-1592 (1994).
Stoeckli, F., Dubinin's theory for the volume filling of micropores: An historical approach, Adsorpt. Sci. Technol., 10, 3-16 (1994).
Sun, L.M., et al., Frequency response for nonisothermal adsorption in biporous pellets, Chem. Eng. Sci., 49(3), 373-382 (1994).
Tan, H.K.S., Analysis of cyclic fixed bed sorption processes for second-order irreversible kinetics, Chem. Eng. Sci., 49(8), 1277-1285 (1994).
Tolmachev, A.M., The characteristic curve method in adsorption, Adsorpt. Sci. Technol., 10, 155-164 (1994).
Van Tassel, P.R., et al., Lattice model and simulation of dynamics of adsorbate motion in zeolites, Chem. Eng. Sci., 49(17), 2979-2990 (1994).
Van Tassel, P.R.; Davis, H.T., and McCormick, A.V., New lattice model for adsorption of small molecules in zeolite micropores, AIChE J., 40(6), 925-934 (1994).
Vortmeyer, D.; Giese, M., and Lingg, G., Extension of the van Deemter equation to nonlinear isotherms of the Langmuir type, Chem. Eng. Sci., 49(16), 2593-2598 (1994).
Wei, J., Nonlinear phenomena in zeolite diffusion and reaction, Ind. Eng. Chem. Res., 33(10), 2467-2472 (1994).
Yao, C., and Tien, C., Approximate solution of intraparticle diffusion equations and their application to continuous-flow stirred tank and fixed-bed adsorption calculations, Sep. Technol., 4(2), 67-80 (1994).
Zhou, C.; Hall, F., and Gasem, K.A.M., Predicting gas adsorption using two-dimensional equations of state, Ind. Eng. Chem. Res., 33(5), 1280-1289 (1994).

**1995**

Andrade, J.S., et al., A network model for diffusion and adsorption in compacted pellets of bidisperse grains, Chem. Eng. Sci., 50(12), 1943-1952 (1995).
Arnost, D., and Schneider, P., Dynamic transport of multicomponent mixtures of gases in porous solids, Chem. Eng. J., 57(2), 91-100 (1995).
Arvind, G., and Bhatia, S.K., Determination of concentration-dependent adsorbate diffusivities by numerical inversion, Chem. Eng. Sci., 50(8), 1361-1372 (1995).
Bulow, M., and Micke, A., Determination of transport coefficients in microporous solids, Adsorption, 1(1), 29-48 (1995).
Cheng, L.S., and Yang, R.T., Predicting isotherms in micropores for different molecules and temperatures from a known isotherm by improved Horvath-Kawazoe equations, Adsorption, 1(3), 187-196 (1995).
Choudary, N.V.; Jasra, R.V., and Bhat, S.G.T., An improved Langmuir approach for prediction of binary gas mixture adsorption equilibria, Sep. Sci. Technol., 30(11), 2337-2348 (1995).
Chu, K.H., and Hashim, M.A, Simulated countercurrent adsorption processes: A comparison of modelling strategies, Chem. Eng. J., 56(2), 59-65 (1995).
Coutelieris, F.A.; Burganos, V.N., and Payatakes, A.C., Convective diffusion and adsorption in a swarm of spheroidal particles, AIChE J., 41(5), 1122-1134 (1995).
Cracknell, R.F., and Nicholson, D., Adsorption of gas mixtures on solid surfaces, theory and computer simulation, Adsorption, 1(1), 7-16 (1995).
Critoph, R.E., and Turner, L., Heat transfer in granular activated carbon beds in the presence of adsorbable gases, Int. J. Heat Mass Transfer, 38(9), 1577-1586 (1995).
de Kock, F.P., and van Deventer, J.S.J., Stochastic model for equilibrium adsorption onto activated carbon, Chem. Eng. J., 59(3), 205-220 (1995).
Devotta, I., et al., A new phenomenological model for adsorption-diffusion in polymer solutions: Role of disengagement dynamics, Chem. Eng. Sci., 50(7), 1129-1142 (1995).
Do, D.D., Dynamics of a semi-batch adsorber with constant molar supply rate: A method for studying adsorption rate of pure gases, Chem. Eng. Sci., 50(3), 549-553 (1995).
Do, D.D., and Rice, R.G., Revisiting approximate solutions for batch adsorbers: Explicit half time, AIChE J., 41(2), 426-429 (1995).
Do, H.D., and Do, D.D., Effect of pore size distribution on the surface diffusivity in activated carbon: Hybrid Dubinin-Langmuir isotherm, Adsorption, 1(4), 291-302 (1995).

Edwards, D.A., and Davis, A.M.J., Diffusion and convective dispersion through arrays of spheres with surface adsorption, diffusion, and unequal solute partitioning, Chem. Eng. Sci., 50(9), 1441-1454 (1995).
Forbert, R., and Schlunder, E.U., A model for the prediction of the adsorption equilibria of arbitrary organic vapours on active carbon, Chem. Eng. Process., 34(1), 49-59 (1995).
Garbacz, J.K., and Dabrowski. A., Partially mobile adsorption of a single gas on a microporous adsorbent, Adsorpt. Sci. Technol., 12(4), 353-363 (1995).
Giona, M., and Giustiniani, M., Size-dependent adsorption models in microporous materials: Thermodynamic consistency and theoretical analysis, Ind. Eng. Chem. Res., 34(11), 3848-3855 (1995).
Giona, M.; Giustiniani, M., and Viola, A., Two-step adsorption models in molecular sieves, Chem. Eng. J., 58(1), 21-32 (1995).
Giustiniani, M.; Giona, M., and Ludlow, D.K., Size-dependent adsorption models in microporous materials: Comparison with experimental data, Ind. Eng. Chem. Res., 34(11), 3856-3864 (1995).
Grenier, P., et al., Single-step thermal method to measure intracrystalline mass diffusion in adsorbents, AIChE J., 41(9), 2047-2057 (1995).
Gruber, J., Waves in a two-component system: The oxide surface as a variable charge adsorbent, Ind. Eng. Chem. Res., 34(8), 2769-2781 (1995).
Hajji, A., and Khalloufi, S., Theoretical and experimental investigation of a constant-pressure adsorption process, Int. J. Heat Mass Transfer, 38(18), 3349-3358 (1995).
Hu, X., and Do, D.D., Validity of isothermality in adsorption kinetics of gases in bidispersed solids, AIChE J., 41(6), 1581-1584 (1995).
Hu, X., and Do, D.D., Comparing various multicomponent adsorption equilibrium models, AIChE J., 41(6), 1585-1592 (1995).
Huang, R.T.; Chen, T.L., and Weng, H.S., Binary adsorption in a fixed-bed column packed with an ion-exchange resin, Sep. Sci. Technol., 30(13), 2731-2746 (1995).
Hwang, K.S.; Jun, J.H., and Lee, W.K., Fixed-bed adsorption for bulk component system: Non-equilibrium, non-isothermal and non-adiabatic model, Chem. Eng. Sci., 50(5), 813-826 (1995).
Keller, J.U., Theory of measurement of gas-adsorption equilibria by rotational oscillations, Adsorption, 1(4), 283-290 (1995).
Koh, C.A.; Nooney, R.I., and Gubbins, K.E., Spectroscopic and simulation studies of adsorption and removal of hydrocarbons using novel materials, Ind. Eng. Chem. Res., 34(12), 4170-4173 (1995).
Kruglov, A.V., amd Aris, R., Adsorption kinetics for the case of step and s-shape isotherms, AIChE J., 41(11), 2393-2398 (1995).
Kruszewski, S., Influence of surface roughness variation on the surface-enhanced Raman scattering (SERS) signal, Adsorpt. Sci. Technol., 11(4), 217-224 (1995).
Leitao, A., and Rodrigues, A., Adsorptive processes using "large-pore" materials: Analysis of a criterion for equivalence of diffusion-convection, "apparent" diffusion and "extended" linear driving force models, Chem. Eng. J., 60(1), 81-88 (1995).
Lerch, V.; Haul, R., and Hease, D., Differential pressure-jump method for routine measurement of sorption kinetics, Adsorption, 1(3), 265-282 (1995).
Malek, A., et al., Effect of velocity variation due to adsorption-desorption on equilibrium data from breakthrough experiments, Chem. Eng. Sci., 50(4), 737-740 (1995).
Mann, R., et al., Interpretation of water isotherm hysteresis for an activated charcoal using stochastic pore networks, Adsorption, 1(3), 253-264 (1995).
Mendes, A.M.M.; Costa, C.A.V., and Rodrigues, A.E., Linear driving force approximation for isothermal non-isobaric diffusion/convection with binary Langmuir adsorption, Gas Sep. Purif., 9(4), 259-270 (1995).
Millot, S.; Loubet, J.L., and Georges, J.M., Variation of the hydrodynamic polymer layer thickness on solid surfaces with molecular weight and concentration, Adsorpt. Sci. Technol., 12(3), 259-265 (1995).
Nivarthi, S.S.; Davis, H.T., and McCormick. A.V., Effectiveness of window blocking in zeolite NaY by strongly coadsorbed molecules, Chem. Eng. Sci., 50(20), 3217-3230 (1995).
Olivares, J.M., and Peppas, N.A., In-situ penetrant transport in thin coal sections, Chem. Eng. Commun., 132, 91-106 (1995).
Palkar, S.A., and Lenhoff, A.M., Effect of taper on diffusion and adsorption in cylindrical pore, Ind. Eng. Chem. Res., 34(10), 3551-3555 (1995).
Peck, J.D., and Yang, R.T., Effect of binary cross-term diffusivities in molecular sieve on adsorber dynamics, Chem. Eng. Sci., 50(21), 3487-3491 (1995).

Rangarajan, B.; Lira, C.T., and Subramanian, R., Simplified local density model for adsorption over large pressure ranges, AIChE J., 41(4), 838-845 (1995).

Rudzinski, W., et al., On the theoretical origin and applicability of the potential theory approach to predict mixed-gas adsorption on solid surfaces from single-gas adsorption isotherms, Chem. Eng. Sci., 50(16), 2641-2660 (1995).

Sikavitsas, V.I., and Yang, R.T., Predicting multicomponent diffusivities for diffusion on surfaces and in molecular sieves with energy heterogeneity, Chem. Eng. Sci., 50(19), 3057-3066 (1995).

Sircar, S., Influence of adsorbate size and adsorbent heterogeneity on IAST, AIChE J., 41(5), 1135-1145 (1995).

Subramanian, R.; Pyada, H., and Lira, C.T., An engineering model for adsorption of gases onto flat surfaces and clustering in supercritical fluids, Ind. Eng. Chem. Res., 34(11), 3830-3837 (1995).

Tondeur, D., Paradigms and paradoxes in modeling adsorption and chromatographic separations, Ind. Eng. Chem. Res., 34(8), 2782-2788 (1995).

van den Broeke, L.J.P., Simulation of diffusion in zeolitic structures, AIChE J., 41(11), 2399-2414 (1995).

Various, Advances in membrane separation and adsorption (topic issue), Dev. Chem. Eng. Mineral Process., 3(3), 122-234 (1995).

Watson, J.S., Simplified predictions of breakthrough fronts for constant-pattern adsorption and ion exchange, Sep. Sci. Technol., 30(7), 1351-1372 (1995).

Yang, R.T., and Sikavitsas, V.I., Kinetic theory for predicting multicomponent diffusivities from pure-component diffusivities for surface diffusion and diffusion in molecular sieves, Chem. Eng. Sci., 50(20), 3319-3322 (1995).

## 1996

Ballamudi, R.K., and Bitsanis, I.A., Structural transitions at solid-liquid interfaces, Adsorption, 2(1), 69-76 (1996).

Barbari, T.A.; Kasargod, S.S., and Fieldson, G.T., Effect of unequal transport rates and intersolute solvation on the selective batch extraction of a dilute mixture with a dense polymeric sorbent, Ind. Eng. Chem. Res., 35(4), 1188-1194 (1996).

Bojan, M.J., et al., Topologies of capillary condensation, Adsorption, 2(1), 51-58 (1996).

Bourdin, V., et al., Application of the thermal frequency response method and of pulsed field gradient NMR to study water diffusion in zeolite NaX, Adsorption, 2(3), 205-216 (1996).

Bourdin, V., et al., Thermal frequency response method for the study of mass-transfer kinetics in adsorbents, AIChE J., 42(3), 700-712 (1996).

Brandani, S., Analytical solution for ZLC desorption curves with bi-porous adsorbent particles, Chem. Eng. Sci., 51(12), 3283-3288 (1996).

Brandani, S., and Ruthven, D.M., Analysis of ZLC desorption curves for gaseous systems, Adsorption, 2(2), 133-144 (1996).

Brandani, V.; Giona, M., and Giustiniani, M., Comments and reply on "Size-dependent adsorption models in microporous materials: Thermodynamic consistency and theoretical analysis", Ind. Eng. Chem. Res., 35(8), 2813-2815 (1996).

Chaiken, R.F., Absorption/desorption model of diffusion for heterogeneous rate processes, AIChE J., 42(7), 2087-2089 (1996).

Chiang, A.S.T.; Lee, C.K., and Wu, F.Y., Theory of adsorbed solutions: Analysis of one-dimensional systems, AIChE J., 42(8), 2155-2161 (1996).

Czepirski, L., et al., Apparatus for the investigation of adsorption equilibria and kinetics at elevated pressures, Adsorpt. Sci. Technol., 14(2), 77-82 (1996).

Devotta, I., and Mashelkar, R.A., Competitive diffusion-adsorption of polymers of differing chain lengths on solid surfaces, Chem. Eng. Sci., 51(4), 561-570 (1996).

Dogu, T., et al., One-sided single-pellet technique for adsorption and intraparticle diffusion, AIChE J., 42(2), 516-523 (1996).

Esposito, F., et al., A generalized form of the Langmuir isotherm for gas adsorption in glassy polymers, Ind. Eng. Chem. Res., 35(9), 2939-2945 (1996).

Fichthorn, K.A., Diffusion of short-chain molecules on metal surfaces, Adsorption, 2(1), 77-88 (1996).

Filippov, L.K., Multicomponent adsorption dynamics of polymers in a fixed bed, Chem. Eng. Sci., 51(16), 4013-4024 (1996).

Foeth, F., et al., Equilibrium adsorption data from temperature-programmed desorption experiments, Adsorption, 2(4), 279-286 (1996).

Foeth, F., et al., Equilibrium adsorption data from breakthrough curves with variable velocity and pressure, Sep. Sci. Technol., 31(1), 21-38 (1996).

Garbacz, J.K.; Kopkowski, A., and Dabrowski, A., Isosteric heat of single gas adsorption on an energetically homogeneous adsorbent surface, Adsorpt. Sci. Technol., 13(2), 105-114 (1996).

Garbacz, J.K.; Kowalczyk-Dembinska, H., and Dabrowski, A., Partially mobile adsorption of single gases on graphitized carbon black surfaces, Adsorpt. Sci. Technol., 14(1), 69-76 (1996).

Georgiou, A., and Kupiec, K., Nonlinear driving force approximations of intraparticle mass transfer in adsorption processes: The effect of pellet shape, Dev. Chem. Eng. Mineral Process., 4(3), 235-244 (1996).

Giona, M., and Giustiniani, M., Thermodynamics and kinetics of adsorption in the presence of geometric roughness, Sep. Technol., 6(2), 99-110 (1996).

Giona, M., et al., Exact solution of linear transport equations in fractal media: Adsorption and chemical reaction, Chem. Eng. Sci., 51(22), 5065-5076 (1996).

Goncharuk, V.V.; Gorchev, V.F., and Karakhim, S.A., A new approach for modelling heterogeneous surfaces, Adsorpt. Sci. Technol., 14(6), 375-382 (1996).

Grechko, L.G., et al., Some aspects of the absorption of electromagnetic radiation by heterogeneous systems, Adsorpt. Sci. Technol., 14(6), 363-374 (1996).

Gusev, V., et al., Theory for multicomponent adsorption equilibrium: Multispace adsorption model, AIChE J., 42(10), 2773-2783 (1996).

Gutsche, R., and Hartmann, K., Generalized criteria for predicting the dynamics of continuous-flow chemical systems: Application to adsorber systems, Chem. Eng. Sci., 51(13), 3519-3536 (1996).

Hanagandi, V.; Ploehn, H., and Nikolaou, M., Solution of the self-consistent field model for polymer adsorption by genetic algorithms, Chem. Eng. Sci., 51(7), 1071-1078 (1996).

Heslop, M.J.; Buffham, B.A., and Mason, G., A test of the polynomial-fitting method of determining binary-gas-mixture adsorption equilibria, Ind. Eng. Chem. Res., 35(4), 1456-1466 (1996).

Hinderer, J., and Keil, F.J., Three-dimensional Monte Carlo calculations of diffusion and reaction phenomena in zeolites, Chem. Eng. Sci., 51(11), 2667-2672 (1996).

Holysz, L., Surface free energy interactions of a 'Thermisil' glass surface: A comparison of the thin layer wicking and contact angle techniques, Adsorpt. Sci. Technol., 14(2), 89-100 (1996).

Kaczmarski, K., and Antos, D., Fast finite difference method for solving multicomponent adsorption-chromatography models, Comput. Chem. Eng., 20(11), 1271-1276 (1996).

Keffer, D.; Davis, H.T., and McCormick, A.V., The effect of nanopore shape on the structure and isotherms of adsorbed fluids, Adsorption, 2(1), 9-22 (1996).

Kim, D.H., A new linear formula for cyclic adsorption in a particle, Chem. Eng. Sci., 51(17), 4137-4144 (1996).

Lang, R., et al., Enhancement of the heat and mass transfer in compact zeolite layers, Adsorption, 2(2), 121-132 (1996).

Lavanchy, A., et al., Binary adsorption of vapours in active carbons described by the Dubinin equation, Adsorpt. Sci. Technol., 13(6), 537-545 (1996).

Ma, Z.; Whitley, R.D., and Wang, N.H.L., Pore and surface diffusion in multicomponent adsorption and liquid chromatography systems, AIChE J., 42(5), 1244-1262 (1996).

Maddox, M.W.; Sowers, S.L., and Gubbins, K.E., Molecular simulation of binary mixture adsorption in buckytubes and MCM-41, Adsorption, 2(1), 23-32 (1996).

Martinez, G.M., and Basmadjian, D., Towards a general gas adsorption isotherm, Chem. Eng. Sci., 51(7), 1043-1054 (1996).

Monson, P.A. (ed.), Molecular modeling of adsorption (topic issue), Adsorption, 2(1), 5-110 (1996).

Nan, H.S.; Dias, M.M., and Rodrigues, A.E., Measurements of effective diffusivity in large-pore permeable pellets with various geometries using the chromatographic method, Chem. Eng. Commun., 146, 201-220 (1996).

Ngai, S., and Gomes, V.G., Nonlinear sorption isotherm of zeolites by frequency response analysis, Ind. Eng. Chem. Res., 35(4), 1475-1479 (1996).

Niemeyer, B.; Feilenreiter, T., and Tiltscher, H., Theoretical studies on biospecific adsorption for large-scale affinity separations, Chem. Eng. Sci., 51(24), 5263-5272 (1996).

Ogenko, V.M., Surface chemistry in modern nanotechnologies, Adsorpt. Sci. Technol., 14(5), 295-300 (1996).

Park, I.S.; Do, D.D., and Rodrigues, A.E., Measurement of the effective diffusivity in porous media by the diffusion cell method, Catal. Rev. Sci. Eng., 38(2), 189-248 (1996).

Prochaska, K., Adsorption of extractants and modifiers in mixed binary model systems, Solvent Extr. Ion Exch., 14(6), 1057-1076 (1996).

Ray, M.S., Adsorption and adsorptive separations: A review and bibliographical update (1994), Adsorption, 2(2), 157-178 (1996).

Ray, M.S., Adsorptive and membrane-type separations: A bibliographical update (1994), Adsorpt. Sci. Technol., 13(1), 49-69 (1996).

Rosner, D.E., and Papadopoulos, D.H., Jump, slip, and creep boundary conditions at nonequilibrium gas/solid interfaces, Ind. Eng. Chem. Res., 35(9), 3210-3222 (1996).

Rudzinski, W.; Charmas, R., and Piasecki, W., Nature of energetic surface heterogeneity in ion adsorption at electrolyte/oxide interface: Theoretical studies of correlations between binding-to-surface energies of various surface complexes, Adsorpt. Sci. Technol., 14(1), 25-38 (1996).

Rudzinski, W.; Dominko, A., and Wojciechowski, B.W., Mixed-gas adsorption on real solid surfaces: Lack of correlations between adsorption energies of various components related to wide applicability of generalized Langmuir-Freundlich isotherm, Chem. Eng. J., 64(1), 85-98 (1996).

Russell, B.P., and LeVan, M.D., Group-contribution theory for adsorption of gas mixtures on solid surfaces, Chem. Eng. Sci., 51(16), 4025-4038 (1996).

Scharfenberg, R.; Meyerhoff, K., and Hesse, D., Problems in the determination of pore connectivity by digital image processing, Chem. Eng. Sci., 51(10), 1889-1898 (1996).

Shumyatsky, Y.I., and Alekhina, M.S., The effects accompanying passing current through a granular bed of activated carbon, Adsorption, 2(4), 317-322 (1996).

Silva, J.A.C., and Rodrigues, A.E., Analysis of ZLC technique for diffusivity measurements in bidisperse porous adsorbent pellets, Gas Sep. Purif., 10(4), 207-224 (1996).

Sircar, S., Data representation for binary and multicomponent gas adsorption equilibria, Adsorption, 2(4), 327-335 (1996).

Somers, S.A., et al., Binary fluids in planar nanopores: Adsorptive selectivity, heat capacity and self-diffusivity, Adsorption, 2(1), 33-40 (1996).

Sun, L.M., and Do, D.D., Frequency response analysis of a closed diffusion cell with two resonators, Adsorption, 2(4), 265-278 (1996).

Talbot, J., Time dependent desorption: A memory function approach, Adsorption, 2(1), 89-94 (1996).

Taqvi, S.M., and LeVan, M.D., Role of convection and diffusion in a single pore with adsorptive walls, Adsorption, 2(4), 299-310 (1996).

Tondeur, D., et al., Multicomponent adsorption equilibria from impulse response chromatography, Chem. Eng. Sci., 51(15), 3781-3800 (1996).

Various, Proceedings of the 1st Polish-Ukrainian Symposium on Theoretical and Experimental Studies of Interfacial Phenomena and their Technological Application (Kazimierz Dolny, Poland; 8-10 June 1995), Adsorpt. Sci. Technol., 14(1), 1-76 (1996).

Vega, L.F., et al., Effect of surface active sites on adsorption of associating chain molecules in pores: A Monte Carlo study, Adsorption, 2(1), 59-68 (1996).

Xiu, G.H., Modeling breakthrough curves in a fixed bed of activated carbon fiber: Exact solution and parabolic approximation, Chem. Eng. Sci., 51(16), 4039-4041 (1996).

Yang, M., et al., A neural network model for prediction of binary adsorption using single solute and limited binary solute adsorption data, Sep. Sci. Technol., 31(9), 1259-1266 (1996).

Yi, X.; Shing, K.S., and Sahimi, M., Molecular simulation of adsorption and diffusion in pillared clays, Chem. Eng. Sci., 51(13), 3409-3426 (1996).

**1997**

Bereznitski, Y.; Jaroniec, M., and Gadkaree, K.P., Influence of analysis conditions on low pressure adsorption measurements and its consequences in characterization of energetic and structural heterogeneity of microporous carbons, Adsorption, 3(4), 277-282 (1997).

Bhatia, S.K., Transport in bidisperse adsorbents: Significance of the macroscopic adsorbate flux, Chem. Eng. Sci., 52(8), 1377-1386 (1997).

Bradley, R.H., Polar interactions at carbon surfaces described by the Dubinin equations and by an XPS-based model, Adsorpt. Sci. Technol., 15(7), 477-484 (1997).

Braeuer, P.; Salem, M., and Harting, P., Calculation of single adsorption isotherms from gravimetrically measured binary gas mixture adsorption isotherms on activated carbon at high pressures, Sep. Purif. Technol., 12(3), 255-263 (1997).

Burganos, V.N.; Coutelieris, F.A., and Payatakes, A.C., Sherwood number for a swarm of adsorbing spheroidal particles at any Peclet number, AIChE J., 43(3), 844-846 (1997).
De Kock, F.P., and van Deventer, J.S.J., An evaluation of isotherms in the description of competitive adsorption equilibria, Chem. Eng. Commun., 160, 35-58 (1997).
Do, D.D., and Do, H.D., A new adsorption isotherm for heterogeneous adsorbent based on the isosteric heat as a function of loading, Chem. Eng. Sci., 52(2), 297-310 (1997).
Garbacz, J.K., and Dabrowski, M., A non-analytical solution of the description of single gas adsorption on a microporous adsorbent, Adsorpt. Sci. Technol., 15(2), 99-108 (1997).
Garbacz, J.K., and Kopkowski, A., Adsorbate-adsorbate association in a one-component mobile adsorption phase on energetically homogeneous adsorbents, Adsorpt. Sci. Technol., 15(9), 695-706 (1997).
Hutson, N.D., and Yang, R.T., Theoretical basis for the Dubinin-Radushkevitch (D-R) adsorption isotherm equation, Adsorption, 3(3), 189-196 (1997).
Jaroniec, M., Fifty years of the theory of the volume filling of micropores, Adsorption, 3(3), 187-188 (1997).
Kaczmarski, K., et al., Modeling fixed-bed adsorption columns through orthogonal collocations on moving finite elements, Comput. Chem. Eng., 21(6), 641-660 (1997).
Kaneko, K., and Murata, K., An analytical method of micropore filling of a supercritical gas, Adsorption, 3(3), 197-208 (1997).
Khamis, E.; Moussa, M.N., and Ibrahim, A.K., Application of the kinetic-thermodynamic model to the adsorption of organophosphorus compounds on to a steel surface, Adsorpt. Sci. Technol., 15(3), 173-188 (1997).
Kim, D.H., A new linear approximation formula for cyclic adsorption in a biporous adsorbent, Chem. Eng. Sci., 52(20), 3471-3482 (1997).
Malek, A., and Farooq, S., Effect of velocity variation on equilibrium calculations from multicomponent breakthrough experiments, Chem. Eng. Sci., 52(3), 443-448 (1997).
Maurer, R.T., Multimodel approach to mixed-gas adsorption equilibria prediction, AIChE J., 43(2), 388-397 (1997).
Mota, J.P.B.; Rodrigues, A.E., and Tondeur, D., Charge dynamics of a methane adsorption storage system: Intraparticle diffusional effects, Adsorption, 3(2), 117-126 (1997).
Myers, A.L.; Calles, J.A., and Calleja, G., Comparison of molecular simulation of adsorption with experiment, Adsorption, 3(2), 107-116 (1997).
Prisciandaro, M., and Pepe, F., Adsorption with zero and pseudo-zero order chemical reaction, Can. J. Chem. Eng., 75(2), 362-368 (1997).
Ray, M.S., Adsorptive and membrane-type separations: A bibliographical update (1996), Adsorpt. Sci. Technol., 15(9), 627-648 (1997).
Russell, B.P., and LeVan, M.D., Nonlinear adsorption and hydrodynamic dispersion in self-similar networks, Chem. Eng. Sci., 52(9), 1501-1510 (1997).
Rutherford, S.W., and Do, D.D., Permeation time lag with non-linear adsorption and surface diffusion, Chem. Eng. Sci., 52(5), 703-714 (1997).
Rynders, R.M.; Rao, M.B., and Sircar, S., Isotope exchange technique for measurement of gas adsorption equilibria and kinetics, AIChE J., 43(10), 2456-2470 (1997).
Serbezov, A.S., and Sotirchos, S.V., Mathematical modeling of the adsorptive separation of multicomponent gaseous mixtures, Chem. Eng. Sci., 52(1), 79-92 (1997).
Sheng, P., and Costa, C.A.V., Modified linear driving force approximations for cyclic adsorption-desorption processes, Chem. Eng. Sci., 52(9), 1493-1500 (1997).
Siddiqui, F.A., and Franses, E.I., Dynamic adsorption and tension of nonionic binary surfactant mixtures, AIChE J., 43(6), 1569-1578 (1997).
Stoeckli, F.; Wintgens, D.; Lavanchy, A., and Stockli, M., Binary adsorption of vapours in active carbons described by the combined theories of Myers-Prausnitz and Dubinin(II), Adsorpt. Sci. Technol., 15(9), 677-684 (1997).
Suen, S.Y., An isotherm model describing concave-down Scatchard curve for heterogeneous affinity adsorption, J. Chem. Technol. Biotechnol., 70(3), 278-286 (1997).
Talbot, J., Analysis of adsorption selectivity in a one-dimensional model system, AIChE J., 43(10), 2471-2478 (1997).
Taqvi, S.M., and LeVan, M.D., A simple way to describe nonisothermal adsorption equilibrium data using polynomials orthogonal to summation, Ind. Eng. Chem. Res., 36(2), 419-423 (1997).

Taqvi, S.M., and LeVan, M.D., Virial description of two-component adsorption on homogeneous and heterogeneous surfaces, Ind. Eng. Chem. Res., 36(6), 2197-2206 (1997).

Taqvi, S.M.; Vishnoi, A., and LeVan, M.D., Effect of macropore convection on mass transfer in a bidisperse adsorbent particle, Adsorption, 3(2), 127-136 (1997).

Vaidya, D.S., et al., Convection-diffusion of solutes in dynamic media, Adsorption, 3(1), 41-54 (1997).

Valladares, D., and Zgrablich, G., A test of the Horvath-Kawazoe method by Monte-Carlo simulation, Adsorpt. Sci. Technol., 15(1), 15-24 (1997).

van Zee, G.; Bisschops, M., and de Graauw, J., Internal Sherwood numbers for diffusion in sorption operations, Chem. Eng. Process., 36(3), 201-208 (1997).

Various, Adsorption in micropores (topic issue), Adsorption, 3(3), 187-250 (1997).

Watson, J.S., Constant-pattern adsorption fronts with mass transfer resistances in both solid and fluid films, Sep. Sci. Technol., 32(16), 2607-2621 (1997).

Xui, G.H.; Nitta, T.; Li, P., and Jin, G., Breakthrough curves for fixed-bed adsorbers: Quasi-lognormal distribution approximation, AIChE J., 43(4), 979-985 (1997).

Zhang, R., and Ritter, J.A., New approximate model for nonlinear adsorption and diffusion in a single particle, Chem. Eng. Sci., 52(18), 3161-3172 (1997).

Ziolkowska, D., and Garbacz, J.K., Adaptation of single gas adsorption equations for the description of adsorption from non-aqueous liquid solutions of iodine onto active carbons, Adsorpt. Sci. Technol., 15(3), 155-164 (1997).

Zumkeller, H.J., and Bart, H.J., Influence of residual loadings on mass transfer and efficiency in sorption cycles, Chem. Eng. Process., 36(6), 459-468 (1997).

## ADSORPTION DESIGN METHODS AND DATA

### 1967

Haydel, J.J., and Kobayashi, R., Adsorption equilibria in the methane-propane-silica gel system at high pressures, Ind. Eng. Chem. Fund., 6(4), 546-554 (1967).

Lerch, R.G., and Ratkowsky, D.A., Optimum allocation of adsorbent in stagewise adsorption operations, Ind. Eng. Chem. Fund., 6(2), 308-310; 6(3), 480 (1967).

Schumacher, W.J., and York, R., Separation of hydrocarbons in fixed beds of molecular sieves, Ind. Eng. Chem. Process Des. Dev., 6(3), 321-327 (1967).

Todes, O.M., and Lezin, Y.S., Dynamics of continuous adsorption and desorption in fluidized bed, Int. Chem. Eng., 7(3), 447-450; 7(4), 577-581 (1967).

### 1968

Carter, J.W., Some aspects of the prediction of performance of a large air-dryer, Trans. IChemE, 46, T222-T224 (1968).

Chen, J.W.; Buege, J.A.; Cunningham, F.L., and Northam, J.I., Scale-up of a column adsorption process by computer simulation, Ind. Eng. Chem. Process Des. Dev., 7(1), 26-31 (1968).

Collins, J.J., Where to use molecular sieves, Chem. Eng. Prog., 64(8), 66-71 (1968).

Dayan, J., and Levenspiel, O., Longitudinal dispersion in packed beds of porous adsorbing solids, Chem. Eng. Sci., 23(11), 1327-1334 (1968).

Fukunaga, P.; Hwang, K.C.; Davis, S.H., and Winnick, J., Mixed-gas adsorption and vacuum desorption of carbon dioxide on molecular sieve, Ind. Eng. Chem. Process Des. Dev., 7(2), 269-275 (1968).

Masukawa, S., and Kobayashi, R., Correlations for adsorption of a binary gas mixture on a heterogeneous adsorbent: Methane-ethane-silica gel system, AIChE J., 14(5), 740-746 (1968).

Rimpel, A.E.; Camp, D.T.; Kostecki, J.A., and Canjar, L.N., Kinetics of physical adsorption of propane from helium on fixed beds of activated alumina, AIChE J., 14(1), 19-24 (1968).

Shen, J., and Smith, J.M., Adsorption isotherms for benzene-hexane mixtures, Ind. Eng. Chem. Fund., 7(1), 100-105 (1968).

Shen, J., and Smith, J.M., Rates of adsorption in the benzene-hexane system, Ind. Eng. Chem. Fund., 7(1), 106-114 (1968).

Shulman, H.L.; Youngquist, G.R.; Marsh, J.L., and Mehta, V.K., Development of a continuous countercurrent fluid-solids adsorber, Ind. Eng. Chem. Process Des. Dev., 7(4), 493-496 (1968).

Tikonova, N.M., Gas flow distribution in packed beds, Int. Chem. Eng., 8(1), 135-138 (1968).

**1969**

Asher, W.J.; Campbell, M.L.; Epperly, W.R., and Robertson, J.L., Desorb n-paraffins from molecular sieves with ammonia, Hydrocarbon Process., 48(1), 134-138 (1969).

Avery, D.A., and Boiston, D.A., The recovery of solvents from gaseous effluents, Chem. Eng. (Rugby, Engl.), January, CE8-11 (1969).

Carter, J.W., Scale-up in the design of fixed-bed adsorption plant, Brit. Chem. Eng., 14(3), 303-307 (1969).

Danner, R.P., and Wenzel, L.A., Adsorption of CO-nitrogen, CO-oxygen, and oxygen-nitrogen mixtures on synthetic zeolites, AIChE J., 15(4), 515-520 (1969).

Emery, D.L., Drying natural gas with alumina, Hydrocarbon Process., 48(3), 130-132 (1969).

Layton, L., and Youngquist, G.R., Sorption of sulfur dioxide by ion exchange resins, Ind. Eng. Chem. Process Des. Dev., 8(3), 317-324 (1969).

Lucas, J.P., and Ratkowsky, D.A., Optimization in various multistage adsorption operations, Ind. Eng. Chem. Fund., 8(3), 576-581 (1969).

Mehta, D.S., and Calvert, S., Calculating actual plates in continuous sorptions, Brit. Chem. Eng., 14(11), 1563 (1969).

Romberg, E., and Shorrock, J.C., Determination of adsorption equilibria for the helium-methane-krypton system on active charcoal using a dynamic method, Trans. IChemE, 47, T3-T10 (1969).

Symoniak, M.F., Correlation for sizing adsorption systems, Chem. Eng. (N.Y.), 6 October, 172-174 (1969).

Zwiebel, I., Fixed-bed adsorption with variable gas velocity due to pressure drop, Ind. Eng. Chem. Fund., 8(4), 803-807 (1969).

**1970**

Allen, T., and Patel, R.M., Adsorption of alcohols on finely divided powders, J. Applied Chem., 20, 165-171 (1970).

Kehat, E., and Heineman, M., Desorption of normal paraffins from 5A molecular sieve, Ind. Eng. Chem. Process Des. Dev., 9(1), 72-78 (1970).

Kidnay, A.J., and Hiza, M.J., Purification of helium gas by physical adsorption at 76K, AIChE J., 16(6), 949-954 (1970).

Kotb, A.K., Adsorption of sulphur dioxide on coal, J. Applied Chem., 20, 147-152 (1970).

LaPlante, L.J., and Symoniak, M.F., Molecular sieves for protein-from-paraffins process, Hydrocarbon Process., 49(12), 77-82 (1970).

Marcussen, L., Kinetics of water adsorption on porous alumina, Chem. Eng. Sci., 25(9), 1487-1500 (1970).

Phillips, J.R., Desorption of normal paraffins from 5A molecular sieves, Ind. Eng. Chem. Process Des. Dev., 9(3), 484-485 (1970).

Snyder, C.F., and Chao, K.C., Heat of adsorption of light hydrocarbons and their mixtures on activated carbon, Ind. Eng. Chem. Fund., 9(3), 437-443 (1970).

Tomassi, W., Effect of temperature variations on the state of chlorine in adsorption layer on activated carbon, Int. Chem. Eng., 10(3), 372-374 (1970).

**1971**

Alonso, J.R.F., Simple breakthrough curves yield more information about fixed-bed adsorbers, Brit. Chem. Eng., 16(7), 602-603 (1971).

Gonzalez, A.J., and Holland, C.D., Adsorption equilibria of light hydrocarbon gases on activated carbon and silica, AIChE J., 17(2), 470-475 (1971).

Hales, G.E., Drying reactive fluids with molecular sieves, Chem. Eng. Prog., 67(11), 49-53 (1971).

Hales, G.E., Gas dehydration using molecular sieves, Hydrocarbon Process., 50(6), 151-154 (1971).

Kondis, E.F., and Dranoff, J.S., Kinetics of isothermal sorption of ethane on 4A molecular sieve pellets, Ind. Eng. Chem. Process Des. Dev., 10(1), 108-114 (1971).

McCarthy, W.C., Adsorption of light hydrocarbons from nitrogen with activated carbon, Ind. Eng. Chem. Process Des. Dev., 10(1), 13-18 (1971).

Reikert, L., Rates of sorption and diffusion of hydrocarbons in zeolites, AIChE J., 17(2), 446-454 (1971).

Ruthven, D.M., and Loughlin, K.F., Sorption and diffusion of n-butane in Linde 5A molecular sieve, Chem. Eng. Sci., 26(8), 1145-1154 (1971).

Thomas, W.J., and Lombardi, J.L., Binary adsorption of benzene-toluene mixtures, Trans. IChemE, 49, T240-T250 (1971).

Thomas, W.J., and Qureshi, A.R., Adsorption of toluene in fixed beds of active carbons, Trans. IChemE, 49, T60-T69 (1971).

**1972**

Alonsono, J.R.F., Design of fixed bed adsorbers, Brit. Chem. Eng., 17(4), 326; 17(6), 517-522 (1972).

Carter, J.W., and Husain, H., Adsorption of carbon dioxide in fixed beds of molecular sieves, Trans. IChemE, 50, T69-T75 (1972).

Chen, J.W.; Cunningham, F.L., and Buege, J.A., Computer simulation of plant-scale multicolumn adsorption processes under periodic countercurrent operation, Ind. Eng. Chem. Process Des. Dev., 11(3), 430-434 (1972).

Colley, C.R., Adsorbent selection for gas drying, Brit. Chem. Eng., 17(3), 229-233 (1972).

Fernbacher, J.M., and Wenzel, L.A., Adsorption equilibria at high pressures in the helium-nitrogen-activated carbon system, Ind. Eng. Chem. Fund., 11(4), 457-465 (1972).

Friederich, R.O., and Mullins, J.C., Adsorption equilibria of binary hydrocarbon mixtures on homogeneous carbon black at 25°C, Ind. Eng. Chem. Fund., 11(4), 439-445 (1972).

Jeffreson, C.P., Prediction of breakthrough curves in packed beds, AIChE J., 18(2), 409-420 (1972).

Johnston, W.A., Designing fixed-bed adsorption columns, Chem. Eng. (N.Y.), 27 November, 87-92 (1972).

Joithe, W.; Bell, A.T., and Lynn, S., Removal and recovery of NOx from nitric acid plant tail gas by adsorption on molecular sieves, Ind. Eng. Chem. Process Des. Dev., 11(3), 434-439 (1972).

King, P.J., and Denizman, E., Channelling in porous beds, Chem. Eng. (Rugby, Engl.), May, 177-181 (1972).

Kowler, D.E., and Kadlec, R.H., Optimal control of a periodic adsorber, AIChE J., 18(6), 1207-1219 (1972).

Lavie, R., and Reilly, M.J., Limit cycles in fixed adsorption beds operated in alternating modes, Chem. Eng. Sci., 27(10), 1835-1844 (1972).

Loughlin, K.F., and Ruthven, D.M., Sorption and diffusion of ethane in type A zeolites, Chem. Eng. Sci., 27(7), 1401-1408 (1972).

Ma, Y.H., and Mancel, C., Diffusion studies of carbon dioxide, nitric oxide, nitrogen dioxide, and sulfur dioxide on molecular sieve zeolites by gas chromatography, AIChE J., 18(6), 1148-1153 (1972).

Malkin, L.S., et al., Mass transfer during drying of oils by NaA synthetic zeolites, Int. Chem. Eng., 12(2), 331-335 (1972).

Maslan, F., and Aberth, E.R., Low-temperature adsorption of hydrogen on activated carbon, Sep. Sci., 7(5), 601-605 (1972).

Rimmer, P.G., and Bowen, J.H., Design of fixed bed sorbers using a quadratic driving force equation, Trans. IChemE, 50, T168-T175 (1972).

Webber, D.A., Adsorptive removal of carbon dioxide from air at intermediate low temperatures, Chem. Eng. (Rugby, Engl.), January, 18-23 (1972).

**1973**

Carter, J.W., and Barrett, D.J., Comparative study for fixed bed adsorption of water vapour by activated alumina, silica gel, and molecular sieve adsorbents, Trans. IChemE, 51, T75-T81 (1973).

Garg, D.R., and Ruthven, D.M., Theoretical prediction of breakthrough curves for molecular sieve adsorption columns, Chem. Eng. Sci., 28(3), 791-806 (1973).

Gregg, S.J.; Nashed, S., and Malik, M.T., Adsorption of water vapour on a microporous carbon black, Powder Technol., 7, 15-19 (1973).

Hutchins, R.A., Design of activated-carbon systems, Chem. Eng. (N.Y.), 20 August, 133-138 (1973).

Jury, S.H., and Horng, J.S., Molecular sieve 4A water vapor sorption therm, AIChE J., 19(2), 371-372 (1973).

Lukchis, G.M., Design of adsorption systems, Chem. Eng. (N.Y.), 11 June, 111-116; 9 July, 83-87; 6 August, 83-90 (1973).

Ma, Y.H., and Roux, A.J., Multicomponent rates of sorption of sulfur dioxide and carbon dioxide on sodium mordenite, AIChE J., 19(5), 1055-1059 (1973).

Zanker, A., Space rates for fixed-bed adsorption columns, Chem. Eng. (N.Y.), 26 November, 102 (1973).

**1974**

Carter, J.W., and Husain, H., Simultaneous adsorption of carbon dioxide and water vapour by fixed beds of molecular sieves, Chem. Eng. Sci., 29(1), 267-274 (1974).

Chiu, H.M.; Hashimoto, N., and Smith, J.M., Chromatographic studies of adsorption of nitric oxide on activated carbon, Ind. Eng. Chem. Fund., 13(3), 282-285 (1974).

Donnet, J.B.; Kobel, L., and Sevenster, A., Helium adsorption on solid surfaces, Ind. Eng. Chem. Fund., 13(1), 83-86 (1974).
Garg, D.R., and Ruthven, D.M., Performance of molecular sieve adsorption columns: Systems with micropore diffusion control, Chem. Eng. Sci., 29(2), 571-582 (1974).
Garg, D.R., and Ruthven, D.M., Performance of molecular sieve adsorption columns: Macropore diffusion control, Chem. Eng. Sci., 29(9), 1961-1968 (1974).

**1975**

Aharoni, C.; Neuman, M., and Notea, A., Adsorption of lead chloride vapors, Ind. Eng. Chem. Process Des. Dev., 14(4), 417-421 (1975).
Bunke, G., and Gelbin, D., Effects of cyclic operation on adsorber performance, Chem. Eng. Sci., 30(10), 1301-1304 (1975).
Carter, J.W., Regeneration of fixed adsorber beds, AIChE J., 21(2), 380-382 (1975).
Garg, D.R., and Ruthven, D.M., Sorption of carbon dioxide in Davison 5A molecular sieve, Chem. Eng. Sci., 30(4), 436-437 (1975).
Garg, D.R., and Ruthven, D.M., Performance of molecular sieve adsorption columns: Combined effects of mass transfer and longitudinal diffusion, Chem. Eng. Sci., 30(9), 1192-1195 (1975).
Papadatos, K.; Svrcek, W.Y., and Bergougnou, M.A., Holdup dynamics of single-stage gas-solid fluidized bed adsorber, Can. J. Chem. Eng., 53, 686-695 (1975).
Ruthven, D.M.; Garg, D.R., and Crawford, R.M., Performance of molecular sieve adsorption columns: Non-isothermal systems, Chem. Eng. Sci., 30(8), 803-810 (1975).
Subbotin, A.I., et al., Dynamics of adsorption of methanol-containing air-vapor mixture, Int. Chem. Eng., 15(3), 392-394 (1975).

**1976**

Kiovsky, J.R.; Koradia, P.B., and Hook, D.S., Molecular sieves for sulfur dioxide removal, Chem. Eng. Prog., 72(8), 98-103 (1976).
Mostafa, H.A., and Said, A.S., Theoretical-plate concept for fixed-bed adsorption and ion-exchange, Trans. IChemE, 54, T132-T134 (1976).
Neretnieks, I., Analysis of some adsorption experiments with activated carbon, Chem. Eng. Sci., 31(11), 1029-1036 (1976).
Ruthven, D.M., Sorption of oxygen, nitrogen, carbon monoxide, methane, and binary mixtures of these gases in 5A molecular sieve, AIChE J., 22(4), 753-759 (1976).
Ruthven, D.M., and Doetsch, I.H., Diffusion of hydrocarbons in 13X zeolite, AIChE J., 22(5), 882-886 (1976).

**1977**

Ammons, R.D.; Dougherty, N.A., and Smith, J.M., Adsorption of methylmercuric chloride on activated carbon, Ind. Eng. Chem. Fund., 16(2), 263-269 (1977).
Bye, G.C., and Chigbo, G.O., Analysis of some adsorption isotherms on calcium silicate hydrates, J. Applied Chem. Biotechnol., 27, 48-54 (1977).
Koradia, P.B., and Kiovsky, J.R., Drying chlorinated hydrocarbons using zeolites, Chem. Eng. Prog., 73(4), 105-106 (1977).

**1978**

Allen, T., and Burevski, D., Models of adsorption of sulphur dioxide on powdered adsorbents, Powder Technol., 21, 91-96 (1978).
Bunke, G., and Gelbin, D., Breakthrough curves in cyclic steady state for adsorption systems with concave isotherms, Chem. Eng. Sci., 33(1), 101-108 (1978).
Larsen, J.W.; Kennard, L., and Kuemmerle, E.W., Thermodynamics of adsorption of organic compounds on the surface of Bruceton coal measured by gas chromatography, Fuel, 57(5), 309-313 (1978).
Ruthven, D.M., Prediction of desorption times for fixed-bed adsorbers, AIChE J., 24(3), 540-542 (1978).

**1979**

Danner, R.P., and Choi, E.C.F., Mixture adsorption equilibria of ethane and ethylene on 13X molecular sieves, Ind. Eng. Chem. Fund., 17(4), 248-253 (1978); 18(3), 300 (1979).
Ikeda, K., Performance of nonisothermal fixed-bed adsorption column with nonlinear isotherms, Chem. Eng. Sci., 34(7), 941-950 (1979).
Martin, M.; Cuellar, J., and Galan, M.A., Adsorption characteristics of n-propanol on silica gel by gas-chromatography, Chem. Eng. Sci., 34(5), 691-696 (1979).

Ruthven, D.M., and Kumar, R., Chromatographic study of diffusion of nitrogen, methane and their binary mixtures in 4A molecular sieve, Can. J. Chem. Eng., 57, 342-352 (1979).

Scamehorn, J.F., Removal of vinyl chloride from gaseous streams by adsorption on activated carbon, Ind. Eng. Chem. Process Des. Dev., 18(2), 210-217 (1979).

Sung, E.; Han, C.D., and Rhee, H.K., Optimal design of multistage adsorption-bed systems, AIChE J., 25(1), 87-100 (1979).

Yang, R.T., and Liu, R.T., Preferential adsorption of methane over hydrogen on certain coals, Ind. Eng. Chem. Fund., 18(3), 299-300 (1979).

**1980**

Moharir, A.S.; Kunzru, D., and Saraf, D.N., Effect of adsorbent particle size distribution on breakthrough curves for molecular sieve columns, Chem. Eng. Sci., 35(8), 1795-1802 (1980).

Robinson, K.S., and Thomas, W.J., The adsorption of methane/ethane mixtures on a molecular sieve, Trans. IChemE, 58, T219-T227 (1980).

Rolniak, P.D., and Kobayashi, R., Adsorption of methane and methane-carbon dioxide mixtures at elevated pressures and ambient temperatures on 5A and 13X molecular sieves by tracer perturbation chromatography, AIChE J., 26(4), 616-625 (1980).

Skalsky, M., and Farrell, P.C., Uptake characteristics of selected biochemicals on coated and uncoated activated charcoal, Trans. IChemE, 58, T91-T97 (1980).

Snape, E., and Lynch, F.E., Metal hydrides adsorb hydrogen, Chemtech, September, 578-583; December, 768-773 (1980).

Wankat, P.C., and Partin, L.R., Process for recovery of solvent vapors with activated carbon, Ind. Eng. Chem. Process Des. Dev., 19(3), 446-451 (1980).

Weber, W.J., and Liu, K.T., Determination of mass-transport parameters for fixed-bed adsorbers, Chem. Eng. Commun., 6(1), 49-60 (1980).

Weiss, A.; Freund, T., and Biron, E., A method to determine the residual capacity of an adsorber, Chem. Eng. (Rugby, Engl.), April, 213-215 (1980).

Werling, K., and Wimmerstedt, R., A kinetic study of the adsorption and desorption process of two commercial molecular sieves, Chem. Eng. Sci., 35(8), 1783-1786 (1980).

**1981**

Al-Sahhaf, T.A.; Sloan, E.D., and Hines, A.L., Application of the modified potential theory to the adsorption of hydrocarbon vapors on silica gel, Ind. Eng. Chem. Process Des. Dev., 20(4), 658-662 (1981).

Andrieu, J., and Smith, J.M., Adsorption rates for sulfur dioxide and hydrogen sulfide in beds of activated carbon, AIChE J., 27(5), 840-842 (1981).

Basmadjian, D., and Wright, D.W., Non-isothermal sorption of ethane-carbon dioxide mixtures in beds of 5A molecular sieves, Chem. Eng. Sci., 36(5), 937-940 (1981).

Coppola, A.P., and LeVan, M.D., Adsorption with axial diffusion in deep beds. Chem. Eng. Sci., 36(6), 967-972 (1981).

Costa, E.; Sotelo, J.L.; Calleja, G., and Marron, C., Adsorption of binary and ternary hydrocarbon gas mixtures on activated carbon, AIChE J., 27(1), 5-14 (1981).

Fraenkel, D., Hydrogen adsorption by zeolites, Chemtech, January, 60-65 (1981).

Gala, H.B.; Kara, M.; Sung, S.; Chiang, S.H., and Klinzing, G.E., Hydrogen permeation through nickel in gas and liquid phase, AIChE J., 27(1), 159-162 (1981).

Kincal, S., and Culfaz, A., Sorption of sulfur dioxide on hydrogen-form and aluminum-deficient mordenites, Sep. Sci. Technol., 16(3), 229-236 (1981).

Knoblauch, K.; Richter, E., and Jungten, H., Application of active coke in flue gas purification, Fuel, 60(9), 832-838 (1981).

Reich, R.; Ziegler, W.T., and Rogers, K.A., Adsorption of methane, ethane, and ethylene gases and their binary and ternary mixtures and carbon dioxide on activated carbon (212-301K and pressures to 35 atmospheres), Ind. Eng. Chem. Process Des. Dev., 19(3), 336-344 (1980); 20(1), 175 (1981).

Resnick, W., and Golt, M., Particle-to-gas mass-transfer measurements and coefficients in fixed beds at low Reynolds numbers, Int. J. Heat Mass Transfer, 24(3), 387-394 (1981).

Sandrock, G.D., and Huston, E.L., How metals adsorb hydrogen, Chemtech, December, 754-762 (1981).

**1982**

Cavalletto, G., and Smith, J.M., Adsorption of ethylene in beds of activated carbon, AIChE J., 28(6), 1039-1040 (1982).

Ghezelayagh, H., and Gidaspow, D., Micro-macropore model for sorption of water on silica gel in a dehumidifier, Chem. Eng. Sci., 37(8), 1181-1198 (1982).

Hunt, L., Adsorptive gas-generators, Process Eng. (London), August, 30-31 (1982).

Kumar, R., Effect of Freon-12 exposure on sieving property of 4A zeolite, Can. J. Chem. Eng., 60, 577-580 (1982).

Kumar, R.; Duncan, R.C., and Ruthven, D.M., Chromatographic study of diffusion of single components and binary mixtures of gases in 4A and 5A zeolites, Can. J. Chem. Eng., 60, 493-499 (1982).

Liapis, A.I., and Crosser, O.K., Comparison of model predictions with non-isothermal sorption data for ethane-carbon dioxide mixtures in beds of 5A molecular sieves, Chem. Eng. Sci., 37(6), 958-962 (1982).

Liberti, L., and Passino, R., Simplified method for calculating cyclic exhaustion-regeneration operations in fixed-bed adsorbers, Ind. Eng. Chem. Process Des. Dev., 21(2), 197-203 (1982).

Marcussen, L., and Vinding, C., Comparison of experimental and predicted breakthrough curves for adiabatic adsorption in fixed beds, Chem. Eng. Sci., 37(2), 299-318 (1982).

Matsuda, S.; Kamo, T.; Imahashi, J., and Nakajima, F., Adsorption and oxidative desorption of hydrogen sulfide by molybdenum oxide-titanium dioxide, Ind. Eng. Chem. Fund., 21(1), 18-22 (1982).

**1983**

Balzli, M.W.; Liapis, A.I., and Rippin, D.W.T., Applications of mathematical modelling to simulation of multicomponent adsorption in activated carbon columns, Trans. IChemE, 56, T145-T156 (1978); Chem. Eng. Res. Des., 61, 393 (1983).

Milewski, M., et al., Efficiency of dynamic tests used to evaluate adsorbents for separation of C8 isomers, Sep. Sci. Technol., 18(2), 187-194 (1983).

Miura, K.; Nakanishi, A., and Hashimoto, K., Methyl bromide vapor adsorption by activated carbon, Ind. Eng. Chem. Process Des. Dev., 22(3), 469-477 (1983).

Ray, M.S., Separation and purification of gases using solid adsorbents (review paper), Sep. Sci. Technol., 18(2), 95-120 (1983).

Sacco, A.; Bac, N.; Hammarstrom, J.L., and Chung, B., Prediction of residual capacity in thin adsorbers, Can. J. Chem. Eng., 61(5), 665-671 (1983).

Santangelo, J.G., and Chen, G.T., Metal hydrides for hydrogen gas purification, Chemtech, October, 621-623 (1983).

Sorial, G.A.; Granville, W.H., and Daly, W.O., Adsorption equilibria for oxygen and nitrogen gas mixtures on 5A molecular sieves, Chem. Eng. Sci., 38(9), 1517-1524 (1983).

Umehara, T.; Harriott, P., and Smith, J.M., Regeneration of activated carbon, AIChE J., 29(5), 732-741 (1983).

**1984**

Abd El-Salaam, K.M., et al., Physical adsorption studies on mixed vanadium oxide catalysts, Adsorpt. Sci. Technol., 1(2), 169-176 (1984).

Carrott, P.J.M., and Sing, K.S.W., Adsorption of nitrogen on precipitated and pyrogenic silicas, Adsorpt. Sci. Technol., 1(1), 30-40 (1984).

Galan, M.A., Adsorption characteristics of nitric oxide on various adsorbents, Chem. Eng. J., 28(2), 105-114 (1984).

Glanz, P., and Findenegg, G.H., Adsorption of gas mixtures of propene and propane on graphitized carbon black, Adsorpt. Sci. Technol., 1(1), 41-50 (1984).

Hymore, K., and Laguerie, C., Analysis and modelling of operation of counterflow multistage fluidized bed adsorber for drying moist air, Chem. Eng. Process., 18(5), 255-268 (1984).

Ishikawa, T., et al., Adsorption of organic halides on silver-impregnated silica gels, Adsorpt. Sci. Technol., 1(1), 93-102 (1984).

Lee, T.V., et al., Adsorption of binary mixtures of ethane and acetylene on activated carbon, Sep. Sci. Technol., 19(1), 1-20 (1984).

Lopez-Garzon, F.J., et al., High-temperature adsorption of hydrocarbons prepared from olive stones, Adsorpt. Sci. Technol., 1(1), 103-110 (1984).

Ruthven, D.M.; Tezel, F.H.; Devgun, J.S., and Sridhar, T.S., Adsorptive separation of krypton from nitrogen, Can. J. Chem. Eng., 62(4), 526-534 (1984).

Tan, H.K.S., Programs for fixed-bed sorption, Chem. Eng. (N.Y.), 24 December, 57-61 (1984).
Urano, K., and Yamamoto, E., Adsorption of organic vapors on activated carbon, Ind. Eng. Chem. Process Des. Dev., 23(4), 665-669 (1984).
Yue, P.L., and Olaofe, O., Kinetic analysis of the catalytic dehydration of alcohols over zeolites, Chem. Eng. Res. Des., 62, 81-91 (1984).
Yue, P.L., and Olaofe, O., Molecular sieving effects of zeolites in the dehydration of alcohols, Chem. Eng. Res. Des., 62, 167-172 (1984).

**1985**

Baba, Y.; Nakao, T.; Inoue, K., and Nakamori, I., Adsorption mechanism of iodine on activated carbon, Sep. Sci. Technol., 20(1), 21-32 (1985).
Bottani, E.J., and Cascarini, L.E., Physical adsorption of carbon dioxide on graphite: Monolayer and multilayer regions, Adsorpt. Sci. Technol., 2(4), 253-262 (1985).
Carlson, N.W., and Dranoff, J.S., Adsorption of ethane by 4A zeolite pellets, Ind. Eng. Chem. Process Des. Dev., 24(4), 1300-1302 (1985).
Casey, J.T., and Liapis, A.I., Fixed bed sorption with recycle, Chem. Eng. Res. Des., 62, 315-320, 344-350 (1984); 63, 398-402 (1985).
Ching, C.B., and Ruthven, D.M., Experimental study of a simulated countercurrent adsorption system, Chem. Eng. Sci., 40(6), 877-892; 40(8), 1411-1418 (1985).
Costa, E.; Calleja, G., and Domingo, F., Adsorption of gaseous hydrocarbons on activated carbon: Characteristic kinetic curve, AIChE J., 31(6), 982-991 (1985).
Emesh, I.T.A., and Gay, I.D., Adsorption and NMR studies of ethene, ethane and carbon monoxide on Zn- and Cd-exchanged A-zeolites, J. Chem. Technol. Biotechnol., 35A(3), 115-120 (1985).
Friday, D.K., and LeVan, M.D., Hot purge-gas regeneration of adsorption beds with solute condensation: Experimental studies, AIChE J., 31(8), 1322-1328 (1985).
Jean, G., et al., Selective removal of nitrogeneous-type compounds from fuels using zeolites, Sep. Sci. Technol., 20(7), 541-564 (1985).
Joshi, J.B.; Mahajani, V.V., and Juvekar, V.A., Adsorption of NOx gases (review paper), Chem. Eng. Commun., 33(1), 1-92 (1985).
Kaguei, S.; Yu, Q., and Wakao, N., Thermal waves in an adsorption column: Parameter estimation, Chem. Eng. Sci., 40(7), 1069-1076 (1985).
Lao, M.Z., and Ye, Z.H., Phase equilibrium of adsorption separation of xylene isomers with type-Y zeolites, Chem. Eng. Commun., 35(1), 89-100 (1985).
Lee, T.V.; Madey, R., and Huang, J.C., Adsorption equilibria for ethane and propane gas mixtures on activated carbon, Sep. Sci. Technol., 20(5), 461-480 (1985).
Morbidelli, M., et al., Role of the desorbent in bulk adsorption separations, Chem. Eng. Sci., 40(7), 1155-1168 (1985).
Rudzinski, W., et al., Simple adsorption equation for adsorption of nonionic surfactants on hydrophilic surfaces of silica, Adsorpt. Sci. Technol., 2(4), 207-218 (1985).
Santacesaria, E.; Gelosa, D.; Danise, P., and Carra, S., Separation of xylenes on Y-zeolites in the vapor phase, Ind. Eng. Chem. Process Des. Dev., 24(1), 78-92 (1985).
Sircar, S., and Myers, A.L., Gas adsorption operations: Equilibrium, kinetics, column dynamics and design, Adsorpt. Sci. Technol., 2(2), 69-88 (1985).
Yang, R.T., and Saunders, J.T., Adsorption of gases on coals and heat-treated coals at elevated temperature and pressure, Fuel, 64(5), 616-626 (1985).
Yumura, M., and Furimsky, E., Comparison of adsorbents for hydrogen sulfide removal at high temperatures, Ind. Eng. Chem. Process Des. Dev., 24(4), 1165-1168 (1985).
Zagorevskaya, E.V., et al., Adsorption of polychlorocarbons on carbon blacks by gas chromatography, Adsorpt. Sci. Technol., 2(4), 219-228 (1985).

**1986**

Bulow, M.; Schlodder, H., and Struve, P., Sorption uptake of molecular mobility of n-paraffins in ZSM-5 type zeolite, Adsorpt. Sci. Technol., 3(4), 229-232 (1986).
Furlong, D.N.; Sing, K.S.W., and Parfitt, G.D., Adsorption of water vapour on rutile and silica-coated rutile, Adsorpt. Sci. Technol., 3(1), 25-32 (1986).

Ghosh, A.K., and Agnew, J.B., Adsorption of acetylene, hydrogen chloride and vinyl chloride on activated carbons: Kinetics and thermodynamics, using transient response technique, Chem. Eng. Commun., 40, 169-182 (1986).

Heiti, R.V., and Thodos, G., Water sorption from air by celite-supported calcium chloride, AIChE J., 32(7), 1169-1175 (1986).

Janchen, J., and Stach, H., n-Decane adsorption on silicon dioxide adsorbents, Adsorpt. Sci. Technol., 3(1), 3-10 (1986).

Rajniak, P., and Ilavsky, J., Mathematical and experimental modelling of adsorption in fixed bed, Adsorpt. Sci. Technol., 3(4), 233-252 (1986).

Roethe, K.P., et al., Adsorption of aromatics on silica gel, Adsorpt. Sci. Technol., 3(2), 65-74 (1986).

Schoellner, R., and Mueller, U., Influence of mono- and bivalent cations in 4A zeolites on adsorptive separation of ethene and propene from crack-gases, Adsorpt. Sci. Technol., 3(3), 167-172 (1986).

Van Deventer, J.S.J., Competitive equilibrium adsorption of metal cyanides on activated carbon, Sep. Sci. Technol., 21(10), 1025-1038 (1986).

**1987**

Al-Damkhi, A.M.; Al-Ameeri, R.S.; Jeffreys, G.V., and Mumford, C.J., Optimal separation of n-paraffins from Kuwait kerosene using a molecular sieve adsorbent, J. Chem. Technol. Biotechnol., 37(4), 215-228 (1987).

Aracil, J., Use of factorial design of experiments in the determination of adsorption equilibrium constants (methyl iodide on charcoals), J. Chem. Technol. Biotechnol., 38(3), 143-152 (1987).

Arnold, M., Hydrogen adsorption by metals, Process Eng. (London), August, 24-25 (1987).

Beschmann, K.; Kokotailo, G.T., and Riekert, L., Kinetics of sorption of aromatics in zeolite ZSM-5, Chem. Eng. Process., 22(4), 223-230 (1987).

Bottani, E.J., et al., Argon and nitrogen physical adsorption on boron nitride, Adsorpt. Sci. Technol., 4(1), 121-130 (1987).

Braslaw, J.; Golovoy, A., and Beckwith, E.C., Adsorption of automotive paint solvents on activated carbon, Chem. Eng. Commun., 51, 321-333 (1987).

Carton, A.; Gonzalez, G.; Torre, A.I., and Cabezas, J.L., Separation of ethanol-water mixtures using 3A molecular sieve, J. Chem. Technol. Biotechnol., 39(2), 125-132 (1987).

Chou, C.L., Dynamic modelling of water vapor adsorption by activated alumina, Chem. Eng. Commun., 56, 211-228 (1987).

Chow, D.K., Optimal process parameters for fixed bed adsorption, Can. J. Chem. Eng., 65(5), 871-876 (1987).

Costa, E.; Calleja, G., and Marijuan, L., Adsorption of phenol and p-nitrophenol on activated carbon, Adsorpt. Sci. Technol., 4(1), 59-78 (1987).

Dolesjs, V., and Machac, I., Pressure drop in flow of fluid through fixed bed of particles, Int. Chem. Eng., 27(4), 730-737 (1987).

Kler, S.C., and Lavin, J.T., Computer simulation of gas distribution in large-shallow-packed adsorbers, Gas Sep. Purif., 1(1), 55-61 (1987).

McAllister, J., and Parsons, P., Meeting hydrocarbon dewpoint requirements: Moisture removal using adsorbents, Chem. Eng. (Rugby, Engl.), April, 24-26 (1987).

McKay, G., and Bino, M.J., Adsorption of pollutants on to activated carbon in fixed beds, J. Chem. Technol. Biotechnol., 37(2), 81-94 (1987).

Miller, G.W.; Knaebel, K.S., and Ikels, K.G., Equilibria of nitrogen, oxygen, argon, and air in molecular sieve 5A, AIChE J., 33(2), 194-201 (1987).

Mohilla, R.; Argyelan, J., and Szolcsanyi, P., Rapid method for calculating breakthrough curve of gas-cleaning adsorbers, Int. Chem. Eng., 27(4), 723-730 (1987).

Pesaran, A.A., and Mills, A.F., Moisture transport in silica gel packed beds, Int. J. Heat Mass Transfer, 30(6), 1037-1060 (1987).

Richter, E.; Knoblauch, K., and Juntgen, H., Mechanisms and kinetics of sulphur dioxide adsorption and NOx reduction on active coke, Gas Sep. Purif., 1(1), 35-43 (1987).

Simons, G.A.; Garman, A.R., and Boni, A.A., Kinetic rate of sulfur dioxide sorption by CaO, AIChE J., 33(2), 211-217 (1987).

Singh, V.S., and Pandey, B.P., Adsorption of petroleum sulfonate TRS10-80 on a mineral sand, J. Chem. Technol. Biotechnol., 37(1), 45-58 (1987).

## 1988

Abdallah, K.; Grenier, P.; Sun, L.M., and Meunier, F., Nonisothermal adsorption of water by synthetic NaX zeolite pellets, Chem. Eng. Sci., 43(10), 2633-2644 (1988).

Awum, F.; Narayan, S., and Ruthven, D.M., Measurement of intracrystalline diffusivities in NaX zeolite by liquid chromatography, Ind. Eng. Chem. Res., 27(8), 1510-1515 (1988).

Banerjee, B.D., Spacing of fissuring network and rate of desorption of methane from coals, Fuel, 67(11), 1584-1586 (1988).

Chiang, P.C.; Lou, J.C., and Tseng, S.K., Development of a procedure for selecting the optimum adsorbent for ABS removal, Can. J. Chem. Eng., 65(6), 913-920 (1988).

De-Xin, Z.; Zhi-Jing, X., and Zhen, F., Prediction of breakthrough curves of oxygen-nitrogen coadsorption system on molecular sieves, Gas Sep. Purif., 2(4),184-189 (1988).

Dexin, Z., and Youfan, G., Prediction of breakthrough curves in system of methane, ethane, and carbon dioxide coadsorption on 4A molecular sieve, Gas Sep. Purif., 2(1), 28-33 (1988).

Filippov, L.K., Modelling of adsorption separation of gas mixtures with allowance for thermal effects accompanying adsorption, Gas Sep. Purif., 2(3), 144-150; 2(4), 190-195 (1988).

George, N., and Davies, J.T., Parameters affecting adsorption of microorganisms on activated charcoal cloth, J. Chem. Technol. Biotechnol., 43(3), 173-186 (1988).

Henning, K.D., et al., Impregnated activated carbon for mercury removal, Gas Sep. Purif., 2(1), 20-22 (1988).

Jedrzejak, A., and Paderewski, M., Adsorption-desorption cycles in fixed bed of adsorbent, Int. Chem. Eng., 28(4), 707-713 (1988).

Joshi, S., and Fair, J.R., Adsorptive drying of toluene, Ind. Eng. Chem. Res., 27(11), 2078-2084 (1988).

Kim, H.C., and Woo, S.I., Preparation of oxygen-enriched air by selective adsorption with platinum crystallites supported on alumina, Ind. Eng. Chem. Res., 27(11), 2135-2139 (1988).

Kuo, S.L., and Hines, A.L., Adsorption of chlorinated hydrocarbon pollutants on silica gel, Sep. Sci. Technol., 23(4), 293-304 (1988).

Niklasson, C., and Anderson, B., Adsorption and reaction of hydrogen and deuterium on nickel/silicon dioxide catalyst, Ind. Eng. Chem. Res., 27(8), 1370-1376 (1988).

Nirdosh, I.; Tremblay, W.B.; Muthuswami, S.V., and Johnson, C.R., Adsorption-desorption studies on the radium-silica system, Can. J. Chem. Eng., 65(6), 927-933 (1988).

Rybolt, T.R., and English, K.J., Virial analysis of methane adsorption in 5A zeolite, AIChE J., 34(7), 1207-1210 (1988).

Shadman, F., and Dombek, P.E., Enhancement of sulfur dioxide sorption on lime by structure modifiers, Can. J. Chem. Eng., 66(6), 930-935 (1988).

Shah, D.B., et al., Sorption and diffusion of benzene in HZSM-5 and silicalite crystals, AIChE J., 34(10), 1713-1717 (1988).

Sowerby, B., and Crittenden, B.D., Experimental comparison of type-A molecular sieves for drying the ethanol-water azeotrope, Gas Sep. Purif., 2(2), 77-83 (1988).

Sowerby, B., and Crittenden, B.D., Vapour-phase separation of alcohol-water mixtures by adsorption onto silicalite, Gas Sep. Purif., 2(4),177-183 (1988).

Suckow, M., Modelling of large-scale adsorptive gas separation processes, Gas Sep. Purif., 2(3), 151-158; 2(4), 196-204 (1988).

Tan, C.S., and Liou, D.C., Desorption of ethyl acetate from activated carbon by supercritical carbon dioxide, Ind. Eng. Chem. Res., 27(6), 988-991 (1988).

van der Merwe, P.F., and van Deventer, J.S.J., Influence of oxygen on adsorption of metal cyanides on activated carbon, Chem. Eng. Commun., 65, 121-138 (1988).

Yan, T.Y., Effects of moisture in separation of C8-aromatics using medium-pore zeolites, Ind. Eng. Chem. Res., 27(9), 1665-1668 (1988).

## 1989

Carrott, P.J.M., Effect of temperature on neopentane isotherms determined on non-porous non-graphitized carbon black (Elftex), Adsorpt. Sci. Technol., 6(2), 103-109 (1989).

Carrott, P.J.M., and Sing, K.S.W., Virial analysis of the adsorption of trifluorochloromethane and methane by microporous carbons and zeolites, Adsorpt. Sci. Technol., 6(3), 136-146 (1989).

Carrott, P.J.M.; Brotas de Carvalho, M., and Sing, K.S.W., Gas chromatographic study of specific physisorption by activated carbons, Adsorpt. Sci. Technol., 6(2), 93-102 (1989).

Chevreau, T., et al., Sorption of benzene by Y-zeolites exchanged with chromic ions and protons, Adsorpt. Sci. Technol., 6(3), 119-135 (1989).
Ching, C.B., and Ruthven, D.M., Sorption and diffusion of some amino acids in KX zeolite crystals, Biochem. Eng. J., 40(1), B1-B6 (1989).
Ching, C.B.; Hidajat, K., and Uddin, M.S., Evaluation of equilibrium and kinetic parameters of smaller molecular size amino acids on KX zeolite crystals, Sep. Sci. Technol., 24(7), 581-598 (1989).
Choi, C.; Bloomquist, C.A.A., and Dyrkacz, G.R., Adsorption of surfactants on coals and macerals, Energy Fuels, 3(1), 38-42 (1989).
Choudhary, V.R.; Akolekar, D.B., and Singh, A.P., Single- and multicomponent sorption/diffusion of hydrocarbons from their iso-octane solution in H-ZMS-5 zeolite, Chem. Eng. Sci., 44(5), 1047-1060 (1989).
Davini, P., Adsorption of sulphur dioxide on thermally treated active carbon, Fuel, 68(2), 145-148 (1989).
Frimpong, S.; Plank, C.A., and Laukhuf, W.L.S., Evaluation of sorption and transport of sulfur dioxide in polycarbonate, Chem. Eng. J., 42(1), 25-36 (1989).
Gaicobbe, F.W., Use of physical adsorption to facilitate the production of high-purity oxygen, Gas Sep. Purif., 3(3), 133-138 (1989).
George, M.A.; Kiefer, J.H., and Hessler, J.P., Removal of simple hydrocarbons from a rare gas by a 70%Zr-25%V-5%Fe getter, Gas Sep. Purif., 3(2), 50-55 (1989).
Gray, P.G., and Do, D.D., Adsorption and desorption of gaseous sorbates on a bidispersed particle with Freundlich isotherm: Experimental study of sulphur dioxide sorption on activated carbon particles, Gas Sep. Purif., 3(4), 201-208 (1989).
Guo, C.J.; Talu, O., and Hayhurst, D.T., Phase transition and structural heterogeneity: Benzene adsorption on silicalite, AIChE J., 35(4), 573-578 (1989).
Hall, C.R., and Holmes, R.J., Observations and comments on the displacement of pre-adsorbed water from BPL activated carbon by chloropicrin vapour, Adsorpt. Sci. Technol., 6(2), 83-92 (1989).
Helmy, A.K.; de Bussetti, S.G., and Ferreiro, E.A., Adsorption of 1.10-phenanthroline and quinoline on charcoal from mixed solutions, Adsorpt. Sci. Technol., 6(1), 1-8 (1989).
Hills, J.H., and Pirzada, I.M., Analysis and prediction of breakthrough curves for packed bed adsorption of water vapour on corn-meal, Chem. Eng. Res. Des., 67(5), 442-450 (1989).
Horstmann, B.J., and Chase, H.A., Modelling the affinity adsorption of immunoglobulin-G to protein-A immobilised to agarose matrices, Chem. Eng. Res. Des., 67(3), 243-254 (1989).
Huang, T.C., and Cho, L.T., Adsorption of p-nitrophenol on anion exchange resin at various temperatures, Can. J. Chem. Eng., 67(6), 1030-1033 (1989).
Jaroniec, M., et al., Use of a polynomial equation for analyzing low-concentration adsorption measurements of ethane on activated carbons, Sep. Sci. Technol., 24(15), 1355-1362 (1989).
Kapoor, A., and Yang, R.T., Correlation of equilibrium adsorption data of condensible vapors on porous adsorbents, Gas Sep. Purif., 3(4), 187-192 (1989).
Kapoor, A., and Yang, R.T., Kinetic separation of methane-carbon dioxide mixture by adsorption on molecular sieve carbon, Chem. Eng. Sci., 44(8), 1723-1734 (1989).
Kirchgessner, D.A., and Jozewicz, W., Enhancement of reactivity in surfactant-modified sorbents for sulfur dioxide control, Ind. Eng. Chem. Res., 28(4), 413-418 (1989).
Kubota, K., et al., Adsorption characteristics of propylene on molecular sieve 4A, Sep. Sci. Technol., 24(9), 709-720 (1989).
Kwon, K.C., Determination of heats of adsorption on activated carbons by flow microcalorimetry, Gas Sep. Purif., 3(1), 13-18 (1989).
Kyotani, T., et al., Removal of hydrogen sulphide from hot gas in presence of Cu-containing sorbents, Fuel, 68(1), 74-79 (1989).
Mohring, W.R., and Hawley, M.C., Gravimetric analysis of hydrogen fluoride vapour adsorption by Bigtooth Aspen wood, Ind. Eng. Chem. Res., 28(2), 237-243 (1989).
Newton, G.H.; Chen, S.L., and Kramlich, J.C., Role of porosity loss in limiting sulfur dioxide capture by calcium-based sorbents, AIChE J., 35(6), 988-994 (1989).
Osei-Owusu, A., and Aharoni, C., Adsorption of sulfur compounds by hydrodesulfurization catalysts, Adsorpt. Sci. Technol., 6(3), 155-168 (1989).
Punjak, W.A.; Uberoi, M., and Shadman, F., High-temperature adsorption of alkali vapors on solid sorbents, AIChE J., 35(7), 1186-1194 (1989).

Rohner, M.C.; Sharma, V.K., and Richarz, W., NMR study of chlorine adsorption on alumina, Can. J. Chem. Eng., 67(3), 513-516 (1989).

Starzewski, P., and Grillet, Y., Thermochemical studies of adsorption of helium and carbon dioxide on coals at ambient temperature, Fuel, 68(3), 375-380 (1989).

Tan, C.S., and Liou, D.C., Regeneration of activated carbon loaded with toluene by supercritical carbon dioxide, Sep. Sci. Technol., 24(1), 111-128 (1989).

Tan, C.S., and Liou, D.C., Supercritical regeneration of activated carbon loaded with benzene and toluene, Ind. Eng. Chem. Res., 28(8), 1222-1226 (1989).

Tarasevich, Y.I., et al., Non-specific and specific interactions of hydrocarbons with the surface of a non-swelling organo-substituted layer silicate, Adsorpt. Sci. Technol., 6(3), 147-154 (1989).

Valsaraj, K.T., Partitioning of hydrophobic nonpolar volatile organics between the aqueous and surfactant aggregate phases on alumina, Sep. Sci. Technol., 24(14), 1191-1206 (1989).

Yan, T.Y., Separation of p-xylene and ethylbenzene from C8-aromatics using medium-pore zeolites, Ind. Eng. Chem. Res., 28(5), 572-577 (1989).

Yoshida, H., and Kataoka, T., Adsorption of BSA on cross-linked chitosan: Equilibrium isotherm, Biochem. Eng. J., 41(1), B11-B15 (1989).

Zabasajja, J., and Savinell, R.F., Electrosorption of n-alcohols on graphite particles, AIChE J., 35(5), 755-763 (1989).

Zarkanitis, S., and Sotirchos, S.V., Pore structure and particle size effects on limestone capacity for sulfur dioxide removal, AIChE J., 35(5), 821-830 (1989).

**1990**

Abdul-Rehman, H.B.; Hasanain, M.A., and Loughlin, K.F., Quaternary, ternary, binary, and pure component sorption on zeolites: Light alkanes at moderate to high pressures, Ind. Eng. Chem. Res., 29(7), 1525-1546 (1990).

Ahmed, K.; Kershenbaum, L.S., and Chadwick, D., Adsorption of thiophene on nickel/alumina catalysts, Ind. Eng. Chem. Res., 29(2), 150-156 (1990).

Ajisegiri, E.S., and Sopade, P.A., Moisture sorption isotherms of Nigerian millet at varying temperatures, J. Food Eng., 12(4), 283-292 (1990).

Alishusky, J.J., and Fournier, R.L., Partitioning and adsorption of aromatic compounds on microporous silica gel, AIChE J., 36(10), 1605-1609 (1990).

Asanuma, H., et al., Adsorption of nitrogen monoxide by the chelate resin-immobilized iron(II) complex and its application for simultaneous removal of nitrogen monoxide and sulfur dioxide, Ind. Eng. Chem. Res., 29(11), 2267-2272 (1990).

Barthomeuf, D., and de Mallmann, A., Adsorption of aromatics in NaY and $AlPO_4$: Correlation with the sorbent properties in separations, Ind. Eng. Chem. Res., 29(7), 1435-1438 (1990).

Bhatia, S.K., et al., Effective diffusivity of phenol in activated carbon, Chem. Eng. Commun., 98, 139-154 (1990).

Braibanti, A.; Fisicaro, E., and Palmia, F., Water activity and pseudo-activity coefficient of sorbed water, J. Food Eng., 12(4), 307-324 (1990).

Campbell, G.M., The irreversible adsorption of plutonium hexafluoride, Chem. Eng. Commun., 98, 223-230 (1990).

Carli, R., et al., Mathematical modelling of gas purification with getter-based purifiers, Gas Sep. Purif., 4(3), 171-176 (1990).

Cen, P.L., Adsorption uptake curves of ethylene on Cu(I)-NaY zeolite, AIChE J., 36(5), 789-793 (1990).

Chmelka, B.F., et al., Transport of aromatic molecules in NaY zeolite powders, AIChE J., 36(10), 1562-1568 (1990).

Choudhary, V.R., and Mamman, A.S., Diffusion of cumene in H-ZSM-8 and modified H-ZSM-8 zeolites, AIChE J., 36(10), 1577-1580 (1990).

Crawshaw, J.P., and Hills, J.H., Sorption of ethanol and water by starchy materials, Ind. Eng. Chem. Res., 29(2), 307-309 (1990).

Eiden, U., and Schlunder, E.U., Adsorption equilibria of pure vapors and their binary mixtures on activated carbons, Chem. Eng. Process., 28(1), 1-22 (1990).

Farooq, S., and Ruthven, D.M., Heat effects in adsorption column dynamics, Ind. Eng. Chem. Res., 29(6), 1076-1090 (1990).

Forsythe, R.; Madey, R., and Photinos, P.J., Transport of methane, acetylene, and acetaldehyde through activated carbon, Sep. Sci. Technol., 25(6), 659-672 (1990).

Gail, E., and Kast, W., Kinetics of sulphur dioxide sorption by single pellets of activated carbon, Chem. Eng. Sci., 45(2), 403-410 (1990).

Ghosh, T.K., and Hines, A.L., Adsorption of acetaldehyde, propionaldehyde, and butyraldehyde on silica gel, Sep. Sci. Technol., 25(11), 1101-1116 (1990).

Grant, T.M., and King, C.J., Mechanism of irreversible adsorption of phenolic compounds by activated carbons, Ind. Eng. Chem. Res., 29(2), 264-272 (1990).

Harrison, D.P., et al., Reaction between hydrogen sulfide and zinc oxide-titanium oxide sorbents: Single-pellet kinetic studies and sulfidation modeling, Ind. Eng. Chem. Res., 29(7), 1160-1172 (1990).

Hasatani, M.; Watanabe, F., and Marumo, C., Separation of air at low temperature by structural molecular-sieving carbon, Int. Chem. Eng., 30(1), 134-141 (1990).

Hidajat, K.; Aracil, J., and Kenney, C.N., Adsorption of methyl iodide on charcoal: Mass transfer rates in a recycle adsorber, Process Safety Environ. Prot., 68(1), 57-66 (1990).

Hidajat, K.; Aracil, J., and Kenney, C.N., Fixed bed adsorption of methyl iodide on charcoal, Process Safety Environ. Prot., 68(1), 67-74 (1990).

Hinckley, C.C., et al., Carbon monoxide detection of chemisorbed oxygen in coal and other carbonaceous materials, Fuel, 69(1), 103-109 (1990).

Ho, Y.S., and Yeh, C.T., The adsorption isobar of dioxygen on rhodium powders, Adsorpt. Sci. Technol., 7(1), 1-8 (1990).

Jaroniec, M., et al., Comparative studies of adsorption of ethane and benzene on microporous activated carbons, Chem. Eng. Sci., 45(6), 1539-1546 (1990).

Keener, T.C., and Zhou, D., Prediction of activated carbon adsorption performance under high relative humidity conditions, Environ. Prog., 9(1), 40-46 (1990).

Maeda, S., et al., Selective adsorption of arsenic(V) ion by use of iron(III) hydroxide-loaded coral limestone, Sep. Sci. Technol., 25(5), 547-556 (1990).

Mehdizadeh, S., and Durning, C.J., Prediction of differential sorption kinetics near $T_g$ for benzene in polystyrene, AIChE J., 36(6), 877-884 (1990).

Nigam, S.C.; Siahpush, A.R., and Wang, H.Y., Analysis of bioproduct separation using gel-enclosed adsorbents, AIChE J., 36(8), 1239-1248 (1990).

Ramakrishnan, T.S., and Wasan, D.T., Effect of adsorption on the optimal displacement of acidic crude oil by alkali, AIChE J., 36(5), 725-737 (1990).

Ray, M.S., Adsorptive and membrane separations: A bibliographical guide (1985-1989), Adsorpt. Sci. Technol., 7(1), 28-64 (1990).

Siemieniewska, T., et al., Application of the Dubinin-Astakhov equation to evaluation of benzene and cyclohexane adsorption isotherms on steam gasified humic acid chars from brown coal, Energy Fuels, 4(1), 61-69 (1990).

Uberoi, M., and Shadman, F., Sorbents for removal of lead compounds from hot flue gases, AIChE J., 36(2), 307-309 (1990).

Uddin, M.S.; Hidajat, K., and Ching, C., Liquid chromatographic evaluation of equilibrium and kinetic parameters of large molecule amino acids on silica gel, Ind. Eng. Chem. Res., 29(4), 647-651 (1990).

Van den Bulck, E., Isotherm correlation for water vapor on regular-density silica gel, Chem. Eng. Sci., 45(5), 1425-1429 (1990).

Yu, J.W., and Neretnieks, I., Single-component and multicomponent adsorption equilibria on activated carbon of methylcyclohexane, toluene, and isobutyl methyl ketone, Ind. Eng. Chem. Res., 29(2), 220-232 (1990).

**1991**

Anstice, P.J.C.; Halliday, N., and Alder, J.F., Adsorption of chloropicrin onto humidified activated carbon surfaces, Adsorpt. Sci. Technol., 7(3), 107-113 (1991).

Bhatt, B.L.; Golden, T.C., and Hsiung, T.H., Adsorptive removal of catalyst poisons from coal gas for methanol synthesis, Sep. Sci. Technol., 26(12), 1559-1574 (1991).

Davini, P.; DeMichele, G., and Bertacchi, S., Reaction between calcium-based sorbents and sulphur dioxide: A thermogravimetric investigation, Fuel, 70(2), 201-204 (1991).

Do, D.D.; Hu, X., and Mayfield, P.L.J., Multi-component adsorption of ethane, n-butane and n-pentane in activated carbon, Gas Sep. Purif., 5(1), 35-48 (1991).

Doong, S.J., and Ho, W.S.W., Sorption of organic vapors in polyethylene, Ind. Eng. Chem. Res., 30(6), 1351-1361 (1991).

Friedrich, M., et al., Adsorption of water on activated carbon: A thermodynamic study by immersion calorimetry, Adsorpt. Sci. Technol., 7(3), 133-139 (1991).

Frimpong, S.; Plank, C.A., and Laukhuf, W.L.S., Sorption and transport of sulfur dioxide in polycarbonate-polyarylate blends, Chem. Eng. J., 47(2), 63-74 (1991).

Fujita, I., et al., Adsorption of nonionic surfactants on chemically modified styrene-divinylbenzene copolymers, Sep. Sci. Technol., 26(10), 1395-1402 (1991).

Ghosh, T.K., and Hines, A.L., Adsorption of acetaldehyde, propionaldehyde, and butyraldehyde on molecular sieve 13X, Sep. Sci. Technol., 26(7), 931-946 (1991).

Giacobbe, F.W., Adsorption of very low level carbon dioxide impurities in oxygen on a 13X molecular sieve, Gas Sep. Purif., 5(1), 16-20 (1991).

Golden, T.C.; Hsiung, T.H., and Snyder, K.E., Removal of trace iron and nickel carbonyls by adsorption, Ind. Eng. Chem. Res., 30(3), 502-507 (1991).

Gray, P.G., and Do, D.D., Dynamics of carbon dioxide sorption on activated-carbon particles, AIChE J., 37(7), 1027-1034 (1991).

Green, T.K.; Ball, J.E., and Conkright, K., Rate of benzene sorption by o-alkylated Illinois No.6 coal, Energy Fuels, 5(4), 609-610 (1991).

Hasany, S.M.; Chughtai, F.A., and Ghaffar, A., Sorption behavior of microamounts of zinc on titanium oxide from aqueous solutions, Sep. Sci. Technol., 26(8), 1131-1146 (1991).

Hassan, N.M., et al., Adsorption of water vapor on molecular sieve 13X, Chem. Eng. Commun., 105, 241-254 (1991).

Hassan, N.M., et al., New apparatus for measuring radon adsorption on solid adsorbents, Ind. Eng. Chem. Res., 30(9), 2205-2211 (1991).

Hradil, J., et al., Sorption and desorption of organic compounds by synthetic polymeric sorbents, Ind. Eng. Chem. Res., 30(8), 1926-1931 (1991).

Kobayashi, S., et al., Adsorption behavior of chlorofluorocarbons in zeolitic pores: Adsorption isotherm, Ind. Eng. Chem. Res., 30(10), 2340-2344 (1991).

Krammer, G., and Staudinger, G., Sulphur dioxide removal from flue gas with dry limestone, Gas Sep. Purif., 5(4), 259-260 (1991).

Kuo, S.L.; Hines, A.L., and Dural, N.H., Correlation of methyl chloride, methylene chloride, chloroform, and carbon tetrachloride adsorption data on silica gel, Sep. Sci. Technol., 26(8), 1077-1092 (1991).

Lin, W.F., and Tan, C.S., Separation of m-xylene and ethylbenzene in gaseous carbon dioxide, Sep. Sci. Technol., 26(12), 1549-1558 (1991).

Llewellyn, P.L., and Theocharis, C.R., A diffuse reflectance Fourier transform infra-red study of carbon dioxide adsorption on silicalite-I, J. Chem. Technol. Biotechnol., 52(4), 473-480 (1991).

Martyniuk, H., and Augustyn, D., Sorption of metal cations on sulphonated brown coals, Fuel, 70(4), 551-556 (1991).

Mayfield, P.L.J., and Do, D.D., Measurement of the single-component adsorption kinetics of ethane, butane, and pentane onto activated carbon using a differential adsorption bed, Ind. Eng. Chem. Res., 30(6), 1262-1270 (1991).

Meijer, T., et al., The interaction of water, carbon dioxide, hydrogen, and carbon monoxide with the alkali-carbonate/carbon system: A thermogravimetric study, Fuel, 70(2), 205-214 (1991).

Mochida, I., and Kawano, S., Capture of ammonia by active carbon fibers further activated with sulfuric acid, Ind. Eng. Chem. Res., 30(10), 2322-2327 (1991).

Nishioka, M., Strong adsorption of unreacted maleic anhydride in coal during Diels-Alder reaction, Energy Fuels, 5(3), 523-525 (1991).

Niwa, M.; Yamazaki, K., and Murakami, Y., Separation of oxygen and nitrogen due to the controlled pore-opening size of chemical vapor deposited zeolite-A, Ind. Eng. Chem. Res., 30(1), 38-42 (1991).

Otto, K., et al., Adsorption of hydrocarbons and other exhaust components on silicalite, Ind. Eng. Chem. Res., 30(10), 2333-2340 (1991).

Panagiotis, G.; Smirniotis, G., and Ruckenstein, E., Optimum dilution profiles of composite zeolites in packed beds, Chem. Eng. Commun., 106, 119-138 (1991).

Rees, L.V.C.; Bruckner, P., and Hampson, J., Sorption of nitrogen, methane and carbon dioxide in silicalite-1, Gas Sep. Purif., 5(2), 67-76 (1991).

Rojas, J.A.; Voilley, A., and Thomas, C., Study of adsorption of 1-octen-3-ol on porous hydrophobic polymers, Gas Sep. Purif., 5(3), 187-193 (1991).

Selles-Perez, M.J., and Martin-Martinez, J.M., Classification of alpha plots obtained from nitrogen/77K adsorption isotherms of activated carbons, Fuel, 70(7), 877-882 (1991).

Tezel, H.O., and Ruthven, D.M., The effect of dehydration conditions on the adsorption and diffusion of propane and nitrogen in Linde 5A zeolite, Can. J. Chem. Eng., 69(1), 371-376 (1991).

Timmermann, E.O., and Chirife, J., The physical state of water sorbed at high activities in starch in terms of the GAB sorption equation, J. Food Eng., 13(3), 171-180 (1991).

Tobis, J., and Vortmeyer, D., Scale-up effects due to near-wall channelling in isothermal adsorption columns: On the limitations in the use of plug flow models, Chem. Eng. Process., 29(3), 147-154 (1991).

Unal, S.; Wood, D.G., and Harris, I.J., Propensity of raw Victorian brown coal for physical adsorption of nitrogen, oxygen, carbon dioxide and water, Fuel, 70(12), 1481-1482 (1991).

Vitaya, V.B., and Toda, K., Kinetics and mechanism of the adsorption of Sulfolobus acidocaldarius on coal surfaces, Biotechnol. Prog., 7(5), 427-433 (1991).

Wu, Y.Y.; Wong, D.S.H., and Tan, C.S., Thermodynamic model for the adsorption of toluene from supercritical carbon dioxide on activated carbon, Ind. Eng. Chem. Res., 30(11), 2492-2496 (1991).

**1992**

Ahmad, S.; Mannan, A., and Qureshi, I.H., Adsorption studies of radioactive cobalt on a minerals mixture, Sep. Sci. Technol., 27(4), 523-534 (1992).

Allen, S., et al., An evaluation of single resistance transfer models in the sorption of metal ions by peat, J. Chem. Technol. Biotechnol., 54(3), 271-276 (1992).

Baksh, M.S.A.; Kikkinides, E.S., and Yang, R.T., Lithium type-X zeolite as a superior sorbent for air separation, Sep. Sci. Technol., 27(3), 277-294 (1992).

Bardacki, T., and King, F.G., Measurements of argon, nitrogen and carbon dioxide diffusion through random assemblies of small spheres, Gas Sep. Purif., 6(1), 43-48 (1992).

Bhaskar, G.V., and Bhamidimarri, R.S.M., Adsorption of 2,4-D onto activated carbon: Application of *n*th order approximation, J. Chem. Technol. Biotechnol., 53(3), 297-300 (1992).

Bhutani, M.M.; Santosh, A., and Mitra, A.K., Abnormal mode of chromate adsorption on ignited sulphates as carrier surfaces, Adsorpt. Sci. Technol., 8(1), 1-12 (1992).

Bojan, M.J.; van Slooten, R., and Steele, W., Computer simulation studies of the storage of methane in microporous carbons, Sep. Sci. Technol., 27(14), 1837-1856 (1992).

Boyer, P.M., and Hsu, J.T., Effects of ligand concentration on protein adsorption in dye-ligand adsorbents, Chem. Eng. Sci., 47(1), 241-252 (1992).

Bruckner, P., et al., Adsorption and immersion of benzene in active carbons, Adsorpt. Sci. Technol., 8(1), 57-68 (1992).

Buroni, M., et al., Behaviour of calcium-based sorbents towards sulphur-dioxide capture in a high-pressure thermobalance, Fuel, 71(8), 919-924 (1992).

Carrasco-Marin, F., et al., Adsorption of sulphur dioxide in flowing air onto activated carbons from olive stones, Fuel, 71(5), 575-578 (1992).

Crawshaw, J.P., and Hills, J.H., Experimental determination of binary sorption and desorption kinetics for the system ethanol, water, and maize at 90°C, Ind. Eng. Chem. Res., 31(3), 887-892 (1992).

Desai, R.; Hussain, M., and Ruthven, D.M., Adsorption of water vapour on activated alumina: Equilibrium and kinetic behaviour, Can. J. Chem. Eng., 70(4), 699-715 (1992).

Economy, J., et al., Tailoring carbon fibres for adsorbing volatiles, Chemtech, October, 597-603 (1992).

Eiden, U., and Schlunder, E.U., Inert gas desorption of single organic vapours from activated carbon, Chem. Eng. Process., 31(2), 63-76 (1992).

Gonzalez-Pradas, E.; Villafranca-Sanchez, M., and Gallego, A.C., Effects of experimental variables on phosphate adsorption on bentonite, J. Chem. Technol. Biotechnol., 54(3), 291-296 (1992).

Gow, A.S., and Phillips, J., Calorimetric-study of oxygen adsorption on a high surface area polymer-derived carbon, Energy Fuels, 6(2), 184-188 (1992).

Gray, P.G., and Do, D.D., Modelling of the interaction of nitrogen dioxide with activated carbon: Adsorption dynamics at the single particle scale, Chem. Eng. Commun., 117, 219-240 (1992).

Gusev, V.Y., and Fomkin, A.A., Thermodynamics of high-pressure Tian-Calvet adsorption calorimetry and the heats of adsorption of Xe on Na-X zeolite over the temperature range 229-467K, Adsorpt. Sci. Technol., 8(2), 75-85 (1992).

Hall, C.R.; Holmes, R.J., and Lawston, I.W., Further observations on the displacement of pre-adsorbed water from BPL activated carbon by chloropicrin vapour, Adsorpt. Sci. Technol., 8(2), 69-74 (1992).
Hassan, N.M., et al., Adsorption of radon from a humid atmosphere on activated carbon, Sep. Sci. Technol., 27(14), 1955-1968 (1992).
Hedges, S.W., and Yeh, J.T., Kinetics of sulfur dioxide uptake on supported cerium oxide sorbents, Environ. Prog., 11(2), 98-103 (1992).
Hu, X., and Do, D.D., Multicomponent adsorption kinetics of hydrocarbons onto activated carbon: Effect of adsorption equilibrium equations, Chem. Eng. Sci., 47(7), 1715-1726 (1992).
Hylton, T.D., Evaluation of the TCE catalytic oxidation unit at Wurtsmith Air Force base, Environ. Prog., 11(1), 54-57 (1992).
Irabien, A.; Cortabitarte, F., and Ortiz, M.I., Kinetics of flue gas desulfurization at low temperatures: Nonideal surface adsorption model, Chem. Eng. Sci., 47(7), 1533-1543 (1992).
Ismail, H.M.; Fouad, N.E., and Zaki, M.I., Nitrogen and pyridine adsorption on chromia-coated silica and alumina catalysts: Probing the chromia dispersity, Adsorpt. Sci. Technol., 8(1), 34-43 (1992).
Jasra, R.V., and Bhat, S.G.T., Thermal desorption of linear alkenes from zeolite molecular sieve 5A, Adsorpt. Sci. Technol., 8(4), 174-183 (1992).
Kaguei, S., and Wakao, N., Relationships between surface diffusivity and pore diffusivity in batch adsorption: Measurements of the diffusivities for n-hexane and n-decane in 5A molecular sieves, Chem. Eng. Sci., 47(8), 2109-2113 (1992).
Kaur, P., et al., Studies on the sorption behaviour of some amino acids on silica gel pretreated with alkalies in relation to chromatography, Adsorpt. Sci. Technol., 8(3), 157-173 (1992).
Klobucar, J.M., and Pilat, M.J., Continuous flow thermal desorption of VOCs from activated carbon, Environ. Prog., 11(1), 11-17 (1992).
Kocjan, R., Additional purification of some salts by using silica gel modified with Calmagit as a sorbent, Sep. Sci. Technol., 27(3), 409-418 (1992).
Kodama, K.; Kaguei, S., and Wakao, N., Batch adsorption of trichlorotrifluoroethane (Freon-113) onto activated carbon: Surface diffusivity and pore diffusivity, Can. J. Chem. Eng., 70(2), 244-249 (1992).
Lee, C.K.; Chiang, A.S.T., and Wu, F.Y., Lattice model for the adsorption of benzene in silicalite-1, AIChE J., 38(1), 128-135 (1992).
Ludmany, A., and Nagy, L.G., Investigation on the factors influencing the mechanism of the alkaline sorption on the surface of titanium phosphate sorbent, Period. Polytech. Chem. Eng., 34(1), 31-36 (1992).
Maeda, S., et al., Iron(III) hydroxide-loaded coral limestone as an adsorbent for arsenic(III) and arsenic(V), Sep. Sci. Technol., 27(5), 681-690 (1992).
Matranga, K.R.; Myers, A.L., and Glandt, E.D., Storage of natural gas by adsorption on activated carbon, Chem. Eng. Sci., 47(7), 1569-1579 (1992).
Mentasty, L.; Faccio, R.J., and Zgrablich, G., High-pressure methane adsorption in 5A zeolite and the nature of gas-solid interactions, Adsorpt. Sci. Technol., 8(2), 105-113 (1992).
Nassar, M.M., Equilibrium studies on the adsorption of glycine on resin, Adsorpt. Sci. Technol., 8(2), 86-94 (1992).
Preiss, H., and Kant, W., Swelling and sorption of pyrolysed halogenated mesophase pitches, Fuel, 71(6), 635-640 (1992).
Raghuram, S., and Wilcher, S.A., The separation of n-paraffins from paraffin mixtures, Sep. Sci. Technol., 27(14), 1917-1954 (1992).
Rajakovic, L.V., The sorption of arsenic onto activated carbon impregnated with metallic silver and copper, Sep. Sci. Technol., 27(11), 1423-1434 (1992).
Ramesh, R., et al., Isosteric and calorimetric heats of adsorption of methanol on coal, Energy Fuels, 6(3), 239-241 (1992).
Ray, M.S., Adsorptive and membrane-type separations: A bibliographical guide (1991), Adsorpt. Sci. Technol., 8(3), 114-133 (1992).
Rives, V., A computer program for analyzing nitrogen adsorption isotherms on porous solids, Adsorpt. Sci. Technol., 8(2), 95-104 (1992).
Ruthven, D.M., Diffusion of oxygen and nitrogen in carbon molecular sieve, Chem. Eng. Sci., 47(17), 4305-4308 (1992).
Sato, T., et al., Uptake of benzenecarboxylate ions by magnesium aluminium oxides, J. Chem. Technol. Biotechnol., 55(4), 385-390 (1992).

Shah, G.C., Improve activated carbon bed adsorber operations, Hydrocarbon Process., 71(11), 61-63 (1992).

Sifton, J.B.; Eic, M., and Ruthven, D.M., Development of an alternative technique for sampling stack gas by using zeolite 3A, Gas Sep. Purif., 6(4), 179-184 (1992).

Sircar, S., Estimation of isosteric heats of adsorption of single gas and multicomponent gas mixtures, Ind. Eng. Chem. Res., 31(7), 1813-1819 (1992).

Sobolik, J.L.; Ludlow, D.K., and Hessevick, W.L., Parametric sensitivity comparison of the BET and Dubinin-Radushkevich models for determining char surface area by carbon dioxide adsorption, Fuel, 71(10), 1195-1202 (1992).

Suckow, M., et al., Calculation of the hydrothermal long-term stability of zeolites in gas-desulphurization and gas-drying processes, Gas Sep. Purif., 6(2), 101-108 (1992).

Susarla, S.; Bhaskar, G.V., and Rao Bhamidimarri, S.M., Competitive adsorption and desorption of 2,4-D and PCOC in a volcanic soil, Adsorpt. Sci. Technol., 8(4), 184-195 (1992).

Tan, C.S.; Liao, C.L., and Chiang, S.T., Separation of diethylbenzene isomers on silicalite in the presence of high pressure carbon dioxide and propane, Adsorpt. Sci. Technol., 8(4), 226-234 (1992).

Todorovic, M., et al., Adsorption of radioactive ions $^{137}Cs^{-}$, $^{85}Sr^{2+}$, and $^{60}Co^{2+}$ on natural magnetite and hematite, Sep. Sci. Technol., 27(5), 671-680 (1992).

Xiao, J., and Wei, J., Diffusion mechanism of hydrocarbons in zeolites: Theory and analysis of experimental observations, Chem. Eng. Sci., 47(5), 1123-1160 (1992).

**1993**

Ahlbeck, J., et al., A method for measuring the reactivity of absorbents for wet flue gas desulfurization, Chem. Eng. Sci., 48(20), 3479-3484 (1993).

Aitani, A.M., Sour natural gas drying, Hydrocarbon Process., 72(4), 67-73 (1993).

Aksenenko, E.V., and Tarasevich, Y.I., Non-specific adsorption of hydrocarbons on microporous surfaces: A comprehensive molecular-statistical/chromatographic approach, Adsorpt. Sci. Technol., 9(2), 54-71 (1993).

Akubuiro, E.C., Potential mechanistic routes for the oxidative disintegration of ketones on carbon adsorbents, Ind. Eng. Chem. Res., 32(12), 2960-2968 (1993).

Ching, C.B.; Hidajat, K., and Liu, X., Sorption and diffusion of cresols on bonded beta-cyclodextrin-silica stationary phase, Ind. Eng. Chem. Res., 32(11), 2789-2793 (1993).

Choudary, V.N.; Jasra, R.V., and Bhat, T.S.G., Adsorption of a nitrogen-oxygen mixture in NaCaA zeolites by elution chromatography, Ind. Eng. Chem. Res., 32(3), 548-552 (1993).

Choudhary, V.R., and Mayadevi, S., Adsorption of methane, ethane, ethylene, and carbon dioxide on X, Y, L, and M zeolites using a gas chromatography pulse technique, Sep. Sci. Technol., 28(8), 1595-1608 (1993).

Ciembroniewicz, A., and Marecka, A., Kinetics of carbon dioxide sorption for two Polish hard coals, Fuel, 72(3), 405-408 (1993).

Eissmann, R.N., and LeVan, M.D., Coadsorption of organic compounds and water vapor on BPL activated carbon: 1,1,2-trichloro-1,2,2-trifluoroethane and dichloromethane, Ind. Eng. Chem. Res., 32(11), 2752-2757 (1993).

Eldridge, R.B., Olefin/paraffin separation technology: A review, Ind. Eng. Chem. Res., 32(10), 2208-2212 (1993).

Ghosh, T.K.; Lin, H.D., and Hines, A.L., Hybrid adsorption-distillation process for separating propane and propylene, Ind. Eng. Chem. Res., 32(10), 2390-2399 (1993).

Goto, A., et al., Desorption of uranium from amidoxime fiber adsorbent, Sep. Sci. Technol., 28(13), 2229-2236 (1993).

Gray, P.G., and Do, D.D., Modelling of the interaction of nitrogen dioxide with activated carbon: Kinetics of reaction with pore evolution, Chem. Eng. Commun., 125, 109-120 (1993).

Hawash, S., et al., Useful adsorption equilibriums by means of natural clay, Adsorpt. Sci. Technol., 9(4), 231-243 (1993).

Hawash, S.; Farah, J.Y., and El-Geundi, M.S., Investigation of nickel ion removal by means of activated clay, Adsorpt. Sci. Technol., 9(4), 244-257 (1993).

Helmy, A.K.; Ferreiro, E.A., and de Bussetti, S.G., Apparent and partial specific sorption of phosphate by binary mixtures of hydrated Al oxides, Adsorpt. Sci. Technol., 9(2), 72-83 (1993).

Hershkowitz, F., and Madiara, P.D., Simultaneous measurement of adsorption, reaction, and coke using a pulsed microbalance reactor, Ind. Eng. Chem. Res., 32(12), 2969-2974 (1993).

Hu, X., and Do, D.D., Multicomponent adsorption kinetics of hydrocarbons onto activated carbon: Contribution of micropore resistance, Chem. Eng. Sci., 48(7), 1317-1324 (1993).

Hu, X.; Rao, G.N., and Do, D.D., Multicomponent sorption kinetics of ethane and propane in activated carbon: Simultaneous adsorption, Gas Sep. Purif., 7(1), 39-46 (1993).
Hu, X.; Rao, G.N., and Do, D.D., A mathematical model for multicomponent adsorption, desorption and displacement kinetics of ethane, propane and n-butane on activated carbon, Gas Sep. Purif., 7(4), 197-206 (1993).
Hufton, J.R., and Danner, R.P., Chromatographic study of alkanes in silicalite: Equilibrium properties, AIChE J., 39(6), 954-961 (1993).
Hufton, J.R., and Danner, R.P., Chromatographic study of alkanes in silicalite: Transport properties, AIChE J., 39(6), 962-974 (1993).
Illan-Gomez, M.J., et al., NO reduction by activated carbons: The role of carbon porosity and surface area, Energy Fuels, 7(1), 146-154 (1993).
Jarvelin, H., and Fair, J.R., Adsorptive separation of propylene-propane mixtures, Ind. Eng. Chem. Res., 32(10), 2201-2207 (1993).
Jha, S., et al., Chromatographic utilization of the sorption behaviour of some nitrophenols on acid-treated alumina, Adsorpt. Sci. Technol., 9(2), 92-108 (1993).
Kabay, N., and Egawa, H., Kinetic behavior of lightly crosslinked chelating resins containing amidoxime groups for batchwise adsorption of $UO_2^{2+}$, Sep. Sci. Technol., 28(11), 1985-1994 (1993).
Kapoor, A.; Krishnamurthy, K.R., and Shirley, A., Kinetic separation of carbon dioxide from hydrocarbons using carbon molecular sieve, Gas Sep. Purif., 7(4), 259-263 (1993).
Kats, B.M., and Kutarov, V.V., Adsorption of the vapour of low-molecular substances by swelling polymers, Adsorpt. Sci. Technol., 9(1), 30-35 (1993).
Kats, B.M.; Kutarov, V.V., and Chagodar, A.A., Applications of non-linear diffusion equations to the kinetics of water vapour adsorption by polymeric fibres, Adsorpt. Sci. Technol., 9(4), 269-275 (1993).
Kawamura, Y., et al., Adsorption of metal ions on polyaminated highly porous chitosan chelating resin, Ind. Eng. Chem. Res., 32(2), 386-391 (1993).
Kikkinides, E.S., and Yang, R.T., Gas separation and purification by polymeric adsorbents: Flue gas desulfurization and sulfur dioxide recovery with styrenic polymers, Ind. Eng. Chem. Res., 32(10), 2365-2372 (1993).
Krings, U.; Kelch, M., and Berger, R.G., Adsorbents for the recovery of aroma compounds in fermentation processes, J. Chem. Technol. Biotechnol., 58(3), 293-299 (1993).
Loughlin, K.F., et al., Rate and equilibrium sorption parameter for nitrogen and methane on carbon molecular sieve, Gas Sep. Purif., 7(4), 264-273 (1993).
Lu, G.Q., and Do, D.D., Retention of sulfur dioxide as sulfuric acid by activated coal reject char, Sep. Technol., 3(2), 106-110 (1993).
Madras, G.; Erkey, C., and Akgerman, A., Supercritical fluid regeneration of activated carbon loaded with heavy molecular weight organics, Ind. Eng. Chem. Res., 32(6), 1163-1168 (1993).
Mehandjiev, D.R., and Nickolov, R.N., Dependence of the C-constant in the Brunauer-Emmett-Teller equation on water pre-adsorbed on activated carbon, Adsorpt. Sci. Technol., 9(1), 48-53 (1993).
Miyake, Y., and Suzuki, M., Removal of trichloroethylene from air stripping off-gas by adsorption on activated carbon fibre, Gas Sep. Purif., 7(4), 229-234 (1993).
Mushtaq, A., Sorption behavior of carrier-free technetium-99m on zinc dust, Sep. Sci. Technol., 28(9), 1743-1752 (1993).
Niswander, R.H., et al., A more energy efficient product for carbon dioxide separation, Sep. Sci. Technol., 28(1), 565-578 (1993).
Park, S.W.; Lee, W.K., and Moon, H., Adsorption and desorption of gaseous methyl iodide in a triethylenediamine-impregnated activated carbon bed, Sep. Technol., 3(3), 133-142 (1993).
Perona, J.J.; Byers, C.H., and Prazniak, J.K., Vacuum sorption pumping studies of argon and oxygen on 4A molecular sieves, Sep. Sci. Technol., 28(1), 595-614 (1993).
Rajniak, P., and Yang, R.T., A simple model and experiments for adsorption-desorption hysteresis: Water vapor on silica gel, AIChE J., 39(5), 774-786 (1993).
Reed, B.E., and Matsumoto, M.R., Modeling cadmium adsorption by activated carbon using the Langmuir and Freundlich isotherm expressions, Sep. Sci. Technol., 28(13), 2179-2196 (1993).
Ruhl, M.J., Recover VOCs via adsorption and activated carbon, Chem. Eng. Prog., 89(7), 37-41 (1993).
Ruthven, D.M., and Kaul, B.K., Adsorption of aromatic hydrocarbons in NaX zeolite: Equilibrium, Ind. Eng. Chem. Res., 32(9), 2047-2052 (1993).

Ruthven, D.M., and Kaul, B.K., Adsorption of aromatic hydrocarbons in NaX zeolite: Kinetics, Ind. Eng. Chem. Res., 32(9), 2053-2057 (1993).

Ruthven, D.M., and Xu, Z., Diffusion of oxygen and nitrogen in 5A zeolite crystals and commercial 5A pellets, Chem. Eng. Sci., 48(18), 3307-3312 (1993).

Sadowski, Z., The influence of the sodium oleate adsorption on the behavior of calcite suspensions, Chem. Eng. Sci., 48(2), 305-312 (1993).

Schweiger, T.A.J., and LeVan, M.D., Steam regeneration of solvent adsorbers, Ind. Eng. Chem. Res., 32(10), 2418-2429 (1993).

Sharma, P.K., and Hickey, G.S., A comparison of the oxygen uptake characteristics of copper-exchanged zeolite with copper dispersed on a silica support, Gas Sep. Purif., 7(3), 141-146 (1993).

Stankiewicz, Z., and Schreiner, H., Temperature-vacuum process for the desorption of activated charcoal, Process Safety Environ. Prot., 71(B2), 134-140 (1993).

Stenger, H.G.; Hu, K., and Simpson, D.R., Competitive adsorption of NO, sulphur dioxide and water onto mordenite synthesized from perlite, Gas Sep. Purif., 7(1), 19-26 (1993).

Stenzel, M.H., Remove organics by activated carbon adsorption, Chem. Eng. Prog., 89(4), 36-43 (1993).

Stradella, L., Heats of adsorption of different gases on polycrystalline transition metals, Adsorpt. Sci. Technol., 9(3), 190-198 (1993).

Strauss, H.; Heegn, H., and Strienitz, I., Effect of PAA adsorption on stability and rheology of titanium dioxide dispersions, Chem. Eng. Sci., 48(2), 323-332 (1993).

Takeuchi, Y., and Itoh, T., Removal of ozone from air by activated carbon treatment, Sep. Technol., 3(3), 168-175 (1993).

Takeuchi, Y.; Shigeta, A., and Iwamoto, H., Adsorption of solvent vapor mixture in air by activated carbon fiber bed, Sep. Technol., 3(1), 46-52 (1993).

Teng, H., and Suuberg, E.M., Chemisorption of nitric oxide on char: Irreversible carbon oxide formation, Ind. Eng. Chem. Res., 32(3), 416-423 (1993).

Tezel, F.H., and Apolonatos, G., Chromatographic study of adsorption for nitrogen, CO and methane in molecular sieve zeolites, Gas Sep. Purif., 7(1), 11-18 (1993).

Various, Activated carbon and carbon molecular sieves (topic issue), Gas Sep. Purif., 7(4), 195-284 (1993).

Wakker, J.P.; Gerritsen, A.W., and Moulijn, J.A., High temperature hydrogen sulfide and COS removal with MnO and FeO on gamma-alumina acceptors, Ind. Eng. Chem. Res., 32(1), 139-149 (1993).

Webb, O.F.; Phelps, T.J., and Bienkowski, P.R., Multicomponent adsorption of polycyclic aromatic hydrocarbons in manufactured gas plant soils, Sep. Sci. Technol., 28(1), 873-894 (1993).

Weber, G., et al., t-Curves for n-hexane, Adsorpt. Sci. Technol., 9(4), 258-268 (1993).

Westgate, P.J., and Ladisch, M.R., Sorption of organics and water on starch, Ind. Eng. Chem. Res., 32(8), 1676-1680 (1993).

**1994**

Afzal, A.; Khan, M., and Ahmad, H., Kinetics of the sorption of organic vapours in chromatographic silica gel, Adsorpt. Sci. Technol., 11(2), 113-122 (1994).

Aguayo, A.T., et al., Isotherms of chemical adsorption of bases on solid catalysts for acidity measurement, J. Chem. Technol. Biotechnol., 60(2), 141-146 (1994).

Anthony, R.G.; Dosch, R.G., and Philip, C.V., Use of silicotitanates for removing cesium and strontium from defense waste, Ind. Eng. Chem. Res., 33(11), 2702-2705 (1994).

Biernat, J.F.; Konieczka, P., and Izatt, R.M., Complexing and chelating agents immobilized on silica gel and related material and their application for sorption of inorganic species, Sep. Purif. Methods, 23(2), 77-120 (1994).

Cal, M.P.; Larson, S.M., and Rood, M.J., Experimental and modeled results describing the adsorption of acetone and benzene onto activated carbon fibers, Environ. Prog., 13(1), 26-30 (1994).

Calleja, G., et al., Multicomponent adsorption equilibrium of ethylene, propane, propylene and carbon dioxide on 13X-zeolite, Gas Sep. Purif., 8(4), 247-256 (1994).

Chanda, M., and Rempel, G.L., Quaternized poly(4-vinylpyridine) gel-coated on silica: Fast kinetics of diffusion-controlled sorption of organic sulfonates, Ind. Eng. Chem. Res., 33(3), 623-630 (1994).

Chatzopoulos, D.; Varma, A., and Irvine, R.L., Adsorption and desorption studies in the phase for the toluene/activated carbon system, Environ. Prog., 13(1), 21-25 (1994).

Chen, Y.D.; Yang, R.T., and Uawithya, P., Diffusion of oxygen, nitrogen and their mixtures in carbon molecular sieve, AIChE J., 40(4), 577-585 (1994).

Choudhary, V.R.; Mayadevi, S., and Kamble, K.R., Adsorption of oxygen and nitrogen on $AlPO_4$-5 and SaPO-5 at moderate pressures using a novel adsorption/desorption method, Ind. Eng. Chem. Res., 33(5), 1319-1323 (1994).
Chung, T.W., Predictions of moisture removal efficiencies for packed-bed dehumidification systems, Gas Sep. Purif., 8(4), 265-268 (1994).
Do, H.D., and Do, D.D., Structural heterogeneity in the equilibrium data for hydrocarbons and carbon oxides on activated carbons, Gas Sep. Purif., 8(2), 77-94 (1994).
Eligwe, C.A., and Okolue, N.B., Adsorption of iron(II) by a Nigerian brown coal, Fuel, 73(4), 569-572 (1994).
Fitch, F.R.; Bulow, M., and LaCava, A.I., Investigation of the mechanism for the separation of nitrogen-oxygen mixtures on carbon molecular sieves, Gas Sep. Purif., 8(1), 45-51 (1994).
Foeth, F.; Andersson, M., and Bosch, H., Separation of dilute carbon dioxide/methane mixtures by adsorption on activated carbon, Sep. Sci. Technol., 29(1), 93-118 (1994).
Gao, G.N.; Hu, X., and Do, D.D., Multicomponent sorption of hydrocarbons in activated carbon: Simultaneous desorption and displacement, Gas Sep. Purif., 8(2), 67-76 (1994).
Ghosh, D.R., and Keinath, T.M., Effect of clay minerals present in aquifer soils on the adsorption and desorption of hydrophobic organic compounds, Environ. Prog., 13(1), 51-59 (1994).
Ghosh, M.M.; Cox, C.D., and Yuan-Pan, J., Adsorption of selenium on hydrous alumina, Environ. Prog., 13(2), 79-88 (1994).
Golden, T.C., and Sircar, S., Equilibrium and kinetics of adsorption of Freon-12 at infinite dilution, AIChE J., 40(6), 935-943 (1994).
Gonzalez-Martin, M.L., et al., Adsorption of sodium dodecyl sulphate on cassiterite surface, Can. J. Chem. Eng., 72(3), 551-554 (1994).
Goubkina, M.L.; Polyakov, N.S., and Tatarinova, L.I., Immersion heats of active carbons in water and organic substances, Adsorpt. Sci. Technol., 10, 27-34 (1994).
Green, T.K., and Selby, T.D., Pyridine sorption isotherms of Argonne premium coals: Dual-mode sorption and coal microporosity, Energy Fuels, 8(1), 213-218 (1994).
Hassan, N.M.; Al-Ameeri, R.S., and Oweysi, F., Separation of n-paraffins from kerosene-range feedstock by adsorption on fixed-bed urea, Sep. Sci. Technol., 29(3), 401-414 (1994).
Hu, X., and Do, D.D., Effect of energy distribution shape on the sorption equilibrium and dynamics of sulphur dioxide in activated carbon, Chem. Eng. Sci., 49(6), 919-923 (1994).
Hu, X.; Do, D.D., and Rao, G.N., Experimental concentration dependence of surface diffusivity of hydrocarbons in activated carbon, Chem. Eng. Sci., 49(13), 2145-2152 (1994).
Hu, X.; King, B., and Do, D.D., Ternary adsorption kinetics of gases in activated carbon, Gas Sep. Purif., 8(3), 175-186 (1994).
Hu, X.; King, B., and Do, D.D., Ternary desorption and displacement kinetics of gases in activated carbon, Gas Sep. Purif., 8(3), 187-190 (1994).
Huang, R.T.; Chen, T.L., and Weng, H.S., Adsorption of o-cresol and benzoic acid in an adsorber packed with an ion-exchange resin: A comparative study of diffusional models, Sep. Sci. Technol., 29(15), 2019-2034 (1994).
Huang, Y.H., et al., Experimental study of binary equilibrium adsorption/desorption of propane-propylene mixtures on 13X sieves by differential sorption bed system and investigation of their equilibrium expressions, Sep. Technol., 4(3), 156-166 (1994).
Hwang, K.S., and Lee, W.K., The adsorption and desorption breakthrough behavior of carbon monoxide and carbon dioxide on activated carbon: Effect of total pressure and pressure-dependent mass transfer coefficients, Sep. Sci. Technol., 29(14), 1857-1892 (1994).
Iwai, Y.; Uchida, H., and Mori, Y., Separation of isomeric dimethylnaphthalene mixture in supercritical carbon dioxide by using zeolite, Ind. Eng. Chem. Res., 33(9), 2157-2160 (1994).
Izumi, Y., and Iwasawa, Y., CO-breathing ruthenium and rhodium clusters, Chemtech, 24(7), 20-27 (1994).
Kusakabe, K.; Isumi, N., and Morooka, S., Desorption of uranium recovered with fibrous amidoxime adsorbent shaped into balls, Sep. Sci. Technol., 29(11), 1501-1508 (1994).
Lee, J.K., et al., Sulfur dioxide adsorption over activated lignite char prepared from fluidized bed pyrolysis, Chem. Eng. Sci., 49(24A), 4483-4490 (1994).
Li, G.Q., Separation of oxygen from air using coordination complexes: A review, Ind. Eng. Chem. Res., 33(4), 755-783 (1994).

Li, J., and Talu, O., Adsorption equilibrium of benzene/para-xylene vapour mixture on silicalite, Chem. Eng. Sci., 49(2), 189-197 (1994).
Mahle, J.J.; Buettner, L.C., and Friday, D.K., Measurement and correlation of the adsorption equilibria of refrigerant vapors on activated carbon, Ind. Eng. Chem. Res., 33(2), 346-354 (1994).
Mariwala, R.K., and Foley, H.C., Calculation of micropore sizes in carbogenic materials from the methyl chloride adsorption isotherm, Ind. Eng. Chem. Res., 33(10), 2314-2321 (1994).
Masthan, S.K.; Rao, K.S.R., and Rao, P.K., The influence of metal loading on the structure-sensitive chemisorption of nitrogen on Ru/alumina catalysts, Adsorpt. Sci. Technol., 11(3), 175-186 (1994).
Mentasty, L.; Woestyn, A.M., and Zgrablich, G., High-pressure methane adsorption on natural and synthetic zeolites, Adsorpt. Sci. Technol., 11(2), 123-133 (1994).
Milewska-Duda, J.; Ceglarska-Stefanska, G., and Duda, J., A comparison of theoretical and empirical expansion of coals in the high pressure sorption of methane, Fuel, 73(6), 975-979 (1994).
Mishra, S.P., and Chaudhury, G.R., Kinetics of zinc adsorption on charcoal, J. Chem. Technol. Biotechnol., 59(4), 359-364 (1994).
Moller, K.P.; Kojima, M., and O'Connor, C.T., Diffusion and adsorption in zeolite-Y and mordenite deactivated by propene oligomerization and hexane cracking, Chem. Eng. J., 54(3), 115-124 (1994).
Niwa, M.; Yamazaki, K., and Murakami, Y., Separation of lower olefins by chemical vapor deposited zeolite-A, Ind. Eng. Chem. Res., 33(2), 371-374 (1994).
Olguin, M.T., et al., $UO_2^{2+}$ sorption in natural Mexican erionite and Y-zeolite, Sep. Sci. Technol., 29(16), 2161-2178 (1994).
Pan, Z.J., et al., Pore structure alteration of a carbon molecular sieve for the separation of hydrogen sulfide from methane by adsorption, Adsorpt. Sci. Technol., 10, 193-202 (1994).
Payne, G.F.; Patel, A.R., and Sun, W.Q., Tyrosinase reaction/chitosan adsorption: Potential for removing polymerization storage inhibitors, Ind. Eng. Chem. Res., 33(9), 2168-2173 (1994).
Pehlivan, E., et al., Sorption of heavy metal ions on new metal-ligand complexes chemically derived from Lycopodium clavatum, Sep. Sci. Technol., 29(13), 1757-1768 (1994).
Polyakov, N.S.; Larin, A.V., and Goubkina, M.L., Elution adsorption dynamics of organic substances on carbon adsorbents, Adsorpt. Sci. Technol., 10, 173-180 (1994).
Reiss, G., Status and development of oxygen generation process on molecular sieve zeolites, Gas Sep. Purif., 8(2), 95-100 (1994).
Salas-Peregrin, M.A., et al., Adsorption of carbon dioxide on activated carbons from diluted ambient environments (see errata in 9(2), 390, 1995), Energy Fuels, 8(1), 239-243 (1994).
Saleem, M., et al., Adsorption characteristics of organic dyes on alumina powder from aqueous solutions, Adsorpt. Sci. Technol., 11(2), 95-104 (1994).
Salem, A.B.S.H., Naphtha desulfurization by adsorption, Ind. Eng. Chem. Res., 33(2), 336-340 (1994).
San, J.Y., and Jiang, G.D., Modeling and testing of a silica gel packed-bed system, Int. J. Heat Mass Transfer, 37(8), 1173-1180 (1994).
Sengupta, A.K., and Zhu, Y., Selective and reversible ligands sorption through a novel regeneration scheme, Ind. Eng. Chem. Res., 33(2), 382-386 (1994).
Sharma, D.C., and Forster, C.F., Compost as an adsorbent for the treatment of hexavalent chromium, Process Safety Environ. Prot., 72(B4), 234-240 (1994).
Shoubary, Y.E.; Woodmansee, D.E., and Shilling, N.Z., Sorption and desorption of contaminants from different host matrices, Environ. Prog., 13(1), 37-44 (1994).
Singh, B.K., and Rawat, N.S., Comparative sorption kinetic studies of phenolic compounds on fly ash and impregnated fly ash, J. Chem. Technol. Biotechnol., 61(1), 57-66 (1994).
Singh, B.K., and Rawat, N.S., Comparative sorption equilibrium studies of toxic phenols on flyash and impregnated flyash, J. Chem. Technol. Biotechnol., 61(4), 307-318 (1994).
Smirniotis, P.G., and Ruckenstein, E., Comparison of the performance of ZSM-5, beta-zeolite-Y, USY, and their composites in the catalytic cracking of n-octane, 2,2,4-trimethylpentane, and 1-octene, Ind. Eng. Chem. Res., 33(4), 800-813 (1994).
Sorrento, L., The proven process of carbon adsorption, Chem. Eng. (N.Y.), July, 94-95 (1994).
Sun, W., and Costa, C.A.V., Activated diffusion of permanent gases in 4A zeolite at low coverage, Chem. Eng. Sci., 49(12), 1913-1924 (1994).
Theis, T.L.; Iyer, R., and Ellis, S.K., Parameter estimation for trace element sorption on a new granular iron oxide, Environ. Prog., 13(1), 72-77 (1994).

Tsai, W.T., and Chang, C.Y., Adsorption of methylene chloride vapor on activated carbons, J. Chem. Technol. Biotechnol., 61(2), 145-152 (1994).

Tung, L.A., and King, C.J., Sorption and extraction of lactic and succinic acids at pH > pKa1: Factors governing equilibria, and regeneration and process conditions, Ind. Eng. Chem. Res., 33(12), 3217-3229 (1994).

Vega, L.F.; Panagiotopoulos, A.Z., and Gubbins, K.E., Chemical potentials and adsorption isotherms of polymers confined between parallel plates, Chem. Eng. Sci., 49(17), 2921-2930 (1994).

Wang, K.; Hawley, M.C., and Reath, M.L., Adsorption model of HF on wood chips in a packed-bed reactor, Chem. Eng. Sci., 49(14), 2321-2330 (1994).

Yan, T.Y., A novel process for Hg removal from gases, Ind. Eng. Chem. Res., 33(12), 3010-3014 (1994).

Yoshida, H.; Kishimoto, N., and Kataoka, T., Adsorption of strong acid on polyaminated highly porous chitosan: Equilibria, Ind. Eng. Chem. Res., 33(4), 854-859 (1994).

You, Y.H.; Chiang, H.L., and Chiang, P.C., Comparison of adsorption characteristics for VOCs on activated carbon and oxidized activated carbon, Environ. Prog., 13(1), 31-36 (1994).

**1995**

Acevedo S., et al., Adsorption of asphaltenes and resins on organic and inorganic substrates and their correlation with precipitation problems in production well tubing, Fuel, 74(4), 595-598 (1995).

Alkandary, J.A.M.; Al-Ammeri, R., and Salem, A.B.S.H., Adsorption equilibria of normal paraffins on 5A molecular sieve, Sep. Sci. Technol., 30(16), 3195-3210 (1995).

Allen, S.J., and Brown, P.A., Isotherm analyses for single component and multi-component metal sorption onto lignite, J. Chem. Technol. Biotechnol., 62(1), 17-24 (1995).

Andreev, B.M.; Kruglov, A.V., and Selivanenko, Y.L., Continuous isotope separation in systems with solid phase: Gas-phase separation of isotopes of the light elements, Sep. Sci. Technol., 30(16), 3211-3228 (1995).

Arrou, D., and Baboulene, M., Enantioselective separation of basic amino acids on talc, J. Chem. Technol. Biotechnol., 63(1), 92-96 (1995).

Besedova, E., and Bobok, D., Adsorption of acetone and cumene on activated carbon: Single-component adsorption equilibria of acetone and cumene on supersorbon HS-4, Adsorpt. Sci. Technol., 12(1), 39-50 (1995).

Biswas, P.; Owens, T.M., and Wu, C.Y., An equilibrium analysis for reaction of metal compounds with sorbents in high temperature systems, Chem. Eng. Commun., 133, 31-52 (1995).

Branton, P.J.; Hall, P.G., and Sing, K.S.W., Physisorption of alcohols and water vapour by MCM-41, a model mesoporous adsorbent, Adsorption, 1(1), 77-82 (1995).

Buss, E., Gravimetric measurement of binary gas adsorption equilibria of methane-carbon dioxide mixtures on activated carbon, Gas Sep. Purif., 9(3), 189-198 (1995).

Cavalcante, C.L., and Ruthven, D.M., Adsorption of branched and cyclic paraffins in silicalite: Equilibrium, Ind. Eng. Chem. Res., 34(1), 177-184 (1995).

Cavalcante, C.L., and Ruthven, D.M., Adsorption of branched and cyclic paraffins in silicalite: Kinetics, Ind. Eng. Chem. Res., 34(1), 185-191 (1995).

Chang, C., and Ku, Y., The adsorption and desorption characteristics of EDTA-chelated copper ion by activated carbon, Sep. Sci. Technol., 30(6), 899-916 (1995).

Cheng, L.S., and Yang, R.T., Monolayer cuprous chloride dispersed on pillared clays for olefin-paraffin separations by pi-complexation, Adsorption, 1(1), 61-76 (1995).

Cho, S.Y., and Lee, Y.Y., Equilibria of 1,1,2-trichloro-1,2,2-trifluoroethane on activated carbons, Ind. Eng. Chem. Res., 34(7), 2468-2472 (1995).

Chue, K.T.; Kim, J.N., and Yang, R.T., Comparison of activated carbon and zeolite 13X for carbon dioxide recovery from flue gas by pressure swing adsorption, Ind. Eng. Chem. Res., 34(2), 591-598 (1995).

Chung, T.W., et al., Efficiency of a liquid desiccant dehumidification system regenerated by using solar collectors/regenerators with photovoltaic fans, Sep. Sci. Technol., 30(6), 1039-1045 (1995).

Deng, S.G., and Lin, Y.S., Sulfur dioxide sorption properties and thermal stability of hydrophobic zeolites, Ind. Eng. Chem. Res., 34(11), 4063-4070 (1995).

Dutta, N.N., and Patil, G.S., Developments in CO separation, Gas Sep. Purif., 9(4), 277-284 (1995).

El-Nabarawy, T., et al., Benzene hydrogenation activities of supported nickel catalysts in relation to their chemisorption properties, Adsorpt. Sci. Technol., 12(2), 151-160 (1995).

Farooq, S., Sorption and diffusion of oxygen and nitrogen in molecular sieve RS-10, Gas Sep. Purif., 9(3), 205-212 (1995).

Fedorov, A.N., Investigation and improvement of cryogenic adsorption purification of argon from oxygen, Gas Sep. Purif., 9(2), 137-145 (1995).
Fenouil, L.A., and Lynn, S., Study of calcium-based sorbents for high-temperature hydrogen sulfide removal: Kinetics of hydrogen sulfide sorption by calcined limestone, Ind. Eng. Chem. Res., 34(7), 2324-2333 (1995).
Hassan, N.M., et al., Adsorption of radon and water vapor on commercial activated carbons, Sep. Sci. Technol., 30(4), 565-584 (1995).
Heesink, A.B.M., and Van Swaaij, W.P.M., The adsorption of hydrogen sulphide on sulphided limestone, Chem. Eng. Sci., 50(22), 3651-3656 (1995).
Hong, A., et al., Adsorbate shape selectivity: Separation of the HF/134a azeotrope over carbogenic molecular sieve, Ind. Eng. Chem. Res., 34(3), 992-996 (1995).
Huang, Y.H., et al., Binary adsorption and desorption rates of propylene-propane mixtures on 13X molecular sieves, Sep. Technol., 5(1), 1-12 (1995).
Ilic, M.R., et al., Sorption of hydrogen cyanide onto activated carbon cloth impregnated with metallo-organic compounds, Sep. Sci. Technol., 30(13), 2707-2730 (1995).
Illan-Gomez, M.J., et al., NO reduction by activated carbons: Catalytic effect of potassium, Energy Fuels, 9(1), 97-103 (1995).
Illan-Gomez, M.J., et al., NO reduction by activated carbons: Influence of catalyst loading on the catalytic effect of potassium, Energy Fuels, 9(1), 104-111 (1995).
Illan-Gomez, M.J., et al., NO reduction by activated carbons: Catalysis by calcium, Energy Fuels, 9(1), 112-118 (1995).
Illan-Gomez, M.J.; Linares-Solano, A., and de Lecea, C.S.M., NO reduction by activated carbon: Catalysis by transition metals, Energy Fuels, 9(6), 976-983 (1995).
Inoue, K., et al., Adsorption of some platinum group metals on some complexane types of chemically modified chitosan, Sep. Sci. Technol., 30(12), 2477-2490 (1995).
Inui, T., and Pu, S.B., Separation of 2,6-dimethylnaphthalene from a mixture of its isomers using lithium-incorporated zeolite-Y synthesized by rapid crystallization method, Sep. Technol., 5(4), 229-238 (1995).
Jobic, H., et al., Measurement of the diffusivity of benzene in microporous silica by quasi-elastic neutron scattering and NMR pulsed-field gradient technique, Adsorption, 1(3), 197-202 (1995).
Jozewicz, W., and Gullett, B.K., Reaction mechanisms of dry Ca-based sorbents with gaseous HCl, Ind. Eng. Chem. Res., 34(2), 607-612 (1995).
Juang, R.S., and Lin, H.C., Metal sorption with extractant-impregnated macroporous resins: Particle diffusion kinetics, J. Chem. Technol. Biotechnol., 62(2), 132-140 (1995).
Juang, R.S., and Lin, H.C., Metal sorption with extractant-impregnated macroporous resins: Chemical reaction and particle diffusion kinetics, J. Chem. Technol. Biotechnol., 62(2), 141-147 (1995).
Khalil, L.B., and Girgis, B.S., Adsorption of p-nitrophenol on activated carbon prepared from phosphoric acid-treated apricot stone shells, Adsorpt. Sci. Technol., 12(1), 79-92 (1995).
Khamis, E.; Kandil, S., and Ibrahim, A.K., Adsorption action of phosphine inhibitors on the corrosion of steel, Adsorpt. Sci. Technol., 12(4), 279-292 (1995).
Khodakov, A.Y., and Rees, L.V.C., Effect of propane on the kinetics of carbon dioxide adsorption in NaA zeolites, Gas Sep. Purif., 9(4), 253-258 (1995).
Kikkinides, E.S.; Sikavitsas, V.I., and Yang, R.T., Natural gas desulfurization by adsorption: Feasibility and multiplicity of cyclic steady states, Ind. Eng. Chem. Res., 34(1), 255-262 (1995).
Kishimoto, N., and Yoshida, H., Adsorption of glutamic acid on crosslinked chitosan fiber: Equilibria, Sep. Sci. Technol., 30(16), 3143-3164 (1995).
Kong, Y., and Cha, C.Y., NOx abatement with carbon adsorbents and microwave energy, Energy Fuels, 9(6), 971-975 (1995).
Lee, S.Y., et al., Impact of environmental tobacco smoke on the water adsorption capacity of desiccant materials, Gas Sep. Purif., 9(4), 285-291 (1995).
Lim, B.G.; Ching, C.B., and Tan, R.B.H., Determination of competitive adsorption isotherms of enantiomers on a dual-site adsorbent, Sep. Technol., 5(4), 213-228 (1995).
Lodyga, B., et al., A simple method of adsorbent washing, Adsorpt. Sci. Technol., 11(4), 235-242 (1995).
Lopez, F.A., et al., Adsorption of $Pb^{2+}$ on blast furnace sludge, J. Chem. Technol. Biotechnol., 62(2), 200-206 (1995).
Lu, X.C.; Li, F.C., and Watson, A.T., Adsorption measurements in Devonian shales, Fuel, 74(4), 599-603 (1995).

MacNaughton, S.J., and Foster, N.R.,, Supercritical adsorption and desorption behavior of DDT on activated carbon using carbon dioxide, Ind. Eng. Chem. Res., 34(1), 275-282 (1995).
Martin-Martinez, J.M.; Torregrosa, R., and Mittelmeijer, M.C., Mechanisms of adsorption of carbon dioxide in the micropores of activated anthracite, Fuel, 74(1), 111-114 (1995).
Mentasty, L., et al., High-pressure methane adsorption on NaX and NaY zeolites with different Si/Al ratios, Adsorpt. Sci. Technol., 11(4), 209-216 (1995).
Mojtahedi, W., and Abbasian, J., Hydrogen sulfide removal from coal gas at elevated temperature and pressure in fluidized bed with zinc titanate sorbents: Cyclic tests, Energy Fuels, 9(3), 429-434 (1995).
Natarajan, V., and Bishnoi, P.R., Langmuir constant computations for gas hydrate systems, Ind. Eng. Chem. Res., 34(4), 1494-1498 (1995).
Nevskaia, D.M., et al., Interaction of Triton X-100 on silica: A relationship between surface characteristics and adsorption isotherms, J. Chem. Technol. Biotechnol., 63(3), 249-256 (1995).
Nickolov, R.N., and Mehandjiev, D.R., Application of the simplified equation for micropore size distribution to the study of water vapour adsorption on activated carbon, Adsorpt. Sci. Technol., 12(3), 203-210 (1995).
Park, S.W., et al., Effect of water vapor on adsorption of methyl iodide to triethylenediamine-impregnated activated carbons, Sep. Technol., 5(1), 35-44 (1995).
Parker, G.R., Optimum isotherm equation and thermodynamic interpretation for aqueous 1,1,2-trichloroethane adsorption isotherms on three adsorbents, Adsorption, 1(2), 113-132 (1995).
Perona, J.J., et al., A simple model for strontium breakthrough on zeolite columns, Sep. Sci. Technol., 30(7), 1259-1268 (1995).
Rauf, M.A., et al., Adsorption studies of arsenic on manganese dioxide using radiometric techniques, Adsorpt. Sci. Technol., 12(2), 93-100 (1995).
Ray, M.S., Adsorption and adsorptive separations: A bibliographical update (1992-1993), Adsorption, 1(1), 83-97 (1995).
Raychoudhuri, A., and Gaikar, V.G., Adsorptive separations of 2,6-xylenol/cresol mixtures with zeolites, Sep. Technol., 5(2), 91-96 (1995).
Reed, B.E., Identification of removal mechanisms for lead in granular activated carbon (GAC) columns, Sep. Sci. Technol., 30(1), 101-116 (1995).
Salas-Peregrin, M.A., et al., Adsorption of carbon dioxide on activated carbons from diluted ambient environments (errata), Energy Fuels, 9(2), 390 (1995).
Sambeth, J.; Gambaro, L., and Thomas, H., Study of the adsorption/oxidation of methanol over vanadium pentoxide, Adsorpt. Sci. Technol., 12(3), 171-180 (1995).
Samra, S.E.; Youssef, A.M., and Girgis, B.S., Catalytic properties of silica-supported 12-molybdophosphoric acid in the conversion of 2-propanol, Adsorpt. Sci. Technol., 12(3), 191-202 (1995).
Saure, R., and Schlunder, E.U., Sorption isotherms for methanol, benzene and ethanol on poly(vinyl acetate) (PVAc), Chem. Eng. Process., 34(3), 305-316 (1995).
Schilde, U., et al., Selectivity of amidoxime polymers for the sorption of gallate, Sep. Sci. Technol., 30(10), 2245-2250 (1995).
Setoyama, N., and Kaneko, K., Density of He adsorbed in micropores at 4.2K, Adsorption, 1(2), 165-174 (1995).
Sirkecioglu, A.; Altav, Y., and Erdem-Senatalar, A., Adsorption of hydrogen sulfide and sulfur dioxide on bigadic clinoptilolite, Sep. Sci. Technol., 30(13), 2747-2762 (1995).
Swisher, J.H.; Yang, J., and Gupta, R.P., Attrition-resistant zinc titanate sorbent for sulfur, Ind. Eng. Chem. Res., 34(12), 4463-4471 (1995).
Takeuchi, Y., et al., Adsorption of 1-butanol and p-xylene vapor and their mixtures with high silica zeolites, Sep. Technol., 5(1), 23-34 (1995).
Talu, O.; Li, J., and Myers, A.L., Activity coefficients of adsorbed mixtures, Adsorption, 1(2), 103-112 (1995).
Tantet, J.; Eic, M., and Desai, R., Breakthrough study of the adsorption and separation of sulfur dioxide from wet gas using hydrophobic zeolites, Gas Sep. Purif., 9(3), 213-220 (1995).
Thibaud-Erkey, C.; Campagnolo, J.F., and Akgerman, A., Adsorption of volatile organic contaminants on soil and soil constituents, Sep. Purif. Methods, 24(2), 129-200 (1995).
Triebe, R.W., and Tezel, F.H., Adsorption of nitrogen, carbon monoxide, carbon dioxide and nitric oxide on molecular sieves, Gas Sep. Purif., 9(4), 223-230 (1995).
Triebe, R.W., and Tezel, F.H., Adsorption of nitrogen and carbon monoxide on clinoptilolite: Determination and prediction of pure and binary isotherms, Can. J. Chem. Eng., 73(5), 717-724 (1995).

Volesky, B., and Holan, Z.R., Biosorption of heavy metals (review paper), Biotechnol. Prog., 11(3), 235-250 (1995).
Vyas, R.K., and Kumar, S., Estimation of temperature-dependent thermal conductivity of a packed bed of 13X molecular sieves, Ind. Eng. Chem. Res., 34(11), 4058-4062 (1995).
Wang, K.; Furney, T.D., and Hawley, M.C., Modelling the HF adsorption process on woodchips in a packed-bed reactor, Chem. Eng. Sci., 50(18), 2883-2898 (1995).
Yoshida, H.; Kishimoto, N., and Kataoka, T., Adsorption of glutamic acid on polyaminated highly porous chitosan: Equilibria, Ind. Eng. Chem. Res., 34(1), 347-355 (1995).

**1996**

Acharya, D.R., On molecular sieve/water interactions, Sep. Sci. Technol., 31(8), 1179-1183 (1996).
Aguerre, R.J.; Viollaz, P.E., and Suarez, C., A fractal isotherm for multilayer adsorption in foods, J. Food Eng., 30(1), 227-238 (1996).
Al-Lohedan, H.A.; Khamis, E., and Issa, Z.A., Studies on the influence of temperature on the adsorption of some cationic surfactants on to steel, Adsorpt. Sci. Technol., 13(3), 137-152 (1996).
Alpay, E., et al., Adsorption parameters for strongly adsorbed hydrocarbon vapours on some commercial adsorbents, Gas Sep. Purif., 10(1), 25-34 (1996).
Bakker, W.J.W., et al., Sorbent development for continuous regenerative hydrogen sulphide removal in a rotating monolith reactor, Can. J. Chem. Eng., 74(5), 713-718 (1996).
Bart, H.J.; Germerdonk, R., and Ning, P., Two-dimensional non-isothermal model for toluene adsorption in a fixed-bed adsorber, Chem. Eng. Process., 35(1), 57-64 (1996).
Bercic, G.; Pintar, A., and Levec, J., Desorption of phenol from activated carbon by hot water regeneration (desorption isotherms), Ind. Eng. Chem. Res., 35(12), 4619-4625 (1996).
Beristain, C.I.; Garcia, H.S., and Azuara, E., Enthalpy-entropy compensation in food vapor adsorption, J. Food Eng., 30(3), 405-416 (1996).
Boyle, R.W., et al., Chemisorption of carbon monoxide on the iridium(111) surface: In-situ studies of adsorption and desorption kinetics via vibrational spectroscopy, Ind. Eng. Chem. Res., 35(9), 2986-2992 (1996).
Bradford, M.C.J., and Vannice, M.A., Estimation of CO heats of adsorption on metal surfaces from vibrational spectra, Ind. Eng. Chem. Res., 35(9), 3171-3178 (1996).
Brady, T.A.; Rostam-Abadi, M., and Rood, M.J., Applications for activated carbons from waste tires: Natural gas storage and air pollution control, Gas Sep. Purif., 10(2), 97-102 (1996).
Cal, M.P.; Rood, M.J., and Larson, S.M., Removal of VOCs from humidified gas streams using activated carbon cloth, Gas Sep. Purif., 10(2), 117-122 (1996).
Carriere, P., et al., Performance of a virtual adsorber system for removal of lead, Sep. Sci. Technol., 31(7), 965-986 (1996).
Chang, C.Y.; Tsai, W.T., and Lee, H.C., Desorption kinetics of N,N-dimethylformamide vapor from granular activated carbon and hydrophobic zeolite, Sep. Sci. Technol., 31(12), 1675-1686 (1996).
Chen, N., and Yang, R.T., Ab-initio molecular orbital study of adsorption of oxygen, nitrogen, and ethylene on silver-zeolite and silver halides, Ind. Eng. Chem. Res., 35(11), 4020-4027 (1996).
Craknell, R.F., et al., Adsorption and selectivity of carbon dioxide with methane and nitrogen in slit-shaped carbonaceous micropores: Simulation and experiment, Adsorption, 2(3), 193-204 (1996).
Czepirski, L.; Komorowska-Czepirska, E., and Cacciola, G., Adsorption equilibria and kinetics of water vapour on modified chabazite, Adsorpt. Sci. Technol., 14(2), 83-88 (1996).
Da Silva, L.R.D.; Peixoto, C.R.M., and Gushikem, Y., Use of alpha-cellulose coated with titanium(IV) oxide in separation of Cr(VI), Sep. Sci. Technol., 31(8), 1045-1058 (1996).
Dahal, M.P., and Lawrance, G.A., Adsorption of thallium(I), lead(II), copper(II), bismuth(III) and chromium(III) by electrolytic manganese dioxide, Adsorpt. Sci. Technol., 13(4), 231-240 (1996).
Dasmahapatra, G.P.; Pal, T.K., and Bhattacharya, B., Evaluation of thermodynamic parameters for separation of hexavalent chromium by fly ash, Sep. Sci. Technol., 31(20), 2843-2854 (1996).
Davini, P., Investigation of the sulphur dioxide adsorption properties of calcium hydroxide/fly ash systems, Fuel, 75(6), 713-716 (1996).
Do, D.D., A model for surface diffusion of ethane and propane in activated carbon, Chem. Eng. Sci., 51(17), 4145-4158 (1996).
El-Nabarawy, T.; Petro, N.S., and El-Aziz, S.A., Adsorption characteristics of coal-based activated carbons: Adsorption of nitrogen and carbon dioxide, Adsorpt. Sci. Technol., 13(3), 177-186 (1996).

Ferro-Garcia, M.A., et al., Chemical and thermal regeneration of an activated carbon saturated with chlorophenols, J. Chem. Technol. Biotechnol., 67(2), 183-189 (1996).

Funke, H.H., et al., Separation of hydrocarbon isomer vapors with silicalite zeolite membranes, Ind. Eng. Chem. Res., 35(5), 1575-1582 (1996).

Furst, E.M.; Pagac, E.S., and Tilton, R.D., Coadsorption of polylysine and the cationic surfactant cetyltrimethylammonium bromide on silica, Ind. Eng. Chem. Res., 35(5), 1566-1574 (1996).

Gaikar, V.G.; Mandal, T.K., and Kulkarni, R.G., Adsorptive separations using zeolites: Separation of substituted anilines, Sep. Sci. Technol., 31(2), 259-270 (1996).

Ghosh-Dastidar, A., et al., Investigation of high-reactivity calcium carbonate sorbent for enhanced sulfur dioxide capture, Ind. Eng. Chem. Res., 35(2), 598-606 (1996).

Goto, M., et al., Impulse response analysis for adsorption of ethyl acetate on activated carbon in supercritical carbon dioxide, Sep. Sci. Technol., 31(12), 1649-1662 (1996).

Gurgel, J.M., and Kluppel, R.P., Thermal conductivity of hydrated silica-gel, Chem. Eng. J., 61(2), 133-138 (1996).

Hu, X., and Do, D.D., Contribution of concentration-dependent surface diffusion in ternary adsorption kinetics of ethane, propane and n-butane in activated carbon, Adsorption, 2(3), 217-226 (1996).

Hubicki, Z.; Hubicka, H., and Lodyga, B., Studies of the separation of palladium(II) microquantities from macroquantities of salts of other elements on selective ion exchangers, Adsorpt. Sci. Technol., 14(1), 5-24 (1996).

Huddersman, K., Adsorption of hexane isomers on ion exchanged mordenite, AIChE J., 42(10), 2990-2992 (1996).

Huddersman, K., and Klimczyk, M., Separation of branched hexane isomers using zeolite molecular sieves, AIChE J., 42(2), 405-408 (1996).

Inoue, K., et al., Adsorption of metal ions on gallium(III)-templated oxine type of chemically modified chitosan, Sep. Sci. Technol., 31(16), 2273-2286 (1996).

Jasra, R.V.; Choudary, N.V., and Bhat, S.G.T., Correlation of sorption behavior of nitrogen, oxygen, and argon with cation locations in zeolite-X, Ind. Eng. Chem. Res., 35(11), 4221-4229 (1996).

Kim, Y.S., et al., Adsorption of organic phenols onto hexadecyltrimethyl-ammonium-treated Montmorillonite, Sep. Sci. Technol., 31(20), 2815-2830 (1996).

King, B., and Do, D.D., Measurement of multicomponent adsorption kinetics of gases in activated carbon by a batch adsorber FT-IR technique, Chem. Eng. Sci., 51(3), 423-440 (1996).

Kondo, K.; Sumi, H., and Matsumoto, M., Adsorption characteristics of metal ions on chitosan chemically modified by D-galactose, Sep. Sci. Technol., 31(12), 1771-1775 (1996).

Kong, Y., and Cha, C.Y., Microwave-induced regeneration of NOx-saturated char, Energy Fuels, 10(6), 1245-1249 (1996).

Kovbuz, M.O.; Khimyak, Y.Z., and Kesh, A.M., Adsorption of some antitumour compounds on polymer adsorbents, Adsorpt. Sci. Technol., 14(4), 217-228 (1996).

Lai, Y.D., and Liu, J.C., Fluoride removal from water with spent catalyst, Sep. Sci. Technol., 31(20), 2791-2804 (1996).

Lee, J.K.; Hudgins, R.R., and Silveston, P.L., Sulfur dioxide oxidation in a periodically operated trickle-bed: Comparison of activated carbon catalysts, Environ. Prog., 15(4), 239-244 (1996).

Lin, S.H.; Chen, Y.W., and Huang, J.K., Hydrochlorofluorocarbon adsorption by granular and extruded activated carbons, Adsorpt. Sci. Technol., 13(3), 187-196 (1996).

Lizzio, A.A., and DeBarr, J.A., Effect of surface area and chemisorbed oxygen on the sulphur dioxide adsorption capacity of activated char, Fuel, 75(13), 1515-1522 (1996).

Lordgooei, M., et al., Activated-carbon cloth adsorption/cryogenic system to recover toxic volatile organic compounds, Gas Sep. Purif., 10(2), 123-130 (1996).

Lu, G.Q., Preparation and evaluation of adsorbents from waste carbonaceous materials for SOx and NOx removal, Environ. Prog., 15(1), 12-18 (1996).

Lu, G.Q., and Lau, D.D., Characterisation of sewage sludge-derived adsorbents for hydrogen sulfide removal: Surface and pore structural evolution in chemical activation, Gas Sep. Purif., 10(2), 103-112 (1996).

Magalhaes, F.D.; Laurence, R.L., and Conner, W.C., Transport of n-paraffins in zeolite-T, AIChE J., 42(1), 68-86 (1996).

Malek, A., and Farooq, S., Comparison of isotherm models for hydrocarbon adsorption on activated carbon, AIChE J., 42(11), 3191-3201 (1996).

Mathias, P.M., et al., Correlation of multicomponent gas adsorption by the dual-site Langmuir model: Application to nitrogen/oxygen adsorption on 5A-zeolite, Ind. Eng. Chem. Res., 35(7), 2477-2483 (1996).

Mazzotti, M., et al., Vapor-phase SMB adsorptive separation of linear/nonlinear paraffins, Ind. Eng. Chem. Res., 35(7), 2313-2321 (1996).

NcNamara, J.D., and Wagner, N.J., Process effects on activated carbon performance and analytical methods used for low level mercury removal in natural gas applications, Gas Sep. Purif., 10(2), 137-140 (1996).

Nodzenski, A., A method for the measurement of carbon dioxide desorption from coal in the elevated pressure range, Adsorpt. Sci. Technol., 13(2), 71-84 (1996).

Ogino, H., and Oi, T., Extraction of lithium from $LiMn_2O_4$ by ammonium peroxodisulfate and lithium isotope selectivities of adsorbents thus obtained, Sep. Sci. Technol., 31(9), 1215-1232 (1996).

Olguin, M.T., et al., Sorption of $^{239}Np$ and $^{235}U$ fission products by zeolite-Y, Mexican natural erionite, and bentonite, Sep. Sci. Technol., 31(15), 2021-2044 (1996).

Pu, S.B.; Tanaka, Y., and Inui, T., Adsorption and separation of ß,ß-dimethylnaphthalene isomers on various large-pore zeolites having 12-oxygen-member-ring structure, Sep. Technol., 6(3), 189-196 (1996).

Qadeer, R., Adsorption of cerium ions on activated charcoal, Adsorpt. Sci. Technol., 13(6), 519-526 (1996).

Rauf, M.A., et al., Kinetic and thermodynamic aspects of ytterbium adsorption on sand, Adsorpt. Sci. Technol., 13(2), 97-104 (1996).

Rauf, M.A.; Iqbal, M.J., and Ellahi, I., Studies on the adsorption behaviour of lutetium on sand, Adsorpt. Sci. Technol., 13(2), 127-135 (1996).

Ray, M.S., Adsorptive and membrane-type separations: A bibliographical update (1995), Adsorpt. Sci. Technol., 13(6), 433-459 (1996).

Rossi, P.F., and Rossi, P., Heats of adsorption of aliphatic alcohols on alpha-alumina at 25-200°C: Variations with experimental temperature, Adsorpt. Sci. Technol., 13(4), 215-230 (1996).

Rota, R., et al., Adsorption equilibria of dimethylnaphthalene isomers, Ind. Eng. Chem. Res., 35(1), 199-206 (1996).

Rubel, A.M., and Stencel, J.M., Effect of pressure on NOx adsorption by activated carbons, Energy Fuels, 10(3), 704-708 (1996).

Ruthven, D.M., and Kaul, B.K., Adsorption of n-hexane and intermediate molecular weight aromatic hydrocarbons on LaY zeolite, Ind. Eng. Chem. Res., 35(6), 2060-2064 (1996).

Scott, J.A., and Karanjkar, A.M., Enhanced activated carbon surfaces from heat-fixation of biosorbed metals, Process Safety Environ. Prot., 74(B1), 45-51 (1996).

Setoyama, N.; Li, G., and Kaneko, K., Nitrogen adsorption on fluorinated activated carbon fiber, Adsorption, 2(4), 293-298 (1996).

Sheikh, M.A.; Hassan, M.M., and Loughlin, K.F., Adsorption equilibria and rate parameters for nitrogen and methane on Maxsorb activated carbon, Gas. Sep. Purif., 10(3), 161-168 (1996).

Singh, I.J., and Misra, B.M., Studies on sorption of radiocesium on copper-hexacyanoferrate-loaded resins, Sep. Sci. Technol., 31(12), 1695-1706 (1996).

Singh, R.P., The influence of organic cosolvent methanol on the adsorption of carbofuran on three different types of Indian soils, Adsorpt. Sci. Technol., 13(5), 305-316 (1996).

Smirniotis, P.G., and Zhang, W., Effect of the Si/Al ratio and of the zeolite structure on the performance of dealuminated zeolites for the reforming of hydrocarbon mixtures, Ind. Eng. Chem. Res., 35(9), 3055-3066 (1996).

Soltys, M.M., and Yaremko, Z.M., The adsorption of polymethacrylic acid on titanium dioxide, Adsorpt. Sci. Technol., 14(3), 199-207 (1996).

Sun, J., et al., Natural gas storage with activated carbon from a bituminous coal, Gas Sep. Purif., 10(2), 91-96 (1996).

Sun, M.S.; Talu, O., and Shah, D.B., Diffusion measurements through embedded zeolite crystals, AIChE J., 42(11), 3001-3007 (1996).

Talu, O., et al., Measurement and analysis of oxygen/nitrogen/5A-zeolite adsorption equilibria for air separation, Gas. Sep. Purif., 10(3), 149-160 (1996).

Tamon, H.; Kitamura, K., and Okazaki, M., Adsorption of carbon monoxide on activated carbon impregnated with metal halide, AIChE J., 42(2), 422-430 (1996).

Tan, C.S., and Huang, C.H., Separation of trimethylbenzene isomers on molecular sieves 13X in high pressure carbon dioxide, Sep. Sci. Technol., 31(15), 2011-2020 (1996).

Tihmillioglu, F., and Ulku, S., Use of Clinoptilolite in ethanol dehydration, Sep. Sci. Technol., 31(20), 2855-2866 (1996).

Triebe, R.W.; Tezel, F.H., and Khulbe, K.C., Adsorption of methane, ethane and ethylene on molecular sieve zeolites, Gas Sep. Purif., 10(1), 81-84 (1996).

Tsuchiai, H., et al., Removal of sulfur dioxide from flue gas by the absorbent prepared from coal ash: Effects of nitrogen oxide and water vapor, Ind. Eng. Chem. Res., 35(3), 851-855 (1996).

Various, Proceedings of the 1st Polish-Ukrainian Symposium on Theoretical and Experimental Studies of Interfacial Phenomena and their Technological Application (Kazimierz Dolny, Poland; 8-10 June 1995), Adsorpt. Sci. Technol., 14(1), 1-76 (1996).

Wei, J., Adsorption and cracking of n-alkanes over ZSM-5: Negative activation energy of reaction, Chem. Eng. Sci., 51(11), 2995-3000 (1996).

Yan, T.Y., Mercury removal from oil by reactive adsorption, Ind. Eng. Chem. Res., 35(10), 3697-3701 (1996).

Yang, J., and Volesky, B., Intraparticle diffusivity of Cd ions in a new biosorbent material, J. Chem. Technol. Biotechnol., 66(4), 355-364 (1996).

Yang, R.T., et al., Zeolites containing mixed cations for air separation by weak chemisorption-assisted adsorption, Ind. Eng. Chem. Res., 35(9), 3093-3099 (1996).

Youssef, A.M., et al., Adsorption of heavy metals on coal-based activated carbons, Adsorpt. Sci. Technol., 13(2), 115-126 (1996).

Yrjas, K.P.; Zevenhoven, C.A.P., and Hupa, M.M., Hydrogen sulfide capture by limestone and dolomite at elevated pressure: Sorbent performance, Ind. Eng. Chem. Res., 35(1), 176-183 (1996).

Zevenhoven, C.A.P.; Yrjas, K.P., and Hupa, M.M., Hydrogen sulfide capture by limestone and dolomite at elevated pressure: Sorbent particle conversion modeling, Ind. Eng. Chem. Res., 35(3), 943-949 (1996).

Zhou, L., and Zhou, Y., A comprehensive model for the adsorption of supercritical hydrogen on activated carbon, Ind. Eng. Chem. Res., 35(11), 4166-4168 (1996).

**1997**

Abdel Raouf, M.W., and Daifullah, A.A.M., Potential use of bone charcoal in the removal of antimony and europium radioisotopes from radioactive wastes, Adsorpt. Sci. Technol., 15(8), 559-570 (1997).

Allen, S.J., et al., The adsorption of pollutants by peat, lignite and activated chars, J. Chem. Technol. Biotechnol., 68(4), 442-452 (1997).

Anon., Recovering olefins using adsorption, Chem. Eng. (N.Y.), November, 124-125 (1997).

Anstice, P.J.C., and Alder, J.F., The effect of sulphur dioxide on the adsorption properties of activated carbon towards chloropicrin, Adsorpt. Sci. Technol., 15(7), 541-550 (1997).

Asada, H.; Seiyama, H., and Takechi, M., Displacement transition in methane/cyclohexane adsorbed on graphite, Adsorpt. Sci. Technol., 15(4), 271-276 (1997).

Cal, M.P.; Rood, M.J., and Larson, S.M., Gas phase adsorption of volatile organic compounds and water vapor on activated carbon cloth, Energy Fuels, 11(2), 311-315 (1997).

Chanda, M., and Rempel, G.L., Chromium(III) removal by epoxy-cross-linked poly(ethylenimine) used as gel-coat on silica: Sorption characteristics, Ind. Eng. Chem. Res., 36(6), 2184-2189 (1997).

Chanda, M., and Rempel, G.L., Chromium(III) removal by epoxy-cross-linked poly(ethylenimine) used as gel-coat on silica: A new kinetic model, Ind. Eng. Chem. Res., 36(6), 2190-2196 (1997).

Chen, J.H.; Wong, D.S.H., and Tan, C.S., Adsorption and desorption of carbon dioxide onto and from activated carbon at high pressures, Ind. Eng. Chem. Res., 36(7), 2808-2815 (1997).

Chou, S.H.; Wong, D.S.H., and Tan, C.S., Adsorption and diffusion of benzene in activated carbon at high pressures, Ind. Eng. Chem. Res., 36(12), 5501-5506 (1997).

Choudhary, V.R.; Nayak, V.S., and Choudhary, T.V., Single-component sorption/diffusion of cyclic compounds from their bulk liquid phase in H-ZSM-5 zeolite, Ind. Eng. Chem. Res., 36(5), 1812-1818 (1997).

Dalai, A.K.; Tollefson, E.L., and Yang, A., Oxidation of methyl mercaptan over an activated carbon in a fixed-bed reactor, Ind. Eng. Chem. Res., 36(11), 4726-4733 (1997).

DeBarr, J.A.; Lizzio, A.A., and Daley, M.A., Adsorption of sulfur dioxide on bituminous coal char and activated carbon fiber, Energy Fuels, 11(2), 267-271 (1997).

Denayer, J.F.M., and Baron, G.V., Adsorption of normal and branched paraffins in faujasite zeolites NaY, Pt/NaY and USY, Adsorption, 3(4), 251-266 (1997).

Denizli, A., et al., Cibacron Blue F3G-A attached poly(vinyl alcohol) particles for specific albumin adsorption, Sep. Sci. Technol., 32(5), 1003-1016 (1997).

Deorkar, N.V., and Tavlarides, L.L., Zinc, cadmium, and lead separation from aqueous streams using solid-phase extractants, Ind. Eng. Chem. Res., 36(2), 399-406 (1997).
Dutta, M.; Baruah, R., and Dutta, N.N., Adsorption of 6-aminopenicillanic acid on activated carbon, Sep. Purif. Technol., 12(2), 99-108 (1997).
El-Hakim Daifullah, A.; El-Reefy, S., and Gad, H., Adsorption of p-nitrophenol on Inshas incinerator ash and on the pyrolysis residue of animal bones, Adsorpt. Sci. Technol., 15(7), 485-496 (1997).
El-Nabarawy, T., Dehydrogenation of cyclohexane in relation to some textural and catalytic properties of Ni/alumina and Co/alumina catalysts, Adsorpt. Sci. Technol., 15(1), 25-38 (1997).
El-Nabarawy, T.; Petro, N.S., and Abdel-Aziz, S., Adsorption characteristics of coal-based activated carbons: Adsorption of water vapour, pyridine and benzene, Adsorpt. Sci. Technol., 15(1), 47-58 (1997).
Eltekova, N.A., and Eltekov, Y.A., Adsorption equilibria in carbon/hydrocarbon/water system, Adsorpt. Sci. Technol., 15(2), 109-114 (1997).
Fernandez, J.; Renedo, J., and Garea, A., Preparation and characterization of fly ash/hydrated lime sorbents for sulphur dioxide removal, Powder Technol., 94(2), 133-140 (1997).
Furuya, E.G.; Chang, H.T.; Miura, Y., and Noll, K.E., A fundamental analysis of the isotherm for the adsorption of phenolic compounds on activated carbon, Sep. Purif. Technol., 11(2), 69-68 (1997).
Garcia, E., et al., Kinetic study of high-temperature removal of hydrogen sulfide by novel metal oxide sorbents, Ind. Eng. Chem. Res., 36(3), 846-853 (1997).
Garcia-Garcia, A., et al., NO reduction by potassium-containing coal briquettes: Effect of mineral matter content and coal rank, Energy Fuels, 11(2), 292-298 (1997).
Grzegorczyk, D.S., and Carta, G., Adsorption of amino acids on porous polymeric adsorbents: Intraparticle mass transfer (corrigendum), Chem. Eng. Sci., 52(4), 655 (1997).
Hourieh, M.A.; Alaya, M.N., and Youssef, A.M., Carbon dioxide adsorption and decolourizing power of activated carbons prepared from pistacia shells, Adsorpt. Sci. Technol., 15(6), 419-428 (1997).
Hwang, K.S., et al., Adsorption and thermal regeneration of methylene chloride vapor on an activated carbon bed, Chem. Eng. Sci., 52(7), 1111-1124 (1997).
Ishihara, T., et al., Thermostable molecular sieves, silicoaluminophosphate (SAPO)-34, for the removal of NOx with propane in the coexistence of oxygen, water, and sulfur dioxide, Ind. Eng. Chem. Res., 36(1), 17-22 (1997).
Jeong, S.M., and Kim, S.D., Enhancement of the sulfur dioxide sorption capacity of CuO/gamma-alumina sorbent by an alkali-salt promoter, Ind. Eng. Chem. Res., 36(12), 5425-5431 (1997).
Jinzhou, D.; Xinmin, Y., and Xiangke, W., Adsorption study of 47 elements on alumina and bentonite using a multitracer technique, Adsorpt. Sci. Technol., 15(5), 341-348 (1997).
Juang, R.S., and Chen, M.L., Application of the Elovich equation to the kinetics of metal sorption with solvent-impregnated resins, Ind. Eng. Chem. Res., 36(3), 813-820 (1997).
Kobayashi, M.; Shirai, H., and Nunokawa, M., Investigation on desulfurization performance and pore structure of sorbents containing zinc ferrite, Energy Fuels, 11(4), 887-896 (1997).
Kolhapure, N.H., and Venkatesh, K.V., An unsaturated flow of moisture in porous hygroscopic media at low moisture contents, Chem. Eng. Sci., 52(19), 3383-3392 (1997).
Kwon, K.C.; Crowe, E.R., and Gangwal, S.K., Reactivity of metal oxide sorbents for removal of sulfur compounds from coal gases at high temperature and pressure, Sep. Sci. Technol., 32(1), 775-792 (1997).
Lee, J.W.; Park, H.C., and Moon, H., Adsorption and desorption of cephalosporin C on nonionic polymeric sorbents, Sep. Purif. Technol., 12(1), 1-12 (1997).
Lee, S.H.; Vigneswaran, S., and Moon, H., Adsorption of phosphorus in saturated slag media columns, Sep. Purif. Technol., 12(2), 109-118 (1997).
Levy, J.H.; Day, S.J., and Killingley, J.S., Methane capacities of Bowen Basin coals related to coal properties, Fuel, 76(9), 813-820 (1997).
Li, W.B., et al., Selective adsorption of NOx from hot combustion gases by Ce-doped CuO/titanium dioxide, Energy Fuels, 11(2), 428-432 (1997).
Li, Z., and Flytzani-Stephanopoulos, M., Cu-Cr-O and Cu-Ce-O regenerable oxide sorbents for hot gas desulfurization, Ind. Eng. Chem. Res., 36(1), 187-196 (1997).
Lin, S.H., and Lin, C.M., Adsorption characteristics of humic acids by granular activated carbon, Adsorpt. Sci. Technol., 15(7), 507-516 (1997).
Lizzio, A.A., and DeBarr, J.A., Mechanism of sulfur dioxide removal by carbon, Energy Fuels, 11(2), 284-291 (1997).

Lizzio, A.A., and Pollack, N.R., Symposium on production and use of carbon-based materials for environmental cleanup: An introduction, Energy Fuels, 11(2), 249 (1997).
Mahata, N., and Vishwanathan, V., Dependency of phenol hydrogenation activity on hydrogen chemisorption over supported palladium catalysts, Adsorpt. Sci. Technol., 15(3), 165-172 (1997).
Malek, A., and Farooq, S., Kinetics of hydrocarbon adsorption on activated carbon and silica gel, AIChE J., 43(3), 761-776 (1997).
Martyniuk, H., and Wieckowska, J., The effect of coal rank and carbonization temperature on sulphur dioxide adsorption properties of coal chars, Fuel, 76(7), 563-566 (1997).
Maskan, M., and Gogus, F., The fitting of various models to water sorption isotherms of pistachio nut paste, J. Food Eng., 33(3), 227-238 (1997).
Meeyoo, V.; Trimm, D.L., and Cant, N.W., Adsorption-reaction processes for the removal of hydrogen sulphide from gas streams, J. Chem. Technol. Biotechnol., 68(4), 411-416 (1997).
Mekhemer, G.A.H., and Zaki, M.I., Low-temperature IR spectroscopy of CO adsorption on calcined supported ceria: Probing adsorbed species and adsorbing sites, Adsorpt. Sci. Technol., 15(5), 377-390 (1997).
Mochida, I., et al., Initial period of NO-ammonia reaction over a heat-treated pitch-based active carbon fiber, Energy Fuels, 11(2), 307-310 (1997).
Mochida, I., et al., Kinetic study of the continuous removal of SOx on polyacrylonitrile-based activated carbon fibres: Catalytic activity of PAN-ACF heat-treated at 800°C, Fuel, 76(6), 533-536 (1997).
Mochida, I., et al., Kinetic study of the continuous removal of SOx on polyacrylonitrile-based activated carbon fibres: Kinetic model, Fuel, 76(6), 537-542 (1997).
Mutlu, M.; Sag, Y., and Kutsal, T., Adsorption of copper(II) by Z. ramigera immobilized on Ca-alginate in packed bed columns: A dynamic approach by stimulus-response technique and evaluation of adsorption data by moment analysis, Chem. Eng. J., 65(1), 81-86 (1997).
Petersen, F.W., and VanDeventer, J.S.J., Competitive adsorption of gold cyanide and organic compounds onto porous adsorbents, Sep. Sci. Technol., 32(13), 2087-2103 (1997).
Pineda, M., et al., Modelling of performance of zinc ferrites as high-temperature desulfurizing sorbents in a fixed-bed reactor, Fuel, 76(7), 567-574 (1997).
Predescu, L.; Tezel, F.H., and Chopra, S., Adsorption of nitrogen, methane, carbon monoxide, and their binary mixtures on aluminophosphate molecular sieves, Adsorption, 3(1), 7-26 (1997).
Radwan, G.M.H., and Nassar, M.M., Desulphurization of Um-al-Nar refinery straight-run kerosene gas oil using palm fruit kernel activated charcoal: A locally-made adsorbent, Adsorpt. Sci. Technol., 15(4), 311-322 (1997).
Rao, S.V.S.; Lal, K.B., and Ahmed, J., Separation of cobalt from synthetic intermediate and decontamination radioactive wastes using polyurethane foam, Sep. Sci. Technol., 32(18), 3007-3026 (1997).
Rawat, J.P.; Uraqi, S.M.U., and Singh, R.P., Effect of various factors on the sorption of endosulphan on two Indian soils, Adsorpt. Sci. Technol., 15(9), 649-660 (1997).
Rege, S.U., and Yang, R.T., Limits for air separation by adsorption with LiX zeolite, Ind. Eng. Chem. Res., 36(12), 5358-5365 (1997).
Rinkus, K.; Reed, B.E., and Lin, W., NaOH regeneration of Pb- and phenol-laden activated carbon: Batch study results, Sep. Sci. Technol., 32(14), 2367-2384 (1997).
Rossi, P.F., Heats of adsorption of aliphatic alcohols on $\alpha$-alumina at 25-200°C: Variations with chain length, Adsorpt. Sci. Technol., 15(1), 69-78 (1997).
Rubel, A.M., and Stencel, J.M., The effect of low-concentration sulphur dioxide on the adsorption of NO from gas over activated carbon, Fuel, 76(6), 521-526 (1997).
Russell, B.P., and LeVan, M.D., Coadsorption of organic compounds and water vapor on BPL activated carbon: Ethane, propane, and mixing rules, Ind. Eng. Chem. Res., 36(6), 2380-2389 (1997).
Rynders, R.M.; Rao, M.B., and Sircar, S., Isotope exchange technique for measurement of gas adsorption equilibria and kinetics, AIChE J., 43(10), 2456-2470 (1997).
Sambeth, J.; Juan, A.; Gambaro, L., and Thomas, H., Adsorption/oxidation of methanol on vanadia: A theoretical and experimental study, Adsorpt. Sci. Technol., 15(7), 517-530 (1997).
Shmidt, J.L.; Pimenov, A.V., and Lieberman, A.I., Kinetics of adsorption with granular, powdered, and fibrous activated carbon, Sep. Sci. Technol., 32(13), 2105-2114 (1997).
Shu, H.T.; Li, D.; Scala, A.A., and Ma, Y.H., Adsorption of small organic pollutants from aqueous streams by aluminosilicate-based microporous materials, Sep. Purif. Technol., 11(1), 27-36 (1997).

Silva, J.A.C., and Rodrigues, A.E., Sorption and diffusion of n-pentane in pellets of 5A zeolite, Ind. Eng. Chem. Res., 36(2), 493-500 (1997).
Silva, J.A.C., and Rodrigues, A.E., Fixed-bed adsorption of n-pentane/isopentane mixtures in pellets of 5A zeolite, Ind. Eng. Chem. Res., 36(9), 3769-3777 (1997).
Silva, J.A.C., and Rodrigues, A.E., Equilibrium and kinetics of n-hexane sorption in pellets of 5A zeolite, AIChE J., 43(10), 2524-2534 (1997).
Singh, R.P., and Singh, D., Effect of cosolvent (acetone) on the adsorption and movement of cypermethrin in Indian soils, Adsorpt. Sci. Technol., 15(2), 135-145 (1997).
Suer, M.G.; Dardas, Z., and Lu, Y., In-situ CIR-FTIR study of the diffusion of supercritical hydrocarbons in zeolite-L, AIChE J., 43(7), 1717-1726 (1997).
Taulbee, D.N.; Graham, U.; Rathbone, R.F., and Robl, T.L., Removal of carbon dioxide from multicomponent gas streams using dry-FGD wastes, Fuel, 76(8), 781-786 (1997).
Tiwari, D.P.; Saksena, D.N., and Singh, D.K., Kinetics of adsorption of Pb(II) on used tea leaves and Cr(VI) on acacia arabica bark, Dev. Chem. Eng. Mineral Process., 5(1), 79-88 (1997).
Uchida, H., et al., Adsorption behaviors of 2,6- and 2,7-dimethylnaphthalenes in supercritical carbon dioxide using NaY-type zeolite, Ind. Eng. Chem. Res., 36(2), 424-429 (1997).
van Zee, G., and de Graauw, J., Efficient sorption processes by fibrous sorbents, Ind. Eng. Chem. Res., 36(6), 2346-2352 (1997).
Various, Production and use of carbon-based materials for environmental cleanup (symposium papers), Energy Fuels, 11(2), 249-353 (1997).
Wang, N.; Sasaki, M., and Yoshida, T., Flow microcalorimetric study of methanol adsorption from n-hexane on coals, Energy Fuels, 11(6), 1293-1298 (1997).
White, D.A., and Rautiu, R., The sorption of anionic species on hydrous tin dioxide, Chem. Eng. J., 66(2), 85-90 (1997).
Xie, Y., et al., Zeolites modified by CuCl for separating CO from gas mixtures containing carbon dioxide, Adsorption, 3(1), 27-32 (1997).
Yang, J.; Lee, C.H., and Chang, J.W., Separation of hydrogen mixtures by a two-bed pressure swing adsorption process using zeolite 5A, Ind. Eng. Chem. Res., 36(7), 2789-2798 (1997).
Yehia, A.; Atia, A.A., and Ateya, B.G., The adsorption of sulphonate on fluorite single crystals as examined by in-situ FT-IR internal reflectance spectroscopy, Adsorpt. Sci. Technol., 15(9), 685-694 (1997).
Yeo, S.D.; Tuncer, E., and Akgerman, A., Adsorption of volatile organic compounds on soil and prediction of desorption breakthroughs, Sep. Sci. Technol., 32(15), 2497-2512 (1997).
Zapfel, W.; Marr, R., and Siebenhofer, M., Adsorptive NOx-reduction, Sep. Sci. Technol., 32(1), 617-640 (1997).

## ADSORBENT MATERIALS

### 1967

Mathur, B.C., and Banerjee, P.K., Solid desicants for vapour-phase dehydration, Brit. Chem. Eng., 11(11), 1388-1389 (1966); 12(3), 381-384 (1967).

### 1969

Gribkova, L.V.; Sarakhov, A.I.; Tverdokhleb, N.A., and Ryabikova, A.I., Adsorption and mechanical properties of molded zeolites containing clay, Int. Chem. Eng., 9(3), 503-505 (1969).
Hawtin, P.; Dawson, R.W., and Roberts, J., The diffusion of gases through graphite, Trans. IChemE, 47, T109-T113 (1969).

### 1970

Koshelev, V.S., and Byk, S.S., Regeneration of NaA and KA synthetic zeolites by propylene, Int. Chem. Eng., 10(1), 21-23 (1970).

### 1971

Green, D.W.; Hardy, R.G.; Beri, P., and Vickburg, C.D., Make activated carbon from coke, Hydrocarbon Process., 50(1), 105-108 (1971).
Ruthven, D.M., and Loughlin, K.F., Effect of crystallite shape and size distribution on diffusion measurements in molecular sieves, Chem. Eng. Sci., 26(5), 577-584 (1971).

**1973**

Borthakur, P.C., et al., Quality assessment of molecular sieve zeolites by thermal analysis, J. Applied Chem. Biotechnol., 23, 415-418 (1973).

**1974**

Barrer, R.M., and Harding, D.A., Optimization and modification of Offretite sorbents, Sep. Sci., 9(3), 195-210 (1974).

Burwell, R.L., Modified silica gel as adsorbents and catalysts, Chemtech, June, 370-377 (1974).

Parkash, S., Chemistry of activated carbon, Chem. Ind. (London), 1 June, 445-449 (1974).

**1975**

DeJohn, P.B., Granular activated carbons from lignite and coal, Chem. Eng. (N.Y.), 28 April, 113-116 (1975).

Matsumura, Y., Production of acidified active carbon by wet oxidation, and its carbon structure, J. Applied Chem. Biotechnol., 25, 39-56 (1975).

Various, Activated carbon regeneration (topic issue), Chem. Eng. Prog., 71(5), 80-91 (1975).

**1976**

Hancil, V., Regeneration of granulated active carbon, Int. Chem. Eng., 16(4), 663-668 (1976).

Matsumoto, Z., and Numasaki, K., Regenerate granular carbon, Hydrocarbon Process., 55(5), 157-160 (1976).

Trimm, D.L., The deactivation and reactivation of nickel-based catalysts, Trans. IChemE, 54, T119-T123 (1976).

**1977**

Allen, T., and Burevski, D., Adsorption of gases on microporous carbons, Powder Technol., 18, 139-148 (1977).

**1978**

Burfield, D.R.; Gan, G.H., and Smithers, R.H., Molecular sieves: Desiccants of choice, J. Applied Chem. Biotechnol., 28, 23-30 (1978).

Gorbaty, M.L., Effect of drying on the adsorptive properties of subbituminous coal, Fuel, 57(12), 796-797 (1978).

Robson, H., Synthesizing zeolites, Chemtech, March, 176-180 (1978).

Suzuki, M.; Misic, D.; Koyama, O., and Kawazoe, K., Thermal regeneration of spent activated carbons, Chem. Eng. Sci., 33(3), 271-280 (1978).

Zanitsch, R.H., and Lynch, R.T., Selecting a thermal regeneration system for activated carbon, Chem. Eng. (N.Y.), 2 January, 95-100 (1978).

**1979**

Durie, R.A., and Schafer, H.N.S., Production of active carbon from brown coal in high yields, Fuel, 58(6), 472-476 (1979).

Roques, M., and Bastick, M., Some structural characteristics of microporous carbons deduced from adsorption and diffusion data, Fuel, 58(8), 561-564 (1979).

Yang, R.T., and Liu, R.T., Gaseous diffusion in carbon with particular reference to graphite, Ind. Eng. Chem. Process Des. Dev., 18(2), 245-249 (1979).

**1980**

Dixon, D.R., Magnetic adsorbents: Properties and applications, J. Chem. Technol. Biotechnol., 30, 572-578 (1980).

Linares-Solano, A., et al., Active carbons from almond shells as adsorbents in gas and liquid phases, J. Chem. Technol. Biotechnol., 30, 65-72 (1980).

Seko, M.; Miyake, T., and Inada, K., Sieves for mixed xylenes separation, Hydrocarbon Process., 59(1), 133-138 (1980).

**1981**

Barrer, R.M., Sorption in porous crystals: Equilibria and their interpretation, J. Chem. Technol. Biotechnol., 31, 71-85 (1981).

Chihara, K.; Smith, J.M., and Suzuki, M., Regeneration of powdered activated carbon, AIChE J., 27(2), 213-225 (1981).

Juntgen, H.; Knoblauch, K., and Harder, K., Carbon molecular sieves: Production from coal and application in gas separation, Fuel, 60(9), 832-838 (1981).

Ordoyno, N.F., and Rowan, S.M., Adsorptive properties of carbons prepared by the pyrolysis of binary mixtures of textile fibres, J. Chem. Technol. Biotechnol., 31, 415-423 (1981).

Wilson, J., Active carbons from coals, Fuel, 60(9), 823-831 (1981).

### 1982

Idiculla, R., and Seshadri, S.K., Desiccant characteristics of coconut pith, J. Chem. Technol. Biotechnol., 32, 1065-1072 (1982).

Mahajan, P.O.; Youssef, A., and Walker, P.L., Surface-modified carbons for drying of gas streams, Sep. Sci. Technol., 17(8), 1019-1026 (1982).

### 1983

Bac, N.; Sacco, A., and Hammarstrom, J.L., Measurement of residual adsorption capacity of charcoal filters under conditions of variable humidity, Chem. Eng. Commun., 24(4), 205-214 (1983).

Wilkins, C.S.H., New uses for activated carbon, Chem. Eng. (Rugby, Engl.), October, 15, 23 (1983).

### 1984

Gobet, J., and Kovats, E., Preparation of hydrated silicon dioxide for reproducible chemisorption experiments, Adsorpt. Sci. Technol., 1(1), 77-92 (1984).

Gobet, J., and Kovats, E., Specific surface area of hydrated silicon dioxide measured by chemisorption, Adsorpt. Sci. Technol., 1(2), 111-122 (1984).

Goodboy, K.P., and Fleming, H.L., Trends in adsorption with aluminas, Chem. Eng. Prog., 80(11), 63-68 (1984).

Miura, K., and Hashimoto, K., Model representing change of pore structure during activation of carbonaceous materials, Ind. Eng. Chem. Process Des. Dev., 23(1), 138-145 (1984).

Various, Zeolites for industry (topic issue), Chem. Ind. (London), 2 April, 237-272 (1984).

### 1985

Bohra, J.N., and Sing, K.S.W., Micropore structure of carbonized rayon yarn by nitrogen adsorption and nonane preadsorption, Adsorpt. Sci. Technol., 2(2), 89-96 (1985).

Goworek, J., Adsorption properties of hydrated and thermally modified silica gel, Adsorpt. Sci. Technol., 2(3), 195-206 (1985).

Lopez-Peinado, A., et al., Porous texture characterization of coals and chars, Adsorpt. Sci. Technol., 2(1), 31-38 (1985).

Shah, D.B., and Hayhurst, D.T., Research on molecular sieve technology, Chem. Eng. Educ., 19(4), 198-202 (1985).

### 1986

Hudec, P., et al., Possibility of using t-plots obtained from nitrogen adsorption for valuation of zeolites, Adsorpt. Sci. Technol., 3(3), 159-166 (1986).

Lohse, U., et al., Dependence of adsorption capacity and thermal stability of Y-zeolites upon Si/Al ratio, Adsorpt. Sci. Technol., 3(3), 149-158 (1986).

Lohse, U., et al., Adsorption properties of zeolites prepared by different processes, Adsorpt. Sci. Technol., 3(3), 173-180 (1986).

Lohse, U.; Noack, M., and Jahn, E., Adsorption properties of the $AlPO_4$-5 molecular sieve, Adsorpt. Sci. Technol., 3(1), 19-24 (1986).

Lopez-Gonzalez, J.D., et al., Preparation of active carbons from olive wood, Adsorpt. Sci. Technol., 2(4), 263-270 (1985); 3(1), 41-48 (1986).

Stach, H., et al., Influence of pore diameter on adsorption behaviour of nonpolar molecules on silica adsorbents, Adsorpt. Sci. Technol., 3(4), 261-270 (1986).

Stanley-Wood, N.G.; Sadeghnejad, G.R., and York, P., Adsorption potential characterisation of modified celluloses, Powder Technol., 46(2), 195-200 (1986).

Thamm, H., et al., Calorimetric investigation of adsorption properties of microporous alumino-phosphate, Adsorpt. Sci. Technol., 3(4), 217-220 (1986).

Tipnis, P.R., and Harriott, P., Thermal regeneration of activated carbons, Chem. Eng. Commun., 46, 11-28 (1986).

### 1987

Allen, S.J., Equilibrium adsorption isotherms for peat, Fuel, 66(9), 1171-1175 (1987).

Buczek, B., and Czepirski, L., Improvement of capacity for active carbons, Adsorpt. Sci. Technol., 4(4), 217-223 (1987).

Dubinin, M.M., and Kadlec, O., Microporous adsorbents with limiting development of micropores, Adsorpt. Sci. Technol., 4(1), 45-52 (1987).

Weisz, P.B., Zeolites: The remarkable active site, Chemtech, June, 368-373 (1987).

## 1988

Chen, N.Y., and Degnan, T.F., Industrial catalytic applications of zeolites, Chem. Eng. Prog., 84(2), 32-41 (1988).

Chiang, P.C., and You, J.H., Use of sewage-sludge for manufacturing adsorbents, Can. J. Chem. Eng., 65(6), 921-926 (1988).

Costa, E.; de Lucas, A.; Uguina, M.A., and Ruiz, J.C., Synthesis of 4A zeolite from calcined kaolins for use in detergents, Ind. Eng. Chem. Res., 27(7), 1291-1296 (1988).

Harriott, P., and Cheng, A.T.Y., Kinetics of spent activated carbon regeneration, AIChE J., 34(10), 1656-1662 (1988).

Iovtchev, K.; Nikolov, L., and Elenkov, D., Experimental study of new mechanically resistant adsorbent, Can. J. Chem. Eng., 66(6), 1013-1016 (1988).

Rabo, J.A., New advances in molecular sieve science and technology, Period. Polytech. Chem. Eng., 32(4), 211-234 (1988).

Ramdas, S., Computer graphics and zeolite chemistry, Chem. Eng. Prog., 84(2), 68 (1988).

Ruthven, D.M., Zeolites as selective adsorbents, Chem. Eng. Prog., 84(2), 42-50 (1988).

Schork, J.M., and Fair, J.R., Steaming of activated carbon beds, Ind. Eng. Chem. Res., 27(8), 1545-1548 (1988).

Van Vliet, B.M., and Weber, W.J., Particle surface roughness effects on interfacial mass transfer dynamics of microporous adsorbents, Chem. Eng. Commun., 68, 165-176 (1988).

Vaughan, D.E.W., Synthesis and manufacture of zeolites, Chem. Eng. Prog., 84(2), 25-31 (1988).

## 1989

Burganos, V.N., and Sotirchos, S.V., Fragmentation in random pore structures, Chem. Eng. Commun., 85, 95-112 (1989).

Carniglia, S.C., and Ping, W.L., Alumina desiccant isosteres: Thermodynamics of desiccant equilibria, Ind. Eng. Chem. Res., 28(7), 1025-1030 (1989).

Constantinides, G.N., and Payatakes, A.C., Three-dimensional network model for consolidated porous media: Basic studies, Chem. Eng. Commun., 81, 55-82 (1989).

Herron, N., Toward Si-based life: Zeolites as enzyme mimics, Chemtech, September, 542-548 (1989).

Inoue, M.; Kitamura, K., and Inui, T., Synthesis of wide-pore alumina support from Gibbsite, J. Chem. Technol. Biotechnol., 46(3), 233-248 (1989).

Kasaoka, S., et al., Preparation of activated fibrous carbon from phenolic fabric and its molecular-sieve properties, Int. Chem. Eng., 29(1), 101-115 (1989).

Kirchgessner, D.A., and Jozewicz, W., Structural changes in surfactant-modified sorbents during furnace injection, AIChE J., 35(3), 500-506 (1989).

LaCava, A.I.; Koss, V.A., and Wickens, D., Non-Fickian adsorption rate behavior of some carbon molecular sieves: Slit-potential rate model, Gas Sep. Purif., 3(4), 180-186 (1989).

Martin-Martinez, J.M., et al., Application of a reference material to characterization of porous carbons, Fuel, 68(2), 204-208 (1989).

Morales, M.I.; Blanco, M.N., and Thomas, H.J., Molybdenum concentration profiles in thin layers of porous supports, Adsorpt. Sci. Technol., 6(4), 169-181 (1989).

Patrick, V., et al., High-temperature sulfidation-regeneration of CuO-alumina sorbents, Ind. Eng. Chem. Res., 28(7), 931-940 (1989).

Pierre, A.; Mercier, R., and Lamarche, J.M., The adsorption of cement superplasticizers on to mineral dispersions, Adsorpt. Sci. Technol., 6(4), 219-231 (1989).

Ragai, J., Characterization of hydrous zirconia gels: Adsorption and related studies, Adsorpt. Sci. Technol., 6(1), 9-17 (1989).

Rawtani, A.V.; Rao, M.S., and Gokhale, K.V., Synthesis of ZSM-5 zeolite using silica from rice husk ash, Ind. Eng. Chem. Res., 28(9), 1411-1414 (1989).

Rixey, W.G., and King, C.J., Fixed-bed multisolute adsorption characteristics of nonwet adsorbents, AIChE J., 35(1), 69-74 (1989).

Rodriguez, R.A.E., et al., Adsorption parameters and surface species in the nickel-molybdenum alumina system, Adsorpt. Sci. Technol., 6(4), 192-211 (1989).

Shukla, D.B., and Pandya, V.P., Estimation of crystalline phase in ZSM-5 zeolites by infrared spectroscopy, J. Chem. Technol. Biotechnol., 44(2), 147-154 (1989).

## 1990

Anon., New large-pore molecular sieve, Chem. Eng. (Rugby, Engl.), 26 July, 17 (1990).

Baksh, M.S.A.; Kapoor, A., and Yang, R.T., A new composite sorbent for methane-nitrogen separation by adsorption, Sep. Sci. Technol., 25(7), 845-868 (1990).

Balkose, D., Effect of preparation pH on properties of silica gel, J. Chem. Technol. Biotechnol., 49(2), 165-172 (1990).

Bhat, R.N., and Kumar, R., Synthesis of zeolite-beta using silica gel as a source of silicon dioxide, J. Chem. Technol. Biotechnol., 48(4), 453-466 (1990).

Dagani, R., No organics needed to synthesize alumino-phosphate molecular sieves, Chem. Eng. News, 10 December, 19-20 (1990).

Duda, J.L., A random walk in porous media, Chem. Eng. Educ., 25(3), 136-144 (1990).

Hu, H.C., and Lee, T.Y., Synthesis kinetics of zeolite-A, Ind. Eng. Chem. Res., 29(5), 749-754 (1990).

Irving-Monshaw, S., New zip in activated carbon, Chem. Eng. (N.Y.), February, 43-47 (1990).

Jaroniec, M., et al., Use of argon adsorption isotherms for characterizing microporous activated carbons, Fuel, 69(4), 516-518 (1990).

Jaroniec, M., Evaluation of the fractal dimension of microporous activated carbons, Fuel, 69(12), 1573-1575 (1990).

Kenny, M.B., and Sing, K.S.W., The hydrophobicity of silicalite and ZSM-5, Chem. Ind. (London), 15 January, 39-40 (1990).

Kosmulski, M., et al., Influence of the leaching process on adsorption properties of porous glasses, Sep. Sci. Technol., 25(7), 953-960 (1990).

Milne, C.R., et al., Calcination and sintering models for application to high-temperature, short-time sulfation of calcium-based sorbents, Ind. Eng. Chem. Res., 29(2), 139-150 (1990).

Mondragon, F., et al., New perspectives for coal ash utilization: Synthesis of zeolitic materials, Fuel, 69(2), 263-266 (1990).

## 1991

Abbasian, J.; Rehmat, A., and Banerjee, D.D., Sulfation of partially sulfided calcium-based sorbents, Ind. Eng. Chem. Res., 30(8), 1990-1994 (1991).

Ackley, M.W., and Yang, R.T., Adsorption characteristics of high-exchange clinoptilolites, Ind. Eng. Chem. Res., 30(12), 2523-2530 (1991).

Baksh, M.S.A., and Yang, R.T., Model for spherical cavity radii and potential functions of sorbates in zeolites, AIChE J., 37(6), 923-930 (1991).

Bar-Ziv, E.; Longwell, J.P., and Sarofim, A.F., Determination of the surface area of single particles from high-pressure carbon dioxide adsorption-desorption measurements in an electrodynamic chamber, Energy Fuels, 5(1), 227-228 (1991).

Cacciola, G., et al., Preformed zeolite products to be used in adsorption heat pumps: Preparation methodology and adsorption measurements, Adsorpt. Sci. Technol., 7(3), 163-171 (1991).

Cox, M.A.A., Mercury injection measurements used in the prediction of rock pore dimensions employing a crystalline lattice of capillaries, Chem. Eng. Sci., 46(1), 167-172 (1991).

Davis, M.E., Zeolites and molecular sieves: Not just ordinary catalysts (review paper), Ind. Eng. Chem. Res., 30(8), 1675-1683 (1991).

Do, D.D., and Rice, R.G., A simple method of determining pore and surface diffusivities in adsorption studies, Chem. Eng. Commun., 107, 151-162 (1991).

El-Wakil, A.M.; Youssef, A.M., and Tollan, K.A., Coal-based carbons with molecular sieve properties, Sep. Sci. Technol., 26(3), 445-450 (1991).

Faux, D.A.; Hall, C.K., and Sundaresan, S., Diffusion in zeolites: Effect of directional bias, Chem. Eng. Sci., 46(9), 2359-2362 (1991).

Floess, J.K.; Lee, K.J., and Oleksy, S.A., Kinetics of oxygen chemisorption on microporous carbons, Energy Fuels, 5(1), 133-137 (1991).

Fuda, K., et al., Chemical changes occurring during sulphidation and regeneration of iron containing sorbents, Fuel, 70(1), 100-106 (1991).

Guha, O.K.; Roy, J., and Choudhury, A., Non-polar carbon adsorbents from coal: Effect of coal rank studied by molecular probe chromatography, Fuel, 70(1), 9-12 (1991).

Haggin, J., Titanium silicate molecular sieves developed, Chem. Eng. News, 30 September, 31-32 (1991).

Kennedy, C.R., et al., Effect of crystallite size on the activity and poison resistance of a shape-selective zeolite, Ind. Eng. Chem. Res., 30(1), 12-18 (1991).

Kenny, M.B.; Sing, K.S.W., and Theocharis, C.R., On the stability of VPI-5: A very large pore molecular sieve, Chem. Ind. (London), 18 March, 216-217 (1991).

Kirton, P.J.; Ellis, J., and Crisp, P.T., Investigation of adsorbents for sampling compounds found in coke oven emissions, Fuel, 70(1), 4-8 (1991).

Lafyatis, D.S.; Tung, J., and Foley, H.C., Poly(furfuryl alcohol)-derived carbon molecular sieves: Dependence of adsorptive properties on carbonization temperature, time, and poly(ethylene glycol) additives, Ind. Eng. Chem. Res., 30(5), 865-873 (1991).

Laine, J.; Simoni, S., and Calles, R., Preparation of activated carbon from coconut shell in a small scale concurrent flow rotary kiln, Chem. Eng. Commun., 99, 15-24 (1991).

Nakano, T., et al., Control of the micropores in molecular sieving carbon (MSC) by impregnation of hydrocarbons and heat treatment, Int. Chem. Eng., 31(2), 342-350 (1991).

Neimark, A.V., Calculating surface fractal dimensions of adsorbents, Adsorpt. Sci. Technol., 7(4), 210-219 (1991).

Okieimen, F.E.; Okundia, E.U., and Ogbeifun, D.E., Sorption of cadmium and lead ions on modified groundnut (Arachis hypogea) husks, J. Chem. Technol. Biotechnol., 51(1), 97-104 (1991).

Pollard, S.J.T.; Sollars, C.J., and Perry, R., A low cost adsorbent from spent bleaching earth: The selection of an activation procedure, J. Chem. Technol. Biotechnol., 50(2), 265-276 (1991).

Pollard, S.J.T.; Sollars, C.J., and Perry, R., A low cost adsorbent from spent bleaching earth: Optimisation of the zinc chloride activation procedure, J. Chem. Technol. Biotechnol., 50(2), 277-292 (1991).

Ruplis, A., Pore structure of crystalline ferric hydroxides and oxides, Adsorpt. Sci. Technol., 7(4), 239-246 (1991).

Schreifels, J.A., et al., Adsorption of a metal deactivator addditive onto metal surfaces, Energy Fuels, 5(2), 263-268 (1991).

Seaton, N.A., Determination of the connectivity of porous solids from nitrogen sorption measurements, Chem. Eng. Sci., 46(8), 1895-1910 (1991).

Seifert, J., and Emig, G., Studies of the microstructure of porous solids by physisorption measurements, Int. Chem. Eng., 31(1), 29-41 (1991).

Stoeckli, F., and Ballerini, L., Evolution of microporosity during activation of carbon, Fuel, 70(4), 557-559 (1991).

Tsetsekou, A.; Androutsopoulos, G., and Mann, R., Mercury porosimetry hysteresis and entrapment predictions based on a corrugated random pore model, Chem. Eng. Commun., 110, 1-30 (1991).

Tsiao, C., and Botto, R.E., $^{128}$Xe-NMR investigation of coal micropores, Energy Fuels, 5(1), 87-92 (1991).

White, D.A., and Labayru, R., Synthesis of a manganese dioxide-silica hydrous composite and its properties as a sorption material for strontium, Ind. Eng. Chem. Res., 30(1), 207-210 (1991).

White, W.E., et al., Changes in surface area, pore structure and density during formation of high-temperature chars from representative US coals, Adsorpt. Sci. Technol., 7(4), 180-209 (1991).

Wildman, J., and Derbyshire, F., Origins and functions of macroporosity in activated carbons from coal and wood precursors, Fuel, 70(5), 655-662 (1991).

Yang, R.T., and Baksh, M.S.A., Pillared clays as a new class of sorbents for gas separation, AIChE J., 37(5), 679-686 (1991).

## 1992

Akubuiro, E.C., and Wagner, N.J., Assessment of activated carbon stability toward adsorbed organics, Ind. Eng. Chem. Res., 31(1), 339-346 (1992).

Baksh, M.S.A.; Kikkinides, E.S., and Yang, R.T., Characterization by physisorption of a new class of microporous adsorbents: Pillared clays, Ind. Eng. Chem. Res., 31(9), 2181-2189 (1992).

Baksh, M.S.A.; Kikkinides, E.S., and Yang, R.T., Lithium-type-X zeolite as a superior sorbent for air separation, Sep. Sci. Technol., 27(3), 277-294 (1992).

Baksh, M.S.A., and Yang, R.T., Unique adsorption properties and potential energy profiles of microporous pillared clays, AIChE J., 38(9), 1357-1368 (1992).

Bhattachrya, A., et al., Studies on the synthesis of SAPO-5 molecular sieves, J. Chem. Technol. Biotechnol., 54(4), 399-407 (1992).

Brown, S., New interest in activated carbon, Process Eng. (London), May, 139-140 (1992).

Costello, L.M., and Koros, W.J., Temperature dependence of gas sorption and transport properties in polymers: Measurement and applications, Ind. Eng. Chem. Res., 31(12), 2708-2714 (1992).
Davies, M.E., Large-and extra-large-pore molecular sieves, Chem. Ind. (London), 17 February, 137-139 (1992).
de Lucas, A., et al., Synthesis of 13X zeolite from calcined kaolins and sodium silicate for use in detergents, Ind. Eng. Chem. Res., 31(9), 2134-2140 (1992).
Ernstson M.L., and Rasmuson, A., Field and laboratory measurements of the air permeability of chipped forest-fuel materials, Fuel, 71(8), 963-970 (1992).
Fan, L.T.; Boateng, A.A., and Walawender, W.P., Surface fractal dimension of rice hull-derived charcoal from a fluidized-bed reactor, Can. J. Chem. Eng., 70(2), 387-390 (1992).
Gray, P.G., and Do, D.D., A graphical method for determining pore and surface diffusivities in adsorption systems, Ind. Eng. Chem. Res., 31(4), 1176-1182 (1992).
Guy, P.J., and Perry, G.J., Victorian brown coal as a source of industrial carbons (review paper), Fuel, 71(10), 1083-1086 (1992).
Haggin, J., Molecular sieves have controlled pore volume, Chem. Eng. News, 2 November, 28 (1992).
Ho, C.S., and Shih, S.M., Calcium hydroxide/fly ash sorbents for sulfur dioxide removal, Ind. Eng. Chem. Res., 31(4), 1130-1135 (1992).
Kago, T., et al., Preparation and performance of amidoxime fiber adsorbents for recovery of uranium from seawater, Ind. Eng. Chem. Res., 31(1), 204-209 (1992).
Kocjan, R., Additional purification of some salts by using silica gel modified with Calmagit as a sorbent, Sep. Sci. Technol., 27(3), 409-418 (1992).
Liu, P.K.T., Surface and porous characterization of two commercial activated carbons with polymer and petroleum coke as precursors, Ind. Eng. Chem. Res., 31(9), 2216-2222 (1992).
Lu, G.Q., and Do, D.D., Physical structure and adsorption properties of coal washery reject, Fuel, 71(7), 809-814 (1992).
Maeda, S., et al., Iron(III) hydroxide-loaded coral limestone as an adsorbent for arsenic(III) and arsenic(V), Sep. Sci. Technol., 27(5), 681-690 (1992).
Mann, R., and Yousef, H.N.S., Quantification of random pore structures of porous adsorbents, Adsorpt. Sci. Technol., 8(4), 196-216 (1992).
Merchant, A.A., and Petrich, M.A., Preparation and characterization of activated carbons from scrap tires, almond shells, and Illinois coal, Chem. Eng. Commun., 118, 251-264 (1992).
Munoz-Guillena, M.J., et al., Activated carbons from Spanish coals: Two-stage carbon dioxide activation, Energy Fuels, 6(1), 9-15 (1992).
Oger, L., et al., Heterogeneities and characteristic lengths in porous media, Int. Chem. Eng., 32(4), 674-688 (1992).
Portsmouth, R.L., and Gladden, L.F., Mercury porosimetry as a probe of pore connectivity, Chem. Eng. Res. Des., 70(1), 63-70 (1992).
Todorovic, M., et al., Adsorption of radioactive ions $^{137}Cs^{-}$, $^{85}Sr^{2+}$, and $^{60}Co^{2+}$ on natural magnetite and hematite, Sep. Sci. Technol., 27(5), 671-680 (1992).

**1993**

Apel, M.L., and Torma, A.E., Determination of kinetics and diffusion coefficients of metal sorption on Ca-alginate beads, Can. J. Chem. Eng., 71(4), 652-656 (1993).
Barton, S.S., and Evans, M.J.B., A phenomenological study of the reduction potential of porous and non-porous carbons, Adsorpt. Sci. Technol., 9(2), 84-91 (1993).
Basaldella, E.I.; Bonetto, R., and Tara, J.C., Synthesis of NaY zeolite on preformed kaolinite spheres: Evolution of zeolite content and textural properties with the reaction time, Ind. Eng. Chem. Res., 32(4), 751-752 (1993).
Blocki, S.W., Hydrophobic zeolite adsorbent: A proven advancement in solvent separation technology, Environ. Prog., 12(3), 226-230 (1993).
Boniface, H.A., and Ruthven, D.M., Selectivity of some zeolites for adsorption of atmospheric gases, Gas Sep. Purif., 7(3), 183-184 (1993).
Bradley, R.H., and Rand, B., Activated carbon adsorbents for organic vapours, Fuel, 72(3), 389-394 (1993).
Browne, T.E., and Cohen, Y., Polymer-grafted silica: A screening system for polymeric adsorption resin development, Ind. Eng. Chem. Res., 32(4), 716-725 (1993).
Chanda, M., and Rempel, G.L., Poly(4-vinylpyridine) gel-coated on silica: High capacity and fast kinetics in uranyl sulfate recovery, Ind. Eng. Chem. Res., 32(4), 726-732 (1993).

de Lucas, A., et al., Use of Spanish natural clays as additional silica sources to synthesize 13X zeolite from kaolin, Ind. Eng. Chem. Res., 32(8), 1645-1650 (1993).

Gergova, K., et al., Evolution of the active surface of carbons produced from various raw materials by steam pyrolysis/activation, J. Chem. Technol. Biotechnol., 58(4), 321-330 (1993).

Gergova, K.; Eser, S., and Schobert, H.H., Preparation and characterization of activated carbons from anthracite, Energy Fuels, 7(5), 661-668 (1993).

Gergova, K.; Petrov, N., and Minkova, V., A comparison of adsorption characteristics of various activated carbons, J. Chem. Technol. Biotechnol., 56(1), 77-82 (1993).

Gladden, L.F., Nuclear magnetic resonance studies of porous media, Chem. Eng. Res. Des., 71(6), 657-674 (1993).

Guzel, F., and Tez, Z., The characterization of the micropore structures of some activated carbons of plant origin by nitrogen and carbon dioxide adsorptions, Sep. Sci. Technol., 28(8), 1609-1628 (1993).

Haggin, J., Progress in design of zeolite synthesis, Chem. Eng. News, 13 September, 41-43 (1993).

Henning, K.D., and Schafer, S., Impregnated activated carbon for environmental protection, Gas Sep. Purif., 7(4), 235-240 (1993).

Ho, C.S., and Shih, S.M., Characteristics and sulphur dioxide capture capacities of sorbents prepared from products of spray-drying flue gas desulphurization, Can. J. Chem. Eng., 71(6), 934-939 (1993).

Knoblauch, K., Activated carbon and carbon molecular sieves in gas separation and purification, Gas Sep. Purif., 7(4), 195-196 (1993).

Li, R.J., et al., Transport of gases in miscible polymer blends above and below the glass transition region, AIChE J., 39(9), 1509-1518 (1993).

Merchant, A.A., and Petrich, M.A., Pyrolysis of scrap tires and conversion of chars to activated carbon, AIChE J., 39(8), 1370-1376 (1993).

Miyabe, K., and Suzuki, M., Adsorption characteristics of octadecylsilyl-silica gel in gaseous systems, AIChE J., 39(11), 1791-1798 (1993).

Otowa, T.; Tanibata, R., and Itoh, M., Production and adsorption characteristics of MAXSORB: High-surface-area active carbon, Gas Sep. Purif., 7(4), 241-246 (1993).

Patrick, V.; Gavalas, G.R., and Sharma, P.K., Reduction, sulfidation, and regeneration of mixed iron-aluminum oxide sorbents, Ind. Eng. Chem. Res., 32(3), 519-532 (1993).

Rees, L.V.C., and Shen, D., Characterization of microporous sorbents by frequency-response methods, Gas Sep. Purif., 7(2), 83-90 (1993).

Rossi, P.F., Adsorption heat-flow microcalorimetry applied to coal surface properties, Adsorpt. Sci. Technol., 9(3), 148-189 (1993).

Schroter, H.J., Carbon molecular sieves for gas separation processes, Gas Sep. Purif., 7(4), 247-252 (1993).

Spaull, A.J.B., Cahn's perfect wetting theory in determining monolayer capacity, J. Chem. Technol. Biotechnol., 57(1), 87-92 (1993).

Tomasko, D.L., et al., Pilot scale study and design of a granular activated carbon regeneration process using supercritical fluids, Environ. Prog., 12(3), 208-217 (1993).

Tsuji, M., and Komarneni, S., Selectivity study of alkaline earth and divalent transition metal ions on [$Al^{3+}$/$Na^{+}$]-substituted tobermorites, Sep. Sci. Technol., 28(11), 2061-2072 (1993).

Various, Activated carbon and carbon molecular sieves (topic issue), Gas Sep. Purif., 7(4), 195-284 (1993).

Vyas, S.N., et al., Synthesis of carbon molecular sieves by activation and coke deposition, Fuel, 72(4), 551-556 (1993).

Yenkie, M.K.N., and Natarajan, G.S., Determination of specific surface area of granular activated carbon by aqueous phase adsorption of phenol and from pore size distribution measurements, Sep. Sci. Technol., 28(5), 1177-1190 (1993).

Zhong, S., and Hepworth, M.T., Thermodynamic studies of iron oxysulfide as a sulfur sorbent, Energy Fuels, 7(6), 1073-1078 (1993).

**1994**

Anon., Advanced imaging techniques provide insights about flow in porous solids, Chem. Eng. Prog., 90(7), 22-23 (1994).

Arriagada, R.; Garcia, R., and Reyes, P., Steam and carbon dioxide activation of Eucalyptus globulus charcoal, J. Chem. Technol. Biotechnol., 60(4), 427-433 (1994).

Balci, S.; Dogu, T., and Yucel, H., Characterization of activated carbon produced from almond shell and hazelnut shell, J. Chem. Technol. Biotechnol., 60(4), 419-426 (1994).

Bardina, I.A., et al., Adsorption properties and structure of some active carbons, Adsorpt. Sci. Technol., 10, 211-220 (1994).
Barton, S.S.; Evans, M.J.B., and MacDonald, J.A.F., Oxidized BPL carbons, Adsorpt. Sci. Technol., 10, 75-84 (1994).
Basta, N., New sorbents and adsorption processes, Chem. Eng. (N.Y.), November, 39-43 (1994).
Bhatia, S.; Ghosh, B., and Agrawal, D.C., Synthesis of zeolite-A from calcined diatomaceous clay: Optimization studies, Ind. Eng. Chem. Res., 33(9), 2107-2110 (1994).
Botha, F.D., and McEnaney, B., Chemical activation of a South African coal using phosphoric acid, Adsorpt. Sci. Technol., 10, 181-192 (1994).
Cheng, L.S., and Yang, R.T., Improved Horvath-Kawazoe equations including spherical pore models for calculating micropore size distribution, Chem. Eng. Sci., 49(16), 2599-2610 (1994).
Davis, M.E., Zeolites: Can they be synthesized by design? Chemtech, 24(9), 22-26 (1994).
Dubinin, M.M.; Polyakov, N.S., and Petukhova, G.A., Porous structure and surface chemistry of active carbons, Adsorpt. Sci. Technol., 10, 17-26 (1994).
Girgis, B.S.; Khalil, L.B., and Tawfik, T.A.M., Activated carbon from sugar cane bagasse by carbonization in the presence of inorganic acids, J. Chem. Technol. Biotechnol., 61(1), 87-92 (1994).
Goldstein, T.P., and Schmitt, K.D., Carbon centre types as determined by MAS-CP/NMR on frozen fossil fuels, Fuel, 73(3), 397-404 (1994).
Gomez-Serrano, V., et al., Oxidation of activated carbon by hydrogen peroxide: Study of surface functional groups by FT-IR, Fuel, 73(3), 387-396 (1994).
Goworek, J.; Stefaniak, W., and Swiatkowski, A., The porosity of carbon materials as estimated by thermogravimetric analysis, Adsorpt. Sci. Technol., 11(2), 105-112 (1994).
Guijarro, M.I.; Mendioroz, S., and Munoz, V., Impact of the preparation conditions in the sulfur distribution of a new sulfurized porous adsorbent, Ind. Eng. Chem. Res., 33(2), 375-381 (1994).
Haggin, J., Molecular sieve with unique structural features synthesized, Chem. Eng. News, 4 July, 23 (1994).
Haggin, J., Carbogenic sieves improve catalyst action, Chem. Eng. News, 19 December, 36-37 (1994).
Kabay, N., Preparation of amidoxime-fiber adsorbents based on poly(methacrylonitrile) for recovery of uranium from seawater, Sep. Sci. Technol., 29(3), 375-384 (1994).
Khalil, L.B.; Mourad, W.E., and Girgis, B.S., Changes in the surface area of silica-supported 12-molybdophosphoric acid in relation to the degree of loading and the calcination temperature, Adsorpt. Sci. Technol., 11(2), 83-94 (1994).
Kocjan, R., and Garbacka, M., Silica gel modified with methylthymol blue for separation and preconcentration of trace amounts of heavy metals from some salts, Sep. Sci. Technol., 29(6), 799-807 (1994).
Ku, B.J.; Lee, J.K., and Rhee, H.K., Treatment of activated carbon to enhance catalytic activity for reduction of nitric oxide with ammonia, Ind. Eng. Chem. Res., 33(11), 2868-2874 (1994).
Kumar, M., and Gupta, R.C., Correlation of reactivity and properties of wood chars, Fuel, 73(11), 1805-1806 (1994).
Leboda, R., et al., Effect of catalytic gasification of active carbon from plum stones on its mesoporous structure, Adsorpt. Sci. Technol., 10, 221-240 (1994).
Liu, H., and Seaton, N.A., Determination of the connectivity of porous solids from nitrogen sorption measurements: Solids containing large mesopores, Chem. Eng. Sci., 49(11), 1869-1878 (1994).
Lu, G.Q., and Do, D.D., Development of carbonaceous adsorbents from coal reject for acidic gases removal, Gas Sep. Purif., 8(1), 17-30 (1994).
Mariwala, R.K., and Foley, H.C., Evolution of ultramicroporous adsorptive structure in poly(furfuryl alcohol)-derived carbogenic molecular sieves, Ind. Eng. Chem. Res., 33(3), 607-615 (1994).
Pan, Z.J., et al., Pore structure alteration of a carbon molecular sieve for the separation of hydrogen sulfide from methane by adsorption, Adsorpt. Sci. Technol., 10, 193-202 (1994).
Park, T.G., and Hoffman, A.S., Estimation of temperature-dependent pore size in poly(n-isopropylacrylamide) hydrogel beads, Biotechnol. Prog., 10(1), 82-86 (1994).
Petrov, N.; Gergova, K., and Eser, S., Effect of water vapour on the porous structure of activated carbon from lignite, Fuel, 73(7), 1197-1201 (1994).
Pielaszek, J., X-ray powder diffractometry in studies of solid carbons, Fuel, 73(11), 1792-1796 (1994).
Polyakov, N.S., et al., The hydrophobization of active carbons, Adsorpt. Sci. Technol., 10, 165-172 (1994).
Razvigorova, M., et al., On the composition of volatiles evolved during the production of carbon adsorbents from vegetable wastes, Fuel, 73(11), 1718-1722 (1994).

Rudzinski, W.; Nieszporek, K., and Dabrowski, A., Application of the Dubinin-Astakhov equation to the study of the geometric structure, energetic surface heterogeneity and surface topography of carbonaceous adsorbents, Adsorpt. Sci. Technol., 10, 35-62 (1994).
Ruplis, A., and Bumans, R., The influence of acid treatment on the sorption properties of Latvian clays, Adsorpt. Sci. Technol., 10, 137-146 (1994).
Segarra, E.I., and Glandt, E.D., Model microporous carbons: Microstructure, surface polarity and gas adsorption, Chem. Eng. Sci., 49(17), 2953-2966 (1994).
Sharma, D.C., and Forster, C.F., Compost as an adsorbent for the treatment of hexavalent chromium, Process Safety Environ. Prot., 72(B4), 234-240 (1994).
Siriwardane, R.V.; Poston, J.A., and Evans, G., Spectroscopic characterization of molybdenum-containing zinc titanate desulfurization sorbent, Ind. Eng. Chem. Res., 33(11), 2810-2818 (1994).
Sorrento, L., The proven process of carbon adsorption, Chem. Eng. (N.Y.), July, 94-95 (1994).
Stoeckli, F., and Huguenin, D., Pretreatment and physical activation of acetylene cokes, Fuel, 73(12), 1929-1930 (1994).
Tarasevich, Y.I., et al., Study of the structure of a surface-porous adsorbent using the gas chromatographic version of the molecular probe method, Adsorpt. Sci. Technol., 10, 147-154 (1994).
Xu, Y.; Koga, Y., and Watkinson, A.P., Pore size distribution of coals and chars from Western Canada, Fuel, 73(11), 1797-1801 (1994).
Zhong, S., and Hepworth, M.T., Kinetic studies of iron oxysulfide as a sulfur sorbent, Energy Fuels, 8(1), 276-283 (1994).

**1995**

Allen, S.J., et al., Contrasting adsorption exhibited by lignite-based activated carbons, J. Chem. Technol. Biotechnol., 64(3), 261-268 (1995).
Amarasekera, G.; Scarlett, M.J., and Mainwaring, D.E., Micropore size distributions and specific interactions in coals, Fuel, 74(1), 115-118 (1995).
Amin, A., Surface characteristics of hardened slag-lime pastes, Adsorpt. Sci. Technol., 12(4), 323-334 (1995).
Bailey, A.; Lawrie, G.A., and Williams, M.R., A novel approach for determining the position of primary water-adsorbing centres within the porous structure for active carbons, Adsorpt. Sci. Technol., 11(4), 193-200 (1995).
Basaldella, E.I.; Kikot, A., and Tara, J.C., Effect of pellet pore size and synthesis conditions in the in-situ synthesis of low-silica X-zeolite, Ind. Eng. Chem. Res., 34(9), 2990-2992 (1995).
Burban, J.H.; He, M., and Cussler, E.L., Organic microporous materials made by bicontinuous microemulsion polymerization, AIChE J., 41(4), 907-914 (1995).
Chanda, M., and Rempel, G.L., Poly(ethylene imine)-based granular sorbents by a new process of templated gel-filling: High capacity and selectivity of copper sorption in acidic and alkaline media, Ind. Eng. Chem. Res., 34(8), 2574-2583 (1995).
Cheng, L.S., and Yang, R.T., A new class of non-zeolitic sorbents for air separation: Lithium ion exchanged pillared clays, Ind. Eng. Chem. Res., 34(6), 2021-2028 (1995).
Davis, W.T.; Hood, C.C., and Dever, M., Analysis of a thin activated carbon loaded adsorption medium, Sep. Sci. Technol., 30(7), 1309-1324 (1995).
Deng, S.G., and Lin, Y.S., Sol-gel preparation and properties of alumina adsorbents for gas separation, AIChE J., 41(3), 559-570 (1995).
DiPanfilo, R., and Egiebor, N.O., Structural features of activated carbon produced from oil sands coke, Dev. Chem. Eng. Mineral Process., 3(1), 3-13 (1995).
El-Geundi, M.S.; Ismail, H.M., and Attyia, K.M.E., Activated clay as an adsorbent for cationic dyestuffs, Adsorpt. Sci. Technol., 12(2), 109-118 (1995).
El-Nabarawy, T., et al., Surface and catalytic properties of alumina, CuO/alumina and ZnO/alumina systems, Adsorpt. Sci. Technol., 12(2), 161-170 (1995).
El-Shobaky, G.A.; Ghozza, A.M., and Hammad, S., Effect of sodium oxide doping on the surface and catalytic properties of the manganese oxide/alumina system, Adsorpt. Sci. Technol., 12(2), 119-128 (1995).
Fisher, C., Building large-pore molecular sieves, Chem. Ind. (London), 19 June, 461-464 (1995).
Gergova, K., et al., Environmental scanning electron microscopy of activated carbon production from anthracite by one-step pyrolysis-activation, Fuel, 74(7), 1042-1048 (1995).
Gladden, L.F.; Hollewand, M.P., and Alexander, P., Characterization of structural inhomogeneities in porous media, AIChE J., 41(4), 894-906 (1995).

Haggin, J., Salt hydrates hold promise as carbon dioxide adsorbents, Chem. Eng. News, 17 April, 26-28 (1995).
Hairston, D., Deodorizers, Chem. Eng. (N.Y.), March, 65-68 (1995).
Hairston, D., Update on activated carbon, Chem. Eng. (N.Y.), November, 75-78 (1995).
Heschel, W., and Klose, E., On the suitability of agricultural by-products for the manufacture of granular activated carbon, Fuel, 74(12), 1786-1791 (1995).
Hsien, T.Y., and Rorrer, G.L., Effects of acylation and crosslinking on the material properties and cadmium ion adsorption capacity of porous chitosan beads, Sep. Sci. Technol., 30(12), 2455-2476 (1995).
Hu, N., and Scaroni, A.W., Fragmentation of calcium-based sorbents under high heating rate, short residence time conditions, Fuel, 74(3), 374-382 (1995).
Hussein, M.Z., et al., The preparation of activated carbon from chips of oil palm trunk catalysed by zinc chloride/carbon dioxide: Surface area and porosity studies, J. Chem. Technol. Biotechnol., 64(1), 35-40 (1995).
Jagtoyen, M., et al., Relationship between reflectance and structure of high surface area carbons, Fuel, 74(4), 610-614 (1995).
Jarzebski, A.B., and Lorenc, J., Pore network connectivity and effective diffusivity of silica aerogels, Chem. Eng. Sci., 50(2), 357-360 (1995).
Kumar, P.; Jasra, R.V., and Bhat, T.S.G., Evolution of porosity and surface acidity in Montmorillonite clay on acid activation, Ind. Eng. Chem. Res., 34(4), 1440-1448 (1995).
Marinsky, J.A., Gibbs-Donan-based interpretations of the hydrolysis behavior of zeolites, Ind. Eng. Chem. Res., 34(8), 2898-2909 (1995).
McLaughlin, H.S., Regenerate activated carbon using organic solvents, Chem. Eng. Prog., 91(7), 45-53 (1995).
Molinard, A., and Vansant, E.F., Controlled gas adsorption properties of various pillared clays, Adsorption, 1(1), 49-60 (1995).
Nguyen, C.; Ahmadpour, A., and Do, D.D., Effects of gasifying agents on the characterization of nut shell-derived activated carbon, Adsorpt. Sci. Technol., 12(3), 247-258 (1995).
Parra, J.B., et al., Characterization of activated carbons by the BET equation: An alternative approach, Adsorpt. Sci. Technol., 12(1), 51-66 (1995).
Pimenov, A.V., et al., Accelerated adsorption with activated carbon fiber, Sep. Sci. Technol., 30(16), 3183-3194 (1995).
Puziy, A.M., et al., Porous structure and adsorption properties of synthetic active carbons, Adsorpt. Sci. Technol., 12(4), 267-278 (1995).
Qadeer, R.; Hanif, J., and Majeed, A., Characterisation of metal-impregnated charcoal: Nitrogen adsorption, Adsorpt. Sci. Technol., 12(2), 101-108 (1995).
Querol, X., et al., Synthesis of zeolites by alkaline activation of ferro-aluminous fly ash, Fuel, 74(8), 1226-1231 (1995).
Ren, K., and Kagi, D.A., Study of water repellent effect of earth substrates impregnated with water-based silicones, J. Chem. Technol. Biotechnol., 63(3), 237-246 (1995).
Rutherford, S.W.; Nguyen, C., and Do, D.D., Characterisation of through-flowing pores in activated carbon pellets by the method of time-lag, Dev. Chem. Eng. Mineral Process., 3(3), 125-138 (1995).
Shchipko, M.L., and Kuznetsov, B.N., Catalytic pyrolysis of Kansk-Achinsk lignite for production of porous carbon materials, Fuel, 74(5), 751-755 (1995).
Sidorchuk, V.V., et al., Some peculiarities of silica modification under high reagent pressure: Methodological aspects, Adsorpt. Sci. Technol., 12(3), 231-238 (1995).
Sidorchuk, V.V., et al., Some peculiarities of silica modification under high reagent pressure: Influence of chemical nature of modifying reagent, Adsorpt. Sci. Technol., 12(3), 239-246 (1995).
Srinivasan, R.; Auvil, S.R., and Schork, J.M., Mass transfer in carbon molecular sieves: An interpretation of Langmuir kinetics, Chem. Eng. J., 57(2), 137-144 (1995).
Stoeckli, F., et al., Characterization of industrial activated carbons by adsorption and immersion techniques and by STM, Fuel, 74(11), 1582-1588 (1995).
Teng, H.; Serio, M.A., and Solomon, P.R., Reprocessing of used tires into activated carbon and other products, Ind. Eng. Chem. Res., 34(9), 3102-3111 (1995).
Tsuchiai, H., et al., Highly active absorbent for sulfur dioxide removal prepared from coal fly ash, Ind. Eng. Chem. Res., 34(4), 1404-1411 (1995).
Van Weert, G., and Choi, Y., Silver impregnated activated carbon: A new tool to control cyanide in effluents, Chem. Eng. Commun., 134, 87-92 (1995).

Vega, L.F.; Panagiotopoulos, A.Z., and Gubbins, K.E., Chemical potentials and adsorption isotherms of polymers confined between parallel plates (corrigenda), Chem. Eng. Sci., 50(15), 2505 (1995).

Yang, R.T., and Kikkinides, E.S., New sorbents for olefin/paraffin separations by adsorption via pi-complexation, AIChE J., 41(3), 509-517 (1995).

Youssef, A.M., Adsorption properties of activated carbons obtained from polymer wastes, Adsorpt. Sci. Technol., 11(4), 225-234 (1995).

Youssef, A.M., et al., Changes in the adsorption properties of activated carbon due to partial oxidation of the surface, Adsorpt. Sci. Technol., 12(3), 211-220 (1995).

Zhang, B., and Li, S., Determination of the surface fractal dimension for porous media by mercury porosimetry, Ind. Eng. Chem. Res., 34(4), 1383-1386 (1995).

**1996**

Banakh, O.S., et al., Modified zeolites in gas chromatography for the analysis of air pollutants, Adsorpt. Sci. Technol., 14(4), 209-216 (1996).

Barriocanal, C., et al., Reactive-inert interfaces in metallurgical cokes: Effect of added inerts, Fuel, 72(2), 243-245 (1996).

Barthomeuf, D., Basic zeolites: Characterization and uses in adsorption and catalysis, Catal. Rev. Sci. Eng., 38(4), 521-600 (1996).

Belyakova, L.A., Chemical aspects of the formation of highly hydrophobic organosilicas, Adsorpt. Sci. Technol., 14(5), 287-294 (1996).

Bogillo, V.I.; Shkilev, V.P., and Voelkel, A., Chemical heterogeneity of metal oxide surfaces as studied by inverse gas chromatography at finite concentrations, Adsorpt. Sci. Technol., 14(3), 189-198 (1996).

Bolshak, Y.V., et al., Colloid-chemical modification of micellar polyelectrolyte/film-forming agents, Adsorpt. Sci. Technol., 14(6), 393-399 (1996).

Brei, V.V., et al., A study of the Bronsted site acidity of crystalline and amorphous aluminosilicates: Desorption of ammonia, dehydration of 2-propanol and cracking of cumene, Adsorpt. Sci. Technol., 14(6), 349-358 (1996).

Brei, V.V., and Brichka, A.V., A study of the Bronsted site acidity of crystalline and amorphous aluminosilicates: Thermal decomposition of grafted acetyl groups, Adsorpt. Sci. Technol., 14(6), 359-362 (1996).

Buffham, B.A.; Mason, G., and Koash, K., Composition-front scanning of porous media, Chem. Eng. Res. Des., 74(6), 627-634 (1996).

Carrasco-Marin, F.; Alvarez-Merino, M.A., and Moreno-Castila, C., Microporous activated carbons from a bituminous coal, Fuel, 75(8), 966-970 (1996).

Chen, D., et al., Effect of coke deposition on transport and adsorption in zeolites studied by a new microbalance reactor, Chem. Eng. Sci., 51(11), 2687-2692 (1996).

Cook, J.; Sloan, J., and Green, M.L.H., Carbon nanotube chemistry, Chem. Ind. (London), 19 August, 600-603 (1996).

Dagani, R., Carbon nanotube ropes: Moving closer to practical materials, Chem. Eng. News, 29 July, 5-6 (1996).

Dalai, A.K., et al., Preparation of activated carbon from Canadian coals using a fixed-bed reactor and a spouted bed-kiln system, Fuel, 72(2), 227-237 (1996).

Dawidowicz, A.L.; Wasilewska, D., and Rogalski, J., The activation of rigid materials with a surface polymer layer composed of a dextran-polyimine mixture and their application as supports for bioselective adsorption centres, Adsorpt. Sci. Technol., 14(2), 101-112 (1996).

El-Nabarawy, T.; El-Hakam, S.A., and Youssef, A.M., Sorption properties of aluminas in relation to adsorbate-adsorbent interaction, Adsorpt. Sci. Technol., 13(6), 509-518 (1996).

El-Shobaky, G.A., et al., Surface and catalytic properties of gamma-irradiated ferric oxide/alumina solids, Adsorpt. Sci. Technol., 13(3), 153-164 (1996).

Feng, X., et al., Effect of pore size of mesoporous molecular sieves (MCM-41) on Al stability and acidity, Chem. Eng. J., 64(2), 255-264 (1996).

Goworek, J., and Stefaniak, W., Thermal analysis in studies of structural heterogeneity in active carbons, Adsorpt. Sci. Technol., 14(1), 39-46 (1996).

Grzegorczyk, W., and Leboda, R., Controlled gasification of carbonaceous materials in a water vapour atmosphere: Description of the process kinetics, Adsorpt. Sci. Technol., 14(3), 163-172 (1996).

Grzegorczyk, W., and Leboda, R., Controlled gasification of carbonaceous materials in a water vapour atmosphere: Effect of Ca(II) and Fe(III) ions on the gasification kinetics, Adsorpt. Sci. Technol., 14(3), 173-178 (1996).

Guzel, F., The effect of surface acidity upon the adsorption capacities of activated carbons, Sep. Sci. Technol., 31(2), 283-290 (1996).

Guzel, F., Determination of surface areas of active carbons by retention of ethylene glycol and ethylene glycol monoethyl ether, Sep. Sci. Technol., 31(12), 1687-1694 (1996).

Haggin, J., High-capacity oxygen sorbents show promise in air separation applications, Chem. Eng. News, 5 February, 35-36 (1996).

Haggin, J., New large-pore silica zeolite prepared, Chem. Eng. News, 27 May, 5 (1996).

Haggin, J., New model predicts properties of porous solids, Chem. Eng. News, 17 June, 35 (1996).

Haggin, J., Method designs zeolite templates from scratch, Chem. Eng. News, 19 August, 8 (1996).

Hairston, D.W., Desiccants for water removal, Chem. Eng. (N.Y.), August, 55-58 (1996).

Hudek, P., et al., Sorption and diffusion properties of H- and modified-forms of ZSM-5 zeolites, Adsorpt. Sci. Technol., 13(6), 495-508 (1996).

Hussain, S.T., Carbon monoxide adsorption studies on an Ru:Mn bimetallic alloy system for the determination of active metal sites, Adsorpt. Sci. Technol., 13(6), 489-494 (1996).

Illan-Gomez, M.J., et al., Activated carbons from Spanish coals: Chemical activation, Energy Fuels, 10(5), 1108-1114 (1996).

Inoue, M., et al., Synthesis of thermally stable, porous silica-modified alumina via formation of a precursor in an organic solvent, Ind. Eng. Chem. Res., 35(1), 295-306 (1996).

Khalil, K.A., Surface characteristics of Portland Cement/blast furnace slag mixtures, Adsorpt. Sci. Technol., 13(6), 461-468 (1996).

Khalil, L.B., Adsorption characteristics of activated carbon obtained from rice husks by treatment with phosphoric acid, Adsorpt. Sci. Technol., 13(5), 317-326 (1996).

Kimber, G.M., et al., Fabrication of carbon fibre composites for gas separation, Gas Sep. Purif., 10(2), 131-136 (1996).

Kurganov, A.; Reichert, H., and Unger, K., Investigation of the porosity of silica and alumina with chemically bonded polystyrene, Adsorption, 2(4), 287-292 (1996).

Leng, C.C., and Pinto, N.G., An investigation of the mechanisms of chemical regeneration of activated carbon, Ind. Eng. Chem. Res., 35(6), 2024-2031 (1996).

Li, D.; Scala, A.A., and Ma, Y.H., Adsorption and characteristics of base-treated pillared clays, Adsorption, 2(3), 227-236 (1996).

Liou, T.H., and Chang, F.W., The nitridation kinetics of pyrolyzed rice husk, Ind. Eng. Chem. Res., 35(10), 3375-3383 (1996).

Lu, G.Q., and Rostam-Abadi, M., Carbon-based materials for gas separation, purification and clean-up (editorial), Gas Sep. Purif., 10(2), 89-90 (1996).

Luzinov, I.; Voronov, A., and Minko, S., Modification of the surface of a substrate by adsorbed polymer interlayers for the control of adhesion, Adsorpt. Sci. Technol., 14(4), 259-266 (1996).

Marshall, W.E., and Johns, M.M., Agricultural by-products as metal adsorbents: Sorption properties and resistance to mechanical abrasion, J. Chem. Technol. Biotechnol., 66(2), 192-198 (1996).

Masuda, T., et al., Measurement and prediction of the diffusivity of Y-type zeolite, Chem. Eng. Sci., 51(10), 1879-1888 (1996).

McKay, G., Peat for environmental applications: A review, Dev. Chem. Eng. Mineral Process., 4(3), 127-156 (1996).

Minko, S., et al., A kinetic method for investigating the structure of adsorbed polymer layers, Adsorpt. Sci. Technol., 14(4), 251-258 (1996).

Mustafa, S., et al., Cation-exchange properties of aluminium(III) phosphate, Adsorpt. Sci. Technol., 13(4), 261-280 (1996).

Norman, L.M., and Cha, C.Y., Production of activated carbon from coal chars using microwave energy, Chem. Eng. Commun., 140, 87-110 (1996).

Palotas, A.B., et al., Effect of oxidation on the microstructure of carbon blacks, Energy Fuels, 10(1), 254-259 (1996).

Philip, C.A., and Girgis, B.S., Adsorption characteristics of microporous carbons from apricot stones activated by phosphoric acid, J. Chem. Technol. Biotechnol., 67(3), 248-254 (1996).

Philip, C.A.; Amin, A., and Moussa, N.A., Effect of halide impregnation on the structure and surface characteristics of rutile, Adsorpt. Sci. Technol., 13(4), 281-294 (1996).

Pizzio, L.R., et al., Adsorption and impregnation of alumina with molybdenum or tungsten solutions, Adsorpt. Sci. Technol., 13(3), 165-176 (1996).

Rudzinski, W., and Charmas, R., Effects of surface heterogeneity of oxides in simple ion adsorption at oxide/electrolyte interfaces, Adsorption, 2(3), 245-260 (1996).

Sahouli, B., et al., Surface morphology and chemistry of commercial carbon black and carbon black from vacuum pyrolysis of used tyres, Fuel, 75(10), 1244-1250 (1996).

Sidorchuk, V.V., et al., Studies of the effects of aerosilogel modification by organic acids under high pressures, Adsorpt. Sci. Technol., 14(6), 339-348 (1996).

Simpson, E.J.; Koros, W.J., and Schechter, R.S., An emerging class of volatile organic compound sorbents, Friedel-Crafts modified polystyrenes: Synthesis, characterization, and performance in aqueous- and vapour-phase applications, Ind. Eng. Chem. Res., 35(4), 1195-1205 (1996).

Simpson, E.J.; Koros, W.J., and Schechter, R.S., An emerging class of volatile organic compound sorbents, Friedel-Crafts modified polystyrenes: Performance comparison with commercially-available sorbents and isotherm analysis, Ind. Eng. Chem. Res., 35(12), 4635-4645 (1996).

Slimane, R.B.; Stuart, R.W., and Hepworth, M.T., Preparation and testing of value-added sulfur sorbent pellets from in-situ mined Minnesota manganese deposits, Energy Fuels, 10(6), 1250-1256 (1996).

Souchon, I., et al., Trapping of aromatic compounds by adsorption on hydrophobic sorbents, Sep. Sci. Technol., 13(18), 2473-2492 (1996).

Swiatkowski, A.; Trznadel, B.J., and Zietek, S., Description of active carbon micropore size distribution based on the Horvath-Kawazoe equation adapted to benzene adsorption isotherms, Adsorpt. Sci. Technol., 14(1), 59-68 (1996).

Tancredi, N., et al., Activated carbons from Uruguayan eucalyptus wood, Fuel, 75(15), 1701-1706 (1996).

Teng, H., et al., Preparation of activated carbons from bituminous coals with carbon dioxide activation: Effects of oxygen content in raw coals, Ind. Eng. Chem. Res., 35(11), 4043-4049 (1996).

Tokarev, V., et al., Modification of the interface by polymeric peroxides: A new approach to the creation of composites, Adsorpt. Sci. Technol., 14(4), 239-250 (1996).

Tongta, A.; Liapis, A.I., and Hagen, S., Synthesis of anion exchange adsorbent particles by surface functionalization of polystyrene-divinylbenzene based spherical porous matrices, Sep. Technol., 6(1), 77-90 (1996).

Turov, V.V., et al., Surface properties of carbonacious solids as examined by $^1$H-NMR spectroscopy using the bulk freezing procedure, Adsorpt. Sci. Technol., 14(5), 319-330 (1996).

Various, Carbon-based materials (topic issue), Gas Sep. Purif., 10(2), 89-140 (1996).

Venkatraman, A., et al., Surface fractality of wood charcoals through small-angle X-ray scattering, AIChE J., 42(7), 2014-2024 (1996).

Yang, R.T., and Foldes, R., New adsorbents based on principles of chemical complexation: Monolayer-dispersed nickel(II) for acetylene separation by pi-complexation, Ind. Eng. Chem. Res., 35(4), 1006-1011 (1996).

Yazdi, A.V., and Beckman, E.J., Design, synthesis, and evaluation of novel, highly carbon dioxide-soluble chelating agents for removal of metals, Ind. Eng. Chem. Res., 35(10), 3644-3652 (1996).

Youssef, A.M.; El-Nabarawy, T., and Badr, G.E., Effect of gamma-irradiation on the chemisorption and catalytic activity of Ni/alumina and Co/alumina systems, Adsorpt. Sci. Technol., 13(2), 85-96 (1996).

Yue, Y.H., et al., Chemical liquid deposition zeolites with controlled pore-opening size and shape-selective separation of isomers, Ind. Eng. Chem. Res., 35(2), 430-433 (1996).

Zarko, V.I., and Gunko, V.M., Relaxation process of water molecules on surfaces of highly dispersed silica and titania/silica, Adsorpt. Sci. Technol., 14(5), 331-337 (1996).

Zhao, X.S.; Lu, G.Q., and Millar, G.J., Advances in mesoporous molecular sieve MCM-41 (review paper), Ind. Eng. Chem. Res., 35(7), 2075-2090 (1996).

**1997**

Anon., Roll-press briquetters for recycling waste sludge help steelmakers end landfilling and cut scrap costs, Powder Handling Process., 9(1), 64-67 (1997).

Anstice, P.J.C., and Alder, J.F., The effect of sulphur dioxide exposure under controlled environmental conditions on the challenge performance of copper- and chromate-impregnated active carbon against hydrogen cyanide, Adsorpt. Sci. Technol., 15(7), 531-540 (1997).

Attia, A.A., Some surface properties of activated carbons prepared by gasification with different gases, Adsorpt. Sci. Technol., 15(9), 707-715 (1997).

Bandosz, T.J.; Buczek, B., and Grzybek, T., The determination of surface changes in active carbons by potentiometric titration and water vapour adsorption, Fuel, 76(14), 1409-1416 (1997).
Bathen, D.; Schmidt-Traub, H., and Simon, M., Gas adsorption isotherm for dealuminated zeolites, Ind. Eng. Chem. Res., 36(9), 3993-3994 (1997).
Bereznitski, Y.; Jaroniec, M., and Gadkaree, K.P., Influence of analysis conditions on low pressure adsorption measurements and its consequences in characterization of energetic and structural heterogeneity of microporous carbons, Adsorption, 3(4), 277-282 (1997).
Bota, A., et al., Adsorbents from waste materials, Adsorption, 3(1), 81-92 (1997).
Brage, C., et al., Use of amino phase adsorbent for biomass tar sampling and separation, Fuel, 76(2), 137-142 (1997).
Brahim, F.B.; Mgaidi, A., and El Maaoui, M., Exploration of mesoporous structure of Tunisian raw and acid-leached phosphate ore particles, Can. J. Chem. Eng., 75(4), 759-764 (1997).
Choma, J., and Jaroniec, M., Influence of the pore geometry on the micropore size distribution function of active carbons, Adsorpt. Sci. Technol., 15(8), 571-582 (1997).
Dabrowski, A.; Fekner, Z., and Leboda, R., Steam-carbon gasification catalyzed by calcium: Assessment of the porous structure of active carbons from plum stones and synthetic active carbons, Adsorption, 3(3), 233-250 (1997).
Dagani, R., Calixarene container molecule with vast interior, Chem. Eng. News, 6 October, 12 (1997).
Deng, S.G., and Lin, Y.S, Microwave heating synthesis of supported sorbents, Chem. Eng. Sci., 52(10), 1563-1576 (1997).
El-Aal, A.A.; Ghozza, A.M., and El-Shobaky, G.A., Surface and catalytic properties of NiO and ferric oxide solids, Adsorpt. Sci. Technol., 15(8), 593-608 (1997).
El-Fadly, A.M.; Faramawy, S., and El-Naggar, A.Y., Chromatographic characterization of carbowax-modified silica gels: Influence of polymer loading and pore structure, Sep. Sci. Technol., 32(18), 2993-3006 (1997).
El-Nabarawy, T.; Mostafa, M.R., and Youssef, A.M., Activated carbons tailored to remove different pollutants from gas streams and from solution, Adsorpt. Sci. Technol., 15(1), 59-68 (1997).
El-Sharkawy, E.A.; Mostafa, M.R., and Youssef, A.M., Surface and catalytic properties of tin oxide/chromium oxide catalysts, Adsorpt. Sci. Technol., 15(3), 237-249 (1997).
El-Shobaky, G.A.; Ghozza, A.M., and Mohamed, G.M., Surface and catalytic properties of gamma-irradiated chromia/alumina solids, Adsorpt. Sci. Technol., 15(6), 465-476 (1997).
Frere, M.; Jadot, R., and Bougard, J., Determination of the micropore volume distribution function of activated carbons by gas adsorption, Adsorption, 3(1), 55-66 (1997).
Girgis, B.S.; Kader, A.A., and Aly, A.N.H., Development of porosity in bone char during decolourization of sugar syrup, Adsorpt. Sci. Technol., 15(4), 277-288 (1997).
Gomez-Serrano, V.; Acedo-Ramos, M., and Lopez-Peinado, A.J., Study and characterisation of activated carbon treated with sulphuric acid solutions, J. Chem. Technol. Biotechnol., 68(1), 82-88 (1997).
Gomez-Serrano, V.; Acedo-Ramos, M., and Valenzuela-Calahorro, C., Treatment of activated carbon with hydrogen peroxide: Effect on the porous texture, Adsorpt. Sci. Technol., 15(2), 91-98 (1997).
Grafova, I.A.; Khainakov, S.A., and Strelko, V.V., Organoindium precursor purification by new sorbents based on titanium and zirconium phosphates, Adsorpt. Sci. Technol., 15(6), 429-436 (1997).
Hsien, T.Y., and Rorrer, G.L., Heterogeneous cross-linking of chitosan gel beads: Kinetics, modeling, and influence on cadmium ion adsorption capacity, Ind. Eng. Chem. Res., 36(9), 3631-3638 (1997).
Hunger, M., Bronsted acid sites in zeolites characterized by multinuclear solid-state NMR spectroscopy (review paper), Catal. Rev. Sci. Eng., 39(4), 345-393 (1997).
Ishihara, T., et al., Thermostable molecular sieves, silicoaluminophosphate (SAPO)-34, for the removal of NOx with propane in the coexistence of oxygen, water, and sulfur dioxide, Ind. Eng. Chem. Res., 36(1), 17-22 (1997).
Jacoby, M., Zeolite-type crystals incorporate cobalt, Chem. Eng. News, 1 September, 11 (1997).
Jacoby, M., Iron-rich zeolite stands up to water, Chem. Eng. News, 15 September, 8-9 (1997).
Kaminski, M.; Landsberger, S., and Nunez, L., Sorption capacity of ferromagnetic microparticles coated with CMPO, Sep. Sci. Technol., 32(1), 115-126 (1997).
Khalil, K., and Amer, A.A., The effect of fly ash on some surface characteristics and microstructure of beta-$C_2S$, Adsorpt. Sci. Technol., 15(6), 407-418 (1997).
Khokhlova, T.D.; Nikitin, Y.S. and Detistova, A.L., Modification of silicas and their investigation by dye adsorption, Adsorpt. Sci. Technol., 15(5), 333-340 (1997).

Kilduff, J.E., and King, C.J., Effect of carbon adsorbent surface properties on the uptake and solvent regeneration of phenol, Ind. Eng. Chem. Res., 36(5), 1603-1613 (1997).

Kruk, M.; Jaroniec, M., and Choma, J., Critical discussion of simple adsorption methods used to evaluate the micropore size distribution, Adsorption, 3(3), 209-220 (1997).

Kuhr, J.H., et al., Ion exchange properties of a Western Kentucky low-rank coal, Energy Fuels, 11(2), 323-326 (1997).

Leboda, R.; Charmas, B., and Sidorchuk, V.V., Physicochemical and technological aspects of the hydrothermal modification of complex sorbents and catalysts: Modification of porous and crystalline structures, Adsorpt. Sci. Technol., 15(3), 189-214 (1997).

Leboda, R.; Charmas, B., and Sidorchuk, V.V., Physicochemical and technological aspects of the hydrothermal modification of complex sorbents and catalysts: Modification of phase composition and mechanical properties, Adsorpt. Sci. Technol., 15(3), 215-236 (1997).

Li, W., et al., Preparation and characterization of bimodal porous carbons derived from a styrene-divinylbenzene copolymer, Adsorption, 3(1), 67-80 (1997).

Liang, W.; Chen, S., and Peng, S., Difference of diffusivities in zeolites measured by the non-steady-state and the steady-state methods, Ind. Eng. Chem. Res., 36(5), 1882-1886 (1997).

Lizzio, A.A.; DeBarr, J.A., and Kruse, C.W., Production of activated char from Illinois coal for flue gas cleanup, Energy Fuels, 11(2), 250-259 (1997).

Manzi, S.; Valladares, D.; Marches, J., and Zgrablich, G., Characterization of Maxsorb activated carbons and their evaluation for gas storage, Adsorpt. Sci. Technol., 15(4), 301-310 (1997).

Marzec, A., New structural concept for carbonized coals, Energy Fuels, 11(4), 837-842 (1997).

Matynia, T.; Gawdzik, B., and Staszczuk, P., Characterization of carbon beads derived from porous polyimide copolymer, Adsorpt. Sci. Technol., 15(6), 437-444 (1997).

Mehandjiev, D.R.; Nickolov, R.N., and Ioncheva, R.B., Determination of nitrogen structures on activated carbon surfaces by a chemical method, Fuel, 76(5), 381-384 (1997).

Menendez, J.A., et al., Characterization of petroleum coke as an additive in metallurgical cokemaking: Influence on metallurgical coke quality, Energy Fuels, 11(2), 379-384 (1997).

Munoz-Guillena, M.J., et al., CaO dispersed on carbon as a sulphur dioxide sorbent, Fuel, 76(6), 527-532 (1997).

Murakami, K.; Shirato, H., and Nishiyama, Y., In-situ infrared spectroscopic study of the effects of exchanged cations on thermal decomposition of a brown coal, Fuel, 76(7), 655-662 (1997).

Ohkawa, T.; Sasai, T., and Komoda, N., Computer-aided construction of coal molecular structure using construction knowledge and partial structure evaluation, Energy Fuels, 11(5), 937-944 (1997).

Padin, J., and Yang, R.T., Tailoring new adsorbents based on pi-complexation: Cation and substrate effects on selective acetylene adsorption, Ind. Eng. Chem. Res., 36(10), 4224-4230 (1997).

Pino, L., et al., Composite materials based on zeolite 4A for adsorption heat pumps, Adsorption, 3(1), 33-40 (1997).

Querol, X.; Plana, F.; Alastuey, A., and Lopez-Soler, A., Synthesis of Na-zeolites from fly ash, Fuel, 76(8), 793-800 (1997).

Radovic, L.R.; Menon, V.C., and Leon Y Leon, C.A., On the porous structure of coals: Evidence for an interconnected but constricted micropore system and implications for coalbed methane recovery, Adsorption, 3(3), 221-232 (1997).

Rayzman, V.L.; Shcherban, S.A., and Dworkin, R.S., Technology for chemical-metallurgical coal ash utilization (review paper), Energy Fuels, 11(4), 761-773 (1997).

Roshchina, T.M.; Davydov, V.Y., and Khrustaleva, N.M., Adsorption properties of silica with chemically bonded amino and guanidino groups, Adsorpt. Sci. Technol., 15(3), 147-154 (1997).

Rutherford, S.W., and Do, D.D., Review of time lag permeation technique as a method for characterisation of porous media and membranes, Adsorption, 3(4), 283-300 (1997).

Salmones, J.; Garciafigueroa, E., and Mayagoitia, V., Twofold description of the texture of aluminium oxides prepared by the sol-gel method, Adsorpt. Sci. Technol., 15(9), 661-676 (1997).

Sasaoka, E.; Azhar Uddin, M., and Nojima, S., Novel preparation method of macroporous lime from limestone for high-temperature desulfurization, Ind. Eng. Chem. Res., 36(9), 3639-3646 (1997).

Satya Sai, P.M.; Ahmed, J., and Krishnaiah, K., Production of activated carbon from coconut shell char in a fluidized bed reactor, Ind. Eng. Chem. Res., 36(9), 3625-3630 (1997).

Serrano-Talavera, B.; Munoz-Guillena, M., and Linares-Solano, A., Activated carbons from Spanish coals: Preoxidation effect on anthracite activation, Energy Fuels, 11(4), 785-791 (1997).

Shanley, A., Adsorbents keep their edge, Chem. Eng. (N.Y.), December, 61-64 (1997).
Shonija, N.K., and Detistova, A.L., Adsorption properties of native and modified forms of the clay mineral saponite, Adsorpt. Sci. Technol., 15(1), 39-46 (1997).
Staszczuk, P.; Stefaniak, E., and Bilinski, B., Investigations on the adsorption properties and porosity of natural and thermally treated dolomite samples, Powder Technol., 92(3), 253-258 (1997).
Stinson, S.C., Delving into dendrimers, Chem. Eng. News, 22 September, 28-30 (1997).
Stone, B.T.; Costa, V.C., and Bray, K.L., Inorganic and organically modified rare-earth-doped silica gels, AIChE J., 43(11A), 2785-2792 (1997).
Sun, J., et al., Adsorbed natural gas storage with activated carbons made from Illinois coals and scrap tires, Energy Fuels, 11(2), 316-322 (1997).
Tancredi, N.; Cordero, T., and Rodriguez-Mirasol, J., Activated carbons from eucalyptus wood: Influence of the carbonization temperature, Sep. Sci. Technol., 32(6), 1115-1126 (1997).
Tarasevich, Y.I.; Zhukova, A.I., and Aksenenko, E.V., Interaction of water and other polar substances with hydrophilic centres on the surface of hydrophobic adsorbents, Adsorpt. Sci. Technol., 15(7), 497-506 (1997).
Teng, H.; Lin, H.C., and Ho, J.A., Thermogravimetric analysis on global mass loss kinetics of rice hull pyrolysis, Ind. Eng. Chem. Res., 36(9), 3974-3977 (1997).
Tsapatsis, M., and Gavalas, G., Modeling of silica deposition in porous Vycor: Effects of pore network connectivity, AIChE J., 43(7), 1849-1860 (1997).
Various, New adsorbents and ion exchange materials (topic issue), Adsorption, 3(1), 5-105 (1997).
Walker, A., Superabsorbent polymers, Chem. Eng. (Rugby, Engl.), 25 September, 13-14 (1997).
Wang, K.; Ahmadpour, A., and Do, D.D., Equilibria and kinetics characterisation of two different structured nutshell-derived activated carbons, Adsorption, 3(4), 267-276 (1997).
Wang, Z.; Dong, J.; Xu, N., and Shi, J., Pore modification using the supercritical solution infiltration method, AIChE J., 43(9), 2359-2367 (1997).
Wei, S.H.; Mahuli, S.K.; Agnihotri, R., and Fan, L.S., High surface area calcium carbonate: Pore structural properties and sulfation characteristics, Ind. Eng. Chem. Res., 36(6), 2141-2148 (1997).
Whitten, L.J., and Allen, S.J., Development and characterisation of sorbents from Irish sphagnum moss peat, Dev. Chem. Eng. Mineral Process., 5(3), 251-266 (1997).
Wilson, E., Coated mesoporous silica: Supersoaker for heavy metals, Chem. Eng. News, 19 May, 46-47 (1997).
Wu, Z.; Han, S.S., and Yang, R.T., Modification of resin-type adsorbents for ethane/ethylene separation, Ind. Eng. Chem. Res., 36(7), 2749-2756 (1997).
Yamazaki, S., and Tsutsumi, K., Adsorption characteristics of synthesized mordenite membranes, Adsorption, 3(2), 165-172 (1997).
Yang, S.; Vlessidis, A.G., and Evmiridis, N.P., Influence of gel composition and crystallization conditions on the conventional synthesis of zeolites, Ind. Eng. Chem. Res., 36(5), 1622-1631 (1997).
Yildirim, M., and Ozbayoglu, G., Production of ammonium nitrohumate from Elbistan lignite and its use as a coal binder, Fuel, 76(5), 385-390 (1997).
Yoshizawa, N., et al., Coal-based activated carbons prepared with organometallics and their mesoporous structure, Energy Fuels, 11(2), 327-330 (1997).
Zhang, L., and Ying, J.Y., Synthesis and characterization of mesoporous niobium-doped silica molecular sieves, AIChE J., 43(11A), 2793-2801 (1997).

## LIST OF JOURNALS SURVEYED

| | ***Abbreviation*** |
|---|---|
| Adsorption Journal | Adsorption |
| Adsorption Science and Technology | Adsorpt. Sci. Technol. |
| Advances in Chemical Engineering | Adv. Chem. Eng. |
| American Institute of Chemical Engineers Journal | AIChE J. |
| Ammonia Plant Safety | Ammonia Plant Safety |
| Biotechnology Progress | Biotechnol. Prog. |
| Canadian Journal of Chemical Engineering | Can. J. Chem. Eng. |
| Catalysis Reviews in Science and Engineering | Catal. Rev. Sci. Eng. |
| Chemical Engineering (McGraw-Hill, New York) | Chem. Eng. (N.Y.) |
| The Chemical Engineer (IChemE, UK) | Chem. Eng. (Rugby, Engl.) |

| | |
|---|---|
| Chemical Engineering in Australia | Chem. Eng. Aust. |
| Chemical Engineering Communications | Chem. Eng. Commun. |
| Chemical Engineering Education | Chem. Eng. Educ. |
| Chemical Engineering Journal | Chem. Eng. J. |
| (including Biochemical Engineering Journal) | (Biochem. Eng. J.) |
| Chemical and Engineering News | Chem. Eng. News |
| Chemical Engineering and Processing | Chem. Eng. Process. |
| Chemical Engineering Progress | Chem. Eng. Prog. |
| Chemical Engineering Research and Design | Chem. Eng. Res. Des. |
| Chemical Engineering Science | Chem. Eng. Sci. |
| Chemistry and Industry | Chem. Ind. (London) |
| Chemtech | Chemtech |
| Computers in Chemical Engineering | Comput. Chem. Eng. |
| Developments in Chemical Engineering and Mineral Proceesing | Dev. Chem. Eng. Mineral Process. |
| Energy and Fuels | Energy Fuels |
| Energy World | Energy World |
| Environmental Progress | Environ. Prog. |
| Food and Bioproducts Processing | Food Bioprod. Process. |
| Fuel | Fuel |
| Gas Separation and Purification | Gas Sep. Purif. |
| Hydrocarbon Processing | Hydrocarbon Process. |
| Industrial and Engineering Chemistry Process Design and Development | Ind. Eng. Chem. Process Des. Dev. |
| Industrial and Engineering Chemistry Research | Ind. Eng. Chem. Res. |
| International Journal of Heat and Mass Transfer | Int. J. Heat Mass Transfer |
| Journal of Applied Chemistry | J. Appl. Chem. |
| Journal of Chemical Technology and Biotechnology | J. Chem. Technol. Biotechnol. |
| Journal of Food Engineering | J. Food Eng. |
| Journal of Institute of Energy | J. Inst. Energy |
| Journal of Institute of Fuel | J. Inst. Fuel |
| Journal of Loss Prevention in the Process Industries | J. Loss Prev. Process Ind. |
| Powder Handling and Processing | Powder Handling Process. |
| Powder Technology | Powder Technol. |
| Plant/Operations Progress | Plant/Operations Prog. |
| Process Engineering | Process Eng. (London) |
| Processing | Processing (Sutton, Engl.) |
| Process Safety and Environmental Protection | Process Safety Environ. Prot. |
| Process Safety Progress | Process Safety Prog. |
| Separation and Purification Methods | Sep. Purif. Methods |
| Separation Science | Sep. Sci. |
| Separation Science and Technology | Sep. Sci. Technol. |
| Separation Technology | Sep. Technol. |
| Solvent Extraction and Ion Exchange | Solvent Extr. Ion Exch. |
| Transactions of IChemE | Trans. IChemE |

## Subject Index

### A

## B



## C

## D



## E

## F



## G



## H

## I



## K



## L



## M



## N

## O



## P



## Q



## R

# T

## U



## V



## W



## X



## Z

## STUDIES IN SURFACE SCIENCE AND CATALYSIS



---

Volume 1 **Preparation of Catalysts I.** Scientific Bases for the Preparation of Heterogeneous Catalysts. Proceedings of the First International Symposium, Brussels, October 14–17, 1975
edited by **B. Delmon, P.A. Jacobs and G. Poncelet**

Volume 2 **The Control of the Reactivity of Solids.** A Critical Survey of the Factors that Influence the Reactivity of Solids, with Special Emphasis on the Control of the Chemical Processes in Relation to Practical Applications
by **V.V. Boldyrev, M. Bulens and B. Delmon**

Volume 3 **Preparation of Catalysts II.** Scientific Bases for the Preparation of Heterogeneous Catalysts. Proceedings of the Second International Symposium, Louvain-la-Neuve, September 4–7, 1978
edited by **B. Delmon, P. Grange, P. Jacobs and G. Poncelet**

Volume 4 **Growth and Properties of Metal Clusters.** Applications to Catalysis and the Photographic Process. Proceedings of the 32nd International Meeting of the Société de Chimie Physique, Villeurbanne, September 24–28, 1979
edited by **J. Bourdon**

Volume 5 **Catalysis by Zeolites.** Proceedings of an International Symposium, Ecully (Lyon), September 9–11, 1980
edited by **B. Imelik, C. Naccache, Y. Ben Taarit, J.C. Vedrine, G. Coudurier and H. Praliaud**

Volume 6 **Catalyst Deactivation.** Proceedings of an International Symposium, Antwerp, October 13–15, 1980
edited by **B. Delmon and G.F. Froment**

Volume 7 **New Horizons in Catalysis.** Proceedings of the 7th International Congress on Catalysis, Tokyo, June 30–July 4, 1980. Parts A and B
edited by **T. Seiyama and K. Tanabe**

Volume 8 **Catalysis by Supported Complexes**
by **Yu.I. Yermakov, B.N. Kuznetsov and V.A. Zakharov**

Volume 9 **Physics of Solid Surfaces.** Proceedings of a Symposium, Bechyňe, September 29–October 3, 1980
edited by **M. Láznička**

Volume 10 **Adsorption at the Gas–Solid and Liquid–Solid Interface.** Proceedings of an International Symposium, Aix-en-Provence, September 21–23, 1981
edited by **J. Rouquerol and K.S.W. Sing**

Volume 11 **Metal-Support and Metal-Additive Effects in Catalysis.** Proceedings of an International Symposium, Ecully (Lyon), September 14–16, 1982
edited by **B. Imelik, C. Naccache, G. Coudurier, H. Praliaud, P. Meriaudeau, P. Gallezot, G.A. Martin and J.C. Vedrine**

Volume 12 **Metal Microstructures in Zeolites.** Preparation - Properties – Applications. Proceedings of a Workshop, Bremen, September 22–24, 1982
edited by **P.A. Jacobs, N.I. Jaeger, P. Jirů and G. Schulz-Ekloff**

Volume 13 **Adsorption on Metal Surfaces.** An Integrated Approach
edited by **J. Bénard**

Volume 14 **Vibrations at Surfaces.** Proceedings of the Third International Conference, Asilomar, CA, September 1–4, 1982
edited by **C.R. Brundle and H. Morawitz**

Volume 15 **Heterogeneous Catalytic Reactions Involving Molecular Oxygen**
by **G.I. Golodets**

Volume 16 **Preparation of Catalysts III.** Scientific Bases for the Preparation of Heterogeneous Catalysts. Proceedings of the Third International Symposium, Louvain-la-Neuve, September 6–9, 1982
edited by **G. Poncelet, P. Grange and P.A. Jacobs**

Volume 17 **Spillover of Adsorbed Species.** Proceedings of an International Symposium, Lyon-Villeurbanne, September 12–16, 1983
edited by **G.M. Pajonk, S.J. Teichner and J.E. Germain**

Volume 18 **Structure and Reactivity of Modified Zeolites.** Proceedings of an International Conference, Prague, July 9–13, 1984
edited by **P.A. Jacobs, N.I. Jaeger, P. Jirů, V.B. Kazansky and G. Schulz-Ekloff**

Volume 19 **Catalysis on the Energy Scene.** Proceedings of the 9th Canadian Symposium on Catalysis, Quebec, P.Q., September 30–October 3, 1984
edited by **S. Kaliaguine and A. Mahay**

Volume 20 **Catalysis by Acids and Bases.** Proceedings of an International Symposium, Villeurbanne (Lyon), September 25–27, 1984
edited by **B. Imelik, C. Naccache, G. Coudurier, Y. Ben Taarit and J.C. Vedrine**

Volume 21 **Adsorption and Catalysis on Oxide Surfaces.** Proceedings of a Symposium, Uxbridge, June 28–29, 1984
edited by **M. Che and G.C. Bond**

Volume 22 **Unsteady Processes in Catalytic Reactors**
by **Yu.Sh. Matros**

Volume 23 **Physics of Solid Surfaces 1984**
edited by **J. Koukal**

Volume 24 **Zeolites: Synthesis, Structure, Technology and Application.** Proceedings of an International Symposium, Portorož-Portorose, September 3–8, 1984
edited by **B. Držaj, S. Hočevar and S. Pejovnik**

Volume 25 **Catalytic Polymerization of Olefins.** Proceedings of the International Symposium on Future Aspects of Olefin Polymerization, Tokyo, July 4–6, 1985
edited by **T. Keii and K. Soga**

Volume 26 **Vibrations at Surfaces 1985.** Proceedings of the Fourth International Conference, Bowness-on-Windermere, September 15–19, 1985
edited by **D.A. King, N.V. Richardson and S. Holloway**

Volume 27 **Catalytic Hydrogenation**
edited by **L. Cervený**

Volume 28 **New Developments in Zeolite Science and Technology.** Proceedings of the 7th International Zeolite Conference, Tokyo, August 17–22, 1986
edited by **Y. Murakami, A. Iijima and J.W. Ward**

Volume 29 **Metal Clusters in Catalysis**
edited by **B.C. Gates, L. Guczi and H. Knözinger**

Volume 30 **Catalysis and Automotive Pollution Control.** Proceedings of the First International Symposium, Brussels, September 8–11, 1986
edited by **A. Crucq and A. Frennet**

Volume 31 **Preparation of Catalysts IV.** Scientific Bases for the Preparation of Heterogeneous Catalysts. Proceedings of the Fourth International Symposium, Louvain-la-Neuve, September 1–4, 1986
edited by **B. Delmon, P. Grange, P.A. Jacobs and G. Poncelet**

Volume 32 **Thin Metal Films and Gas Chemisorption**
edited by **P. Wissmann**

Volume 33 **Synthesis of High-silica Aluminosilicate Zeolites**
edited by **P.A. Jacobs and J.A. Martens**

Volume 34 **Catalyst Deactivation 1987.** Proceedings of the 4th International Symposium, Antwerp, September 29–October 1, 1987
edited by **B. Delmon and G.F. Froment**

Volume 35 **Keynotes in Energy-Related Catalysis**
edited by **S. Kaliaguine**

Volume 71 **Catalysis and Automotive Pollution Control II.** Proceedings of the 2nd International Symposium (CAPoC 2), Brussels, Belgium, September 10–13, 1990
edited by **A. Crucq**

Volume 72 **New Developments in Selective Oxidation by Heterogeneous Catalysis.** Proceedings of the 3rd European Workshop Meeting on New Developments in Selective Oxidation by Heterogeneous Catalysis, Louvain-la-Neuve, Belgium, April 8–10, 1991
edited by **P. Ruiz and B. Delmon**

Volume 73 **Progress in Catalysis.** Proceedings of the 12th Canadian Symposium on Catalysis, Banff, Alberta, Canada, May 25–28, 1992
edited by **K.J. Smith and E.C. Sanford**

Volume 74 **Angle-Resolved Photoemission. Theory and Current Applications**
edited by **S.D. Kevan**

Volume 75 **New Frontiers in Catalysis, Parts A-C.** Proceedings of the 10th International Congress on Catalysis, Budapest, Hungary, 19–24 July, 1992
edited by **L. Guczi, F. Solymosi and P. Tétényi**

Volume 76 **Fluid Catalytic Cracking: Science and Technology**
edited by **J.S. Magee and M.M. Mitchell, Jr.**

Volume 77 **New Aspects of Spillover Effect in Catalysis. For Development of Highly Active Catalysts.** Proceedings of the Third International Conference on Spillover, Kyoto, Japan, August 17–20, 1993
edited by **T. Inui, K. Fujimoto, T. Uchijima and M. Masai**

Volume 78 **Heterogeneous Catalysis and Fine Chemicals III.**
Proceedings of the 3rd International Symposium, Poitiers, April 5 - 8, 1993
edited by **M. Guisnet, J. Barbier, J. Barrault, C. Bouchoule, D. Duprez, G. Pérot and C. Montassier**

Volume 79 **Catalysis: An Integrated Approach to Homogeneous, Heterogeneous and Industrial Catalysis**
edited by **J.A. Moulijn, P.W.N.M. van Leeuwen and R.A. van Santen**

Volume 80 **Fundamentals of Adsorption.** Proceedings of the Fourth International Conference on Fundamentals of Adsorption, Kyoto, Japan, May 17–22, 1992
edited by **M. Suzuki**

Volume 81 **Natural Gas Conversion II.** Proceedings of the Third Natural Gas Conversion Symposium, Sydney, July 4–9, 1993
edited by **H.E. Curry-Hyde and R.F. Howe**

Volume 82 **New Developments in Selective Oxidation II.** Proceedings of the Second World Congress and Fourth European Workshop Meeting, Benalmádena, Spain, September 20–24, 1993
edited by **V. Cortés Corberán and S. Vic Bellón**

Volume 83 **Zeolites and Microporous Crystals.** Proceedings of the International Symposium on Zeolites and Microporous Crystals, Nagoya, Japan, August 22–25, 1993
edited by **T. Hattori and T. Yashima**

Volume 84 **Zeolites and Related Microporous Materials: State of the Art 1994.** Proceedings of the 10th International Zeolite Conference, Garmisch-Partenkirchen, Germany, July 17–22, 1994
edited by **J. Weitkamp, H.G. Karge, H. Pfeifer and W. Hölderich**

Volume 85 **Advanced Zeolite Science and Applications**
edited by **J.C. Jansen, M. Stöcker, H.G. Karge and J.Weitkamp**

Volume 86 **Oscillating Heterogeneous Catalytic Systems**
by **M.M. Slin'ko and N.I. Jaeger**

Volume 87 **Characterization of Porous Solids III.** Proceedings of the IUPAC Symposium (COPS III), Marseille, France, May 9–12, 1993
edited by **J.Rouquerol, F. Rodriguez-Reinoso, K.S.W. Sing and K.K. Unger**

Volume 88 **Catalyst Deactivation 1994.** Proceedings of the 6th International Symposium, Ostend, Belgium, October 3–5, 1994
edited by **B. Delmon and G.F. Froment**

Volume 89 **Catalyst Design for Tailor-made Polyolefins.** Proceedings of the International Symposium on Catalyst Design for Tailor-made Polyolefins, Kanazawa, Japan, March 10–12, 1994
edited by **K. Soga and M. Terano**

Volume 90 **Acid-Base Catalysis II.** Proceedings of the International Symposium on Acid-Base Catalysis II, Sapporo, Japan, December 2–4, 1993
edited by **H. Hattori, M. Misono and Y. Ono**

Volume 91 **Preparation of Catalysts VI.** Scientific Bases for the Preparation of Heterogeneous Catalysts. Proceedings of the Sixth International Symposium, Louvain-La-Neuve, September 5–8, 1994
edited by **G. Poncelet, J. Martens, B. Delmon, P.A. Jacobs and P. Grange**

Volume 92 **Science and Technology in Catalysis 1994.** Proceedings of the Second Tokyo Conference on Advanced Catalytic Science and Technology, Tokyo, August 21–26, 1994
edited by **Y. Izumi, H. Arai and M. Iwamoto**

Volume 93 **Characterization and Chemical Modification of the Silica Surface**
by **E.F. Vansant, P. Van Der Voort and K.C. Vrancken**

Volume 94 **Catalysis by Microporous Materials.** Proceedings of ZEOCAT'95, Szombathely, Hungary, July 9–13, 1995
edited by **H.K. Beyer, H.G.Karge, I. Kiricsi and J.B. Nagy**

Volume 95 **Catalysis by Metals and Alloys**
by **V. Ponec and G.C. Bond**

Volume 96 **Catalysis and Automotive Pollution Control III.** Proceedings of the Third International Symposium (CAPoC3), Brussels, Belgium, April 20–22, 1994
edited by **A. Frennet and J.-M. Bastin**

Volume 97 **Zeolites: A Refined Tool for Designing Catalytic Sites.** Proceedings of the International Symposium, Québec, Canada, October 15–20, 1995
edited by **L. Bonneviot and S. Kaliaguine**

Volume 98 **Zeolite Science 1994: Recent Progress and Discussions.** Supplementary Materials to the 10th International Zeolite Conference, Garmisch-Partenkirchen, Germany, July 17–22, 1994
edited by **H.G. Karge and J. Weitkamp**

Volume 99 **Adsorption on New and Modified Inorganic Sorbents**
edited by **A. Dąbrowski and V.A. Tertykh**

Volume 100 **Catalysts in Petroleum Refining and Petrochemical Industries 1995.** Proceedings of the 2nd International Conference on Catalysts in Petroleum Refining and Petrochemical Industries, Kuwait, April 22-26, 1995
edited by **M. Absi-Halabi, J. Beshara, H. Qabazard and A. Stanislaus**

Volume 101 **11th International Congress on Catalysis - 40th Anniversary.** Proceedings of the 11th ICC, Baltimore, MD, USA, June 30-July 5, 1996
edited by **J. W. Hightower, W.N. Delgass, E. Iglesia and A.T. Bell**

Volume 102 **Recent Advances and New Horizons in Zeolite Science and Technology**
edited by **H. Chon, S.I. Woo and S. -E. Park**

Volume 103 **Semiconductor Nanoclusters - Physical, Chemical, and Catalytic Aspects**
edited by **P.V. Kamat and D. Meisel**

Volume 104 **Equilibria and Dynamics of Gas Adsorption on Heterogeneous Solid Surfaces**
edited by **W. Rudziński, W.A. Steele and G. Zgrablich**

Volume 105 **Progress in Zeolite and Microporous Materials**
Proceedings of the 11th International Zeolite Conference, Seoul, Korea, August 12-17, 1996
edited by **H. Chon, S.-K. Ihm and Y.S. Uh**